TO THE STUDENT: Three helpful supplemental study aids for the textbook are available:

- **Student Guide and Review Manual** by John K. Harris and Dudley W. Curry contains the following four sections for each chapter in the textbook:

1. *Main Focus and Objectives:* Introduces the chapter, provides the overall learning objective, refers to particular learning objectives in the text, and specifies concepts and techniques that deserve special study.
2. *Review of Key Ideas:* A comprehensive outline which references specific textbook pages and exhibits, as well as Practice Test Questions and Problems, that reinforce important concepts.
3. *Practice Test Questions and Problems:* Includes approximatlely 30–40 objective questions, 2–4 problems, plus 3–5 carefully selected CPA and CMA questions.
4. *Solutions to Practice Test Questions:* Provides keyed explanations for most of the objective answers as well as easy-to-follow solutions for all of the problems.
   (ISBN 0–13–179854–5)

- **Student Solutions Manual** by Charles T. Horngren and George Foster provides worked out solutions to all of the even-numbered problems. (ISBN 0–13–179870–7)

- **Applications in Cost Accounting Using Lotus 1-2-3** by David M. Buehlman and Dennis P. Curtin is a book/disk package that consists of a series of case problems that parallel the concepts in the textbook. The templates on the disk run with Lotus 1-2-3 and assist in solving the problems. (ISBN 0–13–040015–7)

If these supplements are not in your college bookstore, ask the manager for details on how to obtain them.

PRENTICE HALL SERIES IN ACCOUNTING

Charles T. Horngren, Consulting Editor

AUDITING: AN INTEGRATED APPROACH, 5/E	ARENS/LOEBBECKE
KOHLER'S DICTIONARY FOR ACCOUNTANTS, 6/E	COOPER/IJIRI
FINANCIAL STATEMENT ANALYSIS, 2/E	FOSTER
GOVERNMENTAL AND NONPROFIT ACCOUNTING: THEORY AND PRACTICE, 3/E	FREEMAN/SHOULDERS/LYNN
FINANCIAL ACCOUNTING: PRINCIPLES AND ISSUES, 4/E	GRANOF/BELL
COST ACCOUNTING: A MANAGERIAL EMPHASIS, 7/E	HORNGREN/FOSTER
ACCOUNTING	HORNGREN/HARRISON
INTRODUCTION TO FINANCIAL ACCOUNTING, 4/E	HORNGREN/SUNDEM
INTRODUCTION TO MANAGEMENT ACCOUNTING, 8/E	HORNGREN/SUNDEM
ADVANCED MANAGEMENT ACCOUNTING, 2/E	KAPLAN/ATKINSON
INTRODUCTORY FINANCIAL ACCOUNTING, 3/E	MUELLER/KELLY
AUDITING PRINCIPLES, 5/E	STETTLER
BUDGETING, 5/E	WELSCH/HILTON/GORDON

Charles T. Horngren
Stanford University

George Foster
Stanford University

PRENTICE HALL
A Division of Simon & Schuster
Englewood Cliffs, New Jersey 07632

Cost Accounting

A Managerial Emphasis

Seventh Edition

Library of Congress Cataloging-in-Publication Data

HORNGREN, CHARLES T.,
Cost accounting : a managerial emphasis / Charles T. Horngren. George Foster. — 7th ed.
p. cm.
Includes bibliographical references and indexes.
ISBN 0-13-179813-8
1. Cost accounting. 2. Costs, Industrial. I. Foster, George. II. Title.
HF5686.C8H59 1990
658.15·11—dc20 90-48362
CIP

To Professor William J. Vatter

Production editor: *Esther S. Koehn*
Development editor: *Stephen Deitmer*
Acquisition editor: *Terri Peterson*
Interior and cover design: *Kenny Beck*
Cover photo: San Francisco © 1990 *Barrie Rokeach*
Photo editor: *Rona Tuccillo*
Photo researcher: *Tobi Zausner*
Manufacturing buyers: *Trudy Pisciotti/Robert Anderson*

COST ACCOUNTING: A Managerial Emphasis, 7th Edition
Charles T. Horngren/George Foster

A Division of Simon & Schuster
Englewood Cliffs, New Jersey 17632

Printed in the United States of America
10 9 8 7 6 5 4 3 2 1

ISBN 0-13-179813-8

Prentice-Hall International (UK) Limited, *London*
Prentice-Hall of Australia Pty. Limited, *Sydney*
Prentice-Hall Canada Inc., *Toronto*
Prentice-Hall Hispanoamericana, S.A., *Mexico City*
Prentice-Hall of India Private Limited, *New Delhi*
Prentice-Hall of Japan, Inc., *Tokyo*
Simon & Schuster Asia Pte. Ltd., *Singapore*
Editora Prentice-Hall do Brasil, Ltda., *Rio de Janiero*

About the Authors

Charles T. Horngren is the Edmund W. Littlefield Professor of Accounting at Stanford University. A graduate of Marquette University, he received his MBA from Harvard University and his Ph.D. from the University of Chicago. He is also the recipient of honorary doctorates from Marquette University and De Paul University.

A Certified Public Accountant, Horngren served on the Accounting Principles Board for six years, the Financial Accounting Standards Board Advisory Council for five years, and the Council of the American Institute of Certified Public Accountants for three years. For six years, he served as a trustee of the Financial Accounting Foundation, which oversees the Financial Accounting Standards Board and the Government Accounting Standards Board.

In 1990 Horngren was elected to the Accounting Hall of Fame.

A member of the American Accounting Association, Horngren has been its President and its Director of Research. He received the Outstanding Accounting Educator Award in 1973, when the association initiated an annual series of such awards.

The California Certified Public Accountants Foundation gave Horngren its Faculty Excellence Award in 1975 and its Distinguished Professor Award in 1983. He is the first person to have received both awards.

In 1985 the American Institute of Certified Public Accountants presented its first Outstanding Educator Award to Horngren.

Professor Horngren is also a member of the National Association of Accountants, where he was on its research planning committee for three years. He was a member of the Board of Regents, Institute of Management Accounting, which administers the Certified Management Accountant examinations.

Horngren is the coauthor of three other books published by Prentice Hall: *Introduction to Financial Accounting*, Fourth Edition, 1990 (with Gary L. Sundem); *Introduction to Management Accounting*, Eighth Edition, 1990 (with Gary L. Sundem), and *Accounting*, 1989 (with Walter T. Harrison, Jr.).

Charles T. Horngren is the Consulting Editor for the Prentice Hall Series in Accounting.

George Foster is the Paul L. and Phyllis Wattis Professor of Accounting at Stanford University. He graduated with a university medal from the University of Sydney and has a Ph.D. from Stanford University. He has been awarded honorary doctorates from the University of Ghent, Belgium and from the University of Vaasa, Finland.

Foster has received the Distinguished Teaching Award at Stanford University and the Faculty Excellence Award from the California Society of Certified Public Accountants.

Research awards Foster has received include the Manuscript Competition Award of the American Accounting Association, the Notable Contribution to Accounting Literature Award of the American Institute of Certified Public Accountants, and the Citation for Meritorious Contribution to Accounting Literature Award of the Australian Society of Accountants.

He is the author of *Financial Statement Analysis,* published by Prentice Hall, and coauthor of *Security Analyst Multi-Year Earnings Forecasts and The Capital Market,* published by the American Accounting Association. Journals publishing his articles include *Abacus, The Accounting Review, Harvard Business Review, Journal of Accounting and Economics, Journal of Accounting Research, Journal of Cost Management,* and *Management Accounting.*

Foster works actively with many companies including Apple Computer, ARCO, Digital Electric Corp., Frito-Lay Corp., Hewlett-Packard, McDonald's Corp., Octel Communications, Santa Fe Corp., and Wells Fargo. He also has worked closely with Computer Aided Manufacturing-International (CAM-I) in the development of a framework for modern cost management practices. Foster has presented seminars on new developments in cost accounting in North America, Asia, Australasia, and Europe.

PHOTO CREDITS

1 Allen Green/Photo Researchers **24** Stephen Wilkes/The Image Bank **58** Tibor Bognar/The Stock Market **90** Susan Van Horne, © Kaufman and Broad Home Corporation **136** Hewlett-Packard **170** Boise Cascade Corporation **210** Gabe Palmer/The Stock Market **252** Allen-Bradley Co., Inc. **287** Bryan Peterson/The Stock Market **331** Randy Duchaine/The Stock Market **365** Craig Hammell/The Stock Market **395** Charles Gupton/Uniphoto **425** Charles Gupton/The Stock Market **454** Frank Pedrick/The Image Works **491** Ted Horowitz/The Stock Market **526** Nicholas De Sciose/Photo Researchers **556** Miller Brewing Company **590** Chuck O'Rear/West Light **618** Harley-Davidson, Inc. **646** CBS/Geraldine Overton **671** NASA/Science Source/Photo Researchers **704** Pacific Gas and Electric Company **739** Comstock **762** Mark Snyder/The Stock Market **782** Burlington Northern Railroad **817** Blair Seitz/Photo Researchers **851** © 1990 The ARCO Photography Collection, Los Angeles **880** Holiday Inns, Inc. **911** Comstock

Contents

4

Job Costing in Manufacturing 90

5

Job Costing for Services, Process Costing, and Activity-Based Accounting 136

PART TWO
BUDGETS AND STANDARDS AS KEYS TO PLANNING AND CONTROL

6

Master Budget and Responsibility Accounting 170

7

Flexible Budgets and Standards: I 210

Preface

Cost accounting provides data for various purposes, including planning, controlling, and product costing. We stress our major theme of "different costs for different purposes" throughout this book. The favorable reaction to previous editions is evidence that cost accounting courses can be enriched and relieved of drudgery by broadening the course from coverage of procedures alone to a full-fledged coverage of concepts, analyses, and procedures that pays more than lip service to accounting as a management tool.

STRENGTHS OF SIXTH EDITION RETAINED AND ENHANCED

Reviewers of the sixth edition cited the following features that have been retained and strengthened:

- Clarity and understandability of the text
- Coverage of important topics, including current developments in actual practice
- Numerous pedagogical aids (such as exhibits and examples) to assist student learning
- Excellent quantity, quality, and range of assignment material
- Helpful Problems for Self-Study for each chapter
- Flexible organization through a modular approach

The first fifteen chapters provide the essence of a one-term (quarter or semester) course. There is ample text and assignment material in the book's twenty-nine chapters for a two-term course. This book can be used immediately after the student has had an introductory course in financial accounting. Alternatively, this book can build on an introductory course in managerial accounting.

Deciding on the sequence of chapters in a textbook is a challenge. Every instructor has a favorite way of organizing his or her course. Hence, we present a modular, flexible organization that permits a course to be custom-tailored. *Our loosely constrained sequence of chapters facilitates diverse approaches to teaching and learning.*

As an example of the book's flexibility, consider our treatment of process costing. We introduce process costing in Chapter 5—immediately after our chapter on job-order costing—to fill out the student's perspective of costing systems. A more-detailed treatment of process costing appears in Chapters 17 and 18. Indeed, instructors may move from process costing in Chapter 5 immediately to Chapter 17 without interruption in the flow of material, as noted in the text itself. Other instructors may want their students to delve into budgeting and more decision-oriented topics early in the course. These instructors may find the Chapter 5 coverage of process costing sufficient.

CHANGES IN CONTENT AND PEDAGOGY OF SEVENTH EDITION

In keeping with the many changes in cost accounting during the past three years, this edition has been revised more heavily than any of the past editions. Each chapter has been scrutinized by several knowledgeable critics before a final draft was reached.

Real-World Illustrations

Students become more highly motivated to learn cost accounting if they can relate the subject matter to the real world. We have spent considerable time interacting with the management community, investigating new uses of cost accounting data and gaining insight into how changes in technology are affecting the roles of cost accounting information. Illustrations of the practices of actual companies permeate the book's narrative. Also, many new problems are based on an actual organization's situation or on legal cases. In addition, the results of surveys of company practices appear in appropriate places throughout the chapters. Our aim has been to interweave company illustrations to underscore particular concepts and procedures—not merely to tack on company names and examples in stand-alone fashion (and often only tangentially related to the learning process).

A partial list of real-world organizations used as illustrations follows. For a complete list, see the Company Index.

Alcoa	FASB	Stanford University
American Standard	GASB	Toyota
ARCO	Harley-Davidson	U.S. General Accounting Office
Boeing	Hewlett-Packard	U.S. Post Office
Boise Cascade	IBM	Wells Fargo
CASB	Motorola	Weyerhaeuser
Chrysler	PepsiCo.	Yoplait
Emery	Reckitt & Colman	

Changes in Content

Changes in the field of cost accounting have led us to make major changes in the content of the seventh edition.

1. *Three new chapters:*

- Chapter 5: Job Costing for Services, Process Costing, and Activity-Based Accounting
- Chapter 12: Pricing Decisions, Product Profitability Decisions, and Cost Information
- Chapter 29: Strategic Control Systems

2. *Emphasis on all business-function costs, including pervasive coverage of nonmanufacturing costs.* All the major business functions are introduced in Chapter 1 (see Exhibit 1-1). Chapter 2 points out that product costs are not confined to the manufacturing function (see Exhibit 2-11). Chapter 4 introduces the value chain of management functions (research and development, product design, manufacturing, marketing, distribution, and customer service), which is reinforced in subsequent chapters in both the text and assignment material. For example, see Exhibit 5-5. Life-cycle costing (in Chapter 12) presents the comprehensive reporting of revenues and all costs in the value chain on a product-by-product basis (see Exhibit 12-5). Also see Chapter 13, Management Control Systems; Chapter 21, Capital Budgeting and Cost Analysis; and Chapter 29, Strategic Control Systems.

3. *Cost drivers.* Chapter 2 introduces the cost driver concept, which is systematically integrated in the text. For examples, see Chapter 3, Cost-Volume-Profit Relationships; Chapter 10, Determining How Costs Behave; Chapter 12, Pricing Decisions; and Chapter 25, Cost Behavior and Regression Analysis. Nonvolume-related cost drivers (such as number of components in a product or the number of setups) are given more attention (see Exhibit 25-9). New problems reinforce this key concept.

4. *Direct–indirect cost distinction.* Chapter 2 emphasizes the basic distinction between direct and indirect costs, stressing that the higher the proportion of direct costs, the more accurate and believable to managers will be the resulting costs of the cost objects. The importance of this distinction is reinforced in later chapters. For example, Chapter 5 explores the implications of different classifications of direct and indirect costs for setting fees for professional services.

5. *Variable–fixed cost distinction.* Chapter 2 emphasizes the importance of the time horizon in classifying a cost as variable or fixed. Costs that are fixed in the short run may be variable in the long run (see Exhibit 10-2). The phrase "one-time-only special order" is now used in Chapter 11 to highlight the importance of considering the long-run revenue implications of a special order when deciding whether short-run variable costs are the only relevant costs to consider.

6. *Activity-based accounting.* Chapter 5 explains in detail this well-publicized approach to cost accounting. Several subsequent chapters (for example, Chapter 10, Determining How Costs Behave [see pp. 344–45]; Chapter 12, Pricing Decisions; Chapter 15, Cost Allocation [see pp. 492–502]; and Chapter 29, Strategic Control Systems) extensively discuss activity-based accounting. A variety of new problems reinforce this topic as it relates to planning and controlling as well as to product costing.

7. *Service sector.* The service sector is given ample coverage in many chapters as an integral part of the value chain. For example, Chapter 5 has a major section on job costing for the service sector, using an audit engagement as an illustration. Another example is Chapter 13, Systems Choice, which discusses the significant aspects of planning and controlling services, especially the roles of nonfinancial measures. Problems relating to high-profile service sector areas such as television, film making, and professional sports have been added to the assignment material.

8. *Modern cost management.* The thrusts of how companies manage costs are interwoven throughout the book. Examples include:

- More accurate product costing (see Exhibit 12-6)
- Increased prominence of nonfinancial measures for planning, controlling, and measuring performance (see Exhibit 21-2)
- Activities as a pivotal focus for planning and control, as distinct from the costing of products or services
- Quality as a competitive advantage—for example, through cost of quality programs and through quality-based performance measures
- Time as a strategic variable—for example, through the use of breakeven time in capital budgeting (see Chapter 21) and through time-based performance measures (see Chapter 29)
- Implications of modern cost management practices (such as longer contracts with suppliers) for accounting-performance measures.
- Desire for streamlined cost accounting procedures and systems. Chapter 19 covers backflush costing, which is a simplified standard-costing system

9. *Professional ethics.* The seventh edition focuses on ethics for management accountants in Chapter 1 and maintains a perspective on ethics throughout the book. For example, see the section on "cooking the books" in Chapter 28.

Improvements in Pedagogy and Clarity

Several improvements assist student learning:

1. The beginning of each chapter contains a photograph to spur interest, presents a road map of the major topics to be covered, and lists learning objectives, which are cross-linked with the specific topics as they appear in the chapter.
2. Visual aids are used more often and in a consistent format throughout the text. For example, consider the overviews of product costing that emphasize the distinction between direct and indirect costs in Exhibits 4-8, 5-1, 5-7, 14-1, 15-1, 15-7, 17-1, and 18-1. Exhibits 5-5 and 19-2 show the interrelationships among product costing systems and business functions, departments, and activities.
3. Terminology is consistent throughout the text. For example, *operating income* and *net income* are defined early in Chapter 3 and used consistently thereafter.
4. To increase clarity, we have improved in our use of (a) step-by-step expositions of individual topics, (b) numerical examples, and (c) formats of exhibits. For example, the introduction to job costing in Chapter 4 contains a brand new step-by-step set of journal entries and explanations. Moreover, process costing in Chapters 5, 17, and 18 uses identical formats for the exhibits. In addition, the first-in, first-out and weighted-average computations of equivalent units are no longer combined in a single exhibit.
5. The presentation of overhead variances has been thoroughly overhauled and includes a new exhibit that displays the relationships of 4-variance, 3-variance, 2-variance, and 1-variance analyses in one place (together with a class-tested block format that distinguishes among overhead variances, pp. 266–70).
6. To ease learning, Chapters 5, 7, 9, 19, and 26 have been subdivided into easily identifiable independent parts. Each part ends with a Problem for Self-Study. The chapter's assignment material is similarly subdivided. In this way, instructors can readily assign or omit individual parts.

Assignment Material

This edition continues the widely applauded tight linkage between text and assignment material formed in previous editions. The assignment material has two major headings: "Questions" and "Exercises and Problems." We have taken extra care to be sure that the terminology used in the assignment material coincides with that in the chapter. The *Solutions Manual* provides ample suggestions for how to use the material. The *Solutions Manual* also classifies the problems as to their place in the value chain (for example, manufacturing, marketing, customer service). Problems covering the nonprofit sector or the international sector are so identified, as are problems that highlight evolving changes in cost management. The *Solutions Manual* also lists the problems that originally appeared in professional examinations.

Major Changes in Sequence

The sequence of topics in Chapters 1–9 in the seventh edition is unchanged from that of Chapters 1–8 in the sixth edition with the exception that new Chapter 5 covers job costing for services and activity-based accounting. Significant modifications beyond Chapter 9 follow:

- Old Chapter 9 (Relevance) is now divided into two chapters, Chapter 11 (The Decision Process) and Chapter 12 (Pricing Decisions). Old Chapter 10 (Cost Behavior) is still Chapter 10, but it precedes the chapters covering relevant costs.
- Old Chapter 11 (Systems Choice) is now Chapter 13.
- Old Chapters 12–16 (Cost Allocation, Joint Costs, and Process Costing) retain the same sequence, but they are now Chapters 14–18.
- Old Chapter 17 (JIT Costing) is now Chapter 19 (Backflush Costing).
- Old Chapters 18–26 are now Chapters 20–28.

The *Solutions Manual* contains a chapter-by-chapter list of noteworthy changes from the sixth edition.

SUPPLEMENTS TO THE SEVENTH EDITION

A complete package of supplements is available to assist students and instructors in using this book. Supplements available to students are:

- *Student Guide and Review Manual* by John K. Harris and Dudley W. Curry. This is a chapter-by-chapter learning aid for students. It reviews key ideas and contains practice test questions and problems, including an average of four CPA/CMA questions per chapter. Solutions and explanations are also included.
- *Student Solutions Manual* by Charles T. Horngren and George Foster. Designed for student use, this supplement contains solutions for all of the even-numbered questions and problems in the textbook.
- *Lotus Templates for Selected Problems* by C. Rajagopal.
- *Microsoft Excel Topical Templates*
- *Applications in Cost Accounting Using Lotus 1-2-3* by David M. Buehlmann and Dennis P. Curtin. Personal computer applications are keyed to selected chapters of the textbook.

Brand new supplements for instructors are:

- *Annotated Instructor's Edition* with annotations by Linda S. Bamber. This is a great teaching tool. It is the regular textbook amplified by comments in the margins, including points to stress, teaching tips, real-world examples, discussion questions, typical students' misconceptions, class exercises, cross-discipline tie-ins, and cross-references to reinforcing problems in the assignment material.
- *Instructor's Manual* by William O. Stratton. This supplement provides a chapter overview, chapter outline, examples, alternate means of presenting materials, quiz/demonstration exercises, and suggested readings. Transparencies of key exhibits in the manual are also available to instructors.
- *Prentice Hall Course Manager* A three-hole punched *Annotated Instructor's Edition*, packaged in a binder, provides maximum teaching flexibility.
- *Cases and Extended Problems in Cost Accounting* by Charles T. Horngren and George Foster. This supplement contains additional, often longer, problems and cases for nearly every chapter.

Additional supplements available to instructors are:

- *Solutions Manual* by Charles T. Horngren and George Foster. Includes comments on alternative teaching approaches as well as solutions to all assignment material.
- *Test Item File* by John K. Harris and Dudley W. Curry. Includes quiz and examination material and is available in both hard copy and diskette form.
- Transparencies for solutions and key exhibits in the textbook.

ACKNOWLEDGMENTS

We are indebted to many for their ideas and assistance. The acknowledgments in the six previous editions contain a long list of our creditors. Our primary obligation is to Professor William J. Vatter, to whom this book is dedicated. For those who knew him, no words are necessary; for those who did not know him, no words will suffice.

Professor John K. Harris aided us immensely at all stages in the development and production of this book. He critiqued the sixth edition, gave a detailed review of the manuscript of this edition, and assisted us in the essential task of proofing. We are grateful to all the authors of the supplements described earlier.

The following professors influenced this edition by reviewing the preceding edition: L. Bamber, M. Frizzell, A. Hazera, P. Holmes, D. Keys, R. Maschmeyer, M. Oliverio, J. Schiff, K. Sevigny, J. Stancil, and H. Traugh.

Many comments on chapters in draft form were provided by A. Atkinson, L. Bamber, P. Brown, G. Castro, S. Datar, M. Gupta, C. Rajagopal, L. Sjoblom, K. Snowden, V. Srinivasin, and D. Then. Their care and attention to detail are greatly appreciated.

Incisive comments were also received from S. Barty, R. Bernheim, G. Blankenbeckler, A. Corr, M. Ghosal, G. Harwood, R. Hauser, H. Hoverland, M. Iqbal, E. Schwarz, T. Shevlin, G. Staubus, D. Suttles, J. Temmerman, and L. Yuen.

In addition, we have received helpful suggestions by mail from many users, unfortunately too numerous to be mentioned here. The seventh edition is much improved by their feedback.

Many students have read the manuscript and worked the new problems to insure that they are as error-free as possible. They have also contributed ideas and material for revising chapters and preparing new problems. Particular thanks go to G. Castro, S. Kovzan, T. Mongkolcheep, L. Scott, and J. Van Sickle.

Our Stanford colleagues have continually stimulated our thinking on many topics in this book.

A special note of gratitude is extended to T. Bush, I. Cossio, A. Danielson, S. Garg, J. Ochoa, E. Young, and L. Yujuico for their skillful typing of much material in syllabus form.

We thank the people at Prentice Hall: L. Albelli, C. Ciancia, J. Heider, E. Koehn, M. Lines, R. McCarry, R. Mullaney, T. Peterson, T. Pisciotti, A. Rohra, and J. Schmid. In particular, we thank S. Deitmer, whose skillful editorial work helped clarify our book enormously.

Appreciation also goes to the American Institute of Certified Public Accountants, the National Association of Accountants, The Institute of Certified Management Accountants, the Society of Management Accountants of Canada, the Certified General Accountants' Association of Canada, the Financial Executive Institute of America, and to many other publishers and companies for their generous permission to quote from their publications. Problems from the Uniform CPA Examinations are designated (CPA); problems from the Certificate in Management Accounting examinations are designated (CMA); problems from the Canadian examinations administered by the Society of Management Accountants are designated (SMA); problems from the Certified General Accountants' Association are designated (CGA). Many of these problems are adapted to highlight particular points.

We are grateful to the professors who contributed assignment material for this edition. Their names are indicated in parentheses at the start of their specific problems.

Comments from users are welcome.

CHARLES T. HORNGREN AND GEORGE FOSTER

The Accountant's Role in the Organization

Purposes of management accounting and financial accounting

Elements of management control

Cost-benefit approach

The pervading duties of the management accountant

Professional ethics

CHAPTER 1

Managers at all levels of a business—in service and in manufacturing industries—are customers of accounting information. Informed decisions have cost accounting at their roots.

Learning Objectives

When you have finished studying this chapter, you should be able to

1. Identify the three broad purposes of an accounting system
2. Distinguish between financial accounting and management accounting
3. Distinguish between planning and control
4. Identify the five major elements of a management control system
5. Describe the cost-benefit approach to choosing among accounting systems
6. Describe the work of the controller, the chief management accountant
7. Describe the role of professional ethics in management accounting

As we write this chapter, former accountants are the top executives in many large companies, including Nike and PepsiCo. Accounting duties have played a key part in their rise to the management summit. Accounting cuts across all facets of the organization; the management accountant's duties are intertwined with executive planning and control.

The study of modern cost accounting yields insight and breadth regarding both the accountant's role and the manager's role in an organization. How are these two roles related? Where may they overlap? How can accounting help managers? This book offers answers to these questions. In this chapter we look at where the accountant fits in the organization, which gives us a framework for studying the succeeding chapters.

PURPOSES OF MANAGEMENT ACCOUNTING AND FINANCIAL ACCOUNTING

Managers as Customers

Modern cost accounting is often called *management accounting*. Why? Cost accountants regard managers as the primary users of accounting information, as their customers. Around the globe, managers are becoming increasingly aware of the importance of the quality, timeliness, and service extended to their customers. In turn, the accountants who work with managers are becoming increasingly sensitive to the quality, timeliness, and usefulness of accounting numbers.

As Exhibit 1-1 indicates, cost accountants find customers throughout the entire organization, starting with managers doing strategic planning and extending to the designing of products and services, to manufacturing, to marketing, and to follow-up servicing. Management accounting serves diverse management customers. It is a major means of helping managers administer an entire enterprise. Management accounting information provides an organizationwide view that can promote teamwork among all managers.

Managers' long-run success depends on pleasing their customers. Therefore, managers have a continuing responsibility to know their customers' opinions of company products and services. Similarly, management accountants have a continuing responsibility to obtain feedback regarding managers' uses of accounting information. Management accountants' success depends on whether managers' performance is helped by the accounting information.

EXHIBIT 1-1
Modern Cost Accounting or Management Accounting

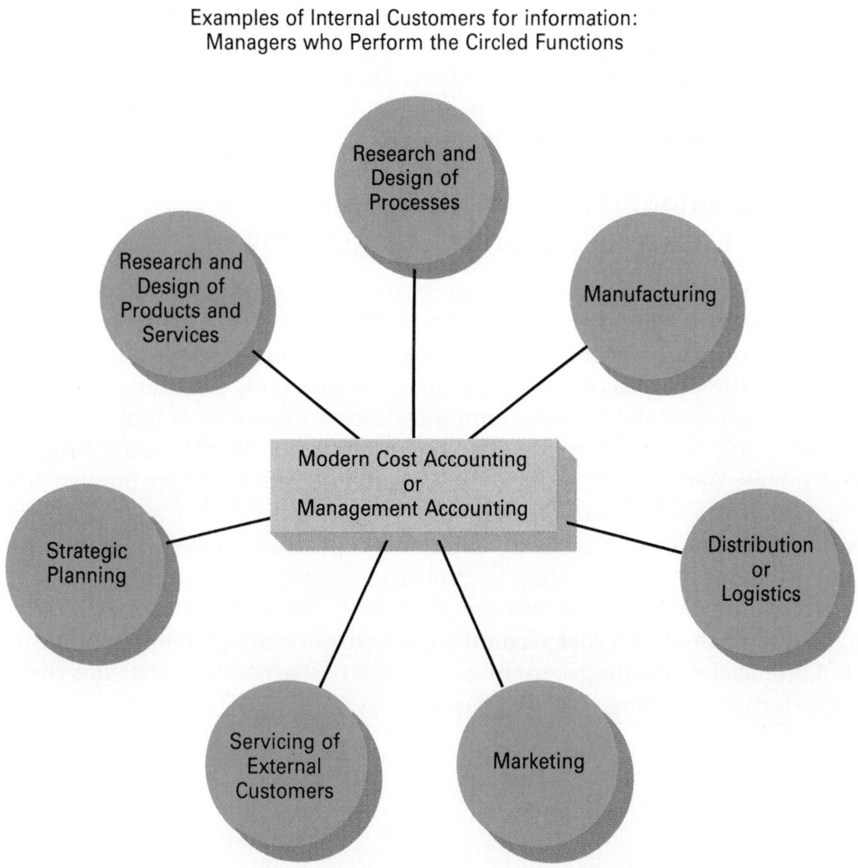

The Major Purposes of the Accounting System

Objective 1

Identify the three broad purposes of an accounting system

The accounting system is the principal quantitative information system in almost every organization. This system should provide information for three broad purposes:

1. ***Internal routine reporting to managers*** to provide information and influence behavior regarding cost management and the planning and controlling of operations
2. ***Internal nonroutine, or special, reporting to managers*** for strategic and tactical decisions on matters such as pricing products or services, choosing which products to emphasize or de-emphasize, investing in equipment, and formulating overall policies and long-range plans
3. ***External reporting*** through financial statements to investors, government authorities, and other outside parties

Both management (internal parties) and external parties share an interest in all three purposes. External users focus on the third purpose. In contrast, internal users concentrate on the first two purposes, which deal with transmitting information and influencing management behavior.

Each broad purpose of accounting may require a different way of aggregating or reporting the data. An ideal underlying data base will collect finely granulated bits of information. In turn, accountants combine or adjust these data to serve the desires of particular internal or external customers. For instance, the costs of design, manufacturing, transportation, and marketing might be combined to guide decisions regarding what products to make domestically and what products to make abroad.

Ultimately, all accounting information is accumulated to help individuals make decisions. However, all the matters to be decided and questions to be answered cannot be foreseen. Consequently, systems are designed to fulfill the broadest set of uses that are common among managers. A single, general-purpose accounting system usually serves the three broad purposes of accounting just discussed.

Distinctions among Financial Accounting, Management Accounting, and Cost Accounting

Objective 2

Distinguish between financial accounting and management accounting

Financial accounting focuses on how accounting can serve external decision makers. In contrast, management accounting is concerned mainly with its internal customers for accounting information. **Management accounting**, focusing on internal customers, is the process of identification, measurement, accumulation, analysis, preparation, interpretation, and communication of information that assists executives in fulfilling organizational goals. A synonym is **internal accounting**.

How do we define cost accounting? We might draw fine distinctive lines in definition, but modern cost accounting is generally used as a near synonym for management accounting. To satisfy external purposes, however, businesses must use cost accounting for measuring income and inventory valuations in accordance with the generally accepted accounting principles that guide financial accounting. When viewed in this way, **cost accounting** is a part of management accounting plus a part of financial accounting—to the extent that cost accounting satisfies the requisites of external reporting. The overlap is large:

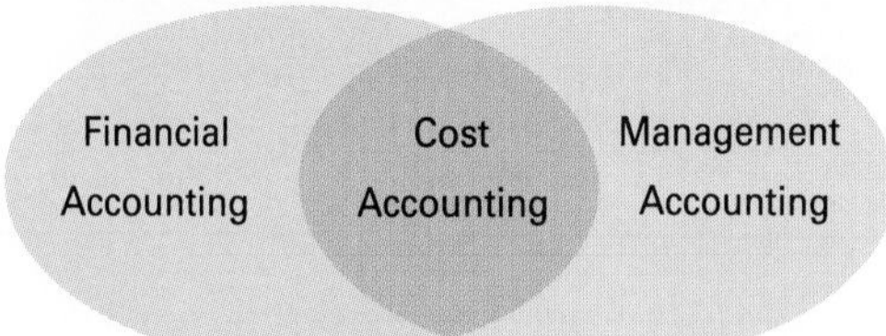

We need not be greatly concerned with the boundaries of cost accounting and management accounting. The major point is that a modern cost accounting system helps managers deal with both the immediate and the distant future. The system's concern with the past is justified only insofar as it helps prediction and satisfies external reporting requirements.

Cost Management and Accounting Systems

The term *cost management* has become widely used in recent years. Unfortunately, no uniform definition exists. We use **cost management** to describe the performance by executives and others in the cost implications of their short-run and long-run planning and control functions. For example, managers may make decisions regarding plant rearrangements or changes in product lines. Accounting systems may help managers in making these decisions. But the systems by themselves are not cost management; instead, the systems assist cost management.

Accounting systems should offer cost management more than a mere collection of financial data for income statements and balance sheets. Well-designed systems provide pertinent information for a wide variety of internal decisions and tasks—such as forecasting, estimating, and collecting nonfinancial measures like spoilage rates, on-time delivery rates, and number of customer complaints. In this way, the systems help management the most.

ELEMENTS OF MANAGEMENT CONTROL

Is Shell's management control system better than Exxon's? Unilever's better than Nestlé's? This book develops a method for answering such questions, and this section provides an overview of management control systems.

Planning and control

Objective 3

Distinguish between planning and control

There are countless definitions of planning and control. Study the left side of Exhibit 1-2. We define **planning** (the top box) as choosing goals, predicting potential results under various ways of achieving goals, and deciding how to attain the desired results. *Control* is usually distinguished from planning. **Control** (the next two boxes) is (a) action that implements the planning decision and (b) performance evaluation that provides feedback on the results.

Budgets and performance reports are two major accounting tools that help managers. Study the right side of Exhibit 1-2. A **budget** is a quantitative expression of a plan of action and an aid to coordination and implementation. Source documents and ledgers compile the results of actions. **Performance reports** measure activities. These reports usually consist of comparisons of budgets with actual results. The deviations of actual results from budget are called **variances**. Understanding the reasons for variances is an important part of **management by exception**, which is the practice of concentrating on areas that deserve attention and ignoring areas that are presumed to be running smoothly.

Exhibit 1-3 shows a simple performance report for a marketing consulting firm. Such reports spur investigation of exceptions. Operations are then brought into

EXHIBIT 1-2
Accounting Framework for Planning and Control

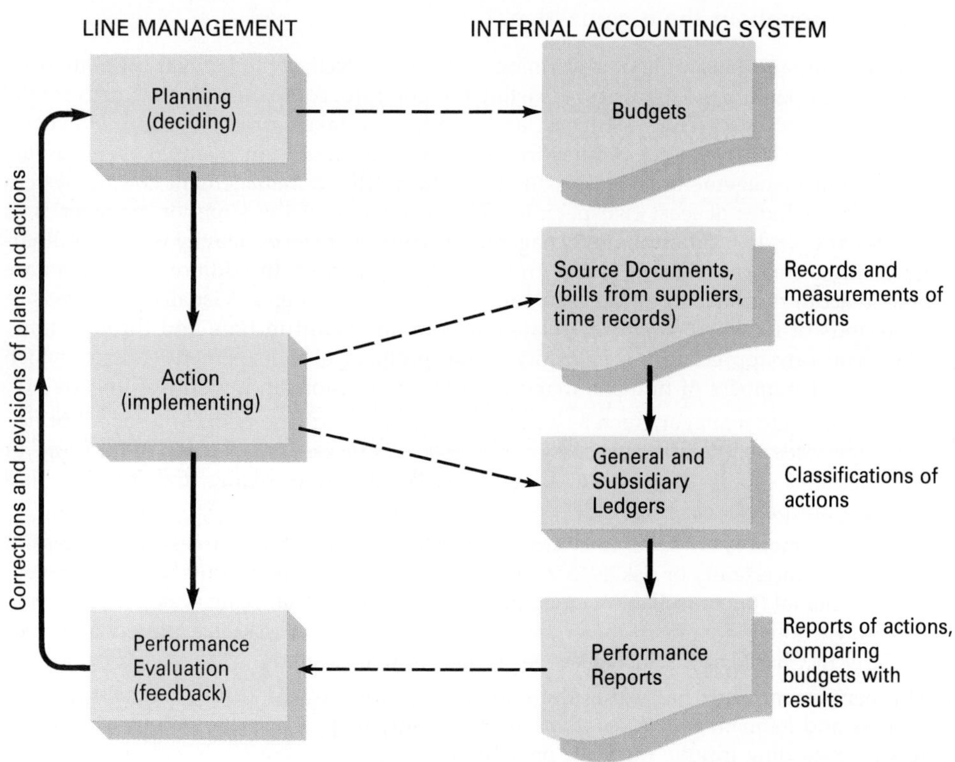

EXHIBIT 1-3
Performance Report

	Budgeted Amounts	Actual Amounts	Deviations or Variances	Explanation
Revenue from fees	xxx	xxx	xx	—
Various expenses	xxx	xxx	xx	—
Net income	xxx	xxx	xx	—

conformity with the plans, or the plans are revised. This is an example of management by exception.

Well-conceived plans include enough flexibility or discretion so that the manager may feel free to seize any unforeseen opportunities. That is, in no case should control mean that managers cling to a preexisting plan when unfolding events indicate the desirability of actions not encompassed in the original plan.

Planning and control are so strongly interlocked that it seems artificial to draw rigid distinctions between them. Managers certainly do not spend time drawing such rigid distinctions. In this book the narrow idea of control will not be used. Unless otherwise stated, we use control in its broadest sense to denote the entire management process of *both* planning and control. For example, instead of referring to a *management planning and control system,* we will refer to a *control system.* Similarly, we will refer to a *control purpose* of accounting instead of the awkward *planning and control purpose* of accounting.

Overview of Control Systems

Objective 4

Identify the five major elements of a management control system

Exhibit 1-4 demonstrates the basic elements of a **management control system**. This system is a means of gathering data to aid and coordinate the process of making decisions throughout the organization.

1. A commonly used example of a planning and control cycle is the familiar room thermostat: Set temperature goal, measure existing temperature, compare with goal, activate the operating process, produce feedback, and, if required, take corrective action. However, the automatic control of a room-heating or -cooling process is an example of *engineering control,* not management control. Exhibit 1-4 shows that a management control system generally includes at least two persons. The major role of the superior manager is to oversee the work of subordinate managers. The control of *human* activity is what ordinarily distinguishes *management control* from *engineering control.* In addition, managers are able to act in response to unforeseen contingencies and changes. Visualize hundreds or thousands of Exhibit 1-4, all interrelated at both an instant in time and through time. Then you can appreciate the enormity of the problem of management control, which entails vast numbers of humans making countless decisions in a world of uncertainty.
2. The subordinate manager, such as a sales manager, an airplane pilot, or a school principal, supervises an operating process or processes. A **process** is a collection of decisions or activities that should be aimed at some ends. For example, selling, transporting, and teaching are processes.
3. The **environment** is the set of uncontrollable factors that affect the success of a process. It is a level of uncertainty or risk as to whether predicted results or outputs will be achieved. Environmental (uncontrollable) variables might include absenteeism, weather, and competitor reactions. The role of the environment (uncertainty or risk) is explored later, particularly in Chapter 20, which may be studied now if desired.
4. The performance box highlights the regular evaluation of the results of the operating process and its managers. Executives are frequently helped here by various accounting reports regarding inputs, outputs, or both.
5. The feedback loops indicate how managers learn to improve the unending sequence of predictions and decisions that are embedded in the management control system.

EXHIBIT 1-4
Basic Elements of a Management Control System

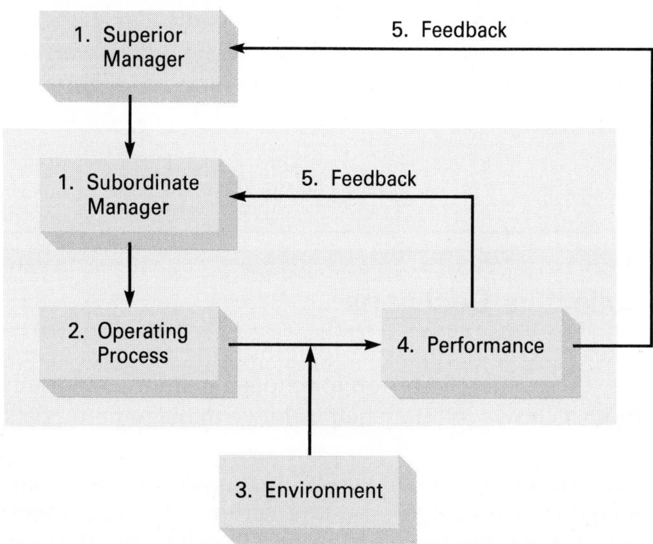

Source: Adapted from H. Itaml, Adaptive Behavior: Management Control and Information Analysis (Sarasota, FL: American Accounting Association), p. 9.

Feedback: A Major Key

In control systems, **feedback** often consists of a comparison of the budget with actual results—that is, with historical information. ***In particular, note that feedback consists of historical information.*** Feedback may have a variety of uses, including the following:

Use of Feedback	Example
• Changing goals	• Based on evaluation of results, General Electric abandons the appliance business.
• Searching for alternative means	• To save costs, California reissues driver's licenses by mail.
• Changing methods for making decisions	• To save costs over the long run, the University of Illinois bases its maintenance decisions on a comparison of expected repair costs over a five-year period rather than over a one-year period.
• Making predictions	• Lockheed uses some new ways to predict costs of material and labor because feedback indicates a poor prediction record, which has adversely affected efforts aimed at winning defense contracts.
• Changing the operating process	• Hewlett-Packard has materials delivered directly to the factory floor instead of to the receiving room or the storeroom.
• Changing the measuring and rewarding of performance	• IBM marketing managers may change their past behavior if their performance is measured by total profits rather than by total sales.

Feedback is obviously important; yet many control systems are weakened by managers not taking advantage of feedback. *Managers are part of the system.* The feedback is usually available at low cost, but it is often ignored by managers. On the other hand, many successful managers actively support their control system by using feedback in a highly visible way. Both accountants and managers should periodically remind themselves that management control systems are not confined exclusively to technical matters such as data processing. ***Management control is primarily a human activity that should focus on how to help individuals do their jobs better.***

COST-BENEFIT APPROACH

Improving Collective Decisions

Objective 5

Describe the cost-benefit approach to choosing among alternative accounting systems

This textbook takes a general approach to accounting referred to as a **cost-benefit approach**. That is, the primary criterion for choosing among alternative accounting systems or methods is how well they help achieve management goals in relation to their costs.

Accounting systems are economic goods. They cost money, just like bread and milk. The old adage says that if you build a better mousetrap, people will beat a path to your door. But many sellers of mousetraps will testify that when the buyers get to the door, they say, "What is the price?" If the price is too high, buyers may be unwilling to pay, and the maker of better mousetraps may face financial ruin.

The costs of buying a new accounting system include the usual clerical and data-processing activities plus educational programs. ***The costs of educating users of the system is frequently substantial, particularly when managers must invest much time in learning the new system and when the users have been reasonably satisfied with the system.***

As customers, managers buy a more elaborate management accounting system when its perceived expected benefits exceed its perceived expected costs. ***Although the benefits may take many forms, they can be summarized as collective sets of decisions that will better attain top-management goals.***

Consider the installation of a company's first budgeting system. The company had probably been using some historical recordkeeping and little formal planning. A major benefit from purchasing a budgeting system is to compel managers to plan and thus make a *different*, more profitable set of decisions than would have been generated by using only a historical system. Thus, the expected benefits exceed the expected costs of the new budgeting system.

Admittedly, the measurement of these costs and benefits is seldom easy. Therefore, you may want to call this approach an abstract theory rather than a practical guide. Nevertheless, the cost-benefit approach provides a starting point for analyzing virtually all accounting issues. Moreover, it is directly linked to a vast theoretical structure of information economics (the application of the microeconomic theory of uncertainty to questions of buying information.)

The cost-benefit way of thinking is widely applicable even if the cost and benefits defy precise measurement. As stated by a motto in the System Analysis Office of the Department of Defense, "It is better to be roughly right than precisely wrong." For example, if two methods of curing the same disease are available, the less costly is preferable. Similarly, if two proposals have equal costs, the proposal that is perceived to yield more benefits is preferable. This judgment can be achieved without a numerical measurement of the levels of benefits.[1]

---

[1]See R. Anthony and D. Young, *Management Control in Nonprofit Organizations*, 4th ed. (Homewood, IL: Irwin, 1988), pp. 409–16.

The Decision's Dependence on Circumstances

A key question asked in applying the cost-benefit approach is, ***How much would we be willing to pay for one system versus another?*** For example, the same concert ticket at a given price may be a "good buy" for one person but a "bad buy" for another person under different circumstances. Similarly, a particular cost accounting system or method may be a good buy for General Motors but a bad buy for Honda. After all, General Motors and Honda have different plants, processes, and managers. The cost-benefit approach takes a skeptical view of such sweeping generalizations as "This budgeting technique is a vast improvement over other techniques. Every company needs it."

The choice of a technique or system inherently depends on specific circumstances. Therefore, this book will concentrate on describing alternative techniques and systems and on how to go about making the choices between them. It will not present one management accounting technique as being innately superior to another. Again, the human element is significant. The same system may work well in one organization but not in another. Why? Because the collective personalities and traditions differ between the two. Systems do not exist in a vacuum. Managers and accountants are integral parts of management control systems. Costs and benefits cannot be evaluated apart from the managers and accountants who will use and prepare systems and reports.

THE PERVADING DUTIES OF THE MANAGEMENT ACCOUNTANT

Overview of an Organization

Exhibit 1-5 illustrates the general organizational relationships in Hewlett-Packard, a company that has many internal customers for accounting information, including managers in research, engineering, and quality assurance. However, we focus here on how accountants serve the manufacturing function. The chief financial officer, who often has the title of financial vice president or vice president–finance, is an integral part of top management. Two key accounting executives, the treasurer and the controller, usually report to the financial vice president.

The manufacturing function has production departments and service departments. The primary purpose of a factory is to produce goods. Therefore, the production-line departments are usually termed *production* ***or*** *operating departments.* ***To facilitate production, most plants also have*** *service departments,* ***also called*** *support departments,* ***which exist to help the production departments.***

Line and Staff Relationships

Most businesses have the production and sale of goods or services as their basic objectives. Line managers are directly responsible for attaining these objectives as efficiently as possible. Staff activities of organizations exist because the scope of the line managers' responsibility and duties generally expands to where they need specialized help to operate effectively. When a department's primary task is that of advice and service to other departments, it is a staff department.

Except for exerting line authority over their own departments, the chief financial officers generally fill a staff role in their companies, as contrasted with the line roles of sales and production executives. This staff role includes advice and help in the areas of budgeting, controlling, pricing, and special decisions. The accounting officers do not exercise direct authority over line departments.

EXHIBIT 1-5
Partial Chart of Hewlett-Packard Organization

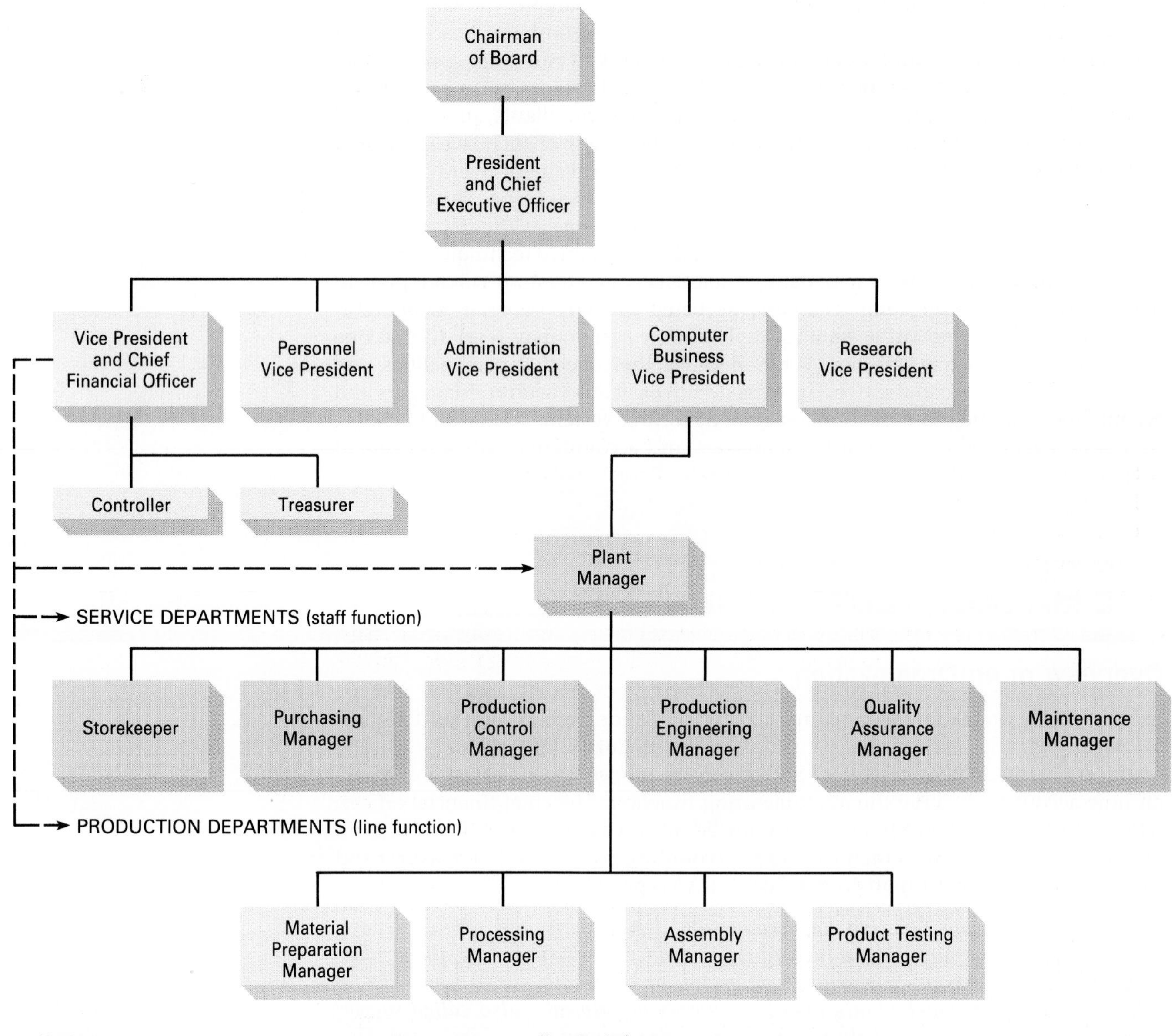

(Solid lines represent line authority; dashed lines represent staff authority)

Management literature is hazy on these distinctions, and we will not belabor them here. For example, some writers distinguish among three types of authority: line, staff, and functional. **Line authority** is exerted downward over subordinates. **Staff authority** is the authority to *advise* but not command others; it may be exercised laterally or upward. **Functional authority** is the right to *command* action laterally and downward with regard to a specific function or specialty.

Uniformity of accounting and reporting is often acquired through the delegation of authority regarding accounting procedures to the chief management accountant, the controller, by the top line management. Note carefully that when the controller prescribes the line department's role in supplying accounting information, he or she is speaking for top line management—not as the controller, a staff person. The president authorizes the uniform accounting procedure; the controller installs it.

Theoretically, the controller's decisions regarding the best accounting procedures to be followed by line people are transmitted to the president. In turn, the president communicates these procedures through a manual of instructions that comes down through the line chain of command to all people affected by the procedures.

Practically, the daily work of the controller is such that face-to-face relationships with the production superintendent or shipping manager may call for directing how production records should be kept or how time records should be completed.[2] The controller usually holds delegated authority from top line management over such matters.

Distinctions Between Controller and Treasurer

Many people confuse the responsibilities of controller and treasurer. The chief financial officer, vice president–finance, typically oversees both the controllership and treasurership functions. Their functions have been distinguished as follows:

Treasurership	Controllership
1. Provision of capital	1. Planning and control
2. Investor relations	2. Reporting and interpreting
3. Short-term financing	3. Evaluating and consulting
4. Banking and custody	4. Tax administration
5. Credits and collections	5. Government reporting
6. Investments	6. Protection of assets
7. Insurance	7. Economic appraisal

The **treasurer** is the financial executive who is primarily responsible for obtaining investment capital and managing cash. We will not dwell on the treasurer's functions. As the seven points indicate, treasurers are concerned mainly with financial, as distinguished from operating, problems. The exact division of various accounting and financial duties varies from company to company.

The **controller** is the financial executive primarily responsible for both management accounting and financial accounting. The controller has been compared to a ship's navigator. The navigator, with the help of specialized training, assists the captain. Without the navigator, the ship may founder on reefs or miss its destination entirely, but the captain exerts the right to command. The navigator guides and informs the captain as to how well the ship is being steered. This navigator role is especially evident in points 1 through 3 of the controller's functions. Note how managerial cost accounting is the controller's primary *means* of implementing the first three functions of controllership. In small companies, one individual may perform tasks of both treasurer and controller. Exhibit 1-6 shows how the controller's role at Hewlett-Packard is organized.

The Controller: The Chief Management Accountant

Objective 6

Describe the work of the controller, the chief management accountant

The world *controller* is applied to various accounting positions. The stature and duties of the controller vary from company to company. In some firms the controller is little more than a glorified bookkeeper who compiles data primarily for conventional balance sheets and income statements. In other firms—for example, General Electric—controllers are key executives who aid management planning and control in about two hundred subdivisions. In most firms the controller's status is somewhere between these two extremes. For example, controllers' opinions

[2]According to some writers, this would be exercising the *functional authority* described above.

EXHIBIT 1-6
Partial Chart of Hewlett-Packard Controller Organization

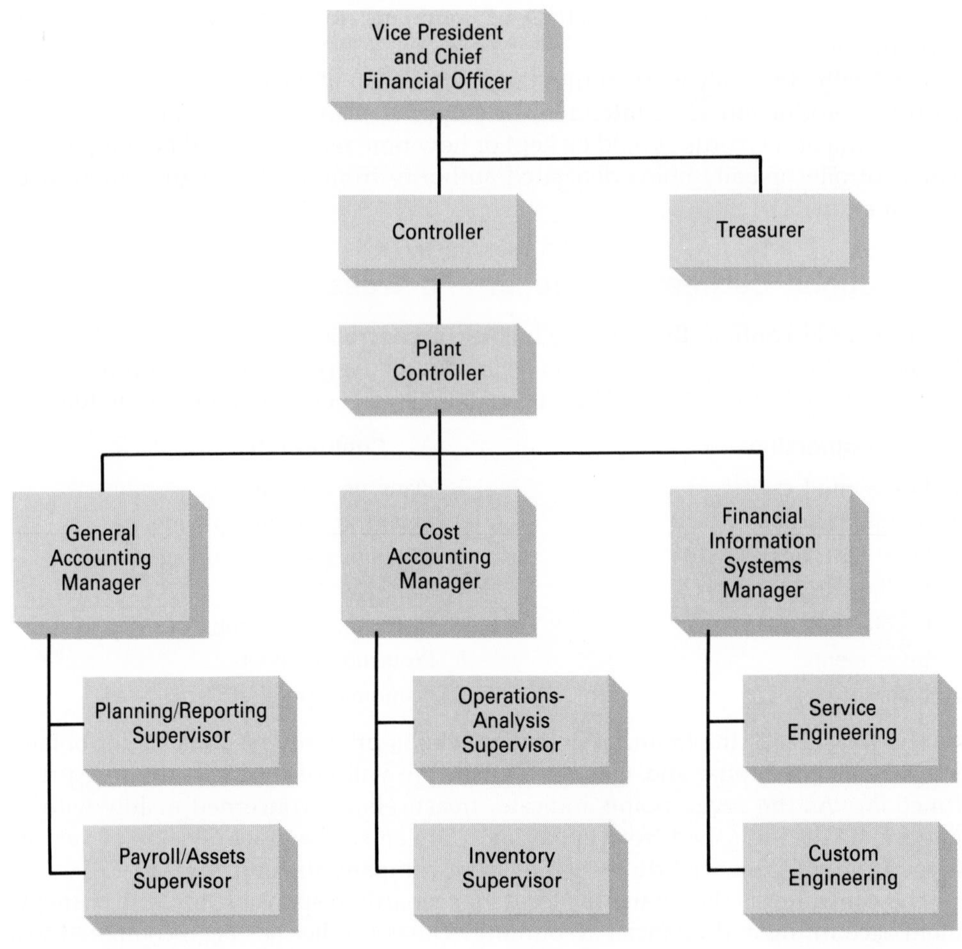

on the tax implications of certain managerial decisions may be carefully weighed, yet their opinions on the other aspects of these decisions may not be sought.

In this book we consider the controller as the chief management accounting executive. The modern controller does not do any controlling in terms of line authority except over his or her own department. Yet the modern concept of controllership maintains that the controller does control in a special sense. That is, by reporting and interpreting relevant data, the controller exerts a force or influence that impels management toward logical decisions consistent with its objectives.

Division of Duties

Accountants often face a dilemma because they are supposed to fulfill two conflicting roles simultaneously. First, they are seen as watchdogs for top managers. Second, they are seen as helpers for all managers. The watchdog role is usually fulfilled through the "scorekeeping" task of accumulating and reporting results to all levels of management. The helper role is usually fulfilled by directing managers' attention to problems (attention directing) and by assisting managers in solving problems.[3]

[3]H. A. Simon, H. Guetzkow, G. Kozmetsky, and G. Tyndall, *Centralization vs. Decentralization in Organizing the Controller's Department* (New York: Controllership Foundation, Inc.). This perceptive study is much broader than its title implies.

The sheer volume of the scorekeeping task is often overwhelming; the day-to-day routines and endless deadlines may shunt the helper role into the background and often into oblivion. If accountants shirk their attention-directing and problem-solving roles, other information systems usually arise to fulfill such demands by internal customers. To prevent the helper role from falling by the wayside, many organizations deliberately split the accountants' duties. For example, Yoplait Company's accounting organization has special positions such as

- operations analyst
- manager—marketing and sales analysis
- director—budgets analysis and reporting

The attention-directing and problem-solving tasks improve cooperation and increase understanding of the budgets and performance reports if a member of the controller's staff personally explains and interprets the information presented to line managers. This attention-directing role (for example, explaining the differences between budgeted and actual performance) is often performed by experienced accountants who, at least to some degree, can talk the line manager's language. Indeed, the interpreters are the individuals who will establish the status of the controller's department in the company. Close, direct contacts between accountants and line managers usually instill confidence in the reliability of performance reports, which are the major measuring devices of most businesses.

Many companies deliberately rotate their newly hired accountants through scorekeeping, attention-directing, and problem-solving posts. In this way, accountants are more likely to appreciate the decision makers' viewpoint and are thus prone to keep the accounting system tuned to the users.

In sum, the modern controller plays a "two-count" role in organizations. The first count is responsibility to top management for the integrity (reliability) of the performance reports of the subunits. The second count is responsibility for helping the subunit managers in planning and controlling operations. Controllers must balance their independence and objectivity against their necessary involvement in assisting line managers.

Roles in Global Business

Financial officers face widening responsibilities in the 1990s. Consider a survey of multinational corporations conducted by a large public accounting firm.[4] Of the 235 respondents, 76% were chief financial officers (CFOs) and 24% were treasurers or controllers. The survey focused on the changing role of the CFO and the qualities needed for success.

The survey results indicate that CFOs spend 40% of their time assisting the chief executive officer (CEO) and other senior executives with management issues. Two-thirds of the CFOs said their influence over operating decisions is increasing.

Survey respondents identified three major forces responsible for these expanded roles: corporate restructuring, globalization of business and financial markets, and the needs of the CEO. A consultant observed, "Today the CFO's focus is on global strategy that synthesizes financial and operational experience."

As part of the survey, respondents were asked to choose three to five personal characteristics from a list of eleven that they considered most important for a CFO in helping a CEO manage the company. The leading five are:

- Good understanding of company operations (84%)
- Ability to solve problems (50%)
- Integrity (49%)
- Ability to view "the big picture" (43%)
- Innovativeness (43%)

[4] *Building Global Profitability and Competitiveness: The New Role of Finance* (Montvale, NJ: KPMG Peat Marwick, 1989).

EXHIBIT 1-7

Standards of Ethical Conduct for Management Accountants

Management accountants have an obligation to the organizations they serve, their profession, the public, and themselves to maintain the highest standards of ethical conduct. In recognition of this obligation, the Institute of Certified Management Accountants and the National Association of Accountants have adopted the following standards of ethical conduct for management accountants. Adherence to these standards is integral to achieving the objectives of management accounting. Management accountants shall not commit acts contrary to these standards nor shall they condone the commission of such acts by others within their organizations.

Competence

Management accountants have a responsibility to:

- Maintain an appropriate level of professional competence by ongoing development of their knowledge and skills.
- Perform their professional duties in accordance with relevant laws, regulations, and technical standards.
- Prepare complete and clear reports and recommendations after appropriate analysis of relevant and reliable information.

Confidentiality

Management accountants have a responsibility to:

- Refrain from disclosing confidential information acquired in the course of their work except when authorized, unless legally obligated to do so.
- Inform subordinates as appropriate regarding the confidentiality of information acquired in the course of their work and monitor their activities to assure the maintenance of that confidentiality.
- Refrain from using or appearing to use confidential information acquired in the course of their work for unethical or illegal advantage either personally or through third parties.

Integrity

Management accountants have a responsibility to:

- Avoid actual or apparent conflicts of interest and advise all appropriate parties of any potential conflict.
- Refrain from engaging in any activity that would prejudice their ability to carry out their duties ethically.
- Refuse any gift, favor, or hospitality that would influence or would appear to influence their actions.
- Refrain from either actively or passively subverting the attainment of the organization's legitimate and ethical objectives.
- Recognize and communicate professional limitations or other constraints that would preclude responsible judgment or successful performance of an activity.
- Communicate unfavorable as well as favorable information and professional judgments or opinions.
- Refrain from engaging in or supporting any activity that would discredit the profession.

Objectivity

Management accountants have a responsibility to:

- Communicate information fairly and objectively.
- Disclose fully all relevant information that could reasonably be expected to influence an intended user's understanding of the reports, comments, and recommendations presented.

Source: Statement on Management Accounting, Standards of Ethical Conduct for Management Accountants (Montvale, NJ: National Association of Accountants, 1983).

PROFESSIONAL ETHICS

Objective 7

Describe the role of professional ethics in management accounting

The distinction between financial accounting and management accounting became institutionalized in the United States in 1972 when the **National Association of Accountants (NAA)**, the largest association of internal accountants in the United States, established a program leading to the Certificate in Management Accounting (CMA).[5] The **Certified Management Accountant (CMA)** is the professional designation for management accountants and financial executives. It is the internal accountant's counterpart of the CPA (Certified Public Accountant). Just like certified public accountants, management accountants have a code of professional ethics.

Public opinion surveys have consistently ranked accountants highly regarding their professional ethics. CPAs and CMAs adhere to codes of ethical conduct. Professional accounting organizations have procedures for reviewing behavior not consistent with these standards. In addition, many companies and government agencies have codes of ethical conduct. Key employees must sign a statement annually indicating that they have complied with the codes.[6]

Line managers and management accountants share responsibility for both external and internal financial reports. The management accountant must ensure that the underlying accounting systems, procedures, and compilations are reliable and free of manipulation.

Exhibit 1-7 contains the Standards of Ethical Conduct for Management Accountants. The exhibit describes the overall obligation of management accountants to maintain the highest standards. Then it lists responsibilities regarding competence, confidentiality, integrity, and objectivity.

Our economic system is based on an orderly flow of honest information for decision making. Accountants are a vital link in the chain of information. What value does accounting information have for decision makers if that information has not been accumulated, processed, and presented honestly? Chapter 28 explores ethical issues in more detail, particularly issues regarding falsifying performance reports.

[5] A key objective of this program is to establish management accounting as a recognized profession. For more information about the CMA program and other programs, see Appendix C at the end of the book.

[6] S. Landekich, *Corporate Codes of Conduct* (Montvale, NJ: National Association of Accountants, 1989).

PROBLEMS FOR SELF-STUDY

(Try to solve these problems before examining the solutions that follow.)

PROBLEM 1

Reexamine the basic elements of a management control system in Exhibit 1-4, p. 7. Describe how the major accounting tools mesh with the ideas of planning and control.

SOLUTION 1

Planning entails choosing goals, predicting potential results, and deciding how to attain goals. The budget is a major accounting tool that helps managers plan. Control entails action, which is the second left-hand box in Exhibit 1-2. It also includes performance evaluation, often using an accounting report that compares budgets with actual results. This report provides feedback, which is the essence of control.

PROBLEM 2

Answer the following questions. The Hewlett-Packard organization charts in this chapter (Exhibits 1-5 and 1-6, pp. 10 and 12) will help answer several of them.

a. Do the following have line or staff authority over the assembly manager: maintenance manager, computer business vice president, plant manager, purchasing manager, storekeeper, personnel vice president, cost accounting manager?
b. What is the general role of the service departments in an organization? How are they distinguished from operating or production departments?
c. Does the controller have line or staff authority over cost accountants? Over accounts-receivable clerks?
d. The chapter mentioned that the Yoplait Company accounting organization has special positions such as manager—marketing and sales analysis. Why are these special positions demanded by management?

SOLUTION 2

a. The only executives having line authority over the assembly manager are the computer business vice president and the plant manager.
b. A typical company's major purpose is to produce and sell goods or services. Unless a department is directly concerned with producing or selling, it is called a service or support department. Service departments exist only to help the production and sales departments with their major tasks: the efficient production and sale of goods or services.
c. The controller has line authority over all members of his or her own department, shown in the controller's organization chart (Exhibit 1-6).
d. The chief financial officer and the controller should be concerned with all three accounting duties: scorekeeping, attention directing, and problem solving. However, there is a perpetual danger that day-to-day pressures will emphasize scorekeeping. Yoplait uses special accounting positions to emphasize the usefulness of other duties besides scorekeeping. As Yoplait illustrates, accountants and managers should constantly see that attention directing and problem solving are also stressed. Otherwise the major management benefits of an accounting system may be lost.

SUMMARY

In its fullest sense, management accounting is well named. It ties management with accounting. Managers are the customers of the management accountant. To maximize their value, accountants must focus on the managers' concerns and attitudes as much as on the technical aspects of accounting measurement.

This chapter stressed the interrelationship of accounting information and management decisions. The first major part of the chapter provided a conceptual overview of this interrelationship. The second major part described how accountants fit into typical organizational settings.

The cost-benefit approach is a major theme of this book. In a most fundamental sense, the question of what accounting system to buy must focus on how different systems would affect collective decisions (and resulting benefits) and at what costs.

Many readers tend to skim the first chapter. After all, it contains neither financial statements nor discussions of various cost terms. Nevertheless, the chapter sets an important tone.

1. Cost accounting is vital to management control, which entails human control of other human beings. Cost accounting is far more than merely a technical subject. It often has an enormous impact on human behavior.
2. Cost accounting systems come in various shapes and sizes. ***Managers buy one system or another, depending on perceptions of relative costs and benefits.***
3. Cost accounting systems are usually the responsibility of the controller, a staff officer. The controller must worry about the system's human and technical aspects.
4. Both external and internal accountants should adhere to standards of ethical conduct.

TERMS TO LEARN

Each chapter will include this section. Like all technical subjects, accounting contains many terms with precise meanings. To learn cost accounting with relative ease, pin down the definitions of new terms when you initially encounter them.

Before proceeding to the assignment material or to the next chapter, be sure you understand the following words or terms. Their meaning is explained in the chapter and also in the Glossary at the end of this book.

budget *(p. 5)* Certified Management Accountant (CMA) *(15)* control *(5)* controller *(11)* cost accounting *(4)* cost-benefit approach *(8)* cost management *(4)* environment *(6)* feedback *(7)* financial accounting *(4)* functional authority *(10)* internal accounting *(4)* line authority *(10)* management accounting *(4)* management by exception *(5)* management control system *(6)* National Association of Accountants (NAA) *(15)* performance reports *(5)* planning *(5)* process *(6)* staff authority *(10)* treasurer *(11)* variances *(5)*

ASSIGNMENT MATERIAL

QUESTIONS

1-1 The accounting system should provide information for three broad purposes. Describe the three purposes.

1-2 Distinguish between *financial accounting* and *management accounting*.

1-3 Explain the meaning of the letters *NAA* and *CMA* as used in accounting.

1-4 "Accounting systems by themselves are not cost management." Do you agree? Explain.

1-5 "Cost accounting is part of management accounting, plus a part of financial accounting." Explain.

1-6 Define *planning*. Distinguish it from *control*.

1-7 "Planning is really more vital than control." Do you agree? Why?

1-8 Identify the basic elements of a management control system.

1-9 Give at least three examples of the environment when it is regarded as a basic element of management control systems.

1-10 Management control has been compared to a room thermostat. What is the major distinction between them?

1-11 Feedback may be used for a variety of purposes. Identify at least five.

1-12 As a new controller, reply to this comment by a factory superintendent: "As I see it, our accountants may be needed to keep records for stockholders and Uncle Sam—but I don't want them sticking their noses in my day-to-day operations. I do the best I know how. No pencil-pusher knows enough about my responsibilities to be of any use to me."

1-13 "We need to record replacement costs because they are more accurate approximations of economic reality." How would an advocate of the cost-benefit approach react to this statement?

1-14 Which two major executives usually report to the vice president–finance?

1-15 "The controller is both a line and a staff executive." Do you agree? Why?

1-16 "The modern concept of controllership maintains that the controller *does* control in a special sense." Explain.

1-17 What are some common causes of friction between line and staff executives?

1-18 How is cost accounting related to the concept of controllership?

1-19 Distinguish among line, staff, and functional authorities.

1-20 "The modern controller plays a two-count role in organizations." Explain.

EXERCISES AND PROBLEMS

1-21 Elements of control system. Examine Exhibit 1-4. Consider the operations of a manufacturing department. The department manager is the superior, and the supervisor is the subordinate. Give at least one example of each of the elements in the exhibit: operating process, environment, performance, and feedback.

1-22 Uses of feedback. A separate section in the chapter, page 7, identified six uses of feedback and provided an example of each:

1. Changing goals
2. Searching for alternative means
3. Changing methods for making decisions
4. Making predictions
5. Changing the operating process
6. Changing the measuring and rewarding of performance

Match the numbers with the appropriate letters:

A. The California State University system adopts a WATS (Wide Area Telephone Service) method for making long-distance telephone calls.
B. Sales commissions are to be based on gross profit instead of total revenue.
C. The Ford Motor Company adjusts its elaborate way of forecasting demand for its cars by including the effects of expected changes in the price of crude oil.
D. The hiring of new sales personnel will include an additional step: an interview and evaluation by the company psychiatrist.
E. Quality inspectors at General Motors are now being used in the middle of the process in addition to the end of the process.
F. Procter & Gamble enters the telecommunications industry.

1-23 Role of the accountant in the organization: line and staff functions.

1. Of the following, who has line authority over a budgetary accountant: manager of accounting for current planning and control; manager of general accounting; controller; storekeeper; manufacturing vice president; president; production-control manager?
2. Of the following, who has line authority over an assembler: stamping manager; assembly manager; plant manager; production-control manager; storekeeper; manufacturing vice president; engineering vice president; president; controller; budgetary accountant; cost-record clerk?

1-24 Draw an organization chart. Draw an organization chart for a company that has the following positions:

Vice president, controller and treasurer
Chief product designer
Receiving and stores superintendent
Branch sales manager
Production superintendent
Chief of finished stockroom
Shipping-room head
Chief of cost accumulation
Maintenance superintendent
Foundry superintendent
Head of job evaluation
Vice president, personnel
Head of general accounting
Budget director
Tool-room superintendent
Chief purchasing agent
Head of cost analysis
Materials inspection superintendent
Employment manager
Welding and assembly superintendent
Machining superintendent
Vice president, manufacturing
Finishing-department superintendent
Vice president, chief engineer
Stamping superintendent
Head of research
President
Head of production control
Vice president, marketing

1-25 Scorekeeping, attention directing, and problem solving. (Alternate is 1-26.) For each of the activities listed below, identify the *major function* (scorekeeping, attention directing, or problem solving).

1. Preparing a monthly statement of Australian sales for the IBM marketing vice president.
2. Interpreting variances on the Stanford University purchasing department's performance report.
3. Preparing a schedule of depreciation for forklift trucks in the receiving department of a Hewlett-Packard factory in Scotland.
4. Analyzing, for a Mitsubishi international manufacturing manager, the desirability of having some auto parts made in Korea.
5. Interpreting why a Birmingham foundry did not adhere to its production schedule.
6. Explaining the stamping department's performance report.
7. Preparing, for the manager of production control of a U.S. Steel plant, a cost comparison of two computerized manufacturing control systems.
8. Preparing a scrap report for the finishing department of a Toyota parts factory.
9. Preparing the budget for the maintenance department of Mount Sinai Hospital.
10. Analyzing, for a General Motors production superintendent, the impact on costs of some new drill presses.

1-26 Scorekeeping, attention directing, and problem solving. (Alternate is 1-25.) For each of the following, identify the major function the accountant is performing—that is, scorekeeping, attention directing, or problem solving.

1. Interpreting variances on a machining manager's performance report at a Nissan plant.
2. Preparing the budget for research and development at a DuPont division.
3. Preparing adjusting journal entries for depreciation on the personnel manager's office equipment at Citibank.
4. Preparing a customer's monthly statement for a Sears store.
5. Processing the weekly payroll for the Harvard University maintenance department.
6. Explaining the welding manager's performance report at a Chrysler factory.
7. Analyzing the costs of several different ways to blend raw materials in the foundry of a General Electric factory.
8. Tallying sales, by branches, for the sales vice president of Unilever.
9. Analyzing, for the General Motors president, the impact of a contemplated new product on net income.
10. Interpreting why a branch did not meet its IBM sales quota.

1-27 Financial and management accounting. David Colhane, an able electrical engineer, was informed that he was going to be promoted to assistant factory manager. David was elated but uneasy. In particular, his knowledge of accounting was sparse. He had taken one course in "financial" accounting but had not been exposed to the "management" accounting that his superiors found helpful.

Colhane planned to enroll in a management accounting course as soon as possible. Meanwhile, he asked Susan Hansley, an assistant controller, to state three or four of the principal distinctions between financial and management accounting, including some concrete examples.

As the assistant controller, prepare a written response to Colhane.

1-28 Cost accounting in nonprofit institutions. The bulk of the revenues of U.S. hospitals do not come directly from patients. Instead the revenues come through third parties such as insurance companies and governmental agencies. Until the early 1980s, these payments were based on the hospital's costs of serving patients. However, such payments are now based on flat fees for specified services. For example, the hospital might receive $4,000 for an appendectomy or $19,000 for heart surgery—no more, no less.

Required
Would the change in the method of payment change the cost accounting practices of hospitals? Explain.

1-29 Responsibility for analysis of performance. Susan Phillipson is the new controller of a multinational company that has just overhauled its organization structure. The company is now decentralized. Each division is under an operating vice president who, within wide limits, has responsibilities and authority to run the division like a separate company.

Phillipson has a number of bright staff members, one of whom, Bob Garrett, is in charge of a newly created performance-analysis staff. Garrett and staff members prepare monthly divisional performance reports for the company president. These reports are divisional income statements, showing budgeted performance and actual performance, and are accompanied by detailed written explanations and appraisals of variances. Each of Garrett's staff members had a major responsibility for analyzing one division; each consulted with divisional line and staff executives and became generally acquainted with the division's operations.

After a few months, Bill Whisler, vice president in charge of Division C, stormed into the controller's office. The gist of his complaint follows:

"Your staff is trying to take over part of my responsibilities. They come in, snoop around, ask hundreds of questions, and take up plenty of our time. It's up to me, not you and your detectives, to analyze and explain my division's performance to central headquarters. If you don't stop trying to grab my responsibilities, I'll raise the whole issue with the president."

Required

1. What events or relationships may have led to Whisler's outburst?
2. As Phillipson, how would you answer Whisler's contentions?
3. What alternative actions can Phillipson take to improve future relationships?

1-30 Accountant's role in planning and control. Dick Victor has been president of Sampson Company, a multinational textile company, for ten months. The company has an industry reputation as being conservative and having average profitability. Previously, Victor was associated with a very successful company that had a heavily formalized accounting system, with elaborate budgets and effectively used performance reports.

Victor is contemplating the installation of a formal budgetary program. To signify its importance, he wants to hire a new vice president for planning and control. This person would report directly to Victor and would have complete responsibility for implementing a system for budgeting and reporting performance.

Required

If you were the controller of Sampson Company, how would you react to Victor's proposed move? What alternatives are available to Victor for installing his budgetary program? In general, should all figure specialists report to one master figure expert, who in turn is responsible to the president?

1-31 Professional ethics, quality control.(CMA) FulRange Inc. produces complex printed circuits for stereo amplifiers. The circuits are sold primarily to major component manufacturers, and any production overruns are sold to small manufacturers at a substantial discount. The small manufacturer market segment appears very profitable.

A common product defect that occurs in production is a "drift" caused by failure to maintain precise heat levels during the production process. Rejects from the 100% testing program can be reworked to acceptable levels if the defect is drift. However, in a recent analysis of customer complaints, George Wilson, the cost accountant, and the quality control engineer have ascertained that normal rework does not bring the circuits up to standard. Sampling shows that about one-half of the reworked circuits will fail after extended, high-volume amplifier operation. The incidence of failure in the reworked circuits is projected to be about 10% over one to five years' operation.

Unfortunately, there is no way to determine which reworked circuits will fail because testing will not detect this problem. The rework process could be changed to correct the problem, but the cost-benefit analysis for the suggested change in the rework process indicates that it is not feasible. FulRange's marketing analyst has indicated that this problem will have a significant impact on the company's reputation and customer satisfaction if the problem is not corrected. Consequently, the board of directors would interpret this problem as having serious negative implications on the company's profitability.

Wilson has included the circuit failure and rework problem in his report that has been prepared for the upcoming quarterly meeting of the board of directors. Due to the potential adverse economic impact, Wilson has followed a longstanding practice of highlighting this information.

After reviewing the reports to be presented, the plant manager and his staff were upset and indicated to the controller that he should control his people better. "We can't upset the board with this kind of material. Tell Wilson to tone that down. Maybe we can get it by this meeting and have some time to work on it. People who buy those cheap systems and play them that loud shouldn't expect them to last forever."

The controller called Wilson into his office and said, "George, you'll have to bury this one. The probable failure of reworks can be referred to briefly in the oral presentation, but it should not be mentioned or highlighted in the advance material mailed to the board."

Wilson feels strongly that the board will be misinformed on a potentially serious loss of income if he follows the controller's orders. Wilson discussed the problem with the quality control engineer, who simply remarked, "That's your problem, George."

Required

1. Discuss the ethical considerations that George Wilson should recognize in deciding how to proceed in this matter.
2. Explain what ethical responsibilities should be accepted in this situation by the (a) controller, (b) quality control engineer, and (c) plant manager and staff.
3. What should George Wilson do in this situation? Explain your answer.

1-32 Professional ethics and disclosure of information.(CMA) EraTech Corporation, a developer and distributor of business applications software, has been in business for five years. The company's main products include programs used for list management, billing, and accounting for the mail-order shopping business. EraTech's sales have increased steadily to the current level of $25 million per year, and the company has 250 employees.

Andrea Nolan joined EraTech approximately one year ago as accounting manager. Nolan's duties include supervision of the company's accounting operations and preparation of the company's financial statements. Nolan has noticed that in the past six months EraTech's sales have ceased to rise and have actually declined in the two most recent months. This unexpected downturn has resulted in cash shortages. Compounding these problems, EraTech has had to hold back the introduction of a new product line due to delays in documentation preparation.

EraTech contracts most of its printing requirements to Web Graphic Inc., a small company owned by Ron Borman. Borman has dedicated a major portion of his printing capacity to EraTech's requirements because EraTech's contracts represent approximately 50% of Web Graphic's business. Andrea Nolan has known Borman for many years. As a matter of fact, she learned of EraTech's need for an accounting manager through Borman.

While preparing EraTech's most recent financial statements, Nolan became concerned about the company's ability to maintain steady payments to its suppliers. She estimated that payments to all vendors, normally made within 30 days, could exceed 75 days. Nolan is particularly concerned about payments to Web Graphic. She knows that EraTech has recently placed a large order with Web Graphic for printing the new product documentation, and she knows that Web Graphic will soon be placing an order for the special paper required for EraTech's documentation. Nolan is considering telling Borman about EraTech's cash problems. However, she is aware that a delay in the printing of the documentation would jeopardize EraTech's new product.

Required

1. Describe Andrea Nolan's ethical responsibilities in the situation described above. Refer to specific standards of Exhibit 1-7, Standards of Ethical Conduct for Management Accountants, to support your answer.
2. Without prejudice to your answer to requirement 1 above, assume that Andrea Nolan learns that Ron Borman of Web Graphic has decided to postpone the special paper order required for EraTech's printing job. Nolan believes Borman must have heard rumors about EraTech's financial problems from some other source because she has not talked to Borman. Should Nolan tell the appropriate EraTech officials that Borman has postponed the paper order? Explain your answer using Exhibit 1-7 for support.

3. Without prejudice to your answers to requirements 1 and 2 above, assume that Ron Borman has decided to postpone the special paper order because he has learned of EraTech's financial problems from some source other than Nolan. In addition, Nolan realizes that Jim Grason, EraTech's purchasing manager, knows of her friendship with Borman. Now Nolan is concerned that Grason may suspect she told Borman of EraTech's financial problems when Grason finds out that Borman postponed the order. Describe the steps that Andrea Nolan should take to resolve this situation. Use Exhibit 1-7 to support your answer.

1-33 Professional ethics and reporting divisional performance. Susan Miller is division controller and George Maloney is division manager of the Ramses Shoe Company. Miller has line responsibility to Maloney, but she also has staff responsibility to the company controller.

Maloney is under severe pressures to achieve budgeted divisional profit for the year. He has asked Miller to book $200,000 of sales on December 31. The customers' orders are firm, but the shoes are still in process. They will be shipped on or about January 4. Maloney said to Miller, "The key event is getting the sales order, not the shipping of the shoes. You should support me, not obstruct my reaching divisional goals."

Required

1. Describe Miller's ethical responsibilities.
2. What should Miller do if Maloney gives her a direct order to book the sales?

CHAPTER 2

To guide decisions, managers need to know the cost of something. This cost object may be an activity, a product, a service, a department, a program, or a project like the construction of a skyscraper.

An Introduction to Cost Terms and Purposes

Costs in general

Direct and indirect costs

Cost drivers

Variable costs and fixed costs

Unit costs and total costs

Manufacturing costs

Costs as assets and expenses

Some cost accounting language

The many meanings of product costs

Classifications of costs

Learning Objectives

When you have finished studying this chapter, you should be able to

1. Define and explain: *cost objects*, *direct costs*, and *indirect costs*
2. Explain the relationships of cost drivers, variable costs, and fixed costs
3. Explain the uses and limitations of unit costs
4. Define and identify each of the three categories of a manufactured product's cost
5. Construct model income statements of a manufacturing company
6. Differentiate between inventoriable costs and period costs
7. Understand some troublesome cost accounting language
8. Provide three different meanings of product costs

We usually do not decide to buy a commodity without some idea of its makeup or characteristics. Similarly, if we know the composition and uses of cost data and systems, we can decide what cost data and what system a manager should "buy" in a particular situation. To acquire this knowledge, we need to speak the accountant's language. What does *cost* mean? We will quickly see that there are different costs for different purposes.

In this chapter you will learn some basic terminology, the jargon that every technical subject seems to have. The chapter contains several widely recognized cost concepts and terms. They are sufficient to demonstrate the multiple purposes of cost accounting systems that we will stress throughout the book. Many other types of costs exist, but we postpone our discussion of them to later chapters.

COSTS IN GENERAL

Cost Objects

Objective 1

Define and explain: *cost objects*, *direct costs*, and *indirect costs*

Accountants usually define **cost** as a resource sacrificed or forgone to achieve a specific objective. For now, consider costs as being measured in the conventional accounting way, as monetary units (for example, dollars) that must be paid for goods and services.

To guide decisions, managers want data pertaining to a variety of purposes. They want the cost of *something*. We call this something a **cost object** and define it as any activity or item for which a separate measurement of costs is desired.[1] A synonym is **cost objective**.

The cost object is a key feature of management accounting. It may be an activity or operation in which resources are consumed or received (repairing automobiles, responding to inquiries for information, testing circuit boards, or reconciling bank accounts). The cost object may be a product or service (manufacturing a personal computer, renting a room, or flying a passenger from Los Angeles to London). The cost object may be a project (constructing a house, building a ship, or designing a

[1]G. Staubus, *Activity Costing and Input-Output Accounting* (Homewood, IL: Irwin), p. 1, stresses that in essence we are determining the cost of an activity or action:

"Costing is the process of determining the cost of doing something, e.g., the cost of manufacturing an article, rendering a service, or performing a function. . . . We may, however, find ourselves speaking of the cost of a product as an abbreviation for the cost of acquiring or manufacturing that product. . . ."

missile). The cost object may be a department (legal departments, shipping departments, or design departments). The cost object may be a program (a drug control program or an athletic program). To summarize, the following are examples of cost objects:

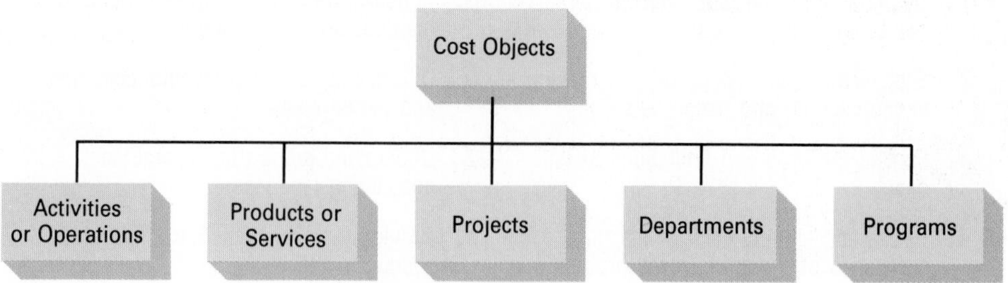

By itself, the term *cost* is meaningless. Cost measurement must be tied to at least one cost object. Of course, the same cost may pertain to many cost objects simultaneously. For instance, the cost of materials may become part of the cost of a product and part of the cost of running a department.

Cost Accumulation and Allocation

A cost system typically accounts for costs in two broad stages: (1) it *accumulates* costs by some "natural" classification such as raw materials used, fuel consumed, or advertising placed; and then (2) it *allocates* (traces) these costs to cost objects. **Cost accumulation** is the collection of cost data in some organized way through an accounting system. **Cost allocation** is a general term that refers to identifying accumulated costs with or tracing accumulated costs to cost objects such as departments, activities, or products.

Cost objects are chosen not for their own sake but to help decision making. The most economically feasible approach to the design of a cost system is typically to assume some common classes of decisions (for example, inventory control and labor control) and to choose cost objects (for example, products and departments) that relate to those decisions. Nearly all systems at least accumulate **actual costs**, which are amounts determined on the basis of costs incurred (historical costs), as distinguished from predicted or forecasted costs. The relationships are shown in Exhibit 2-1.

EXHIBIT 2-1
Relationships of Cost Accumulation to Cost Objects

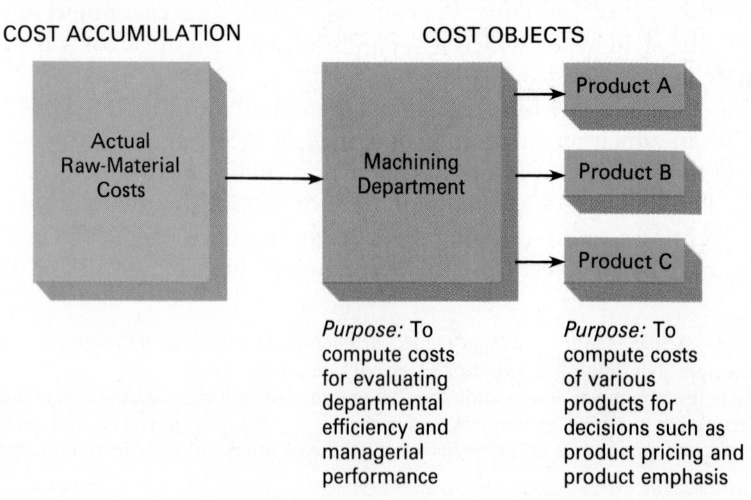

Avoid the widely held belief that cost accounting's sole purpose is to measure income and inventories. The best systems collect data in a form suitable for a variety of purposes. These systems provide a reliable basis for predicting the economic consequences of various decisions such as the following:

- Which products should we continue to make? Discontinue?
- Should we manufacture a product component or should we acquire it from another company?
- What prices should we charge?
- Should we buy the proposed equipment?
- Should we change our manufacturing methods?
- Should we promote a particular manager?
- Should we expand a particular department?

DIRECT AND INDIRECT COSTS

Classification Depends on Cost Objects

A major question concerning costs in both manufacturing and nonmanufacturing functions is whether the costs have a direct or an indirect relationship to a particular cost object:

- **Direct costs**: cost that can be identified specifically with or traced to a given cost object in an economically feasible way
- **Indirect costs**: costs that cannot be identified specifically with or traced to a given cost object in an economically feasible way

"Economically feasible" means "cost effective." Managers do not want cost accounting to be too expensive in relation to expected benefits. The costs of tracing relatively inexpensive items may exceed the benefit of having the resulting information. For example, it may be economically feasible to trace specifically the exact cost of steel and fabric (direct costs) to a specific lot of desk chairs, but it may be economically infeasible to trace specifically the exact cost of rivets or thread (indirect costs) to the chairs.

Managers prefer to classify costs as direct rather than indirect. Why? Because they have greater confidence in the accuracy of the reported costs of products and services. When is a cost considered direct or indirect? The answer depends on the particular cost object. Consider a supervisor's salary in a maintenance department of a telephone company. If the cost object is the department, the supervisor's salary is a direct cost. In contrast, if the cost object is a service (the "product" of the company), such as a telephone call, the supervisor's salary is an indirect cost. In general, many more costs are direct regarding a department as a cost object than regarding a service (a telephone call) or a physical product (a telephone) as a cost object.

This book will explore many problems of how to relate costs to cost objects. For now, be aware that a particular cost may be both direct and indirect. How? The direct-indirect classification depends on the choice of the cost object. As Exhibit 2-1 indicates, the same ***costs are inevitably allocated to more than one cost object.*** For example, managers want to know both the costs of running departments and the costs of products and services. As we have just seen, a supervisor's salary is both direct (with respect to his or her department) and indirect (with respect to the department's individual products or services).[2]

---

[2]Some writers confine the use of the term *allocation* to the assignment of indirect costs only. In contrast, we use allocation to encompass the assignment of both direct and indirect costs.

Accuracy in Determining Costs

Before proceeding, ponder how the direct and indirect costs generally relate to various cost objects. The cost object may be a broadly defined activity (total operation of an airplane repair facility) or a narrowly defined activity (the painting operation of an airplane repair facility). It may be a broadly defined project (designing a new automobile) or a narrowly defined project (designing a knob for a radio in a new automobile). It may be a broadly defined group of products (mass-produced women's suit jackets) or a narrowly defined product (an especially tailored woman's suit jacket).

A useful rule of thumb is that the broader the definition of the cost object, the higher the proportion of its total costs are its direct costs—and the more confidence management has in the accuracy of the resulting cost amounts. The narrower the definition of the cost object, the lower the proportion of its total costs are its direct costs—the less confidence management has in the accuracy of resulting cost amounts.

Consider an example of how direct product costs can differ, depending on what types of costs are classified as direct or indirect. One public accounting firm might find it worthwhile to trace the following as direct costs of an audit engagement (the cost object): compensation of its professional staff and payment for secretarial work based on time specifically devoted to the engagement, and costs for relevant computer time, photocopying, telephone use, and postage. A second firm might classify the compensation of its professional staff as its only direct costs. All other costs are spread broadly over all the engagements as a percentage of direct costs. Thus, the composition of direct costs of an identical product or service may differ between firms. In turn, the total costs of the identical product may also differ.

In this example, accountants would consider the first firm's cost of the audit engagement as more accurate. Why? Because the firm traced many support costs directly, and the second firm lumped the same support costs as indirect costs. The second firm relied on a cruder, less-accurate method for assigning costs.

COST DRIVERS

Objective 2

Explain the relationships of cost drivers, variable costs, and fixed costs

A **cost driver** is any factor whose change causes a change in the total cost of a related cost object. Drivers are causal factors whose effects are increases in total costs. There are many possible cost drivers. For example, in a factory setting, the total cost of materials used may be driven not only by the production volume but also by the quality of the materials, the skills of the workers, the number of parts in a finished product, and the condition of the applicable machines.

EXHIBIT 2-2
Examples of Cost Drivers

Total Costs of Activities	Cost Drivers
Product design	Number of products, number of parts
Engineering	Number of parts, number of engineering-change orders
Purchasing of materials	Number of purchase orders, number of suppliers, negotiating time, expediting time
Manufacturing	Production volume in units of product, number of setups, number of parts
Distribution	Type of transportation, miles driven, weight, number of stops, density of traffic, speed

We can measure cost drivers in miles traveled, production volume, hours worked, patients treated, payroll checks processed, lines typed, sales dollars, and so on—whatever may be appropriate to the related cost object. Exhibit 2-2 presents additional examples. We discuss various cost drivers much more in later chapters, particularly in Chapters 4, 5, 10, 14, 25, and 29. For ease of exposition now and for demonstrating some useful simplifications, we will concentrate on the effects of volume measurements as cost drivers. Examples of measures of volume include units produced, hours worked, tons used, and sales dollars.

VARIABLE COSTS AND FIXED COSTS

Let us now consider two basic types of costs—variable costs and fixed costs. We may define each in terms of whether its total cost changes in response to changes in a related cost driver. A **variable cost** is a cost that does change in total in direct proportion to changes of a cost driver. A **fixed cost** is a cost that does not change in total despite changes of a cost driver. Consider two illustrations.

1. If General Motors buys one type of special clamp at $1 for each of its Buick cars, then the total cost of clamps should be $1 times the number of cars produced. This is an example of a variable cost, a cost that is unchanged per unit of cost driver but changes *in total* in direct proportion to changes in the cost driver. Examples include most materials and parts, many types of assembly labor, sales commissions, and some factory supplies.

   Variable-cost behavior can be plotted graphically. Exhibit 2-3 shows the relationship between direct material costs and units produced; Exhibit 2-4 shows the relationship between sales commissions and sales dollars.

2. General Motors may incur $100 million in a given year for a factory's property taxes, executive salaries, rent, and insurance. These are examples of fixed costs, costs that are unchanged *in total* over a wide range of the cost driver during a given time span but become progressively smaller on a *per unit* basis as the cost driver increases.

EXHIBIT 2-3
Direct Material Costs—
$1.00 Per Unit

EXHIBIT 2-4
Sales Commissions—
10% of Sales

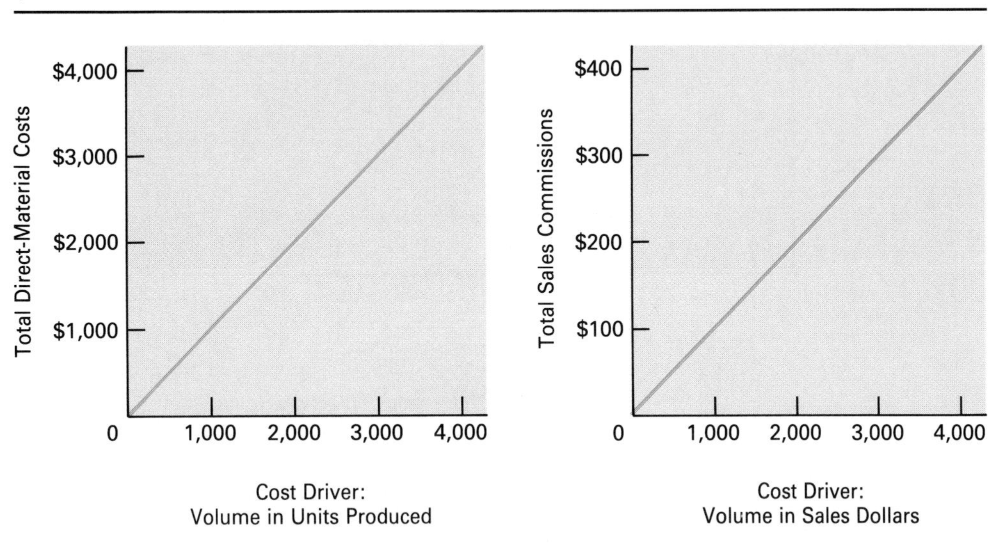

Major Assumptions

The definitions of variable costs and fixed costs have important underlying assumptions:

1. The cost object must be specified. Examples are activities, products, services, projects, departments, territories, and whole companies.
2. The time span must be specified. Examples are months, quarters, years, and product life cycles.
3. Costs are linear. That is, when plotted on ordinary graph paper, a total cost in relation to the cost driver will appear as an unbroken straight line.
4. For the time being, all costs are either variable or fixed. In practice, of course, classification is difficult and nearly always necessitates some simplifying assumptions.
5. There is only one cost driver. The influences of other possible cost drivers on the total cost are held constant or deemed to be insignificant. Volume, often expressed in measures of units produced or sold, is often specified as the lone cost driver.
6. The relevant range of fluctuations in the cost driver must be specified.

We now define and discuss the relevant range.

Relevant Range

A **relevant range** is the band of the cost driver in which a specific relationship between cost (revenue) and a cost (revenue) driver is valid. A fixed cost is fixed only in relation to a given relevant volume range (usually large) and a given time (usually a particular budget period). For example, Exhibit 2-5 shows that a fixed-cost level of $600,000 applies to a relevant range of 30,000 to 95,000 machine-hours—the cost driver—per year. The exhibit also shows that operations on either side of the relevant range would result in different fixed costs. For example, if volume levels fall beneath 30,000 machine-hours, fixed costs would be reduced drastically. With output in the range from shutdown (zero hours) to 30,000 machine-hours, service and executive personnel would likely be laid off. An increase in volume

EXHIBIT 2-5

Total Annual Fixed Costs—Conceptual Analysis

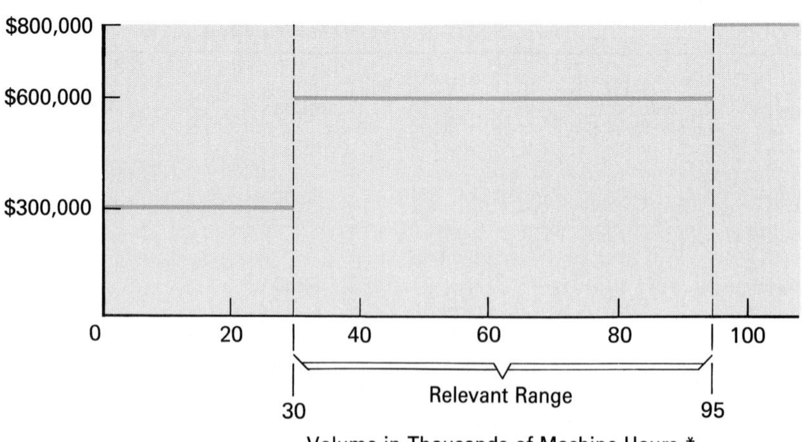

* *$600,000 level between 30,000 and 95,000 hours.*
$800,000 level in excess of 95,000 hours: hiring of additional supervision.
$300,000 level from shutdown (zero hours) to 30,000 hours: laying off of supervision.

above 95,000 machine-hours would increase fixed costs. The business might hire additional personnel to help the increased operations. Fixed costs may differ from one year to the next wholly because of changes in items other than the volume cost driver, such as rent terms, salary levels, and property tax rates.

Exhibit 2-6 illustrates how fixed costs are usually graphed in practice. The likelihood of volume being outside a particular relevant range is usually slight, so $600,000 becomes the fixed-cost level. That is, the three-level refinement in Exhibit 2-5 is not usually graphed. Instead, the $600,000 total is extended back to the zero-volume axis. Such plotting causes no particular harm as long as operating decisions are limited to a relevant range.

The basic assumption of a relevant range also applies to variable costs. That is, outside a relevant range, some variable costs, such as raw materials used, fuel consumed, or labor used, may behave differently per unit of volume. For example, more raw materials and labor-hours may be wasted as techniques are learned at low-volume levels or when crowding or fatigue occurs at high-volume levels. Exhibit 2-7 shows how variable costs may be affected outside a relevant range. The top graph shows the likely behavior of total variable costs for this example. The bottom graph shows how the same costs are graphed in practice. Again, as in the case of fixed costs, such plotting does no harm as long as decisions are confined to a relevant range.

Relationships of Types of Costs

We have introduced two major classifications of costs: direct-indirect and variable-fixed. Costs may simultaneously be variable and direct, fixed and indirect, or some other combination. Examples follow:

Traceability of Cost to Cost Object

Cost Behavior Patterns		Direct	Indirect
	Variable	Tires used in assembly of an automobile where the cost object is each individual automobile assembled	Power costs where power is metered only to the assembly department and the cost object is each individual automobile assembled
	Fixed	Marketing department's supervisor's salary where the cost object is the marketing department	Board of directors' fees where the cost object is the marketing department

UNIT COSTS AND TOTAL COSTS

Using Averages and Unit Costs

Objective 3

Explain the uses and limitations of unit costs

The preceding section concentrated on the behavior patterns of total costs in relation to volume levels. Generally, the decision maker should take a straightforward analytical approach by thinking in terms of total costs rather than average costs. As we will see momentarily, average costs must be interpreted cautiously. Nevertheless, their use is essential in many decision contexts. For example, the chairman of the social committee of a fraternity may be trying to decide whether to hire a musical group for a forthcoming party. The total fee may be predicted with certainty at $1,000. This knowledge is helpful for the decision, but it may not be enough.

EXHIBIT 2-6
Total Annual Fixed Costs as Plotted in Practice

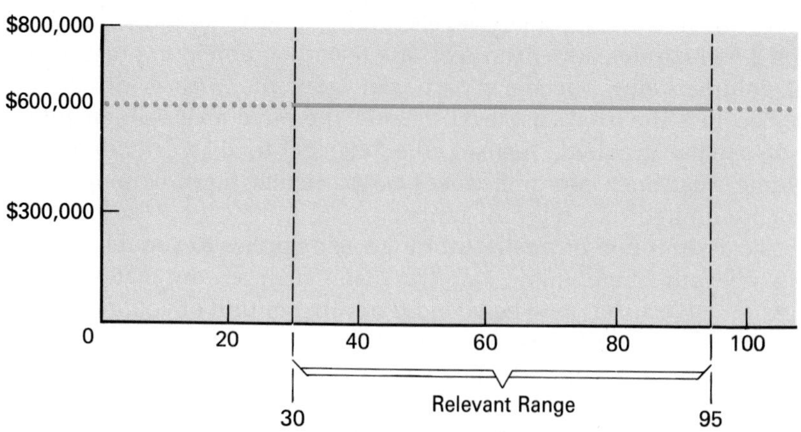

* *$600,000 level between 30,000 and 95,000 hours.*
$800,000 level in excess of 95,000 hours: hiring of additional supervision.
$300,000 level from shutdown (zero hours) to 30,000 hours: laying off of supervision.

EXHIBIT 2-7
Total Variable Costs and a Relevant Range

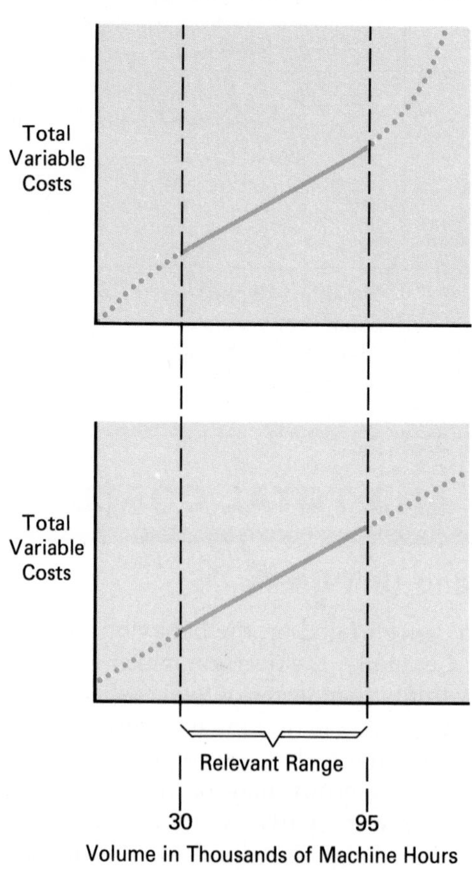

Before a decision can be reached, the chairman must predict both the total cost and the probable number of persons who will attend. Without knowledge of both, he cannot decide intelligently on a possible admission price or even on whether to have a party at all. So he computes an average cost by dividing the total cost by the expected number of persons who will attend. If 1,000 people attend, the average cost is $1 per person; if 100 attend, the average cost soars to $10.

Unless the total cost is averaged (that is, "unitized" with respect to the cost object), the $1,000 cost is difficult to interpret; the average cost (or unit cost) combines the total cost and the number of persons in a handy, communicative way.

Meaning of Average Cost

An **average cost** is computed by dividing some total cost (the numerator) by some denominator. Often the denominator is a measure of volume that is most closely related to the total cost incurred. Examples of denominators include units of product, hours of service, student credit hours, pounds handled in a shipping department, and the number of invoices processed or lines billed in a billing department. Generally, average costs are expressed in terms most informative to the people incurring the costs.

The average cost of making a finished good is frequently computed by accumulating manufacturing costs and then dividing the total by the number of units produced. For example:

Total manufacturing costs of 1,000 units (numerator)		$980,000
Divided by the number of units produced (denominator)	÷	10,000
Equals a unit (average) manufacturing cost	=	$98

Suppose 8,000 units are sold and 2,000 units remain in ending inventory. The average-cost (unit-cost) idea helps the assignment of a total cost to various accounts:

Cost of goods sold, 8,000 units × $98 =	$784,000
Ending inventory of manufactured goods, 2,000 units × $98 =	196,000
	$980,000

Use Unit Costs Cautiously

Unit costs are averages, and they must be interpreted with caution. For example, what does it mean to say that the unit cost for the musicians is $1.00 if 1,000 persons attend the fraternity party? In this case of fixed costs, the *total* cost of $1,000 is unaffected by the volume level, the size of the denominator. But the *unit* cost is strictly a function of the size of the denominator; it would be $1,000 per unit if one person attended, $1.00 per unit if 1,000 persons attended, and $.10 per unit if 10,000 persons attended.

In contrast, assume that the musicians agreed to perform for $1.00 per person. Then we would have a variable-cost situation. If one person attended, the *total* cost would be $1.00; if 1,000 attended, $1,000; if 10,000 attended, $10,000.

Note that for decision purposes the fixed cost per unit must be distinguished from the variable cost per unit. A common mistake is to regard all average costs indiscriminately—as if all costs were variable costs. ***Changes in volume will affect* total *variable costs but not* total *fixed costs.*** In our fraternity example, the committee chairman could use the $1.00 unit variable cost to predict the total costs. But using a $1.00 unit fixed cost to predict the total costs would be perilous. His prediction would be correct if, and only if, 1,000 persons attended—total fixed costs would be $1,000 regardless of the number attending. The moral is: ***Average costs are often useful, but they should be interpreted with extreme caution, especially if they are in the form of fixed costs per unit.***

These relationships are summarized below:

Behavior as Volume Changes Within a Relevant Range

	Total Cost	Average Cost per Unit*
Variable cost	Change	No change
Fixed cost	No change	Change†

*For example, product units, passenger-miles, sales dollars.
†When using data for making predictions, think of fixed costs as a total and variable costs as an amount per unit.

MANUFACTURING COSTS

Objective 4

Define and identify each of the three categories of a manufactured product's cost

Components of Costs

Some type of cost accounting applies to any entity, including manufacturing companies, airlines, retail stores, insurance companies, accounting firms, advertising agencies, government units, hospitals, and other organizations, regardless of whether they operate for a profit goal.

Historically, accounting techniques for planning and control arose in conjunction with manufacturing rather than nonmanufacturing. Why? Because the measurement problems were less imposing and environmental factors such as economic conditions, customer reactions, and competitor activity were generally less influential in manufacturing. However, the basic concepts of planning and control apply equally well to both manufacturing and nonmanufacturing activities.

Consider a company like IBM. Many people would describe IBM as a manufacturing company. More accurately, it should be described as a company that does manufacturing. Why? Because manufacturing is only one of its major business functions. Indeed, IBM's marketing costs exceed its manufacturing costs. The functions in which its main costs arise are:

- Research and development
- Product design
- Manufacturing
- Distribution (logistics)
- Marketing
- Customer service
- General administration for the entire company

A nonmanufacturer's costs are probably a subset of this list; so by focusing on a company that does manufacturing—as we do in this chapter—we can develop a completely general framework of cost accounting for ready application to any organization. In later chapters we will consider nonmanufacturing areas in detail.

Manufacturing is the transformation of materials into other goods through the use of labor and factory facilities. **Merchandising** is the marketing of goods without changing their basic form. For example, assume that Jane Nentlaw wants to make shampoo and sell it directly to retailers. She may purchase a factory and equipment, buy certain oils and containers, hire some workers, and manufacture thousands of units of finished product. This is her manufacturing function. But to persuade retailers to buy her shampoo, Nentlaw will have to convince the ultimate consumer that this product is desirable. This means advertising, including the development of a sales appeal, the selection of a brand name, the choice of media, and so forth. To maximize her success, Nentlaw must effectively manage both manufacturing and merchandising functions.

Instead of handling the merchandising herself, however, Nentlaw may sell her shampoo to a merchandiser (Crump's) that will resell her shampoo to the final consumer. The difference between manufacturer and merchandiser shows up in the income statement. Note in Exhibit 2-8 that Nentlaw's income statement includes a cost of goods manufactured line, which presents costs arising from the manufacture of shampoo. The details of the cost of goods manufactured appear in the separate supporting schedule. Crump's income statement has no cost of goods manufactured line. Instead, it includes a purchases line, which presents costs arising from its buying items—the shampoo from Nentlaw is but one of its purchases—for resale to customers.

Objective 5

Construct model income statements of a manufacturing company

Three Manufacturing Cost Categories

Many companies recognize three major categories of the cost of a manufactured product:

1. **Direct materials costs**. The acquisition costs of all materials that are identified as part of the cost object and that may be traced to the cost object in an economically feasible way. Examples of cost objects are manufactured goods such as sheet steel and subassemblies for an automobile company. Acquisition costs of direct materials include inward delivery charges, sales tax, and custom duties. Direct materials often do *not* include minor items such as glue or tacks. Why? Because the costs of tracing insignificant items do not seem worth the possible benefits of having more accurate product costs. Such items are called *supplies* or *indirect materials* and are classified as part of the indirect manufacturing costs, described below.
2. **Direct labor costs**. The compensation of all labor that can be identified in an economically feasible way with a cost object. Examples in manufacturing are the labor of machine operators and assemblers. **Indirect labor costs** are all factory labor compensation other than direct labor compensation. These are labor costs that are impossible or impractical to trace to a specific product. They are classified as part of the indirect manufacturing costs described below. Examples are wages of janitors and plant guards.
3. **Indirect manufacturing costs**. All manufacturing costs that cannot be identified specifically with or traced to the cost object in an economically feasible way. Other terms describing this category include **factory overhead**, **factory burden**, **manufacturing overhead**, and **manufacturing expenses**. The term *indirect manufacturing costs* is a clearer descriptor than *factory overhead*,[3] but the latter will often be used throughout this book because it is briefer. Examples of factory overhead when products are cost objects include power, supplies, indirect labor, factory rent, insurance, property taxes, depreciation, and factory supervisory compensation.

As mentioned earlier, accountants and managers design their cost accounting systems in light of perceived costs and benefits. How detailed is the tracking of costs? Where the costs of any single category or item become relatively insignificant, separate tracking may no longer be desirable. For example, in highly automated factories direct labor is often less than 5% of total manufacturing costs. Many of these factories no longer track direct labor costs separately as one of the major cost categories.

As an illustration, consider Hewlett-Packard. Several of its plants include all

[3]The term *overhead* is peculiar; its origins are unclear. Some accountants have wondered why such costs are not called "underfoot" rather than "overhead" costs. The answer probably lies in the organization chart. Lower departments ultimately bear all costs, including those coming from over their heads.

EXHIBIT 2-8 *(Place a clip on this page for easy reference.)*

Comparison of Income Statements (in thousands)

Nentlaw (a manufacturer)			**Crump's (a retailer)**		
Income Statement for the Year Ended December 31, 19_2			**Income Statement for the Year Ended December 31, 19_2**		
Sales		$210,000	Sales		$1,500,000
Deduct cost of goods sold:			Deduct cost of goods sold:		
Finished goods, December 31, 19_1	$ 22,000		Merchandise inventory, December 31, 19_1	$ 95,000	
Cost of goods manufactured (see schedule)	104,000		Purchases	1,100,000	
Cost of goods available for sale	$126,000		Cost of goods available for sale	$1,195,000	
Finished goods, December 31, 19_2	18,000		Merchandise inventory, December 31, 19_2	130,000	
Cost of goods sold		108,000	Cost of goods sold		1,065,000
Gross margin (or gross profit)		$102,000	Gross margin (or gross profit)		$ 435,000
Deduct marketing and administrative expenses		80,000	Deduct marketing and administrative expenses		315,000
Operating income*		$ 22,000	Operating income		$ 120,000

Nentlaw

Schedule of Cost of Goods Manufactured†

Direct materials		
Inventory, December 31, 19_1	$11,000	
Purchases of direct materials	73,000	
Cost of direct materials available for use	$84,000	
Inventory, December 31, 19_2	8,000	
Direct materials used		$ 76,000
Direct labor		18,000
Factory overhead (indirect manufacturing costs):		
Indirect labor	$ 4,000	
Supplies	1,000	
Heat, light, and power	1,500	
Depreciation—plant building	1,500	
Depreciation—equipment	2,500	
Miscellaneous	500	11,000
Manufacturing costs incurred during 19_2		$105,000
Add work in process inventory, December 31, 19_1		6,000
Total manufacturing costs to account for		$111,000
Deduct work in process inventory, December 31, 19_2		7,000
Cost of goods manufactured† (to Income Statement)		$104,000

*Note that operating income is determined before the deduction of income taxes. Net income is operating income plus other income and minus other expenses and income taxes.

†Note that the term cost of goods manufactured refers to the cost of goods brought to completion (finished) during the year, whether they were started before or during the current year. Some of the manufacturing costs incurred are held back as costs of the ending work in process inventory; similarly, the costs of the beginning work in process inventory become part of the cost of goods manufactured for 19_2. Note too that this schedule can become a Schedule of Cost of Goods Manufactured and Sold simply by including the opening and closing finished goods inventory figures in the supporting schedule rather than directly in the body of the income statement.

labor costs as just another subpart of factory overhead. Toyota uses the same approach in its factory in Kentucky. In these and other cases, conversion costs are equal to factory overhead.

Prime Costs, Conversion Costs, and the Direct Labor Category

Two of the three major categories of the cost of a manufactured product are sometimes combined in cost terminology as follows: **Prime costs** are all direct manufacturing costs. **Conversion costs** are all manufacturing costs other than direct materials costs.

Think about why some factories have stopped considering direct labor as a separate cost category. Direct labor costs have become less and less important in relation to total manufacturing costs. An important point emerges: The direct-indirect distinction between costs depends heavily on the underlying manufacturing processes. Many cost accounting systems have retained the classic threefold cost categories, but other systems use twofold categories:

Threefold Category	Twofold Category
Direct materials	Direct materials
Direct labor	Conversion costs
Factory overhead	

As the twofold category indicates, conversion costs envelop all manufacturing costs other than direct materials. Many companies use the label *conversion costs* for these indirect costs, but other companies use *factory overhead.* In these companies, direct labor has disappeared as a major cost category; instead, it has merely become another element of factory overhead.

To recapitulate, the classic major distinctions among manufacturing product costs are direct materials, direct labor, and factory overhead. However, be aware that various companies may have other major categories. Some may have only two: direct materials and conversion costs. As information technology improves and some indirect costs become more significant, some companies may add categories by no longer classifying selected costs as indirect. For example, power costs might be metered in various specific areas of a factory. Then a company might have the following: direct materials, direct labor, other direct costs (such as the specifically metered power), and factory overhead. In short, the time-honored threefold category still dominates, but not as much as before.

COSTS AS ASSETS AND EXPENSES

Inventoriable Costs

Objective 6

Differentiate between inventoriable costs and period costs

The rules of financial accounting have a major influence on accounting for manufacturing costs. For example, under generally accepted accounting principles, the manufacturing costs of a product are initially regarded as measures of assets. They are **inventoriable costs,** which are all costs of a product that are regarded as an asset for financial reporting under generally accepted accounting principles. Such costs become expenses (in the form of *cost of goods sold*) only when the units in inventory are sold. Such sales may occur in the same accounting period as manufacture or in a subsequent period.

Other costs, often called **period costs**, are regarded as immediate expenses. These costs are always expensed in the same period in which they are incurred; they are not considered inventoriable costs. Examples are research, marketing, and administrative costs.

Exhibit 2-9 helps clarify the differences between inventoriable and period (non-inventoriable) costs. Study the top of the exhibit. A retailer or wholesaler buys goods for resale without changing their basic form. The *only* inventoriable cost is the cost of merchandise. Unsold goods are held as merchandise inventory whose

EXHIBIT 2-9

Relationships of Inventoriable Costs and Period Costs

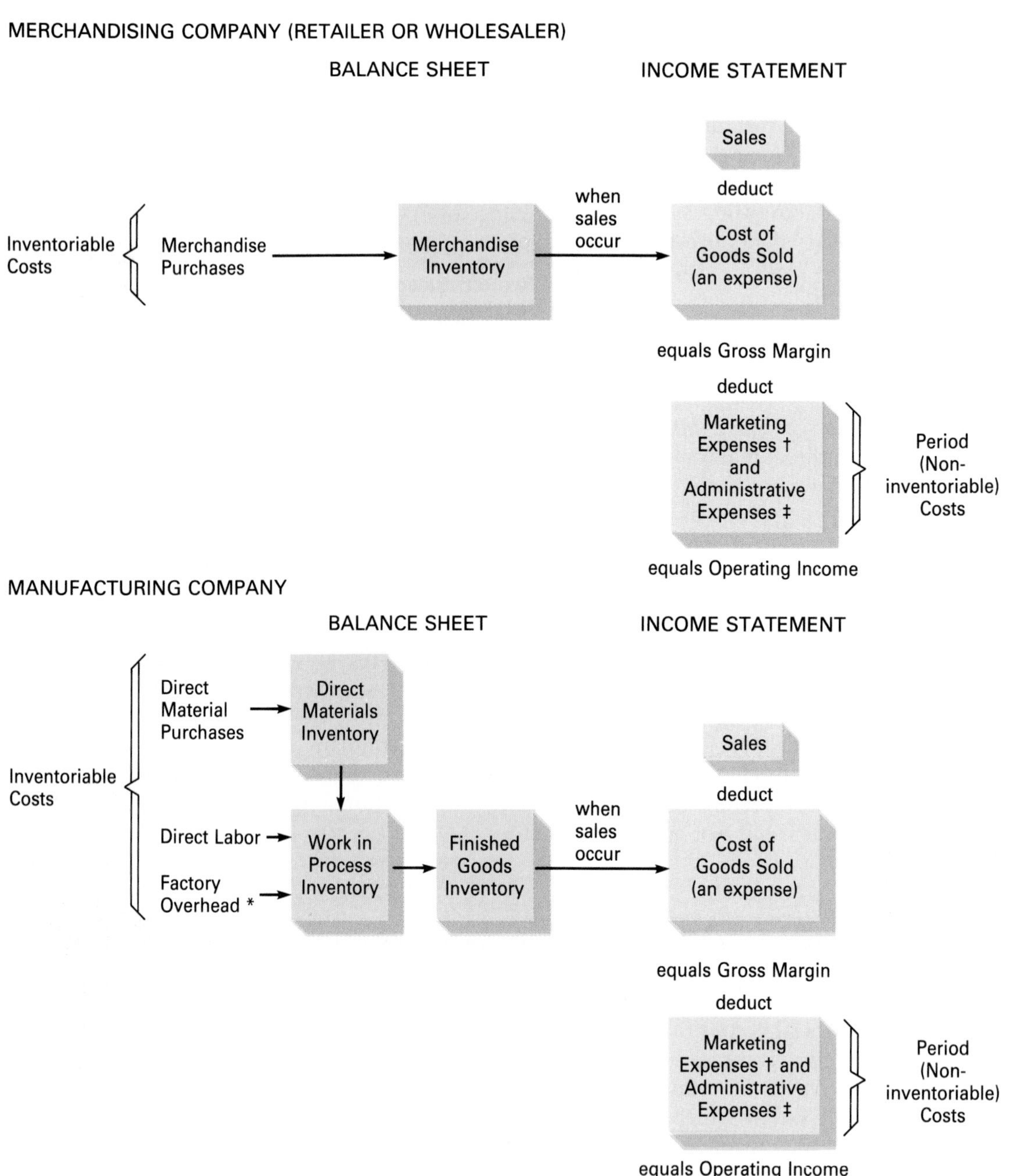

* *Examples: Indirect labor, factory supplies, insurance and depreciation on plant. (Note particularly that where insurance and depreciation relate to the manufacturing function, they are inventoriable; but where they relate to marketing and administration, they are not inventoriable.)*

† *Examples: Insurance on salespersons' cars, depreciation on salespersons' cars, salespersons' salaries, advertising.*

‡ *Examples: Insurance on corporate headquarters building, depreciation on office equipment, clerical salaries.*

cost is shown as an asset on the balance sheet. As the goods are sold, their costs become expenses in the form of cost of goods sold.

A retailer or wholesaler also has a variety of marketing and administrative expenses, which are the main examples of period costs (noninventoriable costs). In the income statement they are deducted from revenue as expenses without ever having been regarded as part of inventory.

A manufacturer, as the bottom half of Exhibit 2-9 shows, transforms direct materials into salable form with the help of direct labor and factory overhead. All these costs are inventoriable costs because they are assigned to inventory until the goods are sold. As in accounting for retailers and wholesalers, the manufacturer's marketing and administrative expenses are regarded as period costs.

Effect on the Balance Sheet

As Exhibit 2-9 shows, balance sheets of manufacturers and merchandisers differ with respect to inventories. The merchandise inventory account is supplanted in a manufacturing company by three inventory categories, each depicting a stage in the production process:

1. **Direct materials inventory**. Direct materials on hand and awaiting use in the production process.
2. **Work in process inventory**. Also sometimes called **work in progress** or **goods in process**. Goods undergoing the production process but not yet fully completed. Costs include the three major manufacturing costs (direct materials, direct labor, and factory overhead).
3. **Finished goods inventory**. Goods fully completed but not yet sold.

The only essential difference between the structure of the balance sheet of a manufacturer and that of the balance sheet of a retailer or wholesaler would appear in their respective current asset sections (numbers are assumed):

Current Asset Sections of Balance Sheets

Manufacturer			Retailer or Wholesaler	
Cash		$ 4,000	Cash	$ 30,000
Accounts receivable		5,000	Accounts receivable	70,000
Direct materials	$ 3,000			
Work in process	2,000			
Finished goods	12,000			
Total inventories		17,000	Merchandise inventories	100,000
Prepaid expenses		1,000	Prepaid expenses	3,000
Total current assets		$27,000	Total current assets	$203,000

Perpetual and Periodic Inventories

There are two fundamental ways of accounting for inventories: perpetual and periodic. The **perpetual inventory method** requires a continuous record of additions to and reductions in materials, work in process, and finished goods, thus measuring on a continuous basis not only these three inventories but also the cumulative cost of goods sold. Such a record helps managerial control and preparation of interim financial statements. Physical inventory counts are usually taken at least once a year to check on the validity of the clerical records.

Companies using perpetual inventory methods frequently have computer-based information-tracking systems. Consider a manufacturer of television sets. Key component parts may have a bar code that is machine-read into the computer as

EXHIBIT 2-10

Summary Comparison of Periodic and Perpetual Inventory Methods
(Figures from Exhibit 2-8)

Periodic Method		Perpetual Method	
Beginning inventories (by physical count)	$ 22,000	Cost of goods sold (kept on a continuous basis rather than being determined periodically)*	$108,000
Add: Manufacturing costs (direct materials used, direct labor, factory overhead)	104,000		
Cost of goods available for sale	126,000		
Deduct ending inventories (by physical count)	18,000		
Cost of goods sold	$108,000		

*Such a condensed figure does not preclude the presentation of a supplementary schedule showing details of manufacturing costs similar to that in Exhibit 2-8.

they are used on the assembly line. Through tracking by bar codes, the manufacturer keeps a continuous record of inventory levels for each component part. Advances in information-gathering technology are now making it more cost effective for companies to use perpetual inventory systems.

The **periodic inventory method** does not require a continuous record of inventory changes. Costs of materials used or costs of goods sold cannot be computed accurately until ending inventories, determined by physical count, are subtracted from the sum of the beginning inventory, purchases, and other purchasing costs. Costs are recorded by natural classifications, such as Material Purchases, Freight In, and Purchase Discounts. See Exhibit 2-10 for a comparison of perpetual and periodic inventory methods.

SOME COST ACCOUNTING LANGUAGE

Objective 7

Understand some troublesome cost accounting language

Troublesome Terms

Many terms have very special meanings in accounting. The meanings often differ from company to company; each organization seems to develop its own distinctive and extensive accounting language. You will save much confusion and wasted time if you find out the exact meaning of any strange jargon that you encounter.

Before proceeding, reflect on some commonly misunderstood terms. Consider the misnomer *manufacturing expenses*, which is often used to describe the factory overhead. ***Factory overhead is not an expense. It is part of inventoriable cost and will funnel into the expenses stream only when the inventoriable costs are released as cost of goods sold.***

Also, *cost of goods sold* is a widely used term that is somewhat misleading when you try to pin down the meaning of *cost*. Cost of goods sold is every bit as much an *expense* as salespersons' commissions. Cost of goods sold is also often called *cost of sales*.

Distinguish clearly between the merchandising accounting and the manufacturing accounting for such costs as wages, depreciation, and insurance. As Exhibit 2-9 demonstrates, in merchandising accounting all such items are period costs (expenses of the current period). ***In manufacturing accounting, many of such items are related to production activities and thus, as factory overhead, are inventoriable costs (and become expenses only when the inventory is sold).***

For reporting in balance sheets and income statements, both merchandising accounting and manufacturing accounting regard marketing and general administrative costs as period costs. The inventoriable cost of a manufactured product excludes sales salaries, sales commissions, advertising, legal, public relations, and the president's salary. ***Manufacturing overhead is traditionally regarded as part of the cost of finished goods inventory, but marketing expenses and general administrative expenses are not.*** The underlying ideas is that the finished goods inventory should *include* all costs of manufacturing functions necessary to get the product to a completed state but should *exclude* all nonmanufacturing costs.

Study Exhibit 2-8, page 36, in its entirety. The following T-accounts for a manufacturing inventory system may help relate some key terms (in thousands):

Work in Process Inventory

Bal. December 31, 19_1	6	Cost of goods manufactured	104
Direct materials used	76		
Direct labor	18		
Factory overhead	11		
To account for	111		
Bal. December 31, 19_2	7		

Finished Goods Inventory

Bal. Dec. 31, 19_1	22	Cost of goods sold	108 - - -→
- - - - - - - - - - - - - →	104		
Available for sale	126		
Bal. Dec. 31, 19_2	18		

Cost of Goods Sold

108	

In particular, the cost of goods manufactured is the cost of all goods completed during the reporting period. Such goods are usually transferred to finished goods inventory. They become cost of goods sold when sales occur, which depends on the nature of the product, types of customers, and business conditions.

Subdivisions of Labor Costs

Labor-cost classifications vary among companies, but the following distinctions are generally found:

- Direct labor (already defined)
- Factory overhead (examples of prominent labor components of this factory overhead follow):
  - Indirect labor (compensation)
    - Forklift truck operators (internal handling of materials)
    - Janitors
    - Plant guards
    - Rework labor (time spent by direct laborers redoing defective work)
    - Overtime premium paid to *all* factory workers
    - Idle time
  - Managers' salaries
  - Payroll fringe costs (for example, health care premiums, pension costs)

All factory labor compensation, other than that for direct labor and managers' salaries, is usually classified as *indirect labor costs,* a major component of factory overhead. The term *indirect labor* is usually divided into many subsidiary classifications. The wages of forklift truck operators are generally not commingled with janitors' wages, for example, although both are regarded as indirect labor.

Managers' salaries are usually not classified as part of indirect labor. Instead, the compensation of supervisors, department heads, and all others who are regarded as part of manufacturing management is placed in a separate classification of factory overhead.

Overtime Premium

Costs are classified in a detailed fashion primarily to associate a specific cost with its specific cause or reason for incurrence. Two classes of indirect labor need special mention. **Overtime premium** consists of the wages paid to all workers (for both direct labor and indirect labor) in *excess* of their straight-time wage rates. Overtime premium is usually considered a part of overhead. If a lathe operator, George Flexner, gets $12 per hour for regular time and gets time and one-half for overtime, his *premium* would be $6 per overtime hour. If he works forty-four hours, including four overtime hours, in one week, his gross earnings would be classified as follows:

Direct labor: 44 hours × $12	$528
Overtime premium (factory overhead): 4 hours × $6	24
Total earnings for 44 hours	$552

Why is overtime premium of direct labor usually considered an indirect rather than a direct cost? After all, it can usually be traced to specific batches of work. It is usually not considered a direct charge because the scheduling of production jobs is generally *random*. For example, assume that Jobs 1 through 5 are scheduled for a specific workday of ten hours, including two overtime hours. Each job requires two hours. Should the job scheduled during hours 9 and 10 be assigned the overtime premium? Or should the premium be prorated over all the jobs? The latter approach does not "penalize"—add to the cost of—a particular batch of work solely because it happened to be worked on during the overtime hours. ***Instead, the overtime premium is considered to be attributable to the heavy overall volume of work, and its cost is thus regarded as part of factory overhead, which is borne by all units produced.***

Sometimes overtime is not random. For example, a special or rush job may clearly be the sole source of the overtime. In such instances, the overtime premium is regarded as a direct cost of the products made for that job.

Another subsidiary classification of indirect labor is the **idle time** of both direct and indirect factory labor. This typically represents wages paid for unproductive time caused by machine breakdowns, material shortages, sloppy production scheduling, and the like. For example, if the lathe operator's machine broke down for three hours, earnings would be classified as follows:

Direct labor: 41 hours × $12	$492
Overtime premium (factory overhead): 4 hours × $6	24
Idle time (factory overhead): 3 hours × $12	36
Total earnings for 44 hours	$552

Payroll Fringe Costs, Direct Labor, and Definitions

We cannot overemphasize the value of quickly obtaining a thorough understanding of the classifications and cost terms introduced in this chapter and later in this book. Managers, accountants, suppliers, and other people will avoid many misunderstandings if they share the meaning of the same technical term.

Consider the classification of factory *payroll fringe costs* (for example, employer contributions to employee benefits such as social security, life insurance, health insurance, and pensions). Most companies classify these costs as factory overhead. In some companies, however, the fringe benefits related to direct labor are charged as an additional direct labor cost. For instance, a direct laborer, such as a lathe operator or an auto mechanic whose gross wages are computed on the basis of a nominal or stated wage rate of $12 an hour, may enjoy fringe benefits totaling, say,

$5 per hour. Most companies classify the $12 as direct labor cost and the $5 as factory overhead. Other companies classify the entire $17 as direct labor cost. The latter approach is conceptually preferable because these costs are a fundamental part of acquiring labor services.

The warning here is to pinpoint what direct labor includes and excludes in a particular situation. Achieving clarity may preclude disputes regarding cost reimbursement contracts, income tax payments, and labor union matters. For example, some countries offer substantial income tax savings to certain companies that locate factories there. To qualify, the "direct labor" costs of these companies in that country must equal at least a specified percentage of the total manufacturing costs of their products. Disputes have arisen regarding how to calculate the direct labor percentage for qualifying for such tax benefits. For instance, are payroll fringe benefits on direct labor an integral part of direct labor costs, or are they part of factory overhead? Depending on how companies classify costs, you can readily see how firms may show "direct labor" as different percentages of total manufacturing costs. Consider a company with $5 million of payroll fringe costs (figures are assumed, in millions):

Classification A			Classification B		
Direct materials	$ 40	40%	Direct materials	$ 40	40%
Direct labor	20	20	Direct labor	25	25
Factory overhead	40	40	Factory overhead	35	35
Total manufacturing costs	$100	100%	Total manufacturing costs	$100	100%

Classification A assumes that payroll fringe costs are part of factory overhead. In contrast, Classification B assumes that payroll fringe costs are part of direct labor. If a country sets the minimum percentage of direct labor costs at 25%, the company would receive a tax break using Classification B, but not using Classification A. In addition to fringe benefits, other debated items are compensation for training time, idle time, vacations, sick leave, and extra compensation for overtime. To prevent these disputes, contracts and laws should be as specific as feasible regarding definitions and measurements.[4]

THE MANY MEANINGS OF PRODUCT COSTS

Objective 8

Provide three different meanings of product costs

The distinction we drew earlier between inventoriable costs and period costs has a long tradition for both internal reporting to management and external reporting to shareholders. During the late 1980s, new U.S. income tax requirements forced companies to account for many selling, general, and administrative costs as inventoriable costs instead of period costs. For example, distribution costs for warehousing are no longer period costs. For another example, legal department costs must be allocated between those related to manufacturing activities (inventoriable costs) and those not so related (period costs). However, these special requirements are confined to reporting to income tax authorities only.

[4]The National Association of Accountants (NAA) has issued a series of *Statements on Management Accounting* that discuss objectives, ethics, terminology, and definitions. They are available from the NAA, Montvale, NJ 07645–0433. For example, Statement 4C is "Definition and Measurement of Direct Labor Cost." This statement favors including as many related fringe benefits as feasible as a part of direct labor costs.

EXHIBIT 2-11
Different Product Costs for Different Purposes

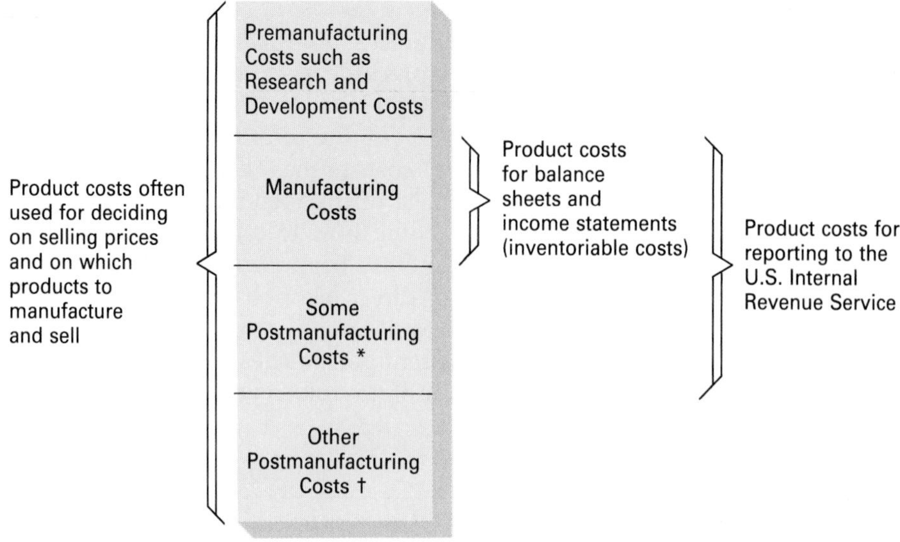

* Includes many administrative costs.
† Includes marketing and customer-service costs.

Accountants frequently use the term *product costs* to describe those costs allocated to units of product. **Product costs** is a general term that denotes different costs allocated to products for different purposes. As Exhibit 2-11 shows, product costs may be inventoriable costs for reporting on balance sheets and income statements, they may be a broader set of costs for purposes of reporting to tax authorities, and they may be a still broader set of costs for purposes of deciding on selling prices and on which products to manufacture and sell.

Multiple purposes of cost accounting are illustrated by the choice of the cost object (product) and the choices of what costs to allocate to the cost object. For balance sheets and income statements, only manufacturing costs are allocated to the products. For reporting to the Internal Revenue Service, some selling and administrative costs are additionally allocated. For decisions regarding choices and pricing of products, managements want the costs of products to include all (or nearly all) costs of running the business.

For external reporting to shareholders, companies with manufacturing operations tend to follow the widely accepted accounting concept that all manufacturing costs are inventoriable, including variable and fixed factory overhead. In practice, however, some companies do not inventory all factory overhead. Instead, they charge some to expense immediately. The most noteworthy example is depreciation on factory equipment. External auditors tend to accept such practices for a variety of reasons. A prominent justification is that consistent application of such depreciation accounting is coupled with undramatic changes in beginning and ending inventory levels. Hence, the final reported net income will not be materially affected. Furthermore, depreciation is often not a significant part of the total manufacturing costs, so inventories are not materially understated.[5]

[5]For a thorough discussion of concepts and practices regarding what costs are inventoriable for both shareholder and tax reporting, see E. Noreen and R. Bowen, "Tax Incentives and the Decision to Capitalize or Expense Manufacturing Overhead," *Accounting Horizons (March 1989).*

CLASSIFICATIONS OF COSTS

This chapter has merely hinted at the vast number of classifications of costs that have proved useful for various purposes. Classifications can be made on the basis of:

1. Ease of traceability to cost object
   a. Direct costs
   b. Indirect costs
2. Behavior in relation to changes of a cost driver
   a. Variable costs
   b. Fixed costs
3. Averaging
   a. Total costs
   b. Unit costs
4. Management function
   a. Manufacturing costs
   b. Marketing costs
   c. Administrative costs
5. Assets or expenses
   a. Inventoriable costs
   b. Period costs
6. Time when computed
   a. Historical costs
   b. Budgeted or predicted costs

PROBLEM FOR SELF-STUDY

(Try to solve this problem before examining the solution that follows.)

PROBLEM

Consider the following data of the Laimon Company for the year 19_1:

Sandpaper	$ 2,000	Depreciation—equipment	$ 40,000
Material handling	40,000	Factory rent	50,000
Lubricants and coolants	5,000	Property taxes on equipment	4,000
Overtime premium	20,000	Fire insurance on equipment	3,000
Idle time	10,000	Direct materials purchased	460,000
Miscellaneous indirect labor	40,000	Direct materials, 12/31/_1	50,000
Direct labor	300,000	Sales	1,260,000
Direct materials, 12/31/_0	40,000	Sales commissions	60,000
Finished goods, 12/31/_1	150,000	Sales salaries	100,000
Finished goods, 12/31/_0	100,000	Shipping expenses	70,000
Work in process, 12/31/_0	10,000	Administrative expenses	100,000
Work in process, 12/31/_1	14,000		

Required

1. Prepare an income statement with a separate supporting schedule of cost of goods manufactured. For all items except sales, purchases of direct materials, and inventories, indicate by "V" or "F" whether each is basically a variable or a fixed cost (where the cost object is a product unit). If in doubt, decide on the basis of whether the total cost will fluctuate substantially over a wide range of production volume.
2. Suppose that both the direct material and rent costs are tied to the manufacturing of the equivalent of 900,000 units. What is the average unit cost for the direct materials assigned to those units? What is the average unit cost of the factory rent? Assume that the rent is a fixed cost.
3. Repeat the computation in requirement 2 for direct materials and factory rent, assuming that the costs are being predicted for the manufacturing of the equivalent of 1,000,000 units next year. Assume that the implied cost behavior patterns persist.
4. As a management consultant, explain concisely to the president why the unit costs for direct materials did not change in requirements 2 and 3, but the unit costs for rent changed.

SOLUTION

1. LAIMON COMPANY

Income Statement
for the Year Ended December 31, 19_1

Sales		$1,260,000
Deduct cost of goods sold:		
Finished goods, December 31, 19_0	$ 100,000	
Cost of goods manufactured (see schedule below)	960,000	
Cost of goods available for sale	$1,060,000	
Finished goods, December 31, 19_1	150,000	
Cost of goods sold		910,000
Gross margin		$ 350,000
Deduct marketing and administrative expenses:		
Sales commissions	$ 60,000 (V)	
Sales salaries	100,000 (F)	
Shipping expenses	70,000 (V)	
Administrative expenses	100,000*	330,000
Operating income		$ 20,000

*Probably a mixture of fixed and variable items.

LAIMON COMPANY

Schedule of Cost of Goods Manufactured
For the Year Ended December 31, 19_1

Direct materials:		
Inventory, December 31, 19_0		$ 40,000
Purchases of direct materials		460,000
Cost of direct materials available for use		$500,000
Inventory, December 31, 19_1		50,000
Direct materials used		$450,000 (V)
Direct labor		300,000 (V)
Indirect manufacturing costs:		
Sandpaper	$ 2,000 (V)	
Lubricants and coolants	5,000 (V)	
Material handling	40,000 (V)	
Overtime premium	20,000 (V)	
Idle time	10,000 (V)	
Miscellaneous indirect labor	40,000 (V)	
Factory rent	50,000 (F)	
Depreciation—equipment	40,000 (F)	
Property taxes on equipment	4,000 (F)	
Fire insurance on equipment	3,000 (F)	214,000
Manufacturing costs incurred during 19_1		$964,000
Add work in process inventory, December 31, 19_0		10,000
Total manufacturing costs to account for		$974,000
Deduct work in process inventory, December 31, 19_1		14,000
Cost of goods manufactured (to Income Statement)		$960,000

2. Direct material unit cost = Direct materials used ÷ Units produced
= $450,000 ÷ 900,000 = $.50

Factory-rent unit cost = Factory rent ÷ Units produced
= $50,000 ÷ 900,000 = $.0556

3. The direct material costs are variable, so they would increase in total from $450,000 to 1,000,000 × $.50 = $500,000. However, their unit costs would be unaffected:

Direct materials used = $500,000 ÷ 1,000,000 units = $.50

In contrast, the factory rent is fixed, so it would not increase in total. However, if the rent is assigned to units produced, the unit costs would decline from $.0556 to $.05:

Factory-rent unit cost = $50,000 ÷ 1,000,000 = $.05

4. The explanation would begin with the answer to requirement 3. As consultant, you should stress that the averaging (unitizing) of costs having different behavior patterns can be misleading. A common error is to assume that a total unit cost, which is often a sum of variable unit costs and fixed unit costs, is an indicator that *total* costs change in a wholly variable way as volume fluctuates. The next chapter demonstrates the necessity for distinguishing between cost behavior patterns. Above all, the user must be wary about unit fixed costs. Too often, unit fixed costs are erroneously regarded as being indistinguishable from variable costs.

SUMMARY

Accounting systems should serve multiple decision purposes, and there are different measures of cost for different purposes. The most economically feasible approach to designing a management accounting system is to assume some common wants for a variety of decisions and choose cost objects for routine data accumulation in light of these wants.

This chapter has concentrated on definitions and explanations of many widely used cost accounting terms. The most basic distinction we make is between direct and indirect costs. The same cost may be direct regarding one cost object and indirect regarding other cost objects.

Be sure you understand Exhibit 2-9, page 38, including the footnotes and the new terms. Newcomers to cost accounting are used to assuming that such costs as power, telephone, and depreciation are expenses unconnected with inventories. However, if these costs are related to manufacturing, they are factory overhead costs that most accounting systems regard as inventoriable.

TERMS TO LEARN

This chapter contains more basic terms than any other in this book. Do not proceed before you check your understanding of the following terms. Both the chapter and the Glossary at the end of the book contain definitions.

actual costs *(p. 26)* average cost *(33)* conversion costs *(37)* cost *(25)*
cost accumulation *(26)* cost allocation *(26)* cost driver *(28)*
cost object *(25)* cost objective *(25)* direct costs *(27)* direct labor costs *(35)*
direct materials costs *(35)* direct materials inventory *(39)* factory burden *(35)*
factory overhead *(35)* finished goods inventory *(39)* fixed cost *(29)*
goods in process *(39)* idle time *(42)* indirect costs *(27)*
indirect labor costs *(35)* indirect manufacturing costs *(35)*
inventoriable costs *(37)* manufacturing *(34)* manufacturing expenses *(35)*
manufacturing overhead *(35)* merchandising *(34)* overtime premium *(42)*
period costs *(37)* periodic inventory method *(40)*
perpetual inventory method *(39)* prime costs *(37)* product costs *(44)*
relevant range *(30)* variable cost *(29)* work in process inventory *(39)*
work in progress *(39)*

ASSIGNMENT MATERIAL

QUESTIONS

2-1 Give three examples of cost objects.

2-2 Explain "economically feasible," or "cost effective," as applied to cost accounting.

2-3 What costs are considered direct? Indirect?

2-4 Give an example of a cost that is both direct and indirect.

2-5 "The composition of direct costs of an identical product or service might differ between firms." Do you agree? Explain.

2-6 Give three examples of cost drivers.

2-7 Define *variable cost, fixed cost,* and *relevant range.*

2-8 Give three examples of fixed factory overhead.

2-9 "Fixed costs are really variable. The more you produce, the less fixed costs you have." Do you agree? Explain.

2-10 Costs may be simultaneously variable and direct or fixed and indirect. Give an example of each case when the cost object is units produced.

2-11 Consider a company like IBM. Identify the business functions in which its main costs arise.

2-12 Distinguish between *manufacturing* and *merchandising.*

2-13 What are the three major categories of the inventoriable cost of a manufactured product?

2-14 Define the following: *direct materials costs, direct labor costs, indirect materials costs, indirect labor costs, factory overhead costs, conversion costs.*

2-15 Give at least four terms that may be substituted for the term *factory overhead.*

2-16 Distinguish among *direct labor, indirect labor, overtime premium,* and *idle time.*

2-17 What is the major difference between the balance sheets of manufacturers and merchandisers?

2-18 "Fixed costs decline as production increases." Do you agree? Explain.

2-19 "For purposes of income determination, insurance, depreciation, and wages should always be treated alike." Do you agree? Explain.

2-20 Why is the term *manufacturing expenses* a misnomer?

2-21 "Cost of goods sold is an expense." Do you agree? Explain.

2-22 Why is the unit-cost concept helpful in accounting?

2-23 Distinguish between costing for *control* and costing for *inventory valuation.*

2-24 Why is overtime premium usually not considered a direct cost of units produced?

2-25 "*Product costs* is a general term that denotes different costs allocated to products for different purposes." Describe three purposes.

EXERCISES AND PROBLEMS

2-26 Average costs and total costs. A fraternity has hired a musical group for a party. The cost will be a fixed sum of $1,000.

Required

1. Suppose 500 persons attend the party. What will be the total cost of the musical group? The cost per person?
2. Suppose 2,000 persons attend. What will be the total cost of the musical group? The cost per person?
3. For prediction of total costs, should the manager of the party use the unit costs in requirement 1? In requirement 2? What is the major lesson of this problem?

2-27 Periodic or perpetual inventory methods. (SMA) The terms *periodic* and *perpetual inventories* are referred to frequently in presenting the accounting procedures followed by businesses in recording their business transactions in any given period of their operations. Discuss the difference between periodic and perpetual inventory procedures.

2-28 Reporting current assets of a manufacturer. The following selected year-end account balances of El Monte Metal Goods are listed in alphabetical order (in thousands)

Accounts receivable	$ 93,000	Factory overhead	$26,000
Cash	9,000	Finished goods inventory	58,000
Cost of goods manufactured	94,000	Marketing expense	39,000
Cost of goods sold	101,000	Direct materials inventory	14,000
Direct labor	10,000	Prepaid expenses	5,000
Direct materials used	25,000	Work in process inventory	32,000

Required

Prepare a schedule of El Monte's current assets. *Hint:* Not all accounts should be used.

2-29 Cost of goods manufactured. Prepare a schedule of cost of goods manufactured for the Canseco Company from the following account balances (in thousands):

	End of 19_1	End of 19_2
Direct materials inventory	$22,000	$26,000
Work in process inventory	21,000	20,000
Finished goods inventory	18,000	23,000
Purchases of direct materials		75,000
Direct labor		25,000
Indirect labor		15,000
Factory insurance		9,000
Depreciation—factory building and equipment		11,000
Repairs and maintenance—factory		4,000
Marketing expenses		93,000
General and administrative expenses		29,000

2-30 Manufacturer's income statement. Prepare an income statement for the company in the preceding problem, assuming sales of $300 million.

2-31 Computing cost of goods manufactured and cost of goods sold. Compute cost of goods manufactured and cost of goods sold from the following account balances relating to 19_2 (in thousands):

Property tax on factory building	$ 3,000
Marketing expenses	37,000
Finished goods inventory, Dec. 31, 19_1	27,000
Factory utilities	17,000
Work in process inventory, Dec. 31, 19_2	26,000
Depreciation of factory building	9,000
Nonfactory administrative expenses	43,000
Direct materials used	87,000
Finished goods inventory, Dec. 31, 19_2	34,000
Depreciation of factory equipment	11,000
Factory repairs and maintenance	16,000
Work in process inventory, Dec. 31, 19_1	20,000
Direct labor	34,000
Indirect labor	23,000
Indirect materials used	11,000
Miscellaneous factory overhead	4,000

2-32 Gross margin for a manufacturer. Supply the missing amounts (in thousands) from the following computation of gross margin:

Sales				$495,000
Cost of goods sold:				
Beginning finished goods inventory			$25,000	
Cost of goods manufactured:				
Beginning work in process inventory		$ 57,000		
Direct materials used	$84,000			
Direct labor	X			
Factory overhead	91,000	220,000		
Total manufacturing costs to account for		$277,000		
Deduct ending work in process inventory		40,000		
Cost of goods manufactured			X	
Goods available for sale			X	
Deduct ending finished goods inventory			37,000	
Deduct cost of goods sold				X
Gross margin				$ X

2-33 Statement of cost of goods manufactured and sold. Suppose Dana Corporation supplies Ford Motor Company with auto parts. Use the accompanying data to answer the following questions regarding Dana (in millions).

	Inventories	
	12/31/_3	12/31/_4
Direct materials	$ 7	$ 5
Work in process	4	3
Finished goods	10	13

Manufacturing Costs Incurred during 19_4		
Direct materials used		$30
Direct labor		10
Indirect manufacturing costs:		
Indirect labor	$ 5	
Utilities	2	
Depreciation—plant and equipment	3	
Other	6	16
Manufacturing costs incurred during 19_4		$56

Required

1. Prepare a statement of cost of goods manufactured and sold.
2. Compare your statement with those in Exhibit 2-8, page 36. How does your statement differ from the Schedule of Cost of Goods Manufactured in Exhibit 2-8?
3. Compute the prime costs and the conversion costs incurred during 19_4.
4. What is the difference in meaning between the terms "total manufacturing costs incurred in 19_4" and "cost of goods manufactured in 19_4"?
5. Draw a T-account for Work in Process. Enter the amounts shown on your financial statement into the T-account as you think they might logically affect Work in Process, assuming that the account is kept on a perpetual inventory basis. Use a single summary number of $16 million for entering the indirect manufacturing costs rather than entering the four individual amounts.

2-34 Income statement and schedule of cost of goods manufactured. (Alternate is 2-36.) The Howell Corporation has the following accounts (in millions):

For Specific Date		For Year 19_2	
Finished goods, Dec. 31, 19_1	$ 70	Purchases of direct materials	$325
Direct materials, Dec. 31, 19_1	15	Direct labor	100
Work in process, Dec. 31, 19_1	10	Depreciation—factory building and equipment	80
Direct materials, Dec. 31, 19_2	20	Factory supervisory salaries	5
Finished goods, Dec. 31, 19_2	55	Miscellaneous factory overhead	35
Work in process, Dec. 31, 19_2	5	Sales	950
		Marketing and administrative expenses (total)	240
		Factory supplies used	10
		Factory utilities	30
		Indirect labor	60

Required

Prepare an income statement and a supporting schedule of cost of goods manufactured for the year ended December 31, 19_2. (For additional questions regarding these facts, see the next problem.)

2-35 Interpretation of statements. Refer to the preceding problem.

Required

1. How would the answer to the preceding problem be modified if you were asked for a schedule of cost of goods manufactured and sold instead of a schedule of cost of goods manufactured? Be specific.
2. Would the sales manager's salary be accounted for any differently if the Howell Corporation were a merchandising company instead of a manufacturing company? Using the boxes in Exhibit 2-9, page 39, describe how an assembler's wages would be accounted for in this manufacturing company.
3. Factory supervisory salaries are usually regarded as indirect manufacturing costs. When might some of these costs be regarded as direct costs? Give an example.

4. Suppose that both the direct materials used and the depreciation were related to the manufacture of one million units of product. What is the unit cost for the direct materials assigned to those units? For depreciation? Assume that yearly depreciation is computed on a straight-line basis.
5. Assume that the implied cost behavior patterns in requirement 4 persist. That is, direct material costs behave as a variable cost and depreciation behaves as a fixed cost. Repeat the computations in requirement 4, assuming that the costs are being predicted for the manufacture of 1.2 million units of product. How would the total costs be affected?
6. As a management accountant, explain concisely to the president why the unit costs differed in requirements 4 and 5.

2-36 Income statement and schedule of cost of goods manufactured. (Alternate is 2-34.) The following items pertain to Chan Corporation (in millions):

For Specific Date		For Year 19_2	
Work in process, Dec. 31, 19_1	$10	Factory utilities	$ 5
Direct materials, Dec. 31, 19_2	5	Indirect labor	20
Finished goods, Dec. 31, 19_2	12	Depreciation—factory building	
Accounts payable, Dec. 31, 19_2	20	and equipment	9
Accounts receivable,		Sales	350
Dec. 31, 19_1	50	Miscellaneous factory overhead	10
Work in process, Dec. 31, 19_2	2	Marketing and administrative	
Finished goods, Dec. 31, 19_1	40	expenses (total)	90
Accounts receivable,		Direct materials purchased	80
Dec. 31, 19_2	30	Direct labor	40
Accounts payable, Dec. 31, 19_1	40	Factory supplies used	6
Direct materials, Dec. 31, 19_1	30	Property taxes on factory	1

Required

Prepare an income statement and a supporting schedule of cost of goods manufactured. (For additional questions regarding these facts, see the next problem.)

2-37 Interpretation of statements. Refer to the preceding problem.

Required

1. How would the answer to the preceding problem be modified if you were asked for a schedule of cost of goods manufactured and sold instead of a schedule of cost of goods manufactured? Be specific.
2. Would the sales manager's salary be accounted for any differently if the Chan Corporation were a merchandising company instead of a manufacturing company? Using the boxes in Exhibit 2-9, page 38, describe how an assembler's wages would be accounted for in this manufacturing company.
3. Factory supervisory salaries are usually regarded as indirect manufacturing costs. When might some of these costs be regarded as direct costs? Give an example.
4. Suppose that both the direct materials used and the depreciation were related to the manufacture of 1 million units of product. What is the unit cost for the direct materials assigned to those units? For depreciation? Assume that yearly depreciation is computed on a straight-line basis.
5. Assume that the implied cost behavior patterns in requirement 4 persist. That is, direct material costs behave as a variable cost and depreciation behaves as a fixed cost. Repeat the computations in requirement 4, assuming that the costs are being predicted for the manufacture of 1.5 million units of product. How would the total costs be affected?
6. As a management accountant, explain concisely to the president why the unit costs differed in requirements 4 and 5.

2-38 Compute cost of goods sold. (SMA) A manufacturing company had the following inventories at the beginning and end of its most recent fiscal period:

	Beginning	End
Direct materials	$22,000	$30,000
Work in process	40,000	48,000
Finished goods	25,000	18,000

During this period, the following costs and expenses were incurred:

Direct materials purchased	$300,000
Direct labor cost	120,000
Indirect labor cost (factory)	60,000
Taxes, utilities, and depreciation on factory building	50,000
Sales and office salaries	64,000

Cost of goods sold during the period was (choose one): (1) $521,000, (2) $514,000, (3) $522,000, (4) $539,000. Show computations.

2-39 Finding unknown balances. An auditor for the Internal Revenue Service is trying to reconstruct some partially destroyed records of two taxpayers. For each of the cases in the accompanying list, find the unknowns designated by capital letters.

	Case 1	Case 2
	(in thousands)	
Accounts receivable, 12/31	$ 6,000	$ 2,100
Cost of goods sold	A	20,000
Accounts payable, 1/1	3,000	1,700
Accounts payable, 12/31	1,800	1,500
Finished goods inventory, 12/31	B	5,300
Gross profit	11,300	C
Work in process, 1/1	–0–	800
Work in process, 12/31	–0–	3,000
Finished goods inventory, 1/1	4,000	4,000
Direct material used	8,000	12,000
Direct labor	3,000	5,000
Factory overhead	7,000	D
Purchases of direct material	9,000	7,000
Sales	32,000	31,800
Accounts receivable, 1/1	2,000	1,400

2-40 Finding unknown balances. An insurance investigator in Birmingham, England, is reconstructing records of the Graff Company. Some data were destroyed in a fire, but the following data (in thousands of British pounds) for the fiscal year survived: cost of goods sold, 32,000; purchases of direct materials, 8,000; factory overhead, 13,000; sales commissions, 2,000; direct labor, 4,000; direct materials used, 7,600; finished goods inventory, beginning, 7,800; gross profit, 12,000; accounts payable, beginning, 1,700 and end, 1,500; work in process, beginning, 1,300 and end, 800.

Required

To evaluate a claim for a fire loss, compute the sales and the ending inventory of finished goods.

2-41 Fire loss, computing inventory costs. A distraught employee, Fang W. Arson, put a torch to a factory on a blustery February 26. The resulting blaze completely destroyed the plant and its contents. Fortunately, certain accounting records were kept in another building. They revealed the following for the period from December 31, 19_1 to February 26, 19_2:

Direct materials purchased, $160,000
Work in process, 12/31/_1, $34,000
Direct materials, 12/31/_1, $16,000
Finished goods, 12/31/_1, $30,000
Factory overhead, 40% of conversion costs
Sales, $500,000
Direct labor, $180,000

Prime costs, $294,000
Gross profit percentage based on net sales, 20%
Cost of goods available for sale, $450,000

The loss was fully covered by insurance. The insurance company wants to know the historical cost of the inventories as a basis for negotiating a settlement, which is really to be based on replacement cost, not historical cost.

Required
Calculate the cost of

1. Finished goods inventory, 2/26/_2
2. Work in process inventory, 2/26/_2
3. Direct materials inventory, 2/26/_2

2-42 Unknowns, T-accounts, schedules. (CPA, adapted) Mat Company's cost of goods sold for the month ended March 31, 19_4, was $345,000. Ending work in process inventory was 90% of beginning work in process inventory. Factory overhead was 50% of direct labor cost. Other information pertaining to Mat Company's inventories and production for the month of March is as follows:

Beginning inventories—March 1	
Direct materials	$ 20,000
Work in process	40,000
Finished goods	102,000
Purchases of direct materials during March	110,000
Ending inventories—March 31	
Direct materials	26,000
Work in process	?
Finished goods	105,000

Required
(*Hint:* Use T-accounts.)

1. Prepare a schedule of cost of goods manufactured for the month of March.
2. Compute the prime costs incurred during March.
3. Compute the conversion costs charged to work in process during March.

2-43 Classification of costs. (Alternate is 2-44.) Classify each of the following as direct or indirect (D or I) with respect to units of production, and as variable or fixed (V or F) with respect to whether the cost changes in total as production volume changes. If in doubt, select on the basis of whether the item will vary over a wide range of activity. You will have two answers, D or I and V or F, for *each* of the ten items:

1. Factory rent
2. Salary of a factory storeroom clerk
3. Manager training program
4. Abrasives (sandpaper, etc.)
5. Cutting bits in a machinery department
6. Workmen's compensation insurance in a factory
7. Cement for a roadbuilder
8. Steel scrap for a blast furnace
9. Paper towels for a factory washroom
10. Food for a factory cafeteria

2-44 Classification of costs. (Alternate is 2-43.) Classify each of the following as direct or indirect (D or I) with respect to a specific product or service, and as variable or fixed (V or F) with respect to whether the cost changes in total as production volume changes. If in doubt, select on the basis of whether the item will vary over a wide range of activity. You will have two answers, D or I and V or F, for *each* of the ten items:

1. Idle time, assembly department
2. Property taxes
3. Coolant for operating machines
4. Supervisory salaries, assembly department
5. Fuel pumps purchased by Ford assembly plant
6. Straight-line depreciation on machinery
7. Factory picnic
8. Fuel for forklift trucks
9. Welding supplies
10. Sheet steel for General Electric's manufacture of refrigerators

2-45 Many meanings of product costs. Consider the following costs of a U.S. corporation:

a. Indirect materials
b. Marketing costs
c. Factory rent
d. Advertising
e. Direct labor
f. Factory administration
g. Direct materials
h. Small tools
i. Quality control
j. Legal department costs related to manufacturing
k. General and administrative (incident to production)
l. Engineering (product development)
m. Insurance (incident to production)
n. Research and development
o. Distribution costs for warehousing
p. Officers' salaries (overall activities)
q. Legal department costs not related to manufacturing
r. Distribution costs related to customer delivery

Required
Identify your answers by letter:

1. Which of these costs are product costs often used for deciding on selling prices and on which products to manufacture and sell?
2. Which of these costs are product costs for balance sheets and income statements (inventoriable costs)?
3. Which of these costs are product costs for reporting to the U.S. Internal Revenue Service?

2-46 Global manufacturing and classification of direct labor. Assume that Apple Computer has manufacturing facilities in Puerto Rico. Suppose manufacturers obtain substantial income tax savings if a company's local direct labor is at least 25% of the total manufacturing cost of products produced there. For tax purposes, the company must follow the same accounting classifications overseas as in the United States. Apple's manufacturing facilities are highly automated. Consequently, the company's nominal direct labor worldwide has averaged less than 6% of total manufacturing costs. However, the following are the assumed figures regarding Apple's products manufactured in Puerto Rico:

Direct materials and parts	$15,000,000	50%
Direct labor	6,000,000	20
Factory overhead	9,000,000	30
Cost of goods manufactured	$30,000,000	100%

Required
1. Suppose Apple accounts for direct labor at nominal rates. Fringe benefits associated with direct labor average 35% of nominal rates. These benefits are classified as part of factory overhead. If Apple accounted for these benefits as part of direct labor rather than factory overhead, what percentage of cost of goods manufactured would direct labor represent?
2. Should Apple classify the associated fringe benefits as direct labor on a worldwide basis? Explain.

2-47 Overtime premium and payroll fringe costs. A city planning department had a flurry of work during a particular week. The printers' labor contract provided for payment to workers at a rate of 150% of the regular hourly wage rate for all hours worked in excess of 8 per day. Anthony Bardo worked 8 hours on Monday through Wednesday, 10 hours on Thursday, and 9 hours on Friday. His regular pay rate is $12 per hour.

Required

1. Suppose that the printing department works on various jobs. All costs of the jobs are eventually allocated to the users, whether the consumers be the property-tax department, the city hospital, the city schools, or individual citizens who purchase some publications processed by the department. Compute Bardo's wages for the week. How much, if any, of Bardo's wages should be classified as direct costs of particular printing jobs. Why?
2. The city pension plan provides for the city to contribute to an employee pension fund for all employees at a rate of 30% of gross wages (not considering pension benefits). How much, if any, of Bardo's retirement benefits should be classified as direct costs of particular printing jobs? Why?

2-48 Overtime and fringe costs. Direct labor is often accounted for at the nominal or stated wage rate. The related "fringe costs," such as employer payroll taxes and employer contributions to pensions and to other employee benefit plans, are accounted for as part of overhead. Suppose that the $20 nominal pay per hour made to Mary Maloney, a direct laborer, caused related fringe benefits payments of $8 per hour.

Required

1. Suppose Maloney works 40 hours during a particular week as an auditor for a public accounting firm, 30 hours for client Garcia and 10 for client Guliani. Compute the costs of direct labor and general overhead.
2. The audit of each client is a cost object. What would be the cost of "direct labor" on the Garcia audit? On the Guliani audit?
3. How would you allocate general overhead to the Garcia audit? The Guliani audit?
4. An overtime premium is paid at 50% of the nominal wage rate for hours worked in excess of 40 per week. Suppose Maloney works a total of 48 hours (30 on the Garcia audit and 18 on the Guliani audit). What would be the cost of direct labor on the Garcia audit? The Guliani audit? What would be the addition to general overhead for fringe benefits and overtime premium?

2-49 Comprehensive problem on unit costs, product costs. The Schramka Company, a small business, makes metal shelving. Costs are as follows (V stands for variable; F stands for fixed):

Cost incurred in 19_1:	
Direct materials costs	$140,000 V
Direct labor costs	30,000 V
Power costs	5,000 V
Indirect labor costs	10,000 V
Indirect labor costs	16,000 F
Other indirect manufacturing costs	8,000 V
Other indirect manufacturing costs	24,000 F
Marketing expenses	122,850 V
Marketing expenses	40,000 F
Administrative expenses	50,000 F

Variable manufacturing costs are variable with respect to units of production. Variable marketing expenses are variable with respect to units sold.

Inventory data are:

	Beginning, January 1, 19_1	Ending, December 31, 19_1
Direct materials	0 lbs.	2,000 lbs.
Work in process	0 units	0 units
Finished goods	0 units	? units

Production in 19_1 was 100,000 units. Two pounds of direct material are used to make one unit of finished product.

Sales in 19_1 were $436,800. The selling price per unit and the purchase price per pound of direct material were stable throughout the year. The company's ending inventory of finished goods is carried at the average unit manufacturing costs for 19_1. Finished goods inventory at December 31, 19_1, was $20,970.

Required

Compute:

1. Direct material inventory, total cost, December 31, 19_1.
2. Finished goods inventory, total units, December 31, 19_1.
3. Selling price per unit, 19_1.
4. Operating income, 19_1. Show computations.

For an additional question regarding these facts, see the next problem.

2-50 Budgeted income statement. This is a continuation of the preceding problem. Assume that in 19_2 the selling price per unit and each variable cost per unit are predicted to be the same as in 19_1. Fixed manufacturing, marketing, and administrative costs in 19_2 are also predicted to be the same as in 19_1. Sales in 19_2 are forecast to be 122,000 units. The desired ending inventory of finished goods, December 31, 19_2, is 12,000 units. Assume zero ending inventories of both direct materials and work in process. The company's ending inventory of finished goods is carried at the average unit manufacturing cost for 19_2. The company uses the first-in, first-out inventory method. Management has asked that you prepare a budgeted income statement for 19_2.

Required

Compute:

1. Unit production of finished goods in 19_2.
2. Budgeted income statement for 19_2.

CHAPTER 3

Computer software companies consider the contribution margin per software package and fixed booth costs when planning their exhibitions at computer shows.

Cost-Volume-Profit Relationships

Cost drivers and revenue drivers: The general case

The breakeven point

Cost-volume-profit assumptions

Interrelationships of cost, volume, and profit

Comparison of contribution margin and gross margin

The P/V chart

Effects of sales mix

Role of income taxes

Measuring volume

All data in dollars

Nonprofit institutions and cost-volume-revenue analysis

CVP models, personal computers, and spreadsheets

Learning Objectives

When you have finished studying this chapter, you should be able to

1. Distinguish between the general and specific concerning cost-volume-profit relationships
2. Demonstrate three methods for calculating breakeven point and target operating income
3. Specify the limiting assumptions of cost-volume-profit analysis
4. Distinguish between contribution margin and gross margin
5. Construct and interpret a P/V chart
6. Explain the effects of sales mix on operating income
7. Compute cost-volume-profit relationships on an after-tax basis

Cost-volume-profit analysis naturally appeals to most business students. Why? Because it provides a sweeping overview of the planning process. It provides a widely used tool that helps answer such questions as How will costs and revenues be affected if we sell 1,000 more units? If we raise or lower our prices? If we increase our occupancy levels by 3% in our hotel or our hospital or our airplanes?

These questions have a common theme: What will happen to financial results if a specified level of volume fluctuates? Convincing answers are not easy to obtain. Managers usually resort to some simplifying assumptions, particularly about cost behavior—the response of costs to a variety of influences. Managers must somehow predict revenue and cost levels. If these predictions are not reasonably accurate, the financial results may be undesirable or even disastrous.

COST DRIVERS AND REVENUE DRIVERS: THE GENERAL CASE

Objective 1

Distinguish between the general and specific cases concerning cost-volume-profit relationships

Chapter 2 defined *cost driver* as any factor whose change causes a change in the total cost of a related cost object. Similarly, a **revenue driver** is any factor whose change causes a change in the total revenue of a related product or service. Chapter 2 emphasized that there are many cost drivers besides volume. For example, the weight, dimensions, and quality of materials may simultaneously affect total costs in addition to the quantity used. Similarly, there are many revenue drivers besides the volume of units sold. For example, changes in selling price, changes in advertising outlays, and changes in quality are underlying factors that might affect total revenues.

The general case for predicting total revenues and total costs would include analyses of how combinations of several revenue drivers and several cost drivers affect the aggregate levels of total revenues and total costs. This general case is described in Chapters 10 and 25. For now, we focus on the specific case where revenue and cost behaviors are linear and where volume is assumed to be the only cost and revenue driver.

We focus on the specific case of straightforward cost-volume-profit relationships for two major reasons. First, many companies have found such relationships helpful in decision making. Second, the straightforward relationships provide an excellent base for learning the more complex relationships that exist in the general case:

General Case	Specific Case
Many revenue drivers	Volume is the only revenue driver
Many cost drivers	Volume is the only cost driver
Various time spans for decisions (short run, long run, product life cycles)	Short-run decisions (time span no longer than a year)

The term cost-volume-profit (CVP) analysis is widely used as representing the specific case. The general case is more accurately labeled as "cost and revenues—cost and revenue drivers—profit analysis." Indeed, some accountants refer to the general case as a major example of "strategic cost analysis," which is a tool for strategic planning.[1]

Accountants and managers should always be on guard to ensure that the simplified relationships of the specific case generate sufficiently accurate predictions of how total revenues and total costs behave. Otherwise managers may be misled into making unwise short-run and long-run decisions. At the outset, remember that we will be considering simplified versions of the real world. Are these simplifications justifiable? The answer depends on the facts in a particular organization. The manager has a method for deciding among courses of action, often called a *decision model*. Models of cost-volume-profit relationships are examples of decision models. The simpler model is always preferable provided that management decisions would not be improved by using a more complicated model. That is, a more complicated model is attractive only when the resulting decisions would be improved in a net benefit sense. More will be said about these simplifications later in the chapter.

THE BREAKEVEN POINT

We obtain an overview of decision models by examining the interrelationships of changes in costs, volume, and profits—sometimes too narrowly described as breakeven analysis. The breakeven point is often only incidental in these studies. The **breakeven point** is that point of volume where total revenues and total expenses (total costs) are equal; that is, there is neither profit nor loss. ***CVP analysis properly considers a broader question: What is the impact on net income of various decisions affecting sales and costs?***

Before we look at CVP analysis, let us pinpoint our terms. We use *net income* and *net profit* interchangeably. Similarly, we use *costs* and *expenses* as synonyms in CVP analysis. That is, costs refer to the costs expiring and so becoming expenses during the period in question.

For consistency, we use the following terms throughout this book:

- **Operating income** is revenues or sales from operations for the accounting period minus all operating costs or expenses, including cost of goods of sold.
- **Net income** is operating income plus nonoperating revenues minus nonoperating expenses minus income taxes.

Examples of nonoperating revenues and nonoperating expenses are interest revenues and interest expenses. For simplicity throughout this chapter, nonoperating

[1]See Chapter 29. See also J. Shank and V. Govindarajan, *Strategic Cost Analysis* (Homewood, IL: Irwin, 1989).

revenues and nonoperating expenses are assumed to be zero. Therefore, operating income and net income are computed as follows:

Operating income = Revenues or sales − All variable costs − All fixed costs
Net income = Operating income − Income taxes

This section shows three methods for calculating the breakeven point: the equation method, the contribution-margin method, and the graphic method. Consider the following example.

Objective 2

Demonstrate three methods for calculating breakeven point and target operating income

EXAMPLE
Mary Frost plans to sell Do-All, a software package, at a heavily attended computer convention in Chicago. Mary can purchase this software at $120 for each package with the privilege of returning all unsold units. The units will be sold at $200 each. The booth rental is $2,000.

Equation Method

The first solution for computing the breakeven point is the *equation method.* Every income statement can be expressed in equation form, a mathematical model, as follows:

$$\text{Sales} - \text{Variable costs} - \text{Fixed costs} = \text{Operating income}$$

or

$$\left(\begin{matrix}\text{Unit} & & \text{Number} \\ \text{sales} & \times & \text{of} \\ \text{price} & & \text{units}\end{matrix}\right) - \left(\begin{matrix}\text{Unit} & & \text{Number} \\ \text{variable} & \times & \text{of} \\ \text{cost} & & \text{units}\end{matrix}\right) - \begin{matrix}\text{Fixed} \\ \text{costs}\end{matrix} = \begin{matrix}\text{Operating} \\ \text{income}\end{matrix}$$

This equation provides the most general and easy-to-remember approach to any breakeven or profit-estimate situation. For the example above:

Let N = Number of units to be sold to break even, where the breakeven point is defined as zero operating income.

$$\$200N - \$120N - \$2{,}000 = 0$$
$$\$80N = \$2{,}000$$
$$N = \frac{\$2{,}000}{\$80}$$
$$N = 25 \text{ units (or \$5,000 total sales at \$200 per unit)}$$

Contribution-margin Method

A second solution method emphasizes a formula. It is the *contribution-margin* or *marginal-income* method. **Contribution margin** is equal to sales minus *all variable* costs. Sales and costs are analyzed as follows:

1. *Unit contribution margin* to cover fixed costs and target operating income:

$$\text{Unit sales price} - \text{Unit variable costs}$$
$$\$200 - \$120 = \$80$$

2. *Breakeven point* in terms of units sold:

$$\frac{\text{Fixed costs}}{\text{Unit contribution margin}}$$
$$\frac{\$2{,}000}{\$80} = 25 \text{ units}$$

Stop a moment, and relate the contribution-margin method to the equation method:

$$(\text{Unit sales} - \text{Unit variable costs}) \times \text{Units} - \text{Fixed costs} = \text{Operating income}$$

The key calculation is dividing $2,000 by $80. Look again at the third line in the equation solution. It reads:

$$N = \frac{\$2{,}000}{\$80}$$

giving us a general formula for a single product and a volume-based cost driver:

$$\textbf{Breakeven point in units} = \frac{\textbf{Fixed costs}}{\textbf{Unit contribution margin}}$$

The *contribution-margin* method is merely a restatement of the *equation* method in different form. Use either technique; the choice is a matter of personal preference.

A condensed income statement at the breakeven point could be presented as follows:

	Total	Per Unit
Sales, 25 units × $200	$5,000	$200
Variable costs, 25 units × $120	3,000	120
Contribution margin	$2,000	$ 80
Fixed costs	2,000	
Operating income	$ 0	

Graphic Method

We can also graph the CVP relationships in this example by using three building blocks:

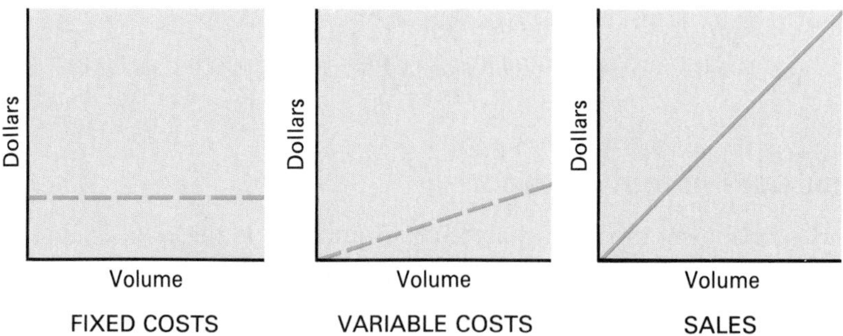

We interpret these graphs in this way: As volume increases, fixed costs remain the same regardless of volume. As volume increases, so do both total variable costs and sales.

Plotting the Graph

These three lines for our software example are plotted as follows on the complete cost-volume-profit (CVP) chart in Exhibit 3-1:

1. To plot fixed costs, measure \$2,000 on the vertical axis (Point A) and extend a line horizontally.
2. To plot variable costs, select a convenient sales volume—say, 30 units. Compute the total variable costs for that volume: 30 × \$120 = \$3,600. Add the \$3,600 to the \$2,000 plot at the 30-unit volume level to get Point B, \$5,600. Using points A and B, draw the total cost "function" to the edge of the graph. The line AB is the sum of the variable costs plus fixed costs.
3. To plot sales, select a convenient sales volume—say, 30 units. Plot Point C for total-sales dollars at that volume: 30 × \$200 = \$6,000. Draw the total-sales line from the origin Point 0 through Point C to the edge of the graph.

The *breakeven point* is where the total-sales line and total-costs line intersect. But note that the graph in Exhibit 3-1 shows the profit or loss outlook for a wide range of volume. The confidence we place in any particular cost-volume-profit chart is naturally a consequence of the relative accuracy of the cost-volume-profit relationships depicted.

Note that total sales and total variable costs fluctuate in direct proportion to changes in physical volume, but fixed costs are the same in total over the entire volume range.

EXHIBIT 3-1
Cost-Volume-Profit Chart

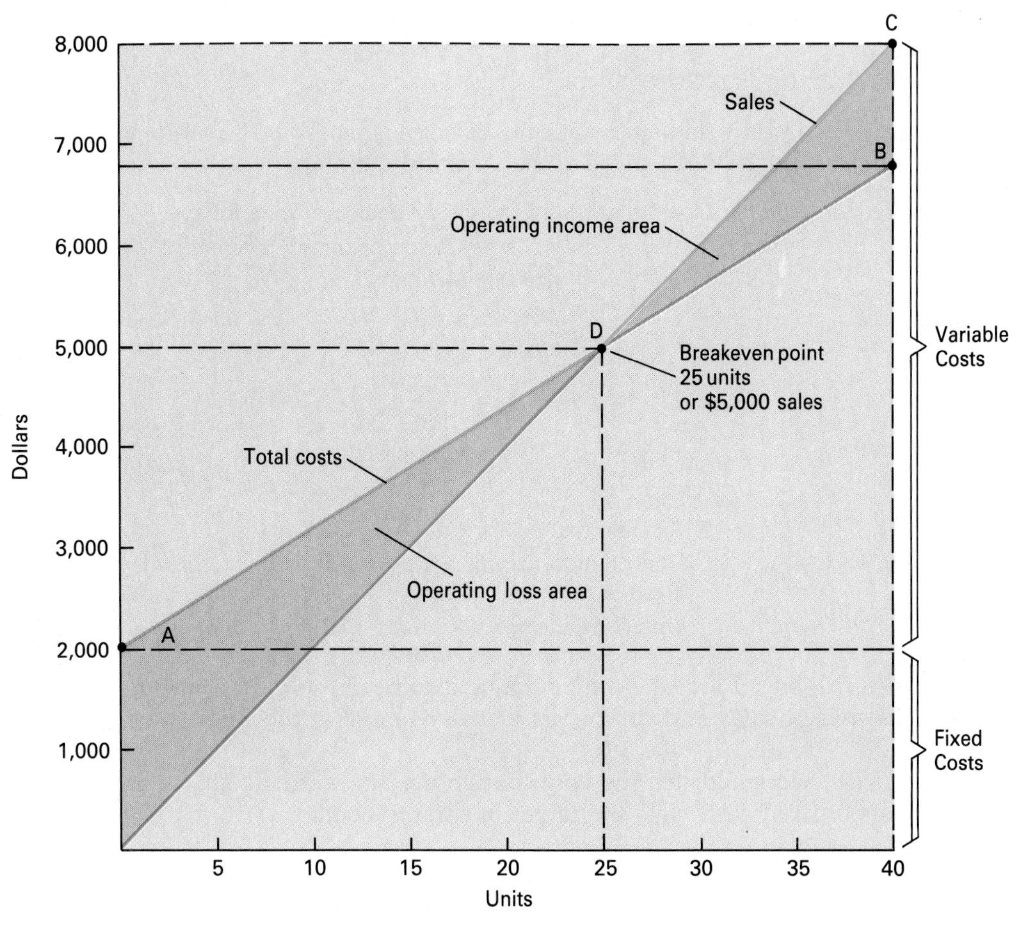

The chart in Exhibit 3-1 can also be depicted as follows:

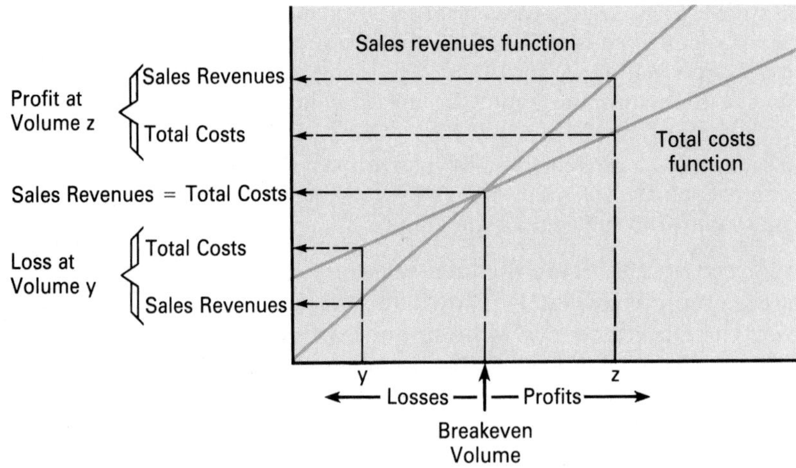

ASSUMED VOLUME LEVEL	Sales	TOTAL COSTS (Variable + Fixed)	OPERATING INCOME
y (10 Units)	\$200 × 10 = \$2,000	(\$120 × 10) + \$2,000 = \$3,200	−1,200
Breakeven (25 Units)	\$200 × 25 = \$5,000	(\$120 × 25) + \$2,000 = \$5,000	0
z (40 Units)	\$200 × 40 = \$8,000	(\$120 × 40) + \$2,000 = \$6,800	\$1,200

Target Operating Income

Let us introduce a profit element by asking, *How many units must be sold to yield an operating income of \$1,200?* We can use the equation method.

$$\text{Let } N = \text{Number of units to be sold to yield target operating income}$$
$$\text{Sales} - \text{Variable costs} - \text{Fixed costs} = \text{Target operating income}$$
$$\$200N - \$120N - \$2{,}000 = \$1{,}200$$
$$\$80N = \$2{,}000 + \$1{,}200$$
$$\$80N = \$3{,}200$$
$$N = \frac{\$3{,}200}{\$80}$$
$$N = 40 \text{ units}$$

Proof:

Sales, 40 × \$200	\$8,000
Variable costs, 40 × \$120	4,800
Contribution margin	\$3,200
Fixed costs	2,000
Operating income	\$1,200

The graph in Exhibit 3-1 indicates an operating income at the 40-unit volume. The difference between sales and total costs at that volume is the \$1,200 operating income.

Alternatively, we could use the contribution-margin method. The numerator now consists of fixed costs plus the target operating income:

$$N = \frac{\text{Fixed costs} + \text{Target operating income}}{\text{Unit contribution margin}}$$
$$N = \frac{\$2{,}000 + \$1{,}200}{\$80}$$

$$\$80N = \$2{,}000 + \$1{,}200$$
$$\$80N = \$3{,}200$$
$$N = 40 \text{ units}$$

COST-VOLUME-PROFIT ASSUMPTIONS

Relevant Range

Recall the concept of relevant range introduced in the preceding chapter. Specific cost and volume relationships—and their graphic representations—may hold for only a limited range of values. The CVP chart shown in Exhibit 3-1 may be unrealistic. The cost and volume relationships valid for the data used to produce the lines may not hold for all values. These relationships would probably not hold if volume fell below a certain level; in that case, the lines extending back to the origin may not be valid. Similarly, if volume rose above to a certain level, relationships may not hold. The lines to the right of that level would not be valid. ***The accompanying modified chart highlights that the sales and cost relationships are valid only within the band of the cost driver called the relevant range.***

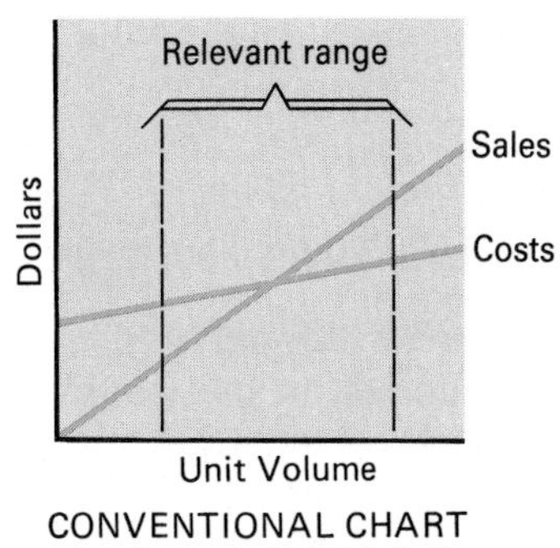

CONVENTIONAL CHART
(as shown in Exhibit 3-1)

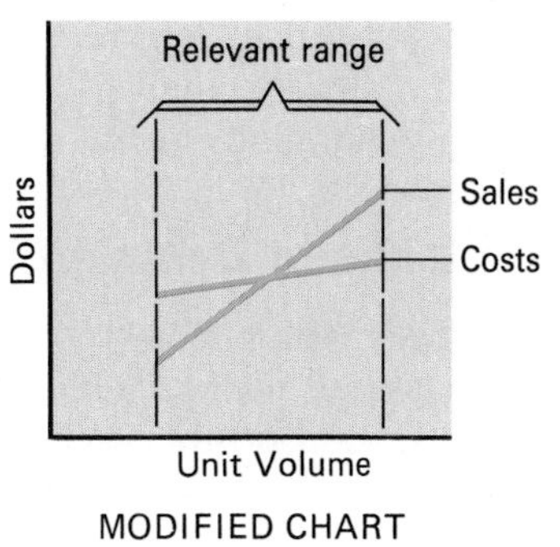

MODIFIED CHART

Limitations of Assumptions

As pointed out earlier, the interplay of a number of factors affects cost behavior. Volume is only one of these factors; others include unit prices of inputs, efficiency, changes in production technology, wars, strikes, legislation, and so forth. Any CVP analysis is based on assumptions about the behavior of revenue, costs, and volume. A change in expected behavior will alter the breakeven point; in other words, profits are affected by changes in factors *besides volume.* A CVP chart must be interpreted in the light of the limitations imposed by its underlying assumptions. ***The real benefit of preparing CVP charts is in the enrichment of understanding of the interrelationships of all factors affecting profits, especially cost behavior patterns over ranges of volume.***

Objective 3

Specify the limiting assumptions of cost-volume-profit analysis

The following underlying assumptions will limit the precision and reliability of a given cost-volume-profit analysis:

1. The behavior of total revenues and total costs has been reliably determined and is linear over the relevant range.[2]
2. Selling prices are constant.
3. All costs can be divided into fixed and variable elements.

[2]Economists do not assume linearity in CVP analysis. For example, they assume that sales price reductions may be needed to spur sales volume.

4. Total fixed costs remain constant.
5. Total variable costs are directly proportional to volume over the relevant range—that is, the variable costs per unit remain constant.
6. Prices of the factors of production inputs (for example, material prices, wage rates) are constant.
7. Efficiency and productivity are constant.
8. The analysis either covers a single product or assumes that a given sales mix will be maintained as total volume changes. (This assumption will be discussed later in this chapter.
9. Revenue and costs are being compared on a single volume base (for example, units produced and sold).
10. Perhaps the most basic assumption of all is that volume is the only driver of costs. Of course, other factors also affect costs and sales. Ordinary cost-volume-profit analysis is a crude oversimplification when these factors are unjustifiably ignored.
11. The production volume equals sales volume, or changes in beginning and ending inventory levels are zero. (The impact of inventory changes on cost-volume-profit analysis depends on what inventory valuation method is used. This complexity is discussed in Chapter 9.)

Business is dynamic, not static. The user of cost-volume-profit analysis must constantly challenge and reexamine assumptions in light of changes in business conditions, such as prices, production inputs, sales mixes, and input mixes. Moreover, cost-volume-profit analysis need not adhere rigidly to the assumptions of linearity and unchanging prices.

The Importance of Decision Situation and Time Horizon

Throughout this chapter, CVP models are described as though costs were neatly and readily classified into variable and fixed categories. In the real world, these classifications depend on the decision situation and the time horizon. Suppose a United Airlines plane will depart from its gate in two minutes. A potential passenger is running down a corridor bearing a transferable ticket from a competing airline. Unless the airplane is held for an extra thirty seconds, the passenger will miss the departure and will not switch to United for the planned trip. What are the variable costs to United of delaying the departure and placing one more passenger in an otherwise empty seat? Variable costs (for example, one more meal) are negligible. Virtually all the costs in that decision situation are fixed.

In contrast, suppose United must decide whether to add another flight, to include another city in its routes, or to acquire another airplane. Many more costs would be regarded as variable and fewer as fixed.

Consider a car owner. The additional cost of driving a car five miles to a store is relatively small. In this instance, the car owner would regard nearly all costs of car ownership as fixed. As the time lengthens and as the volume of miles driven increases, more and more costs, such as tires, that are regarded as fixed in the very short run become variable in the longer run.

The famous economist John Maynard Keynes once said that in the long run we are all dead. Similarly, many managers believe that in the long run all costs are variable. In many short-run situations, however, managers believe all their costs are fixed. Of course, managers and accountants must cope with some short-run decisions and some long-run decisions. Accordingly, distinctions between variable costs and fixed costs have proved useful.

These examples underscore the importance of how the decision situation affects the analysis of cost behavior. Of course, total costs are least likely to be affected by very short time spans and very small increases in volume. In brief, whether costs are really fixed depends heavily on the relevant range, the length of the time horizon in question, and the specific decision situation.

INTERRELATIONSHIPS OF COST, VOLUME, AND PROFIT

Uncertainty and Sensitivity Analysis

Throughout much of this book, we work with single-number "best estimates" in order to emphasize and simplify various important points. For example, our cost-volume-profit models, budget models, and capital-investment models make strong assumptions regarding the levels of variable costs, fixed costs, volume, and other factors. For purposes of introducing and using many decision models, we often conveniently assume a world of certainty.

Obviously, our estimates and predictions are subject to varying degrees of **uncertainty**, which is defined here as the possibility that an actual amount will deviate from an expected amount. How do we cope with uncertainty? There are many complex models available that formally analyze expected values in conjunction with probability distributions, as described in Chapter 20. But the application of *sensitivity analysis* to a basic solution is the most widely used approach.

Sensitivity analysis is a "what-if" technique that essentially asks how a result will be changed if the original predicted data are not achieved or if an underlying assumption changes. In the context of cost-volume-profit analysis, sensitivity analysis answers such questions as "What will operating income be if volume changes from the original prediction?" and "What will operating income be if variable costs per unit increase by 10%?"

A tool of sensitivity analysis is the **margin of safety**, which is the excess of budgeted sales over the breakeven volume. The margin of safety is the answer to this "what-if" question: "If sales drop, how far can they fall below budget before the breakeven point is reached?"

Changes in Variable Costs

Consider an example of sensitivity analysis. Both the unit contribution margin and the breakeven point are altered by changes in unit variable costs. Reconsider our software example:

		Sensitivity Analysis	
	Basic Solution	Increase Variable Cost to $150	Decrease Variable Cost to $75
Sales price per unit	$200	$200	$200
Variable cost per unit	120	150	75
Contribution margin per unit	$ 80	$ 50	$125
Breakeven point in units =	$\frac{\$2,000}{\$80} = 25$	$\frac{\$2,000}{\$50} = 40$	$\frac{\$2,000}{\$125} = 16$

Variable costs are subject to various degrees of control at different volumes. When business is booming, management tends to be preoccupied with the generation of volume "at all costs." When business is slack, management tends to scrutinize costs. Decreases in volume are often accompanied by increases in marketing costs, lower selling prices, lower labor turnover, increases in labor productivity, and decreases in raw-material prices. We see again limitations of a CVP chart; conventional CVP charts assume directly proportional fluctuations of variable costs with volume. This assumption implies adequate and uniform control over costs, but in practice such control is only sometimes attainable.

Changes in Fixed Costs

Fixed costs are not static year after year. Management may deliberately increase them to obtain more profitable combinations of production and distribution. A change in fixed costs may affect revenues, variable costs, or both. For example, a company may increase its sales force to reach markets directly instead of through wholesalers. This strategy usually results in higher unit sales prices and higher fixed marketing costs. To combat increases in labor rates, a firm may invest in automated machinery, which also increases fixed costs but reduces unit variable costs.

In some cases, a company may be wise to reduce fixed costs to obtain a more favorable combination of production inputs. A company may use wholesalers instead of a direct-selling sales force. A company producing ovens may dispose of its foundry if the resulting reduction in fixed costs more than counterbalances increases in the variable costs of purchased castings over the expected volume range.

When a major change in fixed costs is proposed, management uses forecasts of the effect on the targeted net income and the unit contribution margin as a guide toward a wise decision. The management accountant makes continuing analyses of cost behavior and calculates breakeven points periodically. He or she keeps management informed of the cumulative effect of major and minor changes in the company's cost and revenue patterns.

Fixed costs are constant only over a contemplated range of volume for a given time period. The volume range rarely extends from shutdown levels to 100% capacity. When management foresees a radical reduction in volume, it may "jar loose" many fixed costs. Slashing fixed costs lowers the breakeven point and enables the firm to endure a greater decrease in volume before losses occur. In the 1980s, the automobile manufacturers General Motors, Ford, and Chrysler drastically reduced fixed costs. Companies also reduce breakeven points by increasing the contribution margins per unit of product through increases in sales prices, decreases in unit variable costs, or both.

A 1990 news article disclosed that Chrysler's breakeven point had "crept up from 1.1 million vehicles per year to 1.4 million." The story stated that Chrysler planned to reduce costs by starting a selective hiring freeze, limiting merit raises for top executives, and cutting overtime pay. An upward creep of the breakeven point is a signal for management to analyze all its cost-volume-profit relationships more closely.

Managers are also reluctant to add fixed costs. For example, the president of Emery Air Freight commented:

> We would prefer to keep out of the airline business and buy space from the existing airlines because that's a variable cost. If we don't have the need for capacity, we don't buy it, but if we have to own or charter airplanes, the costs become fixed and we're stuck with excess capacity sometimes.

COMPARISON OF CONTRIBUTION MARGIN AND GROSS MARGIN

Objective 4

Distinguish between contribution margin and gross margin

Avoid confusing the terms *contribution margin* and *gross margin* (which is often also called *gross profit*). *Contribution margin* is the excess of sales over *all* variable costs, including variable manufacturing, marketing, and administrative categories. ***In contrast,* gross margin, *or* gross profit, *is the excess of sales over the cost of the goods sold. In a manufacturing company, the manufacturing cost of goods sold includes* fixed *indirect manufacturing costs.***

In a manufacturing company, gross margin and contribution margin would almost always be different amounts. Such amounts would be equal only by the unlikely coincidence that the fixed *manufacturing* costs included in cost of goods sold happened to equal variable *nonmanufacturing* costs. Compare how the same income statement with simplified, assumed figures (in thousands) might be prepared in two basic ways by a company that does manufacturing:

No Distinction between Variable and Fixed Costs

Sales	$900	100%
Manufacturing cost of goods sold	405	45
Gross margin	$495	55%
Nonmanufacturing costs	423	47
Operating income	$ 72	8%

Distinction between Variable and Fixed Costs

Sales	$900	100%
Variable manufacturing costs	$180	20%
Variable nonmanufacturing costs	135	15
Total variable costs	$315	35%
Contribution margin	$585	65%
Fixed manufacturing costs	$225	25%
Fixed nonmanufacturing costs	288	32
Total fixed costs	$513	57%
Total costs	$828	92%
Operating income	$ 72	8%

The gross margin percentage (55%) or ratio (.55) differs from the contribution margin percentage (65%) or ratio (.65). Why? Because fixed manufacturing costs are part of the manufacturing cost of goods sold in computing the gross margin, but no fixed costs are deducted in computing the contribution margin.

Even in retailing, where cost of goods sold consists entirely of the variable costs of merchandise, gross margin is not necessarily equal to the contribution margin. Suppose retail salespersons earn a 10% commission on sales (numbers assumed):

Sales	$100,000
Cost of goods sold	55,000
Gross margin	$ 45,000

Sales		$100,000
Cost of goods sold	$55,000	
Sales commissions	10,000	
Total variable costs		65,000
Contribution margin		$35,000

Both the contribution margin and the gross margin can be expressed as totals, as an amount per unit, or as percentages of sales in the form of ratios. For example, a* contribution-margin ratio *is the total contribution margin divided by the total sales. Similarly, the* variable-cost ratio *is the total variable costs divided by the total sales. Thus, a contribution-margin ratio of 65% means that the variable-cost ratio is 35%.

THE P/V CHART

We can recast Exhibit 3-1 in simpler form as a **P/V chart**, a profit-volume graph showing the impact of changes in volume on operating income. Many managers prefer P/V charts to CVP charts. The first graph in Exhibit 3-2 illustrates the chart, using the data in our example of selling software. Note these points:

Objective 5

Construct and interpret a P/V chart

1. The vertical axis is operating income in dollars. The horizontal axis is volume, which can be expressed in units or in sales dollars.
2. At zero volume, the operating loss is the total fixed costs: $2,000 in this example.
3. The operating-income line will slope upward from the −$2,000 intercept at the rate of the unit contribution margin of $80. The line will intersect the volume axis at the breakeven point of 25 units. Each unit sold beyond the breakeven point will add $80 to operating income. At a volume of 35 units, the operating income would be (35 − 25) × $80 = $800.

EXHIBIT 3-2
P/V Chart

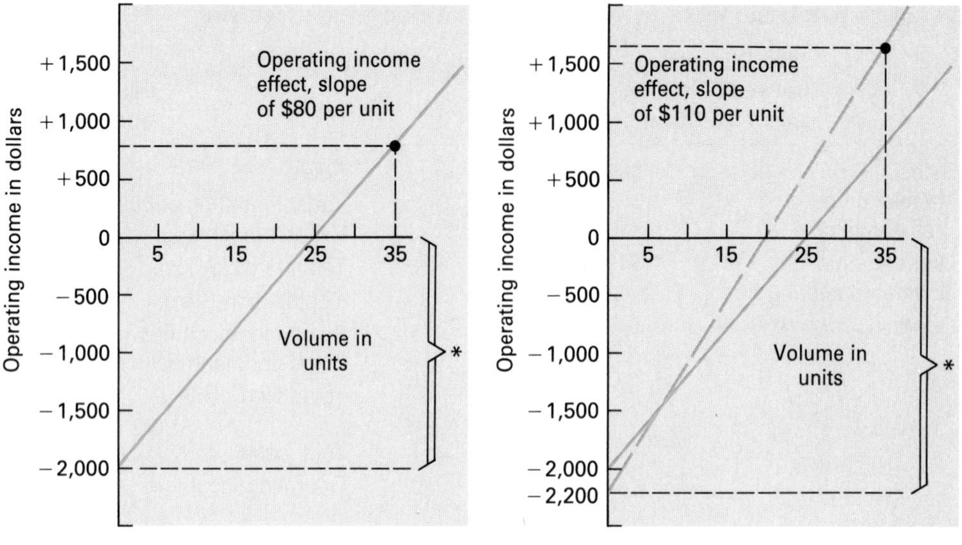

* *Total fixed costs.*

The P/V chart provides a quick, condensed comparison of how alternatives on pricing, variable costs, or fixed costs affect operating income as volume changes. For example, the second graph in Exhibit 3-2 shows how operating income and the breakeven point would be affected by a decrease in unit purchase cost from $120 to $90 and an increase in fixed costs from $2,000 to $2,200. Unit contribution margin would become $110 (recall that the unit sales price is $200), and the breakeven point would fall from 25 units to 20 units:

$$\text{New breakeven point} = \$2{,}200 \div \$110$$
$$= 20 \text{ units}$$

Note also the steeper slope of the new operating-income line, which means that the operating income will increase at a faster rate as volume increases. At a volume of 35 units, the operating income would be $(35 - 20) \times \$110 = \$1{,}650$.

EFFECTS OF SALES MIX

Objective 6

Explain the effects of sales mix on operating income

Sales mix is the relative combination of quantities of products that constitute total sales. If the mix changes, overall sales targets may still be achieved. However, the effects on profits depend on how the original proportions of low-margin or high-margin products have shifted.

Suppose a two-product pen company had the following budget:

	Peerless	Fantastic	Total
Sales in units	120,000	40,000	160,000
Sales @ $5 and $10	$600,000	$400,000	$1,000,000
Variable costs @ $4 and $3	480,000	120,000	600,000
Contribution margins @ $1 and $7	$120,000	$280,000	$ 400,000
Fixed costs			300,000
Operating income			$ 100,000

What would be the breakeven point? The usual answer assumes that the budgeted sales mix will not change. That is, three units of Peerless will be sold for each unit of Fantastic:

$$\text{Let } F = \text{Number of units of Fantastic to break even}$$
$$3F = \text{Number of units of Peerless to break even} = P$$
$$\text{Sales} - \text{Variable costs} - \text{Fixed costs} = \text{Zero operating income}$$
$$\$5(3F) + \$10(F) - \$4(3F) - \$3(F) - \$300{,}000 = 0$$
$$\$25F - \$15F - \$300{,}000 = 0$$
$$\$10F = \$300{,}000$$
$$F = 30{,}000$$
$$3F = 90{,}000 = P$$

The breakeven point is 120,000 units, consisting of 90,000 Peerless pens and 30,000 Fantastic pens. This is the only breakeven point for a sales mix of three Peerless and one Fantastic.

But the breakeven point is not a unique number. It obviously depends on the sales mix. Suppose only Peerless were sold (the unit contribution margin is $1):

$$\$300{,}000 \div \$1.00 = 300{,}000 \text{ units (consisting exclusively of Peerless)}$$

Similarly, suppose only Fantastic were sold (the unit contribution margin is $7):

$$\$300{,}000 \div \$7.00 = 42{,}857 \text{ units (consisting exclusively of Fantastic)}$$

Obviously, for any given sales volume, the higher the proportion of units having relatively high unit contribution margins, the higher the operating income. Suppose in this example that the actual results were sales of 160,000 units—exactly equal to the budget target for total sales volume in units—but the sales mix was 100,000 Peerless and 60,000 Fantastic. Operating income would be $220,000, which is a hefty $120,000 higher than the $100,000 budgeted operating income:

	Peerless	Fantastic	Total
Sales in units	100,000	60,000	160,000
Sales @ $5 and $10	$500,000	$600,000	$1,100,000
Variable costs @ $4 and $3	400,000	180,000	580,000
Contribution margins @ $1 and $7	$100,000	$420,000	$ 520,000
Fixed costs			300,000
Operating income			$ 220,000

Managers naturally want to maximize the sales of all their products. Then why worry about the sales mix? The analysis of a change in mix often clarifies why actual and budgeted sales and profits differ. Coleco Industries provides a notorious example. In 1983, Coleco was budgeting enormous sales and profits from its Adam personal computer. At the same time, Coleco was expecting more modest sales from another major product, the Cabbage Patch doll. The Adam flopped, but the doll soared. This change in sales mix was the principal explanation for how actual profits differed from budgeted profits.

Despite their desire to maximize the sales of all products, managers must frequently cope with limited resources. For instance, additional production capacity may be unavailable. What products should be produced? As Chapters 11 and 24 explain in more detail, the decision is not necessarily to make the product having the higher unit contribution margin. After all, suppose a company can make 1,000 Peerless pens for each hour of capacity instead of 100 Fantastic pens. Then Peerless can generate a contribution margin of 1,000 × $1 = $1,000 per hour, whereas Fantastic can generate only 100 × $7 = $700 per hour.

In sum, cost-volume-profit analysis must be done carefully because so many initial assumptions may not hold. When the assumed conditions change, the breakeven point and the expected operating incomes at various volume levels also change. Of course, the breakeven points are frequently incidental data. Instead, the focus is on the effects on operating income under various production and sales strategies.

ROLE OF INCOME TAXES

When we introduced a profit element in our earlier software example, the following income statement was shown:

Sales, 40 units × \$200	\$8,000
Variable costs, 40 × \$120	4,800
Contribution margin	\$3,200
Fixed costs	2,000
Operating income	\$1,200

Objective 7

Compute the cost-volume-profit relationships on an after-tax basis

Suppose we now want to compute the number of units to be sold to earn net income of \$1,200, assuming federal and state income taxes at a rate of 40%. The only change in the general equation approach is to modify the target operating income to allow for the impact of income taxes. Recall our previous equation approach:

$$\text{Sales} - \text{Variable costs} - \text{Fixed costs} = \text{Operating income}$$

We now have to introduce income tax effects:

$$\text{Target net income} = (\text{Operating income}) - (\text{Tax rate})(\text{Operating income})$$
$$\text{Target net income} = (\text{Operating income})(1 - \text{Tax rate})$$
$$\text{Operating income} = \frac{\text{Target net income}}{1 - \text{Tax rate}}$$

The equation would now be:

$$\$200N - \$120N - \$2,000 = \frac{\text{Target net income}}{1 - \text{Tax rate}}$$
$$\$200N - \$120N - \$2,000 = \frac{\$1,200}{1 - .40}$$
$$\$200N - \$120N - \$2,000 = \$2,000$$
$$\$80N = \$4,000$$
$$N = 50 \text{ units}$$

Proof:		
	Sales, 50 × \$200	\$10,000
	Variable costs, 50 × \$120	6,000
	Contribution margin	\$ 4,000
	Fixed costs	2,000
	Operating income	\$ 2,000
	Income taxes, 40% of \$2,000	800
	Net income	\$ 1,200

Note again that

$$\text{Target operating income} = \frac{\text{Target net income}}{1 - \text{Tax rate}}$$

Suppose the target net income were set at \$1,680, not \$1,200. The needed volume would rise from 50 to 60 units:

$$\$200N - \$120N - \$2{,}000 = \frac{\$1{,}680}{1 - .40}$$

$$\$80N - \$2{,}000 = \$2{,}800$$

$$\$80N = \$4{,}800$$

$$N = 60 \text{ units}$$

The following formula is a shortcut computation of the effects of volume on net income:

$$\text{Change in net income} = \begin{pmatrix}\text{Change}\\ \text{in units}\end{pmatrix} \times \begin{pmatrix}\text{Unit}\\ \text{contribution}\\ \text{margin}\end{pmatrix} \times (1 - \text{Tax rate})$$

In our example:

$$\begin{aligned}\text{Change in net income} &= (60 - 50) \times \$80 \times .60\\ &= 10 \times \$80 \times .60\\ &= \$480, \text{ which would be the increase in target net income from \$1,200 to \$1,680}\end{aligned}$$

In short, each unit beyond the breakeven point adds to net income at the unit contribution margin multiplied by (1 − tax rate).

Note that throughout our illustration, the breakeven point itself is unchanged. Why? Because there is no income tax at a level of zero income.

MEASURING VOLUME

In our examples so far, we have used units of product as a measure of volume. But "volume" is a general term for quantity levels that can be measured in several ways. For example, the volume of a hospital is often expressed in patient-days (number of patients multiplied by the average number of days each patient remained in the hospital). The volume of teaching at a university is often expressed in terms of total student credit hours taken, not number of students enrolled.

The breakeven point is frequently expressed in different forms in different industries. A letter to the shareholders of Trans World Airlines (TWA) stated that personnel reductions and improved scheduling had lowered the company's "breakeven load factor from 52% to 50% (percentage of seats occupied)." Essentially, TWA had improved its breakeven point by lowering fixed costs via personnel reductions and by running the airplanes more intensively.

ALL DATA IN DOLLARS

The most widely encountered aggregate measure of the volume is sales dollars, a monetary rather than a physical calibration of activity. Few companies sell only one product. Apples and oranges can be added together in physical units, but such a sum is harder to interpret than their sales in *dollars.* Similarly, the chief executive of a retailing chain such as Sears or Woolworth would get little help from knowing the total units sold of its thousands of products. Cost-volume-profit analysis is basically the same whether it is anchored to units or to dollars.

Suppose the Ramirez Company has the following income statement:

Sales	$120,000	100%
Variable costs	48,000	40%
Contribution margin	$ 72,000	60%
Fixed costs	60,000	
Operating income	$ 12,000	

The relationships in dollars may be expressed:

$$\text{Contribution-margin ratio or percentage} = \frac{\text{Contribution margin}}{\text{Sales}} = .60 \text{ or } 60\%$$

$$\text{Variable-cost ratio or percentage} = \frac{\text{Variable costs}}{\text{Sales}} = .40 \text{ or } 40\%$$

Using the variable-cost ratio, the breakeven point could be calculated as follows:

$$\begin{aligned}
\text{Let } S &= \text{Total sales dollars to break even} \\
S - (\text{Variable-cost ratio})S - \text{Fixed costs} &= \text{Operating income} \\
S - .40S - \$60{,}000 &= 0 \\
S - .60S &= \$60{,}000 \\
S &= \frac{\$60{,}000}{.60} = \$100{,}000
\end{aligned}$$

Or using the contribution-margin ratio in a formula:

$$\text{Breakeven point} = \frac{\text{Fixed costs}}{\text{Contribution-margin ratio}}$$

$$S = \frac{\$60{,}000}{.60} = \$100{,}000$$

Suppose we want to know the dollar sales needed to earn an operating income of 20% of sales. The equation approach would be:

$$\begin{aligned}
\text{Let } S &= \text{Sales in dollars} \\
\text{Sales} - \text{Variable costs} - \text{Fixed costs} &= \text{Target operating income} \\
S - .40S - \$60{,}000 &= .20S \\
.60S - \$60{,}000 &= .20S \\
.40S &= \$60{,}000 \\
S &= \$150{,}000
\end{aligned}$$

The equation approach just shown could also be recast in a formula that is frequently cited:

$$\text{Sales in dollars} = \frac{\text{Fixed costs} + \text{Target operating income}}{\text{Contribution-margin ratio}}$$

$$\begin{aligned}
S &= \frac{\$60{,}000 + .20S}{.60} \\
.60S &= \$60{,}000 + .20S \\
.40S &= \$60{,}000 \\
S &= \$150{,}000
\end{aligned}$$

NONPROFIT INSTITUTIONS AND COST-VOLUME-REVENUE ANALYSIS

Suppose a social welfare agency has a government budget appropriation for 19_1 of $900,000. The agency's major purpose is to help handicapped persons who are unable to hold jobs. On the average, the agency supplements each person's income by $5,000 annually. The agency's fixed costs are $270,000. There are no other costs. The agency manager wants to know how many persons could be served in 19_1:

$$
\begin{aligned}
\text{Let } N &= \text{Number of persons to be served} \\
\text{Revenue} - \text{Variable costs} - \text{Fixed costs} &= 0 \\
\$900{,}000 - \$5{,}000N - \$270{,}000 &= 0 \\
\$5{,}000N &= \$900{,}000 - \$270{,}000 \\
N &= \$630{,}000 \div \$5{,}000 \\
N &= 126
\end{aligned}
$$

Suppose the manager is concerned that the total budget for 19_1 will be reduced by 15% to a new amount of (1 − .15) ($900,000) = $765,000. The manager wants to know how many handicapped persons will be helped. Assume the same amount of monetary support per person:

$$
\begin{aligned}
\$765{,}000 - \$5{,}000N - \$270{,}000 &= 0 \\
\$5{,}000N &= \$765{,}000 - \$270{,}000 \\
N &= \$495{,}000 \div \$5{,}000 \\
N &= 99
\end{aligned}
$$

Note the two characteristics of the cost-volume-revenue relationships in this nonprofit situation:

1. The percentage drop in service is (126 − 99) ÷ 126, or 21.4%, which is more than the 15% reduction in the budget.
2. If the relationships were graphed, the revenue amount would be a straight horizontal line of $765,000. The manager could adjust operations in one or more of three major ways: (a) cut the volume level, as calculated above; (b) alter the variable cost, the supplement per person; and (c) alter the total fixed costs.

CVP MODELS, PERSONAL COMPUTERS, AND SPREADSHEETS

The CVP model is widely used as a planning model. Moreover, the expanding use of personal computers has led to CVP applications in various organizations and situations. Managers test their plans on the computers, which quickly show changes both numerically and graphically. Managers study assorted combinations

EXHIBIT 3-3

Spreadsheet Analysis of CVP Relationships of Do-All Software Package Example

		Sales Required to Earn Operating Income of		
Fixed Costs	**Variable Costs**	**$1,000**	**$1,500**	**$2,000**
$2,000	50% of Sales	6,000*	7,000	8,000
	60% of Sales	7,500	8,750	10,000
	70% of Sales	10,000	11,667	13,333
$2,500	50% of Sales	7,000†	8,000	9,000
	60% of Sales	8,750	10,000	11,250
	70% of Sales	11,667	13,333	15,000
$3,000	50% of Sales	8,000	9,000	10,000
	60% of Sales	10,000	11,250	12,500
	70% of Sales	13,333	15,000	16,667

*($2,000 + $1,000) ÷ (1.00 − .50) = $6,000
($2,000 + $1,000) ÷ (1.00 − .60) = $7,500
($2,000 + $1,000) ÷ (1.00 − .70) = $10,000

†($2,500 + $1,000) ÷ (1.00 − .50) = $7,000
($2,500 + $1,000) ÷ (1.00 − .60) = $8,750
($2,500 + $1,000) ÷ (1.00 − .70) = $11,667

of changes in selling prices, unit variable costs, fixed costs, and target net incomes. Computerized CVP planning models are used in nonprofit organizations. For instance, hospitals use the models to predict the financial effects of various patient mixes, nursing and other staff requirements, and patient volume.

Computer spreadsheet programs help managers conduct sensitivity analysis. The programs quickly calculate the results of one or more changes in key figures. Managers can enter in different numbers for fixed costs, variable costs, and target operating income in any combination. Using the numbers in our Do-All example, p. 61, Exhibit 3-3 displays how a spreadsheet might appear. The spreadsheet program uses an equation based on CVP relationships. The computer puts the various input amounts in the equation and calculates the 27 operating income amounts that appear in the exhibit. The computer performs these calculations in seconds and without error. In this way, the managers search for their best courses of action.

PROBLEMS FOR SELF-STUDY

PROBLEM 1

A person wants to sell software at a computer convention. Booth rental is $2,000. The unit selling price is $200, and the unit variable cost is $120. How many units must be sold to attain a target operating income of $1,440?

SOLUTION 1

The equation method follows:

Let N = Number of units to be sold to earn target operating income

$$\$200N - \$120N - \$2{,}000 = \$1{,}440$$
$$\$80N = \$3{,}440$$
$$N = 43 \text{ units}$$

PROBLEM 2

Here is the income statement of a British consulting firm (in British £):

Revenues		£ 500,000
Deduct costs:		
Variable	£350,000	
Fixed	250,000	600,000
Operating income (loss)		£(100,000)

Assume that variable costs will remain the same percentage of sales:

a. If fixed costs are increased by £100,000, what amount of sales will cause the firm to break even?

b. With the proposed increase in fixed costs, what amount of sales will yield an operating income of £50,000?

SOLUTION 2

a. Let S = Breakeven sales in British pounds

$$S - \text{Variable costs} - \text{Fixed costs} = \text{Target operating income}$$

$$S - \frac{£350,000}{£500,000} S - (£250,000 + £100,000) = 0$$

$$S - .70S - £350,000 = 0$$

$$.30S = £350,000$$

$$S = £1,166,667$$

b. Let S = Sales needed to earn £50,000 operating income

$$S - .70S - £350,000 = £50,000$$

$$.30S = £400,000$$

$$S = £1,333,333$$

For additional explanation, see the "All Data in Dollars" section, page 73.

SUMMARY

Properly used, cost-volume-profit analysis helps decision making. It offers an overall view of sales and costs as profits are planned for a coming accounting period, and it provides clues to possible changes in strategic plans. CVP models assist in answering what-if questions such as changing a price or conducting an advertising campaign.

Whenever the underlying assumptions of cost-volume-profit analysis do not correspond to a given situation, the limitations of the analysis must be clearly recognized. A single CVP graph is static because it is a picture of relationships that prevail under only one set of assumptions. If conditions change, a different set of cost-volume-profit relationships is likely to appear. The fluid nature of these relationships must be kept uppermost in the minds of executives and accountants.

Check your understanding of basic terms. For example, as page 68 explains, contribution margin and gross margin have different meanings. Contribution margin may be expressed as a total amount, as an amount per unit, as a ratio, or as a percentage.

The contribution margin is the difference between sales and all *variable* costs, including *variable* marketing and *variable* administrative costs. In contrast, gross profit or gross margin is the difference between sales and the *cost of manufacturing goods sold,* including the *fixed* factory overhead.

Changes in sales mix can have a significant impact on the overall breakeven point.

TERMS TO LEARN

This chapter and the Glossary at the end of the book contain definitions of the following important terms:

breakeven point *(p. 60)* contribution margin *(61)*
contribution-margin ratio *(69)* gross margin *(68)* gross profit *(68)*
margin of safety *(67)* net income *(60)* operating income *(60)*
P/V chart *(69)* revenue driver *(59)* sales mix *(70)* sensitivity analysis *(67)*
uncertainty *(67)* variable-cost ratio *(69)*

ASSIGNMENT MATERIAL

QUESTIONS

Special note: *To underscore the basic cost-volume-profit relationships, unless otherwise stated, this assignment material ignores income taxes.*

3-1 "There are many cost drivers besides volume of units produced." Name three.

3-2 Give three examples of revenue drivers besides volume of units sold.

3-3 "Cost-volume-profit analysis represents the specific case." What is an accurate label for the general case?

3-4 Identify three major methods of cost-volume-profit analysis.

3-5 State a general formula for computing the breakeven point in units of a single product and a volume-based cost driver.

3-6 State a general formula for computing the target number of units that must be sold to reach a target operating income for a single product and a volume-based cost driver.

3-7 Name three factors that affect whether costs are really fixed.

3-8 Why is it more accurate to describe the subject matter of this chapter as *cost-volume-profit relationships* rather than as *breakeven analysis?*

3-9 What is a principal difference between the accountant's and the economist's breakeven charts?

3-10 "Advocates of the contribution approach maintain that fixed costs are unimportant." Do you agree? Explain.

3-11 "Gross margin is contribution margin minus fixed factory overhead." True or false? Explain.

3-12 "Even in retailing, where cost of goods sold consists entirely of variable costs, gross margin is not necessarily equal to the contribution margin." True or false? Explain.

3-13 Define *contribution margin, variable-cost ratio, contribution-margin ratio,* and *margin of safety.*

3-14 How is the margin of safety related to sensitivity analysis?

3-15 "The variable-cost ratio plus the contribution-margin ratio will always equal 1.00." True or false? Explain.

3-16 "This breakeven approach is great stuff. All you need to do is worry about variable costs. The fixed costs will take care of themselves." Discuss.

3-17 Give an example of how a manager may decrease variable costs while increasing fixed costs.

3-18 Give an example of how a manager may increase variable costs while decreasing fixed costs.

3-19 Give an example of how a company may reduce its breakeven point by altering its contribution margin per unit of product.

3-20 A lithographic company follows a policy of high pricing each month until it reaches its monthly breakeven point. After this point is reached, the company tends to quote low prices on jobs for the rest of the month. What is your opinion of this policy? As a regular customer, and suspecting this policy, what would you do?

EXERCISES AND PROBLEMS

3-21 Fill in blanks. In the data presented below, fill in the information that belongs in the blank spaces for each of the four unrelated cases.

Sales	Variable Costs	Fixed Costs	Total Costs	Operating Income	Contribution-Margin Ratio
a. $ —	$500	$ —	$ 800	$1,200	—
b. 2,000	—	300	—	200	—
c. 1,000	700	—	1,000	—	—
d. 1,500	—	300	—	—	.40

3-22 Fill in blanks. Fill in the blanks for each of the following independent cases.

(a) Selling Price Per Unit	(b) Variable Cost Per Unit	(c) Total Units Sold	(d) Total Contribution Margin	(e) Total Fixed Costs	(f) Operating Income
1. $30	$20	70,000	$ —	$ —	$15,000
2. 25	—	180,000	900,000	800,000	—
3. —	10	150,000	300,000	220,000	—
4. 20	14	—	120,000	—	12,000

3-23 Changing the relationships. Maria Montez is planning to sell a vegetable slicer-dicer for $15 per unit at a county fair. She purchases units from a local distributor for $6 each. She can return any unsold units for full credit. Fixed costs for booth rental, setup, and cleaning costs are $450.

Required

1. Compute the breakeven point in units.
2. Suppose the unit purchase cost is $5 instead of $6, but the sales price is unchanged. Compute the new breakeven point in units.

3-24 Exercises in cost-volume-profit relationships. The Super Buy Grocers Corporation owns and operates six supermarkets in and around Chicago. You are given the following corporate budget data for next year:

Sales	$10,000,000
Fixed costs	1,700,000
Variable costs	8,200,000

Required

Compute expected operating income for each of the following deviations from budgeted data. (Consider each case independently.)

a. A 10% increase in total contribution margin, holding sales constant
b. A 10% decrease in total contribution margin, holding sales constant
c. A 5% increase in fixed costs
d. A 5% decrease in fixed costs
e. An 8% increase in sales volume
f. An 8% decrease in sales volume
g. A 10% increase in fixed costs and 10% increase in sales volume
h. A 5% increase in fixed costs and 5% decrease in variable costs

3-25 Cost-volume-profit relationships. The Doral Company makes and sells pens. Some pertinent facts follow:
Present sales volume is 5,000,000 units per year at a selling price of 50¢ per unit. Fixed costs are $900,000 per year. Variable costs are 30¢ per unit.

Required
(Consider each case separately.)

1. a. What is the present operating income for a year?
   b. What is the present breakeven point in dollars?

Compute the new operating income for each of the following changes:

2. A 4¢-per-unit increase in variable costs
3. A 10% increase in fixed costs and a 10% increase in sales volume
4. A 20% decrease in fixed costs, a 20% decrease in selling price, a 10% decrease in variable costs per unit, and a 40% increase in units sold

Compute the new breakeven point in units for each of the following changes:

5. A 10% increase in fixed costs
6. A 10% increase in selling price and a $20,000 increase in fixed costs

3-26 Choosing profitable volume level. Shapiro Company manufactures and sells pens at a variable cost of $3 each and a fixed cost of *F*. It can sell 600,000 pens at $5 and $200,000 operating income, or it can sell 350,000 at $6 and another 200,000 at $4 each. Which alternative should Shapiro choose?

3-27 Cost-volume-profit and shoe stores. The Walk Rite Shoe Company operates a chain of rented shoe stores. The stores sell ten different styles of inexpensive men's shoes with identical purchase costs and selling prices. Walk Rite is trying to determine the desirability of opening another store, which would have the following cost and revenue relationships:

	Per Pair
Variable data:	
Selling price	$ 30.00
Cost of shoes	$ 19.50
Sales commissions	1.50
Total variable costs	$ 21.00
Annual fixed costs:	
Rent	$ 60,000
Salaries	200,000
Advertising	80,000
Other fixed costs	20,000
	$360,000

Required
(Consider each question independently.)

1. What is the annual breakeven point in unit sales and in dollar sales?
2. If 35,000 pairs of shoes are sold, what would be the store's operating income (loss)?
3. If the store manager were paid 30¢ per pair as commission, what would be the annual breakeven point in dollar sales and in unit sales?
4. Refer to the original data. If sales commissions were discontinued in favor of an $81,000 increase in fixed salaries, what would be the annual breakeven point in dollars and in unit sales?
5. Refer to the original data. If the store manager were paid 30¢ per pair as commission on each pair sold in excess of the breakeven point, what would be the store's operating income if 50,000 pairs were sold?

3-28 Extension of preceding problem. Refer to requirement 4 of the preceding problem.

1. Calculate the volume level in units where the operating income under a fixed salary plan and a commission plan would be equal. Above that volume level, one plan would be more profitable than the other; below that level, the reverse would occur.
2. Compute the operating income or loss under each plan at volume levels of 50,000 units and 60,000 units.
3. Suppose the target operating income is $168,000. How many units must be sold to reach the target under (a) the commission plan and (b) the salary plan?

3-29 Sensitivity and inflation. Refer to Problem 3-27. As president of Walk Rite, you are concerned that inflation may squeeze your profits. Specifically, you feel committed to the $30 selling price and fear that diluting the quality of the shoes in the face of rising costs would be an unwise marketing move. You expect cost prices of the shoes to rise by 10% during the coming year. You are tempted to avoid the cost increase by placing a noncancelable order with a large supplier that would provide 50,000 pairs of a specified quality for each store at $19.50 per pair. (To simplify this analysis, assume that all stores will face identical demands.) These shoes could be acquired and paid for as delivered throughout the year. However, all shoes must be delivered to the stores by the end of the year.

As a shrewd merchandiser, you foresee some risks. If sales were less than 50,000 pairs, you feel that markdowns of the unsold merchandise would be necessary to sell the goods. You predict that the average selling price of the leftover pairs would be $18.00 each. The regular sales commission of 5% of sales would be paid.

Required

1. Suppose that the actual demand for the year is 48,000 pairs and that you contracted for 50,000 pairs. What is the operating income for the store?
2. If you had had perfect knowledge, you would have contracted for 48,000 rather than 50,000 pairs. What would the operating income have been if you had ordered 48,000 pairs?
3. Given an actual volume of 48,000 pairs, by how much would the average purchase cost of a pair have had to rise before you would have been indifferent between having the contract for 50,000 pairs and not having the contract?

3-30 Miscellaneous relationships, margin of safety. Suppose the breakeven point is sales of $1,000,000. Fixed costs are $400,000.

1. Compute the contribution-margin ratio.
2. Compute the sales price per unit if variable costs are $12 per unit.
3. Suppose 80,000 units are sold. Compute the margin of safety.

3-31 Margin of safety, four alternatives. The Axel Swang Company has the following budget data for selling 300,000 T-shirts to various retailers:

Sales	$930,000	(100%)
Variable costs	325,500	(35%)
Contribution margin	$604,500	(65%)
Fixed costs	520,000	
Operating income	$ 84,500	

Since management is not satisfied with the projected operating income, it is considering four independent possibilities: (1) increasing unit volume 10%; (2) increasing unit selling price 10%; (3) decreasing unit variable costs 10%; or (4) decreasing fixed costs 10%.

Required

1. Without considering any of the independent possibilities, compute the breakeven sales and the margin of safety as a percentage of sales.
2. Rank the four independent possibilities in terms of operating income. Show computations. Are the four possibilities likely to be independent? Why?

3-32 Target prices and operating income (SMA) The Martell Company has recently established operations in a highly competitive market. Management has been aggressive in its attempt to establish a market share.

The price of its product was set at $5 per unit, well below the major competition's selling price. Variable costs were $4.50 per unit, and total fixed costs were $600,000 during the first year.

Required

1. Assume that the firm was able to sell 1,000,000 units in the first year. What was the operating income (loss) for the year?
2. Management has been successful in establishing its position in the market. What price must be set to achieve an operating income of $30,000? Assume that variable costs per unit and total fixed costs do not change in the second year and that output cannot be increased over the first-year level.

3-33 Automation and most profitable volume level. Kelly Company manufactures and sells variations of a basic hand calculator. Kelly is about to expand and faces many combinations of decisions such as whether to make or buy specified components, whether to use few or many robotics, whether to price high or low, and whether to advertise heavily or lightly.

After studying assorted combinations, Kelly has narrowed its choices to the following alternatives:

- Produce and sell up to 3,000,000 units annually at a variable cost of $1.50 per unit and fixed costs of $10,000,000.
- Produce and sell from 3,000,001 units to 6,000,000 units annually at a variable cost of $1.00 per unit and fixed costs of $13,000,000.
- Produce and sell from 6,000,001 units to 10,000,000 units at a variable cost of $.60 per unit and fixed costs of $19,000,000.

Michael Kelly, the president, predicts that 2,500,000 units can be sold for $6 each, or 5,000,000 units for $4 each. If advertising were increased by $1,000,000 and selling costs by $.10 per unit, 8,000,000 units could be sold for $3.50 each. The latter costs are in addition to those already stated for the 6,000,001–10,000,000 unit range.

Required

How many units should Kelly plan to produce and sell: 2.5 million, 5 million, or 8 million?

3-34 Advertising by CPA firms. In late 19_8, a San Francisco CPA firm, Siegel, Sugurman & Seput, decided to advertise in local news publications. First-year billings (revenues) from new clients who called the firm in response to advertisements were about $100,000. These revenues met the firm's goal of attracting clients whose combined annual gross billings are at least double the cost of the campaign. The firm believes that this advertising cost is one-third of the typical cost of acquiring a practice, that is, of buying an existing accounting firm.

Required

Compute the probable cost of acquiring an accounting firm having annual gross billings of $200,000.

3-35 Target operating incomes and contribution margins. The Kaplan Company has a maximum capacity of 200,000 units per year. Variable manufacturing costs are $12 per unit. Fixed factory overhead is $600,000 per year. Variable marketing and administrative costs are $5 per unit, and fixed marketing and administrative costs are $300,000 per year. Current sales price is $23 per unit.

Required

(Consider each situation independently.)

1. What is the breakeven point in (a) units? (b) dollar sales?
2. How many *units* must be sold to earn a target operating income of $240,000 per year?
3. Assume that the company's sales for the year just ended totaled 185,000 units. A strike at a major supplier has caused a materials shortage, so that the current year's sales will

reach only 160,000 units. Top management is planning to slash fixed costs so that the total for the current year will be $85,000 less than last year. Management is also thinking of increasing the selling price, reducing variable costs, or both, in order to earn a target operating income that will be the same dollar amount as last year's. The company has already sold 30,000 units this year at a sales price of $23 per unit with variable costs per unit unchanged. What contribution margin per unit is needed on the remaining 130,000 units in order to reach the target operating income?

3-36 Channels of distribution. Martinez Co., a manufacturer of stationery supplies, has always sold its products through wholesalers. Last year its sales were $2 million and its operating income was $180,000.

As a result of the increase in stationery sales in department stores and discount houses, Martinez is considering eliminating its wholesalers and selling directly to retailers. It is estimated that this would result in a 40% drop in sales, but operating income would be $160,000 as a result of eliminating the wholesaler. Fixed costs would increase from the present figure of $220,000 to $320,000 because of the additional warehouses and distribution facilities required.

Required

1. Would the proposed change raise or lower the breakeven point in dollars? By how much?
2. What dollar sales volume must Martinez attain under the proposed plan to make 10% more operating income than it made last year?

3-37 Effects of size of machines. Chan Pastries, Inc., is planning to manufacture doughnuts for its chain of pastry shops throughout Hong Kong. Two alternatives have been proposed for the production of the doughnuts—use of a semiautomatic machine or use of a fully automatic machine. The shops now purchase their doughnuts from an outside supplier at a cost of 30¢ per doughnut. Projected data follow:

	Semiautomatic	Automatic
Annual fixed cost	$25,000	$45,000
Variable cost per doughnut	$ .10	$ .05

Required

The president has asked for the following information:

1. For each machine, the minimum annual number of doughnuts that must be sold in order to have the total annual costs equal to outside purchase costs
2. The most profitable alternative for 250,000 doughnuts annually
3. The most profitable alternative for 500,000 doughnuts annually
4. The volume level that would produce the same total costs regardless of the type of machine owned

3-38 Choice of production method. The Zuber Company has just been incorporated and plans to produce an orthopedic chair that will sell for $500 per unit. Preliminary market surveys show that demand will be less than 10,000 units per year, but it is not as yet clear how much less.

The company has the choice of arranging a manufacturing process using one of two machines, each of which has a capacity of 10,000 units per year. A special machine will result in total fixed costs for the company of $1,800,000 per year and would yield operating income of $1,200,000 if sales were 10,000 units. A general machine will result in total fixed costs of $1,100,000 per year and would yield operating income of $900,000 if sales were 10,000 units. Variable costs behave linearly for both machines.

Required

1. Compute the breakeven sales for each machine, in dollars and in units.
2. At what sales level are the machines equally profitable?
3. What is the range of sales where one machine is more profitable than the other?

3-39 Choice of production method. (CMA, adapted) Candice Company has decided to introduce a new product. The new product can be manufactured by either a capital-intensive method or a labor-intensive method. The manufacturing method will not affect the quality of the product. The estimated manufacturing costs under the two methods follow.

	Capital Intensive		Labor Intensive	
Raw materials per unit		$5.00		$5.60
Direct labor per unit	.5DLH @ $12	6.00	.8DLH @ $9	7.20
Variable overhead per unit	.5DLH @ $6	3.00	.8DLH @ $6	4.80
Directly traceable additional annual fixed manufacturing costs		$2,440,000		$1,320,000

DLH stands for direct labor hours.

Candice's market research department has recommended an introductory unit sales price of $30. The additional annual marketing expenses are estimated to be $500,000 plus $2 for each unit sold regardless of manufacturing method.

Required

1. Calculate the estimated breakeven point in annual unit sales of the new product if Candice Company uses (a) the capital-intensive manufacturing method or (b) the labor-intensive manufacturing method.
2. Determine the annual unit sales volume at which Candice Company would be indifferent between the two manufacturing methods. Explain the circumstances under which Candice should employ each of the two manufacturing methods.
3. Identify the business factors that Candice must consider before selecting the capital-intensive or labor-intensive manufacturing method.

3-40 Nonprofit institution. A city has a $400,000 lump-sum budget appropriation for a government agency to conduct a counseling program for drug addicts for a year. All of the appropriation is to be spent. The variable costs for drug prescriptions average $400 per patient per year. Fixed costs are $150,000.

Required

1. Compute the number of patients that could be served in a year.
2. Suppose the total budget for the following year is reduced by 10%. Fixed costs are to be unchanged. The same level of service to each patient will be maintained. Compute the number of patients that could be served in a year.
3. As in requirement 2, assume a budget reduction of 10%. Fixed costs are to be unchanged. The drug counselor has discretion as to how much in drug prescriptions to give to each patient. She does not want to reduce the number of patients served. On the average, what is the cost of drugs that can be given to each patient? Compute the percentage decline in the annual average cost of drugs per patient.

3-41 Sales mix, two products. Goldman Company has two products, a standard and a deluxe version of a luggage carrier. The income budget follows.

	Standard Carrier	Deluxe Carrier	Total
Sales in units	150,000	50,000	200,000
Sales @ $20 and $30	$3,000,000	$1,500,000	$4,500,000
Variable costs @ $14 and $18	2,100,000	900,000	3,000,000
Contribution margins @ $6 and $12	$ 900,000	$ 600,000	$1,500,000
Fixed costs			1,200,000
Operating income			$ 300,000

Required

1. Compute the breakeven point in units, assuming that the planned sales mix is maintained.

2. Compute the breakeven point in units (a) if only standard carriers are sold and (b) if only deluxe are sold.
3. Suppose 200,000 units are sold, but only 20,000 are deluxe. Compute the operating income. Compute the breakeven point if these relationships persist in the next period. Compare your answers with the original plans and the answer in requirement 1. That is, what is the major lesson in this problem?

3-42 Sales mix, three products. (Alternate is 3-43.) The Ronowski Company has three product lines of belts—A, B, and C—having contribution margins of $3, $2, and $1, respectively. The president foresees sales of 200,000 units in the coming period, consisting of 20,000 A, 100,000 B, and 80,000 C. The company's fixed costs for the period are $255,000.

Required

1. What is the company breakeven point in units, assuming that the given sales mix is maintained?
2. If the mix is maintained, what is the total contribution margin at a volume of 200,000 units? What is operating income?
3. What would operating income become if 20,000 units of A, 80,000 units of B, and 100,000 units of C were sold? What is the new breakeven point in units if these relationships persist in the next period?

3-43 Sales mix, three products. (Alternate is 3-42.) Mendez Company has three products, tote bags H, J, and K. The president plans to sell 200,000 units during the next period, consisting of 80,000 H, 100,000 J, and 20,000 K. The products have unit contribution margins of $2, $3, and $6, respectively. The company's fixed costs for the period are $406,000.

Required

1. Compute the planned operating income. Compute the breakeven point in units, assuming that the given sales mix is maintained.
2. Suppose 80,000 units of H, 80,000 units of J, and 40,000 units of K are sold. Compute the operating income. Compute the new breakeven point in units if these relationships persist in the next period.

3-44 Product mix effects. (CMA) Kalifo Company manufactures a line of electric garden tools that are sold in general hardware stores. The company's controller, Sylvia Harlow, has just received the sales forecast for the coming year for Kalifo's three products: weeders, hedge clippers, and leaf blowers. Kalifo has experienced considerable variations in sales volumes and variable costs over the past two years, and Harlow believes the forecast should be carefully evaluated from a cost-volume-profit viewpoint. The preliminary budget information for 19_8 is presented below.

	Weeders	Hedge Clippers	Leaf Blowers
Unit sales	50,000	50,000	100,000
Unit selling price	$28.00	$36.00	$48.00
Variable manufacturing cost per unit	13.00	12.00	25.00
Variable selling cost per unit	5.00	4.00	6.00

For 19_8, Kalifo's fixed factory overhead is budgeted at $2,000,000, and the company's fixed selling and administrative expenses are forecast to be $600,000. Kalifo has an effective tax rate of 40%.

Required

1. Determine Kalifo Company's budgeted net income for 19_8.
2. Assuming the sales mix remains as budgeted, determine how many units of each product Kalifo Company must sell in order to break even in 19_8.
3. Determine the total dollar sales Kalifo Company must have in 19_8 in order to earn a net income of $450,000.

4. After preparing the original estimates, Kalifo Company determined that its variable manufacturing cost of leaf blowers would increase 20% and that the variable selling cost of hedge clippers could be expected to increase $1.00 per unit. However, Kalifo has decided not to change the selling price of either product. In addition, Kalifo has learned that its leaf blower has been perceived as the best value on the market, and it can expect to sell three times as many leaf blowers as any other product. Under these circumstances, determine how many units of each product Kalifo Company would have to sell in order to break even in 19_8.
5. Explain the limitations of cost-volume-profit analysis that Sylvia Harlow should consider when evaluating Kalifo Company's 19_8 budget.

3-45 Role of income taxes. Ava Moreno is planning to sell a special kitchen knife for $15 per unit at a county fair. She purchases units from a local distributor for $6 each. She can return any unsold units for full credit. Fixed costs for the booth rental, setup and cleaning costs are $460. Assume an income tax rate of 30%.

Required
How many units must be sold to obtain the target net income of $560?

3-46 Role of income taxes. Here is the income statement of the Wilton Company, a British consulting firm (in British pounds, £):

Revenue		£500,000
Deduct costs		
Variable	£350,000	
Fixed	350,000	700,000
Operating loss		£(200,000)

Assume that variable costs will remain the same percentage of revenue. If fixed costs are increased by £60,000, what amount of revenue will produce a target net income of £60,000? The income tax rate is 40%.

3-47 Role of income taxes. The Bratz Company had fixed costs of $300,000 and a variable-cost ratio of 80%. The company earned net income of $84,000 in 19_3. The income tax rate was 40%.

Required
Compute (1) operating income, (2) contribution margin, (3) total sales, and (4) breakeven point in dollar sales.

3-48 Role of income taxes. The Kleespie Company had an after-tax net income of $150,000 in 19_2. The income tax rate was 25%. Kleespie had a contribution margin of $900,000 and a contribution-margin ratio of 60%.

Required
Compute (1) operating income, (2) total fixed costs, (3) total sales, and (4) breakeven point in dollar sales.

3-49 Restaurant income taxes. The Rapid Meal has two restaurants that are open 24 hours per day. Fixed costs for the two together total $450,000 per year. Service varies from a cup of coffee to full meals. The average sales check for each customer is $8.00. The average cost of food and other variable costs for each customer is $3.20. The income tax rate is 30%. The target net income is $105,000.

Required
1. Compute the total sales needed to obtain the target net income.
2. How many sales checks are needed to earn net income of $105,000? To break even?
3. Compute the net income if the number of sales checks is 150,000.

3-50 Cost-volume relationships, income taxes. (CMA) R. A. Ro and Company, maker of quality handmade pipes, has experienced a steady growth in sales for the past five years. However, increased competition has led Mr. Ro, the president, to believe that an aggressive advertising campaign will be necessary next year to maintain the company's present growth.

To prepare for next year's advertising campaign, the company's accountant has prepared and presented Mr. Ro with the following data for the current year, 19_2:

Variable costs (per pipe):	
Direct labor	$ 8.00
Direct materials	3.25
Variable overhead	2.50
Total variable costs	$13.75
Fixed costs:	
Manufacturing	$ 25,000
Selling	40,000
Administrative	70,000
Total fixed costs	$135,000
Selling price per pipe	$25.00
Expected sales, 19_2 (20,000 units)	$500,000
Tax Rate: 40%	

Required

1. What is the projected net income for 19_2?
2. What is the breakeven point in units for 19_2?
3. Mr. Ro has set the sales target for 19_3 at a level of $550,000 (or 22,000 pipes). He believes an additional selling expense of $11,250 for advertising in 19_3, with all other costs remaining constant, will be necessary to attain the sales target. What will be the net income for 19_3 if the additional $11,250 is spent and the sales goal is met?
4. What will be the breakeven point in dollar sales for 19_3 if the additional $11,250 is spent for advertising?
5. If the additional $11,250 is spent for advertising in 19_3, what is the required sales level in dollar sales to equal 19_2's net income?
6. At a sales level of 22,000 units, what maximum amount can be spent on advertising if an net income of $60,000 is desired?

3-51 ***Review of Chapters 2 and 3.*** For each of the following independent cases, find the unknowns designated by the capital letters.

	Case 1	Case 2
Direct materials used	$ H	$ 40,000
Direct labor	30,000	15,000
Variable marketing and administrative costs	K	T
Fixed manufacturing overhead	I	20,000
Fixed marketing and administrative costs	J	10,000
Gross profit	25,000	20,000
Finished goods inventory, 1/1	0	5,000
Finished goods inventory, 12/31	0	5,000
Contribution margin (dollars)	30,000	V
Sales	100,000	100,000
Direct materials inventory, 1/1	12,000	20,000
Direct materials inventory, 12/31	5,000	W
Variable manufacturing overhead	5,000	X
Work in process, 1/1	0	9,000
Work in process, 12/31	0	9,000
Purchases of direct materials	15,000	50,000
Breakeven point (in dollars)	66,667	Y
Cost of goods manufactured	G	U
Operating income (loss)	L	(5,000)

3-52 Miscellaneous alternatives; contribution income statement. Study the income statement of the Hall Company. Commissions are based on sales dollars. All other variable costs vary in terms of units sold.

The factory has a capacity of 150,000 units per year. The results for 19_1 have been disappointing. Top management is sifting through a number of possible ways to make operations profitable in 19_2.

Hall Company

Income Statement for the Year Ended December 31, 19_1

Sales (90,000 units @ $4.00)				$360,000
Cost of goods sold:				
Direct materials			$90,000	
Direct labor			90,000	
Factory overhead:				
Variable		$18,000		
Fixed		80,000	98,000	278,000
Gross margin				$ 82,000
Marketing costs:				
Variable:				
Sales commissions*	$18,000			
Shipping	3,600	$21,600		
Fixed:				
Advertising, salaries, etc.		40,000	$61,600	
Administrative costs:				
Variable		$ 4,500		
Fixed		20,400	24,900	86,500
Operating income (loss)				$ (4,500)

*Based on sales dollars, not physical units.

Required
Consider each situation independently.

1. Recast the income statement into a contribution format. There will be three major sections: sales, variable costs and fixed costs. Show costs per unit in an adjacent column. Allow adjacent space for entering your answers to requirement 2.
2. The sales manager is torn between two courses of action:
   a. He has studied the market potential and believes that a 15% slash in price would fill the plant to capacity.
   b. He wants to increase prices by 25%, to increase advertising by $150,000, and to boost commissions to 10% of sales. Under these circumstances, he thinks that unit volume will increase by 50%.

   Prepare the budgeted income statements, using a contribution format and two columns. What would be the new net income or loss under each alternative? Assume that there are no changes in fixed costs other than advertising.
3. The president does not want to tinker with the price. How much may advertising be increased to bring production and sales up to 130,000 units and still earn a target operating income of 5% of sales?
4. A mail-order firm is willing to buy 60,000 units of product "if the price is right." Assume that the present market of 90,000 units at $4 each will not be disturbed. Hall Company will not pay any sales commission on these 60,000 units. The mail-order firm will pick up the units at the Hall factory. However, Hall must refund $24,000 of the total sales price as a promotional and advertising allowance for the mail-order firm. In addition, special packaging will increase manufacturing costs on these 60,000 units by 10¢ per unit. At what unit price must the mail-order chain business be quoted for Hall to break even on *total* operations in 19_2?
5. The president suspects that a fancy new package will aid consumer sales and ultimately Hall's sales. Present packaging costs per unit are all variable and consist of 5¢ direct materials and 4¢ direct labor; new packaging costs will be 30¢ and 13¢, respectively. Assume no other changes in cost behavior. How many units must be sold to earn an operating income of $20,000?

3-53 Service business, promoting championship fight. Bob Arum sponsored the World Middleweight Championship fight between Marvelous Marvin Hagler and Sugar Ray Leonard. In an interview, Arum disclosed the following data:

- Leonard will get a lump sum, $11 million.
- Hagler will get $12 million plus 50% of Arum's gross revenues between $24 and $30 million and 70% above $30 million.
- Breakeven point, $24 million.

Arum knew he would not lose money. He was sure of $7 million from Caesars Palace for the privilege of hosting the event, $15 million from closed-circuit television revenues, and $4 million from foreign revenues and other merchandising.

Arum hoped for considerably more, especially from the pay-per-view cable rights, at up to $40 per home. If the fight grosses $50 million, Arum expects a pretax income of $8.5 million.

Required

1. If Arum grosses $50 million and obtains $8.5 million operating income, compute his total costs, which consist of Hagler and other costs.
2. Suppose Arum was not sure of obtaining any revenue. If Arum grossed only $24 million, compute his other costs (exclusive of compensation for Leonard and Hagler).

3-54 Service business, cable TV. Cable Vision has been approached by the city of Mirada (population 800,000) to operate its cable television operations. Mirada city officials have become tired of reporting losses on the cable television company they have operated for the past five years. Cable Vision currently operates cable television facilities for over one hundred other cities or counties.

Cable Vision makes the following assumptions in its planning after negotiations with key parties:

a. A basic set of ten cable television stations will be offered at a rate of $20 per month per subscriber. These ten stations include a sports channel, a news channel, and other general audience channels.

b. The city of Mirada would retain ownership of the physical facilities and would maintain them in working condition. Under a leasing agreement, Cable Vision will pay the city of Mirada $50,000 per month plus 10% of the monthly revenues from the first 10,000 subscribers and 5% of the monthly revenues from additional subscribers.

c. Cable Vision will receive the ten channels in its basic service from Interlink Cable; Interlink acts as an intermediary between cable television stations and companies such as Cable Vision, which sell to individual subscribers. Interlink charges a monthly fixed fee of $20,000 plus a monthly charge of $8 per subscriber for the first 20,000 subscribers and $6 per subsequent subscriber.

d. Cable Vision estimates its own operating costs to include both a fixed and a variable component. The fixed component is $60,000 per month. The variable cost per subscriber is $2 per month (to cover monthly billing, program news mailings, and so on).

Required

1. How does the contribution margin per subscriber behave over the 0 to 30,000 subscriber range?
2. Calculate the breakeven number of subscribers per month for Cable Vision.
3. What is the operating income per month to Cable Vision with (a) 10,000, (b) 20,000, and (c) 30,000 subscribers? Comment on the results.
4. You are hired by Cable Vision to give a second opinion on the reliability of the estimate of a breakeven point in requirement 2 above. What seems most certain in Cable Vision's projected costs? What concerns might you have?

Note: See Chapter 20 for a discussion of cost-volume-profit analysis under uncertainty. Also see Problems 20-17 and 20-22.

CHAPTER 4

In job-order product-costing systems, costs are allocated to specific units or to a small batch of units. Housing construction companies use a job-order product-costing system.

Job Costing in Manufacturing

Choices in cost accounting systems

Overview of two major cost objects: departments and products

Job-order product costing

Illustration of job-order accounting

Applying factory overhead to products

Control by responsibility centers

Manufacturing is only one cost area of the value chain

Appendix: Supplementary description of ledger relationships

Learning Objectives

When you have finished studying this chapter, you should be able to

1. Describe some major choices in designing cost accounting systems
2. Provide an overview of two major purposes of costing systems
3. Distinguish between job-order costing and process costing
4. Prepare summary journal entries for typical transactions of a job-costing system
5. Compute and fully account for factory overhead rates
6. Demonstrate two methods of disposing of year-end underapplied and overapplied overhead
7. Describe how control is achieved by responsibility centers

This chapter examines a general approach to accounting for costs in a multiple-purpose accounting system. Two major cost objects are discussed: departments and products. Regardless of the type of their organization, managers inevitably want to know the costs of operating various departments and the costs of various products or services. In this way, managers are better able to plan and control departmental activities and product strategies.

This chapter will cover a manufacturing setting. The next chapter will cover a service setting and additional topics. Throughout your study, remember that our illustrations offer a general picture of widely held concepts and practices. A specific organization, however, invariably tailors these concepts and practices to fit its particular circumstances.

This chapter stresses techniques because they are an essential part of the accounting function. Equally important, we introduce many terms and fundamental ledger relationships that will aid understanding of the key subjects covered in Chapters 5 through 11.

We wrote this chapter for the reader with little business background. If you have never worked in a factory, please study this chapter and its appendix with care. If you have had some business experience, the appendix to this chapter presents material familiar to you.

CHOICES IN COST ACCOUNTING SYSTEMS

Before exploring the details of a cost or management accounting system, consider the following points.

1. The cost-benefit criterion, or theme, is central in designing management accounting systems and deciding when to change systems. Elaborate systems are expensive in terms of time and money. The costs of educating managers and other personnel must be considered. More-sophisticated systems are installed only if managers believe that collective operations will be sufficiently improved in a net cost-benefit sense.

Objective 1

Describe some major choices in designing cost accounting systems

2. Systems should be tailored to the underlying operations and not vice versa. Any significant change in underlying operations is likely to justify a corresponding

change in the accompanying accounting systems. The best systems design begins with a careful study of how operations are conducted and a resulting determination of what information to gather and report. The worst systems are those that managers perceive as useless or misleading.

3. Most organizations have systems for planning and controlling *departmental* activities. These systems require gathering costs by departments or operations, using budgets, and conducting analyses of variances.

4. *Product-costing* systems differ markedly. They depend on the nature of the industry, the types of manufacturing or service, the variety of products and processes, and other factors. For example, the product-costing systems for a construction company, a suit manufacturer, and a brewery will be quite different.

5. *Cost management* is not a synonym for *cost accounting systems.* Cost management is the performance by executives and others in the cost implications of their short-run and long-run planning and control functions. Cost accounting systems exist to provide information to help executives in performing their cost management duties. Cost systems also exist for other purposes, such as supplying information about inventoriable costs for financial reporting.

6. Management accounting systems are only one source of information for executives. Other sources include (1) personal observation of operations and personnel and (2) nonfinancial performance measures often regarded as production, marketing, or industry data, not as accounting data. Examples of nonfinancial performance measures include setup times, absentee rates, the number of customer complaints, and the number of units sold by an industry.

OVERVIEW OF TWO MAJOR COST OBJECTS: DEPARTMENTS AND PRODUCTS

Objective 2

Provide an overview of two major purposes of costing systems

In the end, all costs are recorded to help individuals make decisions. However, all these decisions cannot be foreseen, so systems are designed to fulfill general purposes that are commonplace among managers. We will frequently distinguish between the *product-costing* purpose of systems and all other purposes. For convenience, we will sometimes refer to all other purposes as *planning and control purposes,* as *budgeting-control purposes,* or, for brevity, as *control purposes.*

Chapter 2 (page 44) pointed out that product costing—that is, using products as cost objects—is conducted for different users. ***The product costs reported as inventoriable costs to shareholders may differ from those reported to tax authorities and may further differ from those reported to managers for guiding pricing and product mix decisions.*** In addition to product costs, managers want *department costs*—using department costs, or the costs of activities within departments, or the costs of groups of departments, divisions, geographic territories, or other parts of the organization as cost objects—for judging the performance of subordinates and the performance of parts of the organization as economic investments.

Management accounting systems usually serve these general planning and control purposes by having various parts of the organization as cost objects. Top management decides how various activities should be conducted and how they should be managed. That is, the work of the organization is divided among **responsibility centers**, which are parts, segments, or subunits of an organization whose managers are accountable for specified sets of activities. Examples are departments, divisions, and territories. Costs are often routinely traced to a cost center, which is another example of a responsibility center. The *cost center* is the smallest segment or area of responsibility for which costs are allocated. Although cost centers are generally the size of departments, in some instances one department may contain several cost centers. For example, a machining department may be under one supervisor, but it may contain various groups of machines such as lathes, punch presses, and milling machines. Each group of machines is sometimes regarded as a separate cost center with its own assistant supervisor.

JOB-ORDER PRODUCT COSTING

Distinction between Job-Order Costing and Process Costing

Objective 3

Distinguish between job-order costing and process-costing systems

Product-costing systems in practice frequently combine elements of one or both of two basic product-costing systems: job-order product-costing systems and process product-costing systems. These systems are best viewed as ends of a spectrum:

For specific units or small batches of custom-made products	← - - - - - - - - →	For mass production of like units
Job order product-costing systems		Process product-costing systems

Product costing tries to identify the resources demanded and consumed by various products. The identification is straightforward if the resources (such as direct materials and direct labor) are unique to a particular product. When resources are used in common (indirect costs like supervision, janitorial services, utilities, and depreciation on equipment) in the manufacturing of products, the challenge is to identify the demands placed by various products on these resources.

In a **job order product-costing system**, product costs are obtained by allocating costs to a specific unit or to a small batch of products or services that proceeds through the production steps as a distinct, identifiable job lot. Synonyms for job-order product-costing systems are **job-order costing systems**, **job-cost systems**, **work-order systems**, or **production-order systems**. These systems arise because of the manufacturing of unlike products. Manufacturing industries that commonly use job-order costing-systems include printing, construction, aircraft, furniture, and machinery. Each job receives varying inputs of direct materials, direct labor, and factory overhead.

In a **process product-costing system** (also called **process-costing system**) product costs are obtained by allocating costs to masses of like units that usually proceed in continuous fashion through a series of uniform production steps. Manufacturing industries that use process costing include chemicals, oil refining, textiles, plastics, pipes, soft drinks, lumber, and coal mining. Process costing is discussed in more detail in the next chapter.

Product costing under both job-order systems and process systems requires averaging. The unit cost of a product is the result of taking some total cost and dividing it by some measure of production. The denominator differs. In job-order costing, it is small (for example, one painting, four printed posters, or two special packaging machines). In process costing, it is large (for example, thousands of cases of soft drinks or thousands of board feet of lumber).

Source Documents

Source documents provide the basic information for managers and accountants. The source document used by job-order costing to compile *product costs* is called the **job-cost record** or **job-order** or **job-cost sheet**. The file of job-cost records for the

EXHIBIT 4-1
Job-Cost Record

SAMPLE COMPANY Job No. ______

For stock ______ Customer ______

Product ______ Date started ______ Date completed ______

Department A – Machining							
Direct Material			Direct Labor			Overhead	
Date	Reference	Amount	Date	Reference	Amount	Date	Amount
	(Materials requisition number)			(Time ticket number)		(Based on budgeted overhead rate)	

Department B – Assembly							
Direct Material			Direct Labor			Overhead	
Date	Reference	Amount	Date	Reference	Amount	Date	Amount

Summary of Costs

	Dept. A	Dept. B	Total
Direct material	XX	XX	XXX
Direct labor	XX	XX	XXX
Factory overhead applied	XX	XX	XXX
Total	XXX	XXX	XXX

uncompleted jobs makes up the subsidiary ledger for Work in Process, a major inventory account. Exhibit 4-1 illustrates a job-cost record.

Job-order manufacturers usually have several jobs passing through the plant simultaneously. Each job typically requires different kinds of materials and department effort. Thus jobs may have different routings, different operations, and different times required for completion. Standardized forms help management keep track of transactions and costs. **Materials requisitions**, sometimes called **stores requisitions** (Exhibit 4-2), are forms used to charge departments and job-cost records for direct materials used. **Time tickets** (Exhibit 4-3) are forms used to charge departments and job-cost records for direct labor used. This time ticket (sometimes called *work ticket* or *time card*) indicates the time spent on a specific job. An em-

EXHIBIT 4-2
Materials Requisition

Job No. 41

Department B Date 2/22

Debit Account WORK IN PROCESS

Authorized by GL

Description	Quantity	Unit Cost	Amount
AT 462 BRACKETS	80	$2.50	$200.00

EXHIBIT 4-3
Time Ticket

Employee No.	741	Date	2/22	Job No.	41
Operation	DRILL	Account	WORK IN PROCESS	Dept.	A
Stop	4:45 P.M.	Rate	$12.00	Pieces: Worked	15
				Rejected	–
Start	4:00 P.M.	Amount	$9.00	Completed	15

ployee who is paid an hourly wage and who operates a drill press will have one **clock card** (Exhibit 4-4), which is a document used as a basis for determining individual earnings, but the worker will also fill out or punch several *time tickets* each day as he or she starts and stops work on particular jobs. Many providers of services (for example, auto mechanics) must account for their time in a similar way.

Of course, all the illustrated source documents may exist only in the form of computer records. As manufacturing and service industries become more automated and as bar codes are used more widely for optical scanning, the time and materials used on jobs are recorded routinely without human intervention. For example, a materials requisition may be entered via a computer terminal, the materials may be picked from shelves and delivered directly to the factory floor by a robot or a conveyor system, and the direct labor time or machine time may be recorded by computer as each machine operation starts and stops.

Responsibility and Control

Management should clearly lay out the department responsibility for usage of direct materials and direct labor. Department heads are usually kept informed of direct material and direct labor performance by hourly, daily, or weekly summaries

EXHIBIT 4-4
Clock Card

Name FRANK YOUNG   Employee Number 741
Department A   Week ending 2/26

Date	AM		PM		Excess Hours		Total Hours
	In	Out	In	Out	In	Out	
2/22	7:58	12:01	1:00	5:01			8
2/23	7:55	12:00	1:00	5:02			8
2/24	8:00	12:02	12:58	5:00	6:00	9:00	11
2/25	7:58	12:02	12:59	5:03			8
2/26	7:56	12:01	12:59	5:01			8

Regular Time	43	hrs @ $12.00	$516.00
Overtime Premium	3	hrs @ $ 6.00	$ 18.00
Gross Earnings			$534.00

of requisitions and time tickets charged to their departments. In addition to this responsibility function, materials requisitions and direct labor time tickets are used to post to job-cost records.

The job-cost records also serve a control function. Comparisons are often made between predictions of job costs and the costs finally applied to the job. Deviations are investigated so that their underlying causes can be discovered.

ILLUSTRATION OF JOB-ORDER ACCOUNTING

General Ledger and Subsidiary Ledgers

The time, costs, and attention devoted by departments to any given job may vary considerably. Manufacturers keep a separate account of each job for inventory purposes and other accounts for department responsibility purposes. In practice, a Work in Process account, supported by a subsidiary ledger of individual job orders, is widely used for inventory-costing purposes. However, practice differs greatly as to the general-ledger accumulation of costs for department responsibility purposes.

Consider a specific example. Suppose Robinson Company manufactures specialized machinery for the paper-making industry. It has two departments, machining and assembly. Exhibit 4-5 (on pages 98–99) shows T-account relationships and relationships between the general and subsidiary ledgers.

The general ledger section of Exhibit 4-5 gives a bird's-eye view of an entire cost accounting system. The amounts shown are developed in the illustration that follows. The subsidiary ledgers and the basic source documents, the section on page 99 of Exhibit 4-5, contain the underlying details—the worm's-eye view. The bulk of the clerical and computer time is spent on these source documents and the subsidiary-ledger accounts; these are the everyday tools for systematically recording operating activities. However, the corresponding general-ledger entries are usually made monthly. They are summaries of the financial effects of perhaps hundreds or thousands of transactions recorded in the subsidiary ledgers and source documents.

Explanations of Transactions

Objective 4

Prepare summary journal entries for typical transactions of a job-costing system

The following transaction-by-transaction summary analysis will explain how product costing is achieved. General ledger entries are usually made monthly:

1. *Transaction:* Direct and indirect materials purchased, $99,000.

   Analysis: The asset Materials is increased. The liability Accounts Payable is increased. A summary of purchases of direct and indirect materials is charged to Materials because the storekeeper is accountable for them. The subsidiary records for Materials would be perpetual inventory records called *materials records.* At a minimum, these records would contain quantity columns for receipts, issues, and balance. Exhibit 4-6 on page 100 presents a materials record.

   Entry: In the journal (explanation omitted):

Materials control	99,000	
Accounts payable		99,000

   Post to the general ledger:

Materials Control*				Accounts Payable	
	99,000				99,000

*The word "control," as used in journal entries and general-ledger accounts, has a narrow bookkeeping meaning. As contrasted with "control" in the management sense, "control" here means that the account in question is supported by an underlying subsidiary ledger. To illustrate: In financial accounting, Accounts Receivable Control is supported by a subsidiary customers' ledger, with one account for each customer. The same meaning applies to the Materials account here; the materials subsidiary ledger consists of individual accounts for the various materials in inventory.

2. *Transaction:* Requisitions of direct materials, $91,000, and indirect materials, $4,000.

Analysis: The assets Work in Process and Factory Department Overhead are increased. The asset Materials is decreased.

Responsibility is fixed by using *materials requisitions* as a basis for charging departments. A materials requisition was shown in Exhibit 4-2. Requisitions are accumulated and journalized monthly.

As they are used, direct materials are charged to jobs, which form a subsidiary ledger to the inventory account Work in Process. Indirect materials (supplies) are charged to individual department overhead cost records (see Exhibit 4-7), which form a subsidiary ledger for Factory Department Overhead. Managers are responsible for monitoring costs, item by item.

In job-cost accounting, a single Factory Department Overhead account may be kept in the general ledger. This account is an asset that is later "cleared" or transferred to other accounts. The overhead is applied to jobs, as will be described later in this illustration.

Entry: In the journal:

Work in process control	91,000	
Factory department overhead control		
(indirect materials)	4,000	
Materials control		95,000

Post to the general ledger:

Materials Control				Work in Process Control			
(1)	99,000	(2)	95,000	(2)	91,000		

Factory Department Overhead Control			
(2)	4,000		

3. *Transaction:* Labor costs incurred, direct ($39,000) and other ($5,000).

Analysis: The assets Work in Process and Factory Department Overhead are increased. Payroll Liability is increased.

Responsibility is fixed by using time tickets (Exhibit 4-3) or individual time summaries as a basis for tracing direct labor to jobs and direct and other labor to departments. Clock cards (Exhibit 4-4) are widely used as attendance records and as the basis for computation of payroll. (Payroll withholdings from employees are ignored in this example.)

Entry: In the journal:

Work in process control	39,000	
Factory department overhead control		
(other labor)	5,000	
Payroll liability		44,000

EXHIBIT 4-5 *(Place a clip on this page for easy reference.)*

Job-Cost System, Diagram of Ledger Relationships

GENERAL LEDGER

① Purchases, $99,000

② Usage of direct materials, $91,000, and indirect materials, $4,000

③ Incurrence of direct labor $39,000, and other labor, $5,000

④ Payment of payroll liability, $44,000

⑤ Incurrence of other factory overhead, $75,000

⑥ Application of factory overhead, $80,000

⑦ Completion of goods $198,800

⑧ Cost of goods sold $190,000

MATERIALS CONTROL	
① 99,000	② 95,000

PAYROLL LIABILITY	
④ 44,000	③ 44,000

ACCOUNTS PAYABLE	
	① 99,000
	⑤ 23,000

ACCUMULATED DEPRECIATION	
	⑤ 50,000

FACTORY DEPARTMENT OVERHEAD CONTROL	
② 4,000	
③ 5,000	
⑤ 75,000	
Bal. 84,000	

FACTORY OVERHEAD APPLIED	
	⑥ 80,000

PREPAID INSURANCE	
	⑤ 2,000

WORK IN PROCESS CONTROL	
② 91,000	⑦ 198,800
③ 39,000	
⑥ 80,000	

CASH	
	④ 44,000

FINISHED GOODS CONTROL	
⑦ 198,800	⑧ 190,000

COST OF GOODS SOLD	
⑧ 190,000	

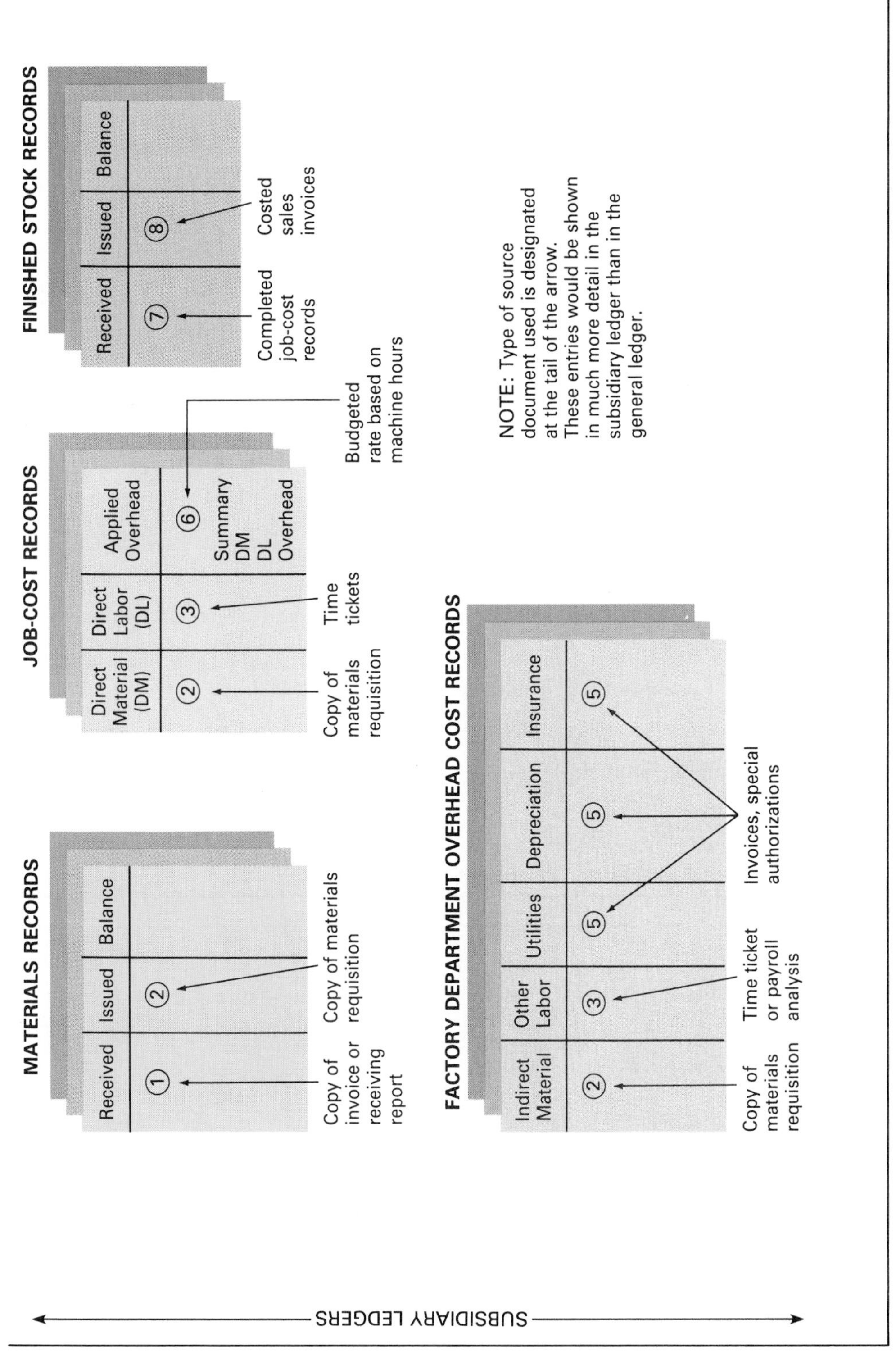

SUBSIDIARY LEDGERS
MATERIALS RECORDS
Received
Issued
Balance
①
②
Copy of invoice or receiving report
Copy of materials requisition
JOB-COST RECORDS
Direct Material (DM)
Direct Labor (DL)
Applied Overhead
②
③
⑥
Summary
DM
DL
Overhead
Copy of materials requisition
Time tickets
Budgeted rate based on machine hours
FINISHED STOCK RECORDS
Received
Issued
Balance
⑦
⑧
Completed job-cost records
Costed sales invoices
FACTORY DEPARTMENT OVERHEAD COST RECORDS
Indirect Material
Other Labor
Utilities
Depreciation
Insurance
②
③
⑤
⑤
⑤
Copy of materials requisition
Time ticket or payroll analysis
Invoices, special authorizations
NOTE: Type of source document used is designated at the tail of the arrow. These entries would be shown in much more detail in the subsidiary ledger than in the general ledger.

EXHIBIT 4-6
Materials Record

Item AF 462 Brackets

		Received			Issued			Balance		
Date	Reference	Quantity	Unit Cost	Total Cost	Quantity	Unit Cost	Total Cost	Quantity	Unit Cost	Total Cost
2/10	V1014	300	2.50	750				300	2.50	750
2/22	R41				80	2.50	200	220	2.50	550

Post to the general ledger:

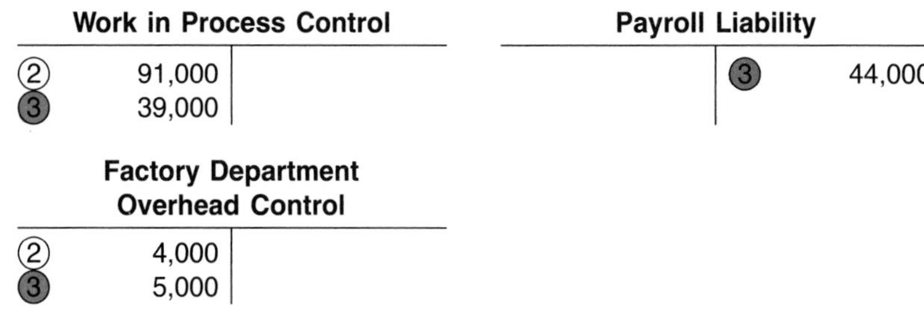

Work in Process Control
(2) 91,000
(3) 39,000
Payroll Liability
(3) 44,000
Factory Department Overhead Control
(2) 4,000
(3) 5,000

4. *Transaction:* Payment of total payroll for the month, $44,000.

Analysis: Payroll Liability is decreased. The asset account Cash is decreased.

Entry: In the journal:

EXHIBIT 4-7
Factory Department Overhead Cost Record, Machining Department

Date	Source Document	Lubricants	Other Supplies	Material Handling	Idle Time	Overtime Premium	Other Labor	Utilities	Insurance	Depr.
	Requisitions	XX	XX							
	Labor recapitulation			XX	XX	XX	XX			
	Invoices							XX		
	Special memos from controller's department on accruals, pre-payments, etc.							XX	XX	XX

Payroll liability	44,000	
Cash		44,000

Post to the general ledger:

Cash

	(4) 44,000

Payroll Liability

(4) 44,000	(3) 44,000

For convenience here, the liability for the month is completely extinguished. However, remember that ***actual payments and entries may be made weekly, even though payroll costs incurred (entry 3) are recorded monthly.*** The reason for this procedure is that paydays seldom coincide with the conventional accounting period (the month), for which costs are accumulated in the general ledger. Thus the Payroll Liability account typically appears as follows:

Payroll Liability

Payments x	Gross earnings x
x	
x	
x	
	Month-end balance represents wages earned but unpaid

5. *Transaction:* Additional factory department overhead costs incurred for the month, \$75,000. These consist of utilities and repairs, \$23,000; depreciation on equipment, \$50,000; and insurance expired, \$2,000.

Analysis: The asset Factory Department Overhead is increased. The liability Accounts Payable is increased and the assets regarding insurance and equipment are decreased.

The detail of these costs is distributed to the appropriate columns of the individual department overhead cost records that make up the subsidiary ledger for Factory Department Overhead. The basic documents for these distributions may be vouchers, invoices, or special memos from the responsible accounting officer.

Entry: In the journal:

Factory department overhead control	75,000	
Accounts payable		23,000
Accumulated depreciation —equipment		50,000
Prepaid insurance		2,000

Post to the general ledger:

Factory Department Overhead Control

(2) 4,000	
(3) 5,000	
(5) 75,000	

Accounts Payable

	(1) 99,000
	(5) 23,000

Accumulated Depreciation—Equipment

	(5) 50,000

Prepaid Insurance

	(5) 2,000

6. *Transaction:* Application of factory overhead to products, $80,000.

Analysis: The asset Work in Process is increased. The asset Factory Department Overhead is decreased by means of its contra account, called Factory Overhead Applied.

Applied factory overhead is factory overhead allocated to products (or services), usually by means of some budgeted (predetermined) rate. The budgeted overhead rate used here is $80 per machine hour. The total amount of overhead applied to a particular job depends on the number of actual machine hours used on that job. It is assumed that 1,000 machine hours were used for all jobs, resulting in a total overhead application of 1,000 × $80 = $80,000. This entry is explained further in a subsequent section of this chapter.

Entry: In the journal:

Work in process control	80,000	
Factory overhead applied		80,000

Post to the general ledger:

Work in Process Control

Debit		Credit
(2)	91,000	
(3)	39,000	
(6)	80,000	

Factory Overhead Applied

Debit	Credit	
	(6)	80,000

7. *Transaction:* Completion and transfer to finished goods of Job Nos. 101–108, $198,800.

Analysis: The asset Finished Goods is increased. The asset Work in Process is decreased.

As job orders are completed, the job-cost records are totaled. Note especially that the totals consist of *actual* direct material, *actual* direct labor, and *applied* factory overhead. Some companies use the *completed* job-cost records as their subsidiary ledger for finished goods. Other companies use separate finished-goods records to form a subsidiary ledger.

Entry: In the journal:

Finished goods control	198,800	
Work in process control		198,800

Post to the general ledger:

Finished Goods Control

Debit		Credit
(7)	198,800	

Work in Process Control

Debit		Credit	
(2)	91,000	(7)	198,800
(3)	39,000		
(6)	80,000		

8. *Transaction:* Cost of goods sold, $190,000.

Analysis: The expense Cost of Goods Sold is increased. The asset Finished Goods is decreased.

Entry: In the journal:

Cost of goods sold (sometimes called cost of sales)	190,000	
Finished goods control		190,000

Post to the general ledger:

Cost of Goods Sold		Finished Goods Control			
 190,000		(7) 198,800		(8) 190,000	

The eight summary entries are usually made monthly. As already emphasized, the biggest share of data accumulated is devoted to compiling the mass of day-to-day details and recording them in subsidiary ledgers. These "ledgers" are usually stored by computers.

At this point, please pause and reexamine all eight entries in the illustration. Be sure to trace each journal entry, step by step, to the accounts in both sections of Exhibit 4-5.

APPLYING FACTORY OVERHEAD TO PRODUCTS

Cost Application or Absorption

In Entry 6 (page 102) in our manufacturing illustration, we used a budgeted overhead rate to apply factory overhead to products. Through requisitions and time tickets, direct materials and direct labor may be traced directly to physical units worked on. But, by its nature, factory overhead cannot be traced directly to specific units. Yet the making of goods would be impossible without incurring overhead costs such as utilities and repairs, depreciation, insurance, janitorial service, and property taxes. These costs accumulate and must be allocated. Different products require different quantities of some of these overhead resources. The objective of the allocation of overhead is to represent the underlying consumption of resources by individual products.

Objective 5

Compute and fully account for factory overhead

Most managers want a close approximation of the manufacturing costs of various products continuously, not just at the end of the year. Managers desire these costs (often together with other costs such as marketing costs) for various ongoing uses, including choosing which products to emphasize or deemphasize, pricing products, producing interim financial statements, and managing inventories. Because management must have immediate access to product costs, few companies wait until the *actual* factory overhead is finally known (at year-end) before allocating overhead costs in computing the costs of products. Instead, a *budgeted* (predetermined) overhead rate is calculated at the beginning of a fiscal year and then applied as products are manufactured.

Before we examine budgeted overhead, let us consider some important terms. *Cost allocation* is a general term that refers to identifying accumulated costs with or tracing accumulated costs to cost objects such as departments, activities, or products. **Cost application**, often called **cost absorption**, is a narrower term that refers to the allocation of costs to *products* as distinguished from the allocation of costs to *departments* or *activities*.

Budgeted Overhead Application Rates

The following steps summarize a widely used approach to accounting for factory overhead:

Step 1. Select a *cost application base* that serves as a common denominator for all products. Examples include direct labor hours, direct labor costs, and machine hours. A **cost application base** (often called a **cost allocation base**) is a factor that is the common denominator for systematically relating a cost or a group of costs, such as factory overhead, to products. The application base or bases should be the best available measure of the cause-and-effect relationships between overhead costs and cost drivers. For example, in a heavily automated department, the choice of machine hours would be preferable to direct labor hours as an application base. Why? Because the total overhead costs will be more heavily affected by machines than by laborers. Costs like depreciation, utilities, and repairs will be most prominent in a heavily automated department.

Step 2. Prepare a factory overhead budget for the planning period, ordinarily a year. The two key items are (a) budgeted total overhead and (b) budgeted total volume of the application base. The company's recent history of actual overhead and actual volume is often used as a starting point for preparing the budget.

Step 3. Compute the **budgeted factory overhead rate** by dividing the budgeted total overhead by the budgeted total volume of the application base.

Step 4. Obtain the actual application base data (such as machine hours) as the year unfolds.

Step 5. Apply the overhead to the jobs by multiplying the budgeted rate times the actual application base data.

Step 6. At the end of the year, account for any differences between the amount of overhead actually incurred and overhead applied to products.

Suppose our illustrative company budgeted its factory overhead for the forthcoming year as $1,056,000. Assume that the forecast is based on a volume of 13,200 machine hours. The budgeted overhead rate would be

$$\text{Budgeted overhead application rate} = \frac{\text{Budgeted total factory overhead}}{\text{Budgeted total volume of application base}}$$

$$= \frac{\$1{,}056{,}000}{13{,}200 \text{ machine hours}} = \$80 \text{ per machine hour}$$

(For ease of learning essential concepts, this example assumes that a single plantwide overhead rate is appropriate. Moreover, a single cost driver [machine hours] is assumed as a cost application base. This calculation is an oversimplification. There are usually different budgeted overhead rates at least for different departments and sometimes for different activities or operations within departments. These are illustrated and explained in Chapters 5 and 14.)

The $80 rate would be used for costing individual jobs. For example, suppose a job-cost record for Job 323 included the following information:

Actual direct materials cost	$700
Actual direct labor cost	$280
Actual machine hours	8

The overhead applied to Job 323 would be: 8 actual machine hours × the budgeted rate of $80 = $640. The total cost of Job 323 would be: $700 + $280 + $640 = $1,620. If actual results for the year exactly conform to the prediction of $1,056,000 overhead costs and 13,200 machine hours, total overhead costs will have been exactly applied to products worked on during the year.

Exhibit 4-8 presents an overview of the job-costing system. A **cost pool** is a grouping of individual costs. The top box is the total "pool" of indirect product costs, factory overhead. It comes down on Job 323 from "overhead" by means of a machine hour application base. The direct costs of materials and labor are specifically identified with Job 323.

Annualized Rate

Should overhead rates be set on the basis of weekly, monthly, or yearly volume of the cost application base? An annualized basis for a budgeted rate is used for two main reasons:

1. To overcome volatility in computed unit costs that would result because of fluctuations in the *volume* of the cost application base from month to month (the denominator reason). This is the principal reason.

EXHIBIT 4-8
Overview of Product Costing for Manufacturing Activities at Robinson Company

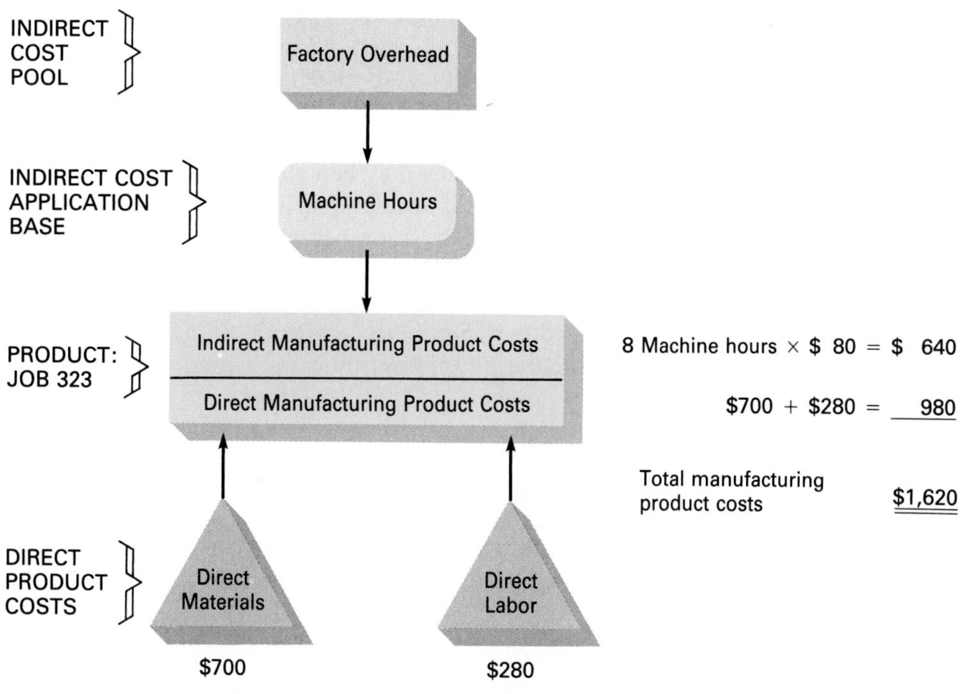

2. To overcome the volatility in computed unit costs that would result because of seasonal, calendar, and other peculiar variations in the *level of total overhead costs* incurred each month (the numerator reason).

The Denominator Reason: Fluctuations in Monthly Volume. Some overhead costs are variable with respect to the cost application base (for example, supplies and some utilities), whereas others are fixed (for example, property taxes, rent, and depreciation). If production fluctuates from month to month, total variable overhead cost incurrence should change in close proportion to variations in production, whereas total fixed overhead will remain unchanged. These relationships mean that overhead rates based on monthly volume may differ greatly from month to month ***solely because of fluctuations in the volume over which fixed overhead is spread.***

Suppose a company schedules its production in harmony with a highly seasonal sales pattern. The total manufacturing overhead has a high proportion of fixed costs. If the company applies actual overhead to products month by month, unit product costs will soar as volume falls and vice versa. Consider the following assumed figures:

	Actual Factory Overhead			Actual Machine Hours	Application Rate per Machine Hour
	Variable	Fixed	Total		
High-volume month	$40,000	$60,000	$100,000	2,000	$ 50
Low-volume month	10,000	60,000	70,000	500	140

The actual overhead rates are obtained by dividing the actual total factory overhead by actual total machine hours. The presence of fixed overhead causes the overhead costs to fluctuate from $50 to $140 per machine hour. The variable element is $20 per machine hour in both months ($40,000 ÷ 2,000 machine hours, and

$10,000 ÷ 500 machine hours). However, the fixed element is $30 per machine hour in the high-volume month ($60,000 ÷ 2,000 machine hours), and $120 per machine hour in the low-volume month ($60,000 ÷ 500 machine hours).

Few people support the contention that an identical product should be charged with a $50 overhead rate during one month and a $140 overhead rate during another. These different overhead rates are not representative of typical, normal production conditions. Management has committed itself to a specific level of fixed costs in light of foreseeable needs far beyond a mere thirty days. Thus, where production fluctuates, monthly overhead rates may be volatile. An average, annualized rate based on the relationship of total annual overhead to total annual volume is more representative of typical relationships between total overhead costs and production volume than is a monthly rate.

The Numerator Reason: Peculiarities of Specific Overhead Items. Fluctuation in monthly volume rather than fluctuation in monthly costs incurred is the principal reason for using an annualized overhead rate. Still, certain costs are incurred in different amounts at various times of the year. If a month's costs alone were considered, the heating cost, for example, could be charged only to winter production.

Typical examples of erratic costs include outlays for repairs, maintenance, and certain indirect materials requisitioned in one month that will be consumed over two or more months. These items may be charged to a department on the basis of monthly repair orders or requisitions. Yet the benefits of such charges may easily extend over a number of months of production. It would be illogical to load products of any single month with costs caused by several months of operations.

The calendar itself has an unbusinesslike design: Some months have twenty workdays, and others have twenty-two or more. Is it sensible to say that a product made in February, the shortest month, should bear a greater share of overhead such as depreciation and property taxes than a product made in March?

Other erratic items that distort monthly overhead rates are vacation and holiday pay, professional fees, extra costs of learning, and idle time related to the installation of a new machine or product line.

All these costs and peculiarities are collected in the annual overhead pool along with the kinds of overhead that do have uniform behavior patterns (for example, supplies). In other words, accountants throw up their hands and say, "We have to start somewhere, so let's pool the year's overhead and develop an annual overhead rate regardless of month-to-month peculiarities of specific overhead costs. ***Such an approach provides a normal manufacturing product cost based on an annual average instead of a so-called actual manufacturing product cost that is affected by month-to-month fluctuations in production volume and by erratic or seasonal behavior of many overhead costs.*** Such a normal cost is used for inventory purposes. It is also often used in combination with marketing and other costs as a point of departure for setting selling prices of products.

Choosing the Application Base: Cost Drivers

Ideally, for wise economic decisions, the cost application base should be the principal *cost driver,* which is the factor(s) that causes the incurrence of costs.[1] Examples of cost drivers are direct labor hours, direct labor costs, machine hours, pounds handled, invoices processed, number of component parts, and number of setups.

Through the years, direct labor has been a significant cost driver in job-costing industries. In many machining departments, however, two or more machines can

[1]A cause-and-effect criterion underlies this recommendation. Chapter 14 discusses alternative criteria for selecting a cost application base.

often be operated simultaneously by a single direct laborer or without any labor at all. In such cases, the use of machines causes most of the overhead costs, such as depreciation and repairs in the machining department. Machine hours, then, are the cost driver and the appropriate base for applying overhead costs. Using machine hours as the cost driver necessitates keeping track of the machine hours used for each job, which creates added data collection costs. Where direct labor and machines are both important, both direct labor costs and machine hours must be compiled for each job.

In contrast, direct labor is often a principal cost driver in some assembly departments of job shops. It is an accurate measure of the relative attention and effort devoted to various jobs. If all workers are paid equal hourly rates, the overhead to apply could be computed simply by multiplying the cost of direct labor, already entered on the job-cost records, by the budgeted percentage overhead rate. No additional job records of the labor hours must be kept. Of course, if the hourly labor rates differ greatly for individuals performing identical tasks, then hours of labor, rather than dollars of labor, should be the application base. For example, more factory overhead would be applied for a worker earning $20 per hour than for a worker earning $18 per hour, even though each worker probably takes nearly the same time and uses the same facilities in doing the same volume of work.

Even if wage rates vary within a department, sometimes direct labor costs are the best overhead application base. For example, better skilled labor is likely higher-paid labor, and these workers may use more costly equipment and require more indirect labor support. The overhead applied to their work ought to be higher than the overhead applied to a lesser-paid worker, even if volume is equal. Also, additional labor costs such as pensions and payroll taxes are often included in factory overhead costs; these costs are generally based on workers' earnings, so they are more closely driven by direct labor costs than by direct labor hours.

No matter what the cost allocation base chosen, the overhead rates are applied day after day throughout the year to cost the various jobs worked on by each department. Suppose management predictions coincide exactly with actual amounts (an extremely unlikely situation). Then the total overhead applied to the year's jobs through these budgeted rates would be equal to the total overhead costs actually incurred. The next section shows how to account for any difference at year-end.

Ledger Procedure for Overhead

When a single overhead cost application base is used for product costing, all overhead items are pooled together, a budgeted annual average overhead rate is computed, and this average rate is used on jobs for costing Work in Process. The use of an annual average results in inventories bearing a normalized share of factory overhead. Because a budgeted overhead rate (such as the $80 per machine hour rate in our earlier illustration) is an average used to apply costs to products, the daily, weekly, or monthly costing of *inventory* is independent of the actual incurrence of overhead costs by *departments.* For this reason, at any given time during the year, the balance in Factory Department Overhead Control is unlikely to coincide with the amount applied to products.

As departments incur overhead costs from month to month, these actual costs are charged in detail to department overhead cost records (the subsidiary ledger) and in summary to the Factory Department Overhead Control account in the general ledger. These costs are accumulated daily, weekly, or monthly without regard to how factory overhead is applied to specific jobs. This ledger procedure helps management control overhead. In reviewing departmental performance reports, management compares actual costs with budgeted amounts. For example, a manager may compare the actual costs of lubricants used against the budget line item called lubricants.

Most accountants confine Factory Department Overhead Control to the accumulation of actual overhead charges incurred. To handle in the ledger the application of these accumulated costs, accountants set up a separate contra account called *Factory Overhead Applied* (sometimes called Factory Overhead *Absorbed*), much as Accumulated Depreciation—Machinery is a contra account for Machinery. To illustrate:

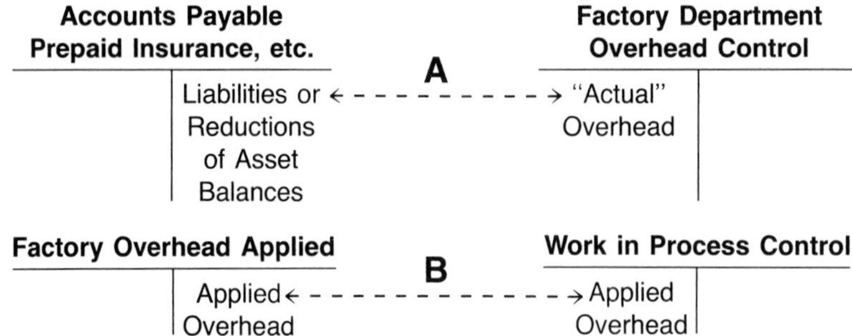

Underapplied or Overapplied Overhead

The workings of the ledger accounts for overhead may be more clearly understood if we pursue our master illustration (as shown in Exhibit 4-5, pp. 98-99). Assume that the month's entries are for January, the first month of the company's year. Postings appear as follows:

Factory Department Overhead Control

Debit		Credit
Jan. 31 (2)	4,000	
Jan. 31 (3)	5,000	
Jan. 31 (5)	75,000	
Jan. 31 Balance	84,000	

Factory Overhead Applied

Debit	Credit	
	Jan. 31 (6)	80,000

The monthly debits to Factory Department Overhead Control will seldom equal the monthly credits to Factory Overhead Applied. In January, for example, we see a $4,000 difference between the two balances. This $4,000 amount is commonly called *underapplied* (or *underabsorbed*) *overhead.* **Overhead** is **underapplied** when the applied balance is less than the incurred (actual) balance; it is **overapplied** when the applied balance exceeds the incurred balance. Although the month-end balances of the two factory overhead accounts rarely coincide, the final year-end balances are ideally not too far apart.

Accounting at the End of the Year

Objective 6

Demonstrate two methods of disposing of year-end underapplied and overapplied overhead

At year-end, most accountants close the incurred and applied accounts against one another. What happens to the difference between them—that is, the underapplied or overapplied overhead? Suppose the amount is small—in comparison to total factory overhead incurred, total cost of goods sold, total operating income, or some other measure of materiality. Then the underapplied or overapplied overhead is regarded as an adjustment of Cost of Goods Sold. This is often called the *immediate write-off method* of disposing of underapplied or overapplied overhead.

Assume that factory overhead incurred is $600,000 and that factory overhead applied is $550,000. The resulting $50,000 underapplication would be accounted for under the immediate write-off method as follows:

Cost of goods sold	50,000	
Factory overhead applied	550,000	
Factory department overhead control		600,000

To close the overhead accounts and to charge underapplied overhead to Cost of Goods Sold.

Under this method, any overapplied overhead would be credited directly to Cost of Goods Sold.

A $50,000 amount may be small to one company but large to another company. When management considers the amount of overapplication or underapplication to be large, accountants favor **proration**, the spreading of underapplied or overapplied overhead among work in process and finished goods inventories and cost of goods sold. In this example, the $50,000 would be spread among Work in Process Control, Finished Goods Control, and Cost of Goods Sold. Assume that the ending balances (before proration) are as shown in accompanying column (a). In column (b) the $50,000 underapplied overhead is prorated over the three pertinent accounts in proportion to their ending balances (before proration), resulting in the ending balances (after proration) in column (c):

	(a) Balance (Before Proration) End of 19_1	(b) Proration of Underapplied Overhead	(c) Balance (After Proration) End of 19_1
Work in process	$ 125,000	125/1,250 × $50,000 = $ 5,000	$ 130,000
Finished goods	500,000	500/1,250 × 50,000 = 20,000	520,000
Cost of goods sold	625,000	625/1,250 × 50,000 = 25,000	650,000
	$1,250,000	$50,000	$1,300,000

The journal entry for this proration follows:

Work in process control	5,000	
Finished goods control	20,000	
Cost of goods sold	25,000	
Factory overhead applied	550,000	
Factory department overhead control		600,000

To close the overhead accounts and to prorate underapplied overhead among the three relevant accounts.

In practical situations, prorating is done only when inventory valuations will be significantly (materially) affected.

Exhibit 4-9 is a schematic comparison of the two methods of disposition of underapplied overhead at year-end. No matter which of the two methods is used, the underapplied overhead is not carried in the overhead accounts beyond the end of the year. That is, the ending balances in Factory Department Overhead Control and Factory Overhead Applied are closed and consequently become zero at the end of each year. (A section in Chapter 9 describes the accounting for underapplied and overapplied overhead on interim financial statements.)

Assumptions in Practice

The proration method may be refined further. Theoretically, the proration should be in proportion to the *applied overhead* component (before proration) in the three accounts described, not their ending balances (before proration), which we used in our illustration. After all, the fundamental objective of proration is to obtain a closer approximation of the "actual" costs of Work in Process, Finished Goods, and Cost of Goods Sold. The proration method illustrated above is defective because it

EXHIBIT 4-9
Year-end Disposition of Underapplied Factory Overhead

METHOD ONE: IMMEDIATE WRITE-OFF

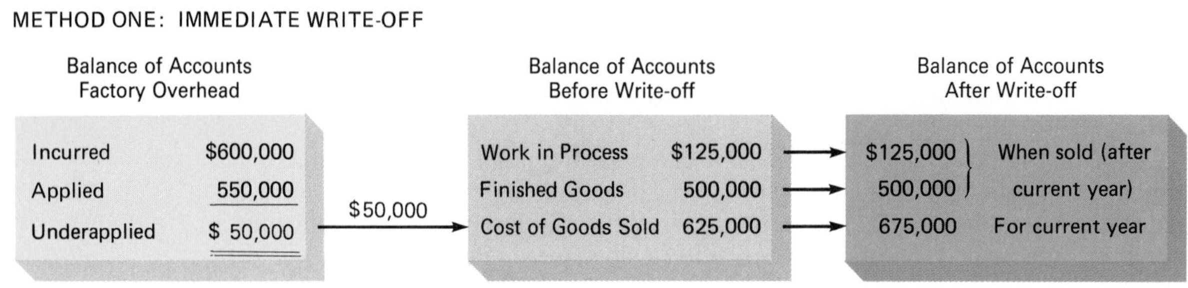

METHOD TWO: PRORATION AMONG INVENTORIES AND COST OF GOODS SOLD

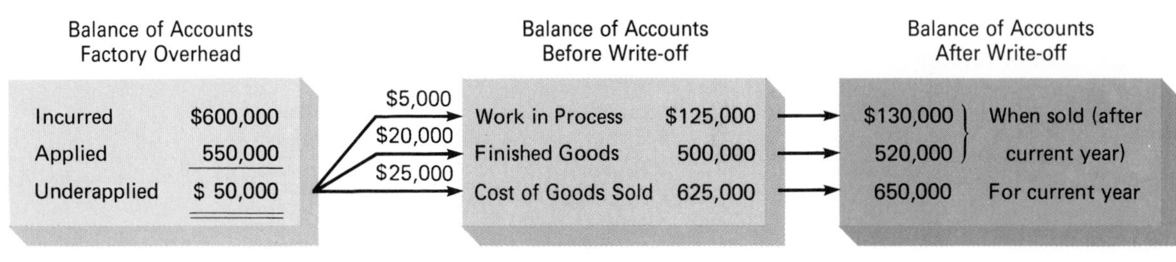

fails to recognize that the proportions of direct materials costs, direct labor costs, and overhead costs are rarely constant among all jobs represented in the three accounts. That is, each of the three accounts may contain different ratios of direct materials, direct labor, and overhead. Overhead is usually applied as in our illustration, using a budgeted rate. Ideally, all applied overhead should be subject to adjustment to "actual" in whatever accounts the applied overhead exists.

Let us illustrate how proration may be faulty. One furniture job may require a high proportion of valuable wood and another furniture job may require inexpensive wood or plastic. The first job completed and charged to cost of goods sold might have an $800 total cost, and the second job in finished goods inventory might have a $200 total cost, but the direct labor hours or machine hours and the related overhead on each job may have been approximately the same. A year-end proration of underapplied overhead based on ending balances (before proration) of the jobs would load $800 ÷ ($800 + $200) = 80% (instead of 50%) of the underapplied overhead to cost of goods sold. The 50% proration would be more accurate. Why? Because it would be in proportion to the equal amounts of applied overhead on each of the two jobs.

Modern companies are becoming increasingly conscious of inventory control, so inventories are lower than in earlier years. Hence, Cost of Goods Sold tends to be higher in relation to inventories of Work in Process and Finished Goods. Also, the inventory balances of job-costing companies are usually relatively low because goods are made in response to specific sales orders. Consequently, writing off underapplied or overapplied overhead instead of prorating it is unlikely to cause significant errors in the financial statements.

If you ponder the practical accounting for overhead, you can easily see the cost-benefit test at work. Averages are used to obtain budgeted rates. In turn, these budgeted rates are used to cost the work in process and finished goods inventories and cost of goods sold. Write-offs of underapplied or overapplied overhead occur only at year-end. Of course, proration of underapplied or overapplied overhead gives more accurate approximations of actual costs. However, most managers and accountants believe that such additional, often costly, attempts at accuracy seldom provide additional useful information. ***Adjusting Cost of Goods Sold for all the underapplied or overapplied overhead is the most widely practiced treatment.***

Actual and Normal Absorption Costing

The following T-accounts illustrate two types of absorption costing—*actual absorption costing* and *normal absorption costing*.

	Chapter 2 Approach, Called Actual Absorption Costing: Work in Process		**Chapter 4 Approach, Called Normal Absorption Costing: Work in Process**	
Direct materials	actual costs		actual costs	
Direct labor	actual costs		actual costs	
Variable factory overhead	actual costs		actual input(s) of the cost application base(s) × budgeted overhead rate(s) (covers both variable and fixed factory overhead)	
Fixed factory overhead	actual costs			

The job-order system described in this chapter is called a *normal-costing* system, not an *actual* costing system, because the overhead is applied to products on an average or "normalized" basis to get representative inventory valuations. Hence the *normal cost* (perhaps more accurately called *normalized cost*) of the manufactured products is composed of *actual* direct materials, *actual* direct labor, and applied overhead [using actual input(s) of the cost application base(s) times budgeted overhead rate(s)].

The Chapter 2 approach does not apply overhead to the products until the end of a fiscal year. Then the actual overhead is used for computing the costs of the products worked on. Reconsider the factory overhead cost in Exhibit 4-5 (pp. 98–99). Under actual costing, the Factory Overhead Applied would be the $84,000 actual factory overhead instead of the $80,000 used under normal costing. All costs actually incurred are precisely offset by costs applied to Work in Process.

Normal costing has replaced actual costing in many companies because the latter fails to provide costs of products as they are manufactured. Of course, it is possible to use a normal-costing system plus year-end adjustments to obtain final results that closely approximate the results under actual costing, as shown earlier by the proration of underapplied overhead. If managers desire yearly "actual" costing results, year-end prorations of underapplied or overapplied overhead may be used in conjunction with the normal costing that occurs throughout the year.

Application Bases in the Future: More Cost Drivers

Throughout the twentieth century, most manufacturers have tended to use broad averages for applying overhead to products. For example, many companies have used a single plantwide overhead rate instead of several departmental rates. Direct labor (costs or hours) has been the single most popular overhead application base. Increasingly, companies use more than one application base. For example, a company may use direct labor costs for applying some factory overhead and machine hours for other factory overhead.

In recent years, manufacturing has become more heavily automated, and overhead has become a much larger proportion of total manufacturing cost. Some companies have seen their factory overhead rates as percentage of direct labor cost soar from 25% in the 1940s to 1,000% and higher in the 1990s. Consequently, many more companies are seriously studying whether to adopt for product costing multiple overhead rates keyed to more than one cost driver.

Advances in information-gathering technology have substantially increased the accuracy of information and decreased the cost of collecting it. As more and more businesses use computerized data bases, we are likely to see closer identification of the costs of departmental activities with the products that flow through them. In some cases, one department will use several overhead rates. There might be an overhead rate related to the direct materials of the product, another rate related to the direct labor, another rate related to metered energy usage, and another rate related to machine usage.

In summary, as managers seek more accurate product costing, overhead application solely on the basis of direct labor hours or direct labor cost is certain to diminish in popularity. Management will use a variety of cost drivers as overhead application bases.[2]

CONTROL BY RESPONSIBILITY CENTERS

Objective 7

Describe how control is achieved by responsibility centers

As mentioned early in this chapter, accountants allocate costs to departments as well as to jobs. Why? To serve the planning and control purpose of cost accounting. Managers control by personally observing activities plus examining accounting reports and other information.

Consider direct materials and direct labor, which are charged directly to jobs for product-costing purposes. The same materials requisitions and time tickets trace departmental responsibility for material and labor usage. These records are used for measuring monthly departmental performance. For example, a computer can easily obtain a sum of materials requisitions and time tickets coded by department for any given month. A departmental performance report can then be prepared that compares budgeted and actual costs for direct materials, direct labor, and factory overhead. If desired, direct material usage might be reported hourly, and direct labor usage might be reported daily, often in physical amounts only. In this way, managers obtain quick feedback about the usage of the major factors of production.

The scope of management accounting extends far beyond and far deeper than general- and subsidiary-ledger bookkeeping. Managers may want quick feedback on important costs like direct materials or direct labor—too quick to await reports based on ledger balances. In such cases, there is little need for keeping a subsidiary multicolumn department cost record for direct material, direct labor, and overhead items. ***Instead the department overhead cost record is usually kept separately, and direct material usage reports and direct labor usage reports are automatically pro-***

[2]Chapters 14 and 15 consider these issues in more detail. Also see R. Howell, J. Brown, S. Soucy, and A. Seed, *Management Accounting in the New Manufacturing Environment* (Montvale, NJ: National Association of Accountants and CAM-I, 1987); M. Sakurai, "The Influence of Factory Automation on Management Accounting Practices: a Study of Japanese Companies," in Robert Kaplan (ed.) *Measures for Manufacturing Excellence* (Boston: Harvard Business School Press, 1990) pp. 39-62; Price Waterhouse, *Survey of Cost Management Practices of Selected Midwest Manufacturers* (1989); NAA Tokyo Affiliate, "Management Accounting in the Advanced Manufacturing Surrounding," October 1988; and Coopers & Lybrand et al., "Management Accounting in Advanced Manufacturing Environments," CAM-I R-87-CMS-08, January 1988.

duced instantly from computer data bases or by frequent summaries of requisitions and time tickets. Thus, the source documents for direct materials and direct labor are often used continuously as bases for control without necessarily having them formally summarized by department in either the subsidiary ledgers or the general ledger.

In many cases, records of physical items—pounds used, hours worked—are the fundamental quantitative means of control. However, managers also want to know the financial impacts of their usage of various resources. This is where the accounting system shines. It provides a financial common denominator (such as dollars) that helps managers to judge performance throughout the entire organization and to concentrate on the most important matters.

MANUFACTURING IS ONLY ONE COST AREA OF THE VALUE CHAIN

This chapter has concentrated on job costing in manufacturing. Companies that do manufacturing also do many other things:

Upstream Cost Areas			Downstream Cost Areas		
Research and Development	Product Design	Manufacturing	Marketing	Distribution	Customer Service

Taken together, these upstream cost areas, manufacturing, and downstream cost areas form a chain of business functions. A **value chain** is the sequence of total business functions in which value is added to a firm's products or services.

We have shown how job costing is used for "building up" product costs of manufacturing. Job-order systems and concepts can be extended to include the upstream and downstream cost areas. Manufacturing product-cost buildups are used

- as inventoriable costs for financial statements.
- as *part* of *full* product costs for pricing and product mix decisions. These full costs extend beyond manufacturing into upstream and downstream areas.

PROBLEMS FOR SELF-STUDY

Restudy the illustration of job-order accounting in this chapter. Then try to solve the following problem, which requires consideration of most of this chapter's important points.

PROBLEM 1

You are asked to bring the following incomplete accounts of a printing plant up to date through January 31, 19_2. Also consider the data that appear after the T-accounts.

Materials Control			
12/31/_1 Balance	15,000		

Payroll Liability			
		1/31/_2 Balance	3,000

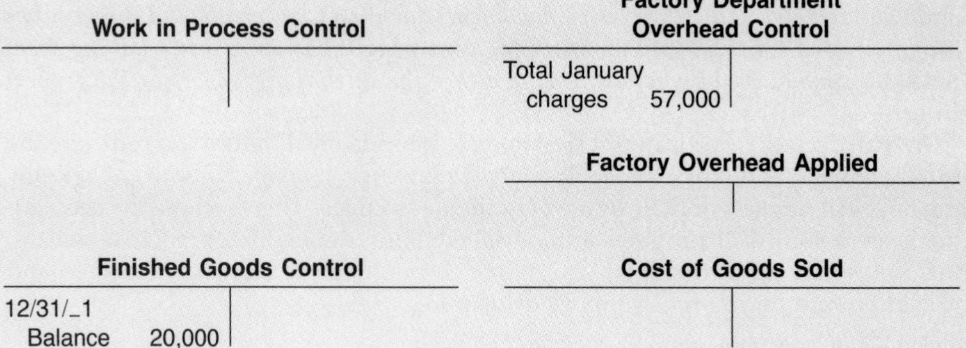

Work in Process Control
Factory Department Overhead Control
Total January charges 57,000
Factory Overhead Applied
Finished Goods Control
12/31/_1 Balance 20,000
Cost of Goods Sold

Additional Information:

1. The overhead is applied using a budgeted rate that is set every December by forecasting the following year's overhead and relating it to forecast direct labor costs. The budget for 19_2 called for $400,000 of direct labor and $600,000 of factory overhead.
2. The only job unfinished on January 31, 19_2, was No. 419, on which total direct labor costs were $2,000 (125 direct labor hours) and total direct materials costs were $8,000.
3. Total materials placed into production during January were $90,000.
4. Cost of goods completed during January was $180,000.
5. Materials inventory as of January 31 was $20,000.
6. Finished goods inventory as of January 31 was $15,000.
7. All factory workers earn the same rate of pay. Direct labor hours for January totaled 2,500. Other labor and supervision totaled $10,000.
8. The gross factory payroll on January paydays totaled $52,000. Ignore withholdings.
9. All "actual" factory overhead incurred during January has already been posted.

Required

a. Materials purchased during January
b. Cost of Goods Sold during January
c. Direct labor costs incurred during January
d. Overhead applied during January
e. Balance, Payroll Liability, December 31, 19_1
f. Balance, Work in Process Control, December 31, 19_1
g. Balance, Work in Process Control, January 31, 19_2
h. Overapplied or underapplied overhead for January

SOLUTION 1

a. Materials purchased: $90,000 + $20,000 − $15,000 = $95,000
b. Cost of Goods Sold: $20,000 + $180,000 − $15,000 = $185,000
c. Direct labor rate: $2,000 ÷ 125 hours = $16 per hour (see 2)
Direct labor costs: 2,500 hours × $16 = $40,000 (see 7)
d. Overhead rate: $600,000 ÷ $400,000 = 150%. Overhead applied: 150% × $40,000 = $60,000
e. Payroll Liability, December 31: $52,000 + $3,000 − $40,000 − $10,000 = $5,000
f. Work in Process Control, December 31: $180,000 + $13,000 − $90,000 − $40,000 − $60,000 = $3,000
g. Work in Process Control, January 31: $8,000 + $2,000 + 150% of $2,000 = $13,000
h. Overapplied overhead: $60,000 − $57,000 = $3,000.

Entries in T-accounts are numbered in accordance with the "additional information" in the problem and are lettered in accordance with the amounts required to be determined.

Materials Control

Debit			Amount	Credit	Amount
12/31/_1 Bal. (given)			15,000		
		(a)	95,000*	(3)	90,000
1/31/_2 Bal.	(5)		20,000		

Work in Process Control

Debit			Amount	Credit	Amount
12/31/_1 Bal.		(f)	3,000*	(4)	180,000
Direct materials	(3)		90,000		
Direct labor	(2) (7)	(c)	40,000		
Overhead	(7) (1)	(d)	60,000		
1/31/_2 Bal.	(2)	(g)	13,000		

Finished Goods Control

Debit		Amount	Credit		Amount
12/31/_1 Bal. (given)		20,000			
	(4)	180,000	(6)	(b)	185,000
1/31/_2 Bal.	(6)	15,000			

Payroll Liability

Debit	Amount	Credit		Amount
(8)	52,000	12/31/_1	(e)	5,000*
		(7)		{ 40,000
				{ 10,000
		1/31/_2 Bal. (given)		3,000

Factory Department Overhead Control

Debit	Amount	Credit	Amount
Total January charges (given)	57,000		

Factory Overhead Applied

Debit	Amount	Credit	Amount
		(7) (1) (d)	60,000

Cost of Goods Sold

Debit		Amount	Credit	Amount
(6)	(b)	185,000		

*Can be computed only after all other postings in the account have been found, so (g) must be computed before (f) in the Work in Process Control computations.

PROBLEM 2

A letter to the shareholders of Marantz, Inc., described many competitive troubles, including the following: "The reduced level of orders . . . resulted in increased unabsorbed costs . . . which further adversely affected the Company's overall competitive position."

Required

Using one or more terms introduced in this chapter, prepare a precise explanation of the quotation.

SOLUTION 2

"Increased unabsorbed costs" means that the *applied* or *absorbed* factory overhead was less than the *actual* factory overhead. In other words, the underapplied or underabsorbed overhead becomes greater and greater if the company falls shorter and shorter of the volume level originally budgeted.

Marantz was confronted with a common difficulty. If sales drop precipitously, production must be cut back. At the same time, the fixed costs of factory facilities, such as rent, depreciation, and property taxes, are not easily reduced. These costs tend to be the bulk of the unabsorbed costs referred to in the quotation. Of course, the most fundamental cause of trouble was the reduction in sales orders.

SUMMARY

Job costing and process costing are two extreme types of product costing. Job-costing systems are expensive because of the details required in operating the system. The cost-benefit criterion is the principal guide for judging whether a job-costing system is appropriate in a particular company.

This chapter focuses on how costs are allocated to two major cost objects: departments and products. In manufacturing, the major direct cost is direct materials. Indirect costs (overhead) are applied to jobs by using some cost application base common to all products. To help managers make better-informed decisions, the application base will be the major cost driver, the factor that causes the incurrence of costs. Direct labor costs, direct labor hours, and machine hours are the most popular cost application bases for applying indirect manufacturing costs to products.

Many companies apply overhead at budgeted rates. The resulting product cost consists of "actual" direct materials, "actual" direct labor, and overhead applied using budgeted rates. This total manufacturing product cost is referred to as a *normal* cost rather than as an *actual* cost. A given product-costing system can properly be called an *actual cost* system or a *normal cost* system (depending on whether actual rates or budgeted rates are used to apply overhead).

In normal costing, throughout the fiscal year Work in Process, Finished Goods, and Cost of Goods Sold are carried at actual direct materials, actual direct labor, and factory overhead applied. Underapplied or overapplied overhead usually accumulates in the accounts from month to month; its final amount is typically added to or subtracted from Cost of Goods Sold at the end of the year. However, if the amount of underapplied or overapplied overhead is significant, it must be prorated over Work in Process, Finished Goods, and Cost of Goods Sold.

As data processing becomes less expensive, more direct costs are likely to be traced to individual jobs. Moreover, multiple overhead application bases will be used.

CHAPTERS 5, 17 AND 18 DESCRIBE PROCESS COSTING AND OTHER FORMS OF PRODUCT COSTING. THEY CAN BE STUDIED NOW, IF DESIRED, WITHOUT BREAKING CONTINUITY.

APPENDIX: SUPPLEMENTARY DESCRIPTION OF LEDGER RELATIONSHIPS

This appendix explains some of the work that underlies the general-ledger relationships described in this chapter. A factory setting illustrates some of the basic manufacturing transactions and shows how source documents and auxiliary records facilitate the accumulation of data. Exhibit 4-10 summarizes sample accounting entries for job costing for the Alvarez Printing Company.

Subsidiary Ledgers and General Ledger

The general ledger is a summary device. Postings are made to it from totals and subtotals of underlying transactions. For example, the balance of Materials Control may be supported by a voluminous file of materials records. Postings to the debit side of Materials Control may be made from the Materials column in a special journal, such as a purchases journal or a voucher register. But the specific materials record in the subsidiary ledger is posted from a copy of a voucher or an invoice.

Source documents, such as materials requisitions, time tickets, clock cards, and other memoranda, are the primary means of recording accounting transactions. They are vital; all subsequent accounting classifications and reports depend on source documents. These documents are increasingly being kept on computers.

Direct Material Usage Reports

A supervisor may prepare a multicopy materials requisition. Separate copies may serve as follows:

- Copy 1—Kept by storekeeper in materials storeroom
- Copy 2—Used by clerk or computer to post to job-cost record
- Copy 3—Used by accounting department as a basis for a summary of requisitions

This summary is the support for the general-ledger entry:

Work in process control	xx	
Materials control		xx

This entry is usually made monthly, although it can be made more frequently if desired.

- Copy 4—Used as a basis for department material usage reports. If these reports are to prove useful for control, they must typically be prepared more often than once a month. Stale, month-old reports concerning major costs are not helpful. Daily or weekly reports are common. The reports for control are needed before formal postings can be made. *We see that the general ledger is oriented toward product costing rather than toward costing for control.*
- Copy 5—Retained by supervisor. Used as a cross-check against the usage reports that the accounting department sends out.

The accounting department ordinarily uses requisitions that become part of a data base for computers. These requisitions may be sorted in many ways. For example:

Requisition Number	Job Number	Department Number	Amount
501	1414	26	$ 32.00
502	1415	27	51.00
503	1408	26	204.00
504	1414	26	19.00
505	1409	28	101.00

Computers accumulate and tabulate data for classification, reclassification, summarization, and resummarization to provide the specific information by management. Thus a material usage report can be submitted to the supervisor of Department 26 on a daily, weekly, or monthly basis:

Department 26
Direct Material Usage
for the week ended ______

Requisition Number	Job Number	Amount
501	1414	$ 32.00
503	1408	204.00
504	1414	19.00
		$255.00

EXHIBIT 4-10

Job-Order System—Sample Entries for Alvarez Printing Company

Transaction	General-Ledger Effects	Subsidiary Ledgers	Source Documents	Explanatory Comments
1. Purchases of materials or supplies	Materials control Accounts payable	Dr. Materials records "Received" column	Approved invoices	
2. Issuance of direct materials	Work in process control Materials control	Dr. Job-cost records Cr. Materials records "Issued" column	Materials requisitions	Requisitions are summarized and classified by department (for hourly, daily, weekly, or monthly, direct material usage reports)
3. Issuance of indirect materials	Factory department overhead control Materials control	Dr. Department overhead cost records, appropriate columns Cr. Materials records "Issued" column	Materials requisitions	
4. Distribution of labor costs	Work in process control Factory department overhead control Payroll liability	Dr. Job-cost records Dr. Department overhead cost records, appropriate columns for various classes of other labor	Summary of time tickets or daily time analyses. This summary is sometimes called a labor cost distribution summary or a payroll recapitulation.	
5. Payment of payroll	Payroll liability Withholdings payable Cash		Summary of clock cards and individual withholdings as shown on payroll records	This entry is usually made weekly, but the cost distribution (entry 4) is not necessarily made at the same time
7. Employer payroll taxes	Factory department overhead control Employer payroll taxes payable	Dr. Department overhead cost records, appropriate columns	Accrual memoranda from accounting officer	

8. Utilities	Factory department overhead control Accounts payable or Accrued utilities	Dr. Department overhead cost records, appropriate columns	Approved invoices or accrual memoranda	
9. Depreciation on factory equipment	Factory department overhead control Accumulated depreciation—equipment	Dr. Department overhead cost records, appropriate columns	Depreciation schedule	
10. Factory insurance expired	Factory department overhead control Prepaid insurance	Dr. Department overhead cost records, appropriate columns	Insurance register or memoranda from accounting officer	
11. Application of overhead to products	Work in process control Factory overhead applied	Dr. Job-cost records	Budgeted overhead rate computed by using overhead budget	
12. Transfer completed units to finished goods	Finished goods control Work in process control	Dr. Finished goods records, "Received" column Cr. Job-cost records	Production reports	Sometimes the completed job order serves as a finished goods record
13. Sales	Accounts receivable control Sales	Dr. Customers' accounts	Copy of sales invoices	
14. Cost of goods sold	Cost of goods sold Finished goods control	Dr. Cost of goods sold record (optional) Cr. Finished goods records	Copy of sales invoices plus costs as shown on finished goods records	
15. Yearly closing of overhead accounts	Factory overhead applied Factory department overhead control Cost of goods sold (cr. if overhead is overapplied; dr. if overhead is underapplied)		General-ledger balances	This illustrates the immediate write-off method.

Dr. = debit; Cr. = credit

Direct Labor Cost Recapitulation

Similar analysis can be applied to the sorting of direct labor costs, using the time ticket as the source document. Production departments may have their labor classified by operations as well as by jobs. For example, the machining department may perform one or more of the following operations: milling, cleaning, grinding, and facing. Time tickets may be summarized as follows:

Time Ticket Number	Employee ID Number	Job Number	Dept. Number	Amount
14	49	1410	26	$20.00
15	49	1410	26	6.00
16	52	1410	27	19.00
17	53	1410	27	16.00
18	30	1411	25	30.00
19	61	1409	28	24.60
20	52	1409	27	9.75

This labor summary can be used as a basis for the general-ledger entry that charges direct labor to products:

Work in process control	xx	
Payroll liability		xx

This entry is usually made monthly, although it can be made more frequently if desired.

The recapitulation also supplies the information for daily, weekly, or monthly usage reports to the department supervisor. These reports may be broken down by jobs or operations to suit the supervisor.

Time tickets may also be used in figuring idle time (for example, caused by machine breakdowns or shortages of material), overtime premium, material moving, and so forth. A computer or timekeeper may prepare a daily reconciliation of employee clock cards with individual time tickets to see that all clock-card time is accounted for as direct labor, idle time, overtime premium, and so forth.

Sample Entries

Exhibit 4-10 on pages 118–19 summarizes the accounting entries for job costing.

TERMS TO LEARN

This chapter and the Glossary at the end of the book contain definitions of the following important terms:

applied factory overhead *(p. 102)* budgeted factory overhead rate *(104)* clock card *(95)* cost absorption *(103)* cost allocation base *(103)* cost application *(103)* cost application base *(103)* cost pool *(104)* job-cost record *(93)* job-cost sheet *(93)* job-cost systems *(93)* job order *(93)* job-order costing systems *(93)* job-order product-costing systems *(93)* materials requisitions *(94)* overapplied overhead *(108)* process-costing system *(93)* process product-costing system *(93)* production-order systems *(93)* proration *(109)* responsibility centers *(92)* stores requisitions *(94)* time tickets *(94)* underapplied overhead *(108)* value chain *(113)* work-order systems *(93)*

ASSIGNMENT MATERIAL

QUESTIONS

4-1 "Product-costing systems differ markedly." Do you agree? Explain.

4-2 "Management accounting systems are only one source of information for executives." Name two other sources.

4-3 "*Cost management* is not a synonym for *cost accounting systems.*" Do you agree? Explain.

4-4 What are the two major cost objects of a cost accounting system?

4-5 Give at least one synonym for the control purpose of a cost accounting system.

4-6 Give two examples of responsibility centers.

4-7 Give two uses of product costs.

4-8 What is the principal difference between *job-cost* and *process-cost* product-costing systems?

4-9 Give two synonyms for a *job-order product-costing system.*

4-10 Give two synonyms for *time ticket.*

4-11 Distinguish between a *clock card* and a *time ticket.*

4-12 Distinguish between *cost allocation* and *cost application.*

4-13 Explain the role of an application base in factory overhead accounting.

4-14 Why is direct labor likely to become a lower proportion of total manufacturing costs?

4-15 "Through the years, direct labor has been a significant cost driver in job-costing industries." Name at least three other cost drivers.

4-16 What are the limitations of the general ledger as a cost accounting device?

4-17 What is a *normal product* cost?

4-18 What is the purpose of a *department overhead cost record?*

4-19 "The term *manufacturing company* is inaccurate. A company may do manufacturing, but it may also do other things." Do you agree? What are the other things?

4-20 "Product costs for manufacturing are used as inventoriable costs, nothing more." Do you agree? Explain.

EXERCISES AND PROBLEMS

4-21 Identifying manufacturing transactions. Describe the transactions indicated by letters in the following accounts:

Materials Control	
(a)	(b)
	(d)

Work in Process Control	
(b)	(g)
(c)	
(f)	

Finished Goods Control	
(g)	(h)

Department Factory Overhead Control	
(c)	
(d)	
(e)	

Factory Overhead Applied	
	(f)

Cost of Goods Sold	
(h)	

Payroll Liability	
	(c)

4-22 Journal entries. (Alternate is 4-23.) The Lee Company uses a job-order cost system for manufacturing missile parts. The following transactions relate to the month of March:

1. Direct materials issued to production, $98,000
2. Direct labor analysis, $50,000
3. Manufacturing overhead is applied to production on the basis of $30 per machine hour. There were 2,000 machine hours incurred
4. Total manufacturing overhead incurred for the month was $64,000
5. Job orders that cost $210,000 were completed during the month
6. Job orders that cost $200,000 were shipped and invoiced to customers during the month at a gross margin of 20% based on manufacturing cost

Required

The beginning inventory of work in process was $40,000. Prepare the general-journal entries required to record this information. What is the ending balance of work in process?

4-23 Journal entries. (Alternate is 4-22.) The Koski Company uses a job-order system for manufacturing parts for diesel engines. Transactions for 19_5 included (in millions):

1. Acquisitions of direct materials, $10
2. Requisitions of direct materials from the storeroom for use in manufacturing, $9
3. Direct labor used, $1
4. Miscellaneous factory overhead incurred (credit Various Liabilities), $5
5. Factory overhead applied, 100,000 machine hours at $40 per hour
6. Costs of orders completed, $13
7. Cost of goods sold, $12

Required

1. Prepare journal entries.
2. Compute the ending balance of Work in Process Control. Assume that the beginning balance was $1 million.

4-24 Basic entries. (Alternate is 4-26.) The University of Chicago Press is wholly owned by the university. A job-order system is used for printing. The bulk of the work is done for other university departments, which pay as though the Press were an outside business enterprise. The Press also publishes and maintains a stock of books for general sale.

The following data pertain to 19_2 (in thousands):

Direct materials and supplies purchased on account	$ 800
Direct materials issued to the production departments	710
Supplies issued to various production departments	100
Labor used directly in production	1,300

Indirect labor incurred by various departments	$ 900
Depreciation, buildings and factory equipment	400
Miscellaneous factory overhead* incurred by various departments (ordinarily would be detailed as repairs, photocopying, utilities, etc.)	550
Factory overhead applied at 160% of direct labor cost	?
Cost of goods manufactured	4,120
Sales	8,000
Cost of goods sold	4,020
Inventories, December 31, 19_1 (not 19_2):	
Materials control	100
Work in process control	60
Finished goods control	500

*The term factory overhead is not used uniformly. Other terms that are often encountered in printing companies include job overhead and shop overhead.

Required

1. Prepare general journal entries to summarize 19_2 transactions. As your final entry, dispose of the year-end overapplied or underapplied factory overhead as a direct adjustment to Cost of Goods Sold. Number your entries. Explanations for each entry may be omitted.
2. Show posted T-accounts for all inventories, Cost of Goods Sold, Factory Department Overhead Control, and Factory Overhead Applied.
3. Sketch how the subsidiary ledger would appear for Factory Department Overhead Control. Assume that there are three departments: art, photo, and printing. You need not show any numbers.

For more details concerning these data, see Problem 4-25.

4-25 Journal entries and source documents. Refer to Problem 4-24. For each journal entry, (a) indicate the source document most likely generating the entry, and (b) give a description of the entry into the subsidiary ledgers, if any.

4-26 Journal entries and T-accounts. (Alternate is 4-24.) The following data relate to operations of the Donnell Printing Company for the year 19_5 (in millions):

Materials control, December 31, 19_4	$ 12
Work in process control, December 31, 19_4	2
Finished goods control, December 31, 19_4	6
Materials and supplies purchased on account	150
Direct materials issued to the production departments	145
Indirect materials (supplies) issued to various production departments	10
Labor used directly in production	90
Indirect labor incurred by various departments	30
Depreciation—plant and factory equipment	19
Miscellaneous factory overhead incurred by various departments (credit Various Liabilities, ordinarily would be detailed as repairs, utilities, etc.)	9
Factory overhead applied, 2,100,000 machine hours at $30 per hour	?
Cost of goods manufactured	294
Sales	400
Cost of goods sold	292

Required

1. Prepare journal entries. Number your entries.
2. Post to T-accounts. What is the ending balance of Work in Process Control?
3. Sketch how the subsidiary ledger would appear for Factory Department Overhead Control, assuming that there are four departments. You need not show any numbers.
4. Show the journal entry for disposing of overapplied or underapplied overhead directly as a year-end adjustment to Cost of Goods Sold. Post the entry to T-accounts.

For more details concerning these data, see Problem 4-27.

4-27 Journal entries and source documents. Refer to Problem 4-26. For each entry, (a) indicate the most likely name of the source documents that would authorize the entry, and (b) give a description of the entry into the subsidiary ledgers, if any.

4-28 Accounting for overhead; budgeted rates. (Alternate is 4-29.) The Solomon Company uses a budgeted overhead rate for applying factory overhead to job orders on a machine hour basis for the machining department and on a direct labor cost basis for the finishing department. The company budgeted the following for 19_1:

	Machining	Finishing
Factory overhead	$10,000,000	$8,000,000
Machine hours	200,000	33,000
Direct labor hours	30,000	160,000
Direct labor cost	$ 900,000	$4,000,000

Required

1. What is the budgeted overhead rate that should be used in the machining department? In the finishing department?
2. During the month of January, the cost record for job order No. 431 shows the following:

	Machining	Finishing
Direct materials requisitioned	$14,000	$3,000
Direct labor cost	$ 600	$1,250
Direct labor hours	30	50
Machine hours	130	10

   What is the total overhead applied to Job 431?
3. Assuming that Job 431 consisted of 200 units of product, what is the unit cost of Job 431?
4. Balances at the end of 19_1:

	Machining	Finishing
Factory overhead incurred	$11,200,000	$7,900,000
Direct labor cost	$ 950,000	$4,100,000
Machine hours	220,000	32,000

   Compute the underapplied or overapplied overhead for each department and for the factory as a whole.
5. Provide reasons why Solomon uses two different overhead application bases.

4-29 Accounting for overhead. (Alternate is 4-28.) The Lynn Company budgeted the following for 19_4:

	Machining	Assembly
Factory overhead	$1,800,000	$3,600,000
Direct labor cost	$1,400,000	$2,000,000
Direct labor hours	100,000	200,000
Machine hours	50,000	200,000

The company uses a budgeted overhead rate for applying overhead to production orders on a machine hour basis in Machining and on a direct labor cost basis in Assembly.

Required

1. Compute the budgeted overhead rate for each department.
2. During February the cost record for Job 494 contained the following:

	Machining	Assembly
Direct materials requisitioned	$45,000	$70,000
Direct labor cost	$14,000	$15,000
Direct labor hours	1,000	1,500
Machine hours	2,000	1,000

Compute the total overhead cost of job 494.

3. At the end of 19_4, the actual factory overhead costs were $2,100,000 in Machining and $3,700,000 in Assembly. Assume that 55,000 actual machine hours were incurred in Machining and actual direct labor cost in Assembly was $2,200,000. Compute the overapplied or underapplied overhead for each department.

4-30 Use of job-cost record. San Diego Tape Company manufactures diskettes for use in reproducing sound. The company uses a job-cost system.

On June 4 the company began production of 20,000 diskettes, assigned Job 471, to be sold to stores for $1.75 each. The following costs pertain to Job 471, which required 20 equipment hours (12 on June 4 and 8 on June 5):

Date	Requisition No.	Description	Amount
6-4	211	50 lbs. polypropylene @ $9	$450
6-4	212	70 lbs. magnetic filament @ $12	840
6-5	217	8 lbs. bucylic acid @ $50	400
	Time Ticket No.	**Description**	**Amount**
6-4	814	12 hours @ $18	$216
6-5	815	25 hours @ $10	250

The company charges overhead to jobs based on the relationship between budgeted factory overhead ($900,000) and budgeted equipment hours (30,000). Equipment hours are recorded automatically by computer. Job 471 was completed on June 5 and transferred to finished goods.

Required

1. Prepare a job-cost record for Job 471.
2. Journalize all costs of Job 471.
3. Journalize the transfer of Job 471 to finished goods.

4-31 Analyzing job-cost data. The Denver Tool Company uses a job-cost system. Consider the following data:

Job No.	Dates: Started	Finished	Sold	Total Cost of Job at April 30	Total Manufacturing Cost Added in May
1	3/26	4/7	4/9	$1,400	
2	4/3	4/12	4/13	6,200	
3	4/3	4/30	5/1	3,600	
4	4/17	5/24	5/27	200	$1,000
5	4/29	5/29	6/3	800	3,200
6	5/8	5/12	5/14		1,600
7	5/23	6/6	6/9		600
8	5/29	6/22	6/26		5,800

Required

1. Compute Denver Tool's cost of (a) work in process at April 30 and May 31, (b) finished goods at April 30 and May 31, and (c) cost of goods sold for April and May.
2. Prepare summary journal entries for the transfer of completed units from work in process to finished goods for April and May.
3. Record the sale of Job 4 for $2,000.

4-32 Subsidiary and general ledgers, journal entries. The Moran Custom Furniture Co. worked on only three jobs during September and October. The job-cost records are summarized as follows:

	410		411		412
	September	October	September	October	October
Direct materials	$19,000	$ —	$12,000	$8,000	$14,000
Direct labor	4,000	2,000	6,000	4,000	1,000
Factory overhead applied	12,000	?	18,000	?	?

Factory overhead is applied as a percentage of direct labor costs. The balances in selected accounts on September 30 were: direct materials control, $31,000; finished goods control, $40,000; cost of goods sold, $900,000; payroll liability, $1,000; and factory overhead applied, $250,000.

Job 410 was completed, transferred to finished goods, and sold along with other finished goods by October 31, the end of the fiscal year. The total cost of goods sold during October was $75,000.

Job 411 was still in process at the end of October. Job 412 was also in process. It had begun on October 23.

Required

1. Taken together, the job-cost records are the subsidiary ledger supporting the general-ledger balance of Work in Process Control. Prepare a schedule of job-cost records showing the balance of Work in Process Control, September 30.
2. Compute the overhead application rate.
3. Prepare summary general journal entries for all costs added to Work in Process Control during October. Also prepare an entry for all costs transferred from Work in Process Control to Finished Goods Control.
4. Post the journal entries to the appropriate T-accounts.
5. Prepare a schedule of job-cost records showing the balance of the Work in Process Control, October 31.

4-33 Accounting for factory overhead. Consider the following selected cost data for the Pittsburgh Forging Company for 19_2:

Budgeted factory overhead cost	$7,000,000
Budgeted machine hours	200,000
Actual factory overhead cost	$6,800,000
Actual machine hours	195,000

Required

1. Compute the budgeted factory overhead rate.
2. Journalize the application of factory overhead.
3. Compute the amount of underapplied or overapplied factory overhead. Is the amount significant? Journalize the disposition of the ending balances in factory overhead accounts.

4-34 Meaning of overapplied overhead. The Umberto Company had budgeted the following performance for 19_4:

Machine hours	30,000
Beginning inventories	None
Sales	$4,000,000
Total variable costs	3,000,000
Total fixed costs	800,000
Operating income	200,000
Factory overhead:	
Variable	300,000
Fixed	600,000

It is now the end of 19_4. A factory overhead rate of $30 per machine hour was used throughout the year for costing products. Total factory overhead incurred was $900,000. Overapplied overhead was $54,000. There is no work in process.

Required
How many machine hours were used in 19_4?

4-35 Overhead balances. (SMA) Budgeted overhead, based on a budgeted volume of 100,000 direct labor hours, was $255,000. Actual overhead costs amounted to $270,000, and actual direct labor hours were 105,000. Overhead overapplied (underapplied) amounted to (choose one): (1) $2,250 overapplied, (2) ($2,250) underapplied, (3) $15,000 overapplied, or (4) ($15,000) underapplied. Show computations.

4-36 Journal entries, year-end disposition of overhead. Spinosa Company uses a job-order cost system. Factory overhead is applied at a rate of $70 per machine hour. Both beginning and ending balances in work in process and finished goods are zero. You are given the following data for 19_4. All goods manufactured are sold.

Machine hours used	21,000
Direct labor hours used	50,000
Direct materials used	$4,000,000
Direct labor	1,000,000
Indirect labor	250,000
Indirect supplies used	100,000
Rent—plant and factory equipment	500,000
Miscellaneous factory overhead	500,000
Cost of goods sold	2,750,000

All underapplied or overapplied overhead is allocated to cost of goods sold at the end of the year.

Required
1. What is factory overhead applied?
2. What is factory overhead incurred?
3. Prepare journal entries to record all the facts above. Include all necessary entries to adjust for overapplied or underapplied overhead.

4-37 Application and proration of overhead. (SMA, heavily adapted) Nicole Limited is a company that produces machinery to customer orders, using skilled labor and a job-order cost system. Manufacturing overhead is applied to production using a budgeted rate. This overhead rate is set at the beginning of each fiscal year by forecasting the coming year's overhead and relating it to direct labor costs. The budget for the company's last fiscal year was:

Direct labor	$280,000
Manufacturing overhead	$168,000

As of the end of the year, two jobs were incomplete. These were No. 1768B—total direct labor charges were $11,000—and No. 1819C—total direct labor costs were $39,000. On these jobs, machine hours were 287 for No. 1768B, and 647 for No. 1819C. Direct materials issued to No. 1768B amounted to $22,000, and $42,000 to No. 1819C.

Total charges to the Manufacturing Overhead Control account for the year were $186,840. Direct labor charges made to all jobs were $400,000, representing 20,000 direct labor hours.

There were no beginning inventories. In addition to the ending work in process described above, the ending finished goods showed a balance of $72,000. Sales for the year totaled $2,700,680; cost of goods sold totaled $648,000; and selling, general, and administrative expenses were $1,857,870.

The amounts for inventories and cost of goods sold were not adjusted for any overapplication or underapplication of manufacturing overhead to production. It is the company's practice to prorate any overapplied or underapplied overhead to inventories and cost of goods sold.

Required
1. Prepare a detailed schedule showing the ending balances in the inventories and cost of goods sold (before considering any underapplied or overapplied manufacturing overhead).

2. Assume that underapplied or overapplied manufacturing overhead is prorated in proportion to the ending balances (before the proration) in Work in Process, Finished Goods, and Cost of Goods Sold. Prepare a detailed schedule showing the proration and the final balances after proration.
3. Assume that all the underapplied or overapplied manufacturing overhead was added to or subtracted from Cost of Goods Sold. Would operating income be higher or lower than the operating income that would result from the prorations in requirement 2 above? By what amount?

4-38 Proration of overhead. The McDonnell Company has commercial and defense contracting business. A contracting officer for the United States Air Force has insisted that underapplied overhead should no longer be written off directly as an adjustment of Cost of Defense Goods Sold for a given year. His insistence arose because $50 million of underapplied overhead was added to the $400 million of unadjusted Cost of Goods Sold in 19_4. There were no beginning inventories.

Factory overhead is applied as a percentage of direct labor costs as contracts are produced. The air force had a large cost-plus-fixed-fee contract representing $300 million of the $400 million of defense production started and sold during 19_4. It had no other contracts pending with McDonnell. An analysis of costs showed (in millions):

	Defense Business		
	Contracts in Progress	Finished Goods Inventory	Cost of Goods Sold
Direct material used	$425	$25	$250
Direct labor cost	45	5	50
Factory overhead applied	90	10	100
Total before adjustment	560	40	400*
Add: Underapplied overhead	—	—	50
Total after adjustment	$560	$40	$450

*Includes $300 million attributable to air force contract.

Required

1. What overhead rate based on direct labor costs would have resulted in factory overhead applied equaling factory overhead incurred?
2. As a judge trying to settle a dispute on the disposition of the underapplied overhead, what position would you favor? Why? Show computations and, assuming your answer would be formally recorded in the general ledger show a journal entry for the proration.
3. As the contracting officer, what basis of prorating of the underapplied overhead would you favor? Why? Show computations.

4-39 Proration of overhead. (Z. Iqbal, adapted) The Zaf Radiator Company has the following data for 19_3:

Budgeted manufacturing overhead	$4,800,000
Overhead application base	Machine hours
Budgeted machine hours	80,000
Manufacturing overhead incurred	$4,900,000
Actual machine hours	75,000

Machine hours data and the ending balances (before proration of underapplied or overapplied overhead) follow:

	Machine Hours	Costs
Cost of goods sold	30,000	$6,250,000
Finished goods	30,000	3,125,000
Work in process	15,000	3,125,000

Required
1. Prorate the underapplied or overapplied overhead by the method commonly used. Compute the ending balances after proration.
2. Repeat requirement 1, except now use the proration method that is conceptually preferable.

4-40 Incomplete data. The Phillips Company uses perpetual inventories and a normal cost system. Balances from selected accounts (in millions) were:

	Balances December 31, 19_1	Balances December 31, 19_2
Factory department overhead control		$ 56
Finished goods control	$50	46
Cost of goods sold		180
Direct materials control	?	20
Work in process control	?	35
Factory overhead applied		72

The cost of direct materials requisitioned for production during 19_2 was $100 million. The cost of direct materials purchased during 19_2 was $90 million. Factory overhead is applied at 200% of direct labor cost.

Required
Before considering any year-end adjustments for overapplied or underapplied overhead, compute:

1. Direct materials control, December 31, 19_1
2. Work in process control, December 31, 19_1

4-41 Overview of general-ledger relationships. The Blakely Company is a small machine shop that uses highly skilled labor and a job-order cost system. The total debits and credits in certain accounts *just before* year-end are:

	December 30, 19_6	
	Total Debits	Total Credits
Direct materials control	$100,000	$ 70,000
Work in process control	320,000	305,000
Factory department overhead control	85,000	—
Finished goods control	325,000	300,000
Cost of goods sold	300,000	—
Factory department overhead applied	—	90,000

Note that "total debits" in the inventory accounts would include beginning inventory balances, if any.

The above accounts *do not* include the following:

a. The labor cost recapitulation for the December 31 working day: direct labor $5,000, and indirect labor $1,000
b. Miscellaneous factory overhead incurred on December 30 and December 31: $1,000

Additional Information:

- Factory overhead has been applied as a percentage of direct labor cost through December 30.
- Direct material purchases during 19_6 were $85,000.
- There were no returns to suppliers.
- Direct labor costs during 19_6 totaled $150,000, not including the December 31 working day described above.

Required
1. Compute the inventories (December 31, 19_5) of direct materials control, work in process control, and finished goods control. Show T-accounts.

2. Prepare all adjusting and closing journal entries for the above accounts. Assume that all underapplied or overapplied overhead is closed directly to Cost of Goods Sold.
3. Compute the ending inventories (December 31, 19_6), after adjustments and closing, of direct materials control, work in process control, and finished goods control.

4-42 Details of job costing. (CMA, adapted) Targon Inc. manufactures lawn equipment. The business uses a job-order system because the products are manufactured in batches rather than on a continuous basis. The balances in selected general-ledger accounts for the eleven-month period ended August 31, 19_2, are presented below.

Materials inventory	$ 32,000
Work in process inventory	1,200,000
Finished goods inventory	2,785,000
Factory overhead control	2,260,000
Cost of goods sold	14,200,000

The work in process inventory consists of two jobs:

Job No.	Units	Items	Accumulated Cost
3005-5	48,000	Estate sprinklers	$ 700,000
3006-4	40,000	Economy sprinklers	500,000
			$1,200,000

The finished goods inventory consists of five items:

Items	Quantity and Unit Cost	Accumulated Cost
Estate sprinklers	5,000 units @ $22 each	$ 110,000
Deluxe sprinklers	115,000 units @ $17 each	1,955,000
Brass nozzles	10,000 gross @ $14 per gross	140,000
Rainmaker nozzles	5,000 gross @ $16 per gross	80,000
Connectors	100,000 gross @ $5 per gross	500,000
		$2,785,000

The factory cost budget prepared for the 19_1–_2 fiscal year is presented below. The company applies factory overhead on the basis of direct labor hours.

The activities during the first eleven months of the year were quite close to the budget. A total of 367,000 direct labor hours have been worked through August 31, 19_2.

Factory Cost Annual Budget
For the Year Ended
September 30, 19_2

Direct materials		$ 3,800,000
Purchased parts		6,000,000
Direct labor (400,000 hours)		4,000,000
Overhead		
Supplies	$190,000	
Indirect labor	700,000	
Supervision	250,000	
Depreciation	950,000	
Utilities	200,000	
Insurance	10,000	
Property taxes	40,000	
Miscellaneous	60,000	2,400,000
Total factory costs		$16,200,000

The September 19_2 transactions are summarized on the next page.

1. All direct materials, purchased parts, and supplies are charged to materials inventory. The September purchases were as follows:

Direct materials	$410,000
Purchased parts	285,000
Supplies	13,000

2. The direct materials, purchased parts, and supplies were requisitioned from materials inventory as shown in the table below.

	Purchased Parts	Materials	Supplies	Total Requisitions
3005-5	$110,000	$100,000	$ —	$210,000
3006-4	—	6,000	—	6,000
4001-3 (30,000 gross of rainmaker nozzles)	—	181,000	—	181,000
4002-1 (10,000 deluxe sprinklers)	—	92,000	—	92,000
4003-5 (50,000 ring sprinklers)	163,000	—	—	163,000
Supplies	—	—	20,000	20,000
	$273,000	$379,000	$20,000	$672,000

3. The payroll summary for September is as follows:

	Hours	Cost
3005-5	6,000	$ 62,000
3006-4	2,500	26,000
4001-3	18,000	182,000
4002-1	500	5,000
4003-5	5,000	52,000
Indirect	8,000	60,000
Supervision	—	24,000
Sales and administration	—	120,000
		$531,000

4. Other factory costs incurred during September were:

Depreciation	$62,500
Utilities	15,000
Insurance	1,000
Property taxes	3,500
Miscellaneous	5,000
	$87,000

5. Jobs completed during September and the actual output were:

Job No.	Quantity	Items
3005-5	48,000 units	Estate sprinklers
3006-4	39,000 units	Economy sprinklers
4001-3	29,500 gross	Rainmaker nozzles
4003-5	49,000 units	Ring sprinklers

6. The following finished products were shipped to customers during September:

Items	Quantity
Estate sprinklers	16,000 units
Deluxe sprinklers	32,000 units
Economy sprinklers	20,000 units
Ring sprinklers	22,000 units
Brass nozzles	5,000 gross
Rainmaker nozzles	10,000 gross
Connectors	26,000 gross

Required

1. a. Calculate the overapplied or underapplied overhead for the year ended September 30, 19_2. Be sure to indicate whether the overhead is overapplied or underapplied.
   b. What is the appropriate accounting treatment for this overapplied or underapplied overhead? Explain your answer.
2. Calculate the dollar balance in the work in process inventory account as of September 30, 19_2.
3. Calculate the dollar balance in the finished goods inventory as of September 30, 19_2, for the estate sprinklers using a FIFO basis.

4-43 Review of job costing: general-ledger relationships. The Weismer Co.'s job-order accounting system is on a calendar-year basis. As of January 31, 19_2, the following information is available:

a. Direct materials used for January totaled $200,000.
b. The cost of goods sold during January was $500,000.
c. Direct materials inventory on January 31, 19_2, was $9,000.
d. The cost of goods completed and transferred to finished goods during January was $600,000.
e. The budgeted factory overhead application rate for 19_2 is 180% of direct labor costs.
f. The finished goods inventory, on December 31, 19_1, was $42,000.
g. Gross factory compensation paid in January totaled $195,000. (Ignore withholdings.)
h. All employees performing direct labor get the same rate of pay. Direct labor hours for January totaled 10,000. Indirect labor, supervision, and miscellaneous factory overhead payroll totaled $30,000.
i. Jobs 480 and 482 were not completed on January 31, 19_2. Together, their total direct labor charges were $6,000 (400 hours). Their total direct material charges were $13,200.
j. The overapplied factory overhead, as of January 31, was $14,000.
k. Direct materials purchased during January totaled $207,000.
l. Balance, Payroll Liability, as of January 31, 19_2, was $4,000.

Required

Compute, showing your work:

1. Balance, Direct Materials Control, on December 31, 19_1
2. Balance, Finished Goods Control, on January 31, 19_2
3. Direct labor costs incurred during January
4. Actual factory overhead incurred during January
5. Balance, Payroll Liability, on December 31, 19_1
6. Balance, Work in Process Control, on January 31, 19_2
7. Balance, Work in Process Control, on December 31, 19_1

4-44 Multiple choice; incomplete data. (Alternate is 4-45.) Some of the general-ledger accounts of the Lucas Manufacturing Company appear as follows on January 31, 19_1.

The accounts are incomplete because the accountant had an emergency operation for ulcers after eating lunch in the company cafeteria on January 31. The treasurer, an old friend of yours, supplied you with the following incomplete accounts and three bits of additional information:

Direct Material Control

Bal. Jan. 1	15,000	
	35,000	

Work in Process Control

Bal. Jan. 1	1,000	40,000
Direct materials requisitioned	20,000	

Finished Goods Control	
Bal. Jan. 1 10,000	30,000

Cost of Goods Sold	

Payroll Liability	
	Bal. Jan. 1 1,000
	Gross earnings of all factory workers 40,000

Additional Information

a. Work tickets for the month totaled 1,650 direct labor hours. All factory workers received $20 per hour.

b. Indirect costs are applied at a rate of $44 per machine hour, and 500 machine hours were used during the month.

After giving you a few minutes to look over the data given, your old friend asks you the following multiple-choice questions. Choose one answer for each question and show computations.

Required

1. Is the January 31 balance of Direct Material Control
   (a) $50,000, (b) $25,000, (c) $20,000, (d) $30,000, (e) $35,000, (f) none of these?
2. Is the amount of total direct labor cost that should have been charged to all the individual production orders worked on during January
   (a) $40,000, (b) $41,000, (c) $33,000, (d) $55,000, (e) $30,000, (f) none of these?
3. Is the total factory overhead cost that should have been applied to production
   (a) $17,000, (b) $75,000, (c) $40,000, (d) $33,000, (e) $24,000, (f) none of these?
4. Is the *total* factory labor cost for the month of January
   (a) $40,000, (b) $41,000, (c) $55,000, (d) $33,000, (e) $56,000, (f) none of these?
5. Is the January 31 balance of Work in Process Control
   (a) $75,000, (b) $77,000, (c) $76,000, (d) $35,000, (e) $36,000, (f) none of these?
6. Is the January 31 balance of Finished Goods Control
   (a) $5,000, (b) $10,000, (c) $20,000, (d) $25,000, (e) $30,000, (f) none of these?
7. Is the Cost of Goods Sold during January
   (a) $40,000, (b) $10,000, (c) $20,000, (d) $30,000, (e) $50,000, (f) none of these?
8. Factory overhead costs actually incurred during the month amount to $24,000. Is the balance in Factory Overhead Control at the end of January
   (a) $24,000, (b) $17,000, (c) $26,000, (d) $41,000, (e) $22,000, (f) $2,000, (g) none of these?
9. Is the January 31 balance of Factory Overhead Applied
   (a) $33,000, (b) $40,000, (c) $24,000, (d) $26,000, (e) $22,000, (f) $30,000, (g) none of these?
10. Is the amount of underapplied (or overapplied) costs for January
    (a) Underapplied by $2,000, (b) Underapplied by $1,000, (c) Overapplied by $1,000, (d) Overapplied by $2,000, (e) Overapplied by $3,000, (f) Neither overapplied nor underapplied, (g) none of these.

4-45 General-ledger relationships; incomplete data. (Alternate is 4-44.) You are asked to bring the following incomplete Dallas Co. accounts up to date through May 19_1. Also, consider the additional information (in dollars) that follows the T-accounts.

Direct Materials Control	
5/31/_1 Balance 20,000	

Accounts Payable	
	4/30/_1 Balance 10,000

Work in Process Control	
4/30/_1 Balance 2,000	

Factory Department Overhead Control	
Total charges for May 55,000	

Factory Overhead Applied	

Finished Goods Control		Cost of Goods Sold	
4/30/_1 Balance 25,000			

Additional Information:

a. The overhead is applied by using a budgeted rate that is set at the beginning of each year by forecasting the year's overhead and relating it to forecasted machine hours. The budget for 19_1 called for a total of 15,000 machine hours and $750,000 of factory overhead.
b. The accounts payable are for direct materials only. The balance on May 31 was $12,000. Payments of $78,000 were made during May.
c. The finished goods inventory as of May 31 was $7,000.
d. The cost of goods sold during the month was $165,000.
e. On May 31 there was only one unfinished job in the factory. Cost records show that $1,000 (40 hours) of direct labor and $2,000 of direct materials had been charged to the job. Thirty machine hours were used on the job.
f. A total of 940 direct labor hours were worked during the month of May. All factory workers earn the same rate of pay.
g. All "actual" factory overhead incurred during May has already been posted.
h. A total of 1,000 machine hours was used during May.

Required

Compute the following:

1. Materials purchased during May
2. Cost of goods completed during May
3. Overhead applied during May
4. Balance, Work in Process Control, May 31, 19_1
5. Direct materials used during May
6. Balance, Direct Materials Control, April 30, 19_1
7. Overapplied or underapplied overhead for May

4-46 Normal job costing, application and proration of overhead. Stylistic Homes assembles motor homes. On December 1, 19_1, Stylistic Homes had zero materials and no work in process or finished goods inventory. During December, 700 homes were sold and 200 homes were finished but not sold. On December 31, 19_1, 100 homes were in varying stages of work in process.

Stylistic Homes uses a job-order costing system. Direct costs include direct materials and direct labor. Factory overhead is applied based on direct materials dollars. Factory overhead control comprises four individual cost accounts:

- Indirect materials
- Indirect labor
- Machining costs (includes depreciation and operation costs)
- Computer costs (includes hardware and software costs related to automated sections of the production line)

The December 31 debit and credit totals before any month-end adjustments are:

	Debits in thousands	Credits in thousands
Materials control	$8,750	$8,550
Payroll liability	2,320	2,320
Accounts payable	8,550	8,750
Factory overhead control—indirect material	950	0
Factory overhead control—indirect labor	480	0
Factory overhead control—machining	2,880	0
Factory overhead control—computer	1,300	0
Factory overhead applied	0	4,560
Work in process control	?	13,320
Finished goods control	?	10,360
Cost of goods sold	?	?

Assume all the above accounts had zero balances as of December 1, 19_1.

Required

1. What is the underapplied or overapplied factory overhead for December?
2. Assume that under- or overapplied factory overhead is directly written off to cost of goods sold. What are the December 31 balances of (a) Materials Control (b) Work in Process Control, (c) Finished Goods Control, and (d) Cost of Goods Sold? Show ledger relationships (T-accounts) underlying your answer.
3. What is the overhead rate per direct material dollar?
4. Assume that underapplied or overapplied factory overhead is prorated in proportion to the December 31 balances (before proration) of work in process, finished goods, and cost of goods sold. After proration of the underapplied or overapplied factory overhead, what are the December 31 balances for (a) Materials Control, (b) Work in Process Control, (c) Finished Goods Control, and (d) Cost of Goods Sold?
5. Provide specific reasons why Stylistic may have selected direct material dollars as the overhead application base.

CHAPTER 5

Printed circuit boards, used in many electronic products, are increasingly being costed using activity accounting systems.

Job Costing for Services, Process Costing, and Activity-Based Accounting

Learning Objectives

When you have finished studying this chapter, you should be able to

1. Describe three alternatives of job costing for a professional service firm
2. Give journal entries for a typical process product-costing system
3. Identify five key steps in process costing
4. Explain the role of equivalent units in process costing
5. Describe activity-based accounting; relate it to control and product costing
6. Give reasons why activity-based accounting is in demand by today's managers
7. Describe the major differences between typical systems and activity-based accounting systems

As its title indicates, this chapter has three major parts. Part One explores how job costing aids planning and control, focusing on services instead of manufacturing. Part Two introduces process costing. Part Three describes activity-based accounting. Each part may be studied independently, and each part concludes with one or more Problems for Self-Study.

PART ONE: JOB COSTING FOR PLANNING AND CONTROL OF SERVICES

GENERAL APPROACH TO PLANNING AND CONTROL

Job-order costing is a valuable management tool, not only in manufacturing companies but in construction companies, service industries, and nonprofit organizations as well. For instance, housebuilders, public accounting firms, consulting firms, law firms, advertising agencies, and hospitals use this costing method. Managers in nonmanufacturing industries, however, may not use the job-costing terminology described in Chapter 4. For example, instead of "jobs," law firms and hospitals have "cases" and consultants have "contracts" or "engagements."

The preceding chapter emphasized inventory costing, not planning and control. However, regardless of the industry setting, job costing helps management plan and control operations. As managers plan each job, they predict the expected quantities of direct resources, such as direct materials or direct labor. The appropriate material prices and labor rates are multiplied by the related physical quantities. Overhead is also included in calculating costs. The result is the budgeted total cost of the job. Comparisons of budgeted and actual costs help managers control the jobs under way and plan future jobs.

Consider a service such as an audit engagement conducted by a certified public accounting firm. A partner would first plan the audit (job). Based on the job's design, the partner predicts the job's costs to the firm (direct costs and indirect costs) and uses this figure in calculating a fee to quote to the client. The partner monitors the job's progress by comparing the hours actually logged at specific dates with the hours budgeted and the hours estimated still to be spent on the audit. The final job records help the partner predict the time and costs of other audits for the same client or similar clients. In this case and in other cases using job costing, accurate time records are critical for planning and control.

Although we will use an illustration of services to show how job costing is used for planning and control, the same general concepts are also used in manufacturing.

Consider the relative importance of direct and indirect costs of jobs in manufacturing and service industries. In manufacturing, direct materials are usually the major cost; sometimes labor is so minor that it is not worth tracking separately to individual jobs. In contrast, in service industries, direct labor is usually the major cost; materials, such as paper or envelopes or postage, are often too minor to be accounted for as direct costs.

BUDGETING AND PRICING: THREE ALTERNATIVES

Objective 1

Described three alternatives of job costing for a professional service firm

Managers face challenging problems in determining how to apply costs to the specific services rendered. The degree of accuracy sought varies in accordance with many factors, including management styles, industry traditions, and competitive pressures.

Professional service organizations use one of three different job-costing approaches:

- *Alternative One*. A single direct cost item (usually direct professional labor), a single indirect cost (overhead) pool, and a single indirect cost (overhead) application rate.
- *Alternative Two*. Multiple direct cost items (direct professional labor, photocopying, computer time, and so on), a single overhead pool, and a single overhead application rate.
- *Alternative Three*. Multiple direct cost items, multiple overhead cost pools, and multiple overhead application rates.

Exhibit 5-1 presents an overview of these alternatives. We will illustrate the first two alternatives. Although desirable for its accuracy, Alternative Three—multiple overhead rates—is costly and is rarely encountered in practice in service industries.

Alternative One: Single Direct Cost, Single Indirect Cost Pool

Like many professional service firms, public accounting firms classify the compensation of their professional personnel who work directly on audits as direct labor. All other costs are usually classified as indirect costs (overhead). We call this procedure *Alternative One*. Suppose a public accounting firm has a condensed budget for audits for 19_1 as follows:

Revenue	$15,000,000	100%
Deduct direct professional labor (for professional hours charged to client engagements)	3,750,000	25
Contribution to overhead and operating income	$11,250,000	75%
Deduct overhead*	10,500,000	70
Operating income	$ 750,000	5%

*All other costs, including unassigned time, training, marketing, and general administration.

As each audit is budgeted, the partner in charge predicts the required number of direct professional hours. *Direct professional hours* are those worked on the audit by partners, managers, and other accountants. The budgeted direct labor cost is the respective hourly labor rates multiplied by the budgeted hours. Partners' time is

EXHIBIT 5-1

Alternative Job-Costing Approaches for a Professional Service Organization (such as an Accounting Firm)

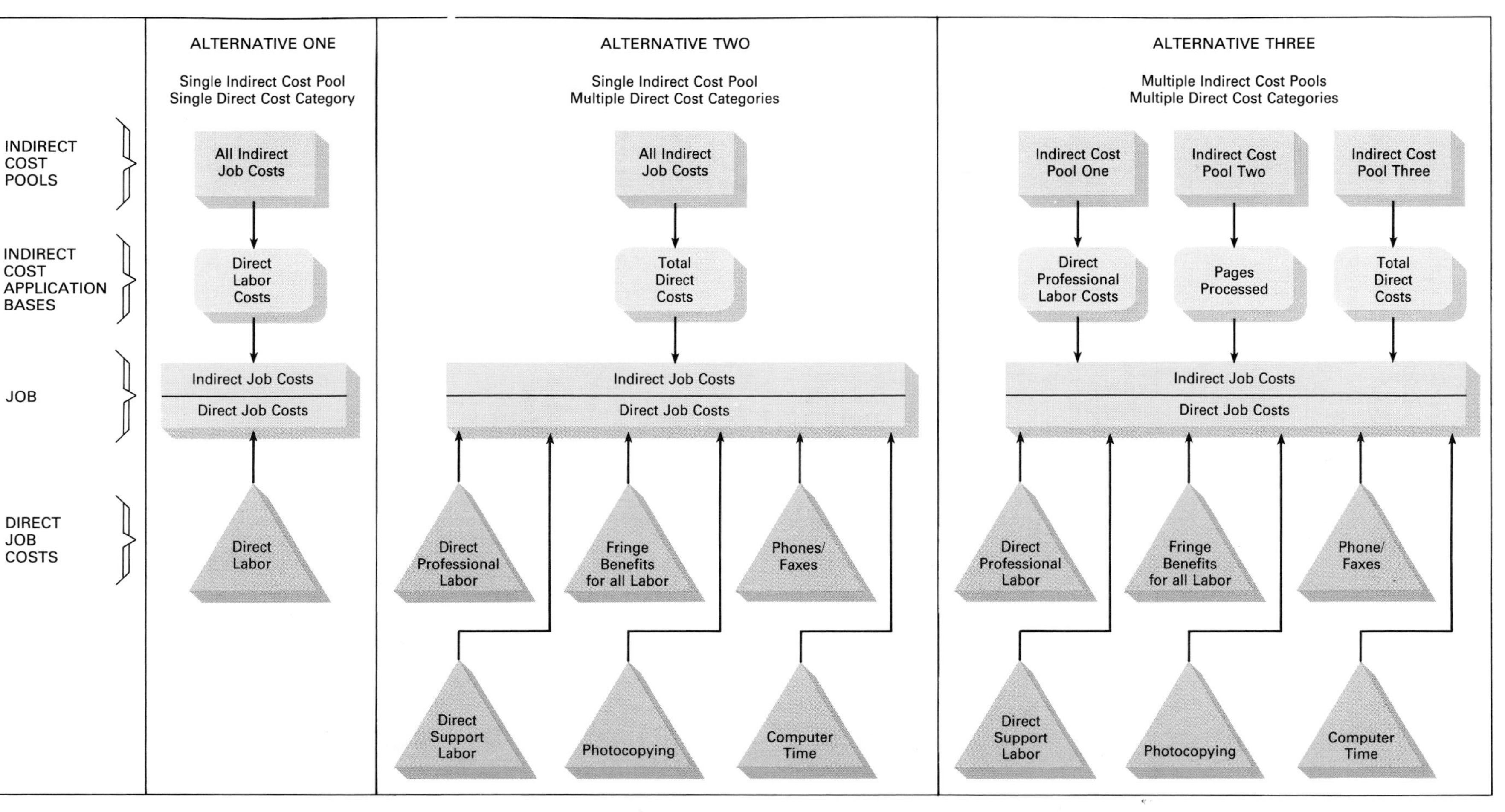

charged to the audit at much higher rates than that of subordinates. Overhead is combined in a single indirect cost pool and is applied as a percentage of direct labor costs. This practice implies that partners require proportionately more overhead support than their subordinates for each of their hours charged. The budgeted total cost of the audit is the direct labor cost plus the applied overhead. In this example, the budgeted overhead rate is 280%, computed as follows:

$$\text{Budgeted overhead rate} = \frac{\text{Budgeted overhead cost}}{\text{Budgeted direct professional labor costs}}$$
$$= \frac{\$10{,}500{,}000}{\$3{,}750{,}000} = 280\%$$

As in all decisions regarding pricing, the revenue for a particular audit depends on competitive factors and the audit partner's assessment of the client's willingness to pay. However, partners frequently use rules of thumb or formulas at least as points of departure for setting a price. In our illustration, the formula might have a markup rate:

$$\text{Markup rate for pricing individual audits} = \frac{\text{Budgeted revenue}}{\text{Budgeted direct professional labor costs}}$$
$$= \frac{\$15{,}000{,}000}{\$3{,}750{,}000} = 400\%$$

That is, the firm will bill a client at four times the direct professional labor costs. If all the partners were able to attain this 400% formula rate consistently, and if the predicted direct labor costs were accurate, the firm would achieve budgeted revenue.

Suppose an audit is expected to have the following summarized costs:

Direct professional labor	$ 40,000
Overhead applied, 280% of $40,000	112,000
Total costs of audit	$152,000

Using the 400% formula rate, the expected revenue and operating income from the audit would be:

Revenue (400% of $40,000)	$160,000
Deduct total costs	152,000
Operating income	$ 8,000

Note that the $8,000 here is 5% of the $160,000 revenue; this is the same percentage of operating income shown earlier in the entire firm's budget ($750,000 ÷ $15,000,000 = 5%).

Specific Control of Work

The partner in charge of the audit usually has a budget that includes detailed scope and steps. For example, the budget for auditing cash or inventories would specify the exact work to be done, the number of hours, and the levels of skilled personnel needed.

As mentioned earlier, the audit partner monitors progress by comparing the actual direct professional hours logged to date with the original budget and with the estimated hours remaining on the engagement. The degree of profitability of each audit heavily depends on whether (a) the audit can be completed within the budgeted time limits, (b) the target revenue can be collected from the client, and (c) the actual costs behaved as expected.

Sometimes the scope of the work changes because of unanticipated early findings or special requests from the client. In these situations, the fees are renegotiated.

Alternative Two: Multiple Direct Costs, Single Indirect Cost Pool

Our audit example described a relatively simple job-costing system for a professional service firm. Only a single direct cost item (labor) and a single overhead rate are used. However, clients have come to demand a more thorough explanation of fees than what Alternative One provides. To supply this explanation in an increasingly competitive business world, professional service firms have strengthened their efforts to track more costs as direct costs rather than as overhead. Many public accounting firms, law firms, and other professional service firms have refined their data processing systems, using computers to help gather the detailed information necessary for providing more direct costs, which we call *Alternative Two*. See Exhibit 5-1.

Assume the following summarized costs for an audit, using the Alternative Two approach:

Direct professional labor	$ 40,000
Direct support labor, such as secretarial costs	10,000
Fringe benefits for all direct labor*	17,500
Photocopying	1,100
Telephone calls	1,000
Computer time (either used inside the firm or acquired outside)	8,000
Total direct costs	$ 77,600
Overhead applied, assumed as 100% of total direct costs	77,600
Total costs of audit	$155,200

*35% assumed rate × ($40,000 + $10,000) = $17,500.

Note that Alternative Two has more direct cost items than Alternative One has. Any item that can be specifically traced and identified with the job might be included as a direct cost. If tracing these costs is economically feasible, items like photocopying, telephone calls, secretarial time, and computer time might be direct costs rather than a part of the indirect costs pooled together as overhead. On average, in comparison with Alternative One, this tracing will benefit some clients and will mean that other clients are charged more.

Our Alternative One illustration used only one class of direct labor, the professional labor worked directly on the audit. Support labor and all fringe benefits were part of overhead. In contrast, our illustration of Alternative Two has two classes of direct labor. Also, it classifies the fringe benefits of all direct labor (professional labor and support labor) as an item separate from direct labor. Another firm might embed fringe benefits in its direct labor costs. Assuming that fringe benefits are 35% of direct labor, this firm would cost its direct professional labor at $40,000 × 135% = $54,000.

How should we classify the compensation costs of the firm's computer employees directly tracked to the audit? Some firms include such costs as direct professional labor; other firms classify them as direct support labor.

Alternative One used direct professional labor as a basis for a 400% markup to obtain revenue of $160,000. Alternative Two could show the same revenue by using the total direct costs as a basis for a percentage markup that would obviously be far lower than 400%. To get the same revenue as in Alternative One, the markup on total direct costs would be $160,000 ÷ $77,600 = 206.2%.

What costs should be included in the total costs used as a basis for markup? Practices differ. There is no single correct answer. For example, some firms will classify computer services purchased from outsiders as not being subject to markups. That is, such computer costs will be billed to clients at 100% of the amounts incurred. Such classification decisions are the result of the preferences of the managing partners of the firms.

Effects of Classifications on Overhead Rates

As might be expected, Alternative Two results in different total costs than does Alternative One ($155,200 compared with $152,000). Depending on how prices are set, these costs may lead to a different total revenue for the audit.

Alternative Two, the more detailed approach, has a lower overhead application rate, 100% of total direct costs instead of the 280% of direct professional labor used in Alternative One. Why? For two reasons. First, there are fewer overhead costs because more costs are traced directly. Second, the application base is broader, including *all* direct costs rather than only direct professional labor.

Some firms prefer to continue to apply their overhead based on all direct professional labor costs, including fringe benefits, rather than on total direct costs. Why? Because the partners believe that overhead is dominantly affected by the amount of direct professional labor costs rather than by other direct costs such as telephone calls.

Whether the overhead application base should be total direct costs, direct professional labor costs or hours, or some other application base is a knotty problem for most professional service firms. Ideally, studies should uncover the principal causes of costs—cost drivers—and these drivers should be used as overhead application bases.

The Alternative Two approach demonstrates a trend in both service and manufacturing industries. That is, as data processing becomes less expensive, more costs than just direct materials and direct labor will be classified as direct costs. Competitive pressures will probably strengthen this trend. Why? Because firms that fail to cost their jobs more accurately will be at a competitive disadvantage.

Alternative Three: Multiple Direct Costs, Multiple Indirect Cost Pools

More than one overhead rate can be used to apply costs to jobs in nonmanufacturing settings, which we call *Alternative Three*. For example, some overhead might be applied on the basis of direct professional labor, and other overhead might be applied on the basis of computer time. Computer time is becoming more widely used as a base as expensive software and programming assume a more prominent role in rendering service to clients or customers.

As multiple overhead application bases are increasingly used, more accurate tracking of costs to specific jobs will result. Managers will then have improved information to guide their decisions. For example, Alternative Three in Exhibit 5-1 shows three indirect cost application bases for a professional service organization: direct professional labor costs, pages processed, and total direct costs. For another example, in manufacturing there may be many cost drivers identified, and various costs may be applied to jobs depending on the number of purchase orders, the quantities of various parts, the number of different parts, the quantities of direct labor, and the quantities of machine hours.

In both manufacturing and service industries, the use of multiple overhead bases will lead to better-informed decisions on pricing, choosing product mixes, and determining whether to buy resources from outside or make the items inside. Thus, it becomes less likely that the multiple-service and multiple-product firms will be outmaneuvered by specialty, niche, or single-product firms.

PROBLEM FOR SELF-STUDY

PROBLEM

"Job costs are important for inventory costing, but they have little to do with planning and control." Do you agree? Explain.

SOLUTION

Job costs are important for both inventory costing and planning and control. Accurate records of direct materials usage and direct labor help managers to monitor and to predict the work remaining on jobs under way and also help managers to predict costs of similar jobs in the future. The chapter illustration of the audit engagement (a job) emphasizes planning and control.

PART TWO: AN INTRODUCTION TO PROCESS-COSTING SYSTEMS

MASS PRODUCTION AND BROAD AVERAGES

Chapter 4 (p. 93) pointed out that job-order product-costing and process product-costing systems are best viewed as ends of a spectrum.

For specific units or small batches of custom-made products ←-------→	For mass production of like units
Job-order product-costing systems	Process product-costing systems

The principal difference between job costing and process costing arises from the types of products that are cost objects. As Chapter 4 emphasized, job-order costing is prevalent where each unit or batch (job) of product tends to be special. In a *process product-costing system* (most often called *process-costing system*), product costs are obtained by allocating total costs to masses of like units that usually proceed in continuous fashion through a series of uniform production steps, such as mixing, cooking, and packaging. Industries that use process costing include chemicals, oil, paper, textiles, plastics, coal mining, glass, and silicon wafers.

All product costing uses averaging to determine manufacturing costs per unit. In job costing, the average is computed using a relatively small number of units, as in the production of a particular printing order. In contrast, process costing uses vast quantities of items—say, a huge number of pencils—in computing the unit average. Accumulated costs for a period—say, a month—are divided by the quantities produced during that period to get the average unit cost. Process costing may also be adopted in nonmanufacturing contexts. For example, we can divide the costs of giving state automobile driver's license tests by the number of tests given, or divide the costs of an X-ray department by the number of X-rays processed.

Exhibit 5-2 shows the major differences between job-order costing and process costing. Several work in process accounts are used in process costing. As goods move from process to process, their costs are transferred accordingly.

The job-order costing sketch in Exhibit 5-2 identifies three major categories of costs: direct materials, direct labor, and factory overhead. In contrast, the process-costing sketch identifies two major categories generally used: direct materials and conversion costs. Direct labor is relatively less significant in process costing, so it is combined with factory overhead in the category frequently called conversion costs. Thus, direct labor is accounted for no differently than are other labor, power, factory rent, and depreciation on equipment.

EXHIBIT 5-2

Comparison of Job-order and Process Costing for Inventory Costing

JOB-ORDER costing: Examples include printing, construction, repairing, and jewelry

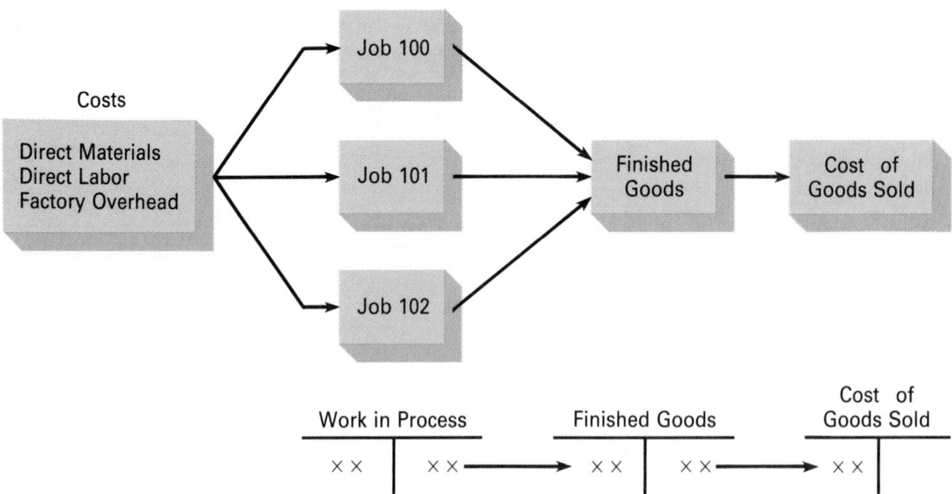

PROCESS COSTING: Examples include chemicals, glass, oil refining, paper, and silicon wafers

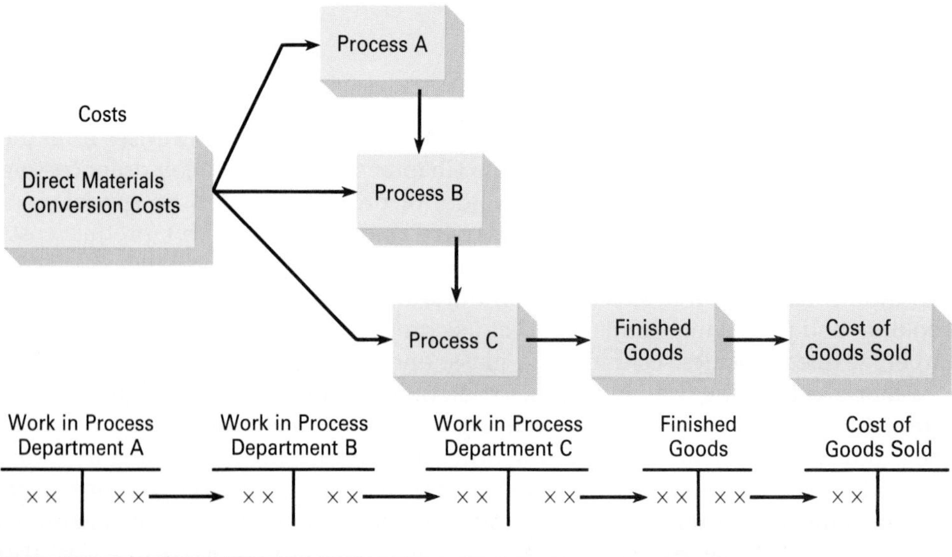

INVENTORY ACCOUNTING AND JOURNAL ENTRIES

Objective 2

Give journal entries for a typical process product-costing system

Process-costing systems are usually simpler and less expensive than job-order costing. Individual jobs do not exist. There are no individual job-cost records. The unit manufacturing cost for inventory purposes is calculated by accumulating the costs of each processing department and dividing the total cost by an appropriate measure of output. For instance, the accumulated cost of a cooking department in a Del Monte tomato-processing plant may be divided by the pounds of tomatoes processed.

The relationships of the inventory accounts are (amounts in millions):

Work in Process—Cooking

Debit		Credit	
Direct materials	30	Transfer cost of goods completed to next department	45
Conversion costs	20		
	50		
Ending inventory	5		

Work in Process—Canning

Debit		Credit	
→ Cost transferred in from Cooking	45	Transfer cost of goods completed to finished goods	50
Additional conversion costs	8		
	53		
Ending inventory	3		

The journal entries are similar to those for the job-order costing system. That is, direct materials and conversion costs are basically accounted for as in Chapter 4. However, there is now more than a single Work in Process account for all units being manufactured. There are two or more Work in Process accounts, one for each processing department. Journal entries follow:

	Debit	Credit
1. Work in process—Cooking	30	
Materials inventory		30
To record direct materials used.		
2. Work in process—Cooking	20	
Materials inventory		2
Payroll liability		8
Accumulated depreciation, equipment		10
To record conversion costs; examples include supplies, all labor, and pertinent depreciation.		
3. Work in process—Canning	45	
Work in process—Cooking		45
To transfer goods to the canning process.		
4. Work in process—Canning	8	
Materials inventory		1
Payroll liability		2
Accumulated depreciation, equipment		5
To record conversion costs; examples as in entry 2.		
5. Finished goods	50	
Work in process—Canning		50
To transfer goods from the canning process.		

The central product-costing problem is how each department should compute the cost of goods transferred out and the cost of goods remaining in the department. If an identical amount of work was done on all units in process, the solution is easy. However, if the units in the inventory are each partially completed, the product-costing system must distinguish between the fully completed units transferred out and the partially completed units not yet transferred.

FIVE KEY STEPS IN PROCESS COSTING

Objective 3

Identify five key steps in process costing

Suppose Davis Company buys wood as the direct material for its forming department. The department processes one type of doll. The dolls are transferred to the finishing department, where hand shaping and fabric are added.

Assume that the forming department manufactured 25,000 units during April. There was no beginning inventory. The department's April costs were:

Direct materials	$ 82,500
Conversion costs	42,500
Costs to account for	$125,000

The unit cost of goods completed would simply be $125,000 ÷ 25,000 = $5.00. An itemization would show:

Direct materials, $82,500 ÷ 25,000	$3.30
Conversion costs, $42,500 ÷ 25,000	1.70
Unit cost of completed unit	$5.00

Suppose not all 25,000 units were completed during April. For example, assume that 5,000 units were still in process at the end of April; only 20,000 were started and fully completed. All direct materials had been placed in process, but on the average only 25% of the conversion costs had been applied to the 5,000 units in ending inventory. How should each department calculate the cost of goods transferred out and the cost of goods remaining in the ending work in process inventory? This is the major inventory-costing problem. The inventory-costing system must distinguish between the cost of the fully completed units and the cost of the partially completed units.

We will describe five key steps in process-cost accounting:

- Step 1: Summarize the flow of physical units.
- Step 2: Compute output in terms of equivalent units.
- Step 3: Summarize the total costs to account for, which are the total debits in Work in Process (that is, the costs applied to Work in Process).
- Step 4: Compute equivalent unit costs.
- Step 5: Apply costs to units completed and to units in ending Work in Process.

Physical Units and Equivalent Units (Steps 1 and 2)

Objective 4

Explain the role of equivalent units in process costing

Step 1, as the first column in Exhibit 5-3 shows, tracks the physical units of production. How should the output for April be measured? Not as 25,000 units. The output was 20,000 *fully* completed units and 5,000 *partially* completed units. A partially completed unit is certainly not a substitute for a fully completed unit. Accordingly, output is usually stated in *equivalent units,* not physical units.

Equivalent units measure the output in terms of the quantities of each of the factors of production that have been applied. That is, an equivalent unit is regarded as the collection of inputs necessary to produce one complete physical unit of product. In our example, as step 2 in Exhibit 5-3 shows, the output would be measured as 25,000 equivalent units of direct materials cost but only 21,250 equivalent units of conversion costs. Only one-quarter of the conversion costs have been added to the ending work in process inventory.

Measures in equivalent units are not confined to manufacturing situations. Universities may measure enrollments in full-time student equivalents. Part-time students would be "measured" in terms of full-time enrollment. For example, two students each taking half the credits of a full-time course of study would be the equivalent of one full-time student "unit."

EXHIBIT 5-3
Davis Company Forming Department
Output in Equivalent Units
For the Month Ended April 30, 19_1

Flow of Production	(Step 1) Physical Units	(Step 2) Equivalent Units	
		Direct Materials	Conversion Costs
Started and completed	20,000	20,000	20,000
Work in process, ending	5,000 (25%)*	5,000	1,250†
Accounted for	25,000		
Work done to date		25,000	21,250

*Degree of completion for conversion costs of this department only at the date of the work in process inventory
†5,000 physical units × .25 degree of completion of conversion costs.

Equivalent units are a popular way of expressing work done in terms of a common denominator. Do there have to be beginning or ending work in process inventories to have equivalent units? No, because as we have just seen, measures of equivalent units are often used to express inputs or outputs in common terms. For example, suppose it is the middle of a torrid summer, and demand for beer has soared. Suppose further that a brewery has no beginning or ending inventories. The brewery's output is expressed in equivalent barrels, even though its products may be sold in bottles, cans, or kegs.

Calculation of Product Costs (Steps 3, 4, and 5)

Exhibit 5-4 is a production cost report. It shows steps 3, 4, and 5. Step 3 summarizes the total costs to account for (that is, the total costs in, or debits to, Work in

EXHIBIT 5-4
Davis Company Forming Department
Production-Cost Report
For the Month Ended April 30, 19_1

	Costs	Totals	Details	
			Direct Materials	Conversion Costs
(Step 3)	Total costs to account for	$125,000	$82,500	$42,500
	Divide by equivalent units		÷25,000	÷21,250
(Step 4)	Equivalent unit costs		$ 3.30	$ 2.00
(Step 5)	Application of costs:			
	Completed and transferred out (20,000 units)	$106,000	(20,000 × $5.30*)	
	Work in process, ending (5,000 units):			
	Direct materials	$ 16,500	5,000 ($3.30)	
	Conversion costs	2,500		1,250 ($2.00)
	Total work in process	$ 19,000		
	Total costs accounted for	$125,000		

*Cost per equivalent whole unit = $3.30 + $2.00 = $5.30

Process—Forming). Step 4 obtains unit costs by dividing total costs by the appropriate measures of equivalent units. The unit cost of a completed unit is $3.30 + $2.00 = $5.30. Why is the unit cost $5.30 instead of the $5.00 calculated earlier in this chapter (p. 146)? Because the $42,500 conversion cost is spread over 21,250 units instead of 25,000 units. Step 5 then uses these unit costs to apply costs to products.

In Exhibit 5-4, notice how the costs are applied to obtain an ending work in process of $19,000. The 5,000 physical units are fully completed regarding direct materials. Therefore, the direct materials applied are 5,000 equivalent units times $3.30, which equals $16,500. In contrast, the 5,000 physical units are 25% completed regarding conversion costs. Therefore, the conversion costs applied are 25% of 5,000 physical units, or 1,250 equivalent units, times $2.00, which equals $2,500.

Journal Entries

The summarized journal entries for the data in our illustration follow:

1. Work in process—Forming	82,500	
Materials inventory		82,500
Direct materials used in production in April.		
2. Work in process—Forming	42,500	
Various accounts		42,500
To record conversion costs; examples include supplies, all labor, and pertinent depreciation.		
3. Work in process—Finishing	106,000	
Work in process—Forming		106,000
Cost of goods completed and transferred in April from Forming to Finishing		

The key T-account would show:

Work in Process—Forming

1. Direct materials	82,500	3. Transferred out to Finishing	106,000
2. Conversion costs	42,500		
Costs to account for	125,000		
Bal: April 30	19,000		

CHAPTERS 17 AND 18 DESCRIBE PROCESS COSTING IN MORE DETAIL. THEY CAN BE STUDIED NOW, IF DESIRED, WITHOUT BREAKING CONTINUITY.

PROBLEM FOR SELF-STUDY

PROBLEM

The Internal Revenue Service must process millions of income tax returns yearly. When the taxpayer sends in a return, documents such as withholding statements and checks are matched against the data on page one. Then various other inspections of the data are conducted. Of course, some returns are more complicated than others, so the expected time allowed to process a return is geared to an "average" return.

Some work-measurement experts have been closely monitoring the processing at a particular branch. They are seeking ways to improve productivity.

Suppose 1 million returns were received on April 15. On April 22, the work-measurement teams discovered that all supplies (computer materials, inspection check-sheets, and so on) had been affixed to the returns, but 40% of the returns still had to undergo a final inspection. The other returns were fully completed.

Required

1. Suppose the final inspection represents 20% of the overall processing time in this process. Compute the total work done in terms of equivalent units of tax returns.
2. The materials and supplies consumed cost $150,000. For these calculations, materials and supplies are regarded just like direct materials. The conversion costs were $1,380,000. Compute the unit costs of materials and supplies and of conversion.
3. Compute the cost of the tax returns not yet completely processed.

SOLUTION

1.

Flow of Production	(Step 1) Physical Units	(Step 2) Equivalent Units: Materials & Supplies	Conversion Costs
Started and completed	600,000	600,000	600,000
Work in process, ending	400,000 (80%)*	400,000	320,000†
Accounted for	1,000,000		
Work done to date		1,000,000	920,000

*Degree of completion for conversion costs of this department only at the date of the work in process inventory

†400,000 × 80% = 320,000

2.

	Costs	Totals	Details: Materials & Supplies	Conversion Costs
(Step 3)	Total costs to account for	1,530,000	$150,000	$1,380,000
	Divide by equivalent units		÷1,000,000	÷920,000
(Step 4)	Equivalent unit costs		$0.15	$1.50

3. (Step 5)			
Application of costs:			
Completed and transferred out (600,000 units)	$990,000	600,000 ($1.65*)	
Work in process, ending (400,000 units):			
Materials and supplies	$ 60,000	400,000 ($.15)	
Conversion costs	480,000		320,000 ($1.50)
Total work in process	$ 540,000		
Total costs accounted for	$1,530,000		

*Cost per equivalent whole unit = $.15 + $1.50 = $1.65

The cost of tax returns not fully processed is $540,000.

PART THREE: ACTIVITY-BASED ACCOUNTING SYSTEMS

SERVING THE PRODUCT-COSTING PURPOSES AND THE PLANNING AND CONTROL PURPOSE

Objective 5

Describe activity-based accounting; relate it to control and product costing

As its name indicates, **activity-based accounting** (or **activity-based costing** or **activity accounting**) is a system that focuses on activities as the fundamental cost objects and uses the costs of these activities as building blocks for compiling the costs of other cost objects. Note that activity-based accounting is generic—that is, it can be a part of a job-order product-costing system or a process product-costing system.

Our descriptions of cost systems in this chapter and in Chapter 4 have focused on product costing. Except for the illustration of the accounting firm, our discussions have centered on how manufacturing costs are applied to products for inventory purposes. However, to maximize long-term profitability, systems should also serve the planning and control purpose.

Many managers plan and control business functions through personal observation and through systems that contain budgets, actual costs, and variances. These business functions are often divided into departments. In this way, managers can better pinpoint responsibilities and costs for various subparts of the business functions. Exhibit 5-5 shows the division of companywide business functions, in Panel A, and then into departments, Panel B.

Recall from Chapter 4 that the general ledger account entitled Factory Department Overhead Control accumulates actual overhead costs. Its subsidiary ledgers contain cost records for each factory department. The factory overhead costs are applied to products as the production departments work on them.[1]

If dividing companywide functions into departments helps managers, wouldn't dividing departments into activities help to sharpen managers' focus further? Panel C of Exhibit 5-5 lists activities common in an assembly department. Note that costs are accumulated for each activity as a separate cost object. The costs are then applied to products as they undergo various activities. ***The final product costs are "built up" from the costs of the specific activities undergone.*** For example, material-handling costs would be the initial activity for assembling an electric motor. If its assembly does not call for any manual insertions of parts, its product costs would have no costs for that activity.

Panel C of Exhibit 5-5 illustrates how the *inventory costs* are built up, depending on the chain of activities used to manufacture a finished product. *Full product costs* could also be built up in a similar fashion by including the costs of various activities in research and development, product design, marketing, distribution, and customer service.

You now have an overview of how costs can be accumulated for planning and control purposes. ***Exhibit 5-5 shows how cost objects become ever more finely granulated, from a particular business function to departments to activities—if that is what managers want.***

[1]Support or service departments (such as repairs and maintenance, purchasing, and power departments) are also a part of the manufacturing function. Their costs are allocated to the production departments. For simplicity, we are not covering these cost allocations here. Chapter 14 explains how to allocate costs from one department to other departments.

EXHIBIT 5-5
Overview of Business Functions, Departments and Activities

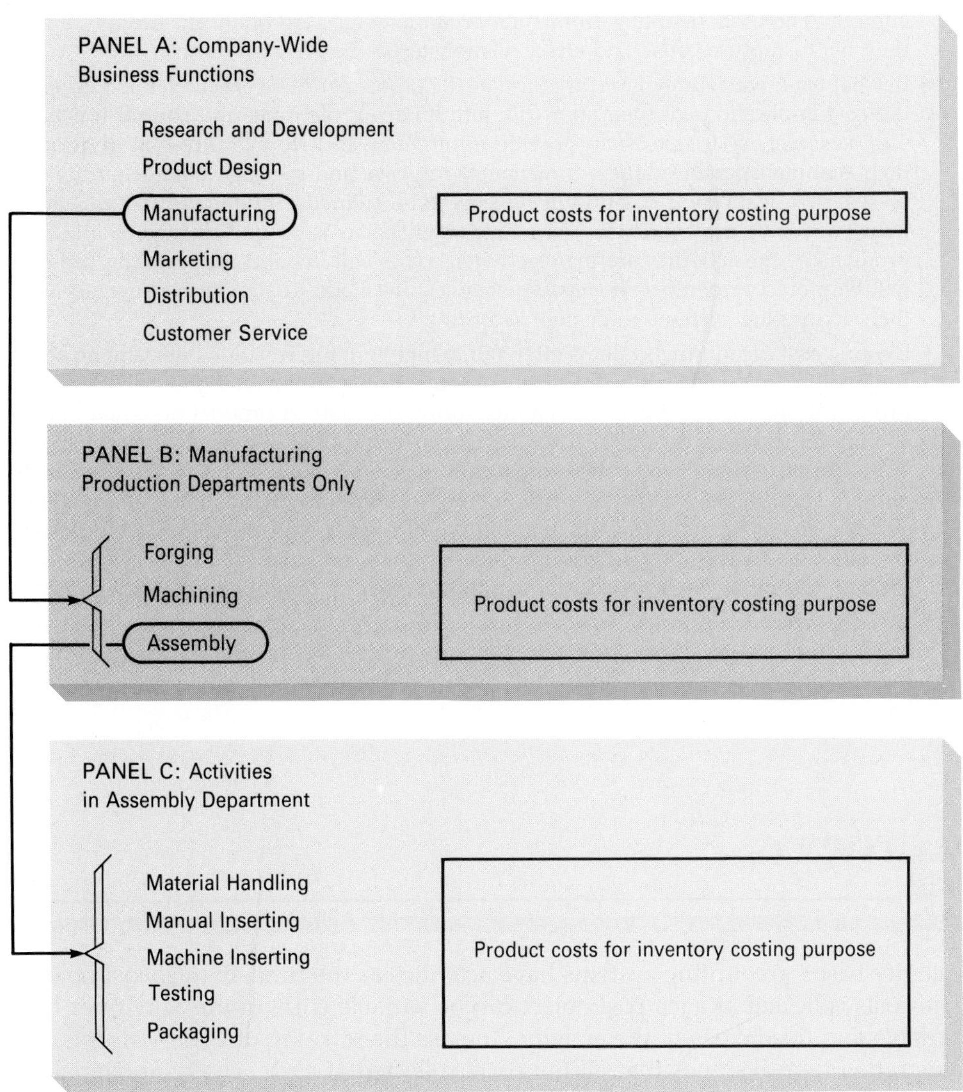

DEMANDS FOR ACTIVITY-BASED ACCOUNTING

Objective 6

Give reasons why activity-based accounting is in demand by today's managers

During the late 1980s, many managers and accountants became discontented with their cost accounting systems. Some companies began to install activity-based accounting systems. Their number is growing. There are several reasons why managers demand activity-based accounting:

- Activities within various departments may be compared or be combined with activities within other departments. For example, the total cost of maintaining quality would encompass many costs. It would be the sum of the inspection costs in the purchasing department, the inspection costs in the production departments, and the customer-service costs in the marketing department. Only if detailed costs are kept by activities can the total company costs of quality (or any other aspect, such as acquiring, storing, and handling materials) be obtained.

- To better manage activities and make wiser economic decisions, executives want to identify the relationships of causes (activities) and effects (costs) in a more detailed, accurate manner. A major concern is whether applications of overhead are sufficiently accurate. Using time-honored single pools of indirect costs and overhead application bases such as direct labor hours or direct labor dollars is no longer considered good enough. The costs resulting from such broad averages are often misleading because they fail to capture cause-and-effect relationships.
- In Chapter 4, we define *cost management* as the performance by executives and others in the cost implications of their short-run and long-run planning and control functions. *Cost accounting systems* exist to provide information to help executives in performing their cost management duties. If managers improve and redesign manufacturing processes, they want their accounting systems to be tailored to the new processes. More than ever, managers' primary cost management focus is on underlying activities, not products. If the activities are managed well, costs will fall and the resulting products will be more competitive. If massive changes are made in activities, managers want their accounting systems to change accordingly.
- Existing cost accounting systems often fail to highlight interrelationships among activities in different departments or functional areas. Consider the relationship between the product design function and the manufacturing function. A product designed to have 40 parts probably causes more costs than a similar product designed to have 10 parts. Why? Because there would be more vendors, more transactions, more inventories, more rework, more inspections, more customer service. Many existing cost accounting systems do not accumulate costs in a way that identifies opportunities to reduce costs. Activity-based accounting induces product designers to reduce the number of parts and prompts operating personnel to reduce the number of vendors and transactions.
- Developments in information-gathering technology—including bar coding, numerically controlled equipment, and hand-held computers—has made practical the gathering and processing of more detailed information demanded by activity-based accounting.

ILLUSTRATION OF ACTIVITY-BASED ACCOUNTING

Activity-based accounting systems have activities as the fundamental cost objects. The costs collected at each cost object can be variable costs of the activity or both variable and fixed costs of the activity. To ease the learning of basic concepts, the illustration here assumes that all the costs collected at each activity are variable.[2]

Our illustration of activity-based accounting summarizes the experience of an actual manufacturing facility, Instruments Inc., that assembles and tests over 800 electronic instrument products, including printed-circuit boards. Prior to the adoption of activity-based accounting, its accounting system was typical of many systems worldwide. Product costing was achieved by summing four categories of manufacturing costs:

- Direct materials
- Direct labor

[2]When fixed costs are included in the computation of the indirect cost application rate of each activity area, issues associated with production volume variances must be considered. Chapters 8 and 9 discuss these variances.

- Indirect manufacturing cost pool 1—applied to products based on their direct material dollars content
- Indirect manufacturing cost pool 2—applied to products based on their direct labor dollars content

Indeed, this product-costing system was more sophisticated than the frequently used system of a single pool for all indirect manufacturing costs.

As business became increasingly competitive, the managers in product product design, manufacturing, and marketing became skeptical about the accuracy of Instruments' cost accounting system. For example, a product designer commented:

> "Why is it when I use a $.20 part (*X*), the materials overhead charge is $.02, but when I use a $100 part (*Y*), the materials overhead charge is $10? Product *Y* does not consume 500 times the resources used to procure and handle *X*."

Managers in manufacturing believed that different factors were causing or driving costs in individual activity areas, but the accounting system did not collect information about these differences. These managers wanted to focus primarily on controlling costs of activity areas. Managers in marketing perceived that the accounting system tended to "overcost" the intensely competitive high-volume products. How? By loading too much indirect manufacturing costs on high-volume products and too little on low-volume products.

Working as a team, representatives of manufacturing and accounting constructed an activity-based accounting system. Manufacturing personnel had the loudest voice in the choice of the individual activity areas and in selecting each area's cost driver for applying costs to products, which we turn to now.

Activity Areas

Every printed-circuit (PC) board of Instruments Inc. has parts (diodes, capacitors, and integrated circuits) inserted on it. The board passes through six activity areas in the manufacturing facility:

1. **Material handling.** All the parts necessary for building the PC board are combined into a kit.
2. **Start station.** Instructions for building the board are entered into a computer. The software program tells the automated equipment which parts to insert where.
3. **Machine insertion of parts.** Automated and semiautomated equipment insert components on the board.
4. **Manual insertion of parts.** Skilled workers insert those components that are not machine-inserted (because of their shape, weight, location on the board, and so on).
5. **Wave soldering.** All parts inserted on the board are simultaneously soldered to ensure that they remain attached.
6. **Quality testing.** Tests are made to check that all components are inserted and in the right place and that the final product performs to specification.

The cost drivers and the cost application rates for computing the indirect manufacturing costs for a PC board follow:

Activity Area	Cost Driver Used as Cost Application Base	Indirect Cost Application Rate
1. Material handling	Number of parts	$1 per part
2. Start station	Number of PC boards	$20 per board
3. Machine insertion of parts	Number of machine-inserted parts	$0.50 per part
4. Manual insertion of parts	Number of manually inserted parts	$4 per part
5. Wave soldering	Number of PC boards	$30 per board
6. Quality testing	Hours of test time	$50 per test hour

Consider the material-handling activity as an illustration of how the indirect cost application rate is computed (using assumed numbers):

$$\text{Application rate for material-handling activity} = \frac{\text{Budgeted material handling}}{\text{Budgeted quantity of parts for the year}}$$

$$= \frac{\$300{,}000}{300{,}000} = \$1 \text{ per part}$$

Each time a raw PC board passes through the material-handling activity, $1 multiplied by its number of parts is added to the board's cost. If a particular type of PC board requires ten parts, its product-cost buildup because of material handling includes $10 of indirect costs; if twenty parts, $20; and so forth.

Direct labor, which is less than 5% of total manufacturing costs, is no longer tracked separately. Instead, it is regarded as an item of indirect costs. Taken together, the manufacturing processes are best described as *machine-paced,* not *labor-paced*. That is, the overall physical production steps and the accompanying manufacturing costs depend more heavily on machines and automation, not on human touching and building products.

Numerical Example

Exhibit 5-6 presents the costing of two PC boards using Instrument Inc.'s activity-based accounting system. An overview of the product-costing system is in Exhibit 5-7. Refer to Exhibits 5-6 and 5-7 throughout this section.

EXHIBIT 5-6

Using Instrument Inc.'s Activity-Based Accounting System
Product Costing of PC Boards A and B

	PC Board A	PC Board B
Direct manufacturing product costs		
• Direct materials	$300	$280
Indirect manufacturing product costs		
• Material handling*		
(A, 81 parts; B, 121 parts) × $1	$81	$121
• Start station		
(A, 1 board; B, 1 board) × $20	20	20
• Machine insertion of parts		
(A, 70 insertions; B, 90 insertions) × $0.50	35	45
• Manual insertion of parts		
(A, 10 insertions; B, 30 insertions) × $4	40	120
• Wave soldering		
(A, 1 board; B, 1 board) × $30	30	30
• Quality testing		
(A, 1.5 hours; B, 2.5 hours) × $50	75	125
Total	281	461
Total manufacturing product costs	$581	$741

*The number of parts includes the raw printed-circuit board (counted as 1 part) plus the number of component parts to be inserted into the board.

EXHIBIT 5-7
Overview of Product Costing at Instruments Inc. Using an Activity-Based Accounting System

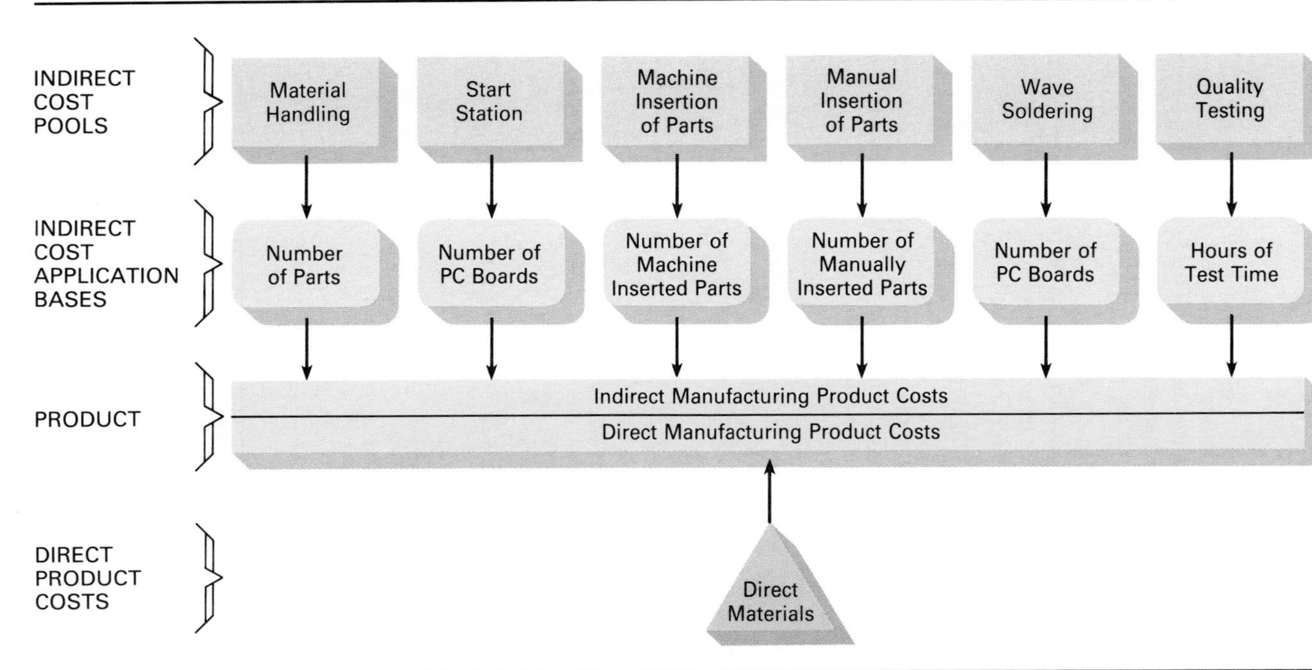

1. The activity-based accounting system pinpoints opportunities for cost reductions. For example, consider the design of products at Instruments Inc. PC Board A has been designed to have
   (i) Fewer parts (81 for A, 121 for B)
   (ii) A higher percentage of insertions by machine (87.5% for A, 75% for B)
   (iii) Less test time (1.5 hours for A, 2.5 hours for B)
   The activity-based accounting system explicitly signals that (i), (ii), and (iii) each reduce the cost of assembling a PC board.
2. Manufacturing personnel at Instruments Inc. are using the activity-based accounting system in their cost-reduction efforts. Each activity area at Instruments Inc. has its own supervisor. The measures used to evaluate the supervisor's performance now include the achievement of cost-reduction targets for the cost driver rate. For example, the supervisor of the machine insertion of components activity area receives a bonus if the actual rate per machine-inserted component is reduced by 10% or more.
3. The reported costs of the standard ("no-frills") PC boards are lower with the activity-based accounting system than with the previous system. PC Board A in Exhibit 5-6 is an example of a standard PC board. In contrast, PC boards that require the manual insertion of a large number of parts or extended test time (as does PC Board B) have higher reported costs under the activity-based accounting system.
4. Each of the indirect cost application bases is a nonfinancial variable (number of parts, hours of test time, and so on). One manufacturing manager used the phrase "Let's get physical" to introduce the application bases used in the activity-based accounting system. ***Controlling physical items such as hours, parts, and defects is the most fundamental way of managing costs.***
5. The differences in the product costs of PC Boards A and B indicate, in part, how these products use different amounts of the resources of each activity area. Consider differences in the relative use of the following four activity areas:

	PC Board A	PC Board B
Material handling	81 parts	121 parts
Machine insertion of parts	70 insertions	90 insertions
Manual insertion of parts	10 insertions	30 insertions
Quality testing	1.5 hours	2.5 hours

A product-costing system that had only one or two indirect cost pools would not recognize this sizable diversity of resource consumption between PC Boards A and B.[3]

MAJOR DIFFERENCES BETWEEN TYPICAL SYSTEMS AND ACTIVITY-BASED ACCOUNTING SYSTEMS

Objective 7

Describe the major differences between typical systems and activity-based accounting systems

Our Instruments Inc. illustration introduced the highlights of an activity-based accounting system. The major differences between typical systems and activity-based accounting systems follow:

Typical Systems	Activity-Based Accounting Systems
• One or a few indirect cost pools for each department or whole plant.	• Many indirect cost pools because of many activity areas. Operating personnel play a key role in designating the areas.

[3]Chapter 12, pp. 409–11, expands this illustration, particularly with respect to the impact of activity-based accounting on target costing and product design.

- Indirect cost application bases may be cost drivers.[4]
- Indirect cost application bases are often financial, such as direct labor costs or direct material costs.

- Indirect cost application bases are more likely to be cost drivers.[4]
- Indirect cost application bases are often nonfinancial variables, such as number of parts in a product or hours of test time.

Typical systems tend to have applications of indirect manufacturing costs that are too crude. They use too few pools of indirect costs leading to cost applications having overly broad averages. The resulting costs may induce erroneous decisions about activities and products. For example, a company may produce too many low-volume products or may make a component part when it should buy. The dangers are especially pronounced when hundreds of diverse products are manufactured in various annual volumes ranging from a few units of, say, one or two kinds of motors or computers to thousands of units of other kinds. Case studies have shown that broad averages can load indirect manufacturing costs too heavily on high-volume products and too lightly on low-volume products.

Most fundamentally, executives manage costs by overseeing activities rather than products. The pooling of costs by activities or activity areas provides information that may help managers to better plan and control costs throughout the chain of business functions, from research and development to customer service.

FUTURE OF ACTIVITY-BASED ACCOUNTING

Whether activity-based accounting is appropriate for a company depends on management judgments: How expensive is activity-based accounting to install and run in comparison with the expected improvements in collective operating decisions?

The companies most likely to benefit from activity-based accounting have some or all of the following characteristics:

- High overhead costs
- Operating personnel with little faith in the accuracy of existing cost information
- A widely diverse set of operating activities
- A widely diverse range of products
- Wide variation in numbers of production runs and costly setups
- Many changes in activities over time, but few corresponding changes in the accounting system
- Improved computer technology

In any event, new information technology is likely to create refinements in cost accounting systems. At a minimum, we can expect wider use of overhead application bases other than direct labor. At a maximum, we can expect extensive use of activity-based accounting throughout a company, including research and development, product design and engineering, manufacturing, marketing, distribution, and customer service.

[4]For example, machine-paced manufacturing departments are more likely to use direct labor hours (rather than machine hours) as a cost application base under a typical costing system than under an activity-based accounting system.

PROBLEMS FOR SELF-STUDY

PROBLEM 1

"Managers should choose which of the following three accounting systems is best for their organization: job-order costing, process costing, or activity-based accounting." Do you agree? Explain.

SOLUTION 1

The statement is misleading. Job-order costing and process costing are two ways to accomplish the purpose of *product costing*. In contrast, activity-based accounting is primarily designed to help *planning and control*. Its focus is on activities as cost objects instead of business functions or departments as cost objects. Therefore, activity-based accounting can be used in conjunction with any form of product costing.

Activity-based accounting can also be useful in improving the accuracy of product costing, whether used in conjunction with job-order or process-costing. For example, more accurate product costs are probable if indirect manufacturing cost application rates are based on individual activities instead of on departments or on the factory as a whole.

PROBLEM 2

The Newark Company has analyzed various activities in its value chain of business functions that extend from research and development through customer service. One of its products, a special diesel engine, has a manufacturing product cost of $13,000 per unit. In addition, the following applicable costs have been identified with the engine: research and development, $800; product design, $1,200; marketing, $5,000; distribution, $900; and customer service, $1,600.

Required

1. Compute the full product cost per unit.
2. Provide a general description of how the $13,000 manufacturing cost would be built up under activity-based accounting.

SOLUTION 2

1.

Business Function in Value Chain	Cost Per Engine
Research and development	$ 800
Product design	1,200
Manufacturing	13,000
Marketing	5,000
Distribution	900
Customer service	1,600
Full product cost	$22,500

In particular, note that manufacturing cost per unit, which is the inventoriable cost, may be far lower than the full product cost.

2. The $13,000 manufacturing cost would be built up by applying costs of the various manufacturing activities undergone as the engine is produced. Among the likely activities are material handling and various machining, painting, and assembly operations. The costs would be applied in a manner similar to that illustrated in Exhibits 5-6 and 5-7, pages 154 and 155. The application bases probably would be nonfinancial variables such as the number of component parts, the hours of machining time, and the square feet of surface painted.

SUMMARY

Job costing in nonmanufacturing settings focuses on obtaining the total cost of a specific service or engagement. The major means of control is the careful budgeting and use of direct professional labor hours.

Although direct labor is becoming less and less prominent as a percentage of total manufacturing costs, it remains the main cost of producing professional services. Control of the hiring of employees and the use of their time is a major challenge to management.

Process costing, which is suitable for the mass production of like units, is the opposite extreme of job costing. An important feature of process costing is the use of equivalent units for measuring the volume of output.

Activity-based accounting refines the focus of cost accounting for planning and control. Beyond business functions or departments, activity-based accounting focuses on activities as the fundamental cost objects and uses the cost of these activities as building blocks for compiling the cost of other cost objects. Activity-based accounting can be used in conjunction with any type of job-order or process product-costing system.

TERMS TO LEARN

This chapter and the Glossary at the end of the book contain definitions of the following important terms:

Activity accounting *(p. 150)*
Activity-based accounting *(150)*
Activity-based costing *(150)*
Equivalent units *(146)*

ASSIGNMENT MATERIAL

QUESTIONS

5-1 "Job-order costing is not confined to manufacturing companies." Give three examples of industries or types of nonmanufacturing companies that use job-order costing.

5-2 Give an example of a rule of thumb or a formula used for pricing by professional service firms.

5-3 Many professional service firms have refined their data processing systems. How have such refinements affected classifications of costs?

5-4 If the total costs are unchanged, what effect will a reclassification of overhead costs as direct costs have on overhead rates?

5-5 "In the production of services, direct materials are usually the major cost." Is this quotation accurate? Explain.

5-6 Give three examples of industries that often use process-costing systems.

5-7 Give three examples of nonprofit organizations that often use process-costing systems.

5-8 Give three examples of equivalent units in various organizations. At least one example should be nonmanufacturing.

5-9 What is the central product-costing problem in process costing?

5-10 "There are five key steps in process-cost accounting." What are they?

5-11 Why does a process-costing system use multiple Work in Process accounts but a job-order system uses only one such account?

5-12 Why might a company have different numbers of equivalent units for direct materials and conversion costs?

5-13 Consider a company's first month of operation. "If there is ending work in process but no beginning work in process, the cost per equivalent unit is greater than if there is no beginning or ending work in process." Do you agree? Explain.

5-14 Name three types of classifications of costs that may be used for planning and control purposes.

5-15 "Cost drivers are cost application bases." Do you agree? Explain.

5-16 "Activity-based accounting is the wave of the present and future. All companies should adopt it." Do you agree? Explain.

5-17 "Activity-based accounting emphasizes nonfinancial variables more than other types of accounting." How and why?

5-18 "Activity-based accounting tends to consider the company as a whole more than other accounting systems do." Do you agree? Explain.

5-19 "Activity-based accounting may be combined with either process costing or job-order costing." Do you agree? Explain.

5-20 Give three examples of technology that have made information gathering less expensive.

EXERCISES AND PROBLEMS

Coverage of Part One of the Chapter

5-21 Cost and price of consulting services. Consider the following data for a consulting firm:

Revenue	$21,000,000
Direct labor for professional services	?
Contribution to overhead and operating income	$16,000,000
Overhead	?
Operating income	$?

The budgeted overhead rate is 300% of direct labor. Fill in the blanks. Also compute the average markup rate of direct labor for the pricing of engagements.

5-22 CPA fees. The partner in charge of an audit engagement has prepared the following budget data for a proposal to a prospective client. These data are for direct professional hours and include all fringe benefits (which total 40% of the basic labor rates):

Partner, 1 × 50 hours × $120	$ 6,000
Manager, 1 × 100 hours × $40	4,000
Senior, 1 × 200 hours × $30	6,000
Assistants, 3 × 180 hours × $20	10,800
	$26,800

The budgeted overhead rate is 250% of direct professional labor dollars. The average markup for the pricing of engagements is 400% of direct professional labor costs.

Required

1. Suppose a normal markup is used to quote a total fixed fee for the audit engagement. Compute the budgeted total revenue, total costs, and operating income.
2. Identify three factors that may cause the actual fee to differ from the budgeted fee.

5-23 Cost and price of services. Consider the following budgeted data for a client case of Finn and Company, a law firm. The client wants a fixed price quotation.

Direct professional labor	$30,000
Direct support labor	9,000
Other operating costs applied at 100% of total direct costs	?
Fringe benefits for all direct labor	13,000
Photocopying	1,000
Telephone calls	1,000
Computer time	6,000

Required

1. Prepare a schedule of the budgeted total costs of the case. Show subtotals for total direct costs and total costs as a basis for markup.
2. Assume that the partner's policy is to quote a fixed fee at 20% above the total costs used as a basis for markup. What fee would she quote?
3. Suppose a competing law firm has the same set of costs. However, its cost accounting system traces only direct professional labor as a direct cost of a case. The firm's policy is to quote a fixed fee at 500% of direct professional labor. What fee would the competitor quote?
4. Compare and comment on the approaches to fee setting described in requirements 2 and 3.

5-24 Job costing in an accounting firm. A public accounting firm has a condensed budget for 19_2 as follows:

Revenue	$20,000,000
Direct labor (for professional hours charged to jobs)	5,000,000
Contribution to overhead and operating income	$15,000,000
Overhead (all other costs)	13,000,000
Operating income	$ 2,000,000

The firm uses a job-costing system for assorted purposes. For example, the partner in charge of an audit will plan the amount of professional time needed, including the time of the partners, managers, senior accountants, and assistants (often called junior accountants). The actual hours will be tabulated, comparisons made, and costs compiled, and the profitability of the job will be measured. The results will be used as a guide for managing future audits.

Required

1. The cost of each job is defined as actual direct labor plus overhead applied as a percentage of direct labor. Compute the budgeted overhead rate.
2. The markup rate for pricing jobs is intended to produce an operating income of 10% of revenue. Compute the markup rate as a percentage of direct labor.
3. The accounting firm's cost per hour of personnel assigned to Job 345, the audit of a local liquor distributor, is: partner, $60; manager, $35; senior, $25; assistant, $15. Last week's time records showed the following:

Partner	3
Manager	15
Senior	40
Assistants	160
Total	218

Compute the total costs added to Job 345 for the week. Suppose the accounting firm is able to realize normal billing rates for this work. Compute the addition to revenue because of this work.

5-25 Job costing in a service industry. Howe and Halling, Certified Public Accountants, use a form of job-costing system. In this respect they are similar to many professional service firms, such as management consulting firms, law firms, and professional engineering firms.

An auditing client may be served by various staff who hold professional positions in the hierarchy from partners to managers to senior accountants to assistants. In addition, there are secretaries and other employees.

Suppose Howe and Halling have the following budget for 19_2:

Compensation of professional staff	$4,000,000
Other costs	2,400,000
Total budgeted costs	$6,400,000

Each professional staff member must submit a weekly time report, which is used to assign costs to jobs. An example is the time report of a senior accountant:

Week of January 8	S	M	T	W	T	F	S	Total
Chargeable Hours								
Client A—Ramirez		8	8	5				21
Client B—Lee				3		4		7
Etc.								
Nonchargeable Hours								
Professional development (attending seminar on computer auditing)					8			8
Unassigned time						4		4
Total	0	8	8	8	8	8	0	40

In turn, these time reports are used for charging hours to a client job-order record, summarized as follows for the Ramirez job:

Employees Charged	Week of Jan. 8	Week of Jan. 15	Total Hours	Billing Rates	Total Billed
Partners	4	4	8	$100	$ 800
Managers	4	4	8	60	480
Seniors	21	30	51	40	2,040
Assistants	48	70	118	20	2,360
Total hours	77	108	185		$5,680

In many cases, these job-cost records bear only a summary of the *hours* charged. Each class of labor is billed at an appropriate hourly rate, so that the job-cost record is the central basis for billing the client.

Required

1. Suppose this firm had a policy of charging overhead to jobs at a budgeted percentage of the salaries charged to the job. The experience of the firm has been that chargeable hours average 80% of available hours for all categories of professional personnel. The non-chargeable hours are regarded as additional overhead. What is the overhead rate as a percentage of the "direct labor," the chargeable professional compensation cost?
2. Compute the *total cost* of the Ramirez job for the two weeks that began January 8. Be as specific as possible. Assume that the average weekly compensation (based on a 40-hour week) of the personnel working on this job is: partners, $2,000; managers, $1,200; seniors, $800; assistants, $400.
3. As the tabulation for Ramirez implies, the job-order record often consists of only the time and no costs. The revenue is computed by multiplying the time by the billing rates. Suppose the partners' operating income objective is 20% of the total costs budgeted. What percentage of the salaries charged to the jobs would be necessary to achieve a total billing that would ultimately provide the income objective? That is, what is the billing rate as a percentage of "direct labor"?
4. In addition to billing, what use might you make of the data compiled on the job-cost records?

5-26 Job costing in a law firm. Serra & Co., a law firm, had the following costs in 19_2:

Direct professional labor	$10,000,000
Overhead	19,000,000
Total costs	$29,000,000

The following costs were included in overhead:

Fringe benefits to direct labor	$1,500,000
Secretarial costs	2,700,000
Telephone call time with clients (estimated but not tabulated)	600,000
Computer time	1,800,000
Photocopying	400,000
	$7,000,000

The firm's data processing capabilities now make it feasible to document and trace these costs to individual cases or jobs. The managing partner is pondering whether more costs than just direct professional labor should be applied directly to jobs. In this way, the firm will be better able to justify billings to clients.

In late 19_2, arrangements were made to trace specified costs to seven client engagements. Two of the case records showed the following:

	Client Cases	
	304	308
Direct professional labor	$20,000	$20,000
Fringe benefits to direct labor	3,000	3,000
Secretarial costs	2,000	6,000
Telephone call time with clients	1,000	2,000
Computer time	2,000	4,000
Photocopying	1,000	2,000
Total direct costs	$29,000	$37,000

Required

1. Compute the overhead application rate based on last year's direct labor costs.
2. Assume that last year's costs were reclassified so that the $7 million would be regarded as direct costs instead of overhead. Compute the overhead application rate as a percentage of direct labor and as a percentage of total direct costs.
3. Using the three rates computed in requirements 1 and 2, compute the total costs of cases 304 and 308.
4. Assume that the billing of clients was based on a 120% markup of total case costs. Compute the billings in requirement 3 for cases 304 and 308.
5. Which method of job costing and overhead application do you favor? Explain.

5-27 Cost-based pricing decisions for a law firm, cross-subsidization. Solve Problem 12-22, page 421.

Coverage of Part Two of the Chapter

5-28 Straightforward process costing. (Alternates are 5-30 and 5-32.) Ming Company produces digital watches in large quantities. For simplicity, assume that the company has two departments, assembly and testing. The manufacturing costs in the assembly department during January were:

Direct materials added	$ 72,000
Conversion costs	76,000
Assembly costs to account for	$148,000

There was no beginning inventory of work in process. Suppose work on 10,000 watches was begun in the assembly department during January, but only 9,000 watches were fully completed. All the parts had been made or placed in process, but only half the labor had been completed for each of the watches.

Required

1. Compute the equivalent units and equivalent unit costs for January.
2. Compute the costs of units completed and transferred to the testing department. Also compute the cost of the ending work in process. (For journal entries, see Problem 5-29.)

5-29 Journal entries. Refer to the data in Problem 5-28. Prepare summary journal entries for the use of direct materials and conversion costs. Also prepare a journal entry to transfer out the cost of goods completed. Show the postings to the Work in Process—Assembly account.

5-30 Straightforward process costing. (Alternates are 5-28 and 5-32.) Mendez Company produces inexpensive radios in large quantities. The manufacturing costs of the assembly department were:

Direct materials added	$1,200,000
Conversion costs	360,000
Assembly costs to account for	$1,560,000

For simplicity, assume that this is a two-department company, assembly and finishing. There was no beginning work in process.

Suppose 1 million units were begun in the assembly department. There were 800,000 units completed and transferred to the finishing department. The 200,000 units in ending work in process were fully completed regarding direct materials but half-completed regarding conversion costs.

Required

1. Compute the equivalent units and equivalent unit costs in the assembly department.
2. Compute the costs of units completed and transferred to the finishing department. Also compute the cost of the ending work in process in the assembly department. (For journal entries, see Problem 5-31).

5-31 Journal entries. Refer to the data in Problem 5-30. Prepare summary journal entries for the use of direct materials and conversion costs. Also prepare a journal entry to transfer out the cost of goods completed. Show the postings to the Work in Process—Assembly account.

5-32 Straightforward process costing. (Alternates are 5-28 and 5-30.) A department produces cotton fabric. All direct materials are introduced at the start of the process. Conversion costs are incurred uniformly throughout the process.

In May there was no beginning inventory. Units started, completed, and transferred, 400,000. Units in process, May 31, 80,000. Each unit in ending work in process was 75% converted. Costs incurred during May: direct materials, $4,320,000; conversion costs, $920,000.

Required

1. Compute the total work done in equivalent units and the equivalent unit costs for May.
2. Compute the cost of units completed and transferred. Also compute the cost of units in ending work in process.

5-33 Process costing: material introduced at start of process. A certain chemical process incurred $37,600 of production costs during a month. Materials costing $22,000 were introduced at the start of processing, and conversion costs of $15,600 were incurred at a uniform rate throughout the production cycle. Of the 40,000 units of product started, 38,000 were completed; 2,000 were still in process at the end of the month, averaging one-half complete. There was no beginning work in process.

Required

In step-by-step fashion, prepare a production-cost report showing cost of goods completed and transferred out and cost of ending work in process.

5-34 Process costs; single department. The following data pertain to the mixing department for July:

Units:	
Work in process, July 1	0
Units started	50,000
Completed and transferred to finishing department	35,000
Costs:	
Chemical P	$250,000
Chemical Q	$ 70,000
Conversion costs	$135,000

Chemical P is introduced at the start of the process, and Chemical Q is added when the product reaches the three-fourths stage of completion. Conversion costs are incurred uniformly throughout the process.

Required

Cost of goods completed and transferred out during July. Cost of work in process as of July 31. Assume that ending work in process is two-thirds completed. (For journal entries, see Problem 5-35.)

5-35 Journal entries. Refer to Problem 5-34. Prepare journal entries without explanations. Assume that the completed goods are transferred to the cooling department.

5-36 Uneven flow. A one-department company manufactured basic hand-held calculators. Various materials were added at different stages of the process. The outer front shell and the carrying case, which represented 10% of the total material cost, were added at the final step of the assembly process. All other materials were considered to be "in process" by the time the calculator reached a 50% stage of completion.

Ninety-one thousand calculators were started in production during 19_1. At year-end, 5,000 calculators were in various stages of completion, but all of them were beyond the 50% stage. On average they were 80% completed. There was no beginning work in process.

The following costs were incurred during the year: direct materials, $271,500; conversion costs, $450,000.

Required

1. Prepare a schedule of physical units and equivalent units.
2. Tabulate the unit costs, cost of goods completed and transferred out, and cost of ending work in process.

5-37 Materials and cartons. A London company manufactures and sells small portable tape recorders. Business is booming. Various materials are added at different stages in the assembly department. Costs are accounted for on a process-cost basis. The end of the process involves conducting a final inspection and adding a cardboard carton.

The final inspection requires 5% of the total processing time. All materials besides the carton are added by the time the recorders reach an 80% stage of completion of conversion.

There were no beginning inventories. One hundred thousand recorders were started in production during 19_3. At the end of the year, which was not a busy time, 4,000 recorders were in various stages of completion. All the ending units in work in process were at the 95% stage. They awaited final inspection and being placed in cartons.

Total direct materials consumed in production, except for cartons, cost £2.0 million. Cartons used cost £172,800. Total conversion costs were £798,400.

Required

1. Present a schedule of physical units, equivalent units, and equivalent unit costs of direct materials, cartons, and conversion costs.
2. Present a summary of the cost of goods completed and transferred out and the cost of ending work in process.

5-38 Computing processing costs and journalizing cost transfers. The following information was taken from the ledger of Miami Polyvinyl Products. Ending inventory is 100% complete as to direct materials but only 25% complete as to conversion costs.

Work in Process—Forming

	Physical Units	*Dollars*		*Physical Units*	*Dollars*
Inventory, November 30	–0–	–0–	Transferred to Painting	72,000	?
Production started:	80,000				
1. Direct materials		680,000			
2. Conversion costs		148,000			
Total to account for	80,000	828,000			

Required

Journalize the transfer of cost to the painting department. Show supporting computations.

Coverage of Part Three of the Chapter

5-39 Business functions, departments, activities. Consider the following items: research and development, customer service, product design, engineering, manufacturing, setups, marketing, engineering change orders, machining, inspection of products, painting, assembly, engineering drawings, distribution.

Required

Arrange these items by business function, by departments, and by activities.

5-40 Activity-based accounting product-cost buildup. The Denver Company uses activity-based accounting. Consider the following information:

Manufacturing Activity Area	Cost Driver Used as Application Base	Conversion Cost per Unit of Application Base
1. Material handling	Number of parts	$ 0.30
2. Machinery	Machine hours	34.00
3. Assembly	Number of parts	1.90
4. Inspection	Number of finished units	20.00

Assume that 50 units of a component for packaging machines have been manufactured. Each unit required 70 parts and 2 machine hours. Direct materials cost $400 per finished unit. All other manufacturing costs were classified into one category, conversion costs.

Required

1. Compute the total manufacturing costs and the unit costs of the 50 units.
2. Suppose upstream activities, such as research and development and product design, were analyzed and applied to this component at $120 per unit. Moreover, similar analyses were conducted of downstream activities, such as distribution, marketing, and customer service. The downstream costs applied to this component were $700 per unit. Compute the full product cost per unit.

5-41 Activity-based accounting, product-cost buildup. The Schramka Company manufactures a variety of chairs. The company's activity areas and related data follow:

Manufacturing Activity Area	Budgeted Conversion Costs for 19_5	Cost Driver Used as Application Base	Conversion Cost per Unit of Application Base
Material handling	$ 200,000	Number of parts	$ 0.25
Cutting	2,000,000	Number of parts	2.50
Assembly	5,000,000	Direct labor hours	25.00
Painting	1,000,000	Number of painted units	22.00

Two styles of chairs were produced in March, the standard chair and an unpainted chair that had fewer parts and required no painting activities. Their quantities, direct material costs, and other data follow:

	Units Produced	Direct Material Costs	Number of Parts	Assembly Direct Labor Hours
Standard chair	5,000	$600,000	100,000	7,500
Unpainted chair	1,000	85,000	15,000	1,100

Required

1. Compute the total manufacturing costs and unit costs of the standard chairs and the unpainted chairs.
2. Suppose upstream activities, such as product design, were analyzed and applied to the standard chairs at $20 each and the unpainted chairs at $15 each. Moreover, similar analyses were conducted of downstream activities, such as distribution, marketing, and customer service. The downstream costs applied were $200 per standard chair and $80 per unpainted chair. Compute the full product cost per unit.

5-42 Activity-based accounting, product-cost buildup. A Hewlett-Packard factory assembles and tests printed-circuit (PC) boards. The factory has a wide variety of products, some high-volume and some low-volume. Using assumed numbers, consider the following data regarding PC Board 82:

Direct materials	$90.00
Indirect manufacturing costs applied	?
Total manufacturing product cost	$?

The activities undergone follow:

Activity Area	Cost Driver	Indirect Manufacturing Costs Applied for Each Activity
1. Start station	No. of raw PC boards	1 × 1.10 = $1.10
2. Axial insertion	No. of axial insertions	45 × .08 = ?
3. Dip insertion	No. of dip insertions	? × .25 = 6.00
4. Manual insertion	No. of manual insertions	11 × ? = 5.50
5. Wave solder	No. of boards soldered	1 × 3.50 = 3.50
6. Backload	No. of backload insertions	6 × .70 = 4.20
7. Test	Standard time board is in test activity	.25 × 90 = ?
8. Defect analysis	Standard time for defect analysis and repair	.10 × ? = 8.00
Total		$?

Required

1. Fill in the blanks in both the opening schedule and the list of activities.
2. How is direct labor identified with products under this product-costing system?
3. Why might managers favor this activity-based accounting system instead of the older system, which applied indirect manufacturing costs based on direct labor?

5-43 Activity-based accounting, six activities. The Lutz Machine Shop produces three types of precisely engineered components for engines. Lutz sells these components to engine manufacturers and to distributors of replacement parts. There are three classifications of products: Quality, Superior, and Superb. The Lutz cost accounting system through 19_4 applied all costs except direct materials to the products based on direct labor hours. During 19_4, several managers and the controller conducted a study of operations and cost drivers that eventually led to the company's dividing its sole manufacturing department into six activities. The 19_5 budgeted costs for these activities follow:

Activity Area	19_5 Budgeted Costs	Cost Driver Used as Application Base
a. Material handling	$ 258,400	Number of parts
b. Production scheduling	114,000	Number of production orders
c. Setup labor used to change machinery and computer configurations for a new production batch	160,000	Number of production setups
d. Automated machinery, including depreciation, repairs, and maintenance	3,510,000	Machine hours
e. Finishing, including related fringe benefits and payroll taxes	1,092,000	Direct labor hours
f. Packaging and shipping room	190,000	Number of orders shipped
Total	$5,324,400	

The company is now preparing a budget for 19_5. The following predicted data are available:

	Quality	Superior	Superb
Units to be produced	10,000	5,000	800
Direct materials, cost per unit	$80	$50	$110
Number of parts per unit	30	50	120
Direct labor hours per unit	2	5	12
Machine hours per unit	7	7	15
Production orders	300	70	200
Production setups	100	50	50
Orders shipped	1,000	2,000	800

Required

1. Prepare a cost application rate for each activity.
2. Prepare a budget that shows total budgeted costs and unit costs of each component, excluding direct materials.
3. Using the old direct-labor-hour-based system, compute the budgeted total and unit costs of each component.
4. Compare your answers in requirements 2 and 3. Give your interpretations of these data for management. Which of these two systems is better? Why?

5-44 Activity-based accounting, unit cost comparisons. The current manufacturing-costing system of Tracy Corporation has two direct product cost categories (direct materials and direct labor). Indirect manufacturing costs are applied to products using a single indirect cost pool. The indirect manufacturing cost application base is direct labor hours; the indirect cost rate is $115 per direct labor hour.

Tracy Corporation is switching from a labor-paced to a machine-paced manufacturing approach at its aircraft components plant. Recently, the plant manager set up five activity areas, each with its own supervisor and budget responsibility. Pertinent data follow:

Activity Area	Cost Driver Used as Indirect Cost Application Base	Cost Per Unit of Application Base
Material handling	Number of parts	$ 0.40
Lathe work	Number of turns	0.20
Milling	Number of machine hours	20.00
Grinding	Number of parts	0.80
Shipping	Number of orders shipped	1,500.00

Information technology has advanced to the point where all the necessary data for budgeting in these five activity areas are automatically collected.

The two job orders processed under the new system at the aircraft components plant in the most recent period had the following characteristics:

	Job Order 410	Job Order 411
Direct materials cost per job	$ 9,700	$59,900
Direct labor cost per job	$ 750	$11,250
Number of direct labor hours per job	25	375
Number of parts per job	500	2,000
Number of turns per job	20,000	60,000
Number of machine hours per job	150	1,050
Number of job orders shipped	1	1
Number of units in each job order	10	200

Required

1. Compute the per unit manufacturing cost of each job under the existing manufacturing-costing system (that is, indirect costs are collected in a single cost pool with direct labor hours as the application base).
2. Assume that Tracy Corporation adopts an activity-based accounting system. Indirect costs are applied to products using separate indirect cost pools for each of the five activity areas (material handling, lathe work, milling, grinding, and shipping). The application base and rate for each activity area are described in the problem. Compute the per unit manufacturing cost of each job under the activity-based accounting system.
3. Compare the per unit cost figures for Job Orders 410 and 411 computed in requirements 1 and 2. Why do they differ? Why might these differences be important to Tracy Corporation?

CHAPTER 6

The budgets, plans, and strategies of a business affect one another. Boise Cascade, a large forest products company, uses key budgets and sales and cost projections as starting points in forming its strategies.

Master Budget and Responsibility Accounting

Learning Objectives

When you have finished studying this chapter, you should be able to

1. Define *master budget*, and describe its major benefits to an organization
2. Distinguish between types of budgets
3. List the major schedules in preparing an operating budget
4. Construct the supporting schedules and main statements for a master budget
5. Describe the uses of financial planning models and computers
6. Describe responsibility accounting
7. Explain how controllability relates to responsibility accounting
8. Construct a cash budget (appendix)

What is the most widely used type of accounting system among well-managed organizations today? A historical cost system? No, although it is often referred to as such. Instead, it is better described as a budgeting system, a system that includes *both* expected results and historical or actual results. A budgeting system builds on historical, or actual, results and expands to include consideration of future, or expected, results. A budgeting system turns managers' perspectives forward.

Budgeting has come of age. It is becoming increasingly important, growing in prominence in the operations of both profit-seeking and nonprofit organizations.

This chapter examines the master, or comprehensive, budget as a planning and coordinating device. The succeeding chapters, especially Chapters 7, 8, and 9, examine various aspects of budgeting decisions and their implementation.

EVOLUTION OF SYSTEMS

Ponder the evolution of control systems. As small organizations begin, personal observation is usually the dominant means of control. A manager sees, touches, and hears the relationship between inputs and outputs; he or she oversees the behavior of various personnel.

As the organization continues to operate, managers add historical records to their personal observations. No fancy cost-benefit analysis is necessary to justify maintaining some historical records for internal purposes. These records help answer essential operating questions: What are the amounts of sales, purchases, cash, inventories, receivables, and payables? Historical records also allow managers to compare current performance with past performance. How did a department perform in 19_3 compared with 19_2? Analyses of past performance may help improve future performance. Managers must deal with a series of periods, not just one. Historical records are the beginning of an accounting system.

As the organization matures, the next step in the growth and improvement of its accounting system is budgeting. A manager would find it helpful to compare actual *performance* in 19_3 with the *plans* that had been drawn up for 19_3. Budgeting systems include this future perspective.

Do budgeting systems meet the cost-benefit test? Evidently they do. Typically, organizations purchase budgeting systems voluntarily rather than as a result of outside forces. Why? Because managers regard budgeting systems as good investments. These systems change human behavior—and decisions—in the ways sought by top management. For example, budgeting may prompt managers to extend their planning horizons much further. Therefore, many prospective difficulties are foreseen and avoided. Without a budgeting system, many managers may veer from one crisis to another with no long-term perspective to guide them.

In sum, managers plan and control with the help of:

1. Personal observation (the basic means)
   plus
2. Historical records
   plus
3. Budgets

(2 and 3: the accounting system means)

MAJOR FEATURES OF BUDGETS

Definition and Role of Budgets

Objective 1

Define *master budget*, and describe its major benefits to an organization

A *budget* is a quantitative expression of a plan of action and an aid to coordination and implemention. Budgets may be formulated for the organization as a whole or for any subunit. The **master budget** summarizes the objectives of all subunits of an organization—sales, production, research, marketing, customer service, and finance. It quantifies management's expectations regarding future income, cash flows, and financial position. These expectations arise from a series of decisions resulting from a careful look at the organization's future.

The accompanying diagram shows how budgets and performance reports help managers. Note the key role that budgets play throughout the process. Our focus in this chapter will be on the planning of operations. Budgets, however, serve a variety of additional functions: coordinating activities, implementing plans, communicating, authorizing actions, motivating, controlling, and evaluating performance. The authorization function seems to predominate in government budgeting and nonprofit budgeting, where budget appropriations serve as approvals and ceilings for management actions.

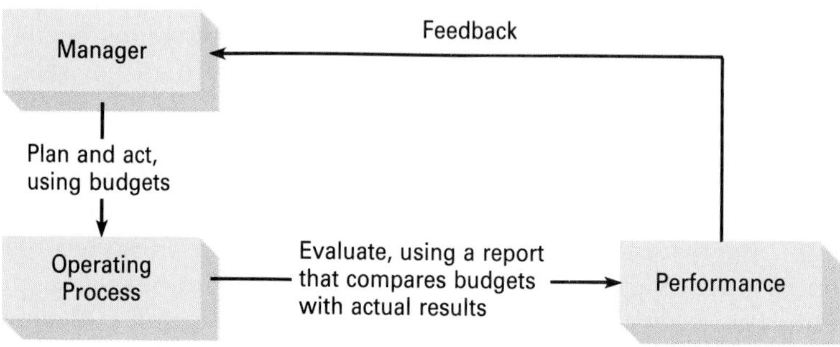

Well-managed organizations usually have the following budgetary cycle:

1. Planning the performance of the organization as a whole as well as its subunits. The entire management team agrees as to what is expected.
2. Providing a frame of reference, a set of specific expectations against which actual results can be compared.
3. Investigating variances from plans. Corrective action follows investigation.
4. Planning again, considering feedback and changed conditions.

The master budget embraces the impact of both *operating* decisions and *financing* decisions. Operating decisions center on the acquisition and use of scarce resources. Financing decisions center on how to get the funds to acquire resources. This book concentrates on how accounting helps the manager make operating decisions; the emphasis in this chapter is on operating budgets. Financing decisions and the role of cash budgets are covered in many finance texts. The chapter appendix explains cash budgets. The leading corporations usually excel in both operating management and financial management. Business failures often arise because of weaknesses in one or the other of these responsibilities.

Wide Use of Budgets

Budgeting systems are more common in larger companies, where formalized techniques often serve management. Still, small concerns also use budgets. Small companies, especially entrepreneurial startups, have a relatively high failure rate. More extensive use of budgets by such concerns would force entrepreneurs to quantify their dreams and directly face the uncertainties of their ventures.[1] For example, a small business with lofty hopes moved into a lush market for school equipment. However, failure to quantify the long collection periods, to forecast a maximum sales potential, and to control costs from the outset resulted in disaster within a year. As one commentator said: "Few businesses plan to fail, but many of those that flop failed to plan."

Managers must grapple with uncertainty, either with a budget or without one. The advocates of budgeting maintain that the benefits from budgeting nearly always exceed the costs. Some budget program, at least, will help almost every organization.

ADVANTAGES OF BUDGETS

Budgets are a major feature of most control systems. When administered intelligently, budgets (a) compel planning, (b) provide performance criteria, and (c) promote communication and coordination.

Strategy and Plans

"Planning" should be a watchword for business managers and for every individual as well. Too often, executives practice "management by crisis." Everyday problems interfere with planning. Operations drift along until the passage of time catches the firms or individuals in undesirable situations that should have been anticipated and avoided. Budgets compel managers to look ahead and be ready for changing conditions. ***This forced planning is by far the greatest contribution of budgeting to management.***

Budgeting is an integral part of strategy and tactics. **Strategy** is a broad term that usually means selection of overall objectives. Tactics are the general means for attaining strategic goals. Strategy analysis may include consideration of such questions as the following:

1. What are the overall goals or objectives of the organization?
2. What trends will affect our markets? How are we affected by the economy, the industry, and our competitors?
3. What are the best ways to invest in our research, design, production, distribution, marketing, and administrative activities?
4. What forms of organizational structure serve us best?
5. What fundamental financial structure is desirable?

[1]Dun & Bradstreet, *Business Failure Record* (1990), reports that the major cause of business failure is the inability to avoid conditions that result in inadequate sales or heavy operating expenses.

6. What are the risks of alternative strategies, and what are our contingency plans if our preferred plan fails?

The most critical strategy questions typically deal with external factors rather than with internal factors. Particularly critical are the current and projected demands of the marketplace, including the behavior of current and projected competitors.

Consider the following diagram:

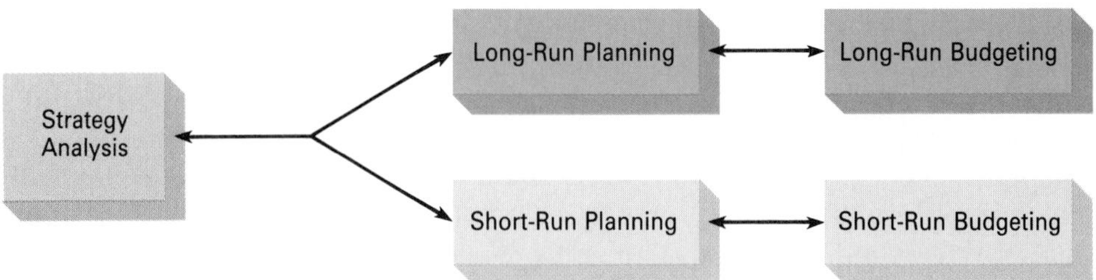

Strategy analysis underlies both long-run and short-run planning.[2] In turn, the plans lead to the formulation of budgets.[3] The arrowheads in the diagram are pointing in two directions. Why? Because strategy, plans, and budgets are interrelated and affect one another. For example, Boise Cascade Corporation, a large forest products company, uses key budgets and elaborate sales and expense projections as the starting points for discussing strategies. In turn, these strategies direct long-run and short-run planning.

Framework for Judging Performance

As a basis for judging actual results, budgeted performance is generally a better criterion than is past performance. For example, sales this year may be higher or direct material costs may be lower, which may be encouraging—but by no means conclusive as a measure of success. Assume that sales rose from 90,000 units to 100,000 units. To keep pace with market growth, however, sales should have risen to 112,000 units this year. A well-designed budgeting system would have forecast this figure, or a figure close to it, and management would likely have set the 112,000 unit sales amount as the year's goal. In subplans to the master budget, individual performance goals would be pegged to the 112,000-unit amount. Management would not greet the news that the business had sold 100,000 units with joy, and personnel responsible for the inadequate sales figure would probably receive a negative evaluation. The key point: Employee evaluation is based on targeted amounts, not management whim, and employees know what is expected of their performance.

A major weakness of using historical data for judging performance is that inefficiences may be buried in the past actions. A good budgeting system, forcing managers to examine the business as they plan, may detect inefficiencies that might otherwise go unnoticed. Management would alert individual employees to these inefficiencies and monitor performance for improvement. Also, the useful-

[2]Not all organizations undertake strategic analysis before or during the development of a budget. For example, some budgets in the nonprofit sector are merely based on last year's expenditures and then are adjusted for an inflation factor.

[3]See A. Ishikawa, *Strategic Budgeting: A Comparison between U.S. and Japanese Companies* (New York: Praeger, 1985), for case studies on linkages between strategic analysis and budgeting. Also see J. Shank and V. Govindarajan, *Strategic Cost Analysis* (Homewood, IL: Irwin, 1989).

ness of comparisons with the past may be hampered by intervening changes in technology, personnel, products, competition, and general economic conditions.

Coordination and Communication

Coordination is the meshing and balancing of all factors of production and of all the departments and functions so that the company can meet organizational objectives.

Coordination implies, for example, that purchasing officers integrate their plans with production requirements and that production officers use the sales budget as a basis for planning personnel needs and machinery use. Top managers want systems designed so that the self-interests of all managers do not conflict with the interests of the organization.

Budgets help management to coordinate in several ways:

1. The existence of a well-laid plan is the major step toward achieving coordination. Executives are forced to think of the relationships among individual operations and the company as a whole.
2. Budgets help to restrain the empire-building efforts of executives. Budgets broaden a manager's thinking to include more than just his or her department and remove bias—conscious or unconscious—in favor of his or her department.
3. Budgets help to search out weaknesses in the organizational structure. The formulation and administration of budgets identify problems in communication, in fixing responsibility, and in working relationships.

The idea that budgets improve coordination and communication may look promising on paper, but it takes plenty of intelligent administration to achieve in practice. Budgets may be a vehicle for communication, but sometimes the vehicle breaks down. For example, the budget staff may find that it is unable to obtain line participation in developing budgets.

Management Support and Administration

Budgets help managers, but budgets need help. ***That is, top management must understand and enthusiastically support the budget and all aspects of the control system.*** Consider a memo from the chief executive officer of the Bank of America: "Operating plans are contracts, and I want them met. If your revenue is off, you should cut your expenses accordingly." Referring to the chief executive officer of Wells Fargo Bank, a news story said: "Expense control is a state of mind around here. Carl is an absolute bear when it comes to meeting budgets."

Despite the quoted words, administration of budgets should not be rigid. Changing conditions call for changes in plans. The budget must receive respect, but it should not prevent a manager from taking prudent action. A department head may commit to the budget, but matters might develop so that some special repairs or a special advertising outlay would best serve the interests of the firm. That manager should feel free to request permission for such outlays, or the budget itself should provide enough flexibility to permit reasonable discretion in deciding how best to get the job done.

To improve acceptance of budgets, companies tend to have budgeting begin at relatively low organizational levels. U.S. managers prefer a participatory, bottom-up approach, as distinguished from an authoritative, top-down approach. In contrast, Japanese controllers and managers (a) prefer less participation, (b) have a more long-term planning horizon, (c) view budgets as more of a communications device, and (d) prefer more budget "padding" than their U.S. counterparts.[4]

---

[4]I. Daley, J. Jiambalvo, G. Sundem, and Y. Kondon, "Attitudes toward Financial Control Systems in the United States and Japan," *Journal of International Business Studies,* Fall 1985.

TYPES OF BUDGETS

Time Coverage

Objective 2

Distinguish between types of budgets

Budgets may span a period of one year or less—or, in cases of plant and product changes that require additional investments, up to five or more years. More and more companies use budgets as essential tools for long-range planning. The usual planning-and-control budget period is one year. The annual budget is often broken down by months for the first quarter and by quarters for the remainder of the year. The budgeted data for a year are frequently revised as the year unfolds. For example, at the end of the first quarter, the budget for the next three quarters is changed in light of new information. Businesses are increasingly using *continuous budgets* (also called *rolling budgets*). A twelve-month forecast is always available by adding a month or quarter in the future as the month or quarter just ended is dropped. Continuous budgets constantly force management to think concretely about the forthcoming twelve months, regardless of the month at hand. Arizona Public Service Co. has a budget that looks ahead two years and is updated every month. The choice of budget periods largely flows from the objectives, uses, and dependability of the budget data.

Classification of Budgets

Various descriptive terms for budgets have arisen. Terminology varies among organizations. For example, budgeted financial statements are sometimes called **pro forma statements**. Some organizations, such as Hewlett-Packard, refer to *targeting* rather than budgeting. Indeed, to give a more positive thrust to budgeting, many organizations do not use the term *budget* at all. Instead, they use the term *profit planning*.

There are countless forms of budgets. Many special budgets and related reports are prepared, including

- Comparisons of budgets with actual performance (performance reports)
- Reports for specific managerial needs—for example, cost-volume-profit projections
- Long-term budgets, often called "capital" or "facilities" or "project" budgets (see Chapter 21)
- Flexible budgets (see Chapter 7)
- Life-cycle budgets (see Chapter 12)

Exhibit 6-1 shows a simplified diagram of the various parts of the *master budget*, the comprehensive plan, a coordinated set of detailed financial statements for short periods, usually a year. As the diagram indicates, many supporting budgets are necessary in actual practice. The bulk of the diagram presents various elements that together are often called the **operating budget**, which is the income statement and its supporting budgets. In contrast, the **financial budget** is that part of the master budget that comprises the capital budget, cash budget, budgeted balance sheet, and budgeted statement of cash flows. It focuses on the impact on cash of operations and other factors, such as planned capital outlays for equipment.

For simplicity, Exhibit 6-1 does not show all functions (for example, research) and the interrelationships among the various budgets. For instance, the amount of interest expense on the budgeted income statement is affected by the cash budget. Moreover, to avoid clutter, the exhibit does not emphasize that once the sales budget is completed, purchasing, production, marketing, and administrative departments can often be working on their budgets simultaneously. Similarly, the various ingredients of the financial budget are often prepared simultaneously.

EXHIBIT 6-1
Master Budget

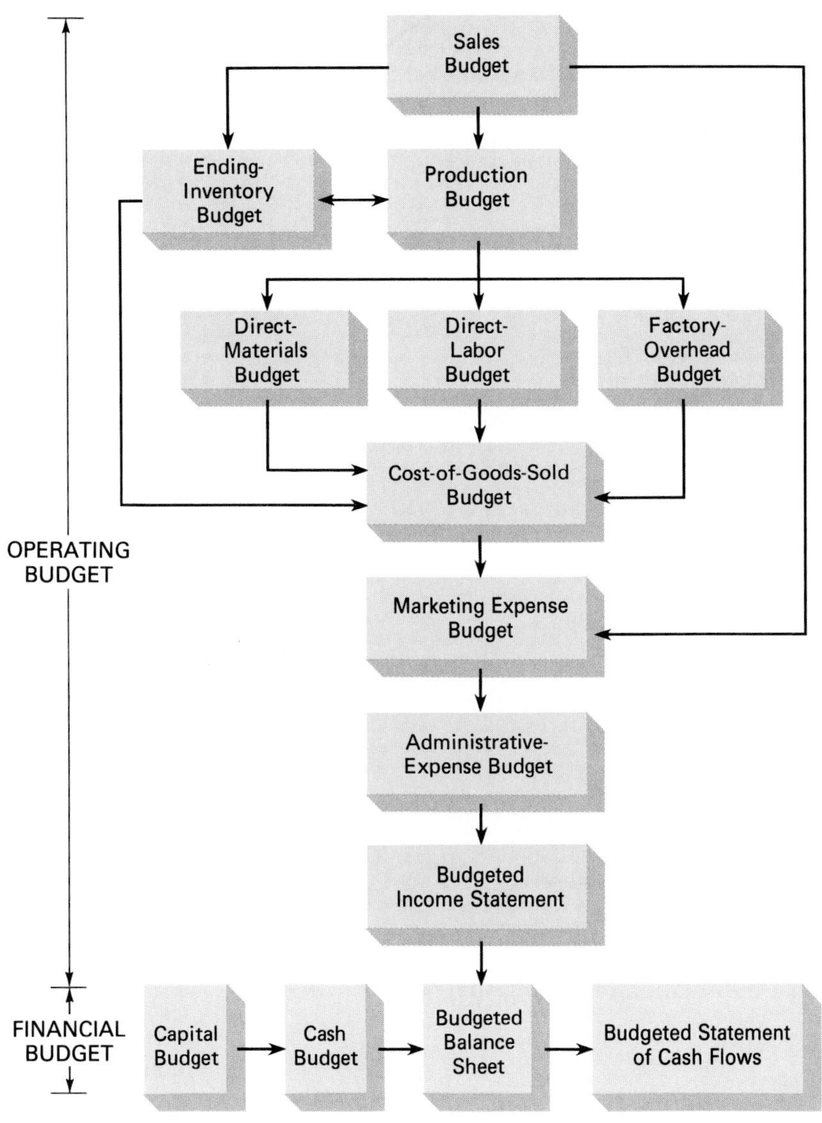

ILLUSTRATION OF MASTER BUDGET

The following illustration is of the final budget itself, but remember that the master-budget process generates key top-management decisions regarding pricing, product lines, production scheduling, capital expenditures, research and development, management assignments, and so on. The first draft of the budget almost always leads to decisions that prompt further drafts before a final budget is chosen.

Basic Data and Requirements

The Alic Company is a machine shop that uses skilled labor and specified metal alloys for manufacturing aircraft replacement parts. Managers are ready to prepare a master budget for the year 19_2. To keep our illustration manageable for clarifying basic relationships, we make the following assumptions:

1. Work in process inventories are negligible and are ignored.
2. Unit prices of direct materials and finished goods remain unchanged.
3. A single cost driver, direct labor hours, is used as an application base for applying all factory overhead. The production steps are labor-paced. That is, worker skills and productivity determine the speed of manufacturing.

Having carefully examined all relevant factors, the executives forecast the following figures:

Direct materials:	
Material 111 alloy	$ 7 per kilogram
Material 112 alloy	$10 per kilogram
Direct labor	$20 per hour
Factory overhead is applied on the basis of direct labor hours	

	Products	
	Product R (regular aircraft parts)	**Product HD (heavy-duty aircraft parts)**
Finished Goods:		
(content of each unit)		
Direct material 111 alloy	12 kilograms	12 kilograms
Direct material 112 alloy	6 kilograms	8 kilograms
Direct labor	4 hours	6 hours

Additional information regarding the year 19_2:

	Products	
	Regular (R)	**Heavy-Duty (HD)**
Expected sales in units	5,000	1,000
Selling price per unit	$ 600	$ 800
Target ending inventory in units*	1,100	50
Beginning inventory in units	100	50
Beginning inventory in dollars	$38,400	$26,200

	Direct Materials	
	111	**112**
Beginning inventory in kilograms	7,000	6,000
Target ending inventory in kilograms*	8,000	2,000

*Target inventories are affected by management policies regarding relations with suppliers and expected sales.

At anticipated volume levels, management believes the following costs will be incurred:

Factory Overhead:	
Supplies	$ 90,000
Indirect labor	200,000
Payroll fringe costs	320,000
Power—variable portion	90,000
Maintenance—variable portion	70,000
Depreciation	230,000
Property taxes	50,000
Property insurance	10,000
Supervision	100,000
Power—fixed portion	20,000
Maintenance—fixed portion	20,000
	$1,200,000

Marketing and Administrative Expenses:	
Sales commissions	$200,000
Advertising	60,000
Sales and customer service salaries	100,000
Travel	90,000
Clerical wages	100,000
Supplies	12,000
Officer salaries	260,000
Miscellaneous	48,000
	$870,000

The budgeting task is to prepare an operating budget (budgeted income statement through operating income) for the year 19_2. The following detailed schedules must be included:

Objective 3

List the major schedules in preparing an operating budget

1. Sales budget
2. Production budget, in units
3. Direct materials usage budget
4. Direct labor budget
5. Factory overhead budget
6. Ending inventory budget
7. Cost-of-goods-sold budget
8. Marketing and administrative expense budget

The preparation of the master budget is fundamentally nothing more than the preparation of familiar financial statements. The major difference is in dealing with expected future data rather than with historical data. Our illustration ends with the major part of the master budget, the income statement. ***Recall that the income statement is part of the operating budget.*** The financial budget, notably the budgeted balance sheet and the cash budget, is discussed in the chapter appendix.

Most organizations have a budget manual, which contains instructions and related information. Although the details differ among organizations, the manual has the following nine basic steps for a manufacturing company. Beginning with the sales budget (sometimes called the sales budget *schedule*), study each budget, step by step. In many cases, computer software is available to speed the computations associated with the preparation of budgets.

Objective 4

Construct supporting schedules and main statements for a master budget

Steps in Preparing Operating Budget

Step 1: *Sales or revenue budget.* *The sales forecast (Schedule 1) is the usual starting point for budgeting.* Why? Because production (and hence costs) and inventory levels are generally geared to the forecasted level of sales.

Sales Budget
For the year Ended December 31, 19_2

Schedule 1	Units	Selling Price	Total Sales
Regular (R)	5,000	$600	$3,000,000
Heavy-duty (HD)	1,000	$800	800,000
Total			$3,800,000

The $3,800,000 is the budgeted sales amount in the income statement. The sales budget is often the result of elaborate information gathering. A later section in this chapter (p. 184) discusses sales forecasts and sales budgets.

In nonprofit organizations, forecasts of revenue or some target level of service are the cornerstone of the master budget. Examples are revenues for hospital patients and police-officer hours for city police protection.

Occasionally, sales are limited by available production capacity. For example, unusually heavy market demands, shortages of personnel or materials, or strikes may cause a company to exhaust its inventory completely. Additional sales cannot be made because no product remains. In such cases, the production capacity—the factor that limits sales—is the starting point for budgeting.

Step 2: *Production budget.* After sales are budgeted, the production budget (Schedule 2) can be prepared. The total number of units to be produced depends on planned sales and also expected changes in inventory levels, as shown in Schedule 2. ***Note that the production budget is stated in finished physical units.***

Production Budget In Units
For the Year Ended December 31, 19_2

	Products	
Schedule 2	**Regular (R)**	**Heavy-Duty (HD)**
Budgeted sales (Schedule 1)	5,000	1,000
Target ending finished goods inventory	1,100	50
Total needs	6,100	1,050
Deduct beginning finished goods inventory	100	50
Units to be produced	6,000	1,000

Many companies try to keep personnel and facilities in steady use. When production is stable throughout the year, inventory will rise and fall with seasonal fluctuations in sales. Management must be sure to have enough product in inventory to meet varying sales demand. (In contrast, as Chapter 19 explains, other companies try to keep inventories extremely low.)

Step 3: *Direct materials usage budget.* The units to be produced (Schedule 2) are the keys to computing the usage of direct materials in quantities and in dollars.

Budgeted Usage of Direct Materials in Kilograms and Dollars
For the Year Ended December 31, 19_2

Schedule 3A	Material 111 Alloy	Material 112 Alloy	Total
Regular (R) (Schedule 2, 6,000 units × 12 and 6 kilograms)	72,000	36,000	
Heavy-duty (HD) (Schedule 2, 1,000 units × 12 and 8 kilograms)	12,000	8,000	
Total direct materials usage	84,000	44,000	
Multiply by price per kilogram	$ 7	$ 10	
Total cost of direct materials used	$588,000	$440,000	$1,028,000

Just like planning the production of finished units, the purchases of direct materials depend on both budgeted usage and inventory levels:

$$\text{Purchases} = \text{Usage} + \frac{\text{Target ending}}{\text{inventories}} - \frac{\text{Beginning}}{\text{inventories}}$$

The computations of budgeted direct material purchases are in Schedule 3B.

Direct Material Purchases Budget
For the Year Ended December 31, 19_2

Schedule 3B	Material 111 Alloy	Material 112 Alloy	Total
Kilograms used in production (Schedule 3A)	84,000	44,000	
Target ending direct material inventory in kilograms	8,000	2,000	
Total needs	92,000	46,000	
Deduct beginning direct material inventory in kilograms	7,000	6,000	
Kilograms to be purchased	85,000	40,000	
Price per kilogram	$ 7	$ 10	
Purchase cost	$595,000	$400,000	$995,000

Step 4: *Direct labor budget.* These costs depend on the types of products, labor rates, and production methods.

Direct Labor Budget
For the Year Ended December 31, 19_2

Schedule 4	Units Produced (Schedule 2)	Direct Labor Hours Per Unit	Total Hours	Total Budget @ $20 Per Hour
Regular (R)	6,000	4	24,000	$480,000
Heavy-duty (HD)	1,000	6	6,000	120,000
Total			30,000	$600,000

Step 5: *Factory overhead budget.* The total of these costs depends on how individual overhead costs behave as production fluctuates.

Factory Overhead Budget*
For the Year Ended December 31, 19_2

Schedule 5	At Budgeted Volume of 30,000 Direct Labor Hours	
Supplies	$ 90,000	
Indirect labor	200,000	
Payroll fringe costs	320,000	
Power—variable portion	90,000	
Maintenance—variable portion	70,000	
Total variable overhead		$ 770,000
Depreciation	$230,000	
Property taxes	50,000	
Property insurance	10,000	
Supervision	100,000	
Power—fixed portion	20,000	
Maintenance—fixed portion	20,000	
Total fixed overhead		430,000
Total factory overhead: ($1,200,000 ÷ 30,000 = $40.00 per direct labor hour)		$1,200,000

*Data are from page 179.

Step 6. *Ending inventory budget.* Schedule 6 shows the calculations of the target ending inventories. This information is required not only for the production budget and the direct material purchases budget but also for detail on a budgeted income statement and a balance sheet.

Ending Inventory Budget
December 31, 19_2

Schedule 6	Kilograms	Price Per Kilogram	Total Amount	
Direct materials:				
111 Alloy	8,000*	$ 7	$ 56,000	
112 Alloy	2,000*	10	20,000	$ 76,000
	Units	**Cost Per Unit**		
Finished goods:				
Regular (R)	1,100†	$384‡	$422,400	
Heavy-duty (HD)	50†	524‡	26,200	448,600
Total				$524,600

*From Schedule 3B, second line
†From Schedule 2, second line
‡Computation of unit costs:

	Cost Per Kilogram or Hour of Input	Products: Regular (R)		Heavy-Duty (HD)	
		Inputs	Amount	Inputs	Amount
Material 111 Alloy	$ 7	12	$ 84	12	$ 84
Material 112 Alloy	10	6	60	8	80
Direct labor	20	4	80	6	120
Factory overhead*	40	4	160	6	240
Total			$384		$524

*Direct labor is the single application base for factory overhead.

Step 7. *Cost-of-goods-sold budget.* The information gathered in Schedules 3 through 6 leads to Schedule 7:

Cost-of-goods-sold Budget
For the Year Ended December 31, 19_2

Schedule 7	**From Schedule**		
Finished goods inventory, December 31, 19_1	Given*		$ 64,600
Direct materials used	3A	$1,028,000	
Direct labor	4	600,000	
Factory overhead	5	1,200,000	
Cost of goods manufactured			2,828,000
Cost of goods available for sale			$2,892,600
Finished goods inventory, December 31, 19_2	6		448,600
Cost of goods sold			$2,444,000

*Given in description of basic data and requirements.

Step 8. *Marketing and administrative expense budget.* Some of these expenses, such as sales commissions, may be directly affected by sales. Other expenses, such as advertising, may be lump-sum appropriations determined by top management as the "correct" amount to spend.

Marketing and Administrative Expense Budget*
For the Year Ended December 31, 19_2

Schedule 8		
Sales commissions	$200,000	
Advertising	60,000	
Sales and customer service salaries	100,000	
Travel	90,000	
Total marketing expenses		$450,000
Clerical wages	$100,000	
Supplies used	12,000	
Officer salaries	260,000	
Miscellaneous	48,000	
Total administrative expenses		420,000
Total marketing and administrative expenses		$870,000

*Data are from page 179.

Step 9. *Budgeted income statement.* Schedules 1, 7, and 8 provide enough information for the income statement, identified in this illustration as Exhibit 6-2. Of course, more details could be included in the income statement, and then fewer supporting schedules would be prepared.[5]

[5]Budgeting questions appear in professional examinations with some regularity. See the supplement to this textbook: John K. Harris and Dudley W. Curry, *Student Guide and Review Manual* (Englewood Cliffs, NJ: Prentice Hall, 1991).

EXHIBIT 6-2

ALIC COMPANY Budgeted Income Statement *For the Year Ended December 31, 19_2*	From Schedule	
Sales	1	$3,800,000
Cost of goods sold	7	2,444,000
Gross margin		$1,356,000
Marketing and administrative expenses	8	870,000
Operating income		$ 486,000

SALES FORECASTING— A DIFFICULT TASK

Factors in Sales Forecasting

The term *sales forecast* is sometimes distinguished from *sales budget* as follows: The forecast is the estimate—the prediction—that may or may not become the sales budget. The forecast often leads to adjustments of managerial plans, so that the final sales budget differs from the original sales forecast. Indeed, companies often take up to four months to complete the forecasting process; sales forecasts are revised an average of five times.[6] The forecast becomes the budget only if management accepts it as an objective.

The marketing vice president usually has direct responsibility for the preparation of the sales budget, the foundation for the quantification of the entire business plan.

A sales forecast is made after consideration of many factors, including the following:

- Past sales volume
- General economic and industry conditions
- Relationship of sales to economic indicators such as gross national product, personal income, employment, prices, and industrial production
- Relative product profitability
- Market research studies
- Pricing policies
- Advertising and other promotion
- Quality of sales force
- Competition
- Seasonal variations
- Production capacity
- Long-term sales trends for various products

Forecasting Procedures

An effective aid to accurate forecasting is to approach the task by several methods. Each forecast acts as a check on the others. The three methods described below are usually combined in some fashion that is suitable for a specific company.

[6]E. A. Imhoff, Jr., *Sales Forecasting Systems* (Montvale, NJ: National Association of Accountants, 1986). Surveys of company practices are often published in the *Journal of Forecasting*.

Sales Staff Procedures As is the case for all budgets, those responsible for sales should have an active role in sales-budget formulation. If possible, the budget data should flow from individual sales personnel or district sales managers upward to the marketing vice president. A valuable benefit from the budgeting process is that discussions occur that generally result in improvements in the budget.

Statistical Approaches Trend, cycle projection, and correlation analysis are useful techniques for forecasting. Correlations between sales and economic indicators help make sales forecasts more reliable, especially if fluctuations in certain economic indicators precede fluctuations in company sales. However, no firm should depend entirely on this approach. Too much reliance on statistical evidence is dangerous because chance variations in statistical data may completely upset a prediction. As always, statistical analysis can provide help but not flawless answers.

Group Executive Judgment All top officers, including production, purchasing, finance, and administrative officers, may use their experience and knowledge to project sales on the basis of group opinion.

It is beyond the scope of this book to give a detailed description of all phases of sales-budget preparation, but its key importance should be kept in mind.

FINANCIAL PLANNING MODELS AND COMPUTERS

A master budget can be a comprehensive planning model for the organization. As the budget is formulated, it is frequently altered as executives exchange views on various aspects of expected activities and ask "what-if" questions. Entering alterations by hand into the budget is cumbersome and time-consuming. Computers greatly ease the job.

Objective 5

Describe the uses of financial planning models and computers

Computer-based **financial planning models** often use the master budget as their structural base. They are mathematical statements of the relationships among all operating activities, financial activities, and the other major internal and external factors that may affect decisions. The models are used for budgeting, for revising budgets, for conducting "what-if" analysis, and for comparing a variety of decision alternatives in how they affect the entire organization. For example, the financial planning model at Ralston Purina Co. tells management that a 1% change in the price of a prime commodity will cause a change in the company's cost models and possibly will mean having to build an entirely new corporate plan and master budget amounts. At Dow Chemical Co., 140 separate cost inputs, constantly revised, are fed into the financial model. Each week, managers monitor such factors as major raw-material costs and prices by country and region.

Computer models have various degrees of sophistication and applicability, depending on how much an organization is willing to pay. Business can acquire unsophisticated general-purpose models from consultants or software firms. These models usually use specific data from the firm as input. The output consists of the conventional financial statements and supporting schedules. Personal-computer users can purchase inexpensive budgeting software programs. At the other extreme, management can have a special-purpose model designed that integrates detailed activities of all subunits of an organization, encompasses internal and external data, and permits "what-if" analyses.

As in all aspects of systems design, managers pursue cost-benefit approaches to using computer-based models. The costs of using general-purpose models continue to decline. The prospects of more widespread use of simple general-purpose models are excellent, but the costs of complex models are still imposing. Most corporations concentrate on the eight to twelve variables most crucial to their industry—rate of inflation, consumer spending on nondurables, interest rates, and so on. These corporations need a sufficiently complex financial model—and the computer capability—to handle these factors.

At least three factors impede the widespread use of the complex financial models that computers make possible:

1. The high costs of developing the models.
2. The high rate of structural change in companies. For example, by the time the model is developed, the business may have dropped a division or may have changed its product line.
3. Unimpressive performance in forecasting. The accuracy of the forecast depends on the quality of the model and its inputs.

RESPONSIBILITY ACCOUNTING

Organization Structure

Objective 6

Describe responsibility accounting

The budget presents the map toward the company goals. Its figures are the company's targets. But how does the company reach these targets? To attain its goals, a company must coordinate all its employees—from the top executive through all levels of management to every supervised worker. Coordinating the company's efforts means assigning responsibility to managers who are accountable for their actions in planning and controlling human and physical resources. This section emphasizes that management is essentially a human activity. Budgets exist not for their own sake, but to help managers.

Organization structure is an arrangement of lines of responsibility within the entity. A company such as Shell may be organized primarily by business functions: exploration, refining, and marketing. Another company, such as Procter & Gamble, may be organized by product lines. If so, the managers of the individual divisions (toothpaste, soap, and so on) would each have decision-making authority concerning all the functions within that division (manufacturing, marketing, and so on).

To improve company performance, top managers often subdivide operating processes into increasingly smaller areas of responsibility. The organization structure that results from these increasing subdivisions is a pyramid. Subordinates make up the broad base. Lower levels of managers report upward to superiors. Top management sits at the peak.

Definition of Responsibility Accounting

Each manager, regardless of level, is in charge of a responsibility center and is accountable for a specified set of activities within a segment or a subunit of the organization. The higher the level of manager, the broader the responsibility center he or she manages. **Responsibility accounting** is a system that measures the plans (by budgets) and actions (by historical records) of each responsibility center. Four major types of responsibility centers are common:

1. **Cost center**—manager accountable for costs (expenses) only
2. **Revenue center**—manager accountable for revenues (sales) only
3. **Profit center**—manager accountable for revenues and costs
4. **Investment center**—manager accountable for investments, revenues, and costs

These four types of responsibility centers are illustrated in various places throughout this book. For example, our Tastee King illustration, which we explain shortly, uses stores, branches, and districts as profit centers. Investment centers are discussed in Chapter 28.

Responsibility accounting affects behavior. For example, consider the following incident:

> The sales department requests a rush production run. The plant scheduler argues that it will disrupt his production and cost a substantial though not clearly determined amount of money. The answer coming from sales is: "Do you want to take the responsibility of losing the X Company as a customer?" Of course the production scheduler does not want to take such a responsibility, and he gives up, but not before a heavy exchange of arguments and the accumulation of a substantial backlog of ill feeling. Analysis of the payroll in the assembly department, determining the costs involved in getting out rush orders, eliminated the cause for argument. Henceforth, any rush order was accepted with a smile by the production scheduler, who made sure that the extra cost would be duly recorded and charged to the sales department—"no questions asked."
>
> As a result, the tension created by rush orders disappeared completely; and, somehow, the number of rush orders requested by the sales department was progressively reduced to an insignificant level.[7]

The responsibility accounting approach traces the costs to either (a) the individual who has the best knowledge about why the cost rose or (b) the activity that caused the cost. In this incident, the cause was the sales activity, and the resulting cost was charged to the sales department. If rush orders occur regularly, the sales department might have a budget for such costs, and the department's actual performance would then be compared against the budget.

Illustration of Responsibility Accounting

The simplified organization chart in Exhibit 6-3 illustrates how companies may use responsibility accounting in a service industry, the fast-food industry. At the top level, a district manager oversees the branch managers, who supervise the manag-

[7]R. Villers, "Control and Freedom in a Decentralized Company," *Harvard Business Review*, XXXII, No. 2, p. 95. Another example of responsibility accounting is Citibank's system for a check-processing department. On any given day, for every $1 million that is not delivered to the Federal Reserve Bank on time, the department is "fined" for the amount the parent company loses in potential interest.

EXHIBIT 6-3
Tastee King
Simplified Partial Organization Chart

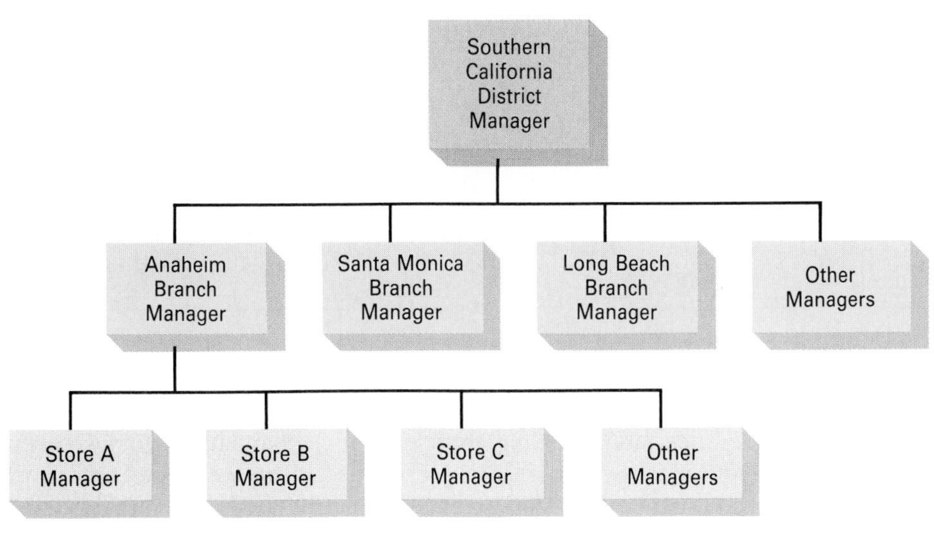

EXHIBIT 6-4
Responsibility Accounting at Various Levels

Tastee King
Responsibility Accounting at Various Levels
(in thousands)

Southern California District Manager Monthly Responsibility Report	**Budget**		**Variance: Favorable (Unfavorable)**	
Operating income of branches and district manager office expense:	**This Month**	**Year to Date**	**This Month**	**Year to Date**
District manager office expense	$ (145)	$ (605)	$ (8)	$ (20)
Anaheim branch	465	1,730	(5)	(4)
Santa Monica branch	500	1,800	19	90
Long Beach branch	310	1,220	31	110
Others	600	2,560	47	130
Operating income	$1,730	$6,705	$84	$306

Anaheim Branch Manager Monthly Responsibility Report	**Budget**		**Variance: Favorable (Unfavorable)**	
Operating income of stores and branch manager office expense:	**This Month**	**Year to Date**	**This Month**	**Year to Date**
Branch manager office expense	$ (20)	$ (306)	$ (5)	$ 4
Store A	48	148	(1)	(5)
Store B	54	228	7	16
Store C	38	160	4	10
Others	345	1,500	(10)	(29)
Operating income	$465	$1,730	$ (5)	$ (4)

Store B Manager Monthly Responsibility Report	**Budget**		**Variance: Favorable (Unfavorable)**	
Sales and expenses:	**This Month**	**Year to Date**	**This Month**	**Year to Date**
Sales	$170	$690	$8	$12
Food expense	50	198	5	14
Paper	15	62	(3)	(2)
Wages	24	98	(4)	(5)
Repairs	5	19	1	(1)
General	12	45	—	(2)
Depreciation	10	40	—	—
Total expenses	116	462	(1)	4
Operating income	$ 54	$228	$7	$16

ers of the stores. Store managers have limited freedom to make operating decisions. They may decide on how to handle local advertising, the number of employees and their schedules, and the store hours. Branch managers oversee several stores, evaluate store managers' performance, and set store managers' compensation levels. In turn, district managers oversee several branches, evaluate branch managers' performance and compensation, and decide on district prices and sales promotions. District managers are accountable to regional managers, who answer to home-office vice presidents.

Exhibit 6-4 provides a more detailed view of how responsibility accounting is used to evaluate profit centers. Examine the lowest level and move to the top. Follow how the reports are related through the three levels of responsibility. All variances may be subdivided for further analysis, either in these reports or in supporting schedules.

Trace the $54,000 operating income from the Store B manager report to the Anaheim branch manager report. The branch manager report summarizes the final results of the stores under his supervision. In addition, charges incurred by the branch manager office are included in this report.

Trace the $465,000 total from the Anaheim branch manager report to the southern California district manager report. The report of the district manager includes data for her own district office plus a summary of the entire district's operating income performance.

Performance Report Format

Exhibit 6-4 stresses variances. This focus is a highlight of management by exception, a management strategy in which executive attention is directed to the important deviations from budgeted amounts. For example, the Anaheim branch's operating income lagged behind the other branches during the current month and for the year to date. The district manager would concentrate her efforts on improving the Anaheim branch. Managers do not waste time investigating smoothly running operations.

The format for reporting operations used in Exhibit 6-4 may be expanded to highlight variances. The expanded performance report for the Store B manager follows:

	Budget		Actual Results		Variance: Favorable (Unfavorable)		Variance: Percent of Budgeted Amount	
	This Month	Year to Date	This Month	Year to Date	This Month	Year to Date	This Month	Year to Date
Sales	$170	$690	$178	$702	$8	$12	4.7%	1.7%

The complete performance report would likely include line-by-line presentations of other data. For example, a report for a restaurant will show the number of customers served and the average sales per customer. In the hotel industry, managers report the percentage of rooms occupied and the average daily rental rate per room as performance measures.

No single format appeals to all users. Some managers prefer the greater detail shown in the eight-column format, but others prefer less detail. The choice is a matter of personal preference.

Feedback and Fixing Blame

Budgets coupled with responsibility accounting provide systematic help for managers, particularly if managers interpret the feedback carefully. Managers, accountants, and students of cost accounting repeatedly tend to "fix the blame"—using variances appearing in the responsibility accounting system to pinpoint fault for operational problems and to provide remedies for these problems. No accounting system—and no variation it detects—can provide answers. Variances can suggest questions or can direct attention to persons who should have answers. ***In looking at revenues, costs, or variances, we should determine whom we should talk to in that specific situation—not whom we should blame.***

RESPONSIBILITY AND CONTROLLABILITY

Definition of Controllability

Objective 7

Explain how controllability relates to responsibility accounting

Controllability is the degree of influence that a specific manager has over costs, revenues, or other items in question. A **controllable cost** is any cost that is primarily subject to the influence of a given *manager* of a given *responsibility center* for a given *time span*. Ideally, responsibility accounting systems either will exclude all uncontrollable costs from a manager's performance report or will segregate such costs from the controllable costs. For example, a machining supervisor's performance report might be confined to quantities (not prices) of direct materials, direct labor, power, and supplies.

In practice, controllability is difficult to pinpoint:

1. Few costs are clearly under the sole influence of one manager. For example, *prices* of direct materials may be influenced by a purchasing manager, and *quantities* used may be influenced by a production manager. Moreover, managers often work in groups or teams. How can individual responsibility be evaluated in a group decision?
2. With a long enough time span, nearly all costs will come under somebody's control. However, performance reports focus on periods of a year or less. A current manager may have inherited inefficiencies from his or her predecessor. For example, the present manager may have to work under undesirable contracts with suppliers or labor unions, contracts negotiated before he or she became manager. How can we separate what the current manager actually controls from the results of decisions made by others? For exactly what is the current manager accountable?

Research on controllability has led to the following observation:

> Holding profit center managers accountable for individual items over which they have only partial control means they bear an extra risk; their rewards depend on factors they cannot control. If the firm fails to compensate them for this risk, it will bear the costs of frustration, lower motivation, and possibly managerial turnover. The key question, then, is: When is the managers' influence significant enough to warrant holding them at risk for something they cannot totally control?[8]

Emphasize Information and Behavior

Beware of overemphasis on controllability. Responsibility accounting is more far-reaching. It focuses on *information and knowledge,* not control. The key question is, Who is the best informed? Put another way, Who is the person who can tell us the most about the specific item, regardless of that person's ability to exert personal control? For instance, purchasing managers may be held accountable for total purchase costs, not because of their ability to affect prices, but because of their ability to predict uncontrollable prices and explain uncontrollable price changes.

In what job would a person have no control but the most information? A store manager may have the most knowledge about storm damage but no control. In what job would a person have full control but little information? This is a more difficult question to answer. However, a store manager might have total control over the sale of chicken and no information that it is contaminated until it is too late.

[8]K. Merchant, *Rewarding Results* (Boston: Harvard Business School Press, 1989), p. 5.

Performance reports for responsibility centers may include uncontrollable items also because such inclusion could change behavior in the directions sought by top management. For example, some companies have changed the accountability of a cost center to a profit center. Why? Because the manager will probably behave differently. In a cost center, the manager may emphasize production efficiency and deemphasize the pleas of sales personnel for faster service and rush orders. In a profit center, the manager is responsible for both costs *and* revenues. Thus, even though the manager still has no control over sales personnel, the manager will now more likely weigh the impact of his or her decisions on costs *and* revenues, rather than solely on *costs.*

Importance of Budget and MBO

The message top executives frequently give to managers is *not* to *maximize* a single measure such as operating income—instead, attain a *budgeted* operating income and simultaneously achieve additional goals such as a targeted share of a market. Focusing on a single amount, whether it be operating income, sales, costs of materials, or some other number, tends to lead managers to neglect other important aspects of their jobs.

Intelligent budgeting can overcome many of the problems of human behavior inherent in commonly used performance measures such as income or return on investment. Many organizations have successfully used some form of **management by objectives (MBO)**. Under this procedure, a subordinate and his or her superior jointly formulate the subordinate's set of goals and the plans for attaining those goals for a subsequent period. For our purposes, *goals* and *objectives* are synonymous.

The plans are frequently crystallized in responsibility accounting and a budget. In addition, various subgoals may be measured, such as levels of product innovation, quality, share of the market, and management training. The subordinate's performance is then evaluated in relation to these agreed-upon budgeted goals.

MBO is also used by nonprofit organizations. For example, in 1988 the Episcopal Diocese of Newark, New Jersey, started paying ministers according to their performance. Under the merit-pay plan, ministers can qualify for salary increases based on such goals as parish growth, education, and quality of sermons.

Responsibility and Uncertainty

Managers live in an uncertain world. Actions and events outside and inside the organization affect the measures of managers' performance. An MBO approach helps managers cope with uncertainty. MBO often reduces complaints regarding responsibility, uncertainties or risks, and lack of control. The focus is on *budgeted results,* given the degree of uncertainties, interdependencies, and limitations of accounting measurements.

The intelligent use of budgets and MBO adds incentives to managers. Executives may be more willing to undertake risks. For example, top management may learn through budgeting that a troubled division will likely lose $50 million. What manager would accept an assignment to such an unsuccessful responsibility center? Top management, aware that the division is facing troubles, will set targets for the new manager that are attainable. Given a set of reasonable goals, an able executive would more likely accept responsibility for the weak division.

In sum, distinctions between controllable and uncontrollable items should be made whenever feasible. However, financial responsibility may be assigned to managers even though controllability may be minimal. The final test is what information and what behavior will be generated by the measurement system. Put another way:

1. What do you want responsibility center managers to worry about? They can't pay attention to costs that aren't assigned to them.
2. How much do you want them to worry? The method of cost assignment may give them more or less direct control over the amount of costs charged to them.[9]

HUMAN ASPECTS OF BUDGETING

Why did we cover the two major topics, master budgets and responsibility accounting, in a single chapter? Primarily to emphasize that human factors are crucial parts of budgeting. Too often, students study budgeting as though it were a mechanical tool.

The budgeting techniques themselves are free of emotion; however, their administration requires education, persuasion, and intelligent interpretation. Many managers regard budgets negatively. To them, the word *budget* is about as popular as, say, *strike* or *layoff*. Top managers must convince their subordinates that the budget is a positive device designed to help managers choose and reach goals. But budgets are not cure-alls. They are not remedies for weak managerial talent, faulty organization, or a poor accounting system.

PROBLEM FOR SELF-STUDY

Before trying to solve the homework problems, review the illustration of the master budget, p. 177.

PROBLEM

Prepare a budgeted income statement, including all necessary detailed supporting schedules. Use the data given in the illustration of the master budget to prepare your own budget schedules. (See the "Basic Data and Requirements" section on pp. 178–79.)

SUMMARY

Comprehensive budgeting is the expression of management's master operating and financial plan—the formalized outlining of company objectives and their means of attainment. When administered wisely, budgets (a) compel management planning, (b) provide definite expectations that are the best framework for judging subsequent performance, and (c) promote communication and coordination among the various segments of the business.

The foundation for budgeting is the sales forecast. Inventory, production, and cost incurrence are generally geared to sales volume.

[9]R. Vancil, *Decentralization: Managerial Ambiguity by Design* (New York: Financial Executives Research Foundation, 1979), p. 21.

Students are frequently overwhelmed by the variety of data required for preparing a master budget. The prediction of sales is almost always the best place to begin.

Responsibility accounting systems identify various decision centers and trace financial data to the individual managers primarily responsible for the data in question. These managers are in the best position to explain their plans and performance.

This chapter began with a section on the evolution of control systems. Ponder the steps:

1. Personal observation
   plus
2. Historical records
   plus
3. Budgets

Budgeting systems are costly. They are bought because managers believe that the systems will help subordinates make better collective operating decisions. If the same collective operating decisions were always made without budgets, then buying a budgeting system would simply be an unnecessary expense.

Some organizations mistakenly tend to use responsibility accounting, budgets, and performance reports to pinpoint blame. The important question is who should be asked, *not* who should be blamed.

APPENDIX: THE CASH BUDGET

Basic Data and Instruction

Objective 8

Construct a cash budget

The major illustration in the chapter featured the operating budget. The other major part of the master budget is the financial budget, which includes the capital budget, cash budget, budgeted balance sheet, and budgeted statement of cash flows. This appendix focuses on the cash budget and the budgeted balance sheet. Capital budgeting is covered in Chapters 21 and 22; coverage of the statement of cash flows is beyond the scope of this book.

Suppose the Alic Company in our chapter illustration had the following balance sheet for the year just ended:

ALIC COMPANY
Balance Sheet
December 31, 19_1

Assets		
Current assets:		
Cash	$ 30,000	
Accounts receivable	400,000	
Direct materials	109,000	
Finished goods	64,600	$ 603,600
Fixed assets:		
Land	$ 200,000	
Building and equipment	2,200,000	
Accumulated depreciation	(690,000)	1,710,000
Total		$2,313,600

Liabilities and Stockholders' Equity		
Current liabilities:		
Accounts payable	$ 150,000	
Income taxes payable	50,000	$ 200,000
Stockholders' equity:		
Common stock, no-par,		
25,000 shares outstanding	$ 350,000	
Retained earnings	1,763,600	2,113,600
Total		$2,313,600

Budgeted cash flows are:

	Quarters			
	1	2	3	4
Collections from customers	$795,000	$900,000	$995,000	$910,000
Disbursements:				
For direct materials	240,000	230,000	260,000	240,000
For other costs and expenses	250,000	240,000	190,000	180,000
For payroll	320,000	380,000	410,000	340,000
For income taxes	80,000	50,000	35,000	55,000
For machinery purchase	—	—	—	84,810

The quarterly data are based on the cash effects of the operations formulated in Schedules 1 through 8 in the chapter.

The company desires to maintain a $35,000 minimum cash balance at the end of each quarter. Money can be borrowed or repaid in multiples of $1,000 at an interest rate of 12% per year. Management does not want to borrow any more cash than is necessary and wants to repay as promptly as possible. In any event, loans may not extend beyond four quarters. By special arrangement, interest is computed and paid when the principal is repaid. Assume that borrowings take place at the beginning and repayments at the end of the quarters in question. Compute interest to the nearest dollar.

An Alic accountant used the above data and the other data contained in the budgets in the chapter (pages 180–184). He was instructed as follows:

1. Prepare a cash budget. That is, prepare a statement of cash receipts and disbursements by quarters, including details of borrowings, repayments, and interest.
2. Prepare a budgeted balance sheet.
3. Prepare a budgeted income statement, including the effects of interest expense and income taxes. Assume income taxes for 19_2 are $185,000.

Preparation of Budgeted Statements

1. The **cash budget** (Exhibit 6-5) is a schedule of expected cash receipts and disbursements. It predicts the effects on cash position at the given levels of operation. The illustrative cash budget is presented by quarters to show the impact of cash-flow timing on bank loan schedules. In practice, monthly—and sometimes weekly—cash budgets are very helpful for cash planning and control. Cash budgets aid in avoiding unnecessary idle cash and unnecessary cash deficiencies. The astute mapping of a financing program keeps cash balances in reasonable relation to needs. Ordinarily, the cash budget has the following main sections:
   a. The beginning cash balance plus cash receipts equals the total cash available for needs, before financing. Cash receipts depend on collections of accounts receivable, cash sales, and miscellaneous recurring sources such as rental or royalty receipts. Studies

of the prospective collectibility of accounts receivable are needed for accurate predictions. Key factors include bad-debt experience and average time lag between sales and collections.

b. Cash disbursements:
   (1) Direct material purchases—depends on credit terms extended by suppliers and bill-paying habits of the buyer.
   (2) Direct labor and other wage and salary outlays—depends on payroll dates.
   (3) Other costs and expenses—depends on timing and credit terms. ***Note that depreciation does not entail a cash outlay.***
   (4) Other disbursements—outlays for fixed assets, long-term investments.
c. Financing requirements depend on how the total cash available for needs, keyed as (a) in Exhibit 6-5, compares with the total cash needed. Needs, keyed as (c), include disbursements, keyed as (b), plus the ending cash balance desired. The financing plans will depend on the relationship of cash available to cash required. If there is an excess, loans may be repaid or temporary investments made. The pertinent outlays for interest expenses are usually shown in this section of the cash budget.
d. The ending cash balance. The total effect of the financing decisions on the cash budget, keyed as (d) in Exhibit 6-5, may be positive (borrowing) or negative (repayment), and the ending cash balance, (a) − (b) + (d).

EXHIBIT 6-5

ALIC COMPANY
Budgeted Statement of Cash Receipts and Disbursements
For the Year Ended December 31, 19_2

	Quarters				For the Year as a Whole
	1	2	3	4	
Cash balance, beginning	$ 30,000	$ 35,000	$ 35,000	$ 35,810	$ 30,000
Add receipts:					
Collections from customers	795,000	900,000	995,000	910,000	3,600,000
(a) Total cash available for needs	825,000	935,000	1,030,000	945,810	3,630,000
Deduct disbursements:					
For direct materials	240,000	230,000	260,000	240,000	970,000
For other costs and expenses	250,000	240,000	190,000	180,000	860,000
For payroll	320,000	380,000	410,000	340,000	1,450,000
For income taxes	80,000	50,000	35,000	55,000	220,000
For machinery purchase	—	—	—	84,810	84,810
(b) Total disbursements	890,000	900,000	895,000	899,810	3,584,810
Minimum cash balance desired	35,000	35,000	35,000	35,000	35,000
(c) Total cash needed	925,000	935,000	930,000	934,810	3,619,810
Cash excess (deficiency) (a − c)*	(100,000)	—	100,000	11,000	10,190
Financing:					
Borrowings (at beginning)	100,000	—	—	—	100,000
Repayments (at end)	—	—	(91,000)	(9,000)	(100,000)
Interest (at 12% per annum)†	—	—	(8,190)	(1,080)	(9,270)
(d) Total effects of financing	100,000	—	(99,190)	(10,080)	(9,270)
Cash balance, ending (a − b + d)	$ 35,000	$ 35,000	$ 35,810	$ 35,920	$ 35,920

*Excess of total cash available over total cash needed before current financing.
†The interest payments pertain only to the amount of principal being repaid at the end of a given quarter. Note that the $9,000 remainder of the loan must be repaid by the end of the fourth quarter. Also note that *depreciation does not necessitate a cash outlay.* The specific computations regarding interest are: $91,000 × .12 × 3/4 = $8,190 and $9,000 × .12 × 4/4 = $1,080.

The cash budget in Exhibit 6-5 shows the pattern of short-term "self-liquidating cash loans." Seasonal peaks of production or sales often result in heavy cash disbursements for purchases, payroll, and other operating outlays as the products are produced and sold. Cash receipts from customers typically lag behind sales. The load is *self-liquidating* in the sense that the borrowed money is used to acquire resources that are combined for sale, and the proceeds from the sale are used to repay the loan. This **self-liquidating cycle** cycle (sometimes called the **working-**

EXHIBIT 6-6

ALIC COMPANY
Budgeted Balance Sheet
December 31, 19_2

Assets			
Current assets:			
Cash (from Exhibit 6-5)		$ 35,920	
Accounts receivable (1)		600,000	
Direct materials (2)		76,000	
Finished goods (2)		448,600	$1,160,520
Fixed assets:			
Land (3)		$ 200,000	
Building and equipment (4)	$2,284,810		
Accumulated depreciation (5)	(920,000)	1,364,810	1,564,810
Total			$2,725,330
Liabilities and Stockholders' Equity			
Current liabilities:			
Accounts payable (6)		$ 305,000	
Income taxes payable (7)		15,000	$ 320,000
Stockholders' equity:			
Common stock, no-par, 25,000 shares outstanding (8)		$ 350,000	
Retained earnings (9)		2,055,330	2,405,330
Total			$2,725,330

Notes:

Beginning balances are used as a start for most of the following computations:

(1) $400,000 + $3,800,000 sales − $3,600,000 receipts = $600,000.
(2) From Schedule 6, page 182.
(3) From beginning balance sheet, page 194.
(4) $2,200,000 + $84,810 purchases = $2,284,810.
(5) $690,000 + $230,000 depreciation from Schedule 5, p. 182.
(6) $150,000 + acquisitions − disbursements = ending balance.
$150,000 + ($995,000 direct material purchases, $600,000 direct labor, $970,000 factory overhead,* $870,000 marketing and administrative expenses) − ($970,000 direct materials, $860,000 other costs and expenses, and $1,450,000 payroll) = $305,000. For your information, the payroll consisted of $600,000 direct labor + $200,000 indirect labor + $100,000 supervision + $200,000 sales commissions + $100,000 clerical wages + $260,000 officer salaries = $1,460,000, of which $1,450,000 was disbursed per Exhibit 6-5.
(7) $50,000 + $185,000 current year − $220,000 payment = $15,000.
(8) From beginning balance sheet.
(9) $1,763,600 + $291,730 net income per Exhibit 6-7 = $2,055,330.

*$1,200,000 from Schedule 5 minus depreciation of $230,000 = $970,000.

capital, cash, or **operating cycle**) is the movement from cash to inventories to receivables and back to cash.

2. The budgeted balance sheet is in Exhibit 6-6. Each item is projected in light of the details of the business plan as expressed in the previous schedules. For example, the ending balance of Accounts Receivable would be computed by adding the budgeted sales (from Schedule 1) to the beginning balance (given) and subtracting cash receipts (given and in Exhibit 6-5).
3. The budgeted income statement is in Exhibit 6-7. It is merely the income statement in Exhibit 6-2, page 184, but now includes interest expense and income taxes.

EXHIBIT 6-7

ALIC COMPANY
Budgeted Income Statement
For the Year Ended December 31, 19_2

	Source	
Sales	Schedule 1	$3,800,000
Cost of goods sold	Schedule 7	2,444,000
Gross margin		$1,356,000
Marketing and administrative expenses	Schedule 8	870,000
Operating income		$ 486,000
Interest expense	Exhibit 6-5	9,270
Income before income taxes		$ 476,730
Income taxes	Assumed	185,000
Net income		$ 291,730

For simplicity, the cash receipts and disbursements were given explicitly in this illustration. Frequently, there are lags between the items reported on an accrual basis in an income statement and their related cash receipts and disbursements. For example, sales may consist of 90% credit and 10% cash sales. In turn, half the total credit sales may be collected in each of the two months subsequent to the sale. Using assumed figures, unrelated to the above illustration, supporting schedules would be constructed as follows:

	May	June	July	August
Schedule A Sales Budget				
Credit sales, 90%	$81,000	$72,000	$63,000	
Cash sales, 10%	9,000	8,000	7,000	
Total sales, 100%	$90,000	$80,000	$70,000	
Schedule B: Cash Collections				
Cash sales this month	$ 9,000	$ 8,000	$ 7,000	
Credit sales last month		40,500*	36,000†	$31,500
Credit sales two months ago			40,500*	36,000†
Total collections	$ (affected by three months of sales‡)			

*.50 × $81,000 = $40,500.
†.50 × $72,000 = 36,000.
‡The data here provide a total for July of $83,500; not enough data are here for obtaining totals for the other months.

Of course, such schedules of cash collections depend on credit terms, collection histories, and expected uncollectible accounts. Similar schedules can be prepared for operating expenses and their related cash disbursements.

TERMS TO LEARN

The chapter, chapter appendix, and Glossary contain definitions of the following important terms:

cash budget *(p. 194)* cash cycle *(197)* controllability *(190)*
controllable cost *(190)* cost center *(186)* financial budget *(176)*
financial planning models *(185)* investment center *(186)* master budget *(172)*
management by objectives (MBO) *(191)* operating budget *(176)*
operating cycle *(197)* organization structure *(186)* profit center *(186)*
pro forma statements *(176)* responsibility accounting *(186)*
revenue center *(186)* self-liquidating cycle *(196)* strategy *(173)*
working-capital cycle *(196)*

ASSIGNMENT MATERIAL

QUESTIONS

6-1 Identify three major steps in the evolution of management control systems.

6-2 "Operating plans are contracts, and I want them met. If your revenue is off, you should cut your expenses accordingly." Do you agree? Explain.

6-3 "*Strategic planning* and *long-term planning* are synonyms." Do you agree? Explain.

6-4 What are the two major features of a budgetary program? Which feature is more important? Why?

6-5 What are the elements of the budgetary cycle?

6-6 "Budgeted performance is a better criterion than past performance for judging managers." Do you agree? Explain.

6-7 "Budgets are wonderful vehicles for communication." Comment.

6-8 "Budgets meet the cost-benefit test. They force managers to act differently." Do you agree? Explain.

6-9 Define *continuous budget, pro forma statements.*

6-10 Distinguish between *sales forecast* and *sales budget.*

6-11 "The sales forecast is the cornerstone for budgeting." Why?

6-12 "Budgets are half-used if they serve only as a planning device." Explain.

6-13 Define *responsibility accounting.*

6-14 Distinguish between *responsibility center, responsibility accounting, cost center, revenue center, profit center,* and *investment center.*

6-15 "A common reporting method shows three sets of dollar figures instead of two sets." Identify the likely three sets.

6-16 "The major purpose of responsibility accounting is to fix blame." True or false? Explain.

6-17 Define *controllable cost.* What two major factors help determine whether a given cost is controllable?

6-18 "An action once taken cannot be changed by subsequent events." What implications does this have for the cost accountant?

6-19 "Responsibility accounting should focus on information or knowledge, not control." Do you agree? Explain.

6-20 "Performance reports by responsibility centers may also include uncontrollable items." Do you agree? Explain.

6-21 "Management by objectives (MBO) prevents the overemphasis on a single goal." Do you agree? Explain.

EXERCISES AND PROBLEMS

6-22 Sales and production budget. The Mendez Company expects 19_2 sales of 100,000 units of serving trays. Mendez's beginning inventory for 19_2 is 7,000 trays; target ending inventory, 11,000 trays. Compute the number of trays budgeted for production.

6-23 Sales and production budget. The Gallo Company had a December 31, 19_2, target inventory of 70,000 four-liter bottles of burgundy wine. Gallo's beginning inventory was 60,000 bottles, and its budgeted production was 900,000 bottles. Compute the budgeted sales in number of bottles.

6-24 Direct materials budget. Inglenook Co. produces wine. The company expects to produce 1,500,000 two-liter bottles of chablis in 19_3. Inglenook purchases empty glass bottles from an outside vendor. Its target ending inventory of such bottles is 50,000; its beginning inventory is 20,000. For simplicity, ignore breakage. Compute the number of bottles to be purchased in 19_3.

6-25 Direct materials budget. Bertelli Co. produces frozen pizzas. The company expects to purchase 910,000 cartons in 19_2. Beginning inventory is 35,000 cartons; target ending inventory, 45,000. For simplicity, ignore damaged cartons. Compute the number of cartons to be used in production in 19_2.

6-26 Budgeted manufacturing costs. The Weber Company has budgeted sales of 100,000 units of its product for 19_1. Expected unit costs, based on past experience, should be $60 for direct materials, $40 for direct labor, and $30 for manufacturing overhead. Assume no beginning or ending inventory in process. Weber begins the year with 30,000 finished units on hand but budgets the ending finished goods inventory at only 10,000 units. Compute the budgeted costs of production for 19_1.

6-27 Budgeting material purchases. The Mahoney Company has prepared a sales budget of 42,000 finished units for a three-month period. The company has an inventory of 22,000 units of finished goods on hand at December 31 and has a target finished goods inventory of 24,000 units at the end of the succeeding quarter.

It takes three gallons of direct materials to make one unit of finished product. The company has an inventory of 90,000 gallons of raw material at December 31 and has a target ending inventory of 110,000 gallons. How many gallons of direct materials should be purchased during the three months ending March 31?

6-28 Responsibility accounting and costs of quality. A St. Regis Co. factory makes grocery bags. A suburban Los Angeles plant produces 8 million to 9 million bags daily, supplying supermarket chains in eleven western states. Machines, which are 25-feet long, turn out bags in batches of 25 or 50. Machine operators inspect their own production by riffling each batch through their hands.

The company has been concerned that as much as 6% of production has been bad, resulting in an intolerable level of sales returns. An employee has suggested signature bags. A plate would be engraved with each employee's signature on it. Workers would slip their nameplate into the bag machines. Each bag would bear the inscription "personally inspected by . . ."

Required

Evaluate the suggestion. Is this an example of responsibility accounting? Explain.

6-29 Budgeting sales, cost of goods sold, and gross profit. Janet Grossman operates Centrum Gift Shop. She expects cash sales of $10,000 for October, a $1,000 increase during November, and a $5,000 increase in December. Credit card sales of $7,000 during October should be followed by $1,000 and $4,000 increases during November and December, respectively. Sales returns can be ignored. Credit card companies like VISA and MasterCard charge 4% on credit card sales, so Centrum net sales will be 96%. Cost of goods sold averages 40% of net sales.

Grossman asks you to prepare a schedule of budgeted sales, cost of goods sold, and gross profit for each month of the last quarter of 19_3. Also show totals for the quarter.

6-30 Sales, production, and purchases budget. The Suzuki Co. in Japan has a division that manufactures two-wheel motorcycles. Its budgeted sales for Model G in 19_3 is 800,000 units. Suzuki's target ending inventory is 100,000 units, and its beginning inventory is 120,000 units. The company's budgeted unit selling price to its distributors and dealers is 400,000 yen (¥).

Suzuki buys all its wheels from an outside supplier. No defective wheels are accepted. (Suzuki's needs for extra wheels for replacement parts are ordered by a separate division of the company.) The company's target ending inventory is 30,000 wheels, and its beginning inventory is 20,000 wheels. The budgeted purchase price is 16,000 yen per wheel.

Required

1. Compute the budgeted sales in yen. (For your general information, during the 1980s and into the 1990s the exchange rate fluctuated between 120 and 280 yen per U.S. dollar. This rate is not necessary to solve this problem.)
2. Compute the number of motorcycles to be produced.
3. Compute the budgeted purchases of wheels in units and in yen.

6-31 Budgeting conversion costs and cost of goods manufactured. Seagate Co. is preparing a budget for 19_4 conversion costs regarding Product LMG, a part used heavily by computer manufacturers. The historical record for 19_3 indicates that an average of 5 painstaking direct labor hours was necessary for each finished unit. The direct labor rate was $20 per hour. The budgeted factory overhead rate was 200% of direct labor cost.

Plans for 19_4 called for more automation, which would reduce direct labor time per unit by 20%. However, increases in repairs, utilities, depreciation, and other factory overhead costs will increase the budgeted factory overhead rate to 500% of direct labor cost. Moreover, wage increases will increase the direct labor rate per hour by 5%.

The company plans to produce 20,000 units of LMG in 19_4. Direct materials will cost $800 per unit.

Required

Prepare a budget of cost of goods manufactured. Show supporting computations. For simplicity, assume that there are no work in process inventories.

6-32 Budgeting purchases, cost of goods sold, and inventory. The sales budget of Galaxy Imports for the nine months ended September 30 follows.

	Quarter Ended			Nine-Month
	March 31	June 30	Sept. 30	Total
Cash sales, 20%	$ 36,000	$ 56,000	$ 42,000	$134,000
Credit sales, 80%	144,000	224,000	168,000	536,000
Total sales, 100%	$180,000	$280,000	$210,000	$670,000

In the past, cost of goods sold has been 65% of total sales. The director of marketing, the production manager, and the financial vice president agree that ending inventory should not go below $30,000 plus 10% of cost of goods sold for the following quarter. Galaxy expects sales of $200,000 during the fourth quarter. The January 1 inventory was $44,000.

Required

Prepare a purchases, cost of goods sold, and inventory budget for each of the first three quarters of the year. Compute the cost of goods sold for the entire nine-month period.

6-33 Budgeting material quantities. (SMA) A sales budget for the first five months of 19_3 is given for a particular product line manufactured by Arthur Guthrie Co. Ltd.

	Sales Budget in Units
January	10,800
February	15,600
March	12,200
April	10,400
May	9,800

The inventory of finished products at the end of each month is to be equal to 25% of the sales estimate for the next month. On January 1, there were 2,700 units of product on hand. No work is in process at the end of any month.

Each unit of product requires two types of materials in the following quantities:

Material A:	4 units
Material B:	5 units

Materials equal to one-half of the next month's requirements are to be on hand at the end of each month. This requirement was met on January 1, 19_3.

Required
Prepare a budget showing the quantities of each type of material to be purchased each month for the first quarter of 19_3.

6-34 Budget for production and direct labor. (CMA, adapted) Roletter Company makes and sells artistic frames for pictures of weddings, graduations, and other special events. Bob Anderson, controller, is responsible for preparing Roletter's master budget and has accumulated the information below for 19_5.

	19_5				
	January	February	March	April	May
Estimated unit sales	10,000	12,000	8,000	9,000	9,000
Sales price per unit	$50.00	$47.50	$47.50	$47.50	$47.50
Direct labor hours per unit	2.0	2.0	1.5	1.5	1.5
Wage per direct labor hour	$8.00	$8.00	$8.00	$9.00	$9.00

Labor-related costs include pension contributions of $.25 per hour, workers' compensation insurance of $.10 per hour, employee medical insurance of $.40 per hour, and social security taxes. Assume that as of January 1, 19_5, the social security tax rates are 7% for employers and 6.7% for employees. The cost of employee benefits paid by Roletter on its employees is treated as a direct labor cost.

Roletter has a labor contract that calls for a wage increase to $9.00 per hour on April 1, 19_5. New labor-saving machinery has been installed and will be fully operational by March 1, 19_5.

Roletter expects to have 16,000 frames on hand at December 31, 19_4, and has a policy of carrying an end-of-month inventory of 100% of the following month's sales plus 50% of the second following month's sales.

Required
Prepare a production budget and a direct labor budget for Roletter Company by month and for the first quarter of 19_5. Both budgets may be combined in one schedule. The direct labor budget should include direct labor hours and show the detail for each direct labor cost category.

6-35 Sales and production budgets. (CPA adapted) The Scarborough Corporation in Australia manufactures and sells two products, Thingone and Thingtwo. In July 19_7, Scarborough's budget department gathered the following data in order to project sales and budget requirements for 19_8:

19_8 Projected Sales:

Product	Units	Price
Thingone	60,000	$ 70
Thingtwo	40,000	$100

19_8 Inventories—in Units:

Product	Expected January 1, 19_8	Target December 31, 19_8
Thingone	20,000	25,000
Thingtwo	8,000	9,000

To produce one unit of Thingone and Thingtwo, the following raw materials are used:

		Amount Used Per Unit	
Raw Material	Unit	Thingone	Thingtwo
A	lbs.	4	5
B	lbs.	2	3
C	each	0	1

Projected data for 19_8 with respect to raw materials are as follows:

Raw Material	Anticipated Purchase Price	Expected Inventories January 1, 19_8	Target Inventories December 31, 19_8
A	$12	32,000 lbs.	36,000 lbs.
B	$ 5	29,000 lbs.	32,000 lbs.
C	$ 3	6,000 units	7,000 units

Projected direct labor requirements for 19_8 and rates are as follows:

Product	Hours Per Unit	Rate Per Hour
Thingone	2	$12
Thingtwo	3	$16

Overhead is applied at the rate of $20 per direct labor hour.

Required
Based on the above projections and budget requirements for 19_8 for Thingone and Thingtwo, prepare the following budgets for 19_8:

1. Sales budget (in dollars)
2. Production budget (in units)
3. Raw-materials purchase budget (in quantities)
4. Raw-materials purchase budget (in dollars)
5. Direct labor budget (in dollars)
6. Budgeted finished goods inventory at December 31, 19_8 (in dollars)

6-36 Responsibility of purchasing agent. (R. Villers, adapted) Richards is the purchasing agent for the Hart Manufacturing Company. Sampson is head of the production planning and control department. Every six months, Sampson gives Richards a general purchasing program. Richards gets specifications from the engineering department. He then selects suppliers and negotiates prices. When he took this job, Richards was informed very clearly that he bore responsibility for meeting the general purchasing program once he accepted it from Sampson.

During Week No. 24, Richards was advised that Part No. 1234—a critical part—would be needed for assembly on Tuesday morning, Week No. 32. He found that the regular supplier could not deliver. He called everywhere, finally found a supplier in the Middle West, and accepted the commitment.

He followed up by mail. Yes, the supplier assured him, the part would be ready. The matter was so important that on Thursday of Week No. 31, Richards checked by phone. Yes, the shipment had left in time. Richards was reassured and did not check further. But on Tuesday of Week No. 32, the part was not in the warehouse. Inquiry revealed that the shipment had been misdirected by the railroad company and was still in Chicago.

Required
What department should bear the costs of time lost in the plant? Why? As purchasing agent, do you think it fair that such costs be charged to your department?

6-37 Responsibility for downtime. Two of several departments at A. O. Smith Co. performed successive operations in the manufacture of automobile frames. The frames were transported from department to department via an overhead conveyor system that was placed in accordance with budgeted time allowances. Each department manager had responsibility for the budgeted costs and budgeted output of his department.

On Tuesday morning some equipment failure in Department D led its manager to ask the manager of Department C to stop the conveyor system. The Department C manager refused, so workers in Department D had to remove the frames, stack them, and return the frames to the conveyor later when production resumed.

The manager of Department D was bitter about the incident and insisted that the manager of Department C should bear the related labor and overhead costs of $11,345. In contrast, the manager of Department C said, "I was just doing my job as specified in the budget."

Required
As the controller, how would you account for the $11,345? Why?

6-38 A study in responsibility accounting. The David Machine Tool Company is in the doldrums. Production volume has fallen to a ten-year low. The company has a nucleus of skilled tool-and-die workers who could find employment elsewhere if they were laid off. Three of these workers have been transferred temporarily to the building and grounds department, where they have been doing menial tasks such as washing walls and sweeping for the past month. They have earned their regular rate of $13 per hour. Their wages have been charged to the building and grounds department. The supervisor of building and grounds has just confronted the controller as follows: "Look at the cockeyed performance report you pencil pushers have given me." The helpers' line reads:

	Budget	Actual	Variance	
Wages of helpers	$4,704	$7,644	$2,940	Unfavorable

"This is just another example of how unrealistic you bookkeepers are! Those tool-and-die people are loafing on the job because they know we won't lay them off. The regular hourly rate for my three helpers is $8. Now that my regular helpers are laid off, my work is piling up, so that when they return they'll either have to put in overtime or I'll have to get part-time help to catch up with things. Instead of charging me at $13 per hour, you should charge about $6—that's all those tool-and-die slobs are worth at their best."

Required
As the controller, what would you do *now?* Would you handle the accounting for these wages any differently?

6-39 Fixing responsibility. (Adapted from a description by Harold Bierman, Jr.) The city of Mountainvale had hired its first city manager four years ago. She favored a "management by objectives" philosophy and accordingly had set up many profit responsibility centers, including a sanitation department, a city utility, and a repair shop.

For many months, the sanitation manager had been complaining to the utility manager about wires being too low at one point in the road. There was barely clearance for large trucks. The sanitation manager asked the repair shop to make changes in the clearance. The repair shop manager asked, "Should I charge the sanitation or the utility department for the $2,000 cost of making the adjustment?" Both departments refused to accept the charge, so the repair department refused to do the work.

Late one day the top of a truck caught the wires and ripped them down. The repair department made an emergency repair at a cost of $2,600. Moreover, the city lost $1,000 of utility revenue (net of variable costs) because of the disruption of service.

Investigation disclosed that the truck had failed to clamp down its top properly. The extra two inches of height caused the catching of the wire.

Both the sanitation and utility managers argued strenuously about who should bear the $2,600 cost. Moreover, the utility manager demanded reimbursement from the sanitation department of the $1,000 of lost utility income.

Required

As the city controller in charge of the responsibility accounting system, how would you favor accounting for these costs? Specifically, what would you do next? What is the proper role of responsibility accounting in determining the blame for this situation?

6-40 Cash receipts and payments, multiple choice. (CMA) Study the appendix. Information pertaining to Noskey Corporation's sales revenue is presented in the following table.

	November 19_5 (actual)	December 19_5 (budget)	January 19_6 (budget)
Cash sales	$ 80,000	$100,000	$ 60,000
Credit sales	240,000	360,000	180,000
Total sales	$320,000	$460,000	$240,000

Management estimates that 5% of credit sales are uncollectible. Of the credit sales that are collectible, 60% are collected in the month of sale and the remainder in the month following the sale. Purchases of inventory each month are 70% of the next month's projected total sales. All purchases of inventory are on account; 25% are paid in the month of purchase, and the remainder are paid in the month following the purchase.

Required

Choose one answer in each case and show computations.

1. Noskey's budgeted cash collections in December 19_5 from November 19_5 credit sales are (a) $144,000, (b) $136,800, (c) $91,200, (d) $96,000, (e) none of these.
2. Noskey's budgeted total cash receipts in January 19_6 are (a) $240,000, (b) $294,000, (c) $239,400, (d) $299,400, (e) none of these.
3. Noskey's budgeted total cash payments in December 19_5 for inventory purchases are (a) $283,500, (b) $405,000, (c) $240,000, (d) $168,000, (e) none of these.

6-41 Collections and disbursements. (CPA) Study the appendix. The following information was available from Montero Corporation's books:

19_2	Purchases	Sales
Jan.	$42,000	$72,000
Feb.	48,000	66,000
Mar.	36,000	60,000
Apr.	54,000	78,000

Collections from customers are normally 70% in the month of sale, 20% in the month following the sale, and 9% in the second month following the sale. The balance is expected to be uncollectible. Montero takes full advantage of the 2% discount allowed on purchases paid for by the tenth of the following month. Purchases for May are budgeted at $60,000, while sales for May are forecasted at $66,000. Cash disbursements for expenses are expected to be $14,400 for the month of May. Montero's cash balance at May 1 was $20,000.

Required

Prepare the following schedules:

1. Expected cash collections during May
2. Expected cash disbursements during May
3. Expected cash balance at May 31

6-42 Identifying amounts in a cash budget. Study the appendix. Hewlett Sales Corporation has completed its cash budget for May and June. The budget is presented with missing amounts identified by a question mark (?). Hewlett's plan for eliminating any cash deficiency is to borrow the exact amount needed from its bank. The current annual interest rate is 12%. Hewlett repays all borrowed amounts less than or equal to $2,000 within one month.

Hewlett Sales Corporation
Cash Budget
May and June

	May	June
Cash balance, beginning	$13,000	$?
Cash collections from customers	65,000	74,800
Sale of plant assets	?	900
(a) Total cash available for needs	92,300	?
Cash disbursements:		
Purchases of inventory	50,400	39,100
Operating expenses	33,900	32,500
(b) Total disbursements	84,300	71,600
Minimum cash balance desired	10,000	?
(c) Total cash needed	?	81,600
Cash excess (deficiency), (a) − (c)	(2,000)	?
Financing of cash deficiency:		
Borrowing (at end of month)	?	—
Principal payments (at end of month)	—	?
Interest expense (at .010 monthly)	—	?
(d) Total effects of financing	?	(2,020)
Cash balance, ending (a) − (b) + (d)	$?	$12,080

Required
Fill in each amount identified by a question mark.

6-43 Preparing a cash budget. Study the appendix. Dulcet Tones, a family-owned stereo store, began October with $10,000 cash. Management forecasts that collections from credit customers will be $90,000 in October and $122,000 in November. The store is scheduled to receive $40,000 cash on a business note receivable in November. Projected cash disbursements include inventory purchases ($102,000 in October and $121,000 in November) and operating expenses ($30,000 each month).

The store's bank requires a $7,500 minimum balance in the store's checking account. At the end of any month when the account balance goes below $7,500, the bank automatically extends credit to the store in multiples of $1,000. Dulcet Tones borrows as little as possible and pays back these loans as rapidly as possible in multiples of $1,000 plus 1.5% monthly interest on the entire unpaid principal. The first payment occurs at the end of the month following the loan.

Required
Prepare the store's cash budget for October and November. Compute the amount owed to the bank on November 30.

6-44 Computing cash receipts and disbursements. Study the appendix. For each of the items a through d, compute the amount of cash receipts or disbursements Fernandez Company would budget for December. A solution to one item may depend on the answer to an earlier item.

a. Management expects to sell 4,000 special radios in November and 4,200 in December. Each radio sells for $60. Cash sales average 20% of the total sales, and credit sales make up the rest. One-third of credit sales are collected in the month of sale, with the balance collected the following month.

b. Management has budgeted inventory purchases of $300,000 for November and $250,000 for December. Fernandez Company pays for 25% of its inventory at the time of purchase in order to get a 2% discount. The business pays the 75% balance the following month, with no further discount.
c. The company pays rent and property taxes of $60,000 each month. Commissions and other selling expenses average 20% of sales. Fernandez Company pays two-thirds of these costs in the month incurred, with the balance paid in the following month.
d. Management expects to sell equipment that cost $141,000 at a gain of $20,000. Accumulated depreciation on this equipment is $60,000.

6-45 Preparing a budgeted balance sheet. Use the following information to prepare a budgeted balance sheet for Fashion Store at July 31, 19_6. Show computations for cash and owners' equity amounts.

a. June 30 cash balance, $8,000.
b. July budgeted sales, $124,000.
c. July 31 accounts receivable balance, one-fourth of July sales.
d. July cash receipts, $155,000.
e. July 31 inventory balance, $90,000.
f. July payments for inventory, $59,000.
g. July payments of June 30 accounts payable and accrued liabilities, $61,000.
h. July 31 accounts payable balance, $49,000.
i. June 30 furnitures and fixtures balance, $348,000; accumulated depreciation balance, $277,000.
j. July capital expenditures of $12,000 budgeted for cash purchase of furniture.
k. July operating expenses, including income tax, total $42,000, half of which will be paid during July and half accrued at July 31.
l. July depreciation, $3,000.
m. Cost of goods sold, 50% of sales.
n. June 30 owners' equity, $194,000.

6-46 Cash budgeting. Study the appendix. On December 1, 19_1, the Itami Wholesale Co. is attempting to project cash receipts and disbursements through January 31, 19_2. On this latter date, a note will be payable in the amount of $100,000. This amount was borrowed in September to carry the company through the seasonal peak in November and December.

The trial balance on December 1 shows in part:

Cash	$ 10,000	
Accounts receivable	280,000	
Allowance for bad debts		$15,800
Inventory	87,500	
Accounts payable		92,000

Sales terms call for a 2% discount if paid within the first ten days of the month after purchase, with the balance due by the end of the month after purchase. Experience has shown that 70% of the billings will be collected within the discount period, 20% by the end of the month after purchase, 8% in the following month, and that 2% will be uncollectible. There are no sales for cash.

The average unit sales price of the company's products is $100. Actual and projected sales are:

October actual	$ 180,000
November actual	250,000
December estimated	300,000
January estimated	150,000
February estimated	120,000
Total estimated for year ended June 30	1,500,000

All purchases are payable within fifteen days. Thus approximately 50% of the purchases in a month are due and payable in the next month. The average unit purchase cost is $70. Target ending inventories are 500 units plus 25% of the next month's unit sales.

Total budgeted selling and administrative expenses for the year are $400,000. Of this amount, $150,000 is considered fixed (and includes depreciation of $30,000). The remainder varies with sales. Both fixed and variable selling and administrative expenses are paid as incurred.

Required

Prepare a columnar statement of budgeted cash receipts and disbursements for December and January. Supply supporting schedules for collections of receivables, payments for merchandise, and selling and administrative expenses.

6-47 Comprehensive budget; fill in schedules. Study the appendix. Throughout the problem, ignore income taxes. Following is certain information relative to the position and business of the Newport Stationery Store.

Current assets as of Sept. 30:	
Cash	$ 12,000
Accounts receivable	10,000
Inventory	63,600
Fixed assets—net	100,000
Current liabilities as of Sept. 30	None
Recent and anticipated sales:	
September	$ 40,000
October	48,000
November	60,000
December	80,000
January	36,000

Credit sales: Sales are 75% for cash and 25% on credit. Assume that credit accounts are all collected within 30 days from sale. The accounts receivable on September 30 are the result of the credit sales for September (25% of $40,000).

Gross margin averages 30% of sales. Newport treats purchase discounts on the income statement as "other income."

Expenses: Salaries and wages average 15% of monthly sales; rent, 5%; all other expenses, excluding depreciation, 4%. Assume that these expenses are disbursed each month. Depreciation is $1,000 per month.

Purchases: Newport keeps a minimum inventory of $30,000. The policy is to purchase each month additional inventory in the amount necessary to provide for the following month's sales. Terms on purchases are 2/10, n/30. Assume that payments are made in the month of purchase and that all discounts are taken.

Fixtures: In October, $600 is spent for fixtures, and in November, $400 is to be expended for this purpose.

Assume that a minimum cash balance of $8,000 must be maintained. Assume also that all borrowings are effective at the beginning of the month and all repayments are made at the end of the month of repayment. Loans are repaid when sufficient cash is available. Interest is paid only at the time of repaying principal. Interest rate is 18% per annum. Management does not want to borrow any more cash than is necessary and wants to repay as soon as cash is available.

Required

On the basis of the facts as given above:

1. Complete Schedule A.

Schedule A
Budgeted Monthly Dollar Receipts

Item	September	October	November	December
Total sales	$40,000	$48,000	$60,000	$80,000
Credit sales	10,000	12,000		
Cash sales				
Receipts:				
Cash sales		$36,000		
Collections on accounts receivable		10,000		
Total		$46,000		

2. Complete Schedule B. Note that purchases are 70% of next month's sales.

Schedule B
Budgeted Monthly Cash Disbursements for Purchases

Item	October	November	December	Total
Purchases	$42,000			
Deduct 2% cash discount	840			
Disbursements	$41,160			

3. Complete Schedule C.

Schedule C
Budgeted Monthly Cash Disbursements for Operating Expenses

Item	October	November	December	Total
Salaries and wages	$ 7,200			
Rent	2,400			
Other expenses	1,920			
Total	$11,520			

4. Complete Schedule D.

Schedule D
Budgeted Total Monthly Disbursements

Item	October	November	December	Total
Purchases	$41,160			
Operating expenses	11,520			
Fixtures	600			
Total	$53,280			

5. Complete Schedule E.

Schedule E
Budgeted Cash Receipts and Disbursements

Item	October	November	December	Total
Receipts	$46,000			
Disbursements	53,280			
Net cash increase				
Net cash decrease	$ 7,280			

6. Complete Schedule F (assume that borrowings must be made in multiples of $1,000).

Schedule F
Financing Required

Item	October	November	December	Total
Beginning cash	$12,000	$ 8,720		
Net cash increase				
Net cash decrease	7,280			
Cash position before financing	4,720			
Financing required	4,000			
Interest payments				
Financing retired				
Ending balance	$ 8,720			

7. What do you think is the most logical means of arranging the financing needed by Newport?
8. Prepare a budgeted income statement for the fourth quarter and a budgeted balance sheet as of December 31. Ignore income taxes.
9. Certain simplifications have been introduced in this problem. What complicating factors would be met in a typical business situation?

6-48 Preparing a cash budget for a loan request. Study the appendix. A Computerland store is requesting a $100,000 three-year loan to finance the remodeling of its store. The bank officer wants to investigate the store's current operations. She requires a statement of budgeted cash receipts and disbursements to support the loan application.

The cash balance at February 28 is $7,000. During March the store expects to collect $200,000 from sales made in January and February. Also, in February the store sold some old display cases for $5,000, and the owners expect to receive that amount during March. During March the business will pay off accounts payable (related to purchases of inventory) of $100,000 and a note payable of $70,000 plus $4,000 interest.

The store expects monthly sales of $120,000 for March, April, and May. Experience indicates that the store will collect 60% of sales in the month of sale, 30% in the month following sale, and 7% in the second month after sale. The remaining 3% is uncollectible.

The February 28 inventory is $50,000. Purchases average one-half of sales. The store owners try to keep a minimum inventory of $25,000 plus 20% of purchases for the next month. The owners pay all accounts payable arising from inventory purchases in time to receive a 2% discount, 80% in the month of purchase and 20% the next month.

Budgeted operating expenses for March are rent (8% of sales), advertising ($5,000), employee compensation ($30,000), depreciation ($2,000), and insurance ($1,000). Depreciation and insurance are recorded as the related assets expire. Half the advertising expense is paid as incurred, and half is accrued at the end of the month. During March Computerland will pay advertising of $1,600 that was accrued at February 28.

To make the loan, the bank requires that Computerland maintain a cash balance of at least $15,000 before any effects of financing.

Required

Prepare a cash budget (a statement of budgeted cash receipts and disbursements) for Computerland for March. As the bank loan officer, what time span for cash budgets would you seek?

CHAPTER 7

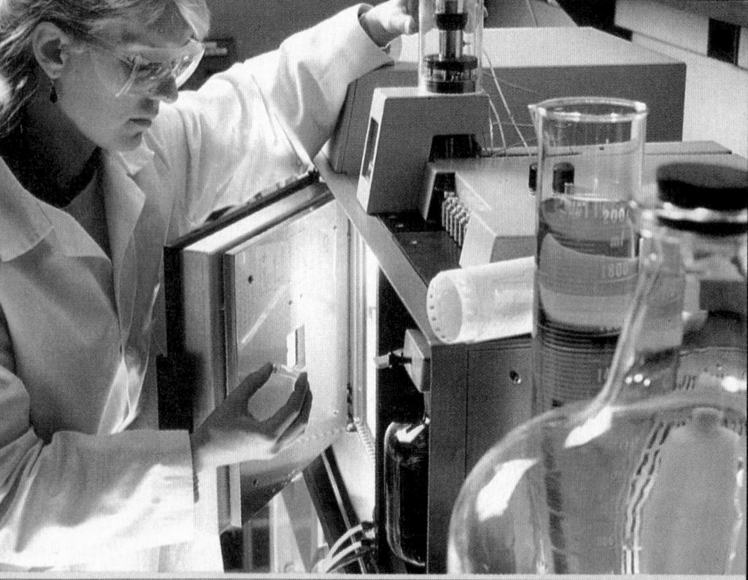

Businesses develop budgets, standards, and variances for planning, controlling, and evaluating performance. Performance standards may be financial or nonfinancial, such as the number of blood tests a medical lab conducts.

Flexible Budgets and Standards: I

Evolution of accounting systems and the cost-benefit approach

PART ONE: FLEXIBLE BUDGETS

Level 0 and Level 1 analysis: Static budget

Level 2 analysis: Flexible budget

Level 3 analysis: Detailed variances

General applicability of expectations

PART TWO: STANDARD COSTS

Standard costs for materials and labor

Impact of inventories

General-ledger entries

Standards used in nonmanufacturing activities

Standards for control

Performance measurement using standards

Controllability and variances

Setup time

When to investigate variances

Learning Objectives

When you have finished studying this chapter, you should be able to

1. Distinguish between static budgets and flexible budgets
2. Use the flexible-budget approach to compute sales volume variances and flexible-budget variances
3. Describe how budgets and nonfinancial (physical) standards are used to control day-to-day activities
4. Distinguish between budget amounts and standard amounts
5. Compute the price and efficiency variances for direct materials and direct labor
6. Prepare journal entries for a typical standard-costing system for direct materials and direct labor
7. Identify the typical responsibilities for controlling variances in material and labor costs

We have learned that managers quantify their plans in the form of budgets. This chapter focuses on how managers use flexible budgets and standard costs as aids for planning and controlling. Recall that a key element of a control system is feedback—the comparison of actual performance with planned performance. This chapter also provides an overview of how budgeting helps performance evaluation.

Our discussion is general. After all, budgets are used in a wide span of organization activities and within various business functions—in research and development, manufacturing, marketing, and customer service.

This chapter has two major parts. Part One describes the flexible budget, which provides an overall look at a company's performance. Part Two describes standard costs, which focus on individual activities. To stress fundamental ideas, we assume throughout this chapter (unless otherwise stated) that there are no beginning or ending inventories of work in process or finished goods. Chapter 9 explores the implications of inventories for the measurement of performance.

EVOLUTION OF ACCOUNTING SYSTEMS AND THE COST-BENEFIT APPROACH

In the last chapter, we traced the evolution of accounting systems, from personal observation to historical records to budgets. In this chapter and the next chapter, we expand the coverage of budgets in general to include static and flexible budgets.

Systems for Planning and Control	Described Primarily In
1. Personal observation (informal system)	General management literature (using the five senses to oversee all facets of operations)
+	
2. Historical records	Chapters 2, 4, and 5
+	
3. Budgets:	
Static budgets	Chapters 6 and 7
+	
Flexible budgets and standard costs	Chapters 7, 8, and 9

Note that these systems build on one another. They are used in combination, not separately. Regardless of the size or complexity of the organization, personal observation continues as the primary means of planning and control.

As we stated, the cost-benefit approach should guide companies in developing their accounting systems. Companies should not invest in a budgeting system unless the extra costs are exceeded by a perceived improvement in collective operating decisions.

PART ONE: FLEXIBLE BUDGETS

LEVEL 0 AND LEVEL 1 ANALYSIS: STATIC BUDGET

Objective 1

Distinguish between static budgets and flexible budgets

Consider the Webb Company. Suppose for simplicity that the company manufactures and sells a single product, a distinctive winter jacket that requires many materials, tailoring, and hand operations. The product comes in several sizes, but the company regards it as essentially a single product having one selling price. Webb bases its budget on last year's performance, general economic conditions, and its expected share of predicted industry sales. Webb makes jackets to order, so that all jackets are manufactured and sold in the same period.

Exhibit 7-1 begins the variance analysis covered throughout this and the next chapter.[1] The top of Exhibit 7-1 presents a *Level 0* analysis of variances. Note that the variance is simply the result of subtracting the budgeted amount from the actual amount. In our example, the variance is $237,000 ($25,000 − $262,000). We do not use a minus sign to denote a negative amount. Instead, a *U* signals an unfavorable variance. An *F* after an amount identifies a favorable variance.

Most managers would not be satisfied with such a skimpy report. Knowing only that a relatively large unfavorable variance exists ($237,000, or 90.5% of budgeted operating income) is not enough. Instead, managers would seek at least the *Level 1* analysis shown in Exhibit 7-1. That is, actual results are compared with the static (master) budget, line by line, for revenue, variable costs, contribution margin, fixed costs, and operating income. If desired, a detailed analysis could be conducted for subclassifications of variable and fixed costs.

The *Level 1* analysis indicates that the *Level 0* analysis stops too soon. The total unfavorable variance of $237,000 in operating income may result from many factors. The *Level 0* analysis does not tell management what those factors are.

Percentages highlight relationships among amounts, as is illustrated by both the *Level 0* and the *Level 1* analysis. In this instance, only 10,000 units were sold instead of the original target of 12,000. This shortfall in sales was not accompanied by a corresponding fall in variable costs, which decreased by only $68,000 and amounted to 60.5% of sales instead of the 55.0% budgeted.[2] Thus, the contribution-margin percentage was only 39.5% instead of the budgeted 45.0%.

---

[1] J. Shank and N. Churchill, "Variance Analysis: A Management-Oriented Approach," *Accounting Review*, LII, No. 4, pp. 950–57. In practice, there are wide varieties of cost systems and variance analyses in both manufacturing and nonmanufacturing and in large and small organizations. For a description of flexible budgeting and variance analysis in a service industry, see H. Sprohge and J. Talboth, "New Applications for Variance Analysis," *Journal of Accountancy*, April 1989, pp. 137–41, which discusses a small thoroughbred boarding farm.

[2] Managers and others tend to watch these percentages regularly. Comparisons are made from period to period and from company to company. For example, consider the U.S. airline industry. In 1989, the labor cost per available seat mile was 3.94 cents for Delta and 2.12 cents for Southwest Airlines.

EXHIBIT 7-1
Webb Company
Overview of Variance Analysis
For the Month Ended April 30, 19_1

Level 0 Analysis		%
Actual operating income	$ 25,000	9.5
Budgeted operating income	262,000	100.0
Static-budget variance of operating income	$237,000 U	90.5

Level 1 Analysis	**(1) Actual Results**	**%**	**(2) Static (Master) Budget**	**%**	**(3) (1)–(2) Static-Budget Variances**
Units sold	10,000	—	12,000	—	2,000 U
Revenue (sales)	$1,850,000	100.0	$2,160,000	100.0	$310,000 U
Variable costs	1,120,000	60.5	1,188,000	55.0	68,000 F
Contribution margin	730,000	39.5	972,000	45.0	242,000 U
Fixed costs	705,000	38.1	710,000	32.9	5,000 F
Operating income	$ 25,000	1.4	$ 262,000	12.1	$237,000 U

$237,000 U
Total static-budget variances*

F = Favorable effect on operating income; U = Unfavorable effect on operating income.
*The static-budget variance shown here is the variance in operating income. The variance can be subdivided, line by line, as shown in Column 3.

The *Level 1* analysis uses the amounts originally formulated in the master budget for April. We say that the master budget is *static* because its amounts don't move; it has a single planned volume level. By definition a **static budget** is not adjusted or altered after it is drawn up, ***regardless of changes in volume, cost drivers, or other conditions during the budget period.***

The combination of percentage analyses and absolute dollar amounts of variances shown in *Level 1* tells managers much more about underlying performance than the *Level 0* analysis. But most managers would consider *Level 1* as not being sufficiently informative.

LEVEL 2 ANALYSIS: FLEXIBLE BUDGET

The chief executive of the Webb Company would probably want a flexible budget to help provide an explanatory trail between the static budget and the actual results. A **flexible budget** (also called a **variable budget**) is a budget that is adjusted for changes in the unit level of the cost (or revenue) driver. The flexible budget is based on a knowledge of how revenue and costs should behave over a range of the driver. In our Webb Company example, the cost driver is sales volume (number of units sold). Exhibit 7-2 shows how the flexible-budget data might appear for the Webb Company for a relevant range of 10,000 to 14,000 units sold.

Because actual sales will probably differ from budgeted sales in the master budget, a flexible budget is better simply because it presents more projections of sales and so has a better chance of "getting it right." The flexible budget provides the data for studying patterns of behaviors of revenues and costs. These patterns may

EXHIBIT 7-2
Webb Company
Flexible-Budget Data
Relevant Range of Product Volume, in Units
For the Month Ended April 30, 19_1

	Budgeted Amount Per Unit	Various Levels of Volume (Units Sold)		
Units sold		10,000	12,000	14,000
Revenue (sales)	$180	$1,800,000	$2,160,000	$2,520,000
Variable costs:				
Direct materials	60	600,000	720,000	840,000
Direct labor	16	160,000	192,000	224,000
Variable factory overhead	12	120,000	144,000	168,000
Variable manufacturing costs	88	880,000	1,056,000	1,232,000
Variable marketing and administrative costs*	11	110,000	132,000	154,000
Total variable costs	99	990,000	1,188,000	1,386,000
Contribution margin	$ 81	810,000	972,000	1,134,000
Fixed costs:				
Manufacturing†		276,000	276,000	276,000
Marketing and administrative‡		434,000	434,000	434,000
Total fixed costs		710,000	710,000	710,000
Total costs		1,700,000	1,898,000	2,096,000
Operating income		$ 100,000	$ 262,000	$ 424,000

*Examples: sales commissions, shipping, photocopying.
†Examples: supervision, depreciation, insurance.
‡Examples: advertising, supervision, accounting.

be of interest to management, but no pattern can emerge from a single-column static budget.

The costs in Exhibit 7-2 can be graphed as in Exhibit 7-3. These exhibits are based on a study of cost behavior patterns that can be expressed as a formula:

$$\text{Total costs per month} = \$710{,}000 + (\$99 \times \text{Units of product sold})$$

Note especially that the graph in Exhibit 7-3 can be transformed to become the familiar cost-volume-profit graph by drawing the sales line, which is not shown in the exhibit. Thus, in its fullest form the flexible budget is an expression of cost-volume-profit relationships, already described in Chapter 3.

Flexible budgets can be useful either before or after the period in question. They can help managers choose a level of operations for planning sales volume. They can also be helpful at the end of the period when managers are trying to analyze actual results. How? As we shall now see, managers can compare actual results against a flexible budget tailored to the volume achieved rather than the original volume in the static budget.

The flexible budget prepared at the end of the period is the budget that *would have been* formulated if all the managers were perfect forecasters of volume. In our example, if the managers could have foreseen that actual units sold would be 10,000 units, a static (master) budget would have been geared to that level.

EXHIBIT 7-3
Webb Company
Graph of Flexible Budget of Costs

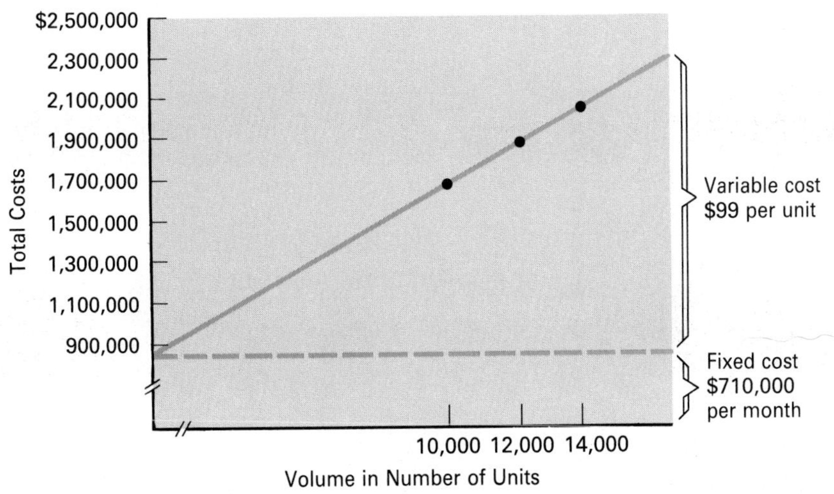

The flexible budget is the path to the more penetrating *Level 2* variance analysis in Exhibit 7-4. The $237,000 static-budget variance of operating income is split into two major categories: (a) sales volume variances and (b) flexible-budget variances. Definitions follow:

- **Sales volume variance**—the difference between the flexible-budget amounts and the static (master) budget amounts. Unit selling prices, unit variable costs, and fixed costs are held constant.
- **Flexible-budget variance**—the difference between actual results and the flexible-budget amounts for the actual output achieved.

Objective 2

Use the flexible-budget approach to compute sales volume variances and flexible-budget variances

Sales Volume Variances

Exhibit 7-4 clearly shows how the variances due to the level of *sales volume* ($162,000 U) are unaffected by any changes in selling prices, unit variable costs, and fixed costs. Why? Because both the flexible budget and the static budget are constructed using the same budgeted prices and budgeted costs. Thus, the final three columns in Exhibit 7-4 focus on effects of sales volume changes only.

The total of the sales volume variances in *Level 2* tells managers that operating income would be $162,000 lower than originally budgeted ($100,000 instead of $262,000.) Why? Solely because of the 2,000-unit decrease in sales volume. The tabulation in Exhibit 7-4, column 4, shows that the underachievement of sales by 2,000 units and $360,000 would decrease contribution margin by $162,000 and so decrease operating income by $162,000.

Level 2 information helps management compute sales volume variance of operating income. The key is the budgeted unit contribution margin. Exhibit 7-2 shows that the budgeted unit contribution margin for our example is $180 − $99 = $81. We use the $81 amount in the following formula:

$$\begin{aligned}\text{Sales volume variance of operating income} &= \left(\text{Flexible-budget units} - \text{Static-budget units}\right) \times \left(\text{Budgeted unit contribution margin}\right)\\ &= (10{,}000 - 12{,}000) \times \$81\\ &= \$162{,}000\text{ U}\end{aligned}$$

EXHIBIT 7-4 *(Place a clip on this page for easy reference.)*

Webb Company
Summary of Performance
For the Month Ended April 30, 19_1

Level 2 Analysis	(1) Actual Results	(2) (1)–(3) Flexible-Budget Variances	(3) Flexible Budget*	(4) (3)–(5) Sales Volume Variances	(5) Static (Master) Budget
Units sold	10,000	—	10,000	2,000 U	12,000
Revenue (sales)	$1,850,000	$ 50,000 F	$1,800,000	$360,000 U	$2,160,000
Variable costs	1,120,000	130,000 U	990,000	198,000 F	1,188,000
Contribution margin	730,000	80,000 U	810,000	162,000 U	972,000
Fixed costs	705,000	5,000 F	710,000	—	710,000
Operating income	$ 25,000	$ 75,000 U	$ 100,000	$162,000 U	$ 262,000

$75,000 U
Total flexible-budget variances

$162,000 U
Total sales volume variances

$237,000 U
Total static-budget variances

*Dollar amounts are computed using the formula for the flexible budget: 10,000 units × $180, $99, and $81, for sales, variable costs, and contribution margin, respectively. Fixed costs are budgeted at $710,000 per month.

For now, we stop at *Level 2* for the sales volume variances. These variances can be analyzed in much more detail. Chapter 26 explores how variances can be compiled for changes in sales mix, market share, and market size.

Who has responsibility for the sales volume variance? Fluctuations in sales are attributable to many factors, but the executive in charge of marketing is usually in the best position to explain why the volume achieved differs from the volume in the static budget.

Flexible-Budget Variances

The first three columns of Exhibit 7-4 compare the actual results (that is, actual total revenue and actual total costs) with the flexible-budget amounts. The flexible-budget variances are the differences between columns 1 and 3. The following summary explanation is offered by *Level 2* analysis:

$$\begin{aligned}\text{Flexible-budget variance of operating income} &= \left(\text{Actual operating income}\right) - \left(\text{Flexible-budget operating income}\right)\\ &= \$25{,}000 - \$100{,}000\\ &= \$75{,}000\ \text{U}\end{aligned}$$

Most managers are not satisfied with such a sparse summary. They want at least a line-by-line listing of various revenues and costs and their flexible-budget variances.

The static-budget variance of revenue for a single-product firm has two causes: volume and price. Sales-volume variances have already been discussed. A **price variance** is defined as the difference between actual unit prices and budgeted unit prices multiplied by the actual quantity of goods or services in question (that is, the actual quantity sold, purchased, or used.) The flexible-budget revenue variance is wholly explained by changes in unit selling prices. Suppose the unit selling price was raised during the period. The *Level 2* analysis in Exhibit 7-4 indicates that selling prices were raised by an average \$5 per unit (from \$180 to \$185, that is, \$1,850,000 ÷ 10,000 units = \$185) after the static (master) budget was formulated:

$$\text{Price variance} = \left(\begin{array}{c}\text{Actual}\\ \text{unit price}\end{array} - \begin{array}{c}\text{Budgeted}\\ \text{unit price}\end{array}\right) \times \left(\begin{array}{c}\text{Actual quantity}\\ \text{of units}\end{array}\right)$$

$$= (\$185 - \$180) \times 10{,}000 \text{ units}$$

$$= \$50{,}000 \text{ F, where } F \text{ means a favorable effect on operating income}$$

A *Level 2* analysis of costs can be conducted for variable and fixed costs on a line-by-line basis. If desired, even more details could be gathered. For example, the actual costs and the flexible-budgeted amounts of subclassifications of direct materials and direct labor could be analyzed.

Compare the *Level 1* analysis on page 213 with the *Level 2* analysis on page 216. Note how the variable costs had a favorable variance when a static budget was used as a basis for comparison. Exhibit 7-1, *Level 1*, shows a favorable variance of \$68,000. But the flexible budget provides a more revealing picture. Exhibits 7-4 and 7-5 pinpoint unfavorable variable-cost variances of \$130,000, which were partially offset by favorable fixed-cost variances of \$5,000. Management now has a clearer evaluation of performance.

Effectiveness and Efficiency

When evaluating performance, some managers like to distinguish between effectiveness and efficiency:

- **Effectiveness**—the degree to which a predetermined objective or target is met
- **Efficiency**—the relationship between inputs used and outputs achieved

Performance may be both effective and efficient, but either condition can occur without the other. For example, Webb Company set a static-budget goal of producing and selling 12,000 units. Only 10,000 units were produced and sold. Performance would be judged as ineffective. Whether performance was efficient is a separate question. The degree of efficiency is measured by comparing actual outputs achieved (10,000 units) with actual inputs (such as quantities of direct materials and direct labor). In Webb's case, as we will see later in this chapter, performance was also inefficient.

How do these ideas of effectiveness and efficiency relate to variances? The sales volume variance of operating income is a measure of effectiveness. The flexible-budget variance of operating income is often a measure of efficiency; however, it is also affected by changes in selling prices and unit costs. Of course, all variances are affected by the care used in formulating credible budgeted or standard amounts.

LEVEL 3 ANALYSIS: DETAILED VARIANCES

In most organizations, top management tends to be satisfied with a *Level 2* depth of variance analysis. Subordinate managers, however, tend to seek more detail. *Level 3* analysis provides additional detailed subdivisions of the *Level 2* variances. For example, marketing managers may want to know sales price variances by sales territory. In our Webb Company illustration, suppose half the unit sales were in each of two territories. The average increase in the originally budgeted \$180 selling price was \$5 per unit, but the price might have been raised in one territory but not

EXHIBIT 7-5 *(Place a clip on this page for easy reference.)*

Webb Company
Overview of Variance Analysis
For the Month Ended April 30, 19_1

Level 0 Analysis

		%
Actual operating income	$ 25,000	9.5
Budgeted operating income	262,000	100.0
Static-budget variance of operating income	$237,000 U	90.5

Level 1 Analysis

	(1) Actual Results	%	(2) Static (Master) Budget	%	(3) (1)–(2) Static-Budget Variances
Units sold	10,000	—	12,000	—	2,000 U
Revenue (sales)	$1,850,000	100.0	$2,160,000	100.0	$310,000 U
Variable costs	1,120,000	60.5	1,188,000	55.0	68,000 F
Contribution margin	730,000	39.5	972,000	45.0	242,000 U
Fixed costs	705,000	38.1	710,000	32.9	5,000 F
Operating income	$ 25,000	1.4	$ 262,000	12.1	$237,000 U

$237,000
Total static-budget variances

Level 2 Analysis

	(1) Actual Results	(2) (1)–(3) Flexible-Budget Variances	(3) Flexible Budget	(4) (3)–(5) Sales Volume Variances	(5) Static (Master) Budget
Units sold	10,000	—	10,000	2,000 U	12,000
Revenue (sales)	$1,850,000	$ 50,000 F*	$1,800,000	$360,000 U‡	$2,160,000
Variable costs	1,120,000	130,000 U†	990,000	198,000 F	1,188,000
Contribution margin	730,000	80,000 U	810,000	162,000 U	972,000
Fixed costs	705,000	5,000 F	710,000	—	710,000
Operating income	$ 25,000	$ 75,000 U	$ 100,000	$162,000 U	$ 262,000

$75,000 U
Total flexible-budget variances

$162,000 U
Total sales volume variances

$237,000 U
Total static-budget variances

Level 3 Analysis

***Detailed Sales Price Variances**

Eastern Territory	$50,000 F
Western Territory	—
Sales price variances	$50,000 F

†Detailed Cost Variances

Price and efficiency variances for materials, labor, and other costs (Covered in this chapter.)

‡Detailed Sales Volume Variances

Sales quantity variances
Sales mix variances
(These aspects are covered in Chapter 26.)

F = Favorable, U = Unfavorable effect on operating income.

in the other. Prices of $190 in the Western Territory and $180 in the Eastern Territory would produce an average price of $185:

Level 3 Analysis	Detailed Sales Price Variances
Eastern Territory	$50,000 F
Western Territory	—
Total sales price variances	$50,000 F

For perspective, you may wish to refer to Exhibit 7-5 occasionally. It provides an overall road map of where we have been and where we are going. *Level 0* and *Level 1* are reproductions of Exhibit 7-1. *Level 2* is a reproduction of Exhibit 7-4. We just discussed *Level 3* regarding revenues.

Please do not attach too much importance to the labels *Level 0, Level 1,* and so forth. By themselves, such labels have no universal meaning in the field of cost accounting. They merely represent increasingly detailed analysis. We will not probe beyond *Level 3* in this chapter; however, many more levels of detail could be specified. For example, sales price variances might be subdivided within territories by customer, by sales personnel, by product color, by week, and by season of the year. Chapter 26 illustrates how analysis beyond *Level 3* can be informative.

The flexible-budget variances for costs are also candidates for further analysis. The remainder of this chapter will show how a *Level 3* analysis uses standards to subdivide these variances into price and efficiency components. Two major categories of costs will be explored: direct materials and direct labor. The next chapter explains variance analysis for factory overhead.

The increasing use of computers in analyzing variances means that managers can readily examine variances with differing levels of detail. To probe a problem situation, a manager may start at, say, *Level 0* and then progressively call to a computer screen *Level 1, Level 2,* and *Level 3.* At another time, the same manager may be content with analyzing only *Levels 0* and *1.*

GENERAL APPLICABILITY OF EXPECTATIONS

Variety of Expectations

Before proceeding with our Webb illustration, we emphasize that the *Level 1, Level 2,* and *Level 3* framework provides a general picture, an overall perspective of how actual results may be analyzed in comparison to a set of expectations. These expectations may be described differently in various organizations as budgets, standards, plans, estimates, forecasts, and the like.

Even companies that do not use budgets or standards will nevertheless always have a set of expected results. For example, companies dedicated to "continual improvement" are simply using their last period's results as an implicit form of a budget or set of standards. All managers have some set of expectations that are essentially "standards" even if that label is not used. These expectations may be developed using data drawn from various sources:

- Internally generated actual costs from the most recent period.
- Actual costs of the most recent period adjusted for expected improvement.
- Internally generated *budgeted cost* (often called *standard cost*) numbers based on an analysis of an efficiently operating manufacturing facility.
- Externally generated *target cost* numbers based on an analysis of the cost structure of the leading competitor in an industry.
- Externally generated *most-efficient* plant costs for a company with multiple plants having the same operations or producing the same product. For example, the company

Yamazaki Mazak compares actual costs at its United States and United Kingdom plants against the actual costs of its Japanese plant that produces similar products.

A small but growing number of companies are experimenting with externally generated target costs (see Chapter 12) or most-efficient plant costs in their formal budgeting system. This experimentation stems from the heightened emphasis on tighter cost management as firms seek ways to remain competitive in the global marketplace.

Activities and Services

Managers do not manage products; they manage activities. If managers skillfully plan and control the activities that underlie business functions, the eventual costs of the products and services sold will be affected favorably. For example, managers try to ensure that research, product design, purchasing, manufacturing, marketing, distribution, and customer-service activities are coordinated and managed effectively and efficiently. Note that various manufacturing activities are only one set among many activities that merit attention.

Consider as an activity customer-service calls made to maintain elevators. Managers, perhaps in conjunction with industrial engineers, might study the steps in the activity, develop some standard times for each step, decide on a major cost driver or drivers, and compute how total cost will change in relation to changes in the levels of the cost drivers. The budgeted cost for the activity would be compiled based on the expected costs and the expected volume of the cost drivers. Suppose two drivers are chosen: number of service calls and service time per call. Performance would be evaluated for a given week (using assumed figures):

	Actual Costs	Flexible Budget	Flexible-Budget Variance
Customer service	$51,700	$43,500*	$8,200 U

*$20,000 per week + $30 per call + $40 per service-call hour, or
$20,000 + $30 (250 service calls) + $40 (400 hours), or
$20,000 + $7,500 + $16,000 = $43,500

The $8,200 variance might prompt further investigation. It provides managers with an overview of the financial effects of a particular activity. However, keep in mind that hour-to-hour and day-to-day control will depend on nonfinancial measures such as a comparison of the number and nature of customer-service calls actually made per hour or day against the budgeted numbers.

Nonfinancial (Physical) Measures

Objective 3

Describe how budgets and nonfinancial standards are used to control day-to-day activities

For the customer-service call activity—as well as for various research and development, product design, engineering, manufacturing, marketing, and distribution activities—planning, controlling, and evaluating performance can be achieved using budgets, standards, and variances. For control purposes, the costs of these activities do not have to be applied to products, and variances do not have to be reported by product or product line.

Each activity may have its individual cost drivers, which are nearly always physical items, such as numbers of phone calls, machine setups, parts per product, and product defects. Monitoring outputs in relation to the cost drivers and conducting personal observations usually enhance control.

Consider how the chief executive of a major manufacturer of aluminum has encouraged managers to base more of their decision making on nonfinancial measures. A news story reported:

> At his request, Alcoa people gathered tons of data about productivity, safety, quality—some 450 nonfinancial measures in all—plus similar information about competitors to give his managers benchmarks.

Let's return to our Webb illustration. The cutting room provides an example of activities in the manufacturing of jackets. Fabric is laid out on long tables and cut using either electric cutters or manual shears. Then the pieces are matched together and assembled. Standards are set for various cutting-room activities. Control is exercised by comparing the total actual outputs of assemblies of cut materials against the expected usage of materials and labor time. Note that such nonfinancial measures are useful even if the data-collection system does not track actual usage to each jacket produced. Indeed, for hour-to-hour or day-to-day control of activities, only physical measures of usage (rather than costs) are usually necessary.

Budgets and actual results show the financial effects of how cost drivers are managed. Nonfinancial (physical) measures are often the primary means for exercising control on a continuing basis. ***Still, the financial reporting system provides the best available way for conveying the economic impacts of the diverse physical activities throughout an organization.***

PROBLEM FOR SELF-STUDY

PROBLEM

Refer to Exhibit 7-4, p. 216. Suppose the actual results pertained to 9,800 units, not 10,000 units. Reconstruct the exhibit.

SOLUTION

Exhibit 7-6 reconstructs Exhibit 7-4. The key change is in the flexible budget. The sales and variable-cost numbers in column 3 are 98% of those in Exhibit 7-4, column 3 (9,800 divided by the 10,000 units in Exhibit 7-4 is 98%). The fixed costs are unchanged.

EXHIBIT 7-6

Webb Company
Summary of Performance
For the Month Ended April 30, 19_1

Level 2 Analysis	(1) Actual Results	(2) (1)–(3) Flexible-Budget Variances	(3) Flexible Budget	(4) (3)–(5) Sales Volume Variances	(5) Static (Master) Budget
Units sold	9,800	—	9,800	2,200	12,000
Revenue (sales)	$1,850,000	$ 86,000 F	$1,764,000	$396,000 U	$2,160,000
Variable costs	1,120,000	149,800 U	970,200	217,800 F	1,188,000
Contribution margin	730,000	63,800 U	793,800	178,200 U	972,000
Fixed costs	705,000	5,000 F	710,000	—	710,000
Operating income	$ 25,000	$ 58,800 U	$ 83,800	$178,200 U	$ 262,000

$58,800 U — Total flexible-budget variances

$178,200 U — Total sales volume variances

$237,000 U — Total static-budget variances

PART TWO: STANDARD COSTS

STANDARD COSTS FOR MATERIALS AND LABOR

Objective 4

Distinguish between budget amounts and standard amounts

Distinction Between Budgets and Standards

Many organizations have budgeting systems that are confined to a static budget. That is, a master budget is prepared and performance is not evaluated beyond the *Level 1* analysis in Exhibit 7-5, p. 218. However, when an organization invests in a flexible budgeting system, it typically develops standards for major costs such as direct materials and direct labor.

Standard costs are carefully predetermined costs that are usually expressed on a per-unit basis. They are costs that should be attained. Standard costs help managers to build budgets, gauge performance, obtain product costs, and save record-keeping costs. Standard costs are the building blocks of a flexible budgeting and feedback system.

A set of standards outlines how a task should be accomplished in nonfinancial terms (minutes, board feet) and how much it should cost. As work is being done, actual costs incurred are compared with standard costs for various tasks or activities to reveal variances. This feedback helps discover better ways of adhering to standards, of altering standards, and of accomplishing objectives.

What is the difference between a standard amount and a budgeted amount? If standards are attainable, as they are assumed to be in this book, there is no conceptual difference. The term *standard cost* usually refers to the cost of a *single finished unit* of output. In contrast, *budgeted cost* usually refers to a *total* amount. Suppose the Webb Company's standard for direct materials is two square yards (often referred to simply as yards) of input at \$30 per yard, or 2 × \$30 = \$60 per unit of output.

The standard cost in Exhibit 7-2, page 214, shows:

	Budgeted Amount Per Unit	Various Levels of Volume (Units Sold)		
Units sold	1	10,000	12,000	14,000
Direct materials	\$60	\$600,000	\$720,000	\$840,000

The standard cost is \$60 per finished unit. The budgeted cost is \$600,000 if 10,000 units are to be produced. Think of a standard as a budget for a single unit. In many companies, the terms *budgeted performance* and *standard performance* are used loosely and often interchangeably as the expectations for an accounting period.

Cost-benefit Approach to Variances

Various surveys of management accounting practices have found that standard costing is widespread. For example, one large survey showed that 85% of manufacturing firms use standard costs.[3] The main purposes of standard costs ranked (1 down to 4) in terms of importance[4] are:

[3]The results of twenty-two surveys are summarized in C. Chow, M. Shields, and A. Wong-Boren, "A Compilation of Recent Surveys and Company-Specific Descriptions of Management Accounting Practices," *Journal of Accounting Education*, (1988), 183–207.

[4]S. Inoue, "Comparative Studies of Recent Development of Cost Management Problems in U.S.A., U.K., Canada, and Japan," Research Paper No. 29, Kagawa University, 1988.

	United States	Canada	Japan	United Kingdom
Cost management	1	1	1	2
Price-making and price-policy	2	3	2	1
Budgetary planning and control	3	2	3	3
Financial statement preparation	4	4	4	4

The uses of standard costs underscore the cost-benefit approach to management accounting systems. Some costs may receive detailed analysis; others, almost no analysis. For example, the flexible-budget variances for direct materials and direct labor are often subdivided into detailed variances, but individual indirect costs may not be. In addition, managers may decide to analyze materials on a daily basis, labor on a weekly basis, and indirect costs on a monthly basis.

In some companies, the control of direct labor is regarded as a key to overall cost control. Why? Because direct labor is a significant cost driver, a cause of many supporting overhead costs. These companies track direct labor in detail as a separate category. Webb Company, like most clothing manufacturers—and most service companies—regards direct labor as worthy of close monitoring. In other companies, particularly those with automated operations, direct labor is a minor cost. No labor has the status of a direct cost. Instead, all types of labor are included as components of indirect costs.

Companies using standard costs frequently track variances for control purposes department-by-department, or activity-by-activity, or operation-by-operation. For example, Webb managers have determined how many sleeves should be sewn in an hour. Some companies also track variances product-by-product or job-by-job—if they believe that the resulting variances are worth calculating from a cost-benefit viewpoint. In any event, the major point here is that standards often help in evaluating the performance of responsibility centers, whether these centers are cutting, assembling, painting, receiving, shipping, cleaning, testing, mailing, baggage handling, clearing checks, or any other activity.

Standard Costs Allowed

Let us continue with our Webb example. Suppose Webb uses over 100 distinct operations to manufacture a quality jacket. The jacket is an assembly of various parts, including sleeves, backs, fronts, collars, and pockets, and comes in various styles, sizes, colors, and so forth. To emphasize the broad concepts under discussion, however, we regard all jackets as a single product. Also, we develop *overall* variances for direct materials, direct labor, and indirect costs. These variances might be subdivided by various activities within the cutting room, such as spreading the cloth and marking the cloth for cutting. Regardless of how detailed the analysis, the underlying concepts used in examining standard costs are the same.

The following are standards for the manufacture of Webb's jackets:

Direct materials	Two square yards of input allowed @ $30 per yard = $60 per unit of output
Direct labor	.8 hour of input allowed @ $20 per hour = $16 per unit of output

In addition to the preceding data, the following data were compiled regarding *actual* performance:

Actual units of output	10,000	Direct labor costs	$171,600
Direct material costs	$701,800	Hours of input	8,800
Yards of input purchased and used	24,200	Labor price per hour	$19.50
Material price per yard	$29		

Our analysis begins with the flexible-budget variances:

	Actual Costs Incurred	Flexible Budget	Flexible-Budget Variance
Direct materials	$701,800	$600,000	$101,800 U
Direct labor	171,600	160,000	11,600 U

The flexible-budget totals for variable costs such as direct materials and direct labor are also sometimes expressed as *total standard costs allowed.* They are computed as follows:

$$\textbf{Units of actual output} \times \textbf{Input allowed per unit of output} \times \textbf{Standard unit price of input} = \textbf{Total standard cost allowed}$$

Direct materials: 10,000 units of actual output × 2 yards × $30 per yard = $600,000

Direct labor: 10,000 units of actual output × .8 hour × $20 per hour = $160,000

Units of output are measures of complete units of products or services generated by the activity or department under discussion. ***Before proceeding, note especially that the flexible-budget amounts (that is, the total standard costs allowed) are keyed to an initial question: What was the* output *achieved?*** To answer that question, some common measure of diverse outputs is often used, such as standard direct labor hours allowed, machine hours, or equivalent units of materials, products, or services.

Definitions of Price and Efficiency Variances

Price variance was defined and discussed earlier in conjunction with revenue. Its definition also holds with respect to costs. The other component of the flexible-budget variance is the efficiency variance. Definitions follow:

- **Price variance**—the difference between actual unit prices and budgeted unit prices multiplied by the actual quantity of goods or services in question (for example, sold, purchased, or used)
- **Efficiency variance**—the difference between the quantity of actual inputs used (such as yards of materials) and the quantity of inputs that *should have been used* (the flexible budget for any quantity of *units of output achieved*), multiplied by the budgeted price

Giving a manager responsibility for the entire flexible-budget variance is often regarded as "unfair" because the manager usually has far more control over the efficiency variance than over the price variance. Moreover, one manager may be in control of efficiency variances and another manager in control of price variances.

Price variances are computed not only for their own sake but to obtain a sharper focus on efficiency. In this way, efficiency can be measured by holding unit prices constant. Thus, managers' judgments about efficiency are unaffected by price changes. Efficiency variances have an important underlying assumption: ***All unit prices are standard prices*** that are sometimes called *budgeted, estimated,* or *predetermined prices.*

Objective 5

Compute the price and efficiency variances for direct materials and direct labor

Price and Efficiency Variance Computations

When calculating the price variance, hold inputs constant at the actual inputs purchased. When calculating efficiency variances, hold price constant at the standard unit price of inputs.

Consider the price variance:

$$\textbf{Price variance} = \left(\textbf{Actual unit price} - \textbf{Standard unit price}\right) \times \textbf{Actual inputs purchased}$$

Direct materials:

$$= (\$29 - \$30) \times 24{,}200 \text{ pounds}$$
$$= \$24{,}200 \text{ Favorable (F)}$$

Direct labor:

$$= (\$19.50 - \$20.00) \times 8{,}800 \text{ hours}$$
$$= \$4{,}400 \text{ Favorable (F)}$$

Consider the efficiency variance: No measurement of efficiency variances can occur without the measurement of *outputs* as well as *inputs*. For any given level of output (that is, actual units produced), the efficiency variance is the difference between the inputs that should have been used and the inputs that were actually used—holding unit input prices constant at the standard unit price:[5]

$$\textbf{Efficiency variance} = \left(\textbf{Inputs actually used} - \textbf{Inputs that should have been used}\right) \times \textbf{Standard unit price of inputs}$$

$$\text{Efficiency variance} = \left(\text{Actual yards or hours used} - \text{Standard yards or hours allowed for output achieved}\right) \times \text{Standard unit price of inputs}$$

Direct materials:

$$= [24{,}200 - (10{,}000 \text{ units} \times 2 \text{ yards})] \times \$30$$
$$= (24{,}200 - 20{,}000) \times \$30$$
$$= \$126{,}000 \text{ U}$$

Direct labor:

$$= [8{,}800 - (10{,}000 \text{ units} \times .8 \text{ hour})] \times \$20$$
$$= [8{,}800 - 8{,}000] \times \$20$$
$$= \$16{,}000 \text{ U}$$

Measuring and Expressing Output

Exhibit 7-7 is a graphical representation of the analysis of direct labor. The cost function is linear, sloping upward at a standard price of $20.00 per hour. We can read the total standard costs that would be allowed for any given quantity of output. Note that quantity is expressed in hours rather than in physical units of output. *This is commonly done because most departments have an assortment of products; hours become a useful common denominator for measuring the total level of all production.*

In highly automated factories, direct labor may be lower than 5% of total manufacturing costs. In these situations, other measures are used for measuring total production or outputs of activity areas. Examples of such measures include total production time and the weighting of various types of products by their relative complexity to arrive at equivalent units of production.

Standard-cost systems frequently do not express output as, say, 10,000 jackets. Instead output is expressed as 10,000 × 0.8 hours = 8,000 **standard hours allowed** (also called **standard hours earned**, **standard hours worked**, or, most accurately, **standard hours of input allowed for good output produced**).

---

[5]Algebraically, these variances are:

$$V_p = (AP - SP) \times AQ$$

and

$$V_e = (AQ - SQ) \times SP$$

V_p = price variance, V_e = efficiency variance, AP = actual unit price of inputs, SP = standard unit price of inputs, AQ = actual quantity of inputs, SQ = standard quantity of inputs allowed for actual output.

EXHIBIT 7-7 *(Place a clip on this page for easy reference.)*
Analysis of Direct Labor

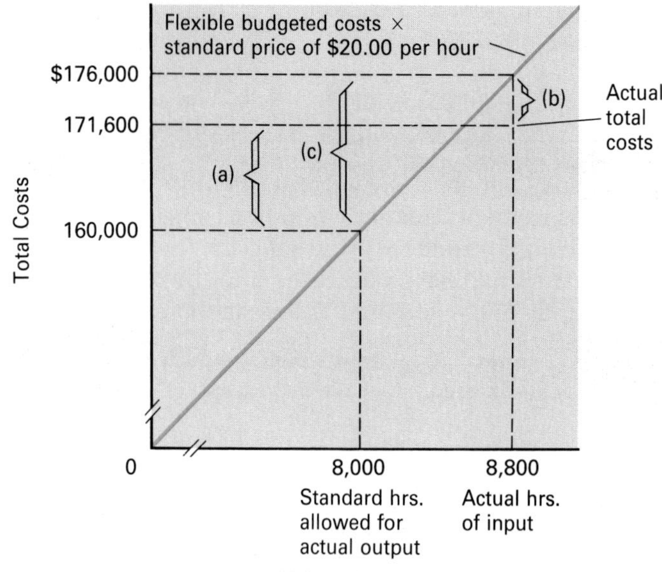

(a) Flexible-budget variance = Actual costs − Flexible budget costs
= $11,600 U, as subdivided in (b) and (c)

(b) Price variance = Difference in actual unit price and standard unit price × Actual inputs
= $4,400 F

(c) Efficiency variance = Difference in actual inputs and standard inputs allowed × Standard price
= $16,000 U

The graphic relationships can also be analyzed as follows:

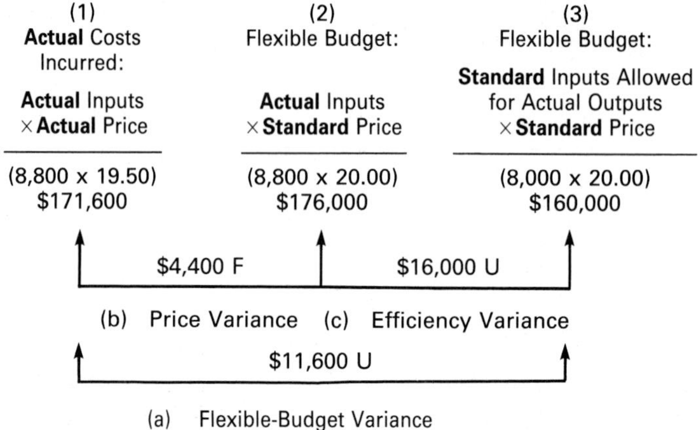

U = Unfavorable; F = Favorable

Standard hours allowed are the number of standard hours that should have been used to obtain any given quantity of output (that is, actual goods produced or actual outputs achieved).

Our extended Webb illustration on flexible budgeting and standards focuses on volume as measured either by units of product or by direct labor hours allowed as a principal cost driver. Nevertheless, as mentioned before, recognize that direct labor hours is only one of many cost drivers. Other cost drivers for various activity areas may be the number of purchase orders issued, machine setups, parts in one finished unit, and lines typed. Flexible budgets and standards can be developed for activities, and variance analysis can be conducted accordingly. The second problem for self-study (p. 237) illustrates the use of variance analysis for labor costs in a medical laboratory that conducts tests of blood samples.

As already explained, times are not necessarily measured in direct labor hours. Another measure is production throughput time (total time from start of production to finish) either in a particular activity area or in a larger set of activities, including the entire factory. Examples of other time measures include setup time and product-testing time.

Relationship of Inputs and Outputs to Flexible Budget

We earlier defined a *flexible budget* as a budget that is adjusted for changes in the unit level of the cost (or revenue) driver. The idea of a flexible budget is completely general. The essential ingredient is the budget formula (for example, outputs × standard price), which may be used to construct a total budget for any given level of a cost (or revenue) driver. In our Webb illustration concerning standards, the cost driver so far has been the number of units produced. That is, the flexible-budget amounts (the total standard costs allowed) are keyed to the actual outputs achieved.

Exhibit 7-7, Column 2, shows that a flexible budget may also be tied to actual inputs, not actual outputs achieved. The Column 2 version of a flexible budget is not as informative as the Column 3 version. Why? Because a manager wants to pinpoint both price and efficiency variances, not just the price variances that would be revealed by comparing Columns 1 and 2 alone. Standard inputs allowed are superior to actual inputs used as a basis for constructing a flexible budget. If the flexible budget is based on actual inputs, then the higher the inefficiency, the higher the flexible-budget allowance. In our example, a department head should not enjoy a more generous flexible budget due to inefficiency simply because the budget was based on actual inputs (direct labor hours used) instead of standard inputs allowed (direct labor hours allowed).

In sum, Columns 2 and 3 are both labeled as flexible budget to remind us of the basic nature and versatility of this budget tool. However, the term flexible-budget variance will always refer to the difference between Columns 1 and 3—that is, between actual costs incurred and a flexible budget based on outputs achieved.

Substitute Terminology

To underscore their basic similarity, the two primary variances for both direct materials and direct labor are designated here as price and efficiency variances. In practice, however, be alert for a variety of terms that are sometimes used to designate these variances. Substitute terminology includes:

- *For price variance:* **Rate variance** (popular term with respect to direct labor)
- *For efficiency variance:* **Quantity variance** or **usage variance** (popular term with respect to direct materials)

In standard-cost systems, the concept of flexible budget, in contrast to static budget, is a key to the analysis of variances. ***For brevity, we frequently use the term budget variance instead of flexible-budget variance. For example, Exhibit 7-7 might have used budget variance instead of flexible-budget variance.***

IMPACT OF INVENTORIES

In our Webb Company illustration, we assumed two points:

1. There were no work in process or finished goods inventories at the beginning or end of the accounting period. All units were produced and sold in the same accounting period.
2. There were no direct materials inventories. All direct materials were purchased and used in the same accounting period.

Suppose production and sales are unequal. The sales volume variance is the difference between the master budget and the flexible budget for *the number of units sold.* Chapter 9 provides details about the effects of inventories on a standard-cost system.

Suppose the quantity of direct materials purchased and the quantity used are unequal. Generally, managers desire rapid feedback for control. They want to pinpoint variances as early as is practical. In the case of material prices, that earliest control point is when materials are purchased, not when they are used. Consequently, direct material price variances are usually based on the quantities purchased, and direct material efficiency variances are always based on the quantities used.[6]

Reconsider our illustration by assuming that the purchasing department acquired 25,000 yards during the month and that the cutting department used only 24,200 yards. The price variance would be 25,000 yards × ($29 − $30) = $25,000, favorable. Note too that actual costs do not have to be maintained as the units flow through the inventory accounts. As long as standards are unchanged, determining actual prices by cost-flow assumptions such as last-in, first-out or first-in, first-out is unnecessary.

GENERAL-LEDGER ENTRIES

Illustrative Entries

Objective 6

Prepare journal entries for a typical standard-costing system for direct materials and direct labor

General-ledger accounting for direct materials and direct labor in earlier chapters has been straightforward. We have used two major inventory accounts:

Materials Control	
Actual costs of direct materials purchased	

Work in Process Control	
Actual costs of direct materials and direct labor used	

The general-ledger procedures in standard-cost systems introduce separate accounts for variances. General-ledger entries in standard-cost systems are usually *monthly* summaries of detailed variance analyses computed from day to day.

Using the data in our earlier illustration, but assuming the 25,000-yard purchases just described, the actual quantities of direct materials purchased are charged to Materials Control at *standard prices.* As Entry 1a shows, this approach isolates (that is, separately identifies) price variances for direct materials as early as is feasible:

[6] M. Laudeman and F. Schaeberle, "The Cost Accounting Practices of Firms Using Standard Costs," *Cost and Management,* July–August 1985, pp. 21–25, indicated that 88% of their surveyed firms base direct material price variances on quantities purchased.

1a.	Materials control (25,000 × $30)	750,000	
	Direct material price variance (25,000 yds. × $1)		25,000
	Accounts payable (25,000 × $29)		725,000
	To record direct materials purchased.		

The direct material efficiency variance is isolated when standard quantities allowed times standard unit prices are debited to Work in Process Control:

1b.	Work in process control (20,000 yds. × $30)	600,000	
	Direct material efficiency variance (4,200 yds. × $30)	126,000	
	Materials control (24,200 yds. × $30)		726,000
	To record direct materials used.		

Note that unfavorable variances are always debits and favorable variances are always credits.

Payroll liability is accounted for at *actual* wage rates. Work in Process Control is charged at standard quantities allowed times *standard rates,* and the direct labor variances are recognized:

2.	Work in process control	160,000	
	Direct labor efficiency variance	16,000	
	Direct labor price variance		4,400
	Payroll liability		171,600
	To record liability for direct labor costs.		

T-accounts would appear as follows:

Materials Control

Debit	Credit
1a. Actual quantity purchased × standard unit price, 750,000	1b. Actual quantity used × standard unit price, 726,000

Direct Material Price Variance

Debit	Credit
	1a. Actual quantity purchased × difference in unit price, 25,000

Work in Process Control

Debit	Credit
Standard quantity allowed × standard unit price 1b. 600,000 2. 160,000	

Direct Material Efficiency Variance

Debit	Credit
1b. Difference in quantity used and allowed × standard unit price, 126,000	

Direct Labor Price Variance

Debit	Credit
	2. Actual quantity purchased × difference in unit price, 4,400

Direct Labor Efficiency Variance

Debit	Credit
2. Difference in quantity used and allowed × standard unit price, 16,000	

The major advantage of this system is its stress on the control feature of standard costs. All variances are isolated as early as possible.

Disposition of Variances

Like the disposition of underapplied or overapplied overhead discussed in Chapter 4, price and efficiency variances are either written off immediately to cost of goods sold or prorated among the inventories and cost of goods sold. A full discussion can be found in Chapter 9, pages 306–8.

Costs of Data Collection

At first glance, it might appear that standard-cost systems would always be more costly to operate than other systems. Obviously, a startup investment must be made to develop the standards. But the ongoing costs can be lower than in so-called actual-cost systems. For example, it is more economical to simply carry all inventories at standard unit prices. Until a set of standards is changed, this system avoids the difficulty, extra data-collection costs, and confusion of making cost-flow assumptions such as first-in, first-out, or last-in, first-out.

An ongoing standard-cost system must be updated for changes in methods, products, and costs. These changes affect not only the accounting department but typically the engineering department, the purchasing department, and the department being monitored, whether in manufacturing, marketing, or some other function.

Of course, standard-cost systems will be more costly if actual direct material costs and actual direct labor costs continue to be traced to individual products. The costs of recordkeeping will increase.

Many standard-cost systems have been adopted with lowering recordkeeping costs as a major purpose. A standard-cost system can be designed with these features:

1. No "actual" costs of materials, work in process, and finished goods are traced to batches of product on a day-to-day basis. Only records of physical counts are kept.
2. Actual consumption of direct materials and direct labor can be totaled by activity, operation, or department for, say, a month without tracing the consumption to products. Feedback on performance can be based on these actual totals of input compared with the total standard inputs allowed for the actual output achieved during the month. A major advantage of this procedure is the simplicity of data collection.
3. Nonfinancial (physical) standards are the bedrock for quick feedback and continuous monitoring of performance. The *financial* impacts of performance need not be accumulated continuously.

STANDARDS USED IN NONMANUFACTURING ACTIVITIES

Avoid the mistaken idea that flexible budgets, standards, and standard costs are confined to inventory costing or to manufacturing. As was shown earlier in this chapter, budgets and standards are found in all management functions and organizations. Standard times are set for repairing automobiles, typing invoices, cleaning hotel rooms, delivering parcels, and cooking hamburgers. Performance is judged accordingly. Flexible budgeting and standards are often used in accounting for these activities, although their use is not as widespread as it is in manufacturing.[7]

[7]A survey of 112 manufacturing firms indicated that 85% used standard costs. See H. Schawarzbach, "The Impact of Automation on Accounting for Indirect Costs," *Management Accounting*, December 1985. W. Cress and J. Pettijohn, in "A Survey of Budget-related Planning and Control Policies and Procedures," *Journal of Accounting Education*, 3, No. 2 (Fall 1985), p. 74, also reported that 85% of the respondents used standard costs. However, only 18% of the respondents used standards for controlling nonmanufacturing costs. A reason frequently given is the perceived difficulty of obtaining accurate input-output relationships.

Note too that standards for control of both manufacturing and nonmanufacturing departments and activities may be confined to nonfinancial measures, such as time taken to respond to a production or service request or to process a mail order. Whether these work measurements are also translated into monetary terms depends on the preferences of the managers who use the data. We discuss work measurement in Chapter 13.

The major point here is that standards may be used for control in a wide variety of organizations that do not necessarily have a full-blown, detailed standard-cost system that tracks actual direct materials and direct labor usage to the individual product level. For example, a department's performance can be judged in the aggregate by comparing total actual costs with total standard costs that are based on standard times allowed for actual outputs achieved.

STANDARDS FOR CONTROL

Perfection Standards

How demanding should standards be? Should they express perfection, or should they allow for the various factors that prevent perfect performance? The answers to these questions differ, depending on the styles and preferences of various managements.[8] Managers seek the standards that provide psychologically productive goals, goals that motivate employees toward optimal production.

Perfection standards (also called *ideal, theoretical,* or *maximum-efficiency standards*) ideally lead to the absolute minimum possible costs for a company operating under the best conceivable conditions using existing specifications and equipment. No provision is made for waste, spoilage, machine breakdowns, and the like. Although not widely used, perfection standards are increasingly being put into practice because they harmonize with recent management thrusts toward continual improvement of quality and operating efficiency.

Currently Attainable Standards

Currently attainable standards generate standard costs that are achievable by a specified level of effort and allow for normal spoilage, waste, and nonproductive time. The designated level of effort varies from company to company, but at least two popular interpretations of currently attainable standards exist.

Under one interpretation, management sets standards that employees regard as goals that they most probably can fulfill. The standards are "expected actuals," predictions of what will likely occur. Taken into account in setting these standards are the inefficiencies that management anticipates. Two major arguments support these "reasonable" standards:

1. "Expected actual" standards serve multiple purposes. They are the expected actual costs that many managers use for costing products, budgeting cash, and budgeting departmental performance. In contrast, perfection standards by themselves cannot be used for cash budgeting or for product costing because these standards will not be achieved. The financial planning resulting from perfection standards will be inaccurate.

---

[8]W. Cress and J. Pettijohn, in "A Survey of Budget-related Planning and Control Policies and Procedures," p. 74, report that 50% of the responding companies use expected actual (but difficult to attain) standards; 42%, standards based on average past performance; and 8%, maximum efficiency (theoretical or perfection) standards. Evidence from this and similar surveys indicates that "difficult to attain" may be a slightly challenging but not overly tight standard. For example, K. Merchant and J. Manzoni, in "The Achievability of Budget Targets in Profit Centers: A Field Study," *Accounting Review*, LXIV, No. 1 (July 1989), 539–58, report that in twelve companies budget targets are set to be achievable an average of eight or nine years out of ten.

2. The "expected actuals" motivate employees. The standards represent reasonable future performance, not fanciful ideal goals, possible but highly improbable goals, or antiquated goals. In many organizations, the "expected actuals" are based on some improvement of average past performance, and the variances should therefore be small. Consequently, unfavorable variances signal performance that requires management's attention.

Another interpretation regards currently attainable standards as tighter than the "reasonable" standards but looser than perfection. Employees regard their fulfillment as possible though unlikely, as ambitious yet perhaps attainable. Only very efficient operations can achieve these standards. Unfavorable variances tend to be larger under this interpretation than under the more "reasonable" currently attainable standards.

Continual Improvement Standard Costs

A *standard cost* is a carefully predetermined cost or a cost that should be attained. Many organizations revise their standards every six or twelve months. Between each revision, the most recently set standard is used in variance analysis.

An alternative approach is a **continual improvement standard cost**, which is a carefully predetermined cost that is successively reduced over succeeding time periods. A synonym is a **moving cost reduction standard cost**. The standard direct materials cost for Webb Company in April 19_1 is $60 per unit. Suppose Webb Company has a 1% monthly continual improvement standard. The standard costs used in variance analysis for subsequent periods would be:

Month	Prior Month's Standard	Reduction in Standard	Revised Standard
April 19_1	—	—	$60.00
May 19_1	$60.00	$0.600 (.01 × $60.00)	59.40
June 19_1	59.40	0.594 (.01 × $59.40)	58.81
July 19_1	58.81	0.588 (.01 × $58.81)	58.22

By using continual improvement standard costs, an organization signals the importance of constantly seeking ways to reduce total costs.

Expected Variances

As we mentioned, financial planning using perfection standards will be inaccurate. Perfection standards may be used for compiling performance reports, but **expected variances** are those specified in the budget for cash planning.

Consider a company that uses a perfection standard for direct materials costs of $16 per unit. Management anticipates an expected unfavorable variance of $1.60 per unit. In the master budget, the total material costs allowed would be $17.60 per unit: $16.00 plus an expected variance of $1.60. The master budget could include the following item:

Direct materials:	
Budget allowance shown on department performance report	$160,000
Expected variance appearing in cash budget	16,000
Total budget allowance for cash planning	$176,000

Developing the Standards

The standard-setting and budget-setting process in an organization often is primarily the responsibility of the line personnel directly involved. The relative tightness of the budget is the result of face-to-face discussion and bargaining between the manager and his or her immediate superior. The management accountants, the industrial engineers, and the market researchers should extend all desired technical assistance and advice, but the final decisions ordinarily should not be theirs. The line manager is the person who is supposed to accept and live with the budget or standard.

The job of the accounting department is (a) to price the physical standards—that is, to express the physical standards in monetary terms—and (b) to report operating performance in comparison with standards.

PERFORMANCE MEASUREMENT USING STANDARDS

Price variances and efficiency variances capture important aspects of performance in many organizations. For example, direct materials costs constitute over 80% of total manufacturing costs for some computer companies. Materials price variances for such companies provide information on a key aspect of management performance. Be careful, however, not to overemphasize a particular single variance.

Assume that a purchasing officer has just negotiated a deal with a favorable price variance. The deal succeeded for three reasons:

1. The purchasing officer's effective bargaining with suppliers
2. Lower-quality materials purchased
3. Higher quantities acquired than necessary in the shortrun, resulting in excessive inventories

If the purchasing officer's performance is evaluated solely on price variances, then only item 1 is considered, and the evaluation will be positive. However, items 2 and 3 will likely result in the company incurring additional costs.

The focus of performance measures is increasingly on reducing the total costs of the company as a whole. In the example of the purchasing officer, the company may ultimately lose more because of items 2 and 3 than it gains from item 1. Similarly, manufacturing costs may be deliberately increased (for instance, by higher material prices or more labor time) in order to obtain better product quality. In turn, the costs of the better product quality may be more than offset by reductions in warranty and servicing costs.

If any single performance measure (for example, a standard-cost variance, a quality standard, or a consumer rating report) receives excessive emphasis, managers tend to make decisions that maximize their own reported performance in terms of that single performance measure. Therefore, managers' actions may conflict with the organization's achieving its overall goals. This faulty perspective on performance arises because top management has designed a performance measurement and reward system that does not emphasize total organization objectives.

CONTROLLABILITY AND VARIANCES

This section explores some underlying features of controlling price and efficiency variances.

Objective 7

Identify the typical responsibilities for controlling variances in material and labor costs

Explanations of Price Variances

Material price variances are often regarded as measures of forecasting ability rather than of failure to buy at specified prices. Still, purchasing officers help control price variances by getting many quotations, buying in economical lot sizes, taking advantage of cash discounts, and selecting the most economical means of delivery.

Companies that use fewer suppliers and longer-term contracts are less likely to have significant price variances. Such contracts tend to specify unchanging prices for fixed time spans.

In most companies, because of union contracts or other predictable factors, management can foresee labor prices with much greater accuracy than it can the prices of materials. Therefore, labor price variances tend to be relatively insignificant.

Labor, unlike materials and supplies, cannot be stored for later use. The purchase and use of labor occur simultaneously. For this reason, labor price variances are usually charged to the same manager who is responsible for labor usage.

Labor price variances may be traceable to faulty predictions of the labor rates. However, the more likely causes include (1) the use of a single average standard labor price for a given activity that is, in fact, performed by individuals earning different rates because of seniority, and (2) the assignment of a worker earning, perhaps, $24 per hour to a given activity that should be assigned to a less-skilled worker earning, say, $16 per hour.

Material Efficiency

Consider a popular method for controlling materials. Exhibit 7-8 is an example of a **standard bill of materials,** which is a record of the standard quantities of materials required for manufacturing a finished unit. In many companies, when production of a product is about to begin, the standard bill becomes the materials requisition. The storeroom then issues the standard amount of materials allowed. As production occurs, any *additional* materials needed may be withdrawn from the materials storeroom only by submitting an **excess materials requisition**. This is a form necessary to obtain any materials needed in excess of the standard amount allowed for the scheduled output. The supervisor must sign the requisition, so he or she is immediately informed of excess usage. A periodic summary of excess materials requisitions provides the cumulative unfavorable efficiency variance. If performance is better than standard, some materials may be unneeded and returned. Special returned-materials forms are used to calculate favorable efficiency variances. Possible reasons for efficiency variances include quality, workmanship, choice of materials, mix of materials, and faulty standards.

EXHIBIT 7-8
Standard Bill of Materials

Assembly No. B — Description TRAY TABLE

Part Number	Number Required	Description
A 1426	4 SQ. FT.	PLASTIC SHEET - PEARL GREY
455	1/8 LB.	ADHESIVE
642	1	TABLE TOP
714	4	STEEL LEGS
961	1	NUT AND BOLT KIT

Labor Efficiency

The control of direct labor is quite important to many companies, less important to others. When deemed important, time standards are usually set for each operation or activity. The source documents for variance reports are usually some form of time ticket showing the actual time used. These tickets are analyzed, and variances are coded and classified. The classifications are almost always by responsibility center (such as, by cost center) and often by operations, activities, products, orders, and causes (machinery breakdowns, rework, faulty materials, use of nonpreferred equipment, use of nonpreferred workers, and so on). Thus, a time ticket may have one number designating departmental responsibility and another number designating the cause of the variance.

Absenteeism can dramatically affect labor efficiency. A General Motors executive has commented:

> When deer season opens, production drops from 70 to 35 cars per hour. Workers are shifted around to handle unfamiliar jobs. Quality declines. And the workers get paid overtime. . . . We have begun computerizing medical excuses signed by local doctors to determine if any definitive pattern exists among the chronic absentees.

SETUP TIME

Setup time is the time required to prepare equipment and related resources for producing a specified number of finished units or operations. Machines and accessory equipment must often be adjusted and "made ready" before a particular operation or job can commence. Managers throughout the world have recently focused on reducing setup time. For example, Toyota has slashed many setup times from hours to minutes.

How should setup time be accounted for? From a cost control viewpoint, setups are separate activities that usually have accompanying expected (budgeted or standard) times. As a minimum, setup time should be distinguished from processing or run time. Hence, setup costs should be coded differently than the run costs.

Setup costs can be important for product-costing purposes. Ordinarily, setup costs should be regarded as direct costs of a product, not as indirect manufacturing costs. Setup costs are easily traceable to an operation or a job, regardless of whether 100 pieces or 2,000 pieces are subsequently processed.

Suppose managers can choose between short and long production runs on components and finished products. Setup costs, then, can significantly affect total costs. If setup costs are regarded as a part of an indirect cost pool, longer-run components and products will probably be burdened with some of the setup costs that are properly regarded as direct costs of the shorter-run products. This impact is particularly large if direct labor is the base for applying factory overhead.

Setup costs underscore the general desirability of classifying as many costs as feasible as direct costs, not as indirect costs. Consider the case of one actual company. Its accounting system classified setup costs as factory overhead. Management had also chosen several short-production-run orders based on cost system data. A new controller doubted the accuracy of the existing product-cost numbers. The cost system was changed so that setup costs were classified as direct costs. The new cost system was used to support modification of selling prices, in some instances to a surprising degree, and modification of the product line. Moreover, future short runs and unique orders were accepted only after considering the direct setup cost.

WHEN TO INVESTIGATE VARIANCES

When should variances be investigated? Frequently the answer is based on subjective judgments, or rules of thumb. ***The most troublesome aspect of feedback is deciding when a variance (either favorable or unfavorable) is significant enough to warrant management's attention.*** For some items, a small deviation may prompt follow-up. For other items, a minimum dollar amount or a certain percentage of deviations from budget may prompt investigations. Of course, a 4% variance in a $1 million material cost may deserve more attention than a 20% variance in a $10,000 repair cost. Therefore, rules such as "Investigate all variances exceeding $5,000, or 25% of standard cost, whichever is lower" are common.

Variance analysis is subject to the same cost-benefit test as other phases of a control system. The trouble with using rules of thumb is that they are too frequently based on subjective assessments, guesses, or hunches. The field of statistics offers tools that can help decisions regarding variance analysis. These tools help answer the cost-benefit question, and they help to separate variances caused by random events from variances that are controllable.

Accounting systems have traditionally implied that a standard is a single acceptable measure. Practically, accountants (and everybody else) realize that the standard is a *band* or *range* of possible acceptable outcomes. Consequently, accountants expect variances to fluctuate randomly within some normal limits.

By definition, a random variance per se is within this band or range. It calls for no corrective action to an existing process. Random variances are attributable to chance rather than to management's implementation decisions. For a further discussion, see Chapter 26.

PROBLEMS FOR SELF-STUDY

PROBLEM 1

Consider the following information for the O'Shea Company for April, when 2,000 finished units of ceramics were produced:

Direct materials used, 4,400 pounds. The standard allowance per finished unit is two pounds at $15 per pound. Six thousand pounds were purchased at $16.50 per pound, a total of $99,000.

Actual direct labor hours were 3,250 at a total cost of $40,300. Standard labor time allowed is 1.5 hours per unit.

Required

1. Calculate the direct material price and efficiency variance and the direct labor price and efficiency variance. The direct materials price variance will be based on a flexible budget for actual quantities purchased, but the efficiency variance will be based on a flexible budget for actual quantities used.
2. Journal entries for a standard-cost system that isolates variances as early as feasible.

SOLUTION 1

1. Exhibit 7-9 presents a general framework for analyzing variances. It may seem awkward at first, but upon review you will discover that it provides perspective and insight. In particular, note the two sets of computations in column 2 for direct materials. The $90,000 relates to the direct materials *purchased*; the $66,000 relates to the direct materials *used*.

EXHIBIT 7-9
Framework for Analysis of Variances

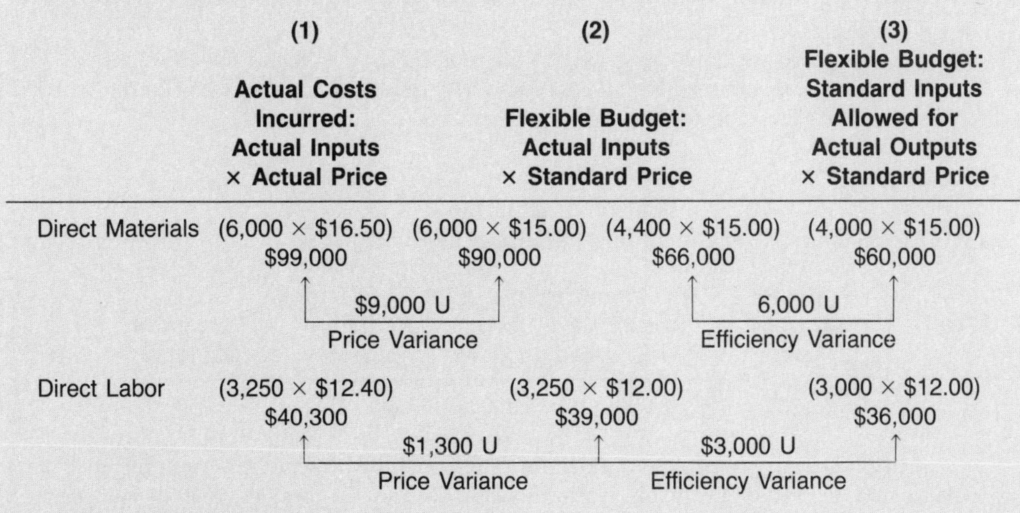

(1)
Actual Costs Incurred: Actual Inputs × Actual Price
(2)
Flexible Budget: Actual Inputs × Standard Price
(3)
Flexible Budget: Standard Inputs Allowed for Actual Outputs × Standard Price
Direct Materials
(6,000 × $16.50) $99,000
(6,000 × $15.00) $90,000
(4,400 × $15.00) $66,000
(4,000 × $15.00) $60,000
$9,000 U
Price Variance
6,000 U
Efficiency Variance
Direct Labor
(3,250 × $12.40) $40,300
(3,250 × $12.00) $39,000
(3,000 × $12.00) $36,000
$1,300 U
Price Variance
$3,000 U
Efficiency Variance

2. **Standard-cost system**

Materials control (6,000 × $15.00)	90,000	
Direct materials price variance	9,000	
Accounts payable		99,000
Work in process control (4,000 × $15.00)	60,000	
Direct materials efficiency variance	6,000	
Materials control (4,400 × $15.00)		66,000
Work in process control (2,000 × $18.00)	36,000	
Direct labor price variance	1,300	
Direct labor efficiency variance	3,000	
Payroll liability		40,300

PROBLEM 2

This problem is more challenging than the preceding problem.

The Rochester Medical Laboratory conducts tests of blood samples. Many physicians use its services because the laboratory has an outstanding reputation for quality, speed, and competitive prices. The laboratory technicians are expected to perform a variety of blood tests requiring various amounts of time.

Production is measured in "weight-units." For example, a red-cell or white-cell blood count might require 5 minutes of time, a cholesterol count might require 10 minutes, and so forth. Thus, the total production is measured not by the unweighted number of blood tests, but by their number weighted by the time allowances (or, in the terminology of process costing, their equivalent units).

Required

1. Assume that five technicians are each expected to have an average output of 8,000 weight-units per 8-hour shift. For a particular shift, 36,000 weight-units were actually processed. Inputs are expressed in terms of equivalent units, 8,000 available weight-units per 8-hour shift. Supply a measure of the technicians' efficiency, where inputs are compared with outputs.
2. Suppose the experienced technicians have a standard labor rate of $25 per hour. (a) Compute the standard labor cost per weight-unit. (b) Suppose two newcomers, working at $20

per hour, replace two experienced technicians. Prepare an analysis showing total direct labor price and direct labor efficiency variances for the shift. For this purpose, the actual inputs expressed in weighted quantities would be 40,000 weight-units. What does your analysis reveal?

3. Suppose there are 22 working days in the month. The results for the entire month were consistent with the results you computed in requirement 2. Prepare a journal entry to show direct labor performance for the month.

SOLUTION 2

1. The technicians were inefficient (data in weight-units):

Inputs Actually Used	Inputs That Should Have Been Used	Efficiency Variance	Percent of Standard
8,000	36,000 ÷ 5 = 7,200	800 U	11%

A major purpose of this requirement is to demonstrate how physical (nonfinancial) standards may be used for evaluating performance. These are the numbers that the manager will watch most closely on a day-to-day basis. Whatever its form or measure, *some* standard is necessary to judge performance.

2a. Inputs can be expressed as 8 hours or sufficient inputs to produce 8,000 weighted-units per technician.

Total labor cost per shift = $25 × 8 = $200
Total equivalent outputs at standard = 8,000 weight-units
Standard labor cost per weight-unit = $200 ÷ 8,000 = $.025

2b.

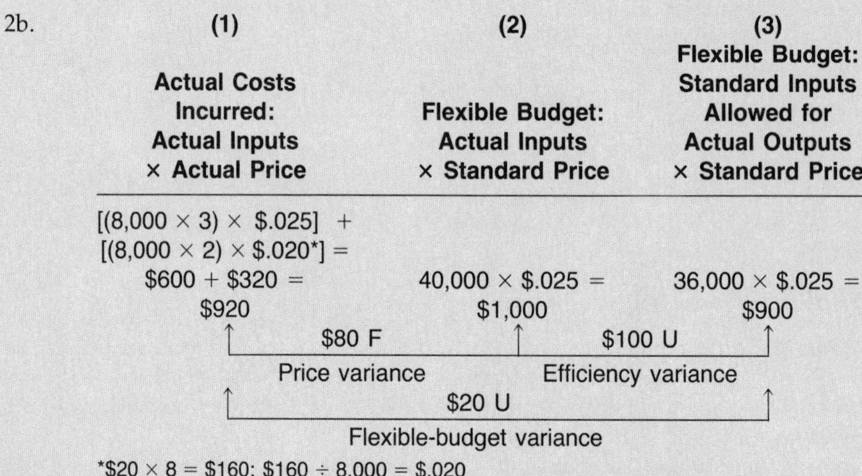

(1) Actual Costs Incurred: Actual Inputs × Actual Price	(2) Flexible Budget: Actual Inputs × Standard Price	(3) Flexible Budget: Standard Inputs Allowed for Actual Outputs × Standard Price
[(8,000 × 3) × $.025] + [(8,000 × 2) × $.020*] = $600 + $320 = $920	40,000 × $.025 = $1,000	36,000 × $.025 = $900

Price variance: $80 F
Efficiency variance: $100 U
Flexible-budget variance: $20 U

*$20 × 8 = $160; $160 ÷ 8,000 = $.020

The analysis shows the financial impact of trading off a lower-priced labor resource ($80 F) against efficiency ($100 U). Perhaps the lower-priced labor worked more slowly because of inexperience.

3. The amounts in requirement 2 would be multiplied by 22. Because there is probably no inventory, the standard costs would be charged to Standard Cost of Services Sold (or similar description):

Standard cost of services sold	19,800	
Direct labor efficiency variance	2,200	
Direct labor price variance		1,760
Payroll liability		20,240

This problem demonstrates the completely general applicability of the concepts in this chapter. For example, outputs are not always measured in terms of physical units produced or in direct labor hours allowed. Moreover, flexible budgets and standards are not confined to manufacturing situations.

SUMMARY

Flexible budgets and standards aid management predictions and provide a framework for judging performance. Actual costs are compared with standard costs to obtain variances. Variances raise questions that lead to improvements in operations.

The flexible-budget variance enables the manager to separate the effects of sales volume from other explanations of why the static (master) budget was not achieved. In turn, the flexible-budget variance is often subdivided into its price and efficiency components. Efficiency is measured by comparing inputs with outputs. Therefore, a common measure of inputs and outputs is necessary to obtain an efficiency variance.

Price variances enhance understanding of the influence of prices on actual results. In addition, price variances permit the exclusion of price effects from *other* variances. The definitions of efficiency variances *and* sales volume variances have a noteworthy common assumption: All unit prices are *standard* or *budgeted* prices.

When standards are currently attainable, there is no conceptual difference between standards and budgets. A standard is a *unit* concept, and a budget is a *total* concept. In a sense, the standard is the budget for one unit. Material and labor variances are primarily divided into two categories: (a) price and (b) efficiency. Price variances are computed by multiplying differences in unit price by actual quantities purchased. Efficiency variances are computed by multiplying differences in actual and standard quantities by standard unit prices.

General-ledger treatments for standard costs vary considerably. Variances should be measured quickly for prompt management attention. Nonfinancial (physical) standards are widely used for day-to-day control. Variances should be assigned to the person primarily responsible for them. Otherwise the investigative follow-ups of reported variances—where the real payoff lies—will be fruitless. Above all, the focus should be on total costs of all factors of production, not on individual variances by themselves.

TERMS TO LEARN

This chapter and the Glossary at the end of the book contain definitions of the following important terms:

continual improvement standard cost *(p. 232)*
currently attainable standards *(231)* effectiveness *(217)* efficiency *(217)*
efficiency variance *(224)* excess materials requisition *(234)*
expected variance *(232)* flexible budget *(213)* flexible-budget variance *(215)*
moving cost reduction standard cost *(232)* price variance *(224)*
quantity variance *(227)* rate variance *(227)* sales volume variance *(215)*
setup time *(235)* standard bill of materials *(234)* standard costs *(222)*
standard hours allowed *(225)* standard hours earned *(225)*
standard hours of input allowed for good output produced *(225)*
standard hours worked *(225)* static budget *(213)* usage variance *(227)*
variable budget *(213)*

ASSIGNMENT MATERIAL

QUESTIONS

7-1 "Flexible budgets can be useful either before or after a specific accounting period in question." How?

7-2 "Performance may be both effective and efficient, but either condition can occur without the other." Give an example of effectiveness. Give an example of efficiency.

7-3 "We expect steadily improving performance, not a static base for comparison. Standards are obsolete." Do you agree? Explain.

7-4 "For hour-to-hour or day-to-day control, nonfinancial (physical) measures are often used." Give three examples of commonly used nonfinancial measures.

7-5 "The idea of comparing performance at one volume level with a plan that was developed at some other volume level is valid in judging the effectiveness of planning and control." Comment.

7-6 "Comparisons of actual costs with standard costs are superior to comparisons of actual data with past data." Why?

7-7 List and briefly describe two different types of standard costs.

7-8 What are the key questions in deciding which variances should be accounted for and analyzed?

7-9 As a minimum, how should setup time be accounted for?

7-10 When can't the terms *budgeted performance* and *standard performance* be used interchangeably?

7-11 List some major factors that affect control procedures for material quantities.

7-12 Define *standard bill of materials* and *excess materials requisition.*

7-13 "Cost control means cost reduction." Do you agree? Why?

7-14 List four common causes of material price variances.

7-15 "Standard costing is OK for big companies that can afford such an elaborate system. But our company is small, and we rely on an 'actual' system because it saves clerical costs." Comment.

7-16 List four purposes of standard costs.

7-17 Who is responsible for developing standards?

7-18 Why do budgeted variances arise?

7-19 When will budgets differ from standards? When will they be the same?

7-20 How does management decide when a variance is large enough to warrant investigation?

EXERCISES AND PROBLEMS

Coverage of Part One of the Chapter

7-21 Flexible budget. Fontana Company made 200,000 units of product in a given year. The total manufacturing costs of $400,000 included $180,000 of fixed costs. Assume that no price changes will occur in the following year and that no changes in production methods are applicable. Compute the total budgeted cost for producing 230,000 units in the following year.

7-22 Flexible budget for services. The head of an army motor pool is trying to estimate the costs of operating a fleet of twenty trucks. Two major costs are fuel, 18¢ per mile, and depreciation per truck per year, $9,000. Prepare the flexible-budget amounts for fuel and depreciation for twenty trucks at levels of 10,000, 20,000, and 30,000 miles per truck.

7-23 Flexible budget. The following data are for April:

	Budgeted Amount per Unit	Various Levels of Volume (Units Sold)		
Units sold		18,000	20,000	22,000
Sales	$30	$?	$?	$?
Variable costs:				
Direct materials	2	360,000	?	?
Fuel	2	?	?	?
Fixed costs:				
Executive salaries		?	?	40,000
Depreciation		?	60,000	?

Required

1. Fill in the unknowns.
2. Draw a freehand graph of the flexible budget for all the items shown here.

7-24 Basic flexible budget. The budgeted prices for direct materials and direct labor per unit of manufactured attaché cases are $20 and $4, respectively. The production manager is pleased with the following data:

	Actual Costs	Static (Master) Budget	Variance
Direct materials	$182,000	$200,000	$18,000 F
Direct labor	39,000	40,000	1,000 F

Required

Actual output was 8,800 units. Is the manager's pleasure justified? Prepare a performance report that uses a flexible budget and a static budget.

7-25 Flexible-budget preparation. The managing partner of Roan Music Box Fabricators has become aware of the disadvantages of static budgets and has asked you to prepare a flexible budget for October for the main style of music box. The following partial data are available for the actual operations in a recent typical month:

Boxes produced and sold	4,500
Direct material costs	$90,000
Direct labor costs	$67,500
Depreciation and other fixed manufacturing costs	$50,700
Average selling price per box	$ 70
Fixed marketing and administrative expenses	$81,350

A 10% increase in the selling price is expected. The only variable marketing expense is a commission of $5.50 per unit paid to the manufacturer's representatives, who bear all their own expenses of traveling, entertaining customers, and so on. The only variable overhead is a patent royalty of $2 per box manufactured. Salary increases that will become effective in October are $12,000 per year for the production superintendent and $15,000 per year for the sales manager. A 10% increase in direct material prices is expected to become effective in October. No changes are expected in direct labor wage rates or in the productivity of the direct labor personnel.

Required

1. Prepare a flexible budget for October, showing expected results at each of three levels of volume: 4,000 units, 5,000 units, 6,000 units.
2. Draw a freehand sketch of a graph for your flexible-budget costs. *Hint:* See Exhibits 7-2 and 7-3, pages 214 and 215.

7-26 Overview of variance analysis. Bank Management Printers, Inc., produces luxury checkbooks with three checks and stubs on each page. Each checkbook is designed for an individual customer and is ordered through the customer's bank. The company's operating budget for September included these data:

Number of checkbooks	15,000
Selling price, per book	$ 20
Variable costs per book	$ 8
Total fixed costs for the month	$145,000

The actual results for September were:

Number of checkbooks produced and sold	12,000
Average selling price per book	$ 21
Variable costs per book	$ 7
Total fixed costs for the month	$150,000

The executive vice president of the company observed that the operating income for September was much less than anticipated, despite a higher-than-budgeted selling price and a lower-than-budgeted variable cost per unit. You have been asked to provide explanations for the disappointing September results.

Required

1. Prepare analyses of September performance. Prepare a *Level 1* and a *Level 2* analysis.
2. Comment briefly on your findings.

7-27 Overview of variance analysis. You have been hired as a consultant by Mary Flanagan, the president of a small manufacturing company that makes automobile parts. Flanagan is an excellent engineer, but she has been frustrated by not having adequate cost data.

You helped install flexible budgeting and standard costs. Flanagan has asked you to consider the following May data and recommend how variances might be computed and presented in performance reports:

Static budget in units	20,000
Actual units produced and sold	23,000
Budgeted selling price per unit	$ 40
Budgeted variable costs per unit	$ 25
Budgeted total fixed costs per month	$200,000
Actual revenue	$874,000
Actual variable costs	$630,000
Favorable variance in fixed costs	$ 5,000

Flanagan was disappointed: although volume exceeded expectations, operating income did not.

Assume that there were no beginning or ending inventories.

Required

1. You decide to present Flanagan with alternative ways to analyze variances so that she can decide what level of detail she prefers. The reporting system can then be designed accordingly. Prepare an analysis similar to *Levels 0, 1,* and *2* in Exhibit 7-5, page 218.
2. What are some likely causes for the variances?

7-28 Flexible and static budgets, service company. Avanti Transportation Company executives have had trouble interpreting operating performance for a number of years. The company has used a budget based on detailed expectations for the forthcoming quarter. For example, the condensed performance report for the recent quarter for a midwestern branch was:

	Budget	Actual	Variance
Revenue	$10,000,000	$9,500,000	$500,000 U
Fuel	1,000,000	986,000	14,000 F
Repairs and maintenance	100,000	98,000	2,000 F
Supplies and miscellaneous	200,000	196,000	4,000 F
Variable payroll	5,700,000	5,500,000	200,000 F
Total variable costs*	7,000,000	6,780,000	220,000 F
Supervision	200,000	200,000	—
Rent	200,000	200,000	—
Depreciation	1,600,000	1,600,000	—
Other fixed costs	200,000	200,000	—
Total fixed costs	2,200,000	2,200,000	—
Total costs charged against revenue	9,200,000	8,980,000	220,000 F
Operating income	$ 800,000	$ 520,000	$280,000 U

U = Unfavorable; F = Favorable.

*For purposes of this analysis, assume that all these costs are completely variable (in relation to revenue dollars). In practice, many are mixed and have to be subdivided into variable and fixed components before a meaningful analysis can be made. Also assume that the prices and mix of services sold remain unchanged.

Although the branch manager was upset about the unfavorable revenue variance, he was happy that his cost performance was favorable; otherwise his operating income would be even worse.

His immediate superior, the vice president for operations, was totally unhappy and remarked: "I can see some merit in comparing actual performance with budgeted performance, because we can see whether actual revenue coincided with our best guess for budget purposes. But I can't see how this performance report helps me evaluate the cost control performance of the department head."

Required

1. Prepare a columnar flexible budget for Avanti at revenue levels of $9,000,000, $10,000,000, and $11,000,000. Use the format of the last three columns of Exhibit 7-2, page 214. Assume that the prices and mix of products sold are equal to the budgeted prices and mix.
2. Express the flexible budget for costs in formula form.
3. Prepare a condensed contribution-format income statement showing the static (master) budget variance, the sales volume variance, and the flexible-budget variance. Use the format of Exhibit 7-4, page 216.

7-29 Overview of variance in service income. Consider the following performance of a division of the Amtrak railroad system during 19_4:

	Actual Results	Static (Master) Budget	Variances
Revenue	$?	$150,000,000	$?
Variable costs	98,000,000	97,500,000*	500,000 U
Contribution margin	?	52,500,000	?
Fixed costs	38,500,000	37,500,000	1,000,000 U
Operating income	$?	$ 15,000,000	$?

*Includes diesel fuel of $45,000,000.

The master budget was based on budgeted revenue per passenger-mile of 10¢. (A passenger-mile equals one paying passenger traveling 1 mile.) To spur an increase in volume, Amtrak decided to cut fares an average of 8%. Actual passenger-miles were 10% in excess of the master budget for the year. The price per gallon of diesel fuel fell below the price used to formulate the master budget. The average price decline of diesel fuel for the year was 6%.

Required

1. As an explanation for the president, prepare a summary performance report that shows for each line item the flexible-budget variance and the revenue volume variance (sales volume variance).
2. Assume that fuel was used at the same level of efficiency as predicted in the master budget. What amount of the flexible-budget variance for variable costs is not attributable to diesel fuel costs? Explain.

7-30 Analysis of service, airline operations. The Carruthers Airline had the accompanying data for 19_2 (in thousands).

	Static (Master) Budget	Actual Results	Variances
Revenue	$100,000	$84,000	$16,000 U
Variable expenses	60,000*	51,600	8,400 F
Contribution margin	40,000	32,400	7,600 U
Fixed expenses	35,000	35,000	0
Operating income	$ 5,000	$(2,600)	$ 7,600 U

*Includes $45 million of wages, meals, and fuel.

You discover that a midyear airfare increase of 9% resulted in a 5% increase in average revenue per passenger mile flown for the entire year. Wage, meal, and fuel increases and various inefficiencies exceeded the pace of the fare increases and averaged 10% for the year. Other price and efficiency effects were minor and can be ignored.

Required

Prepare an analytical tabulation as an explanation for the Carruthers president. Begin with actual results at actual prices and end with the static (master) budget. Show (a) flexible-budget variances and (b) revenue volume variances.

7-31 Overview of variance in service income. Ferraro Legal Services (FLS) was founded by Eugene Ferraro after the legal profession removed its ban on advertising. The firm provides brief legal consultation on a variety of topics at a fixed price. Offices are now open in eighteen major cities. The staff is a mix of attorneys and paraprofessionals. The average client requires one-half hour of time and is billed $75 for the service. Attorneys are paid a percentage of the revenue generated; paraprofessionals are on a straight salary. Budgeted contribution margins averaged 60% of billed sales.

The FLS static (master) budget for 19_2 predicted volume of 700,000 clients, but only 600,000 were served. Fixed costs were $21.5 million, which exceeded the budget by $1.5

million, primarily because of an extra advertising campaign. There were no variances from the average selling prices, but attorneys handled a higher percentage of clients than planned, causing an unfavorable flexible-budget variance for variable costs of $1 million.

Ferraro was disappointed when he learned that operating income for 19_2 was less than half the $11.5 million budgeted. He said, "I knew that profit would be lower than expected. After all, I authorized the extra advertising. I also realize that volume is down a little. But why is operating income *so much* below budget?"

Required

1. Explain why the budgeted operating income was not attained. Use a presentation similar to Exhibit 7-4, page 216. Enough data have been given to permit you to construct the complete exhibit by filling in the known items and then computing the unknowns (in thousands of dollars).
2. Complete your explanation by summarizing what happened, using no more than three sentences.

Coverage of Part Two of the Chapter

7-32 Material and labor variances. Consider the following data:

	Direct Materials	Direct Labor
Costs incurred: actual inputs × actual prices	$200,000	$90,000
Actual inputs × standard prices	214,000	86,000
Standard inputs allowed for actual outputs achieved × standard prices	225,000	80,000

Required

Compute the price, efficiency, and flexible-budget variances for direct materials and direct labor. Use *U* or *F* to indicate whether the variances are unfavorable or favorable.

7-33 Materials variances, nonfinancial measures. Assume that a table manufacturer uses plastic tops. Plastic is purchased in large sheets, cut down as needed, and then glued to the tables. A given-sized table will require a specified amount of plastic. The amount of plastic needed per Type-F TV table is 4 square feet and the cost per square foot is 65¢. A certain production run of 1,000 Type-F TV tables results in purchases and usage of 4,300 square feet at 70¢ per square foot, a total cost of $3,010.

Required

1. Compute the price variance and the efficiency variance for direct materials.
2. What type of data would the production manager probably watch most closely on a day-to-day basis?

7-34 Efficiency variances, nonfinancial measures. Assume that 10,000 units of a particular alloy were produced. Suppose the standard direct materials allowance is two pounds per unit, at a cost per pound of $5. Actually, 21,000 pounds of materials (input) were used to produce the 10,000 units (output).

Similarly, assume that it is supposed to take four direct labor hours to produce one unit and that the standard hourly labor cost is $15. But 42,000 hours (input) were used to produce the 10,000 units.

Required

1. Compute the efficiency variances for direct materials and direct labor.
2. What type of data would the production manager probably watch most closely on a day-to-day basis?

7-35 Material and labor variances, nonfinancial measures. Consider the following data regarding the manufacture of a line of tables:

	Direct Materials	Direct Labor
Actual price per unit of input (board feet and hours)	$14	$ 9
Standard price per unit of input	$12	$10
Standard inputs allowed per unit of output	5	2
Actual units of input	48,000	22,000
Actual units of output (product)	10,000	10,000

Required

1. Compute the price, efficiency, and flexible-budget variances for direct materials and direct labor. Use *U* or *F* to indicate whether the variances are unfavorable or favorable.
2. Prepare a plausible explanation for the performance.
3. What type of data would the production manager probably watch most closely on a day-to-day basis?

7-36 Material and labor variances, nonfinancial measures. Consider some data for the Cobb Company, a manufacturer of raincoats:

Units produced, 10,700
Budgeted or standard amounts per unit

Direct materials (4 square yards @ $5.00)	$20
Direct labor (2 hours @ $8.00)	16

Actual data:

Direct material costs	$270,000	Direct labor costs	$171,600
Square yards of input purchased and used	50,000	Hours of input	22,000
Price per square yard	$ 5.40	Labor price per hour	$ 7.80

Required

1. Compute (a) price and efficiency variances for direct materials and (b) price and efficiency variances for direct labor. Present your answer in a format similar to the analysis that appears in Exhibit 7-9, page 237.
2. Suppose that the Cobb Company control system were designed to isolate material price variances upon purchase rather than upon usage of the materials. Suppose further that 60,000 pounds of material were purchased during April and that only 50,000 pounds were issued to production. Compute the purchase-price variance that would be reported by the control system.
3. What type of data would the production manager probably watch most closely on a day-to-day basis?

7-37 Material and labor variances, nonfinancial measures. (SMA, adapted) The Carberg Co. manufactures and sells a single product for $20. The company uses a standard-cost system, isolating all variances as soon as possible. The standards for one finished unit include:

Direct materials (1 kg. at $2)	$2.00
Direct labor (.6 hr. at $10)	6.00

Actual results for November were:

Units produced	5,100
Direct materials purchased at $2.10 per kg.	5,200 kg.
Direct materials used in production	5,300 kg.
Direct labor at $10.20 per hr.	3,200 hrs.

Required

1. Compute the price and efficiency variances for direct materials and direct labor.

2. What type of data would the production manager probably watch most closely on a day-to-day basis?

7-38 General journal entries for materials. Consider the following data regarding direct materials compiled by the manufacturer of 1,000 television tables:

Actual quantity purchased	6,000 square feet
Standard quantity allowed	4,000 square feet
Actual quantity used	4,300 square feet
Standard price per square foot	$.65
Actual price per square foot	$.62

Required

1. The manager of purchasing wants early monitoring of price variances, and the manager of production wants early detection of efficiency variances. Prepare journal entries so that price variances are isolated on purchase of direct materials and efficiency variances are isolated on the use of excess material requisitions.
2. Another company has the same process, but its managers are less concerned about isolating variances early. Instead, price variances are isolated upon issuance of direct material to the producing departments. In this way, the production department's efficiency can still be measured because price volatility will not affect the efficiency variance. This company does not isolate efficiency variances until after production is completed. Prepare journal entries that would isolate each variance closest to the time management wants that information. Show T-accounts.
3. Suppose journal entries isolate variances later rather than earlier. Does later isolation necessarily mean that the costs of underlying operations are controlled more loosely? Explain.

7-39 Journal entries for direct labor. The Grafton Company had a standard rate (price) of $16.00 per direct labor hour. Actual hours in July were 10,300 at an actual price of $15.80 per hour. The standard direct labor hours allowed for the output achieved were 10,000.

Required

Prepare general-journal entries that isolate price and efficiency variances. Prepare one entry in which Work in Process Control is carried at standard hours allowed times standard prices. Then prepare a second set of entries in which (a) price variances are isolated as labor costs are originally journalized and (b) efficiency variances are isolated as units are transferred from Work in Process Control to Finished Goods Control.

7-40 Analysis of variances. Chemical, Inc., has set up the following standards for direct materials and direct labor:

	Per Finished Batch
Direct materials: 10 lbs. @ $3.00	$30.00
Direct labor: .5 hours @ $20.00	10.00

The number of finished units budgeted for the period was 10,000; 9,810 units were actually produced.

Actual results were:

Direct materials: 98,073 lbs. used	
Direct labor: 4,900 hrs.	$102,900

Required

During the month, purchases amounted to 100,000 lbs, at a total cost of $310,000. Price variances are isolated upon purchase.

1. Prepare journal entries to record the data above.
2. Show computations of all material and labor variances.
3. Comment on each of the variances.

7-41 Comparison of general-ledger entries for direct materials and direct labor. The Lee Co. in Korea has the following data for the month of March when 1,100 finished units were produced:

- Direct materials used, 3,600 pounds. The standard allowance per finished unit is 3 pounds at $3.00 per pound. Five thousand pounds were purchased at $3.25 per pound, a total of $16,250.
- Direct labor, actual hours, was 2,450 hours at a total cost of $19,600. The standard labor cost per finished unit was $15.20. Standard time allowed is 2 hours per unit.

Required

1. Prepare journal entries for a "normal"-cost system.
2. Prepare journal entries for a "standard"-cost system, assuming purchase price variances are isolated at the time of purchase. Support your entries with a detailed variance analysis, using the columnar analytical format illustrated in the chapter.
3. Show an alternative approach, including journal entries, to the way you quantified the material price variance in requirement 2. Which way is better, and why?

7-42 Elementary variance analysis and graph. Consider the following selected data regarding the manufacture of a line of upholstered chairs:

	Standards
Direct materials	Two square yards of input at $5 per square yard, or $10 per unit of output
Direct labor	One-half hour of input at $10 per hour, or $5 per unit of output

The following data were compiled regarding actual performance: actual units produced, 20,000; square yards of input acquired and used, 37,000; price per square yard, $5.10; direct labor costs, $88,200; actual hours of input, 9,000; labor rate per hour, $9.80.

Required

1. Show computations of price and efficiency variances for direct materials and for direct labor. Prepare a plausible explanation of why the variances occurred.
2. Sketch a graphical analysis of the direct labor variance, using the vertical axis for total costs and the horizontal axis for volume in hours. Indicate what vertical distances represent the flexible-budget variance, price variance, and efficiency variance.
3. Suppose 60,000 square yards of materials were purchased (at $5.10 per square yard) even though only 37,000 square yards were used. Suppose further that variances are identified with their most likely control point; accordingly, direct material purchase-price variances are isolated and traced to the purchasing department rather than to the production department. Compute the price and efficiency variances under this approach.

7-43 Journal entries and T-accounts. Prepare journal entries and T-accounts for all transactions in Problem 7-42, including requirement 3. Summarize in three sentences how these journal entries differ from the normal costing entries as described in Chapter 4, p. 111.

7-44 Flexibility in budgets. Refer to Problem 7-42. Suppose the master budget was for 24,000 units of output. The general manager is gleeful about the following report:

	Actual Costs	Master Budget	Variance
Direct materials	$151,700	$240,000	$88,300 F
Direct labor	$ 88,200	$120,000	$31,800 F

Required

Is the manager's glee warranted? Prepare a report that might provide a more detailed explanation of why the master budget was not achieved. Actual output was 20,000 units.

7-45 Developing standard costs per unit (CMA, adapted) Ogwood Company is a small manufacturer of wooden household items. Al Rivkin, corporate controller, plans to implement a standard-cost system for Ogwood. Rivkin has information from several co-workers that will assist him in developing standards for Ogwood's products.

One of Ogwood's products is a wooden cutting board. Each cutting board requires 1.25 board feet of lumber and 12 minutes of direct labor time to prepare and cut the lumber. The cutting boards are inspected after they are cut. Because the cutting boards are made of a natural material that has imperfections, one board is normally rejected for each five that are accepted. Four rubber foot pads are attached to each cutting board. A total of fifteen minutes of direct labor time is required to attach all four foot pads and finish each cutting board. The lumber for the cutting boards costs $3.00 per board foot, and each foot pad costs $.05. Direct labor is paid at the rate of $8.00 per hour.

Required

Develop the standard cost for the direct cost components of the cutting board. The standard cost should identify, for each direct cost component of the cutting board, the standard quantity, standard rate, and standard cost per unit.

7-46 Solving for unknowns. The city of Chicago has a maintenance shop where all kinds of truck repairs are performed. Through the years, various labor standards have been developed to judge performance. However, during a March strike, some labor records vanished. The actual hours of input were 1,000. The direct labor flexible-budget variance was $1,700, favorable. The standard labor price was $14 per hour; however, a recent labor shortage had necessitated using higher-paid workers for some jobs and had produced a labor-price variance for March of $400, unfavorable.

Required

1. Actual labor price per hour
2. Standard hours allowed for output achieved

7-47 Responsibility for purchase price variances. The Chester Company uses standard costs for metal-working operations. The purchasing manager, Amy Strotz, is responsible for material price variances, and the production manager, Juan Morales, is responsible for material efficiency variances and direct labor price and efficiency variances.

The standard price for metal used as a principal raw material was $2 per pound. The standard allowance was six pounds per finished unit of product.

The standard rate for direct labor was $14 per hour. The standard allowance was one-half hour per finished unit of product.

During the past week, 10,000 good finished units were produced. However, labor trouble caused the production manager to use much nonpreferred personnel. Actual labor costs were $78,000 for 6,500 actual hours; 80,000 pounds of metal were purchased for $1.80 per pound; and 71,000 pounds of metal were consumed during production.

Required

1. Compute the material purchase-price variance, material efficiency variance, direct labor price variance, and direct labor efficiency variance.
2. As a supervisor of both the purchasing manager and the production manager, how would you interpret the feedback provided by the computed variances?
3. What are the flexible-budget allowances for the production manager for direct materials and direct labor? Would they be different if production were 7,000 good finished units?
4. Prepare a condensed responsibility-performance report for the production manager for the 10,000 units produced. Show three columns: charges to department, flexible budget, and variance.
5. Describe how the managers probably control their day-to-day operations.

7-48 Variances in food service labor. St. Mary's Hospital uses hourly paid personnel to take serving carts and meals from the kitchen to the patients' rooms. The hospital uses a flexible budget and standards to evaluate performance. The budget for a recent four-week period was 24,000 meals. The server personnel were budgeted at $6,400, computed on the basis of an average wage rate of $8 per hour.

The following data were compiled for the four-week period, when 24,000 meals were actually served. The servers actually worked 960 hours:

	Actual	Budget	Variance
Dietary department:			
Wages, servers	$7,296	$6,400	$896 U

U = Unfavorable.

Required

1. Compute the price and efficiency variances and also the flexible-budget variance. Give some possible reasons for the variances.
2. Suppose 28,800 meals had been served. Repeat requirement 1.
3. What would the food service manager use on a day-to-day basis to exert control?

7-49 Variances in hospital food costs. The Good Samaritan Hospital food service uses a flexible budget and standards to help plan and control its food and labor requirements. A weekly budget and labor schedule is prepared. The budgeted volume for a week in January was for 9,800 meals. The average cost of food per meal was budgeted at $3. The following data were compiled at the end of the week, when 10,300 meals were actually served:

	Actual	Budget	Variance
Dietary department:			
Cost of food	$35,226	$29,400	$5,826 U

U = Unfavorable.

The department manager was upset by the unfavorable variance, especially because the prices paid for food were 5% below the budgeted unit cost. The manager had obtained some attractive bargains in purchases of fresh fruits, vegetables, and meats. However, she had to cope with untrained personnel who wasted food. Excess usage was 20% of the standard allowed. She has asked you to help analyze the situation.

Required

1. Compute the flexible-budget variance and the price and efficiency variances.
2. Provide some possible reasons for the variances.
3. How are costs of food probably managed on a day-to-day, meal-to-meal basis?

7-50 Nonprofit service variances. During the peak season, the Internal Revenue Service (IRS) hires temporary personnel at hourly rates to open mail, check tax returns for completeness, and prepare returns for additional processing. The IRS uses standards to evaluate performance. Suppose the standards are 15 tax returns per hour and a $9 hourly wage rate. Each worker has a workweek of five days at eight hours daily. Severe local labor shortages have forced the IRS to pay an actual wage rate of $10 hourly.

The supervisor has the following data regarding performance for the most recent four-week period, when 20 temporary personnel worked 40-hour workweeks and handled 52,500 tax returns:

	Actual Costs	Flexible Budget	Flexible-Budget Variance
Wages	$32,000	?	?

Required

1. Compute the flexible-budget amount and the flexible-budget variance.
2. Compute the labor price variance and the labor efficiency variance.
3. Identify possible reasons for the variances.
4. What means does the supervisor probably use for managing this tax-processing activity? How do flexible budgets and standards help in managing the total cost of this activity?

7-51 Long-term agreement with supplier. Yamazaki Mazak manufactures large-scale machining systems that are sold to other industrial companies. Each machining system has a sizable direct materials cost, consisting primarily of the purchase price for a metal compound. For its Lexington, Kentucky, manufacturing facility, Mazak has a long-term contract with Fuji Metals. Fuji will supply to Mazak up to 2,400 pounds of metal per month at a fixed purchase price of $120 per pound for each month in 19_6. For purchases above 2,400 pounds in any month, Mazak renegotiates the price for the additional amount with Fuji Metals (or another supplier). The standard price per pound is $120 for each month in the January to December 19_7 period.

Production data, direct materials actual usage in dollars, and direct materials actual price per pound for the January to May 19_7 period are:

	Number of machining systems produced	Total actual direct materials usage	Average actual direct materials purchase price per pound of metal
January	10	$242,400	$120
February	12	286,560	120
March	18	442,260	126
April	16	395,264	128
May	11	253,440	120

The average actual direct materials purchase price is for *all* units purchased in that month. Assume that (a) the direct materials purchased in each month are all used in that month and (b) each machining system is started and completed in the same month.

The Lexington facility is one of three plants that Mazak operates to manufacture large-scale machining systems. The other plants are in Worcester, the United Kingdom, and Tokyo.

Required

1. Assume that Mazak's standard materials input per machining system is 198 pounds of metal. Compute the direct materials price variance and direct materials efficiency variance for each month of the January to May 19_7 period.
2. How does the signing of a long-term agreement with a supplier that includes a fixed purchase-price clause affect the interpretation of a materials purchase-price variance?

7-52 Continual improvement standards. Assume in Problem 7-51 that Mazak uses the following continual improvement standards for the direct materials input per machining system:

January	200 pounds of metal
February	198 pounds of metal
March	196 pounds of metal
April	194 pounds of metal
May	192 pounds of metal

Required

1. Using these standards, compute the direct materials efficiency variance for each month of the January to May 19_7 period.
2. Outline two ways that Mazak might develop continual improvement standards for the direct materials input per machining system.

CHAPTER 8

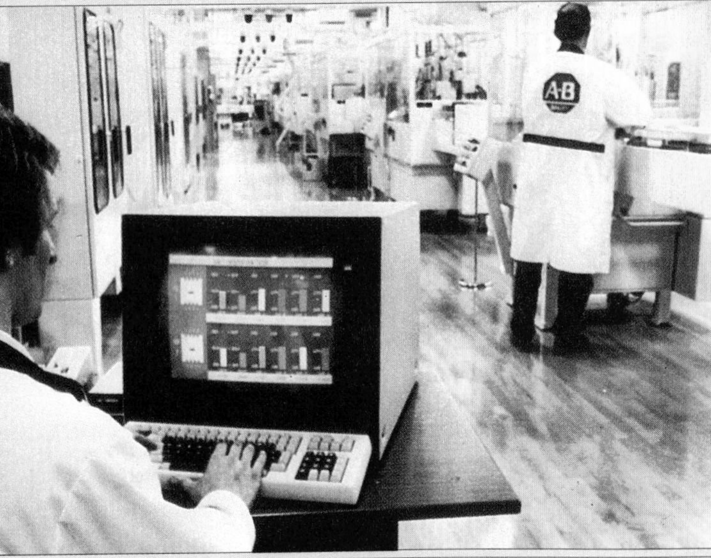

Manufacturing overhead is a major cost category in highly automated plants, such as this Allen-Bradley facility, which produces electronics components. Only six attendants oversee operations that generate about $40 million in sales a year.

Flexible Budgets and Standards: II

Variable factory overhead: Control and inventory costing

Fixed factory overhead: Control and inventory costing

Journal entries for overhead

Standard, normal, and actual costing

Analysis of fixed-factory-overhead variances

4-variance, 3-variance, 2-variance overhead analysis of activity area

Overhead variances in the ledger

Checkpoints for analyzing variances

Performance evaluation and variances

Learning Objectives

When you have finished studying this chapter, you should be able to

1. Describe how to develop standard costs for variable factory overhead
2. Explain the computation and meaning of spending and efficiency variances for variable factory overhead
3. Explain how to compute the budgeted fixed-factory-overhead rate
4. Describe how the unit inventory costs can differ depending on the choice of the denominator level of production volume
5. Prepare journal entries for factory overhead under standard costing
6. Identify actual costing, normal costing, and standard costing
7. Explain four ways to analyze factory-overhead variances
8. Explain the variance effects of a combined-overhead rate
9. Prepare journal entries for factory-overhead variances

This chapter continues the *Level 3* analysis of manufacturing costs introduced in the preceding chapter. We analyze variable and fixed factory overhead. We also compare the effects of the *budgetary control* purpose and the *inventory-costing* purpose of accounting for overhead.

In addition, this chapter covers some widely used methods of budgeting overhead, applying overhead, and analyzing overhead variances. Practices vary, depending on the preferences of the individual managers concerned.

Please proceed slowly as you study this chapter. Pause frequently. Trace the data to the diagrams and graphs in a systematic manner. In particular, ponder how fixed factory overhead is accounted for in one way for the budgetary-control purpose and in a different way for the inventory-costing purpose. The accounting for fixed factory overhead is usually the most puzzling aspect of the study of flexible budgets and standards.

This chapter concentrates on production, not selling, so we focus on units produced, not units sold. Thus, we do not contend with the sales volume variance that was examined in the preceding chapter. (For more on the sales volume variance, see Chapter 26.)

VARIABLE FACTORY OVERHEAD: CONTROL AND INVENTORY COSTING

Direct materials and direct labor have traditionally received more thorough analyses than factory overhead. In recent years, however, factory overhead has become increasingly prominent as a major manufacturing cost category, second only to direct materials. Hence, executives are seeking better ways to manage both variable and fixed overhead.

The control of variable factory overhead requires the identification of the cost drivers for such overhead items as energy, supplies, and repairs. Control increasingly entails monitoring nonfinancial measures such as kilowatts used, quantities of lubricants used, and repair hours employed.

Managers get insights for controlling overhead by using accounting tools such as flexible budgets. We will see how subdividing the flexible-budget variance helps pinpoint the causes of costs.

Standards and Flexible Budgets

Consider a continuation of the Webb Company illustration of the preceding chapter. The following data pertain to the production of 10,000 jackets:

	Actual	Flexible Budget	Flexible-Budget Variance
Variable factory overhead	\$131,000	\$120,000	\$11,000 U

Objective 1

Describe how to develop standard costs for variable factory overhead

The development of standard costs for variable factory overhead per unit of product requires the following steps:

1. Choose the cost driver of variable-factory-overhead costs. This cost driver becomes the application base for inventory-costing purposes. We use direct labor hours in the Webb case because direct labor is the principal overhead cost driver in the manufacturing of jackets. In other cases, different cost drivers may be better choices (for example, machine hours, direct material costs, equivalent units produced, or throughput time).
2. Estimate the inputs that should be used to produce one unit of actual output, which is .8 direct labor hours per jacket in the Webb case.
3. Develop a budgeted variable-overhead rate per direct labor hour, which is \$15 in the Webb case. Because this overhead is variable, the rate could be developed at any level of volume within the relevant range (see Exhibit 7-2, p. 214):

   At 10,000 units: \$120,000 ÷ 10,000 = \$12 per unit, or \$12 ÷ .8 = \$15 per hour
   At 12,000 units: \$144,000 ÷ 12,000 = \$12 per unit, or \$12 ÷ .8 = \$15 per hour
   At 14,000 units: \$168,000 ÷ 14,000 = \$12 per unit, or \$12 ÷ .8 = \$15 per hour

4. Apply the variable overhead to work in process, using the budgeted rate multiplied by the standard hours allowed for the actual outputs. For one unit of output, the standard cost is \$15 × .8 hours, or \$12. For 10,000 units, the flexible budget would be \$15 × (10,000 × .8 hours), or \$120,000.

A summary of the Webb data follows:

Budgeted variable-overhead rate per direct labor hour	\$15.00
Standard direct labor hours allowed per unit of finished product	.8
Budgeted variable-overhead rate per unit of finished product, \$15 × .8 hours	\$12.00
Actual variable-overhead costs	\$131,000
Actual direct labor hours of input	8,800
Standard direct labor hours allowed for actual outputs, .8 × 10,000	8,000
Flexible budget for 10,000 units, 8,000 hours × \$15	\$120,000

Two Major Variances: Spending and Efficiency

Objective 2

Explain the computation and meaning of spending and efficiency variances for variable factory overhead

Exhibit 8-1 displays a useful general format for analyzing variable-overhead variances. Recall the general formula for an efficiency variance:

$$\text{Efficiency variance} = \left(\begin{array}{c}\text{Inputs}\\\text{actually}\\\text{used}\end{array} - \begin{array}{c}\text{Inputs}\\\text{that should}\\\text{have been}\\\text{used}\end{array}\right) \times \begin{array}{c}\text{Standard}\\\text{unit price}\\\text{of inputs}\end{array}$$

The efficiency variance for variable overhead is a measure of the extra overhead incurred (or saved) *solely* because the chosen cost driver's *inputs actually used* differed from *the inputs that should have been used:*

EXHIBIT 8-1
Variable-Factory-Overhead Variances

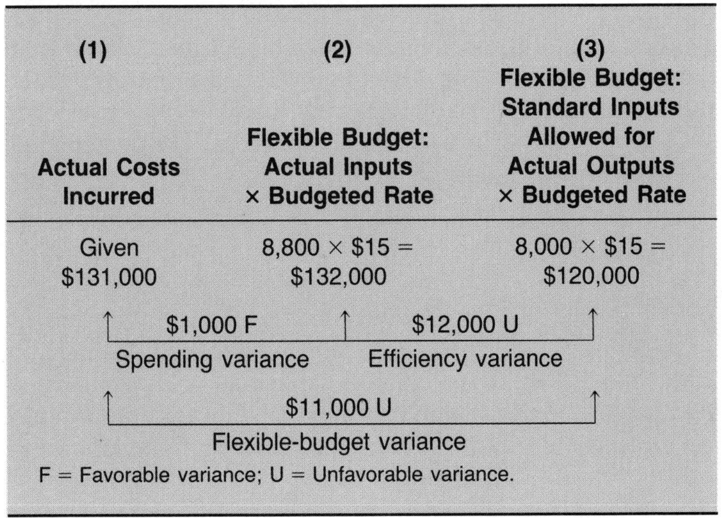

(1)
Actual Costs Incurred
(2)
Flexible Budget: Actual Inputs × Budgeted Rate
(3)
Flexible Budget: Standard Inputs Allowed for Actual Outputs × Budgeted Rate
Given
$131,000
8,800 × $15 =
$132,000
8,000 × $15 =
$120,000
$1,000 F
$12,000 U
Spending variance
Efficiency variance
$11,000 U
Flexible-budget variance
F = Favorable variance; U = Unfavorable variance.

$$\text{Variable-overhead efficiency variance} = \left(\begin{array}{c}\text{8,800 direct labor} \\ \text{hours actually used}\end{array} - \begin{array}{c}\text{8,000 direct labor hours} \\ \text{that should have been used}\end{array} \right) \times \begin{array}{c}\text{\$15 budgeted} \\ \text{rate per hour}\end{array}$$

$$= 800 \times \$15$$

$$= \$12{,}000 \text{ U}$$

In practice, a fragile assumption frequently underlies the efficiency variance computation for variable-overhead items: Variable overhead fluctuates in direct proportion to some measure of production volume. Suppose the variable overhead consisted wholly of supplies such as polishing cloths. Because direct labor was inefficiently used, as the above analysis shows, management could *expect* the related usage of supplies to be proportionately excessive by $12,000. Whether in fact a clear-cut, proportional relationship exists depends on the particular circumstances.

The $1,000 variance labeled in the analytical format shown in Exhibit 8-1 is usually called a variable-overhead **spending variance**. It is defined as the actual amount of overhead incurred minus the expected amount based on the flexible budget for actual inputs. Expressed another way, the spending variance for variable overhead is the flexible-budget variance minus the efficiency variance.

The spending variance is really a composite of price and other factors. It is the part of the flexible-budget variance unexplained by the efficiency variance attributable to the relationship of variable overhead to direct labor. For this reason, most practitioners used the term "spending variance" rather than merely "price variance":

$$\text{Variable-overhead spending variance} = \text{Flexible-budget variance} - \text{Efficiency variance}$$

$$\$1{,}000 \text{ F} = \$11{,}000 \text{ U} - \$12{,}000 \text{ U}$$

Note that the spending variance in this case could result exclusively from favorable unit price changes for polishing cloths. But it could also arise exclusively from careful use of these supplies. Why? Because variable-overhead costs may be affected by specific control of the items themselves, as distinguished from specific control of the related direct labor. Thus, this variance could be partially or completely traceable to the efficient use of supplies, even though it is labeled as a spending variance.

Consider another example. Suppose the Webb direct labor efficiency variance were zero. Then the variable-overhead efficiency variance would also be zero. Suppose further that there were no unit-price changes for polishing cloths. Would this mean no spending variance, too? Not necessarily. A direct laborer might be completely efficient in terms of the time used to produce a given output. That laborer, however, could have used too many cloths, not because of excess production time but because of simple waste. The cost of this excess usage of polishing cloths would be measured by the spending variance.

The Control Purpose

The first graph in Exhibit 8-2 shows how variable factory overhead relates to responsibility budgeting for planning and control purposes. Managers study how individual variable-overhead items should behave as volume changes. Taken together, total variable overhead should change in direct proportion to changes in volume, as is indicated on the first graph. Flexible-budget variances arise when actual costs differ from the budgeted costs depicted in the first graph.

We have presented an overall picture of how variable overhead might behave in its entirety. Of course, variable factory overhead typically consists of many items, including energy costs, repairs, indirect labor, and idle time; employer payroll taxes and fringe benefits such as pensions and health care are also included in variable factory overhead. Managers help control variable-overhead costs by budgeting each line item and then investigating possible causes for any significant flexible-budget variances.

Some variable-manufacturing-overhead costs, such as equipment repairs, are affected by how costs of other business functions are controlled. Thus, cooperation among the design, engineering, manufacturing, and marketing functions is necessary so that the total costs of the business are wisely managed. For example, in the long run, repair costs may be higher or lower not primarily because of how equipment is operated day to day, but because of a combination of strategic decisions regarding product designs, equipment choices, and production schedules aimed at satisfying more customers.

EXHIBIT 8-2

Graphs of Variable-Factory-Overhead Behavior

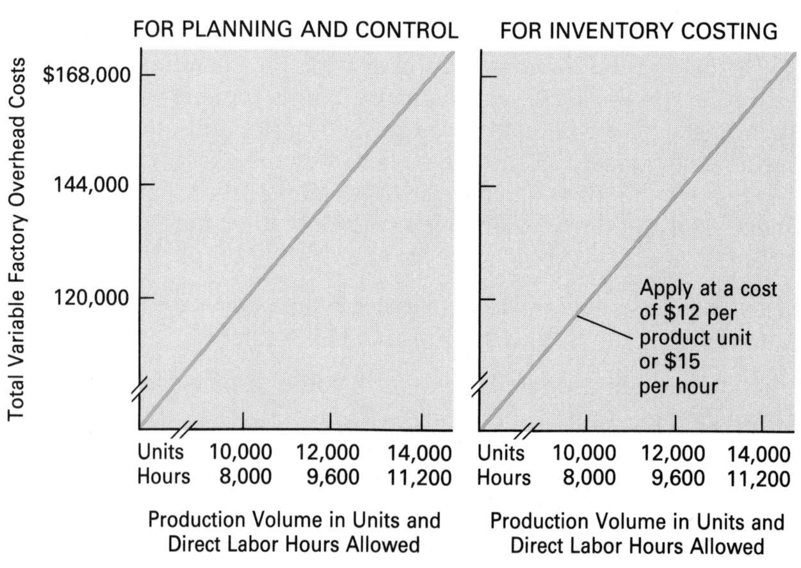

The Inventory-Costing Purpose

The second graph in Exhibit 8-2 shows how total variable factory overhead relates to inventory costing. Recall that inventory costing requires the application of factory overhead to products; inventory costs are the manufacturing costs only. Recall too that, in contrast to the narrower scope of inventory costing, full product costing includes manufacturing costs plus application of costs of other business functions such as marketing and general administration. In this chapter, however, we confine product costing to inventory costing as it relates to the application of factory overhead.

Compare the two graphs in Exhibit 8-2. The lines on the two graphs are identical. Both graphs show how total costs will behave. The total budget for variable factory overhead will increase at the rate of $15 per direct labor hour. Moreover, variable factory overhead will be applied also at $15 per hour.

8,000 × $15 = $120,000
9,600 × $15 = $144,000
11,200 × $15 = $168,000

Managers tend to think of *total costs* when exerting budgetary control over operations at various levels of volume. In contrast, managers tend to think of *unit costs* when applying costs for inventory purposes. For example, managers will say that their budget for 10,000 units is $120,000 and that their cost per unit is $120,000 ÷ 10,000 units = $12.

FIXED FACTORY OVERHEAD: CONTROL AND INVENTORY COSTING

The Control Purpose

Objective 3

Explain how to compute the budgeted fixed-factory-overhead rate

The Webb Company illustration has budgeted fixed manufacturing overhead of $276,000 (see Exhibit 7-2, p. 214). By definition, this is a lump sum that is not expected to vary with changes in the cost driver (direct labor hours) within the relevant range. Fixed costs are frequently a component of a flexible budget; however, the "flex" in the flexible budget arises from the variable costs, not the fixed costs. The flexible-budget formula for Webb Company's factory overhead is:

Flexible budget for factory overhead	= Budgeted fixed factory overhead	+ Budgeted variable factory overhead
	= $276,000 per month	+ $15 per direct labor hour
For 10,000 units	= $276,000	+ $15 (8,000) = $396,000
For 12,000 units	= $276,000	+ $15 (9,600) = $420,000

Budgetary control of fixed factory overhead concentrates on line-by-line, detailed plans for such typical items as supervision, depreciation, insurance, property taxes, and rentals. Fixed factory overhead is generally not subject to as much day-to-day (or month-to-month) managerial influence as is variable overhead. The variances of actual costs from fixed-overhead budgets tend to be relatively small.[1]

[1] Various surveys have shown that from 70% to 82% of the respondent firms identify variable and fixed components of manufacturing overhead. See C. Chow, M. Shields, and A. Wong-Boren, "A Compilation of Recent Surveys and Company-Specific Descriptions of Management Accounting Practices," *Journal of Accounting Education*, Vol. 6 (1988), pp. 183–207, which summarizes twelve surveys of practice.

The Inventory-Costing Purpose

The basic procedure for applying overhead is the same as that used for job-order costing in Chapter 4. The budgeting and computation of the overhead rate are usually done annually. The budgeted rate for applying fixed factory overhead is computed as follows, continuing with data from our Webb example.

$$\text{Budgeted fixed-factory-overhead rate for applying costs to inventory} = \frac{\text{Budgeted total fixed factory overhead}}{\text{Some preselected production volume level of application base for the budget period}}$$

$$= \frac{\$276{,}000 \times 12 \text{ months}}{12{,}000 \text{ units} \times 12 \text{ months}}$$

$$= \frac{\$3{,}312{,}000}{144{,}000 \text{ units}} = \$23 \text{ per unit}$$

or

$$= \frac{\$3{,}312{,}000}{144{,}000 \text{ units} \times .8 \text{ direct labor hour}}$$

$$= \frac{\$3{,}312{,}000}{115{,}200 \text{ direct labor hours}} = \$28.75 \text{ per direct labor hour}$$

The preselected production volume level in this case is based on the originally expected production volume for the year. It can be expressed as either 144,000 units or 115,200 hours per year, or 12,000 units or 9,600 hours per month.

The preselected production volume level of the application base used to set a budgeted fixed-factory-overhead rate for applying costs to inventory is called the **denominator volume.** Synonyms are **denominator activity** and **denominator level**.

For inventory-costing purposes, a $23 fixed-factory-overhead cost will be applied to each unit as it is produced. The first 5,000 units would bear a total inventory cost for fixed factory overhead of $115,000. The next 5,000 units would bear an additional $115,000, as the next graphs demonstrate:

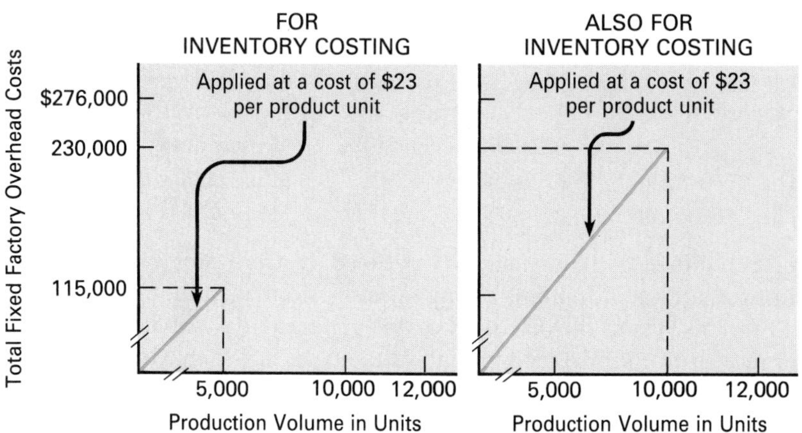

The first of the following graphs shows that the focus for budgetary *control* is on the $276,000 lump-sum total of fixed factory overhead. The second graph shows how, for *inventory-costing* purposes, ***the total costs in the first graph are typically "unitized" by using some denominator volume:***

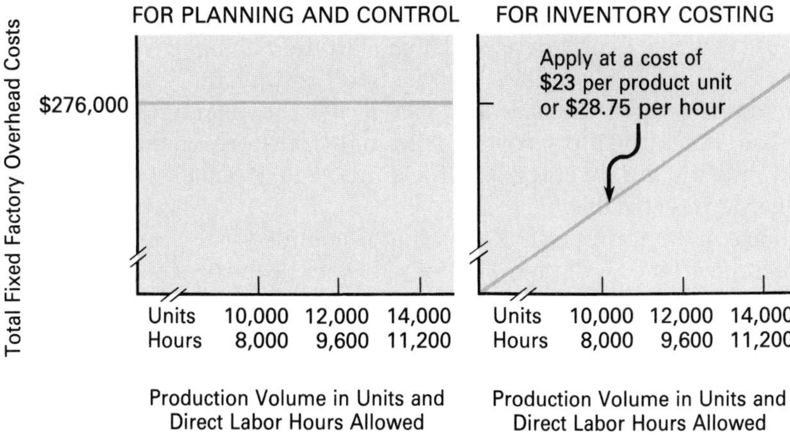

Unit Costs with Different Production Volumes

Objective 4

Describe how the unit inventory costs can differ depending on the choice of the denominator level of production volume

The *total* fixed factory overhead for budgetary control is unaffected by the specific preselected production volume used as a denominator. However, if fixed factory overhead is sizable, the choice of a denominator volume can have a significant effect on *unit* product costs. Assume that the total annual fixed overhead is $3,312,000:

(1) Total Annual Fixed Overhead	(2) Total Production Volume Level for Year	(1) ÷ (2) Budgeted Fixed-Overhead Rate for Inventory Costing
$3,312,000	10,000 × 12 = 120,000	$27.60
3,312,000	12,000 × 12 = 144,000	23.00
3,312,000	14,000 × 12 = 168,000	19.71

Graphs of these data on a monthly basis follow:

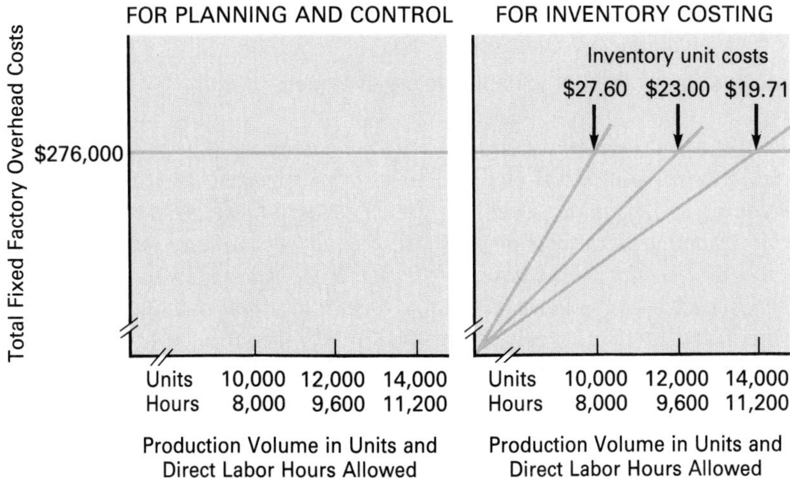

Choosing the Denominator Volume

The right-hand graph clearly demonstrates that the unit cost depends on the production volume level chosen as the denominator in the computation; the higher the denominator volume, the lower the inventory cost per unit. Unit costing is troublesome because managers usually want a single representative unit fixed cost despite month-to-month fluctuations in production volume.

The two graphs underscore an important difference. The planning and control purpose regards fixed costs in accordance with their actual cost behavior pattern, costs that arise in chunks rather than in finely granulated portions. In contrast, inventory costing views these *fixed* costs *as though* they had a *variable* cost behavior pattern. Therefore, inventory costing seemingly transforms a fixed cost into a variable cost, and for the inventory-costing purpose, *all manufacturing costs* are routinely regarded as variable.[2]

The choice of an appropriate denominator volume for the budgeted fixed-overhead rate is a matter of judgment, not of science. The most widely used denominator level is the budgeted volume for the year. The next chapter, p. 301, discusses the choice of the denominator more fully.

JOURNAL ENTRIES FOR OVERHEAD

Objective 5

Prepare journal entries for factory overhead under standard costing

In the general-ledger procedure in Chapter 4, we used a single Factory Department Overhead Control account. In this chapter we use two new accounts, one for variable overhead and another for fixed overhead. Thus, we also use two Applied accounts instead of the single Factory Overhead Applied account used in Chapter 4.

Consider the journal entries for the Webb Company. Assume that in April, actual variable factory overhead was $131,000; actual fixed factory overhead, $271,000.

As described in Chapter 4, the entries for overhead during the accounting period do not isolate overhead variances in separate accounts. Instead actual overhead is accumulated in one account and overhead is applied through a separate account:

a. Variable factory overhead control	131,000	
Accounts payable and other accounts		131,000
To record actual variable factory overhead incurred.		
b. Work in process control	120,000	
Variable factory overhead applied		120,000
To record application of variable factory overhead.		

Note that the variable-overhead variances are not shown in separate variance accounts in these entries, whereas direct material and direct labor variances are commonly isolated in separate accounts as the initial general-ledger entries are made. However, we can readily compute the accumulated variable-overhead flexible-budget variance for the year by taking the difference between the debit balance of the Variable Factory Overhead Control account and the credit balance of the Variable Factory Overhead Applied account.

Budgeted fixed factory overhead was $276,000. Assume that the denominator volume was 12,000 units, or 9,600 direct labor hours. The fixed factory overhead would be applied at $23 per unit or $28.75 per direct labor hour. In April, volume was 10,000 units.

The journal entries would be:

[2]This approach assumes *absorption costing,* which applies fixed factory overhead costs to inventories and is the most widely used inventory method. In contrast, *variable costing* does not apply fixed factory overhead costs to inventories. The next chapter compares the absorption costing and variable costing inventory methods in detail.

a.	Fixed factory overhead control	271,000	
	Payroll liability, accumulated depreciation, etc.		271,000
	To record actual fixed overhead incurred. (Detailed postings of fixed-overhead items such as salaries, depreciation, property taxes, and insurance would be made to the department overhead records in the subsidiary ledger for Fixed Factory Overhead Control.)		
b.	Work in process control	230,000	
	Fixed factory overhead applied		230,000
	To apply overhead at the budgeted rate of $23 per unit or $28.75 per standard direct labor hour allowed for the output achieved. (Note how this total differs from the fixed-overhead budget for this level of volume. The budget is $276,000 for *any* level of volume.)		

Again, note that the fixed-factory-overhead variance (that is, the underapplied fixed factory overhead) is not shown at this point in a separate variance account.

The general-ledger treatment practiced in standard-cost systems is not uniform. However, if you understand the features of standard costs, you can easily adapt to any given accounting system. Differences in general-ledger treatment usually center on (a) the number of detailed variance accounts used and (b) the timing of isolating variances in the ledger.

STANDARD, NORMAL, AND ACTUAL COSTING

Objective 6

Identify actual costing, normal costing, and standard costing

The cost accounting literature often mistakenly divides inventory-costing systems into three categories, as if they were mutually exclusive: job costing, process costing, and standard costing. Such a three-way classification is erroneous. Why? Because standard costs can be used in a wide variety of organizations and in conjunction with any kind of inventory costing, whether it be job-order costing, process costing, or some hybrid inventory costing.

There is a major difference in inventory costing between standard costing and normal or actual costing. Under standard costing, Work in Process is usually carried at what the inventories *should* cost rather than at actual cost. Exhibit 8-3 indicates the following:

- **actual costing:** a costing method that allocates actual direct costs to products and uses actual rates to apply indirect costs based on the actual inputs of the indirect-cost application base
- **normal costing:** a costing method that allocates actual direct costs to products and uses budgeted rates to apply indirect costs based on the actual inputs of the indirect-cost application base
- **standard costing:** a costing method that allocates standard direct costs to products and uses budgeted rates to apply indirect costs based on the standard inputs allowed for actual output

Ponder the major difference in overhead application in *normal* costing and *standard* costing. In normal costing, a budgeted rate is multiplied by *actual* inputs such as machine hours. In standard costing, a budgeted rate is multiplied by *standard inputs allowed for the actual outputs achieved.* In normal costing, data regarding only *actual inputs* are generated. In standard costing, data regarding both *actual inputs* and *standard inputs allowed* are generated.

EXHIBIT 8-3
Comparison of Approaches to Inventory Costing

Cost Category	Chapter 2 Approach, Called Actual Costing: Work in Process Inventory	Chapter 4 Approach, Called Normal Costing: Work in Process Inventory	Chapter 8 Approach, Called Standard Costing: Work in Process Inventory
Direct materials	Actual inputs × Actual prices	Actual inputs × Actual prices	Standard inputs allowed for actual output × Standard prices
Direct labor	Actual inputs × Actual prices	Actual inputs × Actual prices	Standard inputs allowed for actual output × Standard prices
Variable factory overhead	Actual inputs × Actual overhead rates	Actual inputs × Budgeted overhead rates	Standard inputs allowed for actual output × Budgeted overhead rates
Fixed factory overhead	Actual inputs × Actual overhead rates	Actual inputs × Budgeted overhead rates	Standard inputs allowed for actual output × Budgeted overhead rates

ANALYSIS OF FIXED-FACTORY-OVERHEAD VARIANCES

General Framework

The most perplexing and controversial variance in standard-cost accounting arises from the inevitable failure to achieve a production volume exactly equal to the denominator volume. Consider the Webb Company facts:

Denominator production volume in product units	12,000 units
Denominator production volume in hours, 12,000 × .8 =	9,600 hours
Actual production achieved	10,000 units
Standard inputs of direct labor allowed, 10,000 × .8 =	8,000 hours
Actual inputs used	8,800 hours

The framework that we used at the beginning of this chapter to analyze variable overhead is arranged in Exhibit 8-4 so that it may easily be compared with the framework for fixed overhead.

Ponder the differences. For all variable costs, the *flexible-budget* allowance will *always* equal the amount *applied* to products. Thus, the amount of the flexible-budget variance for variable overhead will always be identical to underapplied (or overapplied) variable overhead, $11,000 U in this example. In contrast, except when actual output achieved equals the denominator volume, the *flexible budget* for fixed factory overhead will *never* equal the amount of fixed overhead *applied* to products. Thus, the amount of the flexible-budget variance for fixed overhead will rarely be identical to underapplied (or overapplied) fixed overhead, $5,000 F versus $41,000 U in this example.

Finally, note that an efficiency variance is computed for variable but not fixed overhead. Why? Because efficiency variances are measures to help the short-run control of performance. Efficient use of direct materials, direct labor, and variable factory overhead can affect actual costs, but short-run fixed overhead is unaffected by efficiency.

EXHIBIT 8-4

Framework for Analyzing Variable- and Fixed-Factory-Overhead Variances
4-Variance Analysis

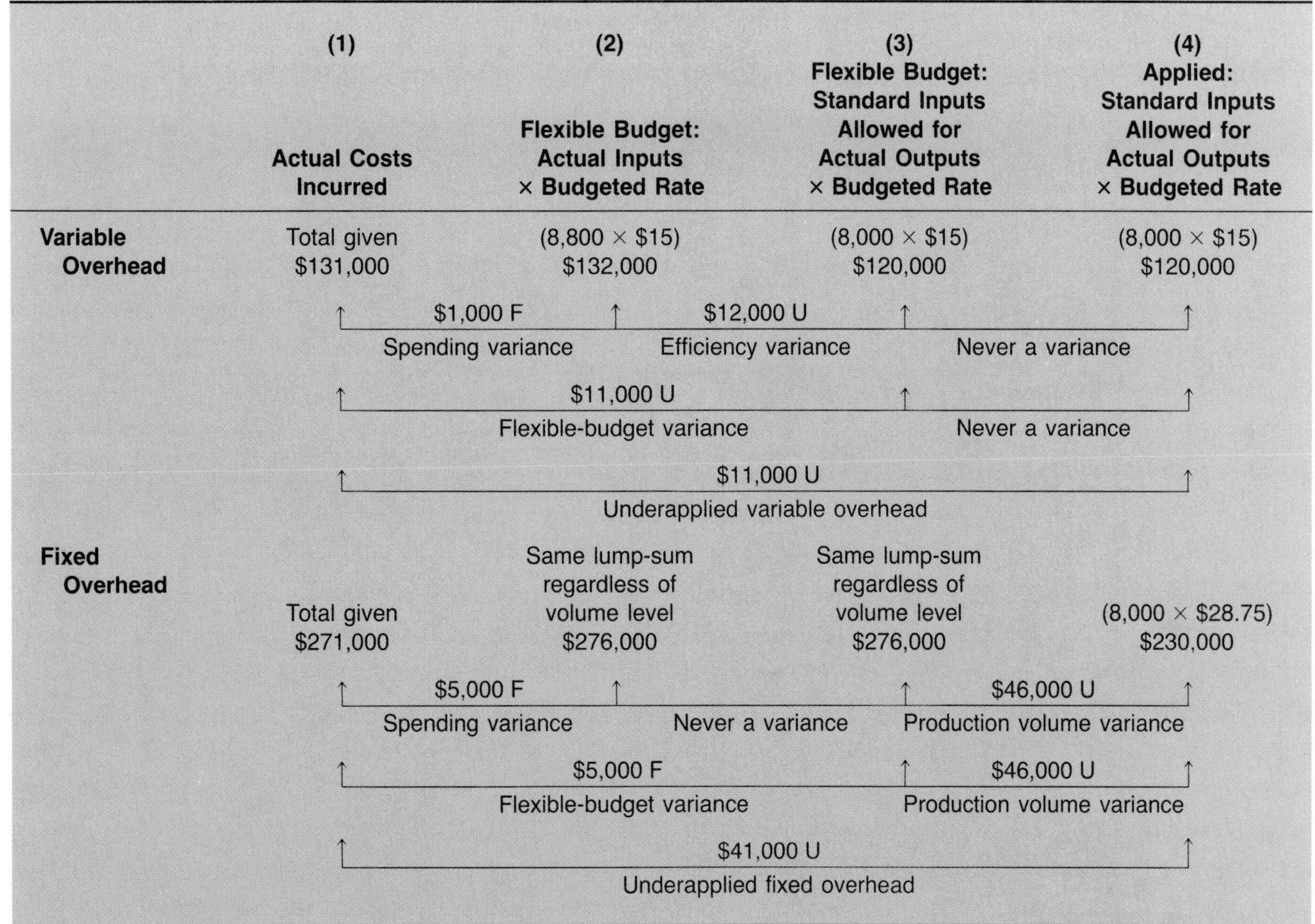

The variance analysis in Exhibit 8-4 is often called a **4-variance analysis** *or* **4-way analysis** *of factory overhead because four variances are computed—two spending variances, one efficiency variance, and one production volume variance. That is, the top sets of arrows for variable and fixed overhead show the most detailed analyses:*

Variable factory overhead	*Spending variance* *Efficiency variance*	*4 variances*
Fixed factory overhead	*Spending variance* *Production volume variance*	

Production Volume Variance

In a standard-costing system, the first step in analyzing fixed overhead is to calculate the underapplied or overapplied amounts:

Actual of $271,000 − Applied of $230,000 = Underapplied fixed overhead of $41,000 U

This $41,000 can be subdivided into two variances: the flexible-budget variance of $5,000 F and a *production volume variance* of $46,000 U. The **production volume variance** is the difference between budgeted fixed overhead and applied fixed overhead.

Exhibit 8-5 is a graph showing the $46,000 production volume variance. ***When actual volume achieved is less than the denominator volume, the production volume***

EXHIBIT 8-5
Production Volume Variance

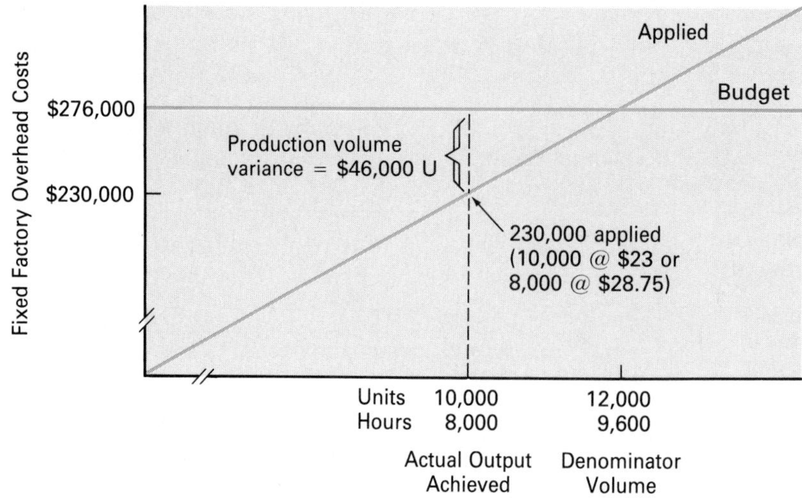

variance is unfavorable. Computations can be expressed using standard inputs, direct labor hours in our example as follows:

$$\begin{aligned}\text{Production volume variance} &= \text{Budgeted fixed overhead} - \text{Applied fixed overhead}\\ &= \$276{,}000 - (8{,}000 \times \$28.75)\\ &= \$276{,}000 - \$230{,}000\\ &= \$46{,}000\ \text{U}\end{aligned}$$

or

$$\begin{aligned}\begin{matrix}\text{Production volume}\\ \text{variance}\end{matrix} &= \left(\begin{matrix}\text{Denominator}\\ \text{volume in}\\ \text{units of}\\ \text{standard inputs}\end{matrix} - \begin{matrix}\text{Units of}\\ \text{standard inputs}\\ \text{allowed for}\\ \text{actual outputs}\end{matrix}\right) \times \left(\begin{matrix}\text{Budgeted fixed-}\\ \text{overhead rate}\\ \text{per unit of}\\ \text{standard inputs}\end{matrix}\right)\\ &= (9{,}600 - 8{,}000) \times \$28.75\\ &= \$46{,}000\ \text{U}\end{aligned}$$

The computation also can be expressed in product units:

$$\begin{aligned}\text{Production volume variance} &= \left(\begin{matrix}\text{Denominator}\\ \text{volume in}\\ \text{product units}\end{matrix} - \begin{matrix}\text{Actual}\\ \text{outputs in}\\ \text{product units}\end{matrix}\right) \times \left(\begin{matrix}\text{Budgeted fixed-}\\ \text{overhead rate}\\ \text{per product unit}\end{matrix}\right)\\ &= (12{,}000 - 10{,}000) \times \$23\\ &= \$46{,}000\ \text{U}\end{aligned}$$

When actual production volume exceeds denominator volume, the production volume variance is favorable. Why? Because managers regard the variance as a benefit of better-than-expected utilization of facilities.

There is no production volume variance for variable overhead. The concept of production volume variance arises for fixed overhead because of the conflict between accounting for planning and control (by budgets) and accounting for inventory costing (by budgeted application rates). Note again that the fixed-overhead budget serves the planning and control purpose, ***whereas the development of the inventory-costing rate results in the treatment of fixed overhead*** **as if** ***it were a variable cost.*** In other words, the applied line in Exhibit 8-5 is artificial in the sense that, for inventory-costing purposes, it seemingly transforms a fixed cost into a variable cost. This bit of magic forcefully illustrates the distinction between accounting for planning and control and accounting for inventory costing.

To summarize, the production volume variance arises because the actual production volume level achieved usually does not coincide with the production level used as a denominator volume for computing a budgeted application rate for inventory costing of fixed factory overhead.

Economic Meaning of Unfavorable Production Volume Variance

The economic meaning of the flexible-budget variance for fixed factory overhead is relatively straightforward. The amounts actually incurred are compared with the lump-sum amounts budgeted.

In contrast, the economic meaning of the production volume variance is less clear. The Webb Company variance in April is (12,000 − 10,000) × \$23 = \$46,000, unfavorable. Remember that the variance is a creature of the "unitizing" of fixed costs. Does the production volume variance measure the economic cost to Webb Company of producing and selling 10,000 rather than 12,000 units in April? The presence of idle facilities has no bearing on the amount of fixed costs currently incurred.

How should the costs of idle capacity be measured? The economic impacts of the inability to reach a target denominator volume are often directly measured by lost contribution margin, even if it must be approximated. In the Webb Company example, Exhibit 7-4 (page 216) revealed that the lost contribution margin was 2,000 units × \$81 = \$162,000. Managers tend to understand the meaning of this number, but they find the production volume variance less informative.

The use of historical fixed-overhead unit costs might be justified in an economic sense on the following grounds: "We cannot regularly maintain measurements of lost contribution margin per unit. Therefore, as a practical but crude substitute, we use historical fixed-overhead unit costs instead."

Impact of Choice of Denominator Volume

The economic meaning of the production volume variance is also clouded because its amount is affected by the choice of the denominator volume. For example, consider Webb's April operations with three different denominator volumes:

	Alternative Denominator Volumes		
(1) Denominator volume in product units	10,000	12,000	14,000
(2) Budgeted fixed factory overhead	\$276,000	\$276,000	\$276,000
(3) Budgeted fixed-overhead rate per unit of product	\$27.60	\$23.00	\$19.71
(4) Applied fixed overhead (10,000 units of actual output × \$27.60, \$23.00, \$19.71)	\$276,000	\$230,000	\$197,100
(5) Production volume variance (2)–(4)	0	\$46,000 U	\$78,900 U

The table illustrates three of many possible denominator volumes. The choice of a denominator volume might reasonably fall anywhere within the relevant range of, say, 9,000 to 15,000 units per month. Thus, the choice of denominator level results in different amounts of applied fixed overhead. Therefore, it affects inventory valuations and, if inventory levels change, operating income—even if all other facts and conditions are the same.

Different Terminology

Production volume variances have a number of widely used synonyms: **capacity variance**, **idle capacity variance**, **activity variance**, **denominator variance**, and just plain **volume variance**. The last term is particularly popular, but we use *production volume variance* here. Why? To distinguish the *production* volume variance, which is unique to inventory costing, from *sales* volume variances. As explained in the preceding chapter, sales volume variances are encountered in *all* accounting systems and in *all* types of organizations, including service industries and nonprofit entities, but production volume variances are associated with manufacturing companies.

4-VARIANCE, 3-VARIANCE, 2-VARIANCE OVERHEAD ANALYSIS OF ACTIVITY AREA

The analysis of overhead is a front-line task that managers should conduct at the activity-area level. We now examine a simplified flexible budget for variable and fixed overhead for a machining activity area of Precision Company. In this illustration, we will review some fundamental concepts and also explore some alternative ways to analyze factory overhead variances.

Data for Illustration

Exhibit 8-6 shows a flexible budget for variable and fixed factory overhead. Exhibit 8-7 shows actual factory overhead costs.

EXHIBIT 8-6
Precision Company
Machining Activity Area
Simplified Flexible Factory-Overhead Budget for Anticipated Monthly Volume Range

Standard machine hours allowed	800	900	1,000	1,100
Variable factory overhead:				
Machine adjustments and repairs	$ 8,000	$ 9,000	$10,000	$11,000
Idle time	800	900	1,000	1,100
Rework	800	900	1,000	1,100
Overtime premium	400	450	500	550
Supplies	3,600	4,050	4,500	4,950
Total	$13,600	$15,300	$17,000	$18,700
Variable-overhead rate, $17 per machine hour				
Fixed factory overhead:				
Supervision	$ 2,700	$ 2,700	$ 2,700	$ 2,700
Depreciation—plant	1,000	1,000	1,000	1,000
Depreciation—equipment	15,000	15,000	15,000	15,000
Property taxes	1,000	1,000	1,000	1,000
Insurance—factory	300	300	300	300
Total	$20,000	$20,000	$20,000	$20,000
Total factory overhead	$33,600	$35,300	$37,000	$38,700

A team of operating personnel and accountants jointly decided that machine hours was the principal cost driver. Using machine hours as the application base, the team computed a budgeted variable overhead application rate of $17 per hour. The

EXHIBIT 8-7
Precision Company
Machining Activity Area
Actual Factory-Overhead Costs For the Month Ended March 31, 19_1

Variable overhead:		Fixed overhead:	
Machine adjustments and repairs	$ 8,200	Supervision	$ 2,700
Idle time	600	Depreciation—plant	1,000
Rework	850	Depreciation—equipment	15,000
Overtime premium	600	Property taxes	1,150
Supplies	4,000	Insurance—factory	350
Total	$14,250	Total	$20,200

Total actual factory overhead: $14,250 + $20,200 = $34,450

team also selected a denominator volume of 1,000 machine hours for setting the budgeted fixed-overhead application rate.

The actual hours of input were 790. The standard time allowed per unit of product was .4 hour. There were 2,000 units of product manufactured, which we can express as 800 (2,000 × .4) standard hours allowed.

4-Variance Analysis of Overhead Variances

Objective 7

Explain four ways to analyze factory-overhead variances

Exhibit 8-8, Panel A, presents an analysis of the overhead variances. The framework is the same one used earlier in the chapter. Note especially that the efficiency variance is 790 − 800 hours, or 10 hours, favorable, in nonfinancial terms. This physical measure is a popular way on the factory floor to judge if operations are under control. The financial impact on variable overhead because of efficient use of machine hours is $170 (10 hours × $17), favorable.

The budgeted fixed-factory-overhead rate is $20 per machine hour ($20,000 ÷ 1,000 hours). Note the unfavorable production volume variance, which can be computed in either of two ways:

Budget amount	$20,000
Applied amount, 800 machine hours × $20	16,000
Production volume variance	$ 4,000 U

Denominator volume in machine hours	1,000
Actual volume of output achieved expressed in machine hours allowed	800
Difference (underutilized capacity)	200 hours
Budgeted rate for applying fixed overhead	× $20
Production volume variance	$ 4,000 U

The detailed analysis in Exhibit 8-8, Panel A, is often called a *4-variance analysis* or a *4-way analysis* because four variances are computed—two spending variances (variable and fixed overhead), one efficiency variance, and one production volume variance.

Combined Rate

Objective 8

Explain the variance effects of a combined-overhead rate

Many companies separate variable overhead and fixed overhead for planning and control purposes and combine them for inventory-costing purposes. These companies use a single budgeted-overhead rate. In Exhibit 8-8, Panel B, this rate would be $37, the variable-overhead rate of $17 plus the fixed-overhead rate of $20. For example, the applied combined (total) single overhead is $29,600 (800 machine hours × $37). The overhead variance analysis in Panel B is conceptually the same as in Panel A but shows less detail.

Consider this summary of the relationships among the variances in Exhibit 8-8. The 4-variance analysis of Panel A is shown below:

4-Variance Analysis	Spending Variance	Efficiency Variance	Production Volume Variance
Variable Factory Overhead	$820 U	$170 F	
Fixed Factory Overhead	$200 U		$4,000 U

The 3-variance analysis in Panel B makes no distinction between variable factory overhead incurred and fixed factory overhead incurred. The two spending variances in the 4-variance analysis are shown as a single spending variance in the 3-variance analysis:

3-Variance Analysis	Spending Variance	Efficiency Variance	Production Volume Variance
Combined Factory Overhead	$1,020 U	$170 F	$4,000 U

The 2-variance analysis is confined to the flexible-budget variance and the production volume variance:

2-Variance Analysis	Flexible-Budget Variance	Production Volume Variance
Combined Factory Overhead	$850 U	$4,000 U

The 1-variance analysis is the single difference between the combined actual costs incurred and the applied costs:

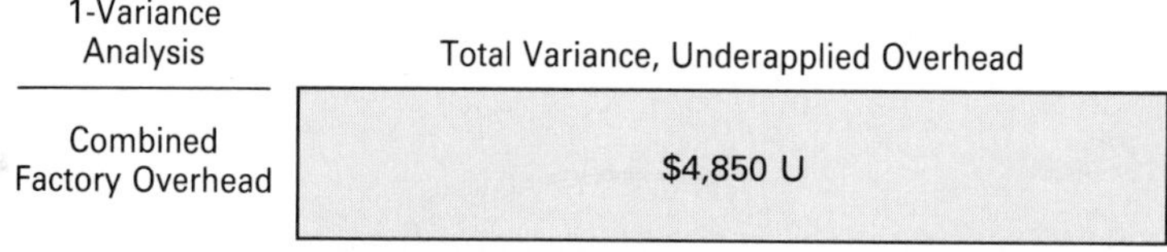

2-Variance and 3-Variance Analysis

Exhibit 8-8, Panel B, provides a comprehensive analysis of all relationships among the combined-overhead analyses and how combined overhead consists of variable

EXHIBIT 8-8 *(Place a clip on this page for easy reference.)*

Framework for Analysis of Factory-Overhead Variances

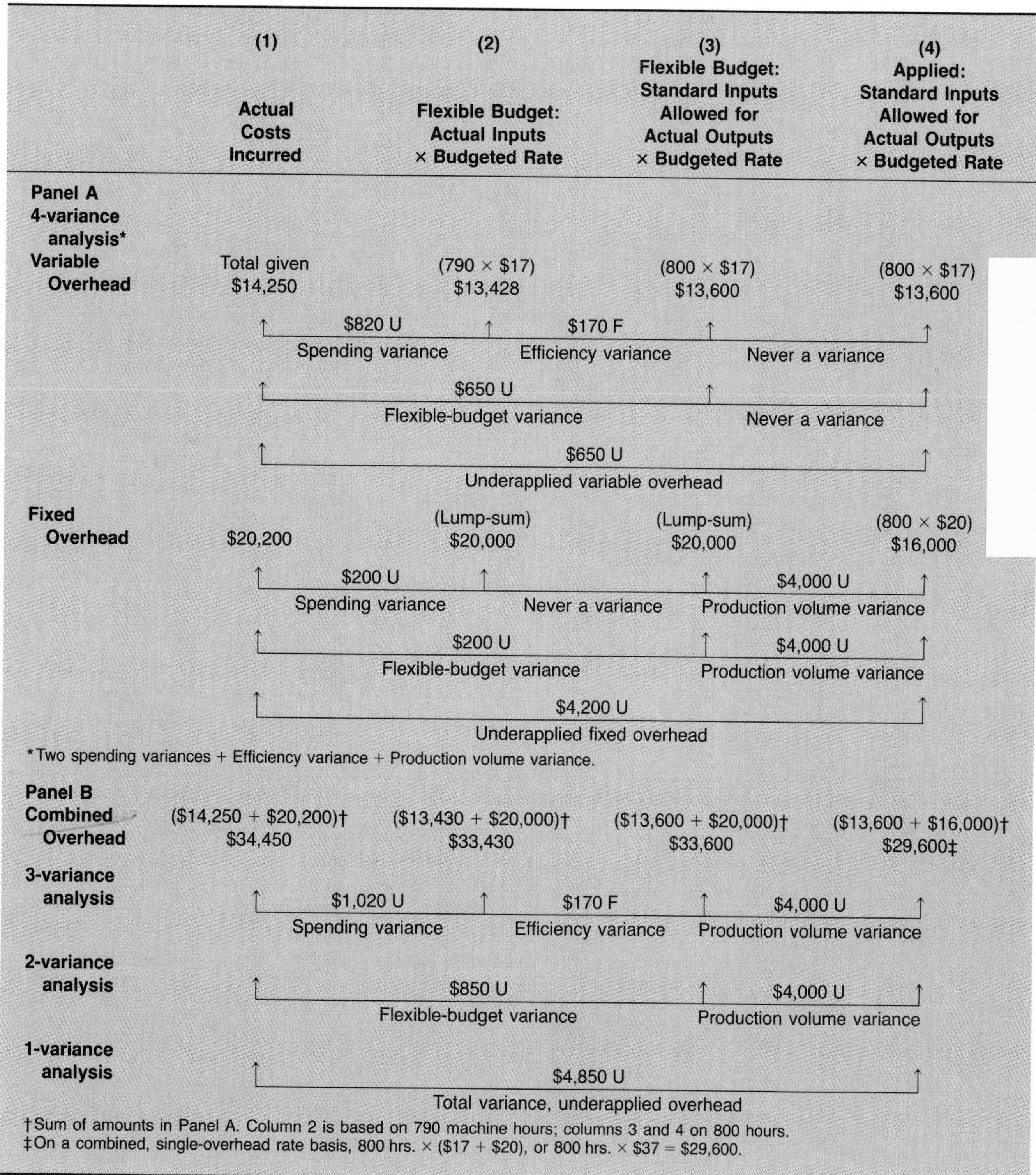

	(1) Actual Costs Incurred	(2) Flexible Budget: Actual Inputs × Budgeted Rate	(3) Flexible Budget: Standard Inputs Allowed for Actual Outputs × Budgeted Rate	(4) Applied: Standard Inputs Allowed for Actual Outputs × Budgeted Rate
Panel A 4-variance analysis*				
Variable Overhead	Total given $14,250	(790 × $17) $13,428	(800 × $17) $13,600	(800 × $17) $13,600
	$820 U Spending variance (1)–(2)	$170 F Efficiency variance (2)–(3)	Never a variance (3)–(4)	
	$650 U Flexible-budget variance (1)–(3)		Never a variance (3)–(4)	
	$650 U Underapplied variable overhead (1)–(4)			
Fixed Overhead	$20,200	(Lump-sum) $20,000	(Lump-sum) $20,000	(800 × $20) $16,000
	$200 U Spending variance (1)–(2)	Never a variance (2)–(3)	$4,000 U Production volume variance (3)–(4)	
	$200 U Flexible-budget variance (1)–(3)		$4,000 U Production volume variance (3)–(4)	
	$4,200 U Underapplied fixed overhead (1)–(4)			

*Two spending variances + Efficiency variance + Production volume variance.

	(1)	(2)	(3)	(4)
Panel B Combined Overhead	($14,250 + $20,200)† $34,450	($13,430 + $20,000)† $33,430	($13,600 + $20,000)† $33,600	($13,600 + $16,000)† $29,600‡
3-variance analysis	$1,020 U Spending variance (1)–(2)	$170 F Efficiency variance (2)–(3)	$4,000 U Production volume variance (3)–(4)	
2-variance analysis	$850 U Flexible-budget variance (1)–(3)		$4,000 U Production volume variance (3)–(4)	
1-variance analysis	$4,850 U Total variance, underapplied overhead (1)–(4)			

†Sum of amounts in Panel A. Column 2 is based on 790 machine hours; columns 3 and 4 on 800 hours.
‡On a combined, single-overhead rate basis, 800 hrs. × ($17 + $20), or 800 hrs. × $37 = $29,600.

and fixed components, as shown in the 4-variance analysis in Panel A. Study it slowly, step by step. Begin at the bottom with the 1-variance analysis, which shows the combined (total) overhead variance only.

Note the distinction between the 2-variance and 3-variance overhead analysis. The 3-variance analysis computes three variances: spending, efficiency, and production volume. The 2-variance analysis computes only two variances: flexible-budget (sometimes called the controllable variance) and production volume variance. The flexible-budget variance, as shown in Exhibit 8-8, is simply the difference between actual costs and the flexible-budget allowance based on standard hours allowed. The 2-variance analysis stops there; it does not subdivide the flexible-budget variance into spending and efficiency variances.[3]

The essential distinction between the 4-variance analysis and all other levels of analysis is the separate analysis of variable overhead and fixed overhead.

Computing a Flexible Budget for Combined Overhead

A close study of Panel B of Exhibit 8-8 also shows that the flexible budget for the combined factory overhead is based on a formula with fixed and variable components: \$20,000 + \$17 per machine hour. Such a formula can be developed by analyzing the budgeted behavior of costs as they change between two levels of volume. For example, in Exhibit 8-8, Panel B:

$$\text{Variable-overhead rate} = \frac{\text{Change in combined total overhead}}{\text{Change in production volume}}$$

$$= \frac{\$33{,}600 - \$33{,}430}{800 - 790} = \frac{\$170}{10} = \$17$$

Choose either level of volume to compute the fixed-overhead component:

$$\text{Fixed-overhead component} = \$33{,}600 - 800\ (\$17) = \$20{,}000$$

or

$$\text{Fixed-overhead component} = \$33{,}430 - 790\ (\$17) = \$20{,}000$$

OVERHEAD VARIANCES IN THE LEDGER

Objective 9

Prepare journal entries for factory-overhead variances

There are several ways of accounting for overhead variances. The easiest way is probably to allow the department-overhead control accounts and applied accounts to accumulate month-to-month postings until the end of the year. Monthly variances would not be isolated formally in the accounts, although monthly variance reports could be prepared. Assume that the data in Exhibit 8-8, Panel A, are for the *year* rather than for the *month*. At year-end, isolating (that is, separately identifying) and closing entries could be made as follows:

1. Variable factory overhead applied	13,600	
Variable-overhead spending variance	820	
Variable-overhead efficiency variance		170
Variable-factory-overhead control		14,250
To isolate variances for the year.		

[3]The approaches to analyzing overhead variances are not as uniform as they are for direct materials and direct labor variances. J. Chiu and Y. Lee, "A Survey of Current Practice in Overhead Accounting and Analysis," *Proceedings of Meeting of American Accounting Association* (O. R. Whittington, San Diego State University) indicated that 40% of respondents used the 2-variance analysis; 37%, the 3-variance; and 23%, some other analysis.

2. Fixed factory overhead applied	16,000	
Fixed-overhead spending (flexible budget) variance	200	
Fixed-overhead production volume variance	4,000	
Fixed-factory-overhead control		20,200
To isolate variances for the year.		
3. Income summary (or Cost of goods sold)	650	
Variable-overhead efficiency variance	170	
Variable-overhead spending variance		820
To close.		
4. Income summary (or Cost of goods sold)	4,200	
Fixed-overhead spending (flexible budget) variance		200
Fixed-overhead production volume variance		4,000
To close.		

If desired, the isolating entries for the variances (1 and 2 above) could be made monthly, although the closing entries (3 and 4 above) are usually confined to year-end.

Of course, rather than being closed directly to the Income Summary or Cost of Goods Sold, in certain cases the overhead variances may be prorated at the year-end, as shown in the next chapter, p. 306.

The general ledger is designed mainly to serve the purpose of inventory costing. Yet management's major purpose, that of planning and control, is aided by using flexible-budget figures, which are not highlighted in general-ledger balances. As is often the case, conventional general-ledger accounting for overhead often provides only minimal information for control.

CHECKPOINTS FOR ANALYZING VARIANCES

The Precision Company activity-area illustration provides a useful means for checking your overall understanding of overhead variances. Consider the following checkpoints:

1. Identify the actual output achieved and the standard inputs allowed for that output before proceeding with any other part of variance computations. The Applied column in the analytical framework is based on either actual product outputs or standard inputs allowed for actual outputs.
2. Clearly distinguish between variable factory overhead, which has an efficiency variance but no production volume variance, and fixed factory overhead, which has a production volume variance but no efficiency variance.
3. Remember that a production volume variance can be computed in either of two ways: (a) Budget − Applied, or (b) Budgeted fixed-overhead rate × (Denominator volume − Actual volume achieved).
4. Remember that the flexible budget for fixed factory overhead is a lump sum that is unaffected by either actual hours of input or standard hours of input allowed for actual output achieved.

The following table is a capsule review of the key variances presented in this chapter and the preceding chapter.

Applicability to a Particular Cost	Direct Materials	Direct Labor	Variable Factory Overhead	Fixed Factory Overhead
Price variance?	Yes	Yes	No	No
Spending variance?	No	No	Yes	Yes*
Efficiency variance?	Yes	Yes	Yes	No
Flexible-budget variance?	Yes	Yes	Yes	Yes
Denominator volume concept?	No	No	No	Yes
Production volume variance?	No	No	No	Yes

*For fixed overhead, the flexible-budget variance and the spending variance are synonymous.

The checkpoints and the table will help analysis. However, to get the big picture of overall cost management, accountants and executives should initially ask three questions:

- How do *individual* costs, including individual overhead cost items, behave?
- Who is responsible for their control?
- How do costs and variances interrelate and affect *total* costs for the organization as a whole?

PERFORMANCE EVALUATION AND VARIANCES

The variances outlined in this chapter capture important aspects of managerial performance. However, like any set of performance measures, they can be inappropriately used. In Chapter 7 (p. 233), we note how it is inappropriate to overemphasize individual price or efficiency variances for direct materials or direct labor. The same caution applies to the manufacturing-overhead variances presented in this chapter.

For cost control purposes, the variances presented in Chapters 7 and 8 are best viewed as attention directors, not problem solvers. Problem solving on the manufacturing floor frequently uses nonfinancial measures reported on a daily, an hourly, or even a continuous basis. Examples of these nonfinancial measures include:

- first-time through yield—the percentage of products manufactured "right" (to specification) the first time, and
- throughput time—the time that a product takes from the first stage of manufacture to its completion.

Many companies use both nonfinancial measures and financial measures when evaluating the performance of their managers. The overhead variances outlined in this chapter highlight the financial consequences of actions that usually initially appear as nonfinancial measures. For example, a worker may consider the hour-to-hour scrap rate (a nonfinancial measure) as the first signal of inferior direct materials. This signal might prompt rapid corrective actions. The cost of excess material usage may appear as an efficiency variance (a financial measure) reported later (daily, weekly, or monthly).

Performance analysis is discussed in many subsequent chapters. For example, Chapters 9 and 28 note potential conflicts between financial and nonfinancial performance measures. There are no easy solutions to designing an appropriate performance measurement system. Focusing only on nonfinancial measures or focusing only on financial measures is nearly always too simplistic.

PROBLEM FOR SELF-STUDY

The following problem reviews both this chapter and the preceding chapter. Although the numbers have deliberately been kept small to simplify the computations, the underlying concepts have widespread use.

PROBLEM

The McDermott Furniture Company has established standard costs for the cabinet department, in which one size of a single four-drawer style of dresser is produced. The standard costs are used in evaluating actual performance. The standard costs of producing one of these dressers are shown below:

Standard-Cost Record
Dresser, Style AAA

Direct materials: Lumber—50 board feet @ $5	$250
Direct labor: 3 hours @ $24	72
Factory overhead:	
Variable costs—3 hours @ $10	30
Fixed costs—3 hours @ $20	60
Total per dresser	$412

The actual costs of operations to produce 400 of these dressers during January are as follows (there were no initial inventories):

Direct materials purchased:	25,000 board feet @ $5.60	$140,000
Direct materials used:	19,000 board feet	
Direct labor:	1,100 hours at $22	24,200
Factory overhead:		
Variable costs		12,760
Fixed costs		29,800

The flexible budget for this department at the monthly volume level used to set the budgeted fixed-overhead rate called for 1,400 direct labor hours of operation. At this level, the variable overhead was budgeted at $14,000, and the fixed overhead at $28,000.

Required
Make all journal entries for January. Compute the following variances from standard cost. Label your answers as *favorable* (F) or *unfavorable* (U).

1. Direct material price, isolated at time of purchase
2. Direct material efficiency
3. (a) Direct labor price
   (b) Direct labor efficiency
4. (a) Variable-overhead flexible-budget variance
   (b) Fixed-overhead spending (flexible-budget) variance
   (c) Fixed-overhead production volume variance
5. (a) Variable-overhead spending variance
   (b) Variable-overhead efficiency variance

SOLUTION

Journal entries are supported by pertinent variance analysis (see Exhibit 8-9):

1. Materials control (25,000 @ $5.00)	125,000	
Direct material price variance (25,000 @ $.60)	15,000	
Accounts payable (25,000 @ $5.60)		140,000

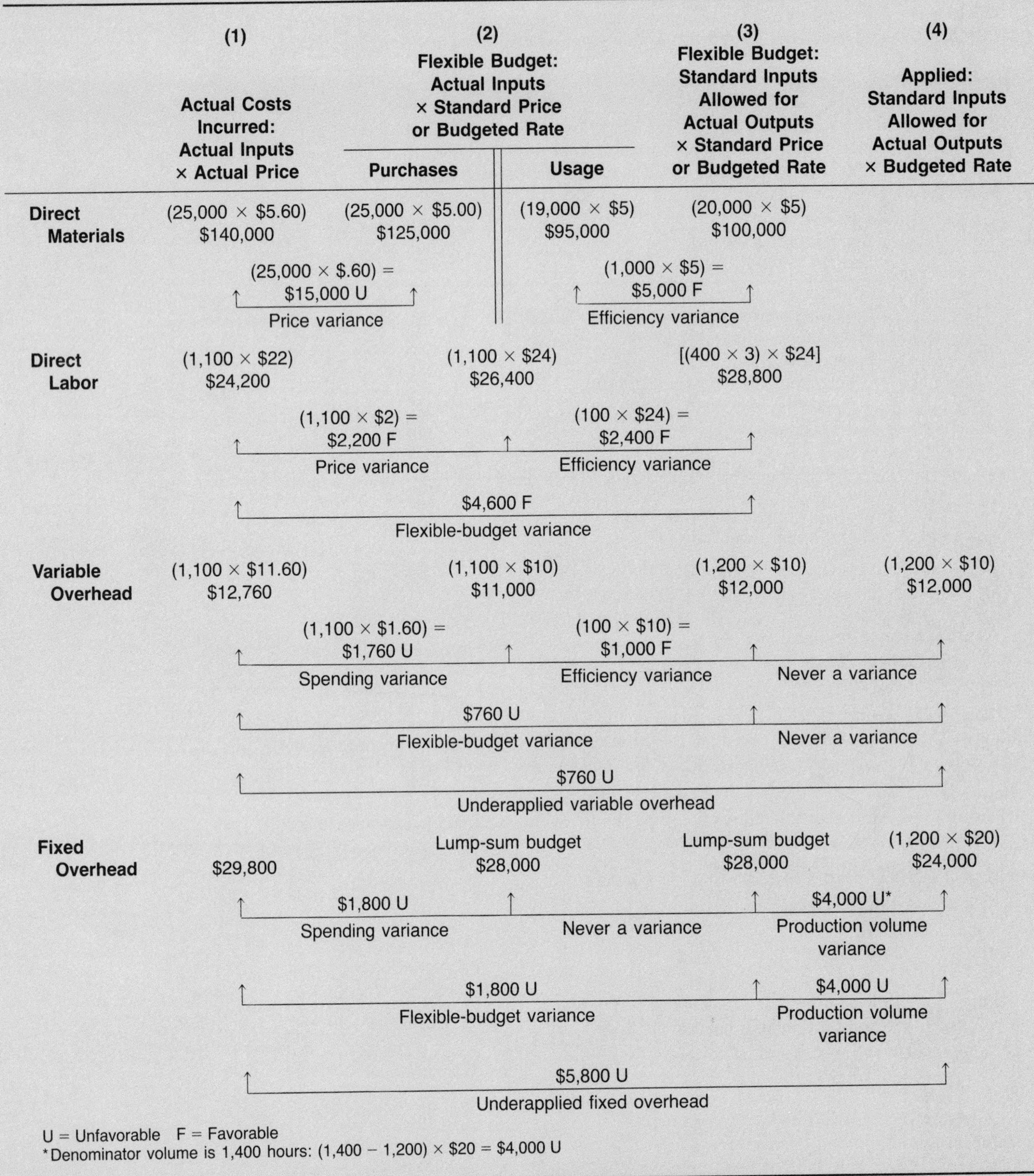

EXHIBIT 8-9 *(Place a clip on this page for easy reference.)*
McDermott Furniture Company
Analysis of Manufacturing Costs

	(1) Actual Costs Incurred: Actual Inputs × Actual Price	(2) Flexible Budget: Actual Inputs × Standard Price or Budgeted Rate — Purchases	(2) Flexible Budget: Actual Inputs × Standard Price or Budgeted Rate — Usage	(3) Flexible Budget: Standard Inputs Allowed for Actual Outputs × Standard Price or Budgeted Rate	(4) Applied: Standard Inputs Allowed for Actual Outputs × Budgeted Rate
Direct Materials	(25,000 × $5.60) $140,000	(25,000 × $5.00) $125,000	(19,000 × $5) $95,000	(20,000 × $5) $100,000	
Direct Labor	(1,100 × $22) $24,200	(1,100 × $24) $26,400		[(400 × 3) × $24] $28,800	
Variable Overhead	(1,100 × $11.60) $12,760	(1,100 × $10) $11,000		(1,200 × $10) $12,000	(1,200 × $10) $12,000
Fixed Overhead	$29,800	Lump-sum budget $28,000		Lump-sum budget $28,000	(1,200 × $20) $24,000

Direct Materials
(25,000 × $.60) = $15,000 U — Price variance
(1,000 × $5) = $5,000 F — Efficiency variance

Direct Labor
(1,100 × $2) = $2,200 F — Price variance
(100 × $24) = $2,400 F — Efficiency variance
$4,600 F — Flexible-budget variance

Variable Overhead
(1,100 × $1.60) = $1,760 U — Spending variance
(100 × $10) = $1,000 F — Efficiency variance
Never a variance
$760 U — Flexible-budget variance
Never a variance
$760 U — Underapplied variable overhead

Fixed Overhead
$1,800 U — Spending variance
Never a variance
$4,000 U* — Production volume variance
$1,800 U — Flexible-budget variance
$4,000 U — Production volume variance
$5,800 U — Underapplied fixed overhead

U = Unfavorable F = Favorable
*Denominator volume is 1,400 hours: (1,400 − 1,200) × $20 = $4,000 U

2. Work in process control (400 units × 50 board feet × $5)	100,000	
Direct material efficiency variance (1,000 × $5)		5,000
Materials control (19,000 × $5)		95,000
3. Work in process control (400 units × $72)	28,800	
Direct labor price variance (1,100 hrs. × $2)		2,200
Direct labor efficiency variance (100 hrs. × $24)		2,400
Payroll liability (1,100 hrs. × $22)		24,200
4. Variable-factory overhead control	12,760	
Accounts payable and other accounts		12,760
Work in process control	12,000	
Variable-factory overhead applied (400 × 3 × $10)		12,000
5. Fixed-factory overhead control	29,800	
Payroll liability, accumulated depreciation, etc.		29,800
Work in process control	24,000	
Fixed-factory overhead applied (400 × 3 × $20)		24,000

The analysis of variances in Exhibit 8-9 summarizes the characteristics of different cost-behavior patterns. The approaches to direct labor and variable overhead are basically the same. Furthermore, there is no fundamental conflict between the planning-and-control and inventory-costing purposes: that is, the applied amounts in column (4) also equal the flexible-budget allowances. In contrast, the cost-behavior pattern and control features of fixed overhead require a different analytical approach. The budget is a lump sum. There is no efficiency variance for fixed factory overhead because short-run performance cannot ordinarily affect the incurrence of fixed factory overhead. Finally, there will nearly always be a conflict between the planning-and-control and inventory-costing purposes because the applied amount in column (4) for fixed overhead usually will differ from the lump-sum budget amount. The latter conflict is highlighted by the production volume variance, which measures the effect of working at a volume level other than the denominator volume used to set the budgeted-fixed-overhead application rate.

SUMMARY

This chapter highlights and contrasts two major purposes that must be served in accounting for overhead—planning and control through the use of flexible budgets and inventory costing through the use of budgeted overhead rates.

When analyzing overhead variances, managers regard the flexible-budget variances as controllable, at least to some degree. Managers generally regard the production volume variance as less subject to control in the short run.

The general-ledger entries in this chapter sharply distinguish between variable and fixed overhead. This treatment is more effective for planning and control because it emphasizes the basic differences in cost behavior. These differences are often important in influencing managerial decisions. The final section demonstrates that these distinctions can be maintained even if a combined-overhead rate is used for inventory costing.

The worksheet analysis illustrated in Exhibit 8-8 provides a useful approach to the analysis of overhead variances. The first step is to obtain the underapplied or overapplied overhead, the total overhead variance. Then any further variance breakdowns can be added algebraically and checked against the total variance.

TERMS TO LEARN

This chapter and the Glossary at the end of the book contain definitions of the following important terms:

activity variance *(p. 265)* actual costing *(261)* capacity variance *(265)*
denominator activity *(258)* denominator level *(258)*
denominator variance *(265)* denominator volume *(258)*
idle capacity variance *(265)* normal costing *(261)*
production volume variance *(263)* spending variance *(255)*
standard costing (261) volume variance *(266)*

ASSIGNMENT MATERIAL

QUESTIONS

8-1 "Control of factory overhead increasingly entails monitoring nonfinancial variables." Name three nonfinancial variables that affect factory overhead.

8-2 Identify three cost drivers that are frequently related to variable factory overhead.

8-3 State the meaning of the efficiency variance for variable factory overhead.

8-4 "The two tidy categories of price and efficiency are a crude split for analyzing variable-factory-overhead variances." Explain.

8-5 "A favorable variable-factory-overhead efficiency variance would be better described as a favorable variable-overhead labor efficiency variance." Do you agree? Explain.

8-6 "The spending variance for variable factory overhead is really affected by several factors." Explain.

8-7 "Managers also control overhead by examining costs across various business functions together." Give an example.

8-8 "Variances of actual costs from budgeted fixed overhead tend to be relatively small." Explain.

8-9 Give two synonyms for *denominator volume.*

8-10 Explain how the accounting for fixed factory overhead differs between the planning and control purpose and the inventory-costing purpose.

8-11 Give three synonyms for *production volume variance.*

8-12 Why is it better to use the term *production volume variance* instead of *volume variance?*

8-13 "The spending variance and the flexible-budget variance for fixed factory overhead will always be identical amounts." Explain.

8-14 What is the essential difficulty in applying fixed overhead to a product?

8-15 "There should be an efficiency variance for fixed overhead. A supervisor can inefficiently use fixed resources." Comment.

8-16 Why is the title "flexible budget" a misnomer?

8-17 What three basic questions should be asked in approaching the control of overhead?

8-18 Describe a useful first step in variance computations.

8-19 Describe two main ways to compute a production volume variance.

8-20 "There is a basic difference in types of variances if they are computed for combined overhead instead of for separate variable and fixed categories." Do you agree? Explain.

EXERCISES AND PROBLEMS

8-21 4-Variance analysis, fill in the blanks. Use the given factory overhead data to fill in the blanks.

	Variable	Fixed
Actual costs incurred	$11,900	$6,000
Applied to product	9,000	4,500
Flexible budget: Standard inputs allowed for actual outputs × budgeted rate	9,000	5,000
Flexible budget: Actual inputs × budgeted rate	10,000	5,000

Use *F* for favorable and *U* for unfavorable:

	Variable	Fixed
1. Spending variance	$_____	$_____
2. Efficiency variance	_____	_____
3. Production volume variance	_____	_____
4. Flexible-budget variance	_____	_____
5. Underapplied (overapplied) factory overhead	_____	_____

8-22 4-Variance analysis, fill in the blanks. Using the given factory overhead data and *F* for favorable and *U* for unfavorable, fill in the blanks:

	Variable	Fixed
Budget for actual machine hours of input	$40,000	$60,000
Applied	36,000	55,000
Budget for standard machine hours allowed for actual outputs	?	?
Actual costs incurred	43,000	59,300
1. Spending variance	$_____	$_____
2. Efficiency variance	_____	_____
3. Production volume variance	_____	_____
4. Underapplied overhead	_____	_____
5. Flexible-budget variance	_____	_____

8-23 Straightforward coverage of factory overhead in standard-cost system. The Singapore division of a Canadian telecommunications company uses a standard-cost system for its machine-paced production of telephone equipment. Data regarding production during June follow:

Variable-factory-overhead costs incurred, $155,100
Variable-factory-overhead costs applied at $12 per machine hour
Fixed-factory-overhead costs incurred, $401,000
Fixed factory overhead budgeted, $390,000
Denominator volume in machine hours, 13,000
Standard machine hours allowed per equivalent unit of output, 0.3
Equivalent units of output, 41,000
Actual machine hours used, 13,300
Ending work in process inventory, zero

Required

1. Prepare journal entries for factory overhead without explanations.
2. Prepare an analysis of all variances. Use the framework illustrated in Exhibit 8-4, p. 263.
3. Describe how variable-overhead items are controlled from day to day. Also, describe how individual fixed-overhead items are controlled.

8-24 Graphs and overhead variances. The Carvelli Company is a manufacturer of housewares. The company uses a standard-cost system. The budget for 19_2 included:

Variable factory overhead	$9 per machine hour
Fixed factory overhead	$72,000,000
Denominator volume based on expected machine hours	4,000,000 hours

Required

1. Prepare four freehand graphs, two for variable overhead and two for fixed overhead. Each pair should display how total costs will fluctuate for (a) planning and control purposes and (b) inventory-costing purposes.
2. Suppose 3,500,000 standard machine hours were allowed for the production actually achieved in 19_2. However, 3,800,000 actual hours of input were used. Actual factory overhead was: variable, $36,100,000; fixed, $72,200,000. Compute (a) variable-overhead spending and efficiency variances and (b) fixed-overhead spending variance and production volume variance. Use the analytical framework illustrated in Exhibit 8-4, p. 263.
3. Prepare a graph for fixed factory overhead that shows a budget line and an applied line. Indicate the production volume variance on the graph.
4. What is the amount of the underapplied variable overhead? Underapplied fixed overhead? Why are the flexible-budget variance and the underapplied overhead always the same amount for variable factory overhead but rarely the same amount for fixed factory overhead?
5. Suppose the denominator volume level was 3,000,000 rather than 4,000,000 hours. What variances in requirement 2 would be affected? What would be their new amounts?

8-25 Journal entries. Refer to the preceding problem, requirement 2. Consider variable factory overhead and then fixed factory overhead. Prepare the journal entries for (1) the incurrence of overhead, (2) the application of overhead, and (3) the isolation and closing of overhead variances to Cost of Goods Sold for the year.

8-26 Straightforward 4-variance overhead analysis. The Lopez Company uses a standard-cost system. Its standard cost of an auto part, based on a denominator volume of 40,000 units per year, included 6 machine hours of variable overhead at $8 per hour and 6 machine hours of fixed overhead at $15 per hour. Actual output achieved was 44,000 units. Actual variable factory overhead was $245,000. Actual machine hours of input were 28,400. Actual fixed overhead was $373,000.

Required

1. Prepare journal entries.
2. Prepare an analysis of all variable-overhead and fixed-overhead variances, using the approach illustrated in Exhibit 8-4, p. 263.
3. Describe how variable-overhead items are controlled from day to day. Also, describe how individual fixed-overhead items are controlled.

8-27 Characteristics of fixed-overhead variances. Fox Company executives have studied their operations carefully and have been using a standard-cost system for many years. They are now formulating standards for 19_1. Total fixed overhead is expected to be $1,200,000. Two hours of machine time is the standard time for finishing one unit of product.

Required

1. Graph the budgeted fixed overhead for 40,000 to 80,000 standard machine hours allowed, assuming that the total budget will not change over that volume level. What would be the appropriate inventory-costing rate per standard machine hour for fixed overhead if denominator volume is 50,000 hours? Graph the applied fixed-overhead line.
2. Assume that 25,000 units of product were produced. How much fixed overhead would be applied to production? Would there be a production volume variance? Why? Assume that 20,000 units were produced. Would there be a production volume variance? Why? Show the production volume variance for 20,000 units on a graph. In your own words, define *production volume variance*. Why does it arise? Can production volume variance exist for variable overhead? Why?
3. Assume that 22,000 units are produced. Fixed-overhead costs incurred were $1,234,000. What is the total underapplied or overapplied fixed overhead? Spending variance? Production volume variance? Use the analytical technique illustrated in Exhibit 8-4, p. 263.
4. In requirement 1 (ignore requirements 2 and 3), what would be the appropriate inventory-costing rate per standard machine hour for fixed overhead if denominator volume is estimated at 40,000 hours? At 60,000 hours? At 80,000 hours? Draw a graph showing budgeted fixed overhead and three "applied" lines, using the three rates just calculated.

If 20,000 units are produced and the denominator volume is 60,000 standard machine hours, what is the production volume variance? Now compare this with the production volume variance in your answer to requirement 2; explain the difference.

5. Specifically, what are the implications in requirements 1 and 4 regarding (a) the setting of the inventory-costing rate for fixed overhead, (b) the meaning of spending and production volume variances for fixed overhead, and (c) the major differences in planning and control techniques for variable and fixed costs?

8-28 Flexible budget, 4-variance analysis. (CMA, adapted) Nolton Products developed its overhead application rate from the current annual budget. The budget is based on an expected actual output of 720,000 units requiring 3,600,000 direct labor hours (DLH). The company is able to schedule production uniformly throughout the year.

A total of 66,000 units requiring 315,000 DLH was produced during May. Actual overhead costs for May amounted to $375,000. The actual costs as compared with the annual budget and one-twelfth of the annual budget are shown below.

Nolton uses a standard-costing system and applies factory overhead on the basis of DLH.

	Annual Budget				
	Total Amount	Per Unit	Per DLH	Monthly Budget	Actual Costs for May 19_3
VARIABLE					
Indirect labor	$ 900,000	$1.25	$ .25	$ 75,000	$ 75,000
Supplies	1,224,000	1.70	.34	102,000	111,000
FIXED					
Supervision	648,000	.90	.18	54,000	51,000
Utilities	540,000	.75	.15	45,000	54,000
Depreciation	1,008,000	1.40	.28	84,000	84,000
Total	$4,320,000	$6.00	$1.20	$360,000	$375,000

Required

Calculate the following amounts for Nolton Products for May 19_3:

1. Applied overhead costs
2. Variable-overhead spending variance
3. Fixed-overhead spending variance
4. Variable-overhead efficiency variance
5. Production volume variance

Be sure to identify each variance as favorable (F) or unfavorable (U).

8-29 4-variance analysis, find the unknowns. Consider each of the following situations—cases A, B, and C—independently. Data refer to operations for a week in April. For each situation, assume a standard-cost system. Also assume the use of a flexible budget for control of variable and fixed overhead based on standard machine hours.

	Cases		
	A	B	C
(1) Actual fixed overhead	$10,600	—	$12,000
(2) Actual variable overhead	7,000	—	—
(3) Denominator volume in hours	500	—	1,100
(4) Standard hours allowed for actual output	—	650	—
Flexible-budget data:			
(5) Fixed factory overhead	—	—	—
(6) Variable factory overhead (per standard hour)	—	8.50	5.00
(7) Budgeted fixed factory overhead	10,000	—	11,000
(8) Budgeted variable factory overhead*	—	—	—
(9) Total budgeted factory overhead*	—	12,525	—

Additional data:

(10) Standard variable overhead applied	7,500	—	—
(11) Standard fixed overhead applied	10,000	—	—
(12) Production volume variance	—	500 U	500 F
(13) Variable-overhead spending variance	950 F	-0-	350 U
(14) Variable-overhead efficiency variance	—	-0-	100 U
(15) Fixed-overhead spending-variance	—	300 F	—
(16) Actual hours of input	—	—	—

*For standard hours allowed for actual output.

Required

Fill in the blanks under each case. Prepare a worksheet similar to that in Exhibit 8-4, p. 263. Fill in the knowns and then solve for the unknowns.

8-30 2-variance and 3-variance analysis. The Wilson Company has a standard-costing system for its machine-paced outputs. A single hourly rate is used to apply factory overhead to production. However, a flexible-budget formula is used for planning and control purposes: $8,000,000 per year plus $7 per machine hour.

For 19_1, the denominator volume for developing an inventory-costing rate was the expected volume of 1,600,000 machine hours. The actual factory overhead was $18,000,000. Machine hours were: standard allowed for actual output achieved, 1,300,000; actual hours of input, 1,500,000.

Required

1. Prepare an analysis of all overhead variances. Use the format of Exhibit 8-8, Panel B, p. 269, to display a 2-variance analysis and a 3-variance analysis.
2. Describe how variable-overhead items are controlled from day to day. Also, describe how individual fixed-overhead items are controlled.

8-31 Disadvantages of standard-cost variances. Critics of standard costs have emphasized that managers are sometimes inclined to focus too narrowly on minimizing individual unfavorable variances. Such behavior may be disadvantageous to a company in other ways.

Required

For each of the following variances, indicate at least one way that minimization of an unfavorable variance may be undesirable for the company as a whole:

1. Variable-overhead spending
2. Variable-overhead efficiency
3. Fixed-overhead spending
4. Production volume variance

8-32 Hospital overhead variances, 4-variance analysis. The Sharon Hospital, a large metropolitan health-care complex has had much trouble in controlling its accounts receivable. Bills for patients, for various government agencies, and for private insurance companies have frequently been inaccurate and late. This has led to intolerable levels of bad debts and investments in receivables.

You were employed by the hospital as a consultant on this matter. After conducting a careful study of the billing operation, you developed some currently attainable standards that were implemented in conjunction with a flexible budget four weeks ago. You had divided costs into fixed and variable categories. You regarded the bill as the product, the unit of output.

You have reasonable confidence that the underlying source documents for compiling the results have been accurately tallied. However, the accountant has had some trouble summarizing the data and has provided the following:

Variable-overhead costs, allowance per standard hour	$ 10
Fixed-overhead budget variance, favorable	200
Combined budgeted overhead costs for the bills produced	22,500
Production volume variance, favorable	900
Variable-cost spending variance, unfavorable	2,000
Variable-cost efficiency variance, favorable	2,000
Standard hours allowed for the bills produced, 1,800	

Required
Compute:

1. Actual hours of input
2. Fixed-overhead budget
3. Fixed overhead applied
4. Budgeted fixed-overhead rate
5. Denominator volume in hours

8-33 Combined overhead, 3-variance analysis. The Wright-Patterson Air Force Base contained an extensive repair facility for jet engines. It had developed standard costing and flexible budgets to account for this activity. Budgeted *variable* overhead at an 8,000 standard monthly direct-labor-hour level was $64,000; budgeted *total* overhead at a 10,000 standard direct-labor-hour level was $197,600. The standard cost applied to repair output included a combined overhead rate of 120% of standard direct labor cost.

Total actual overhead for October was $249,000. Direct labor costs actually incurred were $202,440. The direct labor price variance was $9,640, unfavorable. The direct labor flexible-budget variance was $14,440, unfavorable. The standard labor price was $16.00 per hour. The production volume variance was $14,000, favorable.

Required
1. Compute: Direct labor efficiency variance and the spending, efficiency, and production volume variances for combined overhead. Also, compute the denominator volume. (*Hint:* See Exhibit 8-8, Panel B, p. 269.)
2. Describe how variable-overhead items are controlled from day to day. Also, describe how individual fixed-overhead items are controlled.

8-34 Combined-overhead variances, 2-variance analysis (SMA, adapted) The Weser Company uses a budgeted total overhead rate of $40 per unit based on a denominator volume of 60,000 units a year, or 5,000 units a month.

During October, the company produced 5,200 units and experienced the following combined overhead variances:

Overhead flexible-budget variance	$2,500 unfavorable
Overhead production volume variance	$5,000 favorable

During November, production was 4,900 units and the actual overhead cost incurred was $2,000 less than October's overhead.

Determine the overhead flexible-budget variance and production volume variance for the month of November.

8-35 Flexible-budget, spending, and efficiency variances. John Eastman, the manager of a medical testing laboratory, has been under pressure from the director of a large clinic, Dr. Elvira Coleman, to get more productivity from his employees.

Eastman had been provided with some industry standards for various tests and had compiled a budgeted cost-behavior pattern for direct labor and indirect costs. Some selected variable-cost items for a recent four-week period follow:

	Budgeted Cost-Behavior Pattern per Direct Labor Hour	Budget Based on 4,000 Actual Labor Hours	Actual Costs Incurred	Variance	
				Amount	Percentage of Budget
Direct labor	$12.00	$48,000	$48,000	—	—
Supplies	2.00	8,000	6,400	1,600 F	20%
Indirect labor	1.00	4,000	3,000	1,000 F	25
Miscellaneous	2.00	8,000	6,000	2,000 F	25

Eastman was pleased with his budgetary performance, but Coleman was not happy. She asked you to reconstruct the performance report, saying. "This report seems incomplete. I want a better pinpointing of the degree of basic efficiency in that department."

Your investigation of the industry literature and your discussion with a management consultant in health care revealed that standard times have been developed for these clinical procedures. Although the standard times are open to criticism if examined on an hour-by-hour or day-by-day basis, in the aggregate they provide a fairly reliable benchmark for measuring outputs against inputs. Your study of the output showed that about 2,600 standard direct labor hours should have been allowed for the output actually achieved by the laboratory.

Required

Prepare a performance report based on the flexible budget for standard hours allowed for the actual output achieved. For each line item, show two final columns, one for spending variance and one for efficiency variance. Is this report more informative than the performance report shown above? Explain.

8-36 Price and spending variances. The Favaro Company used a flexible budget and standard costs for controlling customer services. In March, the company serviced 7,000 units of its product. Assume that 15,000 actual hours were used at an actual hourly rate of $16.40. Two direct labor hours is the standard allowance for servicing one unit. The standard labor rate is $16 per hour. The flexible budget for miscellaneous supplies is based on a formula of $2.40 per unit, which can also be expressed as $1.20 per direct labor hour. The actual cost of supplies was $18,800.

Required

1. Compute the "price" and "efficiency" variances for direct labor and for supplies.
2. The "price" variance for supplies is rarely labeled as such. Instead it is often called a spending variance. The plant manager has made the following comments: "I have been troubled by the waste of supplies for months. I know that the prices are exactly equal to the budgeted prices for every supply item, because a two-year stock of supplies was bought a year ago in anticipation of prolonged inflation. Consequently, we used known prices for preparing our budgets. Given these facts, please explain how a spending variance can arise for miscellaneous supplies. Why doesn't all waste appear as an efficiency variance?" Respond to the manager's comments. Be clear and specific; the manager is impatient with muddled explanations, and your next pay raise will be affected by her appraisal of your explanation.
3. If prices rise, the two-year stock of supplies will result in savings of overall purchase costs. Give two arguments against acquiring such a large stock of supplies.

8-37 Detailed analysis of variable overhead. An activity area is scheduled to repair 10,000 units of product in 5,000 standard direct labor hours. However, it has taken 6,000 actual direct labor hours to repair the 10,000 units. The variable-factory-overhead items are assumed to vary in proportion to the actual direct labor hours of input.

	Budget Formula per Standard Direct Labor Hour Allowed	Actual Costs Incurred
Indirect labor	$2.00	$11,700
Maintenance	.20	1,150
Lubricants	.10	600
Cutting tools	.16	1,500
	$2.46	$14,950

The actual direct labor rate is $16.20 per hour; the standard rate is $16.00.

Required

1. Prepare a performance report with separate lines for direct labor, indirect labor, maintenance, lubricants, and cutting tools. Show three columns: actual costs incurred, flexible budget based on actual outputs, and flexible-budget variance.
2. Prepare a more detailed version of your answer to requirement 1, showing six columns: actual costs incurred, flexible budget based on actual inputs, flexible budget based on actual outputs, flexible-budget variance, spending variance, and efficiency variance.
3. Explain the similarities and differences between the direct labor variances and the variable-overhead variances.

8-38 Review of fundamental variance analysis, 3-variance analysis. (CPA, adapted) The Beal Company uses a standard-costing system. At the beginning of 19_6, Beal adopted the following standards:

	Input	Total
Direct materials	3 lbs. @ $2.50 per lb.	$ 7.50
Direct labor	5 hrs. @ $7.50 per hr.	37.50
Factory overhead:		
Variable	$3.00 per direct labor hour	15.00
Fixed	$4.00 per direct labor hour	20.00
Standard cost per unit		$80.00

Denominator volume per month is 40,000 standard direct labor hours. Beal's January 19_6 flexible budget, set up on January 1, was based on denominator volume. Beal's actual January production was 7,800 units. The records for January indicated the following:

Direct materials purchased	25,000 lbs. @ $2.60
Direct materials used	23,100 lbs.
Direct labor	40,100 hrs. @ $7.30
Total actual factory overhead (variable and fixed)	$300,000

Required

1. Prepare a schedule of total standard production costs for the production of 7,800 units in January.
2. For the month of January 19_6, compute the following variances, indicating whether each is favorable (F) or unfavorable (U):
   a. Direct materials price variance, based on purchases
   b. Direct materials efficiency variance
   c. Direct labor price variance
   d. Direct labor efficiency variance
   e. Total-factory-overhead spending variance
   f. Variable-factory-overhead efficiency variance
   g. Production volume variance

8-39 Working backward from given variances. The Mancuso Company uses a flexible budget and standard costs to aid planning and control. At a 60,000-direct-labor-hour level, budgeted variable overhead is $120,000 and budgeted direct labor is $480,000.

The following are some results for August:

Variable-overhead flexible-budget variance	$ 10,500 U
Variable-overhead efficiency variance	20,000 U
Actual direct labor costs incurred	574,000
Material purchase price variance (based on purchases)	16,000 F
Material efficiency variance	9,000 U
Fixed overhead incurred	50,000
Fixed-overhead spending variance	2,000 U

The standard cost per pound of direct materials is $1.50. The standard allowance is one pound of direct materials for each unit of finished product. Ninety thousand units of product were made during August. There was no beginning or ending work in process. In July, the material efficiency variance was $1,000, favorable, and the purchase price variance was $.20 per pound, unfavorable. In August, the purchase price variance was $.10 per pound.

In July, labor troubles caused an immense slowdown in the pace of production. There had been an unfavorable direct labor efficiency variance of $60,000; there was no labor price variance. These troubles persisted into August. Some workers quit. Their replacements had to be hired at higher rates, which had to be extended to all workers. The actual average wage rate in August exceeded the standard average wage rate by $.20 per hour.

Required

Compute for August:

1. Total pounds of direct materials purchased during August

2. Total number of pounds of excess material usage
3. Variable-overhead spending variance
4. Total number of actual hours of input
5. Total number of standard hours allowed for the finished units produced
6. Describe how variable-overhead items are controlled from day to day. Also, describe how individual fixed-overhead items are controlled.

8-40 Variance analysis from fragmentary evidence, 4-variance analysis. Being a bright young person, you have just landed a wonderful job as assistant controller of Gyp-Clip, a new and promising Singapore division of Croding Metals Corporation. The Gyp-Clip Division has been formed to produce a single product, a new-model paper clip. Croding Laboratories has developed an extremely springy and lightweight new alloy, Clypton, which is expected to revolutionize the paper-clip industry.

It is your first day on the job. Gyp-Clip has been in business one month. The controller takes you on a tour of the plant and explains the operation in detail: Clypton wire is received on two-mile spools from the Croding mill at a fixed price of $40 a spool, which is not subject to change. Clips are bent, cut, and shipped in bulk to the Croding packaging plant. Factory rent, depreciation, and all other items of fixed factory overhead are handled by the home office at a set amount of $100,000 per month. Ten thousand tons of paper clips have been produced, but this is only 75% of denominator volume.

The controller has just figured out the month's variances. She is looking for a method of presenting them in clear, logical form to top management at the home office. You say that you know of just the method and promise to have the analysis ready the next morning.

Filled with zeal and enthusiasm, feeling that your future as a rising star in this growing company is secure, you decide to take your spouse out to dinner to celebrate the trust and confidence that your superior has placed in you.

Upon returning home, you are horrified to discover your dog happily devouring the controller's figure sheet. You manage to salvage only the accompanying fragments.

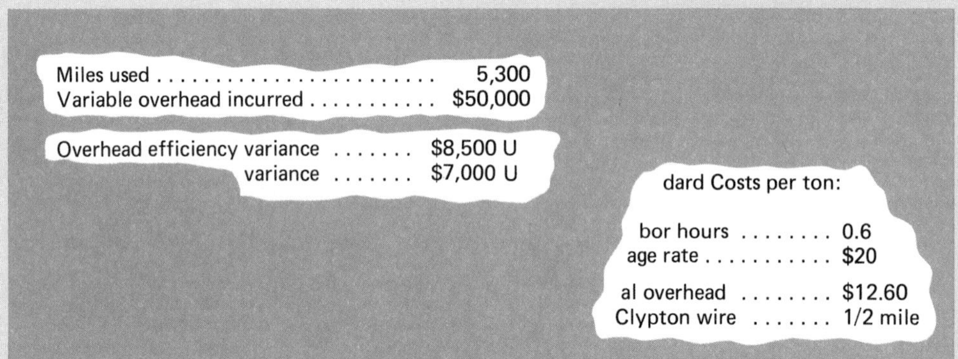

Miles used 5,300
Variable overhead incurred $50,000
Overhead efficiency variance . . . . $8,500 U
variance . . . . $7,000 U
dard Costs per ton:
bor hours 0.6
age rate $20
al overhead $12.60
Clypton wire . . . . 1/2 mile

You remember that the $7,000 variance did not represent the grand total of all variances. You also recall that the company applies overhead based on direct labor hours.

Required

Don't let the controller down. Go ahead and make up your analysis of all variances.

8-41 4-variance analysis, bonuses, and activity areas. Easy Rider assembles prestige motorcycles. Over the last three years, sales have increased despite the heavy advertising of foreign competitors. A key to its success has been dramatic improvements in both the quality and the manufacturing cost of each Easy Rider motorcycle.

Nancy Hyatt, the manufacturing manager of Easy Rider, has played a key role in bringing about improvements in manufacturing. Over this period, however, Hyatt has had an ongoing battle with the president of Easy Rider about the compensation to her manufacturing supervisors. Each supervisor is in charge of a separate activity area in the manufacturing plant. The current compensation scheme rewards each supervisor with a quarterly bonus based on 10% of the sum of the two direct materials variances and the four manufacturing overhead variances when that sum is favorable.

Easy Rider records the following variances for each activity area of the manufacturing plant:

- Direct materials price variance
- Direct materials efficiency variance
- Variable-manufacturing-overhead spending variance
- Variable-manufacturing-overhead efficiency variance
- Fixed-manufacturing-overhead spending variance
- Production volume variance

Easy Rider has only one category of direct costs—direct materials. Several years ago, it stopped tracking direct labor (because it is less than 5% of total manufacturing costs). All labor costs are now included in manufacturing overhead.

Hyatt has just had a meeting with Eddie Gilbert, the supervisor of the machine-welding area of the manufacturing plant. Gilbert believes that the performance of the welding activity area in the January–March 19_6 quarter warrants his receiving a much larger bonus than the existing plan awards him. Summary data for the January–March 19_6 quarter are:

Direct materials:	
Standard welding materials used per bike	2.2 lbs.
Actual welding materials used per bike	2.0 lbs.
Standard price per lb. of welding material used	$20.00
Actual price per lb. of welding material used	$19.20

Other manufacturing data:	
Budgeted variable overhead	= $40 per welding hour
Actual variable overhead	= $135,000
Budgeted fixed overhead	= $440,000
Actual fixed overhead	= $428,000
Budgeted welding time per motorcycle	= 0.25 hours
Actual welding time per motorcycle	= 0.20 hours
Budgeted production volume	= 22,000 motorcycles
Actual production volume	= 16,000 motorcycles
Denominator production volume	= 22,000 motorcycles

Variable and fixed welding manufacturing overhead is applied to each motorcycle that passes through the welding activity area based on the budgeted welding time for each motorcycle. There are no beginning or ending inventories of direct materials, work in process, or finished goods in the welding activity area.

Starting in February 19_6, Easy Rider sales dropped. Tariffs and duties on foreign competitors were abolished in February when a change occurred in the political party that heads the Foreign Trade Committee.

Required

1. Compute the budgeted fixed-manufacturing-overhead rate in the welding activity area for the January–March 19_6 period.
2. Compute the six variances Easy Rider calculates for the machine-welding activity area for the January–March 19_6 period. Show all computations.
3. Compute the bonus paid to Eddie Gilbert, the supervisor of the machine-welding activity area of Easy Rider's manufacturing plant, for the January–March 19_6 period.
4. Outline two advantages and two disadvantages to Easy Rider of including the production volume variance in the bonus calculation for each activity-area supervisor.

8-42 Comprehensive variance analysis, executive bonus. The Charlotte beverage plant of National Goods bottles cola soft drinks. The production manager at the plant receives a monthly bonus that is 10% of the aggregate of the following eight variances if that aggregate is favorable:

- Direct materials price variance
- Direct materials efficiency variance

- Direct labor price variance
- Direct labor efficiency variance
- Variable-manufacturing-overhead spending variance
- Variable-manufacturing-overhead efficiency variance
- Fixed-manufacturing-overhead spending variance
- Fixed-manufacturing-overhead production volume variance

Standard-cost data for the most recent month (March) were:

- Standard Direct Materials
  - 2 lbs. input per barrel of cola bottled
  - \$17.50 per lb. of input
- Standard Direct Labor
  - 30 minutes input per barrel of cola bottled
  - \$18 per hour
- Budgeted Variable Manufacturing Overhead
  - \$5 per lb. of direct materials input

Fixed manufacturing overhead was applied on the basis of direct materials (lbs.) input. Budgeted fixed manufacturing overhead for March was \$350,000. Budgeted production volume was 50,000 barrels of cola.

The following data were reported for March:

- Actual direct materials cost of \$18 per lb. of input
- Actual direct labor cost of \$19 per hour
- Actual variable-manufacturing-overhead cost = \$387,600
- Actual total-manufacturing-overhead cost = \$727,600
- Direct materials price variance = \$38,000 U
- Direct labor efficiency variance = \$36,000 F
- Variable-manufacturing-overhead spending variance = \$7,600 U

Actual March production volume was reported to be 40,000 barrels of cola. There were no beginning or ending inventories of direct materials or bottled cola.

Required

1. What was the actual direct materials input (in lbs.) per barrel of cola beverage bottled?
2. What was the actual direct labor input (in minutes) per barrel of cola beverage bottled?
3. Compute the eight variances used to calculate the production manager's bonus. Present your analysis using the format in Exhibit 8-9, p. 274. What bonus should the manager receive for the month of March?
4. Why did the fixed-manufacturing-overhead production volume variance arise? Give three possible causes of this variance.
5. What changes would you recommend in the plant's existing bonus plan for the production manager?

CHAPTER 9

Income Effects of Alternative Inventory-Costing Methods

PART ONE: VARIABLE AND ABSORPTION COSTING

Role of fixed indirect manufacturing costs

Comparison of standard variable costing and standard absorption costing

Breakeven points in variable costing and absorption costing

Performance measures and absorption costing

PART TWO: ROLE OF VARIOUS DENOMINATOR LEVELS IN ABSORPTION COSTING

Characteristics of capacity

Measurement of capacity

Inventory and income effects

PART THREE: STANDARD-COST VARIANCES AND FINANCIAL STATEMENTS

Effects of prorations

Adjusting inventories for external reporting

Variances and interim reporting

Appendix: Standard-cost variances: a more accurate approach to proration

Managers can increase operating income by increasing inventory levels when absorption costing is used.

Learning Objectives

When you have finished studying this chapter, you should be able to

1. Identify the fundamental feature that distinguishes variable costing from absorption costing
2. Construct an income statement using absorption costing and variable costing
3. Explain differences in operating income under absorption costing and variable costing
4. Describe the effects of absorption costing on computing the breakeven point
5. Understand how absorption costing influences performance-evaluation decisions
6. Explain the role of various denominator levels in absorption costing
7. Describe how standard-cost variances affect financial statements
8. Show how to convert variable-costing inventory valuation to absorption-costing
9. Explain the role of variances in interim reporting

When they design an accounting system, managers and accountants usually choose an inventory-costing method. This decision is crucial because of its effects on reported income. After all, income is a principal measure of managers' performance as viewed from both inside and outside the organization.

This chapter examines and compares the effects of some costing alternatives on the measurement of inventories and income. We divide the chapter into three parts: (1) variable and absorption costing, (2) the role of various denominator levels in absorption costing, and (3) standard-cost variances and financial statements. These three topics are closely enough related to warrant inclusion within a single chapter. *However, they may be studied independently.* For your convenience, each part ends with a Problem for Self-Study. A chapter appendix provides additional discussion of proration of variances.

We concentrate on different approaches to measuring inventories and income. These approaches, however, have implications for many management decisions, including choices of product mix, pricing, and making or buying components. For instance, the computation of the full product cost for guiding pricing includes the manufacturing costs commonly regarded as inventoriable costs in addition to the costs of other business functions such as product design, distribution, and customer service.

PART ONE: VARIABLE AND ABSORPTION COSTING

ROLE OF FIXED INDIRECT MANUFACTURING COSTS

Objective 1

Identify the fundamental feature that distinguishes variable costing from absorption costing

Impact of Fixed Factory Overhead

Two major methods of inventory costing are absorption costing and variable costing. These methods differ in only one conceptual respect: whether fixed manufacturing overhead (indirect manufacturing costs) is an inventoriable cost.

- **Absorption costing** is a method of inventory costing that includes all direct manufacturing costs and all indirect manufacturing costs (both variable and fixed) as inventoriable costs.
- **Variable costing** is a method of inventory costing that includes all direct manufacturing costs and variable indirect manufacturing costs as inventoriable costs; fixed indirect manufacturing costs are excluded from the inventoriable costs.

Throughout this chapter, to emphasize underlying concepts we assume that some measure of overall production volume is the cost application base. As always, remember that "volume" is measured in a variety of ways; examples include machine hours, equivalent units, physical units, direct labor hours, and total value of production. In the most fundamental terms, volume is the number of cost driver inputs required for the total outputs in question.

The basic impact of variable and absorption costing on income is clear-cut. Suppose that fixed factory overhead budgeted and incurred at Radius Company was $2,200 for the year 19_1. Also suppose that 1,100 units were produced, but only 1,000 units were sold. Assuming the denominator volume for absorption costing was 1,100 units, the fixed overhead rate would be $2,200 ÷ 1,100 units = $2.00 per unit. The following chart summarizes the accounting:

	Fixed-Factory-Overhead Cost to Account for	**Inventoriable Costs**	**Expense in 19_1**
		None	**Expires as a period cost when incurred**
Variable Costing	$2,200	——————→	$2,200
Absorption Costing	$2,200 →	Additions to inventory: 1,100 × $2.00 = $2,200	
		Cost of goods sold: 1,000 × $2.00 = $2,000 →	$2,000
		Ending inventory: 100 × $2.00 = $ 200	

The chart contains only one item—fixed factory overhead—because *all* other manufacturing costs are accounted for exactly alike under both variable costing and absorption costing. That is, all direct manufacturing costs and variable factory overhead are always inventoriable. Moreover, all nonmanufacturing costs are accounted for alike under both variable costing and absorption costing.

Trace the $2,200 in Exhibit 9-1, which presents fuller information. The income statement under variable costing deducts the lump sum as an expense in 19_1. In contrast, the income statement under absorption costing regards each finished unit as bearing $2 of fixed factory overhead. The variable manufacturing costs are accounted for the same in both statements.

The general format of these income statements should be familiar. The $200 difference in operating income ($1,100 − $900) arises from the $200 ($800 − $600) difference in ending inventories. Under absorption costing, $200 of the $2,200 fixed manufacturing costs is held back in inventory, but under variable costing the $200 is written off in the period incurred.

Never overlook the heart of the matter. The differences between variable costing and absorption costing center on how to account for fixed factory overhead. If inventory levels change, income will differ because of the different accounting for

EXHIBIT 9-1
Comparison of Variable Costing and Absorption Costing
Radius Company
Income Statements for the Year Ended Dec. 31, 19_1

(Data assumed; there is no beginning inventory)

Variable Costing

Sales: 1,000 units @ $17.00		$17,000
Variable manufacturing cost of goods available for sale: 1,100 units @ $6.00	$6,600*	
Deduct ending inventory: 100 units @ $6.00	600	
Variable manufacturing cost of goods sold	$6,000	
Add variable marketing and administrative costs	2,400	
Total variable costs		8,400
Contribution margin†		$ 8,600
Deduct fixed costs:		
Fixed manufacturing costs	$2,200	
Fixed marketing and administrative costs	5,500	7,700
Operating income		$ 900

Absorption Costing

	Unit Cost	Total	
Sales			$17,000
Cost of goods sold:			
Variable manufacturing costs: 1,100 units	$6.00	$6,600*	
Fixed manufacturing costs	2.00	2,200	
Cost of goods available for sale	$8.00	$8,800	
Deduct ending inventory: 100 units	8.00	800	8,000
Gross margin†			$ 9,000
Deduct total marketing and administrative costs, including $2,400 of variable costs			7,900
Operating income			$ 1,100

*Composed of

	Assumed Unit Cost	Total
Direct materials	$3.00	$3,300
Direct labor	2.00	2,200
Variable overhead	1.00	1,100
Variable manufacturing costs	$6.00	$6,600

†The contribution margin and the gross margin are two important intermediate line items that highlight the conflict of the underlying concepts of variable costing and absorption costing.

fixed factory overhead. Compare sales of 900, 1,000, and 1,100 units. Fixed factory overhead would be included in expense as follows:

	Fixed Factory Overhead Expense in 19_1
Variable costing, whether sales are 900, 1,000, or 1,100 units	$2,200
Absorption costing, where sales are:	
900 units, $400 held back in inventory	$1,800
1,000 units, $200 held back in inventory	$2,000
1,100 units, $0 held back in inventory	$2,200

The term *variable costing* is a reasonably accurate description, but **direct costing** is also widely used to signify the exclusion of fixed factory overhead from inventories. *Direct costing* is an unfortunate choice of terms; after all, the approach includes in inventory not only direct manufacturing costs but also *variable indirect* manufacturing costs. Indeed, the term *variable costing* could be improved to *variable manufacturing costing*. Why? Because the distinction between approaches centers on how to account for manufacturing costs. Variable costs of other business functions—such as marketing, administrative, and customer-service—are written off as period costs under both variable costing and absorption costing.

Income statements prepared using variable costing are sometimes called *contribution* or *contribution-approach* income statements. The contribution approach stresses the lump-sum amount of all fixed costs to be recouped before net income emerges. This highlighting of total fixed costs helps to focus management attention on fixed-cost behavior and control when making both short-run and long-run plans. Keep in mind that advocates of the contribution approach ***do not maintain that fixed costs are unimportant or irrelevant, but they do stress that the distinction between variable and fixed costs is crucial for many decisions.***

Capsule Comparisons of Inventory-Costing Methods

Exhibit 9-2 presents a capsule comparison of six alternative inventory-costing systems using a Work in Process account:

Variable Costing	**Absorption Costing**
1. Actual costing	4. Actual costing
2. Normal costing	5. Normal costing
3. Standard costing	6. Standard costing

The boxes in Exhibit 9-2 represent the debits to Work in Process (that is, the amounts applied to product) under alternative inventory-costing systems. The major systems in use are standard-costing systems, which were introduced in the preceding two chapters, and normal-costing systems, which were introduced in Chapter 4. As the boxes indicate, variable costing or absorption costing may be combined with actual, normal, or standard costing. ***Variable costing signifies that fixed factory overhead is not inventoried.***

Variable costing has been a controversial subject among accountants—not so much because there is disagreement about the need for delineating between variable costs and fixed costs for management planning and control, but because there is a question about using variable costing for *external* reporting. Those favoring variable costing maintain that the fixed portion of factory overhead is more closely related to the capacity to produce than to the production of specific units. Those

EXHIBIT 9-2

Capsule Comparison of Six Alternative Inventory-Costing Systems

			Work in Process Inventory		
			Actual Costing	**Normal Costing**	**Standard Costing**
Absorption Costing	**Variable Costing**	**Direct Costs**	Actual inputs × Actual prices	Actual inputs × Actual prices	Standard inputs allowed for actual output achieved × Standard prices
		Variable Factory Overhead	Actual inputs × Actual overhead rates	Actual inputs × Budgeted overhead rates	Standard inputs allowed for actual output achieved × Budgeted overhead rates
		Fixed Factory Overhead	Actual inputs × Actual overhead rates	Actual inputs × Budgeted overhead rates	Standard inputs allowed for actual output achieved × Budgeted overhead rates

opposing variable costing maintain that inventories should carry a fixed-cost component. Why? Because both variable and fixed manufacturing costs are necessary to produce goods; both types of costs should be inventoriable, regardless of their differences in behavior patterns.

Variety of Practices Concerning Inventoriable Costs

Surveys indicate that 52% of the respondent companies use a variable costing format in reporting to top management (23% as the primary format and 29% as a supplemental format)[1] Various interpretations of variable costing will be found for internal purposes, because management can decide how to apply the basic concepts without worrying about reporting to outsiders.

Regarding internal reporting, some managers advocate "super-variable costing." This interpretation assumes that in the short run, only direct materials are inventoriable costs; all other costs are fixed and not identifiable with products.[2] Conversely, some opponents of variable costing favor "super-absorption costing." They take a long-run view and maintain that almost all costs (including research, design, marketing, administrative, and customer-service costs) are variable and must be identified with the products.[3]

For reporting to the U.S. Internal Revenue Service, all manufacturing costs plus some marketing and administrative costs must be included as inventoriable costs. For example, legal department costs must be allocated between those related to manufacturing activities (inventoriable costs) and those not related to manufacturing activities (period costs).

For external reporting to shareholders, companies around the globe tend to follow the generally accepted accounting principle that all manufacturing overhead is inventoriable. In practice, however, some companies do not inventory all factory overhead. Instead, they charge some factory overhead to expense immediately. A notable example is depreciation on factory equipment. External auditors tend to accept such practices for at least two reasons:

1. The consistent application of such depreciation accounting is coupled with undramatic changes in beginning and ending inventory levels. Hence, the final reported net income will not be materially affected.
2. Depreciation is seldom a significant part of the total manufacturing costs, so inventories are not materially affected.[4]

COMPARISON OF STANDARD VARIABLE COSTING AND STANDARD ABSORPTION COSTING

The following illustration explores the implications of accounting for fixed factory overhead in more depth.

Stassen Company began business on January 1, 19_1. It is now the end of 19_2. The company uses standard absorption costing. The president has asked you to

[1]C. Chow, M. Shields, A. Wong-Boren, "A Compilation of Recent Surveys and Company-Specific Descriptions of Management Accounting Practices," *Journal of Accounting Education*, Vol. 6 (1988), p. 197.

[2]E. Goldratt and J. Cox, *The Goal* (Croton-on-Hudson, N.Y.: North River Press, 1984).

[3]R. Kaplan, "Management Accounting for Advanced Technological Environments," *Science*, August 1989, p. 822.

[4]For a thorough discussion of concepts and practices regarding what costs are inventoriable for both shareholder and tax reporting, see E. Noreen and R. Bowen, "Tax Incentives and the Decision to Capitalize or Expense Manufacturing Overhead," *Accounting Horizons*, March 1989.

prepare comparative income statements for 19_1 and 19_2 under absorption costing and variable costing.

The following simplified unit data are available:

	19_1	19_2
Beginning inventory	—	4,000
Production	6,000	900
Sales	2,000	3,000
Ending inventory	4,000	1,900

Other Data:

Variable manufacturing costs per unit	$ 2
Fixed manufacturing costs	$10,000
Denominator volume in units	10,000
Fixed manufacturing costs per unit	$ 1
Fixed marketing and administrative costs	$ 1,400
Variable marketing and administrative costs per unit sold	$ .50
Selling price per unit	$ 8.50

There were no inventories of work in process or materials. There were no price, efficiency, or spending variances for any costs. Production was far below expected volume in 19_2 because of persistent shortages of raw materials.

The following sections will:

1. Present comparative income statements under variable costing and absorption costing.
2. Explain the difference in operating income between absorption costing and variable costing in each of the two years.
3. Explain the comparative effects of sales and production on operating income.

Comparative Income Statements

Objective 2

Construct an income statement using absorption costing and variable costing

Exhibit 9-3 contains the comparative income statements under variable costing (Panel A) and absorption costing (Panel B).

- Panel A—Actual costs and standard costs are assumed to be equal. There are no variances here. The statement begins with all items valued at standard. If there were 19_1 variances of variable costs, they would appear as follows:

Contribution margin—at standard	$12,000
Variable-cost variances (detailed)*	X,XXX
Contribution margin—at actual	$Y,YYY

*Net unfavorable (favorable) variances are deducted from (added to) contribution margin at standard.

- Panel B—The statement begins with all items valued at standard. A production volume variance is shown immediately after the line item Manufacturing cost of goods sold—at standard. Any other variances of manufacturing costs would be grouped with the production volume variances.

In practice, the presentation of variances differs among companies. Some commonly encountered presentations follow:

- Adjustments to contribution margin—at standard; or adjustments to gross margin—at standard
- Adjustments to operating income—at standard
- Adjustments to individual line items—at standard

For example, direct materials price variance would appear as an immediate adjustment to its line item: direct materials used—at standard.

EXHIBIT 9-3 *(Place a clip on this page for easy reference.)*
Stassen Company
Comparative Income Statements
For the Years 19_1 and 19_2

PANEL A: VARIABLE COSTING	19_1	19_2
Sales 2,000 units and 3,000 units at $8.50	$17,000	$25,500
Beginning inventory at $2	—	$ 8,000
Add variable cost of goods manufactured at $2	12,000	1,800
Available for sale	12,000	9,800
Deduct ending inventory	8,000	3,800
Variable manufacturing cost of goods sold	4,000	6,000
Variable marketing costs at $.50 per unit sold	1,000	1,500
Total variable costs	5,000	7,500
Contribution margin	12,000	18,000
Fixed factory overhead	10,000	10,000
Fixed marketing and administrative costs	1,400	1,400
Total fixed costs	11,400	11,400
Operating income	$ 600	$ 6,600

PANEL B: ABSORPTION COSTING	19_1	19_2
Sales	$17,000	$25,500
Beginning inventory of $3	—	12,000
Add absorption cost of goods manufactured	18,000	2,700
Available for sale	18,000	14,700
Deduct ending inventory	12,000	5,700
Manufacturing cost of goods sold—at standard	6,000	9,000
Production volume variance*	4,000 U	9,100 U
Adjusted manufacturing cost of goods sold	10,000	18,100
Gross margin or gross profit—at actual	7,000	7,400
Marketing and administrative costs†	2,400	2,900
Operating income	$ 4,600	$ 4,500

*Computation of production volume variance based on denominator volume of 10,000 units:

19_1	$ 4,000 underapplied (10,000 − 6,000) × $1
19_2	9,100 underapplied (10,000 − 900) × $1
Two years together	$13,100 underapplied (20,000 − 6,900) × $1

†19_1	$ 1,400 fixed + 2,000 units × $.50 = $2,400
19_2	$ 1,400 fixed + 3,000 units × $.50 = $2,900

Keep these points in mind about absorption costing as you study Exhibit 9-3:

a. The unit product cost is $3, not $2, because variable manufacturing costs ($2) *plus* fixed factory overhead ($1) are applied to product.
b. The $1 fixed-overhead application rate was based on a denominator volume of 10,000 units. Whenever *production* (not sales) deviates from the denominator volume, a production volume variance arises. The measure of the variance is $1 multiplied by the difference between the actual volume of production and the denominator volume.
c. The production volume variance exists only under absorption costing, not variable costing. All other variances exist under both absorption costing and variable costing.
d. The absorption-costing income statement classifies costs primarily by *business function,* such as manufacturing, marketing, and administrative costs. In contrast, the variable-costing income statement features *cost behavior* as the primary classification scheme. Absorption-costing income statements seldom differentiate between the variable and fixed costs. Exhibit 9-3 does not—although such differentiation is possible.

Explaining Differences in Operating Income

Objective 3

Explain differences in operating income under absorption costing and variable costing

In addition to points (a) through (d) in the preceding section, let's consider another important point:

e. If inventories increase during a period, variable costing will generally report less operating income than will absorption costing; when inventories decrease, variable costing will report more operating income than absorption costing. These differences in operating income are due *solely* to moving fixed factory overhead in and out of inventories as they increase and decrease, respectively.[5]

To illustrate, assume that all variances are written off as period costs, no change occurs in work in process inventory, and no change occurs in the budgeted fixed overhead rate (budgeted fixed overhead divided by the denominator volume in units). Then the difference between the operating incomes under absorption and variable costing may be shown by formulas:

FORMULA 1

$$\begin{pmatrix}\text{Absorption-} \\ \text{costing} \\ \text{operating} \\ \text{income}\end{pmatrix} - \begin{pmatrix}\text{Variable-} \\ \text{costing} \\ \text{operating} \\ \text{income}\end{pmatrix} = \begin{pmatrix}\text{Units} \\ \text{produced} \\ \text{minus} \\ \text{units} \\ \text{sold}\end{pmatrix} \times \begin{pmatrix}\text{Budgeted} \\ \text{fixed} \\ \text{overhead} \\ \text{rate}\end{pmatrix}$$

or

FORMULA 2

$$\begin{pmatrix}\text{Absorption-} \\ \text{costing} \\ \text{operating} \\ \text{income}\end{pmatrix} - \begin{pmatrix}\text{Variable-} \\ \text{costing} \\ \text{operating} \\ \text{income}\end{pmatrix} = \begin{pmatrix}\text{Increase or} \\ \text{decrease in} \\ \text{inventory units}\end{pmatrix} \times \begin{pmatrix}\text{Budgeted} \\ \text{fixed overhead} \\ \text{rate}\end{pmatrix}$$

Let's use the data from Exhibit 9-3 to illustrate these formulas:

USING FORMULA 1

19_1 $4,600 − $600 = (6,000 − 2,000 units) × $1
= $4,000

19_2 $4,500 − $6,600 = (900 − 3,000 units) × $1
= $−2,100

USING FORMULA 2

19_1 $ 4,000 = (Increase in inventory of 4,000 units) × $1

19_2 $−2,100 = (Decrease in inventory of 2,100 units) × $1

Effects of Sales and Production on Income

A final important point, in addition to the previous points (a) through (e), deals with cost-volume-profit relationships:

f. Under variable costing, operating income is driven by fluctuations in sales volume. Operating income should rise as sales rise, and vice versa, at the rate of the contribution margin per unit:

$$\begin{matrix}\text{Increase in operating} \\ \text{income from 19\_1 to 19\_2}\end{matrix} = \begin{pmatrix}\text{Contribution margin} \\ \text{of \$8.50 − \$2.50}\end{pmatrix} \times \begin{pmatrix}\text{3,000 units} \\ \text{sold in 19\_2}\end{pmatrix} - \begin{pmatrix}\text{2,000 units} \\ \text{sold in 19\_1}\end{pmatrix}$$

$$= \$6.00 \times 1{,}000 = \$6{,}000$$

[5]The illustrations assume that the fixed-overhead rate per unit is unchanged from one accounting period to the next. If the rate does change significantly, these generalizations may not hold. Also see T. Shevlin, "Reconciliation of the Difference in Operating Income under Absorption Costing and Variable Costing" (Working paper, University of Washington, 1988).

Under absorption costing, operating income is driven by both sales volume and production volume. For example, in 19_1 the $4,600 operating income under absorption costing is higher than the $600 under variable costing because production exceeded sales. The opposite effect occurred in 19_2. Moreover, in 19_2 sales rose over the 19_1 level. Nevertheless, under absorption costing, operating income fell from $4,600 to $4,500. Why? Because 19_2 had to bear some of the fixed overhead carried over in inventory from 19_1. In addition, 19_2 bore the effects of production being below denominator volume. Thus, *production* schedules affect operating income under absorption costing but not under variable costing. The detailed effects of fixed factory overhead on 19_2 operations are shown in Exhibit 9-4.

For absorption costing we have seen that, under some circumstances, operating income could fall even though sales volume rises. ***Heavy reductions of beginning inventory levels might combine with low production and a large production volume variance to result in unusually large amounts of fixed overhead being charged to a single accounting period. Thus, under absorption costing, an increase in sales volume could be accompanied by a lower reported income, which could result in a confusing and misleading portrayal of operating results.***

Exhibit 9-5 is a concise comparison of the preceding six points regarding variable costing and absorption costing.

EXHIBIT 9-4
Stassen Company
Analysis of Fixed Factory Overhead in 19_2

		Inventory		Expense
PANEL A: VARIABLE COSTING				
No fixed factory overhead carried over from 19_1				
Fixed factory overhead incurred in 19_2	$10,000	→		$10,000
PANEL B: ABSORPTION COSTING		**Units**	**Dollars**	
Fixed factory overhead in beginning inventory	$ 4,000	4,000	$4,000	
Fixed factory overhead incurred in 19_2	10,000			
To account for overhead:	$14,000			
Applied to product, 900 × $1 =		900	900	
Available for sale		4,900	$4,900	
Contained in standard cost of goods sold	$ 3,000	3,000	3,000 →	$ 3,000
In ending inventory	1,900	1,900	$1,900	
Not applied, so it becomes production volume variance, (10,000 − 900) × $1	9,100	→		9,100
Fixed factory overhead written off as an expense in 19_2				$12,100
Fixed factory overhead accounted for in 19_2	$14,000			
Difference in 19_2 factory overhead charged to expense, $12,100 − $10,000				$ 2,100
This amount is the same as the difference in 19_2 operating income, $6,600 − $4,500				$ 2,100

EXHIBIT 9-5
Comparative Income Effects

	Question	Variable Costing	Absorption Costing	Comments
a.	Is fixed factory overhead inventoried?	No	Yes	Basic theoretical question of when this cost should become an expense.
b.	Is there a production volume variance?	No	Yes	Choice of denominator volume affects measurement of operating income under absorption costing only.
c.	How are other variances treated?	Same	Same	Highlights that the basic difference is the accounting for fixed factory overhead, not the accounting for any variable manufacturing costs.
d.	Are classifications between variable and fixed costs routinely made?	Yes	No	However, absorption cost can be modified to obtain subclassifications of variable and fixed costs, if desired.
e.	How do changes in inventory levels affect operating income?			Differences are attributable to timing of the transformation of fixed factory overhead into expense.
	Production = sales	Equal	Equal	
	Production > sales	Lower*	Higher†	
	Production < sales	Higher	Lower	
f.	What are the effects on cost-volume-profit relationships?	Driven by sales volume	Driven by production volume *and* sales volume	Management control benefit: Effects of changes in production volume on operating income are easier to understand under variable costing.

*That is, lower operating income than under absorption costing.
†That is, higher operating income than under variable costing.

BREAKEVEN POINTS IN VARIABLE COSTING AND ABSORPTION COSTING

Objective 4

Describe the effects of absorption costing on computing the breakeven point

Chapter 3 introduced cost-volume-profit analysis. The discussion assumed (a) the use of variable costing or (b) the use of absorption costing with no changes in beginning or ending inventory levels.

If variable costing is used, the breakeven point is computed in the usual manner. It is unique; there is only one breakeven point. Moreover, income is a function of sales. As sales volume rises, income rises, and vice versa. In our Stassen illustration for 19_2:

$$\text{Breakeven point in units} = \frac{\text{Total fixed costs incurred during period}}{\text{Unit contribution margin}}$$

$$\text{Let } N_S = \text{Breakeven point in units}$$

$$N_S = \frac{\$10{,}000 + \$1{,}400}{\$8.50 - (\$2.00 + \$.50)} = \frac{\$11{,}400}{\$6} = 1{,}900$$

In contrast, ***if absorption costing is used, operating income is a function of both sales volume and production volume. Changes in inventory levels can dramatically affect income, as our illustration demonstrates. The breakeven point is not unique. There are many combinations of sales and production that produce an operating income of zero.***[6]

We see that variable costing dovetails precisely with cost-volume-profit analysis. The user of variable costing can easily compute the breakeven point or any effects that changes in sales volume may have on short-run income. In contrast, the user of absorption costing must consider both sales volume and production volume before making such computations.

Consider an additional example. Suppose in our illustration that actual production in 19_2 were equal to the denominator volume, 10,000 units. Also suppose that there were no sales and no marketing and administrative costs. All the production would be placed in inventory, and so all the fixed factory overhead would be held back in inventory. There would be no production volume variance. Thus, the company could break even with no sales whatsoever! In contrast, under variable costing the operating loss would be equal to the fixed costs.

Absorption costing enables a manager to affect reported income for a given accounting period by (a) choosing a production schedule and (b) choosing a level of denominator volume. Given such opportunities, some managers may manipulate income. These temptations are not available under variable costing.

PERFORMANCE MEASURES AND ABSORPTION COSTING

Undesirable Buildups of Inventories

Objective 5

Understand how absorption costing influences performance-evaluation decisions

Absorption costing enables managers to increase operating income in the short run by increasing the production schedule independent of customer demands for products. Such a variation in the production schedule can add to the costs of doing

---

[6]A formula for computing the breakeven point under absorption costing is similar to the formula under variable costing. However, the numerator must include the fixed-factory-overhead rate multiplied by the difference between the number of units sold and units produced. In our Stassen Company illustration for 19_2:

$$\text{Breakeven sales in units} = \frac{\text{Total fixed costs incurred during period} + \left[\text{Fixed overhead rate}\left(\text{Breakeven sales in units} - \text{Units produced}\right)\right]}{\text{Unit contribution margin}}$$

Let N_S = Breakeven sales in units

$$N_S = \frac{(\$10{,}000 + \$1{,}400) + \$1.00(N_S - \$900)}{\$8.50 - (\$2.00 + \$.50)}$$

$$\$6.00\ N_S = \$11{,}400 + \$1.00N_S - \$900$$

$$\$5.00\ N_S = \$10{,}500$$

$$N_S = 2{,}100$$

Proof of 19_2 breakeven point:

Gross margin:		
(Sales price − unit cost of goods sold) × units sold,		
($8.50 − $3.00) × 2,100 =		$11,550
Production volume variance (10,000 − 900) × $1.00	$9,100	
Marketing and administrative costs:		
Variable, $.50 × 2,100	1,050	
Fixed	1,400	11,550
Operating income		$ 0

You can readily see that the breakeven point depends on three factors: (1) the units sold, (2) the units produced, and (3) the denominator volume chosen to set the fixed overhead rate. In this case, the numbers are 2,100 sold, 900 produced, and 10,000 denominator volume.

business without any attendant increase in sales. For example, a manager who is evaluated based on absorption-costing income may increase production at the end of a performance-review period solely to make the production volume variance less unfavorable or more favorable. Each extra unit produced absorbs fixed manufacturing overhead that would otherwise have been an expense of the period.

The undesirable effects of such an increase in production at the end of a reporting period may be sizable, and they can arise in several ways. For example:

1. A plant manager may switch production to those orders that absorb the highest amount of overhead, irrespective of their demand by customers (called "cherry picking" the production line). Some difficult-to-manufacture items may be delayed, resulting in missed promises on customer delivery dates.
2. A plant manager may accept a particular order to increase his or her production even though another plant in the same company is better suited to handle that order.
3. To meet increased production, a manager may defer maintenance beyond the current period. Although operating income may increase currently, future income will probably decrease because of increased repairs and less-efficient equipment.

Early criticisms of absorption costing concentrated on whether fixed manufacturing overhead qualified as an asset under generally accepted accounting principles. Criticisms of absorption costing have increasingly emphasized its potential undesirable incentives for managers. Indeed, one critic labels absorption costing as "one of the black holes of cost accounting," in part because it may induce managers to make decisions "against the long-run interests" of the company.[7]

Proposals for Revising Performance Evaluation

Critics of absorption costing are making one or both of two proposals for revising how the performance of managers is evaluated: (a) reduce the types of costs that are inventoried and (b) emphasize nonfinancial performance measures.

By accounting for fixed manufacturing overhead as period costs, variable costing reduces the incentives for managers to build up inventory levels. Some authors argue that a large number of costs should be expensed. For example, some propose "super-variable costing," in which all costs other than direct materials are expensed to the period in which they are incurred.[8] By expensing direct labor and variable manufacturing overhead to the current accounting period, managers will have less incentive to build inventory levels beyond those required to meet customer demand.

Nonfinancial performance measures can be used to monitor key variables such as attaining but not exceeding inventory levels, meeting of promised customer delivery dates, and abiding by plant maintenance schedules. By including such nonfinancial performance measures in performance reviews, managers are given explicit signals about the factors top management views as important.

Designing an effective performance measurement and reward system is a challenging task. Chapter 28 provides further discussion of the potential limitations of using single-performance benchmarks, whether financial or nonfinancial in nature.

---

[7]R. Schmenner, "Escaping the Black Holes of Cost Accounting," *Business Horizons*, January–February 1988, p. 66.

[8]Goldratt and Cox, *The Goal*.

PROBLEM FOR SELF-STUDY

PROBLEM

Reconsider the facts in Exhibit 9-3, page 294.

Required

1. Assume a total of $1,000 in unfavorable price, spending, and efficiency variances for variable manufacturing costs in 19_2. These variances are regarded as adjustments to cost of goods sold. Compute the operating income under variable costing and absorption costing.
2. Ignore requirement 1. Suppose production in 19_2 were 5,900 instead of 900 units and sales were unchanged. Compute the operating income under variable costing and absorption costing.

SOLUTION

1. Variable costing and absorption costing do not differ regarding price, spending, and efficiency variances. Therefore, the operating income under each method would be lowered by $1,000:

	Variable Costing	Absorption Costing
Operating income for 19_2 per Exhibit 9-3	$6,600	$4,500
Price, spending, and efficiency variances for variable costs	1,000 U	1,000 U
Operating income	$5,600	$3,500

The $1,000 in unfavorable variances would appear as a line item in the income statement, ordinarily as an adjustment to cost of goods sold at standard.

Alternatively, the $1,000 variances could be displayed as follows (amounts for detailed variances assumed):

		Variable Costing	Absorption Costing
Operating income for 19_2 per Exhibit 9-3		$6,600	$4,500
Variances for variable costs:			
Direct materials price variance	$ 800 F		
Direct materials efficiency variance	1,200 U		
Direct labor price variance	100 F		
Direct labor efficiency variance	300 U		
Variable-factory-overhead spending variance	300 U		
Variable-factory-overhead efficiency variance	100 U	1,000 U	1,000 U
Operating income		$5,600	$3,500

2. The operating income under variable costing would be unaffected by the units produced, so it would be $6,600, as before. Why? Because operating income under variable costing is a function of units sold.

The operating income under absorption costing would be affected by the units produced. The production volume variance would be $4,100 unfavorable instead of $9,100 unfavorable. Therefore, operating income would be higher by $5,000:

Sales, as before	$25,500
Beginning inventory, as before	12,000
Add absorption cost of goods manufactured, $2,700 before, now 5,900 × $3, or	17,700
Available for sale	29,700
Deduct ending inventory, $5,700 before, now 6,900 × $3, or	20,700

Cost of goods sold—at standard	9,000
Production volume variance, unfavorable (10,000 − 5,900) × $1	4,100
Cost of goods sold—as adjusted	13,100
Gross margin, at actual	12,400
Marketing and administrative costs	2,900
Operating income	$ 9,500

PART TWO: ROLE OF VARIOUS DENOMINATOR LEVELS IN ABSORPTION COSTING

Objective 6

Explain the role of various denominator levels in absorption costing

Chapter 8 pointed out that the choice of a particular production volume level as a denominator in the computation of fixed-overhead rates significantly affects product costs and income. We now study how various alternative levels of volume can affect operating income under absorption costing. Fixed manufacturing costs have become a more prominent part of an organization's total costs. Thus, the choice of a denominator level becomes increasingly important. The choice affects not only income but also decisions regarding pricing, product mix, and whether to make or buy component parts. The denominator is often expressed in terms of plant capacity.

CHARACTERISTICS OF CAPACITY

The term *capacity* means "constraint," "an upper limit." Although the term capacity is usually applied to plant and equipment, it is equally applicable to other resources, such as people and materials. A shortage of machine time, executive time, or materials may be critical in limiting company production or sales.

A company may take steps—paying overtime, subcontracting, paying premium prices for additional materials, and the like—to expand production. But these steps may be unattractive and unacceptable from an economic viewpoint. ***Management specifies the upper limit of capacity for current planning and control purposes after considering engineering* and *economic factors.*** Management, not external forces, usually imposes the upper limit for capacity. In setting capacity size, management considers its decisions regarding the acquisitions of plant and equipment. In turn, managers reach decisions on these long-term assets after studying the expected impact of these capital outlays on operations over a number of years.

MEASUREMENT OF CAPACITY

Measurement of capacity usually begins with **theoretical capacity** (also called *maximum* or *ideal* capacity), which assumes the production of output 100% of the time. Then managers deduct for Sundays, holidays, downtime, changeover time, and similar items to attain a measure of **practical capacity**. The latter term is defined as the maximum level at which the plant or department can operate efficiently. Practical capacity often allows for unavoidable operating interruptions such as repair time or waiting time.

Practical capacity is often used as a denominator volume. Two other levels of capacity commonly used as denominators are:

1. **Normal volume**, the level of capacity utilization (which is less than 100% of practical capacity) that will satisfy average customer demand over a span of time (often five years) that includes seasonal, cyclical, and trend factors.
2. **Master-budget volume**, the anticipated level of capacity utilization for the coming year or other planning period (such as six months). Company practices vary, but surveys indicate that at least 60% of the large firms use master-budget volume as the denominator level.

There are apt to be differences in terms used among companies, so be sure to obtain their exact meaning in a given situation. For example, *master-budget volume* is sometimes called **budgeted volume, expected annual volume, expected annual capacity, expected annual activity,** or **master-budget activity**.

INVENTORY AND INCOME EFFECTS

Consider the data below for comparing the choices of a denominator volume:

	Denominator Based on		
	Master-Budget Volume	Normal Volume*	Practical Capacity
Budgeted fixed factory overhead for 19_1	$4,500,000	$4,500,000	$4,500,000
Volume in standard machine hours allowed	75,000	90,000	100,000
Budgeted fixed-factory-overhead rate per hour	$60	$50	$45

*Expected average volume per year for the next five years.

We will deal with fixed factory overhead only. Why? Because variable overhead fluctuates in direct proportion to changes in volume, but fixed overhead does not. ***The entire problem of choosing among master-budget volume, normal volume, or practical capacity is raised by the presence of fixed overhead.*** Recall that the production volume variance in Chapter 8 was confined to fixed overhead.

Suppose that in 19_1, 70,000 units were produced (in 70,000 actual and standard hours allowed) and 60,000 units were sold. There was no beginning inventory. Exhibit 9-6 compares the effects of choosing different levels of denominator volume for setting the budgeted fixed-factory-overhead rate. There are different inventory costs, different operating incomes. Further, the measure of utilization of facilities—the production volume variance—will differ markedly. In Exhibit 9-6, operating income is lowest where practical capacity is the denominator volume and highest when master-budget volume is the denominator volume. Why? Because a smaller (larger) portion of overhead is held back as an asset in inventory when a lower (higher) overhead rate is used.

Exhibit 9-6 also indicates that the accounting effects of using practical capacity are lower unit costs for inventory purposes and the steady appearance of an unfavorable production volume variance on the income statement.

The journal entry to record the production volume variance at the end of 19_1 would be:

	Denominator Based on					
	Master-Budget Volume		Normal Volume		Practical Capacity	
Fixed factory overhead applied	4,200,000		3,500,000		3,150,000	
Production volume variance	300,000		1,000,000		1,350,000	
Fixed factory overhead control		4,500,000		4,500,000		4,500,000

EXHIBIT 9-6

Income Statement Effects of Using Various Volume Bases as Denominators for Overhead Application

	Master-Budget Volume (75,000 Hours)		Normal Volume (90,000 Hours)		Practical Capacity (100,000 Hours)	
	Using a $60 Fixed Overhead Rate		Using a $50 Fixed Overhead Rate		Using a $45 Fixed Overhead Rate	
Sales		$ xxx		$ xxx		$ xxx
Production costs:						
Direct materials, direct labor, variable overhead	$ xxx		$ xxx		$ xxx	
Fixed overhead applied to product*	4,200,000		3,500,000		3,150,000	
Total production costs, 70,000 units	$ xxx		$ xxx		$ xxx	
Ending inventory, fixed-overhead component, 10,000 units†	600,000		500,000		450,000	
Total fixed-overhead component of cost of goods sold		3,600,000		3,000,000		2,700,000
Production volume variance, unfavorable‡		300,000		1,000,000		1,350,000
Total fixed overhead charged to the period's sales		$3,900,000		$4,000,000		$4,050,000
Operating income		Highest		Middle		Lowest
Recap:						
Overhead charged as follows:						
Cost of goods sold (expense)		$3,600,000		$3,000,000		$2,700,000
Production volume variance (expense)		300,000		1,000,000		1,350,000
Ending inventory (asset)		600,000		500,000		450,000
Overhead accounted for§		$4,500,000		$4,500,000		$4,500,000

*70,000 × $60 and 70,000 × $50 and 70,000 × $45, respectively.
†10,000 × $60 and 10,000 × $50 and 10,000 × $45, respectively.
‡(75,000 − 70,000) × $60 and (90,000 − 70,000) × $50 and (100,000 − 70,000) × $45, respectively. This amount is sometimes called "loss from idle capacity."
§Assumed here that actual and budgeted fixed overhead was $4,500,000.

Practical Capacity

Management often wants to keep running at full capacity, which really means at practical capacity. As the first footnote in Exhibit 9-6 indicates, using practical capacity as the denominator leads to the lowest unit cost ($45 instead of $50 or $60). Where product costs are used as guides for pricing, some managers say that choosing practical capacity as the denominator induces lower prices and thus maximizes production volume.

Using practical capacity as the denominator volume became increasingly popular with U.S. businesses in the 1980s. A major reason is probably the position of the Internal Revenue Service, which between 1975 and 1987 permitted the use of practical capacity as a denominator volume. As compared with normal volume or master-budget volume, the practical-capacity denominator volume results in the faster write-offs of fixed factory overhead as a tax deduction.

There is no requirement that American companies use the same denominator volumes for management purposes and for income tax purposes. Nevertheless, the economies of recordkeeping and the desire for simplicity often lead companies to choose the same denominator for management and tax purposes.

In 1987, the Internal Revenue Service specified that practical capacity could no longer be used for tax purposes. Companies must use master-budget volume (or its equivalent), along with the full proration of variances between inventories and cost of goods sold. Hence, to the extent that income tax laws affect accounting for internal purposes, the use of practical capacity will probably diminish.

Normal Volume Versus Master-Budget Volume

Master-budget volume is the denominator for applying all fixed factory overhead to products on a year-to-year basis. The denominator based on *normal* volume attempts to apply fixed overhead by using a *longer-run* average expected volume. Conceptually, the *normal rate* results in favorable production volume variances in years of above-average volume, which are offset by unfavorable production volume variances in years of below-average volume.

A major reason for choosing master-budget volume over normal volume is the overwhelming forecasting problem that accompanies the determination of normal volume. Sales not only fluctuate cyclically but exhibit trends over the long run. In effect, the use of normal volume implies an unusual talent for accurate long-run forecasting. Many accountants and executives who reject normal volume as a denominator claim that the nature of their company's business precludes sufficiently accurate forecasts beyond one year.

When companies use normal volume, the objective is to choose a period long enough to average out sizable fluctuations in volume and to allow for trends in sales. The uniform rate for applying fixed overhead supposedly provides for "recovery" of fixed costs over the long run. The basic reason in favor of using normal volume is that the resulting inventory cost is more representative of the typical relationship between total costs and production volume beyond one year. As production volume varies over different years, low capacity utilization is considered as a cost of excess capacity. Any given year's fixed costs are not spread over a relatively small number of units.

Conceptually, when normal volume is the denominator, the yearly production volume variances should be carried forward in the balance sheet. Practically, however, the year-end balance is closed to Cost of Goods Sold. Why? Because the accounting profession (and the Internal Revenue Service) generally view the year as the terminal time span for allocation of underapplied or overapplied overhead.

Significance of Denominator Volume for Inventory Costing and Control

Obviously the choice of a denominator volume for inventory costing is a matter of judgment. The selection of a denominator becomes crucial when product costs heavily influence managerial decisions. For example, in a cyclical industry, the use of master-budget volume rather than normal volume as a denominator would tend to cause a company to quote low cost-based prices in boom years and high cost-based prices in recession years—in obvious conflict with good business judgment. That is why normal volume may make more sense as a denominator when there are wide swings in business volume through the years, even though the yearly production volume variance is not carried forward in the balance sheet.

In the realm of planning and control for the *current* year, however, normal volume is an empty concept. Normal volume is used as a basis for *long-range* plans. It depends on the time span selected, the forecasts made for each year, and the weighting of these forecasts. A comparison of the 70,000-hour master-budget volume with the 90,000-hour normal volume in Exhibit 9-6 might be the best basis for auditing long-range planning. ***However, normal volume is an average that has no particular significance with respect to a follow-up for a particular year.*** Attempting to use normal volume as a reference point for judging current performance is an example of misusing a long-range measure for a short-range purpose. The pertinent comparison is a particular year's actual volume with the volume level originally predicted in the authorization for the acquisition of facilities.

The master-budget volume, rather than normal volume or practical capacity, is more germane to the evaluation of current results. The master budget is the principal short-run planning and control tool. Managers feel much more obligated to reach the levels stipulated in the master budget, which should have been carefully set in relation to the maximum opportunities for sales in the current year.

PROBLEM FOR SELF-STUDY

PROBLEM

Suppose Budweiser Company opened a new California brewery with a practical capacity of 5 million barrels of beer per year. The budgeted fixed indirect manufacturing costs are $18,000,000 for 19_1. The master-budget volume is 3 million barrels for 19_1 and 4.5 million barrels for 19_2 through 19_5. Normal volume is specified as relating to 19_2 through 19_5.

Required

1. Compute the fixed indirect manufacturing costs per barrel if Budweiser uses (a) master-budget volume, (b) normal volume, or (c) practical capacity as the denominator volume. Also, compute the production volume variance for each alternative, assuming that actual production will equal budgeted production.
2. Which volume level would you prefer to use for judging management performance during 19_1? Explain.

SOLUTION

1.

	(a)	(b)	(c)
		Denominator Based On	
	Master-Budget Volume	**Normal Volume**	**Practical Capacity**
Budgeted fixed indirect manufacturing cost for 19_1	$18,000,000	$18,000,000	$18,000,000
Volume in barrels	3,000,000	4,500,000	5,000,000
Budgeted fixed indirect manufacturing costs per barrel	$ 6.00	$ 4.00	$ 3.60
Production volume variance	0	$ 6,000,000*	$ 7,200,000†

*(4,500,000 − 3,000,000) × $4.00 = $6,000,000
†(5,000,000 − 3,000,000) × $3.60 = $7,200,000

2. Master-budget volume is preferable for judging 19_1 performance. Managers relate more easily to currently attainable targets as quantified in the current year's budget.

PART THREE: STANDARD-COST VARIANCES AND FINANCIAL STATEMENTS

EFFECTS OF PRORATIONS

Proration to Achieve Actual Costs

Objective 7

Describe how standard-cost variances affect financial statements

Managers and accountants tend to think of "actual" costs as the most accurate costs attainable. That is, they believe that normal costs or standard costs provide account balances that are somehow less accurate if month-end or year-end variances are not prorated among the affected accounts to get corrected amounts that better approximate "actual" costs.[9]

How do companies dispose of their variances at year-end? Of course, in all cases immaterial variances can be written off immediately as adjustments to cost of goods sold, but other options are available. A survey of U.S. companies showed:[10]

Closed to cost of goods sold	53.1%
Closed to income account	10.5
Subtotal affecting current income	63.6
Prorated (apportioned) among	
Work in Process, Finished Goods, and Cost of Goods Sold	33.6
Carried forward to next year	1.2
No response	1.6
Total	100.0%

The disposition of *overhead* variances at year-end was initially described in Chapter 4, pages 108–11. The same concepts and procedures apply to *all* variances. First, decide whether *proration* should occur. This decision usually depends on judgments about whether the variances are material in amount. Second, if proration is desired, find out where the related "standard" costs are now lodged in the general-ledger accounts. Use the sum of these standard costs as the basis for apportioning the variances. For example, suppose that all variances are unfavorable and total \$120,000 at year-end. The proration would be in proportion to the balances (assumed) in the accounts containing the relevant standard costs:

	Work in Process	Finished Goods	Cost of Goods Sold
Balances before proration, \$1,000,000	\$100,000	\$300,000	\$600,000
All variances, \$120,000 unfavorable	12,000*	36,000*	72,000*
Balances after proration, \$1,120,000	\$112,000	\$336,000	\$672,000

*Approach 1: \$120,000 ÷ \$1,000,000 = 12%; 12% × \$100,000 = \$12,000; or
Approach 2: \$100,000 ÷ \$1,000,000 = 10%; 10% × \$120,000 = \$12,000; and so on.

The journal entry would be:

Work in process	12,000	
Finished goods	36,000	
Cost of goods sold	72,000	
Cost variance accounts (would be detailed)		120,000
To prorate variances.		

[9]For example, see Standard #407, "Use of Standard Costs for Direct Material and Direct Labor," of the Cost Accounting Standards Board, Washington, D.C.

[10]Chow, Shields, and Wong-Boren, "A Compilation of Recent Surveys," p. 187. The table here is from the Chiu and Lee survey cited therein. Similar results are reported in the F. Rayburn and A. Stewart survey also cited therein.

Impact of Financial Accounting on Proration of Variances

Generally accepted accounting principles and income tax laws typically require that financial statements show actual costs, not standard costs, of inventories and costs of goods sold. Consequently, variance prorations are required if they result in a material change in inventories or operating income.

Prorations also tend to prevent managers from setting standards aimed at manipulating income. If managers do not have to prorate variances, they can more easily affect a year's operating income by how they set standards.

Consider a company starting up a new process or manufacturing a new product. Management may expect variances. For example, learning occurs during the early stages of conducting a new process or manufacturing a new product, and inefficiencies are commonplace. Nevertheless, management may specify standards under assumptions of steady-state conditions that will not be reached until after the learning has occurred.

How should variances occurring during the first stages of new operations be accounted for? Suppose the standards are in fact standards that will apply under steady-state conditions. Should the variances be written off in the period incurred, or should they be carried forward as assets and written off in future periods? We believe that a strong case can be made for the latter accounting—if the standards are specified as being currently attainable only under ordinary ongoing operating conditions.

The Case Against Proration

Some accountants, industrial engineers, and managers reject the idea that actual costs represent the most accurate costs. Instead, they claim that currently attainable standard costs are the "true costs," the only costs that may be carried forward as assets. They contend that variances are measures of inefficiency or abnormal efficiency. Therefore, variances should be completely written off to the accounting period instead of being prorated among inventories and cost of goods sold. In this way, inventory costs will be more representative of desirable and attainable costs. ***In particular, there is no justification for carrying costs of inefficiency as assets, which is what proration tends to accomplish.***

Varieties of Proration

Variations of proration methods may be desirable under some conditions. For instance, price variances may be viewed as being unavoidable and therefore proratable. In contrast, efficiency variances may be viewed as being currently avoidable and therefore nonproratable. The costs of avoidable inefficiency do not qualify as assets under any economic test.

We believe that variances do not have to be prorated to inventories as long as standards are currently attainable. However, if standards are not up to date, or if they reflect perfection (ideal) performance rather than expected performance under reasonably efficient conditions, then conceptually the variances should be split between the portion that reflects departures from currently attainable standards and the portion that does not. The former should be written off as period costs; the latter should be prorated to inventories and cost of goods sold. For example, assume that an operation has a perfection standard time allowed of 50 minutes, which is reflected in a formal standard-cost system. The currently attainable standard is 60 minutes. If it takes, say, 75 actual minutes to perform the operation, the conceptual adjustment would call for writing off the cost of 15 minutes of the 25-minute variance as a period cost and for treating the cost of the remaining 10-minute variance as an inventoriable cost.

The prorations of variances can have significant effects on the measurement of operating income, particularly when inventories have increased or decreased substantially during a given accounting period and when the variances are relatively large. The Problem for Self-Study (p. 310) demonstrates these effects. Also see the chapter appendix for additional discussion of prorations.

ADJUSTING INVENTORIES FOR EXTERNAL REPORTING

Objective 8

Show how to convert variable-costing inventory valuation to absorption costing

To satisfy external reporting requirements, companies often make adjustments to inventory accounts. Examples include (a) converting a variable-costing inventory valuation to absorption costing and (b) prorating variances. Typically, these companies do *not* alter their ongoing inventory records. Instead, for external reporting they create a separate but related "inventory adjustment" valuation account that bears a host of labels, some short, some long.

To illustrate the use of such an account, assume that a company adjusts its Finished Goods Inventory of $700,000 upward by $100,000. The adjustment may be (a) to convert the inventory from variable costing to absorption costing or (b) to restate the inventory because of a proration of unfavorable variances. The journal entry would be:

a. Finished goods inventory adjustment account	100,000	
Fixed factory overhead		100,000

or

b. Finished goods inventory adjustment account	100,000	
Cost variance accounts (detailed, such as direct material price variance, direct material efficiency variance)		100,000

Under the first-in, first-out assumption, the adjustment account would disappear in the subsequent period when the related inventories are sold. For example:

Cost of goods sold (from inventory adjustment account)	100,000	
Finished goods inventory adjustment account		100,000

VARIANCES AND INTERIM REPORTING

Objective 9

Explain the role of variances in interim reporting

Effects of Objectives

Interim reporting is far from uniform. Some companies write off all variances monthly or quarterly to Cost of Goods Sold. Others prorate the variances among inventories and Cost of Goods Sold. For example, consider the practices of 247 companies in a survey concerning interim accounting for overhead variances (underapplied or overapplied overhead):[11]

[11]Chiu and Lee, cited in Chow, Shields, and Wong-Boren, "A Compilation of Recent Surveys," p. 242.

Closed to cost of goods sold	60.3%
Closed to income account	11.3
Subtotal affecting current income	71.6%
Prorated (apportioned) among Work in Process, Finished Goods, and Cost of Goods Sold	23.1
Carried forward to next period	4.1
No response	1.2
Total	100.0%

Most companies follow the same practices for both interim and annual financial statements. They apparently favor the first of two major conflicting objectives of interim reporting.

Objective 1: The results for each interim period should be computed in the same way as if the interim period were an annual accounting period. For example, the interim underapplied overhead would be written off or prorated just like the annual underapplied overhead.

Objective 2: Each interim period is an integral part of the annual period. For example, management may regard a February repair cost as benefiting the entire year's operations. If the repair is the major cause of the underapplied overhead, then the underapplied overhead should be deferred and spread over the entire fiscal year.

Supporters of the second objective argue that overhead application relates to a year as a whole, not just a part of a year. There are bound to be random month-to-month underapplications or overapplications that may come near to off-setting one another by the end of the year. The most frequent causes of these month-to-month deviations are (a) operations at different levels of volume and (b) the presence of seasonal costs, such as heating, that are averaged in with other overhead items in setting an annual overhead rate.

If underapplied manufacturing overhead is carried forward during the year, it would appear on an interim balance sheet as a current asset, a "prepaid expense." Similarly, overapplied overhead would be a current liability, a "deferred credit."

External Reporting

The rules for external reporting distinguish among the types of standard-cost variances. In general, the same reporting procedures should be used for interim and annual statements. However, Accounting Principles Board Opinion No. 28 pinpoints direct material price variances and factory-overhead production volume variances for special treatment. If these interim variances are expected to be offset by the end of the annual period, they should "ordinarily be deferred at interim reporting dates."

The interim overhead variances are often called *planned variances*. They are especially common in seasonal businesses. Companies favor deferral because such amounts are expected to disappear by the end of the year through the use of averaging as costs are applied to the product. However, "unplanned" or unanticipated underapplied or overapplied overhead should be reported "at the end of an interim period following the same procedures used at the end of a fiscal year."[12]

This approach of Opinion 28 represents a compromise between Objectives 1 and 2 above. Why? Because "unplanned" and "planned" overhead variances are accounted for differently.

[12]Accounting Principles Board Opinion No. 28, *Interim Financial Reporting*, Paragraph 14(d).

PROBLEM FOR SELF-STUDY

PROBLEM

Consider the following standard-cost balances at year-end (before proration):

Work in process	$ 100,000
Finished goods	300,000
Cost of goods sold	600,000
	$1,000,000

Assume that variances were:

Production Volume Variance		Other Variances	
30,000			150,000

Management has decided that the $30,000 unfavorable production volume variance should be written off as an adjustment to cost of goods sold. However, the other variances should be prorated in proportion to the balances in work in process, finished goods, and cost of goods sold (before proration).

Required

1. Prepare a schedule that prorates the "other" variances.
2. Prepare journal entries that close all variance accounts.
3. What is the justification for writing off some variances but prorating others?

SOLUTION

1.

	Work in Process	Finished Goods	Cost of Goods Sold
Balances before proration, $1,000,000	$100,000	$300,000	$600,000
"Other" variances, $150,000, favorable	15,000*	45,000*	90,000*
Balances after proration, $850,000	$ 85,000	$255,000	$510,000†

*$150,000 ÷ $1,000,000 = 15%; 15% × $100,000, $300,000, and $600,000, respectively.
†The $30,000 production volume variance will be added to this balance.

2.

Cost of goods sold	30,000	
Production volume variance		30,000
Other variances	150,000	
Work in process		15,000
Finished goods		45,000
Cost of goods sold		90,000

3. The major justification for writing off the production volume variance is that it does not fit the definition of an asset. That is, it is a "lost cost," not a future benefit. In contrast, the other variances may be dominated by price changes or efficiency factors that should have been embedded in the standard costs when they were set at the beginning of the year. Therefore, these variances may indeed fit the definition of an asset. That is, the variances are inventoriable costs that benefit future operations.

SUMMARY

Absorption costing and variable costing have different effects on operating income when inventory levels fluctuate. Differences occur because fixed factory overhead is inventoried only under absorption costing. Operating income is influenced by sales under variable costing, but by sales *and* production under absorption costing.

Variable costing and absorption costing are completely general in the sense that there may be:

actual variable costing	actual absorption costing
normal variable costing	normal absorption costing
standard variable costing	standard absorption costing

Standard-costing systems have the same cost variances regardless of whether variable costing or absorption costing is used, with one exception: Production volume variance is not present when variable costing is used because all fixed factory overhead is regarded as a period cost rather than as an inventoriable cost.

Master-budget volume is the most popular denominator volume. Practical capacity and normal volume are the second- and third-most popular.

If they are significant, standard-cost variances are usually prorated among various inventory accounts and cost of goods sold. In this way, better approximations of "actual" costs are achieved.

The advocates of currently attainable standard costs for inventory costing maintain that the results are conceptually superior to the results under "actual" or "normal" inventory costing systems. They contend that the costs of inefficiency are not inventoriable.

APPENDIX: STANDARD-COST VARIANCES: A MORE ACCURATE APPROACH TO PRORATION

The approach to proration as described in the body of the chapter is relatively simple. More accurate approaches are explored in this appendix, not for accuracy's sake alone, but to review a standard-costing system in its entirety.

The following facts are the basis for our discussion of the general approach to proration. To keep the calculations manageable, the numbers are deliberately small.

Morales Company uses absorption costing. It has the following results for the year:

Purchases of direct materials (charged to Direct Materials Inventories at standard prices), 200,000 lbs. @ $.50	$100,000
Direct material price variance, 200,000 lbs. @ $.05	10,000
Direct materials—applied at standard prices, 160,000 lbs. @ $.50	80,000
Direct material efficiency variance, 8,000 lbs. @ $.50	4,000
Direct labor incurred	45,000
Direct labor—applied at standard rate, 2,000 hrs. @ $20	40,000
Direct labor price variance, 2,200 hrs @ $.4545	1,000
Direct labor efficiency variance, 200 hrs. @ $20	4,000
Manufacturing overhead applied—at budgeted rate per machine hour	70,000
Manufacturing overhead incurred	75,000
Underapplied manufacturing overhead	5,000
Sales	273,000
Marketing, administrative, and customer-service costs	130,000

Assume Morales makes one uniform product, a specialty plastic container. Assume also that 40% of the production is in the ending inventory of finished goods and that 60% of the production has been sold. There is no ending work in process. All variances are unfavorable.

There were no beginning inventories. The balances (before proration) at the end of the year are based on the data given above:

		Percentage
Work in process	$ 0	0%
Finished goods, 40% of the total standard costs applied for material, labor, and overhead ($80,000 + $40,000 + $70,000): 40% of $190,000	76,000	40
Cost of goods sold, 60% of $160,000	114,000	60
Total	$190,000	100%

An analysis of the data regarding direct materials follows:

	Pounds	Total Costs at $.50 Standard Price Per Pound	Percentage
To account for	200,000	$100,000	100%
Now present in:			
Direct material efficiency variance	8,000	$ 4,000	4%
Finished goods	64,000	32,000	32
Cost of goods sold	96,000	48,000	48
Remainder, in direct materials inventories	32,000	16,000	16
Accounted for	200,000	$100,000	100%

Managers at Morales Company want to study a comparative analysis of the effects on operating income (1) without proration of any variances and (2) with proration of all variances. This analysis appears as Exhibit 9-8, page 315. Let us now discuss the steps that lead up to that analysis and the two guidelines that steer us in proration.

Guideline 1. The assumption is often made that standard costs are present in uniform proportions in Work in Process, Finished Goods, and Cost of Goods Sold. If the assumption is invalid and the amounts involved are significant, then the direct materials, direct labor, and factory-overhead components should be *separately* identified; the related variances should then be prorated in proportion to those three amounts.

We confined our analysis to direct labor and factory overhead. These elements (only) are presented in T-accounts. The following entries are numbered in accordance with the logical flows of the amounts through the accounts:

Work in Process

Debit		Credit	
1. Direct labor	40,000	3. Transferred*	110,000
2. Overhead applied	70,000		

Finished Goods

Debit		Credit	
3.	110,000	4. Sold,* 60% of 110,000, or	66,000

Cost of Goods Sold

Debit		Credit
4.	66,000	

Factory Overhead Control

Debit		Credit
5. Incurred	75,000	

Factory Overhead Applied

Debit	Credit	
	2.	70,000

Direct Labor Price Variance

Debit		Credit
1.	1,000	

*Transferred as a part of the total standard costs transferred

Cash, Current Liabilities, etc.			
		1. Direct labor	45,000
		5. Overhead	75,000

Direct Labor Efficiency Variance			
1.	4,000		

In our example, there is no ending Work in Process. Finished Goods represents 40% of total manufacturing costs; Cost of Goods Sold, 60%. All direct labor and factory-overhead variances may be prorated accordingly. See the final three prorations in Exhibit 9-7 on page 314.

	Total Variance	Finished Goods 40%	Cost of Goods Sold 60%
Direct labor price variance	$ 1,000	$ 400	$ 600
Direct labor efficiency variance	4,000	1,600	2,400
Factory-overhead variance	5,000	2,000	3,000
Totals	$10,000	$4,000	$5,000

For simplicity, the factory-overhead variance of $5,000 has not been subdivided into spending, efficiency, or production volume variances.

Based on the proration of these variances, a journal entry would be made:

Finished goods	4,000	
Cost of goods sold	6,000	
Factory overhead applied	70,000	
Direct labor price variance		1,000
Direct labor efficiency variance		4,000
Factory overhead control		75,000

To prorate variances and to close overhead accounts.

Guideline 2. The direct material variances should be prorated slightly differently than direct labor and factory overhead. Why? Because direct material is inventoried before use, but the other elements of production cannot be inventoried before use.

Some key T-accounts follow regarding the flow of the direct materials cost (only) through the accounts:

Direct Materials Inventory			
1. Purchased	100,000	2. Issued	84,000

Work in Process			
2.	80,000	3. Transferred*	80,000

Finished Goods			
3.	80,000	4. Sold*	48,000
Bal. material cost only	32,000		

Cost of Goods Sold			
4.	48,000		

Accounts Payable			
		1.	110,000

Direct Material Price Variance			
1.	10,000		

Direct Material Efficiency Variance			
2.	4,000		

*Transferred as a part of the total standard costs transferred.

The most complex proration is the direct material price variance. To be most accurate, its proration in our illustration should be traced at $10,000 ÷ 200,000 = $0.05 per pound to wherever the 200,000 pounds have been charged at standard prices. As the analysis in the body of the problem indicates, the pounds are not only in Finished Goods and Cost of Goods Sold. They are also in Direct Materials Inventory and in the Direct Material Efficiency Variance account. Hence, we begin with a proration of the material price variance to four accounts, using the percentages shown for direct materials in the problem data on page 311:

Direct material price variance	$10,000
Allocated to:	
Direct material efficiency variance, 4%	$ 400
Finished goods inventory, 32%	3,200
Cost of goods sold, 48%	4,800
Direct materials inventories, 16%	1,600
Total allocated	$10,000

The following journal entry prorates the price variance:

Direct material efficiency variance	400	
Finished goods	3,200	
Cost of goods sold	4,800	
Direct materials inventory	1,600	
Direct material price variance		10,000

After posting the proration of the direct material price variance, the Direct Material Efficiency Variance account would be:

Direct Material Efficiency Variance

Balance before proration	4,000	
Proration of unfavorable direct material price variance	400	
Balance after proration	4,400	

In turn, the material efficiency variance after proration is allocated to:

Finished goods inventory, 40%	$1,760
Cost of goods sold, 60%	2,640
Total allocated	$4,400

EXHIBIT 9-7
Morales Company
Comprehensive Schedule of Prorations of Variances
(All Variances Are Unfavorable)

Type of Variance	(1) Total Variance	(2) To Direct Materials Inventory	(3) To Direct Material Efficiency Variance	(4) To Finished Goods	(5) To Cost of Goods Sold
Direct material price	$10,000*	$1,600	$ 400	$3,200	$ 4,800
Direct material efficiency					
Balance before proration	4,000		4,000		
Balance after proration			$4,400†	1,760	2,640
Direct labor price	1,000†			400	600
Direct labor efficiency	4,000†			1,600	2,400
Factory overhead	5,000†			2,000	3,000
Total variances prorated	$24,000	$1,600		$8,960	$13,440
*Percentages used for proration	100%	16%	4%	32%	43%
†Percentages used for proration	100%			40%	60%

The following journal entry prorates the efficiency variance:

Finished goods	1,760	
Cost of goods sold	2,640	
Direct material efficiency variance		4,400
To prorate the efficiency variance.		

Exhibit 9-7 is a comprehensive schedule of all the variance prorations explained here. The T-accounts for inventories and cost of goods sold after proration are:

Direct Materials Inventory

Purchased	100,000	Issued	84,000
Proration of direct material price variance	1,600		
Balance	17,600		

Work in Process

Direct materials	80,000	Transferred	190,000
Direct labor	40,000		
Overhead applied	70,000		

Finished Goods

Transferred	190,000	Sold	114,000
Proration of direct labor and overhead variances	4,000		
Proration of direct material price variance	3,200		
Proration of direct material efficiency variance	1,760		
Balance	84,960		

Cost of Goods Sold

Sold	114,000		
Proration of direct labor and overhead variances	6,000		
Proration of direct material price variance	4,800		
Proration of direct material efficiency variance	2,640		
Balance	127,440		

Exhibit 9-8 shows how proration affects operating income. Because all the variances are unfavorable, total variances reduce operating income by $24,000. When prorated, the variances reduce income by only $13,440. The difference between these two approaches—$10,560—shows up in the operating income amounts. Not prorating the variances will result in operating income of $5,000; prorating the variances will result in income of $15,560, a relatively significant difference indeed.

EXHIBIT 9-8
Morales Company
Effects of Disposal of Variances on Operating Income

	Standard Absorption Costing	
	Without Proration	**With Proration**
Sales	$273,000	$273,000
Cost of goods sold—at standard	114,000	114,000
Total variances (from column 1, Exhibit 9-7)	24,000	
Prorated variances (from column 5, Exhibit 9-7)		13,440
Cost of goods sold—after effects of variances	138,000	127,440
Marketing, administrative, and customer-service costs	130,000	130,000
Total charges against sales	268,000	257,440
Operating income	$ 5,000	$ 15,560

TERMS TO LEARN

This chapter and the Glossary at the end of the book contain definitions of the following important terms:

absorption costing *(p. 289)* budgeted volume *(302)* direct costing *(290)*
expected annual activity *(302)* expected annual capacity *(302)*
expected annual volume *(302)* master-budget activity *(302)*
master-budget volume *(302)* normal volume *(302)* practical capacity *(301)*
theoretical capacity *(301)* variable costing *(289)*

ASSIGNMENT MATERIAL

QUESTIONS

9-1 "Differences in operating income between variable costing and absorption costing are due solely to accounting for fixed costs." Do you agree? Explain.

9-2 Why is *direct costing* a misnomer?

9-3 Explain the main conceptual issue between variable costing and absorption costing regarding the proper timing for the release of fixed factory overhead as expense.

9-4 "The main trouble with variable costing is that it ignores the increasing importance of fixed costs in modern business." Do you agree? Why?

9-5 "The depreciation on the paper machine is every bit as much a part of the cost of the manufactured paper as is the cost of the raw pulp." Do you agree? Why?

9-6 "The term *variable costing* could be improved if it were called *variable manufacturing costing*." Do you agree? Why?

9-7 "Advocates of the contribution approach prefer to highlight fixed costs as a lump sum." Do you agree? Explain.

9-8 What is the U.S. Internal Revenue Service's position regarding the inventorying of costs of various business functions?

9-9 Why do external auditors accept some company practices that charge depreciation on factory equipment to expense in the period incurred?

9-10 Give an example of how, under absorption costing, operating income could fall even though sales volume rises.

9-11 List the three factors that affect the breakeven point under absorption costing.

9-12 "If absorption costing is used, operating income is a function of both sales volume and production volume." Do you agree? Explain.

9-13 "A manager who is evaluated based on absorption-costing operating income may increase production solely to make the production volume variance less unfavorable or more favorable." Do you agree? Explain.

9-14 "Disruptive effects of an unwarranted increase in production at the end of an accounting period may be sizable." Give an example regarding absorption costing.

9-15 "Criticisms of absorption costing have increasingly emphasized its potential undesirable incentives for managers." Give an example.

9-16 "Some managers advocate super-variable costing." Explain.

9-17 "Some managers favor super-absorption costing." Explain.

9-18 List three different types of denominator volumes.

9-19 "In the realm of planning and control for the current year, normal volume is an empty concept." Do you agree? Explain.

9-20 How do U.S. rules for external quarterly reporting distinguish among variances?

EXERCISES AND PROBLEMS

Coverage of Part One of the Chapter

9-21 Straightforward variable-costing income statement. Prepare a variable-costing income statement (through operating income) for O'Mara Company. Use the following data: No beginning inventories of work in process or finished goods; no ending inventories of work in process. Production was 500,000 units, of which 400,000 were sold for $50 each. Direct material cost was $6 per unit; direct labor cost was $8 per unit; variable manufacturing cost was $1 per unit; fixed manufacturing cost was $2,000,000; variable marketing and administrative cost was $5 per unit sold; and fixed marketing and administrative cost was $7,500,000.

9-22 Absorption and variable costing. (CMA) Osawa Inc. planned and actually manufactured 200,000 units of its single product in 19_5, its first year of operations. Variable manufacturing costs were $30 per unit of product. Planned and actual fixed manufacturing costs were $600,000. Marketing and administrative costs totaled $400,000 in 19_5. Osawa sold 120,000 units of product in 19_5 at a selling price of $40 per unit. Multiple choice:

1. Osawa's 19_5 operating income using absorption costing is (a) $440,000, (b) $200,000, (c) $600,000, (d) $840,000, (e) none of these.
2. Osawa's 19_5 operating income using variable costing is (a) $800,000, (b) $440,000, (c) $200,000, (d) $600,000, (e) none of these.

9-23 Comparison of actual costing methods. The Rehe Company sells its razors at $3 per unit. The company uses a first-in, first-out actual-costing system. That is, a new fixed-factory-overhead application rate is computed each year by dividing the actual fixed factory overhead by the actual production. The following simplified data relate to its first two years of operation:

	Year 1	Year 2
Sales	1,000 units	1,200 units
Production	1,400 units	1,000 units
Costs:		
Factory—variable	$700	$500
—fixed	700	700
Marketing—variable	1,000	1,200
Administrative—fixed	400	400

Required

1. Income statements for each of the years based on absorption costing.
2. Income statements for each of the years based on variable costing.
3. A reconciliation and explanation of the differences in the operating income for each year resulting from the use of absorption costing and variable costing.
4. Critics have claimed that a widely used accounting system had led to undesirable buildups of inventory levels. (a) Which inventory method, variable costing or absorption costing, is more likely to lead to such buildups? Why? (b) What can be done to counteract undesirable inventory buildups?

9-24 Income statements. (SMA) The Mass Company manufactures and sells a single product. The following data cover the two latest years of operations:

	19_3	19_4
Selling price per unit	$ 40	$ 40
Sales in units	25,000	25,000
Beginning inventory in units	1,000	1,000
Ending inventory in units	1,000	5,000
Fixed manufacturing costs	$120,000	$120,000
Fixed selling and administrative costs	$190,000	$190,000

Standard variable costs per unit:

Materials	$10.50	Variable selling and	
Direct labor	9.50	administrative	$1.20
Variable overhead	4.00		

The denominator volume is 30,000 units per year. Mass Company accounting records produce variable-costing information, and year-end adjustments are made to produce external reports showing absorption-costing data. Any variances are charged to cost of goods sold.

Required

1. Prepare two income statements for 19_4, one under the variable-costing method and one under the absorption-costing method. Ignore income taxes.
2. Explain briefly why the operating income figures computed in requirement 1 agree or do not agree.
3. Give two advantages and two disadvantages of using variable costing for internal reporting.

9-25 Variable versus absorption costing. The Mavis Company uses an absorption-costing system based on standard costs. Variable manufacturing costs, including direct-material costs, were $3 per unit; the standard production rate was ten units per machine hour. Total budgeted and actual fixed factory overhead were $420,000. Fixed factory overhead was applied at $7 per machine hour ($420,000 ÷ 60,000 machine hours of denominator volume). Sales price is $5 per unit. Variable marketing and administrative costs, which are related to units sold, were $1 per unit. Fixed marketing and administrative costs were $120,000. Beginning inventory in 19_2 was 30,000 units; ending inventory was 40,000 units. Sales in 19_2 were 540,000 units. The same standard unit costs persisted throughout 19_1 and 19_2. For simplicity, assume that there were no price, spending, or efficiency variances.

Required

1. Prepare an income statement for 19_2 assuming that all underapplied or overapplied overhead is written off directly at year-end as an adjustment to Cost of Goods Sold.
2. The president has heard about variable costing. She asks you to recast the 19_2 statement as it should appear under variable costing.
3. Explain the difference in operating income as calculated in requirements 1 and 2.
4. Prepare a freehand graph of how *fixed factory* overhead was accounted for under absorption costing. That is, there will be two lines—one for the budgeted fixed overhead (which also happens to be the actual fixed factory overhead in this case) and one for the fixed-overhead applied. Show how the overapplied or underapplied overhead might be indicated on the graph.
5. Critics have claimed that a widely used accounting system has led to undesirable build-ups of inventory levels. (a) Which inventory method, variable costing or absorption costing, is more likely to lead to such buildups? Why? (b) What can be done to counteract undesirable inventory buildups?

9-26 Breakeven under absorption costing. Refer to the preceding problem.

Required

1. Compute the breakeven point (in units) under variable costing.
2. Compute the breakeven point (in units) under absorption costing.

3. Suppose production were exactly equal to the denominator volume, but no units were sold. Fixed factory costs are unaffected. However, assume that *all* marketing and administrative costs were avoided. Compute operating income under (a) variable costing and (b) absorption costing. Explain the difference in your answers.

9-27 The All-Fixed Company in 19_6. (R. Marple, adapted) It is the end of 19_6. The All-Fixed Company began operations in January 19_5. The company is so named because it has no variable costs. All its costs are fixed; they do not vary with output.

The All-Fixed Company is located on the bank of a river and has its own hydroelectric plant to supply power, light, and heat. The company manufactures a synthetic fertilizer from air and river water and sells its product at a price that is not expected to change. It has a small staff of employees, all hired on an annual-salary basis. The output of the plant can be increased or decreased by adjusting a few dials on a control panel.

The following are data regarding the operations of the All-Fixed Company:

	19_5	19_6*
Sales	10,000 tons	10,000 tons
Production	20,000 tons	—
Selling price	$30 per ton	$30 per ton
Costs (all fixed):		
Production	$280,000	$280,000
Marketing and administrative	$ 40,000	$ 40,000

*Management adopted the policy, effective January 1, 19_6, of producing only as the product was needed to fill sales orders. During 19_6, sales were the same as for 19_5 and were filled entirely from inventory at the start of 19_6.

Required

1. Prepare income statements with one column for 19_5, one column for 19_6, and one column for the two years together, using
   a. Variable costing
   b. Absorption costing
2. What is the breakeven point under (a) variable costing and (b) absorption costing?
3. What inventory costs would be carried on the balance sheets at December 31, 19_5 and 19_6, under each method?
4. Comment on the results in requirements 1 and 2. Which costing method appears more useful?
5. Assume that the performance of the top manager of the company is evaluated and rewarded largely based on reported operating income. Which costing method would the manager prefer? Why?

9-28 The Semi-Fixed Company in 19_6. The Semi-Fixed Company began operations in 19_5 and differs from the All-Fixed Company (described in Problem 9-27) in only one respect: It has both fixed and variable production costs. Its variable costs are $7 per ton and its fixed production costs $140,000 per year. Denominator volume is 20,000 tons per year.

Required

1. Using the same data as in Problem 9-27 except for the change in production-cost behavior, prepare income statements with adjacent columns for 19_5, 19_6, and the two years together, under
   a. Variable costing
   b. Absorption costing
2. Why did the Semi-Fixed Company have operating income for the two-year period when the All-Fixed Company in Problem 9-27 suffered an operating loss?

3. What inventory costs would be carried on the balance sheets at December 31, 19_5 and 19_6, under each method?
4. How may the variable-costing approach be reconciled with the definition of an asset as being "economic service potential"?
5. Assume that the performance of the top manager of the company is evaluated and rewarded largely based on reported operating income. Which costing method would the manager prefer? Why?

9-29 Comparison of variable costing and absorption costing. Consider the following data:

Hinkle Company
Income Statements for the Year Ended December 31, 19_4

	Variable Costing	Absorption Costing
Sales	$7,000,000	$7,000,000
Costs of goods sold (at standard)	$3,660,000	$4,575,000
Fixed manufacturing overhead	1,000,000	—
Manufacturing variances (all unfavorable):		
Direct materials	50,000	50,000
Direct labor	60,000	60,000
Variable overhead	30,000	30,000
Fixed overhead:		
Spending	100,000	100,000
Production volume	—	400,000
Total marketing costs	1,000,000	1,000,000
Total administrative costs	500,000	500,000
Total costs	$6,400,000	$6,715,000
Operating income	$ 600,000	$ 285,000

The inventories, carried at standard costs, were:

	Variable Costing	Absorption Costing
December 31, 19_3	$1,320,000	$1,650,000
December 31, 19_4	60,000	75,000

Required

1. Marie Hinkle, president of the Hinkle Company, has asked you to explain why the income for 19_4 is less than that for 19_3, even though sales have increased 40% over last year.
2. At what percentage of denominator volume was the factory operating during 19_4?
3. Prepare a numerical reconcilation and explanation of the difference between the operating incomes under absorption costing and variable costing.
4. Critics have claimed that a widely used accounting system has led to undesirable buildups of inventory levels. (a) Which inventory method, variable costing or absorption costing, is more likely to lead to such buildups? Why? (b) What can be done to counteract undesirable inventory buildups?

9-30 Inventory costing and management planning. It is November 30, 19_4. Consider the income statement for a company division's operations for January through November, 19_4 shown on the next page:

Specialized Industrial Products Divison
Income Statement for Eleven Months Ended November 30, 19_4

	Units	Dollars	
Sales @ $1,000	1,000		1,000,000
Deduct cost of goods sold:			
Beginning inventory, December 31, 19_3, @ $800	50	40,000	
Manufacturing costs @ $800, including $600 per unit for fixed overhead	1,100	880,000	
Total standard cost of goods available for sale	1,150	920,000	
Ending inventory, November 30, 19_4, @ $800	150	120,000	
Standard cost of goods sold*	1,000		800,000
Gross margin			200,000
Other costs:			
Variable, 1,000 units @ $50		50,000	
Fixed, @ $10,000 monthly		110,000	160,000
Operating income			40,000

*There are absolutely no variances for the eleven-month period considered as a whole.

Production in the past three months has been 100 units monthly. Practical capacity is 125 units monthly. To retain a stable nucleus of key employees, monthly production is never scheduled at less than 40 units.

Maximum available storage space for inventory is regarded as 200 units. The sales outlook for the next four months is 70 units monthly. Inventory is never to be less than 50 units.

The company uses a standard absorption-costing system. Denominator production volume is 1,200 units annually. All variances are disposed of at year-end as an adjustment to Cost of Goods Sold—at standard.

Required

1. The division manager is given an annual bonus that is geared to operating income. Assume that the manager wants to maximize the company's operating income for 19_4. How many units should the manager schedule for production in December? Note carefully that you do not have to (*nor should you*) compute the operating income for 19_4 in this or in subsequent parts of this question.
2. Assume that standard variable costing is in use rather than standard absorption costing. Would variable-costing operating income for 19_4 be higher, lower, or the same as standard absorption-costing income, assuming that production for December is 80 units and sales are 70 units? Why?
3. If standard variable costing were used, what production schedule should the division manager set? Why?
4. Assume that the manager is interested in maximizing his performance over the long run and that performance is being judged on the basis of income after taxes. Assume that income tax rates will be halved in 19_5 and that the year-end write-offs of variances are acceptable for income tax purposes. Assume that standard absorption costing is used. How many units should be scheduled for production in December? Why?
5. Assume that the total production and total sales for 19_4 and 19_5, taken together, will be unchanged by the specific decision in requirement 4. Assume also that the standards will be unchanged in 19_5. Suppose the decision in requirement 4 is to schedule 50 units instead of an originally scheduled 120 units. By how much will operating income in 19_5 be affected by the decision to schedule 50 units in December 19_4? (That is, how much operating income is shifted from 19_4 to 19_5?)

9-31 Some additional requirements to problem 9-30; absorption costing and production volume variances. Refer to Problem 9-30.

1. What operating income will be reported for 19_4 as a whole, assuming that the implied cost behavior patterns will continue in December as they did in January through November and assuming—without regard to your answer to requirement 1 in Problem 9-30—that production for December is 80 units and sales are 70 units?
2. Assume the same conditions as in requirement 1 except that a monthly denominator of 125 units (practical capacity) was used in setting fixed-overhead rates for inventory costing throughout 19_4. What production volume variance would be reported for 19_4?

9-32 Executive incentives; relevant costing. The data below pertain to the Fry Company:

	Year 19_1
Selling price per unit	$ 2.00
Total fixed cost—production	$ 8,400,000.00
Total fixed costs—marketing and administrative	$ 600,000.00
Variable cost per unit—marketing and administrative	$ .50
Sales in units	17,000,000
Production in units	17,000,000
Denominator volume in units (based on three- to five-year average demand)	30,000,000
Operating loss	$ 500,000.00
No beginning or ending inventories.	

The board of directors approached a competent outside executive to take over the company. He is an optimistic soul and agreed to become president at a token salary. His contract provides for a year-end bonus amounting to 10% of operating income (before considering the bonus or income taxes). The annual income is to be certified by a public accounting firm.

The new president, filled with rosy expectations, promptly raised the advertising budget by $3,500,000 and stepped up production to an annual rate of 30,000,000 units ("to fill the pipelines," the president said). As soon as all outlets had sufficient inventory, the advertising campaign was launched, and sales for 19_2 increased—but only to a level of 25,000,000 units.

The certified income statement for 19_2 contained the following data:

Sales, 25,000,000 × $2		$50,000,000
Production costs:		
Variable, 30,000,000 × $1	$30,000,000	
Fixed	8,400,000	
Total	$38,400,000	
Inventory, 5,000,000 units (1/6)	6,400,000	
Cost of goods sold		32,000,000
Gross margin		$18,000,000
Marketing and administrative expenses:		
Variable	$12,500,000	
Fixed	4,100,000	16,600,000
Operating income		$ 1,400,000

The day after the statement was certified, the persident resigned to take a job with another corporation having difficulties similar to those that Fry Company had a year ago. The president remarked, "I enjoy challenges. Now that Fry Company is in the black, I'd prefer tackling another knotty difficulty." His contract with his new employer is similar to the one he had with Fry Company.

Required

1. As a member of the board of directors, what comments would you make at the next meeting regarding the most recent income statement? Maximum production capacity is 40,000,000 units per year.

2. Would you change your remarks in requirement 1 if (consider each part independently):
   a. Sales outlook for the coming three years is 20,000,000 units per year?
   b. Sales outlook for the coming three years is 30,000,000 units per year?
   c. Sales outlook for the coming three years is 40,000,000 units per year?
   d. The company is to be liquidated immediately, so that the only sales in 19_3 will be the 5,000,000 units still in inventory?
   e. The sales outlook for 19_3 is 45,000,000 units?
3. Assuming that the $140,000 bonus is paid, would you favor a similar arrangement for the next president? If not, and you were outvoted, what changes in a bonus contract would you try to have adopted?

9-33 Prepare income statement on variable-costing basis. A fire partially destroyed the records of the O'Day Manufacturing Company on December 31, 19_1. You have been asked to prepare a comparative schedule of the master budgeted income statement for the year and the actual results. Even though the company has kept records on an absorption-costing basis, you have decided to prepare a statement on the variable-costing basis.

On an absorption-costing basis, the unfavorable production volume variance for fixed factory overhead was $13,650, and 52,000 units were produced and sold at an average selling price of $20 per unit. The standard contribution margin was $5 per unit. Budgeted and actual fixed costs were the same for both manufacturing ($122,850) and nonmanufacturing ($80,000). The total of price, spending, and efficiency variances was $36,000, unfavorable. On an absorption-costing basis, fixed factory overhead had been applied on the basis of the expected volume in the master budget.

9-34 Difficult comparison of variable and absorption costing. Assume that operating income for 19_2 was $600,000 under variable costing and $800,000 under absorption costing. The end-of-year cost of the inventory under standard variable costing was $60,000. The beginning-of-year cost of the inventory under standard absorption costing was $25,000 higher than the cost of the beginning-of-year inventory under standard variable costing. Compute the end-of-year cost of inventory under standard absorption costing.

9-35 Variable or absorption costing; two products. Preston Machining operates a machining shop that cuts and tests metal blades that are sold to light-aircraft assembly companies. Two lines of blades are cut and tested—Series A and Series B. Series A is the more popular line and was the only product Preston made for many years. Series B is a recently added line made possible by advances in machine-cutting technology. In recent years, there has been a substantial increase in the use of automated equipment. Preston's strategy is to use the most advanced cutting equipment to provide its customers with blades meeting very tight quality specifications.

Summary details on the Series A and Series B blades for the current period, 19_2, are:

	Series A	Series B
Actual direct materials per unit	$60	$100
Actual direct labor time per unit	30 minutes	15 minutes
Actual machining time per unit	45 minutes	150 minutes

Actual direct labor costs were $20 per hour for workers on both the Series A and Series B product lines.

Summary data for beginning inventory, production, and sales in 19_2 are:

	Series A	Series B
Beginning inventory	0	0
Actual production	800	400
Actual sales	700	380

For many years, Preston has used an actual-costing system (as distinguished from a normal- or standard-costing system); actual manufacturing overhead costs (both fixed and variable) were allocated to products on the basis of actual direct labor costs. This costing system was also used in the current period. Total actual manufacturing overhead costs were $84,000, $60,000 of which was fixed manufacturing overhead. The costs associated with the machines (depreciation, fluids, cutting tools, and so on) are included in manufacturing overhead costs. Separate variable and fixed overhead rates, both with respect to actual direct labor costs, were computed.

In 19_2, actual sales price per unit was $231 for Series A and $220.50 for Series B. Variable selling, general, and administrative costs (SG&A) were $14 per unit for Series A and $18 per unit for Series B. These costs varied with the number of units sold. Fixed SG&A costs were $28,000.

Required

1. Compute for 19_2:
   (a) The actual variable-manufacturing overhead rate per direct labor dollar
   (b) The actual fixed-manufacturing overhead rate per direct labor dollar
2. Compute for 19_2 the actual manufacturing cost per unit for the Series A and the Series B blades under
   (a) Variable costing
   (b) Absorption costing
3. Prepare income statements for 19_2, using
   (a) Variable costing
   (b) Absorption costing
4. Explain why the operating income in requirement 3 differs between variable costing and absorption costing.

Coverage of Part Two of the Chapter

9-36 Overhead rates and cyclical business. It is a time of severe business recession throughout the capital-goods industries. A division manager for a large corporation in a heavy-machinery industry is confused and unhappy. The manager is distressed with the controller, whose cost accounting department keeps reporting costs to the manager that are of little comfort because they are higher than ever before. At the same time, the manager has to quote lower prices than before in order to get any business.

Required

1. What level of denominator volume is probably being used for application of overhead?
2. How might the overhead be applied in order to make the cost data more useful in making price quotations?
3. Would the product costs being furnished by the cost accounting department be satisfactory for the costing of the annual inventory?

9-37 Role of denominator levels. The Moroso Company incurs fixed manufacturing overhead of $2,700,000 annually. Master-budget volume is 37,500 machine hours; normal volume, 45,000 hours; and practical capacity, 50,000 hours allowed. In 19_1, 37,500 units were produced and 30,000 units were sold. One standard machine hour is allowed for each unit produced. There was no beginning inventory.

Required

1. Prepare a three-column comparison (similar to Exhibit 9-6) using the three denominator volumes for applying fixed overhead to product. Designate which denominator volume would result in the highest and which in the lowest operating income. For each method, show the amounts that would be charged to

   - Cost of goods sold (expense)
   - Production volume variance (expense or loss)
   - Ending inventory (asset)

2. Why is master-budget volume better than either practical capacity or normal volume for judging current operating performance?
3. If your management bonus for the current period is affected by the selection of the denominator volume, which denominator would you prefer? Why?

9-38 Effects of denominator choice. The Wong Company installed standard costs and a flexible budget on January 1, 19_3. The president had been pondering how fixed manufacturing overhead should be applied to products. She decided to wait for the first month's results before making a final choice of what denominator volume should be used from that day forward.

In January, the company operated at a volume of 70,000 standard machine hours allowed for the actual units of output achieved. If the company had used practical capacity as a denominator volume, the fixed overhead spending variance would have been $10,000, unfavorable, and the production volume variance would have been $36,000, unfavorable. If the company had used normal volume as a denominator volume, the production volume variance would have been $20,000, favorable. Budgeted fixed overhead was $120,000 for the month.

Required

1. Compute the denominator volume, assuming normal volume as the denominator.
2. Compute the denominator volume, assuming practical capacity as the denominator.
3. Suppose you are the executive vice president. You want to maximize your 19_3 bonus, which depends on 19_3 operating income. Assume that the production volume variance is charged or credited to income at year-end. Which denominator volume would you favor? Why?

9-39 Income taxes and denominator volume. In the United States, there is nothing illegal or immoral about keeping multiple sets of accounting records, one to satisfy income tax reporting requirements, one to satisfy investor reporting requirements, one to satisfy internal reporting requirements, and so on. Nevertheless, real or imagined expected costs versus expected benefits lead most companies to have one set or at most two sets of records. Income tax regulations have heavy effects on accounting systems because records must be kept to satisfy tax laws.

For years, the Internal Revenue Service (IRS) was liberal regarding the tendency of companies to write off many indirect manufacturing costs, particularly depreciation on equipment, as charges against income immediately. In 1975, however, new income tax regulations forced industry to hold back more indirect costs as "inventoriable" costs. In short, the regulations came down hard against the variable-costing approach to inventory costing and in favor of the absorption-costing approach.

The regulations now insist that any significant underapplied or overapplied overhead be prorated among the appropriate inventory accounts and cost of goods sold. However, the regulations during 1975–1986 explicitly approved a "practical capacity concept." This position permitted the immediate write-off (rather than proration) of the production volume variance that results from using a fixed-overhead rate based on a practical-capacity denominator volume level. The IRS has disallowed the use of the practical-capacity concept since 1987, but some companies are challenging the IRS position through litigation.

The Missile Corporation operates a stamping plant with a theoretical capacity of 60 units per hour. Theoretical capacity is defined as the level of production that the manufacturer could reach if all machines and departments were operated continuously at peak efficiency for all hours when the plant is actually open. The plant is actually open 1,960 hours per year based on an 8-hour day, 5-day week, and 15 shutdown days for vacations and holidays. A reasonable allowance for downtime (the time allowed for ordinary and necessary repairs and maintenance) is 5% of theoretical capacity.

Required

1. Compute practical capacity in units per year. Assume no loss of production during starting up, closing down, or employee work breaks.
2. Suppose, as some companies still insist and plead, that the IRS permitted practical capacity. Assume that 75,000 units are produced for the year and that budgeted and actual fixed indirect manufacturing costs totaled $14,523,600. Also assume that 7,500 units are on hand at the end of the taxable year and that there were no beginning inventories. Compute the amount of fixed indirect costs that would have been applied to production during the year. What amount will be allowed as a deduction in the computation of income taxes for the year? Label your computations.
3. Practical capacity is no longer permitted for income tax purposes. Assume that master-budget volume is permitted and that it was 75,000 units. Compute the amount of fixed

indirect manufacturing costs that would have been applied to production during the year. What amount would be allowed as a deduction in the computation of income taxes for the year? Label your computations. What is the difference in total tax deductions between your answer here and in requirement 2?

Coverage of Part Three of the Chapter

9-40 Straightforward proration. Consider the following balances of standard costs before proration at the end of the year: work in process, $180,000; finished goods, $720,000; cost of goods sold, $900,000. The production volume variance was $50,000, favorable. All other variances were $330,000, unfavorable. Management has decided to prorate all variances in proportion to the ending balances in work in process, finished goods, and cost of goods sold before proration.

Required

1. Prepare a schedule that prorates the variances.
2. Prepare a journal entry that closes all variance accounts.
3. The major justification for proration is the attempt to approximate the "actual" cost of the units produced. What is the most likely inaccuracy in the proration here? Explain.

9-41 Proration of variances, multiple choice. (CPA, adapted) Study the appendix. Toliver Manufacturing Company uses a standard-cost system in accounting for the cost of production of its only product, Product A. The standards for the production of one unit of Product A are as follows:

- Direct materials: 10 feet of item 1 at $.75 per foot and 3 feet of item 2 at $1.00 per foot
- Direct labor: 4 hours at $15.00 per hour
- Manufacturing overhead: applied at 150% of standard direct labor costs

There was no inventory on hand at July 1, 19_2. Following is a summary of costs and related data for the production of Product A during the year ended June 30, 19_3:

- 100,000 feet of item 1 were purchased at $.78 per foot
- 30,000 feet of item 2 were purchased at $.90 per foot
- 8,000 units of Product A were produced, which required 78,000 feet of item 1; 26,000 feet of item 2; and 31,000 hours of direct labor at $16.00 per hour
- 6,000 units of Product A were sold

At June 30, 19_3, there are 22,000 feet of item 1, 4,000 feet of item 2, and 2,000 completed units of Product A on hand. All purchases and transfers are "charged in" at standard.

Required

Choose the correct answers (show computations):

1. For the year ended June 30, 19_3, the total debits to the direct-materials account for the purchase of item 1 would be (a) $78,000, (b) $58,500, (c) $75,000, (d) $60,000.
2. For the year ended June 30, 19_3, the total debits to the work in process account for direct labor would be (a) $496,000, (b) $512,000, (c) $480,000, (d) $465,000.
3. Before allocation of variances, the balance in the material-efficiency-variance account for item 2 was (a) $2,000 debit, (b) $2,600 debit, (c) $600 debit, (d) $1,000 credit.
4. If all variances were prorated to inventories and cost of goods sold, the amount of material efficiency variance for item 2 to be prorated to direct-materials inventory would be (a) $333 credit, (b) $0, (c) $333 debit, (d) $500 debit.
5. If all variances were prorated to inventories and cost of goods sold, the amount of material price variance for item 1 to be prorated to direct-materials inventory would be (a) $0, (b) $647 debit, (c) $600 debit, (d) $660 debit.

9-42 Proration of variances and income effects of standard costs. Study the appendix. The Moraine Company began business on January 1, 19_1, and uses a standard absorption-costing system. Balances in certain accounts at December 31, 19_1, are as follows (in thousands):

At standard unit costs	
Direct materials inventory	$ 20,000
Work in process	10,000
Finished goods	30,000
Cost of goods sold	60,000
Total	$120,000
Variances (all unfavorable):	
Direct materials efficiency	$ 10,000
Direct materials price	12,000
Direct labor price	2,000
Direct labor efficiency	10,000
Underapplied overhead	5,000
Total	$ 39,000
Sales	$150,000

The executives have asked you to compute the gross margin (after deductions for variances, if any), assuming first that no variances are prorated, and second that all variances are prorated.

Assume that all variances that are not prorated are considered direct adjustments of standard cost of goods sold. Assume that prorations are based on the ending balances of the applicable accounts affected, even though more-refined methods would be possible if additional data were available. There are not enough data about the direct material components in the various accounts to warrant the proration of some of the direct material price variance to the direct material efficiency variance. Therefore, prorate the price variance directly to Direct Materials Inventory, Work in Process, Finished Goods, and Cost of Goods Sold.

Required

1. Prepare a comprehensive schedule of proration of all variances.
2. Prepare a compound journal entry for the proration.
3. Prepare comparative summary income statements through gross margin under each of the two assumptions (no variances are prorated and all variances are prorated) specified by the company's executives.

9-43 Proration of direct materials variances. (D. Kleespie, adapted) Study the appendix. The J/E Lahtinen Company manufactures a single plastic product. The company prorates its direct material price and efficiency variances to the appropriate accounts in proportion to the direct material cost components in those accounts. The price variance is $2,800, unfavorable; efficiency variance, $1,200, unfavorable.

The company had no beginning materials inventory. Purchases were 14,000 pounds. The standard allowance is 2 pounds per finished unit. Units produced, 6,000. Units sold, 5,000. The standard unit price for the direct material is $4 per pound.

Required

1. Compute the material efficiency variance in pounds.
2. Compute the ending balance of Direct Materials Inventory in pounds.
3. Compute the amounts of the price variance prorated to Finished Goods and to Cost of Goods Sold.
4. Compute the amounts of the "adjusted" efficiency variance prorated to Finished Goods and to Cost of Goods Sold.
5. Explain why the price variance is prorated before the efficiency variance is prorated.

9-44 Proration of variances and income effects of standard costs. Study the appendix. The Stefano Company uses a standard absorption-costing system, which shows the following account balances (before proration of any variances) at December 31, 19_2:

Direct materials, ending inventory	$175,000
Work in process, ending inventory	100,000
Finished goods, ending inventory	300,000
Cost of goods sold	600,000
Direct materials price variance	64,000
Direct materials efficiency variance	25,000
Direct labor price variance	5,000
Direct labor efficiency variance	25,000
Factory overhead incurred	210,000
Factory overhead applied, at standard rate	170,000
Sales	900,000
Marketing and administrative costs	180,000

Materials price variances are measured when the material is purchased rather than when it is used. Assume that Work in Process, Finished Goods, and Cost of Goods Sold contain standard costs in uniform proportions of direct material, direct labor, and factory overhead. The direct material component represented 60% of the ending balance in Work in Process, Finished Goods, and Cost of Goods Sold. All variances are unfavorable. There are no beginning inventories.

Required

1. Prepare a comprehensive schedule showing the proration of all variances.
2. Prepare a compound journal entry for the proration.
3. Prepare comparative summary income statements based on a standard absorption-costing system:
   a. Without proration of any variances
   b. With proration of all variances
4. Compute the amount of direct labor cost included in the balance of finished goods, ending inventory, before proration.

9-45 Interim reporting. Alberti Company had the following data for the quarter ended March 31, 19_1:

	Actual	Flexible Budget	Flexible-Budget Variances
Variable factory overhead	$200,000	$178,000	$22,000 U
Fixed factory overhead:			
Applied		300,000	
Production volume variance		50,000	
Total	355,000	350,000	5,000 U
Total factory overhead	$555,000	$528,000	$27,000 U
Operating income	$110,000	$180,000	$70,000 U

For simplicity, assume that there are no beginning or ending inventories.

Required

1. Does the company use variable or absorption costing? Explain your answer.
2. Suppose the company had "planned" production volume variances. That is, production volume fluctuated from month to month, but by the end of the year the total production volume variance was expected to be negligible. Under this approach, what "actual" operating income would be shown for the quarter? Explain your answer.
3. If variable costing is used, what "actual" operating income would be reported? Explain.

Special Review Material for Chapters 7 Through 9

9-46 Comparison of alternative income statements. (By J. March and adapted for use in SMA examination) By applying a variety of cost accounting methods to the operating data of a given year, the controller of a manufacturing company prepares the following alternative income statements:

	A	B	C	D
Sales	$1,000,000	$1,000,000	$1,000,000	$1,000,000
Cost of goods sold	$ 375,000	$ 250,000	$ 420,000	$ 395,000
Variances:				
Direct materials	15,000	15,000	—	—
Direct labor	5,000	5,000	—	—
Manufacturing overhead	25,000	—	—	25,000
Other costs (all fixed)	350,000	475,000	350,000	350,000
	$ 770,000	$ 745,000	$ 770,000	$ 770,000
Operating income	$ 230,000	$ 255,000	$ 230,000	$ 230,000

Required

1. The controller used the following costing methods: (a) actual costing, (b) normal costing, (c) standard absorption costing, and (d) standard variable costing. Match each of these methods with the appropriate income statement, A, B, C, or D above, and explain the basis of your selection.
2. During the year, did inventory quantities increase, decrease, or remain the same? Explain.
3. During the year, was the volume of production higher than, lower than, or equal to the company's denominator level of volume? Explain.
4. During the year, was the variable manufacturing overhead incurred more than, less than, or equal to the budget? Explain.

9-47 Review of absorption costing, variable costing, and analysis of variances. On December 30, 19_1, a bomb blast destroyed the bulk of the accounting records of the Horne Division, a small one-product manufacturing division that uses standard costs and flexible budgets. All variances are written off as additions to (or deductions from) income; none are prorated to inventories. In addition, the chief accountant mysteriously disappeared. You have the task of reconstructing the records for the year 19_1. The general manager has said that the accountant had been experimenting with both absorption costing and variable costing.

The records are a mess, but you have gathered the following data (simplified here to save computations) for 19_1:

a.	Cash, December 31, 19_1	$ 10
b.	Sales	128,000
c.	Actual fixed indirect manufacturing costs	21,000
d.	Accounts receivable, December 31, 19_1	20,000
e.	Standard variable manufacturing costs per unit	1
f.	Variances from standard for all variable manufacturing costs	5,000 U
g.	Operating income, absorption-costing basis	14,400
h.	Accounts payable, December 31, 19_1	18,000
i.	Gross margin, absorption costing at standard (before deducting variances)	22,400
j.	Total liabilities	100,000
k.	Unfavorable spending variance, fixed manufacturing costs	1,000
l.	Notes receivable from chief accountant	4,000
m.	Contribution margin, at standard (before deducting variances)	48,000
n.	Direct material purchases, at standard prices	50,000
o.	Actual marketing and administrative costs (all fixed)	6,000

Required

These do not necessarily have to be solved in any particular order. Ignore income taxes.

1. Operating income on a variable-costing basis.
2. Number of units sold.
3. Number of units produced.
4. Number of units used as the denominator volume to obtain fixed-overhead application rate per unit on absorption-costing basis.
5. Did inventory (in units) increase or decrease? Explain.
6. By how much in dollars did the inventory level change (a) under absorption costing, (b) under variable costing?
7. Variable manufacturing cost of goods sold, at standard cost.
8. Manufacturing cost of goods sold at standard cost, absorption costing.

CHAPTER 10

Determining How Costs Behave

General approach to estimating cost functions

Assumptions underlying cost classifications

Cost estimation approaches

Steps in estimating a cost function

Data collection and adjustment issues

Cost behavior in activity areas

Nonlinearity and cost functions

Learning curves and cost functions

Appendix: Inflation and cost estimation

Airline manufacturers use learning curves to predict the behavior of costs incurred in assembling planes.

Learning Objectives

When you have finished studying this chapter, you should be able to

1. State the two assumptions frequently used in cost behavior estimation
2. Distinguish between a linear cost function and a nonlinear cost function
3. Identify a variable-cost function, a fixed-cost function, and a mixed-cost function
4. Provide examples of variable costs and fixed costs when the cost object is a product and when the cost object is an activity
5. Describe four broad classes of approaches to cost estimation
6. Outline six steps in estimating a cost function based on current or past cost relationships
7. Understand data problems encountered in estimating cost functions
8. Distinguish between cumulative average–time learning model and incremental unit–time learning model

Know your costs. Again and again, we have seen that knowledge of how costs behave is frequently the difference between wise and unwise decisions. Cost information is important in planning and control. For example, some firms attempt to position themselves as producers of low-cost products in their industry and compete on the basis of low price. Knowledge of their own costs and of how they compare with their competitors is a key input for decisions in this strategic area. Cost-behavior information also plays a key role in budgeting and other planning decisions, such as bidding on contracts. Decisions in the control area, such as the interpretation of variances, similarly rely heavily on knowledge of cost behavior. This chapter focuses on how to determine cost behavior.

GENERAL APPROACH TO ESTIMATING COST FUNCTIONS

Objective 1

State two assumptions frequently used in cost behavior estimation

Basic Terms and Assumptions

Cost estimation should be distinguished from cost prediction. **Cost estimation** is the attempt to measure *past* cost relationships; an equation is formulated to describe these past relationships. **Cost prediction** is the forecasting of *future* costs. Frequently managers use the cost estimation equation to predict future costs. Efforts that improve our knowledge of how past costs behave will undoubtedly help managers make more accurate predictions of future costs.

Much of this chapter focuses on cost estimation. This topic is sufficiently important to occupy this chapter and, indeed, a subsequent chapter—Chapter 25, Cost Behavior and Regression Analysis. ***Keep in mind, however, that managers are interested in better cost estimates primarily because these estimates enable them to make more accurate cost predictions.*** Subsequent chapters provide numerous examples of decisions using cost predictions. For example, Chapter 11 discusses make-or-buy decisions, Chapter 12 discusses pricing decisions, and Chapters 21 and 22 discuss equipment-replacement decisions.

There are two assumptions that are frequently used in the estimation of cost functions:

Assumption 1. Cost behavior can adequately be approximated by a linear function within the relevant range.

Assumption 2. Variations in the total cost level can be explained by variations in a single cost driver. A *cost driver* is any factor whose change causes a change in the total cost of a related cost object. Examples include machine hours in manufacturing and the weight of items to be shipped in distribution.

These two assumptions are used throughout much of this chapter. The final sections give examples of nonlinear cost behavior. Chapter 25 discusses how variations in two or more cost drivers can explain variations in the level of total cost. The question of what cost driver affects a particular cost's behavior can be answered only in actual situations, on a case-by-case basis.

Given the assumptions of linearity and a single cost driver, each cost has some underlying cost-behavior pattern, more technically described as a *cost function*. Its expected value, $E(y)$, has the form

$$E(y) = \alpha + \beta x$$

where α and β are the underlying (but unknown) parameters. A **parameter** is a constant, such as α, or a coefficient, such as β, in a model or system of equations.

Working with historical data, the cost analyst can develop a linear formula to estimate the cost function:

$$y' = a + bx$$

where y' is the estimated value (as distinguished from the observed value, y) and a (termed the *constant* or *intercept*) and b (termed the *slope coefficient*) are the estimates of the underlying α and β parameters. The **constant** or **intercept**, a, is the estimated component of total costs that, within the relevant range, does not vary with changes in the level of the cost driver. The **slope coefficient**, b, is the amount of change in total cost (y) for each unit change in the cost driver (x) within the relevant range. The *relevant range* is the band of the cost driver in which a specific relationship between total cost and the level of the cost driver is expected to be valid.

Examples of Cost Functions

Objective 2

Distinguish between a linear cost function and a nonlinear cost function

This section illustrates three types of linear cost functions. A **linear cost function** is a function in which a single constant (a) and a single slope coefficient (b) describe the behavior of costs for all changes in the level of the cost driver.

Assume Cannon Services is negotiating with World Wide Communications (WWC) for exclusive use of a telephone line between New York and Paris. WWC offers Cannon Services three alternative cost structures:

- \$5 per minute of phone use
- \$10,000 per month
- \$3,000 per month plus \$2 per minute of phone use

Graph 1 in Exhibit 10-1 presents the familiar *variable cost*, which is often described as being *proportionately variable* or *strictly variable*. Its total changes in direct proportion to changes in x within the relevant range because the intercept a is zero. Graph 1 illustrates the first cost-structure alternative offered by WWC (a = \$0, b = \$5):

$$y' = \$5x$$

Objective 3

Identify a variable-cost function, a fixed-cost function, and a mixed-cost function

where x is the number of minutes of phone time used (the cost driver) each month. Every additional minute used adds \$5 to total costs.

Graph 2 in Exhibit 10-1 presents the familiar *fixed cost*. The cost function in Graph 2 is the second cost-structure alternative offered by WWC (a = \$10,000; b = \$0):

$$y' = \$10{,}000$$

The total costs will be \$10,000 per month regardless of the number of minutes of phone time used.

EXHIBIT 10-1
Examples of Linear Cost Functions

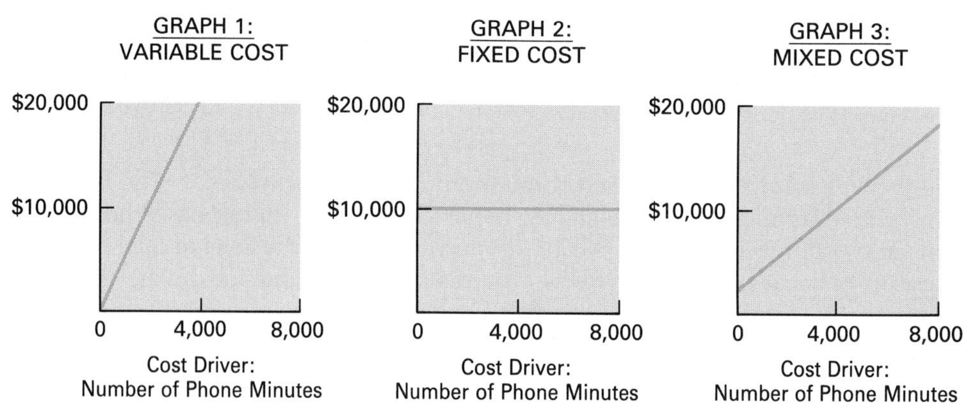

Graph 3 in Exhibit 10-1 presents a **mixed cost,** sometimes called a **semivariable cost**. As the name implies, a mixed cost has both fixed and variable elements. The cost function in Graph 3 is the third cost-structure alternative offered by WWC (a = \$3,000, b = \$2):

$$y' = \$3{,}000 + \$2x$$

The total cost in the relevant range in Graph 3 changes as the number of units of the cost driver change. However, the level of the total cost at two different levels of the cost driver is not proportionally the same as the ratio of the levels of the cost driver.

ASSUMPTIONS UNDERLYING COST CLASSIFICATIONS

The classification of costs into their variable cost or fixed cost components is based on three assumptions:

- The cost object must be specified,
- The time span must be specified, and
- The relevant range for changes in the cost driver must be specified.

Objective 4

Provide examples of variable costs and fixed costs when the cost object is a product, and when it is an activity

Choice of Cost Object

Costs are variable or fixed *with respect to* a chosen cost object. Examples of a variable cost and a fixed cost for a product-cost object and an activity-cost object follow:

	Example of a Cost Object	Example of a Variable Cost	Example of a Fixed Cost
Product-Cost Object	Meal at a restaurant	Food in each meal (meat, fish, vegetables, and so on)	Depreciation on the oven used to cook the meal
Activity-Cost Object	Aircraft maintenance for an airline company	Energy consumed in testing the functioning of each aircraft engine; this energy is metered to each aircraft bay	Leasing cost of the building in which maintenance on many aircraft and aircraft engines is done

Time Span

Whether a cost is variable or fixed is affected by the time span considered in the decision situation. The longer the time span, other things being equal, the higher the proportion of total costs that are variable. Costs that are fixed in the short run may be variable in the long run. Exhibit 10-2 illustrates this general point. The costs in this exhibit are the monthly manufacturing costs at a television-assembly plant of Home Entertainment. Exhibit 10-2 presents three categories of cost drivers:

1. *Volume of product output.* These costs are driven by changes in the volume of television sets assembled.
2. *Product design.* These costs are driven by changes in product design variables. Examples of such variables include the number of different component parts in a television set and the ease of testing each individual component or each television set.
3. *Plant layout.* These costs are driven by changes in the layout of the plant. Examples of such changes include the amount of materials handling and the level of automation.

The most important cost driver at Home Entertainment for short-run (monthly, in our example) changes in total manufacturing costs is volume of product output. Only small changes in product design or plant layout tend to occur in the short

EXHIBIT 10-2

Manufacturing Cost-Behavior Patterns at a Television-Assembly Plant of Home Entertainment: Effect of Differing Time Spans

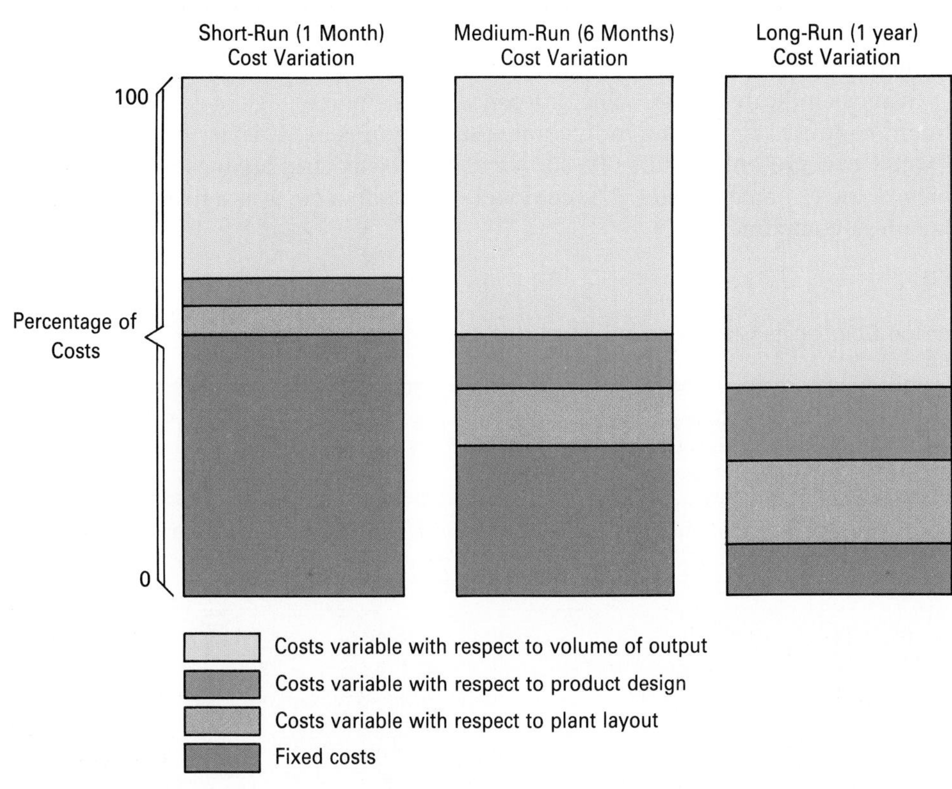

run. In contrast, over a longer time horizon (1 to 2 years), major changes in product design or plant layout may occur. For example, products can be designed to use components that perform multiple functions; this can greatly reduce the number of component parts to be assembled into each television set. Similarly, over a longer time horizon, changes in supplier contracts and redesign of the plant floor can occur; this can significantly reduce the amount of materials inventories and work in process inventory. The result is that, over the long run, product design and plant layout are also important drivers of the level of manufacturing costs at Home Entertainment. Exhibit 10-2 also illustrates how ***fixed manufacturing costs decline as a proportion of total manufacturing costs as the time span is lengthened from the short run to the medium run and then finally to the long run.***

The Relevant Range

Each of the cost-behavior patterns in Exhibit 10-1 (p. 334) has a relevant range within which the specified cost relationship will be valid. The reasons for a specific relevant range are many. For example, it could be based on a labor agreement specifying when overtime premium rates (such as one and a half times the normal hourly rate) go into effect. Alternatively, it could be based on the capacity level of a plant.

Differences Across Organizations and Across Countries

Organizations differ in classifying individual costs. A variable-cost item in one organization can be a fixed-cost item in another organization. Consider labor costs. House construction companies often classify labor costs as a variable cost when determining product costs. These companies frequently make sizable changes in the level of their labor force when there are changes in the demand for housing construction. In contrast, oil-refining companies often classify labor costs as a fixed cost when determining product costs. Few changes in the labor force of oil refineries are made even with sizable changes in the volume or types of oil products being refined.

Surveys indicate considerable differences in the percentage of companies in different countries classifying individual cost categories as variable, fixed, or mixed. Exhibit 10-3 presents results of a survey of U.S.-based and Japanese-based companies. A lower percentage of U.S. companies treat labor costs as a fixed cost than do Japanese companies.

EXHIBIT 10-3

Percentage of U.S. and Japanese Companies Classifying Individual Cost Categories as Variable, Mixed (Semivariable), or Fixed

	U.S.-Based Companies			Japanese-Based Companies		
Cost Category	**Variable**	**Mixed**	**Fixed**	**Variable**	**Mixed**	**Fixed**
Production labor	85.6%	6.0%	8.4%	51.6%	5.4%	43.0%
Setup labor	60.5	24.7	14.8	43.8	6.2	50.0
Material-handling labor	47.6	34.5	17.9	22.8	16.4	60.8
Quality control labor	33.7	36.2	30.1	12.5	12.5	75.0
Tooling	31.6	35.5	32.9	31.5	25.8	42.7
Energy	26.2	45.0	28.8	41.7	30.8	27.5
Building occupancy	1.2	6.0	92.8	0.0	0.0	100.0
Depreciation	1.2	7.0	91.8	0.0	0.0	100.0

NAA Tokyo Affiliate, "Management Accounting in the Advanced Manufacturing Surrounding: Comparative Study on Survey in Japan and U.S.A.," October 1988.

COST ESTIMATION APPROACHES

There are four approaches to cost estimation:

1. Industrial-engineering method
2. Conference method
3. Account analysis method
4. Quantitative analysis of current or past cost relationships

Objective 5

Describe four broad classes of approaches to cost estimation

These approaches differ in the costs of conducting the analysis, the assumptions they make, and the evidence they yield about the accuracy of the estimated cost function. They are not mutually exclusive. Many organizations use a combination of these approaches.

Industrial-Engineering Method

The **industrial-engineering method**, also called the **work-measurement method,** first analyzes the relationship between inputs and outputs in physical terms. For example, inputs for a carpet manufacturer include cotton, wool, dyes, hours of direct labor, hours of machine time, and energy. Output is a measure of square yards of carpet. Time and motion studies may be employed. Then the physical measures are transformed into standard or budgeted costs. For example, 2 bales of cotton and 3 gallons of dye may be required to produce 20 square yards of carpet.

The industrial-engineering method is very time-consuming. For some government contracts, its use is mandatory. Many firms, however, find it too costly for analyzing their entire cost structure. More frequently, firms use this approach for direct cost categories such as materials and labor and not for indirect cost categories such as manufacturing overhead. Physical relationships between inputs and outputs may be difficult to specify for many individual overhead items.

Conference Method

The **conference method** develops cost estimates based on analysis and opinions gathered from various departments of an organization (purchasing, process engineering, manufacturing, employee relations, and so on). One company has a cost-estimating department whose responsibility is to develop product costs based on a consensus of the relevant departments. In another company, representatives of functional areas provide individual cost estimates that are then combined into a product cost estimate.

The advantages of the conference method include the speed at which cost estimates can be developed, the pooling of knowledge from experts in each functional area, and the resulting improved credibility of the cost estimates to all personnel. The accuracy of the cost estimates largely depends on the care and detail taken by those people providing the inputs.[1]

Account Analysis Method

In the **account analysis method,** cost accounts in the ledger are classified as variable, fixed, or mixed. Typically, managers emphasize qualitative rather than quantitative analysis when making cost-classification decisions. The account analysis approach is widely used in practice.[2]

[1]The conference method is further described in W. Winchell, *Realistic Cost Estimating for Manufacturing,* Society for Manufacturing Engineers (Dearborn, MI, 1989).

[2]Survey evidence on the widespread use of the account analysis approach is in M. M. Mowen, *Accounting for Costs as Fixed and Variable,* National Association of Accountants (Montvale, NJ, 1986).

Organizations differ with respect to the care taken in implementing account analysis. In some organizations, individuals thoroughly knowledgeable about the operations make the cost-classification decisions. For example, manufacturing personnel may be called on to classify costs such as machine lubricants and materials-handling labor, and marketing personnel may classify costs such as advertising brochures and sales salaries. In other organizations, only cursory analysis is conducted, sometimes by individuals with limited knowledge of operations, before cost-classification decisions are made.

The account analysis approach may be helpful as a first step in cost classification and estimation. Supplementing this analysis by the conference method improves its credibility.

Quantitative Analysis of Cost Relationships

Data on past cost relationships are often used to estimate cost functions. These data may be time-series data or cross-sectional data. *Time-series data* pertain to the same entity (firm, plant, activity area, and so on) over a sequence of past time periods. For example, monthly observations of manufacturing overhead and machine hours for a particular plant for the most recent year would yield time-series data. *Cross-sectional data* pertain to different entities for the same time period. For example, studies of the personnel costs and loans processed at fifty individual branches of a banking company for March would yield cross-sectional data.

Let us examine indirect manufacturing costs (also called manufacturing overhead or factory overhead costs) at Southern Carpets, which weaves carpets for houses and offices. Its manufacturing plant is highly automated with state-of-the-art weaving machines. Exhibit 10-4 shows monthly (time-series) data for the most recent year. Note that the data are paired. For example, December has indirect manufacturing costs of $275,343 and 2,469 machine hours. The next section uses the data in Exhibit 10-4 to illustrate two different quantitative ways to estimate a cost function: regression analysis and the high-low method.

EXHIBIT 10-4

Southern Carpets: Monthly Indirect Manufacturing Costs and Machine Hours

Month	Indirect Manufacturing Costs	Cost Driver: Machine Hours
January	$341,062	3,467
February	346,471	4,426
March	287,328	3,103
April	262,828	3,625
May	220,843	3,081
June	390,700	4,980
July	337,924	3,948
August	180,000	2,180
September	376,246	4,121
October	295,041	4,762
November	215,121	3,402
December	275,343	2,469

STEPS IN ESTIMATING A COST FUNCTION

Objective 6

Outline six steps in estimating a cost function based on current or past cost relationships

There are six steps in estimating a cost function based on an analysis of current or past cost relationships: (1) choose the dependent variable (the variable to be predicted), (2) choose the cost driver(s), (3) collect data on the dependent variable and the cost driver(s), (4) plot the data, (5) estimate the cost function, and (6) evaluate the estimated cost function. ***Frequently, the cost analyst will proceed through these six steps several times before concluding that an acceptable cost function has been identified.***

Step One: *Choose the dependent variable.* Choice of the **dependent variable** (the variable to be predicted), usually called y, will be guided by the purpose for estimating a cost function. For example, if the purpose is to predict indirect manufacturing costs for a production line, then all costs that are classified as being indirect with respect to the production line should be incorporated into y. Ideally, all the individual items in the dependent variable will have a similar relationship with the cost driver(s), chosen in step 2. Where a single relationship does not exist, the possibility of estimating more than one cost function should be investigated.

Consider several types of benefits paid to employees and their cost drivers:

Benefit	Cost Driver
1. Health benefits	Number of employees
2. Meals in cafeteria	Number of employees
3. Pension benefits	Salaries of employees
4. Life insurance	Salaries of employees

The cost of benefits 1 and 2 can be grouped in one dependent variable, because each has the same cost driver. In contrast, the cost of benefits 3 and 4 should not be included in a dependent variable including 1 and 2, because they are not driven by the number of employees.

Step Two: *Choose the cost driver(s).* Examples of cost drivers where the cost object is a product include the number of parts in a product and the hours of test time for a product. Examples of cost drivers where the cost object is an activity include the number of machine setups and the number of components inserted. Ideally, the chosen cost driver should be economically plausible and accurately measurable.

Step Three: *Collect data on the dependent variable and on the cost driver(s).* This step is usually the most difficult one in cost analysis. The ideal data base would contain numerous observations for a firm whose operations have not been influenced by economic or technological change. Moreover, the time period (for example, daily, weekly, or monthly) used to measure the dependent variable and the cost driver(s) should be identical. A later section of this chapter gives examples of frequently encountered departures from this ideal data base.

Step Four: *Plot the data.* ***This step is critical in estimating cost relationships. The expression "a picture is worth a thousand words" conveys the benefits that come from plotting the data.*** The general relationship between the dependent variable and the cost driver (often called **correlation**) can readily be observed in a plot of the data. Moreover, extreme observations are highlighted. These extreme observations can then be checked to determine whether they arise from an error in recording the data or from an unusual event (such as a labor strike) that would prevent them from being representative of the normal relationship between the dependent variable and the cost driver. Plotting the data can also provide insight into whether a

EXHIBIT 10-5

Southern Carpets: Plot of Monthly Indirect Manufacturing Costs and Machine Hours

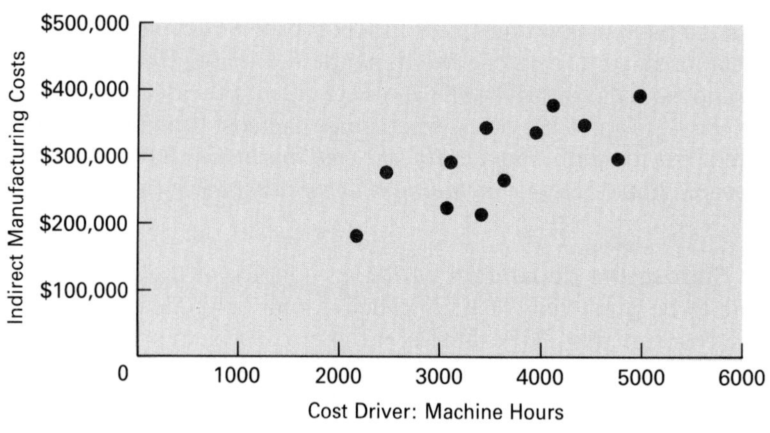

linear function can approximate cost behavior and what the relevant range of the cost function is.

Exhibit 10-5 shows a plot of the monthly data from Exhibit 10-4. There is strong visual evidence of a positive relationship between indirect manufacturing costs and machine hours. There do not appear to be any extreme observations in Exhibit 10-5. The relevant range is from 2,180 to 4,980 machine hours per month.

Step Five: *Estimate the cost function.* The next section of this chapter illustrates the use of both regression analysis and the high-low method to estimate the indirect manufacturing cost function at Southern Carpets.

Step Six: *Evaluate the estimated cost function.* In evaluating a cost function, the analyst should keep two points in mind:

1. *Economic plausibility.* The relationship between the dependent variable and the cost driver(s) should be economically plausible. It should be logical and appeal to common sense.
2. *Goodness of fit.* The closer the actual cost observations are to the values predicted by a cost function, the better the goodness of fit of the cost function. Formal measures of goodness of fit, such as the coefficient of determination (r^2), are available with the regression analysis method described later in this chapter and in Chapter 25.

Economic plausibility and goodness of fit serve as checks on one another. For example, data may show that a clerical overhead cost is more highly related to changes in the cost of electricity than to changes in the number of documents processed. But there may be no logical cause-and-effect relationship that supports such goodness of fit. In this situation, a manager should be reluctant to employ a cost function using electricity costs as the cost driver to predict how clerical overhead costs will behave. ***Managers have greater confidence that an observed statistical relationship will persist in subsequent periods if that relationship is economically plausible.***

Regression Analysis Method

Regression analysis is a statistical model that measures the *average* amount of change in the dependent variable (indirect manufacturing cost in our example) that is associated with a unit change in the amount of one or more cost drivers(s).[3] Regression is discussed at length in Chapter 25.

[3]In a regression model, the x variable(s) in the cost function is called the *independent variable* or the *explanatory variable.* As explained in Chapter 25, only a subset of the independent variables examined will be labeled as cost drivers.

EXHIBIT 10-6
Southern Carpets: Regression Model for Monthly Indirect Manufacturing Costs and Machine Hours

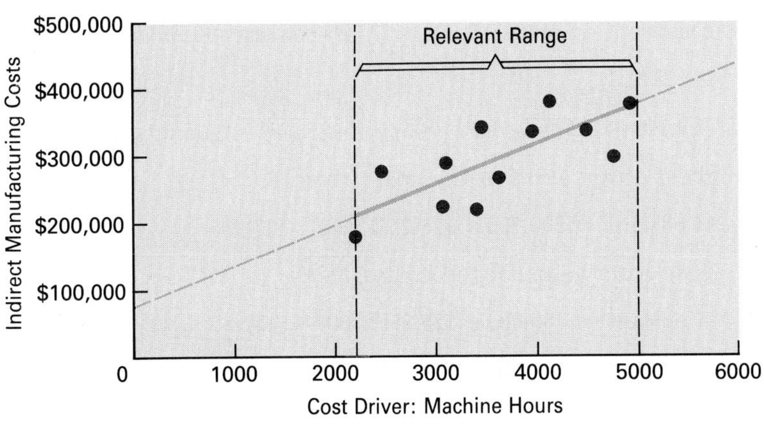

Using regression analysis, Exhibit 10-6 shows the line of best possible fit. The estimated cost function is

$$y' = \$86{,}153 + \$57.27x$$

where y' is the predicted indirect manufacturing cost for any level of machine hours (x). The constant or intercept term of the regression (a) is $86,153, and the slope coefficient (b) is $57.27. (Details on the computation of a and b appear in Chapter 25.)

Management can use this equation for budgeting indirect manufacturing costs. For instance, if 3,000 machine hours were budgeted for the upcoming month, the predicted indirect manufacturing costs would be

$$y' = \$86{,}153 + \$57.27(3{,}000) = \$257{,}963$$

The line in Exhibit 10-6 is deliberately extended to the left and right as a dashed line to emphasize the focus on the relevant range. The manager typically is interested in cost levels *within the relevant range,* not in cost levels outside the relevant range. The $86,153 constant or intercept term is not an estimate of the fixed cost of Southern Carpets. *Instead, it is the constant component of the equation that provides the best available linear approximation of how a cost behaves within the relevant range.*

High-low Method

Very simplified ways of estimating cost functions are occasionally used in practice. An example is the **high-low method,** which entails using only the highest and lowest values of the *cost driver* within the relevant range. The line connecting these two points becomes the estimated cost function.

Using the Exhibit 10-4 data in our illustration:

	Cost Driver: Machine Hours	Indirect Manufacturing Costs
Highest observation of cost driver	4,980	$390,700
Lowest observation of cost driver	2,180	180,000
Difference	2,800	$210,700

$$\text{Slope coefficient} = \frac{\text{Difference between costs associated with highest and lowest observations of the cost driver}}{\text{Difference between highest and lowest observations of the cost driver}}$$

Slope coefficient = \$210,700/2,800 = \$75.25 per machine hour

If $y = a + bx$

then $a = y - bx$

Constant = Total cost − (Slope coefficient × Quantity of cost driver)

At the highest observation of the cost driver:

Constant = \$390,700 − \$75.25(4,980) = \$15,955

At the lowest observation of the cost driver:

Constant = \$180,000 − \$75.25(2,180) = \$15,955

Therefore, the high-low estimate of the cost function is

$y' = a + bx$

= \$15,955 + \$75.25(machine hours)

Compare the high-low equation with the regression equation, which was \$86,153 + \$57.27 per machine hour. For a 2,500 machine-hour level, cost predictions would be:

Regression equation:	\$86,153 + \$57.27(2,500) = \$229,328
High-low equation:	\$15,955 + \$75.25(2,500) = \$204,080

In this illustration, the difference of \$25,248 between these two predictions is 11.0% of the regression prediction. This difference may be significant to a particular decision.

In some cases, the highest (lowest) observation of the cost driver will not coincide with highest (lowest) observation of the dependent variable. Given that causality runs from the cost driver to the dependent variable in a cost function, choosing the highest observation and the lowest observation of the cost driver is appropriate.

There is an obvious danger of relying on only two observations. They may not be representative of all the observations. Always plot all the data. The graph in Exhibit 10-7 illustrates the danger of mechanically applying the high-low method.

EXHIBIT 10-7

Danger of High-Low Method Using Nonrepresentative Observations

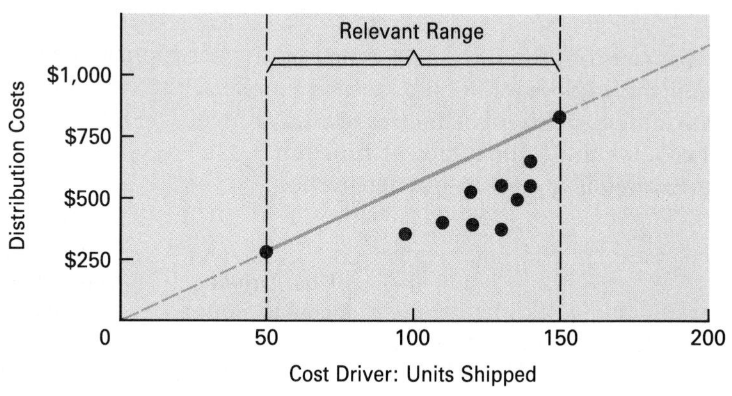

It shows how picking the highest and lowest observations for the units-shipped variable can result in an estimated cost function that poorly describes the underlying cost relationship between distribution costs and units shipped.

Sometimes the high-low method is modified so that the two observations chosen are a "representative high" and a "representative low." The reason is that management wants to avoid having extreme observations that arise from abnormal events affect the cost function. Even with such a modification, this method ignores information on all but two observations when estimating the cost function.

DATA COLLECTION AND ADJUSTMENT ISSUES

The ideal data base to be used in estimating cost functions quantitatively has two characteristics:

1. ***It contains numerous reliably measured observations of the cost driver(s) and the dependent variable, and***
2. ***It includes considerable variation in the values of the cost driver(s).***

Cost analysts typically will not be blessed with a data base having both characteristics. This section outlines some frequently encountered data problems.

Frequently Encountered Data Problems

Objective 7

Understand data problems encountered in estimating cost functions

In most cases, a cost analyst will encounter one or more of the following seven problems.

1. The time period used to measure the dependent variable (for example, overhead costs) is not properly matched with the period used to measure the cost driver(s). This problem often arises when accounting records are not kept on an accrual basis. Consider a cost function with machine-lubricant supplies as the dependent variable and machine hours as the cost driver. Assume that lubricant supplies are purchased sporadically and stored for later use. If records are kept on a cash basis, it will appear that no lubricant supplies are used in many months and that very sizable amounts of lubricant supplies are used in other months, which obviously is an inaccurate picture. Accrual accounting would result in a better matching of costs with the cost driver in this example.
2. Fixed costs are allocated as if they were variable. For example, such costs as depreciation, insurance, or rent may be allocated on a per-unit-of-output basis. *The danger is to regard these costs as variable rather than as fixed. They may seem to be variable because of the allocation methods used.*
3. The same time period is not used for all the items included in the dependent variable and the cost driver(s). For example, labor costs could be accumulated on a monthly basis, whereas volume of output could be accumulated on a weekly basis.
4. Data are either not available for all observations or not uniformly reliable. Missing cost observations often arise from a failure to record a cost or from classifying a cost incorrectly. Data on cost drivers often originate outside the internal accounting system. For example, the accounting department may get data on testing times for medical instruments from the company's manufacturing department and data on the number of items shipped to customers from the distribution department. The reliability of such data varies greatly across organizations. Data are still manually recorded rather than entered electronically in some systems. Manually recorded data typically have a higher percentage of missing observations and incorrectly entered observations than electronically entered data have.

5. Extreme values of observations occur. These extreme values can arise from errors in recording costs (for instance, a misplaced decimal point), from nonrepresentative time periods (for instance, from a period in which a major machine breakdown occurred), or from observations made outside the relevant range.

6. A homogeneous relationship between the individual cost items in the dependent variable and the cost driver(s) does not exist. A homogeneous relationship does exist when each activity whose costs are included in the dependent variable has the same cause-and-effect relationship with the cost driver. Consider materials procurement overhead cost. This overhead cost account can include a diverse set of activities, (for example, new vendor negotiations, materials ordering, incoming inspection, and materials handling). A cost analyst has to decide whether to estimate a separate cost function for each activity or to combine two or more of the cost pools associated with these activities before estimating a cost function for materials procurement overhead.

7. Inflation has occurred in the dependent variable, a cost driver, or both. The appendix to this chapter outlines an approach used to reduce problems arising from inflation. A related problem arises when a variable is an aggregate of several individual cost items and these individual cost items are subject to differing inflation rates. For example, the cost-of-goods-sold figure for a brewing company is a composite of the costs of items such as labor, direct materials (malt, corn, barley, and hops), fuel, and depreciation. Over time, the individual inflation rates of these items have differed significantly.

In many cases, a cost analyst must expend much effort to reduce these seven problems before estimating a cost function based on past data.

COST BEHAVIOR IN ACTIVITY AREAS

Knowledge of how costs behave in activity areas is important in activity-based accounting systems (see Chapter 5) and in cost management (see Chapter 29). Consider the quality-testing area of the Chain Saw Company (CSC). After assembly, each chain saw passes through the quality-testing area. Each chain saw receives a minimum of 10 minutes testing to examine its ability to cut lumber and to operate under differing conditions. Deluxe brands receive a minimum of 15 minutes testing. Several special customers have requested testing that requires over 30 minutes.

Exhibit 10-8 presents monthly data for the quality-testing activity area. The costs in this exhibit are the direct costs of the activity area: the cost of lumber used in testing and the cost of labor working in this area. Personal computers monitor each test and provide the data. Operating personnel believe that the single most important driver of month-to-month cost levels in the quality-testing activity area is hours of test time.

The cost function, based on the high-low method, is estimated as follows:

	Cost Driver: Hours of Testing Time	Costs Incurred at Activity Area
Highest observation of cost driver	986	$80,630
Lowest observation of cost driver	486	45,380
Difference	500	$35,250

$$\text{Slope coefficient} = \frac{\text{Difference between costs associated with highest and lowest observations of the cost driver}}{\text{Difference between highest and lowest observations of the cost driver}}$$

EXHIBIT 10-8

Quality-Testing Activity Area of Chain Saw Company: Monthly Data for Activity-Area Direct Costs and Hours of Testing Time

Month	Activity-Area Direct Costs	Hours of Testing Time in Activity Area
January	$54,235	640
February	59,520	722
March	45,380	486
April	64,000	886
May	59,235	634
June	73,060	812
July	81,625	927
August	80,630	986
September	75,105	958
October	63,970	819
November	67,350	856
December	55,285	546

Visual Analysis

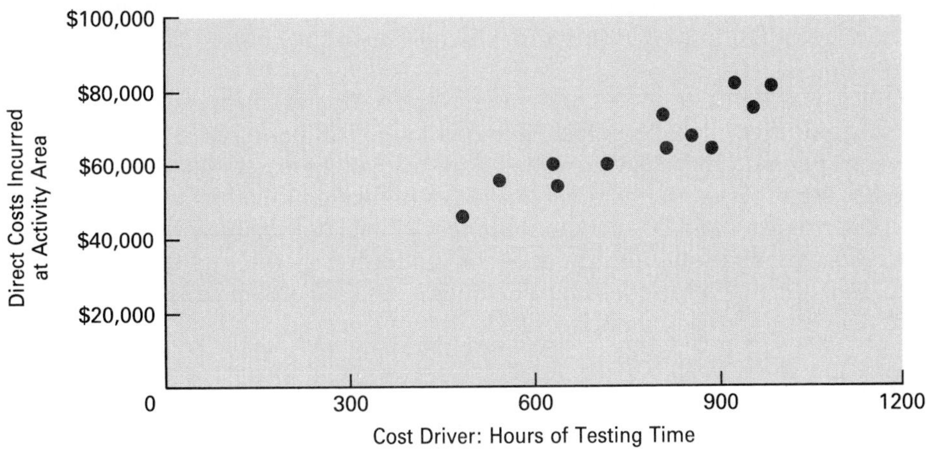

Slope coefficient = \$35,250/500 = \$70.50 per unit

Constant = Total cost − (Slope coefficient × Quantity of cost driver)

At highest observation of the cost driver:

$$\text{Constant} = \$80{,}630 - \$70.50(986) = \$11{,}117$$

At lowest observation of the cost driver:

$$\text{Constant} = \$45{,}380 - \$70.50(486) = \$11{,}117$$

The high-low estimate of the cost function for the quality-testing activity is

$$y' = \$11{,}117 + \$70.50(\text{hours of testing time})$$

With this formula, CSC can examine the effects of changes in testing time on costs. For example, management could investigate what cost reductions would result from designing chain saws that require less testing time. CSC could also predict how the extra testing time some customers require affects costs at the quality-testing area.

NONLINEARITY AND COST FUNCTIONS

A **nonlinear cost function** is a cost function in which a single constant (a) and a single slope coefficient (b) do not describe in an additive manner the behavior of costs for all changes in the level of the cost driver. For example, economies of scale in advertising may mean that an agency can double advertisements for less than a twofold increase in costs. Even direct material costs are not always linear variable costs. Consider the availability of quantity discounts on material purchases. As shown in Exhibit 10-9, the cost per unit falls when the level of each price break is reached—that is, the cost per unit decreases with larger orders. Of course, the cost function continues to rise, as the graph indicates, but it rises more slowly as the cost driver increases. The cost function in Exhibit 10-9 has b = \$20 for 1 to 999 units purchased, b = \$18 for 1,000 to 1,999 units purchased, and b = \$17 for 2,000 plus units purchased (a = \$0 for all ranges of the units purchased).

Step-function Costs

The left graph in Exhibit 10-10 illustrates a purely variable cost for a product where the cost driver is units produced. The right graph in Exhibit 10-10 shows a step-function cost, whereby the cost of the input is constant over various small ranges of the cost driver, but the cost increases by discrete amounts (that is, in steps) as the cost driver moves from one relevant range to the next. This step-pattern behavior occurs when the input is acquired in discrete quantities but is used in fractional quantities.

Vehicle-leasing costs for a package-delivery company illustrate a step-function cost. Each delivery vehicle, which is leased for \$3,000 per month, can make 1,000 deliveries per month. Vehicles are leased for discrete time periods (the minimum is a month) but are used in fractional quantities (to make individual deliveries). Leasing costs increase by \$3,000 per month for each 1,000-delivery step. When planning step costs, the aim is to attain maximum utilization at the highest level for any given step. In the vehicle-leasing illustration, this means making the maximum 1,000 deliveries for each vehicle leased.

Batch Costs

When products are made in batches or production runs, manufacturers often incur a separate set of costs on a per-batch basis. Consider the manufacturing costs of American Cola, a soft-drink company that makes four different flavors of cola (reg-

EXHIBIT 10-9
Effects of Quantity Discounts on Slope of Total Cost Function

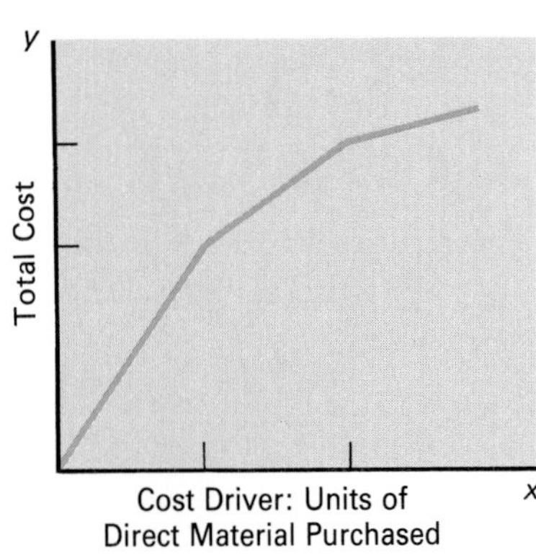

EXHIBIT 10-10
Variable and Step-Variable Cost Functions

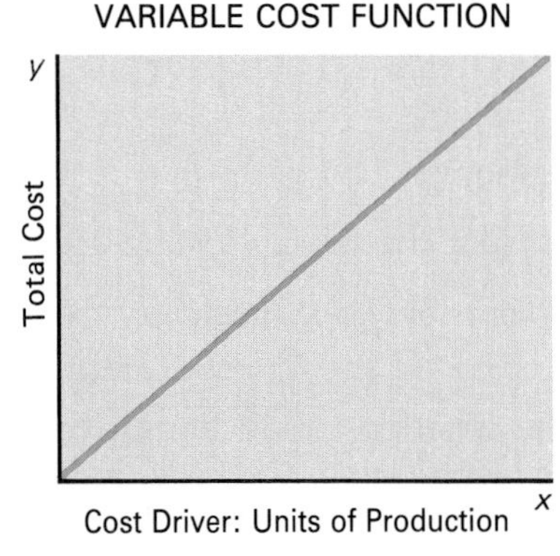

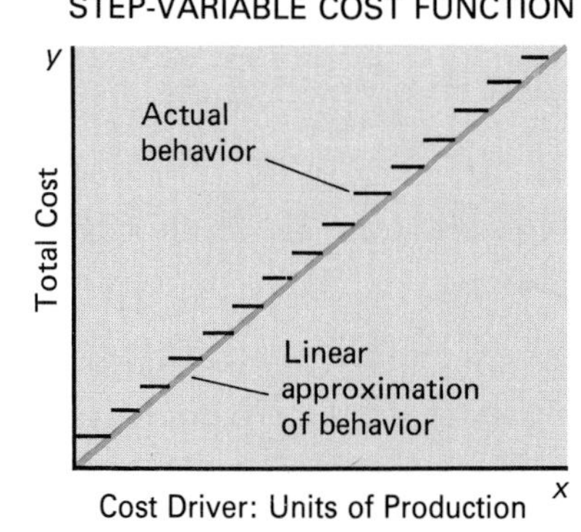

ular, diet, caffeine-free, and cherry). A linear-cost function with the number of cans filled as the cost driver applies to the production of each brand. All flavors have equivalent cost structures. However, before production of a different cola is started, the bottling facility has to be stopped and washed to remove traces of the flavor most recently bottled. These costs are often called changeover, or setup, costs. The company incurs them on a per-batch basis, regardless of the size of each batch. Having batch costs means that total manufacturing costs will not have a linear relationship with the number of cans filled each month. Instead, total costs will be the sum of changeover costs and linear costs:

(cost per setup × number of setups) + (cost per can filled × number of cans filled)

LEARNING CURVES AND COST FUNCTIONS

Some costs do not behave linearly. Consider the assembly of motor vehicles. Over time, the assembly may become more efficient as the people involved become more familiar with the operation. Workers handling repetitive tasks may learn to be more efficient. Managers may learn how to improve the scheduling of work shifts. Plant operators may learn how best to utilize the operating facility.

The effect of learning on output per hour is usually shown by a learning curve. A **learning curve** is a function that shows how labor hours per unit decline as units of output increase. The learning curve helps managers predict how labor hours (or labor costs) will change as more units are produced.

Managers are now extending the learning-curve notion to include other cost areas, such as marketing, distribution, and customer service. The term *experience curve* describes this broader application of the learning curve. An **experience curve** is a function that shows how full costs per unit (including manufacturing, distribution, marketing, and so on) decline as units of output increase.

We now outline two learning-curve models: the cumulative average–time learning model and the incremental unit–time learning model.[4]

- **Cumulative average–time learning model**. The cumulative average time per unit is reduced by a constant percentage each time the cumulative quantity of units produced is doubled.
- **Incremental unit–time learning model**. The incremental unit time (the time needed to produce the last unit) is reduced by a constant percentage each time the cumulative quantity of units produced is doubled.

Objective 8

Distinguish between cumulative average-time learning model and incremental unit-time learning model

Cumulative Average–Time Learning Model

Exhibit 10-11 illustrates the cumulative average–time learning model with an 80% learning curve. The 80% means that when the quantity of units produced is doubled from X to 2X, the cumulative average time *per unit* for the 2X units is 80% of the cumulative average time *per unit* for the X units. The graph on the left side of Exhibit 10-11 shows the average time *per unit* as a function of units produced. The graph on the right side of Exhibit 10-11 shows the total number of labor hours as a

[4]For further discussion, see J. Chen and R. Manes, "Distinguishing the Two Forms of the Constant Percentage Learning Curve Model," *Contemporary Accounting Research,* Spring 1985, pp. 242–52. See also the Northern Aerospace Manufacturing case study in A. A. Atkinson, *Cost Estimation in Management Accounting—Six Case Studies,* Society of Management Accountants of Canada (Hamilton, Ontario, 1987).

EXHIBIT 10-11
Plots for Cumulative Average–Time Learning Model

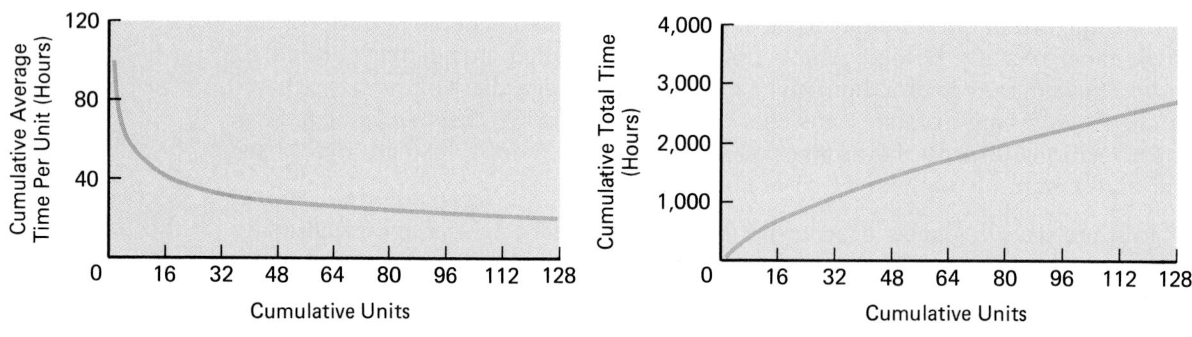

EXHIBIT 10-12
Cumulative Average–Time Learning Model

(1) Cumulative Number of Units	(2) Cumulative Average Time Per Unit (y): Hours	(3) = (1) × (2) Cumulative Total Time: Hours	(4) Individual Unit Time for Xth Unit: Hours
1	100.00	100.00	100.00
2	80.00 (100 × .8)	160.00	60.00
3	70.21	210.63	50.63
4	64.00 (80 × .8)	256.00	45.37
5	59.57	297.85	41.85
6	56.17	337.02	39.17
7	53.45	374.15	37.13
8	51.20 (64 × .8)	409.60	35.45
⋮	⋮	⋮	⋮
16	40.96 (51.2 × .8)	655.36	28.06

NOTE: The mathematical relationship underlying the cumulative average–time learning model is

$$y = pX^q$$

where y = cumulative average time (hours) per unit
X = cumulative number of units produced
p = time (hours) required to produce the first unit, and
q = the index of learning

The value of q is calculated as $q = \dfrac{\ln\text{(\% learning)}}{\ln 2}$

For an 80% learning index $q = \dfrac{-.2231}{.6931} = -.3219$

As an illustration, when $X = 3$, $p = 100$, and $q = -.3219$

$$y = 100 \times 3^{-.3219} = 70.21 \textit{ hours}$$

The cumulative total time when $X = 3$ is $70.21 \times 3 = 210.63$ hours.

The individual unit times in column (4) are calculated using the data in column (3). For example, the individual unit time of 50.63 hours for the third unit is calculated as 210.63 minus 160.00

function of units produced. The observations underlying Exhibit 10-11, and the details of their calculation, are presented in Exhibit 10-12. The cumulative total time is obtained by multiplying the cumulative average time per unit by the cumulative number of units produced.

Incremental Unit–Time Learning Model

Exhibit 10-13 on page 350 illustrates the incremental unit–time learning model with an 80% learning curve. The 80% here means that when the quantity of units produced is doubled from X to $2X$, the time needed to produce the *last unit* at the $2X$ production level is 80% of the time needed to produce the *last unit* at the X production level. The graph on the left side of Exhibit 10-13 shows the average time per unit as a function of units produced. The graph on the right side of Exhibit 10-13 shows the total number of labor hours as a function of units produced. The observations underlying Exhibit 10-13, and the details of their calculation, are presented in Exhibit 10-14 on page 350. We obtain the cumulative total time by summing the individual unit times.

The incremental unit-time model predicts that a higher cumulative total time is required to produce two or more units as compared with the predictions of the cumulative average-time model. Which of these two models is preferable? The one that more accurately approximates the behavior of labor hour usage as output levels increase. The choice can be decided only on a case-by-case approach. Engineers, plant managers, and workers are good sources of information on the amount of learning actually occurring as output increases. This information can be plotted and the appropriate curve selected.

The Problem for Self-Study at the end of this chapter illustrates the cumulative average–time learning model and the incremental unit–time learning model in a job-costing situation.

Setting Budgets and Standards

Predictions of costs should allow for the effects of learning. Consider the data in Exhibit 10-12 for the cumulative average-time model. Suppose the variable costs subject to learning effects consist of direct labor ($20 per hour) and related overhead ($30 per hour). Management could predict the costs shown in Exhibit 10-15 on page 351.

These data on the effects of the learning curve could be used for many different purposes. The data can have a major influence on decisions. For example, a company might set an extremely low selling price on its product in order to generate high demand. As the company's output increases to meet this growing demand, the cost per unit drops. The company "rides the product down the learning curve" as it establishes a higher market share. Although the company may not have earned much on its first sale—it may actually have lost money—the company gains more profit per unit as output increases.

Alternatively, subject to legal and other considerations, the company might set a low price on just the final eight units. After all, the labor and related overhead cost per unit is predicted to be only $12,288 for these final eight units ($32,768 − $20,480). The per unit cost of $1,536 on these final eight units ($12,288 ÷ 8) is much lower than the $5,000 cost per unit of the first unit produced.

Management must take care in using the learning curve in setting standards. Assume that a motor-vehicle company is to begin producing a new car. Management predicts that a learning curve will apply to production of the first 10,000 cars. After this point, no further learning will occur, a condition called steady-state. Standard costs should be lower per unit for the steady-state condition than for the learning-curve period, when workers, schedulers, and operations personnel are becoming familiar with the new assembly. If the steady-state standards are imposed during the learning-curve phase, an unfavorable efficiency variance between

EXHIBIT 10-13
Plots for Incremental Unit–Time Learning Model

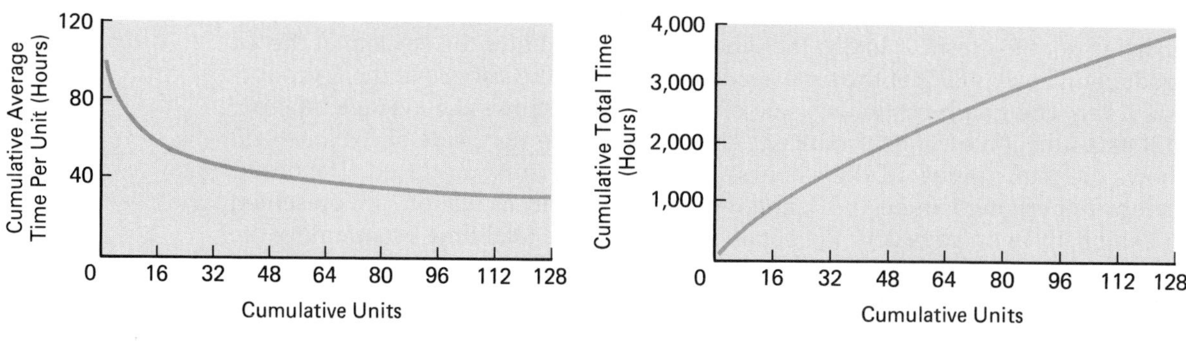

EXHIBIT 10-14
Incremental Unit–Time Learning Model

(1) Cumulative Number of Units	(2) Individual Unit Time for Xth Unit (*m*): Hours	(3) Cumulative Total Time: Hours	(4) = (3) ÷ (1) Cumulative Average Time Per Unit: Hours
1	100.00	100.00	100.00
2	80.00 (100 × .8)	180.00	90.00
3	70.21	250.21	83.40
4	64.00 (80 × .8)	314.21	78.55
5	59.57	373.78	74.76
6	56.17	429.95	71.66
7	53.45	483.40	69.06
8	51.20 (64 × .8)	534.60	66.82
⋮	⋮	⋮	⋮
16	40.96 (51.2 × .8)	892.00	55.75

NOTE: The mathematical relationship underlying the incremental unit–time learning model is

$$m = pX^q$$

where m = time (hours) taken to produce the last single unit,
X = cumulative number of units produced,
p = time (hours) required to produce the first unit, and
q = the index of learning

The value of q is calculated as $q = \frac{\ln(\%\text{ learning})}{\ln 2}$

For an 80% learning curve, $q = -.3219$.

As an illustration, when $X = 3$, $p = 100$, and $q = -.3219$:

$$m = 100 \times 3^{-.3219} = 70.21 \textit{ hours}$$

The cumulative total time when $X = 3$ is 100 + 80 + 70.21 = 250.21 hours.

EXHIBIT 10-15
Predicting Costs Using Learning Curves

Cumulative Number of Units	Cumulative Total* Labor Hours	Cumulative Costs	Additions to Cumulative Costs
1	100	$ 5,000 (100 × $50)	$ 5,000
2	160	8,000 (160 × $50)	3,000
4	256	12,800 (256 × $50)	4,800
8	409.60	20,480 (409.6 × $50)	7,680
16	655.36	32,768 (655.36 × $50)	12,288

*Based on cumulative average–time learning model. See Exhibit 10-12 for computation of cumulative total labor hours.

standards and actual performance may persist, and the employees might reject the standards as unattainable. Conversely, learning-curve cost standards may not be low enough for steady-state conditions.[5]

[5]The learning curve models examined in Exhibits 10-11 to 10-14 assume that learning is driven by a single variable (volume) and is product-related. Current research is examining other drivers of learning, such as product design and how manufacturing processes are configured. See S. Datar, S. Kekre, and E. Svaan, "Cost Drivers of Quality Learning," Working Paper, Carnegie-Mellon University, 1990.

PROBLEM FOR SELF-STUDY

PROBLEM

The Helicopter Division of Aerospatiale is examining helicopter assembly costs at its Marseilles, France, plant. It has received an initial order for eight of its new land-surveying helicopters. Aerospatiale can adopt one of two methods of assembling the helicopters:

Labor-Intensive Assembly Method

1. Direct materials cost of $40,000 per helicopter.
2. Direct labor time for the first helicopter will be 2,000 hours. Labor time per helicopter follows an 85% cumulative average–time learning curve model. (An 85% learning curve is expressed mathematically as $q = -.2345$.) Direct labor costs $30 per hour.
3. Indirect manufacturing costs are predicted using two separate cost functions, each with its own cost driver.
   a. Equipment-related indirect costs at the rate of $12 per direct labor hour.
   b. Materials-handling related indirect costs at the rate of 50% of direct materials cost.

Machine-Intensive Assembly Method

1. Direct materials cost of $36,000 per helicopter.
2. Direct labor time for the first helicopter will be 800 hours. Labor time per helicopter follows a 90% incremental unit–time learning curve model. (A 90% learning curve is expressed mathematically as $q = -.1520$.) Direct labor costs $30 per hour.
3. Indirect manufacturing costs are predicted using two separate cost pools, each with its own cost driver:
   a. Equipment-related indirect costs at the rate of $45 per direct labor hour.
   b. Materials-handling related indirect costs at the rate of 50% of direct materials cost.

The capital-intensive equipment used in this alternative is operated by workers who control the speed of production and so the utilization of the equipment.

Required

1. What is the number of direct labor hours required to assemble the first eight helicopters under
   a. the labor-intensive method?
   b. the machine-intensive method?
2. What is the cost of assembling the first eight helicopters under
   a. the labor-intensive method?
   b. the machine-intensive method?

SOLUTION

1. (a) *Cumulative average–time learning model (85% learning)*

(1) Cumulative Number of Units	(2) Cumulative Average Time Per Unit (*y*): Hours	(3) = (1) × (2) Cumulative Total Time: Hours	(4) Individual Unit Time for Xth Unit: Hours
1	2,000	2,000	2,000
2	1,700 (2,000 × .85)	3,400	1,400
3	1,546	4,638	1,238
4	1,445 (1,700 × .85)	5,780	1,142
5	1,371	6,855	1,075
6	1,314	7,884	1,029
7	1,267	8,869	985
8	1,228.25 (1,445 × .85)	9,826	957

The cumulative average time per unit for the *X*th unit in Column (2) is calculated as $y = pX^q$; see Exhibit 10-12, p. 348. For example, when $X = 3$:

$$y = 2{,}000 \times 3^{-.2345} = 1{,}546 \text{ hours}$$

(b) *Incremental unit–time learning model (90% learning)*

(1) Cumulative Number of Units	(2) Individual Unit Time for Xth Unit (*m*): Hours	(3) Cumulative Total Time: Hours	(4) = (3) ÷ (1) Cumulative Average Time Per Unit: Hours
1	800	800	800
2	720 (800 × .9)	1,520	760
3	677	2,197	732
4	648 (720 × .9)	2,845	711
5	626	3,471	694
6	609	4,080	680
7	595	4,675	668
8	583 (648 × .9)	5,258	657

The individual unit time for the *X*th unit in Column 2 is calculated as $m = pX^q$; see Exhibit 10-14, p. 350. For example, when $X = 3$:

$$m = 800 \times 3^{-.1520} = 677 \text{ hours}$$

2. Costs of assembling the first eight helicopters are:
   (a) With cumulative average–time learning model (85% learning):

Direct materials	8 × $40,000	$320,000
Direct labor	9,826 × $30	294,780
Indirect manufacturing costs:		
Equipment related	9,826 × $12	117,912
Materials related	.50 × $320,000	160,000
		$892,692

   (b) With incremental unit–time learning model (90% learning):

Direct materials	8 × $36,000	$288,000
Direct labor	5,258 × $30	157,740
Indirect manufacturing costs:		
Equipment related	5,258 × $45	236,610
Materials related	.50 × $288,000	144,000
		$826,350

The machine-intensive method has an assembly cost $66,342 lower than the labor-intensive method.

SUMMARY

Many decisions by managers use predictions of future costs. Efforts that improve our knowledge of how past costs behave will undoubtedly help managers make more accurate predictions of future costs. This chapter has examined cost estimation, which is the attempt to measure past cost relationships.

The industrial-engineering method, the conference method, the account analysis method, and quantitative analysis of cost relationships are used in estimating cost functions. Ideally, the cost analyst applies more than one approach; each approach can serve as a check on the others. The chosen cost function should be both economically plausible and meet the goodness-of-fit criterion.

There are six steps in estimating a cost function based on an analysis of current or past cost relationships: (1) choose the dependent variable, (2) choose the cost driver(s), (3) collect data on the dependent variable and the cost driver(s), (4) plot the data, (5) estimate the cost function, and (6) evaluate the estimated cost function. In most applications, the cost analyst will proceed through these steps several times before concluding that an acceptable cost function has been identified. The most difficult task in cost estimation is collecting "high-quality" data on the dependent variable and the cost driver(s).

Not all cost functions are linear. An example of a nonlinear cost function is the learning curve often found in operations such as aircraft or submarine assembly.

Chapter 25 further explores cost estimation, with an emphasis on the use of regression analysis.

APPENDIX: INFLATION AND COST ESTIMATION

One of the frequently encountered data problems listed in this chapter is inflation. Inflation can affect the dependent variable, those cost drivers measured in financial terms, or both.

This appendix illustrates one approach to incorporating inflation into cost estimation. This approach restates the variable(s) subject to inflation using an appropriate price index. Exhibit 10-16 illustrates this approach. The graph on the left side shows quarterly observations at Cybernetics Corporation of indirect manufacturing costs and machine hours for the past nine quarters. These observations appear in the second and third columns of Exhibit 10-17. Although indirect manufacturing costs and machine hours are positively correlated, there is considerable variation around the cost function estimated with a regression model:

$$y' = \$19.22 + \$0.56 \text{ (machine hours)}$$

The fourth column of Exhibit 10-17 presents an index of industry machine construction and maintenance costs. Over these nine quarters, the index has changed from .578 to 1.000, an increase of 73%. The fifth column of Exhibit 10-17 presents the restated indirect manufacturing costs. It is calculated by dividing each indirect manufacturing cost observation in column (2) by the corresponding value of the index in column (4). Thus, the restated cost figure of \$90 in quarter 1 is calculated as \$52 ÷ .578 = \$90. The graph on the right side of Exhibit 10-16 plots the restated

EXHIBIT 10-16
Cost Functions and Inflation Adjustments

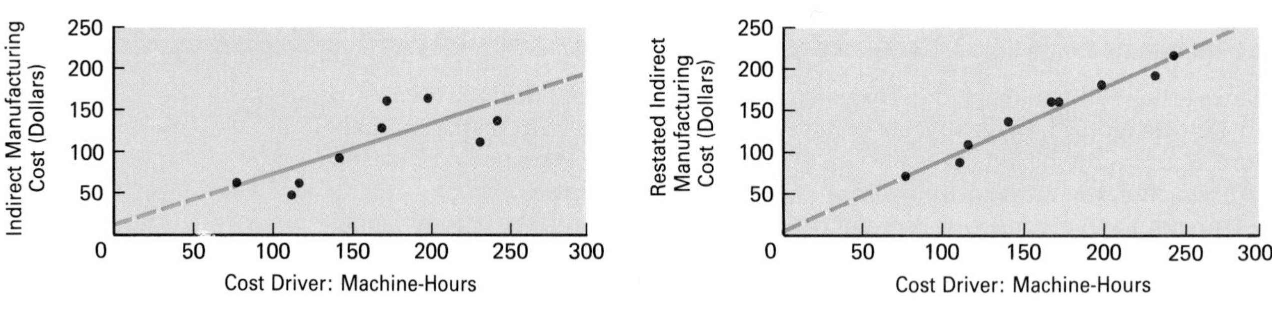

EXHIBIT 10-17
Cost Behavior at Cybernetics Corporation

(1) Quarter	(2) Indirect Manufacturing Costs	(3) Cost Driver: Machine Hours	(4) Index of Industry Machine and Maintenance Costs (Quarter 9 = 1.000)	(5) = (2) ÷ (4) Restated Indirect Manufacturing Costs
1	\$ 52	110	.578	\$ 90
2	113	231	.595	190
3	66	115	.601	110
4	139	242	.647	215
5	95	141	.694	137
6	130	168	.809	161
7	164	198	.913	180
8	67	76	.942	71
9	161	172	1.000	161

indirect manufacturing costs and machine hours; also presented is the cost function estimated with a regression model: $y' = \$10.64 + \$0.84$ (machine hours). The index approach has successfully controlled much of the variation around the fitted cost function observed when the nonrestated indirect manufacturing costs are used.

One limitation of using the index approach in Exhibits 10-16 and 10-17 to control for inflation is that appropriate indexes may not be available for all individual cost items. Readily available indexes (such as the Consumer Price Index) may fail to capture adequately the price changes affecting items included in the costs of individual organizations.

TERMS TO LEARN

This chapter and the Glossary at the end of the book contain definitions of the following important terms:

account analysis method *(p. 337)* conference method *(337)* constant *(333)*
correlation *(339)* cost estimation *(332)* cost prediction *(332)*
cumulative average–time learning model *(347)* dependent variable *(339)*
experience curve *(347)* high-low method *(341)*
incremental unit–time learning model *(347)*
industrial engineering method *(337)* intercept *(333)* learning curve *(347)*
linear cost function *(333)* mixed cost *(334)* nonlinear cost function *(346)*
parameter *(333)* regression analysis *(340)* semivariable cost *(334)*
slope coefficient *(333)* work measurement method *(337)*

ASSIGNMENT MATERIAL

QUESTIONS

10-1 What two assumptions are frequently made when estimating a cost function?

10-2 What is a cost driver? Give two examples of cost drivers.

10-3 Name four approaches to estimating a cost function.

10-4 Describe the conference-method approach to estimating a cost function. What are two advantages of this method?

10-5 List the six steps in estimating a cost function based on an analysis of current or past cost relationships. Which step is typically the most difficult for a cost analyst?

10-6 When using the high-low method, should the high and low observations be based on the dependent variable or the cost driver?

10-7 Discuss four frequently encountered problems when collecting cost data on variables included in a cost function.

10-8 Define *nonlinear cost function.* Give two examples.

10-9 Define *learning curve.* Outline two models that can be used when incorporating learning into the estimation of cost functions.

EXERCISES AND PROBLEMS

10-10 Estimating a cost function. The controller of the Ijiri Co. wants you to estimate a cost function from the following two observations in a general-ledger account called Maintenance:

Monthly Machine Hours	Monthly Maintenance Costs Incurred
4,000	$3,000
7,000	3,900

Required

1. Estimate the cost function for maintenance.
2. Can the constant in the cost function be used as an estimate of fixed maintenance cost per month?

10-11 Identifying variable, fixed, and mixed cost functions. Pacific Corp. operates car rental agencies at over twenty airports. Customers can choose from one of three contracts for car rentals of one day or less:

Contract 1: $50 for the day
Contract 2: $30 for the day plus $0.20 per mile traveled
Contract 3: $1.00 per mile traveled

Required

1. Present separate plots of each of the three contracts with costs on the vertical axis and miles traveled on the horizontal axis.
2. Describe each contract as a linear cost function of the form $y' = a + bx$.
3. Describe each contract as a variable, fixed, or mixed cost function.

10-12 Estimating a cost function, high-low method. Laurie Daley is examining customer-service department costs at the Southern Region of Capitol Products. Capitol Products has over 200 separate electrical products that are sold with a six-month guarantee of full repair or replacement with a new product.

Weekly data for the most recent ten-week period are:

Week	Customer-Service Department Costs	Number of Service Reports
1	$13,845	201
2	20,624	276
3	12,941	122
4	18,452	386
5	14,843	274
6	21,890	436
7	16,831	321
8	21,429	328
9	18,267	243
10	16,832	161

When a product is returned by a customer, a service report is made. This service report includes details of the problem and the time and cost of resolving the problem.

Required

1. Plot the relationship between customer-service costs and number of service reports. Is the relationship economically feasible?
2. Use the high-low method to compute the cost function relating customer-service costs to the number of service reports.
3. What variables, in addition to number of service reports, might be cost drivers of monthly customer-service costs of Capitol Products?

10-13 Linear cost approximation. Terry Lawler, managing director of the Memphis Consulting Group, is examining how overhead costs behave with variations in monthly professional-labor hours billed to clients. Assume the following historical data:

Total Overhead Costs	Professional-Labor Hours Billed to Clients
$340,000	3,000
400,000	4,000
435,000	5,000
477,000	6,000
529,000	7,000
587,000	8,000

Required

1. Compute the linear cost function relating total overhead cost to professional-labor hours, using the representative observations of 4,000 hours and 7,000 hours. Plot the linear cost function.

2. What would be the predicted total overhead costs for (a) 5,000 hours and (b) 8,000 hours using the cost function estimated in requirement 1? Plot the predicted costs and actual costs for 5,000 and 8,000 hours.
3. Lawler had a chance to accept a special job that would have boosted professional-labor hours from 4,000 to 5,000 hours. Suppose Lawler, guided by the linear cost function, rejected this job. It would have brought a total increase in contribution margin of $38,000, before deducting the predicted increase in total overhead cost, $43,000. What is the total contribution margin actually forgone?
4. Does the constant component of the cost function represent the fixed overhead costs of the Memphis Consulting Group? Why?

10-14 Various cost-behavior patterns. (CPA, adapted) Select the graph below that matches the numbered factory-cost data. You are to indicate by letter which of the graphs best fits each of the situations or items described.

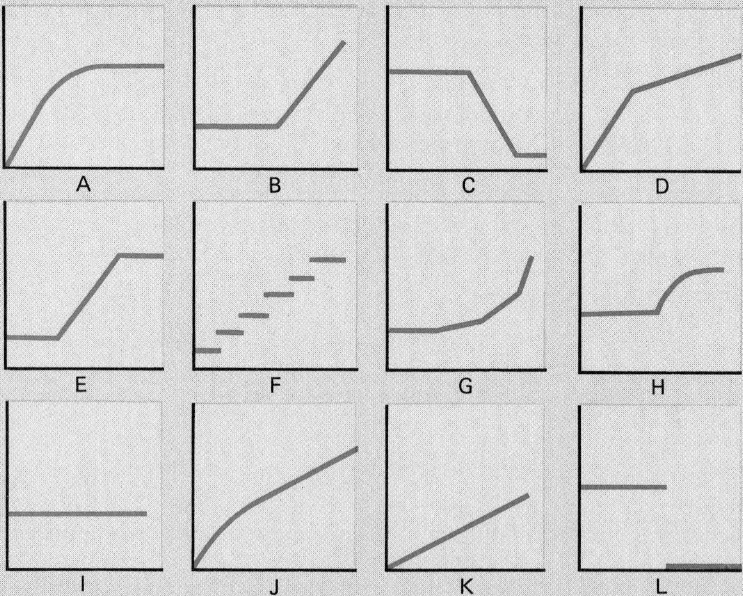

The vertical axes of the graphs represent *total* dollars of cost, and the horizontal axes represent production during a calendar year. In each case, the zero point of dollars and production is at the intersection of the two axes. The graphs may be used more than once.

1. Depreciation of equipment, where the amount of depreciation charged is computed by the machine hours method.
2. Electricity bill—a flat fixed charge, plus a variable cost after a certain number of kilowatt hours are used.
3. City water bill, which is computed as follows:

First 1,000,000 gallons or less	$1,000 flat fee
Next 10,000 gallons	.003 per gallon used
Next 10,000 gallons	.006 per gallon used
Next 10,000 gallons	.009 per gallon used
etc.	etc.

4. Cost of lubricant for machines, where cost per unit decreases with each pound of lubricant used (for example, if one pound is used, the cost is $10; if two pounds are used, the cost is $19.98; if three pounds are used, the cost is $29.94) with a minimum cost per pound of $9.20.
5. Depreciation of equipment, where the amount is computed by the straight-line method. When the depreciation rate was established, it was anticipated that the obsolescence factor would be greater than the wear-and-tear factor.

6. Rent on a factory building donated by the city, where the agreement calls for a fixed-fee payment unless 200,000 labor hours are worked, in which case no rent need be paid.
7. Salaries of repair personnel, where one person is needed for every 1,000 machine hours or less (that is, 0 to 1,000 hours requires one person, 1,001 to 2,000 hours requires two people, and so forth).
8. Cost of direct materials used.
9. Rent on a factory building donated by the county, where the agreement calls for rent of $100,000 reduced by $1 for each direct labor hour worked in excess of 200,000 hours, but minimum rental payment of $20,000 must be paid.

10-15 Matching graphs with descriptions of cost behavior. (D. Green) Given below are a number of charts, each indicating some relationship between cost and a cost driver. No attempt has been made to draw these charts to any particular scale; the absolute numbers on each axis may be closely or widely spaced.

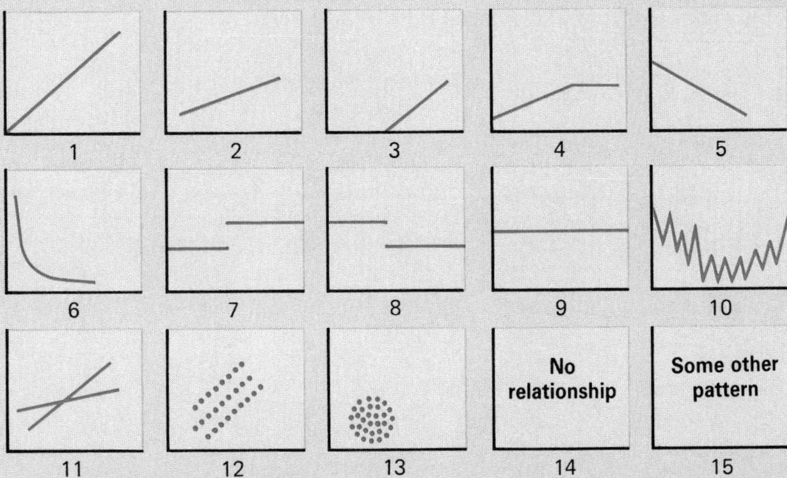

You are to indicate by number which one of the charts best fits each of the situations or items described. Each situation or item is independent of all the others; all factors not stated are assumed to be irrelevant. Some charts will be used more than once; some may not apply to any of the situations. Note that category 14 "No relationship," is not the same as 15, "Some other pattern."

I Taking the horizontal axis as representing the production volume over the year and the vertical axis as representing *total cost* or *revenue,* indicate the one best pattern or relationship for each of the following items:
   a. Direct material cost.
   b. Supervisors' salaries.
   c. A breakeven chart.
   d. Total average unit cost.
   e. Mixed costs—for example, electric power *demand charge* plus usage rate.
   f. Average versus variable cost.
   g. Depreciation of plant, computed on a straight-line basis.
   h. Data supporting the use of a variable cost rate, such as $14 per direct labor hour.
   i. Vacation pay accrued for all workers in a department or plant.
   j. Data indicating that an indirect cost rate based on the given volume measure is spurious.
   k. Variable costs *per unit* of output.
   l. Incentive bonus plan, operating only above some level of production.
   m. Interest charges on money borrowed to finance the acquisition of a plant, before any payments on principal.

II Taking the horizontal axis as representing a *time series of weeks* during a year and the vertical axis as representing *total cost per week,* match the following items with the relationship shown by the charts:

n. Direct labor cost under stable production volume.
o. Direct materials purchased in small quantities during a period of widely fluctuating prices (inventory held at zero).
p. Effect of declining production volume over the year.
q. Result of a shutdown because of vacations, a serious casualty, or complete failure of demand. The shutdown continues to the end of the year.
r. Underapplied or overapplied factory overhead, taken weekly over the year, when volume varies widely and the cost rate is assumed to be correct.
s. Seasonal fluctuation in the use of fuel for heating the plant building over the year.

10-16 Data collection issues, use of high-low method. Trevor Kennedy, the cost analyst at United Packaging's can manufacturing plant, is seeking to develop a cost function(s) that relates engineering support (E.S.) costs to machine hours. These costs have two components: (1) labor (which is paid monthly) and (2) materials and parts (which are purchased from an outside vendor every three months). He collects the following monthly data from the accounting records kept at the factory:

	Engineering Support: Reported Costs			
Month	**Labor**	**Materials and Parts**	**Total**	**Machine Hours**
March	$347	$847	$1,194	30
April	521	0	521	63
May	398	0	398	49
June	355	961	1,316	38
July	473	0	473	57
August	617	0	617	73
September	245	821	1,066	19
October	487	0	487	53
November	431	0	431	42

Required

1. Present plots of the data for the following three cost functions:

	Dependent Variable	Cost Driver
(i)	E.S. labor costs	Machine hours
(ii)	E.S. materials and parts costs	Machine hours
(iii)	E.S. total costs	Machine hours

   Comment on the patterns in these three plots.

2. Compute estimates of each of the three cost functions in requirement 1 using the high-low method.
3. What are two factors that could explain the pattern of monthly E.S. costs for materials and parts? For each factor you cite, explain its implications for examining cost-behavior patterns.

10-17 Data collection issues, use of high-low method. Robin Green, financial analyst at Central Railroad, is examining the behavior of monthly transportation costs for budgeting purposes. Transportation costs at Central Railroad are the sum of two types of costs: (a) operating costs (labor, fuel, and so on), and (b) maintenance costs (overhaul of engines and track, and so on).

Green collects monthly data on (a), (b), and track miles hauled for that month. Track miles hauled are the miles clocked by the engine that pulls the rail carriages. Monthly observations for the most recent year are:

(1) Month	(2) Operating Costs	(3) Maintenance Costs	(4) = (2) + (3) Total Transportation Costs	(5) Track Miles Hauled
January	$471	$437	$ 908	3,420
February	504	338	842	5,310
March	609	343	952	5,410
April	690	347	1,037	8,440
May	742	294	1,036	9,320
June	774	211	985	8,910
July	784	176	960	8,870
August	986	210	1,196	10,980
September	895	282	1,177	4,980
October	651	394	1,045	5,220
November	481	381	862	4,480
December	386	514	900	2,980

Central Railroad earns its greatest revenues carrying agricultural commodities such as wheat and barley.

Required

1. Present plots of the monthly data underlying each of the following cost functions:
   (a) Operating costs = $a + b$ (Track miles hauled)
   (b) Maintenance costs = $a + b$ (Track miles hauled)
   (c) Total transportation costs = $a + b$ (Track miles hauled)
   Comment on the patterns in the three plots.
2. Compute estimates of the three cost functions in requirement 1 using the high-low method. Comment on the estimated cost functions.
3. Outline three limitations of the high-low method for estimating a cost function.

10-18 Cost behavior and budgets. (M. Strecker) Tina Moretti, the president of Shuz, Inc., enlisted the assistance of the vice presidents in charge of sales and of production. Together they constructed two budgets—one optimistic, one pessimistic. These are shown in columns (1) and (2) below. The actual results are shown in column (3). The company's cost accountant was perplexed as to how to present an analysis. Consequently, he produced variances of actual results against both the optimistic projection and the pessimistic projection, columns (4) and (5) respectively.

	(1) Optimistic Budget	(2) Pessimistic Budget	(3) Actual Results	(4) Variance from Optimistic Budget	(5) Variance from Pessimistic Budget
Units sold	200,000	75,000	135,000	65,000	(60,000)
Sales	$2,000,000	$750,000	$1,350,000	$650,000	$(600,000)
Direct materials	$ 200,000	$ 75,000	$ 140,000	$ (60,000)	$ 65,000
Direct labor	400,000	150,000	285,000	(115,000)	135,000
Indirect labor	106,000	43,500	72,000	(34,000)	28,500
Maintenance	20,000	20,000	22,000	2,000	2,000
Supplies	28,000	15,500	21,000	(7,000)	5,500
Power	160,000	60,000	108,000	(52,000)	48,000
Heat	50,000	50,000	53,000	3,000	3,000
Light	7,000	4,500	5,900	(1,100)	1,400
Rent	80,000	80,000	80,000	-0-	-0-
Insurance	20,000	13,750	17,000	(3,000)	3,250

Required

Moretti is baffled by the analysis. She has asked you for a more understandable performance report. In the columns below, in clear and orderly fashion, prepare a new report. Explain your work to the president. Show supporting computations as needed.

Line Item	Actual Results	Sales or Cost Function	Revised Budget	Variance
Units sold	135,000	—	135,000	0
Sales	$1,350,000	y = $10X	$1,350,000	$0
Direct materials	$ 140,000	y = $1X	$ 135,000	$5,000 U
Direct labor	285,000	y = $2X	?	?
Indirect labor	72,000	y = $6,000 + $0.50X	?	?
(and so forth)				

10-19 Cost estimation, cumulative average–time learning curve. The Nautilus Company, which is under contract to the navy, assembles troop deployment boats. As part of its research program, it completes the assembly of the first of a new model (PT109) of deployment boats. The navy is impressed with the PT109. It requests that Nautilus submit a proposal on the cost of producing another seven PT109s.

The accounting department at Nautilus reports the following cost information for the first PT109 assembled by Nautilus:

Direct materials	$100,000
Direct labor (10,000 hours @ $30)	300,000
Tooling cost*	50,000
Variable overhead†	200,000
Other overhead‡	75,000
	$725,000

*Tooling can be reused, even though all of its cost was assigned to the first deployment boat.
†Variable-overhead incurrence is directly affected by direct labor hours; a rate of $20 per hour is used for purposes of bidding on contracts.
‡Other overhead is assigned at a flat rate of 25% of direct labor cost for purposes of bidding on contracts.

Nautilus uses an 85% cumulative average–time learning curve as a basis for forecasting direct labor hours on its assembling operations. (An 85% learning curve implies $q = -.2345$.)

Required

1. Prepare a prediction of the total costs for producing the seven PT109s for the navy. (Nautilus will keep the first deployment boat assembled, costed at $725,000, as a demonstration model for other potential purchasers.)
2. What is the difference between (a) the predicted total costs for producing the seven PT109s in requirement 1 and (b) the predicted total costs for producing the seven PT109s assuming there is no learning curve for direct labor—that is, for (b) assume a linear function for direct labor hours and units produced.

10-20 Cost estimation, incremental unit–time learning curve. Assume the same information for the Nautilus Company as that in requirement 1 of Problem 10-19 with one exception. This exception is that Nautilus uses an 85% incremental unit–time learning curve as a basis for forecasting direct labor hours on its assembling operations.

Required

Prepare a prediction of the total expected costs for producing the seven PT109s for the navy. If you solved requirement 1 of Problem 10-19, compare your cost prediction there with the one you make here.

10-21 Cost estimation, cumulative average–time learning model, competitor cost analysis. United Defense Inc., based in Dallas, Texas, is a large supplier to the U.S. government. Until recently, it had experienced a high success rate in bidding on jet-aircraft components. Last year it lost a bid on the radar system (RS-704) for the new jet fighter of the U.S. Air Force. The successful bidder was a French company (Aerospace Nationale) that also had an existing contract with the French Air Force; this French contract was for a production run of sixteen RS-704 systems. United Defense had manufactured only one RS-704 system to date; 4,000 direct labor hours were required to manufacture this one unit. Based on prior contracts for radar systems, United Defense believed an 85% cumulative average–time learning model for labor hours applied to RS-704 systems.

The cost structure of United Defense for each RS-704 was built up as the sum of direct costs and indirect costs:

- Direct materials cost per RS-704 unit: $90,000
- Direct labor cost: number of direct labor hours per RS-704 unit × $30 an hour
- Indirect procurement costs: 30% of direct materials dollars
- Indirect assembly costs: number of direct labor hours × $40 an hour

Analysts within United Defense pinpointed two areas where Aerospace Nationale had an advantage when bidding on the contract:

1. Aerospace had already assembled sixteen units for another contract. United Defense analysts believed that Aerospace had an 85% cumulative average–time learning model for labor hours, with 4,000 direct labor hours required for the first RS-704 system to be assembled.
2. Aerospace had a direct labor cost of $25 per hour. United Defense analysts believed that Aerospace had a procurement indirect cost rate of 30% of direct materials dollars and an assembly indirect cost rate of $40 per direct labor hour.

Required

1. Estimate the number of direct labor hours required by United Defense to assemble the first sixteen RS-704 systems. (This first sixteen includes the one system already manufactured.)
2. What cost figure did United Defense use in developing its bid for a production run of sixteen RS-704 systems? (Assume it based its bid on the initial production run of sixteen RS-704 systems.)
3. What cost figure would the United Defense analysts have estimated for their competitor Aerospace assembling the additional sixteen units for the U.S. Air Force, given (a) that Aerospace had completed the contract of the sixteen RS-704 systems for the French Air Force and (b) that Aerospace had the lower direct labor cost per hour?
4. Assume you are an analyst at Aerospace Nationale. What evidence would you have gathered to examine whether the assumed learning curve model was appropriate for bidding on the U.S. Air Force contract? What problems may have arisen in this examination?

10-22 Inflation adjustments, use of high-low approach. (See the appendix) Paul Sellinger is examining cost-behavior patterns at the Joseph Schlitz Brewing Company. He collects the following information covering the most recent ten-year period.

Year	Cost of Goods Sold (millions)	Barrels Sold (millions)	Wholesale Price Index of Beer (Year 10 = 1.000)
1	$ 594	16.7	.611
2	686	18.9	.612
3	783	21.3	.623
4	922	22.7	.719
5	1,047	23.3	.759
6	1,102	24.2	.769
7	1,073	22.1	.794
8	1,047	19.6	.848
9	1,050	16.8	.933
10	998	15.0	1.000

The cost of sales is an aggregate of cost items such as labor, direct materials (malt, corn, barley, hops, and so on), and depreciation. The wholesale price index of beer is calculated at the brewing industry level and is based on the price at which wholesalers sell to retail outlets.

Required

1. Plot the relationship between cost of goods sold and barrels sold over this period. What patterns are apparent in the data?
2. Use the high-low method to compute the cost function relating cost of goods sold to barrels sold. Comment on the representativeness of the two points used to estimate the cost function with the high-low method.
3. Use the wholesale price index of beer to restate cost of goods sold of Schlitz each year. Plot the relationship between restated cost of goods sold and barrels sold over this period.
4. Use the high-low method to compute the cost function relating restated cost of goods sold to barrels sold. Compare the cost function with that calculated in requirement 2.
5. What alternatives to the use of the wholesale price index of beer might Sellinger consider to minimize any problems that inflation causes when estimating cost functions?

Relevance, Costs, and the Decision Process

Information and the decision process

The meaning of relevance

Illustration of relevance: Choosing volume levels

Other illustrations of relevance

Opportunity costs, relevance, and accounting records

Irrelevance of past costs

How managers behave

Appendix: Cost terms used for different purposes

CHAPTER 11

Should a fabrics manufacturer accept or reject a one-time-only special order when it has idle production capacity? A relevant cost approach to cost analysis can provide managers with the information essential for making such a decision.

Learning Objectives

When you have finished studying this chapter, you should be able to

1. Describe the five-step sequence in a decision process
2. Outline the meaning of relevant cost; describe its two key aspects
3. Distinguish between quantitative factors and qualitative factors in decisions
4. Identify two ways in which per-unit cost data can mislead decision makers
5. Describe the opportunity cost concept; explain why it is used in decision making
6. Explain why the book value of equipment is irrelevant in equipment-replacement decisions
7. Explain how conflicts can arise between the decision model used by a manager and the performance model used to evaluate that manager

Cost data play an important role in many internal decisions by managers. This chapter illustrates how a solid understanding of cost behavior can help guide decision making. We focus on decisions such as accepting or rejecting a one-time-only special order, manufacturing or purchasing component parts (the make-or-buy decision), and replacing or keeping equipment. Subsequent chapters will discuss pricing decisions (Chapter 12) and capital budgeting decisions (Chapters 21 and 22).

The accountant's role in decisions by line personnel is that of a technical expert. Managers should be supplied with relevant data for guiding decisions. The abilities to distinguish relevant from irrelevant items and to analyze cost-behavior patterns (see Chapter 10) together form the basis for making many decisions.

INFORMATION AND THE DECISION PROCESS

The manager has a method for deciding among courses of action, often called a decision model. A **decision model** is a formal method for making a choice, frequently involving quantitative analysis. For now, let us focus on accounting information as an input into decision models.

Predictions and Models

Consider a decision Home Appliances faces: Should it rearrange a production assembly line to reduce operating labor costs? Assume that the only alternatives are "do not rearrange" and "rearrange." The rearrangement will substantially reduce manual handling of materials and work in process inventories. The current production line uses 20 workers. If the rearrangement is made, 15 workers will be required. The rearrangement is predicted to cost $80,000. The predicted output of 25,000 units for the next year will be unaffected by the decision. Also unaffected by the decision are the predicted selling price per unit of $250, direct materials cost per unit of $50, manufacturing overhead of $30 per unit, and marketing costs of $80 per unit. The cost driver is units of production.

Objective 1

Describe the five-step sequence in a decision process

Study Exhibit 11-1. It outlines a five-step sequence that highlights the role of accounting information in predicting the labor-cost savings. The historical labor cost of $22 per unit of output is the starting point for predicting the labor costs per

EXHIBIT 11-1
Accounting Information and the Decision Process

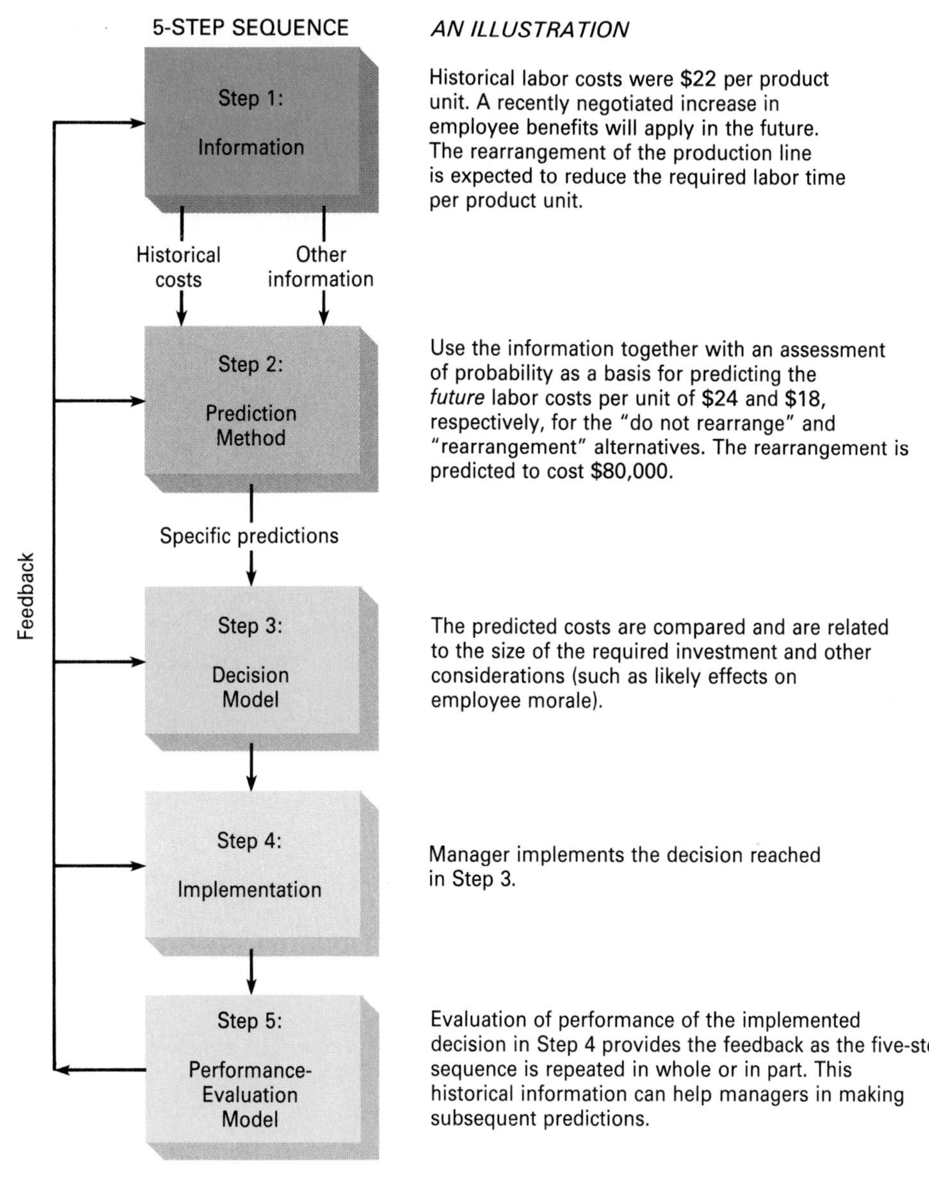

unit of output under both alternatives. The prediction under the "do not rearrange" alternative is $24 per unit, incorporating a recently negotiated increase in employee benefits. The prediction under the "rearrange" alternative is $18 per unit because of the reduced labor requirements.

5-STEP SEQUENCE
AN ILLUSTRATION
Step 1:
Information
Historical labor costs were $22 per product unit. A recently negotiated increase in employee benefits will apply in the future. The rearrangement of the production line is expected to reduce the required labor time per product unit.
Historical costs
Other information
Step 2:
Prediction Method
Use the information together with an assessment of probability as a basis for predicting the future labor costs per unit of $24 and $18, respectively, for the "do not rearrange" and "rearrangement" alternatives. The rearrangement is predicted to cost $80,000.
Specific predictions
Step 3:
Decision Model
The predicted costs are compared and are related to the size of the required investment and other considerations (such as likely effects on employee morale).
Step 4:
Implementation
Manager implements the decision reached in Step 3.
Step 5:
Performance-Evaluation Model
Evaluation of performance of the implemented decision in Step 4 provides the feedback as the five-step sequence is repeated in whole or in part. This historical information can help managers in making subsequent predictions.
Feedback

Models and Feedback

Assume that management chooses the "rearrange" alternative. This decision is implemented, and the subsequent evaluation of actual performance provides feedback. In turn, the feedback might affect future predictions, the prediction method itself, the decision model, or the implementation.

In our illustration, the actual results of the plant rearrangement may show that the new labor costs are $21 per product unit (due to, say, lower than expected labor productivity) rather than the predicted $18. This feedback may lead to better implementation (step 4 in Exhibit 11-1)—through a change in supervisory behavior, employee training, or personnel, for example—so that the $18 target is achieved. On the other hand, the feedback may convince the decision maker that the prediction method, rather than the decision implementation, was faulty. Perhaps the prediction method for similar decisions in the future should be modified to allow for worker training or learning time.

To highlight and simplify various points throughout this chapter, we assume that dollar amounts of future revenues and future costs are certain. In practice, the forecasting of these figures is generally the most difficult aspect of the decision process. Chapter 20 discusses uncertainty and ways to incorporate it into decision making.

THE MEANING OF RELEVANCE

Historical Data and Predictions

Exhibit 11-1 shows that every decision deals with the future—whether it be twenty seconds ahead (the decision to adjust a dial) or twenty years ahead (the decision to plant and harvest pine trees). ***A decision (step 3) always involves a prediction (step 2). Therefore, the function of decision making is to select courses of action for the future. Nothing can be done to alter the past.***

Relevant costs are those *expected future costs* that differ among alternative courses of action. Note the two aspects to this definition:

Objective 2

Outline the meaning of relevant cost; describe its two key aspects

1. expected future costs that
2. differ among alternative courses of action.

Relevant revenues are those expected future revenues that differ among alternative courses of action.

In Exhibit 11-1, the $24 and $18 per unit labor costs are relevant costs—they are expected future costs that differ in the two alternatives. The $22 per unit past labor cost is not relevant, although it may play a role in preparing the $24 and $18 per unit forecasts. ***Historical costs in themselves are irrelevant to a decision, although they may be the best available basis for predicting future costs.***

The quantitative data underlying the choice between the "do not rearrange" and the "rearrange" alternatives are presented in Exhibit 11-2. The first two columns present all data. The last two columns present only expected future costs or revenues that differ between the two alternatives. The revenues, direct materials, manufacturing overhead, and marketing items can be ignored. Why? Because although they are expected future costs, they do not differ between the alternatives. The data in Exhibit 11-2 indicate that rearranging the production line will decrease next year's total costs by $70,000.

The difference in total cost between two alternatives is an **incremental cost**. The incremental cost between Alternative 2 and Alternative 1 in Exhibit 11-2 is $70,000. Synonyms for incremental cost are **differential cost** and **net relevant cost**.

The definition of relevance is the major conceptual lesson in this chapter. The remainder of this chapter will show how to apply the concept of relevance to some commonly encountered decisions. A commentator expressed the importance of *relevance* by saying: "If accounting had been a profession soon after Adam and Eve left the Garden of Eden, it could have placed relevance above all else and even might have gotten it included in the Ten Commandments."

EXHIBIT 11-2
Determining Relevant Revenues and Relevant Costs

	All Data		Relevant Data	
	Alternative 1: Do Not Rearrange	Alternative 2: Rearrange	Alternative 1: Do Not Rearrange	Alternative 2: Rearrange
Revenues*	$6,250,000	$6,250,000	$ —	$ —
Costs:				
Direct materials†	1,250,000	1,250,000	—	—
Direct labor	600,000‡	450,000§	600,000	450,000
Manufacturing overhead‖	750,000	750,000	—	—
Marketing#	2,000,000	2,000,000	—	—
Rearrangement costs	—	80,000	—	80,000
Total costs	$4,600,000	$4,530,000	$600,000	$530,000
Operating income	$1,650,000	$1,720,000		
	$70,000 Difference		$70,000 Difference	

*25,000 × $250 ‡25,000 × $24 ‖25,000 × $30
†25,000 × $ 50 §25,000 × $16 #25,000 × $80

Time Value of Money and Income Taxes

Exhibit 11-2 does not take into account the time value of money or income taxes. We defer discussion of these factors to Chapters 21 and 22. When interest costs associated with the time value of money are recognized, expected future costs and future revenues with the same magnitude but differing timing are relevant. Similarly, when tax issues are introduced, expected future pretax costs and future pretax revenues with the same magnitude may affect income tax costs differently; hence, the income tax costs would be relevant.

Can Qualitative Costs Be Relevant?

Objective 3

Distinguish between quantitative factors and qualitative factors in decisions

We can divide the consequences of alternatives into two broad categories: *quantitative* and *qualitative*. **Quantitative factors** are measured in numerical terms. Examples include costs of direct materials, direct labor, and marketing. **Qualitative factors** cannot be measured in numerical terms. Employees' morale is an example.

Just because qualitative factors cannot be measured in numerical terms does not make them unimportant. Managers must at times give more weight to a qualitative factor than to a quantitative factor. For example, Home Appliances may use a single supplier for a variety of important subassemblies. To preserve good business relationships—a qualitative factor—Home Appliances might decide to continue to buy all subassemblies from that supplier even though it could manufacture one of these subassemblies itself at a cost—a quantitative factor—lower than it pays to the supplier.

ILLUSTRATION OF RELEVANCE: CHOOSING VOLUME LEVELS

Managers often make decisions that affect volume levels. Many such decisions are essentially short run in nature, but they have long-run consequences that should never be overlooked. The decision alternative that managers seek maximizes the chosen objective of the organization (typically operating income in our illustrations).

Cost analysis is an essential step in determining volume levels. Intelligent analysis of costs often depends on explicit distinctions between cost behavior patterns. In the following example, we assume that all costs can be classified as either variable with respect to a single driver (units of output) or fixed. This assumption provides a setting where a contribution-margin approach can provide key information for decisions about choice of the volume of output. The example also illustrates how reliance on fully allocated unit-cost numbers can yield erroneous conclusions about the effect that increasing volume has on operating income.

The One-Time-Only Special Order

Management sometimes faces the decision of accepting or rejecting one-time-only special orders when there is idle production capacity and where the order has no long-run revenue implications.

EXAMPLE

Fancy Fabrics manufactures quality bath towels at its highly automated Charlotte plant. The plant has a capacity of 48,000 towels each month. Current monthly production is 30,000 towels. All existing sales are made through retail department stores. Expected results for the coming month (August) are in Exhibit 11-3. (Note that these amounts are predictions.) The manufacturing cost per unit of $12 comprises direct material $6 (all variable), direct labor $2 ($0.50 of which is variable), and manufacturing overhead $4 ($1 of which is variable). The marketing cost per unit is $7 ($5 of which is variable). Manufacturing costs at Fancy Fabrics also include research and development costs and product design costs. Marketing costs also include distribution costs and customer service costs.

A four-star hotel chain offers to buy 5,000 towels per month at $11 a towel at most for each of the next three months. No subsequent sales to this customer are anticipated. No marketing costs will be necessary for the 5,000 unit special order. The acceptance of this special order is not expected to have any impact on the price or the volume of regular sales. Should the offer from the hotel chain be accepted?

Exhibit 11-3 presents data in an absorption-costing format. The unit manufacturing cost is $12.00, which is above the $11.00 price offered by the hotel chain. Using the $12.00 cost as a guide in deciding, a manager might unwisely reject the offer.

Exhibit 11-4 presents data in a contribution-margin format. The relevant costs are the expected future costs that differ among the alternatives—the variable manufacturing costs of $7.50 per unit. The fixed manufacturing costs and all marketing costs are irrelevant in this case; they will not change in total whether or not the special order is accepted. Therefore, the only relevant items here are sales and variable manufacturing costs. Given the $11.00 relevant revenue per unit (the special order price) and the $7.50 relevant cost per unit, Fancy Fabrics gains an additional $17,500 in operating income per month by accepting the special order. This

EXHIBIT 11-3

Budgeted Income Statement for August, Absorption-Costing Format

	Total	Per Unit
Sales—30,000 towels @ $20	$600,000	$20
Manufacturing cost of goods sold	360,000	12
Gross profit or gross margin	$240,000	$ 8
Marketing costs	210,000	7
Operating income	$ 30,000	$ 1

EXHIBIT 11-4
Comparative Income Statements for August, Contribution-Margin Format

	Without One-Time-Only Special Order, 30,000 Units		With One-Time-Only Special Order, 35,000 Units	Difference
	Per Unit*	Total	Total	
Sales	$20.00	$600,000	$655,000	$55,000†
Variable costs:				
Manufacturing	$ 7.50	$225,000	$262,500	$37,500‡
Marketing	5.00	150,000	150,000	— §
Total variable costs	$12.50	$375,000	$412,500	$37,500
Contribution margin	$ 7.50	$225,000	$242,500	$17,500
Fixed Costs:				
Manufacturing	$ 4.50	$135,000	$135,000	— §
Marketing	2.00	60,000	60,000	— §
Total fixed costs	$ 6.50	$195,000	$195,000	—
Operating income	$ 1.00	$ 30,000	$ 47,500	$17,500

*Analysis of costs per unit:

	Variable Cost	Fixed Cost	Business Function Cost
Direct materials	$ 6.00	$—	$ 6.00
Direct labor	0.50	1.50	2.00
Manufacturing overhead	1.00	3.00	4.00
Manufacturing cost	$ 7.50	$4.50	$12.00
Marketing cost	5.00	2.00	7.00
Product cost	$12.50	$6.50	$19.00

†5,000 × $11.00
‡5,000 × $7.50
§No variable marketing costs will be necessary for the 5,000-unit one-time-only special order. Fixed manufacturing costs and fixed marketing costs are unaffected by the special order.

example illustrates how comparisons based on total amounts (Exhibit 11-4) avoid the misleading implication that the absorption unit-cost data provide (Exhibit 11-3).

The analysis in Exhibit 11-4 assumes that the 5,000-towel special order will use otherwise idle capacity "for at most each of the next three months." Given this assumption, a focus on short-run revenues and short-run costs is appropriate. Chapter 12 explores additional considerations that arise when special order decisions have long-run revenue or cost implications.

Pitfalls in Relevant Cost Analysis

One pitfall in relevant cost analysis is to assume that all variable costs are relevant. In the Fancy Fabrics example, the marketing costs are variable but are not relevant. For the special order decision, Fancy Fabrics incurs no extra marketing costs.

A second pitfall is to assume that all fixed costs are irrelevant. In our example, we assume that the extra 5,000-towel production per month does not affect fixed manufacturing costs. (In terms of the Chapter 3 terminology, we assume that the relevant range is at least from 30,000 to 35,000 towels per month.) In some cases, however, the extra 5,000 towels might increase fixed manufacturing costs. Assume that Fancy Fabrics would have to run three shifts of 16,000 towels per shift to achieve full capacity of 48,000 towels per month. Increasing the monthly production from 30,000 to 35,000 would require a partial third shift because two shifts alone could produce only 32,000 towels. This extra shift would probably increase fixed manufacturing costs.

The best way to avoid these two pitfalls is to focus first and foremost on the relevance concept. Always require each item included in the analysis *both* (1) to be an expected future revenue or cost and (2) to differ among the alternatives.

How Unit Costs Can Mislead

Objective 4

Identify two ways in which per-unit cost data can mislead decision makers

Unit-cost data can mislead decision makers in two major ways:

1. When irrelevant costs are included. Consider the $4.50 per unit allocation of fixed direct labor and manufacturing overhead costs in the special order decision for Fancy Fabrics. This $4.50 per unit cost is irrelevant given the assumptions of our example.
2. When comparisons are made of unit costs not computed using the same volume level. Generally, use total costs rather than unit costs. Then, if desired, the total costs can be unitized. Machinery sales personnel, for example, may brag about the low unit costs of using their new machines. Sometimes they neglect to say that the unit costs are based on outputs far in excess of the current volume of their prospective customer. Unitizing fixed costs over different volume levels can be particularly misleading.

Confusing Terminology

Many different terms are used to describe the costs of specific products and services. Exhibit 11-5 presents several different unit-cost numbers using the data from Column 1 of Exhibit 11-4. **Business function cost** is the sum of all the costs (variable costs and fixed costs) in a particular business function. Manufacturing cost is $12.00 per unit, and marketing cost is $7.00 per unit. For inventory costing purposes, *absorption cost* is often used as a synonym for manufacturing cost.

EXHIBIT 11-5

Variety of Cost Terms*

	Unit Costs			
Variable manufacturing cost	$ 7.50	$ 7.50		$ 7.50
Variable marketing cost	5.00			5.00
Variable product cost	$12.50			
Fixed manufacturing cost		4.50	$4.50	4.50
Manufacturing (absorption) cost		$12.00		
Fixed marketing cost			2.00	2.00
Fixed product cost			$6.50	
Full cost				$19.00

*Fancy Fabrics includes research and development costs and product design costs as components of manufacturing cost. Marketing costs include distribution costs and customer-service costs. Unit-cost data from Exhibit 11-4 are:

	Variable Cost	Fixed Cost	Business Function Cost
Manufacturing cost	$ 7.50	$4.50	$12.00
Marketing cost	5.00	2.00	7.00
Product cost	$12.50	$6.50	$19.00

Full cost refers to the sum of all the costs in all the business functions (research and development, product design, manufacturing, marketing, distribution, and customer service). Full cost in Exhibit 11-5 is $19.00 per unit.

Organizations use terms such as business function cost and full cost differently. In a given situation, be sure to understand their exact meaning.

OTHER ILLUSTRATIONS OF RELEVANCE

Product Emphasis

When a multiproduct plant operates at full capacity, managers must often make decisions regarding which products to emphasize. These decisions frequently have a short-run focus. For example, some food-processing companies continually change their product mix in response to short-run fluctuations in material costs or the selling price of their finished goods.

Analysis of individual product contribution margins provides insight into the product mix that maximizes operating income. Suppose the blades division of Consolidated Engineering has two products: engine blades for cars and engine blades for boats.

	Car Blades	Boat Blades
Selling price per unit	$10	$15
Variable costs per unit	7	9
Contribution margin per unit	$ 3	$ 6
Contribution-margin ratio	30%	40%

At first glance, boat blades look more profitable than do car blades. The product to be emphasized, however, is not necessarily the product with the higher individual unit constribution margin. Rather, managers should aim for the highest contribution margin per unit of the constraining factor—that is, the scarce, limiting, or critical factor. The constraining factor restricts or limits the production or sale of a given product.

Assume the division manager of the blades division has only 1,000 hours of machine capacity available. The constraining factor, then, is machine hours. If the blade division can turn out three car blades per machine hour or one boat blade per machine hour, choosing to emphasize car blades is the correct decision. Producing car blades contributes more margin per machine hour, which is the constraining factor for this example.

	Car Blades	Boat Blades
Units produced per machine hour	3	1
Contribution margin per unit	$ 3	$ 6
Contribution margin per machine hour	$ 9	$ 6
Total contribution margin for 1,000 machine hours	$9,000	$6,000

The constraining factor in this example is machine hours. In a retail department store, it may be cubic feet of display space. The greatest possible contribution margin per unit of the constraining factor yields maximum operating income for a given capacity.

As you can imagine, in many cases a manufacturer or retailer must meet the challenge of trying to maximize total operating income for a variety of products, each with more than one constraining factor. The problem of formulating the most profitable production schedules and the most profitable sales mix is essentially that of maximizing the contribution margin in the face of many constraints. Optimization techniques, such as the linear-programming technique discussed in Chapter 24, help solve these complicated problems.

Make or Buy, and Idle Facilities

Manufacturers often confront the question of whether to make or to buy a product—that is, the question is whether to manufacture their own parts and subassemblies or to buy them from suppliers. Sometimes the manufacture of parts requires special know-how, unusually skilled labor, scarce materials, or the like. These qualitative factors may dictate management's response to the make-or-buy decision. For example, if making the part requires special know-how that the manufacturer lacks, then it must buy the part. Also, many companies make parts only when their facilities are idle and cannot be used to better advantage. Let us assume, however, that quantitative factors predominate in the make-or-buy decision in our example. What quantitative factors are relevant to the make-or-buy decision?

Assume that El Cerrito Company reports the following costs for making Part No. 300:

Cost of Making Part No. 300

	Total Costs for 10,000 Units	Costs per Unit
Direct materials	$ 80,000	$ 8
Direct labor	10,000	1
Variable overhead applied	40,000	4
Fixed overhead applied	50,000	5
Total costs	$180,000	$18

Another manufacturer offers to sell El Cerrito Company the same part for $16 per unit. Should El Cerrito Company make or buy the part?

The unit cost of $18 seemingly indicates that the company should buy. A make-or-buy decision, however, is rarely obvious. The key question for management to ask is, What is the difference in relevant costs between the alternatives? Consider the fixed overhead. The total applied to making the part is $50,000. Assume that $30,000 of this fixed overhead represents costs that will not vary regardless of the decision made. For example, depreciation, property taxes, insurance, and allocated salaries of manufacturing personnel will not change whatever the decision. We see that $30,000, which amounts to $3 per unit for the 10,000 unit volume, is therefore irrelevant and that $20,000 of the fixed costs will be saved if the parts are bought instead of made. In other words, fixed costs that may be avoided in the future are relevant.

For the moment, suppose the capacity now used to make the parts will become idle if the parts are purchased. Exhibit 11-6 presents the relevant-cost computations. Making the part is the preferred alternative. El Cerrito saves $1 per part by making the part rather than buying it from the external manufacturer.

More generally, the choice in our example is not fundamentally whether to make or buy; it is how best to use available facilities. If the component part is bought from the external manufacturer, the released facilities can potentially be used on other jobs. Only if the released facilities are to remain idle are the figures in Exhibit 11-6 valid.

Recognition of how the use of otherwise idle resources can increase profitability occurred at a Chinese machine-repairing factory of the Beijing Engineering Ma-

EXHIBIT 11-6
Relevant Items for Make or Buy Decision

	Total Costs		Per-Unit Costs	
Relevant Items	**Make**	**Buy**	**Make**	**Buy**
Outside purchase of parts		$160,000		$16
Direct materials	$ 80,000		$ 8	
Direct labor	10,000		1	
Variable overhead	40,000		4	
Fixed overhead that can be avoided by not making	20,000		2	
Total relevant costs	$150,000	$160,000	$15	$16
Difference in favor of making	$10,000		$1	

chinery Company. *The China Daily* noted that workers were "busy producing electric plaster-spraying machines" even though the unit cost exceeded the selling price:

> According to the prevailing method of calculating its cost, each sprayer costs 1,230 yuan, but is sold for 985 yuan. Although the factory loses money on products, it continues to put them out to meet market demand. In addition, because workers and machines would otherwise be idle, the production of these sprayers, even at a loss, actually helps cut the factory deficit.
>
> Out of one sprayer's cost of 1,230 yuan, 759 yuan would have to be paid even if nothing had been produced. So because the factory did not have enough work to do, it would have lost even more money, leaving workers and equipment idle, than if it had not produced sprayers.

Relevant Cost Analysis and Contribution-Margin Analysis

Several examples in this chapter illustrate how contribution-margin analysis can assist managers in predicting relevant costs used in decision making. However, do not assume that contribution-margin analysis is synonymous with relevant cost analysis. Relevant cost analysis has a focus on a specific decision with a given set of alternatives. In contrast, contribution-margin analysis can be conducted without a focus on any specific decision. Some companies adopt a contribution-margin format in their internal reporting system.

When using contribution-margin data in relevant cost analysis, examine whether the assumptions underlying the fixed cost and variable cost functions are appropriate for the decision at hand. Chapter 3 noted several key assumptions of contribution-margin analysis:

- a stipulated (typically short) time span,
- a specific relevant range of the cost driver, and
- one cost driver is sufficient to explain variations in the variable cost function.

For decisions such as accepting or rejecting a one-time-only special order or making a short-run change in the product mix, data from a routine contribution-margin income statement can often be helpful. In cases where the decision will affect a relatively long time span or where there are multiple cost drivers for each business-function cost, relevant cost analysis will probably require a detailed special-purpose study.

OPPORTUNITY COSTS, RELEVANCE, AND ACCOUNTING RECORDS

Objective 5

Describe the opportunity cost concept; explain why it is used in decision making

Ideally, a decision maker should be able to make an exhaustive list of alternatives and then compute the expected results under each, giving full consideration to interdependent effects. Practically, the decision maker sifts among the alternatives, quickly discards many as being obviously unattractive, probably overlooks some attractive possibilities, and concentrates on a limited number. ***The idea of an opportunity cost arises because some alternatives are excluded from formal consideration.***

Suppose that at the beginning of the reporting period, a company has 1,000 units in copper inventory, acquired for $110 per unit. The company uses copper as an input for four of its products, A, B, C, and D. The company may decide, however, not to use the 1,000 units on hand as inputs and instead to sell them for salvage. One salvage yard offers $95 per unit, and a second salvage yard offers $90 per unit. Yet another alternative is to throw the units in the city dump. The following table summarizes the "total-alternatives" approach that management could use in analyzing what to do with the copper units (in thousands):

	Choices						
	Use Copper in Product				Salvage		Throw Away
	A	B	C	D	1	2	—
Expected future revenues	$220	$204	$186	$170	$95	$90	—
Expected future costs:							
Labor and overhead	90	70	60	100	—	—	—
Excess of future revenues over future costs	$130	$134	$126	$ 70	$95	$90	—

Using copper as an input to Product B is the preferred alternative.

In many decision-making situations, managers are confronted with too many alternatives to analyze thoroughly. Because managers exclude some alternatives from formal consideration, the idea of an opportunity cost arises. **Opportunity cost** is the maximum available contribution to profit that is forgone (rejected) by using limited resources for a particular purpose.

Suppose in the above analysis that management decides not to analyze the salvage and throw-away alternatives. The alternatives under formal consideration are limited to products A through D. The salvage yard offer of $95 per unit becomes the opportunity cost; that $95 per unit is the maximum available contribution to profit the manufacturer forgoes by limiting the use of the limited resource—the copper—to the specific purpose of manufacturing products.

Under the opportunity-cost approach, the decision maker would organize the analysis of remaining alternatives as follows (in thousands):

	Use Copper in Product			
	A	B	C	D
Expected future revenues	$220	$204	$186	$170
Expected future costs:				
Opportunity cost of salvaging copper	95	95	95	95
Outlay costs: Labor and overhead	90	70	60	100
Total expected future costs	$185	$165	$155	$195
Net advantage of future revenues over future costs	$ 35	$ 39	$ 31	$ (25)

Management would reach the same conclusion: Product B is the preferred alternative.

Note that the *best* of the alternatives excluded from the previous analysis is included as an opportunity cost in the formal comparison. The other excluded alternatives are not in this analysis. The possible uses of the copper in Products A, C, and D also may become excluded alternatives, but they should be excluded only after formal consideration in the decision model.

An opportunity cost is not ordinarily incorporated into formal accounting reports. Such costs represent contributions forgone by rejecting the best alternative excluded from formal consideration; therefore, opportunity costs do not entail cash receipts or disbursements. Accountants usually confine their systematic recording of costs to the outlay costs. An **outlay cost** is a cost that sooner or later requires a cash disbursement. Accountants confine their historical recordkeeping to alternatives selected rather than those rejected, primarily because of the infeasibility of accumulating appropriate data on what might have been.

Carrying Costs of Inventory

The notion of opportunity cost can also be illustrated with a direct-material purchase order decision:

Estimated direct-material requirements for the year	120,000 units
Unit cost, orders below 120,000 units	$ 10.00
Unit cost, orders equal to or greater than 120,000 units; $10.00 minus 2% discount	$ 9.80
Alternatives under consideration:	
A: Buy 120,000 units at start of year	
B: Buy 10,000 units per month	
Average investment in inventory:	
A: (120,000 units × $9.80) ÷ 2	$588,000
B: (10,000 units × $10.00) ÷ 2	$ 50,000

There is a substantial difference in the average investment in inventory under Alternatives A and B: $588,000 − $50,000 = $538,000. At a minimum, $538,000 could be invested in risk-free government bonds and return, say, .06 × $538,000 = $32,280 for the year.

This $32,280 is an opportunity forgone when Alternative A is chosen. This $32,280 is the opportunity cost of the 120,000-unit purchase order. Note that the $32,280, being a *forgone* cost rather than an outlay cost, would not ordinarily be recorded in the accounting system.

IRRELEVANCE OF PAST COSTS

Objective 6

Explain why book value of equipment is irrelevant in equipment-replacement decisions

As defined earlier, a relevant cost is (a) an expected future cost that (b) will differ among alternatives. The illustrations in this chapter have shown that expected future costs that do not differ among alternatives are irrelevant. Now we return to the idea that all past costs are irrelevant.

Consider an example of equipment replacement. The irrelevant cost illustrated here is the **book value** (original cost minus accumulated depreciation) of the existing equipment. Assume that the Toledo Company is considering replacing a metal-cutting machine for aircraft parts with a more technically advanced model. The new machine has an automatic quality-testing capability and is more efficient in using energy. However, the new machine has a shorter life. Toledo company uses the straight-line depreciation method. Revenue from aircraft parts ($1.1 million per year) will be unaffected by the replacement decision. Summary data on the existing machine and the replacement machine follow:

	Existing Machine	Proposed Replacement Machine
Original cost	$1,000,000	$600,000
Useful life in years	5	2
Current age in years	3	0
Useful life remaining in years	2	2
Accumulated depreciation	$ 600,000	$ 0
Book value	$ 400,000	Not acquired yet
Current disposal price (in cash now)	$ 40,000	Not acquired yet
Terminal disposal price (in cash in 2 years)	$ 0	$ 0
Annual cash operating costs (maintenance, energy, repairs, coolants, and so on)	$ 800,000	$460,000

Exhibit 11-7 presents a cost comparison of the two machines. Some managers would not replace the old machine because it would entail recognizing a $360,000 "loss on disposal"; retention would allow spreading the $400,000 book value over the next two years in the form of "depreciation expense" (a term more appealing than "loss on disposal").

We can apply our definition of relevance to four commonly encountered items in equipment-replacement decisions, such as the one facing Toledo Company:

1. *Book value of old equipment.* Irrelevant, because it is a past (historical) cost. All past costs are down the drain. Nothing can change what has already been spent or what has already happened.
2. *Current disposal price of old equipment.* Relevant, because it is an expected future cash inflow that differs between alternatives.
3. *Gain or loss on disposal.* This is the algebraic difference between items 1 and 2. It is a meaningless combination blurring the distinction between the irrelevant book value and the relevant disposal price. It is better to think of each item separately.
4. *Cost of new equipment.* Relevant, because it is an expected future cash outflow that will differ between alternatives.

EXHIBIT 11-7

Cost Comparison—Replacement of Machinery, Including Relevant and Irrelevant Items (in thousands of dollars)

	Two Years Together		
	Keep	**Replace**	**Difference**
Sales	$2,200	$2,200	$—
Costs:			
Operating costs	$1,600	$920	$680
Old machine book value:			
Periodic write-off as depreciation or	400	— }	
Lump-sum write-off	—	400* }	—
Current disposal price of old equipment	—	(40)	40
New machine, written off periodically as depreciation	—	600	(600)
Total costs	$2,000	$1,880	$120
Operating income	$ 200	$ 320	$120

*In a formal income statement, these two items would be combined as "loss on disposal" of $360,000.

EXHIBIT 11-8

Cost Comparison—Replacement of Machinery Relevant Items Only
(in thousands of dollars)

	Two Years Together		
	Keep	Replace	Difference
Operating costs	$1,600	$ 920	$680
Current disposal price of old equipment	—	(40)	40
Depreciation—new equipment	—	600	(600)
Total relevant costs	$1,600	$1,480	$120

Exhibit 11-7 should clarify the above assertions. The difference column in Exhibit 11-7 shows that book value of old equipment is not an element of difference between alternatives and could be completely ignored for decision-making purposes. *No matter what the timing of the charge against revenue, the amount charged is still $400,000 regardless of the available alternative.* Note that the advantage of replacing is $120,000 for the two years together.

In either event, the undepreciated cost will be written off with the same ultimate effect on operating income. The $400,000 enters into the income statement either as a $400,000 offset against the $40,000 proceeds to obtain the $360,000 loss on disposal in the current year or as $200,000 depreciation in each of the next two years. But how it appears is irrelevant to the replacement decision. In contrast, the $600,000 cost of the new equipment is relevant because it can be avoided by deciding not to replace.

Exhibit 11-8 concentrates on relevant items only. Note that the same answer (the $120,000 net difference) will be produced even though the book value is completely omitted from the calculations. The only relevant items are the variable operating costs, the disposal price of the old equipment, and the cost of the new equipment (represented as depreciation in Exhibit 11-8).

Decision makers can vary in their preference between the formats in Exhibits 11-7 and 11-8. Some prefer Exhibit 11-7 because it illustrates why some items are irrelevant to the decision. Other managers prefer Exhibit 11-8 because of its conciseness.

HOW MANAGERS BEHAVE

Consider our equipment-replacement example in the light of the five-step sequence in Exhibit 11-1 (p. 367):

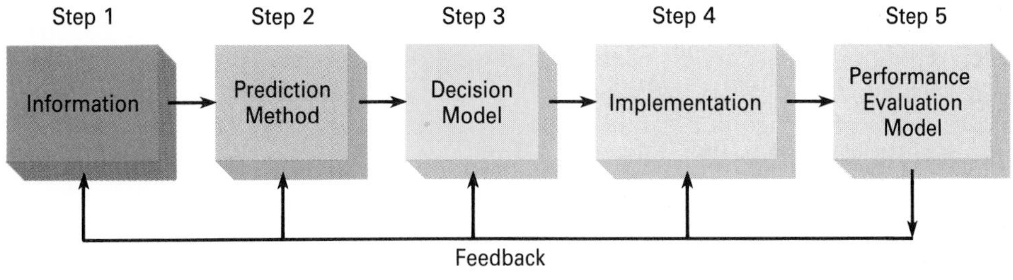

Impact of Reported Loss

If the decision model (step 3) demands choosing the alternative that will minimize total costs over the life span of the equipment, then the analysis in Exhibits 11-7 and 11-8 dictates replacing rather than keeping. In the real world, however, would the manager replace? The answer depends on the manager's perceptions of whether the decision model is consistent with the performance evaluation model (step 5).

Objective 7

Explain how conflicts can arise between the decision model used by a manager and the performance model used to evaluate that manager

Managers tend to favor the alternative that makes their performance look best. They focus on the measures used in the performance-evaluation model. If there is an inconsistency between the decision model and the performance-evaluation model, the latter often prevails in influencing a manager's behavior. Top management may favor following the results of the decision model in Exhibit 11-7, but if the subordinate manager's promotion or bonus hinges on short-run performance, the subordinate's temptation not to replace will be overwhelming. Why? Because the ordinary accrual accounting model for measuring performance will show a first-year operating loss if the old machine is replaced:

	First Year			
	Keep		Replace	
Revenues		$1,100		$1,100
Costs:				
Operating costs	$800		$460	
Depreciation	200		300	
Loss on disposal	—		360	
Total costs		1,000		1,120
Operating income		$ 100		$ (20)

Faced with these data, the subordinate's performance will look much better in the first year by keeping the old machine than by replacing it and showing a loss. Even if top management's goals are long run, the subordinate's concern is more likely with the short run.

Synchronizing the Models

Resolving the conflict between the decision model and the performance-evaluation model is frequently a baffling problem in practice. In theory, resolving the difficulty seems obvious—merely design consistent models. Consider our replacement example. Year-by-year effects on operating income of replacement can be budgeted over the planning horizon of two years. The first year would be expected to be poor, the next year much better. If the subordinate manager knows that top management will evaluate his or her performance over the longer run, the subordinate is more likely to make decisions that have a long-run orientation.

The practical difficulty is that accounting systems rarely track each decision, one at a time. Performance evaluation focuses on responsibility centers for a specific period, not on projects or individual items of equipment for their entire useful lives. Therefore, the impacts of many different decisions are combined in a single performance report. Top management, through the reporting system, is rarely made aware of particular desirable alternatives *not* chosen by subordinates. Chapter 28 further discusses problems of synchronizing decision models and performance-evaluation models.

PROBLEM FOR SELF-STUDY

PROBLEM

Wally Lewis is manager of the engineering development division of Bronco Products, Inc. This division provides contract research for the operating divisions of Bronco Products. Lewis has just received a proposal signed by all ten of his engineers to replace the existing mainframe computing system with 10 workstations. Workstations are minicomputers with extensive memory capacity and work capabilities. Lewis is not enthusiastic about the proposal. The mainframe was purchased only two years ago for $300,000 and has a remaining useful life of three years.

The workstations will cost $13,500 each and have a useful life of three years. Straight-line depreciation is used for all computer equipment at Bronco Products. Given the pace of technology, Lewis believes both the mainframe and the workstations will have zero disposal prices in three years' time. Annual cash-operating costs for the mainframe are $40,000. Annual cash-operating costs for the 10 workstations are $10,000 (10 × $1,000). The current disposal price of the mainframe is $95,000.

Annual revenues of the engineering-development division of $1,000,000 and non-computer-related costs of $880,000 are expected to be unaffected by the computer-equipment replacement decision.

Lewis's annual bonus includes a component based on divisional operating income. He is keen to maintain his track record of three years of increasing divisional operating income. He has a promotion possibility next year that would make him a group vice president of Bronco Products.

Required

1. Summarize the financial data for the two alternatives: (a) keep the mainframe and (b) replace it with workstations.
2. Tabulate a comparison that includes both relevant and irrelevant items. Combine the three years together in a format similar to Exhibit 11-7.
3. Tabulate a comparison of all relevant items for the next three years together.
4. Why might Lewis be reluctant to purchase the 10 workstations?

SOLUTION

1.

	Keep Mainframe	Proposed Workstations
Original cost	$ 300,000	$ 135,000
Useful life in years	5 years	3 years
Current age in years	2 years	0 years
Useful life remaining in years	3 years	3 years
Accumulated depreciation	$ 120,000	Not acquired yet
Current book value	$ 180,000	Not acquired yet
Current disposal price (in cash)	$ 95,000	Not acquired yet
Annual computer-related cash-operating costs	$ 40,000	$ 10,000
Annual revenues	$1,000,000	$1,000,000
Annual non-computer-related operating costs	$ 880,000	$ 880,000

This table includes two types of irrelevant items: (a) past costs (for example, the original cost of the mainframe) and (b) future revenues or costs expected to be the same whether the mainframe is retained or replaced with workstations (for example, the annual revenues).

2.

Relevant and Irrelevant Items	Mainframe	Workstations	Difference
	Three Years Together		
Revenues	$3,000,000	$3,000,000	—
Costs:			
Non-computer-related operating costs	$2,640,000	$2,640,000	—
Computer-related cash-operating costs	120,000	30,000	$90,000
Mainframe book value:			
periodic write-off as depreciation or	180,000	—	
lump-sum write-off	—	180,000	—
Current disposal price	—	(95,000)	95,000
New workstations, written off periodically as depreciation	—	135,000	(135,000)
Total costs	$2,940,000	$2,890,000	$50,000
Operating Income	$ 60,000	$ 110,000	$50,000

This table includes two types of irrelevant items: (a) past costs (for example, the original cost of the mainframe), and (b) future revenues or costs expected to be the same whether the mainframe is retained or replaced with work stations (for example, the annual revenues).

3.

Relevant Items	Mainframe	Workstations	Difference
	Three Years Together		
Computer-related cash-operating costs	$120,000	$ 30,000	$90,000
Current disposal price of mainframe	—	(95,000)	95,000
Depreciation—workstations	—	135,000	(135,000)
Total relevant costs	$120,000	$ 70,000	$50,000

The cost comparison in requirement 3 highlights the three relevant cost items for Bronco Products, Inc. The basic issue is whether an investment of $40,000 now (the $135,000 cost of the workstations minus the $95,000 disposal price of the mainframe) to gain a reduction in annual computer-related operating costs of $30,000 for the next three years is justified.

4. If Bronco Products replaces the mainframe with the workstations, the operating income of Lewis's division will include an $85,000 loss (book value of $180,000 minus $95,000 current disposal price) on the disposal of the mainframe. The net loss for the current year would be $20,000:

Revenues		$1,000,000
Costs:		
Noncomputer-related operating costs	$880,000	
Computer-related cash-operating costs	10,000	
Depreciation—workstations	45,000	
Loss on disposal of mainframe	85,000	
Total costs		1,020,000
Operating income		$ (20,000)

Lewis would probably react negatively to the expected operating loss of $20,000 because it would eliminate the component of his bonus based on operating income. He might also perceive that the $20,000 operating loss would reduce his chances of a promotion to a group vice president.

SUMMARY

Accountants can help managers make better decisions by distinguishing relevant from irrelevant costs. To be relevant to a particular decision, a cost must meet two criteria: (1) it must be an expected *future* cost, and (2) it must be a *difference* between alternatives. The key question is, What difference does it make? If the objective of the decision maker is to maximize long-run profitability, all past (historical) costs are irrelevant to any decision about the future.

The role that past costs play in decision making is an auxiliary one. Past (irrelevant) costs are useful because they provide empirical evidence that often helps sharpen *predictions* of future relevant costs. But the expected future costs are the *only* cost ingredients in any decision model per se.

Use total costs, not unit costs, in cost analysis; unitized fixed costs are often erroneously interpreted as if they behave like unit variable costs. Incremental or differential costs are the differences in total costs between two alternatives.

Cost reports for special decisions may concentrate on relevant items only (Exhibit 11-8), or they may encompass both relevant and irrelevant items (Exhibit 11-7). The best format depends on individual preferences. The shortcut approach concentrates only on the difference column because it summarizes the relevant items.

Top management faces a persistent challenge: making sure that the performance-evaluation model is consistent with the decision model. A common inconsistency is to tell subordinate managers to take a multiple-year view in making decisions but to judge their performance based only on the current year's results.

APPENDIX: COST TERMS USED FOR DIFFERENT PURPOSES

Many companies develop their own distinctive—and sometimes extensive—accounting language. When confronted by unfamiliar terms in practice, you will save much confusion and wasted time if you find out their exact meaning in the given situation. Some terms with which you ought to become familiar follow.

Imputed costs are those costs recognized in particular situations that are not regularly recognized by usual accounting procedures. Imputed costs attempt to make accounting better dovetail with economic reality. Suppose $100,000 is lent to a supplier at simple interest of 2% when loans of comparable risk in the marketplace carry a 16% interest rate. The buyer may extend the 2% loan in consideration for receiving the supplier's products under a purchase contract at prices lower than those prevailing in the market. An accurate accounting requires recognizing that a part of the $100,000 loan is really an additional cost to the buyer for the products purchased during the contract term. This treatment would create an initial loan discount, which would be amortized as interest income over the life of the loan. In this transaction, part of the outlay of $100,000 is imputed as a cost of the product.

Out-of-pocket costs are current or near-future outlays made to meet costs incurred because of a specific decision. For example, accepting a job in order to use facilities that would otherwise be idle would entail costs that would be avoided if the job were not accepted. Examples include purchase of special materials and shipping containers, which are out-of-pocket costs; they would occur only because this specific job had been accepted. Note, however, that depreciation on the machinery and equipment used in this job would be irrelevant; this depreciation would occur whether the job was accepted or not. Depreciation, then, is not an out-of-pocket outlay.

Sunk costs describe past costs that are unavoidable because they cannot be changed no matter what action is taken. If old equipment has a book value of $600,000 and current disposal price of $70,000, what are the sunk costs? The entire $600,000 is sunk because it represents an outlay made in the past that cannot be changed. In our opinion, the term *sunk costs* should not be used at all. Because all past costs are irrelevant, it is fruitless to describe past costs as sunk costs.

Cost management analysts sometimes distinguish between **value-added costs** and **non-value-added costs**. A value-added cost is incurred for an activity that cannot be eliminated without the customer's perceiving a deterioration in the performance, function, or other quality of a product. The cost of a picture tube in a television set is value-added. The costs of those activities that can be eliminated without the customer's perceiving deterioration in the performance, function, or other quality of a product are non-value-added. The costs of handling the materials of a television set through successive stages of an assembly line may be non-value-added. Improvements in plant layout that reduce handling costs may be achieved without affecting the performance, function, or other quality of the television set.

TERMS TO LEARN

This chapter and the Glossary at the end of the book contain definitions of the following important terms:

book value *(p. 377)* business function cost *(372)* decision model *(366)*
differential cost *(368)* full cost *(373)* imputed cost *(383)*
incremental cost *(368)* net relevant cost *(368)* non-value-added cost *(384)*
opportunity cost *(376)* outlay cost *(377)* out-of-pocket cost *(383)*
qualitative factors *(369)* quantitative factors *(369)* relevant costs *(368)*
relevant revenues *(368)* sunk cost *(384)* value-added cost *(384)*

ASSIGNMENT MATERIAL

QUESTIONS

11-1. Define *decision model*.

11-2. Outline the five-step sequence in a decision process.

11-3. Define *relevant cost* as the term is used in this chapter. Why are historical costs irrelevant?

11-4. "All future costs are relevant." Do you agree? Why?

11-5. Distinguish between *quantitative* and *qualitative* factors in decision making.

11-6. Name two ways unit-cost data can mislead a decision maker.

11-7. "Management should always maximize sales of the product with the highest dollar contribution margin." Do you agree? Why?

11-8. Define *opportunity cost*.

11-9. Suppose you are a senior manager with many years of experience. A new employee makes the following comment during a heated disagreement over the purchase of a new machine: "No amount of rhetoric is going to change the fact that all your experience and knowledge is about the past and all your decisions are about the future." How would you respond?

EXERCISES AND PROBLEMS

11-10 Questions on disposal of assets.

1. A company has an inventory of 1,000 assorted missile parts for a line of missiles that has been junked. The inventory cost $80,000. The parts can be either (a) remachined at total additional costs of $30,000 and then sold for a total of $35,000 or (b) scrapped for $2,000. What should be done?
2. A truck, costing $100,000 and uninsured, is wrecked the first day in use. It can be either (a) disposed of for $10,000 cash and replaced with a similar truck costing $102,000 or (b) rebuilt for $85,000 and be brand-new as far as operating characteristics and looks are concerned. What should be done?

11-11 The careening personal computer. (W. A. Paton) An employee in the accounting department of a certain business was moving a personal computer from one room to another. As she came alongside an open stairway, she carelessly slipped and let the computer get away from her. It went careening down the stairs with a great racket and wound up at the bottom in thousands of pieces, completely wrecked. Hearing the crash, the office manager came rushing out and turned rather pale when he saw what had happened. "Someone tell me quickly," he yelled, "if that is one of our fully depreciated items." A check of the accounting records showed that the smashed computer was, indeed, one of those items that had been written off. "Thank God!" said the manager.

Required
Explain and comment on the point of this anecdote.

11-12 Inventory decision, opportunity cost. A manufacturer of lawn mowers predicts that 240,000 spark plugs will have to be purchased during the next year. A supplier quotes a price of $8 per spark plug. The supplier also offers a special discount option: If all 240,000 spark plugs are purchased at the start of the year, a discount of 5% off the $8 price will be available. The manufacturer can invest its cash at 8% per year.

Required
1. What is the opportunity cost of interest forgone from purchasing all 240,000 units at the start of the year instead of in twelve monthly purchases of 20,000 per order?
2. Would this opportunity cost ordinarily be recorded in the accounting system? Why?

11-13 Multiple choice. (CPA) Choose the best answer:

1. Woody Company, which manufactures sneakers, has enough idle capacity available to accept a one-time-only special order of 20,000 pairs of sneakers at $6.00 a pair. The normal selling price is $10.00 a pair. Variable manufacturing costs are $4.50 a pair, and fixed manufacturing costs are $1.50 a pair. Woody will not incur any marketing costs as a result of the special order. What would the effect on operating income be if the special order could be accepted without affecting normal sales? (a) $0, (b) $30,000 increase, (c) $90,000 increase, (d) $120,000 increase.
2. The Reno Company manufactures Part No. 498 for use in its production line. The cost per unit for 20,000 units of Part No. 498 is as follows:

Direct materials	$ 6
Direct labor	30
Variable overhead	12
Fixed overhead applied	16
	$64

   The Tray Company has offered to sell 20,000 units of Part No. 498 to Reno for $60 per unit. Reno will make the decision to buy the part from Tray if there is a savings of $25,000 for Reno. If Reno accepts Tray's offer, $9 per unit of the fixed overhead applied would be totally eliminated. Furthermore, Reno has determined that the released facilities could be used to save relevant costs in the manufacture of Part No. 575. For Reno to have a savings of $25,000, the amount of relevant costs that would have to be saved by using the released facilities in the manufacture of Part No. 575 would have to be (a) $80,000, (b) $85,000, (c) $125,000, (d) $140,000.

11-14 Relevant costs, contribution margin, product emphasis. The Beach Center operates a take-out food center at a popular beach resort. Susan Sexton, owner of the Beach Center, is deciding how much refrigerator space to devote to four different drinks. Pertinent data on these four drinks follow:

	Cola	Lemonade	Natural Punch	Orange-Juice
Selling price per case	$18.00	$19.20	$26.40	$38.40
Variable cost per case	$13.50	$15.20	$20.10	$30.20
Cases sold per foot of shelf space per day	25	24	4	5

Sexton has a maximum front shelf space of 12 feet to devote to the four drinks. She wants a minimum of 1 foot and a maximum of 6 feet of front shelf for each drink.

Required
1. What is the contribution margin per case of each type of drink?
2. A co-worker of Sexton's recommends that she maximize the shelf space devoted to those drinks with the highest contribution margin per case. Evaluate this recommendation.
3. What shelf-space allocation for the four drinks would you recommend for the Beach Center?

11-15 Selection of most profitable product. The Reduction Co. produces two basic types of weight-control equipment, G and H. Pertinent data follow:

	Per Unit	
	G	**H**
Sales price	$100.00	$70.00
Costs:		
Direct materials	$ 28.00	$13.00
Direct labor	15.00	25.00
Variable factory overhead*	25.00	12.50
Fixed factory overhead*	10.00	5.00
Marketing costs (all variable)	14.00	10.00
	$ 92.00	$65.50
Operating income	$ 8.00	$ 4.50

*Applied on the basis of machine hours.

The weight-control craze is such that enough of either G or H can be sold to keep the plant operating at full capacity. Both products are processed through the same production departments.

Required

Which product should be produced? If both should be produced, indicate the proportions of each. Briefly explain your answer.

11-16 Relevance of equipment costs. The Auto Wash Company has just today installed a special machine for polishing cars at one of its several outlets. It is the first day of the company's fiscal year. The machine cost $20,000. Its annual operating costs total $15,000, exclusive of depreciation. The machine will have a four-year useful life and zero terminal disposal price.

After the machine has been used one day, a machine salesperson offers a different machine that promises to do the same job at a yearly operating cost of $9,000, exclusive of depreciation. The new machine will cost $24,000 cash, installed. The "old" machine is unique and can be sold outright for only $10,000, minus $2,000 removal cost. The new machine, like the old one, will have a four-year useful life and zero terminal disposal price.

Sales, all in cash, will be $150,000 annually and other cash costs will be $110,000 annually, regardless of this decision.

For simplicity, assume there is a world of no income taxes and no interest.

Required

1. Prepare a statement of cash receipts and disbursements for each of the four years under both alternatives. What is the cumulative difference in cash for the four years taken together?
2. Prepare income statements for each of the four years under both alternatives. Assume straight-line depreciation. What is the cumulative difference in operating income for the four years taken together?
3. What are the irrelevant items in your presentations in requirements 1 and 2? Why are they irrelevant?
4. Suppose the cost of the "old" machine was $1 million rather than $20,000. Nevertheless, the old machine can be sold outright for only $10,000, minus $2,000 removal cost. Would the net differences in requirements 1 and 2 change? Explain.
5. "To avoid a loss, we should keep the old equipment." What is the role of book value in decisions about replacement of equipment?

11-17 Opportunity cost. (H. Schaefer) Wolverine Corp. is working at full production capacity producing 10,000 units of a unique product, Rosebo. Standard manufacturing costs per unit for Rosebo follow:

Direct materials	$ 2
Direct labor	3
Manufacturing overhead	5
	$10

The unit manufacturing overhead cost is based on a variable cost per unit of $2 and fixed costs of $30,000 (at full capacity of 10,000 units). The nonmanufacturing costs, all variable, are $4 per unit and the sales price is $20 per unit.

A customer, the Miami Co., has asked Wolverine to produce 2,000 units of a modification of Rosebo to be called Orangebo. Orangebo would require the same manufacturing processes as Rosebo, and the Miami Co. has offered to share equally the nonmanufacturing costs with Wolverine. Orangebo will sell at $15 per unit.

Required

1. What is the opportunity cost to Wolverine of producing the 2,000 units of Orangebo? (Assume no overtime is worked.)
2. The Buckeye Corp. has offered to produce 2,000 units of Rosebo for Wolverine so that Wolverine may accept the Orangebo offer. Buckeye would charge Wolverine $14 per unit for the Rosebo. Should Wolverine accept the Buckeye offer? (Support with specific analysis.)
3. Suppose Wolverine had been working at less than full capacity producing 8,000 units of Rosebo at the time the Orangebo offer was made. What is the *minimum* price Wolverine should accept for Orangebo under these conditions? (Ignore the $15 price above.)

11-18 Own or rent, opportunity cost. The U.S. government is running large deficits. The State Department has instructed overseas embassies to conduct "serious belt-tightening" to cut costs. A budget analyst in Washington collects the following information on three properties owned by the U.S. government (in millions):

	In London	In Paris	In Tokyo
Initial purchase cost	$12.000	$3.000	$ 2.000
Selling price now	15.000	9.000	18.000
Outstanding debt on property	10.000	0.200	1.200
Annual interest payment on the outstanding debt on the property	1.400	0.025	0.100
Annual rent of comparable building	1.000	0.720	1.800

The analyst recommends sale of the London property, because this would save the U.S. government $0.400 million ($1.400 million annual interest payment minus $1.000 annual rent) each year. In contrast, he argues that there is no financial gain from selling the Paris or Tokyo properties; the annual rent on a comparable property exceeds the annual interest payment on the outstanding debt on the Paris and Tokyo properties. Currently the State Department is investing money at 12% per year.

Required

1. Critique the reasoning of the budget analyst.
2. What factors would you recommend that the State Department consider in deciding whether to sell the London, Paris, or Tokyo properties?

11-19 Make or buy, unknown level of volume. (A. Atkinson) Oxford Engineering manufactures small engines. The engines are sold to manufacturers who install them in products such as lawn mowers. The company currently manufactures all the parts used in these engines but is considering a proposal from an external supplier who wishes to supply the starter assembly used in these engines.

The starter assembly is currently manufactured in Division 3 of Oxford Engineering. The costs relating to Division 3 for the last twelve months were as follows:

Direct materials	$200,000
Direct labor	150,000
Factory overhead	400,000
Total	$750,000

Over the last twelve-month period, Division 3 manufactured 150,000 engines; the average cost for the starter assembly is computed as $5 ($750,000 ÷ 150,000).

Further analysis of factory overhead revealed the following information. Of the total factory overhead reported, only 25% is considered variable. Of the fixed costs, $150,000 is an allocation of general factory overhead that would remain unchanged if production of the starter assembly is abandoned. A further $100,000 of the fixed overhead is avoidable if self-manufacture of the starter assembly is discontinued. The balance of the current fixed overhead, $50,000, is the division manager's salary. If self-manufacture of the starter assembly is discontinued, the manager of Division 3 will be transferred to Division 2 at the same salary. This will allow the company to save the $40,000 salary that would otherwise be paid to attract an outsider to this position.

Required

1. Tidnish Electronics, a reliable supplier, has offered to supply starter assembly units at $4 per unit. Since this price is less than the current average cost of $5 per unit, the vice president of manufacturing is eager to accept this offer. Should the outside offer be accepted? (*Hint:* Manufacturing volume in the coming year may be different from volume in the last year.)
2. How, if at all, would your response to requirement 1 change if the company could use the vacated factory space for storage and, in so doing, avoid $50,000 of outside storage charges currently incurred? Why is this information relevant or irrelevant?

11-20 Relevant costs, subcontracting of distribution. Geyser Springs bottles and markets sparkling mineral water. It recently upgraded its distribution division through the purchase of a fleet of eight delivery vehicles. The distribution division is a cost center of Geyser Springs. Budgeted costs of the division for the coming year (January 1, 19_2, to December 31, 19_2) follow:

Cash-operating costs	$700,000
Depreciation on fleet of vehicles (8 × $20,000)	160,000
Corporate costs allocated to distribution division	100,000
Total	$960,000

Each delivery vehicle cost $80,000 one year ago (January 1, 19_1) and had an estimated economic life of four years; depreciation is $20,000 per year on a straight-line basis. The estimated terminal disposal price of each vehicle at the end of four years (December 31, 19_4) is $10,000.

The CanAm Express approaches Geyser Springs with an offer to be the sole distributor of all its mineral-water products. CanAm distributes a broad range of food and beverage products using a fleet of over 100 vehicles. It has unused capacity on a sufficient number of vehicles to be able to make daily deliveries of all Geyser Spring products. CanAm offers Geyser Springs a three-year distribution contract for $650,000 each year. The contract will start on January 1, 19_2.

If Geyser Springs signs the $650,000-per-year contract with CanAm, it will close its own distribution division and sell the delivery vehicles. Each of the eight delivery vehicles has a current (January 1, 19_2) disposal price of $25,000. Geyser Springs will avoid all $700,000 of the cash-operating costs of the distribution facility. Of the $100,000 of corporate costs allocated to the distribution division, $25,000 will be avoided if it closes; the remaining $75,000 will remain as a cost of Geyser Springs.

Geyser Springs recently had its stock listed on a national stock exchange. Security analysts have recommended purchase of the stock, in part due to its record of earnings stability. Management of Geyser Springs participates in an annual profit-sharing scheme in which 20% of the reported net income is distributed among the top senior executives. Geyser Springs had a net income of $210,000 in 19_1. Security analysts are forecasting net income of $220,000 for 19_2. This forecast assumes that Geyser Springs continues operation of its own distribution division.

Required

1. Tabulate a comparison of all relevant items for the next three years for the two alternatives—use its own distribution division or use CanAm.

2. Why might Geyser Springs be reluctant to subcontract the distribution of its mineral water to CanAm Express?

11-21 Cost comparisons of airlines, unit costs vs. contribution approach. Charles Smith is chief operating officer for Oceanic United, an international airline operating two roundtrip flights between San Francisco and Nandi, Fiji, each week. Oceanic's capacity level is three roundtrip flights each week. The only other competitor on this 5,600-mile route is South Pacific Express; it operates three roundtrip flights between San Francisco and Nandi each week. Both Oceanic and South Pacific offer only one class of seats (tourist class) on their planes. Assume that no one-way tickets are sold.

Several months ago, Smith hired a consulting firm to compare the costs of Oceanic United and South Pacific. The consultant's report includes the following summary cost comparison to support its "punch line" that Oceanic United was not competitive on cost grounds:

	Oceanic United	South Pacific Express
Total annual costs:		
Variable costs	$ 6,240,000	$17,160,000
Fixed costs	12,480,000	9,360,000
	$18,720,000	$26,520,000
Divide by total passengers carried	÷ 20,800	÷ 34,320
Unit total cost per passenger	$ 900	$ 773

The data analysis underlying these numbers follows:

	Oceanic United	South Pacific Express
Seating capacity per plane	360	310
Roundtrip flights per week	2	3
Roundtrip flights per year	104	156
Average passengers per roundtrip	200	220
Average price per roundtrip	$ 1,000	$ 950
Variable cost per roundtrip	$ 60,000	$ 110,000
Annual fixed costs	$12,480,000	$9,360,000

The variable costs associated with each passenger on a given flight are considered to be close to zero.

Smith is dismayed by the consultant's report and finds it less than informative. He is reminded of the comment that "a consultant is someone who will take your watch off your wrist and tell you what time it is."

Required

1. Would you choose a volume measure different from the number of passengers? Why?
2. Critically evaluate the cost comparison analysis presented by the consultant.
3. Present a more informative financial comparison between Oceanic United and South Pacific Express based on your preferred volume measure. Include in your analysis: differences in the cost structures of the two airlines, volume levels at which total costs (for the two airlines) are equal, contribution margins, and implications of your analysis for pricing decisions.

11-22 Operating decisions for an airline, special order decision, and pricing decision. (Continuation of question 11-21) Assume the same facts as in question 11-21.

A. Oceanic United receives a proposal from a charter group wishing to sell package tours to Fiji. The proposal is to charter a plane once every two weeks to make a roundtrip flight, that is, the charter group will use a separate flight. The tours to Fiji last two weeks. The San Francisco-to-Nandi leg carries a group starting their vacation. The Nandi-to-San Francisco leg carries a group finishing their vacation. The charter group offers to pay Oceanic $700 per roundtrip passenger, with a minimum guarantee of 120 passengers per roundtrip flight. Smith is reluctant to accept the proposal: *$700 is below the $900 unit cost per passenger reported by the consultant.* In your answer to the following questions, assume that there will be no effect on current demand, and ignore the intermediate period of the first two weeks.

Required

1. How much will Oceanic United be better off from each roundtrip charter flight if only 120 passengers travel on a charter roundtrip?
2. What advice would you give Smith concerning the charter group proposal? Include in your answer comment(s) on Smith's concern that the $700 price is below the $900 unit total cost per passenger reported by the consultant.

B. Smith hears that the Fijian government has recently negotiated landing rights in San Francisco for Air Fiji; only one roundtrip flight per week is permitted under the agreement. Air Fiji is managed by Asian Airlines and has a very low cost structure. Smith estimates that its variable cost per roundtrip flight will be $50,000 and that its annual fixed costs will be $3 million a year. Air Fiji plans to offer the one roundtrip flight a week between San Francisco and Nandi and has a seating capacity of 350 per plane. Smith anticipates a price-cutting war on this route and seeks your advice.

Required

1. How low can Oceanic price a roundtrip ticket and still break even, assuming that the demand it faces stays at its existing volume level of 104 roundtrip flights and an average of 200 passengers on each flight?
2. Assume that South Pacific Express does not start a price war. Would you advise Oceanic United to start one? Why?

11-23 Considering three alternatives. (CMA) Auer Company had received an order for a piece of special machinery from Jay Company. Just as Auer Company completed the machine, Jay Company declared bankruptcy, defaulted on the order, and forfeited the 10% deposit paid on the selling price of $72,500.

Auer's manufacturing manager identified the costs already incurred in the production of the special machinery for Jay as follows:

Direct materials used		$16,600
Direct labor incurred		21,400
Overhead applied:		
Manufacturing		
Variable	$10,700	
Fixed	5,350	16,050
Fixed marketing and administrative		5,405
Total cost		$59,455

Another company, Kaytell Corp., would be interested in buying the special machinery if it is reworked to Kaytell's specifications. Auer offered to sell the reworked special machinery to Kaytell as a special order for a net price of $68,400. Kaytell has agreed to pay the net price when it takes delivery in two months. The additional identifiable costs to rework the machinery to the specifications of Kaytell follow:

Direct materials	$ 6,200
Direct labor	4,200
	$10,400

A second alternative available to Auer is to convert the special machinery to the standard model. The standard model lists for $62,500. The additional identifiable costs to convert the special machinery to the standard model are:

Direct materials	$ 2,850
Direct labor	3,300
	$ 6,150

A third alternative for the Auer Company is to sell, as a special order, the machine as is (that is, without modification) for a net price of $52,000. However, the potential buyer of the unmodified machine does not want it for 60 days. The buyer offers a $7,000 down payment with final payment upon delivery.

The following additional information is available regarding Auer's operations:

- Sales commission rate on sales of standard models is 2%, and the sales commission rate on special orders is 3%. All sales commissions are calculated on net sales price (that is, list price minus cash discount, if any).
- Normal credit terms for sales of standard models are 2/10, n/30 (2/10 means a discount of 2% is given if payment is made within 10 days; n/30 means full amount is due within 30 days). Customers take the discounts except in rare instances. Credit terms for special orders are negotiated with the customer.
- The application rates for manufacturing overhead and the fixed marketing and administrative costs are as follows:

Manufacturing:	
Variable	50% of direct labor cost
Fixed	25% of direct labor cost
Marketing and administrative:	
Fixed	10% of the total of direct material, direct labor, and manufacturing overhead costs

- Normal time required for rework is one month.
- A surcharge of 5% of the sales price is placed on all customer requests for minor modifications of standard models.
- Auer normally sells a sufficient number of standard models for the company to operate at a volume in excess of the breakeven point.

Auer does not consider the time value of money in analyses of special orders and projects whenever the time period is less than one year, because the effect is not significant.

Required

1. Determine the dollar contribution that each of the three alternatives will add to the Auer Company's operating income.
2. If Kaytell makes Auer a counteroffer, what is the lowest price Auer Co. should accept for the reworked machinery from Kaytell? Explain your answer.
3. Discuss the influence that fixed-manufacturing-overhead cost should have on the sales prices Auer Company quotes for special orders when (a) the firm is operating at or below the breakeven point; (b) the firm's special orders constitute efficient utilization of unused capacity above the breakeven volume.

11-24 Which school(s) in a school district to close, relevant cost analysis, opportunity costs. Naomi Lance, the superintendent of the Palo Alto School District, faces a difficult decision. Declining student enrollment in the district appears to necessitate closing between one and three elementary or junior-high (middle) schools over the next four years. At present there are seven elementary schools (students of age 6–12 years) and three junior-high schools (students of age 13–14 years). A school in one category can be converted to a school in the other category, although typically at a considerable one-time conversion cost. Eight of the ten schools are on land owned by the school district. Stanford elementary school and Leland middle school are on land leased from Stanford University. The school district pays Stanford University an annual $10,000 lease payment for Stanford elementary school and an annual $15,000 lease payment for Leland middle school.

Lance predicts that over the next four years demand for student places will decline from the current level of 5,000 (also the current capacity) to 4,100. Much of the decline in enrollment will be felt first in the elementary schools. The local press has mentioned the following schools as likely to be closed by the school district:

1. *Addison* elementary school (200-student capacity). Located in the downtown shopping area, this is the smallest of the schools in the district. It was built at a cost of $1.2 million twenty years ago. A shopping mall developer has offered the school district $25 million for the property.
2. *Duveneck* elementary school (500-student capacity). The oldest of the schools in the district, it was built at a cost of $1 million thirty years ago. It is situated in an exclusive residential part of the district. A local developer has offered $8 million for the property.

Two years ago the school district finished a $3 million renovation of Duveneck that made it a showpiece of all elementary schools in California.

3. *Stanford* elementary school (600-student capacity). This school, built at a cost of $5 million ten years ago, is on land leased from Stanford University. There are eighty-nine years remaining on the lease. The lease requires a $10,000 lease payment each year. The land will immediately revert to Stanford University if the school is closed before the lease term expires.
4. *Jane Lathrop* middle school (900-student capacity). This school, built at a cost of $7 million eight years ago, was formerly an elementary school. Last year it was converted to a junior-high school at a cost of $1.5 million. A land developer has offered $9 million for the property. It is especially attractive to the developer because of its open space. It is the only school in the district with six playing fields for athletics.

OTHER INFORMATION

a. If Stanford elementary is not closed, it will require a $2 million capital investment this year to upgrade the classrooms, parking area, and playing fields.
b. If Jane Lathrop is closed, it will cost $1.8 million to convert an existing elementary school to a junior-high school. Three junior-high schools are necessary in the district.
c. Palo Alto School District has a central headquarters' administrative staff of twelve people. The superintendent is paid $80,000 a year (including benefits). The senior assistant is paid $50,000 a year. Each school has one assistant at the central administrative headquarters paid $35,000 a year.

Required

1. Lance believes that the following alternatives warrant further analysis:
   a. Close no schools
   b. Close Addison and Duveneck
   c. Close Addison and Stanford
   d. Close Jane Lathrop

   Prepare a quantitative comparison of these four alternatives. Present your analysis, focusing on relevant costs and other relevant cash inflows and outflows. Distinguish between recurring items and one-time-only items.
2. Lance receives a letter from a concerned parent who lives close to Jane Lathrop. The letter argues strongly for closing both Addison and Stanford elementary. It states that "Stanford is the only school in current need of additional capital investment" and notes that "the school district could avoid a $2 million outlay by immediately closing Stanford elementary school." The letter also points out that "the school district will have to pay Stanford University $890,000 over the next eighty-nine years if it keeps Stanford elementary. In contrast, no cash outflow arises if the school district retains Jane Lathrop." Do you agree with this parent's argument?
3. What cost figures mentioned in the question are irrelevant to the school-closing decision? Explain your reasoning.
4. Other than the costs outlined in requirement 1, what other factors would you recommend that Lance consider in her school-closing decision?

11-25 Multiple choice; comprehensive problem on relevant costs. The following are the Class Company's *unit* costs of manufacturing and marketing a given item at a level of 20,000 units per month:

Manufacturing costs:	
Direct materials	$1.00
Direct labor	1.20
Variable indirect cost	.80
Fixed indirect cost	.50
Marketing costs:	
Variable	1.50
Fixed	.90

The following situations refer only to the data given above; there is *no connection* between the situations. Unless stated otherwise, assume a regular selling price of $6 per unit.

Required

Choose the answer corresponding to the most nearly acceptable or correct answer in each of the seven items. Support each answer with summarized computations.

1. In presenting an inventory of 10,000 items on the balance sheet, the unit cost used is (a) \$3.00, (b) \$3.50, (c) \$5.00, (d) \$2.20, (e) \$5.90.
2. This product is usually sold at the rate of 240,000 units per year (an average of 20,000 per month). At a sales price of \$6.00 per unit, this yields total sales of \$1,440,000, total costs of \$1,416,000, and an operating income of \$24,000, or 10¢ per unit. It is estimated by market research that volume could be increased by 10% if prices were cut to \$5.80. Assuming the implied cost-behavior patterns to be correct, this action, if taken, would (a) decrease operating income by a net of \$7,200; (b) decrease operating income by 20¢ per unit, \$48,000, but increase operating income by 10% of sales, \$144,000 for a net increase of \$96,000; (c) decrease unit fixed costs by 10%, or 14¢, per unit and thus decrease operating income by 20¢ − 14¢, or 6¢ per unit; (d) increase sales volume to 264,000 units, which at the \$5.80 price would give total sales of \$1,531,200; costs of \$5.90 per unit for 264,000 units would be \$1,557,600, and a loss of \$26,400 would result; (e) none of these.
3. A cost contract with the government (for 5,000 units of product) calls for the reimbursement of all costs of production plus a fixed fee of \$1,000. This production is part of the regular 20,000 units of production per month. The delivery of these 5,000 units of product increases operating income from what they would have been, were these units not sold, by (a) \$1,000, (b) \$2,500, (c) \$3,500, (d) \$300, (e) none of these.
4. Assume the same data as in 3 above except that the 5,000 units will displace 5,000 other units from production. The latter 5,000 units would have been sold through regular channels for \$30,000 had they been made. The delivery to the government increases (or decreases) operating income from what they would have been, were the other 5,000 units sold, by (a) \$4,000 decrease, (b) \$3,000 increase, (c) \$6,500 decrease, (d) \$500 increase, (e) none of these.
5. The company desires to enter a foreign market in which price competition is keen. The company seeks a one-time-only order for 10,000 units on a minimum unit-price basis. It expects that shipping costs for this order will amount to only 75¢ per unit, but the fixed costs of obtaining the contract will be \$4,000. Domestic business will be unaffected. The breakeven price is (a) \$3.50, (b) \$4.15, (c) \$4.25, (d) \$3.00, (e) \$5.00.
6. The company has an inventory of 1,000 units of this item left over from last year's model. These must be sold through regular channels at reduced prices. The inventory will be valueless unless sold this way. The unit cost that is relevant for establishing the minimum selling price would be (a) \$4.50, (b) \$4.00, (c) \$3.00, (d) \$5.90, (e) \$1.50.
7. A proposal is received from an outside supplier who will make and ship this item directly to the Class Company's customers as sales orders are forwarded from Class's sales staff. Class's fixed marketing costs will be unaffected, but its variable marketing costs will be slashed 20%. Class's plant will be idle, but its fixed manufacturing overhead would continue at 50% of present levels. How much per unit would the company be able to pay the supplier without decreasing operating income? (a) \$4.75, (b) \$3.95, (c) \$2.95, (d) \$5.35, (e) none of these.

11-26 Make or buy. (Continuation of 11-25) Assume the same facts as in requirement 7 of Problem 11-25 except that if the supplier's offer is accepted, the present plant facilities will be used to make a product whose unit costs will be:

Variable manufacturing costs	\$5.00
Fixed manufacturing costs	1.00
Variable marketing costs	2.00
Fixed marketing costs (new increment)	.50

Total fixed manufacturing overhead will be unchanged, and fixed marketing costs will increase as indicated. The new product will sell for \$9. The minimum desired operating income on the two products taken together is \$50,000 per year.

Required

What is the maximum purchase cost per unit that the Class Company should be willing to pay for subcontracting the old production?

Pricing Decisions, Product Profitability Decisions, and Cost Information

Major influences on pricing

Product-cost categories

Pricing for the short run

Pricing for the long run

Life-cycle product budgeting and costing

Undercosting and overcosting products

Target cost as a pricing guide

Effects of antitrust laws on pricing

CHAPTER 12

Department stores consider customers, competitors, and costs in their decisions on which items to mark down for special sales.

Learning Objectives

When you have finished studying this chapter, you should be able to

1. Discuss the three major influences on pricing decisions
2. Describe six important categories of costs relevant in pricing decisions
3. Distinguish between pricing decisions for the short-run and the long-run
4. Explain how life-cycle product budgeting and costing assist in pricing and product-emphasis decisions
5. Describe how product undercosting/overcosting arises and how it can be reduced
6. Describe the target-costing approach

Managers frequently face decisions on the pricing and the relative profitability of their products. This chapter illustrates the pivotal role that cost data can play in making these decisions.[1] We emphasize how an understanding of cost-behavior patterns and cost drivers can lead to better decisions. This chapter applies the relevant-revenue and relevant-cost framework outlined in Chapter 11.

Pricing decisions differ greatly in both their time horizon and their decision context. Consequently, no single way of computing a product-cost figure is universally relevant for all pricing decisions. This chapter illustrates how managers may find different measures of product cost helpful in making various pricing decisions.

MAJOR INFLUENCES ON PRICING

Objective 1

Discuss the three major influences on pricing decisions

There are three major influences on pricing decisions: customers, competitors, and costs.

Customers. The manager must always examine pricing problems through the eyes of customers. A price increase may cause a customer to reject the company's product and choose one from a competitor. Alternatively, a price increase may drive a customer to choose a substitute product that fits desired specifications in a more cost-effective way. For example, increases in the price of glass drove customers for containers to substitute aluminum cans for bottles.

Competitors. Competitors' reactions influence pricing decisions. At one extreme, a rival's intense drive may force a business to lower its prices to be competitive. At the other extreme, a business without a rival in a given situation can set higher prices. A business with knowledge of its rivals' technology, plant size, and operating policies is better able to estimate its rivals' costs, which is valuable information in setting competitive prices.

Competition spans international borders. For example, overcapacity of firms in their domestic markets can lead to their taking an aggressive pricing policy in their export markets. Increasingly, managers should take a global viewpoint and consider both international and domestic rivals in making pricing decisions. The concept of *target costs,* described later in this chapter, plays a key role in competitor analysis.

[1]For brevity purposes, the term *pricing decision* is used in this chapter to encompass decisions about the relative profitability of products.

Costs. The study of cost-behavior patterns yields insight into the income that results from different combinations of price and volume sold for a particular product. A product consistently priced below its costs can drain sizable amounts of resources from an organization.

Surveys of how executives make decisions reveal that companies weigh customers, competitors, and costs differently. Firms selling commodity-type products in highly competitive markets must accept the price determined by market forces. For example, the gold market has many competitors, each offering the identical product at the same price. The market sets the price, but cost data can assist managers in the gold market in their deciding, say, on the output level that best meets a company's particular objective.

Where managers have some discretion in setting prices, numerous cost-based formulas are available for guiding pricing decisions. These range from variable-cost formulas based on only manufacturing costs to full-cost formulas based on all major cost categories.

PRODUCT-COST CATEGORIES

Objective 2

Describe six important categories of costs relevant in pricing decisions

Exhibit 12-1 presents six basic categories of business-function costs that can be included in the cost buildup of a product or service. For each cost category, there can be both direct product costs and indirect product costs. A *direct product cost* is a cost that can be identified specifically with or traced to a given product in an economically feasible way. An *indirect product cost* is a cost that cannot be identified specifically with or traced to a given product in an economically feasible way. Developing product costs for pricing decisions requires the following four steps:

- Step 1: Decide which business-function cost categories to include in the product-cost buildup. Managers should include only those categories with relevant costs. As described in Chapter 11, relevant costs are those expected future costs that will differ among the decision alternatives under consideration. Not every cost category necessarily includes relevant costs for a specific decision; a manager might use all six categories in Exhibit 12-1 or only a subset of them.
- Step 2: Compute the direct product costs of each of the relevant-cost categories from step 1. Firms can increase the accuracy of product-cost computations by increasing the percentage of direct product costs to total product costs.
- Step 3: Compute the indirect product costs of each of the relevant-cost categories from step 1. Firms can increase the accuracy of product-cost computations by better identifying the drivers of indirect product costs.
- Step 4: Compute the total product costs by summing the direct product costs from step 2 and the indirect product costs from step 3.

EXHIBIT 12-1
Business Function Cost Categories for Product-Cost Buildup

Research and Development	Product Design	Manufacturing	Marketing	Distribution	Customer Service

The Importance of the Time Horizon

Objective 3

Distinguish between pricing decisions for the short-run and the long-run

When computing the relevant costs in a pricing decision (steps 2 and 3), the time horizon of the decision is critical. The two ends of the time-horizon spectrum are:

Short-Run Pricing Decisions	Long-Run Pricing Decisions

Short-run decisions include (1) pricing for a one-time-only special order with no long-term implications and (2) adjusting product mix and volume in a competitive market. The time horizon used to compute those costs that differ among the alternatives for short-run decisions is typically six months or less. Long-run decisions include pricing a main product in a major market where price setting has considerable leeway. A time horizon of one year or longer is often used when computing relevant costs for these long-run decisions. This chapter presents examples of both short-run and long-run pricing decisions.

Short-run pricing decisions and long-run pricing decisions are best viewed as ends of the spectrum. Many pricing decisions will not neatly fall into one or the other of the extreme ends of this spectrum.

PRICING FOR THE SHORT RUN

This section illustrates pricing decisions where the time horizon for computing relevant costs is relatively short. Consider a one-time-only special order for the next four months. Acceptance or rejection of the order will not affect the revenues (volume sold or the price per unit) from existing sales outlets. There are no long-term sales likely from the customer.

EXAMPLE:
American Brewery Company (ABC) operates a brewery with a monthly capacity of 1 million barrels of a nonalcoholic beer product (Champion) that has gained significant market share in recent years. Current production and sales are 600,000 barrels per month. The selling price per barrel is $90. The variable cost per barrel and the fixed cost per barrel (based on a production volume of 600,000 barrels per month) follow:

	Variable Cost Per Barrel	Fixed Cost Per Barrel	Variable and Fixed Cost Per Barrel
Manufacturing costs:			
Direct materials (barley, hops, etc.)	$ 5	$ 0	$ 5
Packaging (bottles, cans, etc.)	22	0	22
Direct labor	2	4	6
Overhead	6	10	16
Marketing costs	24	13	37
Product cost	$59	$27	$86

Canadian Brewery (CB) is constructing a new brewery in Toronto. Brewery operations will not begin for four months. Management, however, wants to start marketing immediately. It decides to buy from another brewery 250,000 barrels of nonalcoholic beer for each of the next four months to sell in Canada. CB has asked ABC and two other brewing companies to bid on this special order. From a production-cost viewpoint, the beer to be brewed for CB is identical to that currently brewed by ABC.

If ABC brews the extra 250,000 barrels, an additional $300,000 in manufacturing costs (material procurement costs of $100,000 and process changeover costs of $200,000) will be required each month. This additional $300,000 is not driven by the volume of the special order; it is a monthly setup cost. No additional costs will be required for research and development, product design, marketing, distribution, or customer service. The 250,000 barrels will be marketed by CB in Canada where ABC does not sell its Champion brand or any other nonalcoholic beers. CB will assume all costs associated with marketing, distribution, and customer service.

ABC's indirect manufacturing costs include variable manufacturing overhead and fixed manufacturing overhead. These costs are allocated on a per-barrel basis. Fixed manufacturing overhead and fixed direct labor costs ($2,400,000 per month) would be unaffected with the additional manufacturing of 250,000 barrels per month.

A vice president of CB notifies each potential bidder that a bid above $45 per barrel will probably be "noncompetitive." ABC knows that one of its competitors, with a highly efficient plant, has sizable idle capacity and will definitely bid for the contract. What price should ABC bid for the 250,000 barrel contract?

The relevant costs for the price-bidding decision can be computed using the four steps outlined earlier in this chapter.

- Step 1: Decide which business-function cost categories to include in the product-cost buildup. Only manufacturing costs are relevant here. Costs associated with research and development, product design, marketing, distribution, and customer service will not change with the manufacturing of the additional 250,000 barrels for CB.
- Step 2: Compute the direct product costs of each of the relevant-cost categories from step 1. The relevant direct-manufacturing costs comprise:
  - Variable manufacturing costs of direct materials (250,000 × $5), packaging (250,000 × $22), and direct labor (250,000 × $2)
  - An additional $300,000 in manufacturing costs (material procurement costs of $100,000 and process changeover costs of $200,000)
- Step 3: Compute the indirect product costs of each of the relevant-cost categories from step 1. The relevant indirect-manufacturing costs are the variable manufacturing-overhead costs (250,000 × $6).
- Step 4: Compute the total product costs by summing the direct product costs from step 2 and the indirect product costs from step 3. Exhibit 12-2 presents the total relevant costs of $9,050,000 per month (or $36.20 per barrel) for the 250,000 special order. Any bid above $9,050,000 per month will increase the profitability of ABC. For example, a successful bid of $40 per barrel, well under CB's ceiling of $45 per barrel, will add $950,000 to the monthly operating income of ABC: 250,000 × ($40 − $36.20) = $950,000.

Cost data, key information in ABC's decision on the price to bid, are not the only inputs. ABC must consider business rivals and their likely bids. The presence of competitors would probably cause ABC to offer a price lower than it would if no competitors were bidding against it.

EXHIBIT 12-2

Relevant Costs for ABC: The 250,000-Barrel Monthly Order

Direct costs:		
Direct materials (250,000 × $5)	$1,250,000	
Packaging (250,000 × $22)	5,500,000	
Direct labor (250,000 × $2)	500,000	
Materials procurement	100,000	
Process changeover	200,000	
Total		$7,550,000
Indirect costs:		
Manufacturing overhead (250,000 × $6)	$1,500,000	
Total		1,500,000
Total relevant costs		$9,050,000

The relevant costs computed in Exhibit 12-2 are developed specifically for price bidding on the one-time-only special order of 250,000 barrels. Exhibit 12-3 presents the June 19_1 monthly income statement of ABC under an absorption-cost format (Panel A) and a contribution-margin format (Panel B) at a volume level of 600,000 barrels. The absorption-cost income statement reports the total manufacturing cost to be $49 per barrel. This income statement erroneously implies that a bid of $45 per barrel to Canadian Breweries will result in ABC sustaining a loss on the contract. Why erroneous? Because the absorption-cost format incorporates a $14 irrelevant cost per barrel amount—the $14 fixed manufacturing cost per barrel ($4 in fixed labor and $10 in fixed manufacturing overhead), which will not be incurred on the 250,000-barrel special order.

The contribution-margin income statement in Exhibit 12-3 (Panel B) highlights the four categories of variable manufacturing costs that are relevant to analyzing the one-time-only special order.

EXHIBIT 12-3

Income Statement for ABC: June 19_1 ($000's)

Panel A: Absorption-Costing Format*

		Total		Per Unit
Revenues		$54,000		$90
Manufacturing cost of goods sold:				
Direct materials	$ 3,000		$ 5	
Packaging	13,200		22	
Direct labor	3,600		6	
Overhead	9,600		16	
Total		29,400		49
Gross margin		$24,600		$41
Marketing costs		22,200		37
Operating income		$ 2,400		$ 4

Panel B: Contribution-Margin Format*

		Total		Per Unit
Revenues		$54,000		$90
Variable costs:				
Manufacturing				
Direct materials	$ 3,000		$ 5	
Packaging	13,200		22	
Direct labor	1,200		2	
Overhead	3,600		6	
Total		21,000		35
Marketing		14,400		24
Total variable costs		$35,400		$59
Contribution margin		$18,600		$31
Fixed costs:				
Manufacturing				
Direct labor	$ 2,400		$ 4	
Overhead	6,000		10	
Total		$ 8,400		$14
Marketing		7,800		13
Total fixed costs		$16,200		$27
Operating income		$ 2,400		$ 4

*Production and sales: in June = 600,000 barrels

Both the absorption-cost format and the contribution-margin format in Exhibit 12-3 exclude the relevant $300,000 additional manufacturing costs each month for material procurement and manufacturing-process changeover.

Setup Costs and Short-Run Pricing Decisions

The relevant manufacturing costs in the special-order decision facing American Brewing Company include both volume-driven costs (the variable manufacturing costs) and the monthly setup cost of $300,000. The buildup of relevant costs for some product-mix decisions may include setup costs incurred on a daily basis or even for a single-shift time period. For example, a mining company may incur setup costs when it processes several different types of minerals rather than one type of mineral on a given shift. Processing 100 tons of lead and then 180 tons of zinc rather than 300 tons of lead alone on the shift may mean that some pieces of equipment have to be cleaned or repositioned. The setup costs of cleaning or repositioning equipment are relevant short-run costs in deciding the relative profitability of alternative combinations of minerals to process.

The nature of the decisions that managers must make should help determine what costs are collected in an internal reporting system on an ongoing basis. For example, when setup costs are important for regularly recurring decisions, separately tracking these costs in the internal reporting system is likely to be cost efficient.

PRICING FOR THE LONG RUN

In many pricing decisions, a long-run time horizon is appropriate. Consider pricing decisions for an automobile company that introduces a new model every three years. Automobile companies recognize that the price set in the initial launch of the new model will have a major influence on the prices set in subsequent years. Suppose a new-model Buick, initially priced at $20,000, is a major success and demand exceeds the initial budgeted production. General Motors, the manufacturer, likely would increase production volume rather than increase the price to, say, $40,000.

An additional factor leads to automobile companies emphasizing the long run in making pricing decisions. Buyers expect predictable differences in the pricing of different models over extended time periods. For example, buyers expect a manufacturer to price its mid-sized family car (such as a Chevrolet) lower than its large-sized luxury car (such as an Oldsmobile). If there is a slowing in demand for the large-sized luxury car, the manufacturer will be reluctant to make substantial short-run decreases in price to a level below the price of its mid-sized family car. Rather, the manufacturer will cut back production of the large-sized luxury car.

Pricing Formulas

When competitive forces set the price for a product, knowledge of long-run product costs can help guide decisions about entering or remaining in the market for that product. When managers have some control over the price charged for a product, long-run product costs can act as a base for setting that price.

Managers can turn to numerous pricing formulas based on cost. The general formula for setting price is adding a markup to the cost:

Cost base	$ *X*
Markup component	*Y*
Prospective selling price	$*X* + *Y*

Many different costs can serve as the cost base in this formula. Consider the cost-based pricing formula that Samlee, Inc., a consumer electronics company, could use in pricing the stereos it assembles. Exhibit 12-4 presents the budgeted cost structure for the coming year based on an annual production volume (and sales volume) of 60,000 units. *Full cost* refers to the sum of all costs of all business functions (R & D, product design, manufacturing, marketing, distribution, and customer service).

Assume Samlee uses a full-cost base with a 40% markup in developing the prospective selling price:

Cost base (full cost)	$400
Markup component (40% × $400)	160
Prospective selling price	$560

Alternative cost bases and markup percentages are as follows:

Cost Base (1)	Unit Cost for Stereo (2)	Markup Percentage (3)	Markup Component for Stereo (4)	Prospective Selling Price for Stereo (5) = (2) + (4)
Variable manufacturing cost	$ 90	500%	$450	$540
Variable product cost	230	150	345	575
Manufacturing function cost	120	400	480	600
Full product cost	400	40	160	560

Note that the selling prices we have computed are *prospective* prices. Other factors beyond the cost base and a markup component enter into managers' pricing decisions. These inputs include expected customer reaction to alternative price levels and the prices of similar products made by competitors.

Survey Evidence

Surveyed managers report a preference for including unit fixed costs as well as unit variable costs in the cost base in their pricing decisions.[2] The advantages cited for this approach include:

1. *Fixed-cost recovery*. In the long run, fixed costs must be recovered to stay in business. Some managers believe that this requirement is best achieved by having every product priced above its full cost (for Samlee, for example, $400 per unit at an annual production volume of 60,000 units).

   Some managers are concerned that if prospective prices are based on variable costs, there will be a temptation to engage in excessive long-run price cutting. The variable product cost of a Samlee stereo is $230. A markup percentage of 150% will yield a $575 prospective price, which more than covers the $400 full cost. However, if Samlee sets long-run prices with a markup percentage of, say, 60% of variable product cost, the resulting $368 prospective price leads to long-run revenues being lower than long-run costs.
2. *Simplicity*. Because of its simplicity, full-cost formula pricing meets the cost-benefit test for some managers. It is expensive for firms to analyze cost-behavior patterns and demand

[2]Unit fixed and variable costs were reported to be used when "setting external prices" by 91% of respondents in one survey of 134 companies. See M. Cornick, W. Cooper, and S. Wilson, "How Do Companies Analyze Overhead?" *Management Accounting*, June 1988, pp. 41–43. Survey evidence has concentrated on the use of variable manufacturing cost information versus fixed and variable manufacturing cost information in pricing decisions. There is little survey evidence on how companies incorporate costs upstream to manufacturing (research & development and product design) or costs downstream to manufacturing (marketing, distribution, and customer service) into their pricing decisions.

EXHIBIT 12-4
Budgeted Annual Cost Structure for Stereo of Samlee, Inc.

Business Function	Variable Costs Per Unit	Fixed Costs Per Unit*	Business Function Cost Per Unit
Research/development	$ 20	$ 60	$ 80
Product design	15	35	50
Manufacturing	90	30	120
Marketing	50	20	70
Distribution	30	15	45
Customer service	25	10	35
Product cost	$230	$170	$400
	Variable product cost	Fixed product cost	Full product cost

*Based on budgeted annual production volume of 60,000 units.

patterns for each individual product. Some managers believe that the benefits from analyzing individual products do not exceed the costs of making the analyses.

3. *Price stability.* Full-cost formula pricing is believed to promote price stability. Managers prefer price stability because it makes planning more programmable.
4. *Price justification.* From a legal standpoint, full-cost formula pricing reduces the likelihood that other parties could prove that the prices are predatory or anticompetitive.

LIFE-CYCLE PRODUCT BUDGETING AND COSTING

Objective 4

Explain how life-cycle product budgeting and costing assist in pricing and product-emphasis decisions

The **product life cycle** spans the time from initial research and development to the time at which sales and support to customers are withdrawn. For motor vehicles, this time span may range from five to ten years. For some pharmaceutical products, the time span may be three to five years. For fashion clothing products, the time span may be less than one year.

Life-cycle budgeting estimates the costs attributable to each product from its initial research and development to its final customer servicing and support in the marketplace. **Life-cycle costing** tracks and accumulates the actual costs attributable to each product from its initial research and development to its final customer servicing and support in the marketplace. The terms "cradle to grave" costing and "womb to tomb" costing convey the attempt to capture fully all costs associated with the product.

Life-Cycle Budgeting and Pricing Decisions

Life-cycle budgeted costs can provide important information for pricing decisions. For some products, the development period is relatively long and many costs are incurred prior to manufacturing. Consider Insight, Inc., a computer software company developing a new accounting package, General Ledger. Assume the following budgeted amounts for the General Ledger software package:

Years 1 & 2			
Research and development	$240,000		
Product design	$160,000		
Years 3 through 6	**One-Time Setup Cost**	+	**Cost Per Package Unit**
Manufacturing	$100,000		$40
Distribution	50,000		16
Marketing	70,000		24
Customer service	80,000		30

A product life-cycle budget highlights to managers the importance of budgeted revenues covering costs in all the categories in Exhibit 12-1 (p. 397) rather than just those costs in a subset of the categories (such as manufacturing). To be profitable, Insight must generate revenue to cover costs in all six categories.

Exhibit 12-5 presents the life-cycle budget for the General Ledger software package of Insight, Inc. Three combinations of the selling price per package and predicted demand are shown. The high premanufacturing and postmanufacturing costs at Insight are readily apparent in Exhibit 12-5. For example, premanufacturing costs (research and development and product design) constitute over 30% of total costs for each of the three combinations of selling price and predicted sales volume. Insight should put a premium on having as accurate a set of revenue and

EXHIBIT 12-5

Budgeted Life-Cycle Revenues and Costs for "General Ledger" Software Package of Insight, Inc.*

	Alternative Selling-Price/ Sales-Volume Combinations		
	1	2	3
Selling price per package	**$400**	**$480**	**$600**
Sales volume	**5,000**	**4,000**	**2,500**
Life-cycle revenue: ($400 × 5,000; $480 × 4,000; $600 × 2,500)	$2,000,000	$1,920,000	$1,500,000
Life-cycle costs:			
Research and development	$ 240,000	$ 240,000	$ 240,000
Product design	160,000	160,000	160,000
Manufacturing ($100,000 +: $40 × 5,000; $40 × 4,000; $40 × 2,500)	300,000	260,000	200,000
Marketing ($70,000 +: $24 × 5,000; $24 × 4,000; $24 × 2,500)	190,000	166,000	130,000
Distribution ($50,000 +: $16 × 5,000; $16 × 4,000; $16 × 2,500)	130,000	114,000	90,000
Customer service ($80,000 +: $30 × 5,000; $30 × 4,000; $30 × 2,500)	230,000	200,000	155,000
Total life-cycle costs	$1,250,000	$1,140,000	$ 975,000
Life-cycle operating income	$ 750,000	$ 780,000	$ 525,000

*The time value of money is not taken into account when summing life-cycle revenue or life-cycle costs in this exhibit. Chapters 21 and 22 outline how this important factor can be incorporated into such summations.

cost predictions for the General Ledger package as is possible, given the high percentage of total life-cycle costs outlayed before any manufacturing begins and before any revenue is received.

Exhibit 12-5 presents the summary life-cycle revenues and life-cycle costs for a product with a six-year time horizon from "cradle to grave." The time value of money is not taken into account when summing life-cycle revenues or costs in Exhibit 12-5. Chapters 21 and 22 outline how the time value of money can affect these amounts.

Developing Life-Cycle Reports

Most accounting systems emphasize reporting on a calendar basis—monthly, quarterly, and annually. In contrast, product-life-cycle reporting does not have this calendar-based focus. Consider four products of a computer software company:

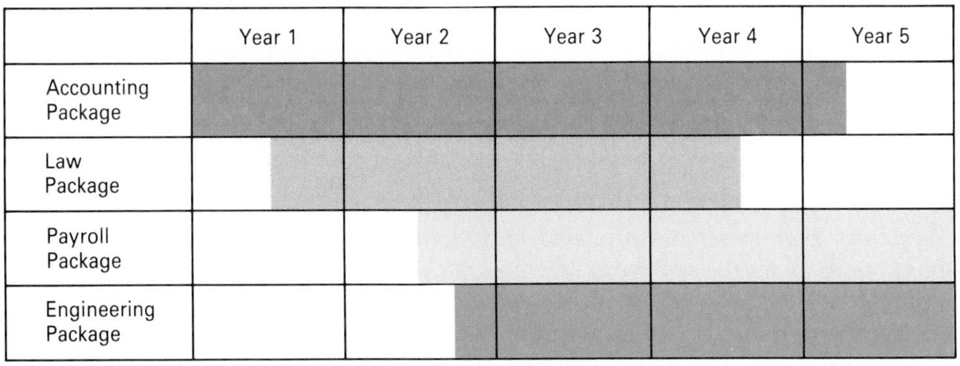

	Year 1	Year 2	Year 3	Year 4	Year 5
Accounting Package					
Law Package					
Payroll Package					
Engineering Package					

Each product spans more than one calendar year, and the Accounting Package product spans five calendar years.

Developing life-cycle reports for each product requires tracking costs and revenues on a product-by-product basis over several calendar periods. The numbers in these life-cycle reports may differ from those in traditional calendar reports. For example, research and development costs are expensed to the period in which they are incurred in many calendar-based accounting systems. The R & D expenses in each period are the aggregate of R & D costs on all products. Individual product-by-product identification of R & D costs typically is not available in calendar reports. In contrast, the R & D costs included in a product-life cost report are often incurred in different calendar years. By tracking R & D costs outlayed over the entire life cycle, the total magnitude of these costs for each individual product can be computed and analyzed.

A product-life-cycle reporting format offers at least three important benefits:

1. The full set of costs associated with each product become more visible. Manufacturing costs are highly visible in most accounting systems. However, the costs associated with upstream areas (for example, research and development) and downstream areas (for example, customer service) are frequently less visible at a product-by-product level in organizations.
2. Differences among products in the percentage of total costs committed at early stages in the life cycle are highlighted. The higher this percentage, the more important it is for managers to develop, as early as possible, accurate predictions of the revenues for that product.
3. Interrelationships across cost categories are highlighted. For example, several companies that have sizably cutback their R & D and product design cost categories have experienced major increases in customer-service-related costs in subsequent years. These costs have arisen from products failing to meet promised quality-performance levels. A life-cycle revenue and cost statement prevents such causally related changes across cost categories from being hidden ("buried") as they are in the calendar income statements of different years.

Exhibit 12-5 presents an example of life-cycle revenues and costs for a software package. The Problem for Self-Study at the end of this chapter presents a second example, using the context of training manuals for company programs. Topics related to life-cycle costing are discussed further in several chapters, especially Chapter 19 on project costing, Chapters 21 and 22 on capital budgeting, and Chapter 29 on strategic control systems.

UNDERCOSTING AND OVERCOSTING PRODUCTS

Objective 5

Describe how product undercosting/overcosting arises and how it can be reduced

A product cost is a measure of the resources consumed (sacrificed or forgone) to produce and deliver the product to the customer. Accurate product-cost information is essential to a manager making a pricing decision, as we will see in this section. Consider Exhibit 12-6, which presents two types of product-cost distortion:

- **Product undercosting**—a product consumes a relatively high level of resources but is reported to have a relatively low product cost. This category is Cell II in Exhibit 12-6.
- **Product overcosting**—a product consumes a relatively low level of resources but is reported to have a relatively high product cost. This category is Cell III in Exhibit 12-6.

Companies that undercost products may actually accept sales that are bringing about losses under the erroneous impression that those sales are profitable. That is, products may be bringing in less in sales than they cost in resources. Companies that overcost products run the danger of allowing competitors to enter a market and take market share. These products actually cost less than what management sees reported. The company could cut the selling price to keep competitors out of the market and still make a profit on the product.

Reasons for Product Undercosting and Overcosting

Undercosting or overcosting of products can happen for several reasons:

1. Cost categories may be inappropriately excluded or included. Exhibit 12-1 outlines six business-function cost categories—research and development, product design, manufacturing/operations, marketing, distribution, and customer service. Undercosting (overcosting) can arise when one or more of these categories is inappropriately excluded (included). For example, a company incurred losses when it undercosted its high-volume

EXHIBIT 12-6
Distortions in Reported Product Costs

Reported product cost	Total resources consumed to produce and deliver a product to the customer: Low	High
Low	**Cell I:** Accurately costed product	**Cell II:** Undercosted product
High	**Cell III:** Overcosted product	**Cell IV:** Accurately costed product

refrigerator product during the initial price-setting period. The company inadvertently omitted all customer-service costs from the cost buildup. Customer-service costs can be 10% or more of full product cost for some consumer products.

2. The time horizon may be inappropriate. Management must select a short time horizon in making short-run pricing decisions and a long time horizon in making long-run pricing decisions. The American Brewing Company example in Exhibits 12-2 and 12-3 illustrates the importance of understanding short-run cost-behavior patterns in pricing decisions with a short-run time horizon.

   Consider this example of an inappropriate time horizon. Management looks only at short-run variable costs when deciding whether to accept a small order from a new customer who will probably become a long-term, large customer. The initial price quoted for the small order may set a ceiling on the price the new customer is willing to pay for much larger orders over a longer time frame.

3. Product-cost cross-subsidization may occur. **Product-cost cross-subsidization** means that at least one miscosted product is resulting in the miscosting of other products in the organization. An overcosted product, in effect, subsidizes one or more undercosted products. Two sources of cross-subsidization are (a) the failure to use direct-cost tracing when it is economically feasible and (b) the use of an inappropriate allocation base when allocating indirect costs to individual products.

The following subsections illustrate two ways of reducing cross-subsidization: increase direct-cost tracing, and use allocation bases that better capture cause-and-effect relationships.

Increase Direct-Cost Tracing. Product-cost cross-subsidization can be reduced by taking a cost that is spread across multiple users without recognition of their different resource demands (the "peanut butter" costing approach) and directly tracing its components to individual users. Consider the costing of a restaurant bill for four colleagues who meet once a month to discuss business developments. Each diner orders separate entrees, desserts, and drinks. The restaurant bill for the most recent meeting is:

	Entree	Dessert	Drinks	Total
Emma	$11	$ 0	$ 4	$ 15
James	20	8	14	42
Jessica	15	4	8	27
Matthew	14	4	6	24
Total	$60	$16	$32	$108
Average	$15	$ 4	$ 8	$ 27

Suppose the restaurant bill is totaled ($108) and the average cost per dinner computed ($27). This costing approach treats each diner the same. Emma would probably object to paying $27. Why? Her meal totals only $15. Indeed, she ordered the lowest-cost entree, had no dessert, and had the lowest drink bill. When costs are averaged across all four diners, Emma is subsidizing those whose total dinner bill exceeds $15. Emma is an example of Cell III (overcosted product) in Exhibit 12-6. James is an example of Cell II (undercosted product) when each diner is charged the average of $27; his directly traceable cost is $42. In this example, the feasibility of direct-cost tracing for each diner makes it possible to eliminate the cross-subsidization that exists with the average-costing approach.

Refine Indirect-Cost Allocation Bases. Product-cost cross-subsidization frequently arises when indirect costs are allocated to products (or product lines) using an allocation base that does not adequately reflect how these different products consume resources. A cause-and-effect relationship is not established. We will consider an example with marketing, distribution, and customer-service costs.

Tasty Snacks manufactures and markets potato chips and other snack items. It sold only to small "Mom and Pop" retail outlets when operations began ten years ago. Tasty rapidly increased sales, in part due to the fresh taste of its products. It invested heavily in a distribution system that guaranteed daily deliveries at all its retail outlets. In recent years, Tasty has expanded its sales base to include large food chain stores. Large food chains negotiate (demand!) much lower prices in return for their high-volume orders.

Exhibit 12-7 (Panel A) presents the product-line income statement for the most recent six-month period. Manufacturing costs at Tasty include the research and product-design categories as well as the costs of manufacturing and packaging the snack products. There is no difference in the manufacturing costs of the products sold in each product line. Marketing, distribution, and customer-service (M, D, & CS) costs are allocated to each product line based on their relative sales revenues ($400/$1,000 = 0.40 for small stores and $600/$1,000 = 0.60 for large stores). Thus, allocation of the $630 million M, D, & CS cost is:

Small stores:	0.40 × $630 = $252
Large stores:	0.60 × $630 = $378

Using the Exhibit 12-7 (Panel A) numbers, Tasty is making money on small store sales and losing money on large store sales.

The assumption underlying the method used to allocate the M, D, & CS indirect costs is that a dollar of sales to small stores consumes the same M, D, & CS resources of Tasty Snacks as does a dollar of sales to large stores. Personnel at Tasty

EXHIBIT 12-7

Product-Line Profitability of Tasty Snacks: June to December, 19_1 (in millions)

Panel A: Reported Income Statement	Small Stores	Large Stores	All Stores
Revenue (percentages)	40%	60%	100%
Revenue			
200 × $2.00; 600 × $1.00	$400	$600	$1,000
Manufacturing costs			
200 × $0.40; 600 × $0.40	80	240	320
Marketing/distribution/customer-service costs			
0.4 × $630; 0.6 × $630	252	378	630
Total costs	$332	$618	$ 950
Operating income	$ 68	$ (18)	$ 50

Panel B: Revised Income Statement	Small Stores	Large Stores	All Stores
Revenue (percentages)	40%	60%	100%
Revenue			
200 × $2.00; 600 × $1.00	$400	$600	$1,000
Manufacturing costs			
200 × $0.40; 600 × $0.40	80	240	320
Marketing/distribution/customer-service costs			
(based on special study)	340	290	630
Total costs	$420	$530	$ 950
Operating income	$ (20)	$ 70	$ 50

Products decided to examine the appropriateness of this assumption. Analysis was made of past data, and interviews were conducted with Tasty personnel. The conclusion was that two variables drive M, D, & CS costs:

- number of separate sales calls to each store, and
- number of separate items purchased per sales call.

Based on a special study of these two cost drivers, management learned that for the most recent six-month period, $340 of $630 million M, D, & CS costs should be allocated to the small stores and $290 to the large stores. Exhibit 12-7 (Panel B) presents the revised income statement.

The previously reported and the revised product-line income figures (from Exhibit 12-7) are:

	Small Stores	Large Stores
Operating income—previously reported	$ 68	$(18)
Operating income—revised	$(20)	$ 70

The revised figures indicate that sales to large stores are profitable and sales to small stores are unprofitable. The cross-subsidization in our example was the overcosting of the large-store product line, which resulted in the undercosting of the small-store product line. The small-store product line is an example of Cell II in Exhibit 12-6, and the large-store product line is an example of Cell III.

The previously reported product-line income figures in Exhibit 12-7 (Panel A) are misleading regarding which product line to emphasize. Managers might erroneously have chosen to push sales to small stores, which would have decreased rather than increased operating income. Given the revised figures, managers should emphasize sales to large stores or try to increase prices to small stores. Further discussion of cost allocation issues and of refinements in cost allocation bases appears in Chapters 14 and 15.

TARGET COST AS A PRICING GUIDE

Objective 6

Describe the target-costing approach

Should a company enter a new market? The prices competitors set for their products are important factors in making this decision. The estimated long-run cost of a product that will enable a company to enter or to remain in the market and compete profitably against its competitors is called the **target cost**. The target-cost concept has an external market focus as its starting point—that is, the competitors and the prices they set. Japanese consumer-product companies often use target costs in their pricing strategies.[3]

Activity-based accounting systems (described in Chapter 5, pp. 150–57) can provide key information for use in target-cost analysis. *Activity-based accounting* is a system that focuses on activities as the fundamental cost objects. It then uses the cost of these activities as building blocks for compiling the costs of other cost objects (such as a product).

EXAMPLE:
Instruments Inc. is considering entering the medical-instruments market and selling heart-monitoring equipment. The prices charged by competitors already selling heart-monitoring instruments provide a ceiling price for Instruments Inc.'s product. Given this

[3]M. Sakurai, "The Concept of Target Costing and Its Effective Utilization," Working paper, Senshu University, 1988.

ceiling price, Instruments Inc. aims to design a heart-monitoring instrument that (1) provides a profit given its long-run costs and (2) competes with its major competitors. The cost amount that satisfies both (1) and (2) is termed a target cost.

Instruments Inc. has one direct-manufacturing cost category (direct materials) and the following six indirect-manufacturing pools in its activity-based accounting system:

Indirect-Manufacturing Cost Pool	Allocation Base	Allocation Rate
1. Materials handling	Number of parts	$1 per part
2. Start station	Number of printed-circuit boards	$20 per board
3. Machine insertion of parts	Number of machine-inserted parts	$0.50 per machine-inserted part
4. Manual insertion of parts	Number of manually inserted parts	$4 per manually inserted part
5. Wave solder	Number of printed-circuit boards	$30 per board
6. Quality testing	Hours of quality testing time	$50 per testing hour

This activity accounting system is described in more detail in Chapter 5 (pp. 152–56).

Suppose product designers at Instruments Inc. draw up a heart-monitoring instrument (product HM-107) by extending the designs of related heart-monitoring products it is currently assembling. This is an internally oriented approach to product design. Exhibit 12-8, column (2), reports that the estimated manufacturing cost of HM-107 is $2,029. The $2,029 figure comes as a shock to the new-product development (NPD) manager. The NPD manager estimates that a target manufacturing cost of approximately $1,600 is essential if the Instruments Inc. product is to be price-competitive with that of its four major competitors. Product designers at Instruments Inc. were then given the challenge to design a product comparable in performance to the major competitors but with a target manufacturing cost of no more than $1,600. This is an externally oriented approach to product design. After

EXHIBIT 12-8

Costing of Alternative Heart Instrument Products at Instruments Inc.

Costs (1)	HM-107 (2)	HM-108 (3)
Direct manufacturing product costs:		
Direct materials	$1,140	$ 960
Indirect manufacturing product costs:		
Materials handling (210; 170 × $1)	210	170
Start station (1; 1 × $20)	20	20
Machine insertion of parts (142; 138 × $0,50)	71	69
Manual insertion of parts (67; 31 × $4)	268	124
Wave solder (1; 1 × $30)	30	30
Quality testing (5.8; 3.6 × $50)	290	180
Total	$ 889	$ 593
Total manufacturing product costs	$2,029	$1,553

much work and after consultation with manufacturing personnel, product designers at Instruments Inc. put forward a proposal for HM-108, which is comparable in performance to competitors and has an estimated manufacturing cost of $1,553. Column (3) of Exhibit 12-8 presents the costing of HM-108.

Differences in key aspects of the product design of the two proposed products show why HM-108 has a sizably lower manufacturing cost than HM-107:

	HM-107	HM-108
Number of parts	210	170
Number of machine-inserted parts	142	138
Number of manually inserted parts	67	31
Hours of quality testing time required	5.8	3.6

Exhibit 12-8 highlights the important role product designers play in a firm's bringing competitively priced products to the marketplace. Product designers can greatly reduce the costs of products by explicitly considering the implications of their decisions for the manufacturing, marketing, distribution, and customer-service cost areas.

The activity accounting system of Instruments Inc. provides explicit guidelines to a product designer on ways to reduce manufacturing cost per product. These include designing products that have fewer parts, using more parts that are machine-inserted rather than manually inserted, and requiring less quality-testing time.

Exhibit 12-8 also highlights how successful efforts by manufacturing personnel to reduce costs at each activity area translate into lower product costs. For example, assume Instruments Inc. is able to reduce costs at the quality-testing activity area from $50 to $30 per hour. This activity-area cost reduction would decrease the cost of HM-107 by $116 (5.8 hours required × $20 cost reduction per hour) and HM-108 by $72 (3.6 hours required × $20 cost reduction per hour).

EFFECTS OF ANTITRUST LAWS ON PRICING

To comply with United States antitrust laws, such as the Sherman Act, the Clayton Act, and the Robinson-Patman Act, pricing must not be predatory.[4] A business engages in **predatory pricing** when it temporarily cuts prices in an effort to restrict supply and then raises prices rather than enlarge demand or meet competition.[5]

Recent court decisions have been influenced by a classic article by Areeda and Turner,[6] which argues that:

- A price at or above reasonably anticipated short-run marginal and average variable costs should be deemed nonpredatory.
- Unless at or above average cost, a price below reasonably anticipated (1) short-run marginal costs or (2) average variable costs should be deemed predatory.[7]

[4]Discussion of the Sherman Act and the Clayton Act is in A. Barkman and J. Jolley, "Cost Defenses for Antitrust Cases," *Management Accounting*, Vol. 67, No. 10, pp. 37–40.

[5]See D. Greer, *Industrial Organization and Public Policy* (New York: Macmillan, 1984), pp. 316–17.

[6]P. Areeda and D. Turner, "Predatory Pricing and Related Practices under Section 2 of the Sherman Act," *Harvard Law Review*, 88 (1975), 697–733. See also F. Scherer, "Predatory Pricing and the Sherman Act: A Comment," *Harvard Law Review*, 89 (1976), 869–903.

For an overview of case law, see ABA *Antitrust Section, Antitrust Law Developments*, 2d ed. (1984). See also the "Legal Developments" section of the *Journal of Marketing* for summaries of court cases.

[7]Areeda and Turner, "Predatory Pricing," p. 733.

The case of *Adjustor's Replace-a-Car* v. *Agency Rent-a-Car* illustrates how the courts are willing to use variable-cost information in decisions concerning predatory pricing.[8] *Agency Rent-a-Car* (the defendant) used selective price cuts to enter the Austin and San Antonio, Texas, car-rental market; cars were rented to customers for extended time periods. *Adjustor's* (the plaintiff) claimed that it was forced to depart from these markets because *Agency* had engaged in predatory pricing. A circuit-court judge reaffirmed a lower-court decision that *Agency* "did not predatorily price its service by underselling a competing service in light of the facts that its charges were above average variable cost and there were no significant entry barriers to the market." Evidence presented by *Adjustor's* included income statements of *Agency* showing that it had operated its outlets in Austin and San Antonio at a "net operating loss"; these statements included an allocated portion of the overhead cost of *Agency's* headquarters.

The circuit-court judge ruled it was sufficient (in regard to cost justification) for *Agency* "to demonstrate that the price it charged for a rental car never dropped below its average variable cost." The judge noted:

> Agency's expert testified that Agency's average variable costs in San Antonio and Austin during the relevant time periods fluctuated between approximately $3.65 and approximately $5.00 (a day). Thus, Agency's price was always at least 40% greater than its average variable cost. The expert also testified that Agency's average variable costs in Austin were $5.23 when Agency went to a $9.00 price; the price was thus 72% above average variable cost. This testimony cut to the quick of plaintiff's predatory pricing claim.

The circuit-court judge rejected *Adjustor's* claim that "a net loss from operations" on an income statement including an allocation of *Agency's* headquarters' overhead was "effectively an admission of predatory pricing."

The variable-cost guidelines proposed by Areeda and Turner, although having the support of several court decisions, are not explicitly incorporated into the statute law. Caution should always be used when generalizing from individual legal cases. Managers and accountants who are concerned with their conformance to antitrust laws would be prudent to have a system that incorporates the following procedures:

1. Collect data in such a manner as to permit relatively easy compilation of variable costs.
2. Review all proposed prices below variable costs in advance, with a presumption of claims of predatory intent.
3. Keep as detailed a set of records as feasible, not only of manufacturing costs but also of (a) upstream costs such as research and development and product design and (b) downstream costs such as marketing, distribution, and customer service.

A company that follows these guidelines will be prepared for inquiries by regulatory agencies.

---

[8]Adjustor's Replace-a-Car, Inc. v. Agency Rent-a-Car, 735 2nd 884 (1984).

PROBLEM FOR SELF-STUDY

PROBLEM

Continuing Education Programs (CEP) markets teaching packages for companies to use in their own internal training programs. It currently has teaching packages in over twenty different business areas (accounting, personnel management, and so on). The existing internal accounting system at CEP emphasizes overall revenues and costs, year by year. There is no systematized reporting of an individual teaching package's profitability over its life cycle.

Jamie West, the recently appointed editor, is evaluating a proposed teaching package on *Successful Commercial Lending* by Terry Funk. Funk was the author of a prior CEP teaching package, *Successful Consumer Lending*. Funk is requesting a development grant of $350,000 for *Successful Commercial Lending*. West is reluctant to recommend this grant. The highest amount West previously gave was the $250,000 grant to Funk for his prior teaching package. An analysis of the annual financial statements of CEP provides few insights to West in her evaluation of Funk's request. Columns (2) to (7) of Exhibit 12-9 present year-by-year revenues and costs associated with the *Successful Consumer Lending* training package over its six-year life cycle.

Required

1. Did CEP make a profit on the *Successful Consumer Lending* package? (Do not adjust the numbers in your answer for the time value of money.)
2. Why might Jamie West find a product-life-cycle report informative in decisions relating to Funk's proposed *Successful Commercial Lending* teaching package?
3. Name two items other than costs that West should consider in her decisions on Funk's proposed package.

SOLUTION

1. Column (8) of Exhibit 12-9 summarizes the life-cycle revenues and life-cycle costs of the *Successful Consumer Lending* training package. Total life-cycle operating income was $1,092,000. (No adjustment is made for the time value of money.)

2. West makes many decisions on individual training packages. The product life cycle of each package extends over several calendar years. Reports that include costs and revenues for the entire life cycle highlight the overall profitability of an individual training package. Such reports also highlight the magnitude and timing of development costs on each training package. Among the costs for the *Successful Consumer Lending* package, development costs were relatively high, but these were more than offset by relatively high revenue in years 3 through 6. Funk could well argue that "you have to spend money to make money" in the training-package development business.

A year-by-year reporting format captures only part of the costs and revenue associated

EXHIBIT 12-9

Life-Cycle Revenue and Costs of Successful Consumer Lending Training Package (in thousands)

(1)	(2)	(3)	(4)	(5)	(6)	(7)	(8)
Revenue/Costs	**Year 1**	**Year 2**	**Year 3**	**Year 4**	**Year 5**	**Year 6**	**Total***
Life-cycle revenue	$—	$—	$300	$1,030	$820	$370	$2,520
Life-cycle costs:							
Development payments to author	100	150	—	—	—	—	250
Development costs of CEP	2	98	40	—	—	—	140
Production costs of CEP	—	—	75	96	135	80	386
Marketing costs of CEP	—	15	120	100	35	28	298
Distribution & customer-service costs of CEP	—	5	40	30	15	12	102
Royalty payments to author	—	—	30	103	82	37	252
Total life-cycle costs	$102	$268	$305	$ 329	$267	$157	$1,428
Life-cycle operating income							$1,092

*The time value of money is not taken into account when summing life-cycle revenues or life-cycle costs in this exhibit. Chapters 21 and 22 outline how this important factor can be incorporated into such summations.

with each package. Moreover, year-by-year income statements typically report aggregate costs and revenue for CEP rather than project-by-project costs and revenue for that year.

From Jamie West's perspective, the issue is not a year-by-year set of reports versus a product-by-product life-cycle set of reports. Rather, the issue is whether, given the existing year-by-year reporting system, her collective decisions will be improved by adding a set of budgets and actual results on a product-by-product life-cycle basis.

3. Cost information is only one item West will consider in her decisions on Funk's proposed package. Two other items are customers and competitors. West will probably encourage Funk to work with bank lending officers (who are the potential customers) in devising the learning modules to include in the *Successful Commercial Lending* package. Training packages for commercial lending officers available from other organizations will also influence many of West's decisions, including her determining the proposed price to charge and the amount to spend on marketing and production.

SUMMARY

Pricing decisions, including product profitability decisions, are among the most challenging decisions facing managers. Cost information is an important input into many pricing decisions. Costs are incurred at all stages of the life cycle of a product. These stages are research and development, product design, manufacturing, marketing, distribution, and customer service. The systematic budgeting and tracking of costs at all these stages provides important information to managers making decisions about product introductions and product pricing and determining relative product profitability.

The time horizon appropriate to a decision on pricing or product emphasis helps dictate what costs are relevant. Management should focus on those future costs that will differ among the decision alternatives under consideration; these costs are the relevant costs. Short-run decisions may have different relevant costs from long-run decisions.

This chapter emphasized the role of cost data in pricing decisions. Always keep in mind two other major influences on pricing decisions: customers and competitors.

TERMS TO LEARN

This chapter and the Glossary at the end of the book contain definitions of the following important terms:

life-cycle budgeting *(p. 403)* life-cycle costing *(403)* predatory pricing *(411)*
product-cost cross-subsidization *(407)* product life cycle *(403)*
product overcosting *(406)* product undercosting *(406)* target cost *(409)*

ASSIGNMENT MATERIAL

QUESTIONS

12-1 What are the three major influences on pricing decisions?

12-2 Name the six business function categories of costs that can be included in the cost buildup of a product or service.

12-3 Outline a four-step procedure for developing product costs for pricing decisions.

12-4 Give two examples of pricing decisions with a short-run focus.

12-5 What is the product life cycle?

12-6 What are three benefits of using a product-life-cycle reporting format?

12-7 Describe three ways that products may be undercosted or overcosted.

12-8 What is product-cost cross-subsidization?

12-9 Outline two ways of reducing product-cost cross-subsidization.

12-10 What is a target cost?

12-11 What are two ways a company can manufacture products that are more competitively priced?

12-12 Define predatory pricing.

EXERCISES AND PROBLEMS

12-13 An argument about pricing. A column in a newspaper contained the following item:

> Dear Ms. Personal: My husband and I are in constant disagreement. He drives 10 miles to work every day (five days a week) and drives a mile in the opposite direction to pick up and deliver the man who rides with him. Therefore, the total daily mileage is 24 for the roundtrip. For this service, the man pays $3 every week toward gas. Bus fare would cost $16 each week. I say this fellow should pay at least $9 each week, which would be one-half the cost for the week's gas. But my husband says he can't just ask him for the money, so he settles for this arrangement month after month. We have three children and are in debt for several thousand dollars, and even this $9 a week would really help, as we live on a very tight budget. Am I reasonable to think the cost of gas should be split 50-50?

Required
As Ms. Personal, write a reply.

12-14 Pricing of hotel rooms on weekends. Paul Diamond is the owner of the Galaxy chain of four-star prestige hotels. These hotels are in Chicago, London, Los Angeles, Montreal, New York, Seattle, San Francisco, and Tokyo. Diamond is currently struggling in setting weekend rates for the San Francisco hotel (the San Francisco Galaxy). From Sunday through Thursday, the Galaxy has an average occupancy rate of 90%. On Friday and Saturday nights, however, average occupancy declines to less than 30%. Galaxy's major customers are business travelers who stay mainly Sunday through Thursday.

The current room rate at the Galaxy is $150 a night for single occupancy and $180 a night for double occupancy. These rates apply seven nights a week. For many years, Diamond has resisted having rates for Friday and Saturday nights that are different from rates for the rest of the week. Diamond has long believed that price reductions convey a "nonprestige" impression to his guests. The San Francisco Galaxy values highly its reputation for treating its guests as "royalty."

Most room costs at the Galaxy are fixed on a short-stay (per night) basis. Diamond estimates the variable cost of servicing each room to be $20 a night per single occupancy and $22 a night per double occupancy.

Many prestige hotels in San Francisco offer special weekend rate reductions (Friday and/or Saturday) of up to 50% of their Sunday-through-Thursday rates. These weekend rates also include additional items such as a breakfast for two, a bottle of champagne, and discounted theater tickets.

Required

1. Would you recommend that Diamond reduce room rates at the San Francisco Galaxy on Friday and Saturday nights? What factors should be considered in his decision?
2. In six months' time the Super Bowl is to be held in San Francisco. Diamond observes that several four-star prestige hotels have already advertised a Friday through Sunday rate for Super Bowl weekend of $300 a night. Should Diamond charge extra for the Super Bowl weekend? Explain.

12-15 Relevant-cost approach to pricing decisions, special order. The following financial data apply to the video-tape production plant of the Dill Company:

October 19_4	Standard Factory Cost Per Video Tape
Direct materials	$1.50
Direct labor	0.80
Variable factory overhead	0.70
Fixed factory overhead	1.00
Total factory cost	$4.00

Variable factory overhead varies with respect to units produced. Fixed factory overhead of $1.00 per tape is based on budgeted fixed factory overhead of $150,000 per month and budgeted production of 150,000 tapes per month. Dill Company sells each tape for $5.00.

Marketing costs have two components:

- Variable marketing costs (sales commissions) of 5% of dollar sales
- Fixed monthly costs of $65,000

During October 19_4, Lyn Randell, a Dill Company salesperson, asked the president for permission to sell 1,000 tapes at $3.80 per tape to a customer not in its normal marketing channels. The president refused this special order on the grounds that the selling price was below the total standard factory cost.

Required

1. What would have been the effect on monthly operating income of accepting the special order?
2. Comment on the president's "below factory cost" reasoning for rejecting the special order?
3. What factors would you recommend that the president consider when deciding whether to accept or reject the special order?

12-16 Relevant-cost approach to short-run pricing decisions. The San Carlos Company is an electronics business with eight product lines. Income data for one of the products (XT-107) for the month just ended (June 19_3) follow:

Sales—200,000 units at average price of $100		$20,000,000
Variable costs:		
Direct materials at $35	$7,000,000	
Direct labor at $10	2,000,000	
Variable factory overhead at $5	1,000,000	
Sales commissions at 15% of sales	3,000,000	
Other variable costs at $5	1,000,000	
Total variable costs at $70		14,000,000
Contribution margin		$ 6,000,000
Fixed costs		5,000,000
Operating income		$ 1,000,000

Abrams Inc., an instruments company, has a problem with its preferred supplier of XT-107 component products. This supplier has had a three-week strike by its employees and will not be able to supply Abrams 3,000 units next month. Abrams approaches the sales representative, Sarah Holtz, of the San Carlos Company about providing 3,000 units of XT-107 at a price of $80 per unit. Holtz informs the XT-107 product manager, Jim McMahon, that she would accept a flat commission of $6,000 if this special order were accepted. San Carlos has the capacity to produce 300,000 units of XT-107 each month, but demand has not exceeded 200,000 in any month in the last year.

Required

1. If Holtz accepts the 3,000-unit order from Abrams Inc., what will be the effect on monthly operating income? (Assume the same cost structure as occurred in June 19_3.)
2. McMahon ponders whether to accept the 3,000-unit special order. He is afraid of the precedent that might be set by cutting the price. He said, "The price is below our full cost of $95 per unit. I think we should quote a full price, or Abrams will expect favored treatment again and again if we continue to do business with them." Do you agree with McMahon? Explain.

12-17 Relevant-cost approach to pricing decisions: contribution-margin vs. absorption-cost income statement. Stardom Inc. cans peaches for sale to food distributors. All costs are classified as either manufacturing or marketing. Stardom prepares monthly budgets. The March 19_3 budgeted absorption-cost income statement follows:

Revenues (1,000 crates at $100 a crate)	$100,000	100%
Manufacturing cost of goods sold	60,000	60
Gross profit	40,000	40%
Marketing costs	30,000	30
Operating income	$ 10,000	10%

Normal markup percentage:
$40,000 ÷ $60,000 = 66.7% of absorption cost

Monthly costs are classified as fixed or variable (with respect to the cans produced for manufacturing costs and with respect to the cans sold for marketing costs):

	Fixed	Variable
Manufacturing	$20,000	$40,000
Marketing	16,000	14,000

Stardom has the capacity to can 1,500 crates per month. The relevant range in which monthly fixed manufacturing costs will be "fixed" is from 500 crates to 1,500 crates per month.

Required

1. Recast the income statement in a contribution-margin format. Indicate the normal markup percentage based on total variable costs.
2. Assume a new customer approaches Stardom to buy 200 crates at $55 per crate. The customer does not require additional marketing effort. Additional manufacturing costs of

$2,000 (for special packaging) will be required. Stardom believes this is a one-time-only special order, as the customer is discontinuing business in six weeks' time. Stardom is reluctant to accept this 200-crate special order because the $55 per-crate price is below the $60 per-crate absorption cost. Do you agree with this reasoning?

3. Assume the new customer decides to remain in business. How would this longevity affect your willingness to accept the $55 per-crate offer?

12-18 Average-cost-based pricing, cross-subsidization. For many years, five former classmates—Steve Armstrong, Lola Gonzales, Rex King, Elizabeth Poffo, and Gary Young—have had a reunion dinner at the annual meeting of the American Accounting Association. The bill for the most recent dinner at the Seattle Spaceneedle Restaurant was broken down as follows:

Diner	Entree	Dessert	Drinks	Total
Armstrong	$27	$8	$24	$59
Gonzales	24	3	0	27
King	21	6	13	40
Poffo	31	6	12	49
Young	15	4	6	25

For at least the last ten annual dinners, King put the total restaurant bill on his American Express card. He then mailed to each person a bill for the same price (the average cost). They shared the gratuity at the restaurant by paying cash. King continued this practice for the Seattle dinner. However, just before he sent out the bill to the other four diners, Young phoned him to complain. He was livid at Poffo for ordering the steak and lobster entree ("She always does this!") and at Armstrong for having three glasses of imported champagne ("What's wrong with domestic beer?").

Required

1. Compute the price charged per diner under the average-cost pricing approach. What is the rationale(s) for using this approach?
2. What is the amount of cross-subsidization among the five diners with the average-cost pricing approach?
3. What are the likely negative behaviors with an average-cost pricing policy for the reunion dinner?

12-19 Life-cycle product costing, product emphasis. Decision Support Systems (DSS) is examining the profitability and pricing policies of its software division. The DSS software division develops software packages for engineers. DSS collects data on three of its more recent packages:

- EE-46: package for electrical engineers
- ME-83: package for mechanical engineers
- IE-17: package for industrial engineers

Summary details on each package over their two-year "cradle to grave" product lives follow:

Package	Selling Price	Units Sold Year 1	Units Sold Year 2
EE-46	$250	2,000	8,000
ME-83	300	2,000	3,000
IE-17	200	5,000	3,000

Assume that no inventory remains on hand at the end of year 2.

DSS is deciding which product lines in its software division to emphasize. In the past two years, the profitability of this division has been mediocre. DSS is particularly concerned with the increase in research and development costs in several of its divisions. An analyst at the software division pointed out that for one of its most recent packages (IE-17), major efforts had been made to cut back research and development costs.

Last week Nancy Sullivan, the software division manager, attended a seminar on product-life-cycle management. The topic of life-cycle accounting was discussed. Sullivan decides to use this approach in her own division. She collects the following life-cycle revenue and cost information for the EE-46, ME-83, and IE-17 packages.

	EE-46		ME-83		IE-17	
	Year 1	Year 2	Year 1	Year 2	Year 1	Year 2
Revenue ($000s)	$500	$2,000	$600	$900	$1,000	$600
Costs ($000s)						
Research & development	700	0	450	0	240	0
Product design	185	15	110	10	80	16
Manufacturing	75	225	105	105	143	65
Marketing	140	360	120	150	240	208
Distribution	15	60	24	36	60	36
Customer service	50	325	45	105	220	388

Required

1. How does a product-life-cycle income statement differ from a calendar-based income statement? What are the benefits of using a product-life-cycle reporting format?
2. Present a product-life-cycle income statement for each software package. Which package is the most profitable, and which is the least profitable?
3. How do the three software packages differ in their cost structure (the percentage of total costs in each cost category)?

12-20 Target costs, activity-based accounting systems. PART A: Executive Power (EP) manufactures and sells computers and computer peripherals to several nationwide retail chains. John Farnham is the manager of the printer division. Its two largest-selling printers are P-41 and P-63.

The manufacturing cost of each printer is calculated using EP's activity accounting system. EP has one direct-manufacturing cost category (direct materials) and the following five indirect-manufacturing cost pools:

Indirect-Manufacturing Cost Pool	Allocation Base	Allocation Rate
1. Materials handling	Number of parts	$1.20 per part
2. Assembly management	Hours of assembly time	$40 per hour of assembly time
3. Machine insertion of parts	Number of machine-inserted parts	$0.70 per machine-inserted part
4. Manual insertion of parts	Number of manually inserted parts	$2.10 per manually inserted part
5. Quality testing	Hours of quality testing time	$25 per testing hour

Product characteristics of P-41 and P-63 follow:

	P-41	P-63
Direct materials cost	$407.50	$292.10
Number of parts	85 parts	46 parts
Assembly time	3.2 hours	1.9 hours
Number of machine-inserted parts	49 parts	31 parts
Number of manually inserted parts	36 parts	15 parts
Hours of quality testing time	1.4 hours	1.1 hours

Required

What is the manufacturing cost of P-41? of P-63?

PART B: Farnham has just received some bad news. A foreign competitor has introduced products very similar to P-41 and P-63. Given their announced selling prices, Farnham estimates the P-41 clone to have a manufacturing cost of approximately $680 and the P-63 clone to have a manufacturing cost of approximately $390. He calls a meeting of product designers and manufacturing personnel at the printer division. They all agree to have the $680 and $390 figures become target costs for EP's P-41 and P-63, respectively. Product designers examine alternative ways of designing printers with comparable performance but lower cost. They came up with the following revised designs for P-41 and P-63 (termed P-41 REV and P-63 REV, respectively):

	P-41 REV	P-63 REV
Direct materials cost	$381.20	$263.10
Number of parts	71 parts	39 parts
Assembly time	2.1 hours	1.6 hours
Number of machine-inserted parts	59 parts	29 parts
Number of manually inserted parts	12 parts	10 parts
Quality testing time	1.2 hours	0.9 hours

Required

1. What is a target cost?
2. Using the activity-based accounting system outlined in Part A, compute the manufacturing cost of P-41 REV and P-63 REV. How do they compare with the $680 and $390 target costs?
3. Explain the differences between P-41 and P-41 REV and between P-63 and P-63 REV.
4. Assume now that the manufacturing manager of the printer plant has achieved major cost reductions in one of the activity areas. As a consequence, the allocation rate in the assembly-management activity area will be reduced from $40 to $28 per assembly hour. How will this activity-area cost reduction effect the manufacturing cost of P-41 REV and P-63 REV? Comment on the results.

12-21 Pricing of refuse services to different customer categories, cross-subsidization. The residents of the city of Ventura are in "open revolt." The Refuse Department has just announced rate increases that will bring the average annual residential charge to over $500 per household. This rate is approximately double the average rate charged by the three nearby cities. You are hired by a citizens' committee to investigate the city of Ventura's pricing policy for refuse collection and disposal. The existing pricing policy is based on pounds collected. The city's stated rationale for this policy is to prevent residential customers, who have a low average pounds of refuse collected, from subsidizing the higher-volume commercial (offices and retail) customers and industrial customers.

The city of Ventura operated its own Sanitation Department in 19_3. The 19_3 rate charged per customer was $1 per pound of refuse collected. (The city of Ventura collects, hauls, and disposes the refuse but charges customers only on a pounds-collected basis.) This $1 figure per pound collected was calculated as follows:

$$\frac{\text{Budgeted costs of Refuse Department in 19\_3}}{\text{Budget annual pounds of refuse collected in 19\_3}} = \frac{\$38{,}000{,}000}{38{,}000{,}000 \text{ pounds}}$$

The actual costs of the Refuse Department in 19_3 were $39,407,632; the actual refuse collected was 38,480,000 pounds. A breakdown of the actual 38,480,000 pounds follows:

Customer Category	Pickups Per Week	Number of Customers	Average Pounds Collected Per Week	Number of Weeks	Total Annual Pounds Collected
Residential	2	38,000	10 pounds	52	19,760,000
Commercial	5	1,600	100 pounds	52	8,320,000
Industrial	5	400	500 pounds	52	10,400,000
					38,480,000

Two years ago, the nearby city of Santa Barbara subcontracted its refuse requirements to Global Waste Management Services (GWMS). GWMS put on public file the basis of its 19_3 charges to the city of Santa Barbara. These charges follow:

Cost Category	Residential	Commercial	Industrial
I. Fixed cost per pickup	$0.50 per pickup	$2.00 per pickup	$10 per pickup
II. Collection and hauling cost per pound	$0.18 per pound	$0.30 per pound	$0.70 per pound
III. Disposal cost per pound	$0.12 per pound	$0.15 per pound	$1.10 per pound

GWMS offered the same set of charges to the city of Ventura at the start of 19_3. However, the city believed it could operate its Sanitation Department more efficiently than GWMS could.

GWMS is the largest waste-management company in the world. It has contracts with over five hundred cities in North America. An official of GWMS informs you that the cost category with the largest increase in recent years is disposal of industrial refuse due to escalating environmental concerns.

Required

1. Compare the city of Ventura's approach to charging its customers with the GWMS approach to charging cities for refuse services.
2. What is the average actual cost per customer in each of the residential, commercial, and industrial categories in 19_3 using the city of Ventura's $1 per-pound charge approach?
3. Assume the city of Ventura had subcontracted its refuse collection and disposal in 19_3 to GWMS. What would have been the average actual cost GWMS would charge the city of Ventura in each of the residential, commercial, and industrial categories in 19_3?
4. Compare the average cost for each customer category in requirements 2 and 3. Comment on the differences in average cost for each customer category.

12-22 Cost-based pricing decisions for professional services firm, cross-subsidization. PART A: Wigan Associates is a recently formed law partnership. Ellery Hanley, the managing partner of Wigan Associates, has just finished a tense phone call with Martin Offiah, president of Widnes Coal. Offiah complained about the price Wigan charged for some conveyancing (drawing up property documents) legal work done for Widnes Coal. Offiah requested a breakdown of the charges. He also indicated to Hanley that a competing law firm, Hull and Kingston, was seeking more business with Widnes Coal and that he was going to ask them to bid for a conveyancing job next month. Offiah ended the phone call by saying that a price bid similar to the one that Wigan charged last month would guarantee Wigan would not be hired for next month's job.

Hanley was dismayed by the phone call. He was also puzzled, as he believed that conveyancing was an area where Wigan Associates had much expertise and were highly efficient. The Widnes Coal phone call was the bad news of the week. The good news was that yesterday Hanley received a phone call from its only other client (St. Helens Glass) saying it was very pleased with both the quality of the work (primarily litigation) and the price charged on its most recent case.

Hanley decides to collect data on the Widnes Coal and St. Helens Glass cases. Wigan Associates uses a cost-based approach to pricing (billing) each legal case. Currently it uses a single direct-cost category (for professional-labor time) and a single indirect-cost pool (all other costs). Indirect costs are allocated to cases on the basis of professional-labor hours per case. The case files showed the following:

	Widnes Coal	St. Helens Glass
Direct professional-labor time	104 hours	96 hours

Wigan Associates bills clients for professional labor at $70 an hour. Indirect costs are allocated to cases at $105 an hour. Total indirect costs in the most recent period were $21,000. A profit component for the partnership is built into the $70 per professional-labor hour billing rate.

Required

1. Why is it important for Wigan Associates to understand the costs associated with individual cases in their pricing decisions?
2. Compute the amount Wigan Associates billed Widnes Coal and St. Helens Glass.

PART B: Hanley speaks to several partners about the relative pricing of the two cases. Several thought that the relative prices charged seemed out of line with their intuition. One partner observed that a useful approach to obtaining more accurate product costs for pricing purposes would be to increase direct-cost tracing.

Hanley asks his assistant to collect details on those costs included in the $21,000 indirect-cost pool that can be traced to each individual case. After further analysis, Wigan is able to reclassify $14,000 of the $21,000 as direct costs:

	Widnes Coal	St. Helens Glass
Other direct costs traceable to cases:		
Support labor	$1,600	$3,400
Computer time	500	1,300
Travel	600	4,400
Phones/Faxes	200	1,000
Photocopying	250	750
Total	$3,150	$10,850

Hanley decides to calculate the price that would have been billed to each case had the business used six direct-cost pools and a single indirect-cost pool. The single indirect-cost pool would have $7,000 of costs and be allocated to each case using the professional-labor hours base.

Required

1. What is the revised indirect cost allocation rate per professional-labor hour for Wigan Associates when total indirect costs are $7,000?
2. Compute the costs that would have been billed to Widnes and St. Helens if Wigan Associates had directly traced professional labor, support labor, computer time, travel, phones/faxes, and photocopying to each case and had used a single indirect-cost pool with professional-labor hours as the allocation base.
3. Compare the costs billed to Widnes and St. Helens in Part B (requirement 2) with those in Part A (requirement 2). Comment.

PART C: Hanley examines the product-costing approaches in Part A and Part B. He questions the use of a single charge-out rate for all professional time of Wigan Associates. Wigan has two classifications of professional staff—partners and associates. Hanley asks his assistant to examine the relative use of partners and associates on the recent Widnes Coal and St. Helens cases. The Widnes case used 24 hours of partner time and 80 hours of associate time. The St. Helens Glass case used 56 hours of partner time and 40 hours of associate time.

Hanley decides to examine how the use of separate direct and indirect cost pools for partners and for associates would have affected the amount billed to Widnes and St. Helens. Indirect costs in each cost pool would be allocated on the basis of labor hours of that category of professional labor. The rates per category of professional labor are:

Category of Professional Labor	Direct Cost Per Hour	Indirect Cost Per Hour
Partner	$100.00	$57.50
Associate	50.00	20.00

These indirect-cost rates are based on a total indirect-cost pool of $7,000; $4,600 of this $7,000 is attributable to the activities of partners, and $2,400 is attributable to the activities of associates. (The $57.50 indirect-cost-per-hour rate is calculated by dividing $4,600 by 80 hours of partner time; the $20 indirect-cost-per-hour rate is calculated by dividing $2,400 by 120 hours of associate time.)

Required

1. Compute the costs billed to Widnes and St. Helens with separate indirect-cost allocation rates for partners and associates and direct tracing of partner professional labor, associate professional labor, support labor, computer time, travel, phones/faxes, and photocopying to each case.
2. For what decisions might Wigan Associates find it more useful to use the product-costing approach in Part C rather than the product-costing approach in Part A?

12-23 Pricing policies for conferences, average costing vs. direct-cost tracing. John Qualye, executive director of the North American Basketball League (NABL), is planning the second NABL Marketing Conference. Each of the 30 teams in the NABL can send two participants to this conference. The conference will run for three days and will look at basketball marketing opportunities and issues. The first conference, held last year in San Diego, California, was viewed, with one exception, to be a major success. This exception arose from the pricing policy Qualye chose.

Qualye selected what he thought was the simplest pricing policy imaginable. The total cost of the conference was $145,290. Twenty-nine teams sent representatives to San Diego. The average price charged for each team was

$$\frac{\$145{,}290}{29} = \$5{,}010 \text{ per team}$$

Qualye chose a conference-planning group—Successful Events—to handle all the planning. It negotiated a 30% discount from the airlines on business-class rates and a 25% discount from the hotel on room rates. All charges were submitted to Successful Events, which then combined them and gave the NABL the $145,290 bill. There were a total of 52 marketing officials from 29 teams attending. Six officials from the NABL headquarters also attended. These six officials each made presentations at the conference. In addition, two outside speakers were hired for the conference. The total bill sent by Successful Events to the NABL had six categories of costs:

Airfares (58 people)	$ 63,336
Ground transportation (58 people)	6,960
Hotel bill (room charges for 58 people)	26,738
Food and drinks	16,366
Conference materials, speakers, and hotel conference charges	20,640
Successful Events management fee	11,250
	$145,290

One team elected not to send any representatives to the San Diego conference. Two months before the conference, Peter Moore, president of the Anaheim Rollers (located in Anaheim, ninety miles from San Diego), raised with Qualye the issue of "price per participant." Qualye responded that it was to be based on the average cost per team attending. Moore objected to this pricing policy, calling it a "peanut butter" approach. He argued that it penalized teams like the Anaheim Rollers, which were interested in sending only one representative. Moreover, that representative would probably drive down the morning of the conference. After several stormy phone calls with Qualye, Moore decided that "as a matter of principle" the Anaheim Rollers would boycott the conference. Moore said that the last thing the Rollers were willing to do was to subsidize the "self-annointed world-champion Chicago Blizzards attending a conference in California."

The second NABL Marketing Conference will be held in Orlando, Florida. This time John Ribot, president of the Miami Surf, is objecting to the average-cost pricing policy. Moore of the Anaheim Rollers notified Qualye that two Roller representatives would attend the Orlando conference.

Qualye decided to examine alternative pricing policies. He asked Successful Events to provide information on the costs of each participant at the San Diego conference for airfares, ground transportation, and hotel bill. No details were kept on individual participant usage for the other three cost categories. Exhibit 12-10 presents the data for four teams Qualye selected to illustrate alternative pricing policies.

EXHIBIT 12-10

Directly Traceable Costs of Selected Participants at San Diego Conference

	Airfares	Ground Transportation	Hotel Bill
Chicago Blizzards			
Brian Adams	$ 947	$120	$383
Debbie Combs	947	146	504
Los Angeles Firebirds			
Alex Perez	185	140	392
Miami Surf			
Melissa Hyatt	1,124	102	405
José Gonzales	1,124	126	522
Montreal Patriots			
Rick Martel	1,642	94	496
Luke Williams	1,642	108	447

Required

1. What is the average cost per team for the airfare, ground transportation, and hotel bill costs categories? Use the directly traceable cost of each team as the "correct" cost of (charge for) the resources consumed by that team. Classify each of the four teams as undercosted or overcosted. Comment on the numbers you calculate and the classifications you make.
2. Qualye decides to examine pricing on an average cost per individual attendee from each team. Is this a better pricing approach than pricing on a per-team basis for the airfare, ground transportation, and hotel bill costs? Repeat requirement 1 using (a) the average cost per 52 attendees (average the costs of the six NABL officials across the 52 attendees) and (b) the directly traceable costs of the attendees in Exhibit 12-10. Comment on the results.
3. Name two advantages and two disadvantages of the average-cost approach to pricing the NABL marketing conference (either on an "average cost per team" or an "average cost per attendee" basis).
4. Recommend an alternative pricing policy for Qualye to adopt for the Orlando NABL conference. Your recommendations should cover all six categories of costs for the conference. What are two important advantages of your recommended alternative?
5. Two teams have approached Qualye about sending a third person to the Orlando conference. This additional person would be an observer only. The team would cover all out-of-pocket costs for the extra person. Both teams argued that there was zero incremental cost to having this person sit at the back of the room. How should Qualye respond to this approach?

CHAPTER 13

Management Control Systems: Choice and Application

Retailers like Talbot, Macy, and others use incentive plans to motivate personnel to increase sales.

Learning Objectives

When you have finished studying this chapter, you should be able to

1. Describe a management control system
2. Identify levels at which data may be gathered in a management control system
3. Describe the primary criterion used to judge a management control system
4. Distinguish how motivation, goal congruence, and effort relate to each other
5. Describe how engineered costs, discretionary costs, and committed costs differ
6. Explain how ordinary incremental budgets, priority incremental budgets, and zero-base budgets differ
7. Define work measurement; describe its use in an engineered-cost approach to controlling costs
8. Outline how financial and nonfinancial indicators can promote effectiveness and efficiency in discretionary-cost centers

Managers and accountants do not choose a particular management control system based on its technical aspects alone. It is also essential to consider the behavior of people who will use that system. This chapter presents an overview of management control systems. We emphasize behavioral issues in this overview. Our emphasis means that the material in this chapter is "softer" (there is less number crunching) than in most other chapters. Often there will not be a pat answer or, in some cases, even a systematic method of studying the issue. Nevertheless, knowing how to identify the central issue(s) when choosing systems is a skill in itself. This chapter provides a method for identifying central issues and weighing how alternative management control systems may help solve associated problems.

MANAGEMENT CONTROL SYSTEMS

Objective 1

Describe a management control system

A *management control system* is a means of gathering data to aid and coordinate the process of making planning and control decisions throughout the organization. The system improves the collective decisions within an organization.

Exhibit 13-1 shows four representative levels at which data can be gathered in a control system:

1. Customer/market level
2. Total-organization level
3. Individual-facility level
4. Individual-activity level

Objective 2

Identify levels at which data may be gathered in a management control system

Consider General Electric (GE). Data at the customer/market level include product quality, the time taken to respond to customer demands, and the cost of products marketed by GE's competitors. Data at the total-organization level (GE as a corporation) include net income, cash flow, return on assets, and total employment. GE has numerous individual facilities (research and development centers, manufacturing plants, distribution centers, customer-service centers, and so on). A management control system can collect both financial data—for example, labor costs and material costs—and nonfinancial data—for example, the number of product recalls—for each facility. Within each facility, there are many activity areas. Activities

EXHIBIT 13-1
Representative Levels of a Management Control System

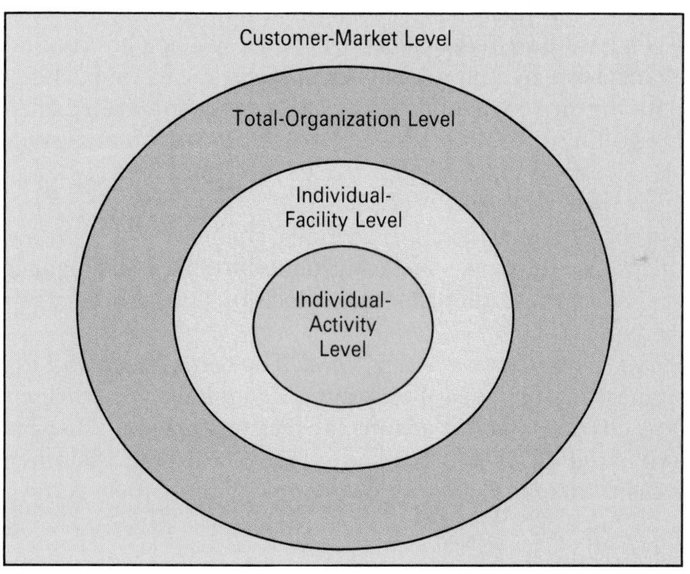

in a warehouse facility, for example, include receiving, storing, assembling, and dispatching. A management control system can collect data on each of these individual activity areas.

Organizations differ with regard to the data collected in their control systems on the four levels distinguished in Exhibit 13-1. Many organizations devote much effort to collecting data at the total-organization level, the individual-facility level, and the individual-activity level in their control systems. Less effort has been devoted to collecting data at the customer/market level. As organizations increasingly adopt a customer-oriented perspective, however, more attention is being given to including customer/market data in management control systems.

EVALUATING MANAGEMENT CONTROL SYSTEMS

What is a useful starting point for evaluating a management control system? *Obtain a specification of top management's goals for the organization as a whole.* Examples of goals include:

- Maximizing five-year reported net income
- Boosting the net-income-to-shareholders'-equity ratio into the top 20% of all publicly traded companies
- Maximizing return on investment to shareholders
- Attaining the largest market share in the industry
- Reaching the highest consumer-satisfaction rating

An observer may disagree with the goals specified by top management, but a control system should be judged in light of how well it helps achieve a given set of goals. For example, the main goal might be maximization of short-run net income, a goal many people regard as unworthy; nevertheless, if top management chooses such a goal, the system should be appraised in relation to it.

Objective 3

Describe the primary criterion used to judge a management control system

Cost, Congruence, Effort

The primary criterion for comparing management control system A against management control system B is how each system promotes the attainment of top management's goals in a cost-beneficial manner. Control systems are commodities. They benefit an organization by helping collective decision making. They cost money. The system with the largest favorable difference between benefits and costs should be chosen. Determining the benefits and costs of individual systems can be an imposing task. Two secondary criteria useful in making the cost-benefit criterion more specific are *goal congruence* and *effort*.

Goal congruence *exists when individuals and groups work toward the organization goals that top management desires.* Goal congruence occurs when managers, working in their own perceived best interests, make decisions that further the overall goals of top management.

Effort *is defined here as exertion toward a goal.* The concept of effort is not confined to its common meaning of physical exertion, such as a worker producing at a faster rate; it embraces all conscientious actions (physical and mental) that accompany the behavior of individuals. This effort can be guided toward the attainment of tangible goals such as cash, cars, and clothing or intangible items such as fun, self-esteem, and power.

Motivation

Considered together, goal congruence and effort are really subparts of *motivation:*

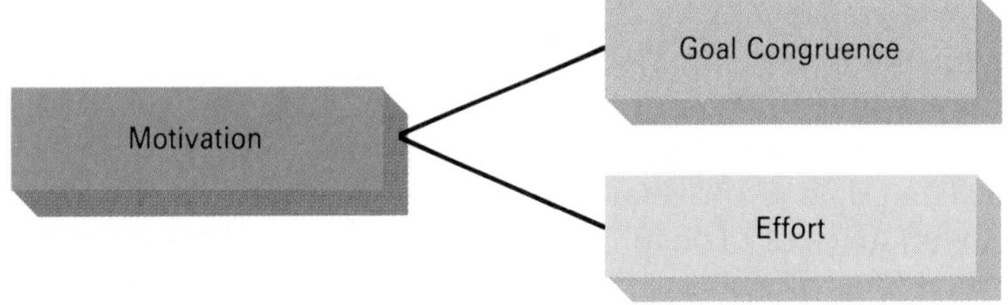

Objective 4

Distinguish how motivation, goal congruence, and effort relate to each other

Motivation is the desire for a selected goal (the goal-congruence aspect) combined with the resulting drive or pursuit toward that goal (the effort aspect).

Goal congruence and effort are often reinforcing aspects of motivation, but each can exist without the other. A manager can share an organizational goal of reaching a target sales volume or a product quality level. But the manager may pursue such goals unenthusiastically. Similarly, a manager may avidly pursue a goal that is incongruent with organizational goals. For example, a manager may increase advertising in an effort to gain sales when the organizational goal is to reduce advertising and instead build up the sales force.

The primary criterion for evaluating a system is how it promotes the attainment of top management's goals in a cost-beneficial manner. To apply that criterion, the motivational effects of system A and of system B deserve center attention. Why? Because the measurement of benefits and costs is based on how each system under consideration motivates individuals.

Formal and Informal Control Systems

The formal control system of an organization includes those explicit rules, procedures, performance-evaluation measures, and incentive plans that guide the behavior of its managers and employees. The management accounting system is one of several information systems that collectively constitute a formal control system. Other information systems pertain to employee relations, product quality, and compliance with environmental regulations.

The informal part of the management control system includes such aspects as shared values, loyalties, and mutual commitments among members of the organization and the unwritten norms about acceptable behavior for promotion. Companies have developed slogans, readily understandable to individuals in diverse parts of the organizations, to reinforce these values and loyalties: "IBM means service" and "At Ford, Quality is Job 1," for example.

Importance of Informal Systems

Many executives, including top managers, often rely on rules of thumb and accumulated experience rather than on the output of formal systems such as accounting reports. This informal method often proves satisfactory because the executives evaluate the information wisely and make decisions that keep their companies competitive.

As companies grow, however, managers become more dependent on formal and less dependent on informal information systems. A formal system provides for the orderly succession of management. Large companies must have such a system; after all, most managers do not have thirty years of experience as a basis for their decisions.

This book emphasizes one part of the formal control system—the cost and management accounting system—in much detail. However, the accounting system is only one of many factors that affect the value of the overall formal control system. Other factors include the stage of development of the organization, the business strategies, and the managerial style of the key senior executives.

Systems Goals and Risk Sharing

What is the role of risk in the design of a management control system? The potential of bearing risk changes behavior. Consider Lewis and Sterling, an accounting firm with offices in many different cities. Assume that the executive committee in New York has decided that the firm's best long-run interest calls for aggressively seeking clients in the banking industry. Suppose a Milwaukee bank seeks competitive bids for its audit. The Milwaukee managing partner of Lewis and Sterling may agree with the accounting firm's goal but be unwilling to make a competitive bid for the audit. The internal cost to the Milwaukee office of preparing a bid is $50,000. The partner views the likelihood of winning the audit engagement to be no more than 25%, even with a billing rate set 10% below normal. How can the executive committee encourage the Milwaukee office to take the 25% chance of winning the new audit client?

The executive committee in New York can take several steps to reduce the risks facing the Milwaukee office. For example:

1. It could fully reimburse the Milwaukee office for the $50,000 cost of preparing the bid, regardless of whether it is awarded the audit engagement.
2. It could award the Milwaukee office its normal billing rate through a subsidy, even though the actual billing rate to the banking client is below normal.

The effect of items 1 and 2 is to shift a major portion of the risk of bidding on the audit engagement from the Milwaukee office to the accounting firm as a whole. Ideally, the control system should encourage the local partner to accept or avoid the risks that top management desires.

General Guideline

A favorite pastime of accountants, managers, and professors regarding management control systems is "fault finding" and "system bashing." But the heart of judging a system is not in assessing its degree of imperfection. The central question should be, *How can the existing system, with all its imperfections, be improved?* A system exists to help managers in their collective decision making in organizations. ***Top managers should predict how system A and system B would affect the collective actions of managers and compare the projected results. In making such predictions, managers must consider the likely motivational effects (goal congruence and managerial effort) of each system.***

AN EXAMPLE OF MOTIVATION: SALES COMPENSATION PLANS

Compensation plans for sales personnel illustrate the importance of motivation, goal congruence, and effort. Organizations that ignore these factors can find themselves losing super performers and retaining poor performers.

Consider Horderns, a major department-store chain with over fifty individual stores. Each store is located in a large shopping mall. The longstanding policy at Horderns has been to pay sales personnel a fixed salary based on years with the company: $1,250 per month for sales personnel in their first year, increasing $125 per year to a maximum of $2,500 per month in the eleventh year. Top management at Horderns emphasizes increasing the sales of each store, but sales growth in recent years has been flat. In fact, the business has begun losing market share to Grace Brothers.

Six months ago, twenty of the top-performing salespeople at the Horderns store in Seattle switched to Grace Brothers. Grace Brothers pays all personnel a base salary of $800 per month and a sales commission of 4% of net sales made by each employee. Horderns decides to examine how the sales plan of Grace Brothers would have compensated two super-performing ex-employees—Stephen Kearny and Nina Landis—who left to join Grace Brothers. The company also wants to know how the sales plan would compensate the average salesperson (total net sales divided by the total number of salespeople).

Panel A of Exhibit 13-2 presents net sales figures for the last three months for Kearney, Landis, and the average Horderns salesperson at Seattle. There are dramatic differences in the sales made by individuals at the Horderns Seattle store. Such differences are common in sales organizations.

Panel B of Exhibit 13-2 shows compensation levels under the current Horderns plan and under the Grace Brothers plan. The current plan at Horderns rewards seniority. It does not highlight the importance of making sales. In contrast, the Grace Brothers plan explicitly rewards successful sales efforts. Consider Nina Landis, whose June sales were more than three times higher than those of the average Horderns salesperson. Under the Grace Brothers plan, Landis would have been paid $3,560 ($800 + 4% of $69,000). Under the existing plan at Horderns, she received $1,250.

There is no upper limit to the sales bonus at Grace Brothers. Superperformers are continually motivated to seek additional sales. It is not surprising that both Stephen Kearney and Nina Landis find the Grace Brothers plan more attractive. At Grace Brothers, goal congruence and effort reinforce each other to promote a high level of motivation toward top management's goal of sales growth.

Advances in Information Technology

An important component of the Grace Brothers management control system is the data on net sales made by each individual salesperson; tracking sales and sales returns requires extensive use of computer-based information systems. Advances

EXHIBIT 13-2

Net Sales by and Compensation Paid to Sales Staff at the Horderns Seattle Store

	April	May	June
Panel A: Net Sales Per Month at Horderns			
Stephen Kearney	$49,300	$56,700	$61,200
Nina Landis	54,200	67,000	69,000
Average Seattle salesperson	22,600	23,200	21,200
Panel B: Compensation			
Current plan at Horderns:			
Stephen Kearney	$1,500	$1,500	$1,500
Nina Landis	1,250	1,250	1,250
Average salesperson	2,100	2,095	2,200
Assuming Grace Brothers plan applied at Horderns:			
Stephen Kearney*	$2,772	$3,068	$3,248
Nina Landis†	2,968	3,480	3,560
Average salesperson‡	1,704	1,728	1,648

	April	May	June
*Stephen Kearney			
Salary base	$ 800	$ 800	$ 800
Commission (4%)	1,972	2,268	2,448
	$2,772	$3,068	$3,248
†Nina Landis			
Salary base	$ 800	$ 800	$ 800
Commission (4%)	2,168	2,680	2,760
	$2,968	$3,480	$3,560
‡Average salesperson			
Salary base	$ 800	$ 800	$ 800
Commission (4%)	904	928	848
	$1,704	$1,728	$1,648

in information technology are enabling retail companies to introduce reward systems that explicitly tie compensation to individual salesperson performance.

These advances are also central to the reward systems being introduced in other industries. For example, one courier company guarantees overnight delivery of packages. Its couriers receive a set bonus in dollars based on how they meet promised delivery schedules:

Percentage of Packages Delivered on Time	Percentage of Bonus Paid
100%	100%
99	80
98	60
97	40
96	20
95	None

This bonus plan clearly signals to the couriers the importance that top management places on meeting the delivery times promised to its customers. Operation of this compensation plan requires an information system that includes pickup and delivery times for every package. Couriers use hand-held recorders to enter this information into the company's computerized information system.

ENGINEERED, DISCRETIONARY, AND COMMITTED COSTS

Objective 5

Describe how engineered costs, discretionary costs, and committed costs differ

Costs can differ in the time horizons used for their planning and in the major accounting technique used in their control. This chapter examines the discretionary-cost type. Exhibit 13-3 summarizes how the discretionary-cost type differs from two other types of cost—engineered cost and committed cost—discussed in more detail elsewhere in this book. This section expands on Exhibit 13-3 and provides examples of engineered, discretionary, and committed costs. Differences among these three cost categories imply that a different mix of approaches to promoting effectiveness and efficiency is appropriate in each category.

Engineered Costs

Engineered costs are costs that result specifically from a clear-cut, measured relationship between inputs and outputs. This relationship is usually personally observable. Examples of inputs include direct material costs, energy costs, and labor costs. Examples of outputs include cars, computers, and telephones.

The major accounting techniques for controlling engineered costs are flexible budgets and standards. Chapters 7 and 8 provide a detailed discussion of these techniques. The feedback time is short, and nonfinancial measures are often used as the foundation of the systems. For example, direct material waste may be monitored as each unit is produced.

Discretionary Costs

Discretionary costs (sometimes called **managed costs** or **programmed costs**) have two important features: (1) they arise from periodic (usually yearly) decisions regarding the maximum outlay to be incurred, and (2) they are not tied to a clear cause-and-effect relationship between inputs and outputs. Examples include advertising, public relations, executive training, teaching, research, health care, and management consulting services. Subsequent sections of this chapter discuss techniques used to control discretionary costs.

The most noteworthy aspect of discretionary costs is that managers are seldom confident that the "correct" amounts are being spent. The founder of Lever Brothers, an international consumer-products company, once noted: "Half the money I spend on advertising is wasted; the trouble is, I don't know which half."

Committed Costs

Committed costs are costs that arise from having property, plant, equipment, and a functioning organization; little can be done in the short run to change committed costs. Capital-expenditure budgeting is the principal accounting technique for con-

EXHIBIT 13-3

Classification of Costs for Planning and Control

Type of Cost	Time Horizon for Planning and Feedback	Major Accounting Techniques for Control
1. Engineered	Short	Flexible budgets and standards
2. Discretionary	Longer	Negotiated static budgets
3. Committed	Longest	Capital-expenditure budgets

trolling committed costs. The planning horizon is long, as is the feedback time. Committed costs include depreciation, property taxes, insurance, and long-term lease rental. There is often a high level of uncertainty about the benefits or cash inflows associated with long-term capital-expenditure decisions.

Chapters 21 and 22 outline the formal decision models (such as discounted cash flow) that managers use in making choices about committed costs. These costs are relatively difficult to influence by short-run actions. Committed costs are affected primarily by long-run sales forecasts that, in turn, indicate the long-run capacity targets. Hence, careful long-range planning, rather than day-to-day monitoring, is the key to managing committed costs.

Once a building has been erected or a telecommunication satellite launched into space, little can be done in day-to-day operations to affect the *total level* of committed costs. From a control standpoint, the objective is usually to increase current utilization of facilities because this will ordinarily increase operating income.

CLASSIFYING COSTS IN FUNCTIONAL AREAS

The following functional-cost categories have been discussed in earlier chapters:

Research and Development	Product Design	Manufacturing	Marketing	Distribution	Customer Service

Engineered costs are most frequently found in manufacturing and, to a lesser extent, in distribution. Discretionary costs are typically found in research and development, product design, marketing, and customer service. Committed costs, in the form of property, plant, and equipment, are found in each of the functional-cost categories.

Exhibit 13-4 compares engineered costs with discretionary costs. The exhibit focuses on four areas: inputs, process, outputs, and the level of uncertainty. Further discussion of committed costs is deferred to Chapters 21 and 22.

Inputs

As Exhibit 13-4 indicates, the primary inputs that affect engineered costs are both physical and human resources. Examples are materials and labor in a manufacturing facility that assembles dishwashing machines. In contrast, the primary inputs that affect discretionary costs are human resources. Examples are salaries paid to research engineers, product designers, and sales representatives.

Process

Exhibit 13-4 shows how processes that are detailed and repetitive are prime candidates for using engineered-cost techniques. An example is the number of deliveries made by drivers from a distribution center. In contrast, discretionary costs are associated with processes that are sometimes labeled as *black boxes*. The latter can be defined as processes that are not easily understood. Knowledge of the precise process may be sketchy, unmeasurable, or unavailable. Examples are research and development and some aspects of product design.

EXHIBIT 13-4
Comparison of Engineered and Discretionary Costs*

	Engineered Costs	Discretionary Costs
1. Primary inputs	Physical and human resources	Human resources
2. Process	a. Detailed and physically observable b. Repetitive	a. Black box (because knowledge of process is sketchy or unavailable) b. Often nonrepetitive or nonroutine
3. Primary outputs	a. Products or quantifiable services b. Value easy to determine c. Quality easy to ascertain	a. Information b. Value difficult to determine c. Quality difficult to ascertain
4. Level of uncertainty	Moderate or small (for example, manufacturing setting)	Great (for example, marketing, lawsuit, research settings)

*This exhibit is a modification of one suggested by H. Itami.

Outputs

The contrast between engineered and discretionary costs is also evident when outputs are considered. *In particular, Exhibit 13-4 points out that the outputs linked with many discretionary-cost centers are types of information.* Examples are written or oral reports from the research, legal, or personnel department. The value and quality of such output are more difficult to measure than are the value and quality of output generated by engineered costs.

Level of Uncertainty

The level of uncertainty also helps to determine whether a cost may be engineered or discretionary. *Uncertainty* is defined here as the possibility that an actual amount will deviate from an expected amount. The higher the level of uncertainty about the relationship between inputs and outputs, the more likely a cost will be classified as a discretionary cost rather than as an engineered cost.

The marketing costs of footwear companies such as Nike and Reebok are typically classified as discretionary costs. There is a high level of uncertainty of the resulting footwear sales when a footwear company signs a multimillion-dollar contract with a sports or recording star. In contrast, material costs of footwear companies are classified as engineered costs. There is a low level of uncertainty regarding the direct materials cost for a particular type of sports shoe.

BUDGETING FOR DISCRETIONARY COSTS

Budgets are an important part of a management control system. They promote effectiveness and efficiency in discretionary-cost centers. Recall that Chapter 7 introduced the following definitions:

- *Effectiveness*—the attainment of a predetermined goal

- *Efficiency*—the relationship between inputs used and outputs achieved. The fewer the inputs used to obtain a given output, the greater the efficiency.

One manager expressed the difference as follows: "Effectiveness is doing the right thing, while efficiency is doing things right." We apply these two terms in the discussion that follows.

The most common accounting technique for controlling discretionary costs is a **negotiated static budget**. In such a budget, a fixed amount of costs is established through negotiations prior to the start of the budget period.

Classifying Negotiated Static Budgets

Objective 6

Explain how ordinary incremental budgets, priority incremental budgets, and zero-based budgets differ

Negotiated static budgets can be classified as ordinary incremental, priority incremental, and zero-base. Each will be discussed in turn.

Ordinary incremental budgets consider the previous period's budget and the actual results. A comparison of these figures may lead to a change to a more realistic budgeted amount for the coming period. Also, expectations for the new period may lead to changes from the previous period's budgeted amounts. For instance, a budget for a research department might be increased because of salary raises, the addition of personnel, or the introduction of a new project.

Priority incremental budgets are similar to incremental budgets. However, priority incremental budgets also include a description of what incremental activities or changes would occur (1) if the budget were increased by, say, 10% and (2) if the budget were decreased by a similar percentage. For example, a university sports department may decide that if its budget is cut by 10%, it will drop scholarships for football, for which it consistently fails to recruit nationally ranked athletes. This plan establishes priorities and forces the manager to think more carefully and concretely about operations. In a way, priority budgeting is a simple and economical compromise between ordinary incremental budgeting and zero-base budgeting.

Zero-base budgeting (ZBB) is budgeting from the ground up, as though the budget were being prepared for the first time. Every proposed expenditure comes under review. ZBB gets at bedrock questions by requiring managers to take the following major steps and to document them when possible:

1. Determine objectives, operations, and costs of all activities under the manager's jurisdiction.
2. Explore alternative means of conducting each activity.
3. Evaluate alternative budget amounts for various levels of effort for each activity.
4. Establish measures of workload and performance.
5. Rank all activities in the order of their importance to the organization.

Chapter 6 discussed many of the issues that arise in using ordinary incremental and priority incremental budgets. We now discuss issues that arise in using ZBB.

Applying Zero-Base Budgeting

Zero-base budgeting typically begins with the **decision units**, which are the lowest units in an organization for which a budget is prepared. A set of *decision packages* is prepared for each unit. These packages describe various levels of operations that the decision unit may perform. For example, a government agency may examine decision packages based on 80%, 100%, and 120% of the current budget appropriation. At Stanford University, the ZBB documentation includes the specification of the objectives, the description of the alternative means of accomplishing objectives, and the financial effects of four different levels of operations, two of which must be below the current level.

ZBB has its boosters and its critics. The strengths that successful adopters of ZBB cite relative to ordinary incremental budgets include the following:

1. Goals have been more sharply established and alternative means of reaching them have been explicitly considered. Justifying each project as if it were a new one forces managers to focus on how each project promotes the goals of the organization.
2. Managers have become more involved in a well-structured process that fosters communication and consensus.
3. Priorities among activities have been better pinpointed.
4. Knowledge and understanding of inputs and outputs have been enhanced. The budget process is generally more rational and less political than most.
5. Resources have been reallocated more effectively and efficiently.

There are many critics of ZBB. Common complaints include the following:

1. The time and cost of preparing the budget are much higher than in less-elaborate budgeting processes. Indeed, in some organizations the heavy paperwork involved in ZBB is referred to as "Zerox-Base Budgeting."
2. The determination of performance measures is difficult.
3. The specification of service levels, especially the minimum level of service, is threatening to many managers.
4. Top managers fail to follow through and use the ZBB information. In short, they do not really support the process.

Most of these weaknesses may result from a rigid application of ZBB. Managers should tailor their chosen budgeting technique to the peculiarities of their specific organization.

What major conclusions should we draw about ZBB in relation to other forms of budgeting? Consider the following:

1. The cost-benefit test will probably not be met if all departments conduct ZBB every year. However, zero-base "review" (a less-intensive form of zero-base budgeting) deserves serious consideration if it is conducted for all departments sequentially every five years or so. Meanwhile, incremental priority budgeting deserves annual use.
2. Unlike incremental budgeting, ZBB forces managers to systematically consider alternative ways of accomplishing objectives (for example, to use an inside cleaning staff or an outside cleaning service), a desirable attribute in any budgeting system.
3. If ZBB acquires a bad name and fades away, it will nevertheless have made one major contribution. Today, more than ever, there has been a decided increase in the prominence of budgeting in general. Inasmuch as intelligent budgeting is a help rather than a hindrance to management, almost any increase in attention to budgeting is a worthwhile contribution.

WORK MEASUREMENT

Objective 7

Define work measurement; describe its use in an engineered-cost approach to controlling costs

Work measurement is the careful analysis of a task, its size, the method used in its performance, and its efficiency. ***The objective of work measurement is to determine the workload in an operation and the number of workers needed to perform that work efficiently. Work measurement is one source of the standards used in cost-management approaches that rely heavily on engineered-cost relationships.***

Work-measurement techniques include:[1]

- *Micromotion study.* Film or videotape records of what a job entails and how long it takes. This technique is used most frequently in studying high-volume settings, such as a work station on an assembly line.

[1]For extensive discussion, see R. Failing, J. Janzen, and L. Blevins, *Improving Productivity Through Work Measurement: A Cooperative Approach* (New York: AICPA, 1988).

- *Work sampling.* A large number of random observations are made of the job. These observations are used to determine the number of steps for the job in its "normal" operating mode.

A **control-factor unit** is the measure of workload used in work measurement. Examples of control-factor units include the following:

Activity	Control-Factor Unit
• Computer software writing	• Lines of computer code written
• Material handling	• Number of incoming material items handled
• Delivery	• Number of individual delivery stops
• Customer service	• Number of customer complaints processed
• Accounts receivable (payable)	• Number of accounts receivable (payable) posted per hour

Dayton-Hudson, a U.S. retailer, is an enthusiastic advocate of work-measurement techniques. This company measures cartons handled, invoices processed, and store items ticketed on a per-hour basis. Productivity programs in its administrative and distribution areas have been aided by detailed work measures. The Target chain division of Dayton-Hudson increased the number of accounts payable processed from 9.7 to 15.0 per hour over a five-year period. (The total number of accounts payable processed each year by the Target chain division exceeds 4 million.)

BUDGETING NONMANUFACTURING COSTS

Alternative approaches exist for controlling many nonmanufacturing costs. A work-measurement approach regards these costs as engineered variable costs. In contrast, a discretionary-cost approach, more often found in practice, regards these costs as fixed.[2]

Assume that Family Farm employs five people (called customer representatives) to handle order processing for its mail-order catalog business. Each employee earns $1,800 per month and should, if operating efficiently, process 1,000 orders per month. In June, 4,700 individual orders were processed. The variances shown using the engineered-cost approach and the discretionary-cost approach are tabulated in the next section.

Essentially, Family Farm has two categories of decisions to make in the planning and control of personnel costs at its customer-order processing department:

1. *Personnel planning.* How many customer-order personnel are required? Should the business hire and lay off as the quantity of work fluctuates?
2. *Control.* How does the business effectively and efficiently control personnel resources on a day-to-day basis?

[2]One survey of controllers of U.S. manufacturing companies (with 223 responses) reports the following data on the approach that "best describes your company's experience with setting standards for nonmanufacturing costs":
a. Extensive use is made = 1.7%
b. Moderate use is made = 16.2%
c. Setting of standards tried, but found not to be of significant benefit = 3.9%
d. Setting of standards not yet attempted = 78.2%
See W. Cress and J. Pettijohn, "A Survey of Budget-related Planning and Control Policies and Procedures," *Journal of Accounting Education*, Vol. 3, No. 2, pp. 61–78.

Engineered-Cost Approach

The engineered approach to control of the customer-order payroll costs bases the budget formula on the unit cost of the individual order processed: \$1,800 ÷ 1,000 orders, or \$1.80 per order. Therefore, the flexible-budget allowance for customer-order personnel would be \$8,460 (4,700 × \$1.80). Assuming that the five people worked throughout the month at \$1,800 per month (giving an actual cost of \$9,000), the following performance report would be prepared:

Actual Cost	Flexible Budget: Total Standard Cost Allowed for Actual Output	Flexible-Budget Variance
\$9,000	\$8,460	\$540 U

The graphic representation of this situation appears in Exhibit 13-5. For simplicity, the flexible-budget variance is not divided into price and efficiency variances here.

The work-measurement approach to day-to-day control assumes a variable-cost budget and the complete divisibility of the workload into small units. Note that the *budget line* on the engineered-cost graph in Exhibit 13-5 is purely variable, even though the costs are actually incurred in steps.

The unfavorable variance of \$540 alerts management to the possibilities of overstaffing or inefficiency; the step of the cost-behavior pattern from \$7,200 to \$9,000 on the graph in Exhibit 13-5 was only partially utilized. The workload capability was 5,000 customer orders, not the 4,700 actually processed.

EXHIBIT 13-5

Comparison of Engineered-Cost Approach and Discretionary-Cost Approach at Family Farm

	Engineered-Cost Approach	Discretionary-Cost Approach
Actual costs incurred	\$9,000	\$6,000
Budget allowance	8,460*	6,000
Variance	540 U	0

*Rate = \$1,800 ÷ 1,000 orders = \$1.80 per order
Total = 4,700 orders at \$1.80 per order = \$8,460

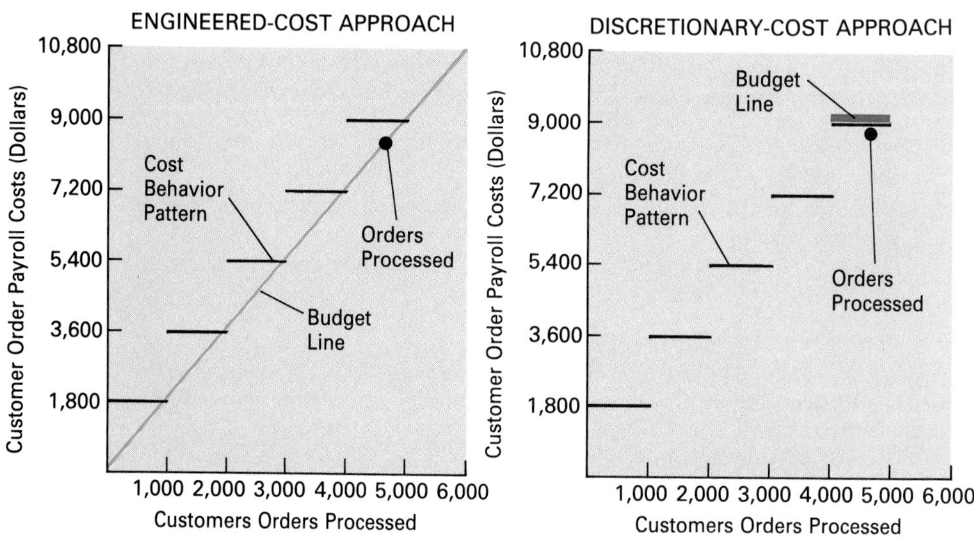

Discretionary-Cost Approach

The budget line in the discretionary-cost approach in Exhibit 13-5 is flat from 4,000 to 5,000. Looking at the right side of Exhibit 13-5 provides no insight into how to control costs between 4,000 and 5,000 customer orders. The primary means of control with a discretionary-cost approach is personal observation. That is, the department manager uses his or her experience to judge the size of the work force needed to carry out the department's functions. The manager is reluctant to overhire because of the corresponding reluctance to discharge or lay off people when volume slackens. Consequently, occasional peak loads are often met by hiring temporary workers or by having the regular employees work overtime.

There is a conflict between common practice and the objective of work measurement, which is to treat most costs as the engineered type and subject them to short-range management control. *The moral is that management's attitudes and its planning and controlling decisions often determine whether a cost is discretionary or engineered. A change in policy can transform a discretionary cost into an engineered cost and vice versa.*

Moreover, management may regard a cost as a discretionary cost for *cash-planning purposes* in the preparation of the master budget but may use the engineered-cost approach for *control purposes* in preparing flexible budgets for performance evaluation. These two views may be reconciled within the overall system. In our example, a master budget for Family Farm could conceivably include the following items:

Customer-order processing personnel:	
Flexible-budget allowance for control	$8,460
Expected unfavorable flexible-budget variance (due to conscious overstaffing)	540
Total budget allowance for cash planning	$9,000

A discretionary-cost approach stresses planning and de-emphasizes daily control through a formal work-measurement system. The company relies more on hiring capable people and less on monitoring their everyday performance.

Whether the engineered-cost approach is "better" than the discretionary-cost approach must be decided on a case-by-case basis, using a cost-benefit approach that focuses on how much operating decisions will be improved. Of course, the consultants who want to install work-measurement systems will typically try to demonstrate that their more costly formal systems will generate net benefits in the form of cost savings and better customer service.

FINANCIAL AND NONFINANCIAL MEASURES FOR A PERSONNEL DEPARTMENT

Objective 8

Outline how financial and nonfinancial indicators can promote effectiveness and efficiency in discretionary-cost centers

This section discusses the use of financial and nonfinancial indicators to promote effectiveness and efficiency in discretionary-cost centers. We use a personnel department to illustrate this approach.

Personnel departments are responsible for activities such as staffing, employee relations, compensation plans, and employee training and development. Salary cost is typically the largest single item in the total costs of the personnel department budget. The benefits from a well-functioning personnel department appear as higher revenues or lower costs in *other* departments and divisions of the organization. These benefits arise from higher-quality and more highly motivated employees being hired and retained by the organization.

What is the optimal amount of resources to spend on the personnel function in an organization? How can top management assess whether a personnel department is increasing the effectiveness or efficiency of its operation? Few organizations ask such questions, in part due to the difficulty in answering them. However, an emerging trend is the use of quantitative measures to monitor (at least) the efficiency of personnel operations.

Efficiency Measures

Exhibit 13-6 illustrates three variables used to monitor the efficiency of the hiring function of a personnel department—cost per hire, acceptance ratio, and average time to fill a vacancy. Consider the cost-per-hire ratio. Total costs of hiring include

1. Cost of advertising vacant positions
2. Fees paid to third parties to locate potential job applicants
3. Cost of the time and resources (for example, phone calls) of the personnel department in interviewing and processing applicants
4. Travel and related costs for staff and applicants
5. Relocation cost (if any) of hired employees
6. Cost of the time and resources used by nonpersonnel department individuals in interviewing job applicants

Once we compute the cost-per-hire, we can use two approaches to evaluate hiring efficiency:

- *Time-series approach:* comparison of the values of the cost-per-hire over time
- *Cross-sectional approach:* comparison of the cost-per-hire of one organization with the values of the same ratio for comparable organizations during the same time period

Exhibit 13-7 illustrates both approaches. We use data from Accounting Professionals Inc. (AP), a medium-size public accounting firm, in its efforts to recruit graduating college students. Exhibit 13-7 includes summary data for comparable accounting firms also recruiting college students. Data for comparisons among organizations may be available from an industry trade association or from a commercial organization.

Using the data in Exhibit 13-7, the personnel department of AP increased the efficiency of its hiring practices in the 19_1 to 19_5 period:

1. AP decreased its cost per hire from $3,800 in 19_1 to $2,800 in 19_5.
2. Comparable firms increased their cost per hire in the same period; for each of the 25, 50, and 75 percentiles, the cost per hire increased each year over the 19_1 to 19_5 period.

EXHIBIT 13-6
Illustrative Variables Used to Monitor the Efficiency of the Hiring Function of a Personnel Department

1. Cost per hire	$\frac{\text{Total hiring costs}}{\text{Total number of hirees}}$
2. Acceptance ratio	$\frac{\text{Number of job offers accepted}}{\text{Number of job offers extended}}$
3. Average time to fill a vacancy	Number of days, on average, between the day the approved vacant position requisition is received and the day the new hire starts work

Source: Saratoga Institute, Saratoga, CA.

EXHIBIT 13-7
Measure of Efficiency (Cost per Hire) at Accounting Professionals Inc. and Comparable Firms

Cost Per Hire	19_1	19_2	19_3	19_4	19_5
Accounting Professionals, Inc.	$3,800	$3,500	$3,300	$3,200	$2,800
Comparable firms*					
25 percentile	1,800	2,000	2,100	2,400	2,900
50 percentile	3,600	3,900	4,400	4,500	4,800
75 percentile	6,400	6,500	6,900	7,300	8,000

*Data for comparable firms are collected by an industry association body. The 25 percentile (50, 75) is that point on the cost-per-hire distribution below which 25% (50%, 75%) of respondents fall. For example, in 19_1, 25% of the firms spent less than $1,800 per hire.

3. In 19_1, the cost per hire for AP was above the 50 percentile of comparable firms. In 19_5, the cost per hire for AP was below the 25 percentile of comparable firms.

The cost-per-hire and the other two measures in Exhibit 13-6 focus only on the cost and speed with which the personnel department carries out its hiring function. None of these measures indicates the "quality" (effectiveness) of the people hired.

Effectiveness Measures

Did the increase in the efficiency of AP's hiring practices come at the expense of a decline in the effectiveness of its hiring practices? (*Effectiveness* means the hiring of "high-quality and strongly motivated" employees.) ***In a discretionary-cost center, such as a personnel department, short-run increases in efficiency indicators need not imply anything about effectiveness.*** For example, the graduating students most likely to accept a job offer from AP are those who are actively seeking employment but have as yet had no alternative job offers. In the short run, AP could reduce many of the individual hiring cost items by interviewing only these students. Such an approach would surely be counterproductive in the long run. The company needs to pursue students who have received offers from other firms. AP must compete successfully against other businesses to hire the highest-quality people, or the business will not compete successfully in the marketplace.

Quantitative measures of effectiveness in discretionary-cost centers often have a long-term focus. Two measures of interest to AP that can be computed only after a considerable lapse of time are:

$$\text{Retention ratio} = \frac{\text{Employees hired as entry-level staff in year 19\_X and still with AP in year 19\_Y}}{\text{Total employees hired as entry-level staff in year 19\_X}}$$

$$\text{Promotion-to-manager ratio} = \frac{\text{Employees hired as entry-level staff in year 19\_X and subsequently promoted to manager by year 19\_Y}}{\text{Total employees hired as entry-level staff in year 19\_X}}$$

The higher these two ratios, the higher the level of effectiveness AP has achieved in its hiring practices. (It is assumed that AP has some absolute standard for promotion to manager.)

Exhibit 13-8 presents these two ratios for AP and for comparable firms in the 19_6 to 19_10 period. Over this time, both AP's retention ratio and its promotion-

EXHIBIT 13-8
Measures of Effectiveness at Accounting Professionals Inc. and Comparable Firms

	19_6	19_7	19_8	19_9	19_10
A. Retention Ratio for 19_5 Hirees					
Accounting Professionals Inc.	.80	.65	.55	.50	.40
Comparable firms*					
25 percentile	.70	.58	.47	.42	.35
50 percentile	.78	.66	.58	.51	.49
75 percentile	.88	.74	.67	.58	.55
B. Promotion-to-Manager Ratio for 19_5 Hirees					
Accounting Professionals Inc.	.00	.02	.07	.10	.12
Comparable firms*					
25 percentile	.00	.01	.02	.03	.08
50 percentile	.00	.02	.06	.09	.11
75 percentile	.02	.06	.10	.14	.17

*Data for comparable firms are collected by an industry association body. The 25 percentile (50, 75) is that point on the distribution below which 25% (50%, 75%) of all respondents fall. For example, for every 10 hirees in 19_5, .80 × 10 = 8 were still with Accounting Professionals in 19_6. In contrast, 25% of comparable firms retained 7 of every 10 hirees (.70 × 10 = 7) for the same time span.

to-manager ratio were similar to the 50 percentile ratios for comparable firms. There is no evidence of a decline in the effectiveness of its recruiting policy in 19_5, which was done at a decreased cost-per-hire (see Exhibit 13-7). Note that the two effectiveness measures in Exhibit 13-8 reflect a combination of many factors. These include:

1. the effectiveness of the hiring decision in 19_5,
2. the actions of the supervisors of each hiree in years subsequent to 19_5,
3. the in-house training and development program of AP, and
4. the competitiveness of the salaries paid by AP.

Interfirm Comparisons

Interfirm comparisons, such as those in Exhibit 13-7 and 13-8, are one means of gaining insight about the relative effectiveness and efficiency of discretionary-cost centers. Several factors, however, limit widespread use of interfirm comparisons:

1. Nonavailability of data on comparable firms. In some cases, comparable data may not exist. In other cases, comparable data exist but are kept confidential.
2. Inconsistencies in the way firms collect and report data. For instance, public accounting firms may differ in how they record the cost of the time that nonpersonnel department individuals spend interviewing potential employees.

Interfirm comparisons provide evidence only on *relative* effectiveness and efficiency. Be careful to consider the importance of any changes in the total set of comparable data when making interfirm comparisons.

Importance of Nonfinancial Performance Measures

Review Exhibit 13-6 (p. 440). Note that only the first measure, cost per hire, is financial. ***Often, only a subset of the variables a company monitors are financial. In the long run, the nonfinancial measures (for example, the acceptance ratio and the average time to fill a vacancy, presented in Exhibit 13-6) may be equally important in giving management a picture of the organization's performance. Also, the accounting system may not include the costs necessary for computing certain financial*** *variables properly.* Consider the cost-per-hire. Many accounting systems do not track the costs of the time and resources of those individuals not in the personnel department who interview applicants. Omitting these costs from the measure means that it will be misleadingly low.

PROBLEM FOR SELF-STUDY

PROBLEM

The Bombay Co. has many small customers. Work measurement of billing activities has shown that each billing clerk can process 4,000 customer accounts each month. The company employs 15 billing clerks at a monthly salary of $1,600. The predicted number of customers for the next month is 56,300; the company had 59,900 in the last month.

Required

1. Assume that management has decided to continue to employ the 15 clerks despite the expected drop in billings. Show two approaches—(a) the engineered-cost approach and (b) the discretionary-cost approach—to budgeting billing-clerk labor. Show how the performance report for the next month would appear under each approach.
2. Some managers favor using tight budgets (based on the engineered-cost approach) as motivating devices for controlling operations. In these cases, managers really expect an unfavorable variance and must allow, in cash planning, for such a variance so that adequate cash will be available as needed. Explain how the flexible-budget variance under the engineered-cost approach could be used in cash planning.
3. Assume that the workers are reasonably efficient. (a) Interpret the flexible-budget variances under the engineered-cost approach and under the discretionary-cost approach. (b) What should management do to exert better control over the cost of billing-clerk labor?

SOLUTION

1. (a) Engineered-cost approach:

Standard unit rate = $1,600 ÷ 4,000 customer accounts
= $0.40 per customer account processed

	Actual Cost	Flexible Budget: Total Standard Cost Allowed for Actual Output	Flexible-Budget Variance
Billing-clerk labor	$24,000 (15 × $1,600)	$22,520 (56,300 × $0.40)	$1,480 U

(b) Discretionary-cost approach:

	Actual Cost	Flexible Budget: Total Standard Cost Allowed for Actual Output	Flexible-Budget Variance
Billing-clerk labor	$24,000	$24,000	$0

2. From 1(a), flexible-budget variance is $1,480 unfavorable. The master budget for financial planning must provide for monthly labor costs of $24,000. Therefore, if the engineered-cost approach were being used for control, the master budget might specify:

Billing-clerk labor	
Flexible-budget allowance	$22,520
Expected unfavorable flexible-budget variance	1,480
Total budget allowance for cash planning	$24,000

3. Management decisions and policies are often crucial in classifying a cost as either fixed or variable. If management refuses, as in this case, to control costs rigidly in accordance with short-run fluctuations in customer-account volume, these costs are discretionary. The unfavorable flexible-budget variance of $1,480 represents the cost that management is willing to pay in order to maintain a stable work force geared to "normal needs."

   Management should be given an approximation of the extra cost of maintaining a stable workforce. There is no single "right way" to keep managers informed on such matters. Two approaches were demonstrated in the previous parts of this problem. The important point is that clerical workloads and capability must be measured before effective control may be exerted. Such measures may be formal or informal. The latter is often achieved through a supervisor's regular observation of how efficiently work is being performed.

SUMMARY

Management control systems exist primarily to improve the collective decisions within an organization. All systems are imperfect. The central question usually is, How can a system, with all its defects, be improved? Using a cost-benefit philosophy, top managers must be conscious of the motivational effects of a particular system. That is, how does system A affect goal congruence and managerial effort in comparison with system B? Behavioral and organizational considerations deserve front and center attention when comparing alternative systems.

For budgetary-control purposes, costs are frequently divided into three categories: engineered, discretionary, and committed. Different accounting control techniques are used for each category. Discretionary costs are especially difficult to control because, by definition, well-specified relationships between inputs and outputs do not exist.

Engineered costs are most frequently found in the manufacturing and distribution areas. Some firms are now investing in efforts to extend control approaches used with engineered costs to areas such as product design, marketing, and customer service. No particular cost item should be viewed as inherently discretionary (or indeed, engineered or committed).

Accounting data play a key role in several approaches to promoting effectiveness and efficiency in discretionary-cost centers; these include the use of budgets and the monitoring of financial performance measures.

APPENDIX: PROMOTING AND MONITORING THE EFFECTIVENESS OR EFFICIENCY OF DISCRETIONARY-COST CENTERS

Exhibit 13-9 summarizes nine approaches used in various combinations to promote the effectiveness and efficiency of cost management in discretionary-cost centers. The text of this chapter discussed the first two approaches in Exhibit 13-9—the use of budgets and the monitoring of financial and nonfinancial performance measures.

Subsequent chapters cover topics related to several other approaches given in Exhibit 13-9. Chapter 27 discusses organization design and the competitive forces

EXHIBIT 13-9

Approaches to Promoting and Monitoring the Effectiveness or Efficiency of Discretionary-Cost Centers

Type of Approach	Example
1. Detailed analysis when preparing financial budgets	• Government departments using zero-base budgeting
2. Monitoring financial and nonfinancial indicators of effectiveness or efficiency on an ongoing and systematic basis	• University departments monitoring input success indicators (for example, the ratio of student acceptances to admission offers) or output success indicators (for example, the starting salaries of graduates)
3. Monitoring customer opinion and satisfaction indicators on an ongoing and systematic basis.	• Wall Street brokerage firm using an independent magazine survey of how investment managers rate the advice of security analysts
4. Organization design	• Keeping corporate (headquarters) functions to a minimum by, for example, allowing each division to have its own computer service center that reports to the president of the division rather than to a vice president of computer services at headquarters
5. Exposing discretionary-cost centers to competitive forces of the marketplace	• Prices charged to divisions for company-run executive-development programs being limited by the prices charged by external programs
6. Rewards for performance	• Running "fraction of the action" programs (for example, R & D staff having a percentage of the operating income associated with the commercial development of their R & D activity).
7. Leadership	• Head of a project team in an advertising agency being a charismatic and creative individual who promotes a high team spirit and a willingness in team members to work long and productive hours
8. Administrative approval mechanisms	• Detailed approval procedures checking the hiring of new personnel, upgrading of classifications, working of overtime, and so on
9. Promotion of organization culture	• High-technology firms promoting shared norms of strong loyalty and a commitment to technological leadership

of the market. Chapter 28 covers rewards for performance. Several of the approaches in Exhibit 13-9—for example, leadership and organization culture—typically use minimal accounting information, but this in no way diminishes their importance. Indeed, leadership is likely to be the single most important approach of the nine listed in Exhibit 13-9. However, it is dangerous to rely on leadership alone. What if a leader resigns to join a competitor? What if the leader is transferred within the company? What if the leader suffers health problems and cannot continue at the job?

TERMS TO LEARN

This chapter and the Glossary contain definitions of the following important terms:

committed costs *(p. 432)* control-factor unit *(437)* decision unit *(435)*
discretionary costs *(432)* effort *(428)* engineered costs *(432)*
goal congruence *(428)* managed costs *(432)* motivation *(428)*
negotiated static budget *(435)* ordinary incremental budget *(435)*
priority incremental budget *(435)* programmed costs *(432)*
work measurement *(436)* zero-base budgeting *(435)* ZBB *(435)*

ASSIGNMENT MATERIAL

QUESTIONS

13-1 What is a *management control system*?

13-2 What are four levels at which data can be gathered in a management control system? Give an example for each level.

13-3 What is the primary criterion for comparing management control system A against management control system B.

13-4 What is the relationship among motivation, goal congruence, and effort?

13-5 "The potential of bearing risk changes behavior." Do you agree? If yes, give an example.

13-6 "Advances in information technology permit the design of sales compensation plans previously not possible." Give an example.

13-7 Give three examples of *engineered costs*.

13-8 Give three examples of *discretionary costs*.

13-9 Give three examples of *committed costs*.

13-10 What are two synonyms for *discretionary costs*?

13-11 What are the most common outputs linked with discretionary costs?

13-12 Distinguish among an *ordinary incremental budget*, a *priority incremental budget*, and a *zero-base budget*.

13-13 What is *work measurement*? Describe two work-measurement techniques.

13-14 Give an example of how a management control system can motivate a manager to behave against the best interests of the company as a whole.

13-15 "I'm majoring in accounting. This study of human behavior is fruitless. You've got to be born with a flair for getting along with others. You can't learn it!" Do you agree? Why?

EXERCISES AND PROBLEMS

13-16 Goals of public accounting firms. All personnel, including partners, of public accounting firms must usually turn in biweekly time reports, showing how many hours were devoted to their various duties. These firms have traditionally looked unfavorably on idle or unassigned staff time. They have looked favorably on heavy percentages of chargeable (billable) time because this maximizes revenue.

Required
What effect is such a policy likely to have on the behavior of the firm's personnel? Can you relate this practice to the problem of goal congruence that was discussed in the chapter? How?

13-17 Budgets and motivation. You are working as a supervisor in a distribution department that has substantial amounts of personnel and equipment costs. You are paid a "base" salary that is actually low for this type of work. The firm has a very liberal bonus plan, which pays you another $1,000 each month you "make the budget" and 2% of the amount you are able to save.

Data on your past experiences follow:

Month	Jan.	Feb.	Mar.	Apr.	May	June
Budget	$40,000	$40,000	$39,000	$36,000	$36,000	$36,250
Actual	41,000	39,500	37,000	37,000	36,500	36,000
Variance	$ 1,000 U	$ 500 F	$ 2,000 F	$ 1,000 U	$ 500 U	$ 250 F

Required

1. What would you do if you were starting the job all over again from January and had the above information?
2. What would you recommend, if anything, be done to the bonus plan if you are now promoted to a higher job in management and required to handle the bonus plan in this department?

13-18 Budgets and motivation. Christine Everette operates a chain of health centers. One of the largest is in the heart of London. Everette uses the services of three managers, Dave, Nick, and Sid. Each is in charge of a group of exercise and apparatus rooms similar in all regards. When she hired them, she offered three methods of weekly payment:

Method X. A base rate of £6 per hour and 30% of all reductions in costs below a "norm" of £600 per week
Method Y. A flat wage of £7 per hour
Method Z. No base rate, but a bonus of £300 for meeting the "norm," plus 10% of all reductions in costs below the norm

The managers chose their method of compensation before starting employment. Assume a 40-hour week. The record for the past six weeks for the three areas follows. All data are in pounds.

	Weeks					
	1	2	3	4	5	6
Dave:						
Utilities	250	250	250	250	250	250
Supplies	180	20	20	260	100	20
Repairs & misc.	305	250	220	265	260	200
Total	735	520	490	775	610	470
Nick:						
Utilities	250	250	250	250	250	250
Supplies	100	100	100	100	100	100
Repairs & misc.	250	200	265	220	260	305
Total	600	550	615	570	610	655
Sid:						
Utilities	250	250	250	250	250	250
Supplies	100	100	100	100	100	100
Repairs & misc.	250	250	250	250	250	250
Total	600	600	600	600	600	600

Required

Which payment method most benefits Dave, Nick, and Sid? Assume that each manager picked a different method. Briefly explain your choices.

13-19 Incentive systems, motivation. Consider the following excerpt from the company employee newsletter of Superior Security Guards. It announces the creation of a Perpetual Trophy, which will be awarded to the branch with the "most outstanding performance" during each fiscal year.

> The Trophy will be several feet high and will move from branch to branch each year. Additionally, a small replica of the trophy will be awarded to the manager of the winning branch.
>
> The Trophy will be awarded to the branch with the best combination of percentage increases in sales and operating income. The actual calculation will be to add the sales percentage increase over the prior year and the operating income percentage increase over the prior year and divide the result by two, which gives equal weight to both sales and operating income percentage growth. Only branches achieving a minimum 15% sales increase will be eligible.

Superior Security Guards has branches in four cities. Details on sales and operating income of the four branches for the most recent four years are (in millions):

	19_4	19_5	19_6	19_7
Sales				
Iowa City branch	$62.4	$75.6	$94.3	$110.4
Kansas City branch	23.7	25.9	28.9	32.8
New Orleans branch	60.5	85.1	122.7	153.8
St. Louis branch	80.7	86.2	99.8	111.3
Operating Income				
Iowa City branch	$ 9.9	$10.5	$15.1	$18.7
Kansas City branch	5.7	5.4	2.6	7.2
New Orleans branch	6.2	8.3	15.9	21.5
St. Louis branch	14.5	14.6	20.9	21.1

Required

1. Which of the four branches won the trophy for the 19_7 year?
2. Comment on the way Superior Security Guards measures "most outstanding performance."

13-20 Incentive systems for used-car sales personnel. Ted Dexter is owner of Regal Motors, a used-car sales company. Regal Motors has sales lots in five different cities. For many years, Dexter has paid his salespeople a monthly salary plus a fixed dollar bonus for each car sold. In March 19_6, the monthly salary was $1,600 and the fixed dollar bonus was $25 per car sold.

In April 19_6, Dexter received information that Fred Truman, one of his best salespersons, was about to resign and switch to Cowdrey Motors. Cowdrey Motors is owned in part by Michael Cowdrey, a recently retired racing driver superstar. Dexter called Truman to discuss his pending resignation. Truman told him the decision was a "tough personal one" but felt Cowdrey offered him a better deal. Salespeople at Cowdrey Motors receive a base monthly salary of $1,200 plus a bonus of 6% of the gross margin on each car sold. *Gross margin* is defined as the selling price minus the purchase cost and all other directly traceable costs incurred before that car is put on the lot for sale.

Dexter collects the following information on Truman, two other salespersons, and the average for all 48 salespeople of Regal Motors for March 19_6:

	Fred Truman	Louise Hutton	Brian Statham	Average for 48 Salespeople
Cars sold	37	16	46	21
Average selling price	$9,000	$10,000	$8,000	$8,400
Average purchase cost and other traceable costs	$7,900	$9,200	$7,800	$7,950

Required

1. Compute the March 19_6 salary and bonus payments Regal Motors paid to Truman, Hutton, Statham, and the average salesperson.

2. Compute the March 19_6 salary and bonus payments that Regal Motors would have paid if it used Cowdrey's payment structure (a base monthly salary of $1,200 plus a bonus of 6% of the gross margin on each car sold).
3. Compare and contrast the Regal Motors and the Cowdrey Motors compensation plans for salespeople.
4. What other factors might be important for Ted Dexter to consider in rewarding the salespeople at Regal Motors?

13-21 Risk sharing, motivation, new clients for advertising agency. The Media Impact Group (MIG) is the second largest advertising agency in Canada with offices in Vancouver, Calgary, Edmonton, Regina, Winnipeg, Toronto, Montreal, Quebec City, and Halifax. The head office is in Toronto. The partners of the firm have agreed to place high priority on developing a larger client base in the lodging industry.

Quality Inns is a major motel chain with over 100 motels across Canada. Its headquarters is in Vancouver. The managing director has called for bids for its advertising account. Quality Inns' current advertising has been described in advertising circles as "uninspiring, nonmemorable, and without focus."

Trent Natham, the partner in charge of MIG's Vancouver office, is unenthusiastic about devoting large amounts of the resources of the Vancouver office to bidding on the Quality Inns account. He views the minimum cost of preparing a credible bid to be $100,000. Moreover, Quality Inns has informed all potential bidders that it will not consider any bid with an average billing rate higher than $60 an hour. The Vancouver office of MIG currently averages $70 an hour on its billing rate to its clients. Moreover, Natham estimates that the Vancouver office would receive only 20% of the revenues of the Quality Inns account. The Toronto and Montreal offices would jointly receive over 50% of the revenues of the Quality Inns account. Competition for the account will be intense with at least six other advertising agencies preparing bids.

The head office views each MIG office as a profit center. Over 50% of Natham's annual bonus is based on profitability for that year at the Vancouver office. The current year has been a highly successful one for the Vancouver office. Many partners have worked 60 hours each week for months at a time.

Required

1. Why is Natham unenthusiastic about making a serious bid for the Quality Inns advertising account?
2. How might Natham be encouraged by the head office in Toronto to make a serious bid for the Quality Inns advertising account?

13-22 Risk sharing, division compensation plans. Steve Simpson is manager of the oil exploration and production division of Ampet. Ampet is 100% owned by its president, Debbie Maceli. Simpson is currently choosing between two ways of spending the remaining $100 million in his 19_4 exploration budget.

(a) Purchase an oil lease in Venezuela, South America, in 19_4. This lease has a high probability of producing commercially viable oil. Extensive exploration work has already been conducted by the current owner. The lease would cost Ampet $100 million cash in 19_4. The cash inflows from the two possible outcomes from this $100 million lease purchase are projected to be (in millions):

	19_4	19_5	19_6	19_7
1. Commercial oil found (90% likelihood)	$70	$60	$40	$20
2. No commercial oil found (10% likelihood)	0	0	0	0

(b) Invest $100 million in exploration in Australia in 19_4. The cash inflows from the two possible outcomes for this $100 million exploration are projected to be (in millions):

	19_4	19_5	19_6	19_7
1. Commercial oil found (30% likelihood)	$10	$340	$390	$200
2. No commercial oil found (70% likelihood)	0	0	0	0

Simpson currently receives a fixed salary and a bonus based on the annual operating income of the oil exploration and production division. He has a low tolerance for bearing risk. His preference is to buy developed oil leases. Much of the uncertainty associated with oil exploration has already been removed from the Venezuela option.

Maceli has three objectives for Ampet: (1) to be a highly profitable oil and gas company over the next five years, (2) to increase the assets Ampet has invested outside of South America (for diversification purposes), and (3) to increase the likelihood that Ampet will become a major oil and gas company in the next decade.

Required

1. Assume you are Simpson. Would you prefer the Venezuelan alternative or the Australian alternative? Why?
2. Assume you are Maceli. You are concerned specifically about the current decision Simpson must make and more generally about ongoing decisions being made in the oil exploration and development division. What changes would you make in the compensation plan for the manager of this division?

13-23 Work measurement, cost control in a warehouse. The manager of a warehouse for a mail-order firm is concerned with the control of her fixed costs. She has recently applied work-measurement techniques and an engineered-cost approach to the staff of order clerks and is wondering if a similar technique could be applied to the workers who collect merchandise in the warehouse and bring it to the area where orders are assembled for shipment.

The warehouse assistant contends that this should not be done, because the present work force of twenty persons should be viewed as a fixed cost necessary to handle the usual volume of orders with a minimum of delay. These employees work a forty-hour week at $15 per hour.

Preliminary studies show that it takes an average of twelve minutes for a worker to locate an article and move it to the order-assembly area and that the average order is for two different articles. At present 1,800 orders are processed per week.

Required

1. For the present volume of orders, prepare the weekly performance report using (a) a discretionary-cost approach and (b) an engineered-cost approach for the weekly performance report.
2. Repeat requirement 1 for volume levels of 1,600 and 1,400 orders per week.
3. What other factors should be compared with the cost variances found in requirements 1 and 2 in order to make a decision on the size of the work force?

13-24 Work measurement, cost control in customer service. Venture Vision is a major cable television operating company. Three months ago (March 19_4) it acquired Cable Galore, the largest cable television operator in California. Cable Galore had 120,000 subscribers, many of whom were not happy with their service. Local newspapers had run several articles in which subscribers voiced their complaints about service under the Cable Galore ownership. One article was titled "Cable Galore Becomes Cable Screw-Up." One customer reported "reading a novel" while waiting for his telephone call to come to the "top of the list."

Venture Vision decided to work fast and effectively on the customer complaint problem. It immediately hired more customer-service representatives. It also sent trouble-shooters to eliminate the causes of the customer complaints and to give personnel additional training.

Venture Vision hired six full-time customer-service representatives, paying each of them a salary of $2,000 a month to handle customer complaints. The two customer-service representatives employed by Cable Galore resigned in March 19_4. The budget for the next three months (April to June) assumed each representative would handle 5 calls (complaints) an hour, work 8 hours a day (8 A.M. to 5 P.M. with 1 hour of breaks) for 20 working days (Monday through Friday) a month. The actual number of calls in the April to June 19_4 period were:

Month	Number of Calls
April	4,900
May	4,200
June	3,300

Required

1. Show the performance report each month (April, May, and June) under (a) the engineered-cost approach, and (b) the discretionary-cost approach. Comment on the different inferences that might be drawn from (a) and (b).
2. What changes should Venture Vision consider in the staffing of its customer-service area?

13-25 Discretionary-cost center, research and development function of a computer software company. Susan Teece, president of Software Advance Inc., has a set of problems many of her competitors would love to have. Teece has a large cash fund of $50 million that could be allocated to research and development. Her twin problems are (a) how much of the $50 million to spend on R & D and (b) how to ensure that the amount budgeted for R & D is effectively and efficiently spent.

The sales, operating income, and R & D expenditures of Software Advance over the last four years (19_1 to 19_4) are (in millions):

	19_1	19_2	19_3	19_4
Sales	$ 2	$12	$47	$86
R & D	0.5	2	8	15
Operating income	(1)	3	11	16

Software Advance's success is based almost exclusively on Insight 1-2-3, a spreadsheet program for personal computers. The Insight program was developed by Jeff Moore, a major shareholder in Software Advance and head of its research and development department. Over 30% of the R & D budget in 19_3 and 19_4 was paid to independent contractors who operate out of their own homes or private offices. Despite $23 million being spent on R & D in 19_3 and 19_4, Software Advance has not been successful in developing a second generation of products to continue its recent high growth rate of sales.

The 19_3 and 19_4 R & D budgets were based on the 19_2 percentage of R & D to sales, the year in which the Insight 1-2-3 program was developed. Teece used this fixed percentage-of-sales approach because it seemed a convenient way to reach a decision quickly. Moreover, the percentage approach allowed Teece to handle her many pressing marketing problems of 19_3 and 19_4.

The electronics industry trade association reports the following 19_4 summary data for 33 software-development companies, classified by firm size:

	Firm Size Category by Sales ($ millions)			
	1–5	5–10	10–20	20–100
Number of firms	18	5	5	5
R & D as a percentage of sales	25.9%	18.4%	15.0%	7.4%

Required

1. Why might the R & D department of Software Advance be viewed as a discretionary-cost center?
2. What limitations arise from use of the fixed percentage-of-sales approach to deciding the size of the annual R & D budget?
3. How might the industry trade association data be useful in the R & D decisions facing Teece?
4. How might Teece and Moore help ensure that the R & D budget of Software Advance is effectively and efficiently spent?

13-26 Discretionary cost center, promoting effectiveness and efficiency in an engineering department. The Sharp Company develops, manufactures, and markets several product lines of low-cost consumer goods. Top management of the company is attempting to evaluate the present method and a new method of charging the different production departments for the services they receive from one of the engineering departments, which is called Manufacturing Engineering Services (MES).

The function of MES, which consists of about thirty engineers and ten drafters, is to reduce the costs of producing the different products of the company by improving machine and manufacturing process design while maintaining the required level of quality. The MES manager reports to the engineering supervisor, who reports to the vice president, manufacturing. The MES manager, George Amershi, may increase or decrease the number of engineers under him. He is evaluated on the basis of several variables, one of which is the difference between the budgeted costs and the actual costs of the department. These costs consist of actual salaries, a share of corporate overhead, the cost of office supplies his department uses, and a cost of capital charge. An individual engineer is evaluated on the basis of the annual savings he or she brings about in relation to annual salary. The salary range of an engineer is defined by personnel classification. There are four classifications, and promotion from one classification to another depends on the approval of a panel that includes both production and engineering personnel.

Production department managers report to a production supervisor, who reports to the vice president, manufacturing. The production department for each product line is treated as a profit center, and engineering services are provided at a cost, according to the following plan. When a production department manager and an engineer agree on a possible project to improve production efficiency, they sign a contract that specifies the scope of the project, the estimated savings to be realized, the probability of success, and the number of engineering hours of each personnel classification required. The charge to the particular production department is the number of hours required multiplied by the "classification rate" for each personnel classification. This rate depends on the average salary for the classification involved and a share of the engineering department's other costs. An engineer is expected to spend at least 85% of his or her time on specific, contracted projects; the remainder may be used for preliminary investigations of potential cost-saving projects or self-improving study. A recent survey showed that production managers have a high degree of confidence in the MES engineers.

A new plan has been proposed to top management. No charge will be made to production departments for engineering services. In all other respects, the new system will be identical to the present. Production managers will continue to request engineering services, as under the present plan. Proponents of the new plan believe that production managers will take a greater advantage of existing engineering talent. Regardless of how engineering services are accounted for, the company is committed to the idea of production departments being profit centers.

Required

Evaluate the strong and weak points of the present and proposed plans. Will the company tend to hire the optimal quantity of engineering talent? Will this engineering talent be used as effectively as possible?

13-27 Incentives, bonus plans for college textbook sales representatives. St. George Press is a publisher of business books for the college market. It employs 30 sales representatives to cover the North American market. These representatives call on professors for two reasons:

- To promote St. George Press books as the textbooks for college courses
- To gather leads on potential authors for new books to be published by St. George Press

Glenda Burgess, the president of St. George Press, recently received complaints from Brian Johnston, a new sales representative for the Illinois region in 19_5. He complained that under the existing bonus plan, he would not receive a single dollar despite delivering what he viewed as a stellar performance. Johnston's predecessor was Michael Beattie, the superstar salesperson in the St. George Press organization. Many Illinois professors affectionately called Beattie "Mr. St. George Press." In his first six months on the job, Johnston was told numerous times by professors that "he had big shoes to fill."

Burgess decides to examine how a proposed change in the existing bonus plan would affect the compensation of sales representatives in the Illinois and the New York regions.

In the existing plan, a bonus is 5% of the net sales gain over the prior year:

$$5\% \times [\text{Net sales in year } t - \text{Net sales in year } t-1]$$

There is no bonus or penalty if net sales of this year are less than last year's. Net sales in year t are defined as gross sales in year t minus sales returns in year t.

The proposed bonus plan has two components:

(a) Sales bonus, calculated as:

Bonus Rate	Net Sales
None	Up to quota
1%	Quota to goal
8%	Over goal

Quota is 80% of last year's net sales in the same territory for existing sales representatives and 70% of last year's net sales for sales representatives in their first year in a territory.

Goal is 105% of last year's net sales in the same territory for existing sales representatives and 95% of last year's net sales for sales representatives in their first year in a territory.

(b) Manuscript signing on bonus, calculated as:

A. Signing an author with a new book proposal in an early development stage: $100

B. Signing an author with a new book in an advanced stage of completion: $200

Burgess collects the following data on the two sales regions over the 19_1 to 19_5 period:

Illinois Sales Region	19_1	19_2	19_3	19_4	19_5
Sales representative	Beattie	Beattie	Beattie	Beattie	Johnston
Gross sales	$600,000	$640,000	$670,000	$690,000	$660,000
Sales returns	$ 30,000	$ 32,000	$ 20,000	$ 30,000	$ 28,000
"Early development stage" books signed	3	2	3	2	1
"Advanced completion stage" books signed	1	1	2	1	0

New York Sales Region	19_1	19_2	19_3	19_4	19_5
Sales representative	Adams	Adams	Noke	Noke	Noke
Gross sales	$400,000	$405,000	$360,000	$390,000	$380,000
Sales returns	$ 20,000	$ 15,000	$ 22,000	$ 18,000	$ 24,000
"Early development stage" books signed	2	2	1	2	1
"Advanced completion stage" books signed	0	1	0	1	0

Required

1. Compute the bonus paid each year (19_2 to 19_5) to the Illinois and New York sales representatives under the existing plan. Compute the bonus that would have been paid under the proposed plan. Comment on the differences in bonuses under the two plans.
2. Critically evaluate the proposed bonus plan. Outline three ways that the proposed bonus plan could be improved. Be specific.
3. Burgess was concerned about sales representatives being excessively generous with the complimentary copies of St. George Press books they give to professors as an inducement to adopt one or more St. George Press books for their courses. Outline two ways the bonus plan could take this factor into account. (Be specific. Include details on how the cost of complimentary books distributed should be estimated when calculating the bonus of a sales representative.)

CHAPTER 14

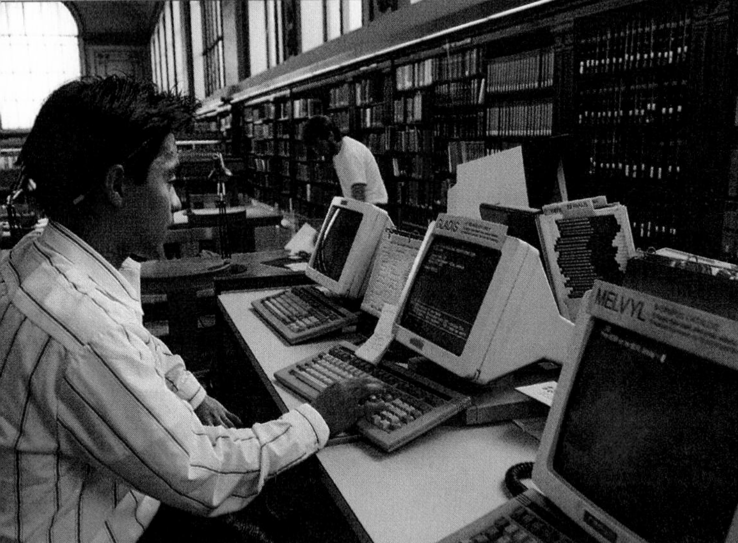

How should university costs, like the costs of a library, be split among departments? Cost allocation provides answers.

Cost Allocation: I

The terminology of cost allocation

Stages I, II, and III cost allocations

Purposes of cost allocation

Criteria to guide cost allocation decisions

The general process of cost allocation

Allocating costs from one department to another

Allocating costs of service departments

Allocation of common costs

Trends in cost allocation practices

Learning Objectives

When you have finished studying this chapter, you should be able to

1. Explain the meaning of key terms
2. Outline four purposes for allocating costs
3. Describe alternative criteria for guiding cost allocation decisions
4. Explain four interdependent decisions for allocating costs
5. Describe cost pool homogeneity and its impact on choosing cost pools
6. Distinguish among direct allocation, step-down, and reciprocal methods of allocating service department costs
7. Outline incremental and stand-alone common cost allocation methods
8. Describe current trends occurring in companies' cost allocation practices

Cost allocation is an inescapable problem in nearly every organization and in nearly every facet of accounting. How should the costs of shared services be allocated among departments? How should university costs be split among undergraduate programs, graduate programs, and research? How should the costs of expensive medical equipment, facilities, and staff be allocated in a hospital? How should manufacturing overhead be allocated to individual products in a multiproduct company?

Finding answers to cost allocation questions is difficult. The answers are seldom clearly right or clearly wrong. Nevertheless, we will try to obtain some insight into cost allocation and to understand the dimensions of the questions, even if the answers seem elusive. You will undoubtedly be faced with cost allocation questions in your career.

THE TERMINOLOGY OF COST ALLOCATION

Objective 1

Explain the meaning of key terms

The terms relating to cost allocation vary throughout the literature and among organizations. Be sure that you and the people with whom you are dealing agree on the meaning of key terms when you confront allocation problems in practice. Important terms used in this chapter include the following:

- *Cost object* (also called *cost objective*) is any activity or item for which a separate measurement of costs is desired.
- *Cost allocation* is a general term that refers to identifying accumulated costs with, or tracing accumulated costs to, cost objects such as activity areas, departments, divisions, territories, or products.[1]
- *Cost application* is a narrower term that refers to a special type of cost allocation, the final assignment of costs to products, as distinguished from activity areas, departments, divisions, or territories.
- *Cost allocation base* is the common denominator for systematically linking a cost or group of costs with a cost object. A *cost application* base describes those cost allocation bases where the cost object is a product (or service).

[1]There are other terms that may be used with assorted shades of meaning: cost reallocation, cost assignment and reassignment, cost apportionment and reapportionment, cost distribution and redistribution, and cost tracing and retracing.

- *Direct cost* is a cost that can be identified specifically with or traced to a given cost object in an economically feasible way.
- *Indirect cost* is a cost that cannot be identified specifically with or traced to a given cost object in an economically feasible way.

Examples of direct and indirect costs for an activity area and a product follow:

Cost Object	Example of a Direct Cost	Example of an Indirect Cost
Activity area: Document-photocopying area of a law firm	Paper and liquids used in photocopying machine	Electricity used to run machine. Electricity metered to firm but not to individual machines
Product: Microwave oven of a home appliance company	Materials assembled in microwave oven	Rent for factory building. Rent is paid for by the company, which manufactures 200 different products

Organizations differ in how they classify costs. A direct cost item in one organization, such as assembly labor or energy, can be an indirect cost item in another. Moreover, changes in classification can occur over time. For example, advances in information-gathering technology, such as bar coding and optical scanning, now enable organizations to change the classification of some items from indirect costs to direct costs. For example, bar codes on individual cutting tools enable some machining shops to trace the costs of these cutting tools directly to products that used them rather than allocating their aggregate costs to products on the basis of (say) machine hours.

STAGES I, II, AND III COST ALLOCATIONS

Managers often distinguish three stages of cost allocation. These three stages help identify different areas where refinements in the costing of cost objects can be made:

- *Stage I Allocation:* Tracing direct costs to cost objects. These cost objects include activity areas, departments, divisions, territories, and products.
- *Stage II Allocation:* Allocating and reallocating costs from one cost object to another cost object (except a product cost object). Examples include (a) the allocation of corporate legal department costs to operating divisions and (b) the allocation of plant maintenance department costs to the machining and assembly production departments.
- *Stage III Allocation:* Allocating indirect costs to products (or services). The term *cost application* is often used to describe Stage III cost allocations. An example is the application of the costs of a quality testing area of an assembly line to products, using hours of testing time as the application base.

Costs may move through all three stages of cost allocation. For example, consider the salary of a plant maintenance supervisor at National Electric Inc. This salary is a direct cost of the plant maintenance department (a Stage I allocation). Plant maintenance department costs are then allocated to the assembly and testing production departments (a Stage II allocation). Finally, costs of each production department are applied to individual products such as a television set (a Stage III allocation). Exhibit 14-1 presents an overview of these three stages using the product-costing format from Chapters 4, 5, and 12. This exhibit is simplified to highlight how Stages I, II, and III relate to the costing of a product. In an actual setting, plant maintenance department costs would be allocated to more than one

EXHIBIT 14-1

Illustration of Stage I, Stage II, and Stage III Cost Allocations at National Electric Inc.

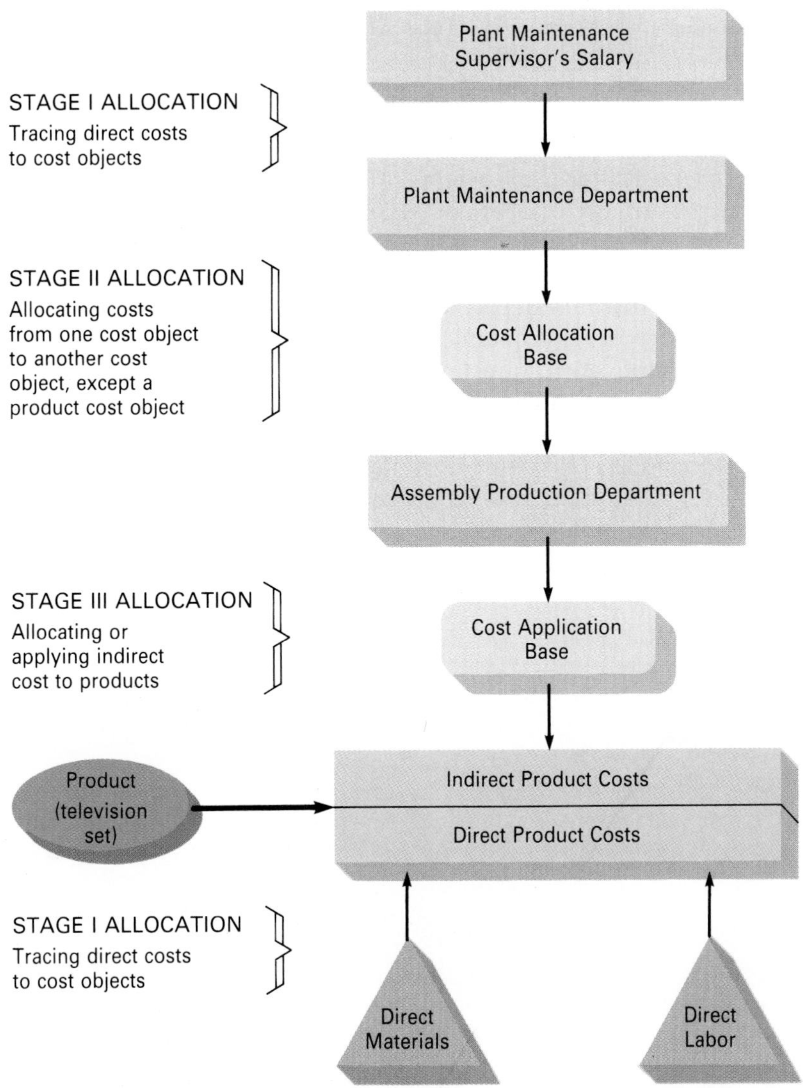

production department, and the assembly production department costs would be allocated to more than one product.

Sometimes Stage II cost allocations are termed *intermediate allocations* and Stage III allocations are called *final allocations*.[2] Although all the topics in cost allocation are heavily interrelated, Chapters 14 and 15 have different focuses. Chapter 14 first examines the purposes served by cost allocation and the general process of cost allocation. It then examines Stage I and Stage II allocations. Chapter 15 discusses topics related to Stage III allocations.

The accuracy of product costs can be increased by increasing the percentage of total costs that are directly traced (Stage I), by refining intermediate allocations (Stage II), and by refining final allocations (Stage III).

[2]Some authors restrict the term *cost allocation* to include only the Stage II and Stage III allocations. See R. Cooper, "The Two-Stage Procedure in Cost Accounting: Part One," *Journal of Cost Management*, Summer 1987, pp. 43–51.

PURPOSES OF COST ALLOCATION

Objective 2

Outline four purposes for allocating costs

Exhibit 14-2 outlines and illustrates four purposes for allocating costs:

1. To make economic decisions for resource allocation
2. To motivate managers and employees
3. To measure income and assets for reporting to internal and external parties
4. To justify costs or compute reimbursement

Ideally, a single-cost allocation would satisfy all purposes simultaneously. Consider the acquisition cost of a direct material such as airplane landing gear by Boeing. All four purposes would require that this acquisition cost be allocated to the cost of assembling the airplane. In contrast, consider the salary of a scientist in a central corporate research department of Boeing. This salary cost may be allocated to satisfy purpose 1 (economic decisions); may or may not be allocated to satisfy purpose 2 (motivation); must not be allocated (under GAAP) to inventory to satisfy purpose 3 (income and asset measurement for reporting to external parties), as it is not an inventoriable cost; and, depending on the laws or regulations, may or may not be allocated to a government contract to justify a cost to be reimbursed to satisfy purpose 4 (cost justification). When all four purposes are unattainable simultaneously, the dominant purpose of making the allocation should guide the cost allocation.

EXHIBIT 14-2
Purposes of Cost Allocation

Purpose	Illustrations
1. To make economic decisions for resource allocation	• To decide whether to add a new airline flight • To decide whether to make a component part of a television set or to purchase it from an outside vendor • To decide what price to charge for a customized product
2. To motivate managers and employees	• To encourage or discourage the use of services such as internal audit, computers, and market research • To decrease the rate of growth of corporate overhead costs • To encourage the design of products that are easier to manufacture or less costly to service
3. To measure income and assets for reporting to internal and external parties	• To cost inventories for financial reporting to stockholders, bondholders, and so on. (Under generally accepted accounting principles, inventoriable costs include manufacturing costs but exclude research, marketing, distribution, and customer service costs.) • To cost inventories for reporting to taxation authorities
4. To justify costs or compute reimbursement	• To cost products as a basis to establish a "fair" price, often done with defense contracting • To compute reimbursement for a consulting firm that is paid a percentage of the cost savings resulting from its recommendations

Different costs for different purposes are a major theme of this book. Exhibit 14-2 emphasizes this theme. Consider product costs in the following business functions in the value-chain:

1. Research and development
2. Product design
3. Manufacturing
4. Marketing
5. Distribution
6. Customer service

For the economic decision purpose (for example, product pricing), costs in all six functions may be included. For the motivation purpose, costs from more than one function are often included to emphasize to managers the importance of how costs in different functions are related to each other. For the purpose of income and asset measurement for reporting to external parties, inventoriable costs under generally accepted accounting principles include only manufacturing costs (and product design costs in some cases). Research and development costs are expensed to the period in which they are incurred, as are marketing, distribution, and customer-service costs. For the cost reimbursement purpose, the particular contract will stipulate whether all six of the above functions or only a subset of them are to be reimbursed.

Managers frequently cite the economic decisions and motivation purposes for allocating costs. For example, executives gave the following responses in a survey on the "primary objective" for allocating costs from one department to other departments (Stage II allocations):[3]

Primary Objective	Respondents (%)
1. To motivate managers and employees	42%
2. To make economic decisions for resource allocation	32
3. Other:	
Cost determination	19
Overhead recovery	5
Fairness/equity	2

CRITERIA TO GUIDE COST ALLOCATION DECISIONS

Role of Dominant Criterion

Objective 3

Describe alternative criteria for guiding cost allocation decisions

Exhibit 14-3 outlines and illustrates four criteria used to guide decisions relating to cost allocations. Managers must first choose the primary purpose that a particular cost allocation is to fulfill and then select the appropriate criterion in implementing the allocation. This text emphasizes the superiority of the cause-and-effect criterion when the purpose behind cost allocation is related to resource allocation or to motivation.

Other Criteria

The benefits-received criterion and fairness-or-equity criterion are sometimes cited in regulations governing U.S. federal government procurement. The Federal Ac-

[3]A. Atkinson, *Intrafirm Cost and Resource Allocations: Theory and Practice,* Society of Management Accountants of Canada and Canadian Academic Accounting Association Research Monograph (1987), p. 9.

EXHIBIT 14-3
Criteria to Guide Cost Allocation Decisions

1. CAUSE AND EFFECT. Using this criterion, managers identify the variable or variables that cause cost objects to incur costs. For example, managers may use hours of testing as the variable when allocating the costs of a quality testing area to products. Cost allocations based on the cause-and-effect criterion are likely to be the most credible to operating personnel.
2. BENEFITS RECEIVED. Using this criterion, managers identify the beneficiaries of the outputs of the cost object. The costs of the cost object are allocated among the beneficiaries in proportion to the benefits each receives. For example, consider a corporatewide advertising program that promotes the general image of the corporation rather than any individual product. The costs of this program may be allocated on the basis of division sales; the higher the sales, the higher the division's allocated cost for the advertising program. The rationale behind this allocation is the belief that divisions with higher sales levels apparently benefited from the advertising more than did divisions with lower sales levels and therefore ought to be allocated more of the advertising's costs.
3. FAIRNESS OR EQUITY. This criterion is often cited in government contracting when cost allocations are the means for establishing a price satisfactory to the government and its client. The allocation here is viewed as a "reasonable" or "fair" means of establishing a selling price in the minds of the contracting parties. For most allocation decisions, however, fairness is a lofty objective rather than an operational criterion.
4. ABILITY TO BEAR. This criterion advocates allocating costs in proportion to the cost object's ability to bear them. An example is the allocation of corporate executive salaries on the basis of divisional operating income; the presumption is that the more profitable divisions have a greater ability to absorb corporate headquarters' costs.

quisition Regulation (FAR) includes the following definition of "allocability" (in FAR 31.201-4):

> A cost is allocable if it is assignable or chargeable to one or more cost objectives on the basis of relative benefits received or other equitable relationship. Subject to the foregoing, a cost is allocable to a Government contract if it—
> (a) Is incurred specifically for the contract;
> (b) Benefits both the contract and other work, and can be distributed to them in reasonable proportion to the benefits received; or
> (c) Is necessary to the overall operation of the business, although a direct relationship to any particular cost objective cannot be shown.[4]

Further discussion of the contract reimbursement purpose of cost allocation is presented in Chapter 15.

The Cost-Benefit Consideration

Many companies place great importance on cost-benefit considerations when designing their cost allocation systems. Companies incur costs not only in gathering data but also in taking the time necessary to educate management about the chosen

[4]See J. Bedingfield and L. Rosen, *Government Contract Accounting* (Washington, DC: Federal Publications Inc., 1985), p. 7.13. Bedingfield and Rosen make the following comment on cost allocation criteria:

> Benefits received, when measurable and traceable, provide guidance for allocation. Benefits received should be interpreted as meaning the receiving of services or goods by the activity represented by the cost objectives to which the costs are being allocated. Where benefits are not reasonably traceable or measurable because of the nature of the cost or its remoteness from the cost objectives to which it is to be allocated, equity must supplant benefit.

system. The more sophisticated the system, generally, the higher these education costs.

The costs of designing and implementing sophisticated cost allocation systems are highly visible, and most companies work to reduce them. In contrast, the benefits from using a well-designed cost allocation system—being able to make more-informed make-or-buy decisions, pricing decisions, cost-control decisions, and so on—are difficult to measure and are frequently less visible. Designers of cost allocation systems, however, should consider these benefits as well as operating costs.

Spurred by rapid reductions in the costs of collecting and processing information, the trend in organizations today is toward more-detailed cost allocation systems. Several companies have developed manufacturing overhead systems that use more than twenty different cost allocation bases. Also, some businesses have state-of-the-art information technology already in place for operating their plants. Applying this existing technology to the development and operation of a manufacturing overhead cost system is less expensive—and thus more inviting—than starting up such a system from scratch.

THE GENERAL PROCESS OF COST ALLOCATION

Managers must make four interdependent decisions in allocating costs:

Objective 4

Explain four interdependent decisions for allocating costs

1. Choose the *cost object*, which is any activity or item for which a separate measurement of costs is desired. A cost object may be:
   - an activity area (such as a road-testing track of an automobile company),
   - a department (such as the personnel department of a retail store),
   - a division (such as the exploration division of an oil and gas company), or
   - a product (such as a kitchen table).
2. Choose the *direct costs* to trace to the chosen cost object. Direct costs for the road-testing track of an automobile company include direct labor and gas.
3. Choose (a) which *indirect costs* to allocate to the chosen cost object and (b) how to aggregate (combine) these costs before allocating them. The accounting system of most organizations collects data in many individual *cost pools*, where a cost pool is any grouping of individual costs. Typically, only some cost pools are classified as indirect costs of the chosen object. Examples of indirect cost pools for the road-testing track area of an automobile company include the general cafeteria where all personnel in the company may eat and the disability insurance policy of the company that covers all personnel. We discuss cost pools in detail in the next section.
4. Choose the *cost allocation base* for each of the indirect cost pools selected in (3). Where a cause-and-effect criterion is used to guide this choice, the cost allocation base should be the driver of costs accumulated at the cost object. An example of a Stage II cost allocation base used to allocate cafeteria department costs to the road-testing track is the number of employees in the testing track area.

Examples of Cost Pools

We use Computer Horizons to illustrate the cost pools found in an organization. Computer Horizons has two manufacturing divisions:

- The Microcomputer Division, which manufactures three computers: (1) the Plum, used mostly by students and in small offices and homes; (2) the Portable Plum, the smaller version of the Plum, which can be carried in a briefcase; and (3) the Super Plum, which has a larger memory and more capabilities than the Plum and is targeted at the business market

- The Peripheral Equipment Division, which manufactures printers and other items that can be used with the Plum line of microcomputers

Exhibit 14-4 illustrates costs typically incurred at the corporate-headquarters level and at the division-manufacturing level. Exhibit 14-4 focuses on indirect costs with respect to the products of the manufacturing divisions.

Service Departments

Many organizations distinguish between production departments and service departments. A **production department** (also called an **operating department**) is one whose efforts are directed toward adding value to a product or service sold to a customer. A **service department** (also called a **support department**) is one whose efforts are directed toward providing services to other departments (production departments and other service departments) in the organization. This section discusses allocating the costs of service departments to production departments (divisions). Two examples of service departments at Computer Horizons are the legal department and the personnel department.

Service departments create special accounting problems when they provide reciprocal services to each other as well as services to production departments. An example of reciprocal services at Computer Horizons would be the legal department providing services to the personnel department (for example, advice on compliance with labor laws) and the personnel department providing services to the legal department (for example, advice about the hiring of lawyers and secretaries).

EXHIBIT 14-4

Indirect Cost Pools (with Respect to Individual Products): Computer Horizons

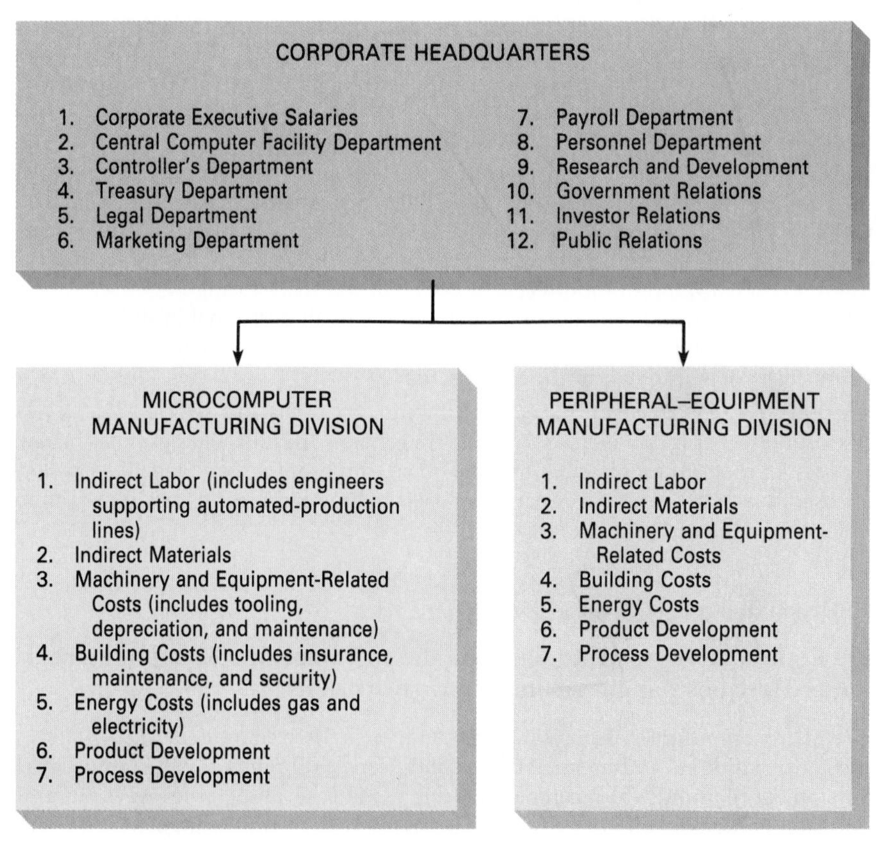

A subsequent section of this chapter provides further discussion of service departments.

Be cautious here. First, organizations differ in the departments located at the corporate and division levels. Some departments located at corporate headquarters of Computer Horizons (for example, research and development) are located at the division level in other organizations. Second, organizations differ in the meaning attached to the terms *production department* and *service department.* Always try to ascertain the precise meaning of these terms when analyzing data that include allocations of "production department" costs and "service department" costs.

Questions in Collecting Costs

Assume the cost object is the Microcomputer Division of Computer Horizons. Direct costs would be those costs that can be identified with or traced specifically to the Microcomputer Division. Indirect costs of this division include costs incurred at Corporate Headquarters (and possibly costs incurred in the Peripheral Equipment Division and other divisions). Decisions required when accumulating and subsequently allocating the indirect costs of the Microcomputer Division relate to the general process of cost allocation (outlined on pp. 461–63) that include:

1. Which cost categories from Corporate Headquarters and the other divisions should be included in the indirect costs of the Microcomputer Division? Should all twelve corporate cost pools in Exhibit 14-4 be allocated, or should only a subset be allocated? (These allocations would be called Stage II allocations.) For example, one company excludes corporate public relations from any corporate cost allocations to the divisions; divisional managers have little say in corporate public relations decisions and were objecting to allocations because of "taxation without representation."
2. How many cost pools should be used when allocating corporate costs to the Microcomputer Division? One extreme is to aggregate all corporate costs into a single cost pool. The other extreme is to have numerous individual corporate cost pools. The concept of homogeneity (described below) is important in making this decision.
3. Which allocation base should be used for each of the corporate cost pools when allocating corporate costs to the Microcomputer Division? Examples follow:

Cost Pool	Possible Allocation Bases
Corporate executive salaries	Sales; Assets employed; Operating income
Legal department	Estimated time or usage; Sales; Assets employed
Marketing department	Sales; Number of sales personnel
Payroll department	Number of employees; Payroll dollars
Personnel department	Number of employees; Payroll dollars; Number of new hires
Treasury department	Sales; Assets employed; Estimated time or usage

These allocation bases are illustrative only. The base preferred by an organization depends on the purpose served by the cost allocation base (see Exhibit 14-2, p. 458), the cause-and-effect relationships between a service department's costs and another department's costs, and the costs of implementing the different allocation bases.

Homogeneity of Cost Pools

Objective 5

Describe cost pool homogeneity and its impact on choosing cost pools

A cost pool is *homogeneous* if each activity whose costs are included in it has the same or a similar cause-and-effect relationship between the cost driver and the costs at that activity. A consequence of using a **homogeneous cost pool** is that the costs allocated using that pool will be the same as the costs that would be made if costs were collected separately for each individual activity in the pool. The greater the degree of homogeneity, the fewer the number of cost pools required to reflect accurately differences in how products consume resources of the organization.

Assume Computer Horizons wishes to use the cause-and-effect criterion to guide cost allocation decisions. ***The company should aggregate only those cost pools that have the same cause-and-effect relationship to the cost object.***[5] For example, if the number of employees in a division is the cause for incurring both corporate payroll costs and corporate personnel costs, the payroll cost pool and the personnel cost pool could be aggregated before determining the combined payroll and personnel cost rate per unit of the allocation base.

ALLOCATING COSTS FROM ONE DEPARTMENT TO ANOTHER

We now move from a general discussion of cost allocation issues to a specific discussion of Stage II allocations (p. 456). Our discussion is of a specific type of Stage II allocation—allocating costs from one department to another. When allocating costs from one department to another, managers must decide on several key issues:

1. Whether to use a single-cost function or have multiple-cost functions for each cost pool. The single-cost function approach is termed the **single-rate method**. Under this method, there is no distinction between fixed-cost and variable-cost functions. The two-cost function approach is termed the **dual-rate method**. Typically, the two functions are a fixed-cost function and a variable-cost function.[6]
2. Whether to use standard prices and standard quantities allowed or actual prices and actual quantities when determining the costs of the department.
3. Whether to use budgeted volume or actual volume as the allocation base for a given cost pool.

The purpose of the cost allocation should guide decisions related to these three issues. This section illustrates how the following guidelines can help achieve the economic-decisions and motivation purposes. These guidelines are presented as starting points for these complex decisions:

- Use multiple-cost functions where a single-cost function does not adequately describe the cost behavior of the cost pool.
- Use standard prices and standard quantities allowed when determining the department costs to be allocated.
- Use budgeted volume as the allocation base for a fixed-cost function and actual volume for the current period as the allocation base for a variable-cost function.

Single-Rate and Dual-Rate Methods

Consider the Central Computer Facility Department of Computer Horizons (see Exhibit 14-4). For simplicity, assume that the only users of this facility are the Microcomputer Division and the Peripheral Equipment Division. The following data apply to the forthcoming budget year:

1. Fixed cost of operating the facility	$300,000
2. Total capacity available	2,000 hours

[5]By the *same* cause-and-effect relationship to the cost object we mean *both* that there is the same cost driver and the same rate per unit of the cost driver.

[6]An example of a cost function with more than two costs is a fixed-cost function, a variable-cost function (variable with respect to units produced), and a per-batch setup-cost function. See Chapters 7 (p. 235) and 10 (p. 347) for discussion of setup costs.

3. Budgeted usage (volume) in hours:

Microcomputer Division	800
Peripheral Equipment Division	400
Total	1,200

4. Estimated standard variable cost per hour in the 1,000- to 1,500-hour relevant range — $200 per hour

Under the single-rate method, the costs of the Central Computer Facility would be allocated as follows (assuming budgeted volume is the allocation base):

$$\text{Total cost pool: } \$300{,}000 + (1{,}200 \times \$200) = \$540{,}000$$

$$\text{Budgeted usage} = 1{,}200 \text{ hours}$$

$$\text{Cost per hour: } \frac{\$540{,}000}{1{,}200} = \$450$$

$$\text{Microcomputer Division} = \$450 \text{ per hour used}$$

$$\text{Peripheral Equipment Division} = \$450 \text{ per hour used}$$

Use of the single-cost function, when combined with the budgeted volume allocation base, transforms what is a fixed cost to the Central Computer Facility (and to Computer Horizons) into a variable cost to users of that facility.

When the dual-rate method is used, allocation bases for each different cost function must be chosen. Assume the allocation bases for fixed costs are budgeted usage and for variable costs are actual usage. The costs allocated to the Microcomputer Division would be:

$$\text{Fixed-cost function: } \left(\frac{800}{1{,}200} \times \$300{,}000\right) = \$200{,}000 \text{ per year}$$

$$\text{Variable-cost function} = \$200 \text{ per hour used}$$

The costs allocated to the Peripheral Equipment Division would be:

$$\text{Fixed-cost function: } \left(\frac{400}{1{,}200} \times \$300{,}000\right) = \$100{,}000 \text{ per year}$$

$$\text{Variable-cost function} = \$200 \text{ per hour used}$$

Assume that during the coming year the Microcomputer Division actually uses 900 hours and the Peripheral Equipment Division actually uses 300 hours. The costs allocated to these two departments would be computed as follows:

Single-Rate Method

$$\text{Microcomputer Division: } 900 \times \$450 = \$405{,}000$$

$$\text{Peripheral Equipment Division: } 300 \times \$450 = \$135{,}000$$

Dual-Rate Method

$$\text{Microcomputer Division: } \$200{,}000 + (900 \times \$200) = \$380{,}000$$

$$\text{Peripheral Equipment Division: } \$100{,}000 + (300 \times \$200) = \$160{,}000$$

The benefit of using a single-cost function is its low cost of implementation. It avoids the often expensive analysis necessary to classify the individual cost items of a department into fixed and variable categories (see Chapter 10). This benefit can be attractive to organizations that are reluctant to invest in fancier accounting systems unless the prospects of better operating decisions are obvious.

Standard Versus Actual Prices and Quantities

Departmental costs can be allocated using standard prices and standard quantities allowed or actual prices and actual quantities. A major benefit of standard prices and standard quantities allowed is that the user department knows the costs in advance. Users are then better equipped to decide about the amount of the service to request, and—if the option exists—whether to use the internal department source or an external vendor.

Standard prices and standard quantities allowed also help motivate the manager of the service department to improve efficiency. During the budget period, the service department bears the risk of any unfavorable flexible-budget variances, not the user department. Why? Because the user department does not pay for any costs that exceed the standard costs.

Some organizations recognize that it may not always be best to impose all the risks of variances from standards completely on the service department (as when costs are allocated using standard quantities allowed and standard prices) or completely on the user department (as when costs are allocated using actual quantities and actual prices). For example, both departments may agree to share the risk (through an explicit formula) of a large uncontrollable increase in the price of a material used by the service department.

Budgeted Versus Actual Usage (Volume) Allocation Bases

The choice between actual usage and budgeted usage for allocating departmental fixed costs can affect a manager's behavior. Consider the budget of $300,000 fixed costs at the Central Computer Facility of Computer Horizons. Assume that actual and budgeted fixed costs are equal. Assume also that the actual usage of the Microcomputer Division is always equal to the budgeted usage. We look at three different cases in Exhibit 14-5 of allocating the $300,000 in total fixed costs.

- Case 1: Actual usage of the Peripheral Equipment Division *equals* budgeted usage.
- Case 2: Actual usage *exceeds* budgeted usage.
- Case 3: Actual usage *is less than* budgeted usage.

The budgeted usage is 800 hours for the Microcomputer Division and 400 hours for the Peripheral Equipment Division. Exhibit 14-5 presents the allocation of total fixed costs of $300,000 to each department for these three variations in actual usage in the Peripheral Equipment Division.

EXHIBIT 14-5

Effect of Variations in Actual Volume on Departmental Cost Allocations

	Actual Volume		Budgeted Volume as Allocation Base		Actual Volume as Allocation Base	
Case	**Micro. Div.**	**Perif. Div.**	**Micro. Div.**	**Perif. Div.**	**Micro. Div.**	**Perif. Div.**
1.	800 hours	400 hours	$200,000*	$100,000†	$200,000*	$100,000†
2.	800 hours	700 hours	$200,000*	$100,000†	$160,000‡	$140,000§
3.	800 hours	200 hours	$200,000*	$100,000†	$240,000‖	$60,000#

* $\frac{800}{(800 + 400)} \times \$300{,}000$ ‡ $\frac{800}{(800 + 700)} \times \$300{,}000$ ‖ $\frac{800}{(800 + 200)} \times \$300{,}000$

† $\frac{400}{(800 + 400)} \times \$300{,}000$ § $\frac{700}{(800 + 700)} \times \$300{,}000$ # $\frac{200}{(800 + 200)} \times \$300{,}000$

With actual usage as the allocation base, the Microcomputer Division will see its fixed-cost allocation differ in each of the three cases relative to what is expected at the start of the year. In Case 1, fixed-cost allocation will equal the expected amount. In Case 2, fixed-cost allocation will be $40,000 less than expected. In Case 3, fixed-cost allocation will be $40,000 more than expected. Consider now Case 3, where there was an increase of $40,000 even though the Microcomputer Division's actual usage is exactly the same as it budgeted. We see that fluctuations in usage in one department can affect the fixed costs allocated to other departments when fixed costs are allocated on the basis of actual usage. When actual usage is the allocation base, user departments will not know their allocated costs until the end of the budget period.

When budgeted usage is the allocation base, user departments will know their allocated costs in advance. This information helps both short-run and long-run planning by user departments. The main justification given for the use of budgeted usage to allocate fixed costs is related to long-run planning. Firms make decisions on committed costs (such as the fixed costs of a service department) using a long-run planning horizon; the use of budgeted usage to allocate these fixed costs is consistent with this long-run horizon.

If fixed costs are allocated on the basis of long-run commitments or plans, some managers will be tempted to underestimate their planned usage. In this way, they will bear a lower fraction of the total costs. Top management can counteract this ploy by systematically comparing actual usage with planned usage. Some organizations offer rewards in the form of salary increases and promotions for managers who make accurate forecasts. Also, some cost allocation methods include penalties for underpredictions of budgeted usage. For instance, a higher variable rate may be charged after a department exceeds its budgeted usage.

Allocations to Influence Behavior

Top managers may resort to cost allocations that are seemingly arbitrary. Why? In some cases, allocation is a means of getting subordinates to behave as top managers desire. Assume that Computer Horizons allocates to its operating divisions the costs incurred at the research and development department at corporate headquarters. Cause-and-effect justifications may be weak, but top managers may argue that this allocation gets division managers to take a needed interest in central research activities. In this way, the formal accounting system provides an important way of influencing behavior through an "arbitrary" cost allocation scheme. This practice may be appropriate to the extent that the desired objective is reached without causing confusion in cost analysis and resentment about the methods of cost allocation.

Cost allocations for motivating desired behavior need not be "full" or "tidy" allocations in the sense that all the costs assigned to a chosen cost object are 100% allocated to other cost objects. For example, partial allocation of some service department costs may encourage use of the service but still remind divisional managers that costs are actually incurred. (Note, however, that for inventory costing purposes under GAAP, complete allocations of costs designated as inventoriable costs are required.)

Allocating costs for motivation purposes is one of several approaches to affecting managers' behavior. Other approaches include top-management edicts and the involvement of all affected managers in the budgeting process. Organizations rarely ask which single approach for motivating managers is best. Rather, they view the challenge as designing a collective set of motivation mechanisms that reinforce each other.

ALLOCATING COSTS OF SERVICE DEPARTMENTS

Objective 6

Distinguish among direct allocation, step-down, and reciprocal methods of allocating service department costs

In many organizations, relationships exist not only between service departments and production departments but also among individual service departments. For example, the personnel department of Computer Horizons provides hiring support services to both the Microcomputer and Peripheral Equipment manufacturing divisions as well as to other service departments (such as the legal department).

We now examine three methods of allocating the costs of service departments: *direct, step-down,* and *reciprocal*. To focus on concepts, we use a single-cost function to allocate the costs of each factory service department. Our justification for this assumption, given the prior discussion in this chapter, is that all costs of both factory service departments are regarded as variable. (The Problem for Self-Study at the end of this chapter illustrates the use of the dual-rate method for allocating service department costs. This problem also illustrates how service-department cost allocation methods can be used to allocate the costs of corporate service departments to operating divisions.)

Consider Wigan Engineering, which manufactures engines used in electric power generating plants. Wigan has two factory service departments and two production departments in its manufacturing facility:

Factory Service Departments	Production Departments
• Plant maintenance	• Machining
• Information systems	• Assembly

Costs are accumulated at each department for planning and control purposes. For inventory-costing purposes, however, the factory service department costs of Wigan must be allocated to the production departments. The data for our example are in Exhibit 14-6.

Direct Allocation Method

The **direct allocation method** (often called the **direct method**) is the most widely used method for allocation of service department costs. This method ignores any service rendered by one service department to another; it allocates each service

EXHIBIT 14-6

Data for Factory Service Department Cost Allocation at Wigan Engineering

	Factory Service Departments		Production Departments		
	Plant Maintenance	Information Systems	Machining	Assembly	Total
Budgeted factory overhead costs before any inter-department cost allocations	$600,000	$116,000	$400,000	$200,000	$1,316,000
Proportions of service furnished:					
By Plant Maintenance					
Budgeted labor hours	—	1,600	2,400	4,000	8,000
Proportion	—	2/10	3/10	5/10	10/10
By Information Systems					
Budgeted computer time	200	—	1,600	200	2,000
Proportion	1/10	—	8/10	1/10	10/10

EXHIBIT 14-7

Direct Method of Allocating Factory Service Department Costs at Wigan Engineering

	Factory Service Departments		Production Departments		
	Plant Maintenance	Information Systems	Machining	Assembly	Total
Budgeted factory overhead costs before any inter-department cost allocations	$600,000	$116,000	$400,000	$200,000	$1,316,000
Allocation of Plant Maintenance: (3/8,5/8)*	(600,000) $ 0		225,000	375,000	
Allocation of Information Systems: (8/9,1/9)†		(116,000) $ 0	103,111	12,889	
Total budgeted factory overhead of production departments			$728,111	$587,889	$1,316,000

*Base is (2,400 + 4,000), or 6,400 hours; 2,400/6,400 = 3/8; 4,000/6,400 = 5/8.
†Base is (1,600 + 200), or 1,800 hours; 1,600/1,800 = 8/9; 200/1,800 = 1/9.

department's total costs directly to the production departments. Exhibit 14-7 illustrates this method using the data in Exhibit 14-6. Note how this method ignores the service rendered by the plant maintenance department to the information systems department and also ignores the service rendered by information systems to plant maintenance. For example, the base used for allocation of plant maintenance is the budgeted total maintenance labor hours worked in the production departments: 2,400 + 4,000 = 6,400 hours. This amount excludes the 1,600 hours of service provided by plant maintenance to the information systems department. Similarly, the base used for allocation of information systems costs is 1,600 + 200 = 1,800 hours of computer time, which excludes the 200 hours of service provided by the information systems to plant maintenance.

The benefit of the direct method is its simplicity. There is no need to predict the usage of service department resources by other service departments.

Step-Down Allocation Method

Some organizations use the **step-down allocation method** (sometimes called the **step** or **sequential allocation method**), which allows for *partial* recognition of services rendered by service departments to other service departments. This method is more complex than the direct method because a sequence of allocations must be chosen. A popular step-down sequence begins with the department that renders the highest percentage of its total services to other service departments. The sequence continues with the department that gives the next-highest percentage of its total services to other service departments, and so on, ending with the service department that renders the lowest percentage of its total services to other service departments.[7]

[7]An alternative approach to selecting the sequence of allocations is to begin with the department that renders the highest dollar amount of services to other service departments. The sequence ends with the allocation of the costs of the department that renders the lowest dollar amount of services to other service departments.

EXHIBIT 14-8
Step-Down Method of Allocating Factory Service Department Costs at Wigan Engineering

	Factory Service Departments		Production Departments		
	Plant Maintenance	Information Systems	Machining	Assembly	Total
Budgeted factory overhead costs before any inter-department cost allocations	$600,000	$116,000	$400,000	$200,000	$1,316,000
Allocation of Plant Maintenance: (2/10,3/10, 5/10)*	(600,000)	120,000	180,000	300,000	
	$ 0				
Allocation of Information Systems: (8/9,1/9)†		(236,000)	209,778	26,222	
		$ 0			
Total budgeted factory overhead of production departments			$789,778	$526,222	$1,316,000

*Base is (1,600 + 2,400 + 4,000), or 8,000 hours; 1,600/8,000 = 2/10; 2,400/8,000 = 3/10; 4,000/8,000 = 5/10.
†Base is (1,600 + 200), or 1,800 hours; 1,600/1,800 = 8/9; 200/1,800 = 1/9.

Let us continue with our Wigan Engineering example, in which we have only two service departments. In Exhibit 14-6, we see that the Plant Maintenance department gives 2/10 of its total service costs to the other service department, Information Systems. The Information Systems department gives 1/10 of its total service to the Plant Maintenance department. We begin, then, with the Plant Maintenance department.

Exhibit 14-8 shows an allocation of $120,000 (2/10 of $600,000) of Plant Maintenance total service costs to Information Systems. The new total of Information Systems, $236,000, is then allocated to the two production departments. The allocation base is the total service hours provided to the departments to which the cost is to be allocated. Once a service department's costs have been allocated, no subsequent service department costs are reallocated or recirculated back to it.

Reciprocal Allocation Method

The **reciprocal allocation method** allocates costs by explicitly including the mutual services rendered among all service departments. The reciprocal method is also called the **cross-allocation method**, the **matrix allocation method**, and the **double-distribution allocation method**. Theoretically, the direct method and the step-down method are not accurate when service departments render services to one another reciprocally. For example, the Plant Maintenance department maintains all the computer equipment in the Information Systems department. Similarly, the Information Systems department provides data-base support for the Plant Maintenance department. The reciprocal allocation method enables us to incorporate fully interdepartmental relationships into the service cost allocations.

Implementing the reciprocal allocation method requires three steps:

Step 1: Express service department costs and service department reciprocal relationships in linear equation form. Let *PM* be the complete reciprocated costs of

Plant Maintenance and *IS* be the complete reciprocated costs of Information Systems. We then express the data in Exhibit 14-6 as follows:

$$(1): PM = \$600{,}000 + .1IS$$
$$(2): IS = \$116{,}000 + .2PM$$

The $.1IS$ term in equation (1) is the percentage of the Information Systems department work that is used by the Plant Maintenance department. The $.2PM$ term in equation (2) is the percentage of the Plant Maintenance work that is used by the Information Systems department.

By **complete reciprocated cost** in equations (1) and (2), we mean the actual incurred cost of the service department plus a part of the costs of the other service departments that provide services to it. This complete reciprocated cost figure is sometimes labeled the **"artificial" cost** of the service department; it is always larger than the actual cost.

Step 2: Solve the system of simultaneous equations to obtain the complete reciprocated cost of each service department. Where there are two service departments, the following substitution approach can be used.

Substituting equation (2) into equation (1):

$$PM = \$600{,}000 + .1(\$116{,}000 + .2PM)$$
$$PM = \$600{,}000 + 11{,}600 + .02PM$$
$$.98PM = \$611{,}600$$
$$PM = \$624{,}082$$

Substituting into equation (2):

$$IS = \$116{,}000 + .2(\$624{,}082) = \$240{,}816$$

Where there are more than two service departments with reciprocal relationships, computer programs can be used to calculate the complete reciprocated cost of each service department.

Step 3: Allocate the complete reciprocated cost of each service department to all other departments (both factory service departments and production departments), using the computed usage proportions (based on total units of service provided to all departments). Consider the Information Systems department, which has a complete reciprocated cost of $240,816. This complete reciprocated cost would be allocated as follows:

To Plant Maintenance (1/10)	=	$ 24,082
Machining (8/10)	=	192,652
Assembly (1/10)	=	24,082
Total		$240,816

Exhibit 14-9 presents summary data pertaining to the reciprocal method.

One source of confusion to some managers using the reciprocal cost allocation method is why the complete reciprocated costs of the service departments ($624,082 and $240,816 in Exhibit 14-9) exceed their budgeted amounts $716,000 ($600,000 and $116,000). This excess of $148,898 ($24,082 for Plant Maintenance and $124,816 for Information Systems) is the total cost that is allocated from one service department to the other service department. The total costs allocated to the production departments under the reciprocal cost allocation are still only $716,000.

Overview of Methods

Assume that the total budgeted overhead of each production department in the example in Exhibits 14-7 to 14-9 is applied to individual products based on budgeted machine hours for the machining department (4,000 hours) and budgeted

EXHIBIT 14-9
Reciprocal Method of Allocating Factory Service Department Costs at Wigan Engineering

	Factory Service Departments		Production Departments		
	Plant Maintenance	Information Systems	Machining	Assembly	Total
Budgeted factory overhead costs before any inter-department cost allocations	$600,000	$116,000	$400,000	$200,000	$1,316,000
Allocation of Plant Maintenance: (2/10,3/10, 5/10)*	(624,082)	124,816	187,225	312,041	
Allocation of Information Systems: (1/10,8/10, 1/10)†	24,082	(240,816)	192,652	24,082	
	$ 0	$ 0			
Total budgeted factory overhead of production departments			$779,877	$536,123	$1,316,000

*Base is (1,600 + 2,400 + 4,000), or 8,000 hours; 1,600/8,000 = 2/10; 2,400/8,000 = 3/10; 4,000/8,000 = 5/10.
†Base is (200 + 1,600 + 200), or 2,000 hours; 200/2,000 = 1/10; 1,600/2,000 = 8/10; 200/2,000 = 1/10.

direct labor hours for the assembly department (3,000 hours). The budgeted overhead application rates associated with each service department allocation method are (rounded to the nearest dollar):

	Total Budgeted Overhead Costs after Allocation of All Service Department Costs		Budgeted Overhead Rate Per Hour for Inventory-Costing Purposes	
Allocation Method	Machining	Assembly	Machining (4,000 machine hours)	Assembly (3,000 labor hours)
Direct	$728,111	$587,889	$182	$196
Step-down	789,778	526,222	197	175
Reciprocal	779,877	536,123	195	179

These differences in budgeted overhead application rates with alternative service-department cost allocation methods can be important to managers. For example, consider a cost-reimbursement contract. The cost allocation method may make a considerable difference.

The reciprocal method, while theoretically the most defensible, is not widely used. The advantage of the direct and step-down methods is that they are relatively simple to compute and understand. However, with the ready availability of computer software to solve systems of simultaneous equations, the extra costs of using the reciprocal method will, in most cases, be minimal. The more likely roadblocks to the reciprocal method being widely adopted are (1) many managers find it difficult to understand, and (2) the numbers obtained by using the reciprocal method differ little, in some cases, from those obtained by using the direct or step-down method.

ALLOCATION OF COMMON COSTS

A **common cost** is a cost of operating a facility that is shared by two or more users. For example, consider the *Financial Accounting Standards Board* (FASB) and the *Governmental Accounting Standards Board* (GASB). The FASB was set up in 1972 to issue standards governing financial reporting by profit-seeking corporations. The GASB was set up in 1984 to issue standards governing financial reporting by public-sector organizations. The FASB and GASB occupy the same building, and a single mailroom facility serves them both. Variable costs of mailing are readily identifiable and kept in separate cost pools that are charged to the user. The fixed costs of the mailroom, however, cannot be identified with each individual user on the basis of a cause-and-effect relationship. How should these fixed costs be allocated between the two users?

Suppose the fixed costs of the mailroom facility for the next year are budgeted at $500,000. If the GASB did not use the mailroom, the fixed operating costs would be $480,000. An outside vendor offers to provide mailroom services to the FASB for a fixed fee of $600,000 plus variable costs. The same vendor offers to provide mailroom services to the GASB for a fixed fee of $200,000 plus variable costs. Two methods have been proposed for allocating common costs to individual users: the incremental method and the stand-alone method.

Incremental Common Cost Allocation Method

Objective 7

Outline incremental and stand-alone common cost allocations methods

The **incremental common cost allocation method** requires that one user be viewed as the primary party and the second user be viewed as the incremental party. The incremental party is allocated that component of the common cost arising because there are two users and not only the primary user. For the mailroom example, the FASB would be allocated $480,000 of the fixed costs and the GASB $20,000. In the extreme case where the fixed costs remain unchanged when the incremental party uses the common facility, the incremental party would receive zero allocation of the common cost. Where there are more than two users of the common facility, the method requires the users to be ranked and the common costs allocated to these users in the ranked sequence.

Under the incremental method, the primary party typically will receive the highest allocation of the common costs. Not surprisingly, most parties in common cost allocation disputes propose themselves as the incremental user. In some cases, the incremental parties are newly formed organizations or new subparts of a corporation, such as a new product line or a new sales territory. Chances for their short-term survival may be enhanced if they bear a relatively low allocation of common costs.

Stand-Alone Common Cost Allocation Method

The **stand-alone common cost allocation method** allocates the common cost on the basis of each user's percentage of the total of the individual stand-alone costs. In the mailroom example, the total of the individual stand-alone costs is $800,000 ($600,000 for FASB and $200,000 for GASB as determined by the outside vendor). The common cost of the internally run mailroom facility is allocated:

$$\text{FASB: } \frac{\$600{,}000}{\$800{,}000} \times \$500{,}000 = \$375{,}000$$

$$\text{GASB: } \frac{\$200{,}000}{\$800{,}000} \times \$500{,}000 = \$125{,}000$$

Advocates of the stand-alone method often emphasize an equity or fairness rationale; they argue that fairness requires that each user pay a percentage of actual common costs based on its share of the total stand-alone costs.

Several of the allocation bases described previously in this chapter have also been used in allocating common costs. For instance, relative revenues are sometimes adopted as an allocation base justified by the ability-to-bear criterion. Also, common fixed costs can be allocated based on the user's estimates of long-run service requirements.

TRENDS IN COST ALLOCATION PRACTICES

Objective 8

Describe current trends in companies' cost allocation practices

Several trends are occurring in the cost allocation practices of companies. Companies are attempting to increase the percentage of total costs that can be classified as a direct cost of the chosen cost object. Advances in information technology are assisting companies in these attempts. For example, many consulting firms now rely on usage meters built into photocopying machines to directly trace photocopying costs to individual jobs. Previously, these costs were included in general overhead cost pools and allocated on the basis of direct professional-labor hours billed. This allocation base overstated the costs of jobs with high labor hours but low usage of photocopying facilities.

Some companies, such as Weyerhaeuser,[8] have adopted a charge-out system for service department usage by divisions (or departments). In a charge-out system, the division using the service pays a charge—known in advance—for a negotiated amount of service. The service costs are thus traced directly to individual users rather than being allocated using a base such as division revenues. In many charge-out systems, the user divisions have the option of choosing between the internal service department and an external vendor of that service. Advocates of charge-out systems maintain that they make service departments more responsive to customer wants—the service department has to provide a service that a customer views as passing a cost-benefit test.

Companies are increasing the number of separate indirect cost pools and using a more diverse set of allocation bases. For example, one company operating with a centralized distribution warehouse previously allocated all distribution costs to different classes of customers on the basis of sales dollars. A detailed study was made of whether this allocation base had a similar cause-and-effect relationship with all the costs being incurred at the distribution center. The study concluded that the aggregate distribution cost pool was not homogeneous and that it should be separated into the following four cost pools, each with a different allocation base:

Distribution Cost Pools	Allocation Base
1. Order generation	Number of separate orders
2. Outbound freight	Number of pounds shipped
3. Shipping	Number of units shipped
4. Customer returns	Percentage of units returned

The company believes that the revised cost allocations enables it to better understand how classes of customers that differ in the number of separate orders made, percentage of units returned, and so on, differ in terms of profitability. The prior single cost pool allocation scheme, based on sales dollars, ignored such factors in profitability computations.

[8]H. T. Johnson and D. A. Loewe, "How Weyerhaeuser Manages Corporate Overhead Costs," *Management Accounting*, August 1987, pp. 20–26.

PROBLEM FOR SELF STUDY

PROBLEM

This problem illustrates how service department cost allocation methods can be used in a setting different from the factory service department example examined in the chapter (Exhibits 14-6 to 14-9). In this problem, the costs of central corporate service departments are allocated to operating divisions. The corporate departments provide services to each other as well as to the operating divisions. Also, this problem illustrates the use of the dual-rate method of allocating service department costs. (The dual-rate method can also be used in factory-department service cost allocations.)

Computer Horizons budgets the following amounts for its two central corporate service departments (legal and personnel) in servicing each other and the two manufacturing divisions—the Microcomputer Division (MCD) and the Peripheral Equipment Division (PED):

To be Supplied by	Budgeted Capacity				
	Legal	Personnel	MCD	PED	Total
Legal—hours	—	250	1,500	750	2,500
—proportions	—	.10	.60	.30	1.00
Personnel—hours	2,500	—	22,500	25,000	50,000
—proportions	.05	—	.45	.50	1.00

Details on actual usage:

Supplied by	Actual Usage by:				
	Legal	Personnel	MCD	PED	Total
Legal—hours	—	400	400	1,200	2,000
—proportions	—	.20	.20	.60	1.00
Personnel—hours	2,000	—	26,600	11,400	40,000
—proportions	.05	—	.665	.285	1.00

The actual costs were:

	Fixed	Variable
Legal	$360,000	$200,000
Personnel	$475,000	$600,000

Fixed costs are allocated on the basis of budgeted capacity. Variable costs are allocated on the basis of actual usage.

Required

What service department costs for legal and personnel will be allocated to MCD and PED using (a) the direct method, (b) the step-down method (allocating the legal department costs first), and (c) the reciprocal method?

SOLUTION

Exhibit 14-10 presents the computations for allocating the fixed and variable service department costs. A summary of these costs follows:

	Microcomputer Division (MCD)	Peripheral Equipment Division (PED)
A. *Direct Method*		
Fixed costs	$465,000	$370,000
Variable costs	470,000	330,000
	$935,000	$700,000

	Microcomputer Division (MCD)	Peripheral Equipment Division (PED)
B. *Step-Down Method*		
Fixed costs	$458,053	$376,947
Variable costs	488,000	312,000
	$946,053	$688,947
C. *Reciprocal Method*		
Fixed costs	$462,513	$372,487
Variable costs	476,364	323,636
	$938,877	$696,123

The simultaneous equations for the reciprocal method are shown on page 477.

EXHIBIT 14-10

Alternative Methods of Allocating Corporate Service Department Costs to Operating Divisions of Computer Horizons: Dual-Rate Method Illustrated

	Corporate Service Departments		Manufacturing Divisions	
Allocation Method	**Legal**	**Personnel**	**MCD**	**PED**
A. Direct Method				
Fixed Costs	$360,000	$475,000		
Legal (6/9, 3/9)	(360,000)		$240,000	$120,000
Personnel (225/475, 250/475)	$ 0	(475,000)	225,000	250,000
		$ 0	$465,000	$370,000
Variable Costs	$200,000	$600,000		
Legal (.25, .75)	(200,000)		$ 50,000	$150,000
Personnel (.7, .3)	$ 0	(600,000)	420,000	180,000
		$ 0	$470,000	$330,000
B. Step-Down Method				
(Legal Department First)				
Fixed Costs	$360,000	$475,000		
Legal (.10, .60, .30)	(360,000)	36,000	$216,000	$108,000
Personnel (225/475, 250/475)	$ 0	(511,000)	242,053	268,947
		$ 0	$458,053	$376,947
Variable Costs	$200,000	$600,000		
Legal (.20, .20, .60)	(200,000)	40,000	$ 40,000	$120,000
Personnel (.7, .3)	$ 0	(640,000)	448,000	192,000
		$ 0	$488,000	$312,000
C. Reciprocal Method				
Fixed Costs	$360,000	$475,000		
Legal (.10, .60, .30)	(385,678)	38,568	$231,407	$115,703
Personnel (.05, .45, .50)	25,678	(513,568)	231,106	256,784
	$ 0	$ 0	$462,513	$372,487
Variable Costs	$200,000	$600,000		
Legal (.20, .20, .60)	(232,323)	46,465	$ 46,465	$139,393
Personnel (.05, .665, .285)	32,323	(646,465)	429,899	184,243
	$ 0	$ 0	$476,364	$323,636

Fixed Costs

$$L = \$360{,}000 + .05P$$
$$P = \$475{,}000 + .10L$$
$$L = \$360{,}000 + .05(\$475{,}000 + .10L) = \$385{,}678$$
$$P = \$475{,}000 + .10(\$385{,}678) = \$513{,}568$$

Variable Costs

$$L = \$200{,}000 + .05P$$
$$P = \$600{,}000 + .20L$$
$$L = \$200{,}000 + .05(\$600{,}000 + .20L) = \$232{,}323$$
$$P = \$600{,}000 + .20(\$232{,}323) = \$646{,}465$$

SUMMARY

Costs are allocated for four major purposes: (1) to make economic decisions for resource allocation; (2) to motivate managers and employees; (3) to measure income and assets for reporting to external parties; and (4) to justify costs or compute reimbursement. When allocation issues arise, a good basic question is, What is the dominant purpose of the cost allocation in this particular instance?

There are four interdependent facets in cost allocation: (1) choosing the cost object, (2) choosing the direct costs to trace to the chosen cost object, (3) choosing the indirect cost pools to be associated with the chosen cost object and deciding how to aggregate these cost pools before allocating them to the cost object, and (4) choosing the cost allocation base for each of the indirect cost pools from (3).

Disputes over cost allocation issues can be frustrating to all parties. There will often be no clearly right or clearly wrong solution. When the purpose of the allocation is related to economic decisions or motivations, the cause-and-effect criterion is a useful guide in choosing an allocation method.

Three major methods are available to allocate service department costs to other departments: direct, step-down, and reciprocal. The last is the most defensible, but the direct and step-down methods are more widely used.

TERMS TO LEARN

This chapter and the Glossary at the end of the book contain definitions of the following important terms:

artificial cost *(p. 471)* common cost *(473)* complete reciprocated cost *(471)*
cross allocation method *(470)* direct allocation method *(468)*
direct method *(468)* double-distribution allocation method *(470)*
dual-rate method *(464)* homogeneous cost pool *(463)*
incremental common cost allocation method *(473)*
matrix allocation method *(470)* operating department *(462)*
production department *(462)* reciprocal allocation method *(470)*
sequential allocation method *(469)* service department *(462)*
single-rate method *(464)* stand-alone common cost allocation method *(473)*
step allocation method *(469)* step-down allocation method *(469)*
support department *(462)*

ASSIGNMENT MATERIAL

QUESTIONS

14-1 What are the three stages of cost allocation?

14-2 What are three ways to increase the accuracy of product costs?

14-3 "Every indirect cost was at one stage a direct cost." Do you agree? Explain.

14-4 "A given cost may be allocated for one or more purposes." List four purposes.

14-5 What criteria might be used to guide cost allocation decisions? Which criterion is the dominant one?

14-6 "There are four interdependent decisions managers make in allocating costs." What are they?

14-7 What is a cost pool? Give an example.

14-8 What is a service department? What is another term for a service department?

14-9 What role does homogeneity of cost pools play in decisions on the number of cost pools to use?

14-10 Why use a dual-rate cost allocation method when it is less costly to use a single-rate method?

14-11 Distinguish among the three methods of allocating the costs of service departments to production departments.

14-12 What is the theoretically most defensible method for allocating service department costs?

14-13 Distinguish between two methods of allocating common costs.

14-14 Describe two trends occurring in the cost allocation practices of companies.

EXERCISES AND PROBLEMS

14-15 Single-rate versus dual-rate cost allocation methods. (W. Crum, adapted) Carolina Company has a power plant designed and built to serve its three factories. Data for 19_1 follow:

	Usage in Kilowatt Hours	
Factory	Budget	Actual
Durham	100,000	80,000
Charlotte	60,000	120,000
Raleigh	40,000	40,000

Actual fixed costs of the power plant were $1 million in 19_1; actual variable costs, $2 million.

Required

1. Using a single-cost pool, compute the amount of power cost that would be allocated to Charlotte. Use two alternative allocation bases.
2. Using multiple-cost pools, compute the amount of power cost that would be allocated to Charlotte. Use two alternative allocation bases.

14-16 Single-rate versus dual-rate allocation methods, service department. The Chicago power plant that services all factory departments of MidWest Engineering has a budget for the forthcoming year. This budget has been expressed in the following terms on a monthly basis:

	Kilowatt Hours	
Factory Departments	Needed at Practical Capacity Production Volume*	Average Expected Monthly Usage
Rockford	10,000	8,000
Peoria	20,000	9,000
Hammond	12,000	7,000
Kankakee	8,000	6,000
Totals	50,000	30,000

*This factor was the most influential in planning the size of the power plant.

The expected monthly costs for operating the department during the budget year are $15,000: $6,000 variable and $9,000 fixed.

Required

1. Assume that a single-cost pool is used for the power plant costs. What dollar amounts will be allocated to each factory department? Use (a) practical capacity and (b) average expected monthly usage as the allocation bases.
2. Assume a dual-rate method; separate cost pools for the variable and fixed costs are used. Variable costs are allocated on the basis of expected monthly usage. Fixed costs are allocated on the basis of practical capacity. What dollar amounts will be allocated to each factory department? Why might you prefer the dual-rate method?

14-17 Allocation of service department costs. Atherton Machining Company has one service department (electric power) and two production departments in its factory. The flexible-budget formula for allocating the service department costs for the next fiscal year is $6,000 monthly plus $0.40 per machine hour in the production departments. Fixed costs are allocated on a lump-sum basis, 60% to Production Department 1 and 40% to Production Department 2. Variable costs are allocated at the budgeted unit rate of $0.40 per machine hour.

Required

1. Assume that the actual costs coincided exactly with the flexible-budget amount. Departments 1 and 2 each worked at 4,000-hour levels. Tabulate the allocations of all costs.
2. Assume the same facts as in requirement 1 except that the fixed costs were allocated on the basis of actual hours rather than capacity available. Tabulate the allocations of all

costs. As the manager of Department 2, would you prefer the method in requirement 1 or in requirement 2? Why?

3. Suppose the service department had inefficiencies at an 8,000-hour level, incurring costs that exceeded budget by $800. How would this change your answers in requirements 1 and 2? Be specific.

14-18 Departmental cost allocation, university computer service center. A computer service center of National University serves two major users, the College of Engineering and the College of Humanities and Sciences (H&S).

Required

1. When the computer equipment was initially installed, the procedure for cost allocation was straightforward. The actual monthly costs were compiled and were divided between the two colleges on the basis of the computer time used by each. In October, the costs were $100,000. H&S used 100 hours and Engineering used 100 hours. How much cost would be allocated to each college? Suppose costs were $110,000 because of various inefficiencies in the operation of the computer center. How much cost would be allocated? Does such allocation seem justified? If not, what improvement would you suggest?
2. Use the same approach as in requirement 1. The actual cost-behavior pattern of the computer center was $80,000 fixed cost per month plus $100 variable cost per hour used. In November, H&S used 50 hours and Engineering used 100 hours. How much cost would be allocated to each college? Use a single-rate method.
3. As the computer service center developed, a committee was formed that included representatives of H&S and Engineering. This committee determined the size and composition of the center's equipment. The committee based its planning on the long-run average utilization of 180 monthly hours for H&S and 120 monthly hours for Engineering. Suppose the $80,000 fixed costs are allocated through a budgeted monthly lump sum based on long-run average utilization. Variable costs are allocated through a budgeted unit rate of $100 per hour. How much cost would be allocated to each college? What are the advantages of this dual-rate allocation method over other methods?
4. What are the likely behavioral effects of lump-sum allocations of fixed costs? For example, if you were the representative of H&S on the facility planning committee, what would your biases be in predicting long-run usage? How would top management counteract the bias?

14-19 Allocation of administrative and marketing costs, revenue as an allocation basis. The Mideastern Transportation Company has had a longstanding policy of fully allocating all costs to its various divisions. Among the costs allocated are general and administrative costs in central headquarters, consisting of office salaries, executive salaries, travel expense, accounting costs, office supplies, donations, rents, depreciation, postage, and similar items.

All these costs are difficult to trace directly to the individual divisions benefited, so they are allocated on the basis of the actual total revenue of each of the divisions. The same basis is used for allocating marketing costs. For example, in 19_3 the following allocations were made (in millions):

	Divisions			
	A	B	C	Total
Revenue	$50.0	$40.0	$10.0	$100.0
Costs allocated on the basis of revenue	6.0	4.8	1.2	12.0

Division A's revenue was expected to rise in 19_4, but the division encountered severe competition. Revenue remained at $50 million. In contrast, Division C enjoyed explosive growth in traffic because of the completion of several huge factories in its area. Its revenue rose to $30 million. Division B's revenue remained unchanged. Careful supervision kept the actual total costs allocated on the basis of revenue at $12 million.

Required

1. What costs will be allocated to each division for 19_4?

2. Using the results in requirement 1, comment on the limitations of using revenue as a basis for cost allocation.

14-20 Allocation of service department costs. For simplicity, suppose Advanced Technologies has one service department (maintenance) plus two production departments (M and N). The flexible budget of the service department is $24,000 monthly plus $1 per machine hour. The expected "long-run" usage of the service department is 75% by Department M and 25% by Department N.

Required

1. Indicate the dual rates for allocating the service department's costs to each production department.
2. Assume that actual costs coincided exactly with the flexible-budgeted amount. Departments M and N each worked at 8,000-hour levels. Tabulate the allocation of all costs.
3. Assume the same facts as in requirement 2 except that the fixed costs were allocated on the basis of actual hours worked by each department rather than on the basis of long-run usage. Tabulate the allocation of all costs. As the manager of Department N, would you prefer the method in requirement 2 or in requirement 3? Why?
4. Suppose the service department had inefficiencies at a 16,000-hour level, incurring costs that exceeded budget by $6,000. How would this change your answers in requirements 2 and 3? Be specific.

14-21 Allocation of travel costs. Joan Ernst, a graduating senior at a university near San Francisco, received an invitation to visit a prospective employer in New York. A few days later she received an invitation from a prospective employer in Chicago. She decided to combine her visits, traveling from San Francisco to New York, New York to Chicago, and Chicago to San Francisco.

Ernst received job offers from both companies. Upon her return, she decided to accept the offer in Chicago. She was puzzled about how to allocate her travel costs between the two employers. She gathered the following data:

Regular roundtrip fares with no stopovers:	
• San Francisco to New York	$1,400
• San Francisco to Chicago	$1,100

Ernst paid $1,800 for her three-leg flight (San Francisco to New York, New York to Chicago, Chicago to San Francisco). In addition, she paid $30 for a limousine from her home to San Francisco Airport and another $30 for a limousine from San Francisco Airport to her home when she returned.

Required

1. How should Ernst allocate the $1,800 airfare between the employers in New York and Chicago? Show the actual amounts you would allocate, and give reasons for your allocations.
2. Repeat requirement 1 for the $60 limousine charges at the San Francisco end of her travels.

14-22 Allocation of common costs, splitting of shared hotel bill. Linda and Mark McGraw are planning a one-week vacation to celebrate their tenth wedding anniversary. For their last two vacations, Linda's sister, Rebecca Miller, has accompanied them. Two months ago, Rebecca graduated from a prestigious MBA program. She recently accepted a job on Wall Street as an investment banker. Linda was very proud of her sister, especially because Rebecca had borrowed money on her own account to put herself through the best (and most expensive) MBA program in the country.

Last year the three of them stayed at the Kokomo Vacation Resort in the Florida Keys for one week. They rented one room with two double beds. The room rate was $120 a night for a single, $140 a double, and $150 for three people in a room. Linda put the $150-a-night bill on her charge account and allocated $140 a night to Mark and herself and $10 to Rebecca. Rebecca was happy with this arrangement and promptly paid her $70 share for the room for the week's vacation.

Linda has just heard that Rebecca would again like to join them for their coming vacation and under the same arrangement as last year. Rebecca was thrilled when Linda told her that

Mark had made reservations at the Kokomo Resort and that the room rates were still $120 a night for a single, $140 a double, and $150 for three people in a room. Mark was less than enthusiastic when Linda told him of the repeat of last year's "cast of characters." Among other complaints, he thought Rebecca could well afford a suite of her own, given her Wall Street salary.

Required

1. What justification is there for the current allocation scheme of $140 to Linda and Mark and $10 to Rebecca?
2. Propose an alternative allocation of the $150 a night rate. Defend your proposal.

14-23 Allocating costs of a central telephone reservation system. Helena Park, the chief operating officer of Happy Inns, faces a difficult problem. The manager of the most prestigious motel in the five-motel chain is upset about her proposed cost allocation scheme for the central telephone reservation system. Until one year ago, each motel handled its own reservations. Then Happy Inns leased a central telephone reservation system for $84,000 per year. Happy Inns pays the variable costs of operating the system. During the first year of operations, Happy Inns allocated none of the $84,000 fixed cost and none of the $56,000 variable costs of operating the system. Summary data for the first year of operating the central reservation system follow:

Motel	Reservations Made Via Central System	Total Number of Room Nights Available for Rental for Year	Total Number of Room Nights Actually Rented During Year	Average Rate Per Room Night	Total Room Revenues for Year
Vancouver	5,000 (12.5%)	28,000 (11.7%)	20,000 (10.8%)	$60	$ 1,200,000 (8.6%)
Halifax	4,000 (10.0)	31,000 (12.9)	25,000 (13.5)	76	1,900,000 (13.6)
Toronto	8,000 (20.0)	37,000 (15.4)	30,000 (16.2)	50	1,500,000 (10.7)
Jasper	3,000 (7.5)	32,000 (13.3)	30,000 (16.2)	100	3,000,000 (21.4)
Quebec City	20,000 (50.0)	112,000 (46.7)	80,000 (43.3)	80	6,400,000 (45.7)
	40,000	240,000	185,000		$14,000,000

The total number of rooms actually rented includes those rooms rented through the central reservation system, those rented through direct dial to the individual motel, and those rented by walk-in clientele. During the first year of operations, approximately 100,000 phone calls were made to the central reservation service.

Park decides to allocate the central telephone costs to each motel. She proposes that the fixed and variable costs be combined in a single pool and be allocated on the basis of actual total room revenues for the year.

Each Happy Inn manager receives a fixed salary plus a percentage of the operating income of the individual motel.

Required

1. What costs would have been allocated to each motel under Park's new proposal (the relative percentage of actual total room revenues) in the first year of operation of the central reservation system? Why might the manager of the Jasper Happy Inn oppose Park's proposal?
2. What limitations are associated with the proposed cost allocation method of Park in requirement 1?
3. Park considers alternative cost allocation systems. She decides to retain a single-cost pool, but to consider alternative allocation bases. What costs would have been allocated to each motel under each of the following bases: (a) reservations made through central system, (b) total number of room nights available for rental, and (c) total number of room nights actually rented during the year?

14-24 Allocation of central corporate costs to divisions. Dusty Rhodes, the corporate controller of Richfield Oil Company, is about to make a presentation to the senior corporate executives and the top managers of its four divisions. These divisions are:

1. Oil & Gas Upstream—the exploration, production, and transportation of oil and gas

2. Oil & Gas Downstream—the refining and marketing of oil and gas
3. Chemical Products
4. Copper Mining

Under the existing internal accounting system, costs incurred at central corporate headquarters are collected in a single pool and allocated to each division on the basis of the actual revenues of each division. The central corporate costs for the most recent year are (in millions):

Interest on debt	$2,000
Corporate salaries	100
Accounting and control	100
General marketing	100
Legal	100
Research and development	200
Public affairs	208
Personnel and payroll	192
	$3,000

Public affairs includes the public relations staff, the lobbyists, and the sizable donations Richfield makes to numerous charities and nonprofit institutions.

Summary data relating to the four divisions for the most recent year are (in millions):

	Oil & Gas Upstream	Oil & Gas Downstream	Chemical Products	Copper Mining	Total
Revenue	$7,000	$16,000	$4,000	$3,000	$30,000
Operating costs	3,000	15,000	3,800	3,200	25,000
Operating income	$4,000	$ 1,000	$ 200	$ (200)	$ 5,000
Identifiable assets	$14,000	$6,000	$3,000	$2,000	$25,000
Number of employees	9,000	12,000	6,000	3,000	30,000

The top managers of each division share in a divisional income bonus pool. *Divisional income* is defined as operating income less allocated central corporate costs.

Rhodes is about to propose a change in the method used to allocate central corporate costs. He favors collecting these costs in four separate pools:

- *Cost Pool 1:* Allocated using identifiable assets of division
  Cost Item: Interest on debt
- *Cost Pool 2:* Allocated using revenue of division
  Cost Items: Corporate salaries, accounting and control, general marketing, legal, research and development
- *Cost Pool 3:* Allocated using operating income (if positive) of division, with only divisions with positive operating income included in the allocation base
  Cost Item: Public affairs
- *Cost Pool 4:* Allocated using number of employees in division
  Cost Item: Personnel and payroll

Required

1. What purposes might be served by the allocation of central corporate costs to each division at Richfield Oil?
2. Compute the divisional income of each of the four divisions when central corporate costs are allocated using revenues of each division.
3. Compute the divisional income of each of the four divisions when central corporate costs are allocated through the four cost pools.
4. What are the strengths and weaknesses of Rhodes's proposal relative to the existing single-cost pool method?

14-25 Division managers' reactions to the allocation of central corporate costs to divisions. (Continuation of Problem 14-24) Dusty Rhodes presents his proposal (outlined in

Problem 14-24) for the use of four separate cost pools to allocate central corporate costs to the divisions. The comments of the top managers of each of the four divisions include the following:

(a) By the top manager of Oil & Gas Upstream Division: "The multiple-pool method of Rhodes is absurd. We are the only division generating a substantial positive cash flow, and this is ignored in the proposed (and indeed the existing) system. We could pay off any debt very quickly if we were not a cash cow for the rest of the dog divisions in Richfield Oil."

(b) By the top manager of Oil & Gas Downstream Division: "Rhodes's proposal is the first sign that the money we spend in the accounting and control function at corporate headquarters is justified. The proposal is fair and equitable."

(c) By the top manager of Chemical Products Division: "I oppose any cost allocation scheme. Last year I was the only major player in the chemical industry to show a positive operating income. We are operating at the bare-bones level. Last year I saved $300,000 by making everyone travel economy class. This policy created a lot of dissatisfaction, but we finally managed to get it accepted. Then at the end of the year we get a charge of $400 million for corporate central costs. What's the point of our division economy drives when they get swamped by allocations of corporate largess?"

(d) By the top manager of Copper Mining Division: "I should probably get concerned, but frankly I view it all as bookkeeping entries. If we were in the black, certain aspects would really infuriate me. For instance, why should corporate research and development costs be allocated to the Copper Division? The only research corporate does for us is how to best prepare our division for divestiture."

Required

How should Rhodes respond to these comments?

14-26 University indirect cost allocation rates, homogeneous cost pools. Tim Sheens, the provost of Ivy League University, is about to announce the 19_7 indirect cost (overhead) rates to be charged on government-sponsored research. This universitywide indirect cost rate is applied to the faculty salaries and fringe benefits (the direct costs) when the total research budget is developed. The 19_6 indirect cost rate of 90% was based on the ratio of 19_6 universitywide budgeted indirect costs of $351 million (equipment costs of $174 million and support salaries and benefits of $177 million) to budgeted direct costs of $390 million (faculty salaries and benefits).

Last month (July 19_6), Sheens had a tense meeting with five professors from the Engineering School. One newly appointed professor (Ralph Randall) reported that to end up with $18,000 in salary and fringe benefits drawn from government-sponsored research funding, he had to apply for a total grant of $34,200 ($18,000 plus 90% of $18,000). Randall did not receive the grant he applied for in 19_6 and believed it was a backlash by the government-funding agency against Ivy League University's 90% indirect cost rate.

The 19_7 budget of Ivy League University for its five schools and in total is (in millions):

	Business School	Engineering School	Humanities & Science School	Law School	Medical School	Total
Indirect Costs						
• Equipment costs	$ 8	$65	$ 32	$ 5	$90	$200
• Support salaries and fringe benefits	16	55	80	12	45	208
Direct Costs						
• Faculty salaries and fringe benefits	25	80	200	20	75	400

The average indirect cost rate for all government-sponsored research of all U.S. universities in 19_6 was 74% of direct costs (faculty salaries and fringe benefits). Sheens believes this average is likely to increase to a 76% indirect cost rate in 19_7.

Required

1. Why does Ivy League University apply an indirect cost (overhead) charge to the research grants its faculty submits to government-funding agencies?
2. What will be the 19_7 universitywide indirect cost rate to be charged on government-sponsored research?

3. Explain any differences between the 19_6 and 19_7 universitywide indirect cost rates.
4. Dave Wilson, a sociology professor in the Humanities and Science School, is planning to submit a proposal that will provide six months of research time ($42,000 salary and fringe benefits) in 19_7. How much will Wilson request in total research support from the government agency?
5. Wilson objects to using a universitywide indirect cost rate. He claims individual schools at Ivy League University are not homogeneous regarding their indirect cost rates. Is there evidence to support this claim?
6. Repeat requirement 4 using the indirect cost rate specific to the Humanities and Science School. Comment on your results.

14-27 Allocating costs of service departments; step-down and direct methods. The Central Valley Company has prepared departmental overhead budgets for normal volume levels before allocations, as follows:

Service departments:		
Building and grounds	$10,000	
Personnel	1,000	
General factory administration	26,090	
Cafeteria—operating loss	1,640	
Storeroom	2,670	
Total		$ 41,400
Production departments:		
Machining	$34,700	
Assembly	48,900	
Total		83,600
Total for both departments		$125,000

Management has decided that the most sensible inventory costs are achieved by using individual departmental overhead rates. These rates are developed after appropriate service department costs are allocated to production departments.

Bases for allocation are to be selected from the following:

Department	Direct Labor Hours	Number of Employees	Square Feet of Floor Space Occupied	Total Labor Hours	Number of Requisitions
Building and grounds		—	—	—	
Personnel*		—	2,000	—	
General factory administration		35	7,000	—	
Cafeteria—operating loss		10	4,000	1,000	
Storeroom		5	7,000	1,000	
Machining	5,000	50	30,000	8,000	2,000
Assembly	15,000	100	50,000	17,000	1,000
	20,000	200	100,000	27,000	3,000

*Basis used is number of employees.

Required

1. Using a worksheet, allocate service department costs by the step-down method. Develop overhead rates per direct labor hour for machining and assembly. Allocate the service departments in the order given in this problem. Use the allocation base for each service department you think is most appropriate.
2. Using the direct method, rework requirement 1.
3. Based on the following information about two jobs, prepare two different total-overhead costs for each job by using rates developed in requirements 1 and 2.

	Direct Labor Hours	
	Machining	Assembly
Job 88	18	2
Job 89	3	17

14-28 Service department cost allocations; single-department cost pools; direct, step-down, and reciprocal methods. The Manes Company has two products. Product 1 is manufactured entirely in Department X. Product 2 is manufactured entirely in Department Y. To produce these two products, the Manes Company has two service departments: A (a material-handling department) and B (a power-generating department).

An analysis of the work done by Departments A and B in a typical period follows:

Supplied by	Used by A	B	X	Y
A	—	100	250	150
B	500	—	100	400

The work done in Department A is measured by the direct labor hours of material-handling time. The work done in Department B is measured by the kilowatt hours of power.

The budgeted costs of the service departments for the coming year are:

	Department A	Department B
Variable indirect labor and indirect material costs	$ 70,000	$10,000
Supervision	10,000	10,000
Depreciation	20,000	20,000
	$100,000	$40,000
	+Power costs	+Material-handling costs

The budgeted costs of the production departments for the coming year are $1,500,000 for Department X and $800,000 for Department Y.

Supervisory costs are salary costs. Depreciation in B is the straight-line depreciation of power-generation equipment in its nineteenth year of an estimated twenty-five-year useful life; it is old but well-maintained equipment.

Required

1. What are the allocations of costs of Service Departments A and B to Production Departments X and Y using the direct method, two different sequences of the step-down method, and the reciprocal method of reallocation?
2. The power company has offered to supply all the power needed by the Manes Company and to provide all the services of the present power department. The cost of this will be $40 per kilowatt hour of power. Should Manes accept? Explain.

14-29 Allocating costs of service departments; dual rates; cost justification. Lindsay Transport Enterprises (LTE) operates an integrated transportation network including both rail operations and road operations. LTE has two service departments and two transportation departments:

Service Departments	Transportation Departments
Equipment and maintenance (EM)	Rail (train) operations
Information systems (IS)	Road (truck) operations

The **budgeted** level of service relationships at the start of the year was:

Supplied By:	Used By EM	IS	Rail	Road
EM	—	.10	.30	.60
IS	.20	—	.50	.30

The **actual** level of service relationships during the year was:

Supplied By:	Used By EM	IS	Rail	Road
EM	—	.20	.40	.40
IS	.25	—	.55	.20

LTE collects fixed costs and variable costs of each service department in separate cost pools. The actual costs in each pool for the year were (in thousands):

	Fixed-Cost Pool	Variable-Cost Pool
EM	$300	$540
IS	80	75

Fixed costs are allocated on the basis of the **budgeted** level of service. Variable costs are allocated on the basis of the **actual** level of service.

LTE monitors the cost per track mile for the rail department and the cost per road mile for the road department. These cost figures include costs allocated from the service departments to the production departments. During the year, the actual transportation miles were:

Rail operations	15,000,000 miles
Road operations	12,000,000 miles

Required

1. Allocate the service department costs to the two transportation departments using the following three methods:
   a. Direct method
   b. Step-down method (allocate EM first)
   c. Reciprocal method
   Show full details of calculations. Present results in a format similar to Exhibit 14-10. Allocate separately the variable and fixed service department costs.
2. Compare the service department total costs per transportation mile for the rail and road operations under each of the three methods in requirement 1. (Round to four decimal places.)
3. The prices charged for rail by LTE are regulated by a government agency and set on a full-cost basis. Full cost includes allocations of service department costs. The road rates set by LTE are unregulated and competition among road transportation operators is intense. What advice would you give the government regulatory agency about how to minimize the ability of rail transportation operators to overstate the service department costs included in their submissions to the agency about the full cost of their rail operations? Be specific.

14-30 Allocating costs of service departments; dual rates; direct, step-down, and reciprocal methods. Magnum T.A. Inc. specializes in the assembly and installation of high-quality security systems for the home and business segments of the market. The four departments at its highly automated state-of-the-art assembly plant are:

Service Departments	Assembly Departments
Engineering Support	Home Security Systems
Information Systems Support	Business Security Systems

The *budgeted* level of service relationships at the start of the year was:

	Used by			
Supplied by	Engineering Support	Information Systems Support	Home Security Systems	Business Security Systems
Engineering Support	—	.10	.40	.50
Information Systems Support	.20	—	.30	.50

The *actual* level of service relationships during the year was:

	Used by			
Supplied by	Engineering Support	Information Systems Support	Home Security Systems	Business Security Systems
Engineering Support	—	.15	.30	.55
Information Systems Support	.25	—	.15	.60

Magnum collects fixed costs and variable costs of each department in separate cost pools. The actual costs in each pool for the year were (in thousands):

	Fixed-Cost Pool	Variable-Cost Pool
Engineering Support	$2,700	$8,500
Information Systems Support	8,000	3,750

Fixed costs are allocated on the basis of the budgeted level of service. Variable costs are allocated on the basis of the actual level of service.

The service department costs allocated to each assembly department are allocated to products on the basis of units assembled. The units assembled in each department during the year were:

Home Security Systems	7,950 units
Business Security Systems	3,750 units

Required

1. Allocate the service department costs to the assembly departments using a dual-rate system and (a) the direct method, (b) the step-down method (allocate Information Systems Support first), (c) the step-down method (allocate Engineering Support first), and (d) the reciprocal method. Present results in a format similar to Exhibit 14-10.
2. Compare the service department costs allocated to each Home Security Systems unit assembled and each Business Security Systems unit assembled under *a*, *b*, *c*, and *d* in requirement 1.
3. What factors might explain the very limited adoption of the reciprocal method by many organizations?

14-31 Cost justification for a contract; analysis of service department cost allocations. Solve Problem 15-31 (P. C. Products), which could have been placed here.

14-32 Service department cost allocation and motivation. Solve Problem 13-26 (Sharp Company), which could have been placed here.

14-33 Common cost allocation, theater facility with multiple users. The Downtown Theater is owned by the city of Los Angeles. It has seating capacity of 2,500. Two companies use the theater.

- *Civic Light Opera Company:* Agreement with the city enables it to use the facilities on Friday, Saturday, and Sunday nights 50 weeks a year. It has used the Downtown Theater each year for the last ten years.
- *Experimental Drama Company:* Agreement with the city enables it to use the facilities on Tuesday, Wednesday, and Thursday nights 50 weeks a year. This company was organized last year and has used the Downtown Theater for one year.

Data for the most recent year are:

	Civic Light Opera	Experimental Drama
Nights theater available	150	150
Nights theater used	120	80
Average attendance per night	2,000	500
Average price per ticket	$15	$10
Revenues	$3,600,000	$400,000
Cost identifiable with each company	$2,200,000	$250,000

The common costs of the Downtown Theater are the $200,000 annual fixed rent payment to the city and the annual fixed operating costs of $800,000.

If the Civic Light Opera Company were the only user of the theater, the fixed costs of the Downtown Theater would be $875,000 ($200,000 fixed rent payment + $675,000 fixed operating costs). Both companies can use the facilities of other theaters providing they sign a one-year lease contract. The Civic Light Opera Company could use the Santa Monica Theater at an annual rent payment of $900,000. The Experimental Drama Company could use the University of Southern California Theater facility for an annual rent payment of $300,000.

Required

1. How will the common cost of $1,000,000 for the Downtown Theater be allocated between the two companies using the following allocation bases: (a) relative capacity (nights theater available), (b) relative use (nights theater used), (c) relative revenues, and (d) relative identifiable costs? What criterion might be invoked to justify methods *a, b, c,* and *d* individually?
2. How will the common cost of $1,000,000 be allocated using (e) the incremental common cost allocation method (assuming Civic Light Opera is the primary user) and (f) the standalone common cost allocation method? What criterion might be invoked to justify methods *e* and *f* individually?
3. What is the operating income of each of the two companies under methods *a, b, c, d, e,* and *f*? *Operating income* is defined as revenues minus the sum of identifiable costs and common cost allocations.
4. The mayor of Los Angeles proposes to use the incremental method as calculated in requirement 2. The manager of the Civic Light Opera Company comments that if the city wants to promote experimental drama, it should provide a direct subsidy to the Experimental Drama Company rather than use a cost allocation method that is unfair to the Civic Light Opera Company. How should the mayor respond to the manager of the Civic Light Opera Company?

14-34 Cost allocation for all cost categories in the value-chain, different costs for different purposes. Laser Technologies develops, assembles, and sells two product lines:

Product Line A:	laser scanning systems
Product Line B:	laser cutting tools

Product Line A is sold exclusively to the Department of Defense under a cost-plus reimbursement contract. Product Line B is sold to commercial organizations.

Laser Technologies classifies costs in each of its six value-chain business functions into two cost pools; direct product-line costs (separately traced to Product Line A or B) and indirect product-line costs. The indirect product-line costs are grouped into a single cost pool for each of the six functions of the value-chain cost structure:

Value-Chain Indirect Product-Line Cost Function	Base for Allocating Indirect Costs to Each Product Line
1. Research and development	Hours of R & D time identifiable with each product line
2. Product design	Number of new products
3. Manufacturing	Hours of machine assembly time
4. Marketing	Number of salespeople
5. Distribution	Number of shipments
6. Customer service	Number of customer visits

Summary data in 19_4 are:

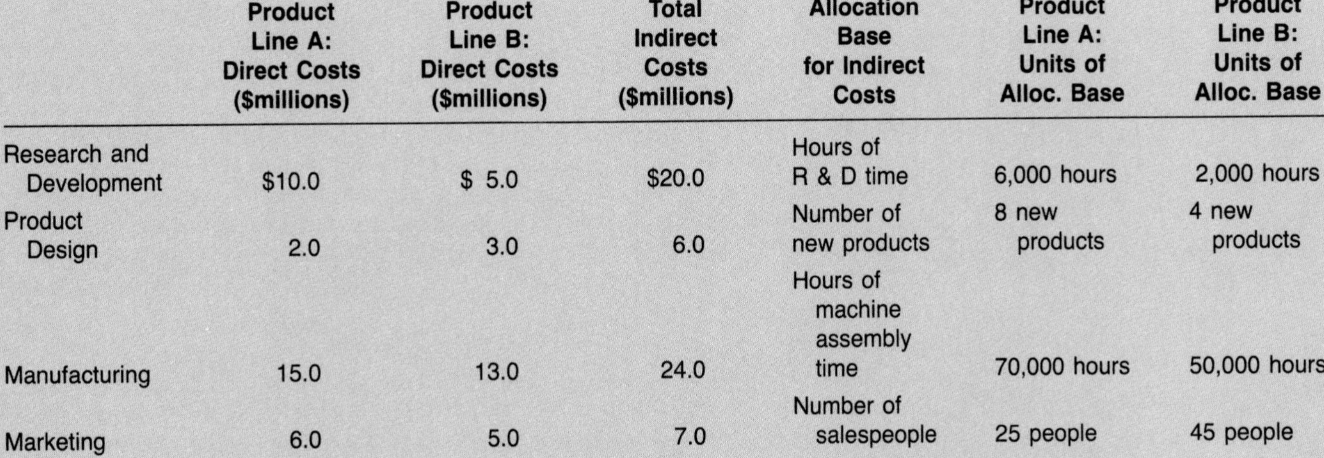

	Product Line A: Direct Costs ($millions)	Product Line B: Direct Costs ($millions)	Total Indirect Costs ($millions)	Allocation Base for Indirect Costs	Product Line A: Units of Alloc. Base	Product Line B: Units of Alloc. Base
Research and Development	$10.0	$ 5.0	$20.0	Hours of R & D time	6,000 hours	2,000 hours
Product Design	2.0	3.0	6.0	Number of new products	8 new products	4 new products
Manufacturing	15.0	13.0	24.0	Hours of machine assembly time	70,000 hours	50,000 hours
Marketing	6.0	5.0	7.0	Number of salespeople	25 people	45 people

	Product Line A: Direct Costs ($millions)	Product Line B: Direct Costs ($millions)	Total Indirect Costs ($millions)	Allocation Base for Indirect Costs	Product Line A: Units of Alloc. Base	Product Line B: Units of Alloc. Base
Distribution	2.0	3.0	2.0	Number of shipments	600 shipments	1,400 shipments
Customer Service	5.0	3.0	1.0	Number of customer visits	1,000 visits	4,000 visits

Required

1. For product pricing on its product Line B, Laser Technologies sets a preliminary selling price of 140% of full cost (made up of both direct costs and the allocated indirect costs for all six of the value-chain cost categories). What is the average full cost per unit of the 2,000 units of Product Line B produced in 19_4?
2. For motivating managers, Laser Technologies separately classifies costs into three groups:

   - Upstream (research & development and product design)
   - Manufacturing
   - Downstream (marketing, distribution, and customer service)

   Calculate the costs (direct and indirect) in each of these three groups for Product Lines A and B.
3. For the purpose of income and asset measurement for reporting to external parties, inventoriable costs under generally accepted accounting principles for Laser Technologies include manufacturing costs and product design costs (both direct and indirect costs of each category). At the end of 19_4, what is the average inventoriable cost for the 300 units of Product Line B on hand? (Assume zero beginning inventories.)
4. The Department of Defense purchases all Product Line A units assembled by Laser Technologies. Laser is reimbursed 120% of allowable costs. *Allowable cost* is defined to include all direct and indirect costs in the research and development, product design, manufacturing, distribution, and customer service functions. Laser Technologies employs a marketing staff that makes many visits to government officials, but the Department of Defense will not reimburse Laser for any marketing costs. What is the 19_4 allowable cost for Product Line A?
5. "Differences in the costs appropriate for different decisions, such as pricing and cost reimbursement, are so great that firms should have multiple accounting systems rather than a single accounting system." Do you agree?

CHAPTER 15

Cost Allocation: II

- Illustrating the general process of application
- Choosing indirect cost pool categories
- Evolving trends in cost pools
- Choosing cost application bases
- Evolving trends in application bases
- The contribution approach to cost allocation
- Cost justification and reimbursement

Choosing cost pools is an important step in allocating costs. To keep pace with the high-technology trend toward robotics, some businesses use a separate machine cost pool for robotic-related costs.

Learning Objectives

When you have finished studying this chapter, you should be able to

1. Outline three factors affecting application of indirect costs to individual products
2. Explain the importance of cost-pool homogeneity in determining the number of indirect cost pools
3. Describe two evolving trends in the number and type of indirect cost pools
4. Evaluate direct labor hours and costs as a cost application base in labor-paced manufacturing and machine-paced manufacturing
5. Outline consequences of using direct labor hours or dollars inappropriately as a cost application base
6. Describe the contribution approach to cost allocation
7. Outline use of the controllability concept in classifying items in a contribution-format income statement
8. Explain importance of explicit agreement between parties when reimbursement is based on costs incurred

Chapter 14 introduced four purposes of cost allocation and discussed the allocation of costs to activity areas or departments. Chapter 15 continues our exploration of cost allocation. We now examine the application of costs to individual products, services, or jobs (such as a government contract). These costs can include the already allocated costs of corporate and other service departments. The final sections of Chapter 15 discuss the contribution approach to cost allocation and specific issues that arise in the cost justification or reimbursement purpose of cost allocation.

In earlier chapters, we noted that *cost allocation* is a general term that refers to identifying accumulated costs with, or tracing accumulated costs to, cost objects such as activity areas, departments, divisions, territories, or products. *Cost application* is a narrower term that refers to a special type of cost allocation, the final assignment of costs to products (called Stage III allocations—see p. 456). Chapter 15 concentrates on issues associated with the choice of *cost application bases.*

ILLUSTRATING THE GENERAL PROCESS OF APPLICATION

Consider an example of cost application for a refrigerator assembled by Consumer Appliances. This example illustrates the four decisions managers make in applying costs to products. It also illustrates the trend among companies to using a diverse series of cost application bases. Consumer Appliances assembles over one hundred consumer electrical products at its single assembly facility. Managers make four decisions:

1. *Choose the cost object.* The cost object is a standard refrigerator model. Consumer Appliances sells refrigerators to retail department stores through its own sales force.
2. *Choose the direct costs to trace to the cost object.* Consumer Appliances traces two categories of direct costs to its standard refrigerator model:

- Direct materials $140
- Direct labor $ 35

3. *Choose (a) the indirect cost categories to be associated with the chosen cost object and (b) the aggregation of these cost categories before applying them to the product cost object.* Consumer Appliances has five indirect cost pools for applying indirect costs to products:

- Procurement
- Production: labor-paced assembly
- Production: machine-paced assembly
- Production: quality testing
- Marketing and distribution

For the economic decision and motivation purposes of cost allocation, all five indirect cost pools are included in the product-cost buildup. When computing the inventoriable product-cost figure that Consumer Appliances reports to external parties, only the first four indirect cost pools are included. (Marketing and distribution costs are (noninventoriable) period costs under generally accepted accounting principles.)

4. *Choose the cost application base for each of the indirect cost categories selected in (3).* Consumer Appliances uses the following application bases and application rates (these rates are revised every six months):

Indirect Cost Pool	Application Base	Application Rate
1. Procurement	Number of parts	$0.10 per part
2. Production: labor-paced assembly	Direct labor hours	$20 per hour
3. Production: machine-paced assembly	Machine hours	$16 per hour
4. Production: quality testing	Testing hours	$30 per hour
5. Marketing and distribution	Number of finished units sold	$150 per unit sold

The application rate for each indirect cost pool is calculated as:

$$\frac{\text{Total budgeted costs allocated to the indirect cost pool}}{\text{Total budgeted amount of application base}}$$

As an illustration, consider the procurement application rate of $0.10 per part. Assume this rate is computed as follows:

$$\frac{\$20{,}000{,}000}{200{,}000{,}000 \text{ parts}} = \$0.10 \text{ per part}$$

Numerator: For the next six months, the total budgeted costs to be allocated to the procurement cost pool are $20,000,000. These costs include personnel in the procurement function, incoming materials handling and inspection, and equipment-related costs for procurement personnel (for example, use of computers) and material handling (for example, use of automated guided vehicles).

Denominator: For the next six months, the total budgeted amount of the application base is 200,000,000 parts. This figure is the budgeted number of parts for all products assembled at the facility in the next six months. It includes a budget of 12,600,000 parts for the standard refrigerator model (420 parts per refrigerator times 30,000 budgeted production volume). The remaining 187,400,000 parts included in the denominator of the procurement application rate are earmarked for other products.

An overview of the product-cost system for Consumer Appliances appears in Exhibit 15-1. The application rate for each indirect cost pool in Exhibit 15-1 is computed in a way similar to that outlined above for the procurement indirect cost

EXHIBIT 15-1
Overview of Product Costing at Consumer Appliances

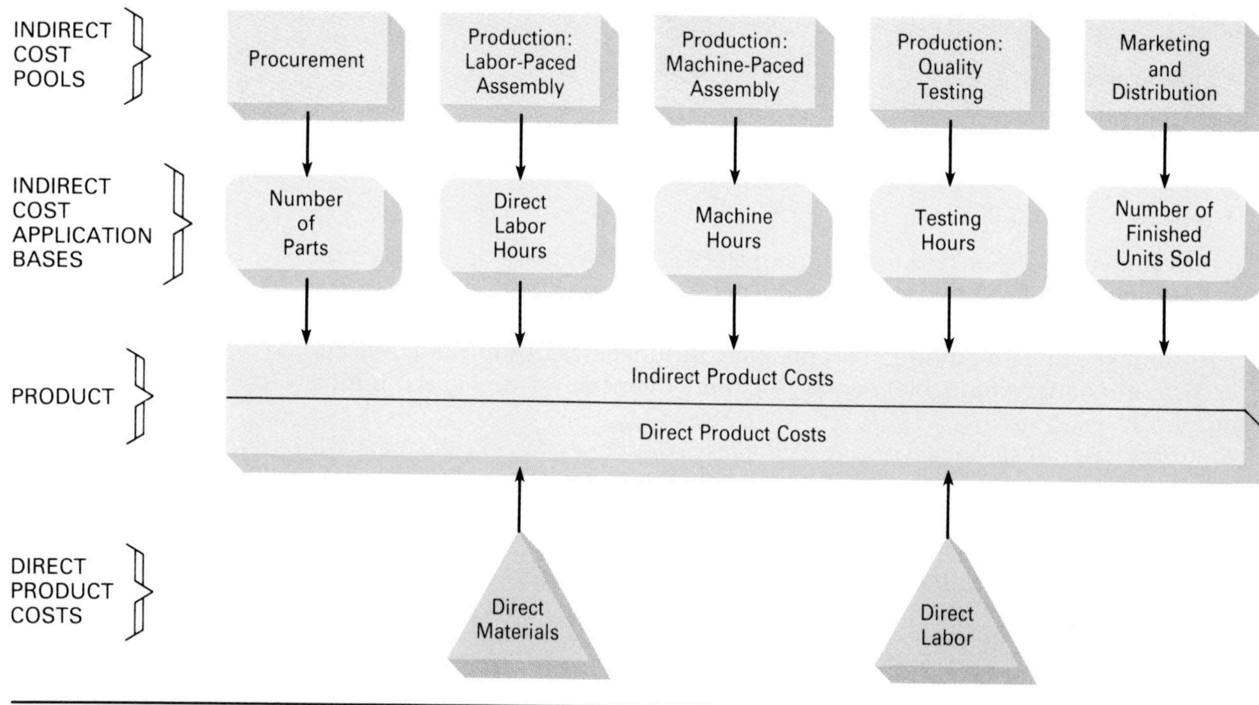

pool. Exhibit 15-2 presents the product-cost buildup for the standard refrigerator model. The full product cost is $455, comprising $175 direct costs and $280 indirect costs. The $280 amount includes $150 for marketing and distribution costs. When computing the inventoriable product cost (for financial reporting to external parties), the $150 amount is excluded. Thus, the inventoriable product cost is $305 per standard refrigerator model.

EXHIBIT 15-2
Costing of Standard Refrigerator Model of Consumer Appliances

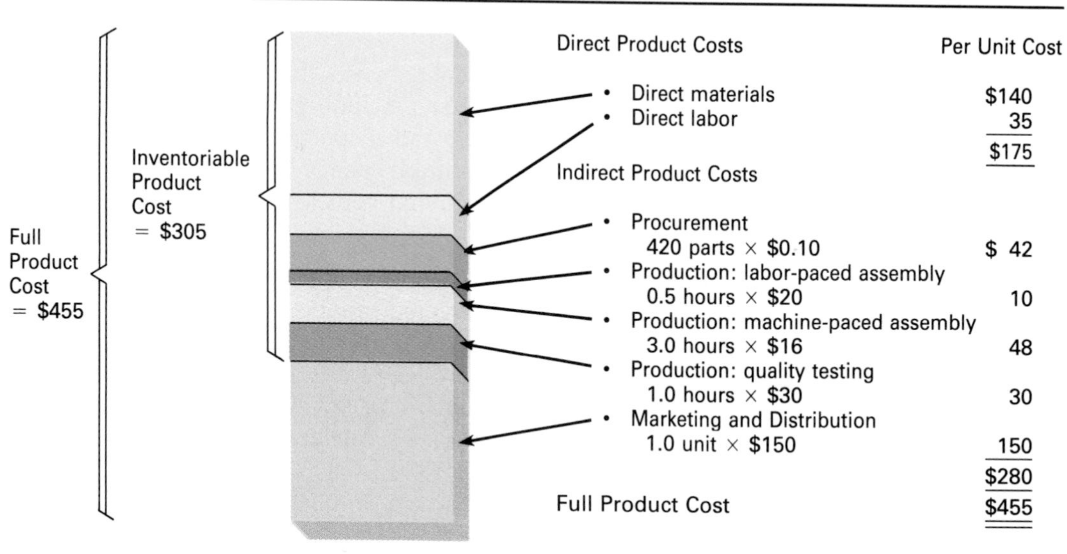

CHOOSING INDIRECT COST POOL CATEGORIES

Objective 1

Outline three factors affecting application of indirect costs to individual products

This section discusses issues related to the choice of cost pools. Subsequent sections discuss the choice of application bases and application rates.

Indirect costs applied to individual products are a function of

- *the aggregate costs in each final (Stage III) indirect cost pool,*
- *the application base used for each indirect cost pool, and*
- *the application rate used for each indirect cost pool.*

The concept of *cost pool homogeneity* discussed in Chapter 14 is central to the issues raised in this section. A cost pool is homogeneous if each activity whose costs are included in it has the same or a similar cause-and-effect relationship between the cost driver and the costs at that activity. The *same* cause-and-effect relationship means there is *both* the same cost driver *and* the same rate per unit of the cost driver. Consider the cost of health benefits paid to employees in our Consumer Appliances example. Health-benefits costs are collected in a single (intermediate) cost pool and then allocated to each of the five (final-stage) indirect cost pools in Exhibit 15-1. The health-benefits cost pool includes health benefits paid to workers in the materials procurement, production, and marketing/distribution areas. Why? Because the same health-benefits rate is paid to a health-insurance group for every employee, the cost driver.

When individual activities within a given cost pool are not homogeneous, more accurate product costs can be computed by dividing that single cost pool into two or more separate cost pools. We now turn to a discussion of how Consumer Appliances uses departmental cost pools instead of a plantwide cost pool.

Plantwide Versus Departmental Cost Pools

Should a company producing many products at the one plant use a single plantwide indirect cost pool or use multiple indirect department cost pools when computing individual product costs? Two important questions arise:

1. Do individual departments *differ* in cause-and-effect relationships between indirect costs and the variables used for application bases?
2. Do individual products *differ* in the way they use individual departments in the plant?

Objective 2

Explain the importance of cost-pool homogeneity in determining the number of indirect cost pools

If the differences found in answering these two questions are sizable, departmental cost pools will provide more accurate product-cost figures. When the differences are minimal, departmental cost pools will result in product-cost numbers similar to those computed using a plantwide cost pool.[1]

Consider product costing at Consumer Appliances. The production plant has three individual departments:

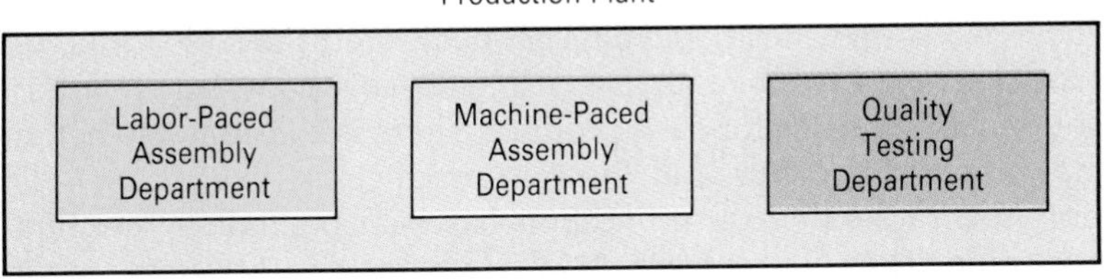

[1]The generic issue being discussed in this section is aggregate versus disaggregate cost pools. The example given discusses a single plantwide cost pool versus departmental cost pools. The same issue arises with the choice of a departmental cost pool versus individual activity cost pools within the one department. Activity-based accounting systems, described in Chapter 5, typically contain multiple indirect cost pools within a single department.

For many years, Consumer Appliances used a single plantwide indirect production cost pool, based on direct labor hours in the labor-paced assembly area. When the plant was first built, labor-paced assembly was the largest area of the plant. In recent years, the size of the labor-paced assembly area has decreased as machine-paced assembly lines have grown. The current product-costing approach has three departmental cost pools for the production plant:

Production Cost Pool	Cost Application Rate
Labor-paced assembly	$20 per direct labor hour
Machine-paced assembly	$16 per machine hour
Quality testing	$30 per testing hour

An indirect cost application rate of $100 per direct labor hour in the labor assembly area would have been applied if the single plantwide cost pool had been used in the current period.

How do the results developed using the three departmental cost pools compare with the results that would have been developed using the previous single plantwide cost pool? Let us examine two products, which use the following resources from three departments:

	Deluxe Refrigerator	Clothes Dryer
Labor-paced assembly department: direct labor hours	0.6 hours	0.8 hours
Machine-paced assembly department: machine hours	4.0 hours	1.5 hours
Quality-testing department: testing hours	3.0 hours	0.4 hours

The indirect production costs applied to these two products follow:

	Deluxe Refrigerator	Clothes Dryer
Using plantwide cost pool	$60	$80
Using departmental cost pools	$166	$52

Exhibit 15-3 summarizes the results.

Deluxe refrigerators have a sizable increase in indirect production costs applied to each product unit when departmental cost pools are used. Why? Because the departmental cost pools capture the relatively high use that the deluxe refrigerator makes of the machine-paced assembly and quality-testing departments. In contrast, clothes dryers have a sizable decrease in indirect production costs applied to each product unit. This product makes relatively high use of the labor-paced assembly area but relatively low use of the machine assembling and testing areas. (Indeed, one of the reasons for Consumer Appliances recently adopting the departmental cost pool system was a complaint from the clothes-dryer products manager. This manager argued that the plantwide cost pool system penalized his product line, making it appear that Consumer Appliances was losing money on a product line he believed was a "winner.")

As Exhibit 15-3 shows, departmental cost pools are preferable to the plantwide cost pool for Consumer Appliances. Departmental cost pools better capture cause-and-effect cost relationships for each product. There is considerable heterogeneity among the three departments and the two products in Exhibit 15-3. The three departments use different cost application bases—labor hours, machine hours, and testing hours. These bases were chosen by operating personnel at Consumer Appliances after analysis of cause-and-effect relationships in each department. The two products differ in their use of each of the three departments; use of a single cost application base does not capture these differences. Many companies have observed similar differences among their departments and their products. These differences are motivating the use of multiple indirect cost pools rather than single plantwide cost pools.

EXHIBIT 15-3

Plantwide versus Departmental Cost Pools at Consumer Appliances: Effect on Indirect Manufacturing Costs Applied to Products

	Deluxe Refrigerator	Clothes Dryer
Plantwide Cost Pool*	$60	$80
Departmental Cost Pools		
Labor-paced assembly department†	$12	$16
Machine-paced assembly department‡	64	24
Quality-testing department§	90	12
	$166	$52

*Direct labor hours per product:	0.6	0.8
Indirect production costs applied $100 × direct labor hours:	$60	$80
†Direct labor hours per product:	0.6	0.8
Indirect production costs applied $20 × direct labor hours:	$12	$16
‡Machine assembly hours per product:	4.0	1.5
Indirect production costs applied $16 × machine hours:	$64	$24
§Testing hours per product:	3.0	0.4
Indirect production costs applied $30 × testing hours:	$90	$12

EVOLVING TRENDS IN COST POOLS

When changes occur in the underlying operations (such as an increase in automation) or in information gathering (such as an increase in the use of bar codes), changes in the set of chosen cost pools should be considered. We now look at two trends in cost pool selection.

Objective 3

Describe two evolving trends in the number and type of indirect cost pools

Elimination of Direct Labor Cost Pool

In some industries, direct labor costs are decreasing relative to direct materials costs and indirect manufacturing costs. When direct labor costs constitute a small percentage of total costs, it may no longer be economically feasible to trace them directly to individual products. Direct tracing requires detailed tracking of labor time and (often) time-consuming attempts to reconcile applied labor time with total labor time available. An increasing number of firms in several industries, such as electronics, are now eliminating direct labor as a direct cost category. They are classifying all labor costs in manufacturing as indirect manufacturing costs.

Harley-Davidson, the motorcycle assembly company, analyzed how its manufacturing product-cost structure compared with the administrative costs required to "collect, inspect, and report" data in its accounting system:[2]

	Manufacturing Product-Cost Structure	Administrative Cost Effort
Materials	54%	25%
Overhead	36	13
Labor	10	62

[2]W. Turk, "Management Accounting Revitalized: The Harley-Davidson Experience," *Journal of Cost Management*, Winter 1990, pp. 28–39.

The administrative costs associated with tracking direct labor as a separate cost category included

- operators' time to fill out labor tickets,
- supervisors' time to review labor tickets,
- timekeepers' time to enter the labor data and review the data output reports for errors, and
- cost accountants' time in reviewing the direct labor and variance data.

Harley-Davidson concluded that tracing direct labor to products did not meet a cost-benefit test. Direct labor costs were only 10% of total manufacturing costs but required 62% of the administrative effort used to track all manufacturing costs. The company now includes all factory labor costs as part of indirect manufacturing costs.

Separate Machining Indirect Cost Pool

Many manufacturing facilities incur high levels of machine-related costs. Some companies are now using a separate machining cost pool when applying indirect costs to products. Machining costs include energy, lubricants, maintenance, repair, and depreciation.

Depreciation is one of the largest cost items in a machining cost pool. Businesses routinely compute depreciation for income taxation. Taxation-based depreciation estimates, however, may not reflect the decline in the economic service potential of machines. To refine estimates of a machine's useful life may require industrial engineers, whose time may be a significant cost to the company.

One company making airplane blades and other metal parts is experimenting with its set of final stage (Stage III) indirect cost pools. The machines used for cutting metal differ considerably in their capital costs, operating costs, and speed; the more expensive machines cut metal at a faster speed. Using three separate machine cost pools instead of a single machine cost pool, the company now applies machining costs on the basis of a rate per machine hour that better measures the cost and speed of the machine being used on a product. Different product costs are now reported for products that require the same total number of machine hours but differ in their relative use of different speed machines.

CHOOSING COST APPLICATION BASES

A *cost application base* is the common denominator for systematically relating a cost or group of costs to products. The choice of the application base should be guided by the major purpose to be served by the cost application and by the cost-benefit perspective, which is ever-present in choices among accounting systems.

Examples of Cost Application Bases

Several application bases are widely used in manufacturing settings. These include (1) direct labor costs or direct labor hours, (2) machine hours, (3) units produced, and (4) material costs or material quantities.

Direct Labor Costs or Direct Labor Hours: The choice between direct labor costs and direct labor hours as the application base depends on which one is more causally related to indirect manufacturing costs. For some indirect cost items, the causal factor is direct labor costs. For example, fringe benefit costs, such as pension benefits, are primarily tied to direct labor cost. For other indirect cost items, direct labor hours is the preferable base. For example, assume that senior and junior workers in an organization are equally efficient and use the same amount of overhead services per hour. In this situation, distorted cost applications will arise with the direct labor cost base. Suppose a senior worker earns $25 per hour and a junior worker earns $18 per hour. With a 200-percent-of-direct-labor-cost rate, the applied overhead cost of a given job requiring one hour of indirect labor would be $50 if the senior worker were used and $36 if the junior worker were used. Yet each worker uses the same amount of overhead services per hour.

Many firms are increasing their investments in machinery. Do not assume, however, that increased use of machines necessarily means direct labor is an inappropriate cost application base. It is useful to distinguish between two environments:

Objective 4

Evaluate direct labor hours and costs as a cost application base in labor-paced manufacturing and machine-paced manufacturing

- **Labor-Paced Manufacturing Environments**: Worker dexterity and productivity determine the speed of production. Machines funtion as tools that aid production workers. In labor-paced manufacturing environments, direct labor costs or direct labor hours may still capture cause-and-effect relationships, even if operations are highly automated.
- **Machine-Paced Manufacturing Environments**: Machines conduct most (or all) phases of production, such as movement of materials to the production line, assembly and other activities on the production line, and shipment of finished goods to the delivery dock areas. Machine workers in such environments may simultaneously operate more than one machine. Workers focus their efforts on supervision of the production line and general trouble-shooting rather than on the actual operation of the machines. Computer specialists and industrial engineers are the real "controllers" of the speed of production.

Machine Hours: In many highly mechanized facilities, the manufacturing operations are machine-paced rather than labor-paced. Here, machine hours frequently capture cause-and-effect relationships for applying indirect costs to products.

The increasing adoption of computerized machine controls is helping managers collect more reliable and detailed data on machine hour usage by individual products. One firm using this application base keeps detailed records on idle time, setup time, and operating time for each machine.

Units Produced: When a production facility manufactures all products in a similar way, a units-of-production application base can capture cause-and-effect relationships. Companies in beverage industries, such as soft drinks and beer, often apply indirect manufacturing costs to products on the basis of gallons or barrels of product processed.

Material Costs or Material Quantities: Materials procurement and handling costs are a sizable indirect cost item in many firms. These costs include purchasing, materials management, and incoming inspection of materials. Bases used to apply materials procurement and handling costs to products include material costs and number of parts. When individual categories of materials procurement costs differ in their cause-and-effect relationship, the use of multiple cost pools (with possibly different application bases) may be appropriate.

Surveys of Practice

Surveys report that direct labor costs or direct labor hours are the most frequently used base for applying indirect manufacturing costs to products. One survey of U.S. midwestern firms reports the following use of each application base:

- Direct labor costs or direct labor hours 74%
- Machine hours 23

- Units produced 21
- Material costs or material quantities 20

Over 70% of all respondents "reported using multiple overhead rates within a single production facility."[3] (This use of multiple overhead rates by individual firms explains why the above percentages exceed 100%.)

Surveys also report differences in the application bases used in labor-paced and machine-paced environments. A survey of U.S.-based and Japanese-based companies reports the following information on applying indirect manufacturing costs to products:[4]

	Labor-Paced Environments		Machine-Paced Environments	
Application Base	**U.S.**	**Japan**	**U.S.**	**Japan**
Standard labor hours	32%	53%	10%	10%
Actual labor hours	25	53	6	18
Standard labor costs	23	20	5	0
Actual labor costs	23	18	6	8
Standard machine hours	7	0	24	38
Actual machine hours	7	0	15	30
Time in machine center	5	3	8	13

Examples of indirect cost application rates are available in some industry surveys. The median application rates for indirect manufacturing costs in segments of the U.S. electronics industry follow.[5]

Segment of the U.S. Electronic Industry	Application Rate Based on Direct Labor Costs	Application Rate Based on Direct Material Costs
Computers (40 firms)	440%	15.5%
Peripherals (39 firms)	277	12.5
Instruments (88 firms)	262	15.0
Telecommunications (54 firms)	233	21.0
Software (56 firms)	68	38.0

Consequences of Inappropriate Use of An Application Base

Objective 5

Outline consequences of using direct labor hours or dollars inappropriately as a cost application base

Many important decisions are based on cost figures. If these figures result from application bases that fail to capture cause-and-effect relationships, managers may make decisions that conflict with maximizing long-run company net income. Consider the use of direct labor costs as an application base in machine-paced manufacturing settings. In this environment, indirect cost rates of 500% of direct labor costs (or more) may be encountered. Possible consequences include the following points.

1. Product managers may make excessive use of external vendors for parts with a high direct labor content. With a 500% indirect cost application rate, every $1 of direct labor spent on a part made internally will result in a $5 indirect cost charge.

[3]Price Waterhouse, *Survey of the Cost Management Practices of Selected Midwest Manufacturers* (1989).

[4]NAA Tokyo Affiliate, Management Accounting in the Advanced Manufacturing Surrounding: Comparative Study on Survey in Japan and U.S.A. (1988).

[5]*1989–90 Operating Ratios Survey* (Santa Clara, CA: American Electronics Association, 1990).

2. Manufacturing managers may pay excessive attention to controlling direct labor hours relative to the attention paid to controlling the more costly categories of materials and machining. By eliminating $1 of direct labor costs, $6 of reported product cost can be eliminated with a 500% of direct labor dollar application base. Managers can control the accounting numbers applied by controlling direct labor use. However, this action does not control the actual incurrence of the larger materials and machining costs.
3. Managers may attempt to classify shop-floor personnel as indirect labor rather than as direct labor, resulting in part of the personnel costs being applied to other products.

EVOLVING TRENDS IN APPLICATION BASES

Increasing Diversity in Application Bases

Some firms are now experimenting with using more indirect cost application bases than typical accounting systems have in the past. The Consumer Appliances example in Exhibits 15-1 and 15-2 illustrates a firm that is increasing the number of application bases from one (direct labor hours) to three (direct labor hours, machine hours, and testing hours) in their production facility.

A second example of increasing diversity in application bases is the marketing, general, and administrative (M, G, & A) cost application system of a mail-order business. These costs are approximately 30% of the total product cost. This company increased the number of application bases from one (revenue dollars) to six. The six M, G, & A cost pools and their application bases are:

M, G, & A Cost Pool	Application Base
Field sales	Units sold
Accounting	Number of transactions
Distribution	Items per order
Sales administration	Items per order
Product marketing	Effort*
New catalog	Number of pages

*Based on estimates of effort made by operating personnel.

This revised cost application scheme has led to sizably larger M, G, & A costs for several products, especially low-priced products that are ordered in small quantities and require extensive new catalog efforts.

Several companies are now considering the use of throughput time as an application base. **Throughput time** is the time from when a product starts on the production line to when it becomes a finished good. The rationale for using this application base is that steps that lengthen throughput time frequently add to the indirect costs at the plant. For example, moving partly assembled products into and out of a work in process inventory area increases throughput time and increases material-handling costs. By using this base, management signals to operating personnel that reported product costs can be reduced by reducing the throughput time of products being assembled.[6] The Problem for Self-Study at the end of this chapter illustrates use of the throughput-time application base for a medical instruments company.

[6]Arguments for using throughput time as an application base are presented in R. Schmenner, "Escaping the Black Holes of Cost Accounting," *Business Horizons*, January–February 1988, pp. 66–72. For example: "C [illegible] ork, and space, and fewer transactions necessary), reduced thr[illegible] ne managers in the factory should concentrate on" (p. 70).

Cost Drivers and Application Bases

When a cause-and-effect criterion is used, the chosen application bases are likely to be cost drivers. Because a change in a cost driver causes a change in the total cost of a related cost object, the use of cost drivers as application bases increases the accuracy of reported product costs. However, not all cost application bases selected by organizations are cost drivers. Consider the following reasons for using bases that are not cost drivers:

1. Improving the accuracy of individual product costs may be less important to an organization than other goals. Consider the goal of restraining the growth in headcount (the number of employees on a company's payroll). Several Japanese companies use direct labor hours as the cost application base. These companies acknowledge that direct labor is not the most important driver of their manufacturing costs, but they want to send a clear signal to all employees that reduction in headcount is a key goal.[7]

   Managers may prefer direct labor hours as an application base even when they are aware that direct labor hours are not a driver of short-run indirect costs. Why? To promote increased levels of automation. Using this application base, managers motivate product designers to decrease the direct labor content of the products they design. Management may view increased automation as a strategic necessity to remain competitive in the long run.
2. Information about cost driver variables may not be reliably measured on an ongoing basis. For example, the number of machine setups is often viewed as a driver of indirect manufacturing costs, but some firms do not systematically record this information.
3. Accounting systems with many indirect cost pools and application bases are more expensive to use than systems with few cost pools and application bases. The investment required to develop, implement, and educate users about a system with many indirect cost pools can be sizable. Some firms place a low priority on investments in their internal accounting systems, given that the benefits from such investments are frequently difficult to quantify.

THE CONTRIBUTION APPROACH TO COST ALLOCATION

Objective 6

Describe the contribution approach to cost allocation

In general, managers using the *contribution approach* to cost allocation distinguish between variable costs and fixed costs in their internal financial reporting and analysis. Exhibit 15-4 stresses cost-behavior patterns in relation to individuals and segments of an organization. A **segment** is any part of an organization for which separate determination of revenues, costs, or both are desired. A synonym for *segment* is **subunit**. Exhibit 15-4 is based on monthly data and the assumption that variable costs are driven by a single volume-related factor such as tons of lumber processed by a lumber mill. Two different types of segments are illustrated in Exhibit 15-4: divisions and products. As you read across Exhibit 15-4, the focus becomes narrower, from the company as a whole, to Divisions A and B, and then to the products of Division B only.

[7]See T. Hiromoto, "Another Hidden Edge—Japanese Management Accounting," *Harvard Business Review*, July–August 1988, pp. 22–26.

EXHIBIT 15-4
The Contribution Approach: Monthly Income Statement by Segments
(In Thousands of Dollars)

	Company as a Whole	Company Breakdown into Two Divisions		Breakdown of Division B by Product				
		Division A	Division B	Not Allocated	Product 1	Product 2	Product 3	Product 4
Net revenues	1,500	500	1,000		300	200	100	400
Variable manufacturing cost of goods sold	780	200	580		120	155	45	260
Manufacturing contribution margin	720	300	420		180	45	55	140
Marketing, service, and administrative costs	220	100	120		60	15	25	20
(1) Contribution margin	500	200	300		120	30	30	120
Fixed costs controllable by segment managers (certain advertising, sales promotion, salespersons' salaries, engineering, research, management consulting, and supervision costs)	190	110	80	45*	10	6	4	15
(2) Contribution controllable by segment managers	310	90	220	(45)	110	24	26	105
Fixed costs controllable by others (such as depreciation, property taxes, insurance, and the division manager's salary)	70	20	50	20	3	15	4	8
(3) Contribution by segments	240	70	170	(65)	107	9	22	97
Unallocated costs (not clearly or practically allocable to any segment except by some questionable allocation base)	135							
(4) Operating income	105							

*Only those costs clearly identifiable with a product line should be allocated.

Revenues, Variable Costs, and Contribution Margins

The allocation of revenues and variable costs is usually straightforward. Each item is directly identifiable with a specific segment of the organization. (We assume no intersegment sales in this section. Issues associated with the transfer pricing of such sales are discussed in Chapter 27.) The contribution margin, item (1) in Exhibit 15-4, is particularly helpful for predicting the impact that short-run changes in volume have on operating income. Changes in operating income can quickly be calculated by multiplying the change in units by the unit contribution margin or by multiplying the increment in dollar sales by the contribution-margin ratio. For example, the contribution-margin ratio of Product 1 is \$120/\$300 = 40%. The increase in operating income resulting from a \$20,000 increase in its sales volume can readily be computed: .40 × \$20,000, or \$8,000. This calculation assumes (a) that no changes in selling prices, operating efficiency, or fixed costs occur, and (b) that costs are variable with respect to a single volume-related factor.

Contributions by Managers and Segments

Item (2) in Exhibit 15-4 is a measure of the manager's controllable contribution, and this item is therefore a measure of the manager's performance. Item (3) describes the performance of the segment as an economic investment.

Objective 7

Outline use of the controllability concept in classifying items in a contribution-format income statement

Controllability is the degree of influence that a specific manager has over the costs, revenues, or other items in question. Distinctions between controllable and uncontrollable items are helpful when feasible. They aid motivation and the analysis of performance. However, the distinction between item (2) and item (3) in Exhibit 15-4 is seldom found in practice. Why? Because it is difficult to draw fine distinctions in identifying which managers control which fixed costs. Nevertheless, distinctions between performances of the manager and the segment may still be established through intelligent budgeting. For example, the manager's objective for the forthcoming year may be to reduce the division's loss from \$5 million to \$3 million. If this objective is attained, the manager may subsequently be judged as successful. However, the division may continue to be regarded as a miserable investment.

The controllable contribution, keyed as item (2) in Exhibit 15-4, should be interpreted in conjunction with the contribution margin. *Managers can often influence some fixed costs.* For example, heavier outlays for maintenance, engineering, or management consulting may reduce repairs, increase machine speeds, or heighten labor productivity. Also, decisions on advertising, research, and sales-promotion budgets are necessarily related to expected impacts on sales volumes. Note, too, that when a time horizon much longer than a month is considered, some items classified as fixed costs in Exhibit 15-4 will become variable costs.

The income statement in Exhibit 15-4 has four numbered measures of performance, ranging from contribution margin to operating income. There is nothing hallowed about these four illustrative measures; some organizations may want to use only two or three such measures. For instance, Reckett & Colman, a large British manufacturer and seller of housewares, has a classification of sales, contribution before marketing costs, marketing costs, contribution after marketing costs, fixed costs, and operating income.

Unallocated Costs

There is a limit to how detailed the allocation of a given cost in an income statement can be. Although many discretionary costs are easily traceable to divisions, they may not all be directly traceable to products. Some advertising costs for Division B may be common to two or more of its products. For example, Products 1, 2, and 3 may be consumer items that all benefit from the same advertisements, and Product 4 may be an item sold to manufacturers by a sales force separate from the sales force selling Products 1, 2, and 3.

Consider the next-to-last line in Exhibit 15-4. An unallocated cost is common to all the segments in question and is not clearly or practically allocable. Examples of unallocated costs include the salaries of the president and other top officers, companywide research and development costs, and some central corporate costs like public relations and corporate-image advertising. Also, income taxes typically are not allocated under the contribution approach because of the difficulty of tracing them to individual segments. Unless the general company costs are clearly traceable to segments, allocations are not made under the contribution approach.[8]

This inability to allocate some costs is the most controversial aspect of the contribution-approach income statement. Accountants and managers are accustomed to the whole being completely subdivided into parts that can be added again to equal the whole. In *traditional* segment income statements, all costs are fully allocated, so that the segments show final operating incomes that can be summed to equal the operating income for the company as a whole.

COST JUSTIFICATION AND REIMBURSEMENT

Examples of the cost-justification or reimbursement purpose of cost allocation are found in diverse settings:

1. A contract between the Department of Defense and a company designing and assembling a new fighter plane; the price paid is based on the contractor's cost plus a preset fixed fee.
2. A contract between two oil companies in a joint venture; the costs of operating a shared oil-refining facility are allocated between the companies on the basis of expected usage of the refinery.
3. A contract between an energy-consulting firm and a hospital; the consulting firm receives a fixed fee plus a share of the energy cost savings arising from the consulting firm's recommendations.

To reduce the areas of dispute between parties to such contracts, the "rules of the game" should be explicit (preferably in written form) and well understood at the time the contract is signed. Such "rules of the game" include the definition of cost items allowed, the cost pools, and the permissible allocation base(s) for each cost pool.

Objective 8

Explain importance of explicit agreement between parties when reimbursement is based on costs incurred

Contracting with the U.S. Government

The U.S. government reimburses most contractors in one of two ways:

1. **Fixed-Priced Contract**. The price the contractor receives is established at the outset. It is not subject to any adjustment due to the actual costs the contractor incurs.
2. **Cost-Reimbursement Contract**. The price the contractor receives is based on the actual costs the contractor incurs. There are many variants on this contract. For example, the contractor may be reimbursed for cost plus a fixed fee or for cost plus an incentive fee.

All contractors doing business with the U.S. government (not just defense contractors) must comply with cost accounting standards issued by the **Cost Accounting Standards Board (CASB)**. The CASB was established in 1970 against a backdrop of systematic cost overruns by contractors on cost-reimbursement contracts

[8]A survey of 79 firms with U.S. headquarters and 26 with European headquarters reported the following percentages of firms *not* allocating costs in the following five categories to divisions:

	Firms With U.S. Headquarters	Firms With European Headquarters
Corporate income taxes	67.1%	71.4%
Interest	60.8	21.4
R & D	34.2	21.4
Headquarters administration	31.6	42.9
Advertising	31.6	28.6

Business International, *Evaluating the Performance of International Operations* (New York: Business International Corporation, 1989), p. 179.

and allegations that contractors were overstating their costs. In 1980, the CASB ceased operations because Congress refused to appropriate money for its annual budget. In 1988, the CASB was recreated as an independent board within the Office of Federal Procurement Policy. It has the exclusive authority to "make, promulgate, amend, and rescind cost accounting standards and interpretations thereof designed to achieve *uniformity* and *consistency* in the cost accounting standards governing measurement, assignment, and allocation of costs to contracts within the United States."

CASB Standards and Interpretations

Exhibit 15-5 lists the titles of selected accounting standards issued by the CASB. Several of the CASB standards (for example, 402) cover general issues relating to the definition of cost items and the prohibition of double-counting. *Double-counting* occurs when a cost item is included both as a direct cost item and as a part of an indirect cost pool allocated to the contract using a budgeted rate. Other standards cover either the allocation of direct costs (for example, 407) or indirect costs (for example, 403) to government contracts.

Standard 403 ("Allocation of Home Office Expenses to Segments") illustrates the CASB approach to determining cost pools and the allocation bases for these pools:

> The allocation of centralized service functions shall be governed by a hierarchy of preferable allocation techniques which represent beneficial or causal relationships. The preferred representations of such relationships is a measure of the activity of the organization performing the function. Supporting functions are usually labor-oriented, machine-oriented, or space-oriented.

Exhibit 15-6 gives examples of the allocation bases presented in Standard 403 for individual home-office expense cost pools.

EXHIBIT 15-5
CASB Standards and Interpretations

CAS 401	Consistency In Estimating, Accumulating And Reporting Costs*
CAS 402	Consistency In Allocating Costs Incurred For The Same Purpose*
CAS 403	Allocation Of Home Office Expenses To Segments*
CAS 404	Capitalization Of Tangible Assets
CAS 405	Accounting For Unallowable Costs
CAS 406	Cost Accounting Period
CAS 407	Use Of Standard Cost For Direct Material And Direct Labor
CAS 408	Accounting For Costs Of Compensated Personal Absence
CAS 409	Depreciation Of Tangible Capital Assets
CAS 410	Allocation Of Business Unit General And Administrative Expense To Final Cost Objectives
CAS 411	Accounting For Acquisition Costs Of Material
CAS 412	Composition And Measurement Of Pension Costs
CAS 413	Adjustment And Allocation Of Pension Cost
CAS 414	Cost Of Money As An Element Of The Cost Of Facilities Capital
CAS 415	Accounting For The Cost Of Deferred Compensation
CAS 416	Accounting For Insurance Costs
CAS 417	Cost Of Money As An Element Of The Cost Of Capital Assets Under Construction
CAS 418	Allocation Of Direct And Indirect Costs
CAS 420	Accounting For Independent Research And Development Costs And Bid And Proposal Costs

*Indicates Standard for which an Interpretation was issued.

EXHIBIT 15-6

Illustrative Allocation Bases Suggested by CASB for Centralized Home-Office Service Functions

Service Rendered	Cost Allocation Bases
1. Personnel administration	1. Number of personnel, labor hours, payroll, number of hires
2. Data-processing services	2. Machine time, number of reports
3. Centralized purchasing and subcontracting	3. Number of purchase orders, value of purchases, number of items
4. Centralized warehousing	4. Square footage, value of material, volume
5. Company aircraft service	5. Actual or standard rate per hour, mile, passenger mile, or similar unit
6. Central telephone service	6. Usage costs, number of telephones

Fairness of Pricing

The entire field of negotiated government contracts is marked by the attempt to use *cost* allocations as a means for establishing a mutually satisfactory *price.* There is a subtle but important distinction here. A *cost* allocation may be difficult to defend on the basis of any cause-and-effect reasoning, but it may be a "reasonable" or "fair" means of establishing a contract price in the minds of the appropriate parties. Various costs become "allowable," but others are "unallowable." An **allowable cost** is a cost that parties to a contract agree to include in the costs to be reimbursed. Some contracts identify cost categories that are nonallowable. For example, marketing costs are often excluded from government contracts. Other contracts specify how allowable costs are to be determined. For example, only economy-class airfares are allowable. Making the allocation rules as explicit as possible (and in writing) reduces argument and litigation when cost allocations are to be used for establishing a contract price.

PROBLEM FOR SELF-STUDY

PROBLEM

Medical Instruments in 19_2 has changed its manufacturing-costing system. The prior system had two direct product-cost categories (direct material and direct labor) and one indirect cost category. Indirect costs were applied to products on the basis of direct labor costs.

The new costing system retains the same two direct product-cost categories. Now, however, indirect manufacturing costs are collected into two cost pools:

1. Materials-handling overhead—applied on the basis of the number of parts in a product. (When each individual part in a product is different, the number of parts and the number of separate part numbers in the product will be equal. When the same part number is used multiple times in a product, the number of parts will exceed the number of separate part numbers in that product.)
2. Production overhead—applied on the basis of the budgeted throughput time for each product. Throughput time is the time from when a product starts on the production line to when it becomes a finished good.

Management made the following assumptions in developing the 19_2 budgeted indirect cost application rates:

Materials-Handling Overhead

Budgeted total materials-handling overhead costs	$8,000,000
Budgeted number of separate part numbers	5,000
Budgeted average usage per separate part number	800
Budgeted total number of parts (5,000 × 800)	4,000,000

$$\text{Budgeted materials-handling overhead cost application rate} = \frac{\$8,000,000}{4,000,000} = \$2 \text{ per part}$$

Production Overhead

Budgeted total production overhead cost	$12,000,000
Budgeted number of individual products	400
Budgeted average production volume per product	100
Budgeted average throughput time per product	6 hours
Budgeted total throughput time (400 × 100 × 6)	240,000 hours

$$\text{Budgeted production overhead cost application rate} = \frac{\$12,000,000}{240,000 \text{ hours}} = \$50 \text{ per hour}$$

Curt Henning is examining how the new costing system affects the reported costs of three products. Details of these products in 19_2 follow:

	Product A	Product B	Product C
Direct materials costs	$1,680	$1,250	$2,070
Direct labor hours	7.2 hours	4.3 hours	6.1 hours
Number of parts	128	86	260
Throughput time in hours	4.8	3.9	18.5

The direct labor rate in 19_2 is $30 per hour. Under the prior product-costing system (with one indirect cost category), an indirect cost application rate of 300% of direct labor costs would have been used in 19_2.

Required

1. What characteristics of a product will lead to its having a much higher cost under the 19_2 costing system than it would have had under the prior costing system?
2. Compute the manufacturing costs of Products A, B, and C using
   (a) the prior product-costing system
   (b) the product-costing system introduced in 19_2
3. Why might there be a cause-and-effect relationship between actual throughput time and production overhead costs?

SOLUTION

1. The characteristics of a product that will lead to its having a much higher cost under the new costing system are (a) low direct labor cost content, (b) high number of parts, and (c) long throughput time.
2. (a)

	Product A	Product B	Product C
Direct manufacturing unit cost			
Direct materials	$1,680	$1,250	$2,070
Direct labor			
(7.2; 4.3; 6.1 × $30)	216	129	183
	$1,896	$1,379	$2,253
Indirect manufacturing unit cost			
($216, $129, $183 × 300%)	648	387	549
Total manufacturing unit cost	$2,544	$1,766	$2,802

(b)

	Product A	Product B	Product C
Direct manufacturing unit cost			
Direct materials	$1,680	$1,250	$2,070
Direct labor			
(7.2; 4.3; 6.1 × $30)	216	129	183
	$1,896	$1,379	$2,253
Indirect manufacturing unit cost			
Material handling			
(128; 86; 260 × $2)	$ 256	$ 172	$ 520
Production			
(4.8; 3.9; 18.5 × $50)	240	195	925
	$ 496	$ 367	$1,445
Total manufacturing unit cost	$2,392	$1,746	$3,698

3. The actions required to reduce throughput time will probably reduce the activities that drive production overhead cost. For example, many firms achieving dramatic reductions in throughput times also achieve

- lower inventory levels (meaning lower materials-handling costs),
- higher quality levels (meaning reduced quality rework activities), and
- reduced complexity in scheduling (meaning lower manufacturing administrative costs).

SUMMARY

Managers use information incorporating cost allocations to make economic decisions for resource allocation, to motivate managers and employees, to measure income and assets for reporting to external parties, and to justify costs or compute reimbursement. Decisions on how costs should be applied to products should be guided by the main purpose for which the product-cost information is desired.

Indirect costs are applied to products through the chosen set of cost pools and the application base used for each cost pool. Two trends in practice include an increase in the number of indirect cost pools used for final (Stage III) allocations and the use of a broader set of application bases (for example, number of part numbers in a product, setup time, and the time to assemble a product). Still, direct labor hours or direct labor costs are the indirect manufacturing cost application bases most frequently used in practice.

When examining the implication of technological change for accounting systems, be careful. Direct labor hours may be an inappropriate application base in machine-paced manufacturing environments but an appropriate application base in labor-paced manufacturing environments. Do not generalize that direct labor hours are an improper application base in all automated plants.

Cost allocations in cost-reimbursement contracts serve mainly as a means of establishing a mutually satisfactory price. In such contracts, both parties may agree to include individual cost items for which cause-and-effect relationships are not easily established. The legal agreement between the contracting parties should be explicit as to the allowable cost items, the cost pools to be used, and the allocation base for each cost pool.

TERMS TO LEARN

This chapter and the Glossary at the end of the book contain definitions of the following important terms:

allowable cost *(p. 507)* CASB *(505)* Cost Accounting Standards Board *(505)*
cost-reimbursement contract *(505)* fixed-price contract *(505)*
labor-paced manufacturing environment *(499)*
machine-paced manufacturing environment *(499)* segment *(502)*
subunit *(502)* throughput time *(501)*

ASSIGNMENT MATERIAL

QUESTIONS

15-1 Name three factors that effect the indirect costs applied to individual products.

15-2 When are departmental overhead rates generally preferable to plantwide overhead rates?

15-3 Describe cost pool homogeneity.

15-4 "To obtain higher homogeneity, have more rather than fewer cost pools." Do you agree? Why?

15-5 "The traditional three cost categories of direct materials, direct labor, and overhead are outdated. It's time all firms recognized that the machining cost component of overhead should be separately reported as a fourth cost category." Do you agree? Why?

15-6 What is the most frequently used base for applying manufacturing overhead costs to products?

15-7 "Manufacturing firms that make extensive use of machines should abandon the use of direct labor cost as an application base." Do you agree? Why?

15-8 What are two consequences of using direct labor hours as an application base in a machine-paced manufacturing environment?

15-9 Why might a firm not use a cost-driver variable as the cost application base?

15-10 List at least three subtotals that might be highlighted in a contribution-approach income statement by segments.

15-11 Name two ways that firms are reimbursed under government contracts.

EXERCISES AND PROBLEMS

15-12 Plant managers and accounting reports. Six months ago, a firm switched from a labor-paced to a machine-paced production line. One of its plant managers makes the following comment on the monthly accounting report on individual product costs: "Nobody at the plant level now understands the unit-cost figures. They are both too high and too volatile. We gave up trying to understand or use them months ago." What advice would you give the firm's accountant?

15-13 Alternative application bases for a professional services firm. The Wolfson Group (WG) provides tax advice to multinational firms. WG charges clients for (a) direct professional time (at an hourly rate) and (b) support services (at 30% of the direct professional costs billed). The three professionals in WG and their rates per professional hour are:

Professional	Billing Rate Per Hour
Myron Wolfson	$500 per hour
Ann Brown	120 per hour
John Anderson	80 per hour

WG has just prepared the May 19_7 bills for two clients. The hours of professional time spent on each client follow:

	Clients	
Professional	Seattle Dominion	Tokyo Enterprises
Wolfson	15	2
Brown	3	8
Anderson	22	30
Total	40	40

Required

1. What amounts did WG bill to Seattle Dominion and Tokyo Enterprises for May 19_7?
2. Suppose support services were billed at $50 per professional-labor hour (instead of 30% of professional-labor costs). How would this change affect the amounts WG billed to the two clients for May 19_7? Comment on the differences between the amounts billed in requirements 1 and 2.
3. How would you determine whether professional-labor costs or professional-labor hours are a more appropriate application base for WG's support services?

15-14 Factory cost application, use of a separate machining cost pool category. Mahitsu Motors is a manufacturer of motorcycles. Production and cost data for 19_1 follow:

	500 CC Brand	1,000 CC Brand
Units produced	10,000	20,000
Direct labor hours per unit	2	4
Machine hours per unit	8	8

A single-cost pool is used for factory overhead. For 19_1, factory overhead was $6,400,000. Mahitsu applies factory overhead costs to products on the basis of direct labor hours per unit.

Mahitsu's accountant proposes that two separate pools be used for factory overhead costs:

- Machining cost pool ($3,600,000 in 19_1)
- General factory overhead cost pool ($2,800,000 in 19_1)

Machining costs are to be applied using machine hours per unit, and general factory costs are to be applied using direct labor hours per unit.

Required

1. Compute the overhead costs applied per unit to each brand of motorcycle in 19_1 using the current single-cost pool approach of Mahitsu.
2. Compute the machining costs and general factory costs applied per unit to each brand of motorcycle assuming that the accountant's proposal for two separate cost pools is used in 19_1.
3. What benefits might arise from the accountant's proposal for separate pools for machining costs and general factory costs?

15-15 Factory cost application, use of a conversion cost pool category, automation. Medical Technology Products manufactures a wide range of medical instruments. Two testing instruments (101 and 201) are produced at its highly automated Quebec City plant. Data for December 19_1 follow:

	Instrument 101	Instrument 201
Direct materials	$100,000	$300,000
Direct labor	$ 20,000	$ 10,000
Units produced	5,000	20,000
Actual direct labor hours	1,000	500

Factory overhead is applied to each instrument product on the basis of actual direct labor hours per unit for that month. Factory overhead cost for December 19_1 is $270,000. The production line at the Quebec City plant is a machine-paced one. Direct labor is made up of costs paid to workers minimizing machine problems rather than actually operating the machines. The machines in this plant are operated by computer specialists and industrial engineers.

Required

1. Compute the cost per unit in December 19_1 for Instrument 101 and Instrument 201 under the existing cost accounting system.
2. The accountant at Medical Technology proposes combining direct labor costs and factory overhead costs into a single conversion cost pool. These conversion costs would be applied to each unit of product on the basis of direct materials costs. Compute the cost per unit in December 19_1 for Instrument 101 and Instrument 201 under the accountant's proposal.
3. What are the benefits of combining direct labor costs and factory overhead costs into a single conversion cost pool?

15-16 Choice of cost pools and cost application bases. (CPA, adapted) You have been engaged to install a cost system for the Martin Company. Your investigation of the manufacturing operations of the business discloses these facts:

1. The company makes a line of lighting fixtures and lamps. The material cost of any particular item ranges from 15% to 60% of total factory cost, depending on the kind of metal and fabric used in making it.
2. The business is subject to wide cyclical fluctuations because the sales volume follows new-housing construction.
3. About 60% of the manufacturing is normally done in the first quarter of the year.
4. For the whole plant, the wage rates range from $12.75 to $25.85 an hour. However, within each of the eight individual departments, the spread between the high and low wage rates is less than 5%.
5. Each product uses all eight of the manufacturing departments but not proportionately.
6. Within the individual manufacturing departments, factory overhead ranges from 30% to 80% of conversion cost.

Required

Based on the information above, you are to write a letter to the president of the company explaining whether in its cost system Martin Company should use

a. A denominator volume tied to normal volume or master-budget volume (see Chapter 9)
b. A plantwide overhead rate or a departmental overhead rate
c. A method of factory overhead application based on direct labor hours, direct labor cost, or prime cost (direct materials plus direct labor)

Include the reasons supporting *each* of your three recommendations.

15-17 Plantwide versus departmental cost pools. (CGA adapted) The Sayther Company manufactures and sells two products, A and B. Manufacturing overhead costs are applied to each product using a plantwide rate of $17 per direct labor hour. This rate is based on budgeted manufacturing overhead of $340,000 and 20,000 of budgeted direct labor hours:

Manufacturing Department	Budgeted Manufacturing Overhead	Budgeted Direct Labor Hours
1	$240,000	10,000
2	100,000	10,000
Total	$340,000	20,000

The number of direct labor hours required to manufacture each product is:

Manufacturing Department	Product A	Product B
1	4	1
2	1	4
Total	5	5

Per-unit costs for the two categories of direct manufacturing costs are:

Direct Manufacturing Costs	Product A	Product B
Direct materials cost	$120	$150
Direct labor cost	$ 80	$ 80

At the end of the year, there was no work in process. There were 200 finished units of Product A and 600 finished units of Product B on hand. Assume that budgeted manufacturing volume was exactly attained.

Sayther sets the listed selling price of each product by adding 120% to its unit manufacturing cost. This 120% markup is designed to cover costs upstream to manufacturing (such as product design) and costs downstream from manufacturing (such as marketing and customer service) as well as to provide a profit.

Required

1. What is the effect on the inventoriable costs for Products A and B of using a plantwide overhead rate instead of departmental overhead rates?
2. What difference in the per-unit selling prices of Product A and Product B would result from the use of a plantwide overhead rate instead of departmental overhead rates?
3. Should Sayther Company prefer plantwide or departmental manufacturing overhead rates?

15-18 Plantwide versus departmental cost pools. (CMA) MumsDay Corporation manufactures a complete line of fiberglass attaché cases and suitcases. MumsDay has three manufacturing departments—molding, component, and assembly—and two service departments—power and maintenance.

The sides of the cases are manufactured in the molding department. The frames, hinges, locks, and so on, are manufactured in the component department. The cases are completed in the assembly department. Varying amounts of materials, time, and effort are required for each of the various cases. The power department and maintenance department provide services to the three manufacturing departments.

MumsDay has always used a plantwide overhead rate. Direct labor hours are used to assign the overhead to its product. The budgeted rate is calculated by dividing the company's total estimated overhead by the total estimated direct labor hours to be worked in the three manufacturing departments.

Whit Portlock, manager of Cost Accounting, has recommended that MumsDay use departmental overhead rates. Portlock has projected operating costs and levels of volume for the coming year. They are presented by department in the accompanying tables (thousands):

	Manufacturing Departments		
	Molding	Component	Assembly
Departmental volume measures:			
Direct labor hours	500	2,000	1,500
Machine hours	875	125	—
Departmental costs:			
Direct materials	$12,400	$30,000	$ 1,250
Direct labor	3,500	20,000	12,000
Variable overhead	3,500	10,000	16,500
Fixed overhead	17,500	6,200	6,100
Total departmental costs	$36,900	$66,200	$35,850
Use of service departments:			
Maintenance			
Estimated usage in labor hours for coming year	90	25	10
Power (in kilowatt hours)			
Estimated usage for coming year	360	320	120
Maximum allotted capacity	500	350	150

	Service Departments	
	Power	Maintenance
Departmental volume measures:		
Maximum capacity	1,000 KWH	Adjustable
Estimated usage in coming year	800 KWH	125 hours
Departmental costs:		
Materials and supplies	$ 5,000	$1,500
Variable labor	1,400	2,250
Fixed overhead	12,000	250
Total service department costs	$18,400	$4,000

Required

1. Calculate the plantwide overhead rate for MumsDay Corporation for the coming year using the same method as used in the past.
2. Whit Portlock has been asked to develop departmental overhead rates for comparison with the plantwide rate. He is to follow these steps in developing the departmental rates:
   a. The maintenance department costs should be allocated to the three manufacturing departments using the direct method.
   b. The power department costs should be allocated to the three manufacturing departments using the dual method, that is, the fixed costs allocated according to long-term capacity and the variable costs according to planned usage.
   c. Calculate departmental overhead rates for the three manufacturing departments using a machine hour base for the molding department and a direct labor hour base for the component and assembly departments.
3. Should MumsDay Corporation use a plantwide rate or departmental rates to apply overhead to its products? Explain your answer.

15-19 Plantwide versus departmental cost pools, automation (CMA) Rose Bach has recently been hired as controller of Empco Inc., a sheet-metal manufacturer. Empco has been in the sheet-metal business for many years and is currently investigating ways to modernize its manufacturing process. At the first staff meeting Bach attended, Bob Kelley, chief engineer, presented a proposal for automating the Drilling Department, Kelley recommended that Empco purchase two robots that would have the capability of replacing the eight direct labor workers in the department. The cost savings outlined in Kelley's proposal included the elimination of direct labor cost in the Drilling Department. Also, Empco charges manufacturing overhead on the basis of direct labor costs using a plantwide rate, so manufacturing overhead cost in the department would be reduced to zero.

The president of Empco was puzzled by Kelley's explanation of cost savings, believing it made no sense. Bach agreed, explaining that as firms become more automated, they should rethink their manufacturing overhead systems. The president then asked Bach to look into the matter and prepare a report for the next staff meeting.

To refresh her knowledge, Bach reviewed articles on manufacturing overhead application for an automated factory and discussed the matter with some of her peers. Bach also gathered historical data on the manufacturing overhead rates Empco had used over the years. Bach also wanted to have some departmental data to present at the meeting and, using Empco's accounting records, was able to estimate averages for each manufacturing department in the 1980s.

Date	Average Annual Direct Labor Cost	Average Annual Manufacturing Overhead Cost	Average Manufacturing Overhead Application Rate
1940s	$1,000,000	$ 1,000,000	100%
1950s	1,200,000	3,000,000	250
1960s	2,000,000	7,000,000	350
1970s	3,000,000	12,000,000	400
1980s	4,000,000	20,000,000	500

Averages for the 1980s

	Cutting Department	Grinding Department	Drilling Department
Direct labor	$ 2,000,000	$1,750,000	$ 250,000
Manufacturing overhead	11,000,000	7,000,000	2,000,000

Required

1. Disregarding the proposed use of robots in the Drilling Department, describe the shortcomings of the system for applying overhead that is currently used by Empco Inc.
2. Explain the misconceptions underlying Bob Kelley's statement that the manufacturing overhead cost in the Drilling Department would be reduced to zero if the automation proposal was implemented.
3. Recommend ways to improve Empco Inc.'s method for applying overhead by describing how it should revise its overhead accounting system:
   a. In the Cutting and Grinding Departments
   b. To accommodate the automation of the Drilling Department

15-20 Application of robotic costs to products; alternative depreciation measures for a robotic cost pool. Consumer Electrics assembles three models of refrigerators (standard, deluxe, and supreme) at its Nashville plant. The plant has 20 robots on its assembly line. There are four factory cost pools: (1) direct materials, (2) direct labor, (3) robotics-related costs, and (4) general factory overhead.

The robotics cost pool for July totaled $900,000:

Depreciation	$380,000
Operating and maintenance	520,000
Total robotic costs	$900,000

Robotic costs are applied to each refrigerator unit using the average actual robot-operating hours per unit for that model in the month times the average actual robotic cost per robot-operating hour in the month. Operating data for July follow:

Model	Total Number of Robot Operating Hours Attributable to Refrigerator Model	Total Units of Model Produced
Standard	10,000	5,000
Deluxe	12,000	4,000
Supreme	8,000	2,000

The 20 robots were purchased eighteen months ago for a total of $12 million. Under an income tax law designed to encourage investment, Consumer Electrics can claim 25% of cost as depreciation in year 1, 38% as depreciation in year 2, and 37% as depreciation in year 3. These tax-based depreciation rates are also used for internal reporting purposes and product-cost accumulation at Consumer Electrics.

Required

1. Compute the total robotic cost applied in July to each unit of (a) standard, (b) deluxe, and (c) supreme refrigerator.
2. An industrial engineer at the plant observes that the robots are superbly maintained. Consequently, the three-year useful life underestimates their economic life to Consumer Electrics. She suggests that "true" depreciation is better measured using the straight-line method, a five-year estimate of useful life (from the initial purchase date). Net disposal proceeds at the end of five years will be $3 million for the 20 robots. Compute how this measure of depreciation would affect the robotic costs applied in July to each unit of (a) standard, (b) deluxe, and (c) supreme refrigerator. (Assume the straight-line method had been used from the time the asset was first purchased.)
3. What are the strengths and weaknesses of relying on tax rules to measure the robotic depreciation cost that is applied to an individual refrigerator unit?

15-21 Allocation of standard costs to products, automation. Photocopy Plus is a manufacturer and distributor of quality photocopying units. Consider its product line at the high-price range. It features color options and reduction/enlargement options. There are three brands in this line of products. The wholesale price of each brand and the units produced in the most recent month follow:

Brand Name	Wholesale Price	Units Produced (August)
Regal	$ 3,000	2,000
Royal	8,000	3,000
Monarch	30,000	5,000

These products are assembled at a highly automated plant in Biloxi, Mississippi.

For internal accounting purposes, Photocopy Plus uses a standard-costing system. Costs at the Biloxi plant are collected into four pools:

A. *Direct materials costs.* Standard costs are based on the standard materials allowed in assembling each photocopying unit times the standard price for each part.

B. *Direct labor costs.* Standard costs are based on standard direct labor hours allowed for each unit (from an engineering study) times the standard rate of $20 per labor hour.

C. *Machining costs.* Standards are based on standard direct machine hours allowed for each unit (from an engineering study) times the standard rate of $50 per machine hour.

D. *General factory overhead costs.* Standard rate is 80% of the total of the direct materials costs, direct labor costs, and machining costs.

Standard unit production costs are calculated as the sum of A, B, C, and D.

The following information underlies the standard unit production costs for each photocopying unit:

	Regal	Royal	Monarch
Direct materials	$800	$2,000	$4,000
Direct labor	5 hours	10 hours	30 hours
Machining	6 hours	20 hours	70 hours

Required

1. Compute the standard unit production costs for the Regal, Royal, and Monarch photocopying units.
2. What percentage is each cost pool of total standard production cost for August?
3. Wendy Reichter, the controller at the Biloxi plant, attends a conference on "Cost Accounting and the Automated Plant." A speaker at the conference proposes eliminating direct labor costs as a separate cost pool and including all labor costs in general factory overhead costs. What factors should Reichter consider in evaluating this proposal for the Biloxi plant?

15-22 Effect of overhead cost application rate changes and product design changes on reported product costs. This problem extends the Problem for Self-Study (pp. 507–509).

In 19_2, Medical Instruments adopted an accounting system with two direct product-cost categories (direct materials costs and direct labor costs) and two indirect manufacturing product-cost categories:

(a) Materials-handling overhead—applied on the basis of actual number of parts in a product

(b) Production overhead—applied on the basis of actual throughput time for each product

It is now 19_3. Curt Henning makes the following assumptions when developing the 19_3 indirect cost rates for materials-handling overhead and production overhead:

Budgeted total materials-handling overhead costs	$7,695,000
Budgeted number of separate part numbers	4,500
Budgeted average usage per separate part number	900
Budgeted total production overhead costs	$12,240,000
Budgeted number of individual products	425
Budgeted average production volume per product	120
Budgeted average throughput time per product	5 hours

Henning is now examining the reported cost of Products A and C in 19_3. (Product B has been discontinued.) Product designers at Medical Instruments made several changes in these two products at the end of 19_2 that reduced both the direct labor hours content and the number of parts in each product in 19_3. Manufacturing has made substantial progress in reducing the throughput time for each product. Details of these two products in 19_2 and 19_3 follow:

	Product A		Product C	
	19_2	19_3	19_2	19_3
Direct material dollars	$1,680	$1,618	$2,070	$2,027
Direct labor hours	7.2	6.9	6.1	5.2
Number of parts	128	116	260	224
Throughput time in hours	4.8	4.2	18.5	14.8

The 19_3 direct labor rate is $32 per hour.

Required

1. Compute the 19_3 budgeted materials-handling overhead cost application rate per part and the 19_3 budgeted production overhead cost application rate per throughput hour.
2. Compute the 19_3 manufacturing cost of Products A and C using the 19_3 indirect cost application rates computed in requirement 1.
3. Compare the manufacturing cost figures for Products A and C in 19_2 (see the Problem for Self-Study) and 19_3. Explain any differences between 19_2 and 19_3 costs for each product.

4. Assume Medical Instruments uses actual rather than budgeted throughput time (at the budgeted rate per throughput hour) when applying production overhead costs to products. The plant is operating at full capacity. A special customer purchases twenty units of Product C on the condition that they be "rushed" (say at 10.8 hours per unit) through the production line. The customer will pay the listed price in the catalog. How should Medical Instruments compute the gross margin (selling price minus manufacturing cost) on this sale of 20 units of Product C? Is the reported product cost of Product C accurate for products sold to this customer?

15-23 Single versus multiple indirect cost pools, behavior change or accuracy in product costing (continuation of 15-22). Medical Instruments uses one indirect cost pool for production overhead. Several companies with production facilities similar to those at Medical Instruments use over ten separate production overhead cost pools, each with a different application base. A manufacturing manager at Medical Instruments made the following observation:

> Our objective in using a single production overhead cost pool based on throughput time is to signal to a broad set of people (in product design, process engineering, and manufacturing) the strategic importance of Medical Instruments reducing throughput time. The system is designed to cause behavioral change at Medical Instruments. A single production overhead rate based on throughput time sends a clear and unambiguous signal to reduce throughput time. Personally, I do not see the need to adopt a system using six or eight production overhead cost pools, each with their own application base. That would be overly complex and complete overkill.

Required
Comment on the manufacturing manager's rationale for using a single indirect cost pool for production overhead costs.

15-24 Application of materials-handling costs to products. Sierra Lumber is a large forest products company. At its plant in Klamath Falls, Oregon, Sierra produces the following three products, each in different production departments:

(1) Lumber for building and housing construction
(2) Wooden poles for electric utility companies
(3) Newsprint for newspaper companies

A fleet of operator-driven forklift trucks performs materials handling. These trucks transport logs from an on-site rail receiving yard to the production lines for each department. The trucks also haul the finished product from each production department to rail cars in the shipping yard. There is a three-stage process for applying costs of materials handling to each product.

- STAGE I: Trace direct costs to the materials-handling department.
- STAGE II: Allocate the materials-handling department costs to each production department on the basis of the sum of (1) tons of raw material (RM) transported from the receiving yard to the production line and (2) tons of finished goods (FG) or product transported from each department to the shipping yard.
- STAGE III: Apply production department costs to individual products on the basis of tons of finished product transported from that department to the shipping yard. (For simplicity, consider only the costs allocated to production departments from Stage II.)

The receiving yard and the shipping yard have a weighing station to record tons carried by each forklift driver. Work in process (WIP), comprising partly finished wood products—mainly odd ends and chips—are also transported from the Lumber and Utility Poles Departments to the Newsprint Department. A record of tons carried of WIP, however, is not maintained.

For the most recent month (March 19_4), directly traceable materials-handling costs were $2 million. Details of recorded tons carried by forklift drivers follow:

Production Department (1)	Tons of Raw Material Transported from Receiving Yard to Department (2)	Tons of Finished Product Transported from Department to Shipping Yard (3)
Lumber	40,000	25,000
Utility poles	60,000	30,000
Newsprint	20,000	25,000

Required

1. Compute the materials-handling costs allocated to each production department.
2. Apply the materials-handling costs allocated in requirement 1 per ton of finished product produced in each production department.
3. What limitations might arise from how Sierra allocates materials-handling costs to each department?

15-25 Application of materials-handling costs to products, changes in allocation bases (continuation of Problem 15-24). Sierra Lumber makes a major change in its materials-handling department in April 19_4. It switches to robotic forklifts. Machine operators in the plant control each robotic forklift. The main motivation for the change was to eliminate the high number of personal injuries occurring to forklift truck operators.

The use of robots has several implications for cost accounting at the Klamath Falls plant. First, a high proportion of the materials-handling cost now consists of machine-related costs. Second, more-detailed information is available regarding differential usage of the materials-handling function by each department. Sierra has decided to allocate materials-handling costs to each department using two separate cost pools:

1. Robotic-related costs, allocated to each production department using hours of robot-operating time
2. Other materials-handling costs, allocated to each production department using tons of raw materials (RM), work in process (WIP), and finished goods (FG) transported by robots

In the most recent month (April 19_4), robotic-related costs were $1.6 million, and other materials-handling costs were $675,000. Operating data relating to the robots for this month are:

Production Department (1)	Hours of Robot-Operating Time Attributable to Department (2)	Tons of Product (RM, WIP, & FG) Transported by Robots and Attributable to Department (3)	Tons of Finished Product Transported from Department to Shipping Yard (4)
Lumber	400	70,000	35,000
Utility poles	250	80,000	35,000
Newsprint	350	50,000	30,000
	1,000	200,000	100,000

Required

1. Compute the materials-handling costs applied per ton of finished product of each production department.
2. Compare and contrast the cost allocation scheme of Sierra Lumber in April 19_4 with that in March 19_4 (described in Problem 15-24).

15-26 Contribution approach to cost allocation: income statement by segments. Stuart Philatelic Sales, Inc., sells postage stamps and supplies to collectors on a retail basis. Stuart also makes up packets of inexpensive stamps that it sells on a wholesale basis to stamp departments of various stores.

Stuart's retail division has two stores, A (downtown) and B (suburban), each of which has a stamp department and an albums and supplies department. Stuart had total net sales in 19_4 of $960,000; cost of merchandise sold was $490,000; and variable operating costs were $120,000. The company's fixed costs, which it fully allocated to its two divisions, were $105,000 of advertising cost and $120,000 of various committed costs. Stuart's common discretionary and committed costs were $35,000.

The costs of merchandise sold and variable operating costs allocated to the retail division were $190,000 and $50,000, respectively. Net sales of the retail division were $390,000, two-thirds of which were Store A's net sales. Sixty percent of the retail division's merchandise costs and 54% of its variable operating costs were allocated to Store A. Advertising costs of $40,000 were allocated to the retail division, which in turn allocated 45% directly to Store A and 5% to Store B; the rest of the $40,000 was unallocated. Of the $120,000 separable committed costs, 50% were allocated to the retail division, which in turn allocated 50% of its committed costs to Store A, had $25,000 of unallocated costs, and allocated the rest to Store B.

Other Information

1. Allocations to Store B—stamp department:

Net sales	$100,000
Cost of merchandise sold	$ 58,000
Variable operating costs	$ 17,000

Allocations to Store B—albums and supplies department:

Net sales	$30,000
Cost of merchandise sold	$18,000
Variable operating costs	$ 6,000

2. One-half of Store B's allocated advertising costs could not be further allocated either to the stamp department or to the albums and supplies department; the other half of B's allocated advertising costs was equally divided between the two departments.
3. Sixty percent of Store B's committed costs were unallocated; three-fourths of the rest were allocated to the stamp department and one-fourth to the albums and supplies department.

Required

1. What was operating income for the company as a whole?
2. Determine the contribution margin, contribution controllable by division managers, and contribution by segments for each of the following:
   a. Company as a whole
   b. Wholesale division
   c. Retail division
   d. Store A of the retail division
   e. Store B of the retail division
   f. Stamp department of Store B
   g. Albums and supplies department of Store B

15-27 Product-line and territorial income statements. The Delvin Company shows the following results for the year 19_1:

Sales	$1,000,000	100.0%
Manufacturing costs of goods sold	$ 675,000	67.5%
Marketing*	220,000	22.0%
Administrative (all fixed)	35,000	3.5%
Total costs	$ 930,000	93.0%
Operating income	$ 70,000	7.0%

*All fixed except for $40,000 freight-out cost.

The sales manager has asked you to prepare statements that will help him assess the company efforts by product line and by territories. You have gathered the following information:

	Product			Territory		
	A	B	C	North	Central	Eastern
Sales*	25%	40%	35%			
Product *A*				50%	20%	30%
Product *B*				15%	70%	15%
Product *C*				14/35	8/35	13/35
Variable manufacturing and packaging costs†	68%	55%	60%			
Fixed separable costs:						
Manufacturing	$15,000	$14,000	$21,000		(not allocated)	
Marketing	40,000	18,000	42,000	$48,000	$32,000	$40,000
Freight out		(not allocated)		13,000	9,000	18,000

Note. All items not directly allocated were considered common costs.
*Percent of company sales.
†Percent of product sales.

Required

1. Prepare a product-line income statement, showing the results for the company as a whole in the first-column, costs not allocated in the second column, and the results for the three products in adjoining columns. Show a contribution margin and a product margin, as well as operating income.
2. Repeat requirement 1 on a territorial basis. Show a contribution margin and a territory margin.
3. Should salespeople's commissions be based on contribution margins, product margins, territorial margins, operating income, or dollar sales? Explain.

15-28 Cost allocation and contribution margin. (R. Anderson) An analogy helps to understand the treatment of costs incident to various types of operations. Consider the following conversation between a restaurant owner (Joe) and his Accountant-Efficiency-Expert about adding a rack of peanuts to the counter in an effort to pick up additional profit in the usual course of business. Some people may consider this conversation an oversimplification, but the analogy highlights some central issues in cost allocation.

EFF EX: Joe, you said you put in these peanuts because some people ask for them, but do you realize what this rack of peanuts is *costing* you?

JOE: It ain't gonna cost. 'Sgonna be a profit. Sure, I hadda pay $250 for a fancy rack to hold bags, but the peanuts cost $0.60 a bag and I sell 'em for $1.00. Figger I sell 50 bags a week to start. It'll take 12½ weeks to cover the cost of the rack. After that I gotta clear profit of $0.40 a bag. The more I sell, the more I make.

EFF EX: That is an antiquated and completely unrealistic approach, Joe. Fortunately, modern accounting procedures permit a more accurate picture, which reveals the complexities involved.

JOE: Huh?

EFF EX: To be precise, those peanuts must be integrated into your entire operation and be allocated their appropriate share of business overhead. They must share a proportionate part of your expenditure for rent, heat, light, equipment depreciation, decorating, salaries for your waitresses, cook———

JOE: The *cook?* What's he gotta do wit'a peanuts? He don' even know I got 'em!

EFF EX: Look Joe, the cook is in the kitchen, the kitchen prepares the food, the food is what brings people in here, and the people ask to buy peanuts. That's why you must charge a portion of the cook's wages as well as a part of your own salary to peanut sales. This sheet contains a carefully calculated cost analysis, which indicates the peanut operation should pay exactly $12,780 per year toward these general overhead costs.

JOE: The peanuts? $12,780 a year for overhead? The nuts?

EFF EX: It's really a little more than that. You also spend money each week to have the windows washed, have the place swept out in the mornings, keep soap in the washroom, and provide free soft drinks to the police. That raises the total to $13,130 per year.

JOE: [Thoughtfully] But the peanut salesman said I'd make money . . . put 'em on the end of the counter, he said . . . and get $0.40 a bag profit . . .

EFF EX: [With a sniff] He's not an accountant. Do you actually know what the portion of the counter occupied by the peanut rack is worth to you?

JOE: Ain't worth nothing . . . no stool there . . . just a dead spot at the end.

EFF EX: The modern cost picture permits no dead spots. Your counter contains 60 square feet and your counter business grosses $150,000 a year. Consequently, the square foot of space occupied by the peanut rack is worth $2,500 per year. Since you have taken that area away from general counter use, you must charge the value of the space to the occupant.

JOE: You mean I gotta add $2,500 a year more to the peanuts?

EFF EX: Right. That raises their share of the general operating costs to a grand total of $15,630 per year. Now then, if you sell 50 bags of peanuts per week for 52 weeks, these allocated costs will amount to approximately $6.00 per bag.

JOE: *What?*

EFF EX: Obviously, to that must be added your purchase price of $0.60 per bag, which brings the total to $6.60. So you see by selling peanuts at $1.00 per bag, you are losing $5.60 on every sale.

JOE: Somethin's crazy!

EFF EX: Not at all! Here are the *figures.* They *prove* your peanuts operation cannot stand on its own feet.

JOE: [Brightening] Suppose I sell *lotsa* peanuts . . . thousand bags a week 'stead of fifty.

EFF EX: [Tolerantly] Joe, you don't understand the problem. If the volume of peanuts sales increases, our operating costs will go up . . . you'll have to handle more bags with more time, more depreciation, more everything. The basic principle of accounting is firm on that subject: "The Bigger the Operation, the More General Overhead Costs That Must Be Allocated." No, increasing the volume of sales won't help.

JOE: Okay, you so smart, *you* tell *me* what I gotta do.

EFF EX: [Condescendingly] Well . . . you could first reduce operating costs.

JOE: How?

EFF EX: Move to a building with cheaper rent. Cut salaries. Wash the windows bi-weekly. Have the floor swept only on Thursday. Remove the soap from the washrooms. Decrease the square-foot value of your counter. For example, if you can cut your costs 50 percent, that will reduce the amount allocated to peanuts from $15,630 to $7,815 per year, reducing the cost to $3.60 per bag.

JOE: [Slowly] That's better?

EFF EX: Much, much better. However, even then you would lose $2.60 per bag if you charge only $1.00. Therefore, you must also raise your selling price. If you want an income of $0.40 per bag you would have to charge $4.00.

JOE: [Flabbergasted] You mean even after I cut operating costs 50 percent I still gotta charge $4.00 for a $1.00 bag of peanuts? Nobody's that nuts about nuts! Who'd buy 'em?

EFF EX: That's a secondary consideration. The point is, at $4.00 you'd be selling at a price based upon a true and proper evaluation of your then reduced costs.

JOE: [Eagerly] Look! I gotta better idea. Why don't I just throw the nuts out . . . put 'em in the ashcan?

EFF EX: Can you afford it?

JOE: Sure. All I got is about 50 bags of peanuts . . . cost about thirty bucks . . . so I lose $250 on the rack, but I'm outa this nutsy business and no more grief.

EFF EX: [Shaking head] Joe it isn't that simple. You are *in* the peanut business! The minute you throw those peanuts out you are adding $15,630 of annual overhead to the rest of your operation. Joe . . . be realistic . . . *can you afford to do that?*

JOE: [Completely crushed] It's unbelievable! Last week I was making money. Now I'm in trouble . . . just because I think peanuts on a counter is gonna bring me some extra profit . . . just because I believe 50 bags of peanuts a week is easy.

EFF EX: [With raised eyebrow] That is the object of modern cost studies, Joe . . . to dispel those false illusions.

Required

1. Is Joe losing $5.60 on every sale of peanuts? Explain.
2. Do you agree that if the volume of peanut sales is increased, operating losses will increase? Explain.
3. Do you agree with the Efficiency-Expert that, in order to make the peanut operation profitable, the operating costs in the restaurant should be decreased and the selling price of the peanuts should be increased? Give reasons.
4. Do you think that Joe can afford to get out of the peanut business? Give reasons.
5. Do you think that Joe should eliminate his peanut operations? Why or why not?

15-29 Overhead disputes. (Suggested by Howard Wright) The Azure Ship Company works on U.S. Navy vessels and commercial vessels. General yard overhead (for example, the cost of the purchasing department) is applied to the jobs on the basis of direct labor costs.

In 19_3, Azure's total $150 million of direct labor cost consisted of $50 million navy and $100 million commercial. The general yard overhead was $30 million.

Navy auditors periodically examine the records of defense contractors. The auditors investigated a nuclear submarine contract, which was based on cost-plus-fixed-fee pricing. The auditors claimed that the navy was entitled to a refund because of double-counting of overhead in 19_3.

The government contract included the following provision:

Par. 15-202. Direct Costs.

(a) A direct cost is any cost which can be identified specifically with a particular cost object. Direct costs are not limited to items which are incorporated in the end product as material or labor. Costs identified specifically with the contract are direct costs of the contract and are to be charged directly thereto. Costs identified specifically with other work of the contractor are direct costs of that work and are not to be charged to the contract directly or indirectly. When items ordinarily chargeable as indirect costs are charged to the contract as direct costs, the cost of like items applicable to other work must be eliminated from indirect costs allocated to the contract.

Azure had formed a special expediting purchasing group, the SE group, to join with the central purchasing group to obtain materials for the nuclear submarine only. Their direct costs, $5 million, had been included as direct labor of the nuclear work. Accordingly, overhead was applied to the contracts in the usual manner. The SE costs of $5 million were not included in the yard overhead costs. The auditors claimed that no overhead should have been applied to these SE costs.

Required

1. Compute the amount of the refund that the navy would claim.
2. Suppose the navy also discovered that $4 million of general yard overhead was devoted exclusively to commercial engine-room purchasing activities. Compute the additional refund that the navy would probably claim. (*Note:* This $4 million was never classified as "direct labor." Furthermore, the navy would claim that it should be reclassified as a "direct cost" but not as "direct labor.")

15-30 Cost justification for a contract; dispute over direct labor and direct materials cost components. P. C. Products specializes in the development of software for accounting applications. The company organizes its activities into four departments:

Service Departments	Manufacturing Departments
Product Development	Inventory Accounting Software Manufacture
Computer Systems	Project-Costing Software Manufacture

P. C. Products is approached by Titan Software, a foreign retailer, which offers to buy 5,000 units of its recently developed Project-Costing Software II package. The contract calls for Titan Software to reimburse P. C. Products on the basis of unit product cost plus 20% gross margin based on product cost. A clause in the contract permits Titan Software to investigate the accounting records of P. C. Products to verify the unit-cost calculations. The contract defines unit product costs as having four components:

A. "Direct materials cost as incurred"

B. "Direct labor cost as incurred"

C. "Service Department costs attributable to the Project-Costing Software Manufacture Department"

D. "Project-Costing Department overhead rate, calculated as 400% of direct labor cost in B"

At the end of the period, P. C. Products submits the following bill to Titan Software:

Unit Product Cost Determination Based on Total Production of 13,250 Units (Packages)

A. Total direct materials	$1,060,000
B. Total direct labor	397,500
C. Total service department costs attributable to Project-Costing Manufacture Department	2,694,393
D. Total overhead applied: 400% × $397,500	1,590,000
Total product costs	$5,741,893
Unit product costs: $5,741,893 ÷ 13,250	$ 433.35

Titan purchases 5,000 of the 13,250 units produced.

Titan Software sends Jessie Ventura to examine the accounting records of P. C. Products. She reports that P. C. Products incorrectly classified $99,375 as a direct labor cost that it should have classified as indirect labor. Indirect labor was to be reimbursed using the 400% of direct labor cost rate included in the contract. P. C. Products recognizes that a misclassification has been made and offers to adjust the price to be charged to Titan Software accordingly.

Ventura also finds that the direct materials cost includes $198,750 attributable to the purchase of a defective batch of iron-oxide material by P. C. Products. Ventura believes that this $198,750 should be excluded from the unit-cost determination for the contract. P. C. Products strongly disagrees and points out that the contract specifies inclusion of "direct materials cost as incurred." Ventura admits her argument would not hold up to a legal challenge.

Required

1. What adjustment in the price should P. C. Products make for the misclassification of the labor-cost item?
2. If the $198,750 item is excluded as an allowable cost of the 5,000-unit contract, what further adjustment in the price would be made?
3. Ventura wonders what can be done in the future to prevent Titan Software's reimbursing P. C. Products for P. C.'s purchasing inefficiencies or materials-usage inefficiencies. What advice would you give her?

15-31 Review of Chapters 14 and 15; cost justification for a contract, analysis of service department cost allocations. Jessie Ventura is investigating the costs charged by P. C. Products to Titan Software under a cost reimbursements contract (see preceding problem). An allowable cost item in the contract is "Service department costs attributable to the Project-Costing Software Manufacture Department." P. C. Products organizes its activities into four departments:

Service departments	Manufacturing departments
Product Development	Inventory Accounting Software Manufacture
Computer Systems	Project-Costing Software Manufacture

P. C. Products reports that the "total service department costs attributable (allocated) to the Project-Costing Software Manufacture Department" were $2,694,393.50:

Fixed costs of service departments allocated	$1,524,706.00
Variable costs of service departments allocated	1,169,687.50
	$2,694,393.50

Ventura discovers that P. C. Products allocates the fixed costs of service departments on the basis of budgeted usage; variable costs of service departments are allocated on the basis of actual usage.

The budgeted usage relationship between the four departments was:

	Used by			
Supplied by	**Product Development**	**Computer Systems**	**Inventory Accounting Software**	**Project-Costing Software**
Product Development	—	20%	40%	40%
Computer Systems	15%	—	25%	60%

The actual usage relationships between the four departments was:

	Used by			
Supplied by	**Product Development**	**Computer Systems**	**Inventory Accounting Software**	**Project-Costing Software**
Product Development	—	15%	45%	40%
Computer Systems	20%	—	30%	50%

The actual costs incurred at the two service departments were:

	Fixed Costs	**Variable Costs**
Product Development	$ 600,000	$ 850,000
Computer Systems	1,700,000	1,200,000

Required

1. What method did P. C. Products use to allocate service department costs to the Project-Costing Software Manufacture Department? (*Hint:* It was not the reciprocal method.)
2. Ventura reads that the reciprocal method of cost allocation is "the most defensible method on theoretical grounds." How would this method affect the costs allocated to the Project-Costing Software Manufacture Department?
3. Ventura views the contract clause that an allowable cost item is "service department costs attributable to the Project-Costing Software Manufacture Department" as being excessively vague. It provides P. C. Products the opportunity to manipulate the costs allocated to the Titan Software contract. In what ways is this clause vague? How might any vagueness be reduced or eliminated?

CHAPTER 16

Dairy companies process raw milk to yield many joint products, including pasteurized milk, skim milk, condensed milk, butter, and cream.

Cost Allocation: Joint Products and Byproducts

- Meaning of terms
- Why allocate joint costs?
- Methods of allocating joint costs
- Irrelevance of joint costs for decision making
- Accounting for byproducts

Learning Objectives

When you have finished studying this chapter, you should be able to

1. Identify the split-off point(s) in a joint-cost situation
2. Distinguish between joint products and byproducts
3. Provide four reasons for allocating joint costs to individual products
4. Describe alternative approaches to costing inventory in joint-cost situations
5. Describe why the sales value at split-off method is widely used
6. Describe the irrelevance of joint costs in deciding to sell or process further
7. Specify alternative accounting methods for byproducts

Joint costs plague the accountant's work. This chapter examines methods for allocating joint costs to products and services. Many of the topics discussed in this chapter are related to issues already covered in Chapters 14 and 15. Before reading on, be sure you are comfortable with pages 455–59 of Chapter 14 (Cost Allocation: I)

MEANING OF TERMS

Consider a single process that yields two or more products or services simultaneously. The distillation of coal, for example, gives us coke, gas, and other products. A **joint cost** is the cost of a single process—distillation, for example—that yields multiple products simultaneously.

The juncture in the process when the products become separately identifiable—when the coal becomes coke, gas, and so on—is called the **split-off point**. **Separable costs** are costs incurred beyond the split-off point that are identifiable with individual products.

Objective 1

Identify the split-off point(s) in a joint-cost situation

Single processes that simultaneously yield two or more products abound in many industries. Exhibit 16-1 presents examples of joint-cost situations in a diverse series of industries. In each example in Exhibit 16-1, no individual product can be produced without the accompanying products appearing, although sometimes the proportions can be varied. A meatpacking company cannot kill a pork chop; it has to slaughter a hog, which supplies hides, bones, fat, and various cuts of dressed meat in addition to pork chops.

The accounting rules for recording the costs and revenues from multiple products are often affected by their being classified as joint or main products, byproducts, or scrap. The distinctions underlying these classifications are based on their relative sales values.

Objective 2

Distinguish between joint products and byproducts

Joint products have relatively high sales value and are not separately identifiable as individual products until the split-off point. When a single process yielding two or more products yields only one product with a relatively high sales value, that product is termed a **main product**. A **byproduct** is a product that has a low sales value compared with the sales value of the main or joint product(s). **Scrap** has a minimal (frequently zero) sales value compared with the sales value of the main or joint product(s).

EXHIBIT 16-1
Examples of Joint-Cost Situations

Industries	Separable Products at the Split-Off Point
1. Agriculture industries	
Raw milk	Cream, liquid skim
Hogs	Pork, hides, bones, fat
2. Extractive industries	
Crude oil	Naphtha, kerosene, jet fuel, gasoline
Copper ore	Copper, silver, lead, zinc
Salt	Hydrogen, chlorine, caustic soda
Coal	Coke, gas, benzole, tar, ammonia
3. Chemical industries	
Naphtha	Ethylene, gasoline, methane, propylene
4. Semiconductor industry	
Fabrication of silicon-wafer chips	Memory chips of different quality as to the number of chips per module, speed, life expectancy, and temperature tolerance

EXHIBIT 16-2
Joint Products, Main Product, Byproduct, and Scrap

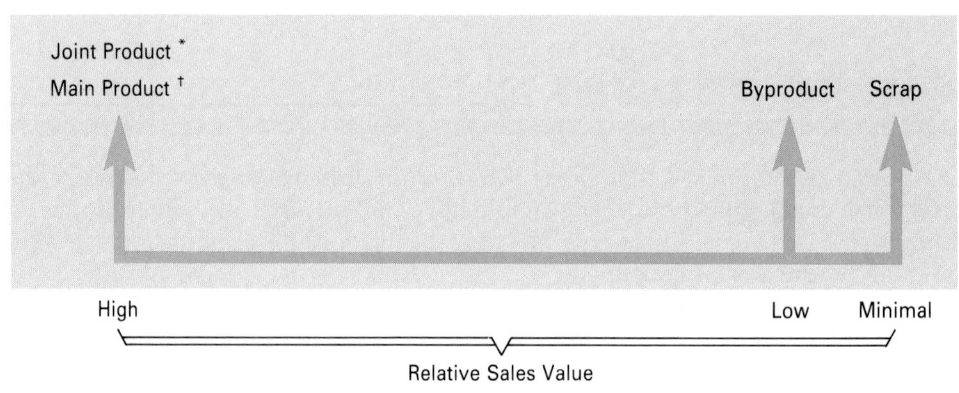

*If multiple products have relatively high sales values.
†If only one product has a relatively high sales value.

Exhibit 16-2 is an overview of these definitions. Be careful. These distinctions are not firm in practice. The variety of terminology and accounting practice is bewildering. Always gain an understanding of the terms as used by the organization if you encounter these terms in an assignment.[1]

WHY ALLOCATE JOINT COSTS?

The purposes for allocating joint costs to individual products or services are similar to the purposes for costs allocation in general (see Exhibit 14-2 on page 458). They include

[1]An extensive overview of the joint-cost literature is in R. Manes and C. Cheng, *The Marginal Approach to Joint Cost Allocation: Theory and Application* (Sarasota, FL: American Accounting Association, 1988).

1. Inventory cost and cost-of-goods-sold computations for external financial statements, including reports for income tax authorities.
2. Inventory cost and cost-of-goods-sold computations for internal financial reporting. Such reports are used in divisional profitability analysis when determining compensation for divisional managers.
3. Cost reimbursement under contracts when only a portion of the separate products or services is sold or delivered to a single customer (such as a government agency).
4. Rate regulation when one or more of the jointly produced products or services are subject to price regulation.[2]

Objective 3

Provide four reasons for allocating joint costs to individual products

METHODS OF ALLOCATING JOINT COSTS

There are three basic approaches to costing inventory (and computing cost of goods sold) in joint-cost situations:

1. Allocate costs using market selling-price data. Three common methods are used in applying this approach:
   - The sales value at split-off method
   - The estimated net realizable value (NRV) method
   - The constant gross-margin percentage NRV method
2. Allocate costs using a physical measure.
3. Do not allocate costs; use market selling-price data to guide inventory costing.

Objective 4

Describe alternative approaches to costing inventory in joint-cost situations

In the simplest situation, the joint products are sold at the split-off point without further processing. We consider this case first. Then we consider situations involving processing beyond the split-off point.

To highlight each joint cost example, this chapter makes extensive use of exhibits. The following notation is used in these exhibits:

Joint Product or Main Product

Byproduct or Scrap

EXAMPLE 1

Farmers' Dairy purchases raw milk from individual farms and processes it up to the split-off point, where two products (cream and liquid skim) are obtained. Exhibit 16-3 presents an overview of the basic relationships. Summary simplified data for the most recent month follow:

- Raw milk processed—110 gallons
- Production—cream 25 gallons
  —liquid skim 75 gallons
- Sales—cream 20 gallons at $8 per gallon
  —liquid skim 30 gallons at $4 per gallon
-

Product	Beginning Inventory	Ending Inventory
Raw milk	0 gallons	0 gallons
Cream	0 gallons	5 gallons
Liquid skim	0 gallons	45 gallons

- Cost of purchasing 110 gallons of raw milk and processing it up to the split-off point to yield 25 gallons of cream and 75 gallons of liquid skim = $400

[2]See J. Crespi and J. Harris, "Joint Cost Allocation under the Natural Gas Act: An Historical Review," *Journal of Extractive Industries Accounting*, Summer 1983, pp. 133–42.

EXHIBIT 16-3
Farmers' Dairy: Example 1 Overview

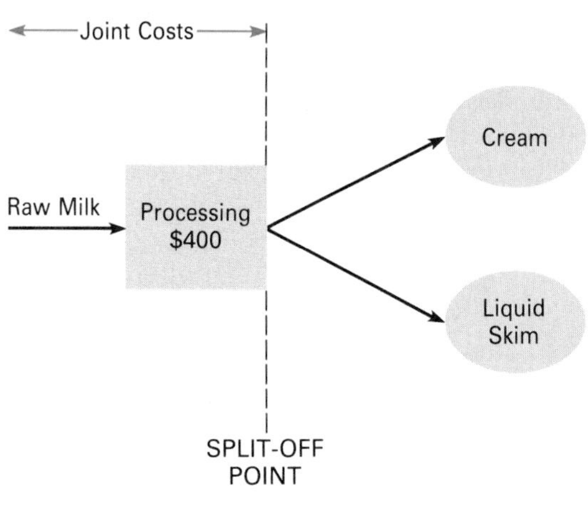

How should the accountant cost the 5 gallons of cream and 45 gallons of liquid skim for inventory purposes? The $400 production costs cannot be uniquely identified with or traced to either product. Why? Because the products themselves were not separated before the split-off point.

The Example 1 data will be used to illustrate the joint-cost allocation methods of sales value at split-off point and physical measure.

Sales Value at Split-Off Method

The **sales value at split-off method** allocates joint costs on the basis of each product's relative sales value at the split-off point. In Example 1, the sales value at split-off is $200 for cream and $300 for liquid skim. We then assign a weighted value to each product as a percentage of total sales value. Using this weighting, we allocate the joint costs to the individual products:

	Cream	Liquid Skim	Total
1. Sales value at split-off point: (Cream, 25 gallons × $8; Liquid skim, 75 gallons × $4)	$200	$300	$500
2. Weighting	$\frac{200}{500} = .40$	$\frac{300}{500} = .60$	
3. Joint costs allocated: (Cream, .40 × $400; Liquid skim, .60 × $400)	$160	$240	$400
4. Production costs per unit: (Cream, $160 ÷ 25 gallons; Liquid skim, $240 ÷ 75 gallons)	$6.40	$3.20	

Note that this method uses the sales value of the entire production including the unsold portion, not just the actual sales. Exhibit 16-4 presents the product-line income statement, using the sales value at split-off method of joint-case allocation. Both cream and liquid skim have gross-margin percentages of 20%.[3]

[3]The equality of the gross-margin percentages for each product is a mechanical result reached with the sales value at split-off method when there are no beginning inventories and all products are sold at the split-off point.

EXHIBIT 16-4
Product-Line Income Statement
Joint Costs Allocated Using Sales Value at Split-off Method

	Cream	Liquid Skim	Total
Sales (Cream, 20 gallons × $8; Liquid skim, 30 gallons × $4)	$160	$120	$280
Joint costs:			
Production costs (Cream, .4 × $400; Liquid skim, .6 × $400)	$160	$240	$400
Deduct ending inventory (Cream, 5 gallons × $6.40; Liquid skim, 45 gallons × $3.20)	32	144	176
Cost of goods sold	$128	$ 96	$224
Gross margin	$ 32	$ 24	$ 56
Gross-margin percentage	20%	20%	20%

An advantage of the sales value at split-off point method is its simplicity. The allocation base (sales value) is expressed in terms of a common denominator (dollars) that is systematically recorded in the accounting system. Many managers cite a second advantage: the costs are allocated in proportion to a measure of the relative revenue-generating power identifiable with the individual products.

Physical Measure Method

The **physical measure method** allocates joint costs on the basis of their relative proportions at the split-off point, using a common physical measure such as weight or volume. In Example 1, the $400 joint cost produced 25 gallons of cream and 75 gallons of liquid skim. Joint costs are allocated as follows using these quantities:

	Cream	Liquid Skim	Total
1. Physical measure of production (gallons)	25	75	100
2. Weighting	$\frac{25}{100} = .25$	$\frac{75}{100} = .75$	
3. Joint costs allocated: (Cream, .25 × $400; Liquid skim, .75 × $400)	$100	$300	$400
4. Production costs per unit: (Cream, $100 ÷ 25 gallons; Liquid skim, $300 ÷ 75 gallons)	$4	$4	

Exhibit 16-5 presents the product-line income statement using this method of joint-cost allocation. The gross-margin percentages are 50% for cream and 0% for liquid skim.

The physical weights used for allocating joint costs may have no relationship to the revenue-producing power of the individual products. Consider a mine that extracts ore jointly containing gold, silver, and lead. Use of a common physical measure (tons) would result in almost all the costs being allocated to the product that weighs the most—that product is lead, which has the lowest revenue-producing power. As a second example, if the joint cost of a hog were assigned to its various products on the basis of weight, center-cut pork chops would have the same cost per pound as pigs' feet, lard, bacon, ham, bones, and so forth. In a product-line income statement, products that have a high sales value per pound (for example, center-cut pork chops) would show a fabulous "profit," and products that have a low sales value per pound (for example, bones) would show consistent losses.

EXHIBIT 16-5
Product-Line Income Statement
Joint Costs Allocated Using Physical Measure Method

	Cream	Liquid Skim	Total
Sales (Cream, 20 gallons × $8; Liquid skim, 30 gallons × $4)	$160	$120	$280
Joint costs:			
Production costs (Cream, .25 × $400; Liquid skim, .75 × $400)	$100	$300	$400
Deduct ending inventory (Cream, 5 gallons × $4; Liquid skim, 45 gallons × $4)	20	180	200
Cost of goods sold	$ 80	$120	$200
Gross margin	$ 80	$ 0	$ 80
Gross-margin percentage	50%	0%	28.6%

Physical measures are sometimes preferred to sales value methods in rate-regulation settings when the objective is to set a fair selling price. Why? Because it is circular reasoning to use selling prices as a basis for setting a selling price, as would be the case under the sales value at split-off method.

EXAMPLE 2
Assume the same situation as in Example 1 except that both cream and liquid skim can be futher processed:

- Cream → Butter cream: 25 gallons of cream are further processed to yield 20 gallons of butter cream at an additional processing (separable) cost of $280. Butter cream is sold for $25 per gallon.
- Liquid skim → Condensed milk: 75 gallons of liquid skim are further processed to yield 50 gallons of condensed milk at an additional processing cost of $520. Condensed milk is sold for $22 per gallon.

Exhibit 16-6 presents an overview of the Example 2 relationships. Inventory information follows:

EXHIBIT 16-6
Farmers' Dairy: Example 2 Overview

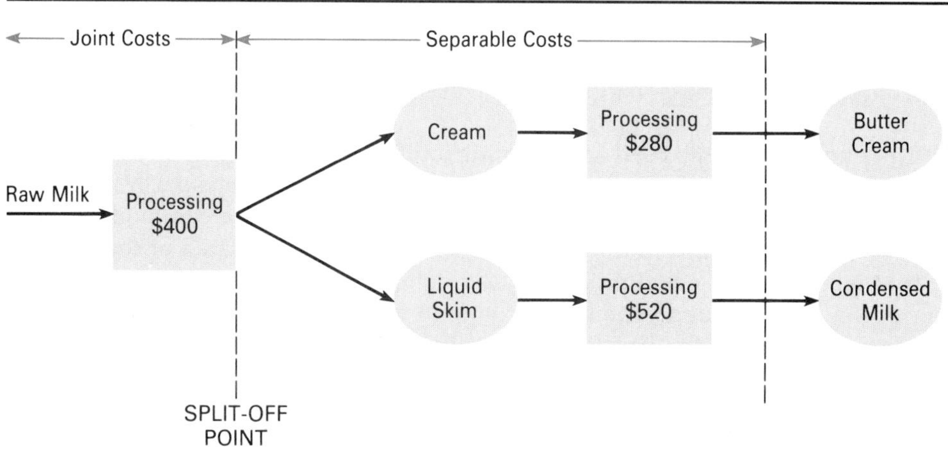

	Beginning inventory	Ending inventory
• Raw milk	0 gallons	0 gallons
• Cream	0 gallons	0 gallons
• Liquid skim	0 gallons	0 gallons
• Butter cream	0 gallons	8 gallons
• Condensed milk	0 gallons	5 gallons

Sales during the period were 12 gallons of butter cream and 45 gallons of condensed milk.

Example 2 will be used to illustrate the estimated net realizable value (NRV) method, the constant gross-margin percentage NRV method, and the no-allocation method.

Estimated Net Realizable Value Method

The **estimated net realizable value (NRV) method** allocates joint costs on the basis of the *relative estimated net realizable value* (expected final sales value in the ordinary course of business minus the expected separable costs of production and marketing). Joint costs would be allocated as follows:

	Butter Cream	Condensed Milk	Total
1. Expected final sales value of production: (Butter cream, 20 gallons × $25; Condensed milk, 50 gallons × $22)	$500	$1,100	$1,600
2. Deduct expected separable costs to complete and sell	280	520	800
3. Estimated net realizable value at split-off point	$220	$ 580	$ 800
4. Weighting	$\frac{220}{800} = .275$	$\frac{580}{800} = .725$	
5. Joint costs allocated: (Butter cream, .275 × $400; Condensed milk, .725 × $400)	$110	$290	$400
6. Production costs per unit: (Butter cream, [$110 + $280]/20 gallons; Condensed milk, [$290 + $520]/50 gallons)	$19.50	$16.20	

Note that** expected **final sales value of the total production of the period is used and not the** actual **final sales of the period. Exhibit 16-7 presents the product-line income statement using the estimated net realizable value method. The gross-margin percentages are 22.0% for butter cream and 26.4% for condensed milk.

Estimating the net realizable value of each product at the split-off point requires information about the subsequent processing steps to be taken (and their expected separable costs).[4] In some plants, there may be many possible subsequent steps. Firms may make frequent changes in further processing to exploit fluctuations in the separable costs of each processing stage or in the selling prices of individual products. Under the estimated net realizable value method, each such change would imply a change in the joint-cost allocation percentages. (In practice, a set of standard subsequent steps is assumed at the start of the accounting period when using this method.)

[4]The estimated net realizable value method is clear-cut when there is only one split-off point. However, when there are multiple split-off points, additional allocations may be required if processes subsequent to the initial split-off point remerge with each other to create a second joint-cost situation.

EXHIBIT 16-7
Product-Line Income Statement
Joint Costs Allocated Using Estimated Net Realizable Value Method

	Butter Cream	Condensed Milk	Total
Sales (Butter cream, 12 gallons × $25; Condensed milk, 45 gallons × $22)	$300	$990	$1,290
Costs of goods sold:			
Joint costs (Butter cream, .275 × $400; Condensed milk, .725 × $400)	$110	$290	$ 400
Separable costs to complete and sell	280	520	800
Cost of goods available for sale	$390	$810	$1,200
Deduct ending inventory (Butter cream, 8 gallons × $19.50; Condensed milk, 5 gallons × $16.20)	156	81	237
Cost of goods sold	$234	$729	$ 963
Gross margin	$ 66	$261	$ 327
Gross-margin percentage	22.0%	26.4%	25.3%

Constant Gross-Margin Percentage NRV Method

The **constant gross-margin percentage NRV method** allocates joint costs so that the overall gross-margin percentage is identical for each individual product. This method entails three steps:

- *Step 1:* Computing the overall gross-margin percentage
- *Step 2:* Using the overall gross-margin percentage and deducting the gross margin from the final sales values to obtain the total costs that each product should bear
- *Step 3:* Deducting the expected separable costs from the total costs to obtain the joint-cost allocation

Exhibit 16-8 presents these three steps for allocating the $400 joint cost between butter cream and condensed milk. The expected final sales value of the total production of the period ($1,600) is used in Exhibit 16-8 and *not* the actual sales of the period.[5]

A product-line income statement for the constant gross-margin percentage NRV method is presented in Exhibit 16-9. The gross-margin percentage for each product (25%) is equal by design.

The tenuous assumption underlying the constant gross-margin percentage NRV method is that every product has the same relationship between sales value and cost. Such a relationship is rarely observed in companies producing multiple products without any joint-cost situations.

No Allocation of Joint Cost

All of these foregoing methods of allocating joint costs to individual products are subject to criticism. Some companies refrain from joint-cost allocation entirely. Instead, they carry all inventories at estimated net realizable value. Income on each

[5]The joint costs allocated to each product need not always be positive under the constant gross-margin percentage NRV method. Some products may receive negative allocations of joint costs to bring their gross-margin percentages up to the overall company average.

EXHIBIT 16-8

Joint Costs Allocated Using Constant Gross-Margin Percentage NRV Method

Step 1:			
Expected final sales value of production:			
(20 gallons × $25) + (50 gallons × $22)		$1,600	
Deduct joint and separable costs ($400 + $280 + $520)		1,200	
Gross margin		$ 400	
Gross-margin percentage ($400 ÷ $1,600)		25%	
Step 2:			
	Butter Cream	**Condensed Milk**	**Total**
Expected final sales value of production: (Butter cream, 20 gallons × $25; Condensed milk, 50 gallons × $22)	$500	$1,100	$1,600
Deduct gross margin, using overall gross-margin percentage of sales (25%)	125	275	400
Cost of goods sold	$375	$ 825	$1,200
Step 3:			
Deduct separable costs to complete and sell	280	520	800
Joint costs allocated	$ 95	$ 305	$ 400

EXHIBIT 16-9

Product-Line Income Statement
Joint Costs Allocated Using Constant Gross-Margin Percentage NRV Method

	Butter Cream	**Condensed Milk**	**Total**
Sales (Butter cream, 12 gallons × $25; Condensed milk, 45 gallons × $22)	$300.0	$990.0	$1,290.0
Costs of goods sold:			
Joint costs (see Exhibit 16-8)	$ 95.0	$305.0	$ 400.0
Separable costs to complete and sell	280.0	520.0	800.0
Cost of goods available for sale	$375.0	$825.0	$1,200.0
Deduct ending inventory (Butter cream, 8 × $18.75*; Condensed milk, 5 × $16.50†	150.0	82.5	232.5
Cost of goods sold	$225.0	$742.5	$ 967.5
Gross margin	$ 75.0	$247.5	$ 322.5
Gross-margin percentage	25%	25%	25%

*$375 ÷ 20 gallons = $18.75
†$825 ÷ 50 gallons = $16.50

product is recognized when production is completed. Industries using variations of this approach include meatpacking, canning, and mining.

Accountants ordinarily criticize carrying inventories at estimated net realizable values. Why? Because income is recognized *before* sales are made. Partly in response to this criticism, some companies using this no-allocation approach carry their inventories at estimated net realizable values minus a normal profit margin.

Exhibit 16-10 presents the product-line income statement with no allocation of joint cost for Example 2. The separable costs are allocated first, which highlights for

EXHIBIT 16-10
Product-Line Income Statement
No Allocation of Joint Costs

	Butter Cream	Condensed Milk	Total
Produced and sold:			
(Butter cream, 12 gallons × $25; Condensed milk, 45 gallons × $22)	$300	$990	$1,290
Produced but not sold:			
(Butter cream, 8 gallons × $25; Condensed milk, 5 gallons × $22)	200	110	310
Total sales value of production	$500	$1,100	$1,600
Separable costs to complete and sell	280	520	800
Contribution to joint costs and profits	$220	$580	$ 800
Joint costs			400
Gross margin			$ 400
Gross-margin percentage			25%

managers the cause-and-effect relationship between individual products and the costs incurred on them. The joint costs are not allocated to butter cream and condensed milk as individual products.

Comparison of the Methods

Objective 5

Describe why the sales value at split-off method is widely used

Which method of allocating joint costs should be chosen? Each one has weaknesses. Because the costs are joint in nature, the cause-and-effect criterion is not helpful in making this choice. The sales value at split-off method is widely used when selling-price data are available (even if further processing is done). Reasons for this practice include the following:

1. *No anticipation of subsequent management decisions.* The sales value at split-off point method does not presuppose the exact number of subsequent steps for which further processing is undertaken.
2. *Availability of a meaningful common denominator to compute the weighting factors.* The denominator of the sales value at split-off method (dollars) is a meaningful one. In contrast, the physical measure method may lack a meaningful common denominator for all the separable products (for example, when some products are liquids and other products are solids).
3. *Simplicity.* The sales value at split-off point is simple. In contrast, the estimated net realizable value method can be very complex in operations with multiple products and multiple split-off points. The total sales value at split-off is unaffected by any change in the production process after the split-off point.

When sales values at the split-off point are not available, other methods must be selected. Individual industries differ in terms of the complexity involved in applying the estimated net realizable value method and in the availability of an informative common physical measure. The same joint-cost allocation method is unlikely to pass the cost-benefit test in all industries. For example, a survey of the joint-cost allocation methods used by United Kingdom chemical and oil-refining companies reports the following:[6]

[6]K. Slater and C. Wootton, *A Study of Joint and By-Product Costing in the U.K.* (London: Institute of Cost and Management Accountants, 1984).

Type of Company	Predominant Joint-Cost Allocation Method Used
Petrochemicals	Sales value at split-off or estimated net realizable value
Coal processing	Physical measure
Coal chemicals	Physical measure
Oil refining	No allocation of joint cost

The authors of the survey note that "it was considered by the majority of oil refineries that the complex nature of the process involved, and the vast number of joint product outputs, made it impossible to establish any meaningful cost apportionment between products."[7] In addition, market prices for many partly processed products at one or more of the split-off points typically are not available.

When external parties are involved, self-interest quite naturally influences perceptions about the propriety of joint-cost allocations. For instance, taxpayers may favor one method, and income tax collectors may favor another method.

IRRELEVANCE OF JOINT COSTS FOR DECISION MAKING

Objective 6

Describe the irrelevance of joint costs in deciding to sell or process further

No technique for allocating joint-product costs should be used for judging the performance of product lines or for making managerial decisions regarding whether a product should be sold at the split-off point or processed beyond split-off. When a product is an inevitable result of a joint process, the decision to process further should not be influenced by either the size of the total joint costs or by the portion of the joint costs allocated to particular products.

Sell or Process Further

The decision to incur additional costs beyond split-off should be based on the incremental operating income attainable beyond the split-off point. Example 2 assumed that it was profitable for both cream and liquid skim to be separately processed into butter cream and condensed milk, respectively. The incremental analysis for these decisions to further process follows:

Further Processing Cream into Butter Cream	
Incremental revenue ($500 − $200)	$300
Incremental costs	280
Incremental operating income	$ 20
Further Processing Liquid Skim into Condensed Milk	
Incremental revenue ($1,100 − $300)	$800
Incremental costs	520
Incremental operating income	$280

The amount of joint costs incurred up to split-off ($400)—and how they are allocated—are irrelevant with respect to the decision to further process cream or liquid skim. Why? Because, by definition, the joint costs of $400 cannot be traced to separate products.

Many manufacturing companies constantly face the decision of whether to process a joint product further. Meat products may be sold as cut or may be smoked, cured, frozen, canned, and so forth. Petroleum refiners are perpetually trying to

[7]Ibid., p. 101.

adjust to the most profitable product mix. The refining process necessitates separating all products from crude oil, even though only one or two may be desired. The refiner must decide what combination of processes to use to get the most profitable quantities of gasoline, lubricants, kerosene, naphtha, fuel oil, and the like. In addition, the manager may occasionally find it profitable to purchase distillates from some other refiner and process them further.

In designing reports for executive decisions, the accountant must concentrate on incremental costs rather than on how historical joint costs are to be split among various products. The only relevant items are incremental revenue and incremental cost. To illustrate the importance of the incremental cost viewpoint, consider another example.

EXAMPLE 3

Fragrance Inc. jointly processes a specialty chemical that yields two cosmetics: 50 ounces of Mystique and 150 ounces of Passion. The sales values per ounce at split-off are $6 for Mystique and $4 for Passion. The joint costs incurred before the split-off point are $880. The manager now has the option of further processing 150 ounces of Passion to yield 100 ounces of Romance. The total additional costs of converting Passion into Romance would be $160, and the selling price per ounce of Romance would be $8. Exhibit 16-11 summarizes the relationships in this example.

The correct approach in deciding whether to further process Passion into Romance is to compare the incremental revenue with the incremental costs:

Incremental revenue of Romance (100 × $8) − (150 × $4)	$200
Incremental costs of Romance, further processing	160
Incremental operating income from converting Passion into Romance	$ 40

The $600 revenue from selling Passion (150 × $4) becomes an opportunity cost if Passion is further processed and sold as Romance.

Another way of looking at the same decision of whether to further process Passion is:

Romance revenue, 100 ounces × $8		$800
Costs:		
Further processing of Romance	$160	
Opportunity cost, forgoing of Passion sales (150 × $4)	600	760
Difference in favor of further processing of Passion into Romance		$ 40

A total income computation of each alternative follows:

EXHIBIT 16-11
Fragrance Inc.: Example 3 Overview

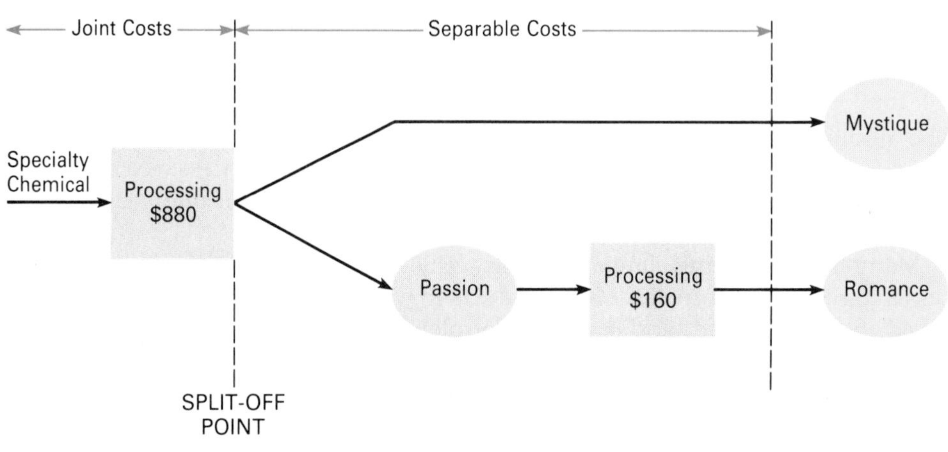

	Alternative 1	Alternative 2	
	Sell Mystique and Passion	Sell Mystique and Romance	Difference
Total revenues	($300 + $600) $900	($300 + $800) $1,100	$200
Total costs	880	($880 + $160) 1,040	160
Operating income	$ 20	$ 60	$ 40

In summary, it is profitable to extend processing or to incur additional costs on a joint product as long as the incremental revenue exceeds incremental costs.

Conventional methods of joint-cost allocation may mislead managers relying on unit-cost data to guide their sell-or-process-further decisions. For example, allocating costs using the physical measure (ounces in our example) method would split the $880 joint cost as follows:

Product	Ounces Produced	Weighting	Allocation of Joint Cost
Mystique	50	50/200 = .25	.25 × $880 = $220
Passion	150	150/200 = .75	.75 × $880 = 660
	200		$880

The resulting product-line income statement for the alternative of selling Mystique and Romance would *erroneously* imply that the business would suffer a loss selling Romance:

	Mystique	Romance
Revenues	$300	$800
Costs		
Joint costs allocated	220	660
Separable costs	—	160
Total cost of goods sold	220	820
Gross margin	$ 80	$(20)

ACCOUNTING FOR BYPRODUCTS

Exhibit 16-2 (p. 528) illustrates that byproducts have relatively low sales value compared with the sales value of the main or joint product(s). We now discuss accounting for byproducts. To simplify the discussion, we deal with a two-product example—a main product and a byproduct. In more complex cases, there could be several joint products and several byproducts as well as scrap. (The accounting alternatives for scrap are discussed in Chapter 18.)

Objective 7

Specify alternative accounting methods for byproducts

EXAMPLE 4

The Meatworks Group processes meat from slaughterhouses. One of its departments cuts lamb shoulders and generates two products:

- Shoulder meat (the main product)—sold for $60 per pack
- Hock meat (the byproduct)—sold for $4 per pack

Both products are sold at the split-off point without further processing, as Exhibit 16-12 shows. Data (number of packs) for this department's most recent period follow:

	Production	Sales	Beginning Inventory	Ending Inventory
• Shoulder meat	500	400	0	100
• Hock meat	100	30	0	70

Total manufacturing costs for these products were $25,000.

EXHIBIT 16-12
Meatworks Group: Example 4 Overview

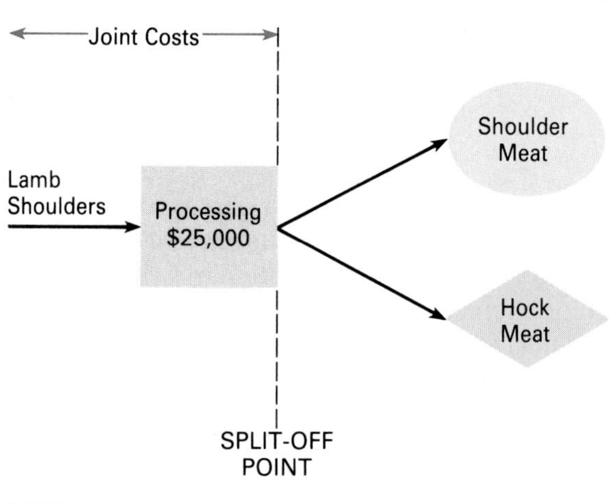

Accounting methods for byproducts address two major questions:[8]

- *When are byproducts first recognized in the general ledger?* The two main answers are (i) at the time of sale and (ii) at the time of production.
- *How do byproducts appear in the income statement?* The two main answers are (i) as a cost reduction of the main or joint product(s) and (ii) as a separate revenue or other income item.

Combining these two aspects gives four possible ways of accounting for byproducts:

Byproduct Accounting Method	Timing of Recognition in General Ledger	Place in Income Statement
A	When sold	As a reduction of cost
B	When produced	As a reduction of cost
C	When sold	As a revenue or other income item
D	When produced	As a revenue or other income item

Exhibit 16-13 presents the income statement figures and inventory figures that the Meatworks Group department would report under each method. Methods A and C are justified primarily on grounds of immateriality or on a cost-benefit criterion. The byproduct is viewed as incidental and therefore does not warrant costly accounting procedures (although procedures may be adopted for control of physical quantities). Methods B and D have more conceptual appeal, as they better match the value of the byproduct with the cost of the products simultaneously produced. Moreover, Methods B and D report costs of byproduct inventories when these inventories exist.[9]

[8]Further discussion on byproduct accounting methods is in C. Cheatham and M. Green, "Teaching Accounting for Byproducts," *Management Accounting News & Views* (Spring 1988), pp. 14–15, and D. Stout and D. Wygal, "Making By-Products a Main Product of Discussion: A Challenge to Accounting Educators," *Journal of Accounting Education* (1989), pp. 219–33.

[9]One version of Method B deducts the estimated net realizable value of the byproduct(s) from the total production cost (joint costs + separable costs). Another version of Method B deducts the estimated net realizable value of the byproduct(s) from the joint costs before they are allocated to individual joint products.

EXHIBIT 16-13
Meatworks Group: Income Statement

	Byproduct Accounting Method			
	A	B	C	D
Timing of Recognition in General Ledger	When Sold	When Produced	When Sold	When Produced
Place in Income Statement	As Reduction of Cost	As Reduction of Cost	Other Revenue Item	Other Revenue Item
Revenues:				
Main product: Shoulder meat (400 × $60)	$24,000	$24,000	$24,000	$24,000
Byproduct: Hock meat (30 × $4)	—	—	120	120
	$24,000	$24,000	$24,120	$24,120
Cost of goods sold:				
Total manufacturing costs	$25,000	$25,000	$25,000	$25,000
Deduct byproduct net revenue (30 × $4)	120	120	—	—
Net manufacturing costs	$24,880	$24,880	$25,000	$25,000
Deduct main product inventory*	4,976	4,976	5,000	5,000
Deduct byproduct inventory (70 × $4)	—	280	—	280
Cost of goods sold	$19,904	$19,624	$20,000	$19,720
Gross margin	$ 4,096	$ 4,376	$ 4,120	$ 4,400
Gross-margin percentage	17.07%	18.23%	17.08%	18.24%
Inventory cost (end of period):				
Main product: Shoulder meat	$ 4,976	$ 4,976	$ 5,000	$ 5,000
Byproduct: Hock meat	$ 0	$ 280	$ 0	$ 280

*(100/500) × net manufacturing costs.

Separable Costs of Byproducts

The byproduct in Example 4 requires no additional (separable) costs to be incurred for processing, marketing, or disposal after the split-off point. When separable costs are incurred, byproduct accounting becomes more complex. Conceptually, these separable costs should be traced to each byproduct. Firms using Method A or C should use the estimated net realizable value of the byproduct: expected revenues minus separable costs incurred (production) and costs yet to be incurred (marketing or disposal). A variant of this approach is to use estimated net realizable value *minus normal profit margin* of byproducts produced. This version is more conservative in that byproducts are valued lower than their estimated net realizable value. It implies, however, that the byproduct is important enough to warrant treatment as a separate product line. In short, the product is treated like a joint product instead of a byproduct.

Byproduct accounting is an area where cost-benefit comparisons often lead to choosing the expedient alternative. When firms recognize byproducts in their general ledger at the time of sale (Methods A and C), the separable byproduct costs typically are expensed to the period in which they are incurred, a procedure justified by the product's immateriality. Although costs of main products are traced to the production or marketing activities associated with those products, it is generally too expensive to trace costs of byproducts in the same way.

PROBLEM FOR SELF-STUDY

PROBLEM

Inorganic Chemicals purchases salt and processes it into more-refined products such as caustic soda, chlorine, and PVC (polyvinyl chloride). In the most recent month (July), Inorganic Chemicals purchased salt for $40,000. Conversion costs of $60,000 were incurred up to the split-off point, at which time two salable products were produced: caustic soda and chlorine. Chlorine can be further processed into PVC.

The July production and sales information follows:

	Production	Sales	Sales Price per Ton
Caustic soda	1,200 tons	1,200 tons	$50
Chlorine	800 tons	—	—
PVC	500 tons	500 tons	$200

All 800 tons of chlorine were further processed, at an incremental cost of $20,000, to yield 500 tons of PVC. There were no byproducts or scrap from this further processing of chlorine. There were no beginning or ending inventories of caustic soda, chlorine, or PVC in July.

There is an active market for chlorine. Inorganic Chemicals could have sold all its July production of chlorine at $75 a ton.

Required

1. Calculate how the joint cost of $100,000 would be allocated between caustic soda and chlorine under each of the following methods: (a) sales value at split-off, (b) physical measure (tons), and (c) estimated net realizable value.
2. What is the gross-margin percentage of caustic soda and PVC under the three methods cited in requirement 1?
3. Lifetime Swimming Pool Products offers to purchase 800 tons of chlorine in August at $75 a ton. This sale of chlorine would mean that no PVC would be produced in August. How would accepting this offer affect August operating income?

SOLUTION

1. (a) Sales value at split-off method

	Caustic Soda	Chlorine	Total
1. Sales value at split-off: (Caustic, 1,200 × $50; Chlorine, 800 × $75)	$60,000	$60,000	$120,000
2. Weighting	$\frac{60}{120} = .5$	$\frac{60}{120} = .5$	
3. Joint costs allocated: (Caustic, .5 × $100,000; Chlorine, .5 × $100,000)	$50,000	$50,000	$100,000

(b) Physical measure method

	Caustic Soda	Chlorine	Total
1. Physical measure	1,200	800	2,000
2. Weighting	$\frac{1,200}{2,000} = .6$	$\frac{800}{2,000} = .4$	
3. Joint costs allocated: (Caustic, .6 × $100,000; Chlorine, .4 × $100,000)	$60,000	$40,000	$100,000

(c) Estimated net realizable value method

	Caustic Soda	Chlorine	Total
1. Expected final sales value of production: (Caustic, 1,200 × $50; PVC from chlorine, 500 × $200)	$60,000	$100,000	$160,000
2. Expected separable costs	—	20,000	20,000
3. Estimated net realizable value at split-off point	$60,000	$ 80,000	$140,000
2. Weighting	$\frac{60}{140} = \frac{3}{7}$	$\frac{80}{140} = \frac{4}{7}$	
3. Joint costs allocated: (Caustic, 3/7 × $100,000; Chlorine, 4/7 × $100,000)	$42,857	$ 57,143	$100,000

2. Caustic Soda

	Sales value at split-off point	Physical measure	Estimated net realizable value
Sales	$60,000	$60,000	$60,000
Joint costs	50,000	60,000	42,857
Gross margin	$10,000	$ 0	$17,143
Gross margin %	16.67%	0%	28.57%

PVC

	Sales value at split-off point	Physical measure	Estimated net realizable value
Sales	$100,000	$100,000	$100,000
Joint costs	50,000	40,000	57,143
Separable costs	20,000	20,000	20,000
Gross margin	$ 30,000	$ 40,000	$ 22,857
Gross margin %	30.00%	40.00%	22.86%

3.

Incremental revenue from further processing of chlorine into PVC [(500 × $200) − (800 × $75)]	$40,000
Incremental costs of further processing chlorine into PVC	20,000
Incremental operating income from further processing	$20,000

The operating income of Inorganic Chemicals would be reduced by $20,000 if it sold 800 tons of chlorine to Lifetime Swimming Pool Products instead of further processing the chlorine into PVC for sale.

SUMMARY

Joint costs permeate accounting. Accountants attempt to allocate joint costs among products having relatively significant sales values. The major purposes for allocating joint costs are inventory costing and the determination of costs for contract reimbursement or rate regulation. Accounting methods available in joint-cost situations include sales value at split-off, estimated net realizable value, constant gross-margin percentage NRV, physical measure, and no-allocation.

The sales value at split-off method has several advantages—simplicity, a meaningful common denominator, and no anticipation of subsequent management decisions. The major obstacle to its widespread adoption is that selling prices for some products at the split-off point are not available.

Only incremental costs and incremental revenues are relevant to decisions on whether to incur additional separable costs beyond the split-off point. As long as the incremental revenue exceeds the incremental costs, further processing is profitable.

TERMS TO LEARN

This chapter and the Glossary at the end of the book contain definitions of the following important terms:

byproduct *(p. 527)* constant gross-margin percentage NRV method *(534)* estimated net realizable value (NRV) method *(533)* joint cost *(527)* joint product *(527)* main product *(527)* physical measure method *(531)* sales value at split-off method *(530)* scrap *(527)* separable costs *(527)* split-off point *(527)*

ASSIGNMENT MATERIAL

QUESTIONS

16-1 What is a *joint cost*?

16-2 Define *separable costs.*

16-3 Give two examples of industries in which joint costs are found. For each example, what are the separable products at the split-off point?

16-4 Distinguish between a *joint product* and a *byproduct.*

16-5 Provide three reasons for allocating joint costs to individual products or services.

16-6 Name four methods of allocating joint costs to joint products.

16-7 Distinguish between the *sales value at split-off method* and the *estimated net realizable value method.*

16-8 Many oil refineries do not allocate joint costs among their products. What reasons could explain why?

16-9 Give three reasons why firms prefer to use the *sales value at split-off point method.*

16-10 "Managers must decide whether a product should be sold at split-off or processed further. The sales value at split-off method of joint-cost allocation is the best method for generating the information managers need." Do you agree? Why?

16-11 Describe two major questions addressed by accounting methods for byproducts.

EXERCISES AND PROBLEMS

16-12 Estimated net realizable value method. Illawara Inc. produces two joint products, A and B, from a single process. In July 19_6, the joint costs of this single process were $24,000. Separable processing costs beyond the split-off point were: A, $30,000; B, $7,500. Product A sells for $50 per unit; Product B sells for $25 per unit. Illawara produced and sold 1,000 units of A and 500 units of B. There are no beginning or ending inventories of A or B.

Required
Allocate the $24,000 joint costs using the estimated net realizable value method.

16-13 Alternative joint-cost allocation methods, further process decision. The Wood Spirits Company produces two products, turpentine and methanol (wood alcohol), by a joint process. Joint costs amount to $120,000 per batch of output. Each batch totals 10,000 gallons, 25% methanol and 75% turpentine. Both products are processed further without gain or loss in volume. Separable processing costs: methanol, $3 per gallon; turpentine, $2 per gallon. Methanol sells for $21 per gallon; turpentine sells for $14 per gallon.

Required
1. What joint costs per batch should be allocated to the turpentine and methanol, assuming that joint costs are allocated on a physical-measure (number of gallons at split-off point) basis?

2. If joint costs are to be assigned on an estimated net realizable value basis, what amounts of joint cost should be assigned to the turpentine and to the methanol?
3. Prepare product-line income statements per batch for requirements 1 and 2. Assume no beginning or ending inventories.
4. The company has discovered an additional process by which the methanol (wood alcohol) can be made into a pleasant-tasting alcoholic beverage. The new selling price would be $60 a gallon. Additional processing would increase separable costs $9 (in addition to the $3 separable cost required to yield methanol). The company would have to pay excise taxes of 20% on the new selling price. Assuming no other changes in cost, what is the joint cost applicable to the wood alcohol (using the estimated net realizable value method)? Should the company use the new process?

16-14 Alternative methods of joint-cost allocation, ending inventories. The Darl Company operates a simple chemical process to reduce a single material into three separate items, here referred to as X, Y, and Z. All three end products are separated simultaneously at a single split-off point.

Products X and Y are ready for sale immediately upon split-off without further processing or any other additional costs. Product Z, however, is processed further before being sold. There is no available market price for Z at the split-off point.

The selling prices quoted below have not changed for three years, and no future changes are foreseen. During 19_3, the selling prices of the items and the total number sold were as follows:

X—120 tons sold for $1,500 per ton
Y—340 tons sold for $1,000 per ton
Z—475 tons sold for $700 per ton

The total joint manufacturing costs for the year were $400,000. An additional $200,000 was spent in order to finish Product Z.

There were no beginning inventories of X, Y, or Z. At the end of the year, the following inventories of completed units were on hand: X, 180 tons; Y, 60 tons; Z, 25 tons. There was no beginning or ending work in process.

Required

1. What will be the cost of inventories of X, Y, and Z for balance sheet purposes and the cost of goods sold for income statement purposes as of December 31, 19_3, using (a) the estimated net realizable value method of joint-cost allocation and (b) the constant gross-margin percentage NRV method of joint-cost allocation?
2. Compare the gross-margin percentages for X, Y, and Z using the two methods given in requirement 1.

16-15 Net realizable value cost allocation method, further process decision. (W. Crum) Tuscania Company crushes and refines mineral ore into three products in a joint-cost operation. Costs and production for 19_8 were as follows:

Department 1: At initial joint costs of $420,000, produces 20,000 pounds of Alco, 60,000 pounds of Devo, 100,000 pounds of Holo
Department 2: Processes Alco further at a cost of $100,000
Department 3: Processes Devo further at a cost of $200,000

Results for 19_8:

Alco: 20,000 pounds completed; 19,000 pounds sold for $20 per pound; ending inventory, 1,000 pounds
Devo: 60,000 pounds completed; 59,000 pounds sold for $6 per pound; ending inventory, 1,000 pounds
Holo: 100,000 pounds completed; 99,000 pounds sold for $1 per pound; ending inventory, 1,000 pounds; Holo required no further processing

Required

1. Use the estimated net realizable value method to allocate the joint costs of the three products.

2. Compute the total costs and unit costs of ending inventories.
3. Compute the individual gross-margin percentages of the three products.
4. Suppose Tuscania receives an offer to sell all of its Devo product for a price of $2 per pound at the split-off point before going through Department 3, just as it comes off the production line in Department 1. Using last year's figures, would Tuscania be better off by selling Devo that way rather than by processing it through Department 3 and selling it as it did? Show computations to support your answer. Disregard all other factors not mentioned in the problem.

16-16 Estimated net realizable value cost allocation method, further process decision. (CPA) The Mikumi Manufacturing Company produces three products by a joint production process. Direct material is put into production in Department A, and at the end of processing in this department three products appear. Product X is sold at the split-off point, with no further processing. Products Y and Z require further processing before they are sold. Product Y is processed in Department B, and Product Z is processed in Department C. The company uses the estimated net realizable value method of allocating joint production costs. A summary of costs and other data for the year ended September 30, 19_2, follows.

	Products		
	X	Y	Z
Pounds sold	10,000	30,000	40,000
Pounds on hand at September 30, 19_2	20,000	–0–	20,000
Sales	$15,000	$81,000	$141,750
	Departments		
	A	B	C
Direct-material cost	$56,000	$ 0	$ 0
Direct labor cost	24,000	40,450	101,000
Manufacturing overhead	10,000	10,550	36,625

There were no inventories on hand at September 30, 19_1, and there was no direct material on hand at September 30, 19_2. All the units of product on hand at September 30, 19_2, were fully complete as to processing.

Required
1. Determine the following amounts for each product: (a) estimated net realizable value as used for allocating joint costs, (b) joint costs allocated, (c) cost of goods sold, and (d) cost of finished goods inventory, September 30, 19_2.
2. Assume that the entire output of Product X could be processed further at an additional cost of $2.00 per pound and then sold at a price of $4.30 per pound. What would be the effect on operating income if all the Product X output for the year ended September 30, 19_2, had been processed further and sold rather than all sold at the split-off point?

16-17 Alternative methods of joint-cost allocation, product-mix decisions. The Sunshine Oil Company buys crude vegetable oil. Refining this oil results in four products at the split-off point: A, B, C, and D. Product C is fully processed at the split-off point. Products A, B, and D can individually be further refined into Super A, Super B, and Super D. In the most recent month (December), the output at the split-off point was:

Product A	300,000 gallons
Product B	100,000 gallons
Product C	50,000 gallons
Product D	50,000 gallons

The joint costs of purchasing the crude vegetable oil and processing it were $100,000.

Sunshine had no beginning or ending inventories. Sales of Product C in December were $50,000. Total output of Products A, B, and D was further refined and then sold. Data relating to December are:

	Separable Processing Cost to Make Super Products	Sales
Super A	$200,000	$300,000
Super B	80,000	100,000
Super D	90,000	120,000

Sunshine had the option of selling Products A, B, and D at the split-off point. This alternative would have yielded the following sales for the December production:

Product A	$50,000
Product B	30,000
Product D	70,000

Required

1. What is the gross-margin percentage for each product sold in December, using the following methods for allocating the $100,000 joint costs: (a) sales value at split-off: (b) physical measure, and (c) estimated net realizable value.
2. Could Sunshine have increased its December operating income by making different decisions about the further refining of Products A, B, or D? Show the effect on operating income of any changes you recommend.

16-18 Comparison of alternative joint-cost allocation methods, further process decision, chocolate products. Roundtree Chocolates manufactures and distributes chocolate products. It purchases cocoa beans and processes these into two intermediate products:

- Chocolate-powder liquor base
- Milk-chocolate liquor base

These two intermediary products become separately identifiable at a single split-off point. Every 500 pounds of cocoa beans yields 20 gallons of chocolate-powder liquor base and 30 gallons of milk-chocolate liquor base.

The chocolate-powder liquor base is further processed into chocolate powder. Every 20 gallons of chocolate-powder liquor base yields 200 pounds of chocolate powder. The milk-chocolate liquor base is further processed into milk chocolate. Every 30 gallons of milk-chocolate liquor base yields 340 pounds of milk chocolate.

An overview of the manufacturing operations at Roundtree Chocolates follows:

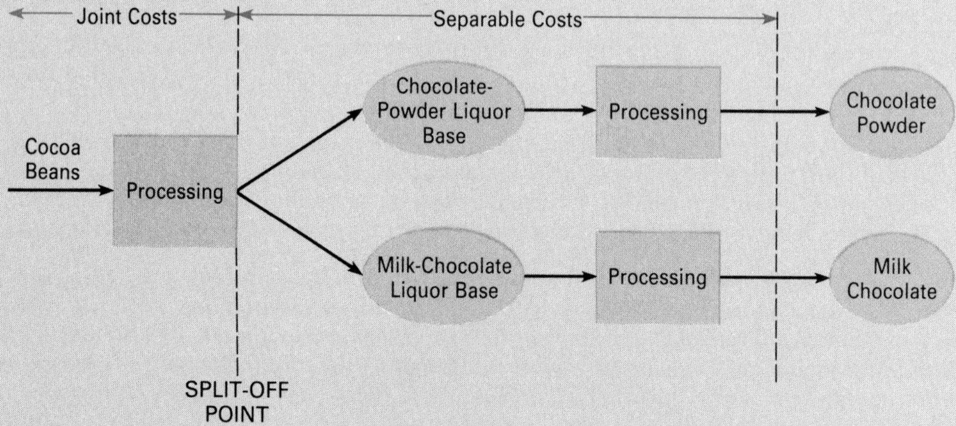

Production and sales data for August 19_4 are:

Cocoa beans processed, 5,000 pounds
Costs of processing cocoa beans to split-off point (including purchase of beans) = $10,000

	Production	Sales	Unit Selling Price
Chocolate powder	2,000 pounds	2,000 pounds	$4 per pound
Milk chocolate	3,400 pounds	3,400 pounds	$5 per pound

The August 19_4 separable costs of processing chocolate-powder liquor base into chocolate powder are $4,250. The August 19_4 separable costs of processing milk-chocolate liquor base into milk chocolate are $8,750.

Roundtree fully processes both of its intermediate products into chocolate powder or milk chocolate. There is an active market for these intermediate products. In August 19_4, Roundtree could have sold chocolate-powder liquor base for $21 a gallon and milk-chocolate liquor base for $26 a gallon.

Required

1. Calculate how the joint costs of $10,000 would be allocated between chocolate-powder liquor base and milk-chocolate liquor base under each of the following methods:
   a. Sales value at split-off
   b. Physical measure (gallons)
   c. Estimated net realizable value
   d. Constant gross-margin percentage NRV method
2. What is the gross-margin percentage of chocolate-powder liquor base and milk-chocolate liquor base under methods *a, b, c,* and *d* in requirement 1?
3. Could Roundtree Chocolate have increased its operating income by a change in its decision to fully process both of its intermediate products?

16-19 Alternative methods of joint-cost allocation, further process decision, memory chips. AMC is a semiconductor firm that specializes in memory chips. In the first stage of the manufacturing operation, raw silicon wafers are photolithographed and then baked at high temperatures. This process yields three individual products at a common split-off point. For each batch of 1,600 raw silicon wafers, these products are:

1. 300 high-density (HD) memory chips
2. 900 low-density (LD) memory chips
3. 400 defective memory chips

The density of a memory chip is based on the number of good memory bits on each chip, with HD chips having more memory bits per chip than LD chips. The 400 defective memory chips from each batch have a zero disposal price. The joint cost of purchasing and processing the 1,600 raw silicon wafers up to the split-off point is $5,000.

AMC has two options for each grade of good memory chip at the split-off point:

1. Sell immediately. The selling price for each HD chip is $10. The selling price for each LD chip is $5.
2. Process further into extended-life memory chips. This processing step further exposes the chips to extreme temperatures, and the chips that survive are sold as extended-life memory chips. Data pertaining to this further processing stage include
   a. Extended-life high-density (EL–HD) chips: From a batch of 300 HD chips, the yield is 200 EL–HD chips. The 100 defective chips from this further processing step have a zero disposal price. The separable costs to further process the 300 HD chips are $1,000. The selling price for each EL–HD chip is $30.
   b. Extended-life low-density (EL–LD) chips: From a batch of 900 LD chips, the yield is 500 EL–LD chips. The 400 defective chips from this further processing step have a zero disposal price. The separable cost to further process the 900 LD chips is $3,000. The selling price for each EL–LD chip is $18.

AMC has consistently followed the policy of further processing the entire output of both the HD and LD chips into their EL–HD and EL–LD forms.

Required

1. Compute how the joint costs of $5,000 would be allocated between HD and LD chips under each of the following methods: (a) sales value at split-off, (b) physical measure (number of good chips at split-off point), (c) estimated net realizable value, and (d) constant gross-margin percentage NRV. Assume that AMC has no beginning or ending inventories.
2. What is the gross-margin percentage of EL–HD and EL–LD chips under methods *a, b, c,* and *d* in requirement 1?

3. Peach Computer Systems offers to buy 900 LD memory chips from AMC at $5 a chip. What would be the effect on operating income of accepting this offer rather than pursuing the current policy of further processing the LD chips into EL–LD form?

16-20 Allocating joint costs, Christmas trees. S. Claus is a retired gentleman who has rented a lot in the center of a busy city for the month of December 19_0 at a cost of $5,000. On this lot he sells Christmas trees and wreaths. He buys his trees by the bundle for $20 each. A bundle is made up of two big trees (average height, seven feet), four regular-sized trees (average height, five feet) and broken branches. The shipper puts in the broken branches merely to make all the bundles of uniform size for shipping advantages. The amount of branches varies from bundle to bundle.

Mr. Claus gets $2 a foot for the trees. He takes home the broken branches. He and Mrs. Claus, Donner, Blitzen, and friends sit around the fire in the evenings and make wreaths, which Mr. Claus sells for $8 each. Except for Christmas Eve, these evenings are a time when there is nothing else to do; therefore, their labor is not a cost.

During the course of the season, Mr. Claus buys 1,000 bundles of trees and 2,000 wreaths are made. In addition to the broken branches, the wreaths contain pine cones (100 pounds, total cost $200), twine (4,000 yards at 10¢ a yard), and miscellaneous items amounting to $400.

In 19_0, Claus sold 1,800 of the seven-foot trees, all the regular-sized ones, and half the wreaths (the local Boy Scouts were also selling wreaths). The department store next door says that it will buy the rest of the wreaths, if Claus will preserve them, for $3 each. The preservative spray costs $2,000 for enough to do the job.

Required

1. What unit cost should Claus assign to each of his items? Use the estimated net realizable value method of allocating joint costs.
2. What is the inventory cost on January 1, 19_1, if he doesn't sell to the department store?
3. Should he sell to the department store?

16-21 Estimated net realizable value method, byproducts. (CMA, adapted) Doe Corporation grows and processes pineapples. It sells three main (joint) pineapple products—sliced pineapple, crushed pineapple, and pineapple juice. The outside skin is cut off in the Cutting Department and processed as animal feed. The skin is treated as a byproduct.

Pineapples are initially processed in the Cutting Department. The pineapples are washed, and the outside skin is cut away. Then the pineapples are cored and trimmed for slicing. The three *main products* (sliced, crushed, juice) and the *byproduct* (animal feed) are recognizable after processing in the Cutting Department. Each product is then transferred to a separate department for final processing.

The trimmed pineapples are forwarded to the Slicing Department, where the pineapples are sliced and canned. The juice generated during the slicing operation is packed in the cans with the slices.

The pieces of pineapple trimmed from the fruit are diced and canned in the Crushing Department. Again, the juice generated during this operation is packed in the can with the crushed pineapple.

The core and surplus pineapple generated from the Cutting Department are pulverized into a liquid in the Juicing Department. There is an evaporation loss equal to 8% of the weight of the good output produced in this department, which occurs as the juice is being heated.

The outside skin, the byproduct, is chopped into animal feed in the Feed Department.

The Doe Corporation uses the estimated net realizable value method to allocate costs of the joint process to its main products. The byproduct is inventoried at its market value. Corporate policy is to subtract the estimated net realizable value of the byproduct produced from the joint costs to be allocated.

A total of 270,000 pounds were entered into the Cutting Department during May. The schedule presented below shows the costs incurred in each department, the proportion by weight transferred to the four final processing departments, and the selling price of each end product.

Department	Costs Incurred	Proportion of Product by Weight Transferred to Departments	Selling Price Per Pound of Final Product
Cutting	$60,000	—	None
Slicing	4,700	35%	$.60
Crushing	10,580	28	.55
Juicing	3,250	27	.30
Animal feed	700	10	.10
Total	$79,230	100%	

Required

1. Calculate:
   a. the pounds of pineapple that result as departmental output for pineapple slices, crushed pineapple, pineapple juice, and animal feed
   b. the estimated net realizable value at the split-off point of the three main products
   c. the amount of the cost of the Cutting Department allocated to each of the three main products and to the byproduct in accordance with corporate policy
   d. the gross margins for each of the three main products
2. Comment on the significance to managerial decisions of the gross-margin information by each main product.

16-22 Joint products, byproducts, sales value at split-off method. (CMA adapted) Multiproduct Corporation is a chemical manufacturer that produces two joint products (Super Pepco-1 and Super Repke-3) and a byproduct (SE-5) from a joint process. If Multiproduct had the proper facilities, it could process SE-5 further into a third joint product.

Multiproduct currently uses the physical measure method of allocating joint costs to the main products. Immediately after it is produced, the byproduct is inventoried at its estimated net realizable value. This amount is subtracted from the joint production costs before the joint costs are allocated to the joint products.

Jim Simpson, Multiproduct's controller, wants to implement the sales value at split-off method of joint-cost allocation. He believes that inventoriable costs should be based on each product's ability to contribute to the recovery of joint-production costs. The estimated net realizable value of the byproduct would be subtracted from the joint-production costs before the joint costs are allocated to the joint products.

Data regarding Multiproduct's operations for November 19_6 follow. The joint costs of production amounted to $2,640,000 for November 19_6.

	Joint Products		
	Super Pepco-1	Super Repke-3	Byproduct SE-5
Finished goods inventory in gallons on Nov. 1, 19_6	0	0	0
November sales in gallons	800,000	700,000	200,000
November production in gallons	900,000	720,000	240,000
Sales value per gallon at split-off point (Pepco-1 & Repke-3)	$2.00	$1.50	$0.55*
Additional processing costs after split-off (separable costs)	$1,800,000	$720,000	—
Final sales value per gallon	$5.00	$4.00	—

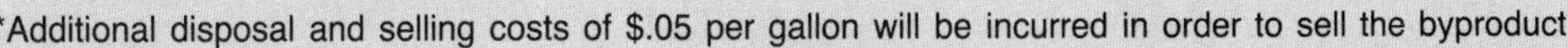

*Additional disposal and selling costs of $.05 per gallon will be incurred in order to sell the byproduct.

Required

1. Allocate the joint costs between Pepco-1 and Repke-3 using the sales value at split-off method.
2. Allocate the joint costs between Pepco-1 and Repke-3 using the estimated NRV method.

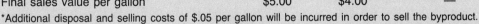

3. Multiproduct Corporation plans to expand its production facilities to enable it to further process SE-5 into a joint product. Discuss how the allocation of the joint-production costs under the sales value at split-off method would change if SE-5 is classified as a joint product.

16-23 Accounting for a main product and a byproduct. (Cheatham and Green, adapted) Bill Dundee is the owner and operator of Louisiana Bottling, a bulk soft-drink producer. A single production process yields two bulk soft drinks—Rainbow Dew (the main product) and Resi-Dew (the byproduct). Both products are fully processed at the split-off point, and there are no separable costs.

Summary data for September 19_8 are:

Cost of soft-drink operations = $120,000
Production and sales data

	Production in Gallons	Sales in Gallons	Selling Price per Gallon
Main product: Rainbow Dew	10,000	8,000	$20.00
Byproduct: Resi-Dew	2,000	1,400	2.00

There were no beginning inventories on September 1, 19_8.

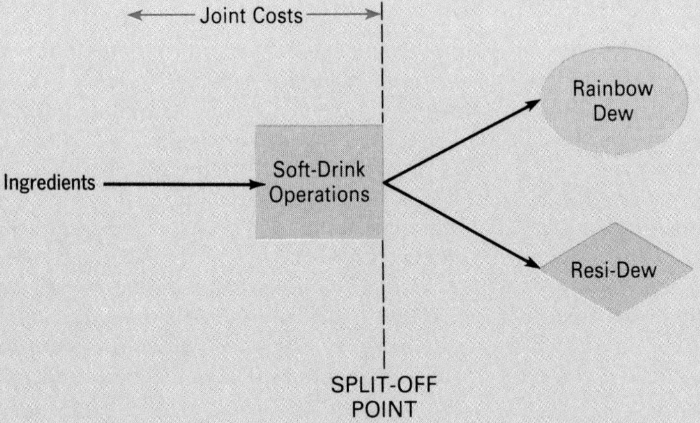

Required

1. What is the gross margin for Louisiana Bottling under the following four methods of byproduct accounting:

Byproduct Accounting Method	Timing of Recognition in General Ledger	Place in Income Statement
A	When sold	As a reduction of cost
B	When produced	As a reduction of cost
C	When sold	As a revenue item
D	When produced	As a revenue item

2. What are the inventoriable costs on September 30, 19_8, for Rainbow Dew and Resi-Dew under each of the four methods of byproduct accounting cited in requirement 1?

16-24 Joint costs and byproducts. (W. Crum) Caldwell Company processes an ore in Department 1, out of which come three products, L, W, and X. Product L is processed further through Department 2. Product W is sold without further processing. Product X is considered a byproduct and is processed further through Department 3. Costs in Department 1 are $800,000 in total; Department 2 costs are $100,000; Department 3 costs are $50,000. Processing 600,000 pounds in Department 1 results in 50,000 pounds of Product L, 300,000 pounds of Product W, and 100,000 pounds of Product X.

Product L sells for $10 per pound. Product W sells for $2 per pound. Product X sells for $3 per pound. The company wants to make a gross margin of 10% of sales on Product X and also allow 25% for marketing costs.

Required

1. Compute unit costs per pound for Products L, W, and X, treating X as a byproduct. Use the estimated net realizable value method for allocating joint costs. Deduct the estimated net realizable value of the byproduct produced from the joint cost of Products L and W.
2. Compute unit costs per pound for Products L, W, and X, treating all three as joint products and allocating costs by the estimated net realizable value method.

16-25 Joint and byproducts, estimated net realizable value method. (CPA) The Harrison Corporation produces three products—Alpha, Beta, and Gamma. Alpha and Gamma are joint products, and Beta is a byproduct of Alpha. No joint costs are to be allocated to the byproduct. The production processes for a given year follow:

A. In Department One, 110,000 pounds of direct material, Rho, are processed at a total cost of $120,000. After processing in Department One, 60% of the units are transferred to Department Two, and 40% of the units (now Gamma) are transferred to Department Three.
B. In Department Two, the material is further processed at a total additional cost of $38,000. Seventy percent of the units (now Alpha) are transferred to Department Four and 30% emerge as Beta, the byproduct, to be sold at $1.20 per pound. Separable marketing costs for Beta are $8,100.
C. In Department Four, Alpha is processed at a total additional cost of $23,660. After this processing, Alpha is ready for sale at $5 per pound.
D. In Department Three, Gamma is processed at a total additional cost of $165,000. In this department, a normal loss of units of Gamma occurs, which equals 10% of the good output of Gamma. The remaining good output of Gamma is then sold for $12 per pound.

Required

1. Prepare a schedule showing the allocation of the $120,000 joint costs between Alpha and Gamma using the estimated net realizable value method. The estimated net realizable value of Beta should be treated as an addition to the sales value of Alpha.
2. Independent of your answer to requirement 1, assume that $102,000 of total joint costs were appropriately allocated to Alpha. Assume also that there were 48,000 pounds of Alpha and 20,000 pounds of Beta available to sell. Prepare an income statement through gross margin for Alpha using the following facts:
   a. During the year, sales of Alpha were 80% of the pounds available for sale. There was no beginning inventory.
   b. The estimated net realizable value of Beta available for sale is to be deducted from the cost of producing Alpha. The ending inventory of Alpha is to be based on the net costs of production.
   c. All other cost and selling price data are listed in A through D above.

16-26 Joint cost allocation in the entertainment industry. Roadshow Ventures—based in Hollywood, California—is a movie production company. Ray Stevens, president of Roadshow Ventures, separately finances each project as a movie syndicate. A *movie syndicate* is a group of investors who provide cash and other resources for a movie project in return for sharing in the revenues or income from that movie. Stevens is also president of Stevens International, which markets outside of North America the movies Roadshow Ventures produces. Stevens owns 100% of Roadshow Ventures and 100% of Stevens International.

Stevens is one of the most recognized names in Hollywood, in large part due to Roadshow Venture's several megasuccessful movies in the past three years. One recent magazine article was titled "Ray Stevens: The Man with the Midas Touch."

Each movie receives revenues from four markets both in North America and outside of North America:

1. Theater distribution market
2. Television market
3. Cable TV market
4. Videocassette market

EXHIBIT 16-14

Summary: First-Year Revenues and Costs to Date of the Movie *Blue Fire Lady* (in millions)

	North American	Non–North American	Total
Revenues			
Theater distribution market	$29.090	$18.600	$47.690
Television market	1.642	0.861	2.503
Cable TV market	2.965	3.263	6.228
Videocassette market	2.643	3.876	6.519
			$62.940
Costs			
Total production costs of movie	—	—	$14.682
Incremental costs:			
Theater distribution market	11.406	5.304	16.710
Television market	0.247	0.103	0.350
Cable TV market	0.261	0.093	0.354
Videocassette market	1.203	0.841	2.044
			$34.140
Operating income			$28.800

Stevens argues that the theater distribution market is the main product and that the other three markets are byproducts.

Each movie syndicate receives all the North American–based revenues from these four markets. Stevens International retains all the non–North American–based revenues from these four markets. Stevens International pays the movie syndicate 20% of the net production costs of the movie. *Net production cost* is defined in the movie syndicate contract with Roadshow Ventures and Stevens International as total production cost minus the incremental income (both North American and non–North American) from the three byproduct markets (television, cable TV, and videocassettes).

Roadshow always releases movies first to the North American theater distribution market and then releases the successful ones to the non–North American markets.

Each movie syndicate that Roadshow Ventures organizes has two classes of investors:

- Roadshow Ventures, the managing partner, contributes 10% of the total production cost and 10% of the incremental costs for the North American theater, television, cable TV, and videocassette markets. Roadshow receives 25% of the operating income of the movie syndicate.
- External investors contribute 90% of the total production cost and 90% of the incremental costs for the North American theater, television, cable TV, and videocassette markets. External investors receive 75% of the operating income of the movie syndicate.

Exhibit 16-14 presents the first year's revenues and the costs to date of *Blue Fire Lady*, a movie produced by Roadshow Ventures.

Required

1. What is the payment Stevens International makes to the *Blue Fire Lady* movie syndicate for 20% of the net production costs of the movie? Use the main product/byproduct classification scheme as set out in the contract.
2. Assume that Stevens International makes two changes in the way it pays the movie syndicate for the rights to non–North American revenues from the four market areas. First, all four market areas will be classified as joint products. Second, the total production cost of the movie will be allocated between the North American and the non–North American markets using the estimated net realizable value method of joint-cost allocation. How much will Stevens International pay the *Blue Fire Lady* movie syndicate for the rights to non–North American revenues from the four market areas?

3. Suppose you are an external investor in the *Blue Fire Lady* syndicate. Would you prefer the requirement 1 or requirement 2 approach to determining how much Stevens International pays for the right to non–North American revenues from the four market areas?
4. You are concerned with the distinction between North American and non–North American revenues and costs. Name two aspects of the movie and entertainment industry that will create problems in making this distinction clear and unambiguous in a contract between a Roadshow Ventures' movie syndicate and Stevens International.

CHAPTER 17

Companies that manufacture units through continuous mass production—for example, Miller Brewing Company—use process product-costing systems to cost inventories.

Process-Costing Systems

Five key steps

Weighted-average method

First-in, first-out method

Comparison of FIFO and weighted-average methods

Standard costs and process costs

Transfers in process costing

Additional aspects of process costing

Learning Objectives

When you have finished studying this chapter, you should be able to

1. Describe five key steps of process costing
2. Demonstrate the weighted-average method of process costing
3. Prepare journal entries for process-costing systems
4. Demonstrate the first-in, first-out (FIFO) method of process costing
5. Show how standard costs affect and simplify process costing
6. Demonstrate how transferred-in costs affect weighted-average process costing
7. Demonstrate how transferred-in costs affect FIFO process costing

Process product-costing systems are used for costing inventory of mass-produced, like units. These units contrast to the custom-made or unique goods costed under job-order product-costing systems. This chapter covers the major *inventory-costing* methods that may be used in process-cost systems. It will be concerned only incidentally with *planning* and *control,* which are discussed in other chapters and are applicable to *all* product-costing systems regardless of whether process costing, job-order costing, or some hybrid system is used.

A basic description of process costing, including the essential concept of equivalent units, is in Chapter 5 (pp. 143–49). ***Please review that material before proceeding.***

FIVE KEY STEPS

Objective 1

Describe five key steps of process costing

Our introduction to process costing in Chapter 5 featured five key steps:

- Step 1: Summarize the flow of physical units
- Step 2: Compute output in terms of equivalent units
- Step 3: Summarize the total costs to account for, which are the total debits in Work in Process (that is, the costs applied to Work in Process)
- Step 4: Compute equivalent unit costs
- Step 5: Apply costs to units completed and to units in the ending work in process

Although shortcuts are sometimes taken, the methodical progression through each of the five steps minimizes errors. The first two steps focus on what is occurring in physical or engineering terms. The dollar impact of the production process is measured in the final three steps.

When beginning inventories are present, these steps are especially useful. Indeed, the main objective of the five-step procedure is to compute the dollar amount of the credit to work in process and the corresponding debit to finished goods (or to work in process in the subsequent department) for work completed during the current period.

We focus mainly on two alternative methods of process costing in this chapter—the weighted-average method and the first-in, first-out method.

Data for Illustration

The easiest way to learn process costing is by example. Let us consider the following scenario.

EXAMPLE 1

A plastics company uses two processes—the first in the Forming Department and the second in the Finishing Department—for making a high-volume children's toy. Direct material is introduced at the *beginning* of the process in Forming, and additional direct material is added at the *end* of the process in Finishing. Conversion costs are applied evenly throughout both processes. As the process in Forming is completed, goods are immediately transferred to Finishing; as goods are completed in Finishing, they are transferred to Finished Goods.

Data for Forming for the month of March 19_1 are:

Work in process,		
Beginning inventory:		10,000 units
Direct materials	$4,000	
Conversion costs (40%)*	1,110	$ 5,110
Units completed and transferred out during March		48,000
Units started during March		40,000
Work in process, ending inventory		2,000 (50%)*
Direct material cost added during March		$22,000
Conversion costs added during March		$18,000

*This means that each unit in process is regarded as being fractionally complete with respect to the conversion costs of the present department only, at the dates of the work in process inventories.

A diagram of how resources are used in our example follows:

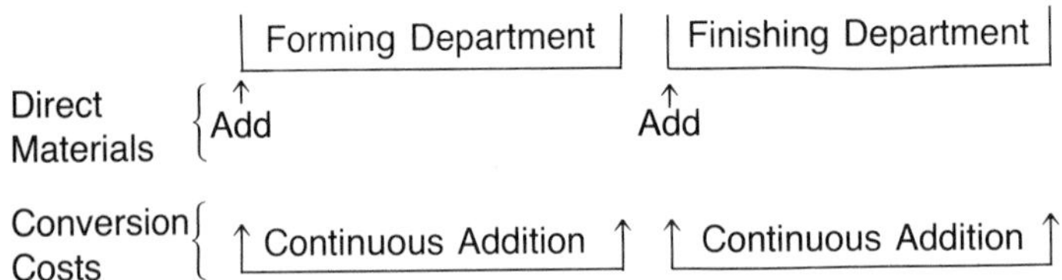

For inventory-costing purposes, the accountant must

- Compute the cost of goods transferred out of a process or a department (in our example, Forming)
- Compute the ending inventory of work in process
- Prepare journal entries summarizing the transactions

Exhibit 17-1 provides an overview of product costing for process-costing activities in an initial process. Note that conversion costs are usually accounted for as indirect costs and that the indirect cost application base is usually expressed as equivalent units of production.

Step 1: Summarize Physical Units

Step 1 traces the physical units of production. Where did units come from? Where did they go? How many units are there to account for? How are they accounted for? Draw flow charts as a preliminary step, if necessary. For example:

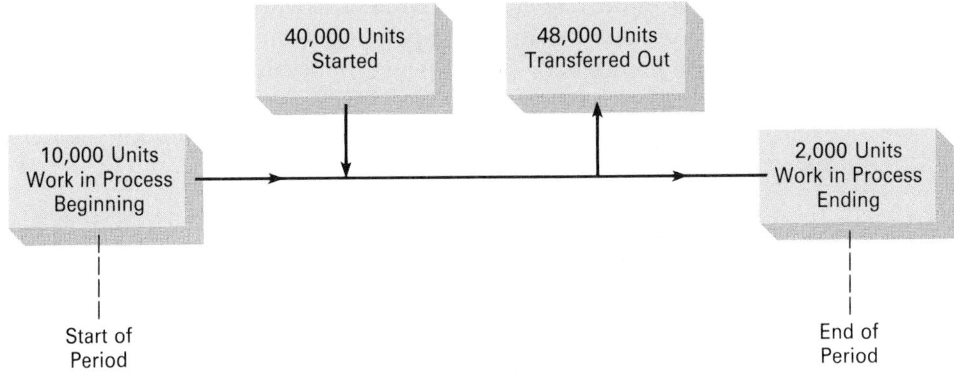

EXHIBIT 17-1

Overview of Product Costing for Process-Costing Activities, Initial Process

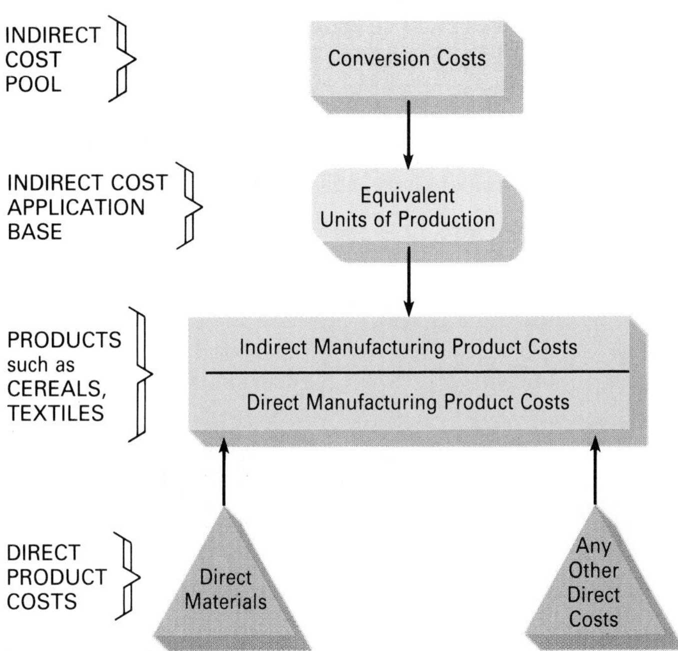

Note: Conversion costs are usually grouped in indirect cost pools. However, if some conversion costs are regarded as direct product costs, they would be accounted for accordingly.

Exhibit 17-2 shows these relationships, which may also be expressed as an equation:

Beginning inventories + Units started = Units transferred out + Ending inventories

The total of the left side of the equation is shown as the units to account for: 10,000 + 40,000 = 50,000 in Exhibit 17-2. The total of the right side is shown as the units accounted for: 10,000 + 38,000 + 2,000 = 50,000 in Exhibit 17-2.

There are 10,000 physical units in process at the start of the period. In addition, 40,000 units were begun during the current period. Of the total of 10,000 +

EXHIBIT 17-2

Step 1: Summarize Physical Units

Work in process, beginning	10,000 (40%)*
Started during current period	40,000
To account for	50,000
Completed and transferred out during current period:†	
From beginning inventory	10,000
Started and completed	38,000
Work in process, ending	2,000 (50%)*
Accounted for	50,000

*Degrees of completion, for conversion costs of this department only, at the dates of the work in process inventories.

†"Current period" is used as a general term. In this example, the current period is one month, March.

40,000 = 50,000 units to account for, 2,000 units remained in process at the end of the period. Therefore, 48,000 units were completed and transferred out during the period, consisting of the 10,000 units from the beginning work in process plus 38,000 units from the 40,000 units started.

All costs incurred are debited to a department account that simultaneously serves as an inventory account, Work in Process—Forming. As goods are completed and transferred out, costs are shifted from Work in Process—Forming to Work in Process—Finishing, usually by monthly entries. For purposes of our example, these accounts will show physical units as well as dollars, which we will enter as we progress. Normally, only dollars are shown in these accounts.

Work in Process—Forming

	Physical Units	Dollars		Physical Units	Dollars
Beginning inventory	10,000	?	Transferred out	48,000	?
Started	40,000	?	Ending inventory	2,000	?
To account for	50,000	?	Accounted for	50,000	?

WEIGHTED-AVERAGE METHOD

Objective 2

Demonstrate the weighted-average method of process costing

Step 1 is the same for both process-costing methods, but Steps 2 through 5 are affected by the choice of method. This section describes the weighted-average method. A subsequent section discusses the first-in, first-out method.

Recall that *equivalent units* measure output in terms of the quantities of each production factor that had been applied. That is, an equivalent unit is a collection of inputs necessary to produce one complete physical unit of output.

Recall also that in process costing, costs are often divided into only two main classifications: direct materials and conversion costs. Direct labor is often not a significant part of total costs, so it is combined with factory overhead costs (such as the costs of energy, repairs, and material handling) in the classification conversion costs.

Step 2: Compute Output in Equivalent Units

Express the physical units in terms of work done. Because direct materials and conversion costs are usually applied to production differently, the equivalent output is divided into direct materials and conversion costs. Instead of thinking of output in terms of physical units, think of output in terms of direct material doses of work and conversion-cost doses of work. ***Disregard dollar amounts until equivalent units are computed.***

In the Forming Department, direct materials are introduced at the beginning of the process. Therefore, both the physical units completed and the physical units in the ending work in process are "fully completed" in terms of equivalent units of work done regarding direct materials. Note especially that the direct material component of work in process is fully completed as soon as work is started. Why? Because in our example all doses of direct material are applied at the initial stage of the process. Specific computations follow:

	Direct Materials
Completed and transferred out, 48,000 × 100%	48,000
Work in process, end, 2,000 × 100%	2,000
Total equivalent units of work done	50,000

EXHIBIT 17-3
Forming Department
Step 2: Computation of Output in Equivalent Units
For the Month Ended March 31, 19_1
Weighted-Average Method

Flow of Production	(Step 1) Physical Units	(Step 2) Equivalent Units Direct Materials	Conversion Costs
Work in process, beginning	10,000 (40%)*	(work done before current period)	
Started during current period	40,000		
To account for	50,000		
Completed and transferred out during current period	48,000	48,000	48,000
Work in process, ending	2,000 (50%)*	2,000	1,000†
Accounted for	50,000		
Work done to date		50,000	49,000†

*Degrees of completion for conversion costs of this department only, at the dates of the work in process inventories.
†2,000 × 50% = 1,000

For conversion costs, the physical units completed and transferred out are fully completed. The physical units in ending work in process are 50% completed (on average); they will be weighted accordingly:

	Conversion Costs
Completed and transferred out, 48,000 × 100%	48,000
Work in process, end, 2,000 × 50%	1,000
Total equivalent units of work done	49,000

Exhibit 17-3 combines Steps 1 and 2. It summarizes the computation of output in terms of equivalent units.

The weighted-average method focuses on the total work done to date regardless of whether that work was done during the preceding period or during the current period. Consequently, equivalent units include the work completed *before* March as well as the work done *during* March. Thus the stage of completion of the March beginning work in process is *not* used in this computation.

Step 3: Summarize Total Costs to Account For

Exhibit 17-4 summarizes the total costs to account for–that is, the total debits in Work in Process. As given in our example data, the debits consist of the beginning balance, $5,110, plus the costs added during March, $22,000 + $18,000 = $40,000.

Step 4: Compute Equivalent Unit Costs

Exhibit 17-5 shows the computation of equivalent unit costs. The weighted-average method has been called a "roll-back" method because the averaging of costs includes the work done in the preceding period(s) on the current period's beginning inventory of work in process. Thus, the total costs and the equivalent units mingle

EXHIBIT 17-4

Step 3: Summarize Total Costs to Account For
(The Total Debits in Work in Process)

Work in Process—Forming					
	Physical Units	Dollars		Physical Units	Dollars
Beginning inventory	10,000	$ 5,110	Transferred out	48,000	?
Started:	40,000		Ending inventory	2,000	?
Direct materials		22,000	Accounted for	50,000	?
Conversion costs		18,000			
To account for	50,000	$45,110			

EXHIBIT 17-5

Step 4: Compute Equivalent Unit Costs
Weighted-Average Method

		Details		
Costs	Totals	Direct Materials	Conversion Costs	Equivalent Whole Unit
Work in process, beginning	$ 5,110	$ 4,000	$ 1,110	
Costs added during current period	40,000	22,000	18,000	
Total costs to account for (Step 3)	$45,110	$26,000	$19,110	
Divide by equivalent units (from Exhibit 17-3)		÷50,000	÷49,000	
Equivalent unit costs		$ .52	$ .39	$.91

the applicable work begun in February with the work done during March. Consequently, the total costs include the cost of the work in process at the beginning of the current period. Note that the equivalent units also include all work done to date, including the work done on beginning work in process before the current period. The division of total costs by equivalent units yields equivalent unit costs.

Step 5: Apply Total Costs

Exhibit 17-6 shows how the equivalent unit costs computed in Step 4 are the basis for applying total costs to units completed and in ending work in process. The 48,000 units completed and transferred out bear a unit cost of \$.52 + \$.39 = \$.91. The 2,000 units in ending work in process bear 2,000 equivalent units of direct materials at \$.52 and 1,000 equivalent units of conversion costs at \$.39. Note how the total costs accounted for can be checked against one another in Steps 3 and 5. The \$45,110 in Exhibit 17-4 agrees with the \$45,110 in Exhibit 17-6.

A **production-cost report** is a report of the units manufactured during a specified period and their costs (Steps 3, 4, and 5). Such a report may be highly summarized or extensively detailed. Supporting schedules are often provided. Exhibit 17-7 is a sample of a production-cost report for the weighted-average method.

EXHIBIT 17-6
Step 5: Apply Costs to Units Completed and in Ending Work in Process
Weighted-Average Method

	Totals	Details: Direct Materials	Details: Conversion Costs
Units completed and transferred out (48,000)	$43,680	48,000 ($.91)	
Work in process, ending (2,000)			
Direct materials	$ 1,040	2,000 ($.52)	
Conversion costs	390		1,000 ($.39)
Total cost of work in process	$ 1,430		
Total costs accounted for	$45,110		

EXHIBIT 17-7
Forming Department
Production-Cost Report
For the Month Ended March 31, 19_1
Weighted-Average Method

	Costs	Totals	Details: Direct Materials	Details: Conversion Costs
	Work in process, beginning	$ 5,110	$ 4,000	$ 1,110
	Costs added during current period	40,000	22,000	18,000
(Step 3)	Total costs to account for	$45,110	$26,000	$19,110
	Divide by equivalent units*		÷50,000	÷49,000
(Step 4)	Equivalent unit costs		$ .52	$ .39
(Step 5)	Application of costs:			
	Completed and transferred out (48,000)	$43,680	48,000 ($.91†)	
	Work in process, ending (2,000):			
	Direct materials	$ 1,040	2,000 ($.52)	
	Conversion costs	390		1,000 ($.39)
	Total work in process	$ 1,430		
	Total costs accounted for	$45,110		

*For work done to date. For more details, see Exhibit 17-3.
†Cost per equivalent whole unit = $.52 + $.39 = $.91

Objective 3

Prepare journal entries for process-costing systems

Journal Entries

Process-costing journal entries are basically like those made in the job-order costing system. That is, direct materials and conversion costs are accounted for as in job-order systems. The main difference is that, in process costing, there is more than one Work-in-Process account.

The data in our weighted-average illustration would be journalized as follows:

EXHIBIT 17-8
Flow of Costs
Process-Costing System

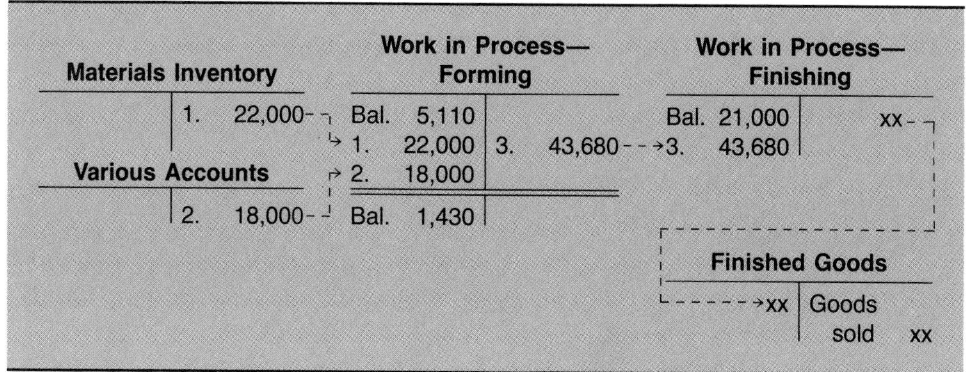

1. Work in process—Forming	22,000	
Materials inventory		22,000
To record requisitions of direct materials for March.		
2. Work in process—Forming	18,000	
Various accounts		18,000
To record conversion costs for March. Examples of various accounts include supplies, all labor, and pertinent depreciation.		
3. Work in process—Finishing	43,680	
Work in process—Forming		43,680
To record cost of goods completed and transferred from Forming to Finishing.		

Exhibit 17-8 shows a general sketch of the flow of costs through the T-accounts. The key T-account, Work in Process—Forming, shows an ending balance of $1,430:

Work in Process—Forming

	Physical Units	Dollars		Physical Units	Dollars
Beginning inventory	10,000	$ 5,110	Transferred out	48,000	$43,680
Started:	40,000		Ending inventory	2,000	1,430
Direct materials		22,000	Accounted for	50,000	$45,110
Conversion costs		18,000			
To account for	50,000	$45,110			
Ending inventory	2,000	$ 1,430			

FIRST-IN, FIRST-OUT METHOD

Objective 4

Demonstrate the first-in, first-out (FIFO) method of process costing

Steps 1 and 2: Physical Units and Equivalent Units

The first-in, first-out (FIFO) method regards the beginning inventory as if it were a batch of goods separate and distinct from the goods started *and* completed in a process during the current period. ***Step 1, the analysis of physical units, is unaffected by this method. However, subsequent steps are affected, as compared to the weighted-average method.***

Step 2, the computation of the output in terms of equivalent units, distinguishes between (a) the goods carried over in beginning work in process and (b) the goods

EXHIBIT 17-9
Forming Department
Step 2: Computation of Output in Equivalent Units
For the Month Ended March 31, 19_1
FIFO Method

Flow of Production	(Step 1) Physical Units	(Step 2) Equivalent Units Direct Materials	Conversion Costs
Work in process, beginning	10,000 (40%)*	(work done before current period)	
Started during current period	40,000		
To account for	50,000		
Completed and transferred out during current period:			
From beginning work in process	10,000	—	6,000†
Started and completed	38,000	38,000	38,000
Work in process, ending	2,000 (50%)‡	2,000	1,000‡
Accounted for	50,000		
Work done in current period only		40,000	45,000

*Degrees of completion for conversion costs of this department only at the dates of the work in process inventories.
†10,000 × (100% − 40%) = 10,000 × 60% = 6,000.
‡2,000 × 50% = 1,000.

started and completed in the current period. What work was done on the beginning inventory during March? None of the direct materials and 60% of the conversion costs (40% of the conversion costs had been done in February). ***The FIFO computations in Exhibit 17-9 confine equivalent units to work done during the current period (March) only.***

Steps 3, 4, and 5: Production-Cost Report

Exhibit 17-10 is the production-cost report for the FIFO method. It presents Steps 3, 4, and 5. Concentrate for now on Steps 3 and 4. The divisor for computing equivalent unit costs is confined to the work done during the current period. Therefore, the costs added during the current period (only) are divided by the equivalent units for work done during the current period (only). Thus, the $5,110 beginning inventory costs and the equivalent units of work done on the beginning inventory in the preceding period are *excluded* from the computation of unit costs.

The bottom half of Exhibit 17-10 shows how the equivalent unit costs computed in Step 4 are the basis for Step 5: applying costs to units completed and in ending work in process:

First, compute the costs of the 48,000 units of goods completed and transferred out:

Work in process, beginning, 10,000 units	$ 5,110
Costs added during the current period to complete the beginning inventory, 6,000 equivalent units of conversion costs × $.40	2,400
Total from beginning inventory	$ 7,510
Started and completed during the current period, 38,000 equivalent whole units × $.95	36,100
Total costs transferred out	$43,610

EXHIBIT 17-10

Forming Department
Production-Cost Report
For the Month Ended March 31, 19_1
FIFO Method

	Costs	Totals	Details: Direct Materials	Details: Conversion Costs
	Work in process, beginning	$ 5,110	(costs of work done before current period)	
	Costs added during current period	40,000	$22,000	$18,000
(Step 3)	Total costs to account for	$45,110		
	Divide by equivalent units*		÷40,000	÷45,000
(Step 4)	Equivalent unit costs		$ .55	$ .40
(Step 5)	Application of costs:			
	Completed and transferred out (48,000):			
	Work in process, beginning (10,000)	$ 5,110		
	Costs added during current period:			
	Direct materials	—		
	Conversion costs	2,400		6,000* ($.40)
	Total from beginning inventory	$ 7,510		
	Started and completed (38,000)	36,100	38,000 ($.95†)	
	Total costs transferred out	$43,610		
	Work in process, ending (2,000):			
	Direct materials	$ 1,100	2,000* ($.55)	
	Conversion costs	400		1,000* ($.40)
	Total work in process	$ 1,500		
	Total costs accounted for	$45,110		

*Equivalent units of work done. (See Exhibit 17-9.)
†Cost per equivalent whole unit = $.55 + $.40 = $.95

Second, compute the cost of the 2,000 units in ending work in process:

Direct materials, 2,000 equivalent units × $.55	$1,100
Conversion costs, 1,000 equivalent units × $.40	400
Total work in process, ending inventory	$1,500

The grand total of costs accounted for is:

Total costs transferred out	$43,610
Total costs of work in process, ending inventory	1,500
Total costs accounted for	$45,110

The average unit cost of goods transferred out is $43,610 ÷ 48,000 units = $.9085. Why does the $.9085 average unit cost of goods transferred out differ from the $.95 unit cost of units started and completed during March? Forming uses FIFO to distinguish between monthly batches of production. Succeeding departments, however, such as Finishing, "cost in" these goods at the *one* average unit cost ($.9085 in this illustration). If the latter averaging were not done, the attempt to trace costs on a pure FIFO basis throughout a series of processes would be too cumbersome.

Only rarely is an application of pure FIFO ever encountered in process costing. It should really be called a *modified* or *departmental* FIFO method. Why? Because FIFO is applied within a department to compile the cost of goods transferred *out*, but the goods transferred *in* during a given period usually bear a single average unit cost as a matter of convenience.

COMPARISON OF FIFO AND WEIGHTED-AVERAGE METHODS

The key difference between FIFO and the weighted-average methods is how equivalent units are computed. Ponder the difference by comparing Exhibit 17-3 (p. 561) with Exhibit 17-9 (p. 565):

- Weighted-average—total work done to date (full weight given to all units completed and transferred out during the current period *plus* partial weights for work done on ending work in process).
- FIFO—work done during current period only (full weight given to all units started and completed currently *plus* partial weights for work done on beginning and ending work in process).

In turn, differences in equivalent units generate differences in equivalent unit costs. Accordingly, there are differences in costs applied to goods completed and those still in process. In our example:

	Weighted Average*	FIFO†
Cost of goods transferred out	$43,680	$43,610
Ending work in process	1,430	1,500
Total costs accounted for	$45,110	$45,110

*From Exhibit 17-7, p. 563.
†From Exhibit 17-10, p. 566.

In this example, the FIFO ending inventory is higher than the weighted-average ending inventory by only $70, or 4.9% ($70 ÷ $1,430 = 4.9%). The difference is attributable to variations in equivalent unit costs of direct materials and conversion costs in different months. The unit cost of the work done in March only was $.55 + $.40 = $.95, as shown in Exhibit 17-10. In contrast, Exhibit 17-7 shows the weighted-average unit cost of $.52 + $.39 = $.91. Therefore, the FIFO method results in a larger cost of work in process inventory, March 31, and a smaller March cost of goods transferred out.

The difference in unit costs from month to month is ordinarily even less significant than the small difference illustrated here. Fluctuations in unit costs are usually caused by volatile direct materials prices, not conversion costs. The latter tend to be relatively stable.

In process-costing industries, the physical inventory levels of work in process tend not to change much from month to month. They also tend to be relatively small in relation to the total number of units completed and transferred out. These conditions also help to explain why unit costs do not differ much between the FIFO and weighted-average methods.

The FIFO measurements of work done during the current period only are essential for judging current-period performance—March in this illustration. A major advantage of FIFO is that managers can judge the performance in March independently from the performance in February. In brief, the "work done during the current period" is vital information for planning and control purposes as well as for FIFO inventory valuation purposes. Standard costing, which is described next in this chapter, uses "work done during the current period" as a basis for comparing actual costs of the current period with budgeted or standard costs of the current period.

In practice, standard costing is especially popular among companies that do not have frequent changes in their basic products. For example, manufacturers of coffee, soap, candy bars, cereals, steel, and glass use standard costing for month-to-month planning, control, and inventory costing. However, FIFO and weighted-average process costing are also widely used, particularly where short product lives reduce the attractions of installing a standard-cost system.[1]

STANDARD COSTS AND PROCESS COSTS

Objective 5

Show how standard costs affect and simplify process costing

This section assumes that you have already studied Chapters 7, 8, and 9. If you have not studied these chapters, proceed to the next major section, "Transfers in Process Costing," page 572.

Previous chapters demonstrated that the use of standard costing is completely general; that is, it can be used in job-costing systems or process-costing systems, and with absorption costing or variable costing. However, standard-cost procedures tend to be most effective when they are adapted to process costing. Mass, continuous, and repetitive production conditions lend themselves rather easily to the setting of appropriate physical standards. Price tags may then be applied to the physical standards to develop standard costs.

Because we have already seen how standard costing aids planning and control, we will concentrate here on its inventory aspects. The intricacies and conflicts between weighted-average and FIFO historical costing methods are eliminated by using standard costs. Further, weighted-average and FIFO methods become very complicated when used in industries that produce a variety of products. Many observers have stressed that standard costing is especially useful where there are various combinations of materials, operations, and product sizes. For example, a steel-rolling mill uses various steel alloys and produces sheets of various sizes and of various finishes. The items of direct materials are not numerous; neither are the operations performed. But used in various combinations, they yield too great a variety of products to justify the broad averaging procedure of historical process costing. Similarly complex conditions are frequently found—for example, in plants manufacturing rubber products, textiles, ceramics, paints, and packaged food products.

Computations under Standard Costing

The facts in Example 2 are basically the same as those we used for the Forming Department in Example 1 except that the following standard costs have been developed for the process.

[1]M. Ghosal, "Include LIFO Process Costing in Cost Accounting Curriculum," Proceedings of American Accounting Association, Southeast Region, April, 1990, reported survey results showing LIFO as the most popular process costing method. The application of LIFO is usually not done on a physical-flow, month-to-month basis. Instead, companies use the process-costing procedures described in this chapter and then subject them to LIFO adjustments at year-end. Incidentally, a major difficulty with mail surveys concerning process costing is the loose use of the term in practice.

EXAMPLE 2

Direct materials, introduced at start of process		$.53 per unit
Conversion costs, incurred evenly throughout process		.37 per unit
Standard cost per unit		$.90 per unit
Work in process, beginning, 10,000 units, 40% completed (direct materials, $5,300; conversion costs, $1,480)	$ 6,780	
Units completed during March	48,000	
Units started during March	40,000	
Work in process, ending	2,000, 50% completed	

For inventory-costing purposes, the accountant must

- Compute the standard cost of goods completed and transferred out of a process or a department (in our example, Forming).
- Compute the ending work in process.
- Compute the total direct-material variance and the total conversion-cost variance. Actual direct-material costs were $22,000. Conversion costs were $18,000.
- Prepare journal entries summarizing the transactions.

A standard-cost system greatly simplifies process-cost computations. As indicated earlier, there is a similarity between the FIFO and standard-cost methods in one important respect. Both use the "work done during current period only"—as shown in Exhibit 17-9, page 565, and repeated in Exhibit 17-11, page 570—as the basis for computing equivalent units.

Steps 1 and 2 are the same for FIFO and standard costing. Exhibit 17-11 shows work done during the current period: direct materials, 40,000 equivalent units, and conversion costs, 45,000 equivalent units.

Steps 3, 4, and 5 are easier for standard costing than for the FIFO method. Why? Because the cost per equivalent unit does not have to be computed, as was done for the FIFO method. Instead the cost per equivalent unit is the standard cost. The latter is the key to computing the total costs to account for, the costs completed and transferred out, and the ending work in process inventory. Exhibit 17-11 illustrates a production-cost report. Exhibit 17-12 summarizes all these calculations.

Accounting for Variances

Standard-cost systems usually accumulate actual costs separately from the inventory accounts. The following is an example. The actual data are recorded in the first two entries, and the variances are recorded in the next two entries. The final entry transfers the completed goods out at standard costs. Exhibit 17-13 shows how the costs flow through the accounts.

1. Forming Department cost control (at actual)	22,000	
Materials inventory		22,000
To record requisitions of direct materials for March. This "cost control" account is debited with actual costs and credited later with standard costs applied to the units worked on.		
2. Forming Department cost control (at actual)	18,000	
Various accounts		18,000
To record conversion costs for March.		
3. Work in process—Forming (at standard)	21,200	
Direct material variances	800	
Forming department cost control		22,000
To record inventory costs and direct material variances.		

EXHIBIT 17-11

Forming Department
Production-Cost Report (at Standard)
For the Month Ended March 31, 19_1
Standard Costs in a Process-Cost System

Flow of Production	Physical Units	Equivalent Units: Direct Materials	Equivalent Units: Conversion Costs
Work in process, beginning	10,000 (40%)*		
Started during current period	40,000		
To account for	50,000		
Units completed and transferred out during current period:			
From beginning inventory	10,000 (60%)*	—	6,000
Started and completed	38,000	38,000	38,000
Work in process, ending	2,000 (50%)*	2,000	1,000
Accounted for	50,000		
Work done in current period only		40,000	45,000

Costs	Totals	Details: Direct Materials	Details: Conversion Costs
Standard cost per equivalent unit (given)		$.53	$.37
Multiply by work done in current period only		×40,000	×45,000
Costs applied at standard prices during current period	$37,850	$21,200	$16,650
Work in process, beginning	6,780	5,300	1,480
Costs to account for	$44,630	$26,500	$18,130
Application of costs:			
Completed and transferred out (48,000)	$43,200	48,000($.90‡)	
Work in process, ending (2,000):			
Direct materials	$ 1,060	2,000 ($.53)	
Conversion costs	370		1,000 ($.37)
Total work in process	$ 1,430		
Total costs accounted for	$44,630		
Summary of variances for current performance:			
Current output in equivalent units		40,000	45,000
Current output at standard costs applied		$21,200	$16,650
Actual costs incurred		$22,000	$18,000
Total unfavorable variance†		$ 800 U	$ 1,350 U

*Degree of completion on conversion costs of present department only, at the dates of the work in process inventories.
†These could be broken down further, depending upon available details.
‡Cost per equivalent whole unit = $.53 + $.37 = $.90

EXHIBIT 17-12
Forming Department
Standard Costs in Process Costing
For the Month Ended March 31, 19_1

Work in Process—Forming (at standard)

	Physical Units	Dollars		Physical Units	Dollars
Beginning inventory	10,000	$ 6,780*	Transferred out	48,000	$43,200§
Started:	40,000		Ending inventory	2,000	1,430‖
Direct materials		21,200†	Accounted for	50,000	$44,630
Conversion costs		16,650‡			
To account for	50,000	$44,630			
Ending inventory	2,000	$ 1,430			

*10,000 × $.53 = $5,300
40% of 10,000 × $.37 = 1,480
$6,780

†40,000 × $.53 = $21,200

‡45,000 × $.37 = $16,650

§48,000 × ($.53 + $.37) = $43,200

‖2,000 × $.53 = $1,060
50% of 2,000 × $.37 = 370
$1,430

EXHIBIT 17-13
Flow of Costs
Standard Process-Costing System

Forming Department Cost Control

Debit	Credit
1. 22,000	3. 22,000 → (WIP—Forming 3.)
2. 18,000	4. 18,000 → (WIP—Forming 4.)

Work in Process—Forming

Debit	Credit
Bal. 6,780	5. 43,200 → (WIP—Finishing 5.)
3. 21,200	
4. 16,650	
44,630	
Bal. 1,430	

Work in Process—Finishing

Debit	Credit
5. 43,200	

Materials Inventory

Debit	Credit
	1. 22,000

Direct Material Variances

Debit	Credit
3. 800	

Various Accounts

Debit	Credit
	2. 18,000

Conversion-Cost Variances

Debit	Credit
4. 1,350	

	Debit	Credit
4. Work in process—Forming (at standard)	16,650	
Conversion-cost variances	1,350	
Forming Department cost control		18,000
To record inventory costs and conversion-cost variances.		
5. Work in process—Finishing	43,200	
Work in process—Forming		43,200
To record cost of goods completed and transferred at standard cost from Forming to Finishing.		

Variances can be measured and analyzed in little or great depth in the same manner as that described in Chapters 7, 8, and 9.

A standard-cost system not only eliminates the intricacies of weighted-average and FIFO inventory methods but also erases the need for burdensome computations of costs per equivalent unit. The standard cost *is* the cost per equivalent unit. In addition, a standard-cost approach helps control operations, as explained in Chapters 7 and 8.

Note how the equivalent units for the current period provide the key measures of the work accomplished during March. The analysis of the physical units and equivalent units in Exhibit 17-11 is vital for evaluating current-period performance. The equivalent units are used to value inventories and to measure variances.

TRANSFERS IN PROCESS COSTING

Many process-costing systems have two or more departments or processes in the production cycle. Ordinarily, as goods move from department to department, related costs are also transferred by monthly journal entries. If standard costs are used, the accounting for such transfers is relatively simple. However, if FIFO or weighted-average is used, the accounting can become more complex.

To make the ideas concrete, we now extend Example 1 to encompass the Finishing Department.

EXAMPLE 1 EXTENDED TO ENCOMPASS FINISHING DEPARTMENT

Recall that the Forming Department transfers its units to the Finishing Department. Here the units receive additional direct materials at the *end* of the process. Conversion costs are applied evenly throughout the Finishing Department's process. As the process in Forming is completed, goods are immediately transferred to Finishing; as goods are completed in Finishing, they are transferred to Finished Goods.

Data for Finishing for the month of March 19_1 are:

Work in process,		
Beginning inventory:		12,000 units
Transferred-in costs	$10,920	
Conversion costs (66⅔%)*	10,080	$21,000
Units completed during March		44,000
Units started during March		?
Work in process, ending inventory (37½%)*		16,000
Direct material cost added during March		$13,200
Conversion costs added during March		$63,000
Costs transferred in during March:		
Weighted-average method		$43,680
FIFO method		$43,610

*This means that each unit in process is regarded as being fractionally complete with respect to the conversion costs of the present department only, at the dates of the work in process inventories.

For inventory-costing purposes, the accountant must

- Compute the cost of goods completed and transferred out of a process or department (in our example, Finishing)
- Compute the ending inventory cost for the goods remaining in the department, using the weighted-average method or first-in, first-out (FIFO) method
- Prepare journal entries for transfers to finished goods

Objective 6

Demonstrate how transferred-in costs affect weighted-average process costing

Transfers and Weighted-Average Method

The five-step procedure described earlier also pertains when accounting for transfers into a department. Exhibit 17-14 shows for Finishing the initial two steps, which analyze physical units and compute equivalent units. Note that direct mate-

EXHIBIT 17-14
Finishing Department
Steps 1 and 2: Computation of Output in Physical Units and Equivalent Units
For the Month Ended March 31, 19_1
Weighted-Average Method

Flow of Production	(Step 1) Physical Units	(Step 2) Equivalent Units Transferred-in Costs	Direct Materials	Conversion Costs
Work in process, beginning	12,000 (66⅔%)*	(work done before current period)		
Transferred in during current period	48,000			
To account for	60,000			
Completed and transferred out during current period	44,000	44,000	44,000	44,000
Work in process, ending	16,000 (37½%)*	16,000	—	6,000†
Accounted for	60,000			
Work done to date		60,000	44,000	50,000

*Degree of completion for conversion costs at the dates of inventories.
†16,000 × 37.5% = 6,000.

rial costs have no degree of completion regarding the ending work in process. Why? Because in Finishing, direct materials are introduced at the *end* of the process. However, the equivalent units for transferred-in costs are, of course, fully completed because they are always introduced at the beginning of the process.

Treatment of transferred-in costs may become complex. **Transferred-in costs** (or **previous department costs**) are costs incurred in a previous department that have been conveyed to a subsequent department. As the units move from the first department to the second department, their costs move with them. Thus, Finishing computations must include transferred-in costs, as well as any additional direct material costs and conversion costs added in Finishing.

The essential difference between applying the weighted-average method to the Forming Department (Exhibit 17-3, p. 561) and applying it to the Finishing Department is the accounting for transferred-in costs in the Finishing Department. In our example, Steps 1 and 2 are unchanged, except for the addition of transferred-in units. Exhibit 17-14 displays the computations of output in equivalent units.

Exhibit 17-15 is a production-cost report. It shows Steps 3, 4, and 5. It presents the total costs to account for. Equivalent unit costs are computed and then applied to the units completed and in ending work in process.

Examine Exhibit 17-15 closely, and note the following points: (1) Under the weighted-average method, unlike under the FIFO method, the costs of beginning work in process must be subdivided into their components. (2) These components are combined with the costs added currently to obtain a total amount incurred to date for each cost element: transferred-in costs, direct materials, and conversion costs. (3) Thus, the unit costs of each cost element will be weighted averages.

Note again that the total costs to account for in each of the three categories are divided by the equivalent units for the "work done to date." For instance, the divisor for conversion costs is 50,000 units. Exhibit 17-16 is a T-account portrayal of the total costs to account for (the total debits in work in process).

Production-cost reports may be presented in a briefer form than that shown in Exhibit 17-15. For example, the data in Exhibit 17-15 could be the supporting computations for a summary production-cost report. The latter would really be a formal

EXHIBIT 17-15

Finishing Department
Production-Cost Report
For the Month Ended March 31, 19_1
Weighted-average Method

	Costs	Totals	Transferred-in Costs	Direct Materials	Conversion Costs
			Details		
	Work in process, beginning	$ 21,000	$10,920	$ —	$10,080
	Costs added during current period	119,880	43,680	13,200	63,000
(Step 3)	Total costs to account for	$140,880	$54,600	$13,200	$73,080
	Divide by equivalent units (from Step 2, Exhibit 17-15)		÷60,000	÷44,000	÷50,000
(Step 4)	Equivalent unit costs		$ .91	$ .30	$1.4616
(Step 5)	Application of costs:				
	Completed and transferred out (44,000)	$117,550		44,000 ($2.6716*)	
	Work in process, ending (16,000):				
	Transferred-in costs	$ 14,560	16,000† ($.91)		
	Direct materials	—		—	
	Conversion costs	8,770			6,000† ($1.4616)
	Total work in process	$ 23,330			
	Total costs accounted for	$140,880			

*Cost per equivalent whole unit = $.91 + $.30 + $1.4616 = $2.6716
†Equivalent units of work done. See Exhibit 17-14 for details.

EXHIBIT 17-16

Finishing Department
Summary of Costs Accounted For
For the Month Ended March 31, 19_1
Weighted-Average Method

Work in Process—Finishing					
	Physical Units	Dollars		Physical Units	Dollars
Beginning inventory	21,000	$ 21,000	Transferred out	44,000	$117,550
Transferred in	48,000		Ending inventory	16,000	23,330
Transferred-in costs		43,680	Accounted for	60,000	$140,880
Direct materials		13,200			
Conversion costs		63,000			
To account for	60,000	$140,880			
Ending inventory	16,000	$ 23,330			

presentation of the effects on the Work in Process—Finishing account (see Exhibit 17-17).

The journal entry for the transfer out to finished goods inventory would be:

Finished goods inventory	117,550	
Work in process—Finishing		117,550
To transfer units to finished goods.		

Sometimes a problem requires that the Work in Process account be split into Work in Process—Direct Materials and Work in Process—Conversion Costs. In

EXHIBIT 17-17
Finishing Department
Summary Production-Cost Report
For the Month Ended March 31, 19_1
Weighted-Average Method

Flow of Production	Physical Units	Total Costs
Work in process, beginning	12,000	$ 21,000
Transferred in during current period	48,000	119,880
To account for	60,000	$140,880
Completed and transferred out	44,000	$117,550
Work in process, ending	16,000	23,330
Accounted for	60,000	$140,880

these cases, the journal entries would contain this greater detail, even though the underlying reasoning and techniques would be unaffected.

Transfers and FIFO Method

Exhibit 17-18 shows the initial two steps for the FIFO method. The method of computation of equivalent units is basically the same as for FIFO in Forming. However, transferred-in costs must now be considered.

Objective 7

Demonstrate how transferred-in costs affect FIFO process costing

Exhibit 17-19, the production-cost report, shows Steps 3, 4, and 5. Note again how the divisor equivalent units for FIFO differ from the divisor equivalent units for the weighted-average method. FIFO uses equivalent units for work done in the current period only.

EXHIBIT 17-18
Finishing Department
Steps 1 and 2: Computation of Output in Physical Units and Equivalent Units
For the Month Ended March 31, 19_1
FIFO Method

		(Step 2) Equivalent Units		
Flow of Production	(Step 1) Physical Units	Transferred-in Costs	Direct Materials	Conversion Costs
Work in process, beginning	12,000 (66⅔%)*	(work done before current period)		
Transferred in during current period	48,000			
To account for	60,000			
Completed and transferred out during current period:				
From beginning work in process	12,000	—	12,000	4,000†
Started and completed	32,000	32,000	32,000	32,000
Work in process, ending	16,000 (37½%)*	16,000	—	6,000‡
Accounted for	60,000			
Work done in current period only		48,000	44,000	42,000

*Degree of completion for conversion costs at the dates of inventories.
†12,000 × 33⅓% = 4,000
‡16,000 × 37.5% = 6,000

EXHIBIT 17-19

Finishing Department
Production Cost Report
For the Month Ended March 31, 19_1
FIFO Method

	Costs	Totals	Details: Transferred-in Costs	Details: Direct Materials	Details: Conversion Costs
	Work in process, beginning	$ 21,000	(costs of work done before current period)		
	Costs added during current period	119,810	$43,610	$13,200	$63,000
(Step 3)	Total costs to account for	$140,810			
	Divide by equivalent units		÷48,000	÷44,000	÷42,000
(Step 4)	Equivalent unit costs		$.90854*	$ .30	$ 1.50
(Step 5)	Application of costs:				
	Completed and transferred out (44,000):				
	Work in process, beginning (12,000)	$ 21,000			
	Costs added during current period:				
	Direct materials	3,600		12,000† ($.30)	
	Conversion costs	6,000			4,000† ($1.50)
	Total from beginning inventory	$ 30,600			
	Started and completed (32,000)	86,673		(32,000 × $2.70854‡)	
	Total costs transferred out	$117,273			
	Work in process, ending (16,000):				
	Transferred-in costs	$ 14,537	16,000† ($.90854)		
	Direct materials	—			
	Conversion costs	9,000			6,000† ($1.50)
	Total work in process	$ 23,537			
	Total costs accounted for	$140,810			

*The unit costs are carried to several decimal places in these exhibits. Of course, they could be rounded with no harm done. However, small discrepancies in totals caused by rounding will occur.
†Equivalent units of work done. See Exhibit 17-18 for details.
‡Cost per equivalent whole unit = $.90854 + $.30 + $1.50 = $2.70854

Exhibit 17-20 summarizes the entries to Work in Process—Finishing. As shown in Exhibit 17-19, the cost of goods completed and transferred out would be $117,273, and the ending inventory would be $23,537.

Remember that in a series of interdepartmental transfers, each department is regarded as a distinct accounting entity. All costs transferred in during a given period are carried at one unit cost, regardless of whether previous departments used weighted-average or FIFO methods.

Common Mistakes

Avoid some common pitfalls when accounting for transferred costs:

1. Remember to include transferred-in costs from previous departments in your calculations. Such costs should be treated as if they were another kind of direct material cost added at the beginning of the process. In other words, when successive departments are involved, transferred goods from one department become all or a part of the direct materials of the next department, although they are called *transferred-in costs*, not direct materials.

EXHIBIT 17-20

Summary of Total Costs to Account For
(The Total Entries to Work in Process)
FIFO Method

Work in Process—Finishing					
	Physical Units	Dollars		Physical Units	Dollars
Beginning inventory	12,000	$ 21,000	Transferred out	44,000	$117,273
Transferred in:	48,000		Ending inventory	16,000	23,537
Transferred-in costs		43,610	Accounted for	60,000	$140,810
Direct materials		13,200			
Conversion costs		63,000			
To account for	60,000	$140,810			
Ending inventory	16,000	$ 23,537			

2. In calculating costs to be transferred on a first-in, first-out basis, do not overlook the costs applied at the beginning of the period to goods that were in process but are now included in the goods transferred. For example, do not overlook the $21,000 in Exhibit 17-20.
3. Unit costs may fluctuate between periods. Therefore, transferred goods may contain batches accumulated at different unit costs (see point 2). These goods, when transferred to the next department, are typically valued by that next department at *one* average unit cost.
4. Units may be expressed in terms of different units of measure in different departments. Consider each department separately. Unit costs could be based on kilograms in the first department and liters in the second. As goods are received by the second department, they must be converted to the liter unit of measure.

ADDITIONAL ASPECTS OF PROCESS COSTING

Estimating Degree of Completion

This chapter's illustrations and almost all process-cost problems blithely mention various degrees of completion for inventories in process. The accuracy of these estimates depends on the care and skill of the estimator and the nature of the process. Estimating the degree of completion is usually easier for direct materials than for conversion costs. The conversion sequence usually consists of a number of standard operations or a standard number of hours, days, weeks, or months for mixing, heating, cooling, aging, curing, and so forth. Thus, the degree of completion for conversion costs depends on what proportion of the total effort needed to complete one unit or one batch has been devoted to units still in process. In industries where no exact estimate is possible, or, as in textiles, where vast quantities in process prohibit costly physical estimates, all work in process in every department is assumed to be complete to some reasonable degree (for example, ⅓ or ½ or ⅔ complete). In other cases, continuous processing entails little change of work in process levels from month to month. In such cases, work in process is safely ignored and monthly production costs are applied solely to goods completed.

We again see the cost-benefit approach. A company may use a simplified system because management decisions will be unaffected by a more elaborate—and more expensive—cost accounting system. Surveys of practice have shown that work in process inventories, as distinguished from finished goods inventories, are typically ignored in process-costing systems.

Businesses use standard costing with process costing far more than they use actual costing (weighted-average or FIFO). The FIFO method is used the least. Process-costing questions appear in professional examinations with some regularity.

Overhead and Budgeted Rates

Managers tend to lump together direct labor and factory overhead applied as conversion costs for process-costing purposes. In many process-cost industries, continuous, steady production results in little fluctuation of total factory overhead from month to month. In such cases, there is no need to use budgeted overhead rates. Of course, when overhead costs and production vary from period to period, budgeted overhead rates are used to compute representative unit costs.

Overhead and Cost Flow

In general, factory overhead is applied using budgeted rates in the same manner as that introduced in Chapter 4. The assumption that all conversion costs are incurred evenly in proportion to the degree of product completion is difficult to justify on theoretical grounds. For example, this assumption implies that a wide variety of factory overhead cost incurrence is driven by direct labor cost. Although such a direct cause-and-effect relationship may not exist, refinements of overhead application beyond this assumption have usually been deemed too costly. Any attempt at more precision generally focuses on developing a budgeted overhead rate to be added to direct material costs to cover indirect costs related to materials, such as purchasing, receiving, storing, issuing, and transferring. In such cases, one overhead rate would be applied along with direct material costs and a separate overhead rate would be applied along with direct labor costs.

Developments in activity-based accounting inevitably refine how conversion costs are applied to work in process. These refinements tend to divide departments and processes into a sequence of activity areas. Chapter 5 provides an illustration (pp. 152-56).

In recent years, companies have revised their manufacturing activities in many ways. Some companies have switched to less detailed process costing. For example, the number of cost centers has shrunk. Consider Omark Industries in Oregon:[2]

	Effect of Continuous Production on Number of Cost Centers	
Product Line	**Before**	**After**
Chain saws	18	4
Sprockets	5	3
Bars	4	1

[2]G. Foster and C. Horngren, "Cost Accounting and Cost Management in a JIT Environment," *Journal of Cost Management for the Manufacturing Industry,* Vol. 1, No. 4 (Winter 1988), p. 10.

PROBLEM FOR SELF-STUDY

PROBLEM

The computation of equivalent units is a key step in process costing. Consider the following June data for a thermoforming process of a plastics company:

Work in process, beginning, 50,000 units, 80% completed for conversion costs	
Units transferred in during current period,	200,000
Units completed and transferred out during current period,	210,000
Work in process, ending, 40% completed for conversion costs	

Additional direct materials are added at the end of the process. Compute the equivalent units for the current period using (1) the weighted-average method and (2) the FIFO method (or the standard-costing method, which would be the same as FIFO).

SOLUTION

1. Weighted-average

Flow of Production	(Step 1) Physical Units	(Step 2) Equivalent Units: Transferred-in Costs	Direct Materials	Conversion Costs
Work in process, beginning	50,000 (80%)*	(Work done before current period)		
Transferred in during current period	200,000			
To account for	250,000			
Completed and transferred out during current period	210,000	210,000	210,000	210,000
Work in process, ending	40,000 (40%)*	40,000	—	16,000†
Accounted for	250,000			
Work done to date		250,000	210,000	226,000

*Degrees of completion for conversion costs at the dates of inventories.
†40,000 × 40% = 16,000.

2. FIFO (or standard costing)

Flow of Production	(Step 1) Physical Units	(Step 2) Equivalent Units: Transferred-in Costs	Direct Materials	Conversion Costs
Work in process, beginning	50,000 (80%)*	(Work done before current period)		
Transferred in during current period	200,000			
To account for	250,000			
Completed and transferred out during current period:				
From beginning work in process	50,000	—	50,000	10,000†
Started and completed during current period	160,000	160,000	160,000	160,000
Work in process, ending	40,000 (40%)*	40,000	—	16,000‡
Accounted for	250,000			
Work done in current period only		200,000	210,000	186,000

*Degrees of completion for conversion costs at the dates of inventories.
†50,000 × 20% = 10,000.
‡40,000 × 40% = 16,000.

SUMMARY

Process product-costing systems are used for costing inventory of mass-produced, like units. The key concept in process costing is that of equivalent units, the expressing of output in terms of the quantities of each of the factors of production that have been applied.

Following five basic steps can help in solving process-cost problems. Process costing is complicated by varying amounts of cost factors, by the presence of beginning inventories, and by the presence of costs transferred in from previous departments.

Two process-costing inventory methods are *weighted-average* and *first-in, first-out*. However, standard costs are widely used, particularly by companies that do not have frequent changes in their basic products. Standard costs are simpler and more practical than other techniques for both product costing and control purposes.

TERMS TO LEARN

This chapter and the Glossary at the end of the book contain definitions of the following important terms:

previous department costs *(p. 573)*
production-cost report *(562)*
transferred-in costs *(573)*

ASSIGNMENT MATERIAL

QUESTIONS

17-1 State the five key steps in process costing.

17-2 What feature of the first two steps of the five-step approach to process costing distinguishes them from the final three steps?

17-3 Name the three inventory methods commonly associated with process costing.

17-4 Name three tasks the accountant must perform for each department regarding inventory costing in process costing.

17-5 Use an equation to express the flow of physical units for a process-costing department.

17-6 "Step 1 of the five key steps is the same for all inventory methods, but Steps 2 through 5 are affected by the choice of method." Do you agree? Explain.

17-7 "In process costing, costs are often divided into only two main classifications." What are they?

17-8 Explain why the stage of completion of the beginning inventory is not used when computing equivalent units under the weighted-average method.

17-9 Why has the weighted-average method been called a roll-back method?

17-10 "A production-cost report has the same format regardless of the process-costing company." Do you agree? Explain.

17-11 Identify the main difference between journal entries in process costing and the ones in job costing.

17-12 Describe the distinctive characteristic of FIFO computations of equivalent units.

17-13 Why is an application of pure FIFO rarely encountered in a series of processes?

17-14 Why should the FIFO method be called a *modified* or *departmental* FIFO method?

17-15 Identify a major advantage of the FIFO method for purposes of planning and control.

17-16 "Standard-cost procedures are particularly applicable to process-costing situations." Do you agree? Why?

17-17 What are some virtues of standard costs as used in process costing?

17-18 Why should the accountant distinguish between *transferred-in costs* and *additional direct material costs* for a particular department?

17-19 "Previous department costs are those incurred in the preceding fiscal period." Do you agree? Explain.

17-20 "There is no need for using budgeted overhead rates for product costing in process-cost industries." Do you agree? Why?

EXERCISES AND PROBLEMS

17-21 Weighted-average equivalent units. Consider the following:

Flow of Production	Physical Units
Work in process, beginning	20,000*
Started in current period	70,000
To account for	90,000
Completed and transferred out during current period	?
Work in process, ending	5,000†
Accounted for	90,000

*Degree of completion: direct materials, 60%; conversion costs, 30%.
†Degree of completion: direct materials, 80%; conversion costs, 60%.

Required

Prepare a schedule of equivalent units for direct materials and conversion costs under the weighted-average method. Show physical units in the first column.

17-22 FIFO equivalent units. The Gagliano Company had computed a portion of the physical units for Department A, for the month of April, as follows:

Units completed:	
From work in process on April 1	10,000
From April production	30,000
	40,000

Direct materials are added at the beginning of the process. Units of work in process at April 30 were 8,000. The work in process at April 1 was 80% complete as to conversion costs, and the work in process at April 30 was 60% complete as to conversion costs. What are the equivalent units of production for the month of April using the FIFO method? Choose one of the following answers and show supporting computations:

	Direct Materials	Conversion Costs
a.	38,000	38,000
b.	48,000	44,800
c.	38,000	36,800
d.	48,000	48,000

17-23 Computing FIFO equivalent units. Insert the missing amounts:

Flow of Production	Physical Units	Equivalent Units: Direct Materials	Equivalent Units: Conversion Costs
Units to account for:			
Work in process, March 31	12,000		
Started during April	?		
To account for	67,000		
Units accounted for:			
Completed and transferred out during current period:			
From beginning inventory	12,000	?*	?*
Started and completed	47,000	?	?
Work in process, April 30	?	?†	?†
Accounted for	67,000		
Work done in current period only		?	?

Costs incurred during April:
*Direct materials: 40%
Conversion costs: 50%

†Direct materials: 30%
Conversion costs: 20%

17-24 Weighted-average and FIFO equivalent units. Consider the following data for May:

	Physical Units
Started in May	50,000
Completed in May	46,000
Ending work in process	12,000
Beginning work in process	8,000

The beginning work in process was 90% complete regarding direct materials and 40% complete regarding conversion costs. The ending work in process was 60% complete regarding direct materials and 30% complete regarding conversion costs.

Required
Prepare schedules of equivalent units for (1) the work done to date (for the weighted-average method) and (2) the work done during May only (for the FIFO method). See Exhibit 17-3, p. 561, and 17-9, p. 565.

17-25 Weighted-average and FIFO equivalent units. This problem is more difficult than Problem 17-24. Consider these September data for physical units in the thermoforming process of a plastics company: beginning work in process, 15,000; transferred in from the Extruding Department during September, 9,000; ending work in process, 5,000. Direct materials are added at the 80% stage of completion of the process in the Thermoforming Department. Conversion costs are incurred evenly throughout the process. The beginning inventory was 60% completed as to conversion costs; the ending inventory was 20% completed as to conversion costs.

Required
Prepare schedules of equivalent units for (1) the work done to date (for the weighted-average method) and (2) the work done during September only (for the FIFO method). Include computations of equivalent units for transferred-in costs as well as for conversion costs and direct materials. See Exhibits 17-14, p. 573, and 17-18, p. 575.

17-26 FIFO equivalent units. Selected production data of the Heating Department of Montez Enterprises follow for February, 19_4:

Flow of Production	Physical Units
Units to account for:	
Work in process, January 31	10,000
Transferred in during February	80,000
Total units to account for	90,000
Units accounted for:	
Completed and transferred out during February	
From beginning inventory	10,000*
Started and completed during February	65,000
Work in process, February 28	15,000†
Total units accounted for	90,000

Costs incurred during February:
*Direct materials: 20% †Direct materials: 50%
Conversion costs: 30% Conversion costs: 40%

Required
Compute equivalent units for goods transferred in, direct materials, and conversion costs. Assume the FIFO method.

17-27 Weighted-average unit cost. (CPA) Barnett Company adds direct materials at the beginning of the process in Department M. Conversion costs were 75% complete as to the 8,000 units in work in process at May 1, 19_3, and 50% complete as to the 6,000 units in work in process at May 31. During May, 12,000 units were completed and transferred out to the next department. An analysis of the costs relating to work in process at May 1 and to production activity for May follows:

	Costs	
	Direct Materials	Conversion
Work in process, 5/1	$ 9,600	$ 4,800
Costs added in May	15,600	14,400

Using the weighted-average method, the total cost per equivalent unit for May was (choose one): (a) $2.47, (b) $2.68, (c) $2.50, (d) $3.16.

17-28 Multiple choice, transfers in, weighted-average. (CPA) On April 1, 19_7, the Collins Company had 6,000 units of work in process in Department B, the second and last stage of their production cycle. The costs attached to these 6,000 units were $12,000 of costs transferred in from Department A, $2,500 of (direct) material costs added in Department B, and $2,000 of conversion costs added in Department B. Materials are added in the beginning of the process in Department B. Conversion was 50% complete on April 1, 19_7. During April, 14,000 units were transferred in from Department A at a cost of $27,000. Material costs of $3,500 and conversion costs of $3,000 were added in Department B. On April 30, 19_7, Department B had 5,000 units of work in process 60% complete as to conversion costs. The costs attached to these 5,000 units were $10,500 of costs transferred in from Department A, $1,800 of material costs added in Department B, and $800 of conversion costs added in Department B.

Required

Choose the best answer for each question:

1. Using the weighted-average method, what were the equivalent units for the month of April?

	Transferred in From Department A	Direct Materials	Conversion Costs
a.	15,000	15,000	15,000
b.	19,000	19,000	20,000
c.	20,000	20,000	18,000
d.	25,000	25,000	20,000

2. Using the weighted-average method, what was the cost per equivalent unit for conversion costs? (a) $4,200 ÷ 15,000, (b) $5,000 ÷ 18,000, (c) $5,800 ÷ 18,000, (d) $5,800 ÷ 20,000.

17-29 Multiple choice, FIFO. (SMA) Choose the best answer for each of the following three multiple-choice questions. Show your supporting computations. The company uses FIFO.

1. A beginning inventory consists of 1,000 units 80% complete as to direct materials and 30% complete as to conversion costs; 11,000 units completed and transferred out; no lost units; and an ending inventory of 900 units, 40% complete as to direct materials and 20% complete as to conversion costs. The total cost of the beginning inventory before completion was $8,000, and current unit costs are $2 for direct materials and $8 for conversion costs. The total direct materials costs amount to (a) $21,800, (b) $22,200, (c) $21,120, (d) $22,720, (e) $23,400.
2. Total conversion costs added during the current period amount to (a) $87,040, (b) $86,680, (c) $86,390, (d) $89,380, (e) $89,440.
3. Costs applied to the goods transferred out amount to (a) $6,000, (b) $14,000, (c) $110,000, (d) $120,000, (e) $114,000.

17-30 Weighted-average method. A toy manufacturer buys wood as its direct material for its Forming Department. The department processes one type of toy. The toys are transferred to the Finishing Department, where hand shaping and metal are added.

Consider the following data:

Units:

Work in process, March 31, 3,000 units, 100% completed for direct materials but only 40% completed for conversion costs

Units started in April, 22,000

Units completed during April, 20,000

Work in process, April 30, 5,000 units, 100% completed for direct materials but only 25% completed for conversion costs

Costs:

Work in process, March 31:		
Direct materials	$7,500	
Conversion costs	2,125	$ 9,625
Direct materials added during April		70,000
Conversion costs added during April		42,500
Total costs to account for		$122,125

Required

Use the weighted-average method. Prepare a schedule of output in equivalent units. Prepare a production-cost report for the Forming Department for April. (For journal entries, see the next problem.)

17-31 Continuation of Problem 17-30, journal entries. Refer to the preceding problem. Prepare a set of summarized journal entries for all April transactions affecting Work in Process—Forming Department. Set up a T-account, Work in Process—Forming Department, and post the entries to it.

17-32 FIFO computations. Repeat Problem 17-30, using FIFO and four decimal places for unit costs.

17-33 Comparing weighted-average and FIFO methods (D. Kleespie) The Estevan Company manufactures blades for saws and uses actual costs in a weighted-average process-cost accounting system. In Department 1, the direct materials are added at the beginning of processing, and conversion costs are considered to be added evenly throughout the processing. Consider the following information regarding the month of May:

	Units
Beginning inventory of work in process ($620 total conversion costs)	100 (60%)*
Completed and transferred out	3,000
Ending inventory of work in process	200 (50%)*
Equivalent unit cost for conversion cost, $11 per unit	

*Degree of completion of each unit

Direct materials cost information has been omitted in order to simplify the problem.

Required

1. How many units were started in production in Department 1 during May?
2. How many equivalent units were used in determining the unit cost for conversion costs?
3. What is the total dollar amount of the conversion-cost portion of the ending work in process inventory?
4. What is the total dollar amount of the conversion-cost portion being transferred out of Department 1?
5. What is the total amount of conversion costs charged to Department 1 during May?
6. Assume instead that the company had used the FIFO method. How many equivalent units would have been used to determine the unit cost for conversion costs?
7. Assume as in requirement 6 that the company had used the FIFO method. What is the total amount of conversion costs charged to Department 1 during May?

17-34 Transfers in, weighted-average. A toy manufacturer has two departments, Forming and Finishing. Consider the Finishing Department, which processes the formed toys through the addition of hand shaping and metal. Although various direct materials might be added at various stages of finishing, for simplicity here suppose all additional direct materials are added at the end of the process.

The following is a summary of the April operations in the Finishing Department:

Units:	
Work in process, March 31, 5,000 units, 60% completed for conversion costs	
Units transferred in during April, 20,000	
Units completed during April, 21,000	
Work in process, April 30, 4,000 units, 30% completed for conversion costs	
Costs:	
Work in process, March 31 (transferred-in costs, $17,750; conversion costs, $7,250)	$ 25,000
Transferred-in costs from Forming Department during April	104,000
Direct materials added during April	23,100
Conversion costs added during April	38,400
Total costs to account for	$190,500

Required

1. Use the weighted-average method. Prepare a schedule of output in equivalent units. Prepare a production-cost report for the Finishing Department for April.
2. Prepare journal entries for April transfers from the Forming Department to the Finishing Department and from the Finishing Department to finished goods.

17-35 FIFO costing. Refer to the preceding problem. Using FIFO costing, repeat the requirements.

17-36 Transferred-in costs, weighted-average and FIFO. Frito-Lay, Inc., manufactures convenience foods, including potato chips and corn chips. Production of corn chips occurs in four steps: cleaning, mixing, cooking, and drying and packaging. Suppose the accounting records of a Frito-Lay plant provided the following information for corn chips in its Drying and Packaging Department during a weekly period:

Cases:	
Work in process, beginning, 5,000 cases, 80% complete for conversion costs	
Units transferred in from the Cooking Department, 20,000	
Units completed, 21,000	
Work in process, ending, 4,000 units, 40% completed for conversion costs	
Costs:	
Work in process, beginning ($29,000 transferred-in costs, $9,060 conversion costs)	$ 38,060
Transferred in from the Cooking Department	96,000
Direct materials added (at the end of the process)	25,200
Conversion costs added	38,400
Total costs to account for	$197,660

Required

1. Compute the equivalent units for transferred-in costs, direct materials, and conversion costs. Use (a) the weighted-average method and (b) the FIFO method.
2. Assume the weighted-average method is used for the Drying and Packaging Department. Prepare a production-cost report.
3. Assume the FIFO method is used for the Drying and Packaging Department. Compute the unit costs for applying the week's costs to products.

17-37 FIFO process costing, transferred-in costs. Pepperell Mills, Inc., manufactures broadloom carpet in seven processes: spinning, dyeing, plying, spooling, tufting, latexing, and shearing.

First, fluff nylon purchased from a company such as Du Pont or Monsanto is spun into yarn that is dyed the desired color. Then two or more threads of the yarn are joined together, or plied, for added strength. The plied yarn is spooled for use in the actual carpet making. Tufting is the process by which yarn is added to burlap backing. After the backing is latexed to hold it together and make it skid resistant, the carpet is sheared to give it an even appearance and feel.

At March 31, before recording the transfer of cost from the Tufting Department to the Latexing Department, the Pepperell Mills general ledger included the following account for one of its lines of carpet:

Work in Process—Tufting Department

February 28 balance	33,900	
Transferred in from Spooling Department	224,000	
Direct materials	24,200	
Conversion costs	28,150	

Work in process inventory of the Tufting Department on February 28 consisted of 75 rolls that were 40% complete as to direct materials and 60% complete as to conversion costs. During March, 560 rolls were transferred in from the Spooling Department. The Tufting Department completed 500 rolls of the carpet in March, and 135 rolls were still in process on March 31. This ending inventory was 100% complete as to direct materials and 80% complete as to conversion costs.

Required

1. Using the FIFO method, compute the equivalent units of production for the Tufting Department for March.
2. Prepare a production-cost report for March.

17-38 Continuation of Problem 17-37, journal entries. Journalize all transactions affecting the Tufting Department during March, including those entries that have already been posted. Omit explanations.

17-39 Continuation of Problem 17-37, standard costs. Assume standard costs per unit for the Tufting Department are $38 for direct materials and $47 for conversion costs. Compute the total March variances for direct materials and conversion costs.

17-40 Basic standard costing. Finest Clock Company uses a standard-cost accounting system for manufacturing some of the parts used in making large clocks. A summary of the June operations of the first process for one of these parts follows:

Beginning inventory of work in process	none
Units started in June	1,300
Ending inventory, 100% completed for direct materials, 70% completed for conversion costs	200
Standard unit costs:	
Direct materials	$ 8
Conversion costs	$ 10
Total actual costs incurred during June:	
Direct materials	$10,900
Conversion costs	$12,960

Required

1. Compute the total standard cost of units transferred out in June and the total standard cost of the June 30 inventory of work in process.
2. Compute the total June variances for direct materials and conversion costs.

17-41 Standard costing with beginning and ending work in process. The Victoria Corporation uses a standard-costing system for its manufacturing operations. Standard costs for the Cooking Process are $6 per unit for direct materials and $3 per unit for conversion costs. All direct materials are introduced at the beginning of the process, but conversion costs are incurred uniformly throughout the process. The operating summary for May included the following data for the Cooking Process:

Work in process inventories:
May 1, 3,000 units, 60% completed
(direct materials $18,000; conversion costs $5,400)
May 31, 5,000 units, 50% completed
Units started in May, 20,000
Units completed and transferred out of Cooking in May, 18,000
Additional actual costs incurred for Cooking during May:
Direct materials, $125,000
Conversion cost, $57,000

Required

1. Compute the total standard cost of units transferred out in May and the total standard cost of the May 31 inventory of work in process.
2. Compute the total May variances for direct materials and conversion costs.

17-42 Process costing for services, weighted-average and FIFO. (CPA, heavily adapted) Webb & Company prepares income tax returns for individuals. Webb uses the weighted-average method and actual costs for financial reporting purposes. The following information pertains to March 19_7:

Inventory data:	
Returns in process, March 1 (25% complete)	200
Returns started in March	825
Returns in process, March 31 (80% complete)	125
Actual cost data:	
Returns in process, March 1:	
Labor	$ 5,700
Overhead	2,600
Labor, March 1 to 31:	
4,000 hours	89,300
Overhead, March 1 to 31	43,700

Required

1. Using the weighted-average method:
   a. Compute the equivalent units of performance for each cost element (labor and overhead).
   b. Compute the actual cost per equivalent unit for each cost element.
   c. Compute the actual cost of returns in process at March 31.
2. Repeat requirement 1 using the first-in, first-out method.

17-43 Continuation of Problem 17-42, standard costing. For internal reporting, Webb uses a standard-cost system. The standards are as follows:

Labor per return	5 hrs. at $20 per hr.
Overhead per return	5 hrs. at $10 per hr.

For March 19_7 performance, budgeted overhead is $49,000 for the standard labor hours allowed.

Required

1. Compute the standard cost per return.
2. Analyze March performance by computing the following variances. Indicate whether each variance is favorable or unfavorable:
   a. Total labor variance
   b. Total overhead variance

17-44 Standard process costing. (N. Melumad and S. Reichelstein, adapted) Enviro-Save Corporation has devised a process for converting raw garbage into liquid fuel. This process is carried out in two departments. In the Extracting Department, processable garbage is extracted from raw garbage; nonprocessable garbage is disposed of at a net cost of zero. In the Fuel Department, processable garbage is processed into liquid fuel. Both operations require conversion costs.

The company uses standard costs in its process-costing system. Variances are treated as period costs and charged monthly to Cost of Goods Sold.

Raw garbage is a direct material, and processable garbage is the transferred-in cost. The company has set the following standard costs and quantities for the different inputs:

	Extracting Department (per 1 ton of processable garbage)	Fuel Department (per 1,000 Liters of Fuel)
Raw garbage	2 tons of raw garbage at $100 per ton	—
Processable garbage	—	1 ton of processable garbage
Conversion costs	5 process hours at $30 an hour	10 process hours at $20 an hour

Of the 100,000 liters of fuel produced during the month of October, 70,000 liters were sold. There were no beginning inventories. At the end of the month, the Extracting Department had no ending inventory of work in process. The Fuel Department had 20,000 liters of fuel 50% completed (with respect to conversion costs).

During the month, the company paid an average hourly wage of $11 and purchased 250 tons of raw garbage. All raw garbage is directly shipped to the Extracting Department and is considered a part of work in process. The actual costs incurred in each department during October were:

	Extracting Department	Fuel Department
Raw garbage	$20,000	—
Conversion costs	17,000	$28,900

Required

1. Compute the total standard costs of the following:
   a. Cost of goods completed (that is, the standard cost of fuel produced) for the month of October
   b. Work in process, October 31
   c. Cost of goods sold for October
2. What will be the actual amount of cost of goods sold in the October income statement?

CHAPTER 18

Electronics companies use extensive quality-testing procedures when determining which products require rework.

Spoilage, Reworked Units, and Scrap

Learning Objectives

When you have finished studying this chapter, you should be able to

1. Distinguish among spoilage, reworked units, and scrap
2. Describe the general accounting procedures for normal and abnormal spoilage
3. Account for spoilage in process costing using the weighted-average method
4. Account for spoilage in process costing using the first-in, first-out method
5. Account for spoilage in job costing
6. Account for reworked units
7. Account for scrap
8. Describe the affects of inspection timing on normal and abnormal spoilage

In recent years, managers have paid more attention to the costs of spoilage, reworked units, and scrap. Accounting systems that record these costs in a timely and detailed way help managers make better-informed decisions, especially concerning production systems. For example, consider investments in cutting-edge production systems such as just-in-time (JIT) and computer-integrated manufacturing (CIM). Managers often cite reductions in costs of spoilage, reworked units, and scrap as a major justification for these investments.

This chapter concentrates on the accounting for internal matters of product quality. We do not contend with external matters such as the impact on attitudes when customers encounter unacceptable or defective products that were originally regarded as having acceptable quality. Chapter 29 discusses these aspects as part of measuring the costs of product quality.

MANAGEMENT EFFORT AND CONTROL

Some causes of spoilage, rework, and scrap are largely beyond managers' control. One example is semiconductor manufacturing, where there is high statistical variability over time in the number of good wafers produced. Another example is a mining company that processes batches of ore containing a varying mix of valuable metals and waste products. The volume of these waste products cannot be controlled. This mining company, however, may employ production workers who pay inadequate attention to detail, and machines may be poorly maintained. Managers could take steps to improve these conditions and could better control quality.

Intensified competition in an expanding, more global marketplace has increasingly focused on improving quality. Executives have learned that a rate of defects regarded as normal in the past is no longer tolerable. Consider these words from a speech by the chief executive officer of Motorola, an electronics manufacturer:

> We want to improve our quality in everything we do by ten times in two years, by a hundred times in four years, and in six years . . . three and a half defects for every million operations, whether typing, manufacturing, or serving a customer.

A Japanese company focused on reducing its 2.64% rate of defective units. The aim was to reduce cost and improve quality by having zero defects. Supervisors asked for suggestions from group leaders. Most of these suggestions seemed unattainable at first glance. Nevertheless, the group leaders actively enforced three suggestions as a part of the company's total quality control activity. Finally, the target defect rate was reduced to 0.23%, a 91% improvement.[1]

TERMINOLOGY

Objective 1

Distinguish among spoilage, reworked units, and scrap

The definitions and accounting used for spoilage, reworked units, and scrap vary considerably from organization to organization. This chapter uses the following definitions:

- **Spoilage.** Unacceptable units of production that are discarded and sold for disposal value. Partially completed or fully completed units may be spoiled. Net spoilage cost is the total of the costs applied to the product up to the point of its rejection (plus its disposal costs or minus net disposal value).
- **Reworked Units.** Unacceptable units of production that are subsequently reworked and sold as acceptable finished goods. Such units may be sold through regular marketing channels or alternative channels, depending on the characteristics of the product and on the available alternatives.
- **Scrap.** Defined in Chapter 16 as a product that has minimal (frequently zero) sales value compared with the sales value of the main or joint product(s). Scrap may be either sold or reused. Examples are shavings and short lengths from woodworking operations and sprues, ingots, and flash from a casting operation in a foundry.

SPOILAGE IN GENERAL

Management Implications and Impact on Quality

The problem of spoilage is important from many aspects, the most important being that of managerial planning and control.[2] Managers must first attempt to select the most economical production method or process. Then they must see that spoilage is controlled within predetermined limits.

Many businesses have investigated whether the costs necessary for great reductions in spoilage are economically justified. Would reduced spoilage affect other related costs? The costs of spoilage really extend beyond the spoiled units themselves. For example, spoilage affects costs of materials handling, storage, and production disruptions. Like many corporations, a division of IBM Corporation compiles an analysis of the "costs of quality." Included are comparisons through time and between manufacturing techniques of the current costs of spoiled goods, reworked units, scrap, material-handling costs, storage costs, and warranty costs. (See Chapter 29 for further discussion of accounting for the costs of quality.)

There is an unmistakable trend in manufacturing to increase quality. Why? Because managers have found that improved quality and intolerance for high spoilage have lowered overall costs and increased sales. The procedures described in this chapter help identify and make visible the costs of spoilage as special management problems. That is, spoilage costs are not ignored or buried as an unidentified part of the costs of good units produced.

[1]M. Sakurai, "Target Costing and How to Use It," *Journal of Cost Management* (Summer 1989), p. 43.

[2]The helpful suggestions of Samuel Laimon, University of Saskatchewan, are gratefully acknowledged.

Normal Spoilage

Working within the selected set of production conditions, management must establish the rate of spoilage that is to be regarded as *normal.* **Normal spoilage** ***is what arises under efficient operating conditions; it is an inherent result of the particular process and so is uncontrollable in the short run.*** Costs of normal spoilage are typically viewed as a part of the costs of *good* units produced, when attaining good units necessitates the simultaneous appearance of spoiled units. In other words, normal spoilage is planned spoilage, in the sense that the choice of a given combination of factors of production entails a spoilage rate that management is willing to accept.

Some managements adhere to a perfection standard as a part of their emphasis on total quality control. Their ideal goal is zero defects. Hence, all spoilage, rework, and even scrap would be treated as abnormal.

Abnormal Spoilage

Abnormal spoilage ***is spoilage that is not expected to arise under efficient operating conditions; it is not an inherent part of the selected production process.*** Most abnormal spoilage is usually regarded as controllable. A first-line supervisor can generally lower the inefficiencies seen in machine breakdowns, accidents, and inferior materials. Costs of abnormal spoilage are the costs of inferior products that should be written off directly as losses for the accounting period in which detection occurs. For the most informative feedback, the Loss from Abnormal Spoilage account should appear on a detailed income statement as a separate line item and not be buried as an indistinguishable part of the cost of goods manufactured.

The normal-abnormal spoilage distinction is an example of the value-added–nonvalue-added cost distinction. Accounting for abnormal spoilage as a period cost rather than as an inventoriable cost is consistent with financial accounting theory. After all, only costs that are reasonably expected to be incurred in getting an asset ready for use for an income-producing purpose are justifiably carried forward as assets. By definition, abnormal spoilage is not reasonably expected to be incurred. It is no more logical to regard abnormal spoilage cost as an asset than it would be to regard as an asset the cost of a speeding ticket incurred by a truck driver hauling new equipment to its destination.

General Accounting Procedures for Spoilage

Objective 2

Describe the general accounting procedures for normal and abnormal spoilage

Before discussing debits and credits for spoiled goods, let us try to relate spoiled goods to two major purposes of cost accounting: control and product costing. Accounting for control is primarily concerned with charging responsibility centers for costs *as incurred.* Manufacturing product costing is concerned with *applying* to inventory or other appropriate accounts the costs *already incurred.* Where does costing for spoiled goods fit into this framework? Spoiled goods incur no cost beyond the amount already incurred before detection of spoilage. Therefore, in accounting for spoiled goods, we are not dealing with any incurrence of new costs. Our objectives are

1. To accumulate data to spotlight the manufacturing cost of spoilage so that management is made aware of its magnitude.
2. To identify the *nature* of the spoilage and distinguish between costs of normal spoilage and abnormal spoilage

Although this discussion of process costing emphasizes accounting for spoilage, the ideas here are equally applicable to waste (shrinkage, evaporation, or lost units). Again we must distinguish between control and inventory costing. For control, most companies use some version of estimated or standard costs that incorporates an allowance for normal spoilage or shrinkage, into the estimate or standard. This section emphasizes inventory costing in actual or normal process-costing systems.[3]

Count All Spoilage

As a general rule, use equivalent units to accumulate the costs of spoilage separately from other inventoriable costs. Then allocate normal spoilage costs to finished goods or work in process, depending on where in the production cycle the spoilage is assumed to take place. Spoilage is typically assumed to occur at the stage of completion where inspection takes place, because spoilage is not detected until this point. ***Normal spoilage need not be allocated to units that have not yet reached this point in the production process, because the spoiled units are included in the units that have passed the inspection point.***

Many writers on process costing advocate ignoring the computation of equivalent units for spoilage or shrinkage. The reason cited in favor of this shortcut is that it automatically spreads normal-spoilage costs over good units through the use of higher equivalent unit costs. However, the results of this shortcut are inaccurate.

Consider the example of direct materials. Suppose $1,800 of direct materials are introduced at the start of a process. Production data are: 1,000 units started, 500 good units completed, 100 units spoiled (all normal spoilage), and 400 units in ending work in process. Spoilage is detected upon completion of the process.

The direct material unit costs would be computed and then applied as shown in Exhibit 18-1. Ignoring the equivalent units for spoilage decreases equivalent units;

EXHIBIT 18-1

Direct Materials Equivalent Units and Spoilage

	Accurate Method: Count Equivalent Units for Spoilage	Inaccurate Method: Ignore Equivalent Units for Spoilage
Costs to account for	$1,800	$1,800
Divide by equivalent units	÷1,000	÷900
Costs per equivalent unit	$ 1.80	$ 2.00
Applied to:		
Good units completed, 500 × $1.80; × $2.00	$ 900	$1,000*
Add: Normal spoilage, 100 × $1.80; × $2.00	180	—
Costs transferred out	$1,080	$1,000
Work in process, ending, 400 × $1.80; × $2.00	720	800†
Costs accounted for	$1,800	$1,800

*500 × $2.00 = $1,000.
†400 × $2.00 = $800.

[3]This section assumes that you have studied Chapter 17.

hence a higher unit cost is computed. A $2.00 equivalent unit cost (instead of a $1.80 equivalent unit cost) is applied to work in process that has not reached the inspection point. Simultaneously, the direct material costs applied to good units completed are too low ($1,000 instead of $1,080). Consequently, the 400 units in ending work in process contain costs of spoilage of $80 ($800 − $720) that do not pertain to such units and that in fact belong with the good units completed. The 400 units in ending work in process undoubtedly include some units that will be detected as spoiled in a subsequent accounting period. In effect, these units will bear two charges for spoilage. The ending work in process is being charged for spoilage in the current period, and it will be charged again when inspection occurs as the units are completed. Such cost distortions would not occur under the accurate method: including spoiled units in the computation of equivalent units.

Base for Computing Normal Spoilage

Normal spoilage rates should be computed using the total good outputs as the base—not the total actual inputs. Why? Because total actual inputs include the abnormal as well as the normal spoilage. For example, consider the spoilage of polio vaccine. Assume total actual inputs of 100,000 cubic centimeters (abbreviated c.c.) and good outputs of 85,500 c.c. Normal spoilage is 5% of good outputs. The relationships are summarized:

Total actual inputs to account for	100,000 c.c.
Output:	
Good units	85,500 c.c.
Normal spoilage in units at 5% × 85,500	4,275
Abnormal spoilage, 100,000 − (85,500 + 4,275)	10,225
Inputs accounted for	100,000 c.c.

The measurement of spoilage or defects varies among industries. For example, several electronics companies in Japan measure defects in parts per million instead of in percentages. Moreover, the worldwide thrusts for improved quality have demonstrated, at least in some cases, that defects thought normal in the past and uncontrollable were really abnormal and controllable.

Data for Illustration

EXAMPLE

One of Anzio Company's plastic products is manufactured in the Processing Department. Direct materials for this product are put in at the beginning of the production cycle. Conversion costs are incurred evenly throughout the cycle. Some units of this product are spoiled as a result of defects not detectable before inspection of finished units. Normally the spoiled units are 10% of the good output.

At January 1, the inventory of work in process of this product was $29,600, representing 2,000 pounds of direct materials ($15,000) and conversion costs of $14,600 representing 80% completion. During January, 8,000 pounds of direct materials ($61,000) were put into production. Conversion costs of $80,400 were applied to the process. The inventory at January 31 consisted of 1,500 pounds, 66⅔% finished. Seventy-two hundred pounds of good units were transferred to finished goods after inspection.

Spoiled units were 2,000 beginning + 8,000 started − 1,500 ending − 7,200 transferred = 1,300. Normal spoilage was 10% of the 7,200 good units completed, or 720. Thus, abnormal spoilage was 1,300 − 720 = 580 units.

Using weighted-average and then FIFO, we will show calculations of

- Equivalent units of production for January
- The dollar and unit amount of the abnormal spoilage during January
- Total inventoriable costs transferred to finished goods inventory

- The cost of work in process inventory at January 31
- Journal entries for units completed and transferred out of the department

The basic five-step approach used in the preceding chapter needs only slight modification to handle spoilage. The following observations pertain to both the weighted-average and FIFO methods:

- *Step 1: Summarize physical units.* Identify both normal and abnormal spoilage.
- *Step 2: Compute output in terms of equivalent units.* Compute equivalent units for spoilage in the same way as for good units. Because inspection occurs upon completion in Example 1, the same amount of work was done on each spoiled unit and each completed good unit.
- *Step 3: Summarize total costs to account for.* The details of this step do not differ from those in the preceding chapter.
- *Step 4: Compute equivalent unit costs.* The details of this step do not differ from those in the preceding chapter. ***However, the divisor includes the equivalent units applicable to the spoiled units.***
- *Step 5: Apply costs to units completed and in process.* This step now includes computation of the cost of spoiled units in the same manner as the cost of good units.

Exhibit 18-2 is an overview of product costing for process costing with spoilage.

Weighted-Average Method and Spoilage

Objective 3

Account for spoilage in process costing using the weighted-average method

Exhibit 18-3 is a production-cost work sheet under the weighted-average method. The costs of abnormal spoilage are assigned to the Loss from Abnormal Spoilage account, 580 units × $17.60 = $10,208. The costs of normal spoilage, $12,672, are added to the costs of their related good units. This illustration assumes inspection

EXHIBIT 18-2

Overview of Manufacturing Product Costing for Process-Costing with Spoilage

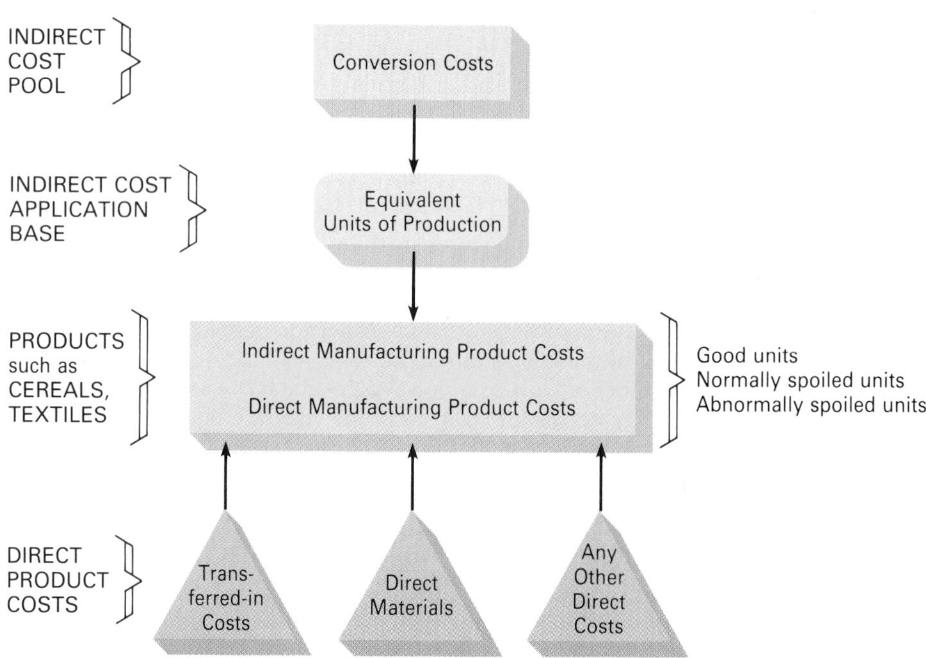

Note: Conversion costs are usually grouped in indirect cost pools. However, if some conversion costs are regarded as direct product costs, they would be accounted for accordingly.

upon completion. In contrast, inspection may take place at some other stage—say, at the halfway point in the production cycle. In such a case, normal spoilage costs would be added to completed goods and to the units in process that are more than half completed.

An alternative way to think about spoilage would be in terms of the normal spoilage cost per good unit passing inspection: $17.60 ÷ 10 units = $1.76. In Exhibit 18-3, the application of normal spoilage costs to the goods completed would be computed as 7,200 units × $1.76 = $12,672.

To provide an overall picture, Exhibit 18-3 shows the results of all five steps in a single exhibit. Obviously, Steps 1 and 2 could be presented in a separate exhibit.

EXHIBIT 18-3

Anzio Company Processing Department Production-Cost Work Sheet
For the Month Ended January 31, 19_1, Weighted-Average Method

	Flow of Production	(Step 1) Physical Units	(Step 2) Equivalent Units: Direct Materials	(Step 2) Equivalent Units: Conversion Costs
	Work in process, beginning	2,000 (80%)*		
	Started during current period	8,000		
	To account for	10,000		
	Good units completed and transferred out during current period	7,200	7,200	7,200
	Normal spoilage (10% × 7,200)	720	720	720
	Abnormal spoilage (1,300 − 720)	580	580	580
	Work in process, ending	1,500 (66⅔%)*	1,500	1,000
	Accounted for	10,000		
	Total work done to date		10,000	9,500

	Costs	Totals	Details: Direct Materials	Details: Conversion Costs
	Work in process, beginning	$ 29,600	$15,000	$14,600
	Costs added during current period	141,400	61,000	80,400
(Step 3)	Total costs to account for	$171,000	$76,000	$95,000
	Divide by equivalent units		÷10,000	÷9,500
(Step 4)	Equivalent unit costs		$ 7.60	$ 10.00
(Step 5)	Application of costs:			
(A)	Abnormal spoilage (580)	$ 10,208	580($17.60†)	
	Completed and transferred out (7,200):			
	Costs before adding spoilage	$126,720	7,200($17.60)	
	Normal spoilage	12,672	720($17.60)	
(B)	Total costs transferred out	$139,392		
	Work in process, ending (1,500):			
	Direct materials	$ 11,400	1,500 ($7.60)	
	Conversion costs	10,000		1,000 ($10.00)
(C)	Total work in process	$ 21,400		
(A) + (B) + (C)	Total costs accounted for	$171,000		

*Degree of completion for conversion costs of this department at the dates of the inventories. Note that direct material costs are fully completed at each of these dates, because in this department materials are introduced at the beginning of the process.
†Cost per equivalent whole unit = $7.60 + $10.00 = $17.60

Managers should try to have inspection done as early in the production process as technically possible. This timing will reduce the direct materials and conversion costs applied to spoiled units. Thus, in the Exhibit 18-3 example, if inspection can occur when units are 80% complete as to conversion costs and 100% complete as to direct materials, the company can avoid incurring the final 20% of conversion costs on the spoiled units.

FIFO Method and Spoilage

Objective 4

Account for spoilage in process costing using the first-in, first-out method

Exhibit 18-4 presents the results of all five steps using the FIFO method. All spoilage costs are assumed to relate to units completed during this period, using the unit costs of the current period.[4] With the exception of accounting for spoilage, the FIFO method is the same as presented in the preceding chapter.

Journal Entries

A production-cost report is shown in Exhibit 18-5. It is confined to physical units and the total costs that flow through the Work in Process account. If managers prefer, the supporting computations could be embodied in a more elaborate production-cost report like the production-cost work sheet in Exhibit 18-4.

The information in Exhibit 18-5 supports the following journal entries:

	Weighted Average		FIFO	
	Debit	**Credit**	**Debit**	**Credit**
Finished goods	139,392		139,060	
Work in process—processing department		139,392		139,060
To transfer good units completed in January.				
Loss from abnormal spoilage	10,208		10,325	
Work in process—processing department		10,208		10,325
To recognize abnormal spoilage in January.				

Assumptions for Allocating Normal Spoilage

Spoilage might actually *occur* at various points or stages of the production cycle but spoilage is typically not *detected* until one or more specific points of inspection. The cost of spoiled units is assumed to be all costs incurred prior to inspection. The unit costs of abnormal and normal spoilage are the same when both are detected simultaneously. However, situations might arise when abnormal spoilage is detected at a different point than normal spoilage. In such cases, the unit cost of abnormal spoilage would differ from the unit cost of normal spoilage.

Various assumptions prevail regarding how to allocate the cost of normal spoilage between completed units and the ending inventory of work in process. A popular approach (applicable to both the weighted-average and FIFO methods) follows.

When normal spoilage is presumed to occur at a specific point in the production cycle, allocate its cost (a unit cost of $17.60 in Exhibit 18-3) over all units that have

[4]If the FIFO method were used in its purest form, normal spoilage costs would be split between the goods started and completed during the current period and those completed from beginning work in process—using the appropriate unit costs of the period in which the units were worked on. The simpler, modified FIFO method, as illustrated in Exhibit 18-4, in effect uses the unit costs of the current period for applying normal spoilage costs to the goods completed from beginning work in process. This modified method assumes that all normal spoilage traceable to the beginning work in process was begun and completed during the current period, an obvious contradiction to a pure FIFO method.

EXHIBIT 18-4
Anzio Company Processing Department
Production-Cost Work Sheet
For the Month Ended January 31, 19_1
FIFO Method

	Flow of Production	(Step 1) Physical Units	(Step 2) Equivalent Units: Direct Materials	(Step 2) Equivalent Units: Conversion Costs
	Work in process, beginning	2,000 (80%)*	(work done before current period)	
	Started during current period	8,000		
	To account for	10,000		
	Good units completed and transferred out during current period:			
	From beginning work in process	2,000	—	400
	Started and completed	5,200	5,200	5,200
	Normal spoilage (10% × 7,200)	720	720	720
	Abnormal spoilage (1,300 − 720)	580	580	580
	Work in process, ending	1,500 (66⅔%)*	1,500	1,000
	Accounted for	10,000		
	Work done in current period only		8,000	7,900

	Costs	Totals	Details: Direct Materials	Details: Conversion Costs
	Work in process, beginning	$ 29,600	(costs of work done before current period)	
	Costs added during current period	141,400	$61,000	$80,400
(Step 3)	Total costs to account for	$171,000		
	Divide by equivalent units		÷8,000	÷7,900
(Step 4)	Equivalent unit costs		$7.625	$10.1772
(Step 5)	Application of costs:			
(A)	Abnormal spoilage (580)	$ 10,325	580 ($17.8022†)	
	Units completed (7,200)			
	Work in process, beginning (2,000)	$ 29,600		
	Costs added during current period	4,071		400 ($10.1772)
	Total from beginning inventory before spoilage	$ 33,671		
	Started and completed before spoilage (5,200)	92,571	5,200 ($17.8022)	
	Normal spoilage	12,818	720 ($17.8022)	
(B)	Total costs transferred out	$139,060		
	Work in process, ending (1,500):			
	Direct materials	$ 11,438	1,500 ($7.625)	
	Conversion costs	10,177		1,000 ($10.1772)
(C)	Total work in process	$ 21,615		
(A) + (B) + (C)	Total costs accounted for	$171,000		

*Degree of completion for conversion costs at the dates of inventories.
†Cost per equivalent whole unit = $7.625 + $10.1772 = $17.8022

EXHIBIT 18-5
Anzio Company Processing Department
Production-Cost Report
For the Month Ended January 31, 19_1

Flow of Production	Physical Units	Total Costs: Weighted-Average Method	Total Costs: FIFO
Work in process, beginning	2,000	$ 29,600	$ 29,600
Started	8,000	141,400	141,400
To account for	10,000	$171,000	$171,000
Good units completed and transferred out	7,200	$139,392	$139,060
Normal spoilage	720		
Abnormal spoilage	580	10,208	10,325
Work in process, ending	1,500	21,400	21,615
Accounted for	10,000	$171,000	$171,000

passed this point. In our example, spoilage is assumed to occur upon completion, so no cost of normal spoilage is allocated to the ending work in process.

Whether the cost of normal spoilage is allocated to the units in ending work in process inventory strictly depends on whether they have passed the point of inspection. For example, if the inspection point is presumed to be the halfway stage of the production cycle, work in process that is 70% completed would be allocated a full measure of normal spoilage cost. But work in process that is 40% completed would not be allocated any normal spoilage cost.

So the point of inspection is the key to the application of spoilage costs. Avoid the idea that normal spoilage costs are applied solely to units completed and transferred out. Why? Because if units in ending work in process have passed inspection, they should have normal spoilage costs added to them. The appendix to this chapter contains additional discussion concerning various assumptions about spoilage.

JOB COSTING AND SPOILAGE

Objective 5

Account for spoilage in job costing

The concepts of normal spoilage and abnormal spoilage are equally applicable to both job-costing systems and process-costing systems. In our illustration of job costing, we show how disposal values are accounted for. When spoiled goods have a disposal value, the net cost of spoilage is computed by deducting disposal value from the costs of the spoiled goods accumulated to the point of inspection.

Spoilage may be considered a normal characteristic of a given production cycle. For example, machines might malfunction at random. Then the causes of spoilage would be attributable to work done on all jobs. Net spoilage cost is budgeted in practice as a part of factory overhead, so that the budgeted overhead rate includes a provision for normal spoilage cost. Therefore, normal spoilage cost is spread, through overhead application, over all jobs rather than loaded on particular jobs only.

In the Chang Machine Shop, assume that 5 pieces out of a job lot of 50 aircraft parts were normally spoiled. Costs applied to the point of inspection were $100 per unit. Disposal value is estimated at $30 per unit. As the normal spoilage is detected, the spoiled goods are inventoried ($30 per unit), the Factory Department

Overhead Control account is debited for the net cost of spoilage ($70 per unit), and work in process is credited for the costs applied to the goods up to the point of inspection ($100 per unit).[5]

Materials control (spoiled goods at disposal value), 5 × $30	150	
Factory department overhead control (normal spoilage), 5 × $70	350	
Work in process control (particular job), 5 × $100		500

Items in parentheses indicate subsidiary postings.

When management finds it helpful for control or for pricing, accountants credit specific jobs with only the disposal value of normally spoiled units, thus forcing the remaining good units in the job to bear net normal spoilage costs. Under this method, the budgeted overhead rate would *not* include a provision for normal spoilage cost because the spoilage would be viewed as directly attributable to the specifications of particular jobs instead of attributable to general factory conditions or processes. The journal entry, with the same data just used, follows:

Materials control (spoiled goods at disposal value)	150	
Work in process control (particular job)		150

REWORKED UNITS

Objective 6

Account for reworked units

Reworked units are unacceptable units of production that are subsequently reworked and sold. Management needs effective control over such actions, because supervisors are tempted to rework rather than to junk unacceptable units. If control is not exercised, supervisors may rework many bad units instead of having them sold outright at a greater economic advantage. Rework should be undertaken only if incremental revenue is expected to exceed incremental costs.

Unless there are special reasons for charging rework to the jobs or batches that contained the bad units, the cost of the extra materials, labor, and so on are in practice usually charged to factory overhead.[6] Thus, once again we see that rework is spread over all jobs or batches as a part of a budgeted overhead rate. Assume that the 5 spoiled pieces used in our Chang illustration are reworked at a total cost of $190 and sold through regular marketing channels. Entries follow:

Original cost applications:	Work in process control	500	
	Materials control		200
	Payroll liability		200
	Factory overhead applied		100
Rework (figures assumed):	Factory department overhead control (rework)	190	
	Materials control		40
	Payroll liability		100
	Factory overhead applied		50
Transfer to finished goods:	Finished goods control	500	
	Work in process control		500

[5]Conceptually, the prevailing treatment just described can be criticized, primarily because *costs already applied to products* are being charged back to Factory Department Overhead Control, which logically should accumulate only *costs incurred,* not both costs incurred and costs applied.

[6]The criticisms of the practical treatment for spoiled goods are also applicable to the treatment described here. The overhead incurred account and the overhead applied account may be padded for amounts that in themselves did not necessitate overhead incurrence. For example, the rework entry illustrated here *applies* $50 of overhead to the account for overhead *incurred.*

ACCOUNTING FOR SCRAP

Objective 7

Account for scrap

Scrap was defined in Chapter 16 in conjunction with the discussion of byproducts. Scrap is a product that has minimal (frequently zero) sales value compared with the sales value of the main or joint product(s).

There are two major aspects of accounting for scrap:

1. Planning and control, including physical tracking
2. Inventory costing, including when and how to affect reported income

Detailed records of physical quantities of scrap are often kept at all stages of production. For example, a survey of manufacturers shows that 71% of the responding companies tracked scrap by specific materials and 37% by specific production operations.[7] Scrap records not only help measure efficiency but also often focus on a tempting source for theft.

Exhibit 18-6 shows a scrap report used in the plastics industry. The amounts are entered in nonfinancial terms such as pounds. Values are determined later after other reports have been compiled. In various industries, items like metal chips, filings, and wood shavings are quantified by weighing, counting, or some other expedient means. Excessive scrap indicates inefficiency. Scrap reports are prepared as source documents for periodic summaries of the amount of actual scrap compared with budgeted norms or standards. Scrap is either quickly sold or disposed of or stored in some routine way for later sale or for reuse.

The tracking of scrap extends into the financial records. For example, of seventy manufacturers responding to a survey, 60% of the firms maintained a distinct cost for scrap somewhere in their cost accounting system.[8] The issues here are similar to those discussed in Chapter 16 regarding the accounting for byproducts:

EXHIBIT 18-6
Scrap Report in Plastics Industry

SCRAP

MODEL # ________ DEPT.: ________ DATE: ________ SHIFT: ________

SCRAP REASON	AMOUNT	SCRAP REASON	AMOUNT
Cracks		Foreign Materials	
Broken		Broken Locator Pins	
Bare Fibers		Broken or Nonfill Screw Holes	
Blisters		Oversanding of Front Area	
OTHER:		OTHER:	
TOTAL SCRAP:		TOTAL SCRAP:	
Disposition:		Approved by: ________	

[7]Price Waterhouse, *Survey of the Cost Management Practices of Selected Midwest Manufacturers*, 1989, p. 21.

[8]Ibid., p. 10.

1. When should any value of scrap be recognized in the accounting records: at the time of production or at the time of sale?
2. How should revenue from scrap be accounted for?

In our Chang example, assume that scrap related to the job lot has a total sales value of $45. Many companies will track the physical amount of scrap (only) and indicate its presence in storage awaiting sale.

When a sale is made, the simplest accounting is to regard scrap sales as a separate line item of other revenues:

Sale of scrap:	Cash or accounts receivable	45	
	Sales of scrap		45

However, many companies will account for the sales as offsets against factory overhead:

Sale of scrap:	Cash or accounts receivable	45	
	Factory department overhead control		45
	Posting made to subsidiary record—"Sales of Scrap" column on department cost record		

This method is both simple and accurate enough in theory to justify its wide use. A normal amount of scrap is an inevitable result of production operations. Basically, this method does not link scrap with any particular physical product; instead, all products bear regular production costs without any particular credit for scrap sales except in an indirect manner. What really happens in such situations is that sales of scrap are considered when budgeted overhead rates are being set. Thus, the budgeted overhead rate is lower than it would be if no credit for scrap sales were allowed in the overhead budget. This accounting for scrap is used in both process-costing and job-costing systems.

In job-costing systems, sometimes sales of scrap are traced to the jobs that yielded the scrap. This method is used only when feasible and economically desirable. For example, the Chang Machine Shop and particular customers, such as the U.S. Department of Defense, may reach an agreement that provides for charging specific, difficult jobs with all rework or spoilage costs and crediting such jobs with all scrap sales that arise from that job. Entries follow:

Scrap returned to storeroom:	No journal entry. (Memo of quantity received and related job is entered on the perpetual record.)		
Sale of scrap:	Cash or accounts receivable	45	
	Work in process control		45
	Posting made to specific job record.		

In these illustrations, we assume that no inventory value is assigned to scrap returned to the storeroom. Scrap, however, sometimes has a significant value, and the time between storing it and selling or reusing it may be quite long. Under these conditions, the company is justified in inventorying scrap at some conservative estimate of net realizable value so that production costs and related scrap recovery may be recognized in the same accounting period.

Some companies tend to delay sales of scrap until the price is most attractive. Volatile price fluctuations are typical for scrap metal. In these cases, if scrap inventory becomes significant, it should be inventoried at some "reasonable" value—a difficult task in the face of volatile market prices.

Scrap is sometimes reused as direct materials rather than sold as scrap. Then it should be debited to Materials Control as a class of direct materials and carried at its estimated net realizable value.[9]

Scrap returned to storeroom:	Materials control	45	
	Factory department overhead control		45
Reuse of scrap:	Work in process control	45	
	Materials control		45

COMPARISON OF ACCOUNTING FOR SPOILAGE, REWORK, AND SCRAP

The basic approach to the accounting for spoilage, rework, and scrap should distinguish among the normal amount that is common to all jobs, the normal amount that is attributable to specific jobs, and abnormal amounts. The following entries recapitulate the preceding examples. Note the parallel approach to the three categories:

Spoilage Cost (net $350 from p. 601)

Normal: (common to all jobs)	Materials control	150	
	Factory department overhead control	350	
	Work in process control		500
Normal: (peculiar to specific jobs)	Materials control	150	
	Work in process control		150
Abnormal:	Materials control	150	
	Loss from abnormal spoilage	350	
	Work in process control		500

Rework Cost ($190 from p. 601)

Normal: (common to all jobs)	Factory department overhead control	190	
	Materials control		40
	Payroll liability		100
	Factory overhead applied		50
Normal: (peculiar to specific jobs)	Same as preceding entry except that the debit of $190 would be to Work in Process Control		
Abnormal:	Same as preceding entry except that the debit of $190 would be to Loss from Abnormal Spoilage		

[9]Kaplan has observed that "it is customary to assign to scrap at least its raw-material value were it to be purchased or sold on the open market. Perhaps, however, the scrap value should be reduced by the increased materials handling and storage costs, plus the cost of disrupting the production schedule, when defective output is produced. This treatment places more of a premium on eliminating defective items in the production process." R. Kaplan, "Measuring Manufacturing Performance: A New Challenge for Managerial Accounting Research," *Accounting Review*, October 1983, p. 701.

Scrap Value Recovered ($45 from p. 604)

Normal: (common to all jobs)	Materials control or cash or accounts receivable	45	
	Factory department overhead control		45
Abnormal: (peculiar to specific jobs)	Same as preceding entry except that the credit would be to Work in Process Control		

Although these journal entries are based on a job-order costing system, similar entries are made in a process-costing system. Of course, process costing, by definition, has no specific jobs for the identification of costs. Practices vary considerably from company to company.

PROBLEM FOR SELF-STUDY

PROBLEM

Suppose the manufacturing cost of some spoiled goods at Burlington Textiles is $4,000. Prepare a journal entry for each of the following conditions under both (a) process costing (Department A) and (b) job costing:

1. Abnormal spoilage of $4,000
2. Normal spoilage of $4,000 related to general factory operations
3. Normal spoilage of $4,000 related to specifications of a particular job

SOLUTION

(a) Process Costing			(b) Job Costing		
1. Loss from abnormal spoilage	4,000		Loss from abnormal spoilage	4,000	
Work in process—Department A		4,000	Work in process control (job)		4,000
2. No entry until goods are transferred. Then the normal spoilage costs are transferred along with the other costs:			Factory department overhead control	4,000	
			Work in process control (job)		4,000
Work in process—Department B	4,000				
Work in process—Department A		4,000			
3. Not applicable			No entry. Let the spoilage cost remain in Work in Process.		

SUMMARY

Despite extensive worldwide efforts to improve quality, nearly every manufacturing company has some problems of spoilage, rework, or scrap. Some of these problems are a normal result of efficient production. Other problems result from inefficient production. Spoilage is a major problem in some industries. Distinguish between normal and abnormal spoilage. Standards or norms must be computed so that performance may be judged and costs properly accounted for. Normal spoilage is spoilage that is unavoidable under a given set of efficient production conditions; abnormal spoilage is spoilage that is not expected to arise under efficient conditions.

When spoilage occurs in process costing, trace the units spoiled as well as the units finished and in process. Compute both normal and abnormal spoiled units separately. Build separate costs of spoiled units. Then add normal spoilage cost to good units passing inspection; charge off abnormal spoilage costs as a loss. Even if no abnormal spoilage exists, it is helpful to compute normal spoilage costs separately. In this way, management will constantly be reminded of the normal spoilage cost of a given process.

The allocation of the cost of normal spoilage usually depends on where spoilage is detected. If spoilage is detected at completion, then no cost of normal spoilage is added to ending work in process inventory. This assumption was used in the illustrations of the weighted-average and FIFO methods. In contrast, inspection may take place at some other stage—say, at the halfway point in the production cycle. In such a case, some normal spoilage cost would also be added to the units in process that are more than half completed.

Accounting for spoilage, rework, and scrap varies considerably in practice. Many companies include net costs in budgeted overhead rates.

APPENDIX: INSPECTION AND SPOILAGE AT INTERMEDIATE STAGES OF COMPLETION IN PROCESS COSTING

Objective 8

Describe the affects of inspection timing on normal and abnormal spoilage

Consider how the timing of an inspection affects the amount of normal and abnormal spoilage. What happens when the inspection occurs when production has reached various stages of completion for conversion costs? Assume that normal spoilage is 10% of the good units passing inspection in the Forging Department of Dana Corporation, a manufacturer of automobile parts. January data follow.

Suppose inspection had occurred at the 50%, the 100%, or the 10% conversion stage. A total of 8,000 units are spoiled in all cases. Note how the number of units of normal spoilage and abnormal spoilage change:

	Physical Units		
	Inspection at Stage of Completion		
Flow of Production	**at 50%**	**at 100%**	**at 10%**
Work in process, beginning (25%)*	11,000	11,000	11,000
Started during January	74,000	74,000	74,000
To account for	85,000	85,000	85,000
Good units completed and transferred out (85,000 − 8,000 spoiled − 16,000 ending)	61,000	61,000	61,000
Normal spoilage	7,700†	6,100‡	6,600§
Abnormal spoilage (8,000 − normal spoilage)	300	1,900	1,400
Work in process, ending (75%)*	16,000	16,000	16,000
Accounted for	85,000	85,000	85,000

*Degree of completion for conversion costs of this department at the dates of the work in process inventories. Direct materials are added at the start of production.
†10% × (11,000 beginning + 50,000 started and completed during current period + 16,000 ending).
‡10% × (11,000 + 50,000).
§10% × (50,000 + 16,000).

EXHIBIT 18-7

Dana Corporation Forging Department
Equivalent Units
For the Month Ended January 31, 19_1
Weighted-Average Method

Flow of Production	(Step 1) Physical Units	(Step 2) Equivalent Units	
		Direct Materials	Conversion Costs
Work in process, beginning	11,000 (25%)		
Started during current period	74,000		
To account for	85,000		
Good units completed and transferred out	61,000	61,000	61,000
Normal spoilage	7,700	7,700	3,850
Abnormal spoilage	300	300	150
Work in process, ending	16,000 (75%)	16,000	12,000
Accounted for	85,000		
Total work done to date		85,000	77,000

Exhibit 18-7 shows the computation of equivalent units under the weighted-average method and assuming inspection at the 50% conversion stage. The calculations depend on how much direct materials and conversion costs were incurred to get the units to the point of inspection. In Exhibit 18-7, the spoiled units have a full measure of direct materials and a 50% measure of conversion costs.

The key computations are those of equivalent units. The computations of equivalent unit costs and the applications of total costs to goods completed and in ending work in process would be similar to those in previous illustrations. In this example, however, note that ending work in process has passed the inspection point. Therefore, these units would bear a normal spoilage cost, just like the units that have been completed and transferred out.[10]

[10]Under the weighted-average method, normal-spoilage equivalent units should be based on all units that ever passed inspection, not just the units that passed inspection during the current period. Suppose in Exhibit 18-7 that beginning work in process had passed inspection in the preceding period. Then computations of equivalent units should include the equivalent units for normal spoilage related to the beginning work in process.

TERMS TO LEARN

This chapter and the Glossary at the end of the book contain definitions of the following important terms:

abnormal spoilage *(p. 593)* normal spoilage *(593)* reworked units *(592)* spoilage *(592)*

ASSIGNMENT MATERIAL

QUESTIONS

18-1 Why is there an unmistakable trend in manufacturing to improve quality?

18-2 "Spoilage costs are only one facet of the costs of quality control." Do you agree? Explain.

18-3 "Management has two major planning and control problems regarding spoilage." What are the two problems?

18-4 "Normal spoilage is planned spoilage." Discuss.

18-5 "Costs of abnormal spoilage are lost costs." Explain.

18-6 "Under some circumstances, all spoilage, rework, and scrap would be treated as abnormal." Explain.

18-7 "Accounting for abnormal spoilage as a period cost is consistent with financial accounting theory." Explain.

18-8 "What has been regarded as normal spoilage in the past is not necessarily acceptable as normal in the present or future." Explain.

18-9 "In accounting for spoiled goods, we are dealing with cost allocation and application rather than cost incurrence." Explain.

18-10 "Total input includes the abnormal as well as the normal spoilage and is therefore irrational as a basis for computing normal spoilage." Do you agree? Why?

18-11 "The point of inspection is the key to the application of spoilage costs." Do you agree? Explain.

18-12 "The unit cost of normal and abnormal spoilage is the same." Do you agree? Explain.

18-13 "The practical treatments of spoilage in job-order costing can be criticized on conceptual grounds." What is the major criticism?

18-14 "In job-order costing, the costs of specific normal spoilage are charged to specific jobs." Do you agree? Explain.

18-15 "The costs of reworking defective units are always charged to the specific jobs where the defects were originally discovered." Do you agree? Explain.

18-16 Describe the general accounting when scrap is reused in production.

18-17 "Abnormal rework costs should be charged to a loss account, not to factory overhead." Do you agree? Explain.

18-18 When is a company justified in inventorying scrap?

18-19 Prepare a journal entry appropriate for the placing of scrap in a storeroom for future use as direct materials. Assume that its estimated net realizable value is $2,000.

EXERCISES AND PROBLEMS

18-20 Reducing normal spoilage rates. The St. Louis Auto Supply Company has been facing increased competition from foreign companies. Customers have complained that St. Louis products have lower-quality parts that frequently require replacements. Company managers have launched a full-scale effort at total quality control, including heavy employee involvement on all aspects of quality—from the receipt of materials through production and customer service.

The production of sparkplug wires in 19_4 through 19_7 had entailed a spoilage rate of 5% of good output, which was considered normal. Employee suggestions, however, had led to a combination of improved materials and better-maintained machines. As a result, by 19_9 a spoilage rate of 1% of good output was considered normal.

The manufacturing cost of good output in 19_9, before considering normal spoilage, was $4,000,000.

Required

1. Compute the manufacturing cost of normal spoilage if the rate is (a) 5%, (b) 1%.
2. Compute the difference between requirement 1a and requirement 1b. Indicate some other possible financial effects of the lower spoilage rate.

18-21 Two ways of accounting for spoilage. (CPA) In manufacturing activities, a portion of the units placed in process is sometimes spoiled and becomes practically worthless. Discuss two ways in which the cost of such spoiled units could be treated in the accounts, and describe the circumstances under which each method might be used.

18-22 Normal and abnormal spoilage in units. A grinding process had the following data in physical units for January:

Work in process, beginning	19,000
Started during current period	150,000
To account for	169,000
Spoiled units	12,000
Good units completed and transferred out (169,000 − 12,000 spoiled − 25,000 ending)	132,000
Work in process, ending	25,000
Accounted for	169,000

Inspection occurs at the 100% conversion stage. Normal spoilage is 5% of the good units passing inspection.

Required

1. Compute the normal and abnormal spoilage in units.
2. Assume that the manufacturing cost of a spoiled unit is $10. Compute the amount of potential savings if all spoilage were eliminated, assuming that all other costs would be unaffected. Comment on your answer.

18-23 Normal and abnormal spoilage in units. Data for a milling process for November follow: work in process, beginning, 20,000; good units completed and transferred out during current period, 90,000; work in process, ending, 17,000. Inspection is at the 100% stage of completion regarding conversion costs, which are incurred evenly throughout the process. Total spoilage is 7,000 units. Normal spoilage is 4% of the good units passing inspection.

Required

1. Compute the normal and abnormal spoilage in units.
2. Assume that the manufacturing cost of a spoiled unit is $1,000. Compute the amount of potential savings if all spoilage were eliminated, assuming that all other costs would be unaffected. Comment on your answer.

18-24 Normal and abnormal spoilage in units. October data for a forging process follow: work in process, beginning, 30,000; good units completed and transferred out during current period, 210,000; work in process, ending, 34,000. Inspection is at the 100% stage of completion regarding conversion costs, which are incurred evenly throughout the process. Total spoilage is 18,000 units. Normal spoilage is 3% of the good units passing inspection.

Required

1. Compute the normal and abnormal spoilage in units for October.
2. Suppose the normal spoilage rate were 5% instead of 3%. Repeat requirement 1.
3. Compute the number of units started in October.

18-25 Unknowns in physical units. An extruding process of a plastics company had the following physical-unit data for September: work in process, beginning, 10,000; good units completed and transferred out, 190,000; work in process, ending, 14,000; total spoilage 15,000. Normal spoilage is 4% of good units completed and transferred out. Inspection occurs at the 100% stage of conversion.

Required

Compute (1) abnormal spoilage and (2) units started during September.

18-26 Physical units, inspection at various stages of completion. (Study the chapter appendix.) Normal spoilage is 6% of the good units passing inspection in a forging process. In March, a total of 10,000 units were spoiled. Other data follow: units started during March, 120,000; work in process, beginning, 14,000 units (20% completed for conversion costs); work in process, ending, 11,000 units (70% completed for conversion costs).

Required

In columnar form, compute the normal and abnormal spoilage in units, assuming inspection at 40%, 100%, and 15% stages of completion.

18-27 Normal and abnormal spoilage. The Van Brocklin Company manufactures one style of long, tapered wax candle, which is used on festive occasions. Each candle requires a two-foot-long wick and one pound of a specially prepared wax. The wick and melted wax are placed in molds and allowed to harden for twenty-four hours. Upon removal from the molds, the candles are immediately dipped in a special coloring mixture that gives them a glossy lacquer finish. Dried candles are inspected, and all defective ones are pulled out. Because the coloring mixture penetrates into the wax itself, the defective candles cannot be salvaged for reuse. They are destroyed in an incinerator. Normal spoilage is regarded as 3% of the number of candles that pass inspection.

Cost and production statistics for a certain week follow:

Direct materials requisitioned (including wicks and wax)	$3,340.00
Direct labor and overhead costs (applied at a constant rate during the hardening process)	1,219.50
Total cost	$4,559.50

During the week, 7,800 candles were completed; 7,500 passed inspection, and the remainder were defective. At the end of the week, 550 candles were still in the molds; they were considered 60% complete. There was no beginning inventory.

Required

1. The cost of the 7,500 candles that passed inspection was (choose one): (a) $4,333.73, (b) $4,125.00, (c) $4,217.85, (d) $4,290.00, (e) $4,248.75. Show computations.
2. The cost of the candles still in the molds at the end of the week was (choose one): (a) $393.25, (b) $269.50, (c) $185.13, (d) $274.72, (e) $300.30. Show computations.

18-28 Weighted-average method, spoilage. Consider the following data for a Cooking Department for the month of January:

	Physical Units
Work in process, beginning	11,000 (25%)*
Started during current period	74,000
To account for	85,000

Good units completed and transferred out during current period:	
From beginning work in process	11,000
Started and completed	50,000
Good units completed	61,000
Spoiled units	8,000
Work in process, ending	16,000 (75%)*
Accounted for	85,000

*Degree of completion for conversion costs of this department at the dates of the work in process inventories. Direct materials are added at the start of production.

Inspection occurs when production is 100% completed. Normal spoilage is 6,600 units for the month.

The following cost data are available:

Beginning inventory:		
Direct materials	$220,000	
Conversion costs	30,000	$ 250,000
Costs added during current period:		
Direct materials		1,480,000
Conversion costs		942,000
Costs to account for		$2,672,000

Required

Prepare a detailed production-cost work sheet. Use the weighted-average method. Distinguish between normal and abnormal spoilage.

18-29. FIFO method, spoilage. Consider the data in Problem 18-28. Using the FIFO method, prepare a detailed production-cost worksheet. Distinguish between normal and abnormal spoilage.

18-30 Weighted-average method, spoilage. The manufacture of convenience foods such as potato chips and corn chips proceeds through several departments. For example, Frito-Lay's departmental processes include cleaning, mixing, cooking, drying, and packaging. At the end of processing in each department, the units of product are inspected; only those units that pass inspection are transferred out. The spoiled units (both normal and abnormal) cannot be salvaged, have no scrap value, and are thrown away. The company has rigid quality standards, so normal spoilage is relatively high.

Consider the Cooking Department. For simplicity, ignore transferred-in costs. Below are data regarding a week's production of barbecue-flavored potato chips:

Costs applied to product:	
Direct materials	At the beginning of processing in the department
Conversion costs	Evenly throughout entire period of processing
Work in process, beginning	
Number of units	600 cases
Percent complete	66⅔%
Accumulated costs:	
Direct materials	$2,844
Conversion costs	$3,120
Work in process, ending:	
Number of units	1,400 cases
Percent complete	50%
Good units completed	2,800 cases
Normal spoilage (detected by inspection of product upon completion of processing)	140 cases
Abnormal spoilage:	
Number of units	60 cases
Percent complete	100%

Costs incurred during current period (exclusive of accumulated cost of work in process at beginning):	
Direct materials	$19,800
Conversion costs	34,260

Required
Indicate your answer by letter. Support your answers with a production-cost work sheet. Assume the weighted-average method. (Round off unit costs to the nearest cent.)

1. The total equivalent units for conversion costs were (a) 3,000, (b) 3,700, (c) 4,400, (d) 2,800, (e) 4,200.
2. The equivalent unit cost for direct materials was (a) $7.71, (b) $8.09, (c) $5.15, (d) $5.39, (e) $5.22.
3. The cost of the ending work in process was (a) $14,280, (b) $18,396, (c) $17,864, (d) $14,616, (e) $14,378.
4. The cost of goods completed and transferred out was (a) $42,700, (b) $43,615, (c) $44,100, (d) $42,010, (e) $44,835.

18-31 Weighted-average method and spoilage. The Alston Company operates under a process-cost system. It has two departments, Cleaning and Milling. For both departments, conversion costs are applied in proportion to the stage of completion. But direct materials are added at the *beginning* of the process in the Cleaning Department, and additional direct materials are added at the *end* of the milling process. The costs and unit production statistics for May follow. All unfinished work at the *end* of May is 25% completed. The beginning inventory (May 1) was 80% completed as of May 1. All completed work is transferred to the next department.

Beginning Inventories	**Cleaning**	**Milling**
Cleaning: $1,000 direct materials, $800 conv. costs	$1,800	
Milling: $6,450 previous dept. cost (transferred-in cost) and $2,450 conversion costs		$8,900
Costs Added During Current Period		
Direct materials	$9,000	$ 640
Conversion costs	$8,000	$4,950
Physical Units		
Units in beginning inventory	1,000	3,000
Units started this month	9,000	7,400
Total units completed and transferred out	7,400	6,000
Normal spoilage	500	400
Abnormal spoilage	500	0

Additional Factors

1. Spoilage is assumed to occur at the *end* of *each* of the two processes when the units are inspected.
2. Assume that there is no shrinkage, evaporation, or abnormal spoilage other than that indicated in the tabulation above.
3. Carry unit-cost calculations to three decimal places where necessary. Calculate final totals to the nearest dollar.

Required
Using the weighted-average method, prepare a production-cost work sheet for the first department only. (Problem 18-33 explores additional facets of this problem.)

18-32 FIFO. Redo Problem 18-31, using FIFO. (Problem 18-34 explores additional facets of this problem.)

18-33 Continuation of Problem 18-31, Second Department. Using the facts in Problem 18-31 and the weighted-average method, prepare a production-cost work sheet for the Milling Department.

18-34 Continuation of Problem 18-32, Second Department. Using the facts in Problem 18-31 and the FIFO method, prepare a production-cost work sheet for the Milling Department.

18-35 FIFO and weighted-average methods, spoilage. (D. Kleespie) The CeeKay Company uses a FIFO process-cost accounting system. Direct materials are added at the beginning of processing in the Mixing Department. Inspection takes place at the end of the process. Conversion costs are added evenly throughout the process. Operating data for August include:

	Pounds
Beginning inventory of work in process, 30% complete as to conversion costs ($13,500 conversion cost)	500
Good units completed and transferred out	2,000
Normal spoilage	60
Abnormal spoilage	40
Ending inventory of work in process, 60% complete as to conversion costs	600
Equivalent unit cost for conversion costs, $100 per pound	

Required

1. How many units were started in production in the Mixing Department during August?
2. How many equivalent units were used in determining the unit cost for conversion?
3. What is the total dollar amount of the conversion-cost portion of the ending work in process inventory?
4. What is the total dollar amount of the conversion-cost portion being transferred out of the Mixing Department?
5. What is the total amount of conversion costs charged to the Mixing Department this period?
6. Assume instead that the CeeKay Company uses the weighted-average method. How many divisor equivalent units are used to determine the average unit cost for conversion?
7. Assume as in requirement 6 that the CeeKay Company uses the weighted-average method. What is the total amount of conversion costs charged to the Mixing Department this period? Assume the weighted-average conversion costs per equivalent unit are $100.

18-36 Process costs, weighted-average, two departments, spoilage. (SMA) The Quebec Manufacturing Company produces a single product. There are two producing departments, Departments 1 and 2, and the product passes through the plant in that order.

There were no work in process inventories at the beginning of the year. In January, direct materials for 1,000 units were issued to production in Department 1 at a cost of $5,000. Direct labor and factory-overhead costs for the month were $2,700. During the month, 800 units were completed and transferred to Department 2. The work in process inventory at the end of the month contained 200 units, complete in materials and one-half complete in labor and overhead.

Direct labor and factory overhead in Department 2 amounted to $6,250 in January. During the month, 500 units were completed and transferred to finished goods. At the end of the month, 200 units remained in process, one-quarter complete. Ordinarily, in Department 2, spoilage is recognized upon inspection at the end of the process, but in January there was an abnormal loss of 50 units when they were one-half complete. The effect of abnormal loss is not to be included in inventory.

Required

Prepare a detailed production-cost work sheet for each department for the month of January. Use the weighted-average method.

18-37 Process costing, inspection at 70% stage. (CMA) APCO Company manufactures various lines of bicycles. Because of the high volume of each product, the company employs a process-cost system using the weighted-average method to determine unit costs. Bicycle parts are manufactured in the Molding Department. The parts are consolidated into a single bike unit in the Molding Department and transferred to the Assembly Department, where they are partially assembled. After assembly, the bicycle is sent to the Packing Department.

Cost per unit data for the 20-inch dirt bike has been completed through the Molding Department. Annual cost and production figures for the Assembly Department are presented in the accompanying schedules.

Defective bicycles are identified at an inspection point when the assembly labor process is 70% complete; all assembly materials have been added prior to this point of the process. The normal rejection percentage for defective bicycles is 5% of the bicycles reaching the inspection point. Any defective bicycles above the 5% quota are considered as abnormal spoilage. All defective bikes are removed from the production process and disposed of with zero net disposal value.

Assembly Department Cost Data

	Transferred in From Molding Department	Assembly Materials	Assembly Conversion Costs	Total Cost of Dirt Bike Through Assembly
Prior period costs	$ 82,200	$ 6,660	$ 11,930	$ 100,790
Current period costs	1,237,800	96,840	236,590	1,571,230
Total costs	$1,320,000	$103,500	$248,520	$1,672,020

Assembly Department Production Data

		Percent Complete		
	Bicycles	Transferred-in Costs	Assembly Materials	Assembly Conversion Costs
Beginning inventory	3,000	100	100	80
Transferred in from molding department during year	45,000	100	—	—
Transferred out to packing department during year	40,000	100	100	100
Ending inventory	4,000	100	50	20

Required

1. Compute the number of
   a. Defective bikes normally spoiled
   b. Defective bikes abnormally spoiled
2. Compute the equivalent units of production for the year for
   a. Bicycles transferred in from the Molding Department
   b. Bicycles produced with regard to assembly materials
   c. Bicycles produced with regard to assembly conversion costs
3. Compute the cost per equivalent unit for the fully assembled dirt bike.
4. Compute the amount of the total production cost of $1,672,020 that will be associated with the following items:
   a. Abnormal defective units
   b. Good units completed in the Assembly Department
   c. Normal defective units
   d. Ending work in process inventory in the Assembly Department
5. Describe how the applicable dollar amounts for the following items would be presented in the financial statements:
   a. Abnormal defective units
   b. Completed units transferred into the Packing Department
   c. Normal defective units
   d. Ending work in process inventory in the Assembly Department

18-38 Different ways of accounting for spoilage. (CPA) The D. Hayes Cramer Company manufactures Product C, whose cost per unit is $1 of (direct) materials, $2 of (direct) labor, and $3 of factory-overhead costs. During the month of May, 1,000 units of Product C were spoiled. These units could be sold for 60¢ each.

The accountant said that the entry to be made for these 1,000 lost or spoiled units could be one of the following four:

Entry No. 1

Spoiled goods inventory	600	
Work in process		600

Entry No. 2

Spoiled goods inventory	600	
Factory department overhead control	5,400	
Work in process		6,000

Entry No. 3

Spoiled goods inventory	600	
Loss on spoiled goods	5,400	
Work in process		6,000

Entry No. 4

Spoiled goods inventory	600	
Accounts receivable	5,400	
Work in process		6,000

Required

Indicate the circumstance under which each of the four entries above would be appropriate. Assume a job-order costing system.

18-39 Scrap in job-order costing. The Lorenzo Company has an extensive job-costing facility that uses a variety of metals. Scrap from Job 372 has a total sales value of $490.

Required

Consider each requirement independently. Unless otherwise stated, assume that the physical amount of scrap (only) is accounted for in storage. That is, no journal entry is made until a sale occurs. Omit explanations of the journal entries.

1. The scrap is sold for $490 cash. Sales of scrap are regarded as a separate line item of other revenues. Prepare the journal entry.
2. The scrap is sold for $490 cash. Sales of scrap are regarded as offsets against factory overhead. Prepare the journal entry.
3. The scrap is sold for $490 cash. Sales of scrap are credited to the jobs that generated the scrap. Prepare the journal entry.
4. The scrap is returned to the storeroom, and an entry is made at its $490 sales value. A month later, the scrap is reused as direct material on a subsequent job. Prepare the journal entries.

18-40 Job-cost spoilage and scrap. (F. Mayne) Santa Cruz Metal Fabricators, Inc., has a large job, No. 2734, which calls for producing various ore bins, chutes, and metal boxes for enlarging a copper concentrator. The following charges were made to the job in November 19_9:

Direct materials	$26,951
Direct labor	15,076
Factory overhead	7,538

The contract with the customer called for the total price to be based on a cost-plus approach. The contract defined *cost* to include direct materials, direct labor, and factory overhead to be applied at 50% of direct labor. The contract also provided that the total costs of all work spoiled were to be removed from the billable cost of the job and that the benefits from scrap sales were to reduce the billable cost of the job.

Required

1. In accordance with the stated terms of the contract, record the following two items in general journal form:
   a. A cutting error was made in production. The up-to-date job-cost record for the batch of work involved showed materials of $650, direct labor of $500, and applied overhead of $250. Because fairly large pieces of metal were recoverable, the company believed that the salvage value was $600 and that the materials recovered could be used on other jobs. The spoiled work was sent to the warehouse.
   b. Small pieces of metal cuttings and scrap in November 19_9 amounted to $1,250, which was the price quoted by a scrap dealer. You may assume that there were no entries made with regard to the scrap until the price was quoted by the scrap dealer. The scrap dealer's offer was immediately accepted.
2. Consider normal and abnormal spoilage. Suppose the contract described above had contained the clause "a normal spoilage allowance of 1% of the job cost will be included in the billable cost of the job."
   a. Is this clause specific enough to define exactly how much spoilage is normal and how much is abnormal? *Hint:* Consider the inputs versus outputs distinction that was presented on page 595.
   b. Repeat requirement 1a with this "normal spoilage of 1%" clause in mind. You should be able to provide two slightly different entries.

18-41 Spoilage and job costing. (L. Bamber) Bamber Kitchens produces a variety of items in accordance with special job orders of hospitals, factories, and food manufacturers. The following $6 cost per case pertains to an order for 2,500 cases of mixed vegetables: direct materials, $3; direct labor, $2; and factory overhead applied, $1. The factory-overhead rate includes a provision for normal spoilage.

Required

1. Assume that a laborer dropped and totally destroyed 200 cases. Prepare a journal entry to record this event. What is the unit cost of the remaining 2,300 cases?
2. Reconsider requirement 1. Suppose part of the 200 cases could be salvaged and sold to a nearby prison for $200 cash. How would your answer to requirement 1 change?
3. Refer to the original data. (The facts given in requirement 1 are not pertinent to requirement 3.) Tasters at the company reject 200 of the 2,500 cases, and the 200 rejected cases are destroyed. Assume this rejection rate is considered normal. Prepare a journal entry to record this event, and calculate the unit cost if
   a. The rejection is attributable to exacting specifications of this particular job
   b. The rejection is characteristic of the production process and is not attributable to this specific job.
4. Reconsider requirement 3. Suppose part of the 200 cases could be salvaged and sold to a nearby welfare agency for $400. How would your answers to requirement 3 change?
5. Refer to the original data. Tasters rejected 200 cases that had insufficient salt. The product can be placed in a vat, salt added, and reprocessed into jars. This operation will cost $200; this step is taken regularly because of the difficulty in seasoning. Prepare a journal entry to record this event. What is the average unit cost of all the cases?
6. How would your answer to requirement 5 change if the rejection occurred because of the exacting specifications of this particular job?
7. How would your answer to requirement 5 change if the rejection occurred because a worker simply forgot to add the salt?

18-42 Weighted-average, inspection at 80% completion. (A. Atkinson) (Study the chapter appendix.) Ottawa Manufacturing produces a plastic toy in a two-stage manufacturing operation. The company uses a weighted-average process-costing system. During the month of June, the following data were recorded for the Finishing Department:

Units of beginning inventory	10,000
% completion of beginning units	25%
Cost of direct materials in beginning work in process	$ 0
Units started	70,000
Units completed	50,000
Units in ending inventory	20,000
% completion of ending units	95%
Spoiled units	10,000
Costs added during current period:	
Direct materials	$655,200
Direct labor	$635,600
Factory overhead	$616,000
Work in process, beginning:	
Conversion costs	$ 42,000
Transferred-in costs	$ 82,900
Cost of units transferred in during current period	$647,500

Conversion costs are incurred evenly throughout the process. Direct material costs are incurred when production is 90% complete. Inspection occurs when production is 80% complete. Normal spoilage is 10% of all good units that pass inspection.

Required
Prepare a production-cost work sheet for the month of June. Show supporting computations.

CHAPTER 19

Harley-Davidson uses backflush costing to simplify its accounting system at its motorcycle assembly plants.

Operation Costing, Backflush Costing, and Project Control

Learning Objectives

When you have finished studying this chapter, you should be able to

1. Explain how hybrid costing systems develop in relation to production systems
2. Describe the major features of just-in-time production systems
3. Relate employee involvement and total quality control to just-in-time production
4. Provide an overview of operation costing
5. Prepare journal entries for an operation costing system
6. Prepare journal entries for three versions of backflush costing systems
7. Identify the four critical success factors of project costing
8. Compute project variances

This chapter has two major parts. After describing some features of production systems—particularly just-in-time, or JIT—Part One focuses on operation costing. We illustrate how underlying production systems affect the design of product costing systems. Part One then describes backflush costing, which fits especially well with just-in-time production systems. Part Two, which can be studied separately, provides an overview of the control of projects.

PART ONE: OPERATION COSTING AND BACKFLUSH COSTING

VARIETIES OF PRODUCTION SYSTEMS

Tailoring Cost Accounting to Production Systems

Objective 1

Explain how hybrid costing systems develop in relation to production systems

Hybrid costing systems are blends of characteristics from both job costing and process costing. Recall that job-order product costing systems and process-product costing systems are best viewed as ends of a continuum:

For specific units or small batches of custom-made products ←		→ For mass production of like units
↑ Job-order product-costing systems	Hybrid costing systems	↑ Process product-costing systems

As we have seen, job costing usually accompanies the custom-order manufacturing of relatively heterogeneous (different) products or services (for example, printing a few posters, hand-tailoring a suit, constructing a homeowner's sidewalk). At its extreme, custom-order manufacturing entails issuing a specific job order, work order, or production order that entails unique direct materials and a set of production steps that differs for each job order.

In contrast, process costing usually accompanies the mass production and continuous-flow manufacturing of homogeneous (uniform) products (for example, printing rolls of wallpaper, stitching shirts, and bagging cement). Continuous-flow manufacturing usually entails mass production of standardized products.

Obviously, a product costing system should be tailored to the underlying production system. Hybrid costing systems develop because there are hybrid production systems, which are blends of custom-order manufacturing and continuous-flow manufacturing. Manufacturers of a relatively wide variety of closely related standardized products probably use a hybrid system. Examples of such products include Ford automobiles, Jantzen swim wear, and Anheuser-Busch beers. Ford automobiles may be manufactured in a continuous flow, but each automobile may be customized with a special combination of motor, transmission, radio, and so on. Each company develops its own hybrid costing system to meet its individual needs.

Just-in-Time (JIT) Production Systems

Objective 2

Describe the major features of just-in-time production systems

Just-in-time (JIT) production is a system in which each component on a production line is produced immediately as needed by the next step in the production line. JIT systems are usually forms of continuous-flow production systems with minimal levels of work in process. Ideally, JIT operates with zero inventories. Many related terms describe the JIT approach, including MAN (materials as needed), MIPS (minimum inventory production system), and ZIPS (zero inventory production system).

In a JIT production line, manufacturing activity at any particular workstation is spurred by the use of that station's output at downstream stations. Ideally, the sale of a unit of finished goods triggers the completion of a unit in final assembly, and so forth, backward in the sequence of manufacturing all the way to purchasing raw material. This characteristic is often called the "demand-pull" feature of a JIT production line. At Hewlett-Packard (HP), workers and managers frequently quote the following demand-pull slogan: "Never build nothing, nowhere, for nobody, unless they *ask* you for it."

Major Features of JIT

In contrast to other production systems, JIT production systems have four major features:

- Inventory is regarded as an evil because of its "non-value added" nature.
- Production is stopped if parts are absent or defects are discovered; corrective actions are taken before production is resumed.
- Emphasis is placed on reducing production cycle time.
- Focus is centered on simplifying all production activities.

Advocates of JIT often talk about their desire to eliminate "non-value-added" activities—that is, activities that do not affect how customers perceive a product. How do we distinguish between value-added and non-value-added characteristics? Consider a person who buys a television set. She probably assigns values based on the set's performance and quality. The non-value-added elements pertaining to the television set may include inventory costs. It matters not at all to this person what costs the manufacturer incurred in handling its television inventory. Eliminating non-value-added activities, then, will save the manufacturer money and do no harm to customer relations. Chapter 29 (pp. 918–19) presents further discussion of the value-added/nonvalue-added approach to cost management.

Ideally, a customer wants a perfect product and an order filled instantly. If purchasing and production are managed flawlessly, a customer order should trigger immediate delivery of direct materials, followed by immediate production and by immediate delivery of the finished goods. Hence, there would be no need for non-value-added activities such as holding any inventories or reworking defective goods. Inventory is regarded as an evil, and rigid limits are therefore imposed on all inventories, from direct materials and parts through all stages of production.

If parts are absent or defects are discovered, production is stopped. In contrast, many traditional manufacturing systems commonly (1) have extra parts and subassemblies held at workstations in anticipation of shortages or production breakdowns or (2) build around a missing part, adding it later when the part becomes available.

In JIT systems, management emphasizes reducing *throughput time,* which is the time from when a product starts on the production line until it becomes a finished good. Throughput time is also often called *production cycle time.* Reduced throughput time enables a company to respond better to changes in customer demand.

JIT production tries to simplify activities on the production line so that non-value-added activities are spotlighted and reduced or eliminated. For example, production layouts make non-value-added activities more visible (such as the accumulation of unwanted inventory items, which are regarded as undesirable rocks in a smooth-flowing stream). Smaller space is allotted, so there is an upper limit to inventories. There are fewer inventory items, so workers can easily count them. Material handling between successive workstations is streamlined. For example, forklift trucks are regarded as non-value-added fossils.

Employee Involvement and Total Quality Control

Objective 3

Relate employee involvement and total quality control to just-in-time production

Factories that have switched to the JIT approach for manufacturing use *employee involvement* as a major means of control. The managers and all other employees, working as a team, are expected to be sufficiently familiar with daily operations to take prompt action to keep production on schedule and to control costs.

The pursuit of *total quality control* (TQC) also tends to get more attention in a JIT system. Each worker assumes responsibility for his or her own work. Vendors guarantee that materials and parts are free from defects. Although JIT production is usually accompanied by an emphasis on both employee involvement and total quality control, these ideas are conceptually distinct. For example, TQC programs can be used in any production system. See Chapter 29 for more discussion of quality issues.

Good relationships with one or a few vendors are a critical component of JIT. Vendors inspect their own goods and guarantee their quality. These vendors provide high-quality items and make frequent deliveries to the factory. In turn, these materials and parts are immediately placed into production. This procedure completely eliminates incoming inspection, the storeroom, and the corresponding Materials Inventory account. The JIT company saves money by using reliable vendors.

If problems arise with materials in non-JIT production systems, the manufacturing of defective products often continues. In contrast, JIT creates an urgency for solving problems immediately.

Financial Benefits of JIT

JIT, employee involvement, and TQC should lead to many financial benefits, including

1. Lower investments in inventories
2. Reductions in carrying and handling costs of inventories
3. Reductions in risks of obsolescence of inventories
4. Lower investments in factory space for inventories and production
5. Reductions in total manufacturing costs

    - Direct materials: Long-term contracts with one or a few vendors often lead to price reductions; improved quality

- Other costs: Lower labor costs from increased overall efficiency despite increases in downtime; reductions of scrap and rework; reductions in paper work

Examples of reductions in paper work include the issuance of blanket long-term orders to suppliers instead of purchase orders. Reductions can be dramatic. For example, a Taiwanese factory was processing 10,000 purchase orders, receiving reports, and materials requisitions per week. After switching to JIT, a simpler accounting system emerged. Ideally, there are no individual purchase orders, no receiving reports, no materials requisitions, no work orders, and no tracking of direct labor as a separate cost element.

JIT tends to focus broadly on the control of *total manufacturing costs* instead of individual costs such as direct labor. For example, idle time may rise because production lines are starved for materials more frequently than before. Still, many overhead costs will decline. Consider probable reductions in the costs of material handling and special inspectors for quality control.

Savings in total manufacturing costs under JIT have been substantial. The Vancouver division of Hewlett-Packard reported the following benefits two years after the adoption of JIT:

Work in process inventory dollars	Down 82%
Space used	Down 40%
Scrap/rework	Down 30%
Production time:	
Impact printers	Down from 7 days to 2 days
Thermal printers	Down from 7 days to 3 hours
Labor efficiency	Up 50%
Shipments	Up 20%

Chapter 23 expands the discussion of JIT. In particular, see Exhibit 23-8, page 754, for a summary of the beneficial effects of JIT production.

OPERATION COSTING

We will now explore two costing systems in some detail. The first, operation costing, is a hybrid that has existed for many years. The second, backflush costing, is a simplified standard costing system that began to receive attention in the 1980s.

Overview of Operation Costing

Objective 4

Provide an overview of operation costing

Operation costing is a hybrid costing system applied to batches of homogeneous products. Each batch uses the same resources of each activity to the same extent. That is, a single batch of products proceeds through a sequence of selected activities or operations. Within each operation, each product unit is treated exactly alike, consuming identical amounts of the operation's resources. Batches are also termed production runs, work orders, and production orders.

Each batch is often a variation of a single design and requires a particular sequence of standardized operations. Consider a business that makes suits. Management may select a single basic design for every suit that the company manufactures. Depending on specifications, batches of suits vary from each other. One batch may use wool; another batch, cotton. One batch may require special hand stitching; another batch, machine stitching. Other products that are likewise often manufactured in batches are semiconductors, textiles, and shoes.

Operation is defined as a standardized method or technique that is repetitively performed regardless of the distinguishing features of the finished product. Operations are usually conducted within departments. For instance, a suit maker may have cutting and hemming as operations within a single department. The term *operation,* however, is often used loosely. It may be a synonym for a *department* or *process*. For example, some companies may call their finishing department a finishing process or a finishing operation.

In operation costing, work orders initiate production. Product costs are compiled for each work order. As in job costing, direct materials—different in each work order—are specifically identified with the appropriate order. *Conversion costs*—all manufacturing costs other than direct material costs—are compiled for each operation and then applied to all physical units passing through the operation. A single average unit conversion cost is applied to each operation. Typical application bases are volume-related cost drivers such as the number of units worked on and the minutes used to complete the individual operation. As will be illustrated, the method for applying conversion costs is similar to applying factory overhead. (Our examples assume only two cost categories, direct materials and conversion costs. Of course, operation costing can have more than two categories.)

Exhibit 19-1 illustrates operation costing. Note how various batches move through some, but perhaps not all, operations. The T-accounts show how costs move through the ledger. Of course, the accounts for each operation could be

EXHIBIT 19-1
Overview of Operation Costing

PHYSICAL FLOWS

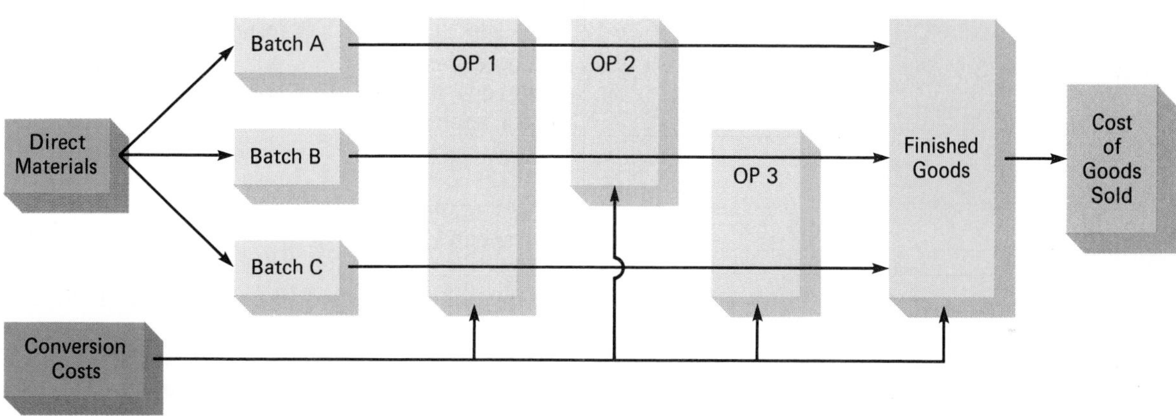

ACCOUNTING ENTRIES

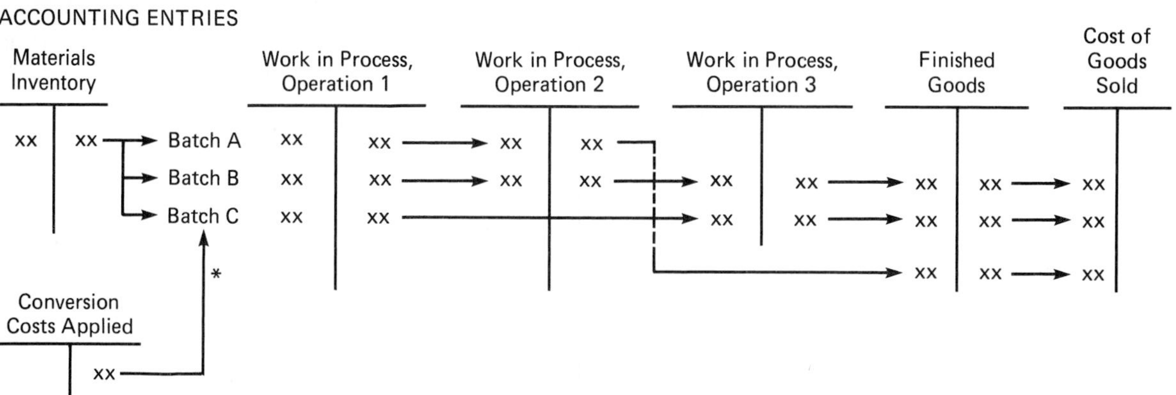

Conversion costs would be applied to the batches in stages as the units moved through the designated operations. Many companies would use the term factory overhead applied *instead of* conversion costs applied. *Moreover, if direct labor is a significant cost, it might be accounted for separately instead of being considered a subpart of factory overhead or conversion costs.*

grouped as a subsidiary ledger or as a single general ledger account, Work in Process.

As always, cost management requires attention to physical processes. For example, in the manufacturing of clothing, managers are concerned with fabric waste, the number of fabric layers that can be cut at one time, and so on. The cost accounting system captures the financial impact of the control of physical processes. Feedback provided by the cost accounting system helps managers improve the physical processes.

Illustration of Operation Costing

An operation costing system uses work orders that specify the needed direct materials and step-by-step operations. Work orders for each batch differ as to the combinations of direct materials and operations to be undergone.

Suppose Baltimore Company, a clothing manufacturer, produces two lines of blazers for Macy's department stores. Wool blazers use better-quality materials and undergo more operations than do polyester blazers. Consider the following condensed operations:

	Work Order 423	Work Order 424
Direct materials	Wool	Polyester
	Full satin lining	Partial rayon lining
	Bone buttons	Plastic buttons
Operations	1. Cutting cloth	1. Cutting cloth
	2. Checking edges	—
	3. Sewing body	3. Sewing body
	4. Checking seams	—
	5. —	5. Sewing collars and lapels by machine
	6. Sewing collars and lapels by hand	6. —

Suppose Work Order 423 is for 100 wool blazers and Work Order 424 is for 200 polyester blazers. The following costs are assumed:

	Work Order 423	Work Order 424
Number of blazers	100	200
Direct materials	$ 6,000	$3,000
Conversion costs:		
Operation 1	580	1,160
Operation 2	400	—
Operation 3	1,900	3,800
Operation 4	500	—
Operation 5	—	875
Operation 6	700	—
Total manufacturing costs	$10,080	$8,835

This operation costing system uses a budgeted or estimated costing rate to apply both direct labor and factory overhead to each operation. As in process costing, direct labor and factory overhead are often combined, and their sum is applied as conversion costs based on the company's budgeted rate for performing each operation. For example, the costs of Operation 5 might be budgeted as follows (amounts assumed):

$$\text{Budgeted rate for applying conversion costs to product in Operation 5} = \frac{\text{Budgeted conversion costs of Operation 5 for the year (direct labor, power, repairs, supplies, depreciation, other factory overhead of this operation)}}{\text{Budgeted machine hours for the year in Operation 5}}$$

$$\text{Budgeted conversion costs rate} = \frac{\$1{,}225{,}000}{35{,}000 \text{ hours}} = \$35 \text{ per machine hour}$$

As goods are manufactured, conversion costs are applied to the work orders by multiplying the $35 hourly application rate times the number of machine hours used in Operation 5. Suppose the 200 polyester blazers require 25 machine hours in Operation 5. The conversion cost of Operation 5 for 200 polyester blazers is 25 × $35 = $875.

Note that each product unit is assumed to consume identical amounts of resources of a particular operation. Consider Operation 3. Work Order 424, containing 200 units, has twice the total cost of the 100 units in Work Order 423 ($3,800 compared with $1,900). If Work Order 424 contained 150 units, its total cost would be 150% rather than 200% of the cost of Work Order 423.

In Operation 3, sewing, Work Order 424 will have twice as much conversion cost applied *only* if the cost application base is the number of output units (or some other factor directly proportional to the number of outputs). Suppose the application base for Operation 3 is direct labor hours. If each jacket in Work Order 423 uses .50 hour of sewing, then the total direct labor hours is .50 × 100 = 50. If each jacket in Work Order 424 uses only .25 hour of sewing—recall that this order is for cheaper jackets and so less time is likely spent on sewing each one—then the total direct labor hours is .25 × 200 = 50. In this case, equal conversion costs would be applied to both Work Orders for Operation 3.

Journal Entries

Objective 5

Prepare journal entries for an operation costing system

Summary journal entries for applying costs to the polyester blazers follow. Entries for the wool blazers would be similar.

The journal entry for the requisition of direct materials, which are traced directly to particular batches, for the 200 polyester blazers is:

1. Work in process, Operation 1	3,000	
Materials inventory		3,000

Actual conversion costs (assumed as $1,400 in Operation 1) are entered into a Conversion Costs account:

2. Conversion costs	1,400	
Various accounts (such as payroll liability and accumulated depreciation)		1,400

The application of conversion costs to products in operation costing uses the budgeted application rate. Assume that the amount applied for Operation 1 is the $1,160 in the preceding tabulation:

3. Work in process, Operation 1	1,160	
Conversion costs applied		1,160

The transfer of the polyester blazers from Operation 1 to Operation 3 (recall that the polyester blazers do not go through Operation 2) would be journalized as follows:

4. Work in process, Operation 3	4,160	
Work in process, Operation 1		4,160

After posting, Work in Process, Operation 1, appears as follows:

Work in Process, Operation 1

1. Direct materials	3,000	4. Transfer to Operation 3	4,160
3. Conversion costs applied	1,160		

As Exhibit 19-1, page 623, implies, the costs of the blazers are transferred through the pertinent operations and then to finished goods in the usual manner. Costs are added throughout the year in the accounts Conversion Costs and Conversion Costs Applied. Any overapplication or underapplication of conversion costs is disposed of in the same way as overapplied or underapplied factory overhead in a job-order costing system.

In sum, operation costing is a hybrid of job costing and process costing:

- Its job-costing feature is the specific tracking of direct materials costs and conversion costs to batches as the batches undergo specific operations.
- Its process-costing feature is that within each operation each product unit is treated exactly alike, consuming identical amounts of the operation's resources.

Operation Costing and Activity-Based Accounting Systems

Exhibit 19-2 presents an overview of the cost accounting systems described in this book. Operation costing is placed on the continuum of product costing between job-order costing and process costing. (It does not necessarily have to be exactly halfway between.) Below the continuum, we see how these three product costing systems can be combined with actual costing, normal costing, and standard costing (which were compared in Chapter 8, p. 261). Our operation costing example above used normal costing, but actual costing or standard costing could also have been used.

The left-hand vertical classification, which is a condensation of a similar presentation in Exhibit 5-5, page 151, displays how an accounting system can accumulate costs for planning and control. Costs can be identified with business functions alone or in more detailed ways with departments or activity areas.

Activity-based accounting systems provide the most detailed or finely granulated accounting data. They emphasize the necessity of using several different cost drivers to obtain accurate data for evaluating the performance of activity areas and

EXHIBIT 19-2

Varieties of Cost Accounting Systems

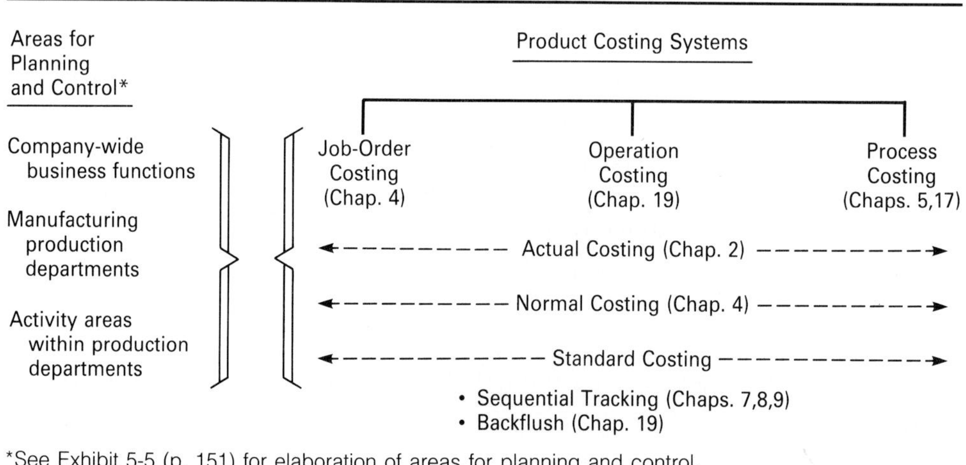

*See Exhibit 5-5 (p. 151) for elaboration of areas for planning and control.

for building up product costs. As Chapter 5 explains, activity-based accounting systems are regarded as the strongest accounting links with intelligent cost management. Activity-based accounting systems may be combined with any form or combination of product costing systems, as Exhibit 19-2 shows.

Exhibit 19-2 emphasizes the variety of choices available to management. The choice of the accounting system depends on the underlying production system and managers' desires regarding accuracy in both product costing and the control of physical processes. Generally, cost accounting systems become simpler and less expensive if (a) there is continuous-flow manufacturing of similar products and (b) there are minor fluctuations in work in process inventories. For example, General Electric's accounting system for the production of light bulbs is simpler than that for the production of aircraft engines.

BACKFLUSH COSTING

Simplified Standard Costing

The traditional standard costing systems (discussed in Chapters 7, 8, and 9) use **sequential tracking**, which is any product costing method that is synchronized with the physical sequences of purchases and production. The traditional standard costing system tracks costs in synchronization with the passage of the products from direct materials, through work in process, to finished goods. But sequential tracking is often expensive, especially if attempts are made to track direct material requisitions and time tickets to individual operations and products. Hence, as indicated in Exhibit 19-2, a simplified standard costing method, *backflush costing*, has been developed.

Backflush costing (also called **delayed costing** or **post-deduct costing**) is a standard costing system for product costing that focuses first on the output and then works backward to apply manufacturing costs to units sold and to inventories. It often accompanies JIT production systems, although it can be coupled with any production system. The term *backflush* probably arose because the trigger points for inventory costing entries can be delayed until as late as the time of sale. Costs are then finally flushed back through the accounting system.

Companies that adopt backflush costing often meet these three conditions:

1. Management wants a simple accounting system. No detailed tracking of actual amounts of direct materials or direct labor through a series of operations step by step to the point of completion is necessary.
2. Each product has a set of standard costs.
3. Backflush costing yields approximately the same financial results as sequential tracking would generate. This agreement in results occurs when total work in process inventory costs and total direct materials inventory costs are low or constant.

If inventories are low, the vast bulk of manufacturing costs will flow into Cost of Goods Sold and not be deferred in inventory. Hence, managers may not want to spend resources sequentially tracking costs through Work in Process, Finished Goods, and Cost of Goods Sold. Backflush costing is especially attractive in companies that have low inventories resulting from JIT. Even if inventory levels are high, if they are relatively stable, sequential tracking and backflush costing should produce approximately the same results. Constant amounts of costs will be deferred in inventory each period.

The following three examples demonstrate backflush costing. To underscore basic concepts, we assume no direct material variances in any of the examples. (We do, however, discuss variances in a separate section following Example 1.) The three examples differ in the number and placement of trigger points for making entries in the accounting system:

Objective 6

Prepare journal entries for three versions of backflush costing systems

	Number of Trigger Points	Location of Trigger Points
Example 1	2	1. Purchase of raw material (direct material) and components 2. Completion of good finished units of product
Example 2	2	1. Purchase of raw material and components 2. Sale of good finished units of product
Example 3	1	1. Completion of good finished units of product

Example 1

This example illustrates how backflushing can eliminate the need for a separate Work in Process account. A hypothetical company, Silicon Valley Computer (SVC), has two trigger points, which are places in their operations at which entries are made in the internal accounting system. SVC produces keyboards for personal computers. For April, there were no beginning inventories of raw materials. Moreover, there is zero beginning and ending work in process.

SVC has only one direct manufacturing cost category (direct materials, which are simply called "raw") and one indirect manufacturing cost category (conversion costs). All labor costs at the manufacturing facility are included in conversion costs. The April direct material cost per keyboard unit is $19; the conversion cost is $12.

SVC has two inventory accounts:

Type	Account Title
Combined materials and work in process	Inventory: Raw and In-process
Finished goods	Finished Goods

Trigger point 1 occurs when materials are purchased. These costs are charged to Inventory: Raw and In-process.

Actual conversion costs are recorded as incurred under backflush costing, just as in other costing systems. Conversion costs are applied to products at trigger point 2—the transfer of units to Finished Goods. This example assumes that underapplied conversion costs are carried forward and not disposed of until year-end.

SVC takes the following steps when applying costs to units sold and to inventories.

Step 1. Record the direct materials purchased in the reporting period. Assume April purchases of $1,950,000:

Entry (a)	Inventory: raw and in-process	1,950,000	
	Accounts payable		1,950,000

Step 2. Record the incurrence of conversion costs during the reporting period. Assume conversion costs were $1,260,000:

Entry (b)	Conversion costs	1,260,000	
	Various accounts (such as accounts payable and payroll liability)		1,260,000

Step 3. Determine the number of finished units manufactured during the reporting period. Assume that 100,000 keyboard units were manufactured in April.

Step 4. Compute the standard cost of each finished unit. This step normally uses a bill of materials (description of the types and quantities of materials) and an operations list (description of operations to be undergone) or similar records. For SVC, the standard cost is $31 ($19 direct material + $12 conversion costs).

Step 5. Record the cost of finished goods completed in the reporting period: (100,000 units × $31 = $3,100,000).

This step gives backflushing its name. At this point in operations, the costs are no longer recorded sequentially with the flow of product along its production route. Instead, the output trigger is going back and pulling costs from Inventory: Raw and In-process.

		Debit	Credit
Entry (c)	Finished goods	3,100,000	
	Inventory: raw and in-process		1,900,000
	Conversion costs applied		1,200,000

Step 6. Record the cost of goods sold in the reporting period. Assume that 99,000 units were sold in April (99,000 units × $31 = $3,069,000).

		Debit	Credit
Entry (d)	Cost of goods sold	3,069,000	
	Finished goods		3,069,000

The April ending inventory balances are:

Inventory: raw and in-process	$50,000
Finished goods, 1,000 units × $31	31,000
Total inventories	$81,000

Exhibit 19-3 provides an overview of backflush costing. The elimination of the typical Work in Process account reduces the amount of detail in the internal accounting system. Units on the production line may still be tracked, but there is no "costs-attach" tracking to specific work orders. There are no work orders or time tickets in the accounting system. "Costs-attach" means sequential tracking in which costs adhere like barnacles to the physical products as they flow along the production stream.

EXHIBIT 19-3

Overview of Backflush Costing, Example 1

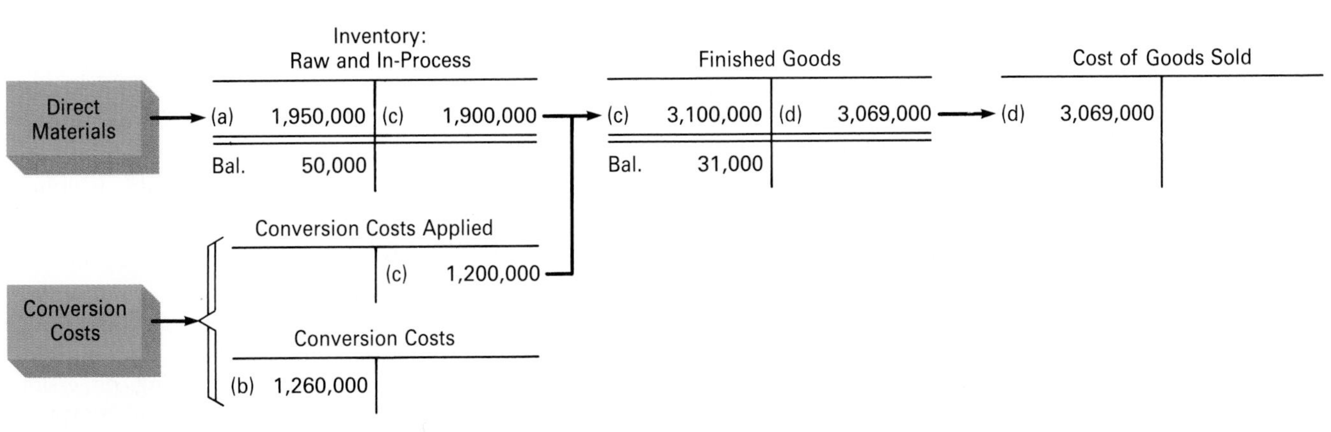

Accounting for Variances

The accounting for variances between actual costs incurred and standard costs applied and the disposition of variances is basically the same under all standard costing systems. This accounting is amply described in Chapters 7, 8, and 9. In our Example 1 of backflush costing, suppose the direct materials purchased had an unfavorable price variance of $42,000. Entry (a) would then be:

Inventory: raw and in-process	1,950,000	
Raw materials price variance	42,000	
Accounts payable		1,992,000

Direct materials are often a large proportion of total manufacturing costs, sometimes over 80%. Consequently, many companies will at least measure direct material efficiency variances in total by physically comparing what is remaining in direct materials inventory against what should be remaining, given the output of finished goods for the accounting period. In our example, suppose such a comparison showed an unfavorable efficiency variance of $90,000. The journal entry would be:

Raw materials efficiency variance	90,000	
Inventory: raw and in-process		90,000

Actual conversion costs may be underapplied or overapplied in any given accounting period. Chapter 9 (pp. 306–9) discussed various ways to account for underapplied or overapplied indirect manufacturing costs. Many companies write off underapplications or overapplications only at year-end; other companies, monthly. In our Example 1, suppose SVC policy was to write off underapplied or overapplied conversion costs to cost of goods sold monthly. The journal entry for this $60,000 difference between actual conversion costs incurred and standard conversion costs applied would be:

Conversion costs applied	1,200,000	
Cost of goods sold	60,000	
Conversion costs		1,260,000

Alternatively, if SVC's policy were not to dispose of underapplied or overapplied conversion costs until year-end, the balances would accumulate throughout the year in Conversion Costs and Conversion Costs Applied. At year-end, an entry like the immediately preceding one would be necessary.

Any direct material variances and underapplied or overapplied conversion costs are usually written off to Cost of Goods Sold instead of being prorated. Companies that use backflush costing typically have low inventories, so proration is less often necessary.

Example 2

Example 2 presents a backflush costing system that is a more dramatic departure from a sequential-tracking inventory costing system than Example 1. The first trigger point in Example 2 is the same as the first trigger point in Example 1, but the second trigger point is the sale—not the manufacture—of finished units. Toyota's cost accounting in its Kentucky factory is similar to Example 2.

There are two justifications for this accounting system:

- To remove the incentive for managers to produce for inventory. Under the typical "costs-attach" tracking assumption, managers can bolster operating income by producing units not sold. Having trigger point 2 as the sale instead of the completion of manufacture, however, reduces the attractiveness of this action by charging conversion costs to expense instead of to inventories. As Chapter 9 explained, under an absorption costing assumption, fixed overhead costs that would otherwise be expenses of the accounting period might be inventoried instead.

- To increase the focus of managers on a plantwide goal (selling units) rather than on an individual subunit goal (for example, increasing machine utilization at an individual work center).

This variation of backflush costing is called super-variable costing. Why? Because it has the same effect on operating income as the immediate expensing of conversion costs. The inventory account is confined solely to direct materials (whether they are in storerooms, in process, or in finished goods). There is only one inventory account:

Type	Account Title
Combined direct materials inventory and any direct materials in work in process and finished goods	Inventory

Entry (a) is prompted by the same trigger point 1 as in Example 1, the purchase of raw materials and components:

Entry (a)	Inventory	1,950,000	
	Accounts payable		1,950,000

The incurrence of conversion costs is recorded in an identical manner as in Example 1:

Entry (b)	Conversion costs	1,260,000	
	Various accounts		1,260,000

Trigger point 2 is the *sale* of good finished units (not their *manufacture*, as in Example 1): 99,000 units sold × \$31 = \$3,069,000, consisting of direct materials (99,000 × \$19 = \$1,881,000) and conversion costs applied (99,000 × \$12 = \$1,188,000).

Entry (c)	Cost of goods sold (99,000 units @ \$31)	3,069,000	
	Inventory (99,000 @ \$19)		1,881,000
	Conversion costs applied (99,000 @ \$12)		1,188,000

No conversion costs are inventoried. The \$1,260,000 in Conversion Costs either is applied to the units sold at standard (\$1,188,000) or is \$72,000 underapplied (\$1,260,000 − \$1,188,000). Suppose SVC, like many companies, writes these underapplied costs off monthly as additions to cost of goods sold:

Entry (d)	Conversion costs applied	1,188,000	
	Cost of goods sold	72,000	
	Conversion costs		1,260,000

The April ending inventory balance is \$69,000 (\$50,000 direct materials still on hand + \$19,000 direct materials embodied in the 1,000 units manufactured but not sold during the period). Exhibit 19-4 provides an overview.

Example 3

This example presents the most extreme and the simplest version of backflush costing. It has only one trigger point for making inventory costing entries. Conversion Costs would be debited as actual costs are incurred. As in Example 1, assume that the trigger point is the manufacture of finished units. Using the same data as in Examples 1 and 2, the summary entry is:

Finished goods	3,100,000	
Accounts payable (for materials)		1,900,000
Conversion costs applied		1,200,000

EXHIBIT 19-4
Overview of Backflush Costing, Example 2

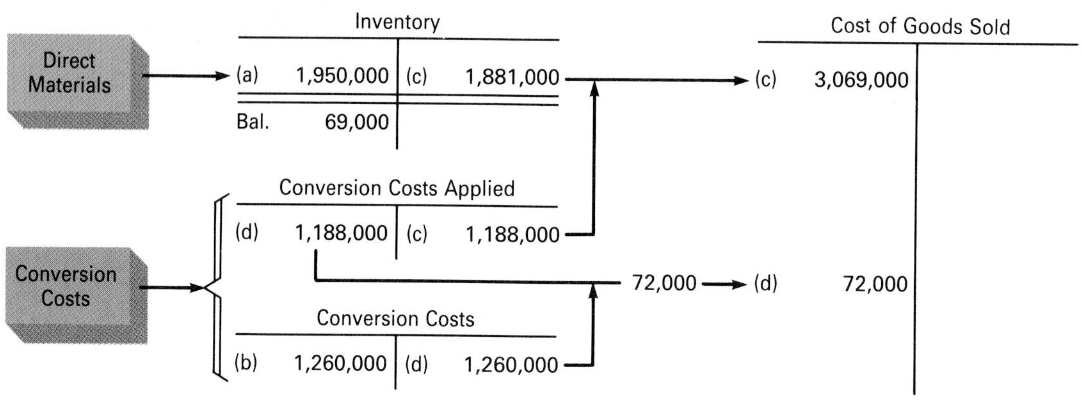

Compare this entry with entries (a) and (c) in Example 1. The simpler version in Example 3 ignores the $1,950,000 purchases of direct materials (entry (a) of Example 1, p. 628). At the end of April, the $50,000 of direct materials purchased has not yet been placed into production ($1,950,000 – $1,900,000 = $50,000), nor has it been entered into the inventory costing system. The Example 3 variation of backflush costing is suitable for a JIT production system with virtually no direct materials inventory and minimum work in process inventories. It is less feasible otherwise.

Difficulties of Backflush Costing

The accounting illustrated in Examples 2 and 3 does not strictly adhere to generally accepted accounting principles of external reporting. Advocates of backflushing, however, cite the materiality concept in support of these versions of backflush accounting. They claim that if inventories are low or not subject to significant change from one accounting period to the next, operating income and inventory costing developed in a backflush costing system will not differ materially from the results generated by a system that does adhere to generally accepted accounting principles.

Suppose material differences in operating income and inventories do exist between the results of a backflush costing system and of a conventional standard costing system. An adjustment can be made to bring the backflush numbers in line with external reporting requirements. For example, the backflush entries in Example 2 would result in expensing all conversion costs as a part of Cost of Goods Sold ($1,188,000 at standard + $72,000 write-off of underapplied conversion costs = $1,260,000). But suppose conversion costs were regarded as sufficiently material in amount to be included in Inventory. Then entry (d), closing the Conversion Cost accounts, would change as follows:

Original entry (d)	Conversion costs applied	1,188,000	
	Cost of goods sold	72,000	
	Conversion costs		1,260,000
Revised entry (d)	Conversion costs applied	1,188,000	
	Inventory (1,000 units × $12)	12,000	
	Cost of goods sold	60,000	
	Conversion costs		1,260,000

A chief attraction of backflush costing is its simplicity. Simple systems, however, generally do not yield as much information as do more complex systems. Criticisms of backflush costing focus mainly on the absence of audit trails—the ability of the accounting system to pinpoint the uses of resources at each step of the production process. Managers, however, keep track of operations by personal observations, computer monitoring, and nonfinancial measures expressed in physical terms. In addition, actual material quantities, conversion costs, and scrap can be identified with individual departments and activity areas. The good units are produced and measured. Their standard costs allowed are also measured. Variances are computed at the department or activity-area level at least monthly and sometimes daily.[1]

[1]For additional descriptions concerning the pros and cons of backflush costing, see W. Turk, "Management Accounting Revitalized: The Harley-Davidson Experience," *Journal of Cost Management for the Manufacturing Industry*, Winter 1990, pp. 28–39; and R. Calvasina, E. Calvasina, and G. Calvasina, "Beware the New Accounting Myths," *Management Accounting*, December 1989, pp. 41–45.

PROBLEM FOR SELF-STUDY

PROBLEM

1. A Dallas company factory uses standard costs and an operation costing system. It has a storeroom and several buffer stocks of parts and in-process inventories at various work centers along the production lines. No separate major cost category for direct labor exists; all factory labor is a part of conversion costs. For simplicity, assume there are no beginning inventories and no standard cost variances of materials. Prepare summary journal entries (without explanations and without disposing of underapplied or overapplied conversion costs) based on the following data for a given month (in thousands):

Raw materials purchased	$35,000
Raw materials used	30,000
Conversion costs incurred	22,000
Conversion costs applied	20,000
Costs transferred to finished goods	47,500
Cost of goods sold	40,000

   For simplicity, you are not given the data to prepare journal entries for each underlying transfer from, say, Work in Process, Operation 1 to Work in Process, Operation 2 to Work in Process, Operation 3. Instead, assume there is only a single account, Work in Process, that is supported by subsidiary Work in Process accounts for each operation.

2. The factory adopts a JIT production system and a backflush costing system with two trigger points: the purchase of materials and the completion of a good finished unit of product. Prepare summary journal entries (without explanations) based on the same data as in requirement 1. Note, however, that the raw materials used and the conversion costs applied would be affected by the goods completed, not the work in process. Assume that 95% of the work placed in process is completed.
3. Post the entries (a) in requirement 1 and (b) in requirement 2 to T-accounts for inventories, conversion costs, conversion costs applied, and cost of goods sold.
4. Compare the inventory balances in requirement 3. Explain any differences.

SOLUTION

1. a. Materials inventory	35,000	
Accounts payable		35,000
b. Conversion costs	22,000	
Various accounts		22,000

		Debit	Credit
	c Work in process	30,000	
	Materials inventory		30,000
	d. Work in process	20,000	
	Conversion costs applied		20,000
	e. Finished goods	47,500	
	Work in process		47,500
	f. Cost of goods sold	40,000	
	Finished goods		40,000
2.	a. Inventory: raw and in-process	35,000	
	Accounts payable		35,000
	b. Conversion costs	22,000	
	Various accounts		22,000
	c. Finished goods	47,500	
	Inventory: raw and in-process (.95 × $30,000)		28,500
	Conversion costs applied (.95 × $20,000)		19,000
	d. Cost of goods sold	40,000	
	Finished goods		40,000

3.

a. *For requirement 1:*

Materials Inventory

(a)	35,000	(c)	30,000
Bal.	5,000		

Work in Process

(c)	30,000	(e)	47,500
(d)	20,000		
Bal.	2,500		

Finished Goods

(e)	47,500	(f)	40,000
Bal.	7,500		

Cost of Goods Sold

(f)	40,000		

Conversion Costs

(b)	22,000		

Conversion Costs Applied

		(d)	20,000

b. *For requirement 2:*

Inventory: raw and in-process

(a)	35,000	(c)	28,500
Bal.	6,500		

Finished Goods

(c)	47,500	(d)	40,000
Bal.	7,500		

Cost of Goods Sold

(d)	40,000		

Conversion Costs

(b)	22,000		

Conversion Costs Applied

		(c)	19,000

4.

	Operation Costing	Backflush Costing
Materials inventory	$ 5,000	$ —
Inventory: raw and in-process	—	6,500
Work in process	2,500	—
Subtotal	7,500	6,500
Finished goods	7,500	7,500
Total inventories	$15,000	$14,000
Underapplied conversion costs	$ 2,000	$ 3,000
Cost of goods sold (at standard)	$40,000	$40,000

The $1,000 difference in inventories is explained by the accounting for conversion costs. Conversion costs applied in operation costing were $1,000 greater than conversion costs applied under backflush costing because the trigger point for application is work in process, not finished goods.

To ease comparisons between the entries for operation costing and backflush costing, the numbers used for the various inventory balances are identical except for the $1,000 difference just explained. Advocates of JIT using backflush costing, however, would insist that inventories would decline substantially under JIT, particularly inventories of raw materials and work in process.

PART TWO: CONTROL OF PROJECTS

Part One covered product costing for products that are ordinarily manufactured within hours or days. Part Two provides an overview of control of projects.

PROJECT FEATURES

Objective 7

Identify the four critical success factors of project costing

What is a job? What is a project? What is an engagement? Distinctions are fuzzy. An accounting firm may use "engagements" to describe its audits for various clients. Still, an accountant may regard the audit of a local country club as a "job" but the audit of a multinational company like Exxon as a "project." In general, a **project** is a complex job that often takes months or years to complete and requires the work of many different departments, divisions, or subcontractors. Examples of long-term projects arise in such fields as construction (for example, bridges, shopping centers, nuclear submarines); developing and introducing a new product model (for example, automobiles, computers, space vehicles); and conducting complex lawsuits (for example, antitrust cases). Note that launching a space vehicle is commonly described as a "mission" rather than a "project."

The planning and controlling of jobs and projects have common characteristics. Projects, however, are more challenging than jobs. Projects are unique and nonrepetitive, have more uncertainties, involve more skills and specialties, and require more coordination over a longer time span. Managers use special control techniques for long-term projects. The projects are often subdivided into a series of work packages, each having its individual time schedules.

Managers' control of projects generally focuses on four critical success factors, the major keys to both profitability and customer satisfaction: (a) scope, (b) quality, (c) time schedule, and (d) costs. *Scope* is the technical description of the final product. Many projects are subjected to "engineering change orders" as the work proceeds, whereby the final product has features different from those originally planned. Obviously, changes in scope usually also affect quality, time schedule, and costs. Trade-offs are frequently necessary. For example, a manager may use more time and incur more costs to obtain a certain level of quality.

In the 1990s, companies are becoming increasingly sensitive to time-based competition. For example, automobile companies struggle to set faster schedules for developing new products. Electronics companies aim at continually reducing schedules for research and development, product design, engineering, manufacturing, and distribution of a new product. Time is a key in project control.

Chapter 12 (pp. 403–6) described life-cycle product budgeting and costing. The focus was on the costs attributable to each product from its initial research and development to its final customer servicing and support in the marketplace. Life-cycle reports for each product require tracking costs and revenues on a product-by-product basis over several calendar periods. Reports for project costing likewise require tracking over longer periods of time.

EXHIBIT 19-5
Cost and Schedule Performance Report

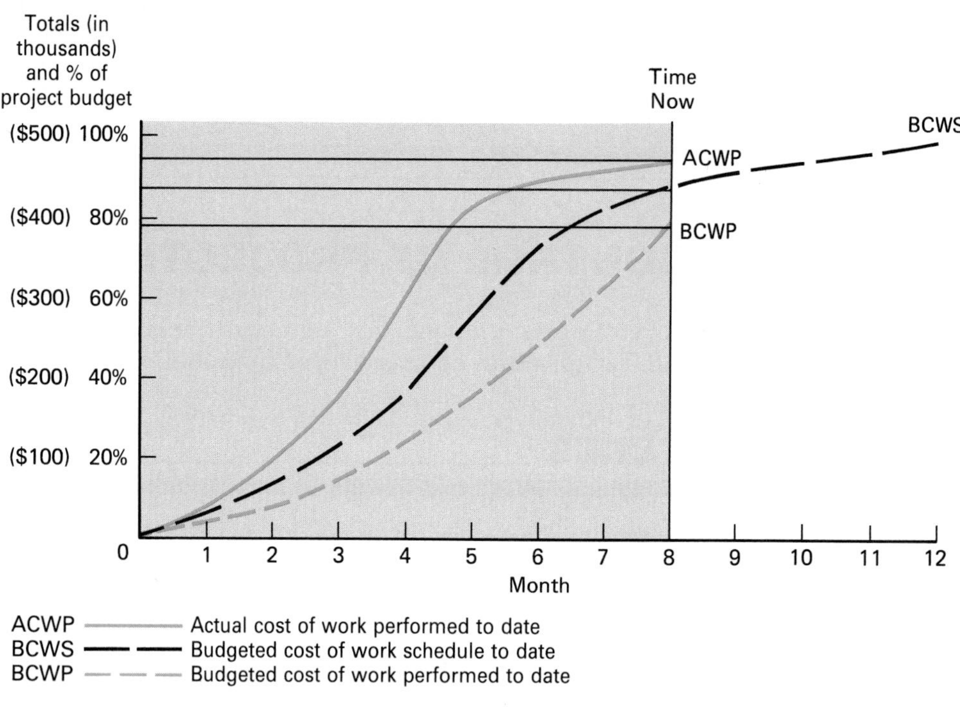

Objective 8

Compute project variances

Project Variances

The U.S. Department of Defense (DOD) spends billions of dollars each year on projects having varying schedules. The DOD requires that these expenditures be monitored by cost performance reporting (CPR). Exhibit 19-5 shows the essence of CPR. Exhibit 19-5 shows a time scale of twelve months, but many projects have time scales that extend for several years.

- BCWS—Budgeted cost of work *scheduled* to date. That is, as of the end of month 8, the project should be 85% completed at a cost of $425,000.
- BCWP—Budgeted cost of work *performed* to date. That is, as of the end of month 8, the project is actually only 80% completed. The output achieved should cost $400,000.
- ACWP—Actual cost of work performed to date, regardless of any budgets or schedules, $480,000.

Many more details are gathered, and many variances are computed. Two variances are especially noteworthy in this CPR system:

- **Project cost variance** is ACWP − BCWP, or $480,000 − $400,000 = $80,000, unfavorable. This unfavorable variance is often called a cost overrun.
- **Project schedule variance** is BCWS − BCWP, or $425,000 − $400,000 = $25,000, unfavorable. This variance is unfavorable because it indicates that the project is behind schedule. When this variance is unfavorable, it is often called schedule slippage.

These two major variances underscore that the cost performance report system would be more accurately labeled as a *cost and schedule performance report system*. The manager can use the unfolding information to predict final costs and completion dates.

Focus on the Future

The graph in Exhibit 19-5 also illustrates another key aspect of project control: a focus on what remains to be accomplished. That is, managers do not want surprises concerning future costs and future time for completion of a project. The "time now" label on the graph is the time for appraisal of where the project has been and where it is going. Such review time may be at scheduled intervals, such as weeks or months, or at designated stages of completion called *milestones*. An example of a milestone is the completion of a building's foundation; a second milestone is the completion of a building's framing; and so forth.

Exhibit 19-5 is labeled as a cost and schedule performance report, but many contractors and consultants refer to it as a *performance report*, an *earned value report*, or an *earned hour report*. The latter two labels are often used when control of hours worked is the major determinant of success.

To illustrate earned hour reporting, suppose a management consultant has an engineering system project:

- Original budget to complete, 1,500 hours.
- BCWS—Budgeted cost (hours) of work scheduled to date, 1,000 hours.
- BCWP—Budgeted cost (hours) of work performed to date, 700 hours. This amount is sometimes called earned value or earned hours.
- ACWP—Actual cost (hours) of work performed to date, 600 hours.

The performance report could be expressed in hours, in dollars, or both. Suppose the budgeted and actual cost is $80 per hour. The variances in dollars would be:

- Project schedule variance is BCWS − BCWP, $80 × (1,000 − 700 hours), $24,000 unfavorable.
- Project cost variance is ACWP − BCWP, $80 × (600 − 700 hours), $8,000 favorable.

In this illustration, the schedule variance indicates that the project has fallen behind schedule, but the cost variance indicates that the work performed to date was done efficiently. Suppose, however, that the budgeted cost per hour of $80 was not met; it was affected by an unfavorable price variance of, say, $5 per hour. Then the cost variance would have two components, an efficiency variance and a price variance (as explained in Chapter 7):

Efficiency variance, as calculated above	$8,000 F
Price variance, actual hours × difference in price per hour, or 600 hours × $5	3,000 U
Project cost variance	$5,000 F*

*or ($85 × 600) − ($80 × 700) = $51,000 − $56,000 = $5,000 F

In view of these variances, the project manager might predict a late completion date but a revised budgeted total final cost of $115,000 (or less), as follows:

Original budget, 1,500 hours × $80	$120,000
Deduct: Favorable project cost variance to date	5,000
Subtotal	$115,000
Deduct: Additional favorable project cost variance (prediction is needed)	?
Revised budget, final cost	$?

SIMILARITY OF CONTROL OF JOBS AND PROJECTS

The control of jobs is like the control of projects, although on a smaller, simpler scale. Moreover, jobs are usually more repetitive. Consider the partner in charge of the audit engagement of a local country club. She must monitor progress and adjust her predictions of actual hours compared with budgeted hours. Perhaps little can be done to alter performance or profitability of the current job. The information gathered on this year's job, however, may be vital to the budgeting and negotiating of fees on next year's job.

Projects such as the development of weapons systems and finding cures for diseases are often undertaken despite high uncertainties and likely changes as work progresses. Should interim performance reports compare progress against the original budget or against a revised budget (much like a flexible budget)? Ideally, management should be provided with both comparisons. In this way, the performance of managers as planners can be assessed by comparing the original budget against the revised budget. Similarly, the performance of managers regarding control of operations can be assessed by comparing the actual results against the revised budget.[2]

[2]For additional discussion, see D. Roman, *Managing Projects: A Systems Approach* (New York: Elsevier, 1986); and N. Augustine, *Managing Projects and Programs* (Boston: HBS Press, 1989). Both of these books provide broad coverage, including the use of network and critical-path analysis.

PROBLEM FOR SELF-STUDY

PROBLEM

Examine Exhibit 19-5, page 636. Suppose actual costs were $380,000 instead of $480,000. Compute the project cost variance and the project schedule variance.

SOLUTION

The project cost variance would be $380,000 – $400,000, or $20,000, favorable. The project schedule variance would be unchanged at $25,000, unfavorable.

SUMMARY

The physical processes of production are the keys to the design of cost accounting systems. Managers can choose among a variety of accounting systems with an assortment of features. Operation costing is a product costing system that falls between the polar extremes of job-order costing and process costing.

Backflush costing is simplified standard costing. It focuses first on output and then works backward to apply manufacturing costs to units sold and to inventories. It often accompanies JIT production systems, but it can be coupled with any production system.

Project costing is more complex than ordinary job costing. Projects tend to have more uncertainties and require coordination over a longer time span.

TERMS TO LEARN

This chapter and the Glossary at the end of this book contain definitions of the following important terms:

backflush costing *(p. 627)* delayed costing *(627)* hybrid costing *(619)* just-in-time (JIT) production *(620)* operation *(622)* operation costing *(622)* post-deduct costing *(627)* project *(635)* project cost variance *(636)* project schedule variance *(636)* sequential tracking *(627)*

ASSIGNMENT MATERIAL

QUESTIONS

19-1 Hybrid costing systems develop because there are hybrid production systems. Do you agree? Explain.

19-2 Distinguish between custom-order manufacturing and continuous-flow manufacturing.

19-3 "JIT is stockless production." Do you agree? Explain.

19-4 List four major features of JIT production systems.

19-5 Give two examples of non-value-added activities that are lessened in JIT production systems.

19-6 "Employee involvement is a key to the success of JIT production systems." Explain.

19-7 "Good relationships with one vendor or a few vendors are a critical component of JIT." Explain.

19-8 Identify three financial benefits that result from a combination of JIT, employee involvement, and TQC.

19-9 Why is operation costing often called hybrid costing?

19-10 "Operation costing means that *conversion costs* and *factory overhead* are synonyms." Do you agree? Explain.

19-11 Give three examples of industries that are likely to use operation costing.

19-12 Identify two application bases that are commonly used to apply conversion costs in operation costing systems.

19-13 Explain how each work order differs in operation costing systems.

19-14 Identify the major job-costing feature of operation costing.

19-15 Identify the major process-costing feature of operation costing.

19-16 Explain how standard costing systems typically track costs.

19-17 Explain the essence of backflush costing.

19-18 "Projects are more challenging than jobs." Explain.

19-19 "Managers' control of projects generally focuses on four critical success factors." Identify those factors.

19-20 What is the relationship between cost and schedule performance reporting and earned hour reporting?

EXERCISES AND PROBLEMS

Coverage of Part One of the Chapter

19-21 Operation costing journal entries. The Omaha Desk Company specializes in making desks of varying shapes, materials, and sizes. It has a cutting operation, an assembly operation, and a staining operation. Some goods are sold as unstained, so they do not go through the staining operation. Consider the following data for November.

Direct materials requisitioned by cutting	$200,000
Direct materials requisitioned by staining	20,000
Conversion costs of all operations (actual)	150,000
Conversions costs applied:	
Cutting	50,000
Assembly	110,000
Staining	30,000

Required
Prepare journal entries (without explanations). Assume that there is no beginning or ending work in process and that all goods are transferred to Finished Goods. The total manufacturing cost of the products not undergoing staining was $60,000.

19-22 Operation costing. Penske Company manufactures a variety of plastic products. The company has an extrusion operation and subsequent operations to form, trim, and finish parts such as buckets, covers, and automotive interior components. Plastic sheets are produced by the extrusion operation. Many of these sheets are sold as finished goods directly to other manufacturers. Additional direct materials (chemicals and coloring) are added in the finishing operation.

The company's manufacturing costs applied to products for October were:

	Extrude	Form	Trim	Finish	Totals
Direct materials	$650,000	$ —	$ —	$ 80,000	$ 730,000
Direct labor	55,000	30,000	20,000	40,000	145,000
Factory overhead	270,000	90,000	40,000	60,000	460,000
	$975,000	$120,000	$60,000	$180,000	$1,335,000

In addition to plastic sheets, two types of automotive products (firewalls and dashboards) were produced:

	Units	Plastic Sheet Direct Materials	Additional Direct Materials
Plastic sheets, sold after extrusion	10,000	$500,000	$ —
Firewalls, sold after trimming	1,000	50,000	—
Dashboards, sold after finishing	2,000	100,000	80,000
	13,000	$650,000	$80,000

For simplicity, assume that each of the items and units produced received the same steps within each operation.

Required

1. Tabulate the conversion costs of each operation, the total units produced, and the conversion cost per unit.
2. Tabulate the total costs, the units produced, and the cost per unit. Be sure that the total costs are all accounted for.

19-23 Operation costing, ending inventory, equivalent units. Gilhooley Products uses three operations in sequence to manufacture an assortment of picnic baskets. In each operation, the same procedures, time, and costs are used to perform that operation for a given quantity of baskets, regardless of the basket style being produced.

During April, a batch of materials for 1,100 baskets of Style X was put through the first operation. This was followed in turn by separate batches of materials for 400 baskets of Style

Y and 1,300 of Style Z. All the materials for a batch are introduced at the beginning of the operation for that batch. The costs as shown below were incurred in April for the first operation:

Direct labor	$29,600
Factory overhead	13,135
Direct materials:	
Style X	18,700
Style Y	8,000
Style Z	13,000

All the units started in April were completed during the month and transferred out to the next operation except 350 units of Style Z, which were only partially completed at April 30. These were 40% completed as to conversion costs and 100% completed as to direct material costs. There were no work in process inventories at the beginning of the month.

Required

1. For each basket style, compute the total cost of work completed and transferred out to the second operation.
2. Compute the total cost of the ending inventory in process.

19-24 Operation costing with ending work in process and journal entries. Galvez Co. produces two models of video recorders. The deluxe units undergo two operations. The super-deluxe units undergo three operations. Consider the following:

	Production Orders	
	For 1,000 Deluxe Units	For 500 Super-Deluxe Units
Direct materials (actual costs)	$50,000	$54,000
Conversion costs (applied on the basis of machine hours used):		
Operation 1	20,000	10,000
Operation 2	?	?
Operation 3	—	5,000
Total manufacturing costs applied	$?	$?

Required

1. Operation 2 is highly automated. Product manufacturing costs depend on a budgeted application rate for conversion costs based on machine hours. The budgeted costs for 19_2 were $100,000 direct labor and $440,000 factory overhead. Budgeted machine hours were 18,000. Each product unit required 6 minutes of time in Operation 2. Compute the total costs of processing the deluxe products and super-deluxe products in Operation 2.
2. Compute the total manufacturing costs and the unit costs of the deluxe and super-deluxe products.
3. Suppose that at the end of the year, 500 deluxe units were in process through Operation 1 only and 300 super-deluxe units were in process through Operation 2 only. Compute the cost of the ending work in process inventory. Assume that no direct materials are applied in Operation 2 but that $4,000 of additional direct materials are to be applied to the 500 units processed in Operation 3.
4. Prepare journal entries that track the costs of all 1,000 deluxe units through operations to Finished Goods.

19-25 Operation costing, journal entries. Pafko Company, a small manufacturer, manufactures a variety of tool boxes. The company's manufacturing operations and their costs for November were:

	Cutting	Assembly	Finishing	Totals
Direct labor	$2,600	$16,500	$4,800	$23,900
Factory overhead	3,000	22,900	3,300	29,200
	$5,600	$39,400	$8,100	$53,100

Three styles of boxes were produced in November. The quantities and direct material costs were:

Style	Quantity	Direct Materials
Standard	1,200	$18,000
Home	600	6,660
Industrial	200	5,400
		$30,060

The company uses actual costing. It tracks direct materials to each style of box. It combines direct labor and factory overhead and applies the conversion costs based on all product units passing through an operation. Each product unit is assumed to receive an identical amount of time and effort in each operation. The Industrial style, however, does not go through the Finishing operation.

Required

1. Compute the total cost and unit cost of each style produced.
2. Prepare summary journal entries for each operation. For simplicity, assume that all direct materials are introduced at the beginning of the cutting operation. Also, assume that all goods were transferred to Finished Goods when completed and that there was no beginning or ending work in process. Prepare one summary entry for all conversion costs incurred, but prepare a separate entry for applying conversion costs in each operation.

19-26 Backflush journal entries and JIT production. The Lee Company has a factory that manufactures transistor radios. The production time is only a few minutes per unit. The company uses a just-in-time production system and a backflush costing system with two trigger points:

- Purchase of raw materials and components
- Completion of good finished units of product

There are no beginning inventories. The following data pertain to April manufacturing (in thousands):

Raw materials purchased	$ 8,800
Raw materials used	8,500
Conversion costs incurred	4,220
Application of conversion costs	4,000
Costs transferred to finished goods	12,500
Cost of goods sold	11,900

Required

1. Prepare summary journal entries for April (without explanations and without disposing of underapplied or overapplied conversion costs).
2. Post the entries in requirement 1 to T-accounts for inventories, finished goods, conversion costs, conversion costs applied, and cost of goods sold.
3. Under an ideal JIT production system, how would the amounts in your journal entries differ from those in requirement 1?

19-27 Backflush costing and JIT production. Ronowski Company produces telephones. For June, there were no beginning inventories of raw materials and no beginning and ending work in process. Ronowski uses a JIT production system and backflush costing with two trigger points for making entries in its accounting system:

- Purchase of raw materials and components
- Completion of good finished units of product

Ronowski's June standard cost per unit of telephone product is: direct materials, $26; conversion costs, $15. There are two inventory accounts:

- Inventory: Raw and In-process
- Finished Goods

The following data apply to June manufacturing:

Raw materials and components purchased	$5,300,000
Conversion costs incurred	$3,080,000
Number of finished units manufactured	200,000
Number of finished units sold	192,000

Required

1. Prepare summary journal entries for June (without explanations and without disposing of underapplied or overapplied conversion costs). Assume no direct material variances.
2. Post the entries in requirement 1 to T-accounts for inventories, conversion costs, conversion costs applied, and cost of goods sold.

Note: Problem 19-28 is a continuation of this problem.

19-28 Backflush, second trigger is sale. We continue with the Ronowski Company factory discussed in Problem 19-27. Now assume, however, that the second trigger point is the sale—rather than the manufacture—of finished units. Also, the inventory account is confined solely to direct materials, whether they would be in a storeroom, in work in process, or in finished goods.

No conversion costs are inventoried. They are applied to the units sold at standard. Any underapplied or overapplied conversion costs are written off monthly to cost of goods sold.

Required

1. Prepare summary journal entries for June, including the disposition of underapplied or overapplied conversion costs (without explanations). Assume no direct material variances.
2. Post the entries in requirement 1 to T-accounts for inventory, conversion costs, conversion costs applied, and cost of goods sold. Explain the composition of the ending balance of Inventory.
3. Suppose conversion costs were sufficiently material in amount to be included in Inventory. Using a backflush system, show how your journal entries would be changed in requirement 1. Explain.

19-29 Backflush, one trigger point. Assume the same facts as in Problem 19-27. Now assume, however, that there is only one trigger point, the completion of finished units.

Required

1. Prepare the journal entry at the trigger point.
2. Compare this entry with your entries (a) and (c) in Example 1, page 628. Explain any differences in accounting results.

19-30 Accounting for variances. Assume the same facts as in Problem 19-27. Suppose the same quantity of raw materials had additional costs as follows:

Unfavorable price variance	$30,000
Unfavorable efficiency variance	70,000

Required

1. Prepare summary journal entries (without explanations) to record the raw-material variances.
2. Assume that underapplied or overapplied conversion costs are written off to cost of goods sold monthly. Prepare the pertinent summary journal entry.

Coverage of Part Two of the Chapter

19-31 Basic project cost control. The marketing, manufacturing, and research and development (R & D) departments of the Perry Company have worked together in deciding which new products and components to develop. The R & D department has launched a project to develop a new compressor for the Perry line of Frosty Refrigerators.

Consider the following data:

• Original budget to complete	30,000 hours
• Budgeted average cost per hour	$ 120
• Price variance to date	$ 0
• Budgeted hours for work scheduled to date	22,100
• Budgeted hours for work performed to date (earned hours)	20,000
• Actual hours of work performed to date	21,200
• Price variance expected	$39,000 unfavorable

Required

1. Compute the project cost variance and the project schedule variance.
2. Prepare a revised budget that is designed to predict actual final cost. Assume that the project cost variance will continue as unfavorable at the same rate as shown to date.

19-32 Basic project cost control. Phoenix Company, a defense subcontractor, has a project to design and produce a subsystem for guided missiles. The original budget to complete is 20,000 hours. The budgeted hours for the work scheduled to date are 12,000 hours. The budgeted hours for the work performed to date (earned hours) were 11,000. The actual hours of work performed to date were 11,440. The budgeted cost is $70 per hour. There was no price variance.

Required

1. Compute the project cost variance and the project schedule variance.
2. Prepare a revised budget that is designed to predict actual final cost. Assume that the project cost variance will continue to be unfavorable at the same rate as shown to date.

19-33 Project cost control. Consider a Grumman research project on a new wing design for a fighter aircraft. The original budget to complete was 30,000 hours at an average cost of $120 per hour. The budgeted hours for the work scheduled to date are 21,000 hours. The budgeted hours for the work performed to date (earned hours) were 22,000 hours. The actual hours of work performed to date were 21,500. Actual costs were $2,795,000.

Required

1. Compute in dollars: the project cost variance, efficiency variance, price variance, and project schedule variance.
2. As the manager, how would you interpret these variances? If the project cost variances persist at the same rate as shown to date, what is the expected actual final cost?

19-34 Project cost control. Wechtel Company, an engineering consulting firm, has a large project: the design and testing of a production control system for an airline's maintenance base. The following data pertain to the project:

Original budget		$600,000
Add (deduct) cost variances to date:		
Unfavorable price variances	$42,000	
Favorable efficiency variances	(60,000)	(18,000)
Subtotal		$582,000
Add (deduct) additional expected variances		?
Revised budget, final cost		$?

The budgeted cost per hour is $100. The budgeted hours for work performed to date were 3,000. The budgeted hours for work scheduled to date were 3,500.

Required

1. Compute the project cost variance and the project schedule variance for the work done to date.
2. Assume that there will be no further price variances. Assume also that the same relative efficiency will persist throughout the remainder of the project. Prepare an estimate of the revised budget of final cost. Will the manager of the consulting firm be pleased? Explain.

CHAPTER 20

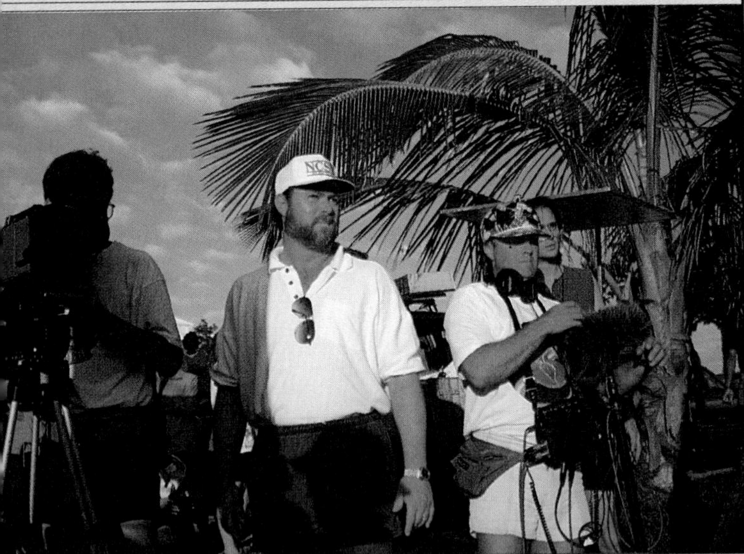

Uncertainty over movie production costs is explicitly recognized in contracts between production companies and movie directors.

Decision Models, Uncertainty, and the Accountant

Coping with uncertainty

Illustrative problem

Buying perfect and imperfect information

Expected monetary value and expected utility

Implementation issues

Learning Objectives

When you have finished studying this chapter, you should be able to

1. Describe five key elements in a decision model
2. Define a probability distribution and explain its role in decision making
3. Describe expected monetary value approach as a special case of the expected value approach
4. Distinguish between a good decision and a good outcome
5. Explain how the expected value of perfect information is useful in purchasing additional information
6. Describe how risk-neutral, risk-averse, and risk-seeking decision makers differ

Predictions and decisions of managers are made in a world of uncertainty. This chapter explores the characteristics of uncertainty and describes how managers can cope with it. We also illustrate the additional insights gained when uncertainty is recognized in accounting models such as cost-volume-profit analysis.

COPING WITH UNCERTAINTY

Role of a Decision Model

Objective 1

Describe five key elements in a decision model

In management and management accounting, as elsewhere, we live in a world of uncertainty. *Uncertainty* is the possibility that an actual amount will deviate from a forecasted amount. For example, marketing costs might be forecast at $400,000 but actually turn out to be $430,000. A *decision model* helps managers deal with uncertainty; it is a formal method for making a choice that often involves quantitative analysis. It usually includes the following elements:

1. A **choice criterion**, which is an objective that can be quantified. This objective can take many forms. Most often it is expressed as a maximization (minimization) of some form of income (cost). The choice criterion, also called an **objective function**, provides a basis for choosing the best alternative action.
2. A set of the alternative actions under explicit consideration.
3. A set of all the relevant **events** that can affect the outcomes (sometimes called **states** or **states of nature**), where an event is a possible occurrence. This set should be mutually exclusive and collectively exhaustive. Two events are mutually exclusive if they have no element in common. Events are collectively exhaustive if, taken together, they make up the entire set of possible events. Rain or no rain, war or peace, sale or no sale are examples. Only one event in a set of mutually exclusive and collectively exhaustive events can actually occur.
4. A set of probabilities, where a **probability** is the likelihood of occurrence of an event.
5. A set of possible **outcomes** (often called **payoffs**) that measures, in terms of the objective function, the predicted consequences of the various possible combinations of actions and events. Each outcome is conditionally dependent on a specific action and specific event.

Exhibit 20-1 presents an overview of the link between a decision model, the implementation of the chosen action, its outcome, and subsequent performance evaluation.

Probabilities

Objective 2

Define a probability distribution; explain its role in decision making

Assigning probabilities is a key aspect of the decision model approach to coping with uncertainty. A **probability distribution** describes the likelihood (or probability) of each of the mutually exclusive and collectively exhaustive set of events. The

EXHIBIT 20-1
Overview of a Decision Model and Its Link to Performance Evaluation

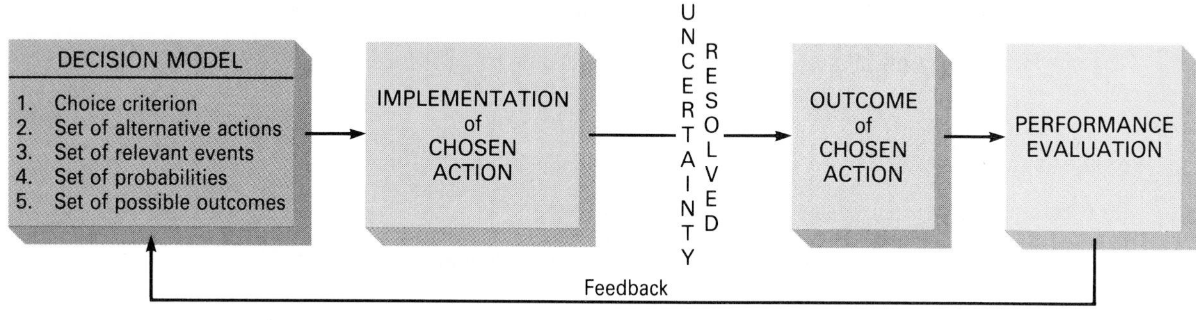

probabilities of these events will add to 1.00 because they are collectively exhaustive. In some cases, there will be much evidence to guide the assignment of probabilities. For example, the probability of obtaining a head in the toss of a fair coin is 1/2; that of drawing a particular playing card from a standard, well-shuffled deck is 1/52. In business, the probability of having a specified percentage of defective units may be assigned with great confidence, based on production experience with thousands of units. In other cases, there will be little evidence supporting estimated probabilities.

The concept of uncertainty can be illustrated by a decision situation facing a book editor. The editor is deciding between publishing a spy novel and publishing a historical novel. Both book proposals require a $200,000 investment at the beginning of the year. (For simplicity, we ignore the time value of money here. See Chapters 21 and 22 for discussion of that topic.) Based on experience, the editor believes that the following *probability distribution* describes the relative likelihood of cash inflows for the next year (assume that the useful sales life of each book is one year):

Proposal A: Spy Novel		Proposal B: Historical Novel	
Probability	**Cash Inflow**	**Probability**	**Cash Inflow**
0.10	$300,000	0.10	$200,000
0.20	$350,000	0.25	$300,000
0.40	$400,000	0.30	$400,000
0.20	$450,000	0.25	$500,000
0.10	$500,000	0.10	$800,000

Exhibit 20-2 shows a graphical comparison of the probability distributions.

Objective 3

Describe expected monetary value as a special case of the expected value approach

Expected Value

An **expected value** is a weighted average of the outcomes with the probabilities of each outcome serving as the weights. Where the outcomes are measured in monetary terms, *expected value* is often called **expected monetary value**. The expected monetary value of the cash inflow from the spy novel—denoted $E(a_1)$—is $400,000:

$$E(a_1) = 0.1(\$300{,}000) + 0.2(\$350{,}000) + 0.4(\$400{,}000) + 0.2(\$450{,}000) + 0.1(\$500{,}000)$$
$$= \$400{,}000$$

The expected monetary value of the cash inflow from the historical novel—denoted $E(a_2)$—is $420,000:

EXHIBIT 20-2
Decisions under Uncertainty: Comparison of Probability Distributions

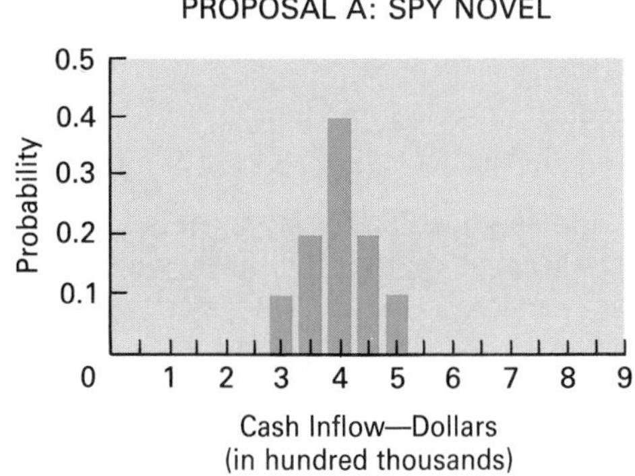

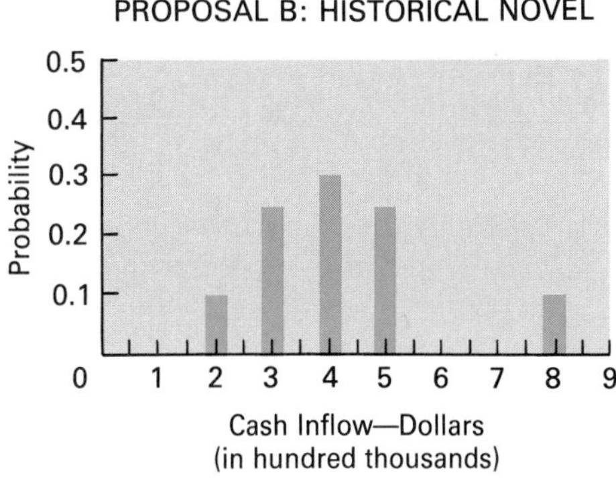

$$E(a_2) = 0.1(\$200{,}000) + 0.25(\$300{,}000) + 0.3(\$400{,}000) + 0.25(\$500{,}000) + 0.1(\$800{,}000)$$
$$= \$420{,}000$$

Expected monetary value is widely used as a decision criterion. For a book editor wanting to maximize the expected monetary value, the historical novel is preferable to the spy novel.

Many statisticians and accountants favor presenting the entire probability distribution directly to the decision maker. Others present information in three categories: optimistic, most likely, and pessimistic. Both of these presentations remind the user that uncertainty exists in the decision at hand.

ILLUSTRATIVE PROBLEM

The following problem demonstrates the expected monetary value approach to uncertainty. Office Machines sells photocopying machines. In addition, it sells an annual service contract for $360 per machine. For the coming year, it has 10,000 machines of its recently introduced model XT-101 under its service contract. Office Machines is considering two alternatives for servicing these machines:

1. Use an external service contractor (Reliable Maintenance Inc.) that will charge Office Machines $75 per service call
2. Use its internal service department with a fixed annual cost of $1,320,000 and a variable cost of $50 per service call

For simplicity, assume that the only uncertainty pertains to the total number of service calls for the coming year. There is a 0.75 probability that the total number of calls will be 40,000 (4 per machine) and a 0.25 probability that it will be 60,000 (6 per machine). These probability assessments are based on existing information in Office Machines' files. Which alternative should Office Machines choose to service the 10,000 machines?

General Approach to Uncertainty

The construction of a model for decision making consists of five steps that are keyed to the five characteristics described at the beginning of this chapter.[1]

[1]The presentations here draw (in part) from teaching notes prepared by R. Williamson.

Step 1. Identify the choice criterion of the decision maker. Assume that Office Machines' choice criterion, or objective function, is to maximize expected operating income for the coming year.

Step 2. Identify the set of actions under consideration. The notation for an action is a. Office Machines has two possible actions:

$$a_1 = \text{Use Reliable Maintenance to service machines}$$
$$a_2 = \text{Use its internal service department}$$

Step 3. Identify the set of relevant events that can occur. The notation for an event is x. Office Machines' only uncertainty is the number of service calls for the 10,000 photocopying machines under contract for servicing:

$$x_1 = 40{,}000 \text{ service calls}$$
$$x_2 = 60{,}000 \text{ service calls}$$

Step 4. Assign probabilities for the occurrence of each event. The notation for the probability of an event is $P(x)$. Analysts at Office Machines examine service-call records for prior models and assess a 75% chance that there will be 40,000 service calls and a 25% chance that there will be 60,000 service calls. Therefore, the probabilities are

$$P(x_1) = 0.75$$
$$P(x_2) = 0.25$$

Step 5. Identify the set of possible outcomes that are dependeent on specific actions and events. The outcomes in this example take the form of four possible operating incomes, which are displayed in a decision table in Exhibit 20-3. A **decision table** (sometimes called a **payoff table** or **payoff matrix**) is a summary of the contemplated actions, events, probabilities, and outcomes.

Office Machines now has the necessary information for making a decision. As shown in Exhibit 20-3, it can compute the expected value of each action. It should select action a_1. The expected operating incomes of each action follow:

$$\text{Use Reliable Maintenance: } E(a_1) = 0.75(\$600{,}000) + 0.25(-\$900{,}000) = \$225{,}000$$
$$\text{Use internal service dept.: } E(a_2) = 0.75(\$280{,}000) + 0.25(-\$720{,}000) = \$30{,}000$$

Expected operating income is higher by $195,000 ($225,000 − $30,000) if Office Machines uses Reliable Maintenance rather than its own service department.

Exhibit 20-4 illustrates how the five steps can be summarized using a decision tree. The following symbols are widely used in drawing a decision tree:

□ represents a decision node

○ represents an event node

EXHIBIT 20-3

Decision Table Presentation of Data for Office Machines

	Probability of Events	
	x_1 = 40,000 service calls $P(x_1) = 0.75$	x_2 = 60,000 service calls $P(x_2) = 0.25$
ACTIONS:		
a_1 = Use Reliable Maintenance	$600,000*	−$900,000†
a_2 = Use internal service department	$280,000‡	−$720,000§

*Operating income = $360 (10,000) − $75(40,000) = $600,000
†Operating income = $360 (10,000) − $75(60,000) = −$900,000
‡Operating income = $360 (10,000) − $50(40,000) − $1,320,000 = $280,000
§Operating income = $360 (10,000) − $50(60,000) − $1,320,000 = −$720,000

EXHIBIT 20-4

Decision-Tree Presentation of Data for Office Machines

Actions (1)	Events (2)	Outcome (3)	Probability of Events (4)	Expected Operating Income (5) = (3) × (4)
a_1	Service calls = 40,000	$600,000	0.75	= $450,000
	Service calls = 60,000	−$900,000	0.25	= −225,000
				$225,000
a_2	Service calls = 40,000	$280,000	0.75	= $210,000
	Service calls = 60,000	−$720,000	0.25	= −180,000
				$ 30,000

a_1 = Use Reliable Maintenance.
a_2 = Use internal service department.

Attached to each decision node will be all the individual actions being considered by the decision marker. Attached to each event node will be the total set of mutually exclusive and collectively exhaustive events.

Chapter 11 explained the relevant-information concept for decision making. Relevant costs (relevant revenues) are those expected future costs (expected future revenues) that will differ among alternative courses of action. Revenues to Office Machines are unaffected by the choice between a_1 and a_2. Only costs are relevant. We could have selected minimizing costs as an alternative approach to that shown in Exhibits 20-3 and 20-4.[2]

Good Decisions and Good Outcomes

Objective 4

Distinguish between a good decision and a good outcome

Always distinguish between a good decision and a good outcome. One can exist without the other. By definition, uncertainty rules out guaranteeing after the fact that the best outcome will always be obtained. It is possible that "bad luck" will produce unfavorable consequences even when "good decisions" have been made.

Suppose you are offered a one-time-only gamble tossing a fair coin. You will win $20 if the event is heads, but you will lose $1 if the event is tails. As a decision maker, you proceed through the logical phases: gathering information, assessing consequences, and making a choice. You accept the bet. Why? Because the expected value is $9.50 [.5($20) + .5(−$1)]. The coin is tossed. You lose. From your viewpoint, this was a good decision but a bad outcome.

A decision can be made only on the basis of information available at the time of the decision. Hindsight is often flawless, but a bad outcome does not necessarily mean that a bad decision was made. Making a good decision is our best protection against a bad outcome.

[2]The expected costs of each alternative are:

$$E(a_1) = 0.75(\$3,000,000) + 0.25(\$4,500,000) = \$3,375,000$$
$$E(a_2) = 0.75(\$3,320,000) + 0.25(\$4,320,000) = \$3,570,000$$

The expected cost of a_1 is $195,000 lower than the expected cost of a_2.

BUYING PERFECT AND IMPERFECT INFORMATION

In many cases, a decision maker faced with uncertainty can gather additional information. This section discusses an approach that aids decisions about how much to invest to obtain more information.

Buying Perfect Information

Objective 5

Explain how the expected value of perfect information is useful in purchasing additional information

Consider again the problem facing Office Machines. It is deciding whether to use an external service contractor (Reliable Maintenance Inc.) or its internal service department for the 10,000 machines under contract for servicing. Although the difference in expected value of the operating income between the two actions is $195,000 ($225,000 – $30,000; see Exhibit 20-3), Office Machines might be concerned. For example, if it uses Reliable Maintenance and 60,000 service calls are required, the operating loss is $900,000.

The only uncertainty facing Office Machines is whether the number of service calls on the XT-101 model photocopying machine will be 40,000 or 60,000 in the coming year. The 0.75 and 0.25 probability assessments are based on the existing information in Office Machines' files, but the company's service-call experience is with other models.

Suppose Office Machines can hire a consultant, Mary O'Leary, to study the service-call situation for the XT-101 model. For illustrative purposes, assume that she has the ability to make perfect predictions in these matters. What is the maximum amount Office Machines should be willing to pay for the consultant's errorless wisdom? This amount is called the *expected value of perfect information* (EVPI). To determine this amount, Office Machines should take five steps:

Step 1. *Identify what the optimal action would be conditional on knowing what event will occur.* If the consultant tells Office Machines that the number of service calls will be 40,000, it will use Reliable Maintenance and obtain an operating income of $600,000. If the number of service calls will be 60,000, Office Machines will use its internal service department and obtain an operating loss of $720,000.

Step 2. *Identify the probabilities of each event occurring—that is, being the event that is perfectly predicted—after the purchase of perfect information.* Office Machines' best estimate of the two levels of service calls being revealed are the current assessments of probabilities, 0.75 and 0.25.

Step 3. *Compute the expected value of a decision made with perfect information.* The perfect information will reveal which event will occur. The expected value of the decision *with* perfect information is the sum of the optimum outcome for each event multiplied by its probability:

$$(0.75 \times \$600{,}000) + (0.25 \times -\$720{,}000) = \$270{,}000$$

Step 4. *Compute the expected value of the preferred action based on existing information.* Using the data in Exhibit 20-4, Office Machines computed an expected value from using Reliable Maintenance (a_1) of $225,000, which is the higher of the expected values computed with existing information for the two actions in this example.

Step 5. *Compute the expected value of perfect information.* The maximum price Office Machines should be willing to pay for perfect advance information would be the difference between the expected value with *perfect* information (from Step 3) and the highest expected value with *existing* information (from Step 4). This *difference* is the **expected value of perfect information (EVPI)**:

Expected value of a decision made with perfect information (from Step 3)	$270,000
Expected value of preferred action made with existing information (from Step 4)	225,000
Expected value of perfect information (EVPI)	$ 45,000

Time for Reflection

In a general sense, the consultant may be considered an *information system*. The decision maker's action will depend on the message or *signal* (the specific prediction) supplied by the system. The decision maker buys the system without knowing the forthcoming signal.

Information gathering is costly, and the cost-benefit test must be met. Indeed, Office Machines would be unwilling to pay for an information system if it did not intend to use the system's signals. It would be unwilling to pay for the system unless the company expected the system to generate an increase in the expected monetary value in excess of its cost. In our example, Office Machines would be willing to pay up to $45,000 for the consultant's perfect predictions.

The sequence of analysis is straightforward but crucial. Its ultimate focus is on changes in the expected monetary value of the possible actions. The model moves from the *system* to the *signals* to the *actions* induced by the expected value of the decision maker's choice criterion. Exhibit 20-5 summarizes this sequence. Office Machines will receive one of two signals prior to its action choice when it hires the consultant. The signal received will perfectly predict whether there will be 40,000 or 60,000 service calls. In contrast, if Office Machines does not hire the consultant, it will receive no signal prior to its action choice.

The choice of an information system is "decision specific." That is, the choice depends on the existing information available to the decision maker, the decision table, the cost of the information system, and the choice criterion of the decision maker.

Buying Imperfect Information

The only uncertainty facing Office Machines is whether x_1 (40,000 calls) or x_2 (60,000 calls) will occur. Obviously, *perfect* information is unobtainable (except in textbooks), but *imperfect* information may still be worth buying. Note that knowledge of EVPI permits outright rejection of any additional information whose cost

EXHIBIT 20-5
Sequence of Analysis in Making Decisions to Purchase Information

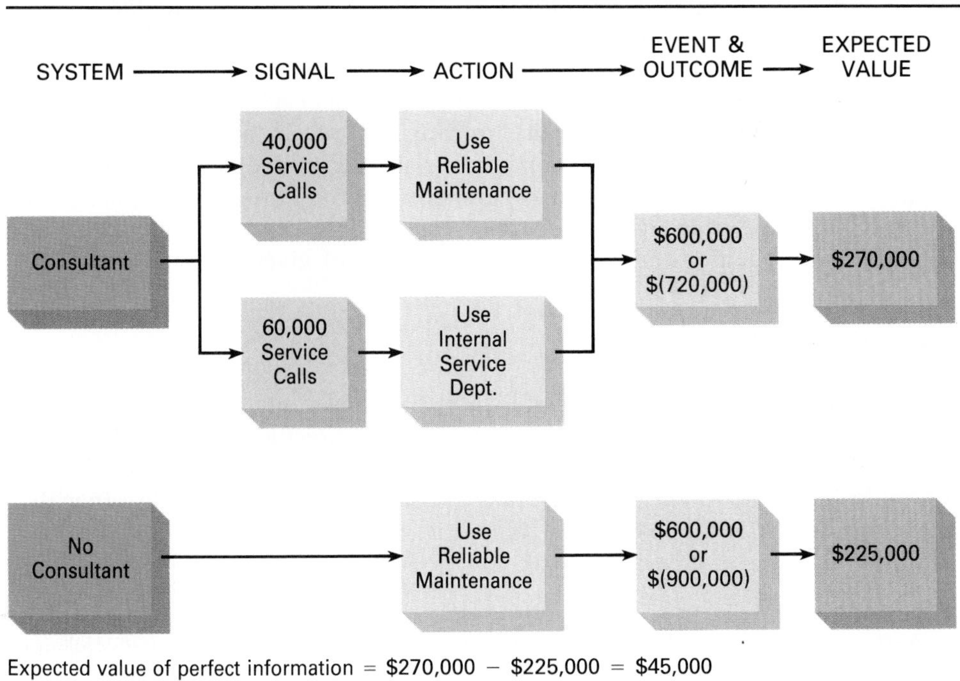

Expected value of perfect information = $270,000 − $225,000 = $45,000

exceeds EVPI. Thus, if the consultant's fee exceeds $45,000, Office Machines should reject the system.

Suppose the consultant's fee is $30,000. Should Office Machines acquire the imperfect information system offered by the consultant? Again, the answer depends on whether the expected value with the new information system exceeds the expected value with existing information.[3]

Cost-Benefit Consideration Is Paramount

This chapter focuses on single-decision situations faced by an individual decision maker. In this context, the cost-benefit perspective can easily be illustrated. In many situations, however, data will be collected for a multitude of decisions. A cost-benefit perspective then focuses on the collective effect of these decisions. Even though the given data may have little or no value in a particular decision situation, the system that provides such data may nevertheless be economically justifiable in a collective sense. For example, an expensive computer-based model for budgeting may be part of an accounting system that supports the planning of operations. For any particular decision, such as whether to acquire 100 or 150 units of material, a simpler computer-based model that provides fewer numbers than are required for planning the company's operations may be cost-effective. The expensive computer-based model, however, may still be economically justifiable when its impact on the entire class of decisions is considered.

EXPECTED MONETARY VALUE AND EXPECTED UTILITY

Objective 6

Describe how risk-neutral, risk-averse, and risk-seeking decision makers differ

The approach used so far in this chapter has had all the computations expressed in expected monetary value terms. This approach is appealing to many managers. Money is the currency in which their reports are expressed. In some cases, however, a more general approach using utility rather than money is appropriate. The more general approach recognizes that individuals differ in how they value the loss of a given dollar amount and how they value the gain of the same dollar amount.

We define **utility value** as the value of a specific outcome to a particular decision maker. A **utility function** is a function that represents each outcome in terms of utility value. A decision maker's utility function depends on attitudes toward risk. There are three basic attitudes: **risk neutral**, **risk averse**, and **risk seeking**. These attitudes are plotted in Exhibit 20-6, where monetary amounts (dollars) are on the horizontal axis and *utils* are on the vertical axis. Utility is arbitrarily measured in **utils**, which are quantifications of the utility value of given monetary amounts.

Attitudes Toward Risk

Risk Neutral. The left-hand panel of Exhibit 20-6 represents the utility function of a risk-neutral decision maker, individual A. Utility is directly proportional to monetary amounts. In other words, the decision maker weighs each dollar as a full dollar (no more, no less). There is equal utility (positive utils) and disutility (negative utils) from a gain and a loss of the same dollar amount.

[3]A structured approach to deciding whether to purchase imperfect information is in C. Holloway, *Decision Making under Uncertainty: Models and Choices* (Englewood Cliffs, NJ: Prentice Hall).

EXHIBIT 20-6
Types of Risk Attitudes and Utility Functions

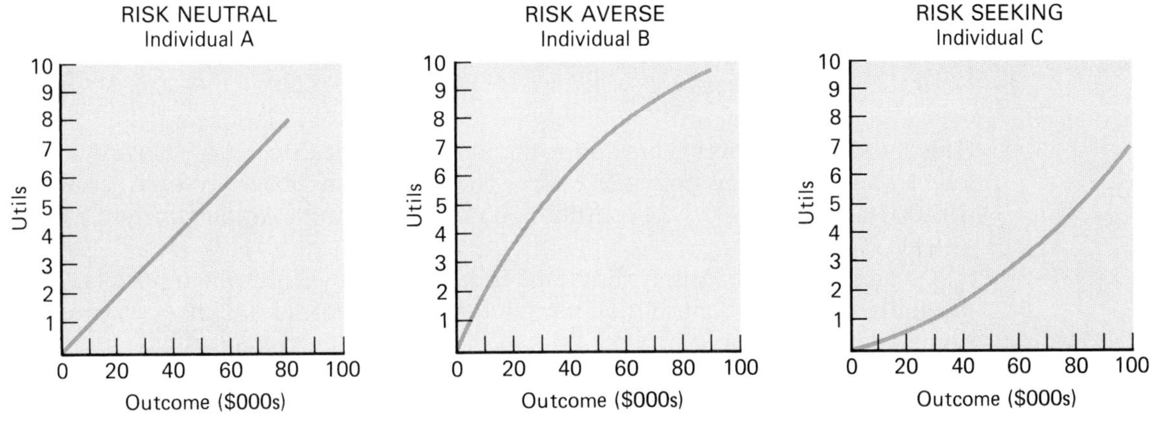

Risk Averse. The middle panel in Exhibit 20-6 represents the utility function of a risk-averse decision maker, individual B. There is more disutility from a loss of a given dollar amount than there is utility from a gain of the same dollar amount.

Risk Seeking. The right-hand panel in Exhibit 20-6 represents the utility function of a risk-seeking decision maker, individual C. There is more utility from a gain of a given dollar amount than there is disutility from the loss of the same dollar amount.

Exhibit 20-6 reveals the nature of the trade-offs made between risks and returns. Individual A may be financially secure and will get as much pleasure from increasing her wealth by $20,000 as pain from decreasing her wealth by $20,000.

Individual B increases his utility from 6 to 8 utils when the payoff increases from $40,000 to $60,000. In contrast, B's utility decreases from 5 to 2 utils when the payoff decreases from $30,000 to $10,000. B puts a large penalty on decision outcomes when high losses may occur. Consequently, risk-averse managers may obtain a widely diversified set of projects to reduce the likelihood of a large overall loss.

Individual C increases her utility from 3 to 6 utils when the payoff increases from $60,000 to $90,000. In contrast, C's utility only decreases from 3 to 1 when the payoff decreases from $60,000 to $30,000. C puts high value on large gains and would be quite willing to invest much of her wealth in a single project that has a chance of achieving a major gain.

Risk and Cost-Volume-Profit Analysis

We will now illustrate how attitudes toward risk can be incorporated into decision making. A decision using cost-volume-profit analysis will be used in this illustration.[4] Suppose Amy Anton is entering the mail-order poster business. Her first offering will be either a football poster or a basketball poster.

Data for each poster follow:

[4]This section considers uncertainty in only the sales volume variable in cost-volume-profit analysis. Uncertainty can also be recognized in other variables, such as selling price, variable costs, and fixed costs. See D. Driscoll, W. Lin, and P. Watkins, "Cost-Volume-Profit Analysis under Uncertainty: A Synthesis and Framework for Evaluation," *Journal of Accounting Literature,* Vol. 3, pp. 85–115.

	Football Poster	Basketball Poster
Selling price per poster	\$16	\$12
Royalty cost per poster	\$ 7	\$ 3
Production and distribution cost per poster	\$ 5	\$ 5
Fixed royalty cost	\$400,000	\$400,000

The contribution margins are \$4 for the football poster (\$16 − \$7 − \$5) and \$4 for the basketball poster (\$12 − \$3 − \$5). The breakeven point for each poster is 100,000 units (\$400,000 ÷ \$4 contribution margin per unit). Anton's current wealth is \$300,000.

Given these facts, Anton should be indifferent between the two posters. Both have the same total contribution margin for any level of sales. However, consider some additional data:

	Probability Distribution	
Sales Volume in Units	Football Poster	Basketball Poster
50,000	0.1	0.2
75,000	0.2	0.3
100,000	0.3	0.2
125,000	0.3	0.1
150,000	0.1	0.1
225,000	0.0	0.1
	1.0	1.0

The expected values of sales volume in units follows.

- For the Football Poster:

$$0.1(50{,}000) + 0.2(75{,}000) + 0.3(100{,}000) + 0.3(125{,}000) + 0.1(150{,}000) = 102{,}500$$

- For the Basketball Poster:

$$0.2(50{,}000) + 0.3(75{,}000) + 0.2(100{,}000) + 0.1(125{,}000) + 0.1(150{,}000) + 0.1(225{,}000) = 102{,}500$$

Both posters have the same expected values for sales volume in units and hence the same expected values for contribution margins of 102,500 × \$4 = \$410,000. With the football poster, there is a 0.3 (0.1 + 0.2) probability that sales volume will fall below the breakeven point of 100,000 units. In contrast, the probability that the sales volume of the basketball poster will fall below 100,000 units is 0.5 (0.2 + 0.3). The basketball poster has a wider range of outcomes (dispersion) around its expected sales and a higher probability of loss. Both posters have the same expected operating income of \$10,000 (\$410,000 − \$400,000).

Anton's choice will depend on her utility function. When using utils in the calculation, it is necessary to know the precise shape of functions such as those presented in Exhibit 20-6. Assume that Amy Anton is risk averse and that her utility can be quantified by a logarithmic utility function:[5]

$$\text{Utility (ending wealth)} = \text{Log}_{10}\text{ (ending wealth)}$$

[5] A logarithmic utility function is one of a broader set of utility functions that can describe a risk-averse decision maker's attitude toward risk.

Utility is defined by the log of ending wealth, not by the log of income, because ending wealth determines Anton's attitude toward risk. The possible outcomes with the football poster and the basketball poster illustrate this utility function:

Units Sold (1)	Income (2)	Ending Wealth (3) = \$300,000 + (2)	Log_{10}(Ending Wealth)* (4)
50,000	−\$200,000	\$100,000	5.00
75,000	−100,000	200,000	5.30
100,000	0	300,000	5.48
125,000	100,000	400,000	5.60
150,000	200,000	500,000	5.70
225,000	500,000	800,000	5.90

*From a table of logarithms not included in this book.

There is a nonlinear relationship between columns (3) and (4). An 800% increase in ending wealth (from \$100,000 to \$800,000) translates to an 18% increase in the log of ending wealth (from 5.00 to 5.90).

Using the logarithmic utility function, the expected utilities from the choice of each poster are:

- Football Poster:

$E(\text{Utility}) = 0.1(5.00) + 0.2(5.30) + 0.3(5.48) + 0.3(5.60) + 0.1(5.70) + 0.0(5.90) = 5.454$

- Basketball Poster:

$E(\text{Utility}) = 0.2(5.00) + 0.3(5.30) + 0.2(5.48) + 0.1(5.60) + 0.1(5.70) + 0.1(5.90) = 5.406$

Given Anton's logarithmic utility function, she would choose the football poster because its expected utility exceeds that of the basketball poster. Note that if Anton were risk neutral, she would be indifferent between the two posters.[6] This example illustrates how the risk preferences of managers can affect their decisions. These decisions include which information system to use within an organization as well as decisions such as what products to make and what prices to charge customers.

IMPLEMENTATION ISSUES

Surveys report several obstacles to the adoption of the decision models outlined in this chapter:[7]

- Difficulty of "selling" the models to managers
- Lack of "good" data
- Lack of time to analyze actual problems with sophisticated models
- Lack of suitable computer software programs

The education efforts required to make managers feel comfortable with decision models that recognize uncertainty can be considerable.

[6]As a general rule, expected utility computations should consider the ending wealth level rather than just the incremental profits or cash flows associated with a specific decision. The intuition behind this general rule is that a loss of (say) \$100,000 can have differing amounts of disutility depending on the wealth level of the decision maker. Only under special conditions (such as risk neutrality) is it valid to make expected utility computations without explicitly considering the wealth level of the decision maker.

[7]See the survey results in W. T. Lin and P. R. Watkins, *The Use of Mathematical Models* (Montvale, NJ: National Association of Accountants, 1986).

Although the analysis in this chapter may appear burdensome in terms of computations, the example situations are relatively simple. We considered settings where there is only one decision maker and where decisions have consequences for only one time period in the future. Consider a more complex case with three parties—the owner of a firm (who makes the decisions about what information system to use) and two operating managers.

The information system the owner chooses can perform several roles in this more complex setting:

1. Improve the quality of decisions made by each operating manager
2. Motivate each operating manager to make decisions that are in the best interest of the owner
3. Assist the owner in evaluating the performance of *each* operating manager and the activities that he or she directs

The first role is illustrated in this chapter. The latter two roles are discussed in Chapters 13, 27, and 28.

The information available to an owner includes that conveyed by the operating managers. These managers may not have incentives to communicate their information in an honest, complete, and timely way. Instead, some operating managers—individually or in collusion—may withhold important information from higher management. They may even distort information if it can be used against them. Moreover, some operating managers may distort or delay the information transmitted to other operating managers. For example, information about capacity levels or cost structures may not be honestly communicated between operating managers.

These complexities do not diminish the importance of a cost-benefit perspective. Instead, they serve to underscore the challenge of applying that perspective in the uncertain world in which owners and managers find themselves.

PROBLEM FOR SELF-STUDY

PROBLEM

Cheri Martel, a cost analyst for the air force, is currently reviewing the cost estimates made by Erland Corp. and Rex Inc., the final two bidders for a jet fighter contract. A committee of four air force generals will make the decision on the contract. Martel must make a presentation to this committee on the expected costs of the jet fighter project for each of the two contractors. The company awarded the contract will be reimbursed on an "actual cost plus a fixed fee" basis. The fixed fee is 8% of the cost bid at the time the contract is awarded.

Erland has made a cost bid of $600 million for the contract. Rex has made a bid of $650 million. Martel decides to use the past experience of both companies on government contracts to assess the probability distribution of actual costs on the jet fighter contract to the air force. She collects data on the cost bid at the time each contract was awarded and the eventual actual cost (before payment of the 8% fixed fee) for all past contracts awarded to Erland and for all past contracts awarded to Rex. The distribution follows:

	Erland	Rex
10% cost underrun	0.05	0.15
No underrun or overrun	0.10	0.25
10% overrun	0.10	0.25
20% overrun	0.15	0.20
30% overrun	0.20	0.10

40% overrun	0.25	0.05
50% overrun	0.10	0.00
60% overrun	0.05	0.00
	1.00	1.00

Required

1. What is the range of outcomes of the actual costs to the air force (including the 8% fixed fee) if the contract is awarded to (a) Erland and (b) Rex? Compute the expected value of the costs to the air force for each contractor.
2. Evaluate the approach used by Martel to compute the probability distribution of actual costs on the jet fighter contract.

SOLUTION

1. A three-step procedure is used to assess the probability distribution of actual costs to the air force:
   a. Determine the actual cost each contractor will report with a 10% underrun, . . . , 50% overrun, and 60% overrun.
   b. Add the fixed fee, which is 8% of the cost bid, and
   c. Use the past distribution for each underrun/overrun category to determine the range of outcomes for the actual costs to the air force for the jet fighter contract.

EXHIBIT 20-7

Probability Distribution of Actual Costs to Air Force with Alternative Contractors (in millions)

A. Erland Corp. Awarded Contract

Cost Underrun/ Overrun Category (1)	Contractor's Actual Cost (2)	Contractor's Actual Cost + 8% of Bid (3) = (2) + $48	Probability (4)	Expected Cost (5) = (3) × (4)
10% cost underrun	$540	$ 588	0.05	$ 29.4
No underrun/overrun	600	648	0.10	64.8
10% overrun	660	708	0.10	70.8
20% overrun	720	768	0.15	115.2
30% overrun	780	828	0.20	165.6
40% overrun	840	888	0.25	222.0
50% overrun	900	948	0.10	94.8
60% overrun	960	1,008	0.05	50.4
			1.00	$813.0

B. Rex Inc. Awarded Contract

Cost Underrun/ Overrun Category (1)	Contractor's Actual Cost (2)	Contractor's Actual Cost + 8% of Bid (3) = (2) + $52	Probability (4)	Expected Cost (5) = (3) × (4)
10% cost underrun	$585	$637	0.15	$ 95.55
No underrun/overrun	650	702	0.25	175.50
10% overrun	715	767	0.25	191.75
20% overrun	780	832	0.20	166.40
30% overrun	845	897	0.10	89.70
40% overrun	910	962	0.05	48.10
50% overrun	—	—	0.00	—
60% overrun	—	—	0.00	—
			1.00	$767.00

Exhibit 20-7 presents the computations, which can be summarized as follows:

Contractor	Current Bid (in Millions)	Expected Value of Cost to Air Force (in Millions)	Range of Outcomes (in Millions)
Erland	$600 + $48 = $648	$813	$588 to $1,008
Rex	$650 + $52 = $702	$767	$637 to $962

Rex has a lower expected value of the actual cost. The magnitude of the largest possible cost overrun with Rex ($962 million) is less than that with Erland ($1,008 million).

2. The approach Martel used to assess probabilities assumes that the past track record of a contractor on all government contracts is a good predictor of likely performance on the jet fighter contract. Martel might consider two factors in evaluating the reasonableness of this assumption:

   a. Explanations for the past cost underruns or overruns for each contractor. These include:

      - Unanticipated inflation,
      - Changes in specifications or quantity mandated by the government after the contract was awarded, and
      - Deliberate underbidding by the contractor (which is frequently alleged, although it is difficult to document).

      If the appropriate explanations are unlikely to persist over the life of the jet fighter contract, a change in the probability distribution for underestimates or overestimates may be appropriate.

   b. The current management of each contractor. Has the management existing at the time of past contracts been replaced? Because of allegations of poor internal control for government contracts, excessive overruns, and so on? Then a reduction in the probability of large overruns may be appropriate.

SUMMARY

Formal decision models are increasingly being used because they are becoming cheaper to implement and because they replace or supplement hunches and implicit rules with explicit assumptions and criteria. The decision maker must choose an appropriate decision model. Both the choice criterion and the complexities of the decision situation affect this choice. The information to compile depends on the chosen prediction method and the decision model.

Managers and accountants often regard quantification as precise. Almost all data however, are subject to uncertainty. Accounting reports for decision making increasingly include the formal, explicit recognition of uncertainty. A knowledge of probability distributions is often essential in providing and evaluating information for decisions. The use of utility values provides a systematic way of measuring trade-offs between risks and returns.

TERMS TO LEARN

This chapter and the Glossary at the end of the book contain definitions of the following important terms:

choice criterion *(p. 647)* decision table *(650)* events *(647)*
expected monetary value *(648)* expected value *(648)*
expected value of perfect information (EVPI) *(652)* objective function *(647)*
outcomes *(647)* payoffs *(647)* payoff matrix *(650)* payoff table *(650)*
probability *(647)* probability distribution *(647)* risk averse *(654)*
risk neutral *(654)* risk seeking *(654)* states *(647)* states of nature *(647)*
utility function *(654)* utility value *(654)* utils *(654)*

ASSIGNMENT MATERIAL

QUESTIONS

20-1 What are five key elements in a decision model?

20-2 Distinguish between *expected value* and *expected monetary value.*

20-3 "A wisdom born after the event is the cheapest of all wisdom." Do you agree? Explain.

20-4 How would you respond to the following statement?

> Why learn about decision making under conditions of uncertainty? When all the computer runs have been digested (if that is possible) and when all the decision trees have shed their leaves, the decision maker has no guarantee that the final outcome will be more desirable than an outcome that would have occurred as the result of a decision made without detailed analysis.

20-5 Why compute the expected value of perfect information when it is unlikely that anyone has perfect information?

20-6 What are the five steps used to compute the expected value of perfect information?

20-7 Distinguish among decision makers who are (a) risk neutral, (b) risk averse, and (c) risk seeking.

20-8 What is the benefit of explicitly recognizing probabilities in decision making?

20-9 Name three obstacles to adoption of the expected value approach to decision making under uncertainty.

EXERCISES AND PROBLEMS

20-10 Decision analysis, definition of a good decision. Consider the following definitions of a good decision:

A. A decision that receives a favorable response from your boss
B. A decision that results in the most favorable outcome when uncertainty is resolved
C. A decision that maximizes expected monetary value at the time of the decision
D. A decision that avoids outcomes that are highly negative

Required

Which of A, B, C, or D is the definition of a good decision that is consistent with the approach outlined in this chapter? Explain.

20-11 Purchase order size, expected monetary value. Once a day Supervalue, a retailer, stocks fresh-cut carnation flowers. Each flower costs 40¢ and sells for $1. Supervalue never reduces the selling price; unsold carnations at the end of each day are given to a nearby church. Supervalue estimates daily demand characteristics as follows:

Demand	Probability
0	0.05
100	0.20
200	0.40
300	0.25
400	0.10
	1.00

Required
How many carnations should Supervalue purchase each day to maximize expected operating income? Why?

20-12 Expected value of perfect information. Refer to Problem 20-11. How much should Supervalue be willing to pay for a perfect prediction concerning the daily demand for carnations?

20-13 Purchase of a new lathe, demand uncertainty. (A. Atkinson) The manager of operations at Purcell's Cove Machine Shop is considering leasing equipment for a new flexible manufacturing system (FMS) to replace existing equipment. The FMS lease will increase annual fixed costs by $900,000 per year and will reduce variable costs by $800 per job.

The manager believes that the annual number of jobs processed by the company will be 900, 1,200, or 1,500. The probabilities of these events occurring are:

Annual Number of Jobs	Probability
900	0.25
1,200	0.45
1,500	0.30
	1.00

Required
Should the company lease the FMS? Support your conclusion with appropriate calculations.

20-14 Setting prices and uncertainty Trent Harris, marketing manager of The Complete Angler, is examining the pricing of its fishing-reel product line. Assume that the unit cost of a fishing reel is known with certainty to be $16. Harris is trying to decide whether to set a selling price of $20 or $22 per unit. The top price has been $20 for the past 30 months. Average monthly sales are forecast as follows:

At a Unit Price of $20

Units	Probability
1,050,000	0.05
1,000,000	0.90
950,000	0.05
	1.00

At a Unit Price of $22

Units	Probability
800,000	0.10
750,000	0.60
700,000	0.30
	1.00

The Complete Angler uses the expected monetary value choice criterion in its decisions.

Required
Determine the optimal selling price of a fishing reel. Show computations.

20-15 Bidding for a contract, uncertainty over amount to bid. (A. Atkinson) Lunenburg Landscaping is considering submitting a bid to provide landscaping services at the site of a new Canadian hotel. The job will require 200 labor hours, $4,000 of material cost, and $2,000 of rental cost for the equipment the company needs to do this job.

The company's regular employees are paid $16 per hour. Normally, these employees work on lawns that the company contracts to maintain on an annual basis. The contribution margin per labor hour devoted to lawn maintenance is $8. The company has learned that it can hire summer students at the rate of $10 per hour to landscape lawns.

Kelly Burns, the owner-manager of Lunenburg Landscaping, figures that the probability is 100% that she can win the contract if she bids $8,500 on the job. Bids must be submitted in multiples of $500. Burns figures that the probability of winning the contract falls by 20% for every increment of $500 over $8,500. For example, she assesses the probability of winning the contract with a $10,000 bid as 40%; that is, there are three multiples of $500 ($10,000 − $8,500) and 3 × 20% = 60%, so 40% results. Burns uses an expected monetary value choice criterion in her decision making.

Required

What bid should Kelly Burns submit on this contract?

20-16 Alternative selling approaches for nonprofits, expected monetary value. Rod Traylor is planning the fifth annual Cobb County Police Officers' Association (CCPOA) benefit concert. The money raised benefits local charities. All the performers donate their services. The concert is to be held at the Omni, which has a maximum capacity of 10,000 people. The Omni offers CCPOA the discounted rate of a $20,000 flat fee plus $3 per ticket sold. Each ticket will have a selling price of $20. The Omni's $3 charge per ticket does not cover any marketing activities.

For each of the last four years, Traylor has organized a team of volunteers to market the concert and sell tickets. Recently, he received a proposal from Nonprofit Support, a company that has successfully assisted many nonprofit organizations in their fundraising. Nonprofit Support offers to handle all marketing functions for the concert in return for 30% of the selling price of each ticket sold.

Traylor makes the following probability assessments of tickets sold:

Tickets Sold	Probability if Volunteers Used	Probability if Nonprofit Support Used
2,000	0.30	0.05
4,000	0.40	0.10
6,000	0.20	0.20
8,000	0.10	0.30
10,000	0.00	0.35

Required

1. Should CCPOA accept Nonprofit Support's proposal if it uses an expected monetary value choice criterion in its decisions?
2. What other factors might Traylor recommend that CCPOA consider in deciding between using volunteers and using Nonprofit Support?

20-17 Cost-volume-profit analysis under uncertainty. (J. Patell) In your recently obtained position as supervisor of new product introduction, you have to decide on a pricing strategy for a specialty product with the following cost structure:

Variable costs per unit = $50
Fixed costs of production = $200,000

The units are assembled upon receipt of orders, and so the inventory levels are insignificant. Your market research assistant is very enthusiastic about probability models and has presented the results of his price analysis in the following form:

a. If you set the selling price at $100/unit, the probability distribution of total sales dollars is uniform between $300,000 and $600,000. Under this distribution, there is a 0.50 probability of equaling or exceeding sales of $450,000.
b. If you lower the selling price to $70/unit, the distribution remains uniform, but it shifts up to the $600,000 to $900,000 range. Under this distribution, there is a 0.50 probability of equaling or exceeding sales of $750,000.

Required

1. This is your first big contract, and, above all, you want to show an operating income. You decide to select the strategy that maximizes the probability of breaking even or earning a positive operating income.
   a. What is the probability of at least breaking even with a selling price of $100/unit?
   b. What is the probability of at least breaking even with a selling price of $70/unit?
2. Your assistant suggests that maximum expected operating income might be a better goal to pursue. Which pricing strategy would result in the higher expected operating income? (Use the expected sales volume under each pricing strategy when making expected operating income computations.)

20-18 Purchasing decision, uncertainty over defect rate. Kim Wolser, the purchasing manager of Crocker Communications, manufactures and markets cable television receiving dishes. For many years, Crocker has used the following two suppliers for a key component (the X407 receiving unit):

1. Silicon Valley Components
2. Sapporo Products

Silicon Valley Components has long been the preferred supplier because of its lower purchase price. Silicon Valley will charge $25 per X407 unit in 19_8, and Sapporo Products will charge $30 per X407 unit in 19_8.

Crocker solders the X407 into the dish and then ships the finished product to its customers. Defects in the X407 come to light when Crocker solders the component into the dish or when customers use the dish. It costs Crocker $40 to repair a defect if it is detected before the dish is shipped to a customer. It costs $280 to repair a defect detected after the dish is shipped to a customer. (This $280 amount includes the cost of customer ill will associated with breakdowns in the receiving dish.)

Based on its experience, Crocker draws up the following probability table:

	Probabilities	
	X407 Supplied by Silicon Valley Components	X407 Supplied by Sapporo Products
1. X407 receiving unit contains no defects	0.90	0.97
2. X407 receiving unit is defective and is repaired before shipping dish to customer	0.07	0.02
3. X407 receiving unit is defective and is repaired after shipping dish to customer	0.03	0.01

Crocker Communications expects to purchase 250,000 X407 receiving units in the coming year, 19_8.

Required

Crocker Communications has decided to use a single supplier in 19_8. Which supplier should Crocker choose? Assume it uses an expected monetary value choice criterion in its decisions?

20-19 Alternative customer service policies for television sets. As an appliance dealer, you are deciding how to service your one-year warranty on the 1,000 color television sets you have just sold to a large local hotel. You have three alternatives:

1. A reputable service firm has offered to service the sets, including all parts and labor, for a flat fee of $18,000.
2. For $15,000, another reputable service firm would furnish all necessary parts and provide up to 1,000 service calls at no charge. Service calls in excess of that number would be $4 each. The number of calls is likely to be:

Event	Chance of Occurrence	Probability of Occurrence	Total Cost
1,000 calls or less	50%	0.5	$15,000
1,500 calls	20	0.2	17,000
2,000 calls	20	0.2	19,000
2,500 calls	10	0.1	21,000
	100%	1.0	

3. You can hire your own labor and buy your own parts. Your past experience with similar work has helped you to formulate the following probabilities and total cost:

Event	Chance of Occurrence	Probability of Occurrence	Total Cost
Little trouble	10%	0.1	$ 8,000
Medium trouble	70	0.7	10,000
Much trouble	20	0.2	30,000
	100%	1.0	

Required
Using the expected monetary value approach, compare the three alternatives. Which plan do you favor? Why?

20-20 Cost reduction uncertainty, compensation plans for hospital consulting firms. Hospital Cost Management Associates (HCMA) provides consulting advice to hospitals. HCMA specializes in identifying areas where major reductions in costs can be made without any reduction in the quality of the service provided to hospital patients.

Calgary Hospital recently received a consulting proposal from HCMA, which impressed the hospital board of advisers. The board is now deciding between two compensation plans for HCMA to provide extensive consulting services during 19_4.

1. Compensation Plan A: Flat payment of $150,000.
2. Compensation Plan B: HCMA receives 40% of the "documented cost savings," defined as the difference between the budgeted costs for the coming year (19_4) and the actual costs.

The cost analyst at Calgary Hospital estimates the following probabilities of events if HCMA is hired:

Events: Cost Difference between 19_4 Budgeted Cost and 19_4 Actual Cost	Probability
$ 50,000	0.2
200,000	0.3
400,000	0.4
600,000	0.1
	1.0

HCMA is willing to work under either compensation plan.

Required
1. Which compensation plan should Calgary Hospital use when hiring HCMA? Assume Calgary Hospital uses the expected monetary value choice criterion in its decision making.

2. A longtime member of Calgary's board of advisers objects strongly to Compensation Plan B. He makes the following comment to the president of HCMA:

   The "documented cost savings" scheme is obscene. We may well pay you $240,000. No consultant is worth this much. When a new heart-monitoring machine costs $120,000, there is no way we can justify $240,000 for the thoughts of some newly minted college graduates.

   How should the president of HCMA respond to this comment?

20-21 Choice of a production-run length. (CMA) Jackston Inc. manufactures and distributes a line of Christmas toys. The company had neglected to keep its dollhouse line current. As a result, sales decreased to approximately 10,000 units per year from a high of 50,000 units. The dollhouse has been redesigned recently, and company officials consider it comparable to its competitors' models. The company plans to redesign the dollhouse each year in order to compete effectively. Joan Blocke, the sales manager, is not sure how many units can be sold next year, but she is willing to place probabilities on her estimates. Blocke's estimates of the number of units that can be sold during the next year and the related probabilities follow:

Estimated Sales in Units	Probability
20,000	0.10
30,000	0.40
40,000	0.30
50,000	0.20
	1.00

The units are to sell for $20 each.

The inability to estimate the sales more precisely is a manufacturing problem for Jackston. Whatever the decision on the number of units to produce, Jackston will schedule a single production run for the entire year's units. If the demand is greater than the number of units manufactured, then sales will be lost. If demand is below supply, the extra units cannot be carried over to the next season and will be given away to various charitable organizations. The production and distribution cost estimates follow:

	Units Manufactured			
	20,000	30,000	40,000	50,000
Variable costs	$180,000	$270,000	$360,000	$450,000
Fixed costs	140,000	140,000	160,000	160,000
Total costs	$320,000	$410,000	$520,000	$610,000

Required

1. Prepare a decision table for the different sizes of production runs required to meet Blocke's four sales estimates. If Jackston relies solely on the expected monetary value approach to make decisions, what size of production run will be selected?
2. Identify the basic steps taken in any decision process. Explain each step by referring to this problem and your answer to requirement 1.

20-22 Uncertainty and cost-volume-profit analysis. (R. Jaedicke and A. Robichek, adapted) The Jaedicke and Robichek Company is considering two new products. Either can be produced by using present facilities. Each product requires an increase in annual fixed costs of $400,000. The products have the same selling price ($10) and the same variable cost per unit ($8).

Management, after studying past experience with similar products, has assessed the following probability distribution:

Event (Units Demanded)	Probability— Product A	Probability— Product B
50,000	—	0.1
100,000	0.1	0.1
200,000	0.2	0.1
300,000	0.4	0.2
400,000	0.2	0.4
500,000	0.1	0.1
	1.0	1.0

Required

1. What is the breakeven point for each product?
2. Which product should be chosen, assuming the objective is to maximize expected operating income? Why? Show computations.
3. Suppose management was absolutely certain that 300,000 units of Product B would be sold. Management still faces the same uncertainty about the demand for Product A as outlined in the problem. Which product should be chosen? Why? What benefits are available to management from the provision of the complete probability distribution instead of just a lone expected value?

20-23 Choice among alternative product designs; demand uncertainty, expected monetary value. (CMA, adapted) Steven Company has been producing component parts and assemblies for use in the manufacture of microcomputers. The company plans to introduce a magnetic tape cartridge backup unit for IBM-compatible microcomputers in the near future.

Steven's Research and Development (R & D) and Market Research departments have been working on this project for an extended period. The development costs of the two departments incurred to date amount to $1,500,000. R & D created several alternative designs for the backup units. Three of the designs were approved for development into prototypes, and from these only one will be manufactured and sold. Market Research has determined that the appropriate selling price would be $540 per unit, regardless of the model selected.

There is uncertainty about demand. Three alternative demand events are possible—light, moderate, and heavy. Steven can meet all demand levels because its production facility is currently operating below full capacity.

Event	Unit Sales	Probability of Event
Light demand	20,000	0.25
Moderate demand	80,000	0.60
Heavy demand	120,000	0.15
		1.00

Variable manufacturing overhead is applied to Steven's products using a plantwide application rate of 250% of direct labor dollars. Steven's engineering and accounting staffs have worked together to develop manufacturing cost estimates for each of the three model designs:

Unit Variable Costs	Model A	Model B	Model C
Direct materials	$150	$100	$114
Direct labor	40	50	48
Manufacturing overhead	100	125	120
Marketing	140	140	140
Total unit variable costs	$430	$415	$422
Other Costs			
Fixed manufacturing overhead	$1,000,000	$1,400,000	$1,300,000
Fixed marketing costs	2,000,000	3,100,000	2,800,000
Development costs (already incurred)	1,500,000	1,500,000	1,500,000

Steven has decided to use an expected monetary value choice criterion in its analysis of which of the three prototypes it will manufacture and sell.

Required

1. Compute the unit contribution margins of Models A, B, and C.
2. Develop a decision-tree presentation of the data for Steven Company.
3. Which prototype should Steven Company manufacture and sell?

20-24 Cost estimation for navy, actual cost plus fixed-fee reimbursement to contractor. Barry Windham, a cost analyst for the navy, is currently reviewing the cost estimates for a nuclear submarine contract. Windham must make a presentation to a selection committee of three admirals. He will compare the costs that the navy will incur if it awards the contract to one of the three following companies:

- Olympus Inc.—has bid $220 million
- Pisces Ventures—has bid $230 million
- Scorpio Corporation—has bid $200 million

The company awarded the contract will be reimbursed on an "actual cost incurred plus fixed-fee" basis. The fixed fee is 5% of the bid.

Windham collects data on the past cost underrun/overrun experience in government contracting of each of the three potential contractors. He will use these data to assess the probability distribution of underruns/overruns on the nuclear submarine project. Windham summarized these data by comparing the cost bid at the time the contract was awarded and the eventual actual cost (before inclusion of the 5% fixed fee). The distribution for each of the three contractors follows:

	Olympus	Pisces	Scorpio
10% cost underrun	0.10	0.00	0.00
No underrun or overrun	0.15	0.60	0.10
10% overrun	0.20	0.40	0.15
20% overrun	0.30	0.00	0.20
30% overrun	0.10	0.00	0.25
40% overrun	0.10	0.00	0.20
50% overrun	0.05	0.00	0.10
	1.00	1.00	1.00

Required

1. What is the probability distribution of the actual costs to the navy (including the 5% fixed fee) if the contract is awarded to (a) Olympus, (b) Pisces, or (c) Scorpio? Compute the expected monetary value of the distribution of actual costs to the navy for each contractor.
2. Windham observes that Pisces has not previously built a nuclear submarine. He also notes that the past projects Pisces has worked on for the government have been technically uncomplicated. How might Windham incorporate this information into his presentation to the contract selection committee?

20-25 Bidding for a contract, expected value of perfect information. (A. Atkinson) Chester Steel Fabrication (CSF) is considering submitting a bid to construct a metal bridge for a new highway. The company controller thinks that a bid of $1,000,000 will cover both the costs associated with this project and the opportunity costs of the equipment and labor that would be used on this project. Julie Dawson, the company president, figures that a CSF bid of $1,000,000 is sure to get the job. Dawson intends to bid that amount, but she wonders about some of the cost estimates included in the controller's report.

After some study, Dawson has determined that the total of all costs associated with this project could actually be $800,000, $900,000, $1,000,000, or $1,200,000 with respective probabilities of 0.10, 0.30, 0.40, and 0.20. CSF uses an expected monetary value choice criterion in its decisions.

Required

1. Given the information provided, what is Dawson's optimal action: bid or not bid?
2. What is the maximum that Dawson should be willing to pay for the controller to undertake a new study to determine precisely the costs associated with this project?

20-26 Expected value of perfect information. As a manager of a post office, you are trying to decide whether to rearrange a production line and facilities in order to save labor and related costs. Assume that the only alternatives are to "do nothing" or "rearrange." Assume also that the choice criterion is that the expected savings from rearrangement must equal or exceed $11,000.

Based on your experience and currently available information, you predict:

a. The "do nothing" alternative will have operating costs of $200,000.
b. The "rearrange" alternative will have operating costs of $100,000 if it is a success and $260,000 if it is a failure. You think there is a 60% chance of success and a 40% chance of failure.

Required

1. Compute the expected value of the costs for each alternative. Compute the difference in expected costs.
2. You can hire a consultant, Joan Zenoff, to study the situation. She would then render a flawless prediction of whether the rearrangement would succeed or fail. Compute the maximum amount you would be willing to pay for the errorless prediction.

20-27 Alternative contracts for a movie producer, uncertainty about production costs and demand. Jillian Armstrong, an independent movie producer, is negotiating with Roadshow Productions on the contract for the production and marketing of her next film, titled *The Revenge of Robocop.* The budget for the film is $10 million. Roadshow offers Armstrong one of three contracts:

Contract A:

1. Roadshow pays all costs of production and marketing.
2. Armstrong receives a fixed fee of $1 million.
3. Armstrong receives 10% of gross revenues for the film in excess of $100 million. (No payment is made for gross revenues up to $100 million.)

Contract B:

1. Roadshow pays 80% of costs of production and marketing up to $10 million and 50% of costs of production and marketing in excess of $10 million.
2. Armstrong receives 10% of *all* gross revenues for the film.

Contract C:

1. Roadshow pays 50% of costs of production and marketing up to $10 million and zero % thereafter.
2. Armstrong receives 30% of *all* gross revenues for the film.

Armstrong assesses the following probabilities for the gross revenues:

P (high demand of $200 million)	0.1
P (medium demand of $50 million)	0.3
P (low demand of $10 million)	0.6
	1.0

She assesses the following probabilities for the costs of production:

P (at budgeted cost of $10 million)	0.6
P (at high cost of $20 million)	0.4
	1.0

Armstrong assesses the following probabilities for six possible events.

Event	Probability
High demand—budgeted cost	0.06
High demand—high cost	0.04
Medium demand—budgeted cost	0.18
Medium demand—high cost	0.12
Low demand—budgeted cost	0.36
Low demand—high cost	0.24
	1.00

Required

1. Compute the expected monetary value to Armstrong under each contract for each of the six possible events.
2. Armstrong will choose the contract that maximizes her expected monetary value from the film. Which contract should she choose? Show calculations.
3. What information might Armstrong use in assessing the probability distribution for the costs of production and marketing of *The Revenge of Robocop* film?

CHAPTER 21

Capital Budgeting and Cost Analysis

Contrast in purposes of cost analysis

Definition and stages of capital budgeting

The discounted cash-flow method

Sensitivity analysis

Analysis of selected items using discounted cash flow

The payback method

The accrual accounting rate-of-return method

Surveys of practice

Breakeven time and capital budgeting for new products

Complexities in capital-budgeting applications

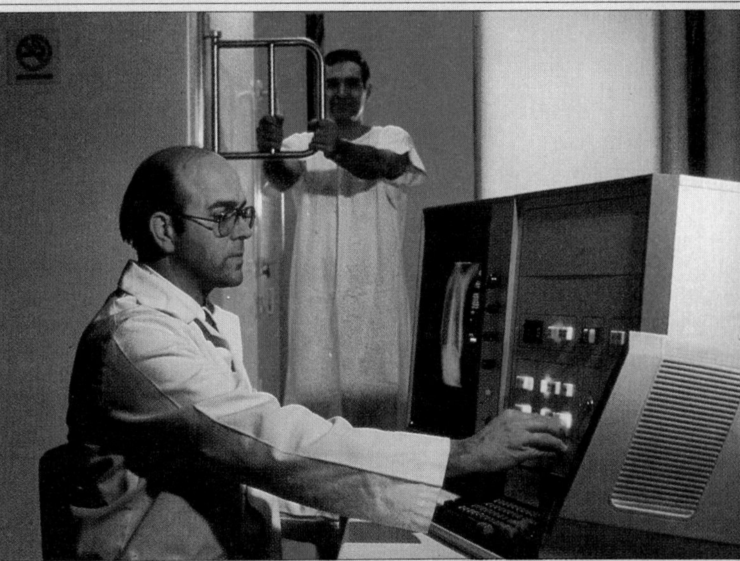

Should a hospital keep its old X-ray machine or buy a new one? Managers can turn to various capital-budgeting decision methods to evaluate alternatives.

Learning Objectives

When you have finished studying this chapter, you should be able to

1. Differentiate between project-by-project orientation of capital budgeting and period-by-period orientation of accrual accounting
2. Identify the four stages of capital budgeting for a project and its predicted outcomes
3. Distinguish among capital-budgeting methods: discounted cash flow, payback, accrual accounting ROR, breakeven time
4. Explain how the two main discounted cash-flow methods (NPV and IRR) differ
5. Explain the differences between the payback and DCF methods
6. Describe conflicts in using DCF for capital budgeting and accrual accounting for performance evaluation
7. Explain the differences between the payback and breakeven-time methods
8. Explain the impact of nonfinancial and qualitative factors in capital-budgeting decisions

Should we invest in a new research and development project, a computer-aided design system for new product development, or a different distribution network for our products? Should a city invest in a large multipurpose hospital or a network of smaller special-purpose hospitals? Such decisions frequently involve large dollar amounts and have long-lasting effects and uncertain outcomes. Accounting data often help managers make these long-range decisions, which are frequently called *capital-budgeting* decisions. This chapter examines the role of accounting data in the popular decision models for capital budgeting. We also study the relationship of the decision models to the performance-evaluation models that managers use to help judge the results of capital-budgeting decisions.

At this stage, we again focus on purpose. Income determination and the planning and controlling of routine operations primarily focus on the *current time period*. Special decisions and long-range planning primarily focus on the *project* or *program* with a much longer time span.

CONTRAST IN PURPOSES OF COST ANALYSIS

Objective 1

Differentiate between project-by-project orientation of capital budgeting and period-by-period orientation of accrual accounting

Exhibit 21-1 illustrates two different dimensions of cost analysis: (a) a project dimension, represented by the vertical axis, and (b) a time dimension, represented by the horizontal axis.

Each project is represented in Exhibit 21-1 as a distinct horizontal rectangle. The life of each project is longer than one year. We focus in this chapter on an individual investment venture throughout its life, recognizing the time value of money. The **time-value of money** takes into account the interest received (paid) from investing (borrowing) money for the time period under consideration.

The distinct vertical rectangle in Exhibit 21-1 illustrates the accounting period focus on income determination and planning and control. This cross-section emphasizes the company's performance for a year (in our example, 19_3). Because the accounting period is a year or less, the time value of money usually is not a primary concern during that time span.

EXHIBIT 21-1
The Project and Time Dimensions of Capital Budgeting

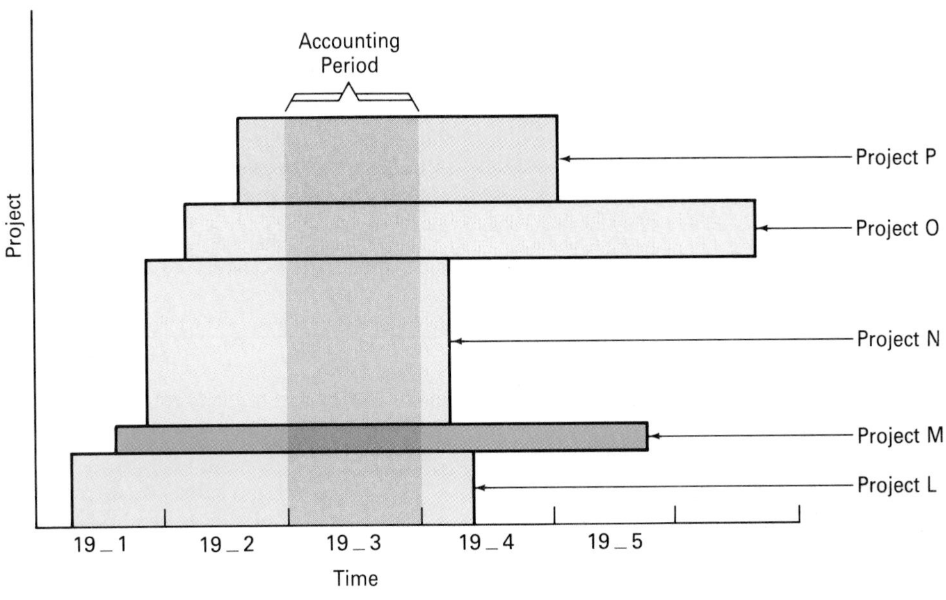

Recall a central theme of this book: different costs for different purposes. Capital-budgeting decisions have a project focus. The revenues and costs specifically associated with the project should be measured over the life of the project. There is a great danger in basing capital-budgeting decisions on an indiscriminate use of the data routinely reported in the current accounting period's income statements. We illustrate this point in several examples in the chapter.

The accounting system that corresponds to the project dimension in Exhibit 21-1 is termed life-cycle costing. This system, described in Chapter 12, accumulates revenues and costs on a project-by-project basis. For example, a life-cycle costing statement for a new-car project at Ford Motor Company could encompass a ten-year period and would accumulate costs in the research and development, product design, manufacturing, marketing, distribution, and customer-service categories. This accumulation, covering many years, differs from the accrual accounting system that measures income on a period-by-period basis.

DEFINITION AND STAGES OF CAPITAL BUDGETING

Objective 2

Identify the four stages of capital budgeting for a project and its predicted outcomes

Capital budgeting is the making of long-term planning decisions for investments and their financing. After a project has been identified and its outcomes have been predicted, there are four stages in capital budgeting. We will illustrate these stages using a decision by Midwest Machinery Inc. (an equipment distributor) of whether to automate its warehouse. The decision is whether to switch from a currently labor-intensive distribution operation to a highly automated distribution operation. Exhibit 21-2 classifies the predicted outcomes of this decision into four categories. Most capital-budgeting decisions involve items in each of these categories.

The four stages in capital budgeting are:

Stage 1: Analyzing the quantitative/financial aspects of the project. This chapter discusses the following methods of summarizing these aspects:

EXHIBIT 21-2

Classification of Predicted Outcomes in a Capital-Budgeting Decision by Midwest Machinery

Category	Example
Quantitative/Financial	Labor-cost reductions
Quantitative/Nonfinancial	Increase in speed with which equipment is delivered to the shipping area
Qualitative/Financial	Increase in financial risk because a higher proportion of total costs is fixed
Qualitative/Nonfinancial	Reduction in workers injured in equipment-handling accidents

Objective 3

Distinguish among capital budgeting methods: discounted cash flow, payback, accrual accounting ROR, breakeven time

- Discounted cash flow (both net present value, and internal rate of return)
- Payback
- Accrual accounting rate of return (ROR)
- Breakeven time

Our discussion emphasizes the discounted cash-flow method. Both the net present value and internal rate-of-return versions of this method explicitly recognize the time value of money as a critical factor in decisions affecting long time spans.

Stage 2: Analyzing the quantitative/nonfinancial and the two qualitative aspects (financial and nonfinancial) of the project.

Stage 3: Financing the project. Sources of financing include internally (within the organization) generated cash, the equity-security capital market, and the debt-security capital market.

Stage 4: Implementing the project and monitoring its performance. In some cases, monitoring may include a post-decision audit in which the predictions made at the time a project was selected are compared with the subsequent outcomes.

The discussion in this chapter focuses on the investment-decision aspects of capital budgeting. However, in understanding the approval or rejection of projects, the influence of individual personalities is often pivotal. Those managers who are best at selling their own projects to the decision maker(s) will often get the lion's share of the available money, and the other managers either get nothing or wait—and then wait some more. Economic considerations can become secondary as individual managers war with words, often citing their impressive operating performance that may or may not be relevant to the capital-budgeting decision at hand.

The relevant cost and revenue framework outlined in Chapter 11 should guide decisions about what items to include in capital-budgeting analysis. Include only future costs and revenues that differ among the alternatives under consideration. Do not assume that only those costs currently classified as variable are relevant. Most capital-budgeting projects have a multiyear time horizon. Costs that are fixed for short time periods are frequently not fixed over longer periods.

THE DISCOUNTED CASH-FLOW METHOD

We now consider the first of four methods for analyzing the quantitative/financial aspects of projects. **Discounted cash flow (DCF)** measures the cash inflows and outflows of a project as if they occurred at a single point in time so that they can be compared in an appropriate way. The *discounted cash-flow* method recognizes that the use of money has a cost—interest forgone. A dollar today is worth more than a dollar five years from today. A dollar can be invested and grow over time as it earns interest. Since the discounted cash-flow method explicitly and routinely weighs the time value of money, it is usually the best (most comprehensive) method to use for long-range decisions.

DCF focuses on *cash* inflows and outflows rather than on *operating income* as used in the conventional accrual accounting method. Cash is invested now with the hope of receiving a greater amount of cash in the future. Try to avoid injecting accrual concepts of accounting into DCF analysis. The compound interest tables and formulas used in DCF analysis are included in Appendix B, pages 930–37. **BEFORE READING ON, BE SURE YOU UNDERSTAND THIS APPENDIX.**

There are two main DCF methods:

1. Net present value (NPV) and
2. Internal rate of return (IRR).

Both DCF methods require as an input the **required rate of return (RRR),** which is the minimum acceptable rate of return on an investment. This rate is also called the **discount rate**, **hurdle rate**, or **cost of capital**. Chapter 22 discusses issues encountered in estimating this rate.

For simplicity, this chapter (along with Chapter 22) assumes that the cash outflows and cash inflows occur at the beginning or end of each period. The following situation will be used to illustrate various capital-budgeting decision methods.

EXAMPLE

The manager of a nonprofit hospital, Lifetime Care, is considering buying a new state-of-the-art X-ray machine that will aid productivity in the X-ray department. The new machine will decrease labor costs and reduce the average number of individual X-rays taken per patient. The machine will cost $379,100 now, and the manager expects it to have a five-year useful life and a zero terminal disposal price. It will result in cash-operating savings of $100,000 annually. The hospital is not subject to income taxes. There are two alternatives:

(1) To continue operation of the X-ray department without change or

(2) To buy the new machine.

Regardless of the decision, revenue will not change, so the only relevant quantitative/financial element in the second alternative is the savings in cash-operating costs.

Compute the net present value of the project. Assume that the required rate of return is 8%. (This relatively low interest rate is not unusual for a nonprofit institution.) Also, compute the internal rate of return on the project.

Net Present Value Method

Net present value (NPV) is a DCF method of calculating the expected net monetary gain or loss from a project by discounting all expected future cash inflows and outflows to the present point in time, using the required rate of return. Projects with higher net present values are preferred to projects with lower net present values, if all other things are equal. Using the NPV method entails the following steps:

1. Drawing a sketch of relevant cash inflows and outflows. The right-hand side of Exhibit 21-3 shows how these cash flows are portrayed. Outflows are placed in parentheses. Although such a sketch is not absolutely necessary, it clarifies relationships. It lets the decision maker organize the data in a systematic way. Note that Exhibit 21-3 includes the outflow at year 0, the time of the new machine acquisition.
2. Choosing the correct compound interest table from Appendix B. Find the discount factor from the appropriate row and column. Table 2 (p. 935) presents discount factors (periods 1 to 5 for the 8% column) for Approach 1 in Exhibit 21-3. Table 4 (p. 937) presents the

EXHIBIT 21-3

Net Present Value Method: Lifetime Care

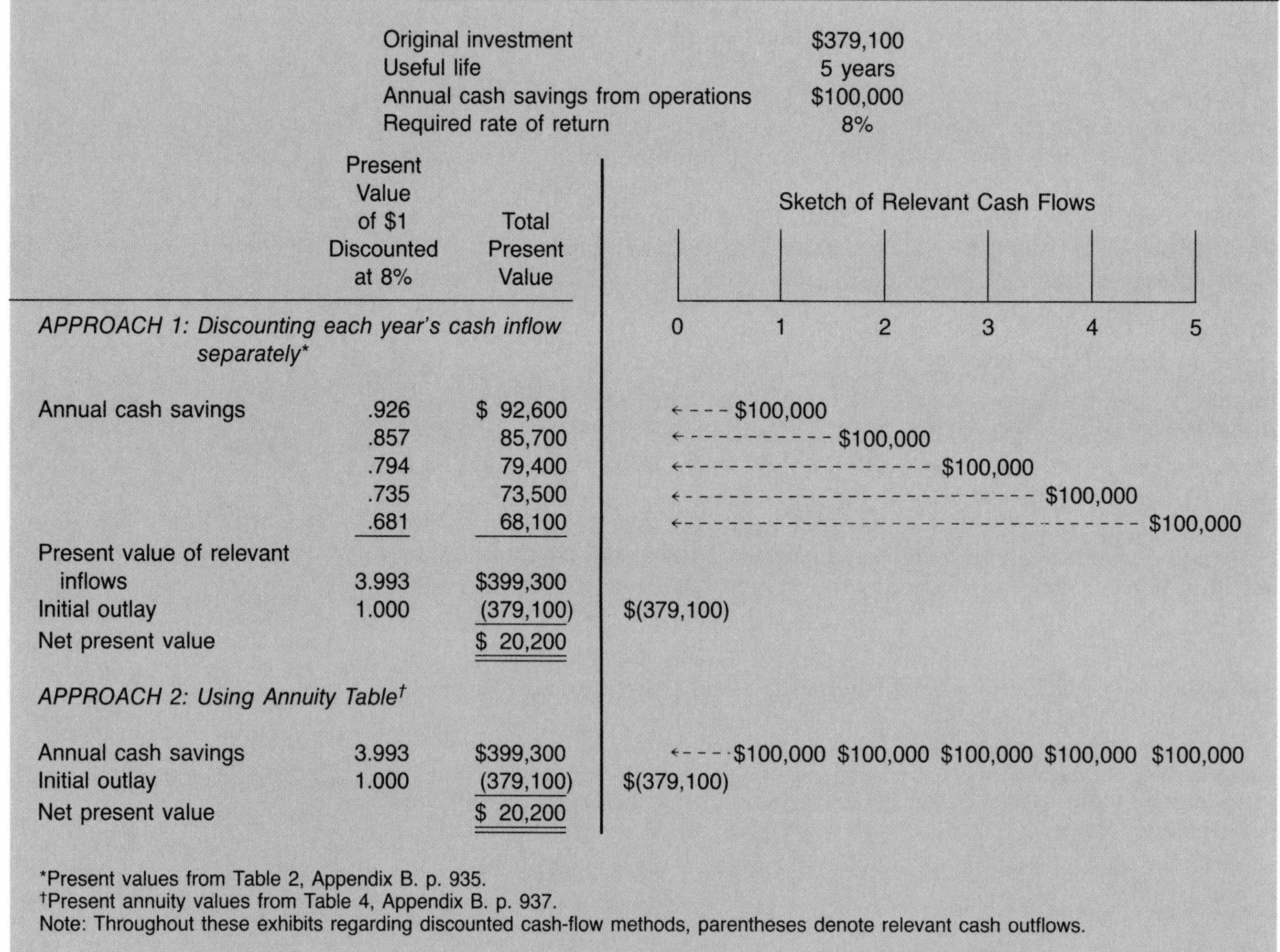

Original investment	$379,100
Useful life	5 years
Annual cash savings from operations	$100,000
Required rate of return	8%

Sketch of Relevant Cash Flows

	Present Value of $1 Discounted at 8%	Total Present Value	0	1	2	3	4	5
*APPROACH 1: Discounting each year's cash inflow separately**								
Annual cash savings	.926	$ 92,600	←	$100,000				
	.857	85,700	←		$100,000			
	.794	79,400	←			$100,000		
	.735	73,500	←				$100,000	
	.681	68,100	←					$100,000
Present value of relevant inflows	3.993	$399,300						
Initial outlay	1.000	(379,100)	$(379,100)					
Net present value		$ 20,200						
APPROACH 2: Using Annuity Table†								
Annual cash savings	3.993	$399,300	←	$100,000	$100,000	$100,000	$100,000	$100,000
Initial outlay	1.000	(379,100)	$(379,100)					
Net present value		$ 20,200						

*Present values from Table 2, Appendix B. p. 935.
†Present annuity values from Table 4, Appendix B. p. 937.
Note: Throughout these exhibits regarding discounted cash-flow methods, parentheses denote relevant cash outflows.

discount factor (period 5 for the 8% column) for Approach 2 in Exhibit 21-3. To obtain the present value, multiply the discount factors by the cash amounts in the sketch in Exhibit 21-3.

3. Summing the present values. If the total is zero or positive, the quantitative/financial aspect of the project indicates it should be accepted, because its expected rate of return equals or exceeds the required rate of return. If the total is negative, the quantitative/financial aspect indicates it is undesirable, because its expected rate of return is below the required rate of return.

Exhibit 21-3 indicates a net present value of $20,200 at the required rate of return of 8%; therefore, the project is desirable based on the quantitative/financial aspect.

The higher the required rate of return, the less willing the manager would be to invest in this project. At a rate of 12%, the net present value would be −$18,600 (3.605, the present value annuity factor from Table 4, × $100,000 = $360,500, which is $18,600 less than the required investment of $379,100). When the required rate is 12%, then, the machine is undesirable at its purchase price of $379,100.

In our example, we have emphasized the quantitative/financial aspect of the decision by Lifetime Care. The manager of the hospital, however, would also weigh other factors. Consider the reduction in the average number of individual X-rays taken per patient with the new machine. This reduction is a qualitative/

nonfinancial benefit of the new machine given the health risks to patients (and employees) of X-rays.

Important: Do not proceed until you thoroughly understand Exhibit 21-3. Compare Approach 1 with Approach 2 in Exhibit 21-3 to see how Table 4 in Appendix B (p. 937) is merely a compilation of the present value factors of Table 2 (p. 935). That is, the fundamental table is Table 2; Table 4 exists merely to reduce calculations when there are a series of equal cash flows at equal intervals.

Internal Rate-of-Return Method

The **internal rate of return (IRR)** is the rate of interest at which the present value of expected cash inflows from a project equals the present value of expected cash outflows of the project. IRR is sometimes called the **time-adjusted rate of return.** Projects with higher IRRs are preferred to projects with lower IRR's, if all other things are equal. Consider the X-ray machine project of Lifetime Care. Exhibit 21-4 presents the cash flows and shows the calculation of the net present value using a 10% discount rate (which is the IRR for this project). The internal rate of return can be described in two ways:

1. The interest rate that makes the net present value of a project equal to zero, as Exhibit 21-4 shows.
2. The maximum interest rate that could be paid to a financial institution for borrowing the money invested over the life of the project to break even. Here "break even" means arriving at the point where the net present value is zero or the internal rate of return equals the required rate of return.

The step-by-step computations of an internal rate of return are not too difficult when the cash inflows are equal. In Exhibit 21-4, the following equation is used:

$379,100 = Present value of annuity of $100,000 at $X\%$ for 5 years, or what factor F in Table 4 (p. 937) will satisfy the following equation:

$$\$379{,}100 = \$100{,}000F$$

$$F = 3.791$$

On the 5-period line of Table 4, find the percentage column that is closest to 3.791. It is exactly 10%. If the factor (F) fell between the factors in two columns, straight-line interpolation would be used to approximate the internal rate of return. (For an illustration, see Part 1 of the Problem for Self-Study, page 690.) In many cases, those people solving capital-budgeting problems have access to a calculator or computer programmed to calculate the internal rate of return.

Note in Exhibit 21-4 that $379,100 is the present value, at a rate of return of 10%, of a five-year stream of cash inflows of $100,000 per year. Ten percent is the rate that equates the amount invested ($379,100) with the present value of the cash inflows ($100,000) per year for five years. In other words, *if* money were borrowed at an interest rate of 10%, the cash inflow produced by the project would exactly repay the loan plus interest over the five years. If the required rate of return is less than 10%, the project will be profitable. If the required rate of return exceeds 10%, the cash inflow will be insufficient to pay interest and repay the principal of the loan. Therefore, 10% is the internal rate of return of this project.

Comparison of Net Present Value and Internal Rate-of-Return Methods

Objective 4

Explain how the two main discounted cash-flow methods (NPV and IRR) differ

In this text, we emphasize the net present value method. It has the important advantage that the end result of the computations is dollars, not a percentage. Therefore, we can add up the net present values of individual independent projects and obtain a valid estimate of the effect of accepting a combination of projects.

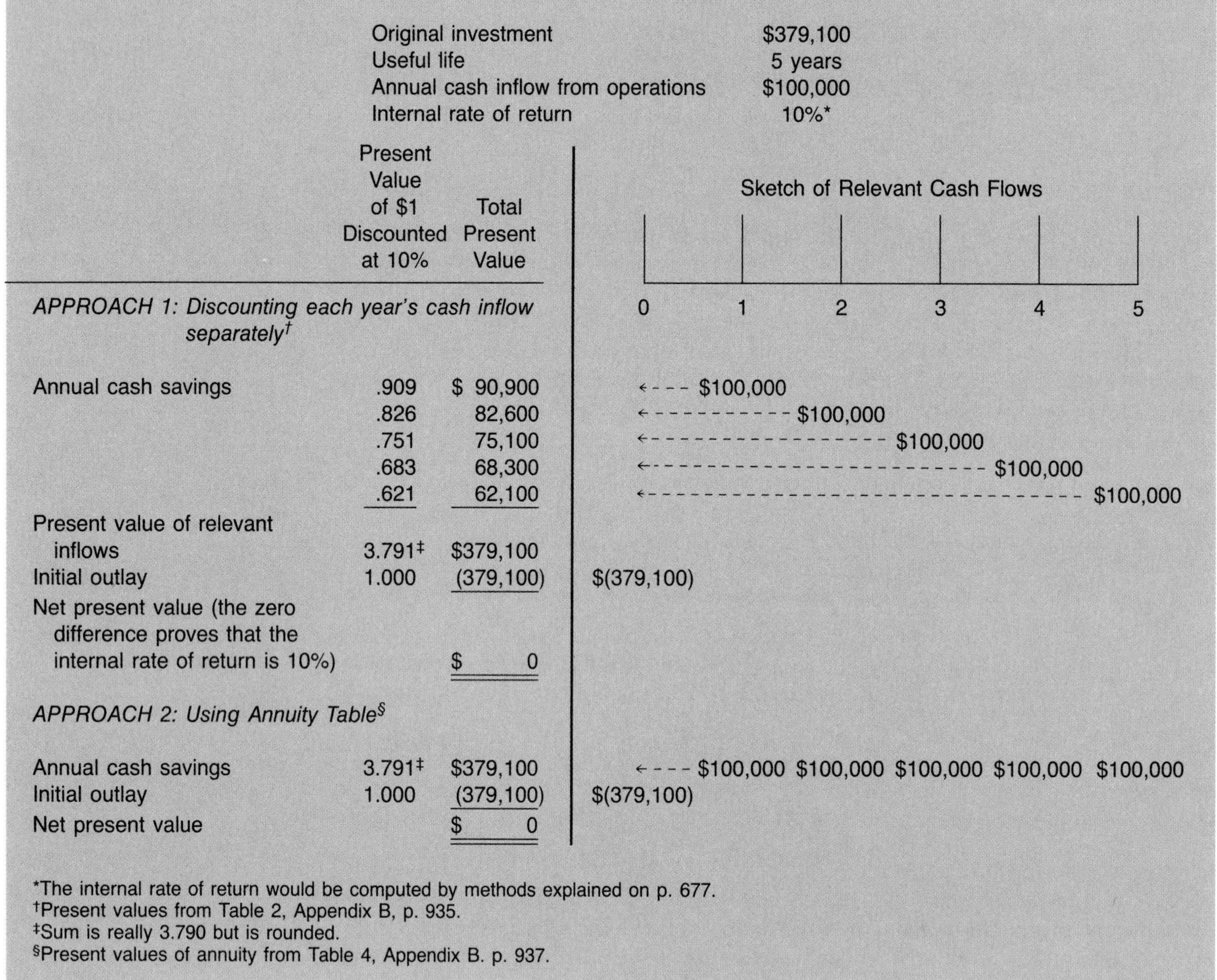

EXHIBIT 21-4
Internal Rate-of-Return Method: Lifetime Care

Original investment	$379,100
Useful life	5 years
Annual cash inflow from operations	$100,000
Internal rate of return	10%*

	Present Value of $1 Discounted at 10%	Total Present Value	Sketch of Relevant Cash Flows
APPROACH 1: Discounting each year's cash inflow separately†			0 1 2 3 4 5
Annual cash savings	.909	$ 90,900	←--- $100,000 (year 1)
	.826	82,600	←--- $100,000 (year 2)
	.751	75,100	←--- $100,000 (year 3)
	.683	68,300	←--- $100,000 (year 4)
	.621	62,100	←--- $100,000 (year 5)
Present value of relevant inflows	3.791‡	$379,100	
Initial outlay	1.000	(379,100)	$(379,100)
Net present value (the zero difference proves that the internal rate of return is 10%)		$ 0	
APPROACH 2: Using Annuity Table§			
Annual cash savings	3.791‡	$379,100	←--- $100,000 $100,000 $100,000 $100,000 $100,000
Initial outlay	1.000	(379,100)	$(379,100)
Net present value		$ 0	

*The internal rate of return would be computed by methods explained on p. 677.
†Present values from Table 2, Appendix B, p. 935.
‡Sum is really 3.790 but is rounded.
§Present values of annuity from Table 4, Appendix B. p. 937.

In contrast, the internal rates of return of individual projects cannot be added to derive the internal rate of return of the combination of projects.

A second advantage is that we can use the net present value method in situations where there is not a constant required rate of return for each year of the project. For example, assume in the X-ray equipment example that Lifetime Care has a required rate of return of 8% in years 1, 2, and 3, and 12% in years 4 and 5. The total present value of the cash inflows is:

Year	Cash Inflow	Required Rate of Return	Present Value of $1 Discounted at Required Rate	Total Present Value of Cash Inflows
1	$100,000	8%	.926	$ 92,600
2	100,000	8	.857	85,700
3	100,000	8	.794	79,400
4	100,000	12	.636	63,600
5	100,000	12	.567	56,700
				$378,000

Given the initial outlay of $379,100, the project is unattractive: It has a negative net present value of $1,100. However, it is not possible to use the internal rate-of-return method to infer that the project should be rejected. The existence of different required rates of return in different years (8% for years 1, 2, and 3 versus 12% for years 4 and 5) means there is not a single benchmark that the internal rate of return (a single figure) must exceed for the project to be acceptable.

SENSITIVITY ANALYSIS

To highlight the basic differences among various decision methods, we assume in this chapter that the expected values of cash flows will occur for certain. Obviously, managers know that predictions are imperfect. Chapter 20 describes how to contend with uncertainty in decision making. For now, we concentrate on sensitivity analysis, which was introduced in Chapter 3. *Sensitivity analysis* is a "what-if" technique that asks how a result will be changed if the original predicted data are not achieved or if an underlying assumption changes.

Sensitivity analysis can take various forms. For instance, management may want to know how far annual cash savings must fall to reach the point of indifference (when the net present value equals zero). For the data in Exhibit 21-3, let ACI = annual cash inflows and let net present value = $0. The initial cash outlay is $379,100, and the present value factor at the 8% required rate for a five-year annuity of $1 is 3.993. Then:

$$\begin{aligned} \text{NPV} &= \$0 \\ 3.993\text{ACI} - \$379{,}100 &= \$0 \\ 3.993\text{ACI} &= \$379{,}100 \\ \text{ACI} &= \$94{,}941 \end{aligned}$$

Thus, the amount by which the annual cash savings can drop before reaching the point of indifference regarding the investment is $100,000 − $94,941 = $5,059.

Computer spreadsheets enable managers to conduct systematic, efficient sensitivity analysis. Exhibit 21-5 shows how the net present value of the X-ray machine

EXHIBIT 21-5
Spreadsheet Analysis of Sensitivity of Net Present Value

Useful Life (in years)	Annual Cash-Operating Savings	Required Rate of Return	Net Present Value
5	$ 80,000	6%	$ (42,140)
5	80,000	8	(59,660)
5	80,000	10	(75,820)
5	90,000	6	(20)
5	90,000	8	(19,730)
5	90,000	10	(37,910)
5	100,000	6	42,100
5	100,000	8	20,200
5	100,000	10	0
5	110,000	6	84,220
5	110,000	8	60,130
5	110,000	10	37,910
5	120,000	6	126,340
5	120,000	8	100,060
5	120,000	10	75,820

project is affected by variations in (a) the cash-operating savings and (b) the required rate of return. The major contribution of sensitivity analysis is that it provides an immediate measure of the financial effect of differences between forecasts and subsequent outcomes. It helps managers focus on those decisions that may be very sensitive indeed, and it eases the manager's mind about those decisions that are not so sensitive. For the X-ray machine project, Exhibit 21-5 documents that variations in either the cash-operating savings or the required rate of return have sizable effects on net present value.

ANALYSIS OF SELECTED ITEMS USING DISCOUNTED CASH FLOW

Managers making capital-budgeting decisions frequently encounter problems when predicting the relevant cash inflows or outflows. We now discuss several of these problems.

1. **Working Capital** Additional investments in plant and equipment, and in sales promotions of product lines, are invariably accompanied by additional investments in working capital in the form of supporting cash, receivables, and inventories. In discounted cash-flow methods, all investments at year 0 are treated as cash outflows at year 0, regardless of how they may be accounted for (capitalized or expensed) in the accrual accounting model. At the end of the useful life of the project, the initial outlays for machines may not be recovered at all, or they may be only partially recovered in the amount of the terminal disposal prices. In contrast, the initial investments in working capital are usually fully recouped when the project is terminated.

Suppose a company buys some new equipment to make a new product. The investment in equipment necessitates an additional investment in working capital in the form of cash, receivables, and inventories. The following example shows the investments (numbers assumed and in millions):

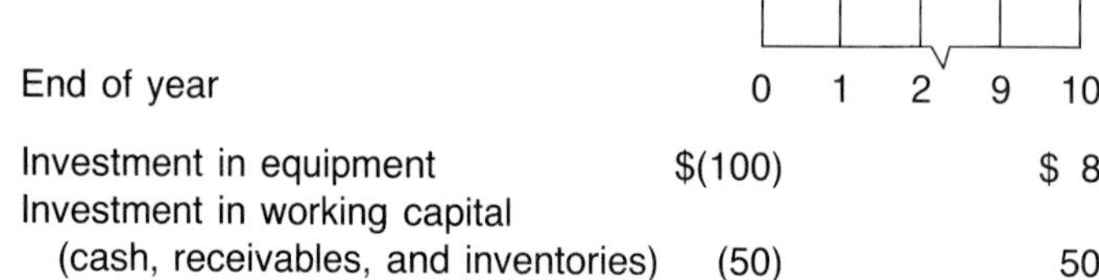

	Sketch of Relevant Cash Flows				
End of year	0	1	2	9	10
Investment in equipment	$(100)				$ 8
Investment in working capital (cash, receivables, and inventories)	(50)				50

As the sketch indicates, the terminal disposal price of the equipment ($8 million at the end of year 10) is relatively small. However, management expects to recover the entire initial investment of $50 million in working capital when manufacture of the new product stops. The difference between the initial outlay for working capital and the present value of its recovery is the present value of the cost of using working capital in the project. Assume a required rate of return of 8%. The present value of $50 million at the end of ten years is $23.15 million—(0.463, the present value factor from Table 2, ×$50 million). Thus, the present value of the cost of using the $50 million working capital for the life of the project is $26.85 million ($50.00 million − $23.15 million).

Some capital-budgeting projects result in reductions of working capital. For example, factory automation projects frequently bring about reduced work in process inventories. In this case, the timing of the cash inflows and outflows is reversed. Assume that a computer-integrated manufacturing project with a seven-year life will reduce working capital by $20 million. This reduction will be represented as a $20 million cash inflow for the project at year 0 and a $20 million cash outflow for the project when it ends in seven years.

2. **Book Value and Depreciation** In using DCF methods, we ignore book value and depreciation. Why? Because DCF is based on inflows and outflows of *cash,* not on the *accrual* concepts of revenues and costs. No adjustments should be made to cash flows for the periodic allocation of the asset cost called depreciation, which is not a cash flow. In DCF methods, the initial cost of an asset is usually regarded as a *lump-sum* outflow of cash at year 0. Deducting depreciation from operating cash inflows would be counting the lump-sum amount twice. (See Chapter 22 for a discussion of how book value and depreciation affect *after-tax* cash flows from operations, as distinguished from *before-tax* cash flows from operations.)

3. **Current Disposal Prices and Required Investment** In deciding whether to keep or to replace equipment, how should the current disposal price of the existing equipment affect the computations? The required investment—the net cash outflow—may be measured as the difference between the gross cost of the new equipment and the current disposal price of the old equipment.

4. **Future Disposal Prices** *The disposal price at the date of termination of a project is an increase in the cash inflow in the year of disposal.* Errors in forecasting the terminal disposal price are seldom crucial because their present value is usually small, especially for amounts to be received in the distant future.

5. **Income Taxes** In practice, comparison between alternatives is best made after considering income-tax effects because the tax impact may alter the picture. (Chapter 22 explains the effects of income taxes.)

THE PAYBACK METHOD

Objective 5

Explain the differences between the payback and DCF methods

We now consider the second of four methods for analyzing the quantitative/financial aspects of projects. The **payback method** measures the time it will take to recoup, in the form of cash inflow from operations, the total dollars invested in a project. Assume that Lifetime Care spent $379,100 for the X-ray machine having a five-year expected useful life and creating a $100,000 annual cash inflow from operations. The payback calculations follow:

$$\text{Payback} = \frac{\text{Incremental amount invested (cash outflow)}}{\text{Uniform annual incremental cash inflow from operations}}$$

$$\text{Payback} = \frac{\$379{,}100}{\$100{,}000} = 3.791 \text{ years}$$

The payback method highlights liquidity, which is often an important factor in business decisions. Projects with shorter paybacks (more liquid) are preferred to projects with longer paybacks, if all other things are equal. This method is easily understood. Like the DCF methods described previously, the payback method is not affected by accrual accounting conventions such as depreciation and depletion. Advocates of payback argue that it is a handy measure (a) when precision in estimates of profitability is not crucial and preliminary screening of many proposals is necessary, and (b) when the predicted cash flows in later years are highly uncertain.

The major weakness of the payback method is its neglect of profitability. The mere fact that Project A has a shorter payback time than Project B does not mean that Project A is preferable. Consider an alternative to the $379,100 machine mentioned earlier. Assume that another machine, with a three-year useful life, requires only a $300,000 investment and will also result in cash savings of $100,000 per year.

Compare the two payback periods:

Project A: \$379,100 ÷ \$100,000 = 3.791 years
Project B: \$300,000 ÷ \$100,000 = 3.000 years

The payback criterion would favor buying the \$300,000 machine. However, the second machine has a useful life of only three years. Ignoring the complexities of compound interest for the moment, we find that the \$300,000 machine results in zero operating income. Its payback period is its useful life. The \$379,100 machine yields operating income for 1.209 years beyond its payback time (a useful life of 5 years − 3.791 years' payback time). A second weakness of the payback method is its neglect of the time value of money.

Nonuniform Cash Flows

The payback formula presented on page 681 is designed for uniform cash inflows. When cash inflows are not uniform, the payback computation takes a cumulative form. Each year's net cash inflows are accumulated until the initial investment has been recovered. Assume a law firm is considering purchase of a \$4,600 facsimile machine for electronically transmitting documents to its clients. This machine is expected to produce a total cash savings of \$13,000 over the next five years (primarily due to a reduction in the use of courier services). The cash savings occur at a nonuniform rate. Payback occurs between the second and third years:

Year	Cash Savings	Accumulated Cash Savings	Cash Investment Yet to be Recovered at End of Year
0	—	—	\$4,600
1	\$1,600	\$1,600	3,000
2	1,800	3,400	1,200
3	2,400	5,800	—
4	3,000	8,800	—
5	4,200	13,000	—

Straight-line interpolation within the third year, which has cash savings of \$2,400, reveals that the final \$1,200 needed to recover the \$4,600 investment (that is, \$4,600 − \$3,400 recovered by the end of year 2) will be saved halfway through year 3:

$$\text{Payback} = 2 \text{ years} + \left(\frac{\$1{,}200}{\$2{,}400} \times 1 \text{ year}\right) = 2.5 \text{ years}$$

The Lifetime Care X-ray machine example (p. 675) has a single cash outflow of \$379,100 at year 0. Where a project has multiple cash outflows, occurring at different points in time, these cash outflows are added to derive a total cash outflow figure for the project. No adjustment is made for the time value of money when adding these cash outflows in payback period computations. Why? Because the payback method ignores the time value of money. (The breakeven time method, discussed on page 686, has several similarities to the payback method but does consider the time value of money.)

The Bailout Factor: An Extension of Payback

The typical payback computation tries to answer the question, How soon can the investment be recouped *if operations proceed as planned?* An additional question to ask is: Which of the competing projects has the best protection if things go wrong? The **bailout payback time** measure—the time when the cumulative cash operating savings plus the disposal price at the end of a particular year equal the original

EXHIBIT 21-6

Comparison of Traditional Payback and Bailout Payback Methods

Machine	Traditional Payback			Bailout Payback			
	Investment	Uniform Cash Savings	Payback	At End of	Cumulative Cash-Operating Savings	Disposal Price	Cumulative Total
General-purpose bottling machine	$100,000	$20,000	5 years	Year 1	$ 20,000 +	$70,000 =	$ 90,000
				Year 2	40,000 +	60,000 =	100,000
				Bailout payback is 2 years.			
Special-purpose bottling machine	$150,000	$40,000	3.75 years	Year 1	$ 40,000 +	$60,000 =	$100,000
				Year 2	80,000 +	45,000 =	125,000
				Year 3	120,000 +	30,000 =	150,000
				Bailout payback is 3 years.			

investment—examines the downside protection if the project is disbanded. Projects with shorter bailout payback times are preferable to those with longer bailout payback times, if all other things are equal.

Disposal prices of general-purpose equipment frequently far exceed those of special-purpose equipment. Consider a decision by Refresh Corp., a soft-drink bottling company. Refresh Corp. is choosing between two competing bottling machines for its cola production line:

	General-Purpose Bottling Machine	Special-Purpose Bottling Machine
Cost	$100,000	$150,000
Annual cash savings (uniform per year)	$ 20,000	$ 40,000
Useful life of machine	10 years	10 years

The general-purpose bottling machine is an "off-the-shelf" model. Management expects its disposal price to be $70,000 at the end of year 1 and to decline $10,000 annually thereafter. The special-purpose bottling machine is specifically designed for Refresh Corp.'s production facilities. Management expects its disposal price to be $60,000 at the end of year 1 and to decline $15,000 annually thereafter.

Exhibit 21-6 compares the two bottling machines using the traditional payback method and the bailout payback method. Using the traditional payback method, the special-purpose machine has a shorter payback period (3.75 years) than does the general-purpose machine (5 years). In contrast, the general-purpose machine has a shorter bailout payback period (2 years) than does the special-purpose machine (3 years).

THE ACCRUAL ACCOUNTING RATE-OF-RETURN METHOD

We now consider the third of four methods for analyzing the quantitative/financial aspects of projects. The **accrual accounting rate of return (ROR)** is an accounting measure of income divided by an accounting measure of investment. It is also called **accounting rate of return.** Advocates of this method prefer projects with higher, rather than lower, accrual accounting rates of return, if all other things are equal. The denominator is frequently the initial increase in required investment:

$$\text{Accrual accounting rate of return} = \frac{\text{Increase in expected average annual operating income}}{\text{Initial increase in required investment}}$$

The facts in our Lifetime Care X-ray machine illustration would yield the following accrual accounting rate of return (average annual depreciation is $379,100 ÷ 5 = $75,820) when the denominator is the initial increase in required investment:

$$\text{ROR} = \frac{\$100{,}000 - \$75{,}820}{\$379{,}100} = \frac{\$24{,}180}{\$379{,}100} = 6.38\%$$

Sometimes the denominator is expressed as the average increase in investment, rather than the initial increase. If we use the average investment, the rate would be doubled to 12.76%.[1]

The label *accrual accounting method* is not universal. Synonyms are *financial statement method, book-value method, rate-of-return-on-assets method, approximate rate-of-return method,* and *accounting return-on-investment method.* Its computations dovetail most closely with accrual accounting models for calculating operating income and required investment.

The accrual accounting rate of return is simple and easily understood. Unlike the payback method, the accounting rate-of-return method considers profitability. However, it ignores the time value of money. Expected future dollars are unrealistically and erroneously regarded as being equal to present dollars. The discounted cash-flow method explicitly allows for interest and considers the exact timing of cash flows. In contrast, the accounting rate-of-return method is based on *annual averages.*

Computing the Investment Base

Practice is not uniform. Should the initial investment or the average investment be used in the denominator of the accounting rate of return? Some companies defend the use of the initial-investment base because it does not change over the life of the investment, which eases comparisons of predicted rates of return with actual rates of return. Such comparisons—on year-to-year, plant-to-plant, and division-to-division bases—are crucial for control and lead to improved capital-budgeting decisions.

In most cases, the rankings of competing projects based on accrual accounting rate of return will not differ regardless of whether the initial or average investment

[1]The cash inflow in the example is $100,000 a year. The amount of average annual depreciation is $379,100 ÷ 5 = $75,820. The terminal disposal price is assumed to be zero. The average investment would be the beginning book value plus the ending book value ($379,100 + $0) divided by 2, or $189,550. Thus, the accrual accounting rate of return on the average investment would be

$$\text{ROR} = (\$100{,}000 - \$75{,}820) \div \$189{,}550$$
$$\text{ROR} = \$24{,}180 \div \$189{,}550 = 12.76\%$$

However, if the project has a nonzero terminal disposal price and straight-line depreciation is used, the *average* investment is computed by *adding* the terminal disposal price to the original cost and dividing by 2. For example, assume that the X-ray machine in Exhibits 21-3 and 21-4 has a terminal disposal price of $60,000 instead of zero. The annual depreciation would be ($379,100 − $60,000) ÷ 5 = $63,820. The average investment is ($379,100 + $60,000) ÷ 2 = $219,550. The accrual accounting rate of return with the denominator calculated as the average investment is

$$\text{ROR} = (\$100{,}000 - \$63{,}820) \div \$219{,}550$$
$$\text{ROR} = \$36{,}180 \div \$219{,}550 = 16.48\%$$

base is used. Using the average base will show substantially higher rates of return; therefore, the desirable rate of return used as a basis for accepting projects should be proportionately higher.

Companies often expand working capital, such as cash, receivables, and inventories, to sustain higher-volume levels. In our Lifetime Care example, if a $40,000 increase in working capital is required, the denominator will be $379,100 plus $40,000, or $419,100. This $40,000 increase in working capital will probably be fully committed for the life of the project. Under the average-investment method, then, the average used for the denominator will be $229,550—that is, $189,550 average investment in equipment plus $40,000 average investment in working capital (alternatively [$419,100 starting + $40,000 terminal recovery of working capital] ÷ 2 = $229,550).

Evaluation of Performance

Objective 6

Describe conflicts in using DCF for capital budgeting and accrual accounting for performance evaluation

The use of the accrual accounting model for evaluating performance is a stumbling block to the implementation of the DCF method for capital-budgeting decisions. Consider John Kent, the manager of a flour-milling division in a multidivision company. Kent took a course in management accounting in an executive program. He became convinced that the discounted cash-flow method would lead to decisions that would better achieve the long-range operating income goals of the company.

Upon returning to his company, Kent was more frustrated than ever. Top management used the accrual accounting rate of return for judging his division's performance. That is, each year divisional operating income was divided by average divisional assets to obtain the rate of return on investment (ROI). Such a measure usually inhibits investments in plant and equipment that might clearly be attractive using the DCF methods. Why? Because a huge investment often boosts depreciation inordinately in the early years under accelerated depreciation methods, thus reducing the numerator in the ROI computation. Also, the denominator is increased substantially by the initial cost of the new assets. As Kent said, "Top management is always giving me hell about my new flour mill, even though I know it is the most efficient we've got regardless of what the figures say."

Obviously, there is an inconsistency between citing DCF methods as being best for capital-budgeting decisions and then applying different concepts in evaluating subsequent performance. As long as such practices continue, management will frequently be tempted to make decisions that may be less than optimal under the DCF methods but optimal, at least over short or intermediate spans of time, under the accrual accounting model. Such temptations become more pronounced when managers are subject to transfers (or promotions), and annual income is important in their evaluations and their compensation plans. Why? Because the time horizon of their concerns is shorter than the time horizon in their capital-budgeting decisions.

SURVEYS OF PRACTICE

Numerous surveys regarding capital-budgeting practices reveal that discounted cash-flow methods are widely used by U.S. firms. One study reports results from 1970, 1975, 1980, and 1988:[2]

[2]T. Klammer, B. Koch, and N. Wilner, "Capital Budgeting Practices—A Survey of Corporate Use," Working Paper, University of North Texas, 1989.

Capital-Budgeting Project Category	Firms Using DCF as the Primary Evaluation Method			
	1970	1975	1980	1988
Replacement projects	28%	45%	56%	60%
Expansion—existing operations	44	62	75	86
Expansion—new operations	41	58	71	87
Foreign operations	45	59	72	79
Abandonment	36	47	55	62
General and administrative	21	29	36	41
Social expenditures	10	14	14	16

There is increased use of DCF in each of these seven categories.

Differences exist across countries in the importance placed on individual capital-budgeting methods. For example, the payback method is more widely used by Japanese companies than by U.S. companies. A survey reports the following percentages for the primary method used by Japanese companies in their capital-budgeting decisions:[3]

- Payback 47%
- Accounting rate of return 23
- Internal rate of return 18
- Net present value 12

Comparative studies of U.S. and Japanese companies report that Japanese companies use a longer payback period as the cutoff point than do U.S. companies. One study of capital budgeting for advanced manufacturing technologies reports that 50% of Japanese companies use a payback period of four years or longer as a cutoff compared with only 12% of U.S. companies that use a similar payback period.[4]

BREAKEVEN TIME AND CAPITAL BUDGETING FOR NEW PRODUCTS

We now consider the last of the four methods we present for analyzing the quantitative/financial aspects of projects—the breakeven time method. Managers are increasingly viewing time as a competitive weapon. Firms that bring a new product to market at a very fast rate are often able to gain sizable shares of total market sales for that product. This increased emphasis on time has led to the use of breakeven time as a capital-budgeting method and as a performance measure. **Breakeven time (BET)** is the time taken from the start of a project (the initial idea date) to when the cumulative present value of the cash inflows of a project equal the present value of the total cash outflows. New-product proposals with shorter BETs are preferred to new-product proposals with longer BETs, if all other things are equal.

Hewlett-Packard (HP) is an enthusiastic advocate of BET:

> One of HP's major objectives is to remain a leader in product development. . . . Some traditional financial methods, e.g., net present value and internal rate of return, were considered for use as HP's primary product-development-process metric. However, within the context of HP's high-technology business, time-to-market is considered to be

[3]N. B. Gultekin and T. Taga, "Financial Management in Japanese Corporations," Working Paper, University of Pennsylvania, 1989.

[4]NAA Tokyo Affiliate, "Management Accounting in the Advanced Manufacturing Surrounding: Comparative Study on Survey in Japan and U.S.A.," 1988.

of utmost importance, and so BET, with its emphasis on how fast these new technologies and products are brought to market, has been chosen to be HP's focal product-development metric.[5]

Companies advocating the use of BET in capital-budgeting decisions emphasize its impact on product development efforts. It promotes aggressive efforts by personnel in different functional areas (such as product design and manufacturing) to speed up the time taken to bring a product to the market and generate sales revenues.

Example of BET Computation

Consider two instrument products (SP-108 and SP-247) being evaluated by the Stanford Park division of Hewlett-Packard. Development work on each product can start at the beginning of year 1. For simplicity, assume all cash flows occur at the end of each year. The estimated cash outflows (covering research and development, product design, manufacturing, marketing, distribution, and customer service) and cash inflows from sales are (in millions):

	Instrument SP-108		Instrument SP-247	
Year	**Cash Outflows**	**Cash Inflows**	**Cash Outflows**	**Cash Inflows**
1	$10	$ 0	$12	$ 6
2	8	16	9	34
3	24	36	19	28
4	15	39	8	10
5	12	24	2	4

Exhibit 21-7 presents the BET computations for both products, assuming Hewlett-Packard uses a 14% required rate of return for discounting cash flows on a before-tax basis.

[5]Corporate Engineering, Hewlett-Packard, *Breakeven Time at Hewlett-Packard*, 1990, p. 2.

EXHIBIT 21-7

Breakeven-Time Computations for Instruments SP-108 and SP-247 (in millions)

Year	PV Factor	Cash Outflows	PV of Cash Outflows	Cumulative PV of Cash Outflows	Cash Inflows	PV of Cash Inflows	Cumulative PV of Cash Inflows
(1)	(2)	(3)	(4) = (2) × (3)	(5)	(6)	(7) = (2) × (6)	(8)
SP-108							
1	0.877	$10.000	$ 8.770	$ 8.770	$ 0.000	$ 0	$ 0
2	0.769	8.000	6.152	14.922	16.000	12.304	12.304
3	0.675	24.000	16.200	31.122	36.000	24.300	36.604
4	0.592	15.000	8.880	40.002	39.000	23.088	59.692
5	0.519	12.000	6.228	46.230	24.000	12.456	72.148
SP-247							
1	0.877	$12.000	$10.524	$10.524	$ 6.000	$ 5.262	$ 5.262
2	0.769	9.000	6.921	17.445	34.000	26.146	31.408
3	0.675	19.000	12.825	30.270	28.000	18.900	50.308
4	0.592	8.000	4.736	35.006	10.000	5.920	56.228
5	0.519	2.000	1.038	36.044	4.000	2.076	58.304

The present value of the total cash outflows for Instrument SP-108 is \$46.230. At the end of year 3, the cumulative present value of the cash inflows is \$36.604. During year 4, the present value of the cash inflows is \$23.088.

$$\text{BET for SP-108} = 3 + \frac{\$46.230 - \$36.604}{\$23.088}$$
$$= 3 + \frac{\$9.626}{\$23.088}$$
$$= 3.42 \text{ years}$$

The present value of the total cash outflows for Instrument SP-247 is \$36.044. At the end of year 2, the cumulative present value of the cash inflows is \$31.408. During year 3, the present value of the cash inflows is \$18.900.

$$\text{BET for SP-247} = 2 + \frac{\$36.044 - \$31.408}{\$18.900}$$
$$= 2 + \frac{\$4.636}{\$18.900}$$
$$= 2.25 \text{ years}$$

SP-247 has a shorter BET (2.25 years) than SP-108 (3.42 years) and thus would be preferred by the Stanford Park division of Hewlett-Packard.

Take considerable care when comparing the BETs of different types of products. A four-year BET for a new-car project may be very short, but a two-year BET for a fashion-dress project may be very long.

The BET computations illustrated in this section use the present value of the total cash outflows for the product over its life cycle. An alternative approach is to consider only the present value of the cash outlays made until the time a product is first available for sale in the market. This approach results in shorter BET's for new products than the approach previously presented.

BET Versus the Payback Method

Objective 7

Explain the differences between the payback and breakeven-time methods

There are two important differences between BET and the payback method. First, BET starts counting time at the genesis of a project, irrespective of when the cash outflows occur. In contrast, payback measures time from when the initial cash outflow is made. Second, BET takes account of the time value of money when cumulating cash inflows and cash outflows. In contrast, payback ignores the time value of money. Both methods are similar in that they ignore cash inflows and cash outflows after the breakeven time or the payback period.

COMPLEXITIES IN CAPITAL-BUDGETING APPLICATIONS

The most challenging aspects of capital budgeting are identifying the project and predicting its outcomes. Textbook examples typically ignore much of the complexity associated with these aspects. This section illustrates some of the complexity often found in practice.

Consider a firm deciding whether to invest in computer-integrated manufacturing (CIM) technology. Applying CIM to its full extent can result in a highly automated factory, where the role of labor is largely restricted to computer programming, engineering support, and maintenance of the robotic machinery. The amounts at stake in CIM decisions can be large—in the billions of dollars for such companies as General Motors and Toyota. Two important factors in capital-budgeting decisions are (a) recognizing the full set of benefits and costs and (b) recognizing the full time horizon of the project.

Recognizing the Full Set of Benefits and Costs

Objective 8

Explain the impact of nonfinancial and qualitative factors in capital-budgeting decisions

The factors that firms consider in making CIM decisions are far broader than the quantitative/financial factors emphasized in many capital-budgeting decisions. Exhibit 21-8 presents examples of the broader set of factors that United States, Japanese, and United Kingdom firms weigh in evaluating CIM technology. The benefits include the following points, which are difficult to quantify.[6]

1. Faster response to marketing changes. An automated factory can, for example, make major design modifications (such as switching from a two-door to a four-door car) relatively quickly. To quantify this benefit requires some notion of consumer demand changes that may occur many years in the future.
2. Increased learning by workers about automation. If workers have a positive experience with CIM, the company can implement other automation projects more quickly and more successfully. Quantifying this benefit requires a prediction of the company's subsequent automation.

Survey evidence emphasizes the importance of linking CIM decisions to the overall competitive strategies of the company.

Recognizing the full set of costs also presents problems. Three classes of costs are difficult to measure and often are underestimated:

1. Costs associated with a reduced competitive position in the industry. If other firms in the industry are investing in CIM, a firm not investing in CIM may suffer a decline in market share. Several firms in the machine-tool industry that continued to use a conventional

---

[6]R. A. Howell, J. D. Brown, S. R. Soucy, and A. H. Seed, *Management Accounting in the New Manufacturing Environment* (Montvale, NJ, and Arlington, TX: National Association of Accountants and CAM-I, 1987); NAA Tokyo Affiliate, *ibid.*, and Coopers & Lybrand, Ernst & Whinney, and Peat Marwick McLintock, "Management Accounting in Advanced Manufacturing Environments" (Arlington, TX: CAM-I, 1988). See also J. A. Hendricks, "Applying Cost Accounting to Factory Automation," *Management Accounting*, December 1988.

EXHIBIT 21-8

Factors Considered in Making Capital-Budgeting Decisions for CIM Projects

Category	Examples
Quantitative/Financial	Lower direct labor cost Lower hourly support labor cost Less scrap and rework Increase in software costs Retraining of personnel
Quantitative/Nonfinancial	Reduction in manufacturing cycle time Increase in manufacturing flexibility Lower product defect rate
Qualitative/Financial	Increase in business risk due to higher fixed-cost structure
Qualitative/Nonfinancial	Improved product delivery and service Improved competitive position Reduction in product development time Faster response to marketing changes Increased learning by workers about automation

manufacturing approach experienced rapid drops in market share after their competitors introduced CIM.

2. Costs of retraining the operating and maintenance personnel to handle the automated facilities.
3. Costs of developing and maintaining the software and maintenance programs to operate the automated manufacturing activities.

Recognizing the Full Time Horizon of the Project

The time horizon of CIM projects can stretch well beyond ten years. Many of the costs are incurred and are highly visible in the early years of adopting CIM. In contrast, important benefits may be realized only many years after the adoption of CIM. Firms that have a fixed upper limit on the time period for computing the benefits of a project (say, five years) are frequently biased against adopting CIM.[7]

[7]Further discussion is in J. A. Brimson, *Activity-based Investment Management* (New York: American Management Association, 1989); and R. S. Kaplan and A. A. Atkinson, *Advanced Management Accounting* (Englewood Cliffs, NJ: Prentice Hall, 1989), Chap. 12.

PROBLEM FOR SELF-STUDY

PROBLEM

Consider again the Lifetime Care X-ray machine project. Assume the expected annual cash savings is $130,000 instead of $100,000. All other facts are unchanged: a $379,100 original investment, a five-year useful life, a zero terminal disposal price, and an 8% required rate of return. Compute the following:

1. Discounted cash flow:
   (a) Net present value
   (b) Internal rate of return
2. Payback period
3. Accrual accounting rate of return on initial investment
4. Breakeven time

Assume (as does the rest of this chapter) that cash outflows and cash inflows occur at the beginning or end of each period (as specified in the problem).

SOLUTION

1. (a) NPV = ($130,000 × 3.993) − $379,100
   NPV = $519,090 − $379,100 = $139,990
   (b) There are several approaches to computing the IRR. One is to use a calculator with an IRR function; this gives a 21.18% IRR. An alternative approach is to use the tables in Appendix B (p. 930–37):

$$\$379{,}100 = \$130{,}000\ F$$

$$F = \frac{\$379{,}100}{\$130{,}000} = 2.916$$

On the 5-period line of Table 4 (Appendix B), the column closest to 2.916 is 22%. To obtain a more accurate number, interpolation can be used:

	Present Value Factors	
20%	2.991	2.991
True rate	—	2.916
22%	2.864	—
Difference	.127	.075

$$\text{Internal rate of return} = 20\% + \frac{.075}{.127}(2\%) = 21.18\%$$

2. $$\text{Payback} = \frac{\text{Incremental amount invested (cash outflow)}}{\text{Uniform annual incremental cash inflow from operations}}$$

$$= \$379{,}100/\$130{,}000 = 2.92 \text{ years}$$

3. $$\text{Accrual accounting ROR} = \frac{\text{Increase in expected average annual operating income}}{\text{Initial increase in required investment}}$$

$$= \frac{\$130{,}000 - \text{Average annual depreciation}}{\$379{,}100}$$

$$= \frac{\$130{,}000 - (\$379{,}100 \div 5)}{\$379{,}100} = \frac{\$54{,}180}{\$379{,}100} = 14.29\%$$

4. The original investment of $379,100 occurs at year 0 and thus has a cumulative present value of $379,100. The cumulative present value of the cash inflows is:

Year	PV Factor	Cash Inflows	Present Value of Cash Inflows	Cumulative Present Value of Cash Inflows
1	.926	$130,000	$120,380	$120,380
2	.857	130,000	111,410	231,790
3	.794	130,000	103,220	335,010
4	.735	130,000	95,550	430,560
5	.681	130,000	88,530	519,090

The cumulative net present value of the cash outflows and cash inflows is:

Year	Cumulative PV of Cash Outflows	Cumulative PV of Cash Inflows	Cumulative Net Present Value
0	$379,100	$ 0	$(379,100)
1	379,100	120,380	(258,720)
2	379,100	231,790	(147,310)
3	379,100	335,010	(44,090)
4	379,100	430,560	51,460
5	379,100	519,090	139,990

$$\text{Breakeven time} = 3 + \frac{\$379{,}100 - \$335{,}010}{\$430{,}560 - \$335{,}010}$$

$$= 3 + \frac{44{,}090}{95{,}550} = 3.46 \text{ years}$$

SUMMARY

Capital budgeting is long-term planning for proposed capital outlays and their financing. Discounted cash-flow (DCF) methods explicitly weigh the time value of money, an important factor in capital-budgeting decisions. Two DCF methods are net present value and internal rate of return.

The payback method is widely used. It is simple and easily understood, but it neglects profitability and the time value of money. The breakeven time method likewise ignores profitability but does incorporate the time value of money into its computation.

The accrual accounting rate-of-return method is also widely used, although it is much cruder than DCF methods. It fails to recognize explicitly the time value of money. Instead, the accrual accounting rate-of-return method depends on averaging techniques that may yield inaccurate answers, particularly when cash flows are not uniform throughout the life of a project.

A serious, practical impediment to adopting DCF methods is the widespread use of accrual accounting models for evaluating the performance of a manager or a division. Frequently, the optimal decision made using a DCF method will not report good "operating income" results in the project's early years based on accrual accounting methods.

The most imposing tasks in capital-budgeting decisions are identifying and defining the project and predicting its outcomes. This chapter described methods of summarizing the quantitative/financial outcomes of a project. In many decisions, managers also give much attention to the nonfinancial and qualitative outcomes of projects.

Chapter 22 discusses special difficulties in relevant costs and capital budgeting, including income-tax factors and inflation.

TERMS TO LEARN

This chapter and the Glossary at the end of the book contain definitions of the following important terms:

Accounting rate of return *(p. 683)* Accrual accounting rate of return *(683)*
Bailout payback time *(682)* Breakeven time (BET) *(686)*
Capital budgeting *(673)* Cost of capital *(675)*
Discounted cash flow (DCF) *(674)* Discount rate *(675)* Hurdle rate *(675)*
Internal rate of return (IRR) *(677)* Net present value (NPV) *(675)*
Payback method *(681)* Required rate of return (RRR) *(675)*
Time-adjusted rate of return *(677)* Time value of money *(672)*

ASSIGNMENT MATERIAL

QUESTIONS

21-1 What are the four stages in capital budgeting after a project has been identified and its outcomes have been predicted?

21-2 What is the essence of the discounted cash-flow method?

21-3 Classify the predicted outcomes of a capital-budgeting decision into four categories.

21-4 List four methods of summarizing the quantitative/financial aspects of a capital-budgeting project.

21-5 What are the two main discounted cash-flow methods? How do they differ?

21-6 What is the payback method? What are its main weaknesses?

21-7 How is the bailout payback time method different from the payback method?

21-8 "The trouble with discounted cash-flow techniques is that they ignore depreciation costs." Do you agree? Why?

21-9 "Let's be more practical. DCF is not the gospel. Managers should not become so enchanted with DCF that strategic considerations are overlooked." Do you agree? Why?

21-10 How is the breakeven time method of capital budgeting different from the payback method?

21-11 Bill Watts, president of Western Publications, accepts a capital-budgeting project advocated by Division X. This is the division in which the president spent his first ten years with the company. On the same day, the president rejects a capital-budgeting project proposal from Division Y. The manager of Division Y is incensed. She believes the Division Y project has an internal rate of return at least ten percentage points above the Division X project. She comments, "What is the point of all our detailed DCF analysis? If Watts is panting over a project, he can arrange to have the proponents of that project "massage" the numbers so that it looks like a winner." What advice would you give the manager of Division Y?

EXERCISES AND PROBLEMS

21-12 Exercises in compound interest. To be sure that you understand how to use the tables in Appendix B at the end of this book, solve the following exercises. Ignore income-tax considerations. The correct answers, rounded to the nearest dollar, are printed on pages 702–3.

1. You have just won $5,000. How much money will you have at the end of ten years if you invest it at 6% compounded annually? At 14%?

2. Ten years from now, the unpaid principal of the mortgage on your house will be $89,550. How much do you have to invest today at 6% interest compounded annually to accumulate the $89,550 in ten years?
3. If the unpaid mortgage on your house in ten years will be $89,550, how much money do you have to invest annually at 6% to have exactly this amount on hand at the end of the tenth year?
4. You plan to save $5,000 of your earnings each year for the next ten years. How much money will you have at the end of the tenth year if you invest your savings compounded at 12% per year?
5. You hold an endowment insurance policy that will pay you a lump sum of $200,000 at age 65. If you invest the sum at 6%, how much money can you withdraw from your account in equal amounts each year so that at the end of ten years there will be nothing left?
6. You have estimated that for the first ten years after you retire you will need an annual cash inflow of $50,000. How much money must you invest at 6% at your retirement age to obtain this annual cash inflow? At 20%?
7. The following table shows two schedules of prospective operating cash inflows, each of which requires the same initial investment of $10,000 now:

	Annual Cash Inflows	
Year	**Plan A**	**Plan B**
1	$ 1,000	$ 5,000
2	2,000	4,000
3	3,000	3,000
4	4,000	2,000
5	5,000	1,000
Total	$15,000	$15,000

The required rate of return is 6% compounded annually. All cash inflows occur at the end of each year. In terms of net present value, which plan is more desirable? Show computations.

21-13 Basic nature of present value. Santa Ynez Products is considering investing in a project with a two-year life and a zero terminal disposal price. Operating cash inflows will be equal payments of $4,000 at the end of each of the two years. How much would the company be willing to invest to earn an internal rate of return of 8%? Use Table 2, Appendix B, to find your answer. Prepare a tabular analysis of each payment. The column headings should be Year, Investment at beginning of year, Operating cash inflow, Return at 8% per year, Amount of investment received at end of year, and Unrecovered investment at end of year.

21-14 Comparison of decision models. The Building Distributors Group is thinking of buying, at a cost of $220,000, some new packaging equipment that is expected to save $50,000 in cash-operating costs per year. Its estimated useful life is ten years, and it will have zero terminal disposal price. The required rate of return is 16%.

Required
Compute:

1. Net present value
2. Payback period
3. Internal rate of return
4. Accrual accounting rate of return based on (a) initial investment and (b) average investment. Assume straight-line depreciation.

21-15 Comparison of approaches to capital budgeting. City Hospital estimates that it can save $28,000 a year in cash-operating costs for the next ten years if it buys a special-purpose machine at a cost of $110,000. A zero terminal disposal price is expected. City Hospital's required rate of return is 14%.

Required
Compute:

1. Net present value
2. Payback period
3. Internal rate of return
4. Accrual accounting rate of return based on (a) initial investment and (b) average investment. Assume straight-line depreciation.

21-16 Capital budgeting with uneven cash flows. Southern Cola is considering the purchase of a special-purpose bottling machine for $28,000. It is expected to have a useful life of seven years with a zero terminal disposal price. The plant manager estimates the following savings in cash-operating costs:

Year	Amount
1	$10,000
2	8,000
3	6,000
4	5,000
5	4,000
6	3,000
7	3,000
Total	$39,000

Southern Cola uses a required rate of return of 16% in its capital-budgeting decisions.

Required
Compute:

1. Net present value
2. Payback period
3. Internal rate of return
4. Accrual accounting rate of return based on (a) initial investment and (b) average investment. Assume straight-line depreciation. Use the average annual savings in cash-operating costs when computing the numerator of the accrual accounting rate of return.

21-17 Equipment purchase for customer-service unit, net present value and payback. Solar Energy Inc. manufactures and markets solar panels for heating water for swimming pools. It currently owns a service van used by its customer-service representatives when making repair and service visits. The van was purchased three years ago for $56,000. The service van has a remaining useful life of five years but will require a $10,000 overhaul two years from now. Its current disposal price is $20,000; in five years its terminal disposal price is expected to be $8,000, assuming that the $10,000 overhaul is done on schedule. The cash-operating costs of this service van are expected to be $40,000 annually.

A salesperson has offered a new van for $51,000 or for $31,000 plus the old service van as a trade-in. The new service van would reduce annual cash-operating costs by $10,000, would not require any overhauls, would have a useful life of five years, and would have a terminal disposal price of $3,000.

Solar Energy has a required rate of return of 14% in its capital-budgeting decisions.

Required
1. Using a net present value criterion, should Solar Energy Inc. purchase the new service van?
2. Compute the payback period for Solar Energy Inc. if it purchases the new service van.

21-18 Sporting contract, net present value and breakeven time. Milano Capri is an Italian soccer team with a long tradition of winning. However, the last three years have been traumatic. The team has not won any major championship, and attendance at games has dropped considerably. Bennetelo Company is the major corporate sponsor of Milano Capri. Rocky Balboa, the president of Bennetelo, is also the president of Milano Capri. Balboa proposes that the team purchase the services of Diego Maradona. Maradona, "a soccer legend in his own lifetime," would create great excitement for Milano's fans and sponsors. Maradona's agent notifies Balboa that terms for the superstar's signing with Milano Capri are a bonus of $3.000 million payable now (start of 19_5) plus the following four-year contract (assume all amounts are in millions and are paid at the end of each year):

	19_5	19_6	19_7	19_8
Salary	$4.500	$5.000	$6.000	$6.500
Living and other expenses	1.000	1.200	1.300	1.400

Balboa's initial reaction is horror. As president of Bennetelo, he has never earned more than $800,000 a year. However, he swallows his pride and decides to examine the expected additions to Milano Capri's cash inflows if Maradona is signed for the four-year contract (assume all cash inflows are in millions and are received at the end of each year):

	19_5	19_6	19_7	19_8
Net gate receipts	$2.000	$3.000	$3.000	$3.000
Corporate sponsorship	3.000	3.500	4.000	4.000
Television royalties	0.000	1.200	1.400	2.000
Merchandise income (net of costs)	0.600	0.600	0.700	0.700

Balboa believes that a 12% required rate of return is appropriate for investments by Milano Capri.

Required

1. For Maradona's proposed four-year contract, compute (a) the net present value, (b) the payback period, and (c) the breakeven time. (Use the present value of the total cash outflows to Milano Capri when computing BET.)
2. What other factors should Balboa consider when deciding whether to sign Maradona to the four-year contract?

21-19 DCF, accrual accounting rate of return, working capital, evaluation of performance. Hammerlink Company has been offered a special-purpose metal-cutting machine for $110,000. The machine is expected to have a useful life of eight years with a terminal disposal price of $30,000. Savings in cash-operating costs are expected to be $25,000 per year. However, additional working capital is needed to keep the machine running efficiently and without stoppages. These working capital items include such items as filters, lubricants, bearings, abrasives, flexible exhaust pipes, and endless belts. These items must continually be replaced, so that an investment of $8,000 must be maintained at all times in them, but this investment is fully recoverable (will be "cashed in") at the end of the useful life. Hammerlink's required rate of return is 14%.

Required

1. Compute the net present value.
2. Compute the internal rate of return.
3. Compute the accrual accounting rate of return (a) on the initial investment and (b) on the average investment. Assume straight-line depreciation.
4. You have the authority to make the purchase decision. Why might you be reluctant to base your decision on the DCF model?

21-20 DCF, accrual accounting rate of return, working capital, evaluation of performance. Jana Wendt is the manager of the local Country West department store. She is considering whether to renovate some space in order to increase sales volume. New display fixtures and equipment will be needed. They will cost $70,000 and are expected to be useful for six years with a terminal disposal price of $4,000. Additional cash-operating inflows are expected to be $25,000 per year.

However, experience has shown that in order to sustain the higher sales volume, similar renovations have required additional investments in current assets, such as merchandise inventories and accounts receivable. An initial investment of $40,000 is needed to finance or "carry" this working capital, and this level must be maintained continuously. If and when she decides to terminate this plan or to use the store space for other purposes, the inventories and receivables can soon be liquidated or "cashed in."

Required

1. Compute (a) net present value, using a required rate of 12%; (b) internal rate of return; (c) accrual accounting rate of return on the initial investment; and (d) accrual accounting rate of return on the average investment. Assume straight-line depreciation.
2. As the store manager, which capital-budgeting method would you prefer for the purposes of making this decision and for evaluating subsequent performance? Give reasons, and compare the methods.

21-21 New-product proposal, NPV, payback, and breakeven time. Detroit Motors is examining a capital-budgeting proposal from its new-product development (NPD) group. The new car, code-named Project Nirvana, has the following projected cash outflows and cash inflows (in millions):

	Years							
	19_1	19_2	19_3	19_4	19_5	19_6	19_7	19_8
Cash Outflows								
Research and development	$20	$40	$ 5	$ 0	$ 0	$ 0	$ 0	$ 0
Product design	10	20	50	10	2	0	0	0
Manufacturing	0	0	3	25	300	200	45	0
Marketing	0	0	0	60	140	80	30	0
Distribution	0	0	0	2	60	40	10	0
Customer service	0	0	0	0	15	60	40	32
Cash Inflows								
Revenues	0	0	0	80	960	640	160	0

For simplicity, assume all cash inflows and cash outflows occur at the end of each year. These projections are viewed as "most likely" outcomes rather than pessimistic or optimistic outcomes.

Projected sales volume is:

Year	Sales Volume (Units)
19_4	5,000
19_5	60,000
19_6	40,000
19_7	10,000

Detroit Motors uses a 12% required rate of return.

Scott Steiner, president of Detroit Motors, has traditionally used discounted cash flow when evaluating new-product proposals. Recently, at a meeting of the President's Forum, he had lunch with Dean Morton, president of Hewlett-Packard. Morton "waxed lyrically" about a new approach Hewlett-Packard was using for evaluating new products: breakeven time (BET). Steiner, intrigued by BET, wants to consider how BET stacks up against the discounted cash-flow method.

Required

1. Compute the net present value of Project Nirvana as of January 1, 19_1.
2. Compute the breakeven time of Project Nirvana from January 1, 19_1. Use the present value of the total cash outflows in your BET computation.
3. What are two ways that the breakeven-time method differs from the traditional payback method?
4. What are two advantages and two disadvantages of the breakeven-time method relative to the discounted cash-flow method when evaluating products or projects?

21-22 New-product proposals, life-cycle costing, breakeven time. Pear Inc. assembles and markets office equipment products. One of its major products is MaxiCalc, a hand-held calculator. This product is near the end of its product life cycle. Several competitors have already introduced new products that are similarly priced to MaxiCalc ($18 per unit) and have more functions and better displays. Pear Inc. is deciding whether to adopt one of the following two investment proposals.

1. MaxiCalc II—an upgraded version of MaxiCalc (using many features of its product design). MaxiCalc II would be priced at $20 per unit. Expected unit sales for the next four years are:

19_3	19_4	19_5	19_6
10,000	50,000	30,000	20,000

2. SuperCalc—the first of a new generation of pocket calculator that will require substantial investments both in research and development and in product design. SuperCalc would be priced at $25 per unit. Expected unit sales for the next four years are:

19_3	19_4	19_5	19_6
0	20,000	60,000	40,000

Exhibit 21-9 presents the budgeted cash outflows for both the MaxiCalc II and SuperCalc proposals for the 19_3 to 19_7 period. These amounts are the expected values of the cash flows for each year for each product. Assume both products will become obsolete at the end of 19_6. The budgeted costs in 19_7 pertain only to handling customer-service complaints on units sold prior to 19_7.

Pear Inc. uses a 12% required rate of return in its new-product investment decisions.

Required

1. Which investment proposal should Pear Inc. adopt if it uses (a) a net present value criterion? (b) a breakeven-time criterion? (Use the present value of the total cash outflows when computing each project's BET.)
2. What factors, other than those examined in requirement 1, should Pear Inc. consider when choosing between the MaxiCalc II and SuperCalc products?

EXHIBIT 21-9

Budgeted Life-Cycle Cash Outflows for MaxiCalc II and SuperCalc (in thousands)

	19_3	19_4	19_5	19_6	19_7
MaxiCalc II					
Research and development	$100	$ 10	$ 0	$ 0	$ 0
Product design	40	15	0	0	0
Manufacturing	80	180	120	60	0
Marketing	90	120	90	30	0
Distribution	5	25	15	10	0
Customer service	7	38	36	23	6
SuperCalc					
Research and development	$190	$ 50	$ 0	$ 0	$ 0
Product design	80	30	10	0	0
Manufacturing	0	120	360	120	0
Marketing	20	260	140	60	0
Distribution	0	10	30	20	0
Customer service	0	20	70	70	20

21-23 Capital budgeting for a sporting franchise. The Midwest Slammers are the leading team in a regional baseball league. Over the past five years, the team has consistently sold out all the 30,000 individual seats and the 100 corporate boxes at its home stadium. The stadium has a current book value of $19 million; annual depreciation using the straight-line method is $1.5 million. The estimated terminal disposal price of the stadium at the end of ten years is $4 million. This $4 million is based on an agreement Midwest signed with the local city authorities. This agreement requires the city to repurchase the stadium for $4 million ten years from the current year. The owner of the stadium must pay the city $1 million each year that the city does not own the stadium.

Jeannette Ochoa, the owner of the Midwest Slammers' sporting franchise, predicts the following amounts for each of the next ten years:

- 10,000 season "A" tickets sold at $300 per year
- 20,000 season "B" tickets sold at $200 per year
- 100 corporate boxes sold at $5,000 per year

Annual operating costs in addition to the $1 million payment to the city will be $5.5 million. These costs include salaries to coaches, players, and administrators, as well as the cost of operating the stadium. Television and radio royalties will be $5 million a year. For simplicity, assume that all the annual cash inflows and annual cash outflows occur at the end of each year.

Ochoa views the sporting franchise as having three assets: the player contracts, the goodwill associated with the Midwest Slammers name, and the stadium, which she believes the city will repurchase in ten years' time. She receives an offer from a local real estate developer to buy the franchise for $80 million. The real estate developer proposes to keep the team at its current stadium. Ochoa would invest the $80 million at 8% per annum. In considering this offer, Ochoa estimates the total value of the Midwest Slammers' sporting franchise at the end of ten years will be $100 million. (This $100 million *includes* the $4 million payment from the city for repurchase of the stadium.)

Required

1. Using the discounted cash-flow method, would you recommend that Ochoa accept the $80 million offer? Her required rate of return on investments is 8% per annum.
2. What other factors would you recommend that Ochoa consider in deciding whether to accept the $80 million offer?
3. Assuming Ochoa decides to keep the sporting franchise, what will be her accrual accounting rate of return on the average investment over the next ten years? (Assume the stadium is the investment base.)

21-24 Capital budgeting for a new sports stadium. (Continuation of Problem 21-23) Jeannette Ochoa, the owner of the Midwest Slammers' sporting franchise, has long viewed the existing home sports stadium as an embarrassment. A national TV sports commentator stated that the stadium "redefined global minimum standards for fan comfort." The city mayor has recently offered Ochoa a new land area on which a modern domed stadium can be built. The land will be given free to the sporting team if it commits itself to construction of the stadium. The cost of constructing the new stadium is $60 million, payable in cash at the start of construction.

The new stadium will be ready by the start of the next season. In addition to its capacity of 60,000 seats, there would be 300 super-deluxe corporate boxes. Ochoa predicts the following amounts for each of the next ten years if the new stadium is constructed:

- 20,000 season "A" tickets sold at $400 per year
- 20,000 season "B" tickets sold at $250 per year
- 10,000 season "C" tickets sold at $200 per year
- 250 corporate boxes sold at $20,000 per year

The annual operating costs of the franchise (including the costs of operating the new stadium) will be $9 million. Television and radio royalties of $5 million a year will be unaffected by the decision concerning the new stadium.

The city authorities require the Midwest Slammers to continue paying $1 million a year for the next ten years on the existing stadium even if it is not used. Ochoa expects the city to repurchase the existing stadium for $4 million ten years from the current year, irrespective of whether the Slammers remain in the existing stadium or move to the new stadium. The estimated selling price of the new stadium at the end of ten years is $30 million.

Required

1. What is the net present value of the proposed move to the $60 million stadium?
2. What is the payback period on the move to the new stadium? (Use the incremental cash inflows from the move as the denominator.)

3. What other factors might be important in deciding whether to construct the new stadium?

21-25 Equipment replacement, sensitivity analysis. A toy manufacturer that specializes in making fad items has just developed a $50,000 molding machine for automatically producing a special toy. The machine has been used to produce only one unit so far. The company will depreciate the $50,000 original cost evenly over four years, after which time production of the toy will be stopped.

Suddenly a machine salesman appears. He has a new machine that is ideally suited for producing this toy. His automatic machine is distinctly superior. It reduces the cost of materials by 10% and produces twice as many units per hour. It will cost $44,000 and will have a zero terminal disposal price at the end of four years.

Production and sales of 25,000 units per year (sales of $100,000) will be the same whether the company uses the molding machine or the automatic machine. The current disposal price of the toy company's molding machine is $5,000. Its terminal disposal price in four years will be $2,600.

With its present equipment, the company's annual costs will be: direct materials, $10,000; direct labor, $20,000; and variable factory overhead, $15,000. Variable factory overhead is applied on the basis of direct labor costs. Fixed factory overhead, exclusive of depreciation, is $7,500 annually, and fixed marketing and administrative costs are $12,000 annually.

Required

1. Assume that the required rate of return is 18%. Using the net present value method, show whether the new equipment should be purchased. What is the role of the book value of the old equipment in the analysis?
2. What is the payback period for the new equipment?
3. As the manager who developed the $50,000 molding machine, you are trying to justify not buying the new $44,000 machine. You question the accuracy of the expected cash-operating savings. By how much must these cash savings fall before the point of indifference—the point where the net present value of the project is zero—is reached?

21-26 Capital budgeting and relevant costs. The city of Los Angeles has been operating a cafeteria for its employees, but it is considering a conversion from this form of food service to a completely automated set of vending machines. If the change is made, the old equipment would be sold now for whatever cash it might bring.

The vending machines would be purchased immediately for cash. A catering firm would take complete responsibility for servicing and replenishing the vending machines and would pay the city a predetermined percentage of the gross vending receipts.

The present cafeteria equipment has ten years remaining of useful economic life. The new vending machine has a ten-year useful economic life. The following data are available (in thousands):

Cafeteria cash revenues per year	$120
Cafeteria cash costs per year	$124
Present cafeteria equipment:	
Net book value	$ 84
Annual depreciation cost	$ 6
Current disposal price	$ 4
Terminal disposal price (10 years from now)	$ 0
New vending machines:	
Purchase price	$ 64
Terminal disposal price	$ 5
Expected annual gross receipts	$ 80
City's percentage share of receipts	10%
Expected annual cash costs (negligible)	
Present values at 14%:	
$1 due in 10 years	$0.27
Annuity of $1 a year for 10 years	$5.20

The city of Los Angeles has a 14% required rate of return.

Required

For the two alternatives, compute the following:

1. Expected increase in net annual operating cash inflows
2. Payback period
3. Net present value
4. Point of indifference (zero NPV) in terms of annual gross vending machine receipts

21-27 Present value analysis, damage assessment in a legal case. Lone Star Oil in 19_4 made a $3.5 billion bid for the 1 billion barrels of oil reserves owned by Sheridan Oil. Sheridan verbally accepted this bid. Two weeks later, the president of Lone Star was shocked to read that Sheridan Oil had signed a written agreement to sell the same 1 billion barrels of oil reserves to Lapton Oil for $3.8 billion. Lone Star sued Lapton Oil for inducing Sheridan to breach its contract with Lone Star.

A trial ensued in 19_6 in which Lapton Oil was found guilty of inducing Sheridan to breach its contract with Lone Star Oil. The court ruled that Lapton Oil should pay Lone Star $6.5 billion (plus interest) in damages. This $6.5 billion amount was based on a submission by Lone Star at the trial that the following expenditures would be necessary over a five-year period to find and develop another 1 billion barrels of oil reserves in (in billions):

19_4	$ 1.6
19_5	1.8
19_6	2.0
19_7	2.2
19_8	2.4
	$10.0

The Lone Star submission argued that it would cost $10 billion to acquire elsewhere 1 billion barrels of oil reserves, which Sheridan Oil had agreed to sell for $3.5 billion. Hence, its estimate of the damage it suffered was $6.5 billion (plus interest). Lapton Oil chose not to dispute this submission during the trial, arguing that it did not breach any contract and so to discuss any possible measure of damage was "defeatist."

The actual amount the judge ordered Lapton Oil to pay Lone Star Oil was $7.865 billion, comprising $6.500 billion plus $1.365 billion forgone interest ($6.5 billion at 10% interest per annum for two years).

Required

Assume you had been hired by Lapton Oil prior to the trial to critique any damage assessments put forward by Lone Star Oil. What criticisms would you make of the $6.5 damage estimated by Lone Star Oil?

21-28 Capital-budgeting approaches, computer-integrated manufacturing. Craig Young, the production manager of Brittania Tools, is concerned about Brittania's ability to maintain its competitive position in the industrial machine-tool market. A recent surge of imports is priced 30% below Brittania's full cost. A major domestic competitor recently switched to a computer-integrated manufacturing (CIM) operation for its machine-tool plant.

Young attends a trade exhibition titled "Automate, Emigrate, or Evaporate" and starts negotiations with a vendor of CIM equipment. This vendor will provide the necessary machines and associated equipment for a cost of $80 million. Young estimates the following annual cost savings from implementing CIM:

1. Reduction in rental payments due to reduced floor space requirements	$ 4.0 million
2. Lower number of product defects and reduced reworking of products	$22.0 million

Another benefit of CIM is reduced levels of working capital. The average combined level of inventories and accounts receivable for Brittania Tools at present is $14 million. Young estimates that following implementation of CIM the average combined level of inventories and accounts receivable would be $4 million. (For simplicity, assume that this reduction in working capital occurs instantaneously at the time the investment in CIM is made.)

The one-time internal costs of implementing CIM are estimated to be \$40 million. These costs include the retraining of operating and maintenance personnel plus any lost production during the changeover. For internal reporting purposes, the \$40 million internal costs are capitalized along with the \$80 million purchase price when determining the investment required for the CIM proposal. (For simplicity, assume that these \$40 million implementation costs are incurred at the same time the \$80 million equipment purchase is made.) The annual costs of maintaining the software programs and of the CIM hardware equipment and machinery are estimated to be \$8 million. The vendor of the CIM equipment maintains that, if properly maintained, a twenty-year useful life may be expected.

The estimated disposal price of the CIM equipment is \$30 million at the end of ten years and \$10 million at the end of twenty years. Brittania uses a required rate of return of 14%. The maximum time period Brittania currently considers for any investment proposal is ten years.

Required

1. Compute the payback period for the CIM proposal.
2. Compute the net present value of the CIM proposal. Should Brittania adopt CIM, given its existing investment criteria?
3. Compute the accrual accounting rate of return based on (a) initial investment and (b) average investment. Assume straight-line depreciation and a ten-year useful life for the investment.
4. Young reads an article that argues that many companies are rejecting CIM proposals because either (a) the discount rate used in DCF analysis is too high or (b) the time period over which the benefits are considered is too short. He believes that Brittania should use an 8% discount rate and consider benefits for a twenty-year period in its CIM analysis. Prepare a report for Young on the effects of making these changes in the DCF calculations.
5. What other factors would you recommend that Brittania Tools consider in deciding whether to adopt the CIM proposal?

Answers to Exercises in Compound Interest (Problem 21-12)

The general approach to these exercises centers on a key question: Which of the four basic tables in Appendix B should be used? No computations should be made until after this basic question is answered with confidence.

1. From Table 1. The \$5,000 is a present value. The value ten years hence is an *amount of future worth*.

$S = P(1 + r)^n$; the conversion factor, $(1 + r)^n$, is on line 10 of Table 1.

Substituting at 6%; $S = 5{,}000(1.791) = \$8{,}955$

Substituting at 14%; $S = 5{,}000(3.707) = \$18{,}535$

2. From Table 2. The \$89,550 is an *amount of future worth*. You want the present value of that amount.

$P = S/(1 + r)^n$; the conversion factor, $1/(1 + r)^n$, is on line 10 of Table 2.

Substituting: $P = \$89{,}550(.558) = \$49{,}969$

3. From Table 3. The \$89,550 is *future worth*. You are seeking the uniform amount (annuity) to set aside annually.

$$S_n = \text{Annual deposit } (F)$$
$$\$89{,}550 = \text{Annual deposit } (13.181)$$
$$\text{Annual deposit} = \frac{\$89{,}550}{13.181} = \$6{,}794$$

4. From Table 3. You are seeking the *amount of future worth* of an annuity of \$5,000 per year.

$S_n = \$5{,}000\ F$, where F is the conversion factor

$S_n = \$5{,}000(17.549) = \$87{,}745$

5. From Table 4. When you reach age 65, you will get $200,000. This is a present value at that time. You must find the annuity that will exactly exhaust the invested principal in ten years.

$$P_n = \text{Annual withdrawal } (F)$$
$$\$200{,}000 = \text{Annual withdrawal } (7.360)$$
$$\text{Annual withdrawal} = \frac{\$200{,}000}{7.360} = \$27{,}174$$

6. From Table 4. You need to find the present value of an annuity for ten years.

$$\text{At 6\%}\begin{cases} P_n = \text{Annual withdrawal } (F) \\ P_n = \$50{,}000(7.360) \\ P_n = \$368{,}000 \end{cases}$$
$$\text{At 20\%}\begin{cases} P_n = \$50{,}000(4.192) \\ P_n = \$209{,}600, \text{ a much lower figure} \end{cases}$$

7. Plan B is preferable. The net present value of Plan B exceeds that of Plan A by $980 ($3,126 − $2,146):

		Plan A		Plan B	
Year	**PV Factor at 6%**	**Cash Inflows**	**PV of Cash Inflows**	**Cash Inflows**	**PV of Cash Inflows**
0	1.000	$(10,000)	$(10,000)	$(10,000)	$(10,000)
1	.943	1,000	943	5,000	4,715
2	.890	2,000	1,780	4,000	3,560
3	.840	3,000	2,520	3,000	2,520
4	.792	4,000	3,168	2,000	1,584
5	.747	5,000	3,735	1,000	747
			$ 2,146		$ 3,126

CHAPTER 22

Income taxes and inflation are important factors electric utility companies consider in their capital budgeting decisions for investments in new plants.

Capital Budgeting: A Closer Look

Learning Objectives

When you have finished studying this chapter, you should be able to

1. Describe two major ways that income taxes effect business decisions
2. Identify three factors that influence the amount of depreciation claimed as a tax deduction
3. Explain why depreciation and the book value of a currently owned asset are not themselves inputs into DCF computations
4. Distinguish between the nominal rate of interest and the real rate of interest
5. Demonstrate the equivalence of the nominal approach and the real approach to incorporating inflation into capital budgeting
6. Describe alternative approaches used to recognize the degree of risk in projects in capital budgeting
7. Explain why the internal rate of return and the net present value decision rules may rank projects differently

Benjamin Franklin said that two things in life are certain: death and taxes. We might add a third: changing prices. This chapter examines how managers analyze income taxes and changing prices in capital budgeting. (We also recognize death in this chapter, although only of projects, not of the individuals who select them.) This chapter also covers the estimation of the required rate of return, capital budgeting in nonprofit organizations, administration of capital budgets, and some issues in using the net present value decision rule and the internal rate-of-return decision rule.

Uncertainty about long-run events makes capital budgeting difficult and challenging. The uncertainty surrounding long-term prediction should prevent any manager from splitting hairs over smaller-scale aspects of the analysis and decisions. A few guideposts, however, help in making intelligent overall decisions. Chapters 20 and 21 discussed some of these guideposts, and this chapter offers more.

No matter where a company does business, income taxes have two major impacts:

Objective 1

Describe two major ways that income taxes effect business decisions

- On the amounts of cash inflow and outflow and
- On the timing of these flows

For simplicity, in this chapter, as in Chapter 21, all cash outflows and cash inflows occur at the beginning or end of each period (as specified).

INCOME TAX FACTORS

The Importance of Income Taxes

Income taxes often have a tremendous influence on decisions. For example, income taxes can sizably reduce the net cash inflows from individual projects and so change their relative desirability. In choosing among projects, managers must get answers to these tax-related questions:

- What income tax rate applies in each year of the project?
- Which cash inflows (outflows) are taxable (tax deductible)?
- What differences exist in tax rates and tax deductions among projects set in different jurisdictions (states, countries, and so on)?

We concentrate on a general approach to understanding income taxes that applies globally. In taking this approach, we focus on income tax provisions affecting depreciation and confine our discussion to corporations (excluding partnerships and individuals).[1]

Many tax rules regarding income measurement are the same as the financial rules regarding reporting to shareholders. Other rules differ. For example, income tax rules frequently permit taxpayers to use shorter useful lives for depreciation than generally accepted accounting principles permit.

Treatment of Depreciation for Taxation

Objective 2

Identify three factors that influence the amount of depreciation claimed as a tax deduction

Taxation rules for depreciation vary considerably among countries. Even within a single country, marked changes can occur over short periods of time. However, taxation rules typically cover three factors that influence depreciation: the amount allowable for depreciation, the time period over which the asset is to be depreciated, and the pattern of allowable depreciation.

Amount Allowable for Depreciation. In many cases, the amount allowable for depreciation is the original purchase cost of the asset. However, sometimes either less than the original cost or more than the original cost is allowable for depreciation. In those countries where corporations have the option of claiming an investment tax credit,[2] the amount allowable for depreciation may be reduced below the original cost of the asset acquired. Some countries permit corporations to write off more than the original cost (as measured by nominal monetary units) for depreciation purposes. For example, when determining the amount allowable for periodic depreciation, Brazilian corporations may use inflation indexes to write up the cost of assets.

Time Period over Which the Asset Is to Be Depreciated. Throughout the years and in various countries, tax authorities have permitted three main methods of determining the depreciation time period:

1. The taxpayer estimates the useful life.
2. The tax authority estimates the useful life.
3. Tax legislation specifies a table of allowable lives. An example is the property-class life categories (recovery periods) used in the Modified Accelerated Cost Recovery System that is applicable in the United States as of this writing.

Other things being equal, the shorter the allowable life, the higher the project's net present value. A short allowable life means that cash savings from tax deductions are in dollars with a relatively high present value.

Pattern of Allowable Depreciation (for a Given Time Period). Tax authorities allow three main depreciation methods:

[1]A general framework for examining income tax factors in business decisions is presented in M. Scholes and M. Wolfson, *Taxes and Corporate Financial Strategy: A Global Planning Approach* (Englewood Cliffs, NJ: Prentice Hall, 1991).

[2]An **investment tax credit (ITC)** is a direct reduction of income taxes arising from the acquisition of depreciable assets. Governments use the ITC to stimulate investments in specific assets or in specific industries. To illustrate: If a firm purchases an asset costing $1,000 and there is a 4% ITC, the firm obtains an immediate tax credit of $40; this credit increases the net present value of an asset purchase by $40. The depreciable amount of the asset would be $960 or $1,000, depending on the specific tax law. In the United States, the ITC option has been made available (and then subsequently withdrawn) several times since 1962. The ITC option is not available in the United States at the time of writing this edition.

1. **Straight-line depreciation (SL),** in which an equal amount of depreciation is taken each year.
2. Accelerated depreciation procedures, such as the double-declining balance (DDB) method.[3] **Accelerated depreciation** is defined as any pattern of depreciation that writes off depreciable assets more quickly than does straight-line depreciation.
3. Depreciation using a table of allowable percentage write-offs as specified by tax legislation.

All other things being equal, the more accelerated the pattern of depreciation, the higher the project's net present value. Accelerated depreciation means that more depreciation occurs in the earlier years of a project when the tax savings from tax deductions are in dollars with a relatively high present value.

EXAMPLE

Martina Enterprises, a corporation, is considering the purchase of testing equipment for a research and development project. The following tax laws apply:

- *Amount allowable for depreciation*—original cost of the equipment minus any predicted terminal disposal price.
- *Time period over which the asset is to be depreciated*—based on an estimate of useful life made by the taxpayer (Martina).
- *Pattern of allowable depreciation*—only the straight-line depreciation method is permitted.

The original cost of the testing equipment is $110,000 payable in cash immediately. Martina predicts that the equipment will have an expected useful life of five years and a terminal disposal price of $20,000. No change in working capital is required if the equipment is purchased. The corporate income tax rate of 40% will apply each year. The company uses a required rate of return of 12% in discounting after-tax cash flows.

Exhibit 22-1 shows the relationship among depreciation, income taxes, and net income. The amount allowable for depreciation is $90,000 (the original cost of $110,000 minus the terminal disposal price of $20,000). The annual depreciation deduction using the straight-line method is $18,000 ($90,000 ÷ 5 years). We can calculate the cash inflow from operations, net of income taxes (see the $30,000 amount in Exhibit 22-1), in three different ways:

1. $S - E - T = \$100{,}000 - \$62{,}000 - \$8{,}000 = \$30{,}000$
2. $NI + D = \$12{,}000 + \$18{,}000 = \$30{,}000$
3. $(S - E)\,(1 - t) + Dt = (\$100{,}000 - \$62{,}000)\,(1 - .4) + \$18{,}000(.4) = \$30{,}000$

where S = sales,
E = expenses excluding depreciation (assumed all paid in cash),
D = depreciation,
T = income taxes,
NI = net income, and
t = income tax rate.

The present value of the tax savings from depreciation to Martina Enterprises is calculated as follows:

[3]**Double-declining balance (DDB) depreciation** is a form of accelerated depreciation in which first-year depreciation is twice the amount of straight-line depreciation when a zero terminal disposal price is assumed. Exhibit 22-2 illustrates DDB.

Year	Income Tax Deduction for Depreciation	Income Tax Savings at 40%	12% Discount Factor	Present Value at 12%
1	$18,000	$7,200	.893	$ 6,430
2	18,000	7,200	.797	5,738
3	18,000	7,200	.712	5,126
4	18,000	7,200	.636	4,579
5	18,000	7,200	.567	4,082
				$25,955

The $25,955 amount is the present value of the tax savings from having the $18,000 depreciation deduction each year for five years. The present value of the tax savings is influenced by the depreciation method used, the applicable tax rate, and the interest rate used for discounting future cash flows.

EXHIBIT 22-1

Basic Analysis of Cash Flow from Operations, Net of Income Taxes, for Martina Enterprises (Data Assumed)

Traditional Income Statement

(*S*)	Sales	$100,000
(*E*)	Deduct: Expenses, excluding depreciation*	$ 62,000
(*D*)	Depreciation (straight-line of $90,000 ÷ 5 years)	18,000
	Total expenses	$ 80,000
	Operating income	$ 20,000
(*T*)	Income taxes at 40%	8,000
(*NI*)	Net income	$ 12,000

Total cash flow from operations, net of income taxes, is:

$S - E - T = \$100{,}000 - \$62{,}000 - \$8{,}000 = \$30{,}000$

or

$NI + D = \$12{,}000 + \$18{,}000 = \$30{,}000$

Item-by-Item Analysis for Capital Budgeting

	Effect of Cash-Operating Items:	
(*S* − *E*)	Cash inflow from operations: $100,000 − $62,000	$38,000
	Income tax cash outflow at 40%	15,200
	After-tax effect of cash-operating items	$22,800
	Effect of Depreciation:	
(*D*)	Straight-line depreciation: $90,000 ÷ 5 = $18,000	
	Income tax savings at 40%	7,200
	Total cash flow from operations, net of income taxes	$30,000

Total cash flow from operations, net of income taxes, can be computed as (letting *t* be the income tax rate):

$$(S - E)(1 - t) + Dt = [(\$100{,}000 - \$62{,}000)(1 - .4)] + \$18{,}000(.4)$$
$$= \$22{,}800 + \$7{,}200 = \$30{,}000$$

*All expenses, other than the depreciation, are assumed to be paid in cash (that is, depreciation is the only accrual expense item).

EFFECTS OF INCOME TAXES ON CASH FLOW

We have just looked at how income tax can affect cash inflow from operations. We turn now to a fuller discussion of how income tax can affect cash inflows and outflows and also how it influences managers' decisions. Our detailed example highlights the effect of the tax deductibility of depreciation on the net present value of a project.

EXAMPLE

Potato Supreme processes potato products for sale to supermarkets and other retail outlets. It is considering replacing an old packaging machine (purchased six years ago) with a new, more efficient packaging machine. The new machine is less labor-intensive and can process a higher volume of potato-chip packs per hour than can the old machine. For simplicity, we assume that:

a. All cash outflows or inflows occur at the start or end of the year,
b. The tax effects of cash inflows and outflows occur at the same time that the inflows and outflows occur,
c. The income tax rate is 30% each year,
d. Gains or losses on the sale of depreciable assets are taxed at the same rate as ordinary income, and
e. Both the old and the new machine have the same working capital requirements.

Our analysis could be refined to account for complications arising when assumptions *a–e* do not hold. However, these complications do not change the basic approach to capital budgeting outlined in Chapter 21 and in the examples presented in this chapter.

The following income tax rules apply to Potato Supreme for the old equipment and the new equipment:

- *Amount allowable for depreciation.* Original cost is the basis for depreciation computations. No account is taken of terminal disposal price when computing depreciation for either the old or the new machine. When an existing asset is sold, any difference between the disposal price and the book value (that is, original cost minus accumulated depreciation at the time of the sale) is treated as ordinary income (or loss) for tax purposes.
- *Time period over which the asset is to be depreciated.* A table of allowable lives specified in tax legislation determines the period for depreciation.
- *Pattern of allowable depreciation.* Straight-line depreciation is required for the old machine. The new machine, however, would qualify for a special tax provision permitting use of the double-declining balance (DDB) depreciation method. A five-year period for depreciating the machine is to be used.

Summary data for the two machines follow:

	Old Machine	New Machine
Original cost	$110,000	$200,000
Accumulated depreciation	$ 60,000	—
Current book value	$ 50,000	—
Current disposal price	$ 26,000	—
Terminal disposal price, 5 years from now	$ 6,000	$ 20,000
Annual cash-operating costs in potato-chip packing area	$250,000	$150,000
Remaining years of useful life (based on industry trade association data)	5 years	5 years
After-tax required rate of return	10%	10%

Present Value of Tax Savings from Depreciation

Depreciation deductions are an important source of tax savings in many projects. We illustrate how to compute the present value of these tax savings for Potato Supreme by reference to the new machine.

Depreciation on the new machine, computed using the double-declining balance method, is:[4]

Year	Book Value at Start of Year	DDB Rate	Income Tax Deduction for Depreciation	Book Value at End of Year
1	$200,000	40%	$80,000	$120,000
2	120,000	40%	48,000	72,000
3	72,000	40%	28,800	43,200
4	43,200	40%	17,280	25,920
5	25,920	—	25,920	—

The present value of the tax savings to Potato Supreme from depreciation of the new machine is calculated as follows:

Year	Income Tax Deduction for Depreciation	Income Tax Rate	Income Tax Savings	10% Discount Factor	Present Value at 10%
1	$80,000	30%	$24,000	.909	$21,816
2	48,000	30%	14,400	.826	11,894
3	28,800	30%	8,640	.751	6,489
4	17,280	30%	5,184	.683	3,541
5	25,920	30%	7,776	.621	4,829
					$48,569

Potato Supreme's income tax rate is assumed to be 30% each year. Predictions of the tax rate applicable for each year of a project are required when this assumption is not appropriate.

Total Project Approach vs. Incremental Approach

Assume Potato Supreme wants to compute the net present value of replacing the old packaging equipment with the new packaging equipment. Consider two approaches:

- **Total project approach** (Exhibit 22-2 on pp. 712–13). Includes all the cash outflows and inflows associated with each project.
- **Incremental approach** (Exhibit 22-3 on pp. 714–15). Includes only those cash outflows and inflows that differ between the two projects. (Exhibit 22-3 includes summary figures from Exhibit 22-2 for the after-tax difference in terminal disposal price and the incremental initial investment at time zero. Stand-alone use of the incremental approach would require more-detailed computations than are included in Exhibit 22-3.)

[4]The DDB depreciation pattern is calculated as follows:
a. Compute the rate (ignoring the terminal disposal price) by dividing 100% by the years of useful life. Then double the rate. In the Potato Supreme example, 100% ÷ 5 years = 20%. The DDB rate would be 2 × 20%, or 40%.
b. To compute the depreciation for any year, multiply the beginning book value at the start of the year (original cost minus any accumulated depreciation) by the DDB rate. Unmodified, this method would never fully depreciate the existing book value. In the Potato Supreme example, for simplicity, the depreciation in the fifth year is the book value at the start of the fifth year.

Both approaches result in a net present value of $141,855 in favor of replacing the old machine with the new machine. Where there are only two alternatives, the incremental approach is faster. The incremental approach, however, rapidly becomes unwieldy when there are more than two alternatives or when computations become intricate.

Five categories of cash flows are included in Exhibits 22-2 and 22-3:

1. Recurring cash-operating costs
2. Income tax cash savings due to depreciation deductions
3. Cost of the new machine
4. Terminal disposal price of the new machine
5. Current disposal price of the old machine

For the replace alternative, items in all these five categories of cash flows are included in the present value computation. For the keep alternative, only items in categories 1, 2, and 5 are included. Having separate categories of cash flows makes possible sensitivity analysis of the amounts within these categories.

Clarification of the Role of Depreciation

Objective 3

Explain why depreciation and book value of a currently owned asset are not themselves inputs into DCF computations

Consider the impact of depreciation on DCF methods of capital budgeting. Study Exhibit 22-4 on page 715, which uses amounts from the Potato Supreme machine replacement example given in Exhibits 22-2 and 22-3. The relevant amounts include the $26,000 current disposal price of the old packaging machine. ***Depreciation and the book value of the old machine are used only when predicting the income tax effects on cash. By themselves, however, they are not inputs to DCF* computations.**

The following points summarize the role of depreciation when considering the replacement of equipment:

1. *Discounted cash-flow method.* Depreciation on the old machine is irrelevant because it is not a future cash outflow. Depreciation on the new machine is irrelevant because the investment should not be counted twice; the investment is usually a one-time cash outlay at time zero, so it should not be deducted again in the form of depreciation.
2. *Relation to income tax cash flows.* Given the definition of *relevance* in Chapter 11, book values and past depreciation are irrelevant in all capital-budgeting methods. The relevant items are the *after-tax cash effects,* not the book values. Using the approach in Exhibit 11-1 (p. 367), the book values and any depreciation charges are essential data for the *prediction method,* but the impacts on expected future cash disbursements for income tax are the relevant data for the *decision model.*

U.S. Taxation Rules

We have concentrated on the general approach to analyzing income tax effects in capital budgeting, an approach that is applicable around the globe. The tax rules in the United States change almost every year. The rules at the time of this writing are called **Modified Accelerated Cost Recovery System (MACRS).** MACRS is a federal income tax regulation that classifies depreciable assets into one of several recovery periods, each of which has a designated pattern of allowable depreciation (double-declining balance, 150% declining balance, or straight-line). MACRS is a modification of tax laws first introduced in 1981. Its specific provisions have been revised several times since then. The two depreciation methods illustrated in the Potato Supreme example in Exhibits 22-2 to 22-4 (and in the exercises and problems for this chapter) are the main alternatives available in the version of MACRS applicable at the time of writing this text. The appendix to this chapter summarizes some key provisions of U.S. tax rules for depreciable assets. This appendix can be read now without any interruption in the flow of this chapter.

EXHIBIT 22-2

Total Project Approach for Potato Supreme: After-Tax Analysis of Equipment Replacement Decision

				Present Value Discount Factors @ 10%	Total Present Value	Sketch of Relevant After-Tax Cash Flows					
End of year						0	1	2	3	4	5
(A) Replace Old Machine											
Recurring cash-operating costs			$150,000								
Income tax savings, @ 30%			(45,000)								
After-tax effect on cash each year for 5 years			$105,000	3.791	$(398,055)	←	$(105,000)	$(105,000)	$(105,000)	$(105,000)	$(105,000)
Income tax savings due to depreciation deductions (DDB):											
Year	Income Tax Deduction	Cash Effects of Income Tax Savings @ 30%									
1	$80,000	$24,000		.909	21,816	←	$24,000				
2	48,000	14,400		.826	11,894	←		$14,400			
3	28,800	8,640		.751	6,489	←			$8,640		
4	17,280	5,184		.683	3,541	←				$5,184	
5	25,920	7,776		.621	4,829	←					$7,776
Terminal disposal price at end of year 5, entirely subject to tax because book value will be zero and so gain equals proceeds			$ 20,000								
Deduct: 30% tax on gain			(6,000)								
Total after-tax effect on cash			$ 14,000	.621	8,694	←					$14,000
Cost of new machine			$200,000	1.000	(200,000)	$(200,000)					

Disposal of old equipment:										
Book value at time zero $(110,000 – $60,000)	$50,000									
Current disposal price	(26,000)	$ 26,000								
Net loss	24,000									
Tax savings	× .30	7,200								
Total after-tax effect on cash		$ 33,200	1.000	33,200	$33,200					
Total present value of all cash flows				$(507,592)						
(B) Keep Old Machine										
Recurring cash-operating costs		$250,000								
Income tax savings, @ 30%		(75,000)								
After-tax effect on cash each year for 5 years		$175,000	3.791	$(663,425)	←----	$(175,000)	$(175,000)	$(175,000)	$(175,000)	$(175,000)
Income tax savings due to depreciation deductions, straight-line $(110,000 ÷ 11)	$10,000									
Income tax savings each year for remaining 5 years at 30%	× .30	$ 3,000	3.790*	11,370	←------	$3,000	$3,000	$3,000	$3,000	$3,000
Terminal disposal price at end of year 5, gain equals proceeds		$ 6,000								
Deduct 30% tax on gain		(1,800)								
Total after-tax effect on cash		$ 4,200	.621	2,608	←------	------	------	------	------	$4,200
Total present value of all cash flows				$(649,447)						
Net Present Value Difference in Favor of Replacement				$141,855						

*Note: Rounded to avoid discrepancies between Exhibits 22-2 and 22-3 due to rounding differences: 3.790 = (.909 + .826 + .751 + .683 + .621)

EXHIBIT 22-3

Incremental Approach for Potato Supreme: After-Tax Analysis of Equipment Replacement Decision

					Present Value Discount Factors @ 10%	Total Present Value	Sketch of Relevant After-Tax Cash Flows					
End of year							0	1	2	3	4	5
Analysis confined to differences between REPLACE and KEEP alternatives in Exhibit 22-2												
A. Recurring cash operating savings, $250,000 − $150,000				$100,000								
Income tax, @ 30%				30,000								
After-tax effect on cash each year for 5 years				$ 70,000	3.791	$265,370	←	$70,000	$70,000	$70,000	$70,000	$70,000
B. Differences in tax savings from depreciation:												
Year	Replace	Keep	Difference	Income Tax Cash Effect @ 30%								
1	$80,000	$10,000	$70,000	$21,000	.909	19,089	←	$21,000				
2	48,000	10,000	38,000	11,400	.826	9,416	←		$11,400			
3	28,800	10,000	18,800	5,640	.751	4,236	←			$5,640		
4	17,280	10,000	7,280	2,184	.683	1,492	←				$2,184	
5	25,920	10,000	15,920	4,776	.621	2,966	←					$4,776
C. After-tax difference in terminal disposal price, end of year 5 (see Exhibit 22-2 for details): $14,000 − $4,200 = $9,800					.621	6,086	←					$9,800
D. Incremental initial investment now (see Exhibit 22-2 for details): $200,000 − $33,200 = $166,800					1.000	(166,800)	$(166,800)					
Net Present Value Difference in Favor of Replacement						$141,855						

EXHIBIT 22-4
Relevant Machine-Related Costs in Capital Budgeting at Potato Supreme

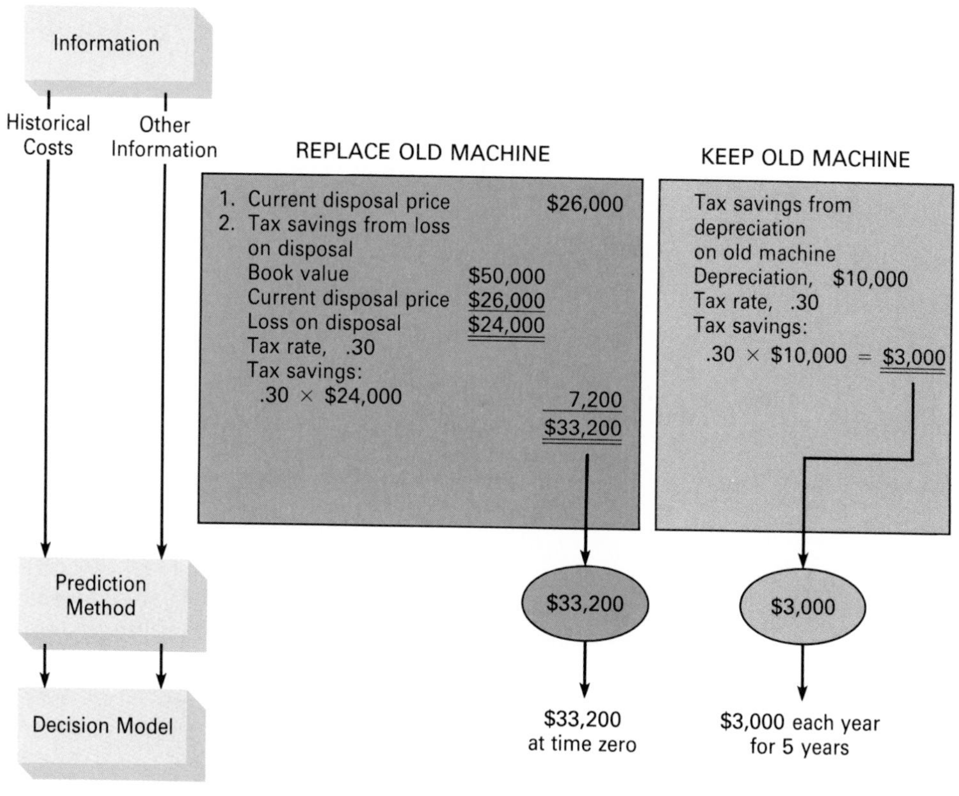

CAPITAL BUDGETING AND INFLATION

Inflation can be defined as the decline in the general purchasing power of the monetary unit. Many countries have inflation rates of 15% or more. Even an annual inflation rate of 5% over, say, a 5-year period can result in sizable losses in purchasing power over that period. We now examine how inflation can be explicitly recognized in capital-budgeting decisions.

Real and Nominal Interest Rates

When analyzing inflation, distinguish between the nominal rate of interest and the real rate of interest:

1. **Nominal rate of interest.** This rate is made up of two elements: (a) the real rate of interest and (b) an inflation element—the premium demanded because of the anticipated decline in the general purchasing power of the monetary unit.
2. **Real rate of interest.** This rate is made up of two elements: (a) a risk-free element—the "pure" rate of interest that is paid on long-term government bonds—and (b) a business-risk element—the "risk" premium above the pure rate that is demanded for undertaking risks.

Objective 4

Distinguish between the nominal rate of interest and the real rate of interest

Assume that the real rate of interest for a high-risk oil exploration project is 20% and an inflation rate of 10% is predicted. The nominal rate of interest is:[5]

$$\begin{aligned}\text{Nominal rate} &= (1 + \text{Real rate})\ (1 + \text{Inflation rate}) - 1\\ &= (1 + .20)\ (1 + .10) - 1\\ &= 1.32 - 1 = .32\end{aligned}$$

Let us now look at applying the nominal rate and the real rate to DCF methods.

Discounted Cash-Flow Methods and Inflation

Objective 5

Demonstrate the equivalence of the nominal approach and the real approach to incorporating inflation into capital budgeting

The watchwords when incorporating inflation into DCF analysis are *internal consistency*. There are two internally consistent approaches:

Nominal Approach:	Predict cash inflows and outflows in nominal monetary units *and* use a nominal discount rate
Real Approach:	Predict cash inflows and outflows in real monetary units *and* use a real discount rate

Many managers find the nominal approach easier to understand. This approach uses the same type of numbers that will be recorded in the accounting system—numbers that include the impact of inflation.

The most frequently encountered error when accounting for inflation in capital budgeting is keeping cash inflows and outflows in* real terms *and using a* nominal *discount rate. This approach is internally inconsistent. It creates a bias against the adoption of many worthwhile capital investment projects. Note that the discounted present value of cash inflows will be understated when this error is made.

Whether a manager uses the nominal rate or the real rate, his or her choice must be consistently applied. Consider a decision by Network Communications to invest in additional testing equipment for its cellular car-phone product line. The testing equipment, which costs $150,000 (payable immediately), will reduce the number of defective car-phones sold to customers and will therefore reduce customer-servicing costs. The equipment is estimated to have a four-year useful life with a zero terminal disposal price. An annual inflation rate of 20% is expected over this four-year period.

The predicted net cash savings (before payment of the $150,000 and any income tax payment) from reduced customer service costs are:

[5] An alternative approach to deriving the nominal rate of interest from the real rate of interest is:

Real rate of interest	.10
Inflation rate	.20
Combination (.20 × .10)	.02
Nominal rate of interest	.32

The real rate of interest can be derived from the nominal rate of interest as follows:

$$\begin{aligned}\text{Real rate} &= \frac{(1 + \text{Nominal rate})}{(1 + \text{Inflation rate})} - 1\\ &= \frac{(1 + .32)}{(1 + .20)} - 1 = .10\end{aligned}$$

Year	Real Dollars	Cumulative Inflation Rate Factor*	Nominal Dollars
(1)	(2)	(3)	(4) = (2) × (3)
1	$80,000	$(1.20)^1$	$ 96,000
2	80,000	$(1.20)^2$	115,200
3	80,000	$(1.20)^3$	138,240
4	60,000	$(1.20)^4$	124,416

*1.20 = 1.00 + .20 inflation rate.

The firm requires an after-tax real rate of return of 10% from this project. The nominal rate of interest for Network Comunications is:

$$\text{Nominal rate} = (1 + \text{Real rate})\ (1 + \text{Inflation rate}) - 1$$
$$= (1 + .10)\ (1 + .20) - 1 = .32$$

The corporate income tax rate is 40%. For tax purposes, the testing equipment will be depreciated using the double-declining balance (DDB) method.[6]

Exhibit 22-5 presents the capital-budgeting approach for predicting cash flows in nominal dollars and using a nominal discount rate.[7] Exhibit 22-6 presents the approach of predicting cash flows in real terms and using a real discount rate.[8] (The present value factors in Exhibits 22-5 and 22-6 have six-decimal digits to show the equivalence of the two approaches.) Both approaches indicate that the project has a net present value of $31,024. Using a net present value criterion, the testing equipment should be purchased.

An often overlooked adjustment to the real approach is necessary in countries (such as the United States) where tax rules restrict the amount allowed for depreciation to the asset's purchase price *in nominal dollars* at the time of the purchase. In these cases, the tax savings each year will be in nominal dollars and will have to be discounted for inflation before they are discounted in real terms. (See the lower portion of Exhibit 22-6.)

[6]Given a four-year useful life, the DDB factor is .5 (2 × .25). Depreciation each year of the machine will be:

Year	Beginning Book Value	DDB Factor	Annual Depreciation
1	$150,000	.5	$75,000
2	75,000	.5	37,500
3	37,500	.5	18,750
4	18,750	—	18.750

[7]The cumulative inflation rate in the example is calculated using six-decimal digits to eliminate doubt about the equivalence of the two approaches. In practice, the cumulative inflation factor (to three-decimal digits) can be obtained using Table 1 (Compound Amount of $1) of Appendix B (p. 934). The Problem for Self-Study at the end of this chapter uses Table 1.

[8]The inflation factors used in Exhibit 22-6 to compute the tax savings in real dollars, given an inflation rate of 20%, are:

Year	Inflation Factor Formula	Inflation Factor
1	$[1 \div (1.20)]$	.833333
2	$[1 \div (1.20)^2]$	.694444
3	$[1 \div (1.20)^3]$	.578704
4	$[1 \div (1.20)^4]$	.482253

EXHIBIT 22-5

Nominal Approach to Inflation: Predict Cash Inflows and Outflows in Nominal Dollars and Use a Nominal Discount Rate*

Cash Savings from Operations

Year	Operating Cash Inflows	Income Tax Outflows (40%)	After-Tax Net Operating Cash Inflows	Present Value Factor (32%)†	Total Present Value	Sketch of Relevant After-Tax Cash Flows (0 1 2 3 4)
0	—	—	—	—	—	
1	$ 96,000	$38,400	$57,600	.757576	$ 43,636	← $57,600 (year 1)
2	115,200	46,080	69,120	.573921	39,669	← $69,120 (year 2)
3	138,240	55,296	82,944	.434789	36,063	← $82,944 (year 3)
4	124,416	49,766	74,650	.329385	24,589	← $74,650 (year 4)
					$ 143,957	

Income Tax Cash Savings from Depreciation Deduction

Year	Depreciation	Tax Savings (40%)		Present Value Factor (32%)†	Total Present Value	Sketch of Relevant After-Tax Cash Flows
0	—	—		—	—	
1	$75,000	$30,000		.757576	$ 22,727	← $30,000 (year 1)
2	37,500	15,000		.573921	8,609	← $15,000 (year 2)
3	18,750	7,500		.434789	3,261	← $7,500 (year 3)
4	18,750	7,500		.329385	2,470	← $7,500 (year 4)
					$ 37,067	
Investment in equipment in year 0					$(150,000)	$(150,000)
Net present value					$ 31,024	

*The nominal discount rate of 32% is made up of the real rate of interest of 10% and the inflation rate of 20%: (1 + .10)(1 + .20) − 1 = .32.
†Present value factors shown to six decimal digits to emphasize that the Exhibit 22-5 and Exhibit 22-6 approaches to inflation are equivalent.

EXHIBIT 22-6

Real Approach to Inflation: Predict Cash Inflows and Outflows in Real Dollars and Use a Real Discount Rate

Cash Savings from Operations

Year	Operating Cash Inflows	Income Tax Outflows (40%)	After-Tax Net Operating Cash Inflows	Present Value Factor (10%)*	Total Present Value	Sketch of Relevant After-Tax Cash Flows: 0	1	2	3	4
0	—	—	—	—						
1	$80,000	$32,000	$48,000	.909091	$ 43,636	←	$48,000			
2	80,000	32,000	48,000	.826446	39,669	←		$48,000		
3	80,000	32,000	48,000	.751315	36,063	←			$48,000	
4	60,000	24,000	36,000	.683031	24,589	←				$36,000
					$ 143,957					

Income Tax Cash Savings from Depreciation Deduction

Year	Depreciation	Tax Savings (40%) in Nominal Dollars	Inflation Factor 20%†	Tax Savings in Real Dollars	Present Value Factor (10%)*	Total Present Value	Sketch of Relevant After-Tax Cash Flows: 0	1	2	3	4
0	—	—	—	—	—						
1	$75,000	$30,000	.833333	$25,000	.909091	$ 22,727	←	$25,000			
2	37,500	15,000	.694444	10,417	.826446	8,609	←		$10,417		
3	18,750	7,500	.578704	4,340	.751315	3,261	←			$4,340	
4	18,750	7,500	.482253	3,617	.683031	2,470	←				$3,617
						$ 37,067					
Investment in equipment in year 0						$(150,000)	$(150,000)				
Net present value						$ 31,024					

*Present value factors shown to six decimal digits to emphasize that the Exhibit 22-5 and Exhibit 22-6 approaches to inflation are equivalent.

†The formula on Table 2 (p. 935) of Appendix B is used to compute the inflation factor.

REQUIRED RATE OF RETURN

Objective 6

Describe alternative approaches used to recognize the degree of risk in projects in capital budgeting

The *required rate of return (RRR)* is a critical variable in discounted cash-flow analysis. It is the minimum desired rate of return on an investment. Alternative names for RRR include the *discount rate*, the *hurdle rate*, and the *cost of capital*. Choosing the RRR for each project is complex and is discussed in finance texts. A safe generalization is: The higher the risk, the higher the required rate of return.

When estimating the RRR, be consistent and use the approach applied in estimating cash inflows and outflows. The options include various combinations of (a) the real rate and the nominal rate and (b) the pretax rate and the after-tax rate. The numerical magnitude of differences among these rates can be sizable, given estimates of inflation that may exceed 20% and corporate tax rates of 30% or more. As noted previously in this chapter, the key words are *internal consistency* in the analysis.

Dealing with Risk

Organizations typically use at least one of the following approaches to recognize the risks associated with projects:

1. *Varying the Payback Time.* Firms that use payback as a project-selection criterion can vary the required payback to reflect differences in project risk. The higher the risk, the shorter the required payback time.
2. *Adjusting the Required Rate of Return.* The higher the risk, the higher the required rate of return. Estimating a precise risk factor for each project is difficult. Some organizations simplify the task by having three or four general-risk categories (for example, very high, high, average, and low). Each potential project is assigned to a specific category. Then a discount rate, which is preassigned by management to that category, is used as the required rate of return.
3. *Adjusting the Estimated Future Cash Inflows.* The estimated future cash inflows of projects viewed as riskier are systematically reduced. One company has a policy of systematically reducing the predicted cash inflows of very high risk projects by 30%, high-risk projects by 20%, and average-risk projects by 10%. It makes no change to the projected cash inflows of low-risk projects.
4. *Sensitivity ("What-If?") Analysis.* This approach involves examining the consequences of changing key assumptions underlying a capital-budgeting project. For example, a copper-mining company might examine what changes would occur if the world price of copper were to increase(decrease) by 10%, 20%, and 30%. How would these changes affect the economic attractiveness of an investment in a new copper mine?
5. *Estimating the Probability Distribution of Future Cash Inflows and Outflows for Each Project.* Chapter 20 discusses this approach to uncertainty. Although the task of determining these distributions is imposing, the results can help managers focus on important issues. For example, suppose a project has a 60% likelihood of very high cash inflows in its early years and a 40% likelihood of minimal cash inflows in its early years. This 40% probability may prompt managers to set up lines of credit with a bank. These lines of credit would enable the company to avoid a short-run cash-flow crisis if the negative outcomes possible with the project occur.

APPLICABILITY TO NONPROFIT ORGANIZATIONS

Discounted cash-flow analysis applies to both profit-seeking and nonprofit organizations. Almost all organizations must decide which fixed assets will accomplish various tasks at the least cost. Moreover, all organizations, including governments, have to bear the cost of money. The required rate of return used in capital budgeting by U.S. federal agencies is 7% for water projects (dams, irrigation, and so on) and 10% for all other projects.

Studies of the capital-budgeting practices of government agencies at various levels (federal, state, and local) and in several countries report similar findings:[9]

1. Urgency is an important factor when allocating funds. For example, capital budgeting for roads often starts with a list of the physical deficiencies in an existing highway rather than with a systematic analysis of whether it would be preferable to build an alternative highway.
2. Systematic biases are often found in project estimates. For example, studies of irrigation projects by the U.S. Bureau of Reclamation report overestimates of the benefits, underestimates of the costs, and underestimates of the time taken to construct dams and other irrigation infrastructure.
3. A tendency exists to cut capital budget projects first when there is a strong push to balance a budget or reduce a deficit.

The General Accounting Office (GAO), in a study of capital-budgeting practices by U.S. federal agencies, gave "low points" to many agencies but "high points" to the U.S. Postal Service (USPS).

> USPS uses return on investment, internal rate of return, and discounted cash flow. The Postal Service also performs a sensitivity analysis. Only the Postal Service conducts follow-up studies to find out if a completed project accomplished its objectives. USPS uses its postaudits to determine the continuing applicability of previous conclusions, to highlight any continuing undesirable trends that warrant management action, to project cost changes through the life of the project, and to compare the results of the project's original economic evaluation.[10]

ADMINISTRATION OF CAPITAL BUDGETS

In profit-seeking organizations, the most important feature of effective capital-budgeting administration is the awareness on the part of all managers that long-run expenditures should be generators of long-run profits. This awareness engenders a constant search for new methods, processes, and products.

Approval of overall capital budgets is usually the responsibility of the board of directors. In many organizations, managers make capital budget requests semiannually or annually. Some of these requests are weeded out very quickly. A subset moves upward through the managerial levels, and successive reviews are made. Some requests finally reach a committee that examines them and makes recommendations to the board of directors. Many organizations have formalized the administration of their capital budgets through the use of a standard set of forms and a specified routine for processing requests.

Two related surveys asked U.S. and Japanese managers about the importance of possible "changes that would improve your business unit's capital-budgeting decisions." The six changes that managers believe would provide the greatest improvement (with their rank in terms of importance) are:[11]

[9]See, for example, two studies: Report to the Congress of the United States by the Comptroller General, *Federal Capital Budgeting: A Collection of Haphazard Practices* (Washington, DC: General Accounting Office, 1981); and Report to the Committee on Environment and Public Works, United States Senate, *Effective Planning and Budget Practices Can Help Arrest the Nation's Deteriorating Public Infrastructure* (Washington, DC: General Accounting Office, 1982).

[10]*Federal Capital Budgeting: A Collection of Haphazard Practices*, p. 67.

[11]See R. Howell, J. Brown, S. Soucy, and A. Seed, *Management Accounting in the New Manufacturing Environment* (Montvale, NJ: National Association of Accountants and CAM-I, 1987); and NAA Tokyo Affiliate, "Management Accounting in the Advanced Manufacturing Surrounding: Comparative Study on Survey in Japan and U.S.A.," 1988.

Change Providing Greatest Improvement	Ranking by U.S. Managers	Ranking by Japanese Managers
Having more accurate forecasts	1	2
Monitoring actual costs and benefits after expenditures are completed*	2	1
Quantifying the impact of not making an investment (for example, considering the financial impact of any degradation of manufacturing capabilities relative to competitors)	3	3
Strengthening long-term orientation	4	6
Emphasizing the rationale of the strategy that the project being analyzed is a component part	5	4
Making a multiyear commitment to the investment plan	6	5

*Most firms currently monitor actual project-specific costs or revenues for only a small percentage of the capital-budgeting projects they implement.

IMPLEMENTING THE NET PRESENT VALUE DECISION RULE

This section discusses problems in using the net present value method when there is a restriction on the total funds available for capital spending. Such constraints may appear irrational in profit-seeking enterprises; after all, they can lead to the rejection of projects with positive net present values. Nevertheless, executives must frequently work within an overall capital-budgeting limit. In nonprofit enterprises, restrictions on the total funds available for capital spending are the norm. For example, an annual government budget typically will provide an upper limit on the funds available to each of the individual government departments.

The **excess present value index** (also sometimes called the *profitability index*) is the total present value of future net cash inflows of a project divided by the total cash outflow. Consider this index for two software graphics packages—Superdraw and Masterdraw—that Business Systems is evaluating. The writers of each package require that Business Systems market only one software graphics package, so accepting one software package automatically means rejecting the other—that is, the packages are mutually exclusive. Summary financial data from the capital budget proposal of each package are:

Project	Present Value at 10% RRR	Cost	Excess Present Value Index	Net Present Value
(1)	(2)	(3)	(2) ÷ (3)	(2) − (3)
Superdraw	$1,400,000	$1,000,000	140%	$400,000
Masterdraw	3,900,000	3,000,000	130%	900,000

Superdraw has an excess present value index of 140% compared with 130% for Masterdraw. (Projects with excess present value indexes of less than 100% are negative net present value projects.) If all other things, such as risk and alternative use of funds, are equal, Superdraw is the preferred project. But "all other things" are rarely equal.

Assume that Business Systems has a total capital budget limit of $5,000,000 for the coming year. It is considering investing in Superdraw or Masterdraw and in any one or more of eight other projects (coded B, C, . . . , H, I). Exhibit 22-7

EXHIBIT 22-7

Allocation of $5,000,000 Capital Budget of Business Systems: Comparison of Two Alternatives

Alternative 1				Alternative 2			
Project	Investment Required	Excess Present Value Index	Total Present Value at 10%	Project	Investment Required	Excess Present Value Index	Total Present Value at 10%
C	$ 600,000	167%	$1,002,000	C	$ 600,000	167%	$1,002,000
Superdraw	1,000,000	140%	1,400,000				
D	400,000	132%	528,000	D	400,000	132%	528,000
				Masterdraw	3,000,000	130%	3,900,000
F	1,000,000	115%	1,150,000	F	1,000,000	115%	1,150,000
					$5,000,000*		$6,580,000‡
E	800,000	114%	912,000	E	$ 800,000	114%	Reject
B	1,200,000	112%	1,344,000	B	1,200,000	112%	Reject
	$5,000,000*		$6,336,000†				
H	$ 550,000	105%	Reject	H	550,000	105%	Reject
G	450,000	101%	Reject	G	450,000	101%	Reject
I	1,000,000	90%	Reject	I	1,000,000	90%	Reject

*Total budget constraint.
†Net present value, $1,336,000.
‡Net present value, $1,580,000.

presents two alternative combinations of these projects. Note that the rationing used in Alternative 2 is superior to that in Alternative 1, despite the greater profitability per dollar obtained by investing in Superdraw compared with Masterdraw. Why? Because the $2,000,000 incremental investment in Masterdraw has an incremental net present value of $500,000. The $2,000,000 would otherwise be invested in Projects E and B, which have a lower combined incremental net present value of $256,000:

	Present Value	Cost	Increase in Net Present Value
Masterdraw	$3,900,000	$3,000,000	
Superdraw	1,400,000	1,000,000	
Increment	$2,500,000	$2,000,000	$500,000
Project E	$ 912,000	$ 800,000	
Project B	1,344,000	1,200,000	
Total	$2,256,000	$2,000,000	$256,000

This example illustrates that managers cannot base decisions involving mutually exclusive investments of different sizes on the excess present value index. The net present value method is the best general guide.

IMPLEMENTING THE INTERNAL RATE-OF-RETURN DECISION RULE

The net present value method will always indicate the project (or set of projects) that will maximize the net present value of future cash flows. However, surveys of practice report widespread use of the internal rate-of-return method. Why? Proba-

EXHIBIT 22-8
Ranking of Projects Using Internal Rate of Return and Net Present Value

Project	Life	Initial Investment Outlay	Annual Net After-Tax Cash Flows	Ranking by Internal Rate of Return		Ranking by Net Present Value		
				IRR	Ranking	PV at 10% RRR	NPV	Ranking
X	5	$286,400	$100,000	22%	1	$379,100	$ 92,700	3
Y	10	419,200	100,000	20	2	614,500	195,300	2
Z	15	509,200	100,000	18	3	760,600	251,400	1

Objective 7

Explain why the internal rate of return and the net present value decision rules may rank projects differently

bly because managers find that method easier to understand and because, in most instances, their decisions would be unaffected by using one method or the other. In some cases, however, the two methods will not indicate the same decision.

Where mutually exclusive projects have unequal lives or unequal investment outlays, the internal rate-of-return method can rank projects differently from the net present value method. Consider Exhibit 22-8.[12] The ranking by the internal rate-of-return method favors Project X, while the net present value method favors Project Z. The projects ranked in Exhibit 22-8 differ in both life (5, 10, and 15 years) and initial investment outlay ($286,400, $419,200, and $509,200).

Managers using the internal rate-of-return method implicitly assume that the reinvestment rate is equal to the indicated rate of return for the shortest-lived project. Managers using the net present value method implicitly assume that the funds obtainable from competing projects can be reinvested only at the company's required rate of return.

Corporate finance textbooks cover, in great detail, the problems of ranking projects with unequal lives or unequal investment outlays. Ideally, there should be a common terminal date for all projects with explicit assumptions as to the appropriate reinvestment rates of funds. The practical difficulties of predicting future profitability on reinvestment are greater than those of predicting profitability of immediate projects. But reinvestment opportunities should be considered when they can be foreseen and measured.

[12]Exhibit 22-8 concentrates on differences in project lives. Similar conflicting results can occur when the terminal dates are the same, but the sizes of the investment outlays differ.

PROBLEM FOR SELF-STUDY

(This is a comprehensive review problem. It illustrates both income tax factors and capital budgeting with inflation.)

PROBLEM

Stone Aggregates operates ninety-two plants across the country. Each plant produces crushed stone aggregate used in many construction projects. Transportation is a major cost item. A scale clerk prepares a delivery ticket (called a dray ticket) using a customer master file. The clerk weighs the products and records details of the product shipped, its weight, its freight charges, and whether or not it is taxed.

Stone Aggregates is considering a proposal to use a computerized ticket-writing system for all of its ninety-two plants. One plant has been a pilot site for the past twelve months, generating cash-operating cost savings of $300,000. These savings arose mainly from a reduction in operating costs at the plant and from a reduction in overshipments (amounts shipped in excess of amounts ordered) to customers. The cost analyst estimates that if the computerized ticket system had been operating at all of the company's plants for the past year, cost savings would have been $25 million (expressed in today's dollars). This cost-savings estimate takes into account the estimated $5 million cash-operating cost that would have been incurred in operating the computerized ticket-writing system at all of the company's plants in the past year.

The cost of the equipment for all ninety-two plants is $45 million, payable immediately. This equipment has an expected useful life of four years and a terminal disposal price of $10 million (expressed in today's dollars). Income tax rules applying to Stone Aggregates are:

- *Amount allowable for depreciation.* Original cost is the basis for depreciation computations. No account is taken of predicted terminal disposal price when computing depreciation. The disposal price in nominal dollars of assets sold will be taxed at the ordinary income tax rate in the year the disposal is made.
- *Time period over which the asset is to be depreciated.* Under a tax law designed to encourage investment, Stone Aggregates can use a three-year write-off period for the asset.
- *Pattern of allowable depreciation.* Straight-line depreciation is required. Given an original cost of $45 million and a three-year write-off period, annual depreciation is $15 million.

Stone Aggregates has a 30% income tax rate in each of the next four years.

Required

1. Does the automated delivery ticket-writing proposal meet Stone Aggregates' 16% after-tax required rate-of-return criterion? This 16% required rate of return includes an 8% inflation prediction. (The real rate of interest is 7.4%.) This 8% inflation prediction applies to both the cost savings and the terminal disposal price of the equipment. Prepare a net present value analysis using nominal dollars and a nominal required rate of return.
2. What factors would you recommend that Stone Aggregates consider in more detail when evaluating the automated delivery ticket-writing proposal?

SOLUTION

1. Exhibit 22-9 shows the net present value analysis. To illustrate an alternative presentation found in practice, the format of Exhibit 22-9 differs from that of Exhibit 22-2. The proposal for an automated delivery ticket-writing system has a net present value of $29.086 million, indicating that—based on quantitative/financial factors—it is an attractive investment. Note especially how the tax law enables Stone Aggregates to fully depreciate the equipment by the end of the third year. No depreciation occurs in year 4.
2. The analysis in Exhibit 22-9 assumes that the cost of operating the system is $5 million each year. However, costs in the year of changeover to the computerized system would probably be much higher. Many companies find that actual implementation and operating costs in the changeover year exceed by a factor of 200% or more the operating costs in subsequent years.

We wrote this problem after reading an article describing Vulcan Materials' adoption of a computerized ticket-writing system.[13] The benefits Vulcan reported included the following:

- "Hauler productivity increased because the new system improved the movement of product across the scales."
- "Scale clerks benefited through job enrichment and a much more predictable work load."

[13]J. Bush and R. Stewart, "Vulcan Materials Automates Delivery Ticket Writing," *Management Accounting*, August 1985, pp. 52–55.

EXHIBIT 22-9

Net Present Value Analysis of Computerized Ticket-Writing System for Stone Aggregates (n.d. = nominal dollars in millions)

	Total Present Value	End of Year 1	End of Year 2	End of Year 3	End of Year 4	Sketch of Relevant After-Tax Cash Flows: 0	1	2	3	4
Operating Savings										
1. Cash-operating savings (real dollars)	—	$25.000	$25.000	$25.000	$25.000					
2. Cumulative inflation factor (from Table 1 on p. 934 for 8%)	—	1.080	1.166	1.260	1.360					
3. Cash-operating savings (n.d.)	—	27.000	29.150	31.500	34.000	←	$27.000	$29.150	$31.500	$34.000
4. Tax payments (30%)	—	8.100	8.745	9.450	10.200					
5. After-tax cash-operating savings (n.d.)	—	18.900	20.405	22.050	23.800					
6. Present value factor (16%)	—	.862	.743	.641	.552					
7. PV of after-tax cash-operating savings (n.d.): 5 × 6	$58.725	16.292	15.161	14.134	13.138					
Depreciation and Disposal Proceeds										
8. Depreciation	—	15.000	15.000	15.000	—					
9. Income tax savings (30%)	—	4.500	4.500	4.500	—	←	$4.500	$4.500	$4.500	
10. After-tax terminal disposal proceeds of equipment*	—	—	—	—	9.520	←				$9.520
11. Total income tax savings + disposal proceeds	—	4.500	4.500	4.500	9.520					
12. Present value factor (16%)	—	.862	.743	.641	.552					
13. PV of income tax savings + disposal proceeds: 11 × 12	15.361	3.879	3.343	2.884	5.255					
PV of total cash inflows: 7 + 13	$74.086	20.171	18.504	17.018	18.393					
PV of cash outflows	(45.000)					$(45.000)				
Net Present Value	$29.086									

*Terminal disposal price of $10 million in year 0 dollars will be $13.600 million ($10 million × 1.360) in year 4 nominal dollars. At a 30% tax rate, the after-tax terminal disposal proceeds in nominal dollars are $9.520 million ($13.600 million × 30%)

- "Communication between the scale-house microcomputers and the division office computers has afforded Vulcan several benefits. The first was the opportunity to eliminate costs associated with transporting the punch cards required under the old system. The second was the elimination of transferring transaction data from punch cards to computer tape. Vulcan also has benefited markedly from the reduction of errors contained in the transaction data and reduced costs associated with making corrections. In addition to improving accuracy, the system has accelerated the issuance of Vulcan's invoices."
- "Overshipments are eliminated because the up-to-date status of a customer's sales order is available at the plant."
- "Cash on delivery customers are identified by the system to prevent unintentional credit sales."
- "Faster flow of data has narrowed the time lag in detecting and correcting problems."

Stone Aggregates should learn from the experience of Vulcan Materials and include analysis of the preceding benefits that will possibly occur with the purchase of a computerized ticket-writing system.

SUMMARY

Income tax factors almost always play an important role in decision making. Income taxes make a major impact on the amounts of cash inflow and outflow and on their timing. Tax rules in many countries are intricate and change often.

We can account for inflation in several ways in capital budgeting. One correct approach is to predict cash inflows and outflows in nominal terms and use a nominal discount rate. A second correct approach is to predict cash inflows and outflows in real terms and use a real discount rate. Both approaches are internally consistent.

The required rate of return should vary according to the riskiness of projects. The higher the risk, the higher the required return.

Use caution when analyzing rankings based on percentages (such as the internal rate of return) to select the total set of investment projects. As a general rule, the net present value method is preferred when selecting investment projects.

APPENDIX: MODIFIED ACCELERATED COST RECOVERY SYSTEM

The tax rules governing depreciation in the United States are collectively called the Modified Accelerated Cost Recovery System (MACRS). Changes in this set of rules are made periodically, as they are in most tax regulations. MACRS is a modification of tax laws first introduced in 1981 termed the Accelerated Cost Recovery System (ACRS). For most fixed assets placed in service in the 1981-86 period, the ACRS system applies. Assets acquired since 1987 use MACRS. Both ACRS and MACRS have more accelerated depreciation schedules than existed with the prior taxation rules. Some highlights of the current version of MACRS follow.

Amount Allowable for Depreciation. Generally, the amount allowable for depreciation is the original cost of the asset. The MACRS uses the phrase "capital recovery" to describe the amount allowable each year as a "depreciation" deduction. Estimates of future disposal prices are ignored under MACRS. Any disposal price of the asset is taxed at the same rate as ordinary income at the time of the sale.

Time Period over Which the Asset Is to Be Depreciated. The time period is specified in a "table of allowable lives" (termed *recovery period*). Eight different recovery periods are possible: 3, 5, 7, 10, 15, 20, 27.5, and 31.5 years. See Exhibit 22-10. These recovery periods do not necessarily reflect the estimated useful life of the assets included in each category.

Pattern of Allowable Depreciation (for a Given Time Period). The depreciation method to be used is a function of the recovery period. The eight different recovery periods and their depreciation method are:

Recovery Period	Depreciation Method
3, 5, 7, 10 years	Double-declining balance (also called 200% declining balance)
15, 20 years	150% declining balance
27.5, 31.5 years	Straight-line

MACRS offers firms the option of using straight-line depreciation (termed the Alternative Depreciation System, or ADS) within any recovery-period class in Exhibit 22-10. This option is attractive for firms expected to suffer tax losses in the early years of an asset's life. Many newly established companies are in this situation. The benefit of the straight-line option is that it defers more of the depreciation tax deductions to years in which the company expects to have taxable income.

Other Tax Considerations

Many items not discussed in this chapter, including investment tax credits, loss carrybacks and carryforwards, state income taxes, short- and long-term capital gains, and distinctions between how capital assets and other assets affect income taxes. Because income tax planning is exceedingly complex, professional tax counsel should be sought.

EXHIBIT 22-10

Accelerated Cost Recovery System: Recovery Periods, Pattern of Allowable Depreciation and Examples

Recovery Period	Pattern of Allowable Depreciation	Examples of Assets in Recovery Period Class
3 years	200% declining balance	Tools for manufacture of selected products Handling equipment for food manufacture
5 years	200% declining balance	Automobiles and trucks Computers, copiers, and typewriters
7 years	200% declining balance	Office furniture, fixtures, and equipment Railroad track
10 years	200% declining balance	Assets used in petroleum refining Assets used in manufacture of tobacco and food products
15 years	150% declining balance	Telephone distribution plants Municipal sewage-treatment plants
20 years	150% declining balance	Multipurpose farm buildings
27.5 years	Straight-line	Residential rental properties
31.5 years	Straight-line	Nonresidential office buildings and warehouses

TERMS TO LEARN

This chapter and the Glossary at the end of the book contain definitions of the following important terms:

accelerated depreciation *(p. 707)*
double-declining balance (DDB) depreciation *(707)*
excess present value index *(722)* incremental approach *(710)* inflation *(715)*
investment tax credit (ITC) *(706)*
modified accelerated cost recovery system (MACRS) *(711)*
nominal rate of interest *(715)* real rate of interest *(715)*
straight-line depreciation *(707)* total project approach *(710)*

ASSIGNMENT MATERIAL

QUESTIONS

22-1 What are the two major impacts that income taxes have on capital-budgeting decisions?

22-2 Describe three factors that influence the amount claimed as a depreciation deduction for tax purposes.

22-3 "It doesn't matter what depreciation method is used. The total dollar tax bills are the same." Do you agree? Why?

22-4 "Accelerated depreciation provides higher cash flows in the early years than does straight-line depreciation." Do you agree? Why?

22-5 Distinguish between the *total* project approach and the *incremental* approach to choosing between two capital-budgeting projects.

22-6 "Depreciation is an irrelevant factor in deciding whether to replace an existing delivery vehicle with a more energy-efficient vehicle." Do you agree?

22-7 What are the main depreciation methods permitted under the Modified Accelerated Cost Recovery System (MACRS)?

22-8 Distinguish between the *nominal* rate of interest and the *real* rate of interest.

22-9 What are the two internally consistent approaches to incorporating inflation into DCF analysis?

22-10 What approaches might be used to recognize risk in capital budgeting?

22-11 "In practice there is no single rate that a given company can use as a guide for sifting among all projects." Why? Explain.

22-12 What three changes do U.S. and Japanese managers think would most improve capital-budgeting decisions?

22-13 How do the reinvestment assumptions implicit in the use of the internal rate-of-return method and the net present value method differ when comparing projects with lives of different lengths?

EXERCISES AND PROBLEMS

22-14 Recapitulation of role of depreciation in Chapters 11, 21, and 22. Antonio Inoki, president of Yokohoma Steel, remarked, "I've read three chapters that have included discussions of depreciation in relation to decisions regarding the replacement of equipment. I'm confused. Chapter 11 said that depreciation on old equipment is irrelevant but that depreciation on new equipment is relevant. Chapter 21 said that depreciation was irrelevant in relation to discounted cash-flow models, but Chapter 22 indicated that depreciation was indeed relevant"

Required
Prepare a clear explanation for the president that would minimize his confusion.

22-15 New equipment purchase, taxation, straight-line and DDB depreciation. Presentation Graphics prepares slides and other aids for individuals making group presentations. It estimates it can save $35,000 a year in cash-operating costs for the next five years if it buys a special-purpose color-slide workstation at a cost of $75,000. The workstation will have a zero terminal disposal price at the end of year 5. Presentation Graphics has a 12% after-tax required rate of return. Its income tax rate is 40% each year for the next five years.

Required

1. Assume straight-line depreciation. Compute (a) net present value, (b) payback period, and (c) internal rate of return.
2. Assume double-declining balance with depreciation for the fifth year being the book value at the start of the fifth year. Compute (a) net present value, (b) payback period, and (c) internal rate of return.

22-16 Multiple choice, including straight-line depreciation. (CPA, adapted) Choose the best answer for each question and show computations. The Apex Company is evaluating a capital-budgeting proposal for the current year. The relevant data follow:

Year	Present Value of an Annuity of $1 in Arrears at 15%
1	$ .870
2	1.626
3	2.284
4	2.856
5	3.353
6	3.785

The initial investment would be $30,000. It would be depreciated on a straight-line basis over six years with a zero terminal disposal price. The before-tax annual cash inflow arising from this investment is $10,000. The income tax rate is 40%, and income tax is paid the same year as incurred. The after-tax required rate of return is 15%. All cash flows occur at year-end.

1. What is the after-tax accrual accounting rate of return on Apex's capital-budgeting proposal? (a) 10%, (b) 16⅔%, (c) 26⅔%, (d) 33⅓%.
2. What is the after-tax payback time (in years) for Apex's capital-budgeting proposal? (a) 5, (b) 3.75, (c) 3, (d) 2.
3. What is the net present value of Apex's capital-budgeting proposal? (a) $(7,290), (b) $280, (c) $7,850, (d) $11,760.
4. How much would Apex have had to invest five years ago at 15% compounded annually to have $30,000 now? (a) $12,960, (b) $14,910, (c) $17,160, (d) cannot be determined from the information given.

22-17 Capital budgeting, income taxes, DDB depreciation, ATM network. Rural Bank is considering whether to invest in an automatic teller machine (ATM) network. Its main competitor (Security Bank) has started installing ATMs at several of its branches. Rick Sterling, a retail banking financial analyst at Rural Bank, estimates the following:

- Initial investment for equipment and software = $30 million. The total $30 million amount would be invested immediately and would be depreciated over a five-year period using the double-declining balance (DDB) method.
- Annual costs of operating the ATMs (for example, equipment and software maintenance) = $3 million.
- Annual reduction in labor and related costs from using the ATMs as compared with the current labor-intensive teller system = $10 million.

Sterling estimates that the useful life of the ATMs is six years. The terminal disposal price of the ATMs at the end of six years is $3 million. For tax purposes, the terminal disposal price is included as an income item in the year of disposal.

Rural Bank has a 14% after-tax required rate of return for retail-banking investments. Rural Bank has a 30% income tax rate, which is expected to remain constant over the foreseeable investment horizon.

Required

1. Sketch the after-tax cash inflows and outflows from Rural Bank's investing in the ATMs. Show all your calculations.
2. Compute the payback period and the net present value of the ATM investment.
3. What questions should Rural Bank consider in deciding whether to invest in an ATM network?

22-18 Comprehensive equipment-replacement decision, income taxes, straight-line depreciation. A manufacturer of automobile parts acquired a special-purpose shaping machine for automatically producing a particular part. The machine has been used for one year. It will have no useful economic life after three more years. The machine is being depreciated on a straight-line basis for income tax purposes. It cost $88,000, has a current disposal price of $29,000, and has a terminal disposal price of $6,000. However, a terminal disposal price of zero was assumed in computing straight-line depreciation for tax purposes.

A new machine has become available and is far more efficient than the present machine. It would cost $63,000, would cut annual cash-operating costs from $60,000 to $40,000, and would have zero terminal disposal price at the end of its useful life of three years. Straight-line depreciation would be used for tax purposes. The applicable income tax is 30%. The after-tax required rate of return is 14%.

Required

Using the net present value method, show whether the new machine should be purchased (a) under a total-project approach and (b) under an incremental approach.

22-19 Automated materials-handling capital project, income taxes, double-declining balance depreciation. Ontime Distributors operates a large distribution network for health-related products. It is considering an automated materials-handling (AMH) proposal for its major warehouse. The before-tax cash-operating savings from the automation are estimated to be $2.5 million a year. These savings arise from reduced storage space requirements, increased labor productivity, less product damage, and higher inventory record-keeping accuracy. This $2.5 million annual savings is calculated as $3.5 million gross cash-operating savings minus the $1.0 million costs of operating and maintaining the AMH equipment. The AMH equipment will cost $6 million, payable immediately. The equipment has a useful life of four years and zero terminal disposal price. The lease on the warehouse expires in four years and is not expected to be renewed. The company has an income tax rate of 40% and after-tax required rate of return of 12%. Under existing tax laws, the $6 million equipment cost qualifies for use of the double-declining balance depreciation method with a four-year useful life. The terminal disposal price of the equipment is included as a taxable income item in the year of its disposal.

Required

1. Compute (a) the net present value and (b) the payback period on the automated materials-handling project.
2. Assume the AMH equipment has a $1 million terminal disposal price at the end of the fourth year instead of a zero terminal disposal price. How will this $1 million terminal disposal price affect the net present value?
3. What other factors should Ontime Distributors consider in its decision?

22-20 Replacement of newspaper printing equipment, income taxes, straight-line depreciation. It is January 1, 19_5, and Anna Murdoch has gained control of the *Morning News* daily newspaper. Her first task is to consider replacing the existing printing machines. The machines were bought secondhand four years ago at a total cost of $10 million. Their book value now (January 1, 19_5) is $2.1 million. If the old machines are not replaced, this $2.1 million will be fully written off as depreciation in the current year (which ends December 31, 19_5). The disposal price on January 1, 19_5, is $1 million. The old machines could be used for another seven years before they are ready to be scrapped, with a zero terminal disposal price. Annual cash-operating costs (before depreciation and before taxes) with the old machines will be $40 million.

Murdoch has already obtained a bid from a manufacturer of printing machines. This manufacturer can immediately install state-of-the-art printing machines for $20 million, payable on January 1, 19_5. The Printers' Union at the *Morning News* strongly opposes the proposed investment. Union members will agree to some reduction in the number of employees at the newspaper, but not to the degree management requests. This partial union cooperation means that annual cash-operating costs (before depreciation and before taxes) with the new machines will be $36 million. The new machines qualify for a five-year recovery period under the existing tax regulations. The straight-line depreciation method is to be used. Any proceeds from selling the machines are classified as taxable income in the year of sale (and ignored in computing annual depreciation deductions). The machines have a useful life of seven years and a terminal disposal price of $3 million. Murdoch uses a 10% required after-tax rate of return in her analysis. The current income tax rate is 35%.

Required

1. Using the net present value method, should Murdoch replace the existing printing machines with the new state-of-the-art printing machines?
2. The head of the Printers' Union has had second thoughts about opposing Murdoch. He agrees to further labor reductions at the *Morning News* that will reduce cash-operating costs (before depreciation and before taxes) with the new machines to $30 million a year. What is the net present value of the new machines to the *Morning News* with this revised union proposal?
3. What other factors should Murdoch consider in deciding whether to replace the existing printing machines?

22-21 Replacement of a machine, income taxes, straight-line depreciation. (CMA, adapted) WRL Company operates a snack-food center at the Hartsfield Airport. On January 2, 19_1, WRL purchased a special cookie-cutting machine, which has been used for three years. WRL is considering purchasing a newer, more efficient machine. If purchased, the new machine would be acquired on January 2, 19_4. WRL expects to sell 300,000 cookies in each of the next four years. The selling price of each cookie is expected to average $0.50.

WRL has two options: (1) continue to operate the old machine or (2) sell the old machine and purchase the new machine. The seller of the new machine offered no trade-in. The following information has been assembled to help management decide which option is more desirable.

	Old Machine	New Machine
Original cost of machine at acquisition	$80,000	$120,000
Terminal disposal price at the end of useful life assumed for depreciation purposes	$10,000	$ 20,000
Useful life from date of acquisition	7 years	4 years
Expected annual cash-operating costs:		
Variable cost per cookie	$0.20	$0.14
Total fixed costs	$15,000	$ 14,000
Depreciation method used for tax purposes	Straight-line	Straight-line
Estimated disposal prices of machines:		
January 2, 19_4	$40,000	$120,000
December 31, 19_7	$ 7,000	$ 20,000

WRL has a 40% income tax rate. Assume that all operating revenues and costs occur at the end of the year. Assume that any gain or loss on the sale of machinery is treated as an ordinary tax item and will affect the taxes paid by WRL at the end of the year in which it occurs. WRL has an after-tax required rate of return of 16%.

Required

1. Use the net present value method to determine whether WRL should retain the old machine or acquire the new machine.
2. Assume that the quantitative differences between the net present values of the two options are so slight that WRL is indifferent between the two proposals. Identify and discuss the nonquantitative factors that WRL should consider.

22-22 Oil and gas project evaluation, depreciation, income taxes, double-declining balance depreciation. Deborah Regent, project analyst for Richfield Oil, is considering a capital-budgeting project to enhance recovery from one of Richfield's existing oil fields. The oil field is in Prudhoe Bay, Alaska. Enhanced recovery will require the construction of a special drilling rig that will recover oil that is deeper in the ground than the current drilling rig can reach. The new drilling rig will be assembled in Anchorage and floated to Prudhoe Bay. The drilling rig project will cost $50 million, payable immediately. The new rig will enable an extra one million barrels of oil to be extracted each year for each of the next seven years. Regent estimates the following operating data:

Revenue per barrel	$21
Variable cost per barrel	$ 5
Fixed operating costs per year for the new drilling rig (excluding depreciation)	$ 1 million

The drilling rig has an expected useful life of seven years with zero terminal disposal price. It will qualify under the taxation guidelines for a five-year recovery period using the double-declining balance depreciation method. A zero terminal disposal price is assumed in computing annual depreciation. Richfield uses an after-tax required rate of return of 8% for low-risk enhanced recovery projects such as the Prudhoe Bay project. Its income tax rate is 30%.

Required

1. Calculate (a) the net present value and (b) the payback period on the enhanced recovery project.
2. Suppose the $21 figure for revenue per barrel is based on the current world price of oil. What is the revenue-per-barrel figure at which the enhanced recovery project is a break-even proposition from a net present value perspective?

22-23 Decision to go to college, net present value, income taxes, inflation. Angela Avila is in her final year of high school. Her best friend, Mary Smith, has decided not to go to college. She has accepted a $16,000-a-year job with a local bank. Mary argues that she will be able to save a fortune by the time Angela has recouped her college tuition fees. Angela, who has never previously considered taking a job straight out of high school, seeks your advice on the financial benefits of going to college. She views herself as similar to Mary in ability, family support, and intelligence. A time period of twenty years from when Angela and Mary leave high school is chosen for analysis. You decide to examine the following two scenarios:

(a) Angela too joins the bank at $16,000 per year. Her salary increases in real terms at 4% per year.

(b) Angela goes to college for four years. The college will cost $12,000 a year for fees, books, and other items (assume paid at the *end* of each year). After college, Angela joins an accounting firm at $30,000 per year. Her salary increases in real terms at 6% per year.

The required real rate of return is 10% per year. (Use 10% in your answers to all four requirements.)

Required

1. Assume zero inflation and zero income taxes. Compare the net present value for the twenty-year period of (a) Angela joining the bank, and (b) Angela going to college and then joining an accounting firm. Comment on your results.
2. Assume zero inflation. Repeat requirement 1 assuming a 28% income tax rate on all income. There are no tax deductions allowed for education costs. Comment on your results.

3. How might a 12% inflation rate affect your answers to requirements 1 and 2? (No computations are required.)
4. What other factors would you recommend that Angela consider in her decision on whether to go to college?

22-24 Capital budgeting for product design, inflation, taxation, straight-line depreciation. Cila Black, president of the Liverpool Product Design Group, is considering an investment in computer-aided design equipment. The equipment will cost £110,000 and will have a five-year useful economic life. It has a zero terminal disposal price and will generate annual cash-operating savings of £36,000 (before income taxes), using 19_0 prices. It is December 31, 19_0.

The after-tax required rate of return is 18% per year.

Required

1. Compute the net present value of the project. Assume a 40% tax rate and straight-line depreciation.
2. Black is wondering if the method in requirement 1 provides a correct analysis of the effects of inflation. The 18% required rate of return incorporates an element attributable to anticipated inflation. For purposes of her analysis, she assumes that the existing rate of inflation, 10% annually, will persist over the next five years. Repeat requirement 1, adjusting the cash-operating savings upward in accordance with the 10% inflation rate.
3. What generalizations about the effects of inflation on capital-budgeting methods and decisions can you develop?

22-25 Capital budgeting, inflation, taxation, straight-line depreciation. (J. Fellingham, adapted) Abbie Young is manager of the customer-service division of an electrical appliance store. Abbie is considering buying a repairing machine that costs $10,000. The machine will last five years. Abbie estimates that the incremental pretax cash savings from using the machine will be $3,000 annually. The $3,000 is measured at current prices and will be received at the end of each year. For tax purposes, she will depreciate the machine straight-line, assuming zero terminal disposal price. Abbie requires a 10% real rate of return (that is, the rate of return is 10% when all cash flows are denominated in 19_0 dollars). Use the 10% rate in your answers to all four requirements.

Required

Treat each of the following cases independently.

1. Abbie lives in a world without income taxes and without inflation. What is the net present value of the machine in this world?
2. Abbie lives in a world without inflation, but there is an income tax rate of 40%. What is the net present value of the machine in this world?
3. There are no income taxes, but the annual inflation rate is 20%. What is the net present value of the machine? The cash savings each year will be increased to reflect the inflation rate.
4. The annual inflation rate is 20%, and the income tax rate is 40%. What is the net present value of the machine?

22-26 Robotics capital project, inflation, income taxes, double-declining balance depreciation. Rustbelt America Inc. purchases secondhand pipeline equipment and "rehabilitates" it for resale. A major problem in its plant is the spot-welding activity. There have been many industrial accidents involving workers at this activity. Rustbelt looks into the possibility of investing in robots. The investment in robots will cost $10 million payable immediately and will reduce labor costs, worker insurance costs, and materials usage costs by a total of $7 million (in 19_0 dollars) a year. The robots require an addition to annual cash-operating costs of $3 million (in 19_0 dollars) a year. Hence the net cash-operating savings from using the robots will be $4 million annually (in 19_0 dollars). Rustbelt believes that using the robots will eliminate industrial accidents involving workers at the spot-welding activity.

The robots have a four-year useful life with a terminal disposal price of $1 million (in 19_0 dollars). The robots qualify for a four-year recovery period using the double-declining balance depreciation method. Any terminal disposal price of the robots is treated as taxable income in the year of the disposal. Rustbelt anticipates inflation in its operating costs and in the terminal disposal price of the robots of 20% per year. It uses a 10% after-tax required rate of return for investments expressed in real dollars. Rustbelt's income tax rate is 40%.

Required

1. What is the nominal after-tax required rate of return of Rustbelt America for investments expressed in nominal dollars?
2. What is the net present value of the $10 million investment in robots? Use the approach of predicting cash inflows and outflows in nominal dollars *and* using a nominal discount rate?
3. What are the advantages of the approach to capital budgeting for inflation in requirement 2 relative to the approach of predicting real cash inflows and outflows *and* using a real discount rate?
4. What factors other than the net present value figure in requirement 2 should Rustbelt America consider in deciding whether to invest in robots?

22-27 Comparison of projects with unequal lives. The manager of the Robin Hood Company is considering two investment projects, which are mutually exclusive.

The after-tax required rate of return of this company is 10%, and the anticipated cash flows are as follows:

Project No.	Investment Required Now	Cash Inflows Year 1	Year 2	Year 3	Year 4
1	$10,000	$12,000	0	0	0
2	$10,000	0	0	0	$17,500

Required

1. Compute the internal rate of return of both projects. Which project is preferable?
2. Compute the net present value of both projects. Which project is preferable?
3. Comment briefly on the results in requirements 1 and 2. Be specific in your comparisons.

22-28 Ranking projects. (Adapted from *N.A.A. Research Report No. 35,* pp. 83–85) Assume that six projects, A through F in the table that follows, have been submitted for inclusion in the coming year's budget for capital expenditures:

	Year	Project A	B	C	D	E	F
Investment	0	$(100,000)	$(100,000)	$(200,000)	$(200,000)	$(200,000)	$(50,000)
	1	0	20,000	70,000	0	5,000	23,000
	2	10,000	20,000	70,000	0	15,000	20,000
	3	20,000	20,000	70,000	0	30,000	10,000
	4	20,000	20,000	70,000	0	50,000	10,000
	5	20,000	20,000	70,000	0	50,000	
Per year	6–9	20,000	20,000		200,000	50,000	
	10	20,000	20,000			50,000	
Per year	11–15	20,000					
Internal rate of return		14%	?	?	?	12.6%	12.0%

Required

1. Compute internal rates of return (to the nearest half percent) for Projects B, C, and D. Rank all projects in descending order in terms of internal rate of return. Show computations.
2. Based on your answer in requirement 1, state which projects you would select, assuming a 10% required rate of return (a) if $500,000 is the limit to be spent, (b) if $550,000 is the limit, and (c) if $650,000 is the limit.
3. Assuming a 16% required rate of return and using the net present value method, compute the net present values and rank all the projects. Which project is more desirable, C or D? Compare your answer with your ranking in requirement 2.

4. What factors other than those considered in requirements 1 through 3 would influence your project rankings? Be specific.

22-29 Ranking of capital budgeting projects, alternative selection methods, capital rationing. (CMA adapted) Franklin Industries has four divisions, each operating in a different industry (tobacco products, cable television, cosmetics, and home appliances). The divisions are currently preparing their capital expenditure budgets for the coming year. The Caledonia Division, located in the northeastern United States, manufactures home appliances that it distributes nationally. The manufacturing and marketing departments of Caledonia have proposed six capital expenditure projects for next year. The division manager must analyze these investment proposals and select those projects that will be included in the capital budget to be submitted to Franklin Industries corporate headquarters for approval. The proposed projects are listed below and are considered to have the same degree of risk.

- *Project A:* Redesign and modification of an existing product that is currently scheduled to be dropped. The enhanced model would be sold for six more years.
- *Project B:* Expansion of a line of cookware that has been produced on an experimental basis for the past year. The expected life of the cookware line is eight years.
- *Project C:* Reorganization of the plant's distribution center, including the installation of computerized equipment for tracking inventory. This project would benefit both administration and marketing.
- *Project D:* Addition of a new product, a combination bread-and-meat slicer. In addition to new manufacturing equipment, a significant amount of introductory advertising would be required. If this project is implemented, Project A would not be feasible due to limited manufacturing capacity.
- *Project E:* Automation of the packaging department, which would result in cost savings over the next six years.
- *Project F:* Construction of a building wing to house offices presently located in an area that could be used for manufacturing. The change would not add capacity for new lines but would alleviate crowded conditions that currently exist, making it possible to improve the productivity of two existing product lines that have been unable to meet market demand.

Franklin Industries has established an after-tax required rate of return of 12% for capital expenditures for all four divisions. Additional information about each of the proposed projects is presented in Exhibit 22-11.

Required

1. Assume that Caledonia Division has no budget restrictions for capital expenditures. Franklin Industries directs the division to identify the capital investment projects that will maximize the value of the company. You are presented with information on the net present value, excess present value index, internal rate of return, and payback period for each project. What projects should be included in the Caledonia Division's capital budget submitted to Franklin Industries corporate headquarters? Explain the basis for your selection.
2. Assume that Franklin Industries specifies that Caledonia Division will be restricted to a maximum of $450,000 for capital expenditures, and Caledonia should select projects that maximize the value to the company. Further, assume that any budget money not spent on any of these projects will be invested at the after-tax required rate of return. Identify the capital investment projects Caledonia should include in its capital expenditures budget submitted to Franklin. Explain the basis for your selection.
3. Should Franklin Industries use the same after-tax required rate of return for all of its four divisions?

22-30 Capital budgeting, income taxes, straight-line depreciation, FMS. Ghent Textiles (GT), located in Belgium, is the largest manufacturer of bath towels in Europe. GT is also a major manufacturer of numerous other textile products, such as tablecloths and bathrobes. GT has a single production plant in Belgium for bath towels that is highly labor-intensive. In recent years, competition from Asian companies has increased substantially. Six months

EXHIBIT 22-11
Caledonia Division
Proposed Capital Projects

Required Investment and Operating After-Tax Cash Flows by Project	**Project A**	**Project B**	**Project C**	**Project D**	**Project E**	**Project F**
Capital investment (cash outflow)	$106,000	$200,000	$140,000	$160,000	$144,000	$130,000
Operating after-tax cash flows:						
Year 1	$ 50,000	$ 20,000	$ 36,000	$ 20,000	$ 50,000	$ 40,000
Year 2	50,000	40,000	36,000	30,000	50,000	40,000
Year 3	40,000	50,000	36,000	40,000	50,000	40,000
Year 4	40,000	60,000	36,000	50,000	20,000	40,000
Year 5	30,000	60,000	36,000	60,000	20,000	40,000
Year 6	40,000	60,000		70,000	11,600	40,000
Year 7		40,000		80,000		40,000
Year 8		44,000		66,000		42,000
Total after-tax cash flows	$250,000	$374,000	$180,000	$416,000	$201,600	$322,000
Additional Project Information						
Present-value of operating cash flows @ 12%	$175,683	$223,773	$129,772	$234,374	$150,027	$199,513
Net present value @ 12%	$ 69,683	$ 23,773	$ (10,228)	$ 74,374	$ 6,027	$ 69,513
Excess present value index (profitability index)	1.66	1.12	.93	1.46	1.04	1.53
Internal rate of return	35%	15%	9%	22%	14%	26%
Payback period	2.2 years	4.5 years	3.9 years	4.3 years	2.9 years	3.3 years
Economic life	6 years	8 years	5 years	8 years	6 years	8 years

ago, Hubert Ooghe, president of GT, visited a competitor's plant in Taiwan. The Taiwan plant uses a production method similar to what the Ghent plant uses, but labor costs per hour for the Taiwanese workers are 20% lower than labor costs for the Belgian workers. Moreover, the Taiwanese are more productive per hour than are the workers in Belgium. Ooghe decides it is time to consider the economic feasibility of adopting a flexible manufacturing system (FMS) at the Belgium plant.

An FMS is a computer-controlled grouping of semi-independent workstations linked by automated material-handling systems. One purpose of an FMS is to manufacture efficiently many different kinds of low- to medium-volume products. An FMS would also enable GT to respond more quickly to customer requests. FMS appears ideally suited to bath-towel production, because GT customers demand numerous variations of bath towels (as to color, length, and design).

Ooghe contacts several companies that manufacture FMSs. The choice quickly becomes either

(a) Retain the existing labor-intensive approach or
(b) Implement an FMS system from a German supplier.

Ooghe and his assistant collect the following data (all amounts are in 100 million Belgian francs, abbreviated F):

a. The projected sales of the European bath-towel market:

19_3	19_4	19_5	19_6	19_7	19_8
400F	420F	440F	460F	480F	500F

b. GT's projected share of the European bath-towel market under the existing labor-intensive approach and using an FMS:

	19_3	19_4	19_5	19_6	19_7	19_8
Labor-intensive	30%	28%	26%	24%	22%	20%
FMS	30%	30%	30%	30%	30%	30%

c. Operating costs (materials, labor, and computer costs) under the two alternatives, as a percentage of GT's sales, are:

	Labor-intensive	FMS
Material	40%	38%
Labor	28%	5%
Computer	2%	7%
	70%	50%

Assume all operating costs are variable with respect to sales francs and are cash outflows at the end of each year.

d. The equipment now used in the plant has a remaining six-year useful life. The disposal price is 8F at the start of 19_3 and 2F at the end of 19_8. The net book value for tax purposes at the start of 19_3 is 20F. GT uses straight-line depreciation for tax purposes, and there are four years remaining before the equipment is fully depreciated. Zero terminal disposal price was assumed in determining the depreciation amount for tax reporting.

e. Costs of implementing the FMS at the start of 19_3 are:

- Equipment (including robots, materials handling, and computer hardware) 96F
- Software and related development 30F

The FMS equipment has an expected useful life of six years. For tax purposes, the original cost of the equipment (96F) minus its terminal disposal price is to be written off over six years using the straight-line method. A terminal disposal price of 6F at the end of 19_8 is used in calculating annual tax depreciation on the equipment. (Ooghe believes the most likely terminal disposal price at the end of 19_8 is also 6F.) For tax purposes, the software development costs (30F) are to be fully expensed in 19_3.

f. The FMS will reduce GT's working capital by 18F starting at the beginning of 19_3. This reduction is due mainly to lower inventory requirements. (For comparing the alternatives, assume working capital will revert to its pre–FMS level at the end of 19_8.)

g. The FMS method will substantially reduce the warehouse and plant space required at the Belgium plant. GT can sell the surplus property to a manufacturer with an adjacent plant for 20F (assume this amount would be received at the start of 19_3). No income tax would be payable on the proceeds from the sale.

h. GT would use internal sources of capital to finance the FMS.

i. GT has an income tax rate of 30% in each year of the 19_3 to 19_8 period. Differences between the terminal disposal price of equipment and its book value at the time of disposal are also taxed at the 30% rate. Assume for discounting purposes that the tax on this difference between terminal disposal price and book value is paid at the end of the year in which the disposal is made.

j. GT uses an after-tax 12% required rate of return for determining the net present value of textile projects.

Required

1. Compute for each year of the 19_3 to 19_8 period: (a) the operating income (income before tax) reported by GT to the tax authorities and the income taxes paid under the existing labor-intensive alternative, and (b) the operating income reported by GT to the tax authorities and the income taxes paid under the FMS purchase alternative.
2. Compute the net present value of (a) the existing labor-intensive alternative and (b) the FMS purchase alternative.
3. Why might Ooghe prefer in 19_3 to capitalize rather than immediately expense the 30F software and related development costs of the FMS method? If capitalized, the 30F amount would initially be treated as an asset and then depreciated over the life of the FMS system. If expensed, the total 30F amount would be an expense of the current period.
4. What factors other than those included in requirements 1 through 3 might Ooghe consider when determining whether to invest in the FMS equipment?

CHAPTER 23

Operations Management and the Accountant: Materials and Inventory

Managing goods for sale in retail organizations

Difficulties with accounting data for managing goods for sale

Just-in-time purchasing

Managing materials in manufacturing organizations

Just-in-time purchasing and production methods are playing a key role in efforts by automobile companies to make major reductions in their manufacturing costs.

Learning Objectives

When you have finished studying this chapter, you should be able to

1. Describe the decisions that operations management typically must make
2. Explain why cost management of materials and inventories is pivotal in many organizations
3. Distinguish among five categories of costs associated with goods for sale
4. Explain the economic-order-quantity (EOQ) model and how it balances ordering costs and carrying costs
5. Compute economic order quantity, reorder point, and level of safety stocks to hold
6. Explain why EOQ models are rarely sensitive to minor variations in cost predictions
7. Describe how just-in-time purchasing and production reduce costs
8. Explain the roles an accountant can play in a materials requirements planning system

The business press is praising the "fresh allure of operations" and the "newfound interest in the shop floor." Global competition in many industries has put a premium on reducing operating costs, improving product quality, and increasing customer service. Managers now give high priority to operations management.

Objective 1

Describe the decisions that operations management typically must make

Operations management is the management of resources to produce or deliver the products or services provided by an organization. Many decisions fall under the operations management umbrella. What is the mix of products or services to sell or manufacture? Which is the best layout of a retail outlet, distribution warehouse, or manufacturing plant? When is the best time to purchase materials? Which is the best way to handle them? These questions are among the many that operations management seeks to answer.

Accounting information can play a key role in operations management. This chapter illustrates the importance of accounting information in two areas:

1. The management of goods for sale in retail organizations
2. The management of materials, work in process, and finished goods in organizations with manufacturing operations

The next chapter examines the role of accounting information in decisions about the mix of products to sell or manufacture. Chapter 29 discusses strategic control systems, an important part of which is managing the quality of materials, work in process, and finished goods.

MANAGING GOODS FOR SALE IN RETAIL ORGANIZATIONS

Objective 2

Explain why cost management of materials and inventories is pivotal in many organizations

Cost of goods sold constitutes the largest single cost item for most retailers. For example, a survey of retail food chains reports the following breakdown of operations.[1]

[1] *Marketing Costs* (Washington,DC: Food Marketing Institute, 1989), p. 1.

Sales		100.0%
Deduct costs:		
Cost of goods sold	75.1%	
Payroll (including fringes)	14.8	
Store operations	5.9	
Other costs	1.6	
Total costs		97.4
Operating income		2.6
Deduct income taxes		1.0
Net income		1.6%

Net income for these surveyed firms averaged only 1.6% of sales, but cost of goods sold averaged 75.1%. This paper-thin net income percentage means that better decisions regarding the purchasing and managing of goods for sale can cause dramatic percentage increases in net income.

Two decisions are central to the management of goods for sale in a retail organization:

1. *How much to order (the economic-order-quantity decision)*
2. *When to order (the reorder decision)*

This chapter discusses the EOQ models that can assist managers in making these decisions.

Costs Associated with Goods for Sale

Objective 3

Distinguish among five categories of costs associated with goods for sale

In managing goods for sale in a retail organization, the following cost categories are important.

1. *Purchase costs:* These costs usually make up the largest single cost category. Discounts for different purchase-order sizes and supplier credit terms affect purchasing costs.
2. *Ordering costs:* **Ordering costs** consist of the costs of preparing a purchase order. Related to the number of orders processed are special processing, receiving, and inspection costs.
3. *Carrying costs:* **Carrying costs** arise when a business holds stocks of goods for sale. These costs consist of the opportunity cost of the investment tied up in inventory (see Chapter 11, p. 377) and the costs associated with storage, such as space rental and insurance.
4. *Stockout costs:* A **stockout** arises when a customer demands a unit of stock and that unit is not readily available. A firm may respond to the shortfall by expediting an order from a supplier. Expediting costs include the additional ordering costs plus any associated transportation costs. Alternatively, the firm may lose a sale due to the stockout. In this case, stockout costs include the lost contribution margin on the sale plus any customer ill will generated by the stockout.
5. *Quality costs:* The **quality** of a product or service is its conformance with a prespecified (and often preannounced) standard. Quality costs are of two kinds: (a) costs incurred to increase the probability that a delivered product is in conformance with its specifications (for example, inspection costs), and (b) costs incurred when a delivered product is not in conformance with its specifications (for example, the cost of a replacement product).

Advances in information-gathering technology are increasing the reliability and timeliness of data in these cost categories. Do not assume, however, that all the relevant data are available in existing accounting systems. Opportunity costs, which are not typically recorded in accounting systems, are an important component in several of these cost categories.

Objective 4

Explain the economic-order-quantity (EOQ) model and how it balances ordering costs and carrying costs

Economic-Order-Quantity Decision Model

The **economic-order-quantity (EOQ)** decision model calculates the optimal quantity of inventory to order. The simplest version of this model incorporates only ordering costs and carrying costs into the calculation. We make the following assumptions when using the simplest version of this model:

1. The same fixed quantity is ordered at each reorder point.
2. Demand, ordering costs, and carrying costs are certain. The **purchase-ordering lead time**—the time between the placement of an order and its delivery—is also certain.
3. Purchasing cost per unit is unaffected by the quantity ordered. This assumption makes purchasing costs irrelevant to determining the optimal EOQ size.
4. No stockouts occur. One justification for this assumption is that the cost of a stockout is prohibitively high. We assume that management, to avoid this potential cost, always maintains inventory so that no stockout can occur.
5. In deciding the size of the purchase order, management considers the costs of quality only to the extent that these costs are components of ordering costs or carrying costs.

Given these assumptions, the purchase costs, stockout costs, and quality costs can be safely ignored. The total relevant costs for determining the optimal EOQ will be the sum of the ordering costs and the carrying costs:

$$\text{Total relevant costs} = \text{Total ordering costs} + \text{Total carrying costs}$$

Example Video Galore sells packages of blank VCR tapes to its customers, in addition to renting out tapes of movies and sporting events. It purchases packages of VCR tapes from a distributor at $20 a package. The distributor pays all incoming freight. No incoming inspection is necessary, as the distributor has a superb reputation for delivering quality merchandise. Annual demand is 10,400 packages, at a rate of 200 packages per week. Video Galore earns 12% on its cash investments. The purchase-order lead time is two weeks. The following cost data are available:

Ordering costs per purchase order		$62.50
Carrying costs per package per year:		
Required annual return on investment, 12% × $20	$2.40	
Relevant insurance, material handling, and so on, per year	2.80	$ 5.20

What is the economic order quantity of packages of VCR tapes?

Solution Exhibit 23-1 deserves careful study. It tabulates the total annual relevant costs of ordering and carrying inventory under various order sizes. The larger the order quantity, the higher the annual carrying costs and the lower the annual ordering costs. The smaller the order quantity, the lower the annual carrying costs and the higher the annual ordering costs. Exhibit 23-2 analyzes the behavior of these two cost functions graphically. The total annual relevant costs will be at a minimum where total purchase-ordering costs and total carrying costs are equal. The EOQ that minimizes the total annual relevant costs in Exhibits 23-1 and 23-2 is 500 packages. The optimal EOQ is most easily found by using the formula given in the next section.

Objective 5

Compute economic order quantity, reorder point, and level of safety stock to hold

EOQ Formula

The formula underlying the EOQ model is

$$\text{EOQ} = \sqrt{\frac{2DP}{C}}$$

EXHIBIT 23-1

Annualized Ordering Costs and Carrying Costs at Various Order Quantities for Video Galore

D: Demand	**10,400**	**10,400**	**10,400**	**10,400**	**10,400**	**10,400**	**10,400**
Q: Order quantity	50	100	400	500	600	1,000	10,400
$Q/2$: Average inventory in units	25	50	200	250	300	500	5,200
D/Q: Number of purchase orders	208	104	26	20.80	17.33	10.40	1
$(D/Q) \times P$: Annual ordering costs for purchase orders	$13,000	$6,500	$1,625	$1,300	$1,083	$ 650	$ 62
$(Q/2) \times C$: Annual carrying costs	130	260	1,040	1,300	1,560	2,600	27,040
Total relevant costs of ordering and carrying inventory	$13,130	$6,760	$2,665	$2,600 (Minimum Cost)	$2,643	$3,250	$27,102

Definition of Symbols
D = demand in units = 10,400 units per year
Q = order quantity
P = ordering cost per purchase order = $62.50
C = carrying cost of one unit in stock = $5.20 per year

EXHIBIT 23-2

Graphic Analysis of Ordering Costs and Carrying Costs for Video Galore

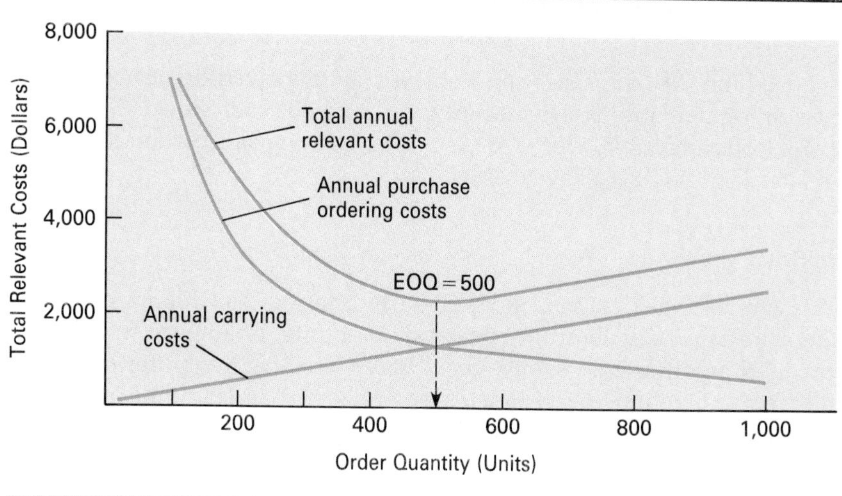

where

EOQ = economic order quantity
D = demand in units
P = ordering costs per purchase order
C = costs of carrying one unit in stock for the time period used for D (one year in this example)

This formula can be illustrated with the data for Video Galore:

$$\text{EOQ} = \sqrt{\frac{2(10{,}400)(\$62.50)}{\$5.20}} = \sqrt{250{,}000} = 500 \text{ packages}$$

The total relevant costs can be calculated using the following formula where Q = the order quantity:

$$TRC = \frac{DP}{Q} + \frac{QC}{2}$$

(In this formula, Q can be any order quantity, not just the EOQ.)

When Q = 500 units:

$$TRC = \frac{10{,}400 \times \$62.50}{500} + \frac{500 \times \$5.20}{2}$$
$$= \$1{,}300 + \$1{,}300 = \$2{,}600$$

The number of deliveries each time period is

$$\frac{\text{Demand}}{\text{EOQ}} = \frac{10{,}400}{500} = 20.8 \text{ deliveries}$$

When to Order, Assuming Certainty

The **reorder point** is the quantity level of the inventory that triggers a new order. The reorder is simplest to compute when both demand and lead time are certain:

Reorder point = Sales per unit of time × Purchase-ordering lead time

Consider our Video Galore example where a week is the chosen unit of time:

Economic order quantity	500 packages
Sales per week	200 packages
Purchase-ordering lead time	2 weeks

Reorder point = Sales per unit of time × Purchase-ordering lead time
= 200 × 2 = 400 packages

The graph in Exhibit 23-3 presents the behavior of the inventory level of VCR packages, assuming demand occurs uniformly throughout each week.[2] If the purchase-ordering lead time is two weeks, a new order will be placed when the inventory level reaches 400 VCR packages.

Safety Stock

Exhibit 23-3 assumes that demand and purchase-ordering lead time are certain. When retailers are uncertain about the demand, lead time, or amount that suppliers can provide, they often hold safety stock. **Safety stock** is the buffer inventory held as a cushion against unexpected increases in demand or lead time and unexpected unavailability of stock from suppliers. Video Galore expects demand to be 200 packages per week, but its managers feel that a maximum demand of 320 packages per week may occur in some weeks. If Video Galore decides that the cost of stockout is prohibitive, it may decide to hold safety stock of 240 packages. This amount is the maximum excess demand of 120 per week for the 2 weeks of purchase-order lead time.

The computation of safety stock hinges on demand forecasts. In our Video Galore example, with uncertainty we used 240 VCR packages as a safety stock. The 240 amount is based on projected demand. Managers will have some notion—usually based on past experience—of the range of weekly demand.

---

[2]This handy but special formula does not apply when the receipt of the order fails to increase stock to the reorder-point quantity (for example, when the lead time is three weeks and the order is a one-week supply). In these cases, there will be overlapping of orders. The reorder point will be average usage during lead time plus safety stock minus orders placed but not yet received. *Safety stock* is the buffer inventory held as a cushion against unexpected increases in demand or lead time and unexpected unavailability of stock from suppliers.

EXHIBIT 23-3
Inventory Level of VCR Packages for Video Galore*

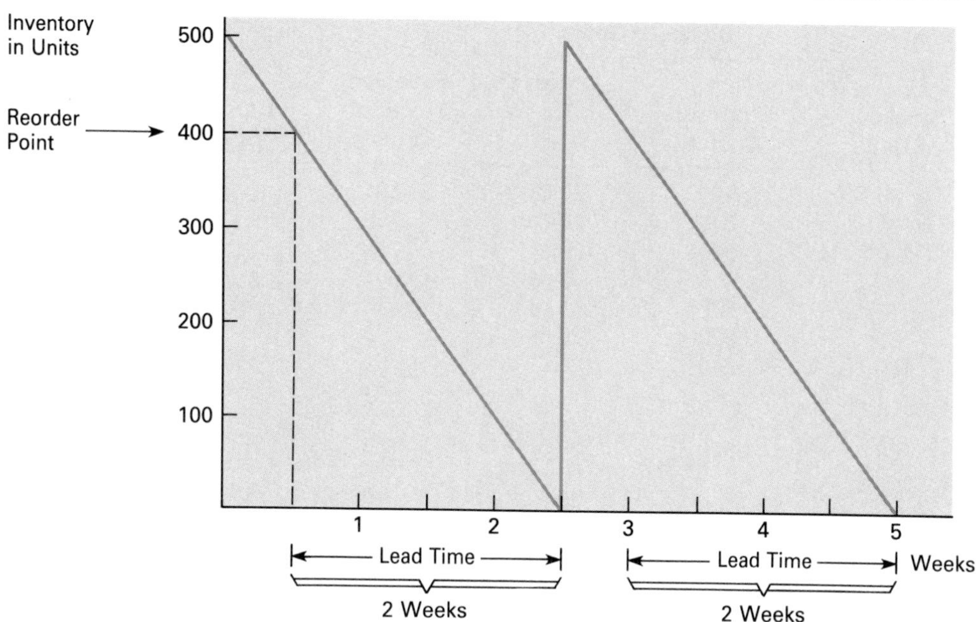

*Exhibit assumes that demand and purchase-ordering lead time are certain.

The major relevant costs in maintaining safety stock are the carrying costs and the stockout costs. Suppose a customer calls Video Galore to buy 5 packages of VCR tapes. The store has none in stock at the moment, but it can supply them within twenty-four hours at an extra payment to the supplier of $4 per package. The customer is willing to wait the twenty-four hours. The stockout cost in this case is a minimum of $4 per package. The optimal safety stock level exists where the costs of carrying an extra unit are exactly counterbalanced by the expected stockout costs. This would be the level that minimizes the total annual stockout costs and carrying costs.

A frequency distribution based on prior daily or weekly levels of demand provides data for computing the associated costs of maintaining safety stock. Assume that one of seven different levels of demand for 2 weeks will occur at Video Galore.

Total demand for 2 weeks	160 units	240 units	320 units	400 units	480 units	560 units	640 units
Probability	0.04	0.06	0.20	0.40	0.20	0.06	0.04

We see that 400 is the most likely level of demand for 2 weeks because it is assigned the highest of the probabilities in the chart. We see also that there is a .30 probability that demand will be between 480 and 640 packages (.20 + .06 + .04 = .30).

If we decide to use a reorder point equal to 400 packages, stockouts would occur only if demand is for 480,560, or 640 units. The following table shows the probability of a stockout with a reorder point of 400 units and safety stocks of 0, 80, 160, and 240 units. For example, suppose actual demand is 640. Subtracting the expected demand provided for (400 units) gives us a stockout of 240 units if the business decides not to keep any safety stock, a stockout of 160 units if 80 units of safety stock are held, and so on.

EXHIBIT 23-4
Computation of Safety Stock for Video Galore

Safety Stock Level in Units	Probability of Stockout	Stockout Costs				Carrying Costs§	Total Costs
		Stockout in Units	Stockout Costs*	Orders Per Year†	Expected Stockout Costs‡		
0	.20	80	$320	20.8	$1,331		
	.06	160	640	20.8	799		
	.04	240	960	20.8	799		
					$2,929	$ 0	$2,929
80	.06	80	$320	20.8	$ 399		
	.04	160	640	20.8	532		
					$ 931	$ 416	$1,347
160	.04	80	$320	20.8	$ 266	$ 832	$1,098
240	.00	0	$ 0	20.8	$ 0	$1,248	$1,248

*Stockout units × stockout cost of $4.00 per unit.
†Annual demand 10,400 ÷ 500 EOQ = 20.8 orders per year.
‡Stockout cost × probability × number of orders per year.
§Safety stock × annual carrying costs of $5.20 per unit (assumes that safety stock is on hand at all times and that there is no overstocking caused by decreases in expected usage).

Probability of stockout	.40	.20	.06	.04
Total actual demand during lead time	400	480	560	640
Deduct expected demand provided for during lead time	400	400	400	400
Stockout if provision is also made for safety stock of:				
0 units	0	80	160	240
80 units	0	0	80	160
160 units	0	0	0	80
240 units	0	0	0	0

Exhibit 23-4 presents the total annual stockout costs and carrying costs assuming the stockout cost is $4 per package and the carrying cost is $5.20 per unit per year. Of the safety-stock levels presented in Exhibit 23-4, the total annual stockout costs and carrying costs are minimized at $1,098, when a safety stock of 160 packages is maintained.

DIFFICULTIES WITH ACCOUNTING DATA FOR MANAGING GOODS FOR SALE

Estimating Cost Parameters

Obtaining accurate estimates of the cost parameters used in the EOQ decision model is a challenging task. For example, the relevant annual carrying costs of inventory consist of *outlay costs* for such items as insurance and property taxes plus the *opportunity cost* of capital. This opportunity cost is the interest forgone by investing capital in inventory. It is calculated as the required rate of return multiplied by those costs per unit that vary with the number of units purchased and that are incurred at the time the units are received. (Examples of these costs per unit are purchase price, freight-in, and incoming inspection.) An opportunity cost must be charged because interest income is forgone when money is tied up in inventory rather than being invested elsewhere (say, in a bank). Most internal reporting systems do not formally record opportunity costs. Users of EOQ models cannot, then, rely exclusively on the accounting system for all the components included in the costs of carrying inventory.

Opportunity costs are also relevant to estimating stockout costs. When a sale is forgone because of an item being out of stock, the opportunity cost includes the lost contribution margin on that sale as well as the lost contribution margin on potential future sales that will not be made to the disgruntled customer.

Cost of a Prediction Error

Objective 6

Explain why EOQ models are rarely sensitive to minor variations in cost predictions

Predicting costs is difficult. Intelligent managers understand that their projections will seldom be perfectly on target. What is the cost of an incorrect prediction when actual costs are different from predicted costs?

Suppose Video Galore's ordering cost per purchase order is $90 instead of the $62.50 prediction used in Exhibits 23-1 and 23-2. We can calculate the cost of this prediction error with a three-step approach.

Step One: Compute the monetary outcome from the best action that could have been taken, given the actual value of the cost input. The appropriate inputs are $D = 10{,}400$ units, $P = \$90$, and $C = \$5.20$. The economic order quantity size is

$$\text{EOQ} = \sqrt{\frac{2DP}{C}}$$

$$\text{EOQ} = \sqrt{\frac{2(10{,}400)(\$90)}{\$5.20}} = \sqrt{\frac{\$1{,}872{,}000}{\$5.20}} = \sqrt{360{,}000}$$

$$= 600 \text{ packages}$$

The total annual relevant cost when EOQ = 600 is

$$\text{TRC} = \frac{DP}{Q} + \frac{QC}{2}$$

$$= \frac{10{,}400 \times \$90}{600} + \frac{600 \times \$5.20}{2}$$

$$= \$1{,}560 + \$1{,}560 = \$3{,}120$$

Step Two: Compute the monetary outcome from the best action based on the incorrect value of the predicted cost input. The planned action when the ordering costs per purchase order are predicted to be $62.50 is to purchase quantities of 500 packages. The total annual relevant cost using this order quantity when $D = 10{,}400$ units, $P = \$90$, and $C = \$5.20$ is

$$\text{TRC} = \frac{DP}{Q} + \frac{QC}{2}$$

$$= \frac{10{,}400 \times \$90}{500} + \frac{500 \times \$5.20}{2}$$

$$= \$1{,}872 + \$1{,}300 = \$3{,}172$$

Step Three: Compute the difference between the monetary outcomes from Step One and Step Two.

	Monetary Outcome
Step One	$3,120
Step Two	3,172
Difference	$ (52)

The cost of the prediction error is only \$52. The total annual relevant cost curve in Exhibit 23-2 is relatively flat over the range of order quantities from 400 to 600. ***A salient feature of the EOQ model is that the total relevant cost is rarely sensitive to minor variations in cost predictions. The square root in the EOQ model reduces the sensitivity of the decision to errors in predicting its inputs.***

Goal-Congruence Issues

Goal-congruence issues can arise when there is an inconsistency between the decision model and the model used for evaluating the performance of the person implementing the model. Consider opportunity costs. Opportunity costs are an important component of the total relevant costs when making decisions with an EOQ model. The absence of recorded opportunity costs in conventional accounting systems, however, raises the possibility of a conflict between the order quantity that the EOQ model indicates as optimal and the order quantity that the purchasing manager regards as optimal.

If annual carrying costs are not included in the measures used to evaluate the performance of store managers, they may favor purchasing larger order quantities than the EOQ decision model implies are optimal. One answer to this conflict is to design the performance evaluation system so that the carrying costs are charged to the appropriate manager. For example, an "imputed interest" charge can be levied against managers for the inventories under their responsibility. This practice inhibits managers from overbuying stocks, a common temptation.

The inventory area provides another illustration of why accountants and managers must be alert to the motivational implications of failing to dovetail a performance-evaluation model with the decision model that top management favors.

JUST-IN-TIME PURCHASING

Objective 7

Describe how just-in-time purchasing and just-in-time production reduce costs

Organizations are giving increased attention to the potential gains from (1) making smaller and more frequent purchase orders, and (2) restructuring their relationship with suppliers. Both (1) and (2) are related to the heightened interest in just-in-time purchasing systems. **Just-in-time (JIT) purchasing** is the purchase of goods or materials such that delivery immediately precedes demand or use. In the extreme, no inventories (goods for sale of a retailer; materials of a manufacturer) would be held.

Organizations using JIT purchasing often stress the "hidden costs" associated with holding high inventory levels. These "hidden costs" include larger amounts of inventory storage space and sizable amounts of spoilage. Retailers have long recognized these costs in relation to perishable goods. For example, daily deliveries of bread and milk to supermarkets have occurred for many years. Retailers using JIT purchasing are now attempting to extend daily deliveries to as many areas as possible so that the goods spend less time in warehouses or on the shelves before sale to customers.

Some vendors are highly cooperative in a business's attempts to adopt JIT purchasing. For example, consider the Frito-Lay company, which has a dominant market share in potato chips and other snack foods. Frito-Lay operates with more frequent deliveries than many of its competitors. Its corporate strategy emphasizes service to retailers and consistency of product quality delivered. Companies moving toward JIT purchasing often argue that the "full cost" of carrying inventories (including the opportunity and other costs not recorded in the accounting system) has been dramatically underestimated in the past.

EXHIBIT 23-5
Sensitivity of EOQ to Variations in Ordering Costs and Carrying Costs

Carrying Costs Per Package Per Year	Ordering Costs Per Purchase Order			
	$62.50	**$50.00**	**$40.00**	**$30.00**
$ 5.20	EOQ = 500	EOQ = 447	EOQ = 400	EOQ = 346
$ 7.00	431	385	345	299
$10.00	361	322	288	250
$15.00	294	263	236	204

*All cells in the matrix assume annual demand is 10,400 units.

EOQ Implications of Just-in-Time Purchasing

Exhibit 23-5 uses sensitivity analysis of the Video Galore example in Exhibit 23-1 to illustrate the economics of smaller and more frequent purchase orders. Reductions in the economic order quantity can be affected by (1) reductions in the ordering costs of each purchase order, and (2) increases in the carrying costs of inventories. The rows of Exhibit 23-5 report how reductions in the ordering costs of Video Galore placing a purchase order (P) are associated with reductions in the EOQ. Consider the case with D = 10,400 and C = $5.20; the EOQ is 500 packages when P = $62.50, 447 packages when P = $50, and 346 when P = $30. The columns axis of Exhibit 23-5 report how increases in the costs of Video Galore carrying a unit of inventory are associated with reductions in the EOQ. Consider the case with D = 10,400 and P = $62.50; the EOQ is 500 when C = $5.20, 431 when C = $7.00, and 294 when C = $15.00.

The analysis that Exhibit 23-5 presents justifies JIT purchasing. Common sense tells us that reduced ordering costs per purchase order and increased carrying costs would lead to reductions in the economic order quantity, and the numbers in Exhibit 23-5 support that reasoning.

Do not assume that a JIT purchasing policy will always be guided by an EOQ decision model. At a minimum, the EOQ model assumes a constant order quantity. If demand fluctuates, a JIT purchasing policy may well require different quantities for each order.

Cost Management

Companies are now giving heightened attention to better management of the costs associated with materials and inventories. Organizations that have moved toward JIT purchasing have made substantial changes in their purchasing practices:

- A reduction in the number of suppliers for each item, with an associated reduction in negotiation time. For example, a division of Xerox reduced its number of suppliers from 5,000 to 300.
- The use of long-term contracts with suppliers, with minimal paperwork involved in each individual transaction. Each purchase transaction may involve only a single telephone call or a single computer entry.
- Minimal checking by purchasers of the quantity and quality of goods received. In the initial negotiations, suppliers are made aware of the premium placed on the delivery of high-quality goods in the exact quantity ordered.
- Payment to suppliers made for batches of deliveries rather than for each individual delivery. For example, the materials-receiving group at one Hewlett-Packard plant receives documents on a daily basis but sends them to Accounts Payable only on a weekly

EXHIBIT 23-6
Ordering Costs per Purchase Order: Conventional and JIT Purchasing Practices

PANEL A: Purchase Order Costs Using Conventional Purchasing Practices

Negotiation	Open-Order Status Paperwork	Expediting	Receiving/ Count	Receiving/ Inspection	Transportation
(1)	(2)	(3)	(4)	(5)	(6)

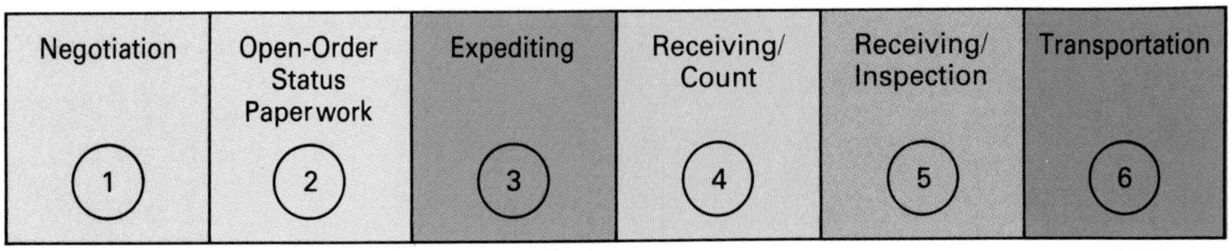

PANEL B: Purchase Order Costs Using Just-in-Time Purchasing

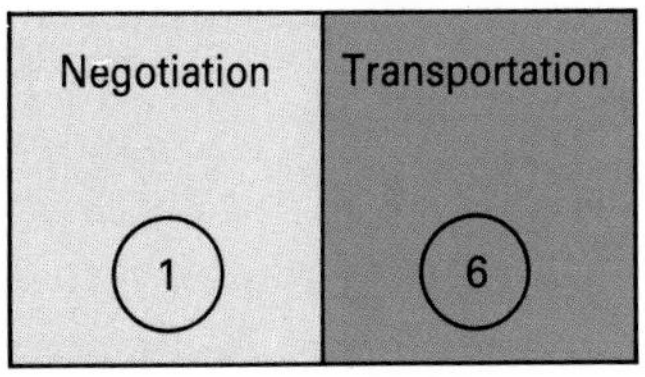

Negotiation	Transportation
(1)	(6)

Source: C. Hahn, P. Pinto, and D. Bragg, "'Just-in-Time' Production and Purchasing," *Journal of Purchasing and Materials Management,* Vol. X, No. 1, p. 8.

basis. Computer software programs "match" each receiving document with the purchase-order number. A computer program then sums all amounts due each supplier, and a single check is written to each supplier.

These changes in purchasing practices can give rise to substantial reductions in the cost of placing each individual purchase order.

Retailers are often highly aggressive in their cost-management activities. For example, Herman's, a sporting-goods store, requires suppliers to use their own salespeople to provide in-store services such as setting up promotion displays and picking up damaged merchandise. Walmart "fines" suppliers for late deliveries and incomplete orders.

Exhibit 23-6 compares the components of ordering costs per purchase order under conventional purchasing practices and under a just-in-time approach. The costs of components (2) to (5) in Panel A of Exhibit 23-6 are sometimes labeled non-value-added costs. A *non-value-added cost* is the cost of an activity that could be eliminated without the customer perceiving a deterioration in the attributes of a product. Chapter 29 discusses the value-added/non-value-added classification of costs and its role in the cost-management practices of several organizations.

MANAGING MATERIALS IN MANUFACTURING ORGANIZATIONS

Managers in companies with manufacturing facilities face the challenging task of producing high-quality products at competitive cost levels. Numerous systems have been developed to assist managers in their planning and implementation activities.

A survey of U.S. companies in the midwest states reports that the following "modern manufacturing systems and practices [were] used in at least some part of their operations":[3]

Objective 8

Explain the roles an accountant can play in a materials requirements planning system

• Materials requirements or resource planning (MRP)	63%
• Just-in-time (JIT)	36
• Computer-integrated manufacturing (CIM)	21
• Flexible manufacturing systems (FMS)	20

A survey of Japanese manufacturing organizations also reported widespread use of MRP and JIT.[4]

The jargon used in describing manufacturing systems in this area can be overwhelming. One production manager commented: "Producing high-quality, low-cost products in competition with the Japanese is challenging but manageable. Understanding how MRP, JIT, CIM, and FMS relate to each other is both unmanageable and a time-sink."

The following section outlines the key features of MRP and JIT production and the role of accounting information in each of these manufacturing systems. We must leave discussion of CIM and FMS to other books. Our descriptions of MRP and JIT are necessarily brief and simple. Be careful when examining the manufacturing systems of individual firms. Businesses vary in the labels they use to describe their systems. Moreover, the vendors of computer software for manufacturing systems use much leeway in labeling their individual products.

Materials Requirements Planning

Materials requirements planning (MRP) is a planning system that focuses first on the amount and timing of finished goods demanded and then determines the demand for materials, components, and subassemblies at each of the prior stages of production.[5] A production structure diagram offers an overview of the production process and its materials requirements. Exhibit 23-7 presents such a diagram that has three end products (FG 1, FG 2, and FG 3). Working backwards, each end product is sequentially exploded (that is, separated) into its necessary components and materials. Exhibit 23-7 illustrates that four direct materials (DM 1, DM 2, DM 3, and DM 4) are purchased for finished goods. For both FG 1 and FG 3, the materials are used to produce components before production of the end product. For FG 2, no intermediary components are produced.[6]

Key components of an MRP system include the following:

1. Master production schedule, which specifies both the quantity and the timing of each item to be produced.
2. Bill of materials file, which outlines the materials, components, and subassemblies for each end product. MRP distinguishes outside purchases from components derived from prior steps in the production process.

---

[3]Price Waterhouse, *Survey of the Cost Management Practices of Selected Midwest Manufacturers,* 1989, p. 3. Another survey also reported MRP to be the most frequently used inventory management system. C. Mecimore and J. Weeks, *Techniques in Inventory Management and Control* (Montvale, NJ: National Association of Accountants, 1987).

[4]S. Inoue, "JIT Production and Its Influence on Cost Management in Japan—Compared with Those of MRP and JOS," Working Paper, Kagawa University, 1989.

[5]MRP is sometimes used to describe materials *resources* planning, which is a planning system that considers such items as capacity planning and labor scheduling as well as materials planning. Common terminology is MRP I for materials requirements planning and MRP II for materials resource planning.

[6]For a case study of MRP, see J. Biggs and E. Long, "Gaining the Competitive Edge with MRP/MRP II," *Management Accounting,* May 1988, pp. 27–32.

EXHIBIT 23-7
Production Structure Diagram Underlying an MRP System

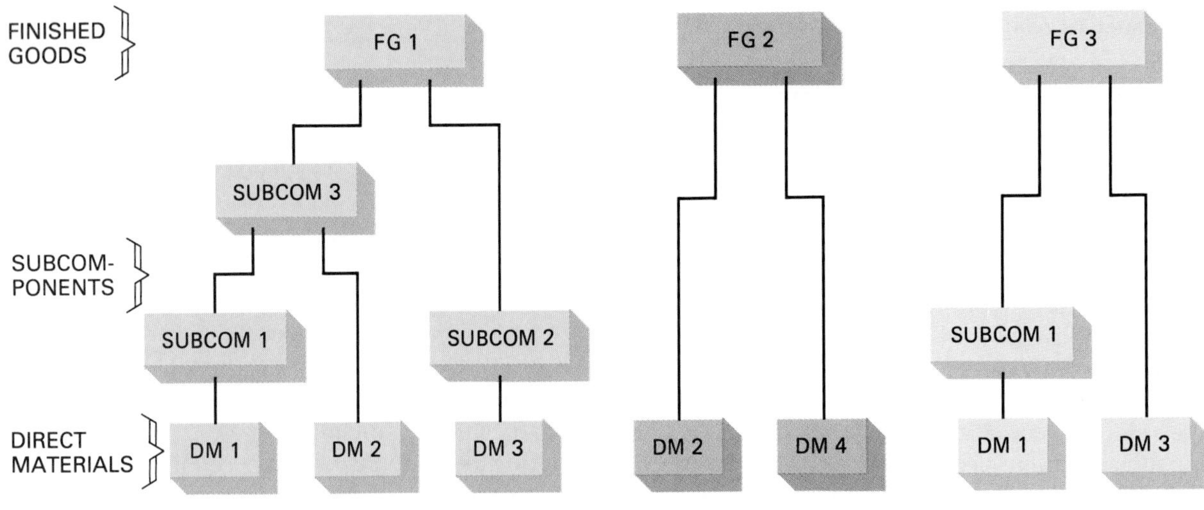

3. Inventory report of each part, component, or subassembly, in which each such item is carried in a separate computer file with details on the number of items on hand and the arrival times and quantities of items scheduled to be received.
4. Lead times of all items to be purchased, and standard construction times for all components and subassemblies produced internally.

Given the information in the MRP system, computer algorithms can determine how changes in the demand for individual finished goods will affect the demand for materials, components, and subassemblies. For instance, when a customer requests that a large order be deferred for one month, an MRP system will help readjust the materials purchases and the production schedules accordingly.

The accountant can play several important roles in the operation of an MRP system. First, MRP requires much detailed information pertaining to materials, work in process, and finished goods. A major cause of unsuccessful attempts to implement MRP has been the problem of collecting and updating inventory records. The chances of MRP operating successfully will be improved by combining the talents of manufacturing personnel and accountants in the early stages of implementing an MRP system. For complex MRP systems, the required records can cover thousands of parts, materials, components, and finished goods. Unless this information is both accurate and timely, many of the benefits of using an MRP approach can be lost.

A second role of the accountant consists of providing estimates of the costs of setting up each production run of a plant, the costs of downtime, and the costs of holding inventory. Some operations managers attempt to combine MRP with decisions about optimal production-run quantities for substages of the production process. When the costs of setting up machines or sections of the production line are high (for example, as with a blast furnace in an integrated steel mill), the benefits from using optimal production-run quantities can be considerable. Similarly, when the costs of downtime are high, there can be sizable benefits from maintaining a continuity of production. One challenge facing the accountant in this context is to recognize interdependencies in the product line. Downtime at one stage (say, *X*) of a production line can have ripple effects at subsequent stages (say, *Y* and *Z*) in the production line. These ripples mean that the cost of downtime is not just the downtime at workstation *X* but also the cost of downtime at workstations *Y* and *Z*.

Just-in-Time Production

Just-in-time (JIT) production is a system in which each component on a production line is produced immediately as needed by the next step in the production line.[7] JIT production includes three key features:

1. The production line is run on a demand-pull basis, so that activity at each workstation is authorized by the demand of downstream workstations.
2. Emphasis is placed on minimizing the throughput time of each unit. *Throughput time* is the interval between the first stage of production and the point at which the finished good comes off the production line.
3. The production line is stopped if parts are absent or defective work is discovered. Stoppage creates an urgency about correcting problems that cause defective units. Each employee puts a premium on minimizing the potential sources of stoppages (such as defective material parts). In contrast, under traditional methods, the inventory of parts and work in process is sufficient in size to enable workers to set aside defective parts and continue their normal operations.

JIT production has as its underlying philosophy the simplification of the production process so that only essential activities are conducted. Several JIT adopters have extended this simplification to their internal accounting system. For example, the product-cost records in several JIT plants have only two entries (based on standard costs). The first is made when direct materials are used at the start of the production line, and the second is made when finished goods leave the production line. The reasons for using simplified product-cost records with JIT include the following:

1. Materials control in JIT plants can best be accomplished by a manager's personal observations. The absence of large amounts of materials and work-in-process inventory means that managers can spend more time observing and monitoring the existing materials and work in process.
2. Work in process constitutes a lower percentage of the total cost of production because of the reductions in lead time and the demand-pull feature of production.
3. There is a reduction in detail associated with rework. Rework can occur at many different stages in a production process. Detailed records are necessary if the cost of rework is to be traced to individual jobs. Reduce the percentage of jobs with rework, and you reduce the need for detailed recordkeeping of them.

Additional examples of simplifications in accounting systems are described in the backflush costing section of Chapter 19 (pp. 627–33.)

Firms adopting JIT production make heavy use of nonfinancial performance measures in the day-to-day control of operations at the plant level.[8] Examples of such measures are:

- Throughput time
- $\dfrac{\text{Total setup time for machines}}{\text{Total production time}}$
- $\dfrac{\text{Number of units requiring rework or scrap}}{\text{Total number of units started and completed}}$

---

[7]Detailed discussion of JIT production is in W. Duncan, *Just-in-Time in American Manufacturing* (Dearborn, MI: Society of Manufacturing Engineers, 1988). For a case study, see W. Turk, "Management Accounting Revitalized: The Harley-Davidson Experience," *Journal of Cost Management*, Winter 1990, pp. 28–39.

[8]See J. Lessner, "Performance Measurement in a Just-in-Time Environment: Can Traditional Performance Measurements Still Be Used? *Journal of Cost Management*, Fall 1989, pp. 22–28.

EXHIBIT 23-8
The Effects of JIT Production

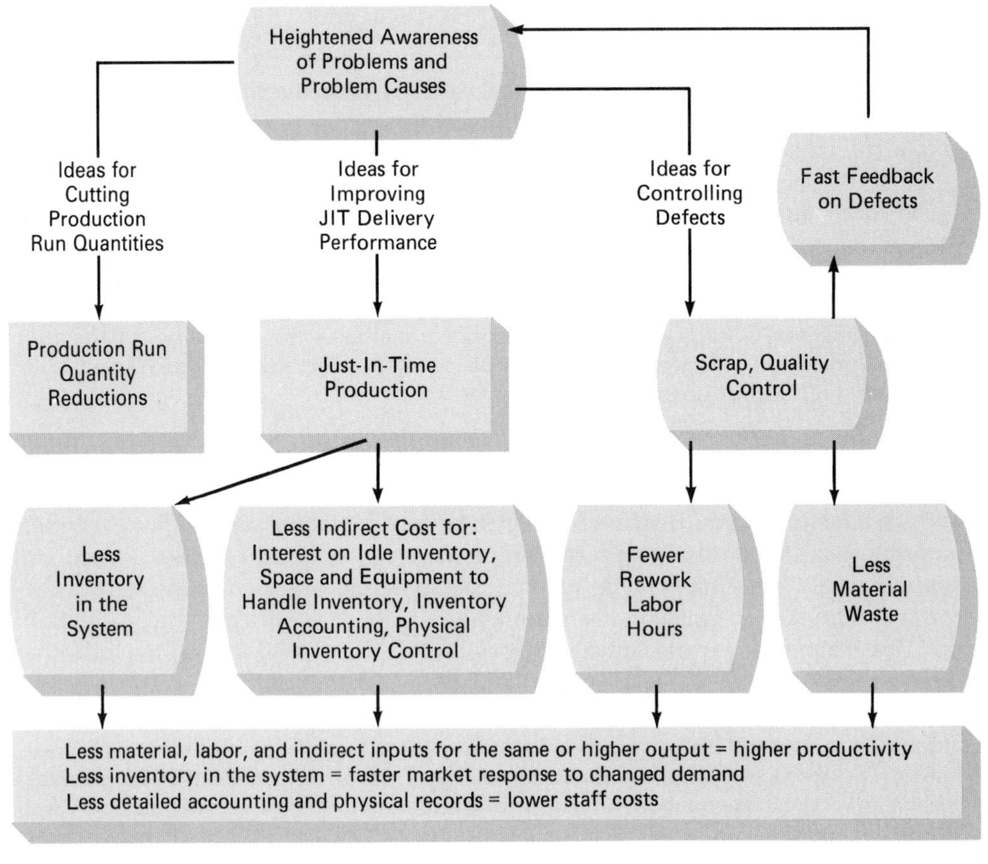

Source: Hewlett-Packard, adapted.

Exhibit 23-8 summarizes the effects Hewlett-Packard reported from adopting JIT at several of its production plants. Early advocates of JIT emphasized the benefits associated with lower inventories. Most firms adopting JIT report even greater benefits from the heightened emphasis on eliminating the root causes of rework and waste and on reducing the production lead time of their products.

PROBLEMS FOR SELF-STUDY

PROBLEM 1

The Complete Gardener (CG) is deciding on the economic order quantity for two brands of lawn fertilizer: Super Grow and Nature's Own. The following information is collected:

	Super Grow	**Nature's Own**
Annual demand	2,000 bags	1,280 bags
Ordering costs per purchase order	\$30	\$35
Annual carrying costs per bag	\$12	\$14

Required

1. Compute the EOQ for Super Grow and Nature's Own.
2. What is the sum of the total ordering costs and total carrying costs for Super Grow and Nature's Own?
3. Compute the number of deliveries per year for Super Grow and Nature's Own.

SOLUTION

1.

Super Grow

$$EOQ = \sqrt{\frac{2(2{,}000)(\$30)}{\$12}}$$
$$= 100 \text{ bags}$$

Nature's Own

$$EOQ = \sqrt{\frac{2(1{,}280)(\$35)}{\$14}}$$
$$= 80 \text{ bags}$$

2.

Super Grow

$$TRC = \frac{2{,}000(\$30)}{100} + \frac{100(\$12)}{2}$$
$$= \$1{,}200$$

Nature's Own

$$TRC = \frac{1{,}280(\$35)}{80} + \frac{80(\$14)}{2}$$
$$= \$1{,}120$$

3.

Super Grow

$$\frac{2{,}000}{100} = 20 \text{ deliveries}$$

Nature's Own

$$\frac{1{,}280}{80} = 16 \text{ deliveries}$$

PROBLEM 2

The Complete Gardener (CG) signs a long-term contract with the Super Grow distributor, and they set up a new procedure for placing purchase orders. A single entry is made into a computer network operated by the Super Grow distributor. CG will make no incoming inspection of bags; the distributor has "guaranteed" to maintain a 100% product quality level in return for CG signing the long-term contract. CG's new ordering cost per purchase order will be $4.50. CG reexamined its materials-handling costs and revised upward its annual carrying cost per bag to $28.80.

Required

1. Consider the new ordering cost per purchase order and the new carrying cost per bag. Compute CG's economic order quantity and the number of deliveries per year for Super Grow.
2. How might your answers to requirement 1 provide insight into a just-in-time purchasing policy?

SOLUTION 2

1. For Super Grow, $D = 2{,}000$, $P = \$4.50$, and $C = \$28.80$:

$$EOQ = \sqrt{\frac{2(2{,}000)(\$4.50)}{\$28.80}}$$
$$= 25 \text{ bags}$$
$$\frac{D}{EOQ} = \frac{2{,}000}{25} = 80 \text{ deliveries}$$

2. A just-in-time purchasing policy involves the purchase of goods or materials such that delivery immediately precedes demand. The decrease in the EOQ for Super Grow from 100 bags to 25 bags increases the number of deliveries from 20 to 80. By restructuring relationships with its supplier, CG has dramatically reduced its ordering costs. This pattern is a familiar one for firms adopting JIT purchasing.

SUMMARY

This chapter examined the management of materials and inventories for two reasons:

- Materials costs (goods for sale in retail) is the largest single cost category in many organizations.
- Accounting information plays a key role in decisions relating to managing the costs of materials and inventories.

Five cost categories important to managing goods for sale in a retail organization are purchasing costs, ordering costs, carrying costs, stockout costs, and quality costs. The economic-order-quantity (EOQ) decision model highlights the implications of ordering costs and carrying costs for the total relevant costs of different purchase-order quantities. Accounting information used to estimate inputs for the EOQ model includes those transactions routinely recorded in the accounting system, and opportunity costs not routinely recorded.

The potential conflict between decision models and performance-evaluation models is exemplified by the EOQ decision model. The opportunity cost of the interest on the investment in inventory plays a key role in the EOQ decision model. Still, many organizations measure managerial performance on the basis of reported costs or total reported operating income, without providing for any opportunity cost of interest.

A major trend in organizations is a reduction in the relative level of inventories held. This trend in part reflects downward revisions in estimates of ordering costs and upward revisions in estimates of carrying costs. The trend also reflects the growing body of evidence that the costs of delivering a high-quality product increase (1) when goods are held in warehouses or sit for a long time on a retailer's sales floor; (2) when materials or finished goods are held in storage areas; and (3) when work in process is allowed to accumulate beyond what is immediately demanded by subsequent stages in the production process.

The important topics of managing product quality and the role of accounting data in these efforts are discussed in Chapter 29.

TERMS TO LEARN

This chapter and the Glossary at the end of the book contain definitions of the following important terms:

carrying costs *(p. 741)* economic order quantity (EOQ) *(742)*
just-in-time (JIT) purchasing *(748)* materials requirements planning (MRP) *(751)*
operations management *(740)* ordering costs *(741)*
purchase-ordering lead time *(742)* quality *(741)* reorder point *(744)*
safety stock *(744)* stockout *(741)*

ASSIGNMENT MATERIAL

QUESTIONS

23-1 Give two examples of decisions that fall under the operations management umbrella.

23-2 Why do better decisions regarding the purchasing and managing of goods for sale frequently cause dramatic percentage increases in net income?

23-3 Name two decisions central to the management of goods for sale in a retail organization.

23-4 Name five cost categories important in managing goods for sale in a retail organization.

23-5 What assumptions are made when using the simplest version of the economic-order-quantity (EOQ) decision model?

23-6 What costs are included in annual carrying costs of inventory when using the EOQ decision model?

23-7 Give three examples of opportunity costs that are typically not recorded in accounting systems, although they are relevant to the EOQ model.

23-8 What are the steps in computing the cost of a prediction error when using the EOQ decision model?

23-9 "The practical approach to determining economic order quantity is concerned with locating a minimum cost range rather than a minimum cost point." Explain.

23-10 Why might goal-congruence issues arise when an EOQ model is used to guide decisions on how much to order?

23-11 Name two cost factors that can explain why an organization finds it cost effective to make smaller and more frequent purchase orders.

23-12 "Accountants have placed inventories on the wrong side of the balance sheet. They are a liability, not an asset." Comment on this statement by a plant manager.

23-13 Give two examples of how organizations that have moved toward just-in-time purchasing have made substantial changes in their purchasing practices.

23-14 What roles can the accountant play in the operation of a materials requirements planning system?

23-15 Describe key features of just-in-time (JIT) production.

EXERCISES AND PROBLEMS

23-16 Economic order quantity for retailer. The Cloth Centre sells fabrics to a wide range of industrial and consumer users. One of its products is denim cloth, used in the manufacture of jeans and carrying bags. The supplier for the denim cloth pays all incoming freight. No incoming inspection of the denim is necessary because of the supplier's past track record of delivering quality merchandise. The purchasing officer of the Cloth Centre has collected the following information:

Annual demand for denim cloth	20,000 yards
Ordering costs per purchase order	$160
Carrying costs per year	20% of purchase cost
Safety stock requirements	None
Cost of denim cloth	$8 per yard

The purchasing lead time is two weeks. The Cloth Centre is open 250 days a year (50 weeks for 5 days a week).

Required

1. Calculate the economic order quantity for denim cloth.
2. Calculate the number of orders that will be placed each year.
3. Calculate the reorder point for denim cloth.

23-17 Economic order quantity for retailer. Tru-Value Hardware (TVH) is deciding the purchase order quantity for its standard line of neon light fittings. Annual demand is 18,000 units. Ordering costs per purchase order are $16. Carrying costs per light-fitting unit are $1.50 per year. TVH uses an economic-order-quantity model in its purchasing decisions. The store is open 360 days a year.

Required

1. Calculate TVH's economic order quantity for neon light fittings.
2. Assume demand is known with certainty and the purchasing lead time is five days. Calculate TVH's reorder point for neon light fittings.

23-18 Effect of different order quantities on ordering costs and carrying costs, EOQ. Koala Blue retails a broad line of Australian merchandise at its Santa Monica store. It sells 26,000 Ken Done linen bedroom packages (two sheets and two pillow cases) each year. Koala Blue pays Ken Done Merchandise Inc. $104 per package. Its ordering costs per purchase order are $72. The carrying costs per package are $10.40 per year.

Pat Carrol, manager of the Santa Monica store, seeks your advice on how ordering costs and carrying costs vary with different order quantities. Ken Done Merchandise Inc. guarantees the $104 purchase cost per package for the 26,000 units budgeted to be purchased in the coming year.

Required

1. For purchase order quantities of 300, 500, 600, 700, and 900 compute the annual ordering costs, the annual carrying costs, and their sum. Use the format of Exhibit 23-1 (p. 743) to present your results. What is the economic order quantity? Comment on your results.
2. Assume that Ken Done Merchandise Inc. introduces a computerized ordering network for its customers. Pat Carrol estimates that Koala Blue's ordering costs will be reduced to $40 per purchase order. How will this ordering-cost reduction affect Koala Blue's economic order quantity for linen bedroom packages?

23-19 Effect of different order quantities on ordering costs and carrying costs, EOQ. Ritchway Supplies retails office equipment. It purchases 10,000 units a year of the Ultimate Executive Desk from Lumber Products at $1,000 per desk. Ritchway's ordering costs per purchase order to Lumber Products are $960. The carrying costs per Ultimate Executive Desk are $120 per year.

David Watson, president of Ritchway Supplies, seeks advice on the economics of purchasing in different order quantities. Lumber Products has guaranteed that the purchase price for the 10,000 units will be $1,000 per desk.

Required

1. For purchase order quantities of 100, 300, 400, 500, and 700 compute the annual ordering costs, the annual carrying costs, and their sum. Use the format of Exhibit 23-1 (p. 743) to present your results. What is the economic order quantity? Comment on your results.

2. Assume that Ritchway Supplies is able to streamline its relationships with Lumber Products. The ordering costs per purchase order drop to $240. How will this ordering-cost reduction affect Ritchway Supplies' economic order quantity for Ultimate Executive Desks?

23-20 Purchase order size for retailer, EOQ, just-in-time purchasing. The *24 Hour Mart* operates a chain of supermarkets. Its best-selling soft drink is Fruitslice. Demand in April 19_2 for Fruitslice at its Memphis supermarket is estimated to be 6,000 cases (24 cans in each case). In March 19_2, the Memphis supermarket estimated the ordering costs per purchase order (*P*) for Fruitslice to be $30. The carrying costs (*C*) of each case of Fruitslice in inventory for a month were estimated to be $1. At the end of March 19_2, the Memphis *24 Hour Mart* reestimated its carrying costs to be $1.50 per case per month to take into account an increase in warehouse-related costs.

During March 19_2, *24 Hour Mart* restructured its relationship with suppliers. It reduced the number of suppliers from 600 to 180. Long-term contracts were signed only with those suppliers that agreed to make product-quality checks before shipping. Each purchase order would be made by linking into the supplier's computer network. The Memphis *24 Hour Mart* estimated these changes would reduce the ordering costs per purchase order to $5. The *24 Hour Mart* is open 30 days in April 19_2.

Required

1. Calculate the economic order quantity in April 19_2 for Fruitslice. Use the EOQ model, and assume in turn:
   (a) $D = 6{,}000$; $P = \$30$; $C = \$1$
   (b) $D = 6{,}000$; $P = \$30$; $C = \$1.50$
   (c) $D = 6{,}000$; $P = \$5$; $C = \$1.50$
2. How does your answer to requirement 1 give insight into the retailer's movement toward just-in-time purchasing policies?

23-21 Purchase order size for retailer, EOQ, just-in-time purchasing. The Family Discount Store (FDS) operates a chain of retail discount stores. Its best-selling brand of baby diapers is Baby Care. Demand in October 19_4 for Baby Care at its Quebec City store is 800 cases (100 one-pound packages of Baby Care are in each case). In October 19_4, the Quebec City store estimated the ordering costs per purchase order (*P*) for Baby Care to be $40. The carrying costs (*C*) of each case of Baby Care diapers in inventory for a month were estimated to be $6. At the end of September 19_4, the Quebec City store reestimated its carrying costs to be $10 per case to take into account an increase in warehouse-related costs.

During September, FDS restructured its relationship with suppliers. It reduced the number of suppliers from 950 to 370. Only those suppliers that agreed to ship in the exact quantities ordered, and with quality-control checks before shipment, were given long-term supply contracts by FDS. Each individual purchase order involved minimal paperwork, and no quality checks were to be made by FDS when each delivery arrived. The Quebec City store estimated its ordering costs per purchase order to be $4.20 after these changes were made.

FDS is open 31 days in October 19_4.

Required

1. Calculate the optimal order quantity in October 19_4 for Baby Care. Use the EOQ model and assume in turn:
   (a) $D = 800$; $P = \$40$; $C = \$6$
   (b) $D = 800$; $P = \$40$; $C = \$10$
   (c) $D = 800$, $P = \$4.20$; $C = \$10$
2. How does your answer to requirement 1 give insight into the retailer's movement toward just-in-time purchasing policies?

23-22 EOQ, cost of prediction error. Ralph Menard is the owner of a truck repair shop. He uses an economic-order-quantity model for each of his truck parts. He initially predicts the annual demand for heavy-duty tires to be 2,000. Each tire has a purchase price of $50. The incremental ordering costs per purchase order are $40. The incremental carrying costs per year are $4 per unit plus 10% of the supplier purchase price.

Required

1. Calculate the EOQ for heavy-duty tires, along with the total sum of ordering costs and carrying costs.
2. Suppose Ralph is precisely correct in all his predictions but is wrong about the purchase price. He ignored a new law that abolished tariff duties on imported heavy-duty tires, which led to lower prices from foreign competitors. If he had been a faultless predictor, he would have foreseen that the purchase price would drop to $30 at the beginning of the year and would be unchanged throughout the year. What is the cost of the prediction error?

23-23 EOQ, uncertainty, safety stock, reorder point. (CMA, adapted) The Starr Company distributes a wide range of electrical products. One of its best-selling items is a standard electric motor. The management of Starr Company uses the economic-order-quantity (EOQ) decision model to determine the optimum number of motors to order. Management now wants to determine how much safety stock to hold.

Starr Company estimates annual demand (300 working days) to be 30,000 electric motors. Using the EOQ decision model, the company orders 3,000 motors at a time. The lead time for an order is five days. The annual carrying costs of one motor in safety stock are $10. Management has also estimated that the stockout cost is $20 for each motor they are short.

Starr Company has analyzed the demand during 200 past reorder periods. The records indicate the following patterns:

Demand During Lead Time	Number of Times Quantity was Demanded
440	6
460	12
480	16
500	130
520	20
540	10
560	6
	200

Required

1. Determine the level of safety stock for electric motors that Starr Company should maintain in order to minimize expected stockout costs and carrying costs. When computing carrying costs, assume that the safety stock is on hand at all times and that there is no overstocking caused by decreases in expected demand. (Consider safety stock levels of 0, 20, 40, and 60 units.)
2. What would be Starr Company's new reorder point?
3. What factors should Starr Company have considered in estimating the stockout costs?

23-24 Just-in-time production, operating efficiency. The Mannheim Group is a major manufacturer of metal-cutting machines. It has plants in Frankfurt and Stuttgart. The managers of these two plants have different manufacturing philosophies.

Richard Stehle, the recently appointed manager of the Frankfurt plant, is a convert to just-in-time production. In September 19_6, he commenced a four-month phase-in period of JIT. By January 19_7, full-scale implementation of JIT had occurred.

Frank Kohl, manager of the Stuttgart plant, has adopted a wait-and-see approach to JIT. He commented to Stehle:

> In my time, I have forgotten more manufacturing acronyms than you have read about in your five-year career. In two years' time, JIT will join the "manufacturing buzzword scrapheap."

Kohl continues with his "well-honed" traditional approach to manufacturing at the Stuttgart plant.

Summary operating data for the two plants in 19_7 follow:

	Jan.–Mar.	Apr.–June	July–Sept.	Oct.–Dec.
Throughput time (days)				
Frankfurt	9.2	8.7	7.4	6.2
Stuttgart	8.3	8.2	8.4	8.1
Total setup time for machines / Total production time				
Frankfurt	52.1%	49.6%	43.8%	39.2%
Stuttgart	47.6	48.1	46.7	47.5
Number of units requiring rework / Total number of units started and completed				
Frankfurt	64.7%	59.6%	52.1%	35.6%
Stuttgart	53.8	56.2	51.6	52.7

Required

1. What are the key features of JIT production?
2. Compare the operating performance of the Frankfurt and Stuttgart plants in 19_7. Comment on any differences you observe.
3. Stehle is concerned about the level of detail on the job-cost records for the cutting machines manufactured at the Frankfurt plant during 19_8. What reasons might lead Stehle to simplify the job-cost records?

CHAPTER 24

Linear-programming models help managers make decisions, often about short-run resource allocations. Petroleum refiners, chemical manufacturers, and food-processing companies frequently take advantage of linear programming.

Operations Management and the Accountant: Linear Programming

The linear-programming (LP) model

Steps in solving an LP problem

Substitution of scarce resources

Implications of LP for managers and accountants

Other uses of linear programming

Building the model

Learning Objectives

When you have finished studying this chapter, you should be able to

1. Understand how the LP model is linked to cost-volume-profit analysis
2. Describe the three steps in solving an LP problem
3. Explain how the trial-and-error and graphic approaches determine the optimal solution
4. Demonstrate why a policy of producing as much as possible of the product with the highest contribution margin per unit need not be optimal
5. Compute the cost of a prediction error for the contribution margin of a product in the objective function
6. Explain the importance of the following three assumptions underlying LP—linearity, certainty, and single cost driver

Accountants play an important role in providing inputs to decision models used by operating managers. This chapter illustrates the role of accounting data in linear-programming decision models. **Linear programming** (often referred to as **LP**) focuses on how best to allocate scarce resources to attain a chosen objective, such as maximization of operating income or minimization of operating costs. All the relationships in the LP model are linear. Industries featuring widespread use of LP include petroleum refining, chemicals, and food processing.

The accountant who is able to interpret the inputs, outputs, assumptions, and limitations of LP can play a vital role in managing the operations of the organization. We begin by examining the LP model. Then we explore LP's implications for accounting systems and measurements. Our cost-benefit theme persists; the worth of formal models and more elaborate systems ultimately depends on whether decisions will be collectively improved.

THE LINEAR-PROGRAMMING (LP) MODEL

The linear-programming model is used for a variety of decisions—such as product mix, production scheduling, and the blending of materials—most frequently as a short-run resource allocation model. LP assumes that a given set of resources is available and that those resources will generate a specified level of fixed costs. In using the LP model, a manager's objective is to choose what types and amounts of products or services to produce or sell. A detailed example concerning the choice of an optimal product mix is presented below.

Linkage to Previous Chapters

Chapter 3 discussed cost-volume-profit relationships. That chapter dealt mostly with a single product or a combination of products being sold without restriction. It focused on maximizing contribution margin, given a level of fixed costs. Chapter 11 (see pp. 373–74) illustrated the importance of considering the scarcity of resources in product-mix decisions. The most profitable product is not necessarily the product with the highest contribution margin per unit. Instead, the most profitable product is the one that produces the highest contribution per unit of the scarce resource or constraining factor, such as total available machine hours.

Objective 1

Understand how the LP model is linked to cost-volume-profit analysis

The basic idea presented in Chapter 11 still holds, but in practice there is usually more than one constraint. Therefore, the problem becomes one of maximizing total contribution margin, given multiple constraints. The LP model is designed to provide a solution to problems where the linear assumptions underlying the model are reasonable.

In applying linear-programming models, we assume that only one factor—volume—causes change in the total cost of products. All other costs are assumed to be fixed. This perspective is called the single cost-driver assumption. For many short-run decisions, this assumption may be reasonable. Where the assumption is not reasonable, other decision models should be considered.[1]

Example: Deriving an Optimal Product Mix

Consider Power Engines, a company that manufactures engines for a broad range of commercial and consumer products. At its Lexington, Kentucky, plant, it assembles and tests two deluxe engine models—a snowmobile engine and an outboard boat engine. Each engine model requires the use of both the assembly department and the testing department. Production data follow:

		Use of Capacity in Hours Per Unit of Product		Daily Maximum Production in Units	
Department	**Available Capacity in Hours**	**Snowmobile Engine (A)**	**Boat Engine (B)**	**Snowmobile Engine**	**Boat Engine**
Department 1: Assembly	300 machine hours	1.5	2.0	200	150
Department 2: Testing	120 testing hours	1.0	0.5	120	240

Suppose a department works exclusively on a single product (engine type). The table indicates that the assembly department can assemble a maximum of 200 snowmobile engines (300 machine hours ÷ 1.5 machine hours per unit = 200 units) or 150 boat engines (300 machine hours ÷ 2.0 machine hours per unit = 150 units). Similarly, the testing department can test 120 snowmobile engines (120 testing hours ÷ 1.0 testing hour per unit = 120 units) or 240 boat engines (120 testing hours ÷ 0.5 testing hour per unit = 240 units).

Exhibit 24-1 summarizes these and other relevant data. Note that snowmobile engines have a contribution margin of $200 per engine and that boat engines have a contribution margin of $250 per engine. Material shortages for boat engines will limit production to 126 engines per day. How many engines of each type should be produced daily to maximize operating income?

STEPS IN SOLVING AN LP PROBLEM

Objective 2

Describe the three steps in solving an LP problem

To solve an LP problem, follow these three steps:

- *Step 1: Determine the objective.* The *objective function* of a linear program expresses the objective or goal to be maximized (for example, operating income) or minimized (for example, operating costs).

[1]Other decision models are described in G. Eppen, F. Gould, and C. Schmidt, *Quantitative Concepts for Management* (Englewood Cliffs, NJ: Prentice Hall, 1988); and S. Nahmias, *Production and Operations Analysis* (Homewood, IL: Irwin, 1989).

EXHIBIT 24-1
Power Engines: Operating Data

Product	Capacities (Per Day) in Product Unit: Department 1: Assembly	Department 1: Testing	Selling Price Per Unit	Variable Costs Per Unit	Contribution Margin Per Unit
If produce snowmobile engines (*A*) only	200	120	$800	$600	$200
If produce boat engines (*B*) only	150	240	$950	$700	$250

- *Step 2: Determine the basic relationships.* These relationships include constraints expressed in linear functions. A **constraint** is a mathematical inequality or equality that must be satisfied by the variables in a mathematical model.
- *Step 3: Compute the optimal solution.* With only two variables in the objective function and a minimal number of constraints, the trial-and-error solution approach or the graphic solution approach can be used. Both approaches are explained in this chapter. An understanding of these two approaches provides insight into LP modeling. In most real-world LP applications, computer software packages provide the optimal solution.

Illustration of Steps

We use the data in Exhibit 24-1 to illustrate the three steps in solving an LP problem. Throughout this discussion, A equals the number of units of snowmobiles produced and B equals the number of units of boat engines produced.

STEP 1: Determine the objective. The objective is to find the combination of products that maximizes the total contribution margin. The linear function expressing this objective is:

$$\text{Total contribution margin (TCM)} = \$200A + \$250B$$

STEP 2: Determine the basic relationships. The relationships can be depicted by the following inequalities:

Department 1 constraint (assembly)	$1.5A + 2.0B \leq 300$
Department 2 constraint (testing)	$1.0A + 0.5B \leq 120$
Material shortage constraint for Product B	$B \leq 126$
Because negative production is impossible	$A \geq 0$ and $B \geq 0$

The coefficients of the constraints are often called technical coefficients. For example, in the assembly department the technical coefficient for product A (snowmobile engines) is 1.5 machine hours, and product B (boat engines) is 2.0 machine hours.

The three solid lines on the graph in Exhibit 24-2 show the existing constraints for Departments 1 (assembly) and 2 (testing) and of the material shortage constraint.[2] The feasible alternatives are those that are technically possible. The "area of feasible solutions" in Exhibit 24-2 shows the boundaries of those product combi-

[2]As an example of how the lines are plotted in Exhibit 24-2, use equal signs instead of inequality signs and assume for Department 1 (assembly) that $B = 0$, then $A = 200$ (300 machine hours ÷ 1.5 machine hours per snowmobile engine). Assume that $A = 0$, then $B = 150$ (300 machine hours ÷ 2.0 machine hours per boat engine). Connect those two points with a straight line.

EXHIBIT 24-2
Power Engines: Linear Programming—Graphic Solution

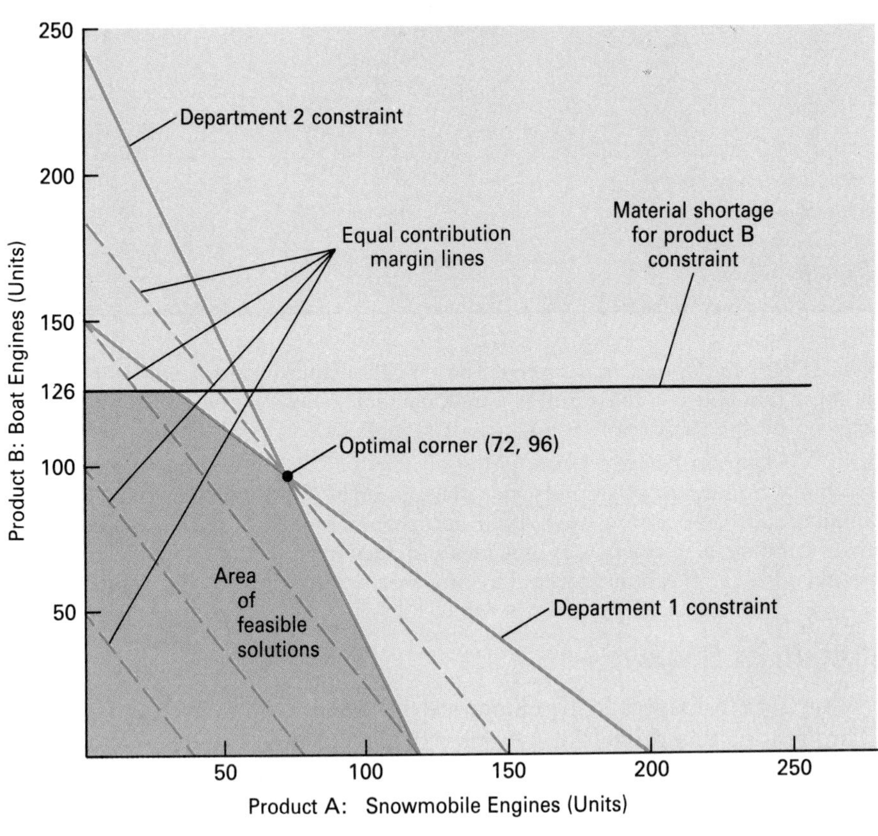

nations that are feasible—that is, combinations of quantities of snowmobile engines and boat engines that satisfy all the constraining factors.

Objective 3

Explain how the trial-and-error and graphic approaches determine the optimal solution

STEP 3: Compute the optimal solution. In step 2, we concentrate on physical relationships. We now return to the economic relationships expressed as the objective function in step 1. The trial-and-error solution approach and the graphic solution approach will be presented.

Trial-and-Error Solution Approach

The optimal solution can be found by trial and error, usually by working with coordinates of the corners of the area of feasible solutions. The approach is simple.

First, select any set of corner points, and compute the total contribution margin. The five corner points appear in Exhibit 24-2. It is helpful to use simultaneous equations to check the graph coordinates. To illustrate, the point ($A = 72; B = 96$) can be derived by solving the two pertinent constraint inequalities as simultaneous equations:

(1)	$1.5A + 2.0B = 300$
(2)	$1.0A + 0.5B = 120$

Multiply (2) by 1.5:

(3)	$1.5A + 0.75B = 180$

Subtract (3) from (1):

$$1.25B = 120$$
$$B = 120 \div 1.25 = 96$$

Substitute for B in (2):

$$1.0A + 0.5\,(96) = 120$$
$$A = 120 - 48 = 72$$

Given $A = 72$ and $B = 96$:

$$\text{TCM} = \$200\,(72) + \$250\,(96) = \$38{,}400$$

Second, move from corner point to corner point, comparing the total contribution margin at that corner point with the total contribution margin at each of the previously examined corner points. These computations, corner by corner, are summarized as follows:[3]

Trial	Corner Point (*A*;*B*)	Product *A*: Snowmobile Engines	Product *B*: Boat Engines	Total Contribution Margin
1	(0;0)	0	0	\$200 (0) + \$250 (0) = \$ 0
2	(0;126)	0	126	\$200 (0) + \$250 (126) = 31,500
3	(32;126)	32	126	\$200 (32) + \$250 (126) = 37,900
4	(72;96)	72	96	\$200 (72) + \$250 (96) = 38,400*
5	(120;0)	120	0	\$200 (120) + \$250 (0) = 24,000

*Indicates optimal solution.

The optimal product mix is 72 snowmobile engines and 96 boat engines.

Graphic Solution Approach

In the graphic approach, the optimal solution must lie on one of the corner points of the "area of feasible solutions." Why? Consider all possible combinations that will produce an equal total contribution margin of, say, \$10,000. That is:

$$\$200A + \$250B = \$10{,}000$$

This set of \$10,000 contribution margins is a straight dashed line through (A = 50;B = 0) and (A = 0;B = 40). Other equal total contribution margins can be represented by lines parallel to this one. In Exhibit 24-2, we show three such lines. The equal total contribution margins increase as the lines get farther from the origin. The optimal line is the one farthest from the origin that has a feasible solution; this happens at a corner (A = 72;B = 96). This solution will become apparent if you put a ruler on the graph and move it outward from the origin and parallel with the \$10,000 line. In general, the optimal solution in a maximization problem lies at the corner where the dashed line intersects an extreme point of the area of feasible solutions.

The slope of the objective function (the dashed line representing the equal total contribution margin, TCM) can be computed from the equation

$$\text{TCM} = \$200A + \$250B$$

[3]Although the trial-and-error and graphic approaches can be useful for two or possibly three variables, they are impractical when many variables exist. Standard computer software packages rely on the simplex method. The **simplex method** is an iterative step-by-step procedure for determining the optimal solution to an LP problem. It starts with a specific feasible solution and then tests it by substitution to see if the result can be improved. These substitutions continue until no further improvement is possible, and thus the optimal solution is obtained.

To find the slope (the rate of change of B for one additional unit of A), divide by the coefficient of B and transfer B to the left-hand side of the equation:

$$\frac{\text{TCM}}{\$250} = \frac{\$200}{\$250}A + B$$

$$B = \frac{\text{TCM}}{\$250} - \frac{\$200}{\$250}A$$

$$B = \frac{\text{TCM}}{\$250} - \frac{4}{5}A$$

The slope of the objective function is a negative \$200/\$250, or −4/5. The graphic solution approach provides insight into the computation of the optimal solution. Its use, however, is restricted to LP problems with two products in the objective function (so that the solution can be represented on a two-dimensional graph).

SUBSTITUTION OF SCARCE RESOURCES

Objective 4

Demonstrate why a policy of producing as much as possible of the product with the highest contribution margin per unit need not be optimal

A common fallacy is to assume that the optimal solution is always to produce as much as possible of the product with the higher contribution margin per unit. In the above example, this would entail maximizing production of boat engines (contribution margin of \$250) and then devoting any remaining capacity to snowmobile engines (contribution margin of \$200). This approach ignores the fact that productive capacity is the scarce resource. ***The key to the optimal solution is exchanging a given contribution margin per unit of scarce resource for some other contribution margin per unit of scarce resource.*** This point becomes clearer if we examine Exhibit 24-2. Moving from corner ($A = 32; B = 126$) to corner ($A = 72; B = 96$) implies that the company is transferring the scarce resource (production capacity) between the products.

In the assembly department, each machine hour devoted to one unit of boat engines (B) may be given up (sacrificed or traded) for 1.33 (2 hours required for 1 boat engine ÷ 1.5 hours required for 1 snowmobile engine) units of snowmobile engines (A). Will this exchange add to profitability? Yes, as shown below:

Total contribution margin at ($A = 32; B = 126$):		
\$200 (32) + \$250 (126)		\$37,900
Added contribution margin from Product A		
by moving to corner ($A = 72; B = 96$):		
72 − 32, or 40 more units @ \$200	\$8,000	
Lost contribution margin from Product B		
by moving to corner ($A = 72; B = 96$):		
126 − 96, or 30 less units @ \$250	7,500	
Net additional contribution margin		500
Total contribution margin at corner		
($A = 72; B = 96$):		
\$200 (72) + \$250 (96)		\$38,400

As we move from corner ($A = 32; B = 126$) to corner ($A = 72; B = 96$), we are contending with the Department 1 (assembly) constraint. There is a net advantage of trading one unit of B (boat engines) for 1.33 units of A (snowmobile engines).

The heart of the substitution between products is a matter of exchanging a given contribution margin per unit of *scarce resource* for some other contribution margin per unit of *scarce resource;* it is not simply a matter of comparing contribution margins per unit of *product.*

IMPLICATIONS OF LP FOR MANAGERS AND ACCOUNTANTS

Many uses of the LP model include both accounting coefficients and technical coefficients. The contribution margins of the snowmobile and boat engines in Exhibit 24-1 are examples of accounting coefficients. Examples of technical coefficients in Exhibit 24-1 are the coefficients of the capacity constraints.

Sensitivity Analysis

What are the implications of uncertainty about the accounting or technical coefficients used in the LP model? ***Both sets of coefficients inevitably affect the slope of the objective function (the equal-contribution margin line in our example) or the area of feasible solutions.*** Questions about uncertainty are frequently explored through sensitivity analysis—that is, by asking "what-if?" questions.

In some cases, the optimal decision about product mix may be unchanged, even though there is a revision in the estimated contribution margin. Errors in approximations of costs and revenues per unit may reduce or increase unit contribution margins but may not tilt the slope of the equal-contribution margin line enough to alter the optimal solution.

An important question is: How much variation in the accounting coefficients or the technical coefficients can exist before the optimal solution will change? Consider the (A = 72;B = 96) optimal solution in Exhibit 24-2. If the slope of the objective function is the same as that of a constraint being examined, any point where the two slopes overlap will be an optimal solution.

Consider how changes in the contribution margin of snowmobile engines might affect the optimal solution. Assume the contribution margin falls from \$200 to \$187.50. The revised objective function will be:

$$\text{TCM} = \$187.50A + \$250B$$

This objective function has the same slope as the constraint for the assembly department:

$$1.5A + 2.0B \leq 300$$

The total contribution margin will be the same for corners (A = 32;B = 126) and (A = 72;B = 96), and for all points joining the two corner points.

Corner Point	Total Contribution Margin
(A = 32;B = 126)	TCM = \$187.50 (32) + \$250 (126) = \$37,500
(A = 72;B = 96)	TCM = \$187.50 (72) + \$250 (96) = \$37,500

In this case, the slope of the constraint and the slope of the objective function are equal. We now consider a case where these slopes are not equal.

Cost of Prediction Error

Objective 5

Compute the cost of a prediction error for the contribution margin of a product in the objective function

Assume the actual contribution margin for the snowmobile engine is \$130 and not the predicted amount of \$200. What is the cost of this prediction error, assuming all other predictions are correct? A three-step procedure answers this question:

- *Step 1:* Compute the monetary outcome of the best action that could have been taken, given the actual value of the predicted variables. Given a \$130 contribution margin for snowmobile engines, the optimal combination is ($A = 32; B = 126$). The total contribution margin is

$$\text{TCM} = \$130\ (32) + \$250\ (126) = \$35{,}660$$

- *Step 2:* Compute the monetary outcome of the best action based on the incorrect value of the predicted variable. The predicted $200 contribution margin for snowmobile engines leads to the selection of the ($A = 72; B = 96$) combination. The total contribution margin from this action, given the $130 actual contribution margin for snowmobile engines, is

$$\text{TCM} = \$130\ (72) + \$250\ (96) = \$33{,}360$$

- *Step 3:* Compute the difference between the monetary outcomes in steps 1 and 2:

	Monetary Outcome
Step 1	$35,660
Step 2	33,360
Difference	$ 2,300

The cost of the prediction error for the contribution margin of snowmobile engines is $2,300.

If prediction errors are costly, more resources should be spent to enhance the accuracy of the predictions before a decision is made. Use of sensitivity analysis, incorporating the possible range of values of the accounting and technical coefficients, will highlight those prediction errors that potentially are very costly. Once the choice is made (such as the product mix), extra efforts may also be made to implement the decision so that the predicted outcomes (for example, costs or production volumes) are achieved. Such endeavors can be economically justifiable when the costs of prediction errors are high.

OTHER USES OF LINEAR PROGRAMMING

LP models have proved useful in a wide variety of decisions, in addition to the product-mix decision illustrated above. LP models also may be used in decisions in the following areas:

Production Scheduling. LP models have helped production planning so that demand requirements are met at minimum cost, subject to constraints concerning production capacity, subcontracting, and inventory holdings. Key cost inputs in this application include the cost per unit of production and the cost of holding each unit of inventory.

Blending of Materials. A typical objective is cost minimization subject to constraints concerning material availability and final end-product quality. Successful applications have been reported in such diverse settings as the mixing of paint, the mixing of meats to make sausages, the blending of chocolate products, and the blending of petroleum products. Key cost inputs include the costs of the individual materials and the processing costs for individual materials and combinations of materials.

Transportation/Distribution. The typical aim of this application is to meet demand for a product from different geographical markets using several geographically dispersed plants, while minimizing transportation and distribution costs. A critical input is the cost of transporting products from each plant to each market area.

Personnel Planning. An example of an LP application in planning personnel is the assignment of nurses to hospital departments that undergo peaks and troughs in demand. Key inputs in these applications include the cost of each type of nurse for normal hours and the cost of overtime and supplemental staff.

With a maximization problem, the solution is found by moving the line representing the objective function away from the origin. In contrast, the line representing the objective function is moved toward the origin in cost minimization problems.

BUILDING THE MODEL

The most challenging aspects of using an LP model are

- Expressing a business problem in a format suitable to LP analysis, and
- Obtaining reliable estimates of the information entered into the model.

Managers and accountants should recognize the types of problems best analyzed by linear programming. These people should also help in constructing the model by specifying the objective function, the constraints, and the variables. Generally, managers and accountants concentrate on building the model and determining its input and leave the technical intricacies to software programmers.

When you encounter applications of LP, be aware that there are several assumptions underlying the optimal LP solution:

Objective 6

Explain the importance of the following three assumptions underlying LP—linearity, certainty, and single cost driver

1. All relationships are linear. This linearity applies both to the objective function and to all constraints. Linearity excludes, for instance, price discounts for volume purchases and cost functions in which learning occurs as production volume increases.

2. All coefficients and constraints in the model are known with certainty. At least one constraint will be binding in the solution to an LP model. For example, the optimal solution to the Power Engines LP model in Exhibit 24-2 uses all 300 machine hours of the capacity in the assembly department. Rarely will a department know that its capacity is exactly 300 machine hours. Typically, there will be uncertainty. Capacity may fluctuate over time because of variations in machine performance, setup times, and so on. One day capacity may be 310 hours of assembly time, and on another day it may be 290 hours. Users of LP who believe this uncertainty is important may decide to make adjustments to the optimal solution reported by an LP model.[4]

3. All costs are either variable with respect to the volume output measure of the objective function or they are fixed. This is the single cost-driver assumption: there is only one factor whose change causes a change in the total cost of the products included in the objective function. For example, the costs of Power Engines are assumed to be either variable with respect to units produced or they are fixed. Most applications of LP are short-run resource allocation decisions. For such short-run decisions, this single cost-driver assumption may be reasonable. For decisions with a longer-term horizon, however, resource allocation models that recognize multiple cost drivers are likely to be appropriate.

[4]Detailed discussion of the effect of recognizing uncertainty is in R. D. Banker, S. M. Datar, and S. Kekre, "Relevant Costs, Congestion and Stochasticity in Production Environments," *Journal of Accounting & Economics*, July 1988.

PROBLEM FOR SELF-STUDY

PROBLEM

Reconsider the Power Engines example. Exhibit 24-2 reports predicted contribution margins of \$200 for snowmobile engines and \$250 for boat engines. After examining the internal accounting data, an analyst discovers that the contribution margin estimate for boat engines should be \$300 rather than \$250. No change in the estimated contribution margin of snowmobile engines is required. What is the cost of incorrectly predicting a contribution margin of \$250 for boat engines instead of \$300?

SOLUTION

The trial-and-error solution approach can be used. Total contribution margin calculations for the five corner product combinations follow:

Trial	Corner Point (*A*;*B*)	Product *A*: Snowmobile Engines	Product *B*: Boat Engines	Total Contribution Margin
1	(0;0)	0	0	\$200 (0) + \$300 (0) = \$ 0
2	(0;126)	0	126	\$200 (0) + \$300 (126) = 37,800
3	(32;126)	32	126	\$200 (32) + \$300 (126) = 44,200*
4	(72;96)	72	96	\$200 (72) + \$300 (96) = 43,200
5	(120;0)	120	0	\$200 (120) + \$300 (0) = 24,000

*Optimal solution.

The cost of a prediction error can be computed as follows:

- Step 1: Compute the monetary outcome of the best action that could have been taken, given the actual value of the predicted variables:

$$TCM = \$200\ (32) + \$300\ (126) = \$44{,}200$$

- Step 2: Compute the monetary outcome of the best action based on the incorrect value of the predicted variable. The incorrect \$250 contribution margin for boat engines leads to the selection of a ($A = 72;B = 96$) combination. The total contribution margin from this action, given the \$300 contribution margin figure for boat engines, is

$$TCM = \$200\ (72) + \$300\ (96) = \$43{,}200$$

- Step 3: Compute the difference between the monetary outcomes in steps 1 and 2:

	Monetary Outcome
Step 1	\$44,200
Step 2	43,200
Difference	\$ 1,000

The cost of the prediction error for the contribution margin of boat engines is \$1,000.

SUMMARY

Linear programming is a popular decision model for determining how to optimize the use of a given set of scarce resources. The LP model is often regarded as a multiple-product extension of the cost-volume-profit model introduced in Chapter 3 and an extension of the resource-allocation model in Chapter 11.

This chapter illustrated both the trial-and-error solution approach and the graphic solution approach. Both approaches provide users with insight about how an LP model selects the optimal solution. Most applications in the real world, having numerous products in the objective function and numerous constraints, rely on readily available computer software programs.

Managers and accountants should be especially alert to the sensitivity of the solution to possible measurement errors in the accounting coefficients and the technical coefficients of the LP model.

TERMS TO LEARN

This chapter and the Glossary at the end of the book contain definitions of the following important terms:

constraint *(p. 765)* linear programming (LP) *(763)* simplex method *(767)*

ASSIGNMENT MATERIAL

QUESTIONS

24-1 What is the most widespread application of the LP model?

24-2 Describe the relationship between cost-volume-profit analysis (Chapter 3) and linear programming.

24-3 What are the three steps in solving a linear-programming problem?

24-4 How might the optimal solution of an LP problem be determined?

24-5 What is the main limitation of the graphic solution approach?

24-6 "The optimal product mix will maximize production of the product with the highest contribution margin." Do you agree? Explain.

24-7 How can sensitivity analysis be used when an LP model guides the choice of product mix?

24-8 Name the three steps in computing the cost of a prediction error in an accounting or technical coefficient of an LP model.

24-9 Name five areas where LP models have proved useful in decision making.

24-10 Name the two most challenging aspects of using the LP model in business decisions.

24-11 What are three assumptions made when using the LP model?

EXERCISES AND PROBLEMS

24-12 Optimal production plan, computer manufacturer. Information Technology, Inc., assembles and sells two products, printers and desktop computers. Customers can purchase either (i) a computer or (ii) a computer plus a printer. The printers are *not* sold without the computer. The result is that the quantity of printers sold is equal to or less than the quantity of desktop computers sold. The contribution margins are $200 per printer and $100 per computer.

Each printer requires 6 hours assembly time on production line 1 and 10 hours assembly time on production line 2. Each computer requires 4 hours assembly time on production line 1 only. (Many of the components of each computer are preassembled by external vendors.) Production line 1 has 24 hours of available time per day. Production line 2 has 20 hours of available time per day.

Let X represent units of printers and Y represent units of desktop computers.

Required

1. Express the relationships in an LP format.
2. Which combination of printers and computers will maximize the operating income of Information Technology? Use both the trial-and-error and the graphic approaches.

24-13 Minimum cost combination, fertilizer mix. The local agricultural center has advised Sam Bowers to spread at least 4,800 pounds of a special nitrogen fertilizer ingredient and at least 5,000 pounds of a special phosphate fertilizer ingredient in order to increase his crop yield. Neither ingredient is available in pure form.

A dealer has offered 100-pound bags of VIM at $10 each. VIM contains the equivalent of 20 pounds of nitrogen and 80 pounds of phosphate. VOOM is also available in 100-pound bags, at $30 each. It contains the equivalent of 75 pounds of nitrogen and 25 pounds of phosphate.

Let X represent bags of VIM and Y represent bags of VOOM.

Required

How many bags of VIM and VOOM should Bowers buy in order to obtain the required fertilizer at minimum cost? Solve graphically.

24-14 Optimal sales mix for a retailer, sensitivity analysis. Always Open, Inc., operates a chain of food stores open twenty-four hours a day. Each store has a standard 40,000 square feet of floor space available for merchandise. Merchandise is grouped in two categories: grocery products and dairy products. Always Open requires each store to devote a minimum of 10,000 square feet to grocery products and a minimum of 8,000 square feet to dairy products. Within these restrictions, each store manager can choose the mix of products to carry.

The manager of the Winnipeg store estimates the following weekly contribution margins per square foot:

Grocery products	$10
Dairy products	3

Required

1. Formulate the decision facing the store manager as an LP model. Use G to represent square feet of floor space for grocery products and D to represent square feet of floor space for dairy products.
2. Why might Always Open set minimum bounds on the floor space devoted to each line of product?
3. Compute the optimal mix of grocery products and dairy products to carry at the Winnipeg store. In the graphic solution approach, use the horizontal axis for grocery products and the vertical axis for dairy products. Use both the graphic solution approach and the trial-and-error solution approach.
4. Suppose the unit contribution margin for dairy products is known to be certain. How much measurement error in the unit contribution margin of grocery products can be tolerated before the optimal solution in requirement 3 will change?

24-15 Optimal floor space allocation, sensitivity analysis. Family Video operates a chain of video stores. Each store has a standard 10,000 square feet of display space for tapes. Merchandise (VHS tapes) is grouped in two categories: new releases and old releases. Family Video requires each store to devote a minimum of 4,000 square feet to new releases and a minimum of 3,000 square feet to old releases. Within these restrictions, each store manager can choose the mix of tapes to carry.

The manager of the Cooperstown store estimates the following monthly contribution margins per square foot:

New releases	$18
Old releases	12

Required

1. Formulate the decision facing the store manager as an LP model. Use A to represent square feet of display space for new releases and B to represent square feet of display space for old releases.
2. Compute the optimal floor space allocation to new releases and old releases at the Cooperstown store. Use both the graphic solution approach and the trial-and-error solution approach. In the graphic solution approach, use the horizontal axis for new releases and the vertical axis for old releases.
3. Suppose the unit contribution margin for old releases is known to be certain. How much reduction in the unit contribution margin of new releases can be tolerated before the optimal solution in requirement 2 will change?

24-16 Formulation of LP problem, assumptions of LP model. (CMA, adapted) The Tripro Company produces and sells three products, A, B, and C. The company is currently changing its short-range planning approach in an attempt to incorporate newer management techniques. The controller and some of his staff have been conferring with a consultant on the feasibility of using a linear-programming model for determining the optimal product mix.

Information available for short-range planning includes expected selling prices and expected direct labor and material costs for each product. Variable-overhead and fixed-overhead costs are assumed to be the same for each product because approximately equal quantities of the products are produced and sold.

Price and Cost Information (Per Unit)

	A	B	C
Selling price	$25.00	$30.00	$40.00
Direct labor	7.50	10.00	12.50
Direct materials	9.00	6.00	10.50
Variable overhead	6.00	6.00	6.00
Fixed overhead	6.00	6.00	6.00

All three products use the same type of direct material, which costs $1.50 per pound. Direct labor is paid at the rate of $5.00 per hour. There are 2,000 direct labor hours and 20,000 pounds of direct materials available in a month.

Required

1. Formulate and label the LP objective function and constraint functions necessary to maximize Tripro's contribution margin. Use Q_A, Q_B, and Q_C to represent units of the three products.
2. What underlying assumptions must be satisfied to justify the use of LP?

24-17 Effect of change in contribution margins, cost of a prediction error. United Foods processes and packages three frozen vegetable products for a chain of restaurants: carrots (C), peas (P), and zucchini (Z). United Foods uses linear programming to determine the optimal product mix. The objective function of this LP in July was

$$\text{Maximize } \$1C + \$2P + \$4Z$$

The optimal product mix for July was C = 100,000 packages, P = 200,000 packages, and Z = 300,000 packages.

Later, investigation of cost behavior revealed that the unit contribution margins should have been $1.10, $2.80, and $4.30 for C, P, and Z, respectively. Although these better cost approximations provide a more accurate estimate of the individual product contribution margins, the optimal decision of 100,000, 200,000, and 300,000 packages for C, P, and Z, respectively, was unchanged. This change in unit contribution margins did not change the corner point solution originally identified by the LP model as optimal.

Required

1. Compute the predicted total contribution margin at the time the July optimal product mix was originally determined.
2. Compute the total contribution margin using the revised estimates of the unit contribution margins per package of C, P, and Z.
3. What is the cost of the error in predicting the July unit contribution margins of C, P, and Z?

24-18 Formulation of LP, multiple choice. (CPA) Gant Company markets two products, Alpha and Gamma. The unit contribution margins per gallon are $5 for Alpha and $4 for Gamma. Both products consist of two ingredients, D and K. Alpha contains 80% D and 20% K. The proportions of the ingredients in Gamma are 40% D and 60% K. The current inventory is 16,000 gallons of D and 6,000 gallons of K. The only company producing D and K is on strike and will neither deliver nor produce them in the foreseeable future. Gant Company wishes to know the numbers of gallons of Alpha and Gamma that it should produce with its present stock of direct materials in order to maximize its total revenue.

$$X_1 = \text{Number of gallons of Alpha}$$

X_2 = Number of gallons of Gamma
X_3 = Number of gallons of D
X_4 = Number of gallons of K

Required

Choose the one correct answer for each requirement. Show computations for items 4 and 5.

1. The objective function for this problem could be expressed as
   a. $f_{max} = 0X_1 + 0X_2 + 5X_3 + 5X_4$
   b. $f_{min} = 5X_1 + 4X_2 + 0X_3 + 0X_4$
   c. $f_{max} = 5X_1 + 4X_2 + 0X_3 + 0X_4$
   d. $f_{max} = X_1 + X_2 + 5X_3 + 4X_4$
   e. $f_{max} = 4X_1 + 5X_2 + X_3 + X_4$
2. The constraint imposed by the quantity of D on hand could be expressed as
   a. $X_1 + X_2 \geq 16{,}000$
   b. $X_1 + X_2 \leq 16{,}000$
   c. $.4X_1 + .6X_2 \leq 16{,}000$
   d. $.8X_1 + .4X_2 \geq 16{,}000$
   e. $.8X_1 + .4X_2 \leq 16{,}000$
3. The constraint imposed by the quantity of K on hand could be expressed as
   a. $X_1 + X_2 \geq 6{,}000$
   b. $X_1 + X_2 \leq 6{,}000$
   c. $.8X_1 + .2X_2 \leq 6{,}000$
   d. $.8X_1 + .2X_2 \geq 6{,}000$
   e. $.2X_1 + .6X_2 \leq 6{,}000$
4. To maximize total contribution margin, the company should produce and market
   a. 106,000 gallons of Alpha only
   b. 90,000 gallons of Alpha and 16,000 gallons of Gamma
   c. 16,000 gallons of Alpha and 90,000 gallons of Gamma
   d. 18,000 gallons of Alpha and 4,000 gallons of Gamma
   e. 4,000 gallons of Alpha and 18,000 gallons of Gamma
5. Assuming that the contribution margins per gallon are $7 for Alpha and $9 for Gamma, the company should produce and market
   a. 106,000 gallons of Alpha only
   b. 90,000 gallons of Alpha and 16,000 gallons of Gamma
   c. 16,000 gallons of Alpha and 90,000 gallons of Gamma
   d. 18,000 gallons of Alpha and 4,000 gallons of Gamma
   e. 4,000 gallons of Alpha and 18,000 gallons of Gamma

24-19 Product mix, optimization, sensitivity analysis. Eveready Batteries (EB) assembles and tests heavy-duty batteries for use in industrial applications. There are two manufacturing departments:

- Assembly department
- Testing department

EB has two products—Eveready Plus and Eveready Triple Duty. The hours required for each product in each department are:

	Assembly Department	Testing Department
Eveready Plus	2.0 hours	1.5 hours
Eveready Triple Duty	4.0 hours	1.0 hours

The assembly department has a maximum of 12,000 hours available per month. The testing department has a maximum of 6,000 hours available per month. Because of the limited availability of component parts for the Eveready Plus battery, the maximum number of Eveready Plus batteries per month is 3,500.

Summary financial data for each product are:

	Eveready Plus	Eveready Triple Duty
Selling price	$220	$350
Direct materials	70	110

Variable assembly department costs are $30 per hour. Variable testing department costs are $20 per hour.

Required

1. What are the unit contribution margins of Eveready Plus and of Eveready Triple Duty?
2. What is the optimal mix of the Eveready Plus and Eveready Triple Duty products per month?
3. Assume that import duties are abolished for products similar to the Eveready Triple Duty. Foreign competition is expected to drive battery prices down. How much can the unit contribution margin for this battery decline before the optimal solution identified in requirement 2 is changed?

24-20 Product-mix decision, scarce productive resources. (CMA, adapted) WoodCo is a wholesale distributor supplying a wide range of equipment and tools to the logging industry. About half of WoodCo's products are purchased from other companies, and the remainder WoodCo manufactures for distribution. One of the items that WoodCo makes is a fiberglass reinforced case used to carry long-bladed chain saws. In a normal year, WoodCo manufactures and sells 5,000 of these cases. This production makes full use of WoodCo's capacity in the Fiberglass Processing Department, which is limited by the amount of direct labor that can be expended at a fixed number of workstations. WoodCo believes it could sell a maximum of 7,000 cases if it had the manufacturing capacity. Presented below are the actual selling price and cost data for manufacturing chain saw cases.

Selling price per case		$80.00
Costs per case		
Fiberglass and resin	$24.00	
Handle, hinges, latch	7.00	
Direct labor ($12.00/hr.)	12.00	
Manufacturing overhead	6.00	
Marketing and administrative costs	16.00	65.00
Operating income per case		$15.00

WoodCo has looked into the possibility of purchasing chain saw cases for distribution. The company learned that Murphy Supply could provide the cases in volume quantities at a price of $60.00 each, delivered to WoodCo. The maximum number of cases that Murphy could supply to WoodCo in any given year would be 6,000, and the cases must be bought in increments of 1,000. Other goods supplied from Murphy in the past have been of very good quality, and WoodCo regards Murphy as a reliable vendor.

The only other product that WoodCo's Fiberglass Processing Department could make is a brightly colored hard hat. At present, WoodCo does not sell hard hats. However, a market analysis indicates there is room for another producer. The market analysis revealed that a selling price of $35.00 per hat would be best. At this price, WoodCo could expect to sell 9,500 hard hats annually. The data gathered indicate that the operating income per hard hat would be considerably less than the operating income per chain saw case. The estimated selling price and manufacturing costs for the hard hats follow:

Selling price per hat		$35.00
Costs per hat		
Fiberglass and resin	$12.00	
Liner, strap, rivets	5.00	
Direct labor ($12.00/hr.)	4.80	
Manufacturing overhead	2.40	
Marketing and administrative costs	3.50	27.70
Operating income per hat		$ 7.30

WoodCo employs departmental rates for applying manufacturing overhead to its products. Direct labor hours are used as the application base for the Fiberglass Processing Department. Included in the manufacturing overhead for the current year is $20,000 of factorywide fixed manufacturing overhead that has been allocated to the Fiberglass Processing Department.

In addition, $80,000 per year of companywide fixed marketing and administrative costs are allocated to the Fiberglass Processing Department using an application base of 20% of sales. WoodCo, to encourage new products, allocates companywide fixed marketing and administrative costs at only 10% of sales rather than the 20% used for established products.

WoodCo's management is, of course, interested in increasing the profitability of the firm. Therefore, the company controller, David Roman, has been asked to review the available data and recommend what combination of products WoodCo should manufacture and purchase in the future.

Required

1. In formulating this recommendation on the nature of WoodCo's future business, (a) explain the decision rule that David Roman should apply and (b) discuss the important factors that David Roman needs to identify when preparing his analysis.
2. Prepare an analysis that will show which product or products WoodCo should manufacture and which it should purchase in order to maximize the company's profitability. Support your answer with calculations.

24-21 Effect of a change in contribution margins, cost of a prediction error. (J. Demski and C. Horngren) Mary Demhorn is the manager of the Tollman Company. Tollman produces two products in two departments. Mary has great confidence in her cost accounting data regarding direct materials and direct labor, but she is less sure about the factory overhead. She has been using a prediction cost function for plantwide total factory overhead in which the budgeted variable factory overhead rate is $4 per direct labor hour (DLH). The plantwide factory overhead cost function is

$9,000 fixed cost per month + $4 variable cost per DLH

Mary collects the following price, cost, and technical data:

	Product	
	X	Y
Selling price per unit	$114	$149
Incremental marketing cost per unit sold	$ 4	$ 8
Direct materials per unit	$ 20	$ 30
DLH per unit:		
Machining	1 hr.	2 hrs.
Finishing	4 hrs.	3 hrs.

The wage rate is $7 per hour in machining and $9 per hour in finishing.

An LP model was used to determine the optimal production combination of products X and Y. The coordinates of the corners of the area of feasible solutions are:

Corner	X	Y
1	0	0
2	200	400
3	0	500
4	250	0

Required

1. Compute the unit contribution margins of products X and Y.
2. Determine the product combination of X and Y that will maximize the total contribution margin of Tollman Company.
3. The day after Mary has made an irrevocable decision about the combination of X and Y to produce (see requirement 2), she is presented with a detailed analysis of factory overhead costs in each department.

Machining overhead = \$6,000 fixed cost per month
+ \$3.00 variable cost per machining department DLH

Finishing overhead = \$3,000 fixed cost per month
+ \$5.00 variable cost per finishing department DLH

Mary is convinced that these departmental cost functions give a "truer" or "more accurate" picture than the plantwide cost function she used. What is the cost of the prediction error associated with using the plantwide cost function for predicting variable factory overhead costs of each product?

24-22 Optimal production plan, food processor, fixed and variable overhead costs. (A. Atkinson, adapted) Coldbrook Manufacturing produces two special concentrates that are derived from processing apples. These two products, Delicious (D) and Tasty (T), are used for flavoring in the food-manufacturing industry. Because of the high quality of these products, Coldbrook Manufacturing can sell all the output it can produce at the existing prices.

The characteristics of the two products are summarized as follows:

	Product D	Product T
Selling price	\$20	\$35
Direct materials used	\$12.40	\$17
Direct labor hours used	0.2 hr.	0.5 hr.
Machine hours used	1 hr.	2 hrs.

The company applies manufacturing overhead to production at the rate of \$18 per direct labor hour worked. The fixed component of the overhead cost per direct labor hour was estimated by dividing the budgeted fixed manufacturing overhead of \$450,000 by the total direct labor hours (45,000) available for production during the year.

Factory workers are paid \$10 per hour worked, and the machine hours available for production amount to 200,000 during the year. All marketing and administrative costs are fixed and are expected to be \$300,000 during the coming year.

Required

1. Determine the optimal production mix for the coming year.
2. Suppose the company controller decides that a cost function with two cost drivers will yield better estimates of variable overhead costs. The new cost function used to estimate variable overhead (V.OVH) will be

   V.OVH per unit = \$1.00 (direct labor hours) + \$2.50 (machine hours)

   How might this approach to variable-overhead cost estimation affect the optimal production mix determined in requirement 1?

24-23 Optimal assembly mix, cost of a prediction error, change in a technical constraint. Video Unlimited assembles and distributes videocassette recorders (VCRs). At its Brisbane plant, Video Unlimited assembles a standard model and a deluxe model. The standard model requires two hours of assembly time in Department 1 and one hour of assembly time in Department 2. The deluxe model requires two hours of assembly time in Department 1 and three hours of assembly time in Department 2. There are multiple-production lines in each department. The total assembly time available in Department 1 is 1,200 hours per month. The total assembly time available in Department 2 is 900 hours per month. A shortage of the microchips used in the deluxe model limits its production to 250 units per month.

The unit contribution margins are \$100 for the standard model and \$160 for the deluxe model.

Required

1. Formulate the Brisbane product-mix decision as an LP model.
2. Compute the mix of standard VCRs (S) and deluxe VCRs (D) that will maximize the total contribution margin. Use both the graphic solution approach and the trial-and-error solution approach.

3. After the production schedule is set, Video Unlimited finds that it must reduce the wholesale price of its standard model by $50 per unit to match unexpected competition from an overseas supplier. (All other data are unchanged.) Compute the cost of the prediction error for the unit contribution margin of the standard model.
4. The microchip supplier informs Video Unlimited that it can supply additional microchip units. Now production is feasible up to 500 deluxe VCRs per month. How will the increased supply of microchip units affect the optimal assembly mix of Video Unlimited? (Assume the unit contribution margin for the standard VCR unit is $50, as explained in requirement 3.)

CHAPTER 25

Regression analysis for railroads frequently requires separate consideration of railroad operating costs and track maintenance costs.

Cost Behavior and Regression Analysis

Learning Objectives

When you have finished studying this chapter, you should be able to

1. Understand why knowledge of both operations and cost analysis is essential in using cost data in regression analysis
2. Explain why exclusive reliance on maximizing goodness of fit in a regression model is dangerous
3. Specify four assumptions underlying regression analysis
4. Choose among alternative regression models in a structured and systematic way
5. Distinguish between simple regression and multiple regression
6. Explain how regression analysis can be used to identify cost drivers
7. Give examples of cost drivers in six business function areas of organizations

Chapter 10 provides an overview of how to determine cost behavior patterns (cost functions) to ensure that predictions of future costs will be as accurate as possible. This chapter extends Chapter 10 by examining the role of regression analysis in identifying cost drivers and in estimating cost functions based on those drivers. We also discuss cost drivers in different functional areas of an organization and the influence of the time horizon on cost behavior patterns. Before reading this chapter, be sure you are comfortable with the contents of Chapter 10.

"Regression analysis" uses a formal model to measure the *average* amount of change in the dependent variable that is associated with unit changes in the amounts of one or more independent variables. When only one independent variable is used, the analysis is called **simple regression**; when more than one independent variable is used, it is called **multiple regression**.

Regression analysis has several attractive features. First, it provides a model for estimating a cost function. Second, it provides measures of probable error for cost estimates. Third, procedures can be performed to examine how well its assumptions describe the data being analyzed. Fourth, regression can be applied where there are several independent variables instead of only one.

We emphasize the interpretation and use of the output from computer software regression analysis programs. Generally available programs (such as SPSS) compute all the statistics (and many more) referred to in this chapter. Appendix 25B presents an introduction to regression analysis.

GUIDELINES FOR REGRESSION ANALYSIS

Objective 1

Understand why knowledge of both operations and cost analysis is essential in using cost data in regression analysis

Knowledge of both operations **and** *cost accounting is required for intelligent application of regression analysis.* Consider repair costs. Repairs may be scheduled when production is at a low level because machines can be taken out of service at these times. In this case, regression analysis would indicate that the higher the level of production, the lower the repair costs, and vice versa. The engineering link of repairs to production, however, is usually clear-cut. Over time there is a cause-and-effect relationship; the higher the level of production, the higher the repair costs. If repair costs were pooled with other overhead costs, the estimated variability of overhead costs in relation to the level of production would be understated. Thus, the true extent of the variation of overhead cost would be masked.

We see that regression analysis, to provide appropriate information, cannot be mechanically applied. Users of regression analysis would do well to follow four guidelines:

- Guideline 1. To the extent that physical relationships or engineering data are available for establishing cause-and-effect links, use them.
- Guideline 2. To the extent that relationships can be implicitly established by logic and knowledge of operations, use them (preferably in conjunction with guidelines 3 and 4). Interviews with the appropriate operating personnel are an essential step in estimating cost functions.
- Guideline 3. Plot the data. The general relationship between costs and other variables can often be readily observed in a plot of the data.
- Guideline 4. To the extent that the relationships in guideline 1 or guideline 2 can be buttressed by an appropriate statistical analysis of data, use that analysis. Regression analysis is often a good check on guidelines 1 and 2.

Above all, avoid mechanically applying regression analysis.

Presentation of Regression Analysis

Regression analysis is a statistical model that measures the average amount of change in the dependent variable that is associated with a unit change in the amount of one or more independent variables. The *dependent variable,* denoted y′, is the variable being estimated using the regression model. The *independent variable,* denoted x, is the variable(s) used to estimate the dependent variable in the regression model. The regression line for a cost function (where y' is the predicted cost) is often presented as follows:

$$y' = a + bx$$

where a is termed the *constant* or *intercept* and b is termed the *slope coefficient.*

We now examine data for Southern Carpets to illustrate regression analysis. This example was introduced in Chapter 10 (p. 338). Southern Carpets weaves carpets for houses and offices. Its manufacturing plant is highly automated with state-of-the-art weaving machines. The role of direct labor in this automated plant is to minimize machine problems rather than to operate the machines. Exhibit 25-1 pres-

EXHIBIT 25-1

Southern Carpets: Monthly Indirect Manufacturing (Overhead) Costs, Machine Hours, Direct Labor Hours, and Number of Production Batches

Month (1)	Indirect Manufacturing (Overhead) Costs (2)	Machine Hours (3)	Direct Labor Hours (4)	Number of Production Batches (5)
January	$341,062	3.467	1,002	451
February	346,471	4,426	1,176	412
March	287,328	3,103	1,193	386
April	262,828	3,625	722	357
May	220,843	3,081	1,553	342
June	390,700	4,980	1,510	428
July	337,924	3,948	1,454	561
August	180,000	2,180	1,261	194
September	376,246	4,121	2,054	373
October	295,041	4.762	1,011	338
November	215,121	3,402	958	192
December	275,343	2,469	1,284	419

EXHIBIT 25-2

Southern Carpets: Plot of Monthly Indirect Manufacturing Costs and Machine Hours

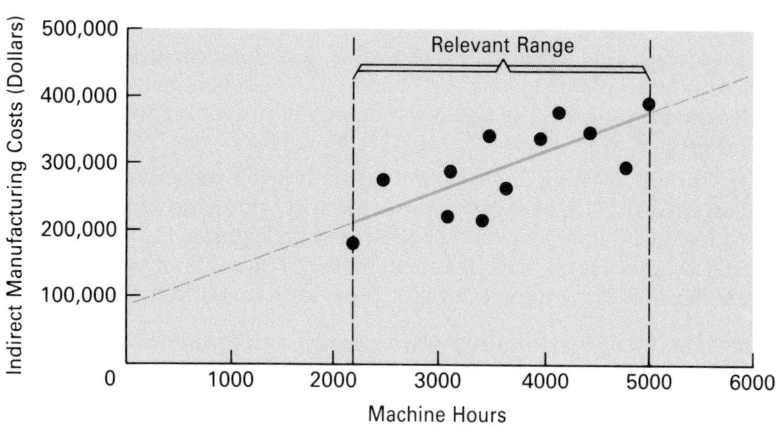

EXHIBIT 25-3

Southern Carpets: Regression Results for Simple Regression with Machine Hours as Independent Variable

Variable (1)	Coefficient (2)	Standard Error (3)	t-Value (4) = (2) ÷ (3)
Constant	86,153	61,152	1.41
Independent variable 1:			
Machine hours	57.27	16.43	3.49
$r^2 = 0.55$; Durbin-Watson statistic = 2.12			

ents data for indirect manufacturing (overhead) costs and for three possible independent variables—machine hours, direct labor hours, and number of production batches.

Exhibit 25-2 depicts the indirect manufacturing costs and machine hours data for Southern Carpets. The cost function estimated with a simple regression model is

$$y' = \$86{,}153 + \$57.27x$$

Exhibit 25-3 presents a convenient format for summarizing the regression results. This regression model, and the results in Exhibit 25-3, will be used in several sections of this chapter.

CHOOSING AMONG COST FUNCTIONS

Managers or accountants must often choose among various cost functions that use different combinations of independent variables. For example, one cost function may use machine hours as the independent variable, a second may use direct labor hours, and a third may use both as independent variables. How do managers select among these alternative cost functions? There are four important selection criteria:

1. *Economic Plausibility.* The basic relationship between the dependent variable and the independent variable(s) should make economic sense and be intuitive to both the operating manager and the accountant.

2. *Goodness of Fit.* The independent variable(s) should explain a considerable percentage of the variation in the dependent variable. The *coefficient of determination* (r^2) is a frequently used measure of goodness of fit. The **coefficient of determination, r^2**, is a measure of the extent to which the independent variable(s) accounts for the variability in the dependent variable.
3. *Significance of Independent Variable(s).* The *t*-value of a slope coefficient (*b*) measures the significance of the relationship between changes in the dependent variable and changes in the independent variable. The *t*-value is computed by dividing the slope coefficient by its standard error.
4. *Specification Analysis.* Testing the assumptions underlying regression analysis is termed **specification analysis**. (These assumptions—linearity within the relevant range, constant variance of residuals, independence of residuals, and normality of residuals—are discussed in this chapter.) If the statistical model satisfies these assumptions, users can place greater confidence in the estimates of cost behavior derived from using the model.

Those independent variables in a regression model satisfying the above criteria can be labeled cost drivers of the dependent variable cost item. Recall from Chapter 2 (p. 28) that a cost driver is any factor whose change causes a change in the total cost of a related cost object. Regression analysis offers a structured approach, based on past data relationships, for identifying cost drivers.

CRITERION 1: ECONOMIC PLAUSIBILITY

The first criterion, economic plausibility, places regression analysis in perspective. Does a high correlation between two variables mean that either is a cause of the other? Correlation, by itself, cannot answer that question: x may cause y, y may cause x, x and y may interact, both may be affected by z, or the correlation may be due to chance. High correlation merely indicates that the two variables move together. No conclusions about cause and effect are warranted.

Consider the high correlation between the conference (league) of the winning team in the Super Bowl and security returns on the New York Stock Exchange (NYSE). Over the past twenty years, in the year when a National Football Conference team (such as the San Francisco 49ers) wins the Super Bowl, there is a high probability that the NYSE index will increase. Conversely, in the year when an American Football Conference team (such as the Miami Dolphins) wins, there is a high probability that the NYSE index will decrease.[1] It is difficult to think of any cause-and-effect reason that makes this high correlation economically plausible. Even though a high correlation has existed in the past between stock prices and the conference of the Super Bowl winner, relying on the correlation existing in subsequent years appears to be very risky. ***Without any economic plausibility for a relationship, it is unlikely that a high level of correlation observed in one set of data will be found in other similar sets of data.***

For the regression reported in Exhibit 25-3, the relationship is economically plausible. Given the nature of the product (woven carpets) and the plant (highly automated state-of-the-art weaving machines), those people knowledgeable about the manufacturing operations would anticipate a positive relationship between indirect manufacturing costs and machine hours. Each additional machine hour used in the relevant range adds $57.27 to the indirect manufacturing costs of Southern Carpets.

[1]See E. Dyl and J. Schatzberg, "Did Joe Montana Save the Stock Market?" *Financial Analysts Journal*, September–October 1989, pp. 4–5.

CRITERION 2: GOODNESS OF FIT

Objective 2

Explain why exclusive reliance on maximizing goodness of fit in a regression model is dangerous

Other things being equal, the preferable regression model has a high goodness of fit, or explanatory power. The r^2 test for goodness of fit is a measure of the extent to which the independent variable explains or accounts for the variability of the dependent variable. The range of r^2 is from 0 (implying no explanatory ability) to 1 (implying perfect explanatory ability).[2] Details of its computation are given on pages 802–4.

Relying exclusively on the goodness-of-fit criterion can be dangerous. It can lead to the indiscriminate inclusion of independent variables that increase r^2 but have no economic plausibility as cost driver(s). The term **data mining** is sometimes used to describe a search for variables that is aimed solely at the maximization of r^2. Determining the dividing line between (a) reasonable data analysis and experimentation and (b) data mining is a difficult judgment call.

The r^2 for the regression reported in Exhibit 25-3 is 0.55, implying that variation in the machine hours variable explains 55% of the variability in indirect manufacturing costs of Southern Carpets. Generally, an r^2 of 0.30 or higher passes the goodness-of-fit test.

CRITERION 3: SIGNIFICANCE OF INDEPENDENT VARIABLE(S)

The coefficient of the chosen independent variable(s) should be significantly different from zero for that independent variable to be considered a possible cost driver. Such a result implies that there is a significant relationship between changes in the dependent variable and changes in the independent variable(s). The t-value of the b coefficient measures significance. A t-value with an absolute value greater than 2.00 is consistent with the b coefficient being significantly different from zero.[3] Details of the computation of the t-statistic are given on pages 804–6.

The major objective of cost estimation is to estimate how costs behave as the independent variable changes within a relevant range, which seldom encompasses

[2]Computer programs frequently report an r^2 and an adjusted r^2. The adjusted r^2 is calculated as follows:

$$\text{Adjusted } r^2 = r^2 - \frac{(k-1)}{(n-k)}(1 - r^2)$$

where k is the number of independent variables and n is the number of observations. The adjusted r^2 reduces the unadjusted r^2 as extra independent variables are added to the regression. The adjusted r^2 measure can in some cases be negative (when there is a low r^2 and many independent variables in the regression). Only r^2 values are reported in this chapter.

[3]The benchmark for inferring that a b coefficient is significantly different from zero is a function of the degrees of freedom in a regression. The benchmark of 2.00 given in the text assumes a sample size of 60 observations. The number of degrees of freedom is calculated as the sample size minus the number of a and b parameters estimated in the regression. For a simple regression, the benchmark values for the t statistic are:

Sample Size	Benchmark
10	$\|t\| > 2.31$
15	$\|t\| > 2.16$
20	$\|t\| > 2.10$
30	$\|t\| > 2.05$
60	$\|t\| > 2.00$

where $|t|$ denotes the absolute value of the t statistic.

zero volume (that is, when $x = 0$). The t-value for a, then, is not usually important because a manager is seldom concerned with estimating the value of y when x is equal to zero. The intercept (a) is the best available starting point for a straight line that approximates how a cost behaves within the relevant range.

For the regression reported for Southern Carpets in Exhibit 25-3, the t-value for the slope coefficient (b) is 3.49, which exceeds the benchmark of 2.00. The coefficient of the machine hours cost variable is therefore significantly different from zero. The relevant range for machine hours in Exhibit 25-1 is from 2,180 to 4,980 hours. Thus, the intercept of $86,153 should be interpreted only as the amount of indirect manufacturing costs that does not vary with machine hours in the 2,180-to 4,980-hour range. The intercept should not be interpreted as a fixed cost unless the relevant range includes zero volume.

CRITERION 4: SPECIFICATION ANALYSIS

Specification analysis is concerned with testing the assumptions of regression analysis. In each context where regression analysis is used, it is essential to examine whether these assumptions are applicable to the data.

Probably the major limitation of regression is the assumption that the relationship in the equation will persist—that is, that there is an ongoing, stable relationship between the dependent variable and the independent variable(s). For example, for many years manufacturing operations at Southern Carpets were heavily labor-intensive with minimal use of automated machinery. It is only in the last two years that Southern Carpets has invested in highly automated weaving machines. The cost function estimated with data from the labor-intensive manufacturing period would be different from that reported in Exhibit 25-3, which is based on data from the automated-machinery period.

Consider a cost function that has the form

$$E(y) = \alpha + \beta x$$

where α and β are the underlying (but unknown) parameters. Assume regression analysis is used to estimate the cost function

$$y' = a + bx$$

Objective 3

Specify four assumptions underlying regression analysis

When the following four key assumptions are met, the sample values of a and b from a regression model are the best available linear, unbiased estimates of the population parameters α and β:

- Linearity within the relevant range
- Constant variance of residuals
- Independence of residuals
- Normality of residuals

Many software regression packages include tests that can be systematically applied to see whether these and other assumptions hold. These tests are often referred to as *specification analysis*.

Linearity Within the Relevant Range

First, linearity is assumed to exist between the dependent variable (y) and the independent variable (x) within the relevant range. The hypothesized relationship is

$$y = \alpha + \beta x + u$$

where u is called the residual term. We assume the expected value of the residual term, $E(u)$, is zero. This leads to

$$y' = a + bx$$

where y' is the estimated value, as distinguished from the actual value of y. The deviation of the *actual* value of y from the regression line is called the **residual term** u (also called the **disturbance term** or **error term**):

$$u = y - y'$$

The average or expected value of u is zero. The residual term includes the effect of all independent variables excluded from the regression model plus the effect of measurement error in the independent variable(s) included in the model.

Where there is one independent variable, the easiest way to check for the presence of linearity is by studying the data on a scatter diagram, a step that often is unwisely skipped. Exhibit 25-2 (p. 785) presents a scatter diagram for the indirect manufacturing costs and machine hours variables of Southern Carpets in Exhibit 25-1. Linearity appears to be a reasonable assumption for this data.

The learning curve model discussed in Chapter 10 (pp. 347–51) is an example of a nonlinear model; costs increase when production volume increases but by lesser amounts than would occur with a linear cost function. A second example of nonlinearity is when costs are "sticky" downwards or upwards. For example, purchasing department costs of a retail store may rise in response to increases in sales dollars but may not decline as rapidly in response to decreases in sales dollars.

Constant Variance of Residuals

The second assumption is that the standard deviation and variance of the residuals (u's) are constant for all values of the independent variable. This assumption implies that there is a uniform scatter or dispersion of the data points about the regression line.[4] Again, the scatter diagram is the easiest way to check for **constant variance**. This assumption is valid for the graph in Panel A of Exhibit 25-4, but not for the graph in Panel B.

The constant variance assumption implies that the distribution of the u's is unaffected by the level of the independent variable. If the assumption does not hold, the reliability of the estimates of the standard errors is reduced; the a and b regression coefficients are still unbiased. **Bias** in statistics occurs when a random sample fails to represent a population because of some systematic error. Violation of the constant variance assumption is likely in many cost-volume-profit relationships. As the graph in Panel B of Exhibit 25-4 indicates, the higher the production volume, the higher the probability of more scatter of the residuals. Why? One explanation is that random usage in energy consumption is a fixed percentage of production volume. Higher dollar variations in energy costs are therefore associated with higher levels of production volume.

With a small number of observations (as with the data in Exhibit 25-1), tests for constant variance are not very powerful. Data sets of at least twenty observations are necessary before even moderately reliable tests of the constant variance of residuals can be made.

Independence of Residuals

The third assumption is that the residuals (u's) are independent of each other. That is, the deviation of one point about the regression line (its u value, where $u = y - y'$) is unrelated to the deviation of any other point. If the u's are not independent,

[4]Constant variance also is known as homoscedasticity. Violation of this assumption is called heteroscedasticity.

EXHIBIT 25-4
Constant Variance of Residuals Assumption

PANEL A: Example of Constant Variance

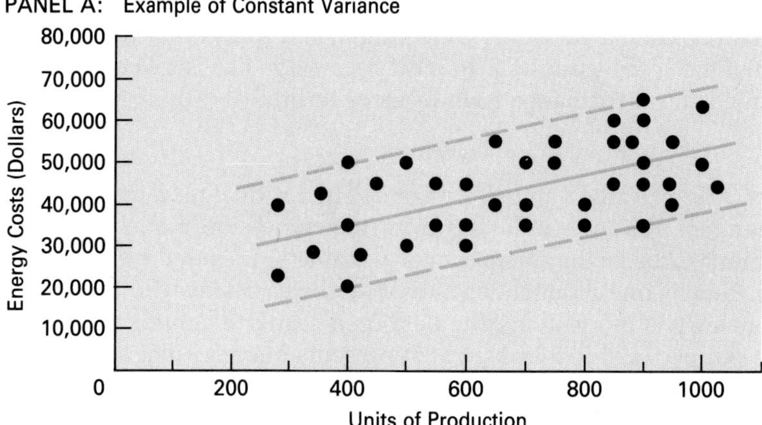

PANEL B: Example of Nonconstant Variance

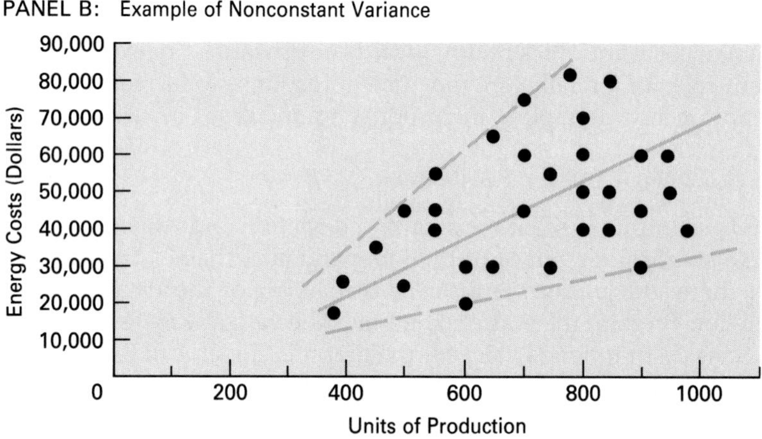

the problem of **serial correlation** in the residuals (also called **autocorrelation**) is present. Serial correlation means that there is a systematic pattern in the residuals such that knowledge of the residual in period t conveys information about the residuals in period $t + 1$, $t + 2$, and so on.

Serial correlation affects the efficiency (but not the unbiasness) of the regression estimates of a and b. The property of efficiency is important in cost accounting applications because with positive (negative) serial correlation, the estimates of the standard errors will be understated (overstated) relative to the underlying population standard errors. Thus, one may infer that the parameter estimates are more (less) precise than they actually are.

The Durbin-Watson statistic is one measure of serial correlation in the estimated residuals. For samples of ten to twenty observations, a Durbin-Watson statistic in the 1.30 to 2.70 range is consistent with the residuals being independent. The 0 to 1.3 range of the Durbin-Watson statistic signals positive serial correlation in the estimated residuals, and the 2.7 to 4.0 range signals negative serial correlation.

For the regression results for Southern Carpets in Exhibit 25-3, the Durbin-Watson statistic is 2.12. An assumption of independence in the estimated residuals seems reasonable for this regression model.

Normality of Residuals

The fourth assumption is that the data points around the regression line are normally distributed. That is, the residuals (u's) are normally distributed. This assumption is necessary in making inferences about y', a, and b. For example, the normality assumption is necessary to make probability statements using the standard error of the residuals.

Tests for normality can be conducted by inspecting plots of the residuals or by examining statistics such as the skewness coefficient and the kurtosis coefficient.[5] As with tests for constant variance, data sets of at least twenty observations are necessary before even moderately reliable tests of the normality of the residuals assumption can be made.

Overview

Appendix 25A presents a convenient overview of the four assumptions examined in specification analysis. ***Many of the items presented in Appendix 25A reinforce two points stressed in Chapter 10: (1) you can benefit from plotting the data, and (2) the most difficult step in cost analysis is collecting reliable data on the dependent variable and the independent variable(s).***

AN EXAMPLE OF CHOOSING AMONG COST FUNCTIONS

Objective 4

Choose among alternative regression models in a structured and systematic way

Let us use the Southern Carpets data to illustrate the previously outlined criterion for choosing among alternative cost functions. Consider two cost functions:

$$y' = a + b \text{ (machine hours)}$$
$$y' = a + b \text{ (direct labor hours)}$$

Exhibits 25-2 and 25-3 present a plot of the data and regression results, respectively, for the cost function using machine hours as the independent variable. Exhibits 25-5 and 25-6 present comparable information for the cost function using direct labor hours as the independent variable.

These two cost functions are compared in Exhibit 25-7. Across several criteria, the cost function based on machine hours is preferable to the cost function based on direct labor hours. The economic plausibility criterion is especially important. Operating personnel at Southern Carpets would probably have greater confidence in machine hours than in direct labor hours as the indirect manufacturing cost driver, given the adoption of highly automated weaving machines. The information in Exhibit 25-7 identifies machine hours as a significant cost driver of monthly indirect manufacturing costs.

Do not always assume that any one cost function will perfectly satisfy all the criteria in Exhibit 25-7. A cost analyst must often make a choice among "imperfect" cost functions, in the sense that the data of any particular cost function will not perfectly meet one or more of the assumptions underlying regression analysis.

---

[5]The skewness coefficient measures whether the distribution of the data departs from the bell-shaped curve of the normal distribution in either the upper or lower parts of the distribution. The kurtosis coefficient measures whether the distribution of the data is more or less fat-tailed than would be expected from the normal distribution.

EXHIBIT 25-5
Southern Carpets: Plot of Monthly Indirect Manufacturing Costs and Direct Labor Hours

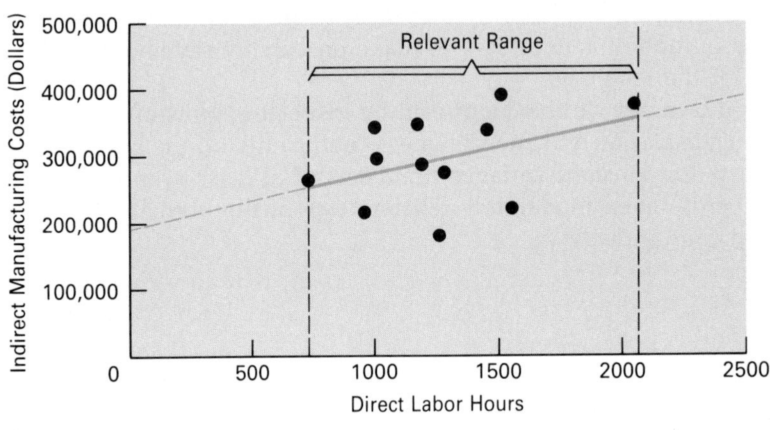

EXHIBIT 25-6
Southern Carpets: Regression Results for Simple Regression with Direct Labor Hours as Independent Variable

Variable (1)	Coefficient (2)	Standard Error (3)	*t*-Value (4) = (2) ÷ (3)
Constant	199,291	72,813	2.74
Independent variable 1:			
Direct labor hours	74.94	55.66	1.35
$r^2 = 0.15$; Durbin-Watson statistic = 2.25			

MULTIPLE REGRESSION

Objective 5

Distinguish between simple regression and multiple regression

In some cases, satisfactory predictions of a cost may be based on only one independent variable, such as machine hours. In many cases, however, accuracy can be improved by basing the prediction on more than one independent variable. The most widely used equations to express relationships between a dependent variable and two or more independent variables are linear in the form

$$y = a + b_1x_1 + b_2x_2 + \cdots + u$$

where

y = the cost variable to be predicted,
$x_1, x_2, \ldots$ = the independent variables on which the prediction is to be based,
$a, b_1, b_2, \ldots$ = the estimated coefficients of the regression model, and
u = the residual term that includes the net effect of other factors.

Example

Consider again the Southern Carpets data in Exhibit 25-1 (p. 784), which include monthly observations on three potential independent variables in a cost function—machine hours, direct labor hours, and number of production batches. The produc-

EXHIBIT 25-7

Southern Carpets: Comparison of Alternative Cost Functions for Indirect Manufacturing Costs Estimated with Simple Regression

Criterion	Cost Function 1: Machine Hours as Independent Variable	Cost Function 2: Direct Labor Hours as Independent Variable
1. Economic Plausibility	Positive relationship between indirect manufacturing costs and machine hours is economically plausible in a highly automated plant.	Positive relationship between indirect manufacturing costs and direct labor hours is economically plausible, but less so than machine hours on a month-to-month basis.
2. Goodness of Fit	$r^2 = 0.55$ Excellent goodness of fit.	$r^2 = 0.15$ Poor goodness of fit.
3. Significance of Independent Variables	*t*-statistic on machine hours of 3.49 is significant.	*t*-statistic on direct labor hours of 1.35 is insignificant.
4. Specification Analysis:		
Linearity within the relevant range	Appears reasonable from a plot of the data.	Some evidence of linearity, but not as strong as with the machine hours variable.
Constant variance of residuals	Appears reasonable, but inferences drawn from only 12 observations are not reliable.	Appears reasonable, but inferences drawn from only 12 observations are not reliable.
Independence of residuals	Durbin-Watson statistic = 2.12 Assumption of independence is not rejected.	Durbin-Watson statistic = 2.25 Assumption of independence is not rejected.
Normality of residuals	Data base too small to make reliable inferences.	Data base too small to make reliable inferences.

tion batch variable measures the number of separate carpet jobs worked on during the month; the same line of carpet worked on multiple times (say, on different days) is counted as multiple batches. Operating personnel at Southern Carpets report sizable changeover costs when production on one carpet batch is stopped and production on another batch is started. For example, materials on the weaving looms must be changed. Management believes there is a positive relationship between the indirect manufacturing costs and the number of production batches.

Exhibit 25-8 presents results for the following multiple regression model, using data in columns 2, 3, and 5 of Exhibit 25-1:

$$y' = \$13{,}749 + \$41.44\, x_1 + \$350.00\, x_2$$

where x_1 is the number of machine hours and x_2 is the number of production batches. It is economically plausible that both machine hours and production batches would help explain variations in indirect manufacturing costs at Southern Carpets. The r^2 of 0.55 for the simple regression using machine hours (Exhibit 25-3) increases to 0.80 with the multiple regression in Exhibit 25-8. The t statistics on both the machine hours and production batch independent variables are significantly different from zero ($t = 3.29$ for the coefficient on machine hours and $t =$

EXHIBIT 25-8

Southern Carpets: Regression Results for Multiple Regression with Two Independent Variables (Machine Hours and Production Batches)

Variable (1)	Coefficient (2)	Standard Error (3)	t-Value (4) = (2) ÷ (3)
Constant	13,749	48,599	0.28
Independent variable 1:			
Machine hours	41.44	12.60	3.29
Independent variable 2:			
Production batches	350.00	106.10	3.30
$r^2 = 0.80$; Durbin-Watson statistic = 2.39			

Objective 6

Explain how regression analysis can be used to identify cost drivers

3.30 for the coefficient on production batches). The multiple regression model in Exhibit 25-8 satisfies both economic and statistical criteria, and it explains much greater variation in indirect manufacturing costs than does the simple regression model using machine hours as the independent variable. Based on the information in Exhibit 25-8, both machine hours and production batches can be considered as important cost drivers of monthly indirect manufacturing costs at Southern Carpets.

In Exhibit 25-8, the slope coefficients, \$41.44 and \$350.00, measure the change in indirect manufacturing costs associated with a unit change in each independent variable (assuming that the other independent variable is held constant). For example, indirect manufacturing costs increase by \$350 when one more production batch is added, assuming that the number of machine hours is held constant.

Multicollinearity

A major concern that arises with multiple regression is multicollinearity. **Multicollinearity** exists when two or more independent variables are highly correlated with each other. Generally, users of regression believe that a coefficient of correlation—that is, r—greater than 0.70 indicates multicollinearity. Multicollinearity has the effect of increasing the standard errors of the coefficients of the individual variables. The result is that there is greater uncertainty about the underlying value of the coefficients of the individual independent variables.

The coefficients of correlation between the potential independent variables for Southern Carpets in Exhibit 25-1 are:

Pairwise Combinations	Coefficient of Correlation
Machine hours and direct labor hours	.12
Machine hours and production batches	.38
Direct labor hours and production batches	.23

These results suggest that multiple regressions using the independent variables in Exhibit 25-1 are not likely to encounter multicollinearity problems.

COST DRIVERS AND DIFFERING TIME SPANS

The Southern Carpets example illustrates how regression analysis can be used to identify two drivers of indirect manufacturing costs—machine hours and production batches. Exhibit 25-9 presents examples of variables found to be important cost drivers in six different business function areas of organizations.

EXHIBIT 25-9
Examples of Cost Drivers in Different Business Function Areas

Business Functions	Examples of Cost Drivers
Research and Development	—number of personnel —number of projects —length of project
Product Design	—number of personnel —number of products —number of engineering-change orders
Manufacturing	—materials costs —materials size or weight —direct labor costs or hours —number of suppliers —number of parts —number of part numbers —number of units produced —number of machine setups —number of production batches —production throughput time —machine setup time —product testing time
Distribution	—dollar value of items distributed —size of items distributed —weight of items distributed —number of total items distributed —number of product types distributed —number of customers —number of distribution outlets
Marketing	—dollar value of marketing expenditures —number of advertisements —number of personnel in marketing
Customer Service	—dollar value of service department expenditures —number of service calls —number of products serviced —time spent servicing products

Objective 7

Give examples of cost drivers in six business function areas of organizations

The time span (short run, intermediate run, or long run) should always be considered when selecting the cost driver(s) for specific decisions. Some cost drivers may be important across all the time spans. For example, the weight of items distributed is likely to be an important predictor of motor-vehicle operating costs for a distributor of industrial equipment in the short, intermediate, and long runs. Not all cost drivers fall into this category. Some drivers that are relatively unimportant in explaining short-run variations in cost behavior can be important determinants of long-run cost behavior. Consider the relationship between (a) the materials procurement and handling costs, (which is part of manufacturing overhead), and (b) the number of suppliers. Included in materials procurement and handling costs are costs for personnel and the inspection of incoming materials. Month-to-month variations in the number of suppliers are not likely to cause corresponding variations in these costs. During a six-month or longer period, however, sizable reductions in the number of suppliers may bring about sizable reductions in materials procurement personnel and material handling costs (such as quality inspection).

PROBLEM FOR SELF-STUDY

PROBLEM

The Clothing Style Center (CSC) operates a mail-order catalog business selling shirts, pants, sweaters, and other clothing items. CSC started business five years ago when it offered a "package deal" of four business shirts. This "package deal" had a single price, which included delivery to the customer. Business boomed. CSC expanded into selling a broad set of clothing items. Customers could order any combination of items in the catalog. Each item had a list price that included delivery. CSC did not charge individual customers a handling cost per order or a separate mailing cost. When analyzing the profitability of individual customers (or classes of customers), management allocated warehouse distribution costs on the basis of sales dollars.

Krystal Carrington decides to examine warehouse distribution cost drivers for CSC. She interviews operating personnel in the warehouse. They argue that the number of customer orders per month is a more important cost driver than the dollar sales of orders per month. Carrington observes more than 20 customer orders being processed and finds that each order has a large fixed amount of time spent on it, irrespective of the number of items purchased or the dollar amount of each item purchased. She collects the following data to use in a regression analysis:

Month	Warehouse Distribution Costs	Total Mail-Order Sales Dollars	Number of Customer Orders
January	$25,300	$324,000	2,730
February	19,000	210,000	1,810
March	12,700	293,000	2,207
April	12,700	201,000	2,200
May	43,800	272,000	3,411
June	40,200	202,000	2,586
July	37,300	342,000	3,364
August	30,700	247,000	2,411
September	49,800	347,000	3,964
October	33,100	328,000	2,897
November	47,100	347,000	3,403
December	35,100	307,000	2,864

She examines two cost functions using regression analysis.

REGRESSION FOR COST FUNCTION 1:
Warehouse distribution costs = $a + b$ (Sales dollars)

Variable	Coefficient	Standard Error	*t*-Statistic
Constant	1,760	17,685	0.10
Independent variable:			
Sales dollars	0.11	0.06	1.83

$r^2 = 0.23$; Durbin-Watson statistic = 1.31

REGRESSION FOR COST FUNCTION 2:
Warehouse distribution costs = $a + b$ (Number of customer orders)

Variable	Coefficient	Standard Error	*t*-Statistic
Constant	−17,077	9,129	−1.87
Independent variable:			
Number of customer orders	17.48	3.17	5.51

$r^2 = 0.75$; Durbin-Watson statistic = 2.23

Required

What can Carrington learn from these regression results for warehouse distribution costs at CSC?

SOLUTION

Exhibit 25-10 presents plots of the data underlying each cost function. Exhibit 25-11 compares regression models for these two cost functions. The preferred cost function, based on the comparison in Exhibit 25-11, uses the number of customer orders as the independent variable. Each customer order, within the relevant range of 1,810 to 3,964 orders per month, adds $17.48 to warehouse distribution costs of CSC. The regression results support the intuition of operating personnel of CSC and Carrington's observations. The number of customer orders is a more important driver of month-to-month variations in warehouse distribution costs than is sales dollars. Prompted by this finding, CSC managers may wish to explore ways of reducing the time spent on handling individual orders. Depending on the competitive situation, managers may also consider charging customers a per-order handling cost.

EXHIBIT 25-10

Clothing Style Center: Plot of Warehouse Distribution Costs and (1) Sales Dollars and (2) Number of Customer Orders

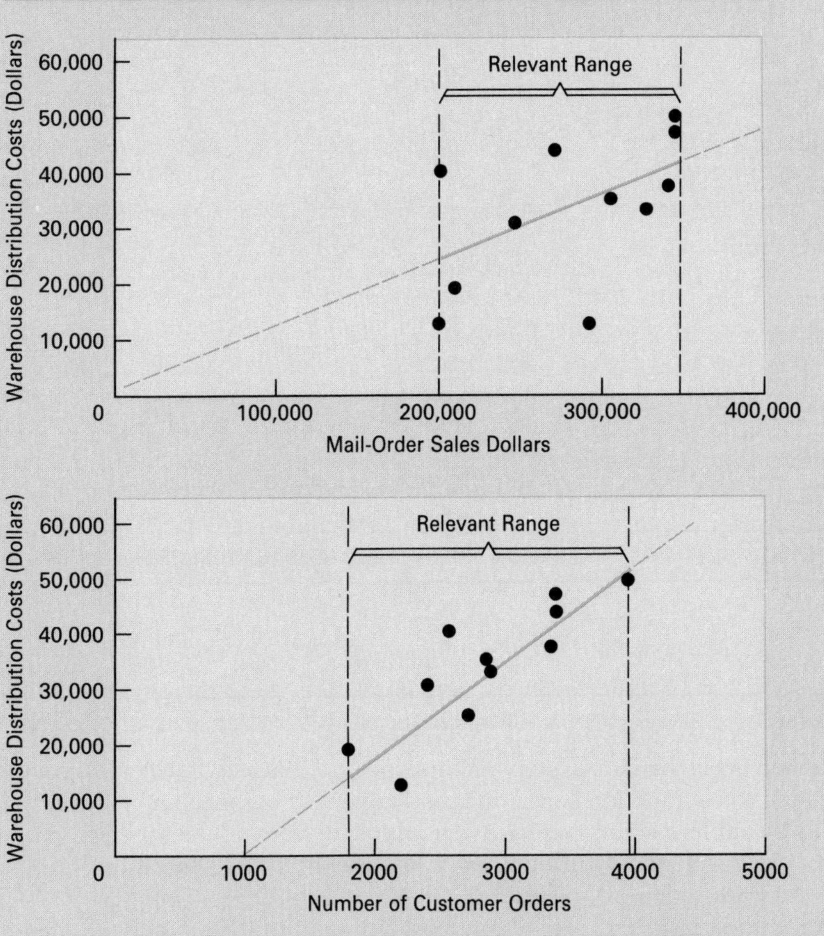

EXHIBIT 25-11
Clothing Style Center: Comparison of Alternative Cost Functions for Warehouse Distribution Costs Estimated with Simple Regression

Criterion	Cost Function 1: Sales Dollars as Independent Variable	Cost Function 2: Number of Customer Orders as Independent Variable
1. Economic Plausibility	Positive relationship between warehouse distribution costs and sales dollars is economically plausible, but operating personnel do not believe it to be the most important driver of costs.	Positive relationship between warehouse distribution costs and the number of customer orders is economically plausible and is supported by beliefs of operating personnel and Carrington's observations at the distribution center.
2. Goodness of Fit	$r^2 = 0.23$ Limited goodness of fit.	$r^2 = 0.75$ Excellent goodness of fit.
3. Significance of Independent Variables	*t*-statistic on sales dollars hours of 1.83 approaches statistical significance.	*t*-statistic on number of customer orders of 5.51 is statistically significant.
4. Specification Analysis:		
Linearity within the relevant range	Not strong evidence of linearity.	Appears reasonable from a plot of the data.
Constant variance of residuals	Appears questionable, but inferences drawn from only 12 observations are not reliable.	Appears reasonable, but inferences drawn from only 12 observations are not reliable.
Independence of residuals	Durbin-Watson statistic = 1.31 Assumption of independence questionable.	Durbin-Watson statistic = 2.23 Assumption of independence is not rejected.
Normality of residuals	Data base too small to make reliable inferences.	Data base too small to make reliable inferences.

SUMMARY

Regression analysis is a systematic approach to identifying cost drivers and to estimating a cost function based on those drivers. It has measures of probable error and can be applied when there are several cost driver variables instead of only one. Ideally, the relationships underlying a regression model should be established by logic and knowledge of operations.

Four criteria to use in identifying cost drivers from the potentially large set of independent variables that can be included in a regression model are economic plausibility, goodness of fit, significance of independent variables, and specification analysis. Those independent variables in a regression model satisfying these four criteria can be labeled as drivers of the dependent variable cost item.

Simple regression analysis is based on only one independent variable. Multiple regression analysis is based on two or more independent variables. The most difficult step in cost analysis is collecting reliable data on the dependent and independent variables. Mechanical use of regression analysis without detailed attempts to collect high-quality data can result in a failure to identify important cost drivers or in a failure to estimate the underlying cost function efficiently.

APPENDIX 25A: SPECIFICATION ANALYSIS

This chapter discusses four assumptions underlying regression analysis: linearity within the relevant range, constant variance of residuals, independence of residuals, and normality of residuals. Exhibit 25-12 covers three topics for each of these assumptions:

1. A description of the assumption and the consequences of its violation,
2. Ways to detect if a violation occurs in the sample of data being examined, and
3. Examples of causes of a violation and their possible solution.

Exhibit 25-12 highlights that successful use of regression analysis in analyzing cost behavior patterns frequently requires inputs from people across different business functions.

APPENDIX 25B: REGRESSION ANALYSIS

Computer software packages are available to estimate the regression statistics discussed in this chapter. This appendix outlines formulas that can be used to estimate several of the commonly used statistics. Columns 2 and 3 of Exhibit 25-13 on page 802 present data on weekly units produced and indirect manufacturing costs for an office-desk manufacturing plant of Executive Desks. The supervisor at the plant believes that units produced is the most important driver of weekly indirect manufacturing costs.

The scatter diagram of these twelve data points in Panel A of Exhibit 25-14 on page 803 indicates that a straight line does provide a reasonable approximation of the relationship that prevailed between indirect manufacturing costs and units produced during the sample period. Using the least-squares technique, means that the sum of the squares of the vertical deviations (distances) from the data points to the straight line is smaller than it would be from any other straight line. *Note especially that, as the graph in Panel B of Exhibit 25-14 shows, the deviations are measured vertically to the regression line.*

The object is to find the values of the a and b coefficients in the predicting equation $y' = a + bx$, where y' is the predicted value as distinguished from the observed value of y. We wish to find the numerical values of the coefficients a and b that minimize $\Sigma\ (y - y')^2$. This calculation is accomplished by using two equations, usually called *normal equations:*

$$\Sigma y = na + b(\Sigma x)$$
$$\Sigma xy = a(\Sigma x) + b(\Sigma x^2)$$

where n is the number of data points; Σx and Σy are, respectively, the sums of the given x values and y values; Σx^2 is the sum of squares of the x values; and Σxy is the sum of the products obtained by multiplying each of the given x values by the associated observed y value.

EXHIBIT 25-12

Specification Analysis for Regressions Using Accounting Data

What Is the Assumption and What Are the Consequences of its Violation?	How to Detect if a Violation Occurs in the Sample of Data Being Examined	Examples of Causes of the Violation and Their Possible "Solution"
A. *LINEARITY WITHIN THE RELEVANT RANGE* Relationship between the dependent variable and the independent variable can be approximated by a linear function. One consequence of violation of this assumption is that the fitted cost function may not describe cost behavior very well; there can be high prediction errors. Different forms of nonlinearity can lead to violations of the constant variance, independence, or normality assumptions, described below.	1. Plot the data and inspect for nonlinearity. 2. Include nonlinear variable in the regression model, such as x^2 or $\sqrt{x}$, and examine its statistical significance. 3. Estimate nonlinear cost functions, such as a learning curve function, and conduct specification analysis on its residuals. (Alternatively, the data may be transformed to examine if the linearity assumption is appropriate.)	1. Data-recording problems. For example, (a) the time period for the dependent variable is weekly, but the time period for the independent variable is monthly; (b) materials costs are reported when materials are purchased and not when materials are used; and (c) pension costs are recorded only in the final pay period. *Possible solution:* Use the same unit of time to measure all variables, and use an accrual accounting basis for recording costs. 2. Learning by workers causes increases in labor productivity. *Possible solution:* Estimate a cost function that explicitly incorporates learning effects. 3. Inflation in input prices. *Possible solution:* Restate the variables affected by inflation using a price index. 4. "Cost stickiness" (costs decline less than predicted when activity declines). *Possible solution:* Estimate a cost function including a dummy variable that assumes a value of 1 if activity declines, and 0 otherwise.
B. *CONSTANT VARIANCE OF RESIDUALS* The variance of the residuals around the "true" regression line should be the same over the entire range of the independent variable. A consequence of violating this assumption is that the standard error used to test the significance of the independent variable(s) is less reliable; it will depend on the range of the independent variable. Estimates of the regression coefficients are still unbiased.	1. Plot the data and inspect for nonconstant variance. 2. Examine the equality of the variance of the estimated residuals in different ranges of the independent variable. 3. Test for specific violations of nonconstant variance. For example, are residuals proportional to the dependent variable?	1. Stoppages, defective materials, and so on, have larger dollar effects at higher activity levels. *Possible solution:* Transform the dependent variable (y) to, say, (log y). 2. Increasing levels of inflation occur as the firm is expanding activity. *Possible solution:* Restate variables affected using a price index. 3. Repairs and maintenance conducted during periods of low activity. *Possible solution:* Estimate a separate cost function for repairs and maintenance.

C. *INDEPENDENCE OF RESIDUALS* Knowledge of the sign or magnitude of a residual should not help predict the sign or magnitude of the residual in the next (or subsequent) periods. Autocorrelation affects the efficiency (but not the unbiasness) of the regression estimates of the *a* and *b* coefficients. Positive (negative) autocorrelation results in the estimated standard errors of the regression coefficients being understated (overstated).	1. Plot the pattern of the residuals around the fitted line, and examine for systematic patterns. 2. Compute the Durbin-Watson statistic, which tests for serial correlation. 3. Compute the autocorrelation function of the residuals, which facilitates testing for serial correlation of any specific lag structure.	1. Underlying cost function is nonlinear. *Possible solution:* See the above section for the linearity assumption. 2. Underlying cost function is linear, but it includes more variables than are currently being analyzed. *Possible solution:* Include additional independent variables in the regression analysis. 3. High autocorrelation is induced by successive activity levels having a high percentage of commonality. *Possible solution:* Estimate the cost function using successive changes in the dependent and independent variables rather than successive values of the variables.
D. *NORMALITY OF RESIDUALS* The distribution of the "true" residuals should be well approximated by a normal distribution. One consequence of violating this assumption is that the standard error may not be a good measure of dispersion, which causes the significance tests to be less reliable.	1. Plot the residuals and inspect whether the distribution approximates the bell-shaped normal curve. 2. Compute distribution statistics for the residuals, such as the skewness coefficient and the kurtosis coefficient.	1. Extreme (nonnormal) observations can arise from errors in recording the data. (This is a very common situation. All extreme values should be checked against their source records.) *Possible solution:* Correct the recording error. 2. Extreme observation arises from an "unusual" event such as a fire or a strike. *Possible solution:* Delete the extreme observation, or make an adjustment to exclude the effect of the "unusual" event. 3. Technological change, such as a marked change in the level of automation, can cause the distribution of residuals to be bimodal (that is, to have separate groupings of observations). *Possible solution:* Restrict the data being analyzed to a period with a low level of technological change.

EXHIBIT 25-13

Executive Desks: Computation for Least-Squares Regression Between Indirect Manufacturing Costs and Units Produced

Week	Units Produced	Indirect Manufacturing Costs				Total Variance of y	Unexplained Variance	Total Variance of x
	x	y	x^2	xy	y′	$(y - \bar{y})^2$	$(y - y')^2$	$(x - \bar{x})^2$
(1)	(2)	(3)	(4)	(5)	(6)	(7)	(8)	(9)
1	34	340	1,156	11,560	284.48	1,958	3,082	3
2	44	346	1,936	15,224	345.48	2,525	0	67
3	31	287	961	8,897	266.18	77	433	23
4	36	262	1,296	9,432	296.68	1,139	1,203	0
5	30	220	900	6,600	260.08	5,738	1,606	34
6	49	416	2,401	20,384	375.98	14,460	1,602	173
7	39	337	1,521	13,143	314.98	1,702	485	10
8	21	180	441	3,780	205.18	13,398	634	220
9	41	376	1,681	15,416	327.18	6,440	2,383	27
10	47	295	2,209	13,865	363.78	1	4,731	125
11	34	215	1,156	7,310	284.48	6,520	4,827	3
12	24	275	576	6,600	223.48	431	2,654	140
Total	430	3,549	16,234	132,211	≈3,549	54,389	23,640	825

We obtain the inputs for our normal equations from Exhibit 25-13. Substituting into the two simultaneous linear equations, we obtain

$$3{,}549 = 12a + 430b$$
$$132{,}211 = 430a + 16{,}234b$$

The solution is $a = \$77.08$ and $b = \$6.10$, which can be obtained by direct substitution if the normal equations are reexpressed symbolically as follows:

$$a = \frac{(\Sigma y)(\Sigma x^2) - (\Sigma x)(\Sigma xy)}{n(\Sigma x^2) - (\Sigma x)(\Sigma x)}$$

$$b = \frac{n(\Sigma xy) - (\Sigma x)(\Sigma y)}{n(\Sigma x^2) - (\Sigma x)(\Sigma x)}$$

For our illustration, we now have

$$a = \frac{(3{,}549)(16{,}234) - (430)(132{,}211)}{(12)(16{,}234) - (430)(430)} = \$77.08$$

$$b = \frac{(12)(132{,}211) - (430)(3{,}549)}{(12)(16{,}234) - (430)(430)} = \$6.10$$

Placing the amounts for a and b in the equation of the least-squares line, we have

$$y' = \$77.08 + \$6.10x$$

where y' is the predicted indirect manufacturing cost for any number of units produced within the relevant range. Using the equation, we would predict, for example, that production of 30 units would have indirect manufacturing costs on average of $\$77.08 + \$6.10(30) = \$260.08$.

Correlation and Goodness of Fit

How much of the total variation of the y values can be attributed to chance? How much can be attributed to the relationship between the two variables x and y? The

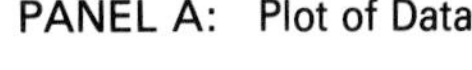

Executive Desks: Plot of Observations and Residuals for Regression Between Indirect Manufacturing Costs and Units Produced

PANEL A: Plot of Data

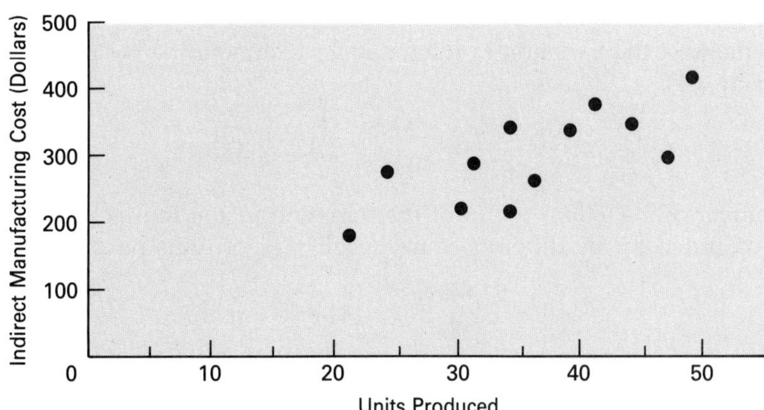

PANEL B: Regression Line and Estimated Residuals

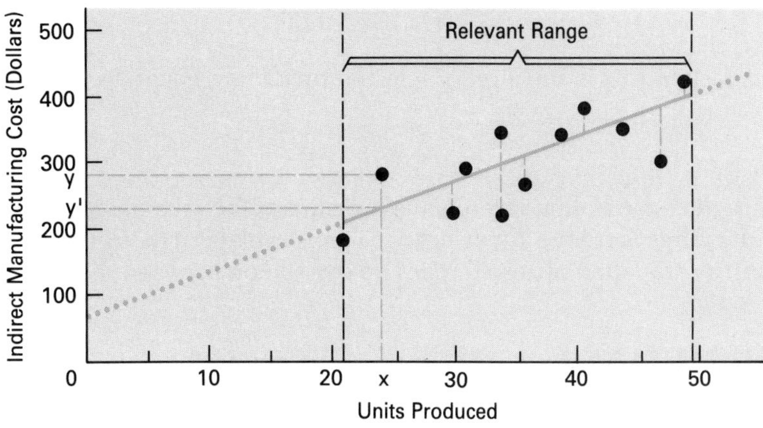

coefficient of determination (r^2) is a measure of the extent to which the independent variable accounts for the variability in the dependent variable.

The coefficient of determination (r^2) indicates the proportion of the variance $(y - \bar{y})^2$ that is explained by the independent variable x; $\bar{y} = (\Sigma y) \div n$. The coefficient of determination is also expressed more informatively as 1 minus the proportion of total variance that is unexplained:

$$r^2 = 1 - \frac{\Sigma(y - y')^2}{\Sigma(y - \bar{y})^2} = 1 - \frac{\text{Unexplained variation}}{\text{Total variation}}$$

The measure of the goodness of fit of a regression line is made by comparing $\Sigma(y - y')^2$ with the sum of the squares of the deviations of y values from their mean $\Sigma(y - \bar{y})^2$. Using our illustration from Exhibit 25-13, $\Sigma y = 3{,}549$ and $\bar{y} = 3{,}549 \div 12 = 295.75$. Therefore, the total squared deviation of the dependent variable from its mean is:

$$\Sigma(y - y)^2 = (340 - 295.75)^2 + (346 - 295.75)^2 + \cdots + (275 - 295.75)^2 = 54{,}389$$

The squared deviation of y from the regression line that is not explained by x is measured by the quantity $\Sigma(y - y')^2$. If all the points actually fell on a straight line, $\Sigma(y - y')^2$ would equal zero. To compute $\Sigma(y - y')^2$ for our illustration, we must predict the values of y' by substituting the given values of x into the least-squares equation. We obtain $y' = \$77.08 + \$6.10(34) = \$284.48$ for the first week's production, $y' = \$77.08 + \$6.10(44) = \$345.48$ for the second week's production, and so on. Substituting these values and the observed y values into $\Sigma(y - y')^2$, we obtain the total squared deviations not explained by x:

$$\Sigma(y - y')^2 = (340 - 284.48)^2 + (346 - 345.48)^2 + \cdots + (275 - 223.48)^2 = 23{,}640$$

The percentage of the variation in the indirect manufacturing costs that cannot be explained by x is:

$$\frac{\Sigma(y - y')^2}{\Sigma(y - \bar{y})^2} = \frac{23{,}640}{54{,}389} = .43, \text{ or } 43\%$$

The remaining 57% (100% − 43%) of the variation in the indirect manufacturing costs is accounted for by differences in the number of units produced:

$$r^2 = 1 - \frac{23{,}640}{54{,}389} = 1 - .43 = .57, \text{ or } 57\%$$

The square root of .57 (that is, the proportion of the total variation in indirect manufacturing costs that is accounted for by differences in the number of units produced) is called the *coefficient of correlation, r:*

$$r = \sqrt{1 - \frac{\Sigma(y - y')^2}{\Sigma(y - \bar{y})^2}}$$

The sign attached to r is the sign of b in the predicting equation:

$$r^2 = .57$$
$$r = +\sqrt{.57} = .755$$

The coefficient of correlation is a relative measure of the relationship between two variables. Its range is from −1 (perfect negative correlation) to +1 (perfect positive correlation); a correlation of zero implies no correlation between the two variables.

STANDARD ERROR OF RESIDUALS

How accurate is the regression line as a basis for prediction? We are using a sample of historical events. If we duplicated that sample using different data, we would not expect to obtain the same line. The values of a and b would differ for each sample taken. What we really want to know is the true regression line of the entire population,

$$y = \alpha + \beta x$$

where α and β are the true coefficients. The values of a and b are estimates based on a sample. Therefore, they are subject to chance variation, as are all sample statistics.

To judge the accuracy of the regression line, we examine the dispersion of the observed values of y around the regression line. A measure of this dispersion assists in judging the probable accuracy of a prediction of the average indirect manufacturing cost of, say, 41 units produced. The measure of the scatter of the actual observations about the regression line is called the **standard error of the residuals**. (It is also called the **standard error of the estimate**.) The standard error of the residuals for the population may be calculated from a sample in a simple linear regression as follows:

$$S_e = \sqrt{\frac{\Sigma(y - y')^2}{n - 2}}$$

where n is the size of the sample.

Using the data in Exhibit 25-13:

$$S_e = \sqrt{\frac{23{,}640}{10}} = 48.62$$

If the assumptions of regression analysis are met (for example, constant variance of residuals and normality of residuals), we can use the standard error to help gauge our confidence in the predictions. Consider the prediction of indirect manufacturing costs when 41 units are produced: $y' = 77.08 + 6.10(41) = \$327.18$. Approximately two-thirds of the points in a normal distribution should be within the band $y' \pm$ one standard error:

$$\$327.18 \pm (1.0)(\$48.62)$$

Thus, the band is \$279 to \$376 (rounded to the nearest dollar). Management can predict that a production level of 41 units will result in indirect manufacturing costs between \$279 and \$376, with approximately two chances out of three that the confidence interval constructed in this manner will contain the true value of y. (For predicting costs beyond the data set used to estimate the regression model, an extra adjustment is necessary that will increase the standard error of prediction.)

SAMPLING ERRORS AND REGRESSION COEFFICIENTS

Does a significant explanatory relationship exist between x and y? For example, the regression coefficient of x of \$6.10 implies an increase in indirect manufacturing cost of \$6.10 for each additional unit produced. The regression coefficient \$6.10 is an estimate of a population parameter. A particular sample may indicate a relationship, even when none exists, by pure chance. If there is no relationship, then the slope β of the true regression line would be zero. A null hypothesis can be set up that $\beta = 0$. If the sample value b is significantly different from zero, we would reject the hypothesis and assert that there is a definite relationship between the variables.

To test this null hypothesis, we need to calculate the standard error of the b coefficient:

$$S_b = \frac{S_e}{\sqrt{\Sigma(x - \bar{x})^2}}$$

For the data in Exhibit 25-13:

$$S_b = \frac{48.62}{\sqrt{825}} = \frac{48.62}{28.72} = 1.69$$

The procedure for deciding whether a positive relationship exists between units produced and indirect manufacturing costs in this:

- Null hypothesis: $\beta = 0$ (no relationship)
- Alternative hypothesis: $\beta \neq 0$ (relationship exists between indirect manufacturing costs and units produced)

The value of b is \$6.10. If the null hypothesis is true, $\beta = 0$, and b is \$6.10 units from β. In terms of its standard error, this is \$6.10 ÷ \$1.69 = 3.61. Therefore, b is 3.61 standard errors from $\beta = 0$. A deviation of more than two standard errors is usually regarded as significant. Therefore, the probability is low that a deviation as large as 3.61 standard errors could occur by chance. Consequently, we reject the null hypothesis and accept the alternative hypothesis that there is a significant relationship between the variables.

The amount of 3.61 standard errors just computed is called the *t*-value of the regression coefficient:

$$t\text{-value} = \frac{\text{Coefficient}}{\text{Standard error of the coefficient}} = \frac{6.10}{1.69} = 3.61$$

High *t*-values enhance confidence in the value of the coefficient as a predictor. Low *t*-values (as a rule of thumb, under 2.00) are indications of low reliability of the predictive power of that coefficient.

TERMS TO LEARN

This chapter and the Glossary at the end of the book contain definitions of the following important terms:

autocorrelation *(p. 790)* bias *(789)* coefficient of determination (r^2) *(786)*
constant variance *(789)* data mining *(787)* disturbance term *(789)*
error term *(789)* multicollinearity *(794)* multiple regression *(783)*
residual term *(789)* serial correlation *(790)* simple regression *(783)*
specification analysis *(786)* standard error of the estimate *(804)*
standard error of the residuals *(804)*

ASSIGNMENT MATERIAL

QUESTIONS

25-1 List four attractive features of using regression analysis in cost estimation.

25-2 What criteria are important when choosing among alternative cost functions?

25-3 "High correlation between two variables means that one is the cause and the other is the effect." Do you agree? Explain.

25-4 What are the four key assumptions examined in specification analysis?

25-5 Give two synonyms for *residual term*. Describe how the residual term is calculated.

25-6 Explain the term *serial correlation of the residuals*.

25-7 "Multicollinearity exists when the dependent variable and the independent variable are highly correlated." Do you agree? Explain.

25-8 "All the independent variables in a cost function estimated with regression analysis are cost drivers." Do you agree?

25-9 Give two examples of cost drivers in each of the following business functions: research and development, product design, manufacturing, marketing, distribution, and customer service.

EXERCISES AND PROBLEMS

25-10 Interpretation of regression coefficient. Gary Wirth, manager of Newcastle Steel Products, learned about linear regression analysis at an evening college course. He decided to apply regression in his study of repair costs at his plant. He plotted 24 points for the past 24 months and fitted a regression line:

$$\text{Total repair cost per month} = \$80{,}000 - \$2.50x$$

where x = number of machine hours worked.

He was baffled because the result was nonsense. Apparently, the more the machines were run, the less the repair costs. He decided that regression was a useless technique.

Required
Why was the puzzling regression coefficient negative? Do you agree with the manager's conclusion regarding regression? Explain.

25-11 Power department cost analysis, fundamentals of regression. (NAA, adapted) Joe Clark, manager of the engineering division of Icarus Parts, collects the following data on monthly power costs and machine hours:

Month	Power Costs (y)	Machine Hours (x)
April	$ 23	22
May	25	23
June	20	19
July	20	12
August	20	12
September	15	9
October	14	7
November	14	11
December	16	14
	$167	129

Required

1. Plot the relationship between monthly power costs *(y)* and machine hours *(x)*.
2. Compute the constant *(a)* and slope coefficient *(b)* of the following cost function using the regression approach:

$$y' = a + bx$$

Comment on the results.

25-12 Maintenance department costs, high-low and regression approaches. (H. Nurnberg, adapted) Bristol Engineering wishes to set flexible budgets for each of its operating departments. A separate maintenance department performs all routine and major repair work on the corporation's equipment and facilities. It has been determined that maintenance cost is primarily a function of machine hours worked in the various production departments. The maintenance cost incurred and the actual machine hours worked during the first four months of 19_1 are as follows:

	Machine Hours in Production Departments (x)	Maintenance Department Costs (y)
January	800	$350
February	1,200	350
March	400	150
April	1,600	550

Required

1. Plot the relationship between monthly maintenance department costs and machine hours in production departments at Bristol Engineering.
2. Compute the constant *(a)* and slope coefficient *(b)* of the following cost function using (i) the high-low approach and (ii) the regression approach:

$$y' = a + bx$$

3. Compute the coefficient of determination, r^2, of the cost function in requirement 2 when the regression approach is used.

25-13 Data collection issues, regression analysis (Extension of Problem 10-17 on page 360). Robin Green, financial analyst at Central Railroad, is examining the behavior of monthly transportation costs for budgeting purposes. Transportation costs at Central Railroad are the sum of two types of costs:

(a) Operating costs, such as fuel and labor, and
(b) Maintenance costs, such as overhaul of engines and track.

Green collects monthly data on (a), (b), and track miles hauled by month. Track miles hauled are the miles clocked by the engine that pulls the railroad cars. Monthly observations for the most recent year follow:

Month (1)	Operating Costs (2)	Maintenance Costs (3)	Total Transportation Costs (4) = (2) + (3)	Track Miles Hauled (5)
January	$471	$437	$908	3,420
February	504	388	842	5,310
March	609	343	952	5,410
April	690	347	1,037	8,440
May	742	294	1,036	9,320
June	774	211	985	8,910
July	784	176	960	8,870
August	986	210	1,196	10,980
September	895	282	1,177	4,980
October	651	394	1,045	5,220
November	481	381	862	4,480
December	386	514	900	2,980

Green examines the results of the following three regressions:

REGRESSION 1:
Operating costs = $a + b$ (Track miles hauled)

Variable	Coefficient	Standard Error	*t*-Value
Constant	$309.19	96.05	3.22
Independent variable 1:			
Track miles hauled	0.054	0.014	3.86

$r^2 = 0.61$; Durbin-Watson statistic = 1.61

REGRESSION 2:
Maintenance costs = $a + b$ (Track miles hauled)

Variable	Coefficient	Standard Error	*t*-Value
Constant	$531.55	46.95	11.32
Independent variable 1:			
Track miles hauled	−0.031	0.007	−4.43

$r^2 = 0.68$; Durbin-Watson statistic = 1.72

REGRESSION 3:
Total transportation costs = $a + b$ (Track miles hauled)

Variable	Coefficient	Standard Error	*t*-Value
Constant	$840.73	80.25	10.48
Independent variable 1:			
Track miles hauled	0.023	0.011	2.09

$r^2 = 0.29$; Durbin-Watson statistic = 2.34

Required

1. Evaluate the three regressions using the economic plausibility, goodness of fit, significance of independent variables, and specification analysis criteria.
2. Name three variables, other than track miles hauled, that could be important cost drivers of railroad operating costs.
3. Describe an alternative data base Green might want to use to examine the cost drivers of railroad maintenance costs.

25-14 Evaluating alternative regression functions, accrual accounting adjustments. (Extension of Problem 10-16 on page 360). Trevor Kennedy, the cost analyst at a can manufacturing plant of United Packaging, used a regression model to examine the relationship between total engineering support costs reported in the plant records and machine hours. After further discussion with the operating manager, Kennedy discovers that the materials and parts numbers reported in the monthly records are on an "as purchased" basis and not on an "as used" or accrual accounting basis. By examining materials and parts usage records, Kennedy is able to restate the materials and parts costs to an "as used" basis. (No restatement of the labor costs was necessary.) The reported and restated costs follow:

Month	Labor: Reported Costs	Materials and Parts: Reported Costs	Materials and Parts: Restated Costs	Total Engineering Support: Reported Costs	Total Engineering Support: Restated Costs	Machine Hours
(1)	(2)	(3)	(4)	(5) = (2) + (3)	(6) = (2) + (4)	(7)
March	$347	$847	$182	$1,194	$529	30
April	521	0	411	521	932	63
May	398	0	268	398	666	49
June	355	961	228	1,316	583	38
July	473	0	348	473	821	57
August	617	0	349	617	966	73
September	245	821	125	1,066	370	19
October	487	0	364	487	851	53
November	431	0	290	431	721	42

The regression results, when total engineering support reported costs (Column 5) are used as the dependent variable, are:

REGRESSION 1:
Engineering Support Reported Costs = *f*(Machine Hours)

Variable	Coefficient	Standard Error	*t*-Statistic
Constant	1,393.20	305.68	4.56
Independent variable 1:			
Machine hours	−14.23	6.15	−2.31

r^2 = .43; Durbin-Watson statistic = 2.26

The regression results, when total engineering support restated costs (Column 6) are used as the dependent variable, are:

REGRESSION 2:
Engineering Support Restated Costs = *f*(Machine Hours)

Variable	Coefficient	Standard Error	*t*-Statistic
Constant	176.38	53.99	3.27
Independent variable 1:			
Machine hours	11.44	1.08	10.59

r^2 = .94; Durbin-Watson statistic = 1.31

Required

1. Present a plot of the data for the cost function relating the *reported costs* for total engineering support to machine hours. Present a plot of the data for the cost function relating the *restated costs* for total engineering support to machine-hours. Comment on the plots.
2. Contrast and evaluate the cost function estimated with regression using restated data for materials and parts with the cost function estimated with regression using the data reported in the plant records. Use the comparison format employed in Exhibit 25-7 (p. 793).
3. What problems might Kennedy encounter when restating the materials and parts costs recorded to an "as used" or accrual accounting basis?

25-15 Choosing one of two regression analyses. (CMA) The Alma Plant manufactures the industrial product line of CJS Industries. Plant management wants to be able to get a good, yet quick, estimate of the manufacturing overhead costs that can be expected each month. The easiest and simplest method to accomplish this task appears to be to develop a flexible-budget formula for the manufacturing overhead costs.

The plant's accounting staff suggested that simple linear regression be used to determine the cost behavior pattern of the overhead costs. The regression data can provide the basis for the flexible-budget formula. Sufficient evidence is available to conclude that manufacturing overhead costs vary with direct labor hours. The actual direct labor hours and the corresponding manufacturing overhead costs for each month of the last three years were used in the linear regression analysis.

The three-year period contained various occurrences common to many businesses. During the first year, production was severely curtailed during two months due to wildcat strikes. In the second year, production was reduced in one month because of material shortages. Overtime was increased during two months to meet a one-time only sales order. At the end of the second year, employee benefits were raised significantly as the result of a labor agreement. Production during the third year was not affected by any special circumstances.

Various members of Alma's accounting staff raised some issues regarding the historical data collected for the regression analysis. These issues follow.

1. Some members of the accounting staff believed that the use of data from all 36 months would provide a more accurate portrayal of the cost behavior. Although they recognized that any of the monthly data could include normal and abnormal amounts, they believed these normal and abnormal amounts would tend to balance out over a longer period of time.
2. Other members of the accounting staff suggested that only those months that were considered normal should be used, so that the regression would not be distorted.
3. Still other members felt that only the most recent 12 months should be used because they were the most current.
4. Some members questioned whether historical data should be used at all to form the basis for a flexible-budget formula.

The accounting department ran two regression analyses of the data—one using the data from all 36 months and the other using the data from only the last 12 months. The information derived from the two linear regressions is:

	Data from all 36 Months	Data from Most Recent 12 Months
Coefficients of the regression equation:		
Constant	\$123,810	\$109,020
Independent variable (DLH)	\$1.6003	\$4.1977
Coefficient of correlation	.4710	.6891
Standard error of the residuals	13,003	7,473
Standard error of the regression coefficient for the independent variable	.9744	1.3959
Calculated *t*-statistic for the regression coefficient	1.6423	3.0072
t-statistic required for a 95% confidence interval		
34 degrees of freedom (36 − 2)	1.960	
10 degrees of freedom (12 − 2)		2.228

Required

1. From the results of Alma Plant's regression analysis that used the data from all 36 months:
   a. Formulate the flexible-budget equation that can be used to estimate monthly manufacturing overhead costs.
   b. Calculate the predicted overhead costs for a month when 25,000 direct labor hours are worked.
2. Using only the results of the two regression analyses, explain which of the two results (12 months versus 36 months) you would use as a basis for the flexible-budget formula.
3. How would the four specific issues raised by the members of Alma's accounting staff influence your willingness to use the results of the statistical analyses as the basis for the flexible-budget formula? Explain your answer.

25-16 Evaluating alternative simple regression models, university business school. Teri Bush, executive assistant to the president of Western University, has been hearing many complaints from university administrators that the nonacademic overhead costs of the business school are out of control. The business school has grown from fourteen full-time faculty to sixty-one full-time faculty in the last twelve years. The nonacademic overhead costs include the salaries of administrators and other nonacademic staff, the costs of handling student applications, and the costs of numerous individual supply items such as photocopying paper.

Bush decides that some data analysis is warranted. She collects the following information pertaining to the business school:

Year	Nonacademic Overhead Costs (thousands)	Number of Nonacademic Staff	Number of Student Applications	Number of Enrolled Students
1	$ 2,200	29	1,010	342
2	4,120	36	1,217	496
3	3,310	49	927	256
4	4,410	53	1,050	467
5	4,210	54	1,563	387
6	5,440	58	1,127	492
7	5,600	88	1,892	513
8	4,380	72	1,362	387
9	5,270	83	1,623	346
10	7,610	73	1,646	487
11	8,070	101	1,870	564
12	10,388	103	1,253	764

She finds the following results for three separate simple regression models.

REGRESSION 1:
Overhead Costs = $a + b$ (Number of Nonacademic Staff)

Variable	Coefficient	Standard Error	*t*-Statistic
Constant	112.04	1,119.40	0.10
Independent variable 1:			
Number of staff	79.68	15.89	5.01

$r^2 = 0.72$; Durbin-Watson statistic = 1.82

REGRESSION 2:
Overhead Costs = $a + b$ (Number of Student Applications)

Variable	Coefficient	Standard Error	*t*-Statistic
Constant	1,147.40	2,710.00	0.42
Independent variable 1:			
Number of applicants	3.10	1.91	1.62

$r^2 = 0.21$; Durbin-Watson statistic = 0.89

REGRESSION 3:
Overhead Costs = $a + b$ (Number of Enrolled Students)

Variable	Coefficient	Standard Error	*t*-Statistic
Constant	−1,382.20	1,350.30	−1.02
Independent variable 1:			
Number of students	14.83	2.84	5.22

$r^2 = 0.73$; Durbin-Watson statistic = 1.09

Required

1. What problems might arise in measuring the nonacademic costs of the business school?

2. Plot the relationship between nonacademic overhead costs and each of the following variables: (a) number of nonacademic staff, (b) number of student applications, and (c) number of enrolled students.
3. Compare and evaluate the three simple regression models estimated by Bush. Use the comparison format employed in Exhibit 25-7 (p. 793).
4. What other variables might Bush collect to explain the behavior of nonacademic overhead costs of the business school?

25-17 Evaluating alternative multiple regression models, university business school (Extension of Problem 25-16). Teri Bush decides that the simple regression analysis reported in Problem 25-16 could be extended to a multiple regression analysis. She finds the following results for several multiple regressions:

REGRESSION 4:
Overhead Costs = $a + b$ (Number of Staff) + c(Number of Applications)

Variable	Coefficient	Standard Error	*t*-Statistic
Constant	1,243.00	1,630.20	0.76
Independent variable 1:			
Number of staff	93.81	21.73	4.32
Independent variable 2:			
Number of applications	−1.50	1.57	−0.96

$r^2 = 0.74$; Durbin-Watson statistic = 1.92

REGRESSION 5:
Overhead Costs = $a + b$ (Number of Staff) + c(Number of Students)

Variable	Coefficient	Standard Error	*t*-Statistic
Constant	−2,114.70	894.79	−2.36
Independent variable 1:			
Number of staff	48.62	12.62	3.85
Independent variable 2:			
Number of students	9.37	2.32	4.04

$r^2 = 0.90$; Durbin-Watson statistic = 1.82

REGRESSION 6:
Overhead Costs = $a + b$ (Number of Applications) + c(Number of Students)

Variable	Coefficient	Standard Error	*t*-Statistic
Constant	−3,178.60	1,714.20	−1.85
Independent variable 1:			
Number of applications	1.68	1.08	1.56
Independent variable 2:			
Number of students	13.70	2.76	4.96

$r^2 = 0.79$; Durbin-Watson statistic = 1.65

The coefficient of correlation (r) between pairwise combinations of the variables is:

	Overhead Costs	Number of Staff	Number of Applications
Number of staff	0.846		
Number of applications	0.455	0.679	
Number of students	0.855	0.610	0.264

Required

1. Compare and evaluate the three multiple regression models estimated by Bush. Use the comparison format employed in Exhibit 25-7 (p. 793).
2. What problems may arise in multiple regressions that do not arise in simple regressions? Is there evidence of such problems in any of the multiple regressions presented in this question?

3. Of what use might the regression results be to the president of Western University in attempting to control the nonacademic overhead costs of the business school?

25-18 Purchasing department cost drivers, simple regression analysis. Fashion Flair operates a chain of ten retail department stores. Each department store makes its own purchasing decisions. Barry Lee, assistant to the president of Fashion Flair, is interested in better understanding the drivers of purchasing department costs at each store. For many years, Fashion Flair has allocated purchasing department costs to products on the basis of the dollar value of merchandise purchased. An item costing $100 is allocated 10 times as much overhead costs associated with the purchasing department as an item costing $10 is allocated.

Lee recently attended a seminar titled "Cost Drivers in the Retail Industry." In a presentation at the seminar, Couture Fabrics, a leading competing firm that has thirty retail outlets, reported the number of purchase orders and the number of suppliers to be the two most important cost drivers of purchasing department costs. The dollar value of merchandise purchased on each purchase order was reported to be not a significant cost driver by Couture Fabrics. Lee interviewed several members of the purchasing department at The Fashion Flair store in Miami. These people told Lee that they believed the Couture Fabrics conclusions also applied to their purchasing department.

Lee collects the following data for the most recent year for the ten retail department stores of Fashion Flair:

Department Store	Purchasing Department Costs (PDC)	Dollar Value of Merchandise Purchased (MP$)	Number of Purchase Orders (# of PO's)	Number of Suppliers (# of S's)
Baltimore	$1,523,000	$68,315,000	4,357	132
Chicago	1,100,000	33,456,000	2,550	222
Los Angeles	547,000	121,160,000	1,433	11
Miami	2,049,000	119,566,000	5,944	190
New York	1,056,000	33,505,000	2,793	23
Phoenix	529,000	29,854,000	1,327	33
Seattle	1,538,000	102,875,000	7,586	104
St.Louis	1,754,000	38,674,000	3,617	119
Toronto	1,612,000	139,312,000	1,707	208
Vancouver	1,257,000	130,944,000	4,731	201

Lee decides to use simple regression analysis to examine whether one or more of three variables (the last three columns in the table) are cost drivers of purchasing department costs. Summary results for these regressions follow:

REGRESSION 1:
PDC = $a + b$(MP$)

Variable	Coefficient	Standard Error	*t*-Statistic
Constant	1,039,376	344,904	3.01
Independent variable 1:			
MP$	0.0032	0.0037	0.86

$r^2 = 0.08$; Durbin-Watson statistic = 2.41

REGRESSION 2:
PDC = $a + b$(# of PO's)

Variable	Coefficient	Standard Error	*t*-Statistic
Constant	718,413	261,938	2.74
Independent variable 1:			
# of PO's	161.63	63.84	2.53

$r^2 = 0.44$; Durbin-Watson statistic = 1.97

REGRESSION 3:
PDC = a + b(# of S's)

Variable	Coefficient	Standard Error	t-Statistic
Constant	830,326	251,869	3.30
Independent variable 1:			
# of S's	3,759	1,710	2.20

$r^2 = 0.38$; Durbin-Watson statistic = 1.97

Required

1. Compare and evaluate the three simple regression models estimated by Lee. Graph each one. Also, use the format employed in Exhibit 25-7 (p. 793) to evaluate the information.
2. Do the regression results support the Couture Fabrics presentation about purchasing department cost drivers?
3. How might Lee gain additional evidence on drivers of purchasing department costs at each store of Fashion Flair?

25-19 Purchasing department cost drivers, multiple regression analysis (Extension of Problem 25-18). Barry Lee decides that the simple regression analysis reported in Problem 25-18 could be extended to a multiple regression analysis. He finds the following results for several multiple regressions:

REGRESSION 4:
PDC = a + b(# of PO's) + c(# of S's)

Variable	Coefficient	Standard Error	t-Statistic
Constant	486,238	257,614	1.89
Independent variable 1:			
# of PO's	128.93	57.94	2.23
Independent variable 2:			
# of S's	2,796	1,465	1.91

$r^2 = 0.63$; Durbin-Watson statistic = 1.88

REGRESSION 5:
PDC = a + b (# of PO's) + c(# of S's) + d(MP$)

Variable	Coefficient	Standard Error	t-Statistic
Constant	495,093	310,362	1.60
Independent variable 1:			
# of PO's	129.71	63.72	2.04
Independent variable 2:			
# of S's	2,827	1,653	1.71
Independent variable 3:			
MP$	−0.0002	.0030	−0.07

$r^2 = 0.63$; Durbin-Watson statistic = 1.87

The coefficient of correlation (r) between pairwise combinations of the variables is:

	PDC	MP$	#PO's
MP$	0.29		
#PO's	0.67	0.27	
#S's	0.62	0.34	0.29

Required

1. Evaluate Regression 4 using the economic plausibility, goodness of fit, significance of independent variables, and specification analysis criteria. Compare Regression 4 with Regressions 2 and 3 in Problem 25-18. Which model would you recommend that Lee use?

2. Compare Regression 5 with Regression 4. Which model would you recommend that Lee use?
3. What difficulties may arise in multiple regressions that do not arise in simple regressions? Is there evidence of such difficulties in either of the multiple regressions presented in this problem?
4. Give two examples of decisions where the regression results reported in this problem (and in 25-18) could be informative.

CHAPTER 26

Variances: Mix, Yield, and Investigation

PART ONE: SALES VARIANCES

Sales volume variances

Sales quantity and sales mix variances

Market size and market share variances

PART TWO: PRODUCTION VARIANCES

Direct material yield and direct material mix variances

Direct labor yield and direct labor mix variances

PART THREE: VARIANCE INVESTIGATION DECISIONS

Sources of variances

Cost-benefit analysis of variance investigation decisions

Difficulties in analysis of variances

Statistical quality control

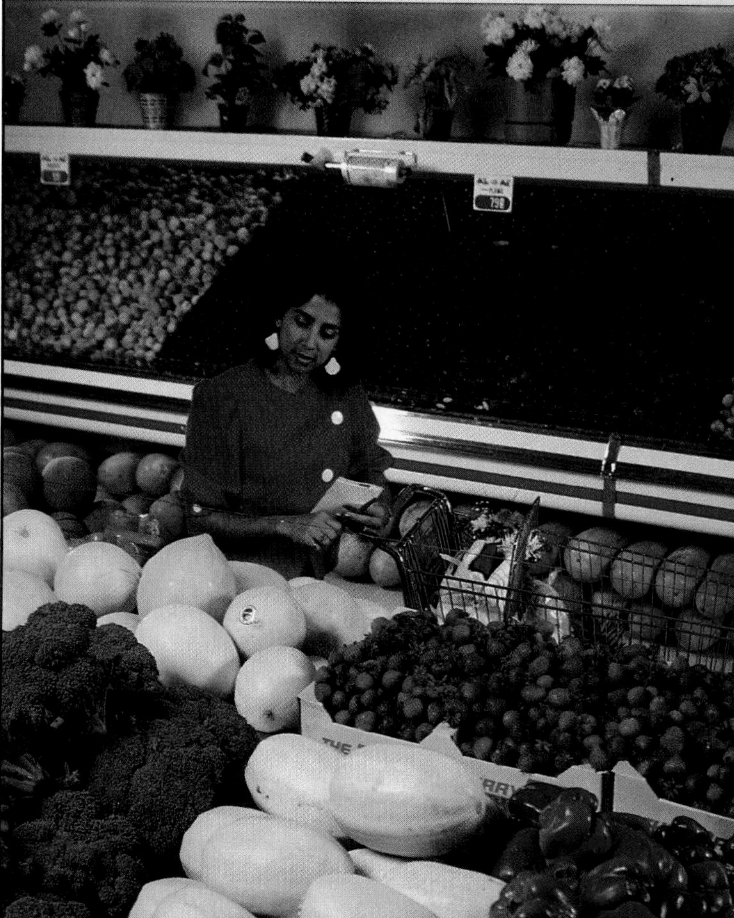

Changes in the mix of products sold can cause sizable changes in the operating income of supermarkets.

Learning Objectives

When you have finished studying this chapter, you should be able to

1. Describe the insight gained from dividing sales volume variance into sales quantity variance and sales mix variance
2. Explain how market size and market share variances provide different explanations for a sales quantity variance
3. Understand how trade-offs among material inputs are highlighted by the material yield and material mix variances
4. Provide explanations for favorable and unfavorable labor mix variances
5. Identify seven sources of variances
6. Explain how decision making under uncertainty can be used to decide whether to investigate a variance
7. Describe statistical quality control and identify problems in using it to analyze cost variances

Comparing results with budgets can help managers evaluate operations. Managers can see the impact of key variables on the actual results and focus on areas that deserve more investigation. Chapters 7 and 8 illustrated various uses of variance information relating to direct materials, direct labor, and factory overhead.

This chapter extends some major areas of the analysis of accounting variances that have been covered only briefly in earlier chapters. Three subjects are discussed. Because each subject can be studied independently, the chapter is divided into three major parts.

1. Part One: *Sales variances.* We examine five variances: sales volume, sales quantity, sales mix, market size, and market share variances.
2. Part Two: *Production variances.* For illustrative purposes, we present a yield and a mix variance for each of two factors of production—direct materials and direct labor. The variances presented can readily be adapted to other inputs, such as energy.
3. Part Three: *Variance investigation decisions.* We examine sources of variances and discuss how the approach to decision making under uncertainty introduced in Chapter 20 can be used in variance investigation decisions.

The examples presented in this chapter involve no more than three products, three direct material inputs, or two direct labor inputs. Computer software packages can greatly reduce the computation burden in calculating variances. When there are thousands of products sold or thousands of direct material or direct labor inputs, the use of computer software packages is essential in variance analysis.

PART ONE: SALES VARIANCES

Most organizations produce more than one product or service. Therefore, budgets for total sales usually specify the quantity of each product to be sold. *Sales mix* is the relative combination of quantities of products that constitute total sales. Managers want help in tracing deviations from budgets. Chapter 7 showed how price, efficiency, and sales volume variances assist managers in single-product situations. We now explore the complications that arise in multiple-product situations when the sales quantity or the sales mix differs from the budget.

The literature on variance analysis in multiple-product situations is sizable and bewildering. It seems that almost every organization has a pet way of defining terms. Be on guard whenever you see the terms *quantity variance, volume variance, activity variance, mix variance,* or *yield variance.* In any discussion, be sure that all parties agree on definitions.

SALES VOLUME VARIANCES

Chapter 7 introduced flexible budgets and standards. Exhibit 7-5 (p. 218) provided an overview of variances. Pause and review that exhibit before reading further.

Managers can make distinctions among variances at various levels of detail. Even in companies with multiple products, the simplest and most informative analysis examines individual products through three variances: *price*, *efficiency*, and *sales volume* variances. Chapter 7 focused on price variances and efficiency variances. We now examine sales volume variances.

The *sales volume variance* is the difference between the amount of the contribution margin in the budget based on actual sales volume (a flexible budget) and that amount in the static (master) budget. Budgeted unit selling prices and unit variable costs are held constant. This variance is measured in terms of contribution margin because fixed costs are the same in a flexible budget and a static budget.

EXAMPLE

Party Wholesaler has an exclusive contract with Bordeaux Winery to import into the United States and sell two brands of wine—jug wine and premium wine. Party Wholesaler sells to liquor stores and other retail outlets. The monthly budget for the Bordeaux Winery contract is based on a combination of last year's performance, a forecast of general industry sales, and the company's expected share of the U.S. market for imported wine.

Assume the following budgeted and actual data for October 19_7 in our example:

	Budgeted		Actual	
	Jug Wine	Premium Wine	Jug Wine	Premium Wine
Selling price per unit	$5	$16	$5.50	$15.50
Variable cost per unit	4	9	4.30	9.50
Contribution margin per unit	$1	$7	$1.20	$6.00
Sales in units	1,200	400	1,100	500

Budgeted fixed costs for October 19_7 are $3,000. Actual fixed costs for October 19_7 are $3,050.

Party Wholesaler prepared the following static (master) budget for October 19_7 for its contract with Bordeaux Winery:

Static (Master) Budget	
Sales	$12,400
Variable costs	8,400
Contribution margin	$ 4,000
Fixed costs	3,000
Operating income	$ 1,000

The budgeted average contribution margin per unit is $2.50 ($4,000 ÷ 1,600 total budgeted units).

Exhibit 26-1 compares the actual results for October 19_7 with the budgeted results. (Exhibit 26-1 is a Level 2 analysis in the terminology used in Chapter 7.) The favorable static-budget variance of $270 is made up of an unfavorable flexible-budget variance of $330 and a favorable sales volume variance of $600.

EXHIBIT 26-1
Variance Analysis Using Flexible Budget: Party Wholesaler, October 19_7

	Actual Results (1)	Flexible-Budget Variance (2)	Flexible Budget* (3)	Sales Volume Variance (4)	Static (Master) Budget (5)
Units sold	1,600	—	1,600	—	1,600
Sales in dollars	$13,800	$300 F	$13,500	$1,100 F	$12,400
Variable costs	9,480	580 U	8,900	500 U	8,400
Contribution margin	$ 4,320	$280 U	$ 4,600	$ 600 F	$ 4,000
Fixed costs	3,050	50 U	3,000	—	3,000
Operating income	$ 1,270	$330 U	$ 1,600	$ 600 F	$ 1,000

$270 F
Static-budget variance (between Actual Results (1) and Static (Master) Budget (5))

*Based on actual units sold times budgeted amounts per unit.
F = Favorable; U = Unfavorable.

The Effects of More Than One Product

Exhibit 26-1 provides a useful overview of the variances, but most managers want more detail. Consider that the final three columns have a puzzling aspect. How can the flexible-budget amounts differ from the static (master) budget amounts if the number of units actually sold equals the 1,600 units originally budgeted? The reason, as explained in this section, is that this is a multiple-product company. In a single-product company, this puzzle would not arise.

In a multiple-product company, each product has its own standards and flexible budget. If the *mix* of products sold changes, the company's overall flexible budget is affected because it is merely a *summation* of the flexible budgets drawn up for the individual products sold.

Exhibit 26-2 shows that the original budgeted contribution margin of $4,000 is the sum of the budgeted contribution margins for 1,200 units of jug wine and 400

EXHIBIT 26-2
Sales Volume Variances by Product: Party Wholesaler, October 19_7

	Jug Wine			Premium Wine			Total		
	Flexible Budget* (1)	Sales Volume Variance (2)	Static (Master) Budget (3)	Flexible Budget* (4)	Sales Volume Variance (5)	Static (Master) Budget (6)	Flexible Budget* (7) = (1) + (4)	Sales Volume Variance (8) = (2) + (5)	Static (Master) Budget (9) = (3) + (6)
Units sold	1,100	100 U	1,200	500	100 F	400	1,600	—	1,600
Sales in dollars†	$5,500	$500 U	$6,000	$8,000	$1,600 F	$6,400	$13,500	$1,100 F	$12,400
Variable costs‡	4,400	400 F	4,800	4,500	900 U	3,600	8,900	500 U	8,400
Contribution margin§	$1,100	$100 U	$1,200	$3,500	$ 700 F	$2,800	$ 4,600	$ 600 F	$ 4,000
Fixed costs							3,000	—	3,000
Operating income							$ 1,600	$ 600 F	$ 1,000

*Based on actual units sold times budgeted amount per unit.
†Jug wine $5 per unit, premium wine $16 per unit.
‡Jug wine $4 per unit, premium wine $9 per unit.
§Jug wine $1 per unit, premium wine $7 per unit.

units of premium wine. But 1,100 units of jug wine and 500 units of premium wine were actually sold. The flexible budget based on actual sales of 1,600 units differs from the static budget because the sales mix used in the standard budget differs from the sales mix that actually occurred. It is only by coincidence that the same number of bottles budgeted—1,600—was the same number of bottles actually sold.

Exhibit 26-2 provides details of the sales volume variance. It shows how the flexible budget based on actual units sold is the sum of individual flexible budgets. The total sales volume variance is $600 favorable:

$$\text{Sales volume variance} = \left(\begin{array}{c}\text{Actual sales}\\ \text{volume in units}\end{array} - \begin{array}{c}\text{Budgeted sales}\\ \text{volume in units}\end{array}\right) \times \left(\begin{array}{c}\text{Budgeted individual product}\\ \text{contribution margin per unit}\end{array}\right)$$

Jug wine:	(1,100 − 1,200) × $1 =	$100 U
Premium wine:	(500 − 400) × $7 =	$700 F
Total		$600 F

Other things being equal, the failure to sell the originally expected 1,200 units of jug wine would cause a decrease in operating income of 100 units × $1 = $100. Selling 100 more units of premium wine than originally expected, however, would increase operating income by 100 units × $7 = $700. Because the sales volume variance depends on how the market reacts to a product, some people prefer to label the sales volume variance as the marketing variance.

How Managers Use Sales Variance Information

The manager in charge of the Bordeaux Winery contract at Party Wholesaler may initially concentrate on the Total columns in Exhibit 26-2 (columns 7, 8, and 9). However, these Total columns provide only a beginning. An inquiring manager would fine-tune perspective by digging more deeply into the available information. The manager should keep the following three points in mind when analyzing the information underlying Exhibit 26-2.

First, the budgeted and actual *aggregated* units are equal. However, is this informative? Adding units of different products together may yield a meaningless amount. (This is often called the "apples and oranges problem.") As we move beyond a one-product analysis, aggregated units fail to provide a common denominator for measuring overall sales volume. To be useful to a manager, units are converted into monetary equivalents such as sales dollars or contribution margins. But we should not forget the fundamental inputs and outputs of organizations—their products or services.

Second, the aggregated dollars given an overall picture of the product line, but managers must often focus on individual products. For example, a manager must decide how best to allocate the advertising budget. Should Party Wholesaler push the premium wine or the jug wine? Without having information on the contribution margins of the individual products that make up the product line, a manager cannot make an informed decision.

Third, there can be confusion regarding the sales volume variances of $1,100 F, $500 U, and $600 F in column 8 of Exhibit 26-2. Taken together, *the total sales volume in units* was unchanged, but the original proportions—the sales mix—changed. The actual volumes of *individual* products deviated from what was expected. If the original budget in total units is attained and a larger proportion of higher-than-average budgeted contribution margin products are sold than was specified in the original mix, higher operating income will result.

The general manager of Party Wholesaler usually receives summary monetary figures rather than a flood of detail. For example, she reviews only the dollar amounts for each contract with a winery or other supplier. Party Wholesaler simply deals with too many outside businesses for the general manager to review the flood of detail for each contract. In our example, the report she receives would present only the information in the Total columns of Exhibit 26-2.

Rapid advances in information technology are making more information more readily available to more people. Marketing managers at Pepsi-Cola, for example, can link into a computer network to seek detailed explanations for individual amounts in their summary sales volume variance reports. Indeed, they may be able to access on-line the same detailed product-by-product variance information typically examined only by individual product managers.

SALES QUANTITY AND SALES MIX VARIANCES

Objective 1

Describe the insight gained from dividing sales volume variance into sales quantity variance and sales mix variance.

Many managers favor probing the sales volume variances further. The analysis in Exhibit 26-2 reveals how two major factors affect total sales volume variances: (1) the actual quantity of units sold and (2) the relative proportions of products bearing different contribution margins. The sales quantity variance captures point 1, and the sales mix variance captures point 2. To simplify the computations, we focus solely on contribution margins, even though these variances could be computed separately for revenues and variable costs.

The **sales quantity variance** is the difference between the amount of contribution margin in the flexible budget based on actual sales volume at budgeted sales mix and that amount in the static (master) budget. Budgeted selling prices and budgeted unit variable costs are held constant. This variance is measured in terms of contribution margin because total fixed costs are the same in a flexible budget and the static budget.

The **sales mix variance** is the difference between the amount of contribution margin in the flexible budget based on actual sales volume at actual mix and that amount in the flexible budget based on actual sales volume at budgeted mix. Budgeted selling prices and budgeted unit variable costs are held constant. This variance is measured in terms of contribution margin because total fixed costs are unchanged in the flexible budgets.[1]

Exhibit 26-3 presents the formulas and computations for jug wine and premium wine for both the sales quantity variance and the sales mix variance:

	Sales Quantity Variance	Sales Mix Variance
Jug wine	$250 U	$150 F
Premium wine	250 F	450 F
Total	$ 0	$600 F

Observe that the sales quantity variance weighs all units at a single overall budgeted average contribution margin per unit. For a given change in actual volume, the total budgeted contribution margin would be expected to change at the $2.50 rate of the budgeted average unit contribution margin. With a positive contribution

[1]The discussion of sales mix variance draws extensively on J. Harris and J. Persons, "Methods of Dividing the Sales Mix Variance Among Individual Products," Working Paper, University of Tulsa, 1990.

EXHIBIT 26-3

Sales Quantity and Sales Mix Variance Analysis: Party Wholesaler, October 19_7

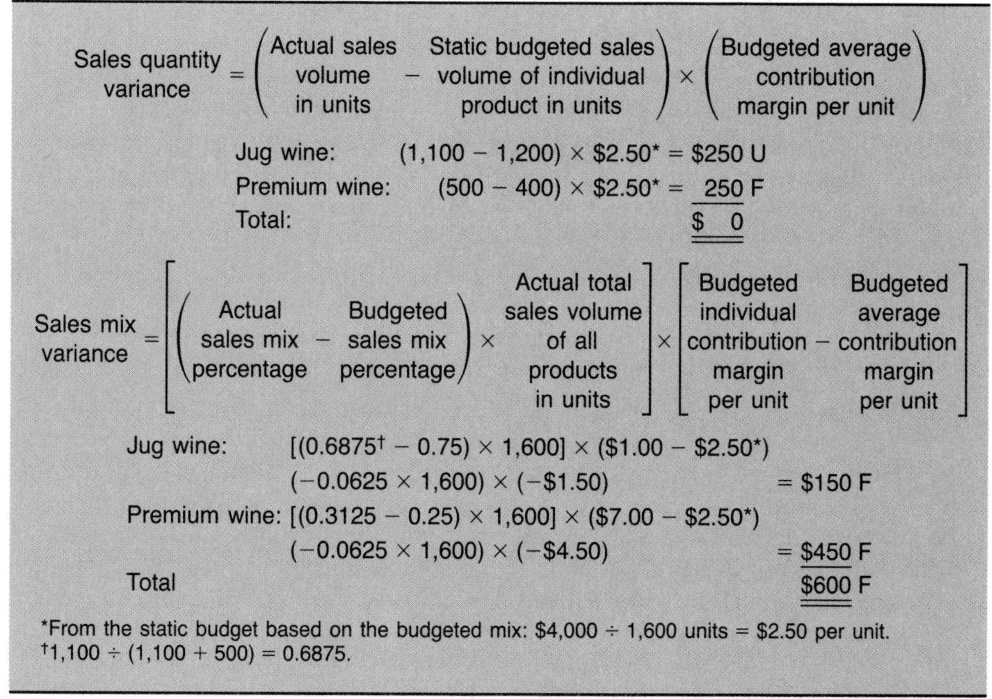

$$\text{Sales quantity variance} = \left(\begin{array}{c}\text{Actual sales}\\ \text{volume}\\ \text{in units}\end{array} - \begin{array}{c}\text{Static budgeted sales}\\ \text{volume of individual}\\ \text{product in units}\end{array}\right) \times \left(\begin{array}{c}\text{Budgeted average}\\ \text{contribution}\\ \text{margin per unit}\end{array}\right)$$

Jug wine:	(1,100 − 1,200) × \$2.50* =	\$250 U
Premium wine:	(500 − 400) × \$2.50* =	250 F
Total:		\$ 0

$$\text{Sales mix variance} = \left[\left(\begin{array}{c}\text{Actual}\\ \text{sales mix}\\ \text{percentage}\end{array} - \begin{array}{c}\text{Budgeted}\\ \text{sales mix}\\ \text{percentage}\end{array}\right) \times \begin{array}{c}\text{Actual total}\\ \text{sales volume}\\ \text{of all}\\ \text{products}\\ \text{in units}\end{array}\right] \times \left[\begin{array}{c}\text{Budgeted}\\ \text{individual}\\ \text{contribution}\\ \text{margin}\\ \text{per unit}\end{array} - \begin{array}{c}\text{Budgeted}\\ \text{average}\\ \text{contribution}\\ \text{margin}\\ \text{per unit}\end{array}\right]$$

Jug wine:	[(0.6875† − 0.75) × 1,600] × (\$1.00 − \$2.50*)	
	(−0.0625 × 1,600) × (−\$1.50)	= \$150 F
Premium wine:	[(0.3125 − 0.25) × 1,600] × (\$7.00 − \$2.50*)	
	(−0.0625 × 1,600) × (−\$4.50)	= \$450 F
Total		\$600 F

*From the static budget based on the budgeted mix: \$4,000 ÷ 1,600 units = \$2.50 per unit.
†1,100 ÷ (1,100 + 500) = 0.6875.

margin, a favorable sales quantity variance arises when the total number of physical units sold exceeds the budgeted amount.

Ponder the role of the budgeted *average* unit contribution margin. It helps managers focus on *overall* results, not *individual product* results. By itself, an individual sales quantity variance for jug wine does not give managers appropriate information if it is based on an average margin. The total sales quantity variance, however, can enhance understanding. Managers can see how income should fluctuate if the budgeted product mix is maintained as total actual volume departs from the total budgeted volume.

The sales mix variance measures the impact of deviations from budgeted percentage mix for each product, taking into account how that product's budgeted contribution margin differs from the budgeted average contribution margin. In the static budget, the budgeted sales mix in units is 75% for jug wine and 25% for premium wine. The actual sales mix in units was 68.75% for jug wine and 31.25% for premium wine. The \$1.00 budgeted contribution margin for jug wine is \$1.50 below the overall budgeted average contribution margin of \$2.50. The \$7.00 budgeted contribution margin for premium wine is \$4.50 above the overall budgeted average contribution margin of \$2.50.

Favorable sales mix variances arise if there is (1) a decrease in the percentage sold of units with below-average budgeted contribution margins, or (2) an increase in the percentage sold of units with above-average budgeted contribution margins. In Exhibit 26-3, jug wine illustrates point 1, and premium wine illustrates point 2.

Unfavorable sales mix variances arise if there is (1) an increase in the percentage sold of units with below-average budgeted contribution margins or (2) a decrease in the percentage sold of units with above-average budgeted contribution margins.

If the original sales mix of 75% of jug wine and 25% of premium wine were maintained and actual volume increased by 5%, there would be no sales mix variance, only a sales quantity variance. The actual sales of jug wine would be 1,260 units (1,600 × 1.05 × 0.75), and actual sales of premium wine would be 420 units

(1,600 × 1.05 × .25). The sales quantity variance would be:

Jug wine:	(1,260 − 1,200) × \$2.50 =	\$150 F
Premium wine:	(420 − 400) × \$2.50 =	50 F
Total		\$200 F

The result is consistent with the idea that, given an unchanging sales mix, a 5% increase in actual unit volume should produce a 5% increase in sales dollars and a 5% increase in contribution margin. The budgeted contribution margin would increase from \$4,000 to \$4,200, which is the \$200 increase calculated above.

MARKET SIZE AND MARKET SHARE VARIANCES

Objective 2

Explain how market size and market share variances provide different explanations for a sales quantity variance.

Many a company regards its fate as significantly affected by overall demand for the industry's products and the company's ability to maintain its share of the market. Statistics for some industries—for example, automobiles, television sets, and soft drinks—are readily available, and so the company can easily monitor its market share. In other industries—for example, management consulting and restaurants—these statistics are difficult to obtain in a timely fashion.

Assume the following information applies to our Party Wholesaler illustration:

	Budget	Actual
Total U.S. market for imported wine (units)	20,000	25,000
Bordeaux Winery's share of total U.S. market for imported wine	8%	6.4%

Party Wholesaler makes all sales of Bordeaux Winery's products in the United States. The sales quantity variance in Exhibit 26-3 can be subdivided into the *market size variance* and the *market share variance*.[2] Exhibit 26-4 presents the formulas and computations for these two variances:

[2]See J. Shank and N. Churchill, "Variance Analysis: A Management-oriented Approach," *Accounting Review,* LII, No. 4, 955.

EXHIBIT 26-4

Market Size and Market Share Variance Analysis: Party Wholesaler, October 19_7

$$\textbf{Market size variance} = \begin{pmatrix}\text{Budgeted market share percentage}\end{pmatrix} \times \begin{pmatrix}\text{Actual industry sales volume in units} - \text{Budgeted industry sales volume in units}\end{pmatrix} \times \begin{pmatrix}\text{Budgeted average contribution margin per unit}\end{pmatrix}$$

= 0.08 × (25,000 − 20,000) × \$2.50*

= \$1,000 F

$$\textbf{Market share variance} = \begin{pmatrix}\text{Actual market share percentage} - \text{Budgeted market share percentage}\end{pmatrix} \times \begin{pmatrix}\text{Actual industry sales volume in units}\end{pmatrix} \times \begin{pmatrix}\text{Budgeted average contribution margin per unit}\end{pmatrix}$$

= (0.064 − 0.080) × 25,000 × \$2.50*

= \$1,000 U

*From the static budget based on the budgeted mix: \$4,000 ÷ 1,600 units = \$2.50 per unit.

Market size variance	$1,000 F
Market share variance	1,000 U
Sales quantity variance	$ 0

Managers are likely to gain some insights from these variances. They measure the additional contribution margin, $1,000, that would be expected because the market size expanded. Unfortunately, the company attained only 6.4% of the industry market. This failure to maintain budgeted market share—the drop from 8% to 6.4%—created an equal offsetting variance. (As the Problem for Self-Study demonstrates, the dollar amount of the market size and the market share variances seldom exactly offset.)

A helpful overview of the five variances we have computed follows:

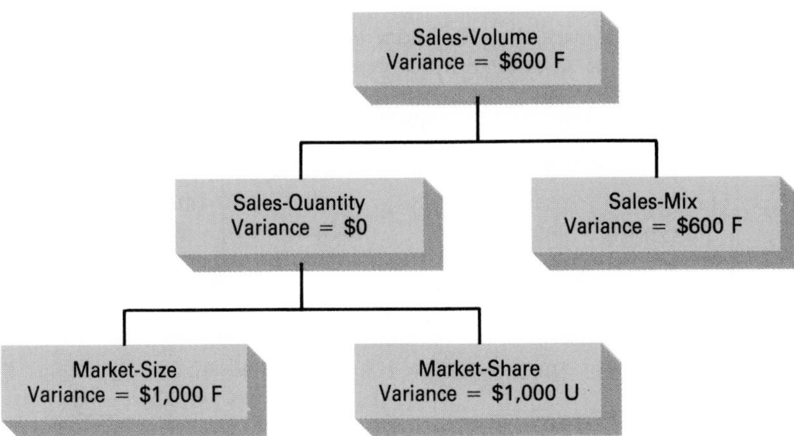

Always consider possible interdependencies among these individual variances.

PROBLEM FOR SELF-STUDY

PROBLEM

Assume the same budget for Party Wholesaler as in Exhibit 26-2 (p. 820). Suppose, however, that actual sales were 1,440 units of jug wine and 360 units of premium wine and that actual total wine imports into the U.S. market were 18,000 units.

Compute the sales volume variances for each product and in total. Subdivide the sales volume variance into sales quantity and sales mix variances. Then divide the sales quantity variance into market size and market share variances. Using your numerical results, comment on the meaning of each variance.

SOLUTION

$$\text{Sales volume variance} = \left(\begin{array}{c}\text{Actual sales}\\ \text{volume}\\ \text{in units}\end{array} - \begin{array}{c}\text{Budgeted sales}\\ \text{volume in units}\end{array}\right) \times \left(\begin{array}{c}\text{Budgeted individual}\\ \text{product contribution}\\ \text{margin per unit}\end{array}\right)$$

Jug wine:	(1,440 − 1,200) × $1 =	$240 F
Premium wine:	(360 − 400) × $7 =	280 U
Total		$ 40 U

This variance is straightforward. The manager can see that selling 240 more units of jug wine increased the total contribution margin by $240. Selling 40 less units of premium wine, however, which is more profitable per unit, more than offsets the favorable effects of jug wine.

$$\text{Sales quantity variance} = \left(\begin{array}{c}\text{Actual sales}\\ \text{volume}\\ \text{in units}\end{array} - \begin{array}{c}\text{Static budgeted sales}\\ \text{volume of individual}\\ \text{product in units}\end{array}\right) \times \left(\begin{array}{c}\text{Budgeted average}\\ \text{contribution}\\ \text{margin per unit}\end{array}\right)$$

Jug wine:	(1,440 − 1,200) × $2.50 =	$600 F
Premium wine:	(360 − 400) × $2.50 =	100 U
Total		$500 F

Given an unchanging mix, exceeding the budgeted unit quantity of 1,600 units by 200 units would produce a favorable sales quantity variance of $500.

$$\text{Sales mix variance} = \left[\left(\begin{array}{c}\text{Actual}\\ \text{sales mix}\\ \text{percentage}\end{array} - \begin{array}{c}\text{Budgeted}\\ \text{sales mix}\\ \text{percentage}\end{array}\right) \times \begin{array}{c}\text{Actual total}\\ \text{sales volume}\\ \text{of all}\\ \text{products}\\ \text{in units}\end{array}\right] \times \left[\begin{array}{c}\text{Budgeted}\\ \text{individual}\\ \text{contribution}\\ \text{margin}\\ \text{per unit}\end{array} - \begin{array}{c}\text{Budgeted}\\ \text{average}\\ \text{contribution}\\ \text{margin}\\ \text{per unit}\end{array}\right]$$

Jug wine:	[(0.80* − 0.75) × 1,800] × ($1.00 − $2.50)	
	= (0.05 × 1,800) × (−$1.50) =	$135 U
Premium wine:	[(0.20 − 0.25) × 1,800] × ($7.00 − $2.50)	
	= (−0.05 × 1,800) × ($4.50) =	405 U
Total		$540 U

*1,440 ÷ (1,440 + 360) = 0.80.

The actual sales mix is 80% of jug wine (compared with 75% in the static budget) and 20% of premium wine (compared with 25% in the static budget). We see an increase in the percentage sold of jug wine, which has a lower-than-average budgeted contribution margin, and a decrease in the percentage sold of premium wine, which has a higher-than-average budgeted contribution margin. Both changes are unfavorable.

$$\text{Market size variance} = \left(\begin{array}{c}\text{Budgeted}\\ \text{market}\\ \text{share}\\ \text{percentage}\end{array}\right) \times \left(\begin{array}{c}\text{Actual}\\ \text{industry}\\ \text{sales}\\ \text{volume}\\ \text{in units}\end{array} - \begin{array}{c}\text{Budgeted}\\ \text{industry}\\ \text{sales}\\ \text{volume}\\ \text{in units}\end{array}\right) \times \left(\begin{array}{c}\text{Budgeted}\\ \text{average}\\ \text{contribution}\\ \text{margin}\\ \text{per unit}\end{array}\right)$$

$$= 0.08 \times (18{,}000 - 20{,}000) \times \$2.50 = \$400\ \text{U}$$

$$\text{Market share variance} = \left(\begin{array}{c}\text{Actual}\\ \text{market}\\ \text{share}\\ \text{percentage}\end{array} - \begin{array}{c}\text{Budgeted}\\ \text{market}\\ \text{share}\\ \text{percentage}\end{array}\right) \times \left(\begin{array}{c}\text{Actual}\\ \text{industry}\\ \text{sales}\\ \text{volume}\\ \text{in units}\end{array}\right) \times \left(\begin{array}{c}\text{Budgeted}\\ \text{average}\\ \text{contribution}\\ \text{margin}\\ \text{per unit}\end{array}\right)$$

$$= (0.10 - 0.08) \times 18{,}000 \times \$2.50 = \$900\ \text{F}$$

There are two offsetting amounts. The total market size declined from 20,000 units to 18,000 units, which is unfavorable to Party Wholesaler's income. The market, however, is smaller, so that by holding its units sold steady, Party Wholesaler gains a larger market share, which is favorable.

A summary of these variances follows:

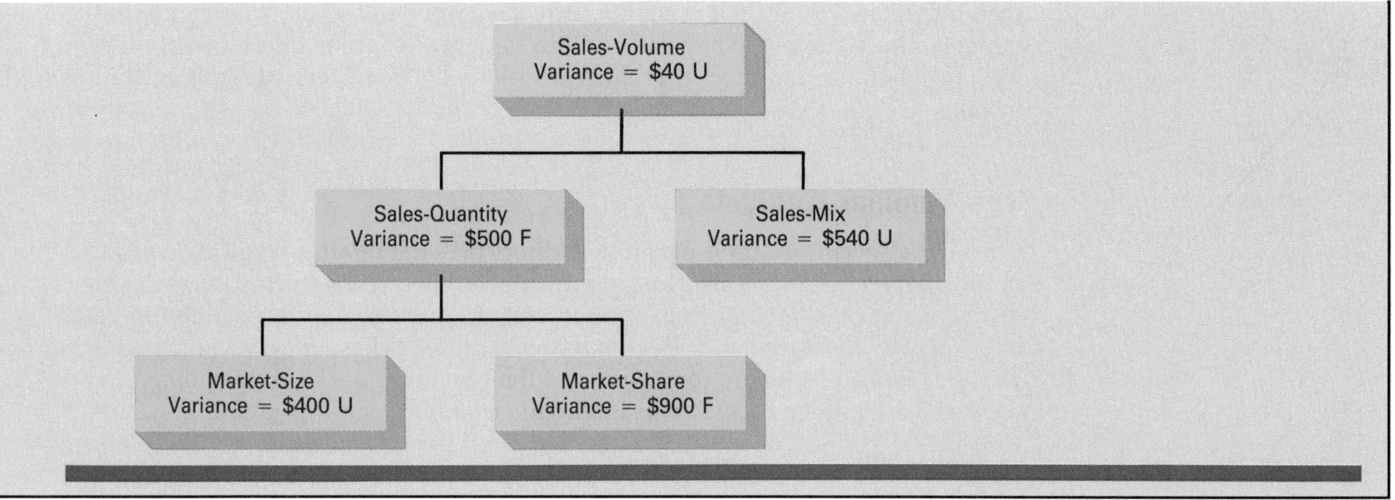

PART TWO: PRODUCTION VARIANCES

Manufacturing processes often require that a number of different direct materials and a number of different direct labor skills be combined to obtain a unit of finished product. For some products, this combination must be exact. The manager of a plant that assembles portable computers has a prespecified list of parts to include in each computer. For other products, a manufacturer has some leeway in combining the materials. The manager of an agricultural fertilizer plant can combine materials (such as elemental phosphorus and acids) in varying proportions.

What do the terms *yield* and *mix* mean in a production setting? *Yield* refers to the quantity of finished output produced from a budgeted or standard mix of inputs. *Mix* refers to the relative proportion or combination of the various inputs required to produce the quantity of finished output.

Recall that a *variance* is the difference between actual results and budgeted figures. The sources of the budgeted figures discussed in this chapter include:

- Internally generated actual costs from the most recent period
- Actual costs of the most recent period adjusted for expected improvement
- Internally generated *budgeted cost* (often called *standard cost*) numbers based on an analysis of efficient operations
- Externally generated *target cost* numbers based on an analysis of the cost structure of the leading competitor in an industry
- Externally generated *most-efficient* actual costs for a company with multiple facilities having the same operations or producing the same product

The interpretation of a variance will be affected by the source of the budgeted figure(s) used in computing the variance.

DIRECT MATERIAL YIELD AND DIRECT MATERIAL MIX VARIANCES

When we initially examined material and labor variances in Chapter 7, we saw that managers sometimes make trade-offs between price and efficiency variances. For example, an orange-juice bottler may use oranges whose juice content is lower than budgeted if their price is significantly lower than the price of oranges with the budgeted juice content. The yield and mix variances computed in this section provide insight into the effect that such decisions by managers have on operating income.

The Familiar Analysis

Consider a specific example of multiple direct-material inputs and a single product output. Perfect Gardens sells bags of prestige garden rocks to customers desiring distinctive color combinations. Three types of "raw" rocks are included in each bag—pebble rocks, black rocks, and white rocks. These raw rock inputs are washed, broken rocks are removed, and the remaining rocks are combined to be packaged as "garden rocks." This is a highly labor intensive operation. The budget calls for 10 pounds of raw rocks to produce 9 pounds of garden rocks. Perfect Gardens guarantees that there are no chips in the rocks and that only the highest-quality rocks are bagged.

Perfect Gardens has the following direct material input standards to produce 9 pounds of garden rocks:

5 pounds of pebble rocks at \$0.70 per pound	\$3.50
3 pounds of black rocks at \$1.00 per pound	3.00
2 pounds of white rocks at \$0.80 per pound	1.60
Total	\$8.10

The average standard cost of garden rock is \$0.90 per pound (\$8.10 ÷ 9 pounds of garden rock).

Suppose, for simplicity, that no inventories of direct materials are kept. Purchases are made as needed, and so all price variances relate to direct materials used. Actual results show that a total of 100,000 pounds of raw rocks were used in a recent period:

45,000 pounds of pebble rocks at actual cost of \$0.80 per pound	\$36,000
33,000 pounds of black rocks at actual cost of \$1.05 per pound	34,650
22,000 pounds of white rocks at actual cost of \$0.85 per pound	18,700
100,000	\$89,350
Finished product output was 92,070 pounds of garden rocks at a standard cost of \$0.90 per pound	82,863
Total variance to be explained	\$ 6,487 U

Given the standard that 10 pounds of raw rock are required to produce 9 pounds of garden rock, 102,300 pounds is the standard input required to produce 92,070 pounds of garden rock: 92,070 ÷ 0.9 = 102,300. The budgeted quantity of direct material inputs of each type of raw rock is:

Pebble rocks: $0.5 \times 102{,}300 = 51{,}150$ pounds
Black rocks: $0.3 \times 102{,}300 = 30{,}690$ pounds
White rocks: $0.2 \times 102{,}300 = 20{,}460$ pounds

Exhibit 26-5 presents the familiar approach to analyzing the flexible-budget direct material variance discussed in Chapter 7. Exhibit 26-6 presents further explanation of the \$6,487 unfavorable flexible-budget variance in Exhibit 26-5. Individual direct material price variances and direct material efficiency variances are computed in Exhibit 26-6.

The analysis in Exhibits 26-5 and 26-6 may suffice when managers control each input on an individual basis and when no discretion is permitted regarding the

EXHIBIT 26-5
Direct Material Price and Direct Material Efficiency Variance Analysis: Perfect Gardens

Direct Material	Actual Costs Incurred: Actual Inputs × Actual Prices		Flexible Budget: Actual Inputs × Standard Prices		Flexible Budget: Standard Inputs Allowed for Actual Outputs × Standard Prices	
Pebble rocks	45,000 × $0.80 =	$36,000	45,000 × $0.70 =	$31,500	51,150 × $0.70 =	$35,805
Black rocks	33,000 × $1.05 =	34,650	33,000 × $1.00 =	33,000	30,690 × $1.00 =	30,690
White rocks	22,000 × $0.85 =	18,700	22,000 × $0.80 =	17,600	20,460 × $0.80 =	16,368
Total	100,000	$89,350	100,000	$82,100	102,300	$82,863

$7,250 U — Direct material price variance

$763 F — Direct material efficiency variance

$6,487 U — Flexible-budget direct material variance

EXHIBIT 26-6
Individual Direct Material Price and Direct Material Efficiency Variance Analysis: Perfect Gardens

Direct Material Price Variances

Direct Material	(1) Difference in Unit Price of Material Inputs	(2) Actual Material Inputs Used	(3) = (1) × (2) Direct Material Price Variance
Pebble rocks	($0.70 − $0.80) = −$0.10	45,000	$4,500 U
Black rocks	($1.00 − $1.05) = −$0.05	33,000	1,650 U
White rocks	($0.80 − $0.85) = −$0.05	22,000	1,100 U
Total			$7,250 U

Direct Material Efficiency Variances

Direct Material	Detailed Computations for Next Column	(1) Standard Units of Material Inputs Allowed for Actual Outputs	(2) Actual Units of Material Inputs Used	(3) = (1) − (2) Difference	(4) Standard Price Per Unit of Material Input	(5) = (3) × (4) Direct Material Efficiency Variance
Pebble rocks	(.5 × 102,300)	51,150	45,000	6,150	$0.70	$4,305 F
Black rocks	(.3 × 102,300)	30,690	33,000	−2,310	1.00	2,310 U
White rocks	(.2 × 102,300)	20,460	22,000	−1,540	0.80	1,232 U
Total		102,300	100,000	2,300		$ 763 F

physical mix of material inputs. For example, there is often a specified mix of parts needed for the assembly of motor vehicles, radios, and washing machines. In these cases, all deviations from the input-output relationships are considered attributable to efficiency; the price and efficiency variances individually computed for each material typically provide all the information necessary for decisions. Why? Because no deliberate substitutions of input are tolerated.

Role of Direct Material Yield and Direct Material Mix Variances

Managers may sometimes substitute materials. For example, Perfect Gardens managers have some leeway in combining the pebble rocks, black rocks, and white rocks used to produce the bags of prestige garden rocks. Are the chosen combinations of rocks based on wise trade-offs between material yield and mix? The variances we now examine help us answer this question.

The standard average price per pound for the three rock inputs is \$0.81:

$$(0.5 \times \$0.70) + (0.3 \times \$1.00) + (0.2 \times \$0.80) = \$0.81$$

where 0.5, 0.3, and 0.2 are the standard proportions of the three direct material inputs required to produce one unit of output. This \$0.81 figure is used when computing both direct material yield and direct material mix variances. Exhibit 26-7 details these two subdivisions of the \$763 favorable direct material efficiency variance—a direct material yield variance of \$1,863 F and a direct material mix variance of \$1,100 U.

The direct material yield variance in Exhibit 26-7 shows that if the combination of ingredients is held constant and 2,300 units less of input used, the savings would be \$1,863. The savings result solely from an improved yield of a given mix of inputs. In effect, the yield variance holds standard average unit price constant and measures all units of input as if they were alike.

As the word *mix* implies, the direct material mix variance in Exhibit 26-7 concentrates on the changes in the percentages (proportions) of the individual inputs used. Favorable direct material mix variances arise if there is (1) an increase in the percentage use of direct material input with below-average standard price per unit or (2) a decrease in the percentage use of direct material input with above-average standard price per unit. The white rocks material input in Exhibit 26-7 illustrates point 1.

EXHIBIT 26-7

Direct Material Yield and Direct Material Mix Variance Analysis: Perfect Gardens

$$\textbf{Direct material yield variance} = \left(\begin{array}{c}\text{Actual units}\\\text{of material}\\\text{inputs}\\\text{used}\end{array} - \begin{array}{c}\text{Standard units}\\\text{of material}\\\text{inputs allowed for}\\\text{actual outputs}\end{array}\right) \times \left(\begin{array}{c}\text{Standard}\\\text{average price}\\\text{per unit of}\\\text{material inputs}\end{array}\right)$$

Pebble rocks: = (45,000 − 51,150) × \$0.81 = −6,150 × \$0.81 = \$4,981.50 F
Black rocks: = (33,000 − 30,690) × \$0.81 = 2,310 × \$0.81 = 1,871.10 U
White rocks: = (22,000 − 20,460) × \$0.81 = 1,540 × \$0.81 = 1,247.40 U
Total \$1,863.00 F

$$\textbf{Direct material mix variance} = \left[\left(\begin{array}{c}\text{Actual}\\\text{material}\\\text{mix}\\\text{percentage}\end{array} - \begin{array}{c}\text{Standard}\\\text{material}\\\text{mix}\\\text{percentage}\end{array}\right) \times \begin{array}{c}\text{Actual total}\\\text{units of}\\\text{material}\\\text{inputs}\\\text{used}\end{array}\right] \times \left[\begin{array}{c}\text{Standard}\\\text{individual}\\\text{price per unit}\\\text{of material}\\\text{inputs}\end{array} - \begin{array}{c}\text{Standard}\\\text{average}\\\text{price per unit}\\\text{of material}\\\text{inputs}\end{array}\right]$$

Pebble rocks: = ((.45 − .50) × 100,000) × (\$0.70 − \$0.81) = −5,000 × −\$0.11 = \$ 550 U
Black rocks: = ((.33 − .30) × 100,000) × (\$1.00 − \$0.81) = 3,000 × \$0.19 = 570 U
White rocks: = ((.22 − .20) × 100,000) × (\$0.80 − \$0.81) = 2,000 × −\$0.01 = 20 F
Total \$1,100 U

Unfavorable direct material mix variances arise if there is (1) an increase in the percentage use of direct material input with above-average standard price per unit or (2) a decrease in the percentage use of direct material input with below-average standard price per unit. The black rocks material input in Exhibit 26-7 illustrates point 1, and pebble rocks illustrates point 2.

Objective 3

Understand how trade-offs among material inputs are highlighted by the material yield and material mix variances

The direct material yield variance in Exhibit 26-7 is $1,863 F, and the direct material mix variance is $1,100 U. There was a trade-off among ingredients that boosted yield but caused the average standard unit cost of the overall direct material inputs to be higher than expected. The trade-off resulted in a net favorable variance of $763. The direct material yield and material mix variances provide additional insights for managers comparing actual results and budgeted amounts.

A summary of the direct material variances computed in this section follows:

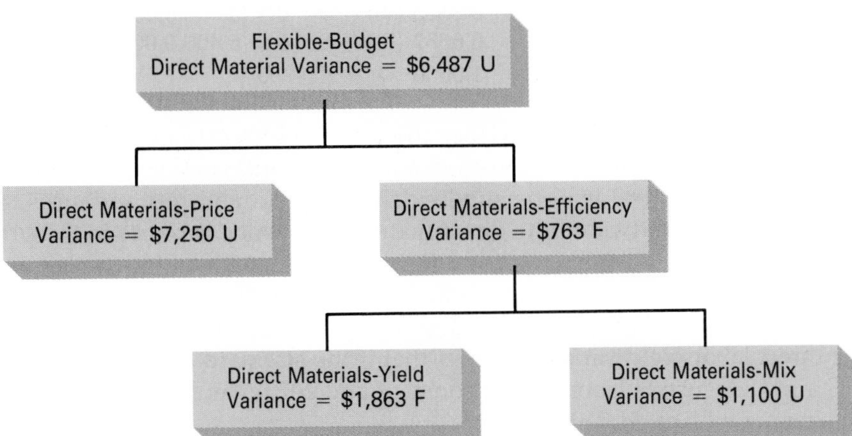

DIRECT LABOR YIELD AND DIRECT LABOR MIX VARIANCES

As we have just seen, managers frequently develop the standard costs of products by using specified combinations of direct materials bearing different individual prices. Managers often use the same approach in developing direct labor standards. Suppose the production process at Perfect Gardens requires considerable labor input. The standard combination of Category G and H workers is 1 hour of Category G worker (at a standard rate of $28 per hour) for every 2 hours of Category H worker (at a standard rate of $16 per hour).

The standard output is 10 pounds of garden rocks per hour. The standard hours of labor allowed for the actual output of 92,070 pounds is:

Category G labor: ⅓ × (92,070/10) =	3,069
Category H labor: ⅔ × (92,070/10) =	6,138
Total standard hours allowed	9,207

Summary data are:

(1) Standard labor price per hour:	
1 hour of Category G workers at $28 per hour	$28
2 hours of Category H workers at $16 per hour	32
Total cost of standard combination of labor	$60
Standard average price per labor hour, $60 ÷ 3 hours	$20

(2) Standard labor price per pound of output at 10 pounds per hour, $20 ÷ 10 — $2

(3) Standard labor cost of 92,070 pounds of output, $2 × 92,070 pounds, or $20 × 9,207 hours — $184,140

(4) Actual inputs, 9,000 hours consisting of

3,600 hours of Category G labor at $30 per hour	$108,000
5,400 hours of Category H labor at $15 per hour	81,000
9,000 hours at $21 per hour	$189,000

The standard mix and actual mix of hours of Category G and Category H workers are:

	Standard Mix	**Actual Mix**
Category G labor	0.3333 (⅓)	0.4 (3,600/9,000)
Category H labor	0.6667 (⅔)	0.6 (5,400/9,000)
	1.0000	1.0

Exhibit 26-8 presents the direct labor price and the direct labor efficiency variances; both of these variances are unfavorable. The unfavorable direct labor efficiency variance of $3,060 may be subdivided into yield and mix variances in the same way that we subdivided the direct material efficiency variance in the preceding section. Exhibit 26-9 presents formulas and computations for the labor yield and labor mix variances. Because we are subdividing the direct labor efficiency variance, all price variances are excluded from these calculations.

The direct labor yield variance shows that if the standard combination of Category G and Category H labor is held constant and 207 fewer labor hours are used, the savings would be $4,140.

Objective 4

Provide explanations for favorable and unfavorable labor mix variances

Favorable direct labor mix variances arise if there is (1) an increase in the percentage use of a labor category with below-average standard price per hour or (2) a decrease in the percentage use of a labor category with above-average standard price per hour. Neither point 1 nor point 2 occurred in our example, as the data in Exhibit 26-9 indicate.

EXHIBIT 26-8

Direct Labor Price and Direct Labor Variance Analysis: Perfect Gardens

Direct Labor Category	**Actual Costs Incurred: Actual Inputs × Actual Prices**		**Flexible Budget: Actual Inputs × Standard Prices**		**Flexible Budget: Standard Inputs Allowed for Actual Outputs × Standard Prices**	
Category G	3,600 × $30 =	$108,000	3,600 × $28 =	$100,800	3,069* × $28 =	$ 85,932
Category H	5,400 × $15 =	81,000	5,400 × $16 =	86,400	6,138† × $16 =	98,208
Total	9,000	$189,000	9,000	$187,200	9,207	$184,140

$1,800 U — Direct labor price variance

$3,060 U — Direct labor efficiency variance

$4,860 U — Flexible-budget direct labor variance

*⅓ × 9,207 = 3,069.
†⅔ × 9,207 = 6,138.

EXHIBIT 26-9

Labor Yield and Labor Mix Variance Analysis: Perfect Gardens

$$\text{Direct labor yield variance} = \left(\text{Actual units of labor inputs used} - \text{Standard units of labor inputs allowed for actual outputs}\right) \times \left(\text{Standard average price per unit of labor inputs}\right)$$

Category G labor: (3,600 − 3,069) × \$20 = 531 × \$20 = \$10,620 U

Category H labor: (5,400 − 6,138) × \$20 = −738 × \$20 = 14,760 F

Total \$ 4,140 F

$$\text{Direct labor mix variance} = \left[\left(\text{Actual labor mix percentage} - \text{Standard labor mix percentage}\right) \times \text{Actual total units of labor inputs used}\right] \times \left[\text{Standard individual price per unit of labor inputs} - \text{Standard average price per unit of labor inputs}\right]$$

Category G labor: [(0.4 − 0.3333) × 9,000] × (\$28 − \$20)

(−0.0667 × 9,000) × \$8 = \$4,800 U

Category H labor: [(0.6 − 0.6667) × 9,000] × (\$16 − \$20)

(−0.0667 × 9,000) × (−\$4) = 2,400 U

Total \$7,200 U

Unfavorable direct labor mix variances arise if there is (1) an increase in the percentage use of a labor category with above-average standard price per hour or (2) a decrease in the percentage use of a labor category with below-average standard price per hour. The Category G labor in Exhibit 26-9 illustrates point 1, and the Category H labor illustrates point 2.

The flexible-budget direct labor variance in Exhibit 26-8 is \$4,860 unfavorable. The trade-off decision the manager made between direct labor yield and mix resulted in lower operating income than was budgeted for at the start of the period. Further analysis revealed that the extra Category G workers employed required substantial initial training. This factor was not considered when the budget was prepared.

A summary of the direct labor variances computed in this section follows:

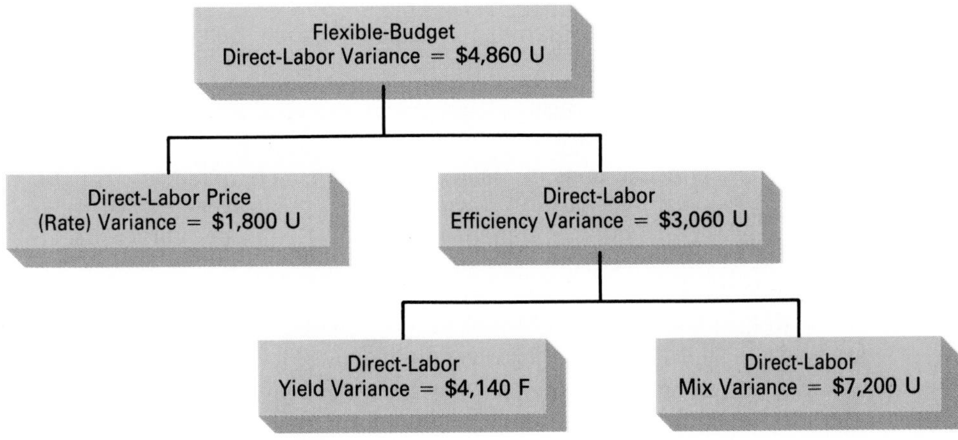

PROBLEM FOR SELF-STUDY

PROBLEM

Review the Perfect Gardens example on direct material variances summarized in Exhibits 26-5 (p. 829) and 26-6 (p. 829). Suppose 46,000 pounds of pebble rocks, 40,000 pounds of black rocks, and 14,000 pounds of white rocks were actually used to produce the 92,070 pounds of garden rocks. Compute the total price and total efficiency direct material variances. Explain why the direct material efficiency variance is no longer $763 favorable. Also compute the direct material yield and direct material mix variances.

SOLUTION

Direct Material	Actual Costs Incurred: Actual Inputs × Actual Prices	Flexible Budget: Actual Inputs × Standard Prices	Flexible Budget: Standard Inputs Allowed for Actual Outputs × Standard Prices
Pebble rocks	46,000 × $0.80 = $36,800	46,000 × $0.70 = $32,200	51,150 × $0.70 = $35,805
Black rocks	40,000 × $1.05 = 42,000	40,000 × $1.00 = 40,000	30,690 × $1.00 = 30,690
White rocks	14,000 × $0.85 = 11,900	14,000 × $0.80 = 11,200	20,460 × $0.80 = 16,368
Total	100,000 $90,700	100,000 $83,400	102,300 $82,863

$7,300 U — Direct material price variance; $537 U — Direct material efficiency variance

$7,837 U — Flexible-budget direct material variance

The unfavorable direct material price variance of $7,300 reflects the increase in the price of each of the three materials relative to their individual standard price per material unit. The favorable direct material efficiency variance of $763 in the original situation has become $537 unfavorable, a difference of $1,300. The new actual input combination is less economical than the old actual input combination:

Direct Material	Direct Material Inputs Used: Old Combination	Direct Material Inputs Used: New Combination	Difference
Pebble rocks	45,000 × $0.70 = $31,500	46,000 × $0.70 = $32,200	$ 700 U
Black rocks	33,000 × $1.00 = 33,000	40,000 × $1.00 = 40,000	7,000 U
White rocks	22,000 × $0.80 = 17,600	14,000 × $0.80 = 11,200	6,400 F
Total	100,000 $82,100	100,000 $83,400	$1,300 U

The direct material yield and direct material mix variances are:

$$\text{Direct material yield variance} = \left(\begin{array}{c}\text{Actual units}\\\text{of material}\\\text{inputs}\\\text{used}\end{array} - \begin{array}{c}\text{Standard units}\\\text{of material}\\\text{inputs allowed for}\\\text{actual outputs}\end{array}\right) \times \left(\begin{array}{c}\text{Standard}\\\text{average price}\\\text{per unit of}\\\text{material inputs}\end{array}\right)$$

Pebble rocks:	= (46,000 − 51,150) × $0.81	= −5,150 × $0.81	= $4,171.5 F
Black rocks:	= (40,000 − 30,690) × $0.81	= 9,310 × $0.81	= 7,541.1 U
White rocks:	= (14,000 − 20,460) × $0.81	= −6,460 × $0.81	= 5,232.6 F
Total			$1,863.0 F

$$\text{Direct material mix variance} = \left[\left(\begin{matrix}\text{Actual}\\\text{material}\\\text{mix}\\\text{percentage}\end{matrix} - \begin{matrix}\text{Standard}\\\text{material}\\\text{mix}\\\text{percentage}\end{matrix}\right) \times \begin{matrix}\text{Actual total}\\\text{units of}\\\text{material}\\\text{inputs}\\\text{used}\end{matrix}\right] \times \left[\begin{matrix}\text{Standard}\\\text{individual}\\\text{price per unit}\\\text{of material}\\\text{inputs}\end{matrix} - \begin{matrix}\text{Standard}\\\text{average}\\\text{price per unit}\\\text{of material}\\\text{inputs}\end{matrix}\right]$$

Pebble rocks:	= [(.46 − .50) × 100,000] × ($0.70 − $0.81) =	−4,000 × −$0.11 =	$ 440 U
Black rocks:	= [(.40 − .30) × 100,000] × ($1.00 − $0.81) =	10,000 × $0.19 =	1,900 U
White rocks:	= [(.14 − .20) × 100,000] × ($0.80 − $0.81) =	−6,000 × −$0.01 =	60 U
Total			$2,400 U

The change in the direct material efficiency variance is attributable solely to the change in the mix of direct materials because there is no change in the direct material yield variance.

This computation of direct material mix and yield variances tells the manager that the trade-off was unwise. That is, yield was better than standard, but the use of higher-cost direct materials more than offset the advantage of yield.

PART THREE: VARIANCE INVESTIGATION DECISIONS

Given feedback in the form of routine of cost variance reports, when should managers investigate variances? Managers typically do not investigate every reported variance line by line. Only some variances lead to investigation, but which ones? How do managers determine which variances require corrective action? Investigation decisions have a cost. What is the relative benefit corresponding to this cost?

The accounting system—a variance report is a part of this system—often plays a key role in the decision of whether to investigate, but managers draw on other sources of information. This section explores the wide dimensions of the decision to investigate variances. We begin by focusing on sources of variances.

SOURCES OF VARIANCES

Objective 5

Identify seven sources of variances

There are seven principal sources of variances: (1) inefficiencies in operations, (2) inappropriate standard (or target), (3) mismeasurement of actual results, (4) implementation breakdown, (5) parameter prediction error, (6) inappropriate decision model, and (7) random variation.[3] Each source may call for its own corrective action, and each corrective action has its costs. In deciding whether to take the corrective steps, managers must compare the costs of the corrective action against any predicted benefits.

Inefficiencies in Operations

A major purpose of using a standard cost system is to pinpoint inefficiencies in operations. Chapter 7 (p. 231) defined *currently attainable standards* as standard costs that are achievable by a specified level of effort and allow for normal spoilage, waste, and nonproductive time. A variance from a standard can arise because there

[3]See also J. Demski, "Optimizing the Search for Cost Deviation Sources," *Management Science*, 16, No. 8, 486–94.

is greater than normal spoilage, waste, or nonproductive time. Correcting this source of a variance entails finding the underlying cause of the inefficiency and eliminating it, which in some cases may not be possible.

Inappropriate Standard

Developing appropriate standards is time-consuming and costly. For example, drawing up standards for an operation with many products and services and many inputs (material, labor, manufacturing overhead, and so on) may be expensive. In some cases, organizations have inaccurate standards because they are unwilling to incur the costs required to develop accurate standards. Also, accurate standards are hard to develop in high-technology organizations, where fast-paced changes in operations may quickly make standards out of date and inappropriate. Inappropriate standards can also arise when individuals undermine the standard-setting process by, say, deliberately working slowly or using more direct materials than necessary during the test period for setting the standards. Correcting this source of a variance requires setting more accurate standards.

Mismeasurement of Actual Results

The *recorded* amounts for costs and revenues can differ from the *actual* amounts. For example, a worker may incorrectly count the ending inventory, which results in an erroneous measure of total costs. Other sources of mismeasurement include improper classification of items (commingling an indirect labor cost item with the direct labor costs, for example) and the improper recording of items (recording an advertising payment as a prepayment rather than a cost of the current period, for example).

Correcting this source of a variance largely depends on getting individual employees to obtain accurate documentation as part of their everyday work habits. Obtaining data with a high level of integrity—that is, accuracy—is an especially troublesome problem in systems design and in employee motivation that permeates all types of organizations. For example, employees sometimes falsify time records so that the true variances from standards on particular operations are masked.

Implementation Breakdown

Employee actions do not always agree with plans. Consider a business that keeps its inventory level within a narrow range to save storage costs. An employee, because of improper motivation or instructions, orders too many inventory units. Why? Because of the attractions of a lower-than-standard price for the larger order. Still, the business will lose more in the additional storage costs for the excessive inventory than it gains from the lower cost.

Correcting this source of a variance often entails restricting the employee's discretion. For example, purchasing officers might not be permitted to order quantity levels above a preset level unless the unit price is less than 85% of the standard price.

Parameter Prediction Error

Planning decisions are based on predictions of future costs, future selling prices, future demand, and so on. In many cases, there will be a difference between the actual value and the predicted value, a difference termed a *prediction error.* For example, one parameter predicted in the economic-order-quantity model is the ordering costs per purchase order. This parameter may be incorrectly predicted because of a failure to recognize a recently negotiated increase in hourly labor rates for purchasing personnel.

Reducing this source of a variance requires the development of a better prediction model. Sensitivity analysis will help managers decide the potential net benefits from maintaining a routine monitoring of possible prediction errors and of devoting resources to reduce these errors.

Inappropriate Decision Model

Decision models differ in their ability to capture reality. Variances can arise when the chosen decision model fails to capture important aspects affecting the decision. Consider the linear-programing (LP) model discussed in Chapter 24. The solution to an LP model can be used when setting standards for direct-material purchase prices. These standards, however, may be inappropriate if the LP solution is not feasible because the LP model fails to recognize a constraint on labor availability or storage capacity. Similar problems can arise if the objective function or the variables in the LP model are incorrectly specified.

A variance arising from an inappropriate decision model is not the same as a variance stemming from a parameter prediction error. With the decision model as the source, the variance pertains to an incorrect functional relationship. The prediction error pertains to an incorrect parameter estimate.

Deciding to correct a decision model source of a variance usually necessitates a comparison of the cost of correction with the benefits over many future time periods. The decision to correct a parameter prediction involves a similar analysis but typically relates to a shorter time period because these predictions are updated more frequently.

Randomness of Operating Processes

Standard cost systems typically represent a standard as a *single* acceptable measure. A more appropriate representation, however, is that the standard is a *band* or *range* of possible acceptable measures. The range of possible measures may be predictable, but the exact amount for any one unit or batch of units is not predictable. For example, at the semiconductor-chip manufacturing facilities of both Hitachi and National Semiconductor, random changes within the predicted tolerable range in yield occur because of the inherent uncertainty of the wafer-fabrication process. By definition, a random deviation source of a variance in itself calls for no corrective action.

Overview

The distinctions among the seven sources of variances illustrate the complexities that decision makers face. The manager should be alert to these seven sources because they focus on assumptions about standard setting, measurement, implementation, prediction methods, and decision models—basic, important management considerations.

The nature of the search prompted by each of the seven sources of variance can differ considerably. For example, when the source is an implementation breakdown, a manager may have to interview a specific worker or even redesign the incentive plan. In contrast, when the source is a decision model, the manager may have to consult with an engineer. Whatever the source or sources of the variance, the manager must compare the expected costs of the investigation with the expected savings that may result from the investigation.

COST-BENEFIT ANALYSIS OF VARIANCE INVESTIGATION DECISIONS

Objective 6

Explain how decision making under uncertainty can be used to decide whether to investigate a variance

Decision Analysis Approach

The approach to decision making under uncertainty explained in Chapter 20 can be used in variance investigation decisions. Exhibit 26-10 presents a simple example involving two events and two actions:

Events: x_1 = In-control process (that is, the process is functioning properly)
x_2 = Out-of-control process (that is, the process is not functioning properly)
Actions: a_1 = Investigate the process
a_2 = Do not investigate the process

Several categories of costs are important in the decision analysis approach summarized in Exhibit 26-10:

C = Cost of investigating = $3,000
D = Cost per period of being out of control = $5,000
M = Cost of correcting, if an out-of-control process is discovered = $2,000
L = Cost of a process being out of control this period and not correcting it this period. This cost includes an estimate of the number of future periods the process will be out of control without a correction and the cost of correcting it when a decision finally is made to correct it = $37,000.

We assume it takes one period to determine if a process is out of control and then to correct it. Our simple example also assumes (1) that a process that goes out of control stays out of control for the planning period and (2) that when an out-of-control process is corrected, it stays in control until the end of the planning period. The length of each time period in Exhibit 26-10 is one day. However, the decision analysis framework is a general one that can be applied to much longer time periods.

The probability of the process being in control is 0.80, and the probability of its being out of control is 0.20:

$$P(x_1) = 0.80 \qquad P(x_2) = 0.20$$

The expected costs from the investigate (a_1) and do not investigate (a_2) actions are:

Investigate: $$E(a_1) = [(P(x_1) \times C)] + [(P(x_2) \times (C + D + M))]$$
$$= (0.80 \times \$3,000) + (0.20 \times \$10,000) = \$4,400$$

EXHIBIT 26-10
Decision Table Presentation of Data

	Probability of Events	
	x_1 = In-control process $P(x_1) = 0.80$	x_2 = Out-of-control process $P(x_2) = 0.20$
Actions		
a_1 = Investigate the process	C = $3,000	$C + D + M$ = $10,000
a_2 = Do not investigate the process	$ 0	L = $37,000

C = Cost of investigation = $3,000
D = Cost per period of being out of control = $5,000
M = Cost of correcting, if out-of-control process discovered = $2,000
L = Cost of a process being out of control this period and not corrected this period = $37,000

$$\text{Do not investigate: } E(a_2) = [(P(x_1) \times \$0)] + [(P(x_2) \times L)]$$
$$= (0.80 \times \$0) + (0.20 \times \$37{,}000) = \$7{,}400$$

The optimal action is investigate the process, because the expected cost is $3,000 less than the do-not-investigate alternative ($4,400 – $7,400).

Role of Probabilities

The preceding example illustrates the critical role of probabilities in a manager's deciding whether to investigate. A lower probability—say, 0.05—of the process's being out of control will change the desirability of conducting an investigation:

$$\text{Investigate: } E(a_1) = (0.95 \times \$3{,}000) + (0.05 \times \$10{,}000) = \$3{,}350$$
$$\text{Do not investigate: } E(a_2) = (0.95 \times \$0) + (0.05 \times \$37{,}000) = \$1{,}850$$

When $P(x_2) = 0.05$, the optimal action is do not investigate.

The probability level at which a manager would be indifferent between investigating and not investigating whether a process is out of control is[4]:

$$P(x_2) = \frac{C}{L - (D + M)}$$

Given $C = \$3{,}000$, $D = \$5{,}000$, $M = \$2{,}000$, and $L = \$37{,}000$, then

$$P(x_2) = \frac{\$3{,}000}{\$37{,}000 - (\$5{,}000 + \$2{,}000)} = 0.10$$

The expected cost when $P(x_2) = 0.10$ for both a_1 and a_2 is the same, $3,700:

$$\text{Investigate: } E(a_1) = (0.90 \times \$3{,}000) + (0.10 \times \$10{,}000) = \$3{,}700$$
$$\text{Do not investigate: } E(a_2) = (0.90 \times \$0) + (0.10 \times \$37{,}000) = \$3{,}700$$

Investigation is desirable in the Exhibit 26-10 example only if the probability of the process being out of control exceeds 0.10.

Estimation of Costs and Benefits

The usefulness of our cost-benefit calculation critically depends on having accurate estimates of $P(x_1)$ and $P(x_2)$ and of the C, M, and L parameters. Much care, then, should be taken in estimating them.

Consider the estimation of C, the cost of investigation. Signals that a process is potentially out of control can often occur well before the cost variance report is prepared. For example, line operators on a soft-drink bottling production line may complain about the quality of the glass bottles if they encounter a greater than expected number of breakages. The line manager may immediately direct that a different batch of bottles be used on the production line. The investigation of the batch currently being used begins. By the time the line manager receives a routine materials yield variance report, the preliminary investigation may already be com-

---

[4]This formula is derived as follows:

Given $P(x_1) + P(x_2) = 1$
then $P(x_1) = 1 - P(x_2)$

Equating the expected costs of a_1 and a_2 and substituting $(1 - P(x_2))$ for $P(x_1)$ yields:

$$[(1 - P(x_2)) \times C] + [P(x_2) \times (C + D + M)] = P(x_2) \times L$$
$$C - [C \times P(x_2)] + [C \times P(x_2)] + [D \times P(x_2)] + [M \times P(x_2)] = [L \times P(x_2)]$$
$$C = P(x_2)[L - (D + M)]$$
$$P(x_2) = \frac{C}{L - (D + M)}$$

pleted. In this case, estimation of C for an investigation prompted by a materials yield variance report should include only the incremental investigation costs beyond those already incurred at the time the report is received. Similarly, estimation of D, M, and L should recognize those steps already taken to correct the glass-bottle breakage problem. This example illustrates how the timeliness of cost accounting data can affect the values of parameters used in cost variance investigation models.

Rules of Thumb

Again and again we have seen that managers use simple decision models or rules of thumb in complex situations. In this area of variance analysis, rules of thumb observed in practice include the following:

1. Investigate all variances. This approach can involve numerous investigations but may be justified if the cost of the process being out of control is extremely high. An example is the manufacture of a space telescope.
2. Investigate all variances greater than a specified percentage (say, 5%) deviation from the standard.
3. Investigate all variances greater than an absolute dollar amount.
4. Investigate all variances greater than a specified point on the frequency distribution of variances. Management often turns to statistics in setting this point, using a measure called the standard deviation (written σ and read "sigma"). For example, management may choose to look into only those variances greater than two standard deviations (2σ) from the budgeted (standard) amount.

Managers often use one (or a combination) of these rules of thumb because, in part, the costs of developing more complex investigation rules is regarded as too high.

DIFFICULTIES IN ANALYSIS OF VARIANCES

Ideally, managers want an information system to report accounting variances that

- Are timely
- Are itemized in sufficient detail, so that important individual variances are not hidden because they cancel each other out
- Pinpoint the source(s) of each variance in one (or more) of the seven categories we discussed
- Indicate the sequence of an investigation that will maximize overall net benefits

Managers face considerable difficulties in their analyses. Consider one study that examined the use of variance investigation models in several cost centers of a grain-processing firm. Operating managers benefitted most from reports of variances in very recent time periods, such as the day before. Weekly or monthly reports of variances were less useful because managers had a harder time locating the cause of problems that had occurred many days earlier. Difficult measurement problems, however, were encountered in computing variances on a daily basis:

> It was difficult to precisely measure inputs (e.g., direct material units) and outputs (finished output units) due to the difficulty of estimating large bin and tank inventory levels. Although a perennial cost accounting problem, it was more acute because of the daily reporting period. The second measurement problem was the difficulty of matching inputs and outputs. This was due to the often lengthy processing times in the grain firm.

For example, a variable supply item was paired with output for the same day even though its purpose may have been to process output for the following day.[5]

Direct material efficiency, yield, and mix variances—all useful for cost management decisions—were affected by the above-noted difficulties in measuring inputs into the computation of these variances.

STATISTICAL QUALITY CONTROL

Objective 7

Describe statistical quality control and identify problems in using it to analyze cost variances

Statistical quality control (SQC) is a formal means of distinguishing between random variation and other sources of variation in an operating process. SQC is an approach used in deciding whether to investigate a variance. A key tool in SQC is a control chart. A **control chart** is a graph of successive observations of a series in which each observation is plotted relative to specified points that present the expected distribution. Only those observations beyond these limits are ordinarily regarded as nonrandom and worth investigating.

Exhibit 26-11 presents control charts for the daily energy usage (in kilowatt hours) of three production lines of an assembly plant. Energy usage in the prior sixty days for each plant was assumed to provide a good basis from which to calculate the distribution of daily energy usage. The arithmetic mean (μ, read "mu") and standard deviation (σ) are the two parameters of the distribution that are used in the control charts in Exhibit 26-11. Based on past experience, the firm decides that any observation outside the $\pm 2\sigma$ range should be investigated.

For production line *A* in Exhibit 26-11, all observations are within the range of $\pm 2\sigma$ from the mean. Management, then, believes no investigation is necessary. For production line *B*, the last two observations signal a possible out-of-control occurrence. Given the $\pm 2\sigma$ rule, both observations would lead to an investigation. Production line C illustrates a process that would not prompt an investigation under the $\pm 2\sigma$ rule but may well be out of control. Note that the last eight observations show a clear direction and that the direction by point 5 (third in the last eight) is away from the mean. Statistical procedures have been developed that consider the trend in recent usage as well as the level of each day's usage.

[5] F. Jacobs, "When and How to Use Statistical Cost Variance Investigation Techniques," *Cost and Management*, LXIV, No. 8, 26–32.

EXHIBIT 26-11

Statistical Quality Control Charts: Daily Energy Use at Three Production Lines

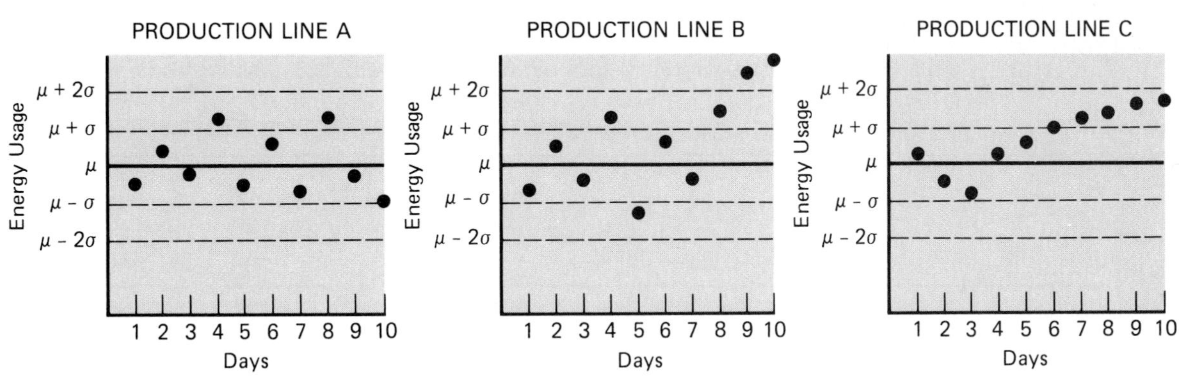

A major benefit of the SQC chart format is that operating personnel find it easy to understand and can eventually become quite adept at recognizing patterns in sequences of observations. Most work on SQC has focused on observations measured in engineering terms (such as labor time, material usage, and number of product rejects) rather than in the monetary terms traditionally found in accounting variance reports.

There are few reported instances where SQC charts have been applied to the analysis of cost variance reports. Several factors could explain this situation. First, adding dollar amounts to engineering items, such as material usage or the number of product rejects, may not provide any greater insight into how such items can be better controlled at the plant floor level. Second, it is often easier to estimate the distribution of items in quantity terms than it is to estimate the distribution of items in price-times-quantity terms. For example, when major changes in oil prices occurred in the last decade, airline companies found that the distribution of fuel cost per mile traveled was changing, which made it difficult to estimate reliably the 1σ and 2σ parameters used in SQC charts. In contrast, the distribution of gallons of fuel consumed per mile traveled was more stable in this period.

PROBLEM FOR SELF-STUDY

PROBLEM

Refer to the example in Exhibit 26-10 (p. 838). Suppose C = \$5,000, D = \$6,000, M = \$4,000, and L = \$30,000. The probabilities as set out in Exhibit 26-10 are unchanged: $P(x_1) = 0.80$ and $P(x_2) = 0.20$. What would be the expected costs of investigate (a_1) and do not investigate (a_2)? At what probability level for x_2 would a manager be indifferent between a_1 and a_2? That is, at what probability level for the process's being out of control would the expected costs of each action be equal?

SOLUTION

Investigate:

$$E(a_1) = (P(x_1) \times C) + [(P(x_2) \times (C + D + M)]$$
$$= (0.80 \times \$5{,}000) + (0.20 \times \$15{,}000) = \$7{,}000$$

Do not investigate:

$$E(a_2) = (P(x_1) \times \$0) + (P(x_2) \times L)$$
$$= (0.80 \times \$0) + (0.20 \times \$30{,}000) = \$6{,}000$$

The optimal action is do not investigate the process.

The probability level for x_2 at which a manager would be indifferent between a_1 and a_2 is

$$P(x_2) = \frac{C}{L - (D + M)} = \frac{\$5{,}000}{\$30{,}000 - (\$6{,}000 + \$4{,}000)} = 0.25$$

If the probability of the process being out of control exceeds 0.25, an investigation should be made.

SUMMARY

Accounting variances are a means of communicating results. Managers use variance reports to help them understand operations and decide which areas warrant investigation.

Quantity, mix, and yield variances are especially useful when sales or production managers have considerable discretion about the mix of products to sell or the way inputs are combined to manufacture and market a given product. Always consider possible interdependencies among individual variances. Standing alone, a variance is rarely sufficient for information for decision making.

In deciding whether to investigate a variance, managers should consider the costs of the investigation, the costs of the process being out of control, and the costs of the correction. These factors are difficult to quantify, and therefore managers often use rules of thumb for deciding when to investigate.

TERMS TO LEARN

This chapter and the Glossary at the end of the book contain definitions of the following important terms:

control chart *(p. 841)* direct labor mix variance *(833)*
direct labor yield variance *(833)* direct material mix variance *(830)*
direct material yield variance *(830)* market share variance *(824)*
market size variance *(824)* sales mix variance *(822)*
sales quantity variance *(822)* statistical quality control (SQC) *(841)*

ASSIGNMENT MATERIAL

QUESTIONS

26-1 "The sales-volume variance might be better labeled as a marketing variance." Explain why such a comment might be made.

26-2 Distinguish between a *sales quantity variance* and a *sales mix variance.*

26-3 Describe two ways that a favorable sales mix variance can arise in a firm selling two products.

26-4 Why might a manager not compute market size variances and market share variances?

26-5 "A company can sell exactly the same number of units as specified in the static (master) budget and still have a flexible budget with numbers that differ from those in the static (master) budget." Do you agree? Explain.

26-6 Name three sources of the standards used in the direct material yield variance and the direct material mix variance.

26-7 Distinguish between a *direct material yield variance* and a *direct material mix variance.*

26-8 The manager of a highly automated plant that assembles desktop computers commented, "Yield and mix variance information is irrelevant to my cost management decisions." Give two possible reasons for the manager making this statement.

26-9 Describe two ways an unfavorable direct labor mix variance can arise in a factory using two categories of direct labor.

26-10 Name five sources of variances.

26-11 Name three rules of thumb used in practice to decide whether to investigate a variance.

26-12 Describe the basic approach of statistical quality control (SQC).

26-13 Describe characteristics of an information system for reporting accounting variances that would be useful to a decision maker.

EXERCISES AND PROBLEMS

Coverage of Part One of the Chapter

26-14 Sales volume, sales quantity, and sales mix variances. The Detroit Penguins play in the American Ice Hockey League. The Penguins play in the Downtown Arena, which has a capacity of 30,000 seats (10,000 in lower-tier seats and 20,000 in the upper-tier seats). All tickets are sold by the Reservation Network. The Penguins' budgeted contribution margin for each ticket in 19_6 is computed as follows:

	Lower-Tier Tickets	Upper-Tier Tickets
Selling price	$35	$14
Downtown Arena fee	10	6
Reservation Network fee	5	3
Contribution margin	$20	$ 5

The budgeted and actual average attendance figures per game in the 19_6 season are:

	Budgeted Seats Sold	Actual Seats Sold
Lower-tier	8,000	6,600
Upper-tier	12,000	15,400
Total	20,000	22,000

There was no difference between the budgeted and actual contribution margins per lower-tier or upper-tier tickets.

The manager of the Penguins was delighted that actual attendance was 10% above budgeted attendance per game, especially given the depressed state of the local economy in the last six months.

Required

1. Compute the sales volume variance for the Detroit Penguins in 19_6.
2. Compute the budgeted average contribution margin per unit (ticket sold) in 19_6.
3. Compute the budgeted and actual sales mix percentages for the lower-tier and the upper-tier seats.
4. Compute the sales quantity variance and the sales mix variance in 19_6.
5. Present a summary of the variances in requirements 1 and 4. Comment on the results.

26-15 Sales volume, sales quantity, and sales mix variances. Debbie's Delight Inc. operates a chain of cookie stores. Budgeted and actual operating data of the downtown store for August 19_2 follow.

Budget for August

	Selling Price per lb.	Variable Cost per lb.	Contribution Margin per lb.	Sales Volume in lbs.
Chocolate chip	$4.50	$2.50	$2.00	45,000
Oatmeal raisin	5.00	2.70	2.30	25,000
Coconut	5.50	2.90	2.60	10,000
White chocolate	6.00	3.00	3.00	5,000
Macadamia nut	6.50	3.40	3.10	15,000
				100,000

Actual for August

	Selling Price per lb.	Variable Cost per lb.	Contribution Margin per lb.	Sales Volume in lbs.
Chocolate chip	$4.50	$2.60	$1.90	57,600
Oatmeal raisin	5.20	2.90	2.30	18,000
Coconut	5.50	2.80	2.70	9,600
White chocolate	6.00	3.40	2.60	13,200
Macadamia nut	7.00	4.00	3.00	21,600
				120,000

Required

1. Compute the individual product and total sales volume variances for August 19_2.
2. Compute the individual product and total sales quantity variances for August 19_2.
3. Compute the individual product and total sales mix variances for August 19_2.
4. Comment on your results in requirements 1, 2, and 3.

26-16 Sales volume, sales quantity, and sales mix variances. Computer Horizons manufactures and sells three related microcomputer products:

1. Plum—sold mostly to college students
2. Portable Plum—smaller version of the Plum that can be carried in a briefcase
3. Super Plum—has a larger memory and more capabilities than the Plum and is targeted at the business market

Budgeted and actual operating data for 19_4 follow.

Budget for 19_4

	Selling Price per Unit	Variable Cost per Unit	Contribution Margin per Unit	Sales Volume in Units
Plum	$1,200	$ 700	$ 500	700,000
Portable Plum	800	500	300	100,000
Super Plum	5,000	3,000	2,000	200,000
				1,000,000

Actual for 19_4

	Selling Price per Unit	Variable Cost per Unit	Contribution Margin per Unit	Sales Volume in Units
Plum	$1,100	$ 500	$ 600	825,000
Portable Plum	650	400	250	165,000
Super Plum	3,500	2,500	1,000	110,000
				1,100,000

During 19_4, competition sparked by overseas suppliers drove the cost of computer chips down, allowing Computer Horizons to buy key components at bargain prices. Computer Horizons had budgeted for a major expansion into the lucrative microcomputer business market in 19_4. Unfortunately, it underestimated the marketing power of its rival, the Big Blue Company.

Required

1. Compute the individual product and total sales volume variances for Computer Horizons in 19_4.
2. Compute the individual product and total sales quantity variances for 19_4.
3. Compute the individual product and total sales mix variances for 19_4.
4. Comment on your results in requirements 1, 2, and 3.

26-17 Market size and market share variances (extension of Problem 26-16). Computer Horizons derived its total unit sales budget for 19_4 from an internal management estimate of a 20% market share and an industry sales forecast by Micro-Information Services of 5,000,000 units. At the end of 19_4, Micro Information reported actual industry sales of 6,875,000 units.

Required

Compute the market size and market share variances for Computer Horizons.

Coverage of Part Two of the Chapter

26-18 Direct material price and efficiency variances, direct material yield and mix variances, food processing. Tropical Fruits Inc. processes tropical fruit into fruit salad mix, which it sells to a food service company. Tropical Fruits has in its budget the following standards for the direct material inputs to produce 80 pounds of tropical fruit salad:

50 pounds of pineapple at $1.00 per pound	$50
30 pounds of watermelon at $0.50 per pound	15
20 pounds of strawberries at $0.75 per pound	15
100	$80

No inventories of direct materials are kept. Purchases are made as needed, so that all price variances relate to direct materials used. The actual direct material inputs used to produce 54,000 pounds of tropical fruit salad for the month of October were:

36,400 pounds of pineapple at $0.90 per pound	$32,760
18,200 pounds of watermelon at $0.60 per pound	10,920
15,400 pounds of strawberries at $0.70 per pound	10,780
70,000	$54,460

Required

1. Compute the direct material price and direct material efficiency variances for each product and for the total output of tropical fruit salad in October.
2. Compute the individual product and total direct material yield variances for October.
3. Compute the individual product and total direct material mix variances for October.
4. Comment on your results in requirements 1, 2, and 3.
5. Why might direct material yield and material mix variances be especially informative to the management of Tropical Fruits Inc.?

26-19 Direct material yield and mix variances, perfume manufacturing. (SMA) The Scent Makers Company produces perfume. To make this perfume, three different types of fluids are used. Dycone, Cycone, and Bycone are applied in proportions of 4/10, 3/10, and 3/10, respectively, at standard, and their standard costs are $6.00, $3.50, and $2.50 per pint, respectively. The chief engineer reported that in the past few months the standard yield has been at 80% on 100 pints of mix. The company maintains a policy of not carrying any direct materials, as storage space is costly. Production has been set at 4,160,000 pints of perfume for the year.

Last week the company produced 75,000 pints of perfume at a total direct materials cost of $449,500. Actual number of pints used and costs per pint for the three fluids follow:

Direct Material	Pints	Cost Per Pint
Dycone	45,000	$5.50
Cycone	35,000	4.20
Bycone	20,000	2.75

Required

1. Compute the price, yield, and mix variances for each of the three direct materials. Reconcile these variances with the total direct material variance.
2. Explain the significance of the price, yield, and mix variances from management's perspective.

26-20 Direct labor price and efficiency variances, direct labor yield and mix variances, externally based standard costs. Choshu Engineering assembles large-scale machining systems at plants in Tokuyama, Manchester, Memphis, and Singapore. Direct labor, comprising assembly-department direct labor and testing-department direct labor, is a major cost category in each plant.

Ricki Choshu, son of the founder, Yoshida Choshu, manages the Memphis plant. Data for the Memphis plant in 19_5 and 19_6 follow:

	19_5 Actual	19_6 Actual
Assembly Department		
Direct labor hours per machining system	52 hours	36 hours
Rate (price) per direct labor hour	$22	$25
Testing Department		
Direct labor hours per machining system	28 hours	24 hours
Rate (price) per direct labor hour	$15	$16

The standard direct labor rates at the Memphis plant in 19_6 were $24 for the assembly department and $17 for the testing department. The Memphis plant assembled 500 machining systems in 19_6.

The 19_6 standard for direct labor hours of input per machining system assembled at the Memphis plant is based on the 19_5 actual results at the most efficient plant operated by Choshu Engineering, in Tokuyama. Results for the Tokuyama plant in 19_5 follow:

Assembly Department	
Direct labor hours per machining system	44 hours
Rate (price) per direct labor hour	$20
Testing Department	
Direct labor hours per machining system	20 hours
Rate (price) per direct labor hour	$18

Thus, the 19_6 standard for direct labor cost for the Memphis assembly department is $24 per hour times 44 hours per machining system. The 19_6 standard for the Memphis testing department is $17 per hour times 20 hours per machining system.

Required

1. Why might Choshu Engineering use 19_5 actual results for direct labor hours of input per machining system at the Tokuyama plant when computing the 19_6 standard direct labor cost per machining system at the Memphis plant?
2. Compute the direct labor price and efficiency variances.
3. Compute the budgeted average price per unit of direct labor input.
4. Compute the actual direct labor mix percentage and the budgeted direct labor mix percentage.
5. Compute the direct labor yield and direct labor mix variances.
6. Present a summary of the variances computed in requirements 2 and 5. Comment on the results.

26-21 Direct labor price and efficiency variances, direct labor yield and mix variances. A supervisor in a sheet metal operation of Mid-West Industries has the following direct labor standard:

(a) Direct labor price per hour:	
2 artisans @ $22	$44
3 helpers @ $12	36
Total cost of standard combination of direct labor	$80
Average price per direct labor hour, $80 ÷ 5 =	$16
(b) Standard direct labor price per unit of output at 8 units per hour, $16 ÷ 8 =	$ 2
(c) Standard direct labor cost of 20,000 units of output, 20,000 × $2, or 2,500 × $16 =	$40,000
(d) Actual inputs, 2,900 hours consisting of:	
900 hours of artisans @ $23	$20,700
2,000 hours of helpers @ $11	22,000
	$42,700

The supervisor had to pay a higher average wage rate to the artisans as the result of a bargained agreement. He tried to save some costs by using more helpers per artisan than usual.

Required

1. Compute the direct labor price and efficiency variances.
2. Divide the direct labor efficiency variance into yield and mix components.
3. What would the actual costs probably have been if the standard direct labor mix had been maintained even though the actual direct labor prices were incurred?

Coverage of Part Three of the Chapter

26-22 Sources of variances. The chapter described seven possible sources of variances: (1) inefficiencies in operations, (2) inappropriate standard, (3) mismeasurement of actual results, (4) implementation breakdown, (5) parameter prediction error, (6) inappropriate decision model, and (7) random variation.

Required

Each of the factors presented in the following list is associated with a separate variance. Use one of the numbers (1) through (7) to identify the most likely source of the variance being described.

(a) A supervisor gets a year-end bonus that is really attributable to overtime work that was performed seven months earlier. The bonus is charged to factory overhead at year-end.
(b) Costs of supplies are charged to factory overhead as acquired rather than as used.
(c) Costs of setting up printing jobs are consistently estimated too low when bids are made.
(d) The disposal price of scrap from production is forecast incorrectly.
(e) Normal spoilage in a food-processing plant amounts to 5% of good output.
(f) The disposal price of scrap from production is completely ignored.
(g) A worker is inefficient because of daydreaming.
(h) A worker is inefficient because of being new at the job and just learning how to do the work.
(i) A worker deliberately used more direct materials than necessary when the initial materials standards were set.

26-23 Cost-benefit analysis of variance investigation decision, timing of variance reports. Southern Beverages bottles mineral waters. Nancy Daus, the newly appointed production manager, has just received the weekly direct material efficiency variance report. Daus's predecessor used the following estimates in his variance investigation decisions:

C = cost of investigating = \$10,000
D = cost per period of being out of control = \$12,000
M = cost of correcting, if an out-of-control process is discovered = \$3,000

The file containing this information also included a report on "L, the cost of a process being out of control this period and not being corrected this period." No estimate on L, however, was included in the file. Daus remembered that her predecessor had been indifferent about conducting an investigation when there was a probability of 0.60 that the process was in control.

Required

1. What is the estimate of L that Daus's predecessor used in deciding whether to investigate a variance?
2. Daus decides to have a daily direct material efficiency report. Would you recommend she use the same estimates of C, D, M, and L as her predecessor in deciding whether to investigate a variance? Explain.

26-24 Cost-benefit analysis of variance investigation decision. You are the manager of a manufacturing process. A material efficiency variance of \$10,000 has been reported for the past week's operations. You are trying to decide whether to investigate this variance. You feel that if you do not investigate and the process is out of control, the cost savings (L) over the planning horizon is \$3,800. The cost to investigate is \$500. The cost per period of being out of control (D) is \$700. If an out-of-control process is discovered, the cost of correcting it (M) is \$300. You assess the probability that the process is out of control at .30.

Required

1. Should the process be investigated? What are the expected costs of investigating and of not investigating?
2. At what level of probability that the process is out of control would the expected costs of each action be the same?
3. If the cost variance is \$10,000, why is L only \$3,800?

26-25 Statistical quality control, airline operations. People's Skyway operates daily round-trip flights on the London–New York route using a fleet of three 747s. These three 747s are the Spirit of Birmingham, the Spirit of Glasgow, and the Spirit of Manchester. The standard quantity of fuel used on each round-trip flight is based on the mean fuel usage over the last twelve months. The distribution of fuel usage per round-trip over this twelve-month period has a mean of 100 gallon units and a standard deviation of 10 gallon units. A gallon unit is 1,000 British gallons.

Cilla Black, the operations manager of People's Skyway, uses a statistical quality control (SQC) approach in deciding whether to investigate variances from the standard fuel usage per round-trip flight. She investigates those flights with fuel usage greater than two standard deviations from the mean. In addition, Black monitors trends in the SQC charts to determine if additional investigation decisions should be made.

Black receives the following report for round-trip fuel usage in October by the three planes operating on the London–New York route:

Flight	Spirit of Birmingham (Gallon Units)	Spirit of Glasgow (Gallon Units)	Spirit of Manchester (Gallon Units)
1	104	103	97
2	94	94	104
3	97	96	111
4	101	107	104
5	105	92	122
6	107	113	118
7	111	99	126
8	112	106	114
9	115	101	117
10	119	93	123

Required

1. Using the $\pm 2\sigma$ rule, what variance investigation decisions would be made?
2. Present SQC charts for round-trip fuel usage by each of the three 747s in October.
3. What inferences can be drawn from the three SQC charts developed for requirement 2?
4. Some managers propose that People's Skyway express its SQC charts in monetary terms rather than in physical quantity terms (gallon units). What are the advantages and disadvantages of using monetary fuel costs rather than gallon units as the unit of analysis in an SQC chart?

CHAPTER 27

Systems Choice: Decentralization and Transfer Pricing

Organization structure and decentralization

Choices about responsibility centers

Transfer pricing

Illustration of transfer pricing

Taxation considerations

Market-based transfer prices

Cost-based transfer prices

A general rule for transfer pricing?

Multinational transfer pricing

Surveys of company practice

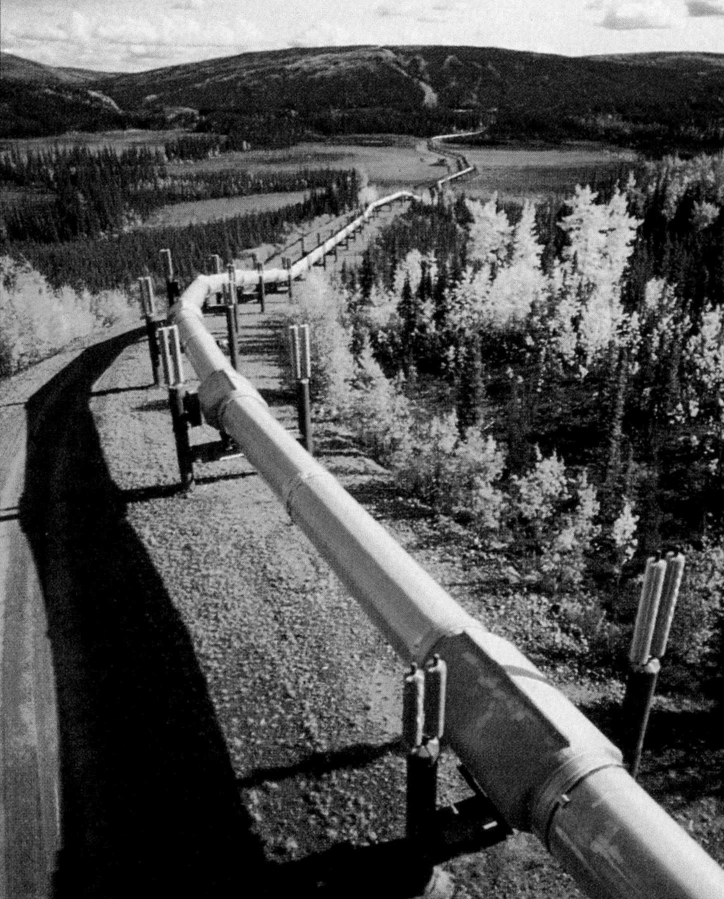

The operating incomes of the production, transportation, and refining divisions of oil companies are heavily influenced by the method used to price transfers of oil among these divisions.

Learning Objectives

When you have finished studying this chapter, you should be able to

1. Describe the benefits and costs of decentralization
2. Explain how geographic, organizational, and product-line reporting units differ
3. Give examples of sourcing decisions and transfer-pricing decisions
4. Identify three general methods for determining transfer prices
5. Understand how a transfer-price method can affect operating income of individual divisions
6. Illustrate how market-based transfer prices generally promote goal congruence in perfectly competitive markets
7. Explain how full-cost-based transfer prices lead to goal congruence problems
8. Describe factors executives consider important when determining transfer prices

Which company has the better management control system: General Motors or Ford? Michelin or Pirelli? What role can accounting information play in management control systems? Should products exchanged between individual profit centers of an organization be transferred at market price or at full cost? Senior managers often ask such questions. Chapters 27, 28, and 29 cover such important topics in this systems choice area. This chapter discusses the benefits and costs of centralized and decentralized organizations and looks at the pricing of products or services transferred between subunits of the same organization. Chapter 28 examines performance evaluation and rewards, and Chapter 29 focuses on strategic control systems.

Chapters 27, 28, and 29 contain a blend of cost accounting (narrowly conceived), strategic management, operations management, economics, and organization behavior. This material is "softer" than that in most other chapters—that is, it presents relatively few numbers. Nevertheless, the concepts in these chapters are important.

ORGANIZATION STRUCTURE AND DECENTRALIZATION

Objective 1

Describe the benefits and costs of decentralization

Top management makes decisions about decentralization that affect day-to-day operations at all levels of the organization. The essence of **decentralization** is the freedom to make decisions. *Decentralization is a matter of degree. Total decentralization means minimum constraints and maximum freedom for managers to make decisions at the lowest levels of an organization. Conversely, total centralization means maximum constraints and minimum freedom for managers to make decisions at the lowest levels.*

Benefits of Decentralization

How should top managers decide on how much decentralization is optimal? Conceptually, they try to choose the degree of decentralization that maximizes the excess of benefits over costs. From a practical standpoint, top managers can seldom quantify either the benefits or the costs. Still, this cost-benefit approach helps them focus on the central issues.

As we discuss the issues of decentralization, we use the term *subunit* to refer to any part of the organization. In practice, a subunit may be a large division (for example, the Chevrolet subunit of General Motors) or a small group (for example, a two-person advertising department of a local boutique).

Advocates of decentralizing decision making and granting responsibilities to managers of subunits claim the following benefits:

1. *Creates greater responsiveness to local needs.* Subunit managers can be more responsive than top managers can to the immediate demands of customers, suppliers, and employees. Eastman Kodak reports that one advantage of decentralization is an "increase in the company's knowledge of the marketplace and improved service to customers." Interlake, a manufacturer of materials-handling equipment, notes this important benefit of increased decentralization: "We have distributed decision-making powers more broadly to the cutting edge of product and market opportunity."
2. *Leads to quicker decision making.* An organization that gives lower-level managers the responsibility for making decisions quickly has a competitive advantage over organizations that drag their heels by sending the decision-making responsibility upward through layer after layer of management.
3. *Increases motivation.* Subunit managers are usually more highly motivated when they can exercise greater individual initiative. Johnson & Johnson, a highly decentralized company, argues that "Decentralization = Creativity = Productivity."
4. *Aids managerial development and learning.* Giving managers more responsibility promotes the development of an experienced pool of managerial talent. The company can draw from this pool of developing talent to fill higher-level management positions. Also, the company learns which people are not managerial material. Tektronix, an instruments company, expressed this benefit as follows: "Decentralized units provide a training ground for general managers, and a visible field of combat where product champions may fight for their ideas."
5. *Sharpens the focus of managers.* In a decentralized setting, the manager of a small subunit has a concentrated focus. A small subunit is more flexible and nimble than a larger subunit and better able to adapt itself quickly to a fast-opening market opportunity. Also, top management, relieved of the burden of day-to-day decision making at the lower level, can focus more time and energy on strategic planning for the entire organization.

Costs of Decentralization

Advocates of more centralized decision making point out the following costs of decentralizing decision making:

1. *Leads to* **suboptimal** (also called **dysfunctional**) **decision making**, *which arises when a decision's benefit to one subunit is more than offset by the costs or loss of benefits to the organization as a whole.* Suboptimal decision making may occur when there is (a) lack of harmony or congruence among the overall organizational goals, the subunit goals, and the individual goals of decision makers or (b) no guidance is given to subunit managers concerning the effects of their decisions on other parts of the organization.

   Suboptimal decision making is most likely to occur when the subunits in the organization are highly interdependent—that is, when decisions affecting one segment of the organization influence the decisions and performance of another segment. Examples of interdependencies are:

   - Subunits competing with each other for the same input factors (such as direct materials) or for the same customers, and
   - Subunits that are vertically integrated so that the end product of one subunit is the direct material of another subunit.

2. *Allows duplication of activities.* Several individual subunits of the organization may undertake the same activity separately. For example, there may be a duplication of staff func-

tions (such as accounting, employee relations, and legal) if an organization is highly decentralized.

3. *Decreases loyalty toward the organization as a whole.* Individual subunit managers may regard the managers of other subunits in the same organization as external parties. Consequently, managers may be unwilling to share significant information or to assist when another subunit faces an emergency situation.
4. *Costs of gathering information may rise.* Managers may take too much time negotiating the prices for internal services on products transferred among subunits.

Sourcing Restrictions

Some organizations restrict what products or services their subunits can purchase from external suppliers when internal subunits can provide the needed products or services. Likewise, some organizations restrict what products or services their subunits can sell to external customers when internal subunits demand those particular products or services. Such restrictions mean that some degree of centralization exists in the organization. Various rationales are given for restricting sources:

1. The attempt to build long-term production plans so that subunits can make capital-budgeting and operating decisions with greater predictability.
2. The attempt to improve long-term supply reliability.
3. The protection of an "infant" segment. A subunit may be in its infancy, and the parent wants to protect it from "market forces" in its development stage by providing a guaranteed market for some of its products.
4. The belief that products or services produced internally are of higher quality.

Comparison of Benefits and Costs

Top managers must compare the benefits and costs of decentralization, often on a function-by-function basis. For example, the controller's function may be highly decentralized for many attention-directing and problem-solving purposes (such as preparing operating budgets and performance reports) but highly centralized for other purposes (such as processing accounts receivables and developing income-tax strategies). Organizations are rarely totally centralized or totally decentralized.

One survey of U.S. and European firms reports that the decisions made most frequently at the profit-center level and least frequently at the corporate level relate to sources of supply, products to manufacture, and product advertising. Decisions relating to the type and source of long-term financing are made least frequently at the profit-center level and most frequently at the corporate level.[1]

CHOICES ABOUT RESPONSIBILITY CENTERS

An organization, whether centralized or decentralized, can measure the performance of its subunits by using one of the four types of responsibility centers presented in Chapter 6:

- Cost Center—manager accountable for costs only
- Revenue Center—manager accountable for revenues only
- Profit Center—manager accountable for revenues and costs
- Investment Center—manager accountable for investments, revenues, and costs

[1]*Evaluating the Performance of International Operations* (New York: Business International, 1989), p. 4.

Centralization or decentralization is not mentioned in these descriptions. Why? Because each of these responsibility units can be found in the extremes of centralized and decentralized organizations.

A common misconception is that the term *profit center* (and in some cases *investment center*) is a synonym for a decentralized subunit. ***Profit centers can be coupled with a highly centralized organization, and cost centers can be coupled with a highly decentralized organization.*** For example, managers in a division organized as a profit center may have little leeway in making decisions. They may need to obtain approval from corporate headquarters for every expenditure over, say, $10,000 and may be forced to accept central-staff "advice." In another company, divisions may be organized as cost centers, but their managers may have great latitude on capital outlays and on where to purchase materials and services. ***In short, the labels "profit center" and "cost center" can be independent of the degree of decentralization in an organization.***

Objective 2

Explain how geographic, organizational, and product-line reporting units differ

Whatever the type of responsibility center, a company may divide itself into reporting units—that is, units that report back to headquarters—in several possible ways. Three frequently encountered alternatives are geographic, organizational, and product line.

A geographic reporting unit is a country, a state or province, a city, or the like. An organizational reporting unit is an R & D division, a manufacturing division, a marketing department, or the like. A product-line reporting unit is the breakfast cereals product line, the candy product line, or the like.

Given four types of responsibility centers and three types of reporting units, twelve possible reporting setups exist:

	Cost Center	Revenue Center	Profit Center	Investment Center
Geographic Reporting Unit				
Organizational Reporting Unit				
Product-Line Reporting Unit				

Product lines are increasingly used as the reporting unit. One survey reports that the percentage of firms using product lines as the reporting unit for their profit centers increased from 31% in the late 1970s to 54% in the late 1980s; 72% of the respondents predicted that they would be using product lines as the reporting unit for their profit centers in the 1990s. Strategic plans of companies typically relate to products or product lines. Managers report that using product lines as the reporting unit helps the development and implementation of these strategic plans.[2]

TRANSFER PRICING

A **transfer price** is the price one segment (subunit, department, division, and so on) of an organization charges for a product or service supplied to another segment of the same organization. Consider News Ltd., which has a Paper Division and a

[2]Ibid., p. 4.

Printing Division. Each division is a profit center. The Paper Division produces rolls of paper that are either transferred to the Printing Division of News Ltd. or sold to other companies that print books, magazines, and newspapers. The Printing Division of News Ltd. prints magazines that are sold in supermarkets and newsstands and through subscriptions.

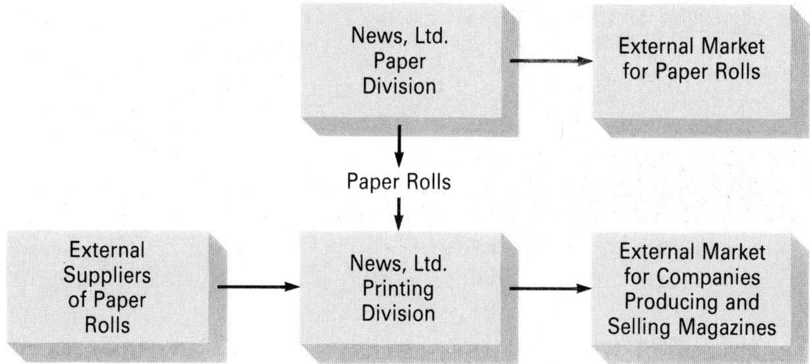

An **intermediate product** is a product transferred from one segment of an organization to another segment of the same organization. This product is further processed and sold to an external customer. Paper rolls are the intermediate product in our News Ltd. example. The price at which paper rolls are transferred from the Paper Division to the Printing Division is termed the *transfer price.*

Objective 3

Give examples of sourcing decisions and transfer-pricing decisions

News Ltd. must make decisions relating to both the sourcing and the pricing of paper rolls:

1. Sourcing decisions. For example, should the Paper Division be able to sell paper rolls to external customers when the paper requirements of the Printing Division have not been met? Should the Printing Division be able to buy paper rolls from external suppliers when the production of the Paper Division is not yet fully sold?
2. Transfer-pricing decisions. For example, what price(s) will be set for internal transfers of paper rolls from the Paper Division to the Printing Division?

Subsequent sections of this chapter discuss issues that News Ltd. and other companies should consider in their sourcing and pricing decisions.

Alternative Transfer-Pricing Methods

Objective 4

Identify three general methods for determining transfer prices

There are three general methods for determining transfer prices:

1. *Market-based transfer prices.* A company may choose to use the price of a similar product or service publicly listed in, say, a trade journal. Also, a company may select for its internal price the external price that a subunit charges to outside customers.
2. *Cost-based transfer prices.* Examples include variable manufacturing costs, absorption costs, and full costs. "Full cost" has various meanings in practice; typically it includes all manufacturing costs as well as costs from some or all of the other business functions (R & D, product design, marketing, distribution, and customer service). The costs can be either actual costs or standard costs.
3. *Negotiated transfer prices.* In some cases, the subunits of a firm are free to negotiate the transfer price between themselves. Information about costs and market prices may be used in these negotiations, but there is no requirement that the chosen transfer price must have any specific relationship to either cost or market-price data.

Ideally, the chosen transfer-price method should lead each subunit manager to make optimal decisions for the organization as a whole. Two specific criteria can help in choosing a transfer-pricing method:

1. *Promotion of goal congruence.* Goal congruence exists when individuals and groups aim at the organization goals that top management sets.

2. *Promotion of a sustained high level of managerial effort. Effort* is defined as exertion toward a goal.

If top management favors a high level of decentralization, a third criterion is also appropriate:

3. *Promotion of a high level of subunit autonomy* in decision making. **Autonomy** is the degree of freedom to make decisions.

These criteria are also explored in Chapter 13 (pp. 427–30).

The next section illustrates the use of market- and cost-based transfer-pricing methods in a petroleum company with production, transportation, and refining divisions. This example shows how the choice of a transfer-pricing method can dramatically affect the operating income of individual divisions. The following sections of the chapter then discuss how the criteria of goal congruence, managerial effort, and subunit autonomy affect the choice of transfer-pricing methods within organizations.

ILLUSTRATION OF TRANSFER PRICING

Horizon Petroleum has three divisions. Each operates as a profit center.

1. Production Division—manages the production of crude oil from an oil field near Tulsa, Oklahoma.
2. Transportation Division—manages the operation of a pipeline that transports crude oil from the Tulsa area to Houston, Texas.
3. Refining Division—manages a refinery at Houston that processes crude oil into gasoline. (For simplicity, assume that gasoline is the only salable product the refinery makes, and that it takes two barrels of crude oil to yield one barrel of gasoline.)

The data in Exhibit 27-1 summarize the variable and fixed costs. Variable costs in each division are assumed to be variable with respect to a single cost driver: barrels

EXHIBIT 27-1
Horizon Petroleum Operating Data

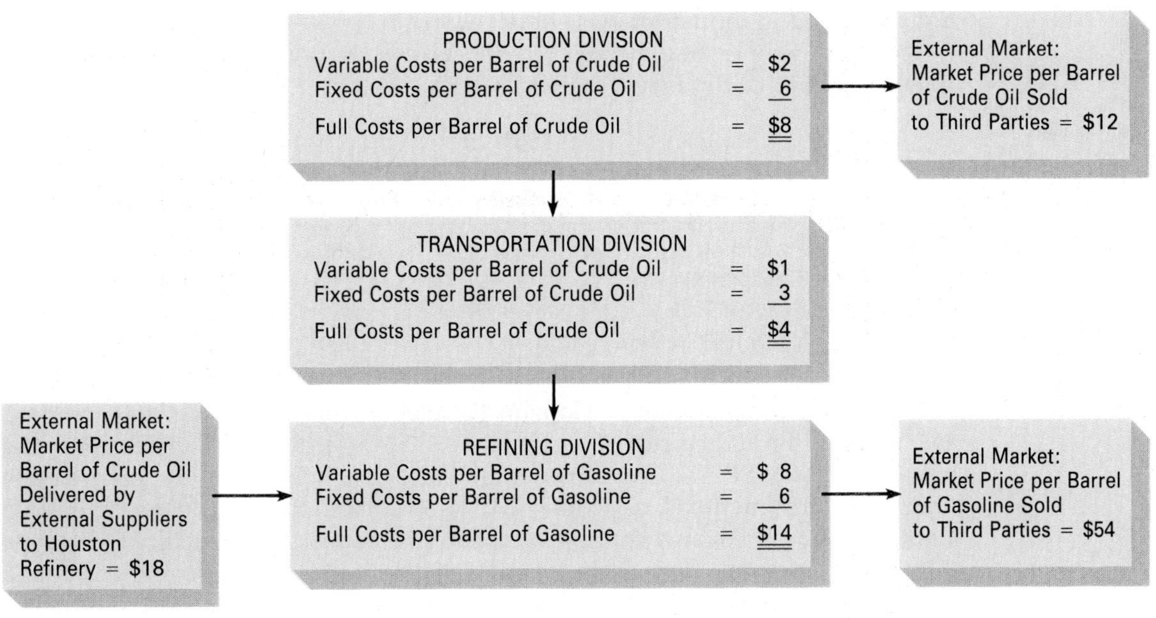

of crude oil produced by the Production Division, barrels of crude oil transported by the Transportation Division, and barrels of gasoline produced by the Refining Division. The fixed costs per unit are based on the budgeted annual volume of crude oil to be produced and transported and gasoline to be produced.

The Production Division can sell crude oil to other parties in the Tulsa area at $12 per barrel. The Transportation Division "buys" crude oil from the Production Division, transports it to Houston, and then "sells" it to the Refining Division. The pipeline from Tulsa to Houston can carry 40,000 barrels of crude oil per day. The Refining Division has been operating at capacity using oil from Horizon's Production Division (on average 10,000 barrels per day) and oil bought from other producers (on average 20,000 barrels per day, at $18 per barrel).

Exhibit 27-2 presents division operating income resulting from three transfer-pricing methods applied to a series of transactions involving 100 barrels of crude oil produced by Horizon's Production Division.

- Method A: 150% of variable costs, where variable costs are the cost of the transferred-in product plus the division's own variable costs.
- Method B: 125% of full costs, where full costs are the cost of the transferred-in product plus the division's own variable and fixed costs.
- Method C: Market price.

The transfer prices per barrel of crude oil under each method follow. The transferred-in cost component in Methods A and B is denoted by the symbol *.

- **Method A: 150% of Variable Costs**
  Production Division to Transportation Division = 1.5($2) = $3
  Transportation Division to Refining Division = 1.5($3* + $1) = $6
- **Method B: 125% of Full Costs**
  Production Division to Transportation Division = 1.25($2 + $6) = $10
  Transportation Division to Refining Division = 1.25($10* + $1 + $3) = $17.50
- **Method C: Market Price**
  Production Division to Transportation Division = $12
  Transportation Division to Refining Division = $18

The $12 and $18 amounts in Method C are market prices per third-party transactions reported in industry newsletters.

Objective 5

Understand how the transfer-price method can affect operating income of individual divisions

The division operating income per 100 barrels of crude oil reported under each transfer-pricing method (see Exhibit 27-2) is:

Division	150% of Variable Costs	125% of Full Costs	Market Price
Production	$ (500)	$200	$400
Transportation	(100)	350	200
Refining	1,400	250	200
Horizon Petroleum	$ 800	$800	$800

The total operating income to Horizon Petroleum from producing, transporting, and refining the 100 barrels of crude oil is $800, regardless of the set of internal transfer prices used. The division operating incomes, however, differ dramatically under the three methods, as Exhibit 27-2 shows. Note that the operating income amounts span a $900 range in the Production Division [$(500) to $400], a $450 range in the Transportation Division [$(100) to $350], and a $1,200 range in the Refining Division ($200 to $1,400). Note also that each division would choose a different

EXHIBIT 27-2

Division Operating Income of Horizon Petroleum for 100 Barrels of Crude Oil under Alternative Transfer-Pricing Methods

	Method A	Method B	Method C
	Internal Transfers at 150% of Variable Costs	Internal Transfers at 125% of Full Costs	Internal Transfers at Market Price
1. PRODUCTION DIVISION			
Revenues:			
$3, $10, $12, × 100 barrels crude oil	$ 300	$1,000	$1,200
Deduct:			
Division variable costs:			
$2 × 100 barrels crude oil	200	200	200
Division fixed costs:			
$6 × 100 barrels crude oil	600	600	600
Division operating income	$ (500)	$ 200	$ 400
2. TRANSPORTATION DIVISION			
Revenues:			
$6, $17.50, $18, × 100 barrels crude oil	$ 600	$1,750	$1,800
Deduct:			
Transferred-in costs:			
$3, $10, $12, × 100 barrels crude oil	300	1,000	1,200
Division variable costs:			
$1 × 100 barrels crude oil	100	100	100
Division fixed costs:			
$3 × 100 barrels crude oil	300	300	300
Division operating income	$ (100)	$ 350	$ 200
3. REFINING DIVISION			
Revenues: $54 × 50 barrels gasoline	$2,700	$2,700	$2,700
Deduct:			
Transferred-in costs:			
$6, $17.50, $18, × 100 barrels crude oil	600	1,750	1,800
Division variable costs:			
$8 × 50 barrels gasoline	400	400	400
Division fixed costs:			
$6 × 50 barrels gasoline	300	300	300
Division operating income	$1,400	$ 250	$ 200

method if its sole criterion were to maximize its own division operating income. Little wonder that division managers take considerable interest in the setting of transfer prices, especially those managers whose compensation or promotion is directly affected by division operating income.

Exhibit 27-2 illustrates that the choice of a transfer-pricing method can affect how the operating income pie is divided among individual divisions. Subsequent sections of this chapter illustrate that the size of the operating income pie itself can also be affected by the choice of a transfer-pricing method.

TAXATION CONSIDERATIONS

Managers should always take into account the taxation implications of alternative transfer-pricing methods. Consider the Horizon Petroleum data in Exhibit 27-2. Assume that the Production Division has to pay the state of Oklahoma income tax (at 5% of operating income) but that both the Transportation and Refining Divisions operate in Texas and do not pay state income tax. Horizon Petroleum would minimize its state income-tax payments with the 150% of variable cost transfer-pricing method:

Transfer-Pricing Method (1)	Operating Income of Production Division (2)	State Income Tax (3) = .05 × (2)
A: 150% of variable cost	$(500)	$ 0
B: 125% of full cost	200	10
C: Market price	400	20

The state of Oklahoma is fully aware of Horizon Petroleum's incentives to minimize income tax and any other taxes it pays. Typically, detailed tax rules constrain the options available to firms when selecting a transfer-pricing method (and other accounting methods). Some firms keep one set of books for tax reporting and a second set for internal management reporting; many firms do not. Using the same transfer-pricing method for both tax reporting and management reporting can help support a company's claim that it is not "manipulating" its reported taxable income to evade paying taxes.

Full consideration of taxation aspects of transfer-pricing decisions is beyond the scope of this book. Tax factors include not only income taxes but also payroll taxes, tariffs, and other government levies on organizations. Our aim here is to highlight the importance of tax factors to managers in their transfer-pricing decisions.

MARKET-BASED TRANSFER PRICES

Objective 6

Illustrate how market-based transfer prices generally promote goal congruence in perfectly competitive markets

When the intermediate market is perfectly competitive[3] and when interdependencies of subunits are minimal, transferring products or services at market prices generally leads to optimal decisions. In using market prices with perfectly competitive markets, a business can meet the three criteria of goal congruence, managerial effort, and subunit autonomy (if desired).

Reconsider the Horizon Petroleum example in Exhibits 27-1 and 27-2. Assume that there is a perfectly competitive market for crude oil in the Tulsa area. Crude oil can be sold or purchased at $12 per barrel. Consider the decisions that Horizon's division managers would make if each had the option of selling or buying crude oil externally. If the transfer price between Horizon's Production Division and Transportation Division is set below $12, the manager of the Production Division will be motivated to sell all production to outside buyers at $12 per barrel. If the transfer price is set above $12, the manager of the Transportation Division will be motivated to purchase all its crude oil requirements from outside suppliers.

When an industry has excess capacity, market prices may drop sizably below their historical average. These low market prices are sometimes called "distress prices." Deciding whether a current market price is a "distress price" is often

[3]A **perfectly competitive market** exists when there is a homogeneous product with equivalent buying and selling prices and no individual buyers or sellers can affect those prices by their own actions.

difficult. Just because market prices precipitously drop from prior levels does not mean the current market price will subsequently increase. The market prices of several agricultural commodities and minerals have stayed for many years at what observers initially believed were temporary "distress levels."

What transfer-pricing method should be used for judging performance if distress prices prevail? Some companies use these distress prices, but others use long-run average prices or "normal" market prices. In the short run, the manager of the supplier division will meet the distress price as long as it exceeds the incremental costs of supplying the product or service. In the long run, the manager must decide whether to dispose of some manufacturing facilities. If long-run average market price is used, managers may be forced to buy internal products at above-current market prices, consequently hurting the short-run performance of the buying division.

COST-BASED TRANSFER PRICES

Sometimes market prices are unavailable, inappropriate, or too costly to obtain for transfer pricing. Many organizations use cost-based measures as transfer prices. This section explores examples of these measures and the problems that may arise with their use.

Objective 7

Explain how full-cost-based transfer prices lead to goal congruence problems

Full-Cost Bases

In practice, companies frequently use transfer prices based on full costs. Using these prices, however, can lead to suboptimal decisions for the company as a whole. Assume that Horizon Petroleum (see Exhibits 27-1 and 27-2) makes internal transfers at 125% of full cost. The purchasing officer at the Houston Refining Division is attempting to reduce the cost of crude oil provided by outside suppliers. At present oil is purchased from a Houston supplier at \$18 per barrel. Assume now that the market for crude oil in Tulsa is not perfectly competitive. The purchasing officer of the Refining Division locates an independent producer in Tulsa that will sell to Horizon Petroleum 20,000 barrels of crude oil per day, at \$13 per barrel, delivered to Horizon's pipeline in Tulsa. Given Horizon's organization structure, the Transportation Division would purchase the 20,000 barrels of crude oil in Tulsa, ship it to Houston, and then sell it to Horizon's Refining Division. The pipeline has excess capacity and can ship the 20,000 barrels at its variable costs of \$1 per barrel without affecting the shipment of crude oil from its own Production Division. Should Horizon Petroleum purchase crude oil from the independent producer in Tulsa? Will the Refining Division show lower costs of purchasing crude oil by using oil from the Tulsa producer or from using its current Houston supplier?

Horizon Petroleum would prefer to purchase oil from the independent producer.

- Alternative 1: Buy 20,000 barrels from Houston supplier at \$18 per barrel.
  Total cost to Horizon Petroleum = 20,000 × \$18 = \$360,000.
- Alternative 2: Buy 20,000 barrels in Tulsa at \$13 per barrel and transport it to Houston at \$1 per barrel variable cost.
  Total cost to Horizon Petroleum = 20,000 × (\$13 + \$1) = \$280,000.

There is a reduction in total costs to Horizon Petroleum of \$80,000 by using the independent producer in Tulsa.

The Refining Division of Horizon Petroleum, however, would see its reported division costs increase if the crude oil is purchased from the independent producer in Tulsa when the transfer price from the Transportation Division to the Refining

Division is 125% of full cost:

$$\text{Transfer price} = 1.25\left(\begin{array}{c}\text{Purchase price}\\ \text{from Tulsa}\\ \text{producer}\end{array} + \begin{array}{c}\text{Unit variable cost}\\ \text{of Transportation}\\ \text{Division}\end{array} + \begin{array}{c}\text{Unit fixed cost}\\ \text{of Transportation}\\ \text{Division}\end{array}\right)$$

- Alternative 1: Buy 20,000 barrels from Houston supplier at $18 per barrel.
  Total cost to Refining Division = 20,000 × $18 = $360,000.
- Alternative 2: Buy 20,000 barrels from the Transportation Division of Horizon Petroleum that are purchased from the independent producer in Tulsa.
  Total cost to Refining Division = 20,000 × $21.25 = $425,000.

Given that the Refining Division is a profit center, it can maximize its short-run division operating income by purchasing from the Houston supplier. If instead it chooses the alternative that maximizes the operating income of the company as a whole, the Refining Division will increase its own division operating costs by $425,000 − $360,000 = $65,000.

This situation is a clear example of goal incongruence that is induced by a transfer price based on full cost. ***The transfer-pricing method has led the Refining Division to regard the fixed cost (and the 25% markup) of the Transportation Division as a variable cost.*** From the viewpoint of the company as a whole, the full-cost-based transfer price can lead to suboptimal decisions in the short run.

Prorating the Overall Contribution

Consider again the example of Horizon Petroleum purchasing crude oil from the independent producer in Tulsa at $13 per barrel. The variable costs to transport each barrel from Tulsa to Houston are $1 per barrel. What transfer price will promote goal congruence and maximum managerial effort for both the Transportation Division and the Refining Division? Recall that the company as a whole benefits from purchasing the crude oil from the independent Tulsa producer. The minimum transfer price is $14 per barrel; a transfer price below $14 does not provide the Transportation Division with an incentive to purchase crude oil from the independent producer in Tulsa. The maximum transfer price is $18 per barrel; a transfer price above $18 will not provide incentives for the Refining Division to purchase crude oil from the Transportation Division.

Companies sometimes impose a variable-cost transfer price and credit each division for a prorated share of the contribution to companywide operating income. The proration of this contribution can be negotiated in many ways. Suppose the proration is made on the basis of standard variable costs incurred by each division. In our example, Horizon Petroleum would prorate the $4 difference ($18 − $14) between the Transportation Division and the Refining Division based on data drawn from Exhibit 27-2 (p. 859):

Variable costs of Transportation Division	$100
Variable costs of Refining Division	400
	$500

The $4-per-barrel overall contribution would be allocated as follows:

To Transportation Division $\quad \frac{\$100}{\$500} \times \$4.00 = \$0.80$

To Refining Division $\quad \frac{\$400}{\$500} \times \$4.00 = \$3.20$

The transfer price between the Transportation Division and the Refining Division would be $14.80 per barrel of crude oil ($13 purchase cost + $1 variable costs + $0.80 allocated contribution). With this transfer price, both divisions will increase their own reported division operating income by purchasing crude oil from the independent producer in Tulsa. Essentially, this approach is a standard-variable-cost plus transfer-pricing system; the "plus" is a portion of the overall contribution to corporate operating income.

To decide on the $0.80 and $3.20 allocation of the $4.00 contribution to total corporate operating income per barrel, the divisions must share information about their variable costs. In effect, each division does not operate (at least for this transaction) in a totally decentralized manner. Because most organizations are hybrids of centralization and decentralization anyway, this approach deserves serious consideration when transfers are significant.

Dual Pricing

There is seldom a *single* transfer price that will simultaneously meet the criteria of goal congruence, managerial effort, and subunit autonomy. Some companies turn to **dual pricing**, which uses two separate transfer-pricing methods to price each interdivision transaction. An example of dual pricing arises when the selling division receives a full-cost plus markup-based price and the buying division pays the market price for the internally transferred products. Assume Horizon Petroleum purchases crude oil from the independent producer in Tulsa at $13 per barrel. One way of recording the transfer between the Transportation Division and the Refining Division is:

1. Credit the Transportation Division (the selling division) with the 125%-of-full-cost transfer price of $21.25 per barrel of crude oil.
2. Debit the Refining Division (the buying division) with the market-based transfer price of $18 per barrel of crude oil.
3. Debit a corporate account for the $3.25 ($21.25 − $18.00) difference between the two transfer prices.

This dual-pricing system essentially gives the Transportation Division a corporate subsidy. The operating income for Horizon Petroleum as a whole will be less than the sum of the operating incomes of the divisions.

Dual pricing is not widely used in practice even though it reduces the goal-congruence problems associated with a pure cost-plus-based transfer-pricing method. One concern of top management is that the manager of the supplying division does not have sufficient incentives to control costs with a dual-price system. A second concern is that the dual-price system does not provide clear signals to division managers about the level of decentralization top management seeks. Above all, dual-pricing tends to insulate managers from the frictions of the marketplace. Managers should know as much as feasible about their subunits' buying and selling markets, and dual pricing reduces the incentives to gain this knowledge.

A GENERAL RULE FOR TRANSFER PRICING?

Is there an all-pervasive rule for transfer pricing that leads toward optimal decisions for the organization as a whole? No. Why? Because the three criteria of goal congruence, managerial effort, and subunit autonomy must all be considered simultaneously. The following general rule, however, has proved to be a helpful *first step* in setting a transfer price:

$$\begin{array}{c}\text{Minimum}\\\text{transfer price}\end{array} = \left(\begin{array}{c}\text{Additional } \textit{outlay costs}\\\text{per unit incurred}\\\text{to the point of transfer}\end{array}\right) + \left(\begin{array}{c}\textit{Opportunity costs}\\\text{per unit to the}\\\text{firm as a whole}\end{array}\right)$$

The term *outlay costs* in this context represents the cash outflows that are directly associated with the production and transfer of the products and services. *Opportunity costs* are defined here as the maximum contribution forgone by the company as a whole if the products or services are transferred internally. The distinction between outlay costs and opportunity costs is made because the accounting system typically records outlay costs but not opportunity costs. Opportunity costs may be forgone contribution margins in some instances or even forgone net proceeds from the sale of facilities in other instances.

If a perfectly competitive intermediate market exists, the opportunity cost is market price minus outlay cost (that is, the cost of the opportunity forgone by the internal supplier). For example, if the outlay cost is \$1 per unit and the market price is \$4 per unit, the transfer price is \$1 + (\$4 − \$1) = \$4, which happens to be the market price. If no market exists for the intermediate product, however, or for alternative products that might utilize the same supplying division's facilities, the opportunity cost may be zero. In this case, outlay costs per unit may be the correct transfer price. In summary, in our example with a perfectly competitive market, if idle capacity exists:

$$\begin{aligned}\text{Minimum transfer price} &= \left(\text{Outlay cost per unit}\right)\\ &= \$1\end{aligned}$$

If no idle capacity exists:

$$\begin{aligned}\text{Minimum transfer price} &= \left(\text{Outlay cost per unit}\right) + \left(\text{Market price per unit} - \text{Outlay cost per unit}\right)\\ &= \$1 + (4 - \$1) = \$4\end{aligned}$$

Opportunity cost to the company as a whole from each potential internal transfer of products or services is the more difficult component to measure in our general rule. Consider a supplying division with idle capacity and an imperfect demand in the intermediate market. Is its opportunity cost zero? Probably not. This division might reduce price to increase demand, hoping to increase overall operating income. But measuring the effect of the price cut is difficult, so measuring the opportunity costs is also difficult. Clearly, the transfer price depends on constantly changing levels of supply and demand. There is not just one transfer price; rather, a transfer-price function yields the transfer price for various quantities supplied and demanded.

MULTINATIONAL TRANSFER PRICING

Many organizations have divisions located in different countries. Deciding on a set of transfer prices to use for exchange between these divisions requires consideration of additional factors that include the following:

1. *International taxation.* Countries have different tax rates, and they enforce their tax codes to different degrees. Also, countries vary in what they allow for deductions. Consider a transfer of motor vehicle parts by Division A of World Motors in Country X to Division B of World Motors in Country Y. Country X has a 20% income tax rate, and Country Y has a 50% income tax rate. World Motors has an incentive to use a high transfer price for transfers from Division A to Division B to maximize the income reported in Country X, which has the lower tax rate. (The tax authorities in Country Y are fully aware of this incentive and will make efforts to maximize the taxes paid by Division B of World Motors.)
2. *Income or Dividend Payment Restrictions.* Some countries restrict the payment of income or dividends to parties outside their national borders. By increasing the prices of goods or

services transferred into divisions in these countries, firms can increase the funds paid out of these countries without "appearing to violate" income or dividend restrictions.

Other additional factors that arise in multinational transfer pricing include tariffs, custom duties, and risks associated with movements in foreign currency exchange rates.

SURVEYS OF COMPANY PRACTICE

Objective 8

Describe factors executives consider important when determining transfer prices

Exhibit 27-3 summarizes the factors that executives consider important when determining transfer prices for both domestic and multinational operations. Exhibit 27-4 summarizes the results of studies of domestic transfer-pricing methods and multinational transfer-pricing methods. Observe the widespread use of both market-based and full-cost-based transfer prices. For companies using cost-based methods, standard costs are more widely used than actual costs. One survey of U.S. companies reports that standard costs were used by 54% of respondents, actual costs by 16%, and some of each by 30%.[4]

Organizations differ in how they handle transfer-pricing disputes. Approaches used in practice include the following:[5]

- *Decentralized bargaining approach.* The parties in the transfer-pricing dispute (for example, the buying division and the selling division) bargain with each other without any directives from or intervention by senior management.
- *Third-party problem-solver approach.* The parties in the dispute air their differences to a third party (such as a senior manager or a "Transfer-Pricing Advisory Panel"), which

[4]*Transfer Pricing Practices of American Industry* (New York: Price Waterhouse, 1984).

[5]R. Eccles, "Analyzing Your Company's Transfer Pricing Practices," *Journal of Cost Management*, Summer 1987.

EXHIBIT 27-3

Factors That Executives Consider Important in Decisions on Transfer Pricing (In Order of Importance)

A. DOMESTIC TRANSFER PRICING*

1. Performance evaluation—to measure the results of each operating subunit
2. Managerial motivation—to provide the company with a "profit-making" orientation throughout each organizational subunit
3. Pricing and product emphasis—to better reflect "costs" and "margins" that must be received from customers
4. External market recognition—to maintain an internal competitiveness so that the company stays in balance with outside market forces

B. MULTINATIONAL TRANSFER PRICING†

1. Overall income to the company
2. The competitive position of subsidiaries in foreign countries
3. Performance evaluation of foreign subsidiaries
4. Restrictions imposed by foreign countries on repatriation of income or dividends‡
5. The need to maintain adequate cash flows in foreign subsidiaries‡
6. Maintaining good relationships with host governments

*Source: Price Waterhouse, *Transfer Pricing Practices of American Industry*, p. 2.
†Source: R. Tang, "Environmental Variables of Multinational Transfer Pricing: A U.K. Perspective," *Journal of Business Finance & Accounting*, Vol. 9, No. 2, p. 182.
‡Ranked equal in importance.

EXHIBIT 27-4
Surveys of Transfer Pricing Methods

A. DOMESTIC TRANSFER-PRICING METHODS

Methods	United States*	Australia†	Canada‡	Japan*	India§	United Kingdom‖
1. Market-price-based	30%	13%	34%	34%	47%	41%
2. Cost-based:						
Variable costs	4	N.D.	6	2	6	6
Absorption or full costs	45	N.D.	37	44	47	19
Other	1	N.D.	3	—	—	4
Total	50%	65%	46%	46%	53%	29%
3. Negotiated	18%	11%	18%	19%	—	24%
4. Other	2%	11%	2%	1%	—	6%
	100%	100%	100%	100%	100%	100%

B. MULTINATIONAL TRANSFER-PRICING METHODS

Methods	United States*	Australia†	Canada‡	Japan*	India§	United Kingdom‖
1. Market-price-based	35%	—	37%	37%	—	31%
2. Cost-based:						
Variable costs	2	—	5	3	—	5
Absorption or full costs	42	—	26	38	—	28
Other	2	—	2	—	—	5
Total	46%	—	33%	41%	—	38%
3. Negotiated	14%	—	26%	22%	—	20%
4. Other	5%	—	4%	—	—	11%
	100%	—	100%	100%	—	100%

SOURCES:
*R. Tang, C. Walter, and R. Raymond, "Transfer Pricing—Japanese vs. American Style," *Management Accounting*, January 1979, pp. 12–16.
†P. Blayney and M. Joye, "Cost Accounting Practices in Australian Manufacturing Companies," Working Paper, University of Sydney, 1989. N.D. means breakdown not disclosed.
‡R. Tang, "Canadian Transfer-Pricing Practices," *CA Magazine*, March 1980, pp. 32–38.
§V. Govindarajan and B. Ramamurthy, "Transfer Pricing Policies in Indian Companies: A Survey," *Chartered Accountant*, November 1983, pp. 296–301.
‖A. Mostafa, J. Sharp, and K. Howard, "Transfer Pricing—A Survey Using Discriminant Analysis," *Omega*, 12, No. 5 (1984), 465–74.

decides the issue. This approach can be very time consuming. Moreover, it does not always yield a solution that all parties view as reasonable.

- *Directive approach.* A group or an individual, such as a senior manager, decides on the transfer price and directs that all parties to the dispute use it.

Case studies and interviews with managers stress the key role that both corporate strategy and performance evaluation play in decisions about transfer pricing. For example, companies that use market-based transfer prices emphasize profit-oriented performance criteria in the evaluation of division managers. Companies that use cost-based transfer prices put less emphasis on these criteria.[6]

[6]R. Eccles, *The Transfer Pricing Problem: A Theory for Practice* (Lexington, MA: Lexington Books, 1985). Companies using cost-based transfer prices often emphasize risk-sharing considerations. See Y. Monden and T. Nagao, "Full Cost-Based Transfer Pricing in the Japanese Auto Industry," in Y. Monden and M. Sakurai, *Japanese Management Accounting* (Cambridge, MA: Productivity Press, 1989).

PROBLEM FOR SELF-STUDY

PROBLEM

Pillercat Corporation is a highly decentralized company. Each division head has full authority for sourcing decisions and sales decisions. The Tractor Division can purchase a key component—the crankshaft—from the Machining Division of Pillercat or from external suppliers:

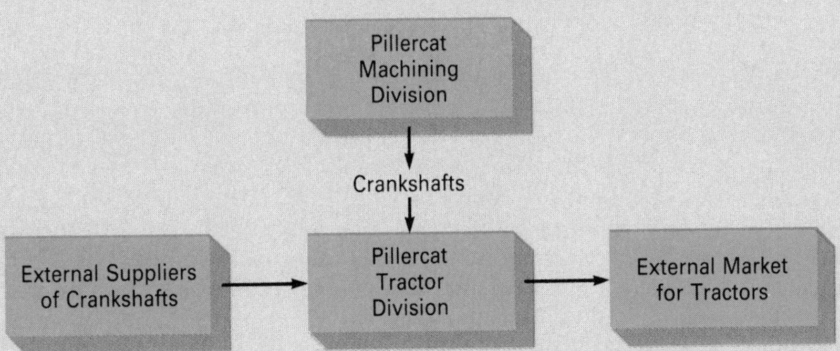

The Machining Division of Pillercat has been the major supplier of crankshafts to the Tractor Division in recent years. The Tractor Division, however, has just announced that it will purchase all its crankshafts in the forthcoming year from two external suppliers at $200 per crankshaft. The Machining Division of Pillercat recently increased its unit price for the forthcoming year to $220 (from $200 in the current year).

Juan Gomez, manager of the Machining Division, felt that the price increase was fully justified. The Machining Division recently purchased some specialized equipment to manufacture crankshafts. The resulting higher depreciation charge as well as an increase in labor cost led to the 10% price increase. Gomez met with the president of Pillercat Corporation and requested that the Tractor Division be directed to buy all its crankshafts from the Machining Division at the $220 price. Gomez supplied the following cost information for the Machining Division: variable cost per crankshaft, $190; fixed cost per crankshaft, $20.

The additional outlay costs per unit that Pillercat incurs to produce each crankshaft is the Machining Division's variable cost of $190. The Tractor Division purchases 2,000 crankshafts per month.

Required

1. Compute the advantage or disadvantage (in terms of monthly operating income) to Pillercat as a whole if the Tractor Division buys crankshafts internally from the Machining Division under each of the following three cases:
   (a) Machining Division has no alternative use for the facilities used to manufacture crankshafts.
   (b) Machining Division can use the facilities for other production operations, which will result in monthly cash-operating savings of $29,000.
   (c) Machining Division has no alternative use for the facilities, and the external supplier drops its price to $185 per crankshaft.
2. As the president of Pillercat, how would you respond to Juan Gomez's request to direct the Tractor Division to purchase all of its crankshafts from the Machining Division? Would your response differ according to the scenarios described in (a), (b), and (c) of requirement 1? Why?

SOLUTION

1.

	(a)	(b)	(c)
Total purchase costs if buying from an external supplier (2,000 × $200, $200, $185)	$400,000	$400,000	$370,000
Total outlay cost if buying from the Machining Division (2,000 × $190)	380,000	380,000	380,000
Total opportunity cost to Pillercat if buying from the Machining Division	—	29,000	—
Total relevant costs	$380,000	$409,000	$380,000
Monthly operating income advantage (disadvantage) to Pillercat if buying from the Machining Division	$ 20,000	$ (9,000)	$ (10,000)

2. Pillercat Corporation is a highly decentralized company. If no forced transfer were made, the Tractor Division would use an external supplier, resulting in an optimal decision for the company as a whole in parts (b) and (c) of requirement 1 but not in part (a).

Suppose that in requirement 1 (a), the Machining Division refuses to meet the price of $200. This decision means that the company will be $20,000 worse off in the short run. Should top management interfere and force a transfer at $200? This interference would undercut the philosophy of decentralization. Many managers would not interfere because they would view the $20,000 as the cost of potential suboptimal decisions that occasionally occur under decentralization. But how high must this cost go before the temptation to interfere would be irresistible? $30,000? $40,000?

Any superstructure that interferes with lower-level decision making weakens decentralization. Of course, such interference may occasionally be necessary to prevent costly blunders. But recurring interference and constraints simply transform a decentralized organization into a centralized organization.

The "general rule" that was introduced in the chapter as a first step in setting a transfer price can be used to highlight the alternatives in requirement 1:

Case	Outlay Costs	+	Opportunity Costs	=	Transfer Price	External Market Price
(a)	$190	+	$0	=	$190	$200
(b)	$190	+	($29,000 ÷ 2,000)	=	$204.50	$200
(c)	$190	+	$0	=	$190	$185

The Tractor Division will maximize monthly operating income of Pillercat as a whole by purchasing from the Machining Division in case (a) and by purchasing from the external supplier in cases (b) and (c).

SUMMARY

A transfer-pricing system should be judged in relation to its impact on (a) goal congruence, (b) managerial effort, and (c) subunit autonomy. Some version of market price as a transfer price will often best motivate managers toward optimal economic decisions. Moreover, the evaluation of performance will then be consistent with the concept of decentralization. When subunits are interdependent, however, and when the markets for intermediate goods are not perfectly competitive, market prices will not always lead to optimal decisions. In such instances, some centralization of control may be desired to prevent suboptimal decisions.

Profit centers or investment centers usually exist in decentralized companies, but profit centers can exist without decentralization. The desirability of a profit center or an investment center versus a cost center should be judged by their predicted relative impacts on collective decisions.

TERMS TO LEARN

This chapter and the Glossary at the end of the book contain definitions of the following important terms:

autonomy *(p. 857)* decentralization *(852)* dual pricing *(853)*
dysfunctional decision making *(853)* intermediate product *(856)*
perfectly competitive market *(860)* suboptimal decision making *(853)*
transfer price *(855)*

ASSIGNMENT MATERIAL

QUESTIONS

27-1 Name three benefits of decentralization.

27-2 Name two costs of decentralization.

27-3 "Organizations typically adopt a consistent decentralization or centralization philosophy across all their business functions." Do you agree? Explain.

27-4 Give an example of each of the following three reporting units: geographic, organizational, and product line.

27-5 Distinguish between *sourcing* restrictions and *transfer-pricing* restrictions.

27-6 "Transfer pricing is confined to profit centers." Do you agree? Why?

27-7 What are the three general methods for determining transfer prices?

27-8 Describe three criteria useful in choosing a transfer-pricing method.

27-9 Why should managers consider income-tax issues when choosing a transfer-pricing method?

27-10 What is the major limitation of transfer prices based on cost?

27-11 In transfer pricing, what is a common conflict between a division and the company as a whole?

27-12 Give two reasons why a dual-pricing approach to transfer pricing is not widely used in practice.

27-13 Name three different ways that firms handle transfer-pricing disputes.

EXERCISES AND PROBLEMS

27-14 Transfer-pricing methods, goal congruence. British Columbia Lumber has a Raw Lumber Division and Finished Lumber Division. The variable costs are:

- Raw Lumber Division: $100 per 100 board feet of raw lumber
- Finished Lumber Division: $125 per 100 board feet of finished lumber

Assume there is no board feet loss in processing raw lumber into finished lumber. Raw lumber can be sold at $200 per 100 board feet. Finished lumber can be sold at $275 per 100 board feet.

Required

1. Should British Columbia Lumber process raw lumber into its finished form?
2. Assume that internal transfers are made at 110% of variable costs. Will each division maximize its division operating income contribution by adopting the action that is in the best interests of British Columbia Lumber?

3. Assume that internal transfers are made at market prices. Will each division maximize its division operating income contribution by adopting the action that is in the best interests of British Columbia Lumber?

27-15 Responding to a transfer-pricing question. The Plastics Company has a separate division that produces a special molding powder (MP). For the past three years, about two-thirds of the output (measured in pounds) has been sold to another division within the company. The remainder has been sold to outsiders. Last year's operating data follow:

	To Other Division		To Outsiders	
Sales	10,000 MP. @ $70*	$700,000	5,000 MP. @ $100	$500,000
Variable costs @ $50		$500,000		$250,000
Fixed costs		150,000		75,000
Total costs		$650,000		$325,000
Gross margin		$ 50,000		$175,000

*The $70 price is ordinarily determined by the outside selling price minus marketing and administrative costs wholly applicable to outside business.

The buying-division manager has a chance to get a firm contract with an outside supplier at $65 per pound for the coming year.

Required

Assume that the molding-powder-division manager asserts that no gross margin can be earned if sales are made at $65 per pound. As the buying-division manager, write a short reply. Assume that the molding-powder division cannot sell the 10,000 pounds to other customers.

27-16 Transfer-pricing dispute. Allison-Chambers Corp., manufacturer of tractors and other heavy farm equipment, is organized along decentralized lines, with each manufacturing division operating as a separate profit center. Each division head has been delegated full authority on all decisions involving the sale of that division's output both to outsiders and to other divisions of Allison-Chambers. Division C has in the past always purchased its requirement of a particular tractor-engine component from Division A. However, when informed that Division A was increasing its selling price to $150, Division C's management decided to purchase the engine component from outside suppliers.

Division C can purchase the component for $135 on the open market. Division A insists that because of the recent installation of some highly specialized equipment and the resulting high depreciation charges A would not be able to make an adequate return on its investment unless it raised its price. A's management appealed to top management of Allison-Chambers for support in its dispute with C and supplied the following operating data:

C's annual purchases of tractor-engine component	1,000
A's variable costs per unit of tractor-engine component	$120
A's fixed costs per unit of tractor-engine component	$ 20

Required

1. Assume that there are no alternative uses for internal facilities. Determine whether the company as a whole will benefit if Division C purchases the component from outside suppliers for $135 per unit.
2. Assume that internal facilities of A would not otherwise be idle. By not producing the 1,000 units for C, A's equipment and other facilities would be used for other production operations that would result in annual cash-operating savings of $18,000. Should C purchase from outside suppliers?
3. Assume that there are no alternative uses for A's internal facilities and that the price from outsiders drops $20. Should C purchase from outside suppliers?

27-17 Transfer-pricing problem. Refer to Problem 27-16. Assume that Division A could sell the 1,000 units to other customers at $155 per unit with variable marketing costs of $5 per unit. If this were the case, determine whether Allison-Chambers would benefit if C purchased the 1,000 components from outside suppliers at $135 per unit.

27-18 Transfer pricing, goal congruence. (NAA, adapted) Nogo Motors, Inc., has several regional divisions, which often purchase component parts from each other. The company is fully decentralized, each division buying and selling to other divisions or in outside markets. Each division makes its decision on where to buy and sell in conformity with divisional goals. Igo Division purchases most of its airbags from Letgo Division. The managers of these two divisions are currently negotiating a transfer price for the airbags for next year, when the airbag will be standard equipment on all Igo vehicles. Letgo Division prepared the following financial information for negotiating purposes:

Costs of airbag as manufactured by Letgo:	
Direct materials costs	$ 40
Direct labor costs	60
Variable manufacturing overhead costs	10
Fixed manufacturing overhead costs	25
Fixed marketing costs	15
Fixed administrative costs	10
Total	$160

Letgo Division is currently working at 80% of its capacity. It is the policy of the division to achieve an operating income that is 20% of sales.

There has been a drop in price for airbags. Their current market price is $130 each.

Required

Consider each of the requirements independently.

1. If Letgo Division desires to achieve its goal of operating income as 20% of sales, what should be the transfer price?
2. Assume that Letgo Division wants to maximize its operating income on sales. What transfer price would you recommend that the Igo Division set to maximize its own operating income?
3. What is the transfer price that you believe Letgo Division should charge if overall company operating income is to be maximized?

27-19 Transfer pricing, media company, captive producer. National Television Network (NTN) is the largest cable-television network in the United States. Its Prime Time channel reaches over 90% of all cable-television homes in North America. The highest-rated program on the Prime Time channel is the one-hour weekly professional wrestling program ("The Power Hour") from the Championship Wrestling Alliance (CWA).

In late 19_1, NTN acquired a 100% interest in the CWA. The CWA had three sources of revenues prior to the acquisition:

a. Arena shows—one-night wrestling shows held in sports auditoriums and other arenas
b. Merchandise—sales of T-shirts, posters, and other items featuring CWA wrestlers such as Misty Blue, The Total Package, and the Bladerunners
c. Television syndication fees—payments by NTN for programs the CWA produced for the Prime Time channel

After NTN's acquisition of the CWA, the issue of pricing CWA's weekly TV program arose. NTN ordered that a transfer price of 110% of actual production cost be used.

Revenue and cost data for the CWA for the 19_2, 19_3, and 19_4 years are (in millions):

	19_2	19_3	19_4
Revenues			
Arena shows	$5.7	$4.3	$3.7
Merchandise	0.6	0.9	1.3
Television syndication fees from NTN	?	?	?
Costs			
Arena shows	$4.3	$4.6	$4.9
Merchandise	0.4	0.6	0.9
Television production for syndicated shows	4.9	6.2	7.4

Jim Ross, president of the CWA, is furious (yet again!). A wrestling newsletter reported in its latest issue (February 19_5) that NTN was unhappy with his management and the financial performance of the CWA. The CWA is organized as a profit center in the NTN group.

The average ratings for "The Power Hour" are currently at an all-time high. The average number of viewers of "The Power Hour" per week has been:

19_2	19_3	19_4
1.6 million	2.1 million	3.1 million

Ross believes NTN would have to pay a fortune to buy a weekly television show with ratings equivalent to those of "The Power Hour." The market price per viewer that NTN was paying to independent television-production companies in the 19_2 to 19_4 period was:

19_2	19_3	19_4
$2.70	$3.20	$4.00

Ross also believes that his major competitor in the wrestling business made 80% of its operating income from syndicated television in 19_4, up from 25% in 19_2. The trend in the industry is that fewer people attend arena shows and more people view TV shows. Within the wrestling industry, Ross's leadership has been described as a "world championship performance." Unfortunately, the CWA bottom line does not back up this rating.

Required

1. Compute the operating income of the CWA each year in the 19_2, 19_3, and 19_4 period using:

   a. The 110% of actual production cost transfer-pricing method, and
   b. The market price transfer-pricing method.

   Comment on the results.

2. What are the limitations to NTN of the 110% of actual production cost transfer-pricing method for the television programs that CWA produces for NTN?

27-20 Pertinent transfer price. Europa Inc., has two divisions, A and B. For one of the company's products, Division A produces a major subassembly and Division B incorporates this subassembly into the final product. There is a market for both the subassembly and the final product. Each division has been designated as a profit center. The transfer price for the subassembly has been set at long-run average market price.

The following data are available to each division:

Estimated selling price for final product	$300
Long-run average selling price for intermediate product	200
Outlay cost for completion in Division B	150
Outlay cost in Division A	120

The manager of Division B has made the following calculation:

Selling price—final product		$300
Transferred-in cost (market)	$200	
Outlay cost for completion	150	350
Contribution (loss) on product		$ (50)

Required

1. Should transfers be made to Division B if there is no excess capacity in Division A? Is market price the correct transfer price?
2. Assume that Division A's maximum capacity for this product is 1,000 units per month and sales to the intermediate market are now 800 units. Should 200 units be transferred to Division B? At what transfer price? Assume for a variety of reasons that A will maintain the $200 selling-price indefinitely; that is, A is not considering cutting the price to outsiders even if idle capacity exists.

3. Suppose A quoted a transfer price of $150 for up to 200 units. What would be the contribution to the company as a whole if the transfer were made? As manager of B, would you be inclined to buy at $150?

27-21 Pricing in imperfect markets. Refer to Problem 27-20.

1. Suppose the manager of Division A has the option of (a) cutting the external price to $195 with the certainty that sales will rise to 1,000 units or (b) maintaining the outside price of $200 for the 800 units and transferring the 200 units to Division B at some price that would produce the same operating income for Division A. What transfer price would produce the same operating income for Division A? Does that price coincide with that produced by the "general rule" in the chapter, so that the desirable decision for the company as a whole would result?
2. Suppose that if the selling price for the intermediate product is dropped to $195, outside sales can be increased to 900 units. Division B wants to acquire as many as 200 units if the transfer price is acceptable. For simplicity assume that there is no outside market for the final 100 units of Division A capacity.
   a. Using the "general rule," what is (are) the transfer price(s) that should lead to the correct economic decision? Ignore performance evaluation considerations.
   b. Compare the total contributions under the alternatives to show why the transfer price(s) recommended lead(s) to the optimal economic decision.

27-22 Effect of alternative transfer-pricing methods on division operating income. Oceanic Products is a tuna-fishing company based in San Diego. It has three divisions:

1. Tuna Harvesting: operates a fleet of 20 trawling vessels
2. Tuna Processing: processes the raw tuna into tuna fillets
3. Tuna Marketing: packages tuna fillets in 2-pound packets that are sold to wholesale distributors at $12 each

The Tuna Processing Division has a yield of 500 pounds of processed tuna fillets from 1,000 pounds of raw tuna provided by the Tuna Harvesting Division. The Tuna Marketing Division has a yield of three hundred 2-pound packets from every 500 pounds of processed tuna fillets provided by the Tuna Processing Division. (The weight of the packaging material is included in the 2-pound weight.) Cost data for each division follow:

Tuna Harvesting Division	
Variable costs per pound of raw tuna	$0.20
Fixed costs per pound of raw tuna	$0.40
Tuna Processing Division	
Variable costs per pound of processed tuna	$0.80
Fixed costs per pound of processed tuna	$0.60
Tuna Marketing Division	
Variable costs per 2-pound packet	$0.30
Fixed costs per 2-pound packet	$0.70

Fixed costs per unit are based on the estimated volume of raw tuna, processed tuna, and 2-pound packets to be produced during the current fishing season.

Oceanic Products has chosen to process internally all raw tuna brought in by the Tuna Harvesting Division. Other tuna processors in San Diego purchase raw tuna from boat operators at $1 per pound. Oceanic Products has also chosen to process internally all tuna fillets into the 2-pound packets sold by the Tuna Marketing Division. Several fish-marketing companies in San Diego purchase tuna fillets at $5 per pound.

Required

1. Compute the overall operating income to Oceanic Products of harvesting 1,000 pounds of raw tuna, processing it into tuna fillets, and then selling it in 2-pound packets.
2. Compute the transfer prices that will be used for internal transfers (i) from the Tuna Harvesting Division to the Tuna Processing Division and (ii) from the Tuna Processing Division to the Tuna Marketing Division under each of the following transfer-pricing methods:

a. 200% of variable cost: Variable cost is the cost of the transferred-in product (if any) plus the division's own variable cost.
b. 150% of full cost: Full cost is the cost of the transferred-in product (if any) plus the division's own variable and fixed costs.
c. Market price.

3. Oceanic rewards each division manager with a bonus, calculated as 10% of division operating income (if positive). What is the amount of the bonus that will be paid to each division manager under each of the three transfer-pricing methods in requirement 2? Which transfer-pricing method will each division manager prefer to use?

27-23 Goal-congruence problems with cost-plus transfer-pricing methods, dual-pricing methods. (Extension of Problem 27-22) Assume that Oceanic Products uses a transfer price of 150% of full cost. Pat Forgione, the company president, attends a seminar on the virtues of decentralization. Forgione decides to implement decentralization at Oceanic Products. A memorandum is sent to all division managers: "Starting immediately, each division of Oceanic Products is free to make its own decisions regarding the purchase of its raw materials and the sale of its finished product."

Required

1. Give two examples of goal-congruence problems that may arise if Oceanic continues use of the 150% of full costs transfer-pricing method and a policy of decentralization is adopted.
2. Forgione is investigating whether a dual transfer-pricing policy will reduce goal-congruence problems at Oceanic Products. Transfers out of each selling division will be made at 150% of full cost; transfers into each buying division will be made at market price. Using this dual transfer-pricing policy, compute the operating income of each division for a harvest of 1,000 pounds of raw tuna that is further processed and marketed by Oceanic Products.
3. Compute the sum of the division operating incomes in requirement 2. Why might this sum not equal the overall corporate operating income from the harvesting of 1,000 pounds of raw tuna and its further processing and marketing?
4. What problems may arise if Oceanic Products uses the dual transfer-pricing system described in requirement 2?

27-24 Multinational transfer pricing, global tax minimization. Industrial Diamonds Inc., based in Los Angeles, has two divisions:

1. Philippine Mining Division—operates a mine containing a rich body of raw diamonds
2. U.S. Processing Division—processes the raw diamonds into polished diamonds used in industrial applications

The costs of the Philippine Mining Division are:

- Variable costs, 2,000 pesos per lb. of raw industrial diamonds
- Fixed costs, 4,000 pesos per lb. of raw industrial diamonds

Industrial Diamonds Inc. has a corporate policy of further processing in Los Angeles all raw diamonds mined in the Philippines. Several diamond-polishing companies in the Philippines buy raw diamonds from other local mining companies at 8,000 pesos per pound. Assume that the current foreign exchange rate is 20 pesos = $1 U.S.

The costs of the U.S. Processing Division are:

- Variable costs, $200 per lb. of polished industrial diamonds
- Fixed costs, $600 per lb. of polished industrial diamonds

Assume that it takes two pounds of raw industrial diamonds to yield one pound of polished industrial diamonds. Polished diamonds sell for $4,000 per pound.

Required

1. Compute the transfer price (in $U.S.) for one pound of raw industrial diamonds transferred from the Philippine Mining Division to the U.S. Processing Division under two methods: (a) 300% of full costs, (b) market price.

2. Assume a world of no income taxes. One thousand pounds of raw industrial diamonds are mined by the Philippine Division and then processed and sold by the U.S. Processing Division. Compute the operating income (in $U.S.) for each division of Industrial Diamonds Inc. under each transfer-pricing method in requirement 1.
3. Assume the corporate income tax rate is 20% in the Philippines and 35% in the United States. Compute the after-tax operating income (in $U.S.) for each division under each transfer-pricing method in requirement 1. (Income taxes are *not* included in the computation of the cost-based transfer price. Industrial Diamonds does not pay U.S. taxes on income already taxed in the Philippines.)
4. Which transfer-pricing method in requirement 1 will maximize the total after-tax net income of Industrial Diamonds?
5. What factors, in addition to global tax minimization, might Industrial Diamonds consider in choosing a transfer-pricing method for transfers between its two divisions?

27-25 Transfer pricing, goal congruence. (NAA, adapted) Frederica Thon, a management accountant, has recently been hired by International Traveler, Inc., a large firm that does business in both the United States and abroad. The company is organized on a divisional basis with considerable vertical integration. Thon is assigned as controller in the Cosmopolitan Division.

Cosmopolitan Division makes several luggage products, one of which is a slim leather portfolio, popular with executives. Sales of the case have been steady, and the marketing department expects continued strong demand. Thon is looking for ways that Cosmopolitan Division can contain its costs and thus boost its operating income from future sales. She discovered that Cosmopolitan Division has always purchased its supply of high-quality tanned leather from another division of International Traveler, the Vagabond Division. Vagabond Division has been providing the three square feet of tanned leather needed for each case for $9 per square foot.

Thon wondered whether Cosmopolitan might purchase leather of comparable quality from a supplier other than Vagabond at a lower price. Top management at International Traveler reluctantly agreed to allow Cosmopolitan Division to consider purchasing outside the company.

Cosmopolitan Division will need leather for 100,000 portfolios during the coming year. Cosmopolitan management has requested bids from several leather suppliers. The following have been received:

(1) Rover, Inc., submitted a bid of $8 per square foot of leather.
(2) Peltry Company offered to supply the leather for $7 per square foot. Thon has been informed that another subsidiary of International Traveler, Conundrum Chemical, supplies Peltry with chemicals that have become an essential ingredient in Peltry's leather preparation. Conundrum Chemical charges Peltry $2 for enough chemicals to prepare three square feet of leather. Conundrum's operating income margin is 30% of sales.
(3) Finally, Vagabond Division of International Traveler has put in writing its desire to supply Cosmopolitan's leather needs. It will continue to charge $9 per square foot as it has done in the past. Bob Trout, Vagabond Division's controller, has made it clear that he believes Cosmopolitan should continue to purchase all its needs from Vagabond. Vagabond's operating income margin is 40% of sales.

You, as International Traveler's vice-president of finance, have called a meeting of the two controllers, Frederica Thon for Cosmopolitan and Bob Trout for Vagabond. Thon is eager to accept Peltry's bid of $7. She points out that Cosmopolitan's operating income will show a significant increase if the division can buy from Peltry.

Trout, however, wants International to keep the business within the company and suggests that you, as International Traveler's vice-president, require that Cosmopolitan purchase its leather from Vagabond. He emphasizes that Vagabond's operating income on the leather should not be lost to the company.

Required

1. What is the cost per square foot of leather for (a) Cosmopolitan Division and (b) International Traveler, Inc., in purchasing the leather from (i) Rover, Inc., (ii) Peltry Company, and (iii) Vagabond Division of International Traveler, Inc.?

2. How should you, as vice-president of finance, resolve the dispute between Thon and Trout?

27-26 Paper company. (Copyright 1986 by the President and Fellows of Harvard College. Reproduced by permission.) "If I were to price these boxes any lower than $480 a thousand," said Mr. Brunner, manager of Birch Paper Company's Thompson division, "I'd be countermanding my order of last month for our sales representatives to stop shaving their bids and to bid full-cost quotations. I've been trying for weeks to improve the quality of our business, and if I turn around now and accept this job at $430 or $450 or something less than $480 I'll be tearing down this program I've been working so hard to build up. The division can't very well show a profit by putting in bids which don't even cover a fair share of overhead costs, let alone give us a profit."

Birch Paper Company was a medium-sized, partly integrated paper company, producing white and kraft papers and paperboard. A portion of its paperboard output was converted into corrugated boxes by the Thompson division, which also printed and colored the outside surface of the boxes. Including Thompson, the company had four producing divisions and a timberland division, which supplied part of the company's pulp requirements.

For several years each division had been judged independently on the basis of its profit and return on investment. Top management had been working to gain effective results from a policy of decentralizing responsibility and authority for all decisions but those relating to overall company policy. The company's top officials felt that in the past few years the concept of decentralization had been successfully applied and that the company's profits and competitive position had definitely improved.

In early 19_7 the Northern Division designed a special display box for one of its papers in conjunction with the Thompson Division, which was equipped to make the box. Thompson's package design and development staff spent several months perfecting the design, production methods, and materials that were to be used. Because of the unusual color and shape these were far from standard. According to an agreement between the two divisions, the Thompson Division was reimbursed by the Northern Division for the cost of its design and development work.

When the specifications were all prepared, the Northern Division asked for bids on the box from the Thompson Division and from two outside companies. Each division manager was normally free to buy from whichever supplier he wished, and, even on sales within the company, divisions were expected to meet the going market price if they wanted the business.

In early 19_7 the profit margins of converters such as the Thompson Division were being squeezed. Thompson, as did many other similar converters, bought its board, liner, or paper and its function was to print, cut, and shape it into boxes. Though it bought most of its materials from other Birch divisions, most of Thompson's sales were to outside customers. If Thompson got the business, it would probably buy the linerboard and corrugating medium from the Southern Division of Birch. The walls of a corrugated box consist of outside and inside sheets of linerboard sandwiching the fluted corrugating medium. About 70% of Thompson's out-of-pocket cost of $400 represented the cost of linerboard and corrugating medium. Though Southern had been running below capacity and had excess inventory, it quoted the market price, which had not noticeably weakened as a result of the oversupply. Its out-of-pocket costs on both liner and corrugating medium were about 60% of the selling price.

The Northern Division received bids on the boxes of $480 a thousand from the Thompson Division, $430 a thousand from West Paper Company, and $432 a thousand from Erie Papers, Ltd. Erie Papers offered to buy from Birch the outside linerboard with the special printing already on it but would supply its own inside liner and corrugating medium. The outside liner would be supplied by the Southern Division at a price equivalent of $90 per thousand boxes and would be printed for $30 a thousand by the Thompson Division. Of the $30, about $25 would be out-of-pocket costs.

Since this situation appeared a little unusual, William Kenton, manager of the Northern Division, discussed the wide discrepancy of bids with Birch's commercial vice-president. He told the vice-president: "We sell in a very competitive market, where higher costs cannot be passed on. How can we be expected to show a decent profit and return on investment if we have to buy our supplies at more than 10% over the going market?"

Knowing that Mr. Brunner had on occasion in the past few months been unable to operate the Thompson Division at capacity, the vice-president found it odd that Mr. Brunner would add the full 20% overhead and profit charge to his out-of-pocket costs. When asked about this, Mr. Brunner's answer was the statement that appears at the beginning of this problem. He went on to say that having done the developmental work on the box, and having received no profit on that, he felt entitled to a good markup on the production of the box itself.

The vice-president explored further the cost structures of the various divisions. He remembered a comment that the controller had made at a meeting the week before to the effect that costs that were variable for one division could be largely fixed for the company as a whole. He knew that in the absence of specific orders from top management, Mr. Kenton would accept the lowest bid, which was that of the West Paper Company for $430. However, it would be possible for top management to order the acceptance of another bid if the situation warranted such action. And though the volume represented by the transactions in question was less than 5% of the volume of any of the divisions involved, other transactions could conceivably raise similar problems later.

Required

1. In the controversy described, which alternative seems best for the company as a whole? Prepare an analysis of the cash flows under each alternative.
2. As the commercial vice-president, what action would you take?

27-27 Transfer pricing for an automotive dealership, goal congruence. European Motors is the largest Mercedes-Benz dealership in the Midwest. The company was started by Stephanie Mortimer, Les Johns, and Kevin Ryan. Mortimer is president. For the first ten years, the company operated with only two divisions:

1. Car Sales Division, with Les Johns as manager. Mercedes-Benz cars were purchased from a German distributor who delivered the cars to the dealership. An independent finance company provided car loans to new car customers. Cars traded in by new car purchasers were sold by an independent auction company for a 10% commission.
2. Car Service Division, with Kevin Ryan as manager. Approximately 40% of the work done in this division was on Mercedes-Benz cars previously sold by the Car Sales Division of European Motors.

The organization operated smoothly. Mortimer stipulated that any work done by the Car Service Division for warranties on cars sold by the Car Sales Division be charged out to the Car Sales Division at 200% of actual cost, the normal charge for work done for external customers. Her rationale was to provide the Car Service Division with the incentive to maintain a high quality of service for the car-purchasing customers of European Motors.

European Motors recently expanded its activities to include the sales of used cars and the provision of financing. The new organization structure was:

1. New Car Sales Division—Les Johns, manager
2. Used Car Sales Division—Terry Lamb, manager (new hire)
3. Car Service Division—Kevin Ryan, manager
4. Car Financing Division—Georgianna Peponis, manager (new hire)

Mortimer ordered that all trade-ins be first worked on by the Car Service Division and then sold by the Used Car Sales Division.

Mortimer labeled each of the four divisions as profit centers. Each division manager was paid a salary plus a bonus equal to 20% of the division operating income. Used cars traded in to the New Car Sales Division were transferred to the Used Car Sales Division at the trade-in price negotiated by the New Car Sales Division; Johns had the final approval on each trade-in price. Work done by the Car Service Division on traded-in used cars before their sale by the Used Car Division was at 200% of actual costs, the normal charge for external work; Ryan had the final approval on the amount of service work done on each trade-in. The Car Financing Division borrowed money from a bank at 14% per year. The Car Financing Division was credited with the difference between the actual rate charged to a customer and the 14% cost of money. The target spread between the rate charged to a customer and the cost of money is 4%.

Problems arose quickly with the current transfer-pricing methods in the new organization structure chosen by Mortimer. Lamb and Peponis used a set of related transactions to illustrate the source of their frustration. The New Car Division recently sold a Mercedes-Benz 190. The list price was $30,000, and the cost from the German distributor was $18,000. A five-year-old Ford Supreme was traded in by the purchaser of the Mercedes 190. The "blue-book" price (an approximation of the market price) for dealers buying a five-year-old Ford Supreme was $12,000.

Sale of Mercedes 190 by New Car Division

1. Sales price of Mercedes 190, $29,500
2. Trade-in price on Ford Supreme, $15,000
3. Cash down payment, $4,500; and a $10,000 one-year loan at 12% per year

Work Done by Car Service Division on Ford Supreme

1. General maintenance, $800 (200% of actual costs of $400)
2. Repair of transmission, $700 (200% of actual costs of $350)

Sale of Ford Supreme by Used Car Division

1. Sale price of Ford Supreme, $16,000
2. Cash down payment, $8,000; and an $8,000 one-year loan at 18% per year

No car was traded in by the buyer of the Ford Supreme.

Lamb argued that the Car Service Division spent more hours than was standard on the Ford Supreme because of their having limited experience with work on non-Mercedes-Benz cars. He also questioned whether the high-quality maintenance and service work approved by Ryan was necessary for the sale of a used car, especially a non-Mercedes-Benz used car. Peponis was livid at Johns' approving financing at 12% per year on the Mercedes 190 $10,000 loan.

Required

1. Compute the operating income of European Motors on the combined sales of the Mercedes-Benz and the Ford Supreme.
2. Compute the individual operating income for *each* of the four divisions on the combined sales of the Mercedes-Benz and the Ford Supreme.
3. What problems may arise with the current transfer-pricing methods in the new organization structure chosen by Mortimer?
4. Assume that Mortimer wishes to implement a policy of decentralization at the division level. Recommend an alternative transfer-pricing system for European Motors.

27-28 Transfer pricing, value-chain cost structure differences across divisions. Global Forest Products (GFP) is a large vertically integrated paper company with operations in twenty different countries. It is one of the three largest forest product companies in the world. GFP recently (March 31, 19_5) acquired Forestry Software (FS), a company developing and selling software programs that operate paper and pulp mills. At the time of the acquisition, GFP promised to retain FS as a separate division that was to continue selling software to other (non-GFP) companies.

FS kept detailed records on a customer-by-customer basis because much of its work was tailored to individual clients. The percentage of FS sales made to GFP in the most recent four years was:

19_1	28%
19_2	30%
19_3	31%
19_4	34%

Revenues and costs associated with sales by FS to GFP over the 19_1 to 19_4 period were (in millions):

	19_1	19_2	19_3	19_4
Revenues	$5.8	$7.0	$8.8	$12.2
Costs:				
Research and development and product design	1.8	2.0	2.4	3.5
Manufacturing	1.3	1.6	1.8	2.8
Marketing, distribution, and customer service	2.2	2.7	3.4	4.2
Full cost	$5.3	$6.3	$7.6	$10.5
Operating income	$0.5	$0.7	$1.2	$ 1.7

Sales between GFP and FS in this four-year period were made at market prices. One reason for FS's growth has been its reputation for both innovative product development and superb customer service.

GFP has a longstanding policy that all internal transfers between its divisions be priced at 140% of actual manufacturing costs (both variable and fixed). Prior to the FS acquisition, all sales between GFP divisions were lumber, pulp, and paper products. (Manufacturing costs in several of the GFP pulp and paper divisions were as high as 90% of the total divisional costs.) After the acquisition, GFP management mandated that this transfer-pricing method would also apply to sales between FS and GFP. GFP divisions, however, were given the option to purchase software from any vendor (rather than from FS only).

Sally Larcker, president and sole owner of FS at the time of the acquisition, received two forms of compensation for relinquishing her 100% ownership:

(a) $17 million in cash and
(b) 25% of operating income of FS for each year of the five years after the acquisition (19_5 to 19_9).

The (b) component of the purchase price is called an earn-out provision.

Key personnel at FS received annual bonuses during 19_1 to 19_4 based on FS's annual operating income. GFP has indicated that these key personnel will retain the same bonus plan for at least 19_5 and 19_6.

Larcker is opposed to the 140% of actual manufacturing cost transfer price between FS and GFP mandated by GFP management. She proposes that all sales be priced at 120% of actual full cost (research and development, product design, manufacturing, marketing, distribution, and customer service); 20% is Larcker's estimate of FS's 19_5 ratio of operating income to revenue if FS had remained an independent company.

Required

1. Compare the operating income for the sales between FS and GFP for each year of the 19_1 to 19_4 period if each of the following transfer-pricing methods had been used:
   (a) 140% of actual manufacturing cost
   (b) 120% of actual full cost
   (c) market price
2. Why might Larcker be opposed to the GFP transfer-pricing method? (Assume that the costs of FS reported in this problem would have been the same if GFP had owned FS since 19_1.)
3. Discuss three problems that may arise at FS if GFP mandates that FS price all internal sales to GFP divisions at 140% of actual manufacturing cost.
4. From GFP's perspective, what are two limitations of the 120% of actual full-cost transfer-pricing method proposed by Larcker?
5. Recommend a transfer-pricing method for use when internal sales are made from FS to GFP. GFP wants to maximize its own overall operating income but also wants to keep incentives high at FS for both new product development and customer service.

CHAPTER 28

Unit managers at Holiday Inns are evaluated on the basis of both financial and nonfinancial performance measures. Nonfinancial measures include attractiveness of front foyer and friendliness of front-desk registration personnel.

Systems Choice: Performance Measurement and Executive Compensation

Financial and nonfinancial measures

Designing an accounting-based performance measure

Different performance measures

Alternative definitions of investment

Measurement alternatives for assets

Goal congruence and performance measures

Nonfinancial performance measures

Executive compensation plans

Ethics and cooking the books

Learning Objectives

When you have finished studying this chapter, you should be able to

1. Provide examples of financial and nonfinancial performance measures
2. Describe the steps in designing an accounting-based performance measure
3. Understand the Du Pont method of profitability analysis
4. Describe the motivation for residual income
5. Identify varieties of financial performance measures used in practice
6. Distinguish between present value, current cost, and historical cost asset-measurement methods
7. Explain goal-congruence problems arising from accrual accounting performance measures
8. Describe how accounting numbers are central to many executive compensation plans
9. Provide examples of management's cooking the books to maximize short-run bonuses

Performance measures are a central component of a management control system. This chapter examines issues in designing performance measures for different levels of an organization and for managers at these different levels. We discuss financial and nonfinancial performance measures.

Performance measurement of an organization's units should be a prerequisite for allocating resources within that organization. When a unit undertakes new activities, projections of revenues, costs, and investments are made. An ongoing comparison of the actual revenues, costs, and investments with the budgeted amounts can help guide top management's decisions about future allocations.

Performance measurement of managers is used in decisions about their salaries, bonuses, future assignments, and status. Moreover, the very act of measuring their performance can motivate managers to strive for the goals used in their evaluation.

FINANCIAL AND NONFINANCIAL PERFORMANCE MEASURES

Chapter 13 discussed four levels at which data can be gathered in a management control system:

1. Customer/market level
2. Total organization level
3. Individual facility level
4. Individual activity level

Exhibit 28-1 presents examples of financial and nonfinancial performance measures at each of these four levels. These four levels illustrate differences in both the source of the performance-measure data and their time-horizon.

Objective 1

Provide examples of financial and nonfinancial performance measures

Source of the data Many performance measures (such as return on investment, or ROI) are reported by the internal accounting system. Other performance measures (such as throughput time) are captured in a firm's manufacturing data base, customer-service data base, or the like.

EXHIBIT 28-1
Examples of Financial and Nonfinancial Performance Measures

Representative Area at Which Data Gathered	Financial Measures	Nonfinancial Measures
Customer/Market Level	Prices of company's products compared with competitors Prices of company's publicly traded securities	Market share held by company's products Third-party quality ratings for all products in the industry
Total-Organization Level	Return on investment Residual income Return on sales	Number of new products introduced Number of new patents filed
Individual-Facility Level	Return on investment Flexible budget variances	Throughput time for products Percentage of times promised delivery dates met (schedule attainment)
Individual-Activity Level	Direct materials variances and direct labor variances Manufacturing overhead variances	Time taken to set up machinery for new production run Number of accounts receivable processed per hour

Time horizon Some performance measures, such as the number of new patents developed, have a long-run time horizon. Other measures, such as a direct materials efficiency variance, have a short-run time horizon.

We now examine accounting-based measures used to evaluate performance over an intermediate- to long-run time horizon. The discussion concentrates on performance measures at the total organization level and at the division level, which may have multiple individual facilities. Chapters 7 and 8 discuss accounting performance measures with a short-run time horizon (such as direct materials and direct labor variances).

DESIGNING AN ACCOUNTING-BASED PERFORMANCE MEASURE

Objective 2

Describe the steps in designing an accounting-based performance measure

Designing an accounting-based performance measure requires the following steps:

Step 1 Choosing a variable(s) that represents top management's financial goal(s). Does operating income, net income, return on investment, or revenues, for example, best measure a division's performance?

Step 2 Choosing definitions of key items included in the variable(s) in step 1 (such as operating income, net income, and investment). For example, should investment be defined as total assets or total assets minus liabilities?

Step 3 Choosing measures for key items included in the variable(s) in step 1. For example, should assets be measured at historical cost? current cost? present value?

Step 4 Choosing a benchmark against which to gauge performance. For example, should all divisions have as a benchmark the same required rate of return on investment?

Step 5 Choosing the timing of feedback. Should reports on the performance of a division be sent to corporate headquarters weekly? monthly? quarterly?

We discuss the first three steps in subsequent sections of this chapter. The fourth step is discussed in Chapter 22 (p. 720), where we point out that different required rates of return should be used for different divisions. The fifth step, the timing of feedback, has been discussed in many places in this book. Timing depends largely on the specific level of management that is receiving the feedback and on the sophistication of the organization's information technology.

These five steps need not be taken sequentially. The issues considered in each step are interdependent, and a decision maker will often proceed through these steps several times before deciding on an accounting-based performance measure(s). The answers to the questions raised at each step depend on top management's beliefs about how each alternative fulfills, in a cost-effective manner, the behavioral criteria of goal congruence, managerial effort, and subunit autonomy discussed in Chapter 27 (p. 856–57).

DIFFERENT PERFORMANCE MEASURES

Hospitality Inns owns and operates three motels, one each in San Francisco, Chicago, and New Orleans. Exhibit 28-2 summarizes data for each of the three motels and for corporate headquarters (located in San Francisco) for the most recent year (19_8). At present, Hospitality Inns does not allocate to the three separate motels the operating costs and assets of corporate headquarters or the total long-term debt of the company. Exhibit 28-2 indicates that the New Orleans motel generates the highest operating income—$480,000. The Chicago motel generates $300,000; the San Francisco motel, $240,000. But is this comparison appropriate? Is the New Orleans motel the most "successful"? Actually, the comparison of operating income ignores potential differences in the size of the investment that corporate headquarters has made in the different motels.

Two approaches incorporate the amount of invested capital into a performance measure: return on investment (ROI) and residual income.

Return on Investment

Return on investment (ROI) is income divided by investment.

$$\text{Return on investment} = \frac{\text{Income}}{\text{Investment}}$$

ROI is the most popular approach to incorporating the investment base into a performance measure. ROI appeals conceptually because it blends all the major ingredients of profitability (revenues, costs, and investment) into a single number. ROI can be compared with opportunities elsewhere, inside or outside the company. Like any single performance measure, however, ROI should be used cautiously and with other performance measures.

ROI is often called the accounting rate of return or the accrual accounting rate of return. Companies vary in how the numerator and the denominator in the ROI are calculated. For example, some firms use operating income for the numerator. Other firms use net income.

EXHIBIT 28-2
Hospitality Inns: Annual Financial Data for 19_8 ($000s)

	San Francisco Motel	Chicago Motel	New Orleans Motel	Corporate Head-Quarters	Total
	(1)	(2)	(3)	(4)	(1) + (2) + (3) + (4)
Motel revenues (sales)	$1,200	$2,000	$3,000	—	$6,200
Motel variable costs	540	750	840	—	2,130
Motel fixed costs	420	950	1,680	—	3,050
Motel operating income	$ 240	$ 300	$ 480	—	$1,020
Central corporate variable costs	—	—	—	$ 80	80
Central corporate fixed costs	—	—	—	120	120
Interest on long-term debt	—	—	—	400	400
Income before income taxes	—	—	—	—	$ 420
Income taxes	—	—	—	—	150
Net income	—	—	—	—	$ 270
Average Book Values for 19_8:					
Current assets	$ 400	$ 500	$ 600	$ 200	$1,700
Fixed assets	600	1,500	2,400	300	4,800
Total assets	$1,000	$2,000	$3,000	$ 500	$6,500
Current liabilities	$ 230	$ 320	$ 350	$ 100	$1,000
Long-term liabilities	—	—	—	4,000	4,000
Stockholders' equity	—	—	—	—	1,500
Total					$6,500

Objective 3

Understand the Du Pont method of profitability analysis

ROI can often provide more insight into performance when it is divided into the following components:

$$\begin{pmatrix}\text{Investment}\\ \text{turnover}\end{pmatrix} \times \begin{pmatrix}\text{Income to revenue}\\ \text{(sales) ratio}\end{pmatrix} = \text{Return on investment}$$

$$\frac{\text{Revenue}}{\text{Investment}} \times \frac{\text{Income}}{\text{Revenue}} = \frac{\text{Income}}{\text{Investment}}$$

This approach is widely known as the Du Pont method of profitability analysis. The components of the Du Pont method lead to the following generalizations: ROI is increased by any action that:

1. increases revenue,
2. decreases costs, or
3. decreases investment

while holding the other two factors constant. Put another way, there are two basic ingredients in profit making: investment turnover and income margin. An improvement in either ingredient without changing the other increases return on investment.

Consider the ROI of each of the three Hospitality Inns motels in Exhibit 28-2. Our definitions here are operating income of each motel for the numerator and total assets of each motel for the denominator:

Motel	Operating Income	÷	Total Assets	=	ROI
San Francisco	$240	÷	$1,000	=	24%
Chicago	300	÷	2,000	=	15%
New Orleans	480	÷	3,000	=	16%

Using these ROI figures, the San Francisco motel appears to make the best use of its total assets.

Assume that the top management at Hospitality Inns adopts a 30% target ROI for the San Francisco motel. How can this return be attained? The Du Pont method illustrates the present situation and three alternatives (in thousands):

	Revenues / Total Assets	×	Operating Income / Revenues	=	Operating Income / Total Assets
Present Situation	$1,200 / $1,000	×	$240 / $1,200	=	$240 / $1,000, or 24%
Alternatives					
A. Increase operating income by increasing revenues	$1,500 / $1,000	×	$300 / $1,500	=	$300 / $1,000, or 30%
B. Increase operating income by decreasing costs	$1,200 / $1,000	×	$300 / $1,200	=	$300 / $1,000, or 30%
C. Decrease total assets	$1,200 / $800	×	$240 / $1,200	=	$240 / $800, or 30%

Alternative A can be achieved by increasing the selling price per unit (the room rate) or by increasing the number of units sold (rooms rented). Alternative B demonstrates the cost-reduction approach to improving performance. Alternative C shows that reducing total assets (such as accounts receivable or inventories) improves ROI.

ROI highlights the benefits that managers can obtain by reducing their investments in current or fixed assets. Some managers are conscious of the need to boost revenues or to control costs but give less attention to reducing their investment base. Investments in cash, accounts receivable, inventory, and fixed assets should be handled carefully at all levels of performance. This approach means investing idle cash, managing credit judiciously, determining proper inventory levels, and spending carefully on fixed assets.

Residual Income

Objective 4

Describe the motivation for residual income

Residual income is income minus an imputed interest charge for the investment base.

Residual income = Income − Imputed interest charge for investment

Assume that Hospitality Inns defines residual income for each motel as motel operating income minus an imputed interest charge of 10% of the total assets of the motel:

Motel	Operating Income	−	Imputed Interest Charge	=	Residual Income
San Francisco	$240	−	$100 (10% of $1,000)	=	$140
Chicago	$300	−	$200 (10% of $2,000)	=	$100
New Orleans	$480	−	$300 (10% of $3,000)	=	$180

Given the 10% imputed interest charge, the New Orleans motel is performing best in terms of residual income.

The objective of maximizing residual income assumes that as long as the division earns a rate in excess of the imputed charge for invested capital, the division should expand. Some firms favor the residual-income approach because managers will concentrate on maximizing an absolute amount (dollars of residual income) rather than a percentage (return on investment).

The objective of maximizing ROI may induce managers of highly profitable divisions to reject projects that, from the viewpoint of the organization as a whole, should be accepted. Assume that Hospitality Inns' required rate of return on investment is 10%. Assume also that an expansion of the San Francisco motel will increase its operating income by $160,000 and increase its total assets by $800,000. The ROI for the expansion is 20% ($160,000 ÷ $800,000), which makes it attractive to Hospitality Inns. By making this expansion, however, the San Francisco manager will see that motel's ROI decrease (in thousands):

$$\text{Preexpansion ROI} = \frac{\$240}{\$1,000} = 24\%$$

$$\text{Postexpansion ROI} = \frac{(\$240 + \$160)}{(\$1,000 + \$800)} = \frac{\$400}{\$1,800} = 22.2\%$$

The annual bonus paid to the San Francisco manager may decrease if ROI is a key factor in the bonus calculation and the expansion option is selected. In contrast, if the annual bonus is a function of residual income, the San Francisco manager will view the expansion favorably:

$$\text{Preexpansion residual income} = \$240 - (10\% \times \$1,000) = \$140$$

$$\text{Postexpansion residual income} = \$400 - (10\% \times \$1,800) = \$220$$

Goal congruence is more likely to be promoted by using residual income rather than ROI as a division manager's performance measure.

Both ROI and residual income represent the results for a single time period (such as twelve months). Managers can take actions that cause short-run increases in ROI or residual income but are in conflict with the long-run interests of the organization. For example, research and development and plant maintenance can be curtailed in the last three months of a fiscal year to achieve a target level of annual operating income.

Return on Sales

The income to revenue (sales) ratio—abbreviated as ROS—is a frequently used key performance measure. The ROS of each motel in the Hospitality Inns chain is:

Motel	Operating Income	÷	Revenues (Sales)	=	ROS
San Francisco	$240	÷	$1,200	=	20%
Chicago	$300	÷	$2,000	=	15%
New Orleans	$480	÷	$3,000	=	16%

The San Francisco motel has a higher ROS than either the Chicago or the New Orleans motel.

ROS performance measures are frequently emphasized by companies with relatively low levels of investment in plant, equipment, and other fixed assets. For example, the major Japanese trading companies, such as Mitsubishi, place greater weight on ROS performance measures than on ROI. These trading companies have annual revenues many times larger than their invested asset base.[1]

Distinction Between Managers and Organization Units

As noted in several chapters, performance evaluation of a manager should be distinguished from performance evaluation of an organization unit, such as a division of the company. The most skillful division manager is often put in charge of the sickest division in an attempt to change its fortunes. Such an attempt may take years, not months. Furthermore, the manager's efforts may merely result in bringing the division up to a minimum acceptable ROI. The division may continue to be a poor profit performer in comparison with other divisions. If top management relied solely on the absolute ROI to judge management, the skillful manager would be foolish to accept such a trouble-shooting assignment. One chairperson expressed the importance of distinguishing between managers and organization units as follows: "In some of our businesses, managers fight unavoidable head winds; in others, they enjoy tail winds not of their own making."

Surveys of Company Practice

Objective 5

Identify varieties of financial performance measures used in practice

Companies vary in the performance measures they emphasize when evaluating their business units. Exhibit 28-3 presents the key financial performance measures used by eight different companies. Note the diversity in the use of absolute dollar amounts and ROS, ROA, and ROI.

Many companies do not fully allocate all corporate costs to business units. In a survey of one hundred multinational companies, only 16.2% of the respondents allocated corporate income taxes, 29.5% allocated interest on debt, and 30.5% allocated headquarters' administrative costs to business units.[2]

---

[1]See M. Sakurai, L. Killough, and R. Brown, "Performance Measurement Techniques and Goal Setting: A Comparison of U.S. and Japanese Practices," in Y. Monden and M. Sakurai ed., *Japanese Management Accounting* (Cambridge, MA: Productivity Press, 1989).

[2]Business International Corporation, *Evaluating the Performance of International Operations* (New York, 1989).

EXHIBIT 28-3

Company Examples of Key Financial Performance Measures

Company Name	Country Headquarters	Product	Key Financial Performance Measures
Dow Chemical	U.S.	Chemicals	Absolute income (profit)
Ford Motor	U.S.	Automotive	ROS and return on assets (ROA)
Guinness	U.K.	Consumer products	Absolute income and ROS
Krones	Germany	Machinery/ Equipment	Absolute sales and income
Mayne Nickless	Australia	Security/ Transportation	ROI and income to sales ratio
Mitsui	Japan	Trading	Absolute sales and income
Pirelli	Italy	Tires/ Manufacturing	Absolute income and cash flow
Swedish Match	Sweden	Consumer products	ROA

SOURCE: Business International Corporation, *Evaluating the Performance of International Operations* (New York, 1989).

ALTERNATIVE DEFINITIONS OF INVESTMENT

Different companies have alternative definitions of investment. The definitions include the following:

1. *Total assets available.* Includes all business assets, regardless of their particular purpose.
2. *Total assets employed.* Defined as total assets available minus idle assets and minus assets purchased for future expansion. For example, if the New Orleans motel in Exhibit 28-2 has unused land set aside for potential expansion, the total assets employed by the motel would exclude the book value of that land.
3. *Working capital (current assets minus current liabilities) plus other assets.* This definition excludes that portion of current assets financed by short-term creditors.
4. *Stockholders' equity.* Use of this definition for each individual motel in Figure 28-2 requires allocation of the long-term liabilities of Hospitality Inns to the three motels, which would then be deducted from the total assets of each motel.

Most firms that employ ROI for performance measurement use either total assets available or working capital plus other assets as the definition of the ROI denominator. However, when top management directs a division manager to carry extra assets, total assets employed can be more informative than total assets available. The most common rationale for using working capital plus other assets is that the division manager often influences decisions on the short-term debt utilized by the division.

Companies need not use the same investment definition for both evaluating the performance of divisions and evaluating the performance of division managers. For evaluating the performance of divisions, the aim is to identify all those assets of the organization (be they division assets or corporate assets) that are attributable to the operation of individual divisions. For evaluating the performance of division managers, many firms are concerned that divisional managers view the chosen definitions as fair.

MEASUREMENT ALTERNATIVES FOR ASSETS

Objective 6

Distinguish between present value, current cost, and historical cost asset-measurement methods

How should the assets included in the investment base be measured? At present value, current cost, current disposal price, or historical cost? Should gross book value or net book value be used for depreciable assets?

Present Value

Chapters 21 and 22 discuss the relevance of discounted cash-flow (DCF) analysis for both asset acquisition and disposal decisions. **Present value** is the asset measure based on DCF estimates. Consider an existing motel with an expected useful life of ten years, expected net cash inflows of $1,200,000 each year, and an expected terminal disposal price of $2,000,000. The required rate of return is 12%. The present value of the motel would be $7.424 million:

Present value of annuity of $1,200,000 for ten years discounted at 12%, $1,200,000 × 5.650	$6,780,000
Present value of $2,000,000 ten years hence discounted at 12%, $2,000,000 × 0.322	644,000
Present value of motel	$7,424,000

For new asset acquisition decisions, the present value of the cash inflows should be compared with the present value of the cash outflows associated with the investment. Assets with a zero or positive net present value should be purchased. For decisions concerning the disposal of assets currently held, the present value of cash inflows from retaining the asset (including its future disposal price) should be compared with the current disposal price of the asset. The very act of continuing to hold an asset means that the decision to hold on to the asset or to dispose of the asset has been made (either explicitly or implicitly).

The relevance of discounted cash-flow analysis to asset acquisition and disposal decisions has led to proposals to use total present value as the investment measure for evaluating the performance of organization units. Annual income under this proposal is the difference between total present values at the beginning and at the end of a year (assuming no cash dividends or additional investments). This present-value-based income number is affected by more than just current operating activities. Income depends on a measurement of total present values at the end of the year, so it is affected also by events influencing future cash flows (such as new inventions, the entry of new competitors, and changes in government regulations) and by changes in the required rate of return on investments.

Several firms (Pirelli in Italy, Cadbury-Schweppes in the United Kingdom, and Westinghouse Electric Corporation in the United States, for example) are experimenting with the use of present value asset valuations in evaluating individual division performance. These firms recognize that the cash-flow estimates underlying the individual asset values are subjective. However, they argue that the sizable problems in implementing the approach should not draw top management's attention from the relevance of information about changes in total present values.

When firms do not systematically incorporate total present value (and disposal price) information into their routine accounting reports, they should make periodic attempts to approximate present values and current disposal prices in judging the desirability of investment in assets. Otherwise managers may overlook investment opportunities.

Current Cost

Current cost is the cost today of purchasing an asset currently held if an identical asset can currently be purchased; it is the cost of purchasing the services provided by that asset if an identical asset cannot currently be purchased. Consider the following comment by American Standard, a manufacturer of household fixtures and an enthusiastic advocate of current-cost measures when evaluating the performance of both its activities and its managers:

> [For some years] American Standard has been modifying its financial measurements to elminate the distortions and inequities of historic cost accounting. [Under historical cost accounting] a company will find that some of its divisions have higher return on book investment than others. This performance criterion is one factor considered in allocating capital resources. A considerable amount of the difference in returns on book assets between units, however, can just be a reflection of when a unit acquired its fixed assets. The unit that acquired its assets some time ago at deflated dollars in today's terms can look very good compared with a unit that acquired its fixed assets recently. It may be that the converse relation is actually true in terms of economic returns on new investments. Then too, the units with high returns may be inhibited from making necessary fresh investments because the investments will depress the historical return on book assets. The internal accounting system should have correct incentives for management to make economic capital investment decisions. By trying to maintain unrealistically high book returns, management can be inhibited from investing in the business. Book returns [under historical cost accounting] can lead management into a trap of "milking the busi-

ness" and shying away from necessary replacements with new and better equipment and facilities.[3]

The American Standard argument can be illustrated using the Hospitality Inns example discussed earlier in this chapter (see Exhibit 28-2). Assume the following information about the fixed assets of each motel (dollar amounts are in thousands):

	San Francisco	Chicago	New Orleans
Age of facility (at end of 19_8)	8 years	4 years	2 years
Gross book value	$1,400	$2,100	$2,800
Accumulated depreciation	$ 800	$ 600	$ 400
Net book value (at end of 19_8)	$ 600	$1,500	$2,400
Depreciation for 19_8	$ 100	$ 150	$ 200

Hospitality Inns assumes a fourteen-year useful life, assumes no terminal disposal price for the physical facilities, and calculates depreciation on a straight-line basis.

An index of construction costs for the eight-year period that Hospitality Inns has been operating (19_0 year-end = 100) is as follows:

Year	19_1	19_2	19_3	19_4	19_5	19_6	19_7	19_8
Construction cost index	110	122	136	144	152	160	174	180

Earlier in this chapter we computed an ROI of 24% for San Francisco, 15% for Chicago, and 16% for New Orleans (see p. 885). One possible explanation of the high ROI for San Francisco is that this motel's fixed assets are expressed in terms of 19_0 construction price levels and that the fixed assets for the Chicago and New Orleans motels are expressed in terms of more recent construction price levels.

Exhibit 28-4 illustrates a step-by-step approach for incorporating current-cost estimates for fixed assets and depreciation into the ROI calculation. The current-cost adjustment dramatically reduces the ROI of the San Francisco motel.

	ROI: Historical Cost	ROI: Current Cost
San Francisco	24%	10.81%
Chicago	15%	11.05%
New Orleans	16%	13.79%

Thus, the 24% ROI of the San Francisco motel can give a highly misleading picture of the returns that Hospitality Inns should expect from subsequent investments in motel facilities.

Obtaining current-cost estimates for some assets can be difficult. The exact assets currently held by an organization may no longer be traded or manufactured. When making current-cost estimates, the company should focus on the expected cash flows from the assets held, not their precise physical or technological features. The aim is to approximate how much it would cost to obtain similar assets that would produce the same expected operating cash inflows as do the assets currently held. Managers rarely replace assets with new assets having identical operating and economic characteristics.

General Price-Level Adjustments

Some companies are exploring the use of historical cost data adjusted for general price-level changes. This approach entails restating *historical* data in terms of current general purchasing power by using a general index, such as a consumer price

[3]This extract is from an internal document ("Management Uses of Inflation-Adjusted Accounting Data: The American Standard Approach") by K. Todd.

EXHIBIT 28-4

Hospitality Inns: ROI Computed Using Current-Cost Estimates as of the end of 19_8 for Depreciation and Fixed Assets

STEP ONE: Restate fixed assets from gross book value at historical cost to current cost as of the end of 19_8.

$$\text{Gross Book Value at Historical Cost} \times \frac{\text{Construction Cost Index in 19\_8}}{\text{Construction Cost Index in Year of Construction}}$$

San Francisco: $1,400 × (180/100) = $2,520
Chicago: $2,100 × (180/144) = $2,625
New Orleans: $2,800 × (180/160) = $3,150

STEP TWO: Derive net book value of fixed assets at current cost as of the end of 19_8. (The estimated total useful life of each motel is 14 years.)

$$\text{Gross Book Value at Current Cost in 19\_8} \times \frac{\text{Estimated Useful Life Remaining}}{\text{Estimated Total Useful Life}}$$

San Francisco: $2,520 × 6/14 = $1,080
Chicago: $2,625 × 10/14 = $1,875
New Orleans: $3,150 × 12/14 = $2,700

STEP THREE: Compute current cost of total assets in 19_8. (Assume current assets of each motel expressed in 19_8 dollars.)

Current Assets (from Exhibit 28-2) + Fixed Assets (From Step Two Above)

San Francisco: $400 + $1,080 = $1,480
Chicago: $500 + $1,875 = $2,375
New Orleans: $600 + $2,700 = $3,300

STEP FOUR: Compute current-cost depreciation expense in 19_8 dollars.

Gross Book Value at Current Cost in 19_8 × 1/14

San Francisco: $2,520 × 1/14 = $180
Chicago: $2,625 × 1/14 = $187.50
New Orleans: $3,150 × 1/14 = $225

STEP FIVE: Compute 19_8 operating income using 19_8 current-cost depreciation.

$$\text{Historical-Cost Operating Income} - \left(\text{Current-Cost Depreciation} - \text{Historical-Cost Depreciation}\right)$$

San Francisco: $240 − ($180 − $100) = $160
Chicago: $300 − ($187.50 − $150) = $262.50
New Orleans: $480 − ($225 − $200) = $455

STEP SIX: Compute ROI using current-cost estimates for fixed assets and depreciation.

$$\frac{\text{Operating Income (From Step Five)}}{\text{Total Assets (From Step Three)}}$$

San Francisco: $160 ÷ $1,480 = 10.81%
Chicago: $262.50 ÷ $2,375 = 11.05%
New Orleans: $455 ÷ $3,300 = 13.79%

index. Companies in Argentina, Brazil, and other high-inflation countries use this approach. As explained more thoroughly in the literature on external reporting, the use of *general* price-level adjustments represents a restatement of historical costs. Conceptually, restated historical costs are *not* the same as current costs, which are sometimes approximated using *specific* price indexes. (A separate line of reasoning, however, claims that general price-level accounting is a practical, cost-effective way of getting crude approximations of results that would be obtained under more expensive current-cost accounting.)

Plant and Equipment: Gross or Net Book Value?

Because historical-cost investment measures are used often in practice, there has been much discussion about the relative merits of using gross book value (original cost) or net book value (original cost minus accumulated depreciation). Those who favor using gross book value claim that it helps comparisons among plants and divisions. If income decreases as a plant ages, the decline in earning power will be made evident. In contrast, if net book value is used, the constantly decreasing base can show a higher ROI in later years; this higher rate may mislead decision makers.

The proponents of using net book value as a base maintain that it is less confusing because (a) it is consistent with the total assets shown on the conventional balance sheet, and (b) it is consistent with net income computations that include deductions for depreciation. If net book value is used in a manner that is consistent with the planning model, it can be useful for auditing past decisions. Surveys of practice report net book value to be the dominant asset measure used by companies in their internal performance evaluations.

GOAL CONGRUENCE AND PERFORMANCE MEASURES

Objective 7

Explain goal-congruence problems arising from accrual accounting performance measures

Individual components of a management control system should be consistent and mutually reinforcing. Goal-congruence problems can arise when the measures used to evaluate a manager's performance conflict with the decision models advocated by top management.

Limitations of Accrual Accounting Performance Measures

As we have discussed, capital investment decisions in many firms are based on discounted cash-flow (DCF) models. These same firms, however, often use accrual accounting models for evaluating the performance of managers. Managers, therefore, may reject capital projects that are justified on a DCF basis because these projects produce dismally low accrual accounting ROI numbers in the first year or two after the initial investment.

Several different approaches promote consistency between the DCF model used to select capital-budgeting projects and the model used for performance evaluation. One approach is to compare directly the cash-flow predictions made in the DCF analysis with the actual cash flows that occur. A second approach is to make predictions of the accrual accounting rate of return for each year of the project at the time the project is selected. The actual accrual accounting rate of return in each year could then be compared with the predicted rate of return.[4]

[4]A third approach is to use the *compound-interest depreciation method*. This method calculates depreciation in such a manner that the year-by-year accounting rate of return on beginning period investment is equal to the DCF internal rate of return on the investment. This depreciation method is rarely adopted in practice.

Comparability Problems with Historical-Cost Measures

The preceding section of this chapter illustrated how historical-cost measures make comparisons difficult. Reconsider the Hospitality Inns example:

	ROI: Historical Cost	Age of Physical Facility (in Years)
San Francisco	24%	8
Chicago	15%	4
New Orleans	16%	2

The managers of the Chicago and New Orleans motels could legitimately claim that the 24% ROI makes the manager of the San Francisco motel appear relatively more effective than would be the case if all three motels had been constructed in the same year.

Hospitality Inns could use one of several approaches to increase the reliability of its performance evaluation of individual motel managers:

1. Restate the fixed assets and depreciation of each motel to a current-cost basis. This approach calculates the ROI figures that would be reported if all three motels had been constructed in the same year. As illustrated earlier in this chapter (p. 890), this approach shows the New Orleans manager to be achieving the highest return on investment.
2. Retain the historical-cost system to compute ROI, but set a budget target for each motel that recognizes the effect of historical cost for motels built at different times. For example, the target ROIs could be 26% for San Francisco, 18% for Chicago, and 16% for New Orleans.

The second alternative is frequently overlooked in the literature. Critics of historical cost have indicated how high rates of return on old assets may erroneously induce a manager not to replace assets. Regardless, the manager's mandate is often "Go forth and attain the budgeted results." The budget, then, should be carefully negotiated with full knowledge of historical-cost accounting pitfalls. The desirability of tailoring a budget to a particular manager and a particular accounting system cannot be overemphasized. For example, many problems of asset valuation and income measurement (whether based on historical cost or current cost) can be satisfactorily solved if top management gets everybody to focus on what is attainable in the forthcoming budget period—regardless of whether the financial measures are based on historical costs or some other measure, such as replacement costs.

NONFINANCIAL PERFORMANCE MEASURES

Companies are giving increased emphasis to using both financial and nonfinancial performance measures when evaluating individual divisions (facilities, activity areas, and so on) or managers of those divisions. Prior sections of this chapter have focused on financial performance measures. We now briefly consider nonfinancial performance measures.

Some organizations have developed explicit and detailed systems to monitor key nonfinancial variables. Consider the "Product Review and Evaluation Summary" system used by the Holiday Inns worldwide chain of hotels and motels. This system is administered by the Quality Assurance Division of Holiday Inns to ensure that a traveler staying at any of its motels has a consistently good experience. Seventeen individual areas of each motel facility (the entrance and lobby, public rest rooms, guest room, guest bathroom, outdoor pool, and so on) are examined for any deviation from the standards set by headquarters. For instance, the items checked in each guest room include cleanliness, lighting, the bed, and the television set. The score on a "Product Review and Evaluation Summary" given to a motel by the Quality Assurance Division is used in determining the compensation of the manager of that motel. The Holiday Inns system provides clear signals to all employees about the specific areas that top management views as critical to the company's success.

The following nonfinancial performance measures were reported in a survey of seventy companies with manufacturing facilities:[5]

• Inventory turns	86%
• Labor efficiency	74
• On-time deliveries	57
• Schedule attainment	53
• Units per hour	44
• Throughput time	16

Many of these measures are reported outside the internal accounting system. Companies attempting to use both financial and nonfinancial performance measures in a single reporting system frequently face the difficult challenge of integrating the organization's diverse data bases.

The design of an effective management control system for many companies requires the joint effort of accountants, engineers, information technologists, marketing personnel, and production or operations managers.

EXECUTIVE COMPENSATION PLANS

Objective 8

Describe how accounting numbers are central to many executive compensation plans

A reward system is an integral part of a management control system. The chosen reward system should strengthen the link between goal congruence and managerial effort. Rewards include monetary items such as base salary, bonus payments, and stock options, as well as nonmonetary items such as an office with a view, a personal secretary, authority, and status.

Executive compensation plans are one component of the reward system. They include a diverse set of short-run and long-run subplans that detail the conditions under which, and the forms in which, compensation is to be paid to executives of the organization. A major challenge in designing an executive compensation plan is achieving the appropriate mix of (1) base salary, (2) annual incentives (for example, cash bonuses based on annual reported income), (3) long-term incentives (for example, stock options based on achieving a set level of ROI by the end of a five-year period), and (4) fringe benefits (for example, life insurance).

Objectives of Executive Compensation

Designers of executive compensation plans cite the following objectives (or subsets of them) as important:

1. To provide incentives for better current and future performance
2. To reward past performance
3. To attract new managers

[5]*Survey of the Cost Management Practices of Selected Midwest Manufacturers* (St. Louis: Price Waterhouse, 1989).

4. To retain superior managers
5. To align the risk attitudes of individual managers with those of corporate stockholders
6. To reduce the income taxes paid by the manager as an individual or by the corporation
7. To promote an "entrepreneurial spirit" within an organization

Simultaneous attainment of all these objectives is unlikely. The challenge facing the designer of an executive compensation package is to choose that mix of rewards (salary, cash bonus, stock options, and so on) and the timing of payment that best achieve the set of objectives top management chooses. Two factors that many designers of compensation plans emphasize are administrative ease and the likelihood that managers affected by the plan will view it as being fair.

Role of Accounting Measures

In many executive compensation plans, accounting performance measures are key components. Short-run plans can make the payment of a bonus contingent on attaining a budgeted income or ROI figure. Some long-run plans, termed **performance plans**, make payment of a definite number of performance units (each assigned a fixed dollar value) contingent on the net income or ROI at the end of an extended time period (say, five years) being at a set level (for example, being in the top 25% of a published list of corporations). By explicitly including a five-year period in the plan, management is encouraged to adopt a long-run focus in decision making.

ROI or residual income need not induce managers to take a short-run focus. The focus on the short run in many compensation plans comes not from the ROI or residual-income measure itself, but from a plan that does not take account of time periods longer than the current year.

Accounting measures can be combined with nonaccounting measures in an executive compensation plan. For example, there are four components of the compensation paid to the senior executives of Holiday Inns: (1) base salary, (2) team profit sharing, (3) individual bonuses, and (4) discretionary incentives. The team profit-sharing component requires attainment of two goals at the company level:

1. Financial Target: "The Hotel Group reaches at least 95% of its planned operating income for that year," and
2. Customer Satisfaction Target: "The Holiday Inns hotel system achieves an established level of customer satisfaction as measured by:
   - Guest inspection scores: The annual number of overall C, D, and F responses on guest inspection cards per 1,000 room-nights sold must be less than 1.670.
   - Guest complaint letters: The annual number of guest complaint letters per 1,000 room-nights sold must be less than 0.457."

The fast-food industry is similar to the lodging industry in its use of both financial targets and customer-service/satisfaction targets in executive compensation plans.

ETHICS AND COOKING THE BOOKS

Objective 9

Provide examples of management's cooking the books to maximize short-run bonuses

Chapter 1 concludes with a discussion of professional ethics for accountants. The same guidelines apply to managers. In particular, the numbers subunit managers report should not be tainted by "cooking the books." That is, managers are responsible for reporting reliable figures uncontaminated by, for example, padded assets, understated liabilities, fictitious sales, and understated costs.

In turn, senior management should unequivocally communicate a positive "tone from the top"—an absolute intolerance for manipulation of accounting transactions and reports. These communications should be both oral and written and repeated often. A culture of integrity can permeate an organization only if senior managers speak and act as good examples.

Division managers often cite enormous pressures "to make the budget" as excuses or rationalizations for not adhering to ethical accounting policies and procedures. The exertion of pressure from senior managers to meet budget targets may have positive motivational effects. A healthy amount of pressure is not bad by itself—as long as the tone from the top simultaneously communicates the absolute need for all managers to behave ethically at all times.

PROBLEM FOR SELF-STUDY

PROBLEM

Budgeted data of the baseball manufacturing division of Home Run Sports for February 19_8 follow:

Current assets	$ 400,000
Fixed assets	600,000
Total assets	$1,000,000

Production volume: 200,000 baseballs per month
Target ROI
(Operating income ÷ total assets): 30%
Fixed costs: $400,000 per month
Variable costs: $4 per baseball

Required

1. Compute the selling price per unit necessary to achieve the 30% target ROI.
2. Using the unit selling price from requirement 1, compute profitability with the Du Pont method.
3. Pamela Stephenson, division manager, receives 15% of the monthly residual income of the baseball manufacturing division as a bonus. What is her bonus for February 19_8, using the selling price per unit from requirement 1? Home Run Sports uses a 12% required rate of return on total division assets when computing divisional residual income.

SOLUTION

1.

$$\text{Target operating income} = 30\% \text{ of } \$1{,}000{,}000 = \$300{,}000$$

Let P = selling price per unit

$$\text{Sales} = \text{Variable costs} + \text{Fixed costs} + \text{Operating income}$$
$$200{,}000P = (200{,}000 \times \$4) + \$400{,}000 + \$300{,}000$$
$$200{,}000P = \$1{,}500{,}000$$
$$P = \$1{,}500{,}000 \div 200{,}000$$
$$P = \$7.50$$

Proof:		
	Sales, 200,000 × $7.50	$1,500,000
	Variable costs, 200,000 × $4	800,000
	Contribution margin	$ 700,000
	Fixed costs	400,000
	Operating income	$ 300,000

2. $$\frac{\text{Revenue}}{\text{Investment}} \times \frac{\text{Income}}{\text{Revenue}} = \frac{\text{Income}}{\text{Investment}}$$

$$\frac{\$1,500,000}{\$1,000,000} \times \frac{\$300,000}{\$1,500,000} = \frac{\$300,000}{\$1,500,000}$$

$$1.5 \times 0.2 = 0.30, \text{ or } 30\%$$

3. $$\begin{aligned} \text{Residual income} &= \text{Operating income} - \text{Imputed interest charge} \\ &= \$300,000 - (0.12 \times \$1,000,000) \\ &= \$300,000 - \$120,000 \\ &= \$180,000 \end{aligned}$$

Stephenson's bonus is $27,000 (15% of $180,000).

SUMMARY

Performance reports should distinguish between the performance of the manager of an organization unit (such as a division or a plant) and the performance of the organization unit as an economic investment.

Choices about top management's financial goals (such as ROI or residual income), about the definitions of components of that goal (such as income and investment), and about the accounting measurements (such as current cost or historical cost) should be made interdependently.

Accounting measures such as ROI or residual income can capture important aspects of both manager performance and organization-unit performance. In many cases, however, they need to be supplemented with nonfinancial measures of performance, such as those relating to customer service, product quality, and productivity. Accounting measures can be an integral part of a long-term executive compensation plan that rewards managers for achieving net income, ROI, or residual-income targets over multiyear periods.

TERMS TO LEARN

This chapter and the Glossary at the end of the book contain definitions of the following important terms:

current cost *(p. 889)* executive compensation plans *(894)*
performance plans *(895)* present value *(888)* residual income *(885)*
return on investment (ROI) *(883)*

ASSIGNMENT MATERIAL

QUESTIONS

28-1 Give an example of a financial performance measure and a nonfinancial performance measure at the total organization level and at the individual facility level.

28-2 What are the steps in designing an accounting-based performance measure?

28-3 Why is return on investment (ROI) a popular performance measure?

28-4 "The accounting rate-of-return tool is so hampered by limitations that we might as well forget it." Do you agree? Why?

28-5 What factors affecting ROI does the Du Pont method highlight?

28-6 "Residual income is identical to ROI. Both measures incorporate income and investment into their computation." Do you agree? Explain.

28-7 Why it is important to distinguish between the performance of a manager and the performance of the organization unit for which the manager is responsible? Use examples.

28-8 Give three definitions of investment used in practice when computing ROI.

28-9 Distinguish between measuring assets based on present value, current-cost, and historical-cost.

28-10 What approaches can promote consistency between the DCF model used to select capital-budgeting projects and the model used for performance evaluation?

28-11 Name four objectives of an executive compensation plan.

28-12 Peta Milano made the following comment when her friend Jake Ali was made manager of the Nuclear Construction Division of General Projects:

> It was like putting a new captain on the bridge of the Titanic after it had hit the iceberg. That division has no prospect of ever being profitable. It is a dinosaur that should have died a long time ago.

How can Jake Ali be motivated to make a sustained effort to achieve General Projects' goal of maximizing companywide accounting rate of return?

EXERCISES AND PROBLEMS

28-13 Return on investment; comparisons of three companies. (*N.A.A.*, adapted.)

1. Return on investment is often expressed as follows:

$$\frac{\text{Income}}{\text{Investment}} = \frac{\text{Revenue}}{\text{Investment}} \times \frac{\text{Income}}{\text{Revenue}}$$

What advantages are there in the breakdown of the computation into two separate components?

2. Fill in the blanks:

	Companies in Same Industry		
	A	B	C
Revenue	$1,000,000	$500,000	$ —
Income	100,000	50,000	—
Investment	500,000	—	5,000,000
Income as a percentage of revenue	—	—	0.5%
Investment turnover	—	—	2
Return on investment	—	1%	—

After filling in the blanks, comment on the relative performance of these companies as thoroughly as the data permit.

28-14 Analysis of return on invested assets, comparison of three divisions. Quality Products, Inc., is a soft drink and food products company. It has three divisions: soft drinks, snack foods, and family restaurants. Results for the past three years follow (in millions):

	Soft Drink Division	Snack Foods Division	Restaurant Division	Quality Products, Inc.
Operating Revenues:				
19_1	$2,800	$2,000	$1,050	$5,850
19_2	3,000	2,400	1,250	6,650
19_3	3,600	2,600	1,530	7,730
Operating Income:				
19_1	$120	$360	$105	$585
19_2	160	400	114	674
19_3	240	420	100	760
Total Assets:				
19_1	$1,200	$1,240	$ 800	$3,240
19_2	1,250	1,400	1,000	3,650
19_3	1,400	1,430	1,300	4,130

Required

Use the Du Pont method (p. 884) to explain changes in the operating income to total assets ratio over the 19_1 to 19_3 period for each division. Comment on the results.

28-15 ROI and residual income. (D. Solomons, adapted) Consider the following data for General Electric Company (in thousands):

	Division A	Division B
Total assets	$1,000	$5,000
Operating income	$200	$750
Return on investment	20%	15%

Required

1. Which is the more successful division? Why?
2. General Electric has used residual income as a measure of management success, the variable it wants a manager to maximize. Using this criterion, what is the residual income for each division if the imputed interest rate is (a) 12%, (b) 14%, (c) 18%? Which division is more successful under each of these imputed interest rates?

28-16 ROI and residual income. (D. Kleespie) The Gaul Company produces and distributes a wide variety of recreational products. One of its divisions, the Goscinny Division, manufactures and sells "menhirs," which are very popular with cross-country skiers. The demand for these menhirs is relatively insensitive to price changes. The Goscinny Division is considered to be an investment center and in recent years has averaged a return on investment of 20%. The following data are available for the Goscinny Division and its product:

Total annual fixed costs	$1,000,000
Variable costs per menhir	$300
Average number of menhirs sold each year	10,000
Average operating assets invested in the division	$1,600,000

Required

1. What is the minimum selling price per unit that the Goscinny Division could charge in order for Mary Obelix, the division manager, to get a favorable performance rating? Management considers an ROI below 20% to be unfavorable.
2. Assume that the Gaul Company judges the performance of its investment center managers based on residual income rather than ROI, as was assumed in requirement 1. The company's minimum acceptable earning rate is considered to be 15%. What is the minimum selling price per unit that the Goscinny Division should charge for Obelix to receive a favorable performance rating?

28-17 Pricing and return on investment. A company assembling motor cycles uses normal volume as the basis for the cost figures used in setting prices. That is, prices are set on the basis of long-run volume predictions. They are then adjusted only for large changes in wage rates or material prices.

You are given the following data:

Materials, wages, and other variable costs	$1,320 per unit
Fixed costs	$300,000,000 per year
Target return on investment	20%
Normal volume	1,000,000 units
Investment (total assets)	$900,000,000

Required

1. What operating income percentage on revenues is needed to attain the target return on investment of 20%?
2. What rate of return on investment will be earned at sales volumes of 1,500,000 and 500,000 units, respectively?
3. The company has a management bonus plan based on yearly divisional performance. Assume that the volume was 1,000,000, 1,500,000, and 500,000 units, respectively, in three successive years. Each of three managers served as division manager for one year before being killed in an automobile accident. As the principal heir of the third manager, comment on the bonus plan.

28-18 ROI, residual income, investment decisions. The Media Group has three major divisions:

(a) Newspapers—owns leading newspapers on four continents
(b) Television—owns major television networks on three continents
(c) Film studios—owns one of the five largest film studios in the world

Summary financial data (in millions) for 19_6 and 19_7 follow:

	Operating Income		Revenues		Total Assets	
	19_6	19_7	19_6	19_7	19_6	19_7
Newspapers	$900	$1,100	$4,500	$4,600	$4,400	$4,900
Television	130	160	6,000	6,400	2,700	3,000
Film studios	220	200	1,600	1,650	2,500	2,600

The manager of each division has an annual bonus plan based on division return on investment (ROI). ROI is defined as operating income divided by total assets. Senior executives from divisions reporting increases in the division ROI from the prior year are automatically eligible for a bonus. Senior executives of divisions reporting a decline in the division ROI have to provide persuasive explanations for the decline to be eligible for any bonus, and they are limited to 50% of the bonus paid to the division managers reporting an increase in ROI.

Ken Kearney, manager of the Newspapers Division, is considering a proposal to invest $200 million in fast-speed printing presses with color-print options. The estimated increment to 19_8 operating income would be $30 million. The Media Group has a 12% required rate of return for investments in all three divisions.

Required

1. Use the Du Pont method to explain differences among the three divisions in their 19_7 division ROI. Use 19_7 total assets as the denominator.

2. Why might Kearney be less than enthusiastic about the fast-speed printing press investment proposal?
3. Rupert Prince, chairman of The Media Group, receives a proposal to base senior executive compensation at each division on division residual income. Compute the residual income of each division in 19_7.
4. Would adoption of a residual income measure reduce Kearney's reluctance to adopt the fast-speed printing press investment proposal?

28-19 Executive compensation (continuation of 28-18). Rupert Prince seeks your advice on revising the existing bonus plan for division managers of the Media Group. He asks you to examine three ideas:

- Adding a more longrun emphasis to the performance measure
- Incorporating companywide (The Media Group) as well as division-based performance measures
- Including goals tailored to specific challenges facing the president of each division

Required

Answer the following questions for each of the three ideas Prince has put forth:

1. How would the proposed change be included in the compensation plan of each division manager?
2. How might the proposed change improve the existing compensation plan?
3. What issues may create difficulties in implementing the changes?

28-20 Various measures of profitability. When the Coronet Company formed three divisions a year ago, the president told the division managers that an annual bonus would be paid to the most profitable division. However, absolute division operating income as conventionally computed would not be used. Instead, the ranking would be affected by the relative investments in the three divisions. Options available include ROI and residual income. Investment can be measured using gross book value or net book value. Each manager has now written a memorandum claiming entitlement to the bonus. The following data are available:

Division	Gross Book Value of Division Assets	Division Operating Income
X	$400,000	$47,500
Y	380,000	46,000
Z	250,000	30,800

All the assets are fixed assets that were purchased ten years ago and have ten years of useful life remaining. A zero terminal disposal price is predicted. Coronet's required return on investment used for computing the imputed interest charge in residual income is 10% of investment.

Required

Which method for computing profitability did each manager choose? Make your description specific and brief. Show supporting computations. Where applicable, assume straight-line depreciation.

28-21 Evaluating managers, ROI, value-chain analysis of cost structure. User Friendly Computer is one of the largest personal computer companies in the world. The board of directors was recently (March 19_5) informed that User Friendly's president, Brian Clay, was resigning to "pursue other interests." An executive search firm recommends that the board consider appointing Peter Diamond (current president of Computer Power) or Norma Provan (current president of Peach Computer). You collect the following financial information on Computer Power and Peach Computer for 19_3 and 19_4 (in millions):

	Computer Power		Peach Computer	
	19_3	19_4	19_3	19_4
Total assets	$360.0	$340.0	$160.0	$240.0
Revenues	$400.0	$320.0	$200.0	$350.0
Costs				
Research and development	$ 36.0	$ 16.8	$ 18.0	$ 43.5
Product design	15.0	8.4	3.6	11.6
Manufacturing	102.0	112.0	82.8	98.6
Marketing	75.0	92.4	36.0	66.7
Distribution	27.0	22.4	18.0	23.2
Customer service	45.0	28.0	21.6	46.4
Total costs	$300.0	$280.0	$180.0	$290.0
Operating income	$100.0	$ 40.0	$ 20.0	$ 60.0

In early 19_5 a computer magazine gave Peach Computer's main product five stars (its highest rating on a 5-point scale). Computer Power's main product was given three stars, down from five stars a year ago because of customer-service problems. The computer magazine also ran an article on new-product introductions in the personal computer industry. Peach Computer received high marks for new products in 19_4. Computer Power's performance was called "mediocre." One "unnamed insider" of Computer Power commented: "Our new-product cupboard is empty."

Required

1. Use the Du Pont method to analyze the ROI of Computer Power and Peach Computer in 19_3 and 19_4. Comment on the results.
2. Compute the percentage of costs in each of the six business-function cost categories for Computer Power and Peach Computer in 19_3 and 19_4. Comment on the results.
3. Rank Diamond and Provan as potential candidates for president of User Friendly Computer.

28-22 ROI and residual income as performance measures, adoption of investment proposals. The budget of World Wide Brands for the coming year (19_9) includes the following items for its five divisions (in millions):

	Operating Income	Identifiable Assets
Tobacco products	$528	$1,650
Hardware and security	40	400
Distilled beverages	66	300
Food products	32	200
Office products	20	400
	$686	$2,950

The manager of each division is paid a base salary plus a bonus that is positively related to the magnitude of the division's return on investment (defined as operating income divided by identifiable assets). World Wide Brands uses a uniform 12% required rate of return for new investments in all divisions.

Each division recently submitted its best investment proposal, based on a discounted cash-flow criterion; each had a positive net present value using the 12% required rate of return. The 19_9 increase in operating income and the increase in identifiable assets of each investment proposal follow:

Investment Proposal From Division	19_9 Increase in Operating Income	19_9 Increase in Identifiable Assets
Tobacco products	$120	$ 500
Hardware and security	26	200
Distilled beverages	38	200
Food products	9	50
Office products	14	100
	$207	$1,050

Required

1. How will adoption of the proposed investment by each division affect the 19_9 ROI of that division? Calculate the ROI of each division before the proposed investment, the ROI of the proposed investment, and ROI of the division including the proposed investment. Comment on the results.
2. How will adoption of the proposed investment by each division affect the 19_9 residual income of that division? In calculating residual income, use a 12% imputed interest charge for identifiable assets. Calculate the residual income of each division before the proposed investment, the residual income of the proposed investment, and the residual income of the division including the proposed investment. Comment on the results.
3. The chief financial officer questions the reasonableness of the uniform required rate of return for all divisions. How might differences among divisions in their required rate of return be incorporated into accounting-based performance measures?

28-23 Alternative measures for the investment base of gasoline stations. ARCO is having trouble in deciding whether to continue to use its old gasoline stations and in evaluating the performance of these stations and their managers in terms of return on investment. Top management has explored various ways of measuring invested capital for the stations:

a. *Historical cost:* original cost of land and buildings minus accumulated depreciation (sometimes called net book value)
b. *Current cost:* cost to currently replace the operating cash inflows provided by the existing gasoline station
c. *Current disposal price:* the net proceeds from selling the gasoline station to another company

Information on three gasoline stations was collected to help clarify the issues:

	Fresno Station	Las Vegas Station	Modesto Station
Operating income	$100,000	$120,000	$ 60,000
Historical cost of investment	$400,000	$200,000	$260,000
Current cost of investment	$640,000	$480,000	$290,000
Current disposal price of investment	$600,000	$2,500,000	$300,000
Age	6 years old	15 years old	2 years old

The Las Vegas station is located next to the largest casino on the Las Vegas Strip and was purchased before the current boom in casinos. The current-cost estimate of the Las Vegas station is for a site one mile away from the existing site with equivalent ability in regard to the generation of operating income. The current-cost estimates for the Fresno and Modesto stations are for the same site as the existing station in each city.

Required

1. Which of the three measures of investment is relevant for deciding whether to dispose of any one or more of the gasoline stations? Why?
2. Compute the ratio of operating income to investment for the Fresno, Las Vegas, and Modesto stations under each of the three measures of investment.
3. Which of the three measures is applicable for judging the performance of a gasoline station as an investment activity?

4. Which of the three measures is applicable for judging the performance of the manager of a gasoline station? Is your answer the same as, or different from, your answers in requirements 1 and 3?
5. What measures of performance, in addition to ROI, might be used to evaluate the performance of a manager of a gasoline station?

28-24 ROI performance measures based on historical cost and current cost. Mineral Waters Ltd. operates three divisions that process and bottle sparkling mineral water. The historical-cost accounting system reports the following data for 1990 (in thousands):

	Calistoga Division	Alpine Springs Division	Rocky Mountains Division
Revenues	$500	$ 700	$1,100
Operating costs			
(excluding depreciation)	300	380	600
Depreciation	70	100	120
Operating income	$130	$ 220	$ 380
Current assets	$200	$ 250	$ 300
Fixed assets—plant	140	900	1,320
Total assets	$340	$1,150	$1,620

Mineral Waters estimates the useful life of each plant to be 12 years with zero terminal disposal price. The straight-line depreciation method is used. The respective age of each plant at the end of 1990 is Calistoga (10 years old), Alpine Springs (3 years old), and Rocky Mountains (1 year old).

An index of construction costs of plants for mineral water production for the ten-year period that Mineral Waters has been operating (1980 year-end = 100) is:

1980	1987	1989	1990
100	136	160	170

Given the high turnover of current assets, management believes that the historical-cost and current-cost measures of current assets are approximately the same.

Required

1. Compute the ROI (operating income to total assets) ratio of each division using historical-cost measures. Comment on the results.
2. Use the approach in Exhibit 28-4 (p. 891) to compute the ROI of each division, incorporating current-cost estimates as of 1990 for depreciation and fixed assets. Comment on the results.
3. What advantages might arise from using current-cost asset measures as compared with historical-cost measures for evaluating the performance of the managers of the three divisions?

28-25 Alternative bonus plans for a multiactivity firm, companywide bonus pool or separate division bonus pools, residual income. Consumer Products Inc. is one of the largest manufacturers and marketers of paper products in the world. Three years ago, as part of a diversification drive, two acquisitions were made:

1. High-Life Beverage, a major beer company
2. Maxwell Products, a major food products company

The table below provides information on the operating revenues, operating income, and invested assets of each of the three lines of business of Consumer Products Inc. for the 19_7 to 19_9 period.

Consumer Products: Operating Data (in millions)

	Paper Division	Beer Division	Food Products Division	Consumer Products Inc.
Operating Revenues:				
19_7	$ 9,000	$3,200	$8,000	$20,200
19_8	10,000	3,000	8,300	21,300
19_9	10,500	2,900	8,600	22,000
Operating Income:				
19_7	$ 1,620	$ 224	$ 608	$ 2,452
19_8	2,100	120	731	2,951
19_9	2,625	87	675	3,387
Invested Assets:				
19_7	$ 5,300	$2,000	$3,800	$11,100
19_8	5,700	1,800	4,300	11,800
19_9	6,000	1,700	4,500	12,200

The current compensation plan for senior executives covers all high-ranking managers (except the president, chief executive officer, and chief operating officer of Consumer Products Inc.). The following table shows the number of senior executives covered by the current plan:

Year	Paper Division	Beer Division	Food Products Division	Consumer Products Inc.
19_7	280	80	140	500
19_8	320	80	150	550
19_9	350	75	175	600

A key component of the current compensation plan is the annual bonus award. All senior executives share equally in a bonus pool that is calculated as 10% of the annual residual income (if positive) of Consumer Products Inc. Residual income is defined as operating income of Consumer Products Inc. minus an imputed interest charge of 14% of invested assets of Consumer Products Inc.

Required

1. Using the Du Pont approach (p. 884), explain differences over the 19_7 to 19_9 period in the profitability of Consumer Products Inc. and in each of its divisions.
2. Compute the size of the annual bonus pool to be shared by the senior executives of Consumer Products Inc. each year of the 19_7 to 19_9 period under the current compensation plan. Compute the annual bonus to be paid to each senior executive in the 19_7 to 19_9 period.
3. The president of the Paper Division proposes that the annual bonus pool be calculated for each division separately. Compute the annual bonus to be paid to each senior executive of each division in the 19_7 to 19_9 period if the bonus pool is defined as 10% of division residual income (if positive). Division residual income is defined as division operating income minus an imputed interest charge of 14% of invested assets of the division. Comment on the results.
4. Discuss the pros and cons to Consumer Products Inc. of basing the bonus pool on the residual income of the company as a whole as compared with the residual income of each division.

28-26 Companywide versus division bonus plans, compensation plans. Andersen, Price and Young (APY) has three divisions: the Auditing Division, Consulting Division, and Tax Division. Gary Schofield, managing director of APY, has just heard that Jonathan Davies and five other senior partners in APY's Consulting Division are about to resign and set up the Davies Consulting Group. The possible resignation of Davies came as no surprise to Schofield, but the departure of the other five partners did. Schofield and Davies had openly clashed in APY partner meetings during the last six months over the current bonus plan. The current bonus per partner is based on 25% of total APY net income divided by the total number of APY partners. All partners receive the same bonus. The salary paid to each partner is based on guidelines set by an executive committee. Currently, the highest-paid partner (Schofield) can receive no more than 300% of the salary of the lowest-paid partner. The financial press recently ran a series of articles on internal personality problems at APY. Both Schofield and Davies were described as having "elephant-sized egos."

Exhibit 28-5 presents the revenues and net income of each division over the 19_1 to 19_5 period. In 19_4, APY had a $100 million out-of-court settlement cost resulting from a banking audit client (Western Bank) going bankrupt in 19_2. Stockholders of the Western Bank sued APY for $800 million. The suit alleged that APY had failed to use appropriate auditing procedures and had failed to detect gross irregularities in loan procedures at Western Bank. The 19_4 Auditing Division net income was based on a $49 million operating income and the $100 million settlement. (APY's insurance company also settled out of court for an additional $200 million beyond the $100 million paid by APY.)

Schofield is in his late fifties. He joined APY thirty years ago. He worked exclusively in the Auditing Division prior to his promotion to APY managing partner. Davies is in his late thirties and does not have an accounting qualification (such as a CPA certificate). He is the author of two books and six articles in the *Harvard Business Review*. Davies has an MBA degree and is on a first-name basis with the senior executives of many corporations. A sizable part of APY's consulting business comes from corporations implementing the strategic frameworks that Davies developed.

The executive committee of APY makes the key decisions concerning compensation structure and resource allocation. There are eleven members of this committee—six from auditing, two from consulting, and three from tax. Davies is not a member of the executive committee, because the APY articles of association require all executive committee members to have an accounting qualification.

Required

1. Compute the bonus paid to each of the APY partners in each year of the 19_1 to 19_5 period.
2. Discuss two advantages and two disadvantages of including the $100 million out-of-court settlement as a 19_4 period cost when computing the total bonus pool for APY.
3. Assume that partner bonuses are separately calculated for each division. All partners in the same division would receive the same bonus. For example, the Consulting Division

EXHIBIT 28-5

Andersen, Price and Young: Summary Data

	19_1	19_2	19_3	19_4	19_5
Revenues (in millions)					
Auditing Division	$601	$610	$ 625	$ 632	$ 648
Consulting Division	110	151	205	278	385
Taxation Division	193	212	248	265	294
Total	$904	$973	$1,078	$1,175	$1,327
Net Income (in millions)					
Auditing Division	$ 60	$ 55	$ 56	$ (51)*	$ 38
Consulting Division	36	45	71	89	121
Taxation Division	38	40	41	42	49
Total	$134	$140	$ 168	$ 80	$ 208
Number of Partners					
Auditing Division	624	614	636	648	628
Consulting Division	68	92	128	170	182
Taxation Division	176	180	188	202	208
Total	868	886	952	1,020	1,018

*Includes $100 million out-of-court settlement cost.

bonus would be based on 25% of the net income of the Consulting Division divided by the number of partners in the Consulting Division. Compute the bonus paid to each of the (a) auditing partners, (b) consulting partners, and (c) tax partners of APY in each year of the 19_1 to 19_5 period under this revised bonus plan. Comment on the differences between the bonuses paid to partners under this approach and the bonuses computed in requirement 1.

28-27 Executive compensation, retention of key employee (continuation of 28-26). The executive committee is committed to retaining Jonathan Davies. You are asked to prepare a memorandum on the key aspects of a "compensation package" you would recommend that APY offer Davies in order to retain him.

28-28 Cooking the books, division managers, internal control. PepsiCo's three main divisions are beverages (main product is the Pepsi soft drink), food products (main products are Frito-Lay chips and other snack foods), and food service (mainly Pizza Hut and Taco Bell).

Over the 19_1 to 19_5 period, the beverage division was organized into Pepsi-Cola Company (U.S. operations) and PepsiCo International. PepsiCo International bottled soft drinks in more than six hundred foreign plants.

In November 19_5, PepsiCo issued the following in a press release:

> Internal auditors at PepsiCo, Inc. recently discovered significant accounting irregularities in certain company-owned foreign bottling operations of its International division. The foreign subsidiaries involved accounted for less than 5% of PepsiCo's operating income in 19_4.
>
> It appears that these irregularities involve the overstatement of assets and understatement of costs over several years, going back at least to 19_1.
>
> The company's investigation, being conducted by a task force that includes special legal counsel and public accountants retained for this purpose, has shown that accounts were falsified by managers of these foreign subsidiaries, principally in Mexico and the Philippines, to improve the apparent performance of their operations. Extensive collusion, creation of false documentation, and the evasion of company internal controls combined to make these misstatements possible. It does not presently appear that these misrepresentations were designed to divert company funds to personal, improper or illegal use.
>
> PepsiCo is terminating and replacing the appropriate individuals, including the U.S.-based manager of the bottling unit of the International division.

In a December 19_5 press release, PepsiCo reported the following details on a restatement of earnings for the 19_1 to 19_5 period based on the investigation (in millions):

	19_1	19_2	19_3	19_4	19_5
Net income as reported	$225.8	$264.9	$291.8	$333.5	$273.0
Net reduction resulting from restatement	2.6	14.5	31.1	36.0	8.1
Net income as restated	223.2	250.4	260.7	297.5	264.9

A former SEC commissioner stated that "numerous techniques were used [by the foreign subsidiaries of PepsiCo] to report false operating income, including falsifying costs, failing to write off broken or unusable bottles and uncollectible accounts receivable, and writing up bottle inventories above cost. To further these schemes, certain individuals made false statements to PepsiCo and its independent auditors concerning the financial condition of certain subsidiaries and, at various times, participated in, or were aware of the falsifications of the books and records of the foreign beverage operations."

The operating income of PepsiCo's three main divisions over the 19_1 to 19_5 period, before any adjustments for the practices disclosed in the investigation, were (in millions):

Division	19_1	19_2	19_3	19_4	19_5
Beverage	$227.0	$254.0	$274.7	$281.9	$217.7
Food Products	158.2	195.4	245.8	298.5	326.4
Food Service	64.1	49.9	59.5	81.9	119.3

Required

1. For each year, compute the percentage that the "restatement of net income" is of (a) net income of PepsiCo as reported and (b) operating income of the beverage division. Comment on the results.
2. What factors may have motivated the Mexico and Philippine senior managers to engage in the unethical practices?
3. One press commentator described the practices reported by the investigation as "business as usual in a company with a high pressure to perform. It is the proverbial storm in a teacup. I don't think the issue is a serious one for PepsiCo's top management." Do you agree? Explain your answer.

28-29 Designing a compensation plan for unit restaurant managers of McDonald's Corporation. The chief operating officer of McDonald's Corporation is considering changes in the compensation package of the manager of each of its company-owned restaurants. McDonald's is the largest chain of fast-food restaurants in the world. A recent annual report included the following comments in a section headed "The McDonald's restaurants: Where technology meets Q.S.C. and V." (*Q.S.C. and V.* stands for quality, service, cleanliness, and value):

> In the tradition of entrepreneurial giants, Ray Kroc [the founder] built a new type of production system. McDonald's took the guesswork out of the foodservice business by applying procedures that geometrically increased productivity while ensuring quality and an enjoyable eating-out experience. The secret of McDonald's success, one expert claimed, was "the rapid delivery of a uniform high-quality mix of prepared foods in an environment of obvious courtesy." The secret, claimed another expert, was McDonald's ability to initiate and maintain its quality control systems through a network of hardworking, dedicated franchisees, and company employees.

McDonald's maintains a year-round training program for all levels of operations. The central training center is Hamburger University at Elk Grove, Illinois. Managers of company-owned restaurants must take an intensive course at Hamburger University at which the Q.S.C. and V. system of values is emphasized (and emphasized, and emphasized . . .).

The options regarding the unit manager compensation plan that the chief operating officer is considering follow.

OPTION A: The existing plan in which the unit manager's annual compensation consists of (1) a base salary and (2) a quarterly bonus that rewards the ability to meet preset objectives in the areas of (i) labor costs, (ii) food and paper costs, (iii) Q.S.C. and V., and (iv) volume projections.

1. The fixed salary: After surveying each market in which it owned restaurants, McDonald's established three salary ranges according to prevailing labor rates and other economic factors. Range I, the highest, usually applied to very large metropolitan areas; Range II applied to somewhat smaller areas where industrial and rural influences on the labor market were about equal; and Range III applied to small-town markets with little industrial influence. In addition, annual merit increases were awarded within each range according to whether an employee was judged superior, satisfactory, or was still in the new employee bracket.

2. The bonus: Meeting the optimal labor crew costs—figured according to projected sales volume and labor crew needs for each month of the quarter—entitled the manager to a bonus of 5% of the base salary.

Together the area supervisor and the unit manager determined the food and paper cost objective based on current wholesale prices, product mix, and other operating factors peculiar to the unit. By meeting the previously agreed objective, the manager earned another 5% bonus.

The exhibit on the facing page is an excerpt from the monthly management visitation report by which each store's Q.S.C. and V. is rated. Based on the average score for the quarter, units were designated "A," "B," or "C." Managers of "A" stores received a bonus of 10% of base salary; "B" store managers, 5%; and "C" store managers, no bonus.

In addition, the manager received a bonus of 2.5% of the increase over the previous year's sales, up to 10% of the base salary. If unit volume was significantly affected by operat-

EXTRACT FROM VISITATION REPORT

Question No.	Section 1 (Outside)	Item Score
1.	Is area within one block of the store free of all litter?	
2.	Are flags being displayed properly and are they in good condition? Are entrance and exit and road signs in excellent condition?	
3.	Are waste receptacles in an excellent state of repair and clean? Is trash being emptied as necessary?	
4.	Is the parking lot and landscaping as clean, litter-free, and well picked up as you could reasonably expect for this business period? Do these areas reflect an excellent maintenance program? Is traffic pattern well controlled?	
5.	Do the sidewalks surrounding the building and the exterior of the building reflect an excellent maintenance program? Were these areas being maintained properly during this visit?	
6.	Were all inside and outside lights which should have been on, on, and were windows clean?	
	SECTION TOTAL	

Question No.	Section II (Inside Store Pre-Purchase of Food)	Item Score
7.	Was the restroom properly maintained? Was the inside lobby and dining area properly maintained?	
8.	Does point of promotion in the store present a unified theme?	
9.	Is menu board in excellent repair and clean? Are napkins and straws available near all registers?	
10.	Is the general appearance of all stations good? Is all stainless steel properly maintained?	
11.	Is there an adequate number of crew and management people working for this business period and are they positioned properly?	
12.	Are all crew members wearing proper McDonald's uniforms, properly groomed, and does their general conduct present a good image?	
13.	Are all counter persons using the Six Step Method and does their serving time per customer meet McDonald's standards?	
	SECTION TOTAL	

ing circumstances beyond the manager's control, the regional manager could grant a semi-annual payout of 5% of base salary.

Therefore, the maximum annual incentive bonus to an "A" store manager who met all the objectives was 20% of the base salary plus an additional 10% of the salary because of the volume gain at the restaurant.

Bonuses for meeting cost objectives were paid quarterly. Those for meeting the Q.S.C. and V. standards and volume increases were paid semiannually.

A group of the unit managers protested that the existing plan is much too complicated. Some managers also complained about its undue subjectivity and the overemphasis on volume increases.

OPTION B: The unit manager's base salary would initially be determined according to the range system described in Option A. Thereafter the manager would be rated monthly by the regional operations staff on six factors: quality, service, cleanliness, training ability, volume, and income. Each factor would be rated 0 for unsatisfactory, 1 for satisfactory, and 2 for outstanding. A manager whose semiannual total is 12 would warrant a bonus of 40% of the base salary for half a year, a score of 11 would warrant a 35% bonus, and so on. At the end of the year, the two semiannual scores would be averaged, and the manager would receive a salary increase of 12% for a score of 12, 11% for a score of 11, and so on, down to a point where the manager would presumably be encouraged to seek a future with a competitor.

OPTION C: The unit manager's base salary would be determined by the range system in Option A. The bonus would be 10% of any sales gain plus 20% of the income (provided that gross margin amounted to at least 10% of the gross sales). The maximum bonus paid would be 50% of the base salary.

A survey of the compensation plans used by competitors of McDonald's Corporation led to several other options being included in the analysis:

OPTION D: The unit manager's compensation plan has a relatively low base salary with a six-month bonus of 10% of any sales gain above the previous highest six-month sales level. There is no restriction on the size of the bonus paid.

OPTION E: The unit manager's compensation plan has a relatively low base salary with a six-month bonus of 20% of the operating income of the restaurant. There is no restriction on the size of the bonus pool.

OPTION F: The unit manager is paid a high base salary, which is based on the prior year's sales of the restaurant. No bonus plan is included in the compensation package. Above-average performers can seek promotions to manage restaurants with higher sales levels.

Required

1. What criteria should the chief operating officer of McDonald's Corporation consider when designing a compensation plan for the managers of its individual restaurants?
2. Using the criteria that you listed in requirement 1, evaluate the six (A–F) compensation plans outlined in this problem.
3. What plan would you recommend that the chief operating officer of McDonald's adopt? (You are not restricted to the six options outlined in this problem.)

Strategic Control Systems

Quality as a competitive weapon

Quality-related management control initiatives

Time as a competitive weapon

Time-related management control initiatives

Cost as a competitive weapon

CHAPTER 29

Overnight couriers promise next-day delivery. On-time delivery is a key performance variable in evaluating managers of courier operations.

Learning Objectives

When you have finished studying this chapter, you should be able to

1. Explain why managers view quality as being strategically important
2. Distinguish among the four cost categories in a cost of quality program
3. Provide examples of nonfinancial quality performance measures
4. Explain why managers view time as being strategically important
5. Describe three ways that control systems can highlight time as a competitive weapon
6. Understand the motivation for selecting throughput time as a cost allocation base
7. Provide examples of value-added and non-value-added costs
8. Describe how customers with equivalent sales dollars can differ in profitability

Quality, time, and cost are three important areas in which companies compete in the marketplace. This chapter examines how management accounting can assist managers in taking strategic initiatives in these areas.

Recall the six business functions that we have discussed in many of the preceding chapters: research and development, product design, manufacturing/operations, marketing, distribution, and customer service. These business functions are interdependent. For example, decisions made by product designers have an impact on subsequent business functions, such as manufacturing and customer service. In each of the quality, time, and cost areas, management accountants can play an important role in promoting greater recognition of these interdependencies. In turn, understanding how a company's business functions work together gives managers a competitive edge.

QUALITY AS A COMPETITIVE WEAPON

Objective 1

Explain why managers view quantity as being strategically important

The **quality** of a product or service is its conformance with a prespecified (and often preannounced) standard. Surveys of both customers and managers rank quality as one of the most important attributes of a product.[1]

Quality differences across companies can be dramatic. Consider one study of eleven U.S. and seven Japanese companies that assemble air-conditioning units. The quality of four variables was measured. The median for companies in each country was:

	U.S. Companies	Japanese Companies
1. Percentage of incoming parts and materials failing to meet specifications	3.30%	0.15%
2. Fabrication: Coil leaks per 100 units	4.4%	0.1%
3. Assembly line: Defects per 100 units	63.5%	0.9%
4. Service-call rate per 100 units under first-year warranty coverage	10.5%	0.6%

D. Garvin, "Japanese Quality Management," *Columbia Journal of World Business*, XIX, No. 3, p. 4.

[1]See KMPG Peat Marwick, *1989 Productivity Management Survey* (American Electronics Association, 1989)—"91% of the survey respondents listed 'quality and reliability of products' among the top three competitive factors in their industry." (p. 1)

Quality differences of this magnitude can give companies at the higher-quality levels a major competitive edge in the marketplace.

A **quality driver** is any factor whose change causes a change in the quality level of a product or service. The more a company understands its quality drivers, the more it can plan its operations to increase quality continually. For example, one company with several different plants assembling printed-circuit boards examined how the defect rate and the level of output differed across the plants. It wanted to determine the quality drivers. It found that plants that used the same production-batch size each production run had the lowest rate of defects. That is, the plants that had the lowest variability in the size of their production runs produced the highest-quality PC boards. This variability in production batches was a quality driver. To promote higher quality, the company encouraged marketing representatives to get orders in multiples of one particular size (8, 16, 24, and so on), which would lower the variability and therefore increase product quality.

QUALITY-RELATED MANAGEMENT CONTROL INITIATIVES

Company initiatives on quality range from comprehensive total quality control (TQC) programs in which all business functions are involved to programs that focus on individual functions (such as the testing function of an aircraft-manufacturing plant).

Cost of Quality Programs

Objective 2

Distinguish among the four cost categories in a cost of quality program

A **cost of quality program** collects product-quality–related costs incurred in the different business functions of an organization and reports them in a comprehensive reporting format. Four categories of costs are often distinguished:

1. **Prevention costs.** Costs incurred in preventing the production of products that do not conform to specification.
2. **Appraisal costs.** Costs incurred in detecting which of the individual products produced do not conform to specification.
3. **Internal failure costs.** Costs incurred when a nonconforming product is detected *before* its shipment to customers.
4. **External failure costs.** Costs incurred when a nonconforming product is detected *after* its shipment to customers.

Exhibit 29-1 presents an extract from the cost of quality program of Texas Instruments.[2] Note the individual line items in each of the prevention, appraisal, internal failure, and external failure categories of quality costs. Each company has its own set of line items.

Each cost line-item in Exhibit 29-1 captures only a small component of the total cost of quality. By pooling costs across different business functions, however, the magnitude of total quality-related costs of Texas Instruments can be determined. The items included in a cost of quality program come from all the major business functions. Any individual in a single business function would probably not be able to observe the magnitude of the total costs of quality unless they are summarized in a comprehensive reporting system. Most existing accounting systems, while reporting a subset of the cost items in Exhibit 29-1 (such as cost of rework—see

[2]"Texas Instruments: Cost of Quality (A)" (Boston, MA: Harvard Business School, Case 9-189-029).

EXHIBIT 29-1
Texas Instruments: Cost of Quality Report

Prevention Costs	Internal Failure Costs
Quality engineering	Scrap
Receiving inspection	Rework
Equipment repair/maintenance	Manufacturing/process engineering on internal failure
Manufacturing process engineering	
Design engineering	**External Failure Costs**
Quality training	Net cost of returned products
Appraisal Costs	Marketing costs on external failure
Technical services laboratory	Manufacturing/process engineering on external failure
Design analysis	Repair
Product acceptance	Travel costs related to quality problems
Manufacturing product inspection	Liability claims

Chapter 18), are not designed to focus comprehensively on cost of quality for the organization as a whole.

The cost of quality report in Exhibit 29-1 can be used to examine interdependencies across the four categories of quality-related costs. For example, Texas Instruments could examine how investments in prevention (say, by designing products that better tolerate rapid temperature changes) are associated with reductions in appraisal, internal failure, or external failure costs. For example, are increased expenditures in product design (a prevention cost category) associated with decreased expenditures on customer service warranty costs (an external failure cost category)?

Nonfinancial Quality Performance Measures

Objective 3

Provide examples of nonfinancial quality performance measures

Many firms monitor both nonfinancial and financial variables for performance evaluations. Exhibit 29-2 illustrates six nonfinancial quality measures monitored by Texas Instruments. These measures include those with an internal orientation (for example, first-pass calibration yield) and those with an external orientation (for example, competitive rank). Operations managers often find nonfinancial measures useful for day-to-day control purposes.

Companies in different countries vary in how they use quality-related variables in their performance measures of managers. Consider manufacturing managers in the Japanese and U.S. electronics industry. A survey reports that 70% of Japanese

EXHIBIT 29-2
Texas Instruments: Examples of Nonfinancial Performance Measures

- First-pass calibration yield (percentage of products passing quality tests first time)
- Outgoing quality level for each product line (defective parts per million as assessed by TI Quality Control)
- Returned merchandise (percentage of shipments returned from customers because of poor quality)
- Customer report card (sample of customers interviewed to get feedback on quality)
- Competitive rank by TI marketing and field sales personnel (ranking of product relative to a number of competitors in that product line)
- On-time delivery (percentage of shipment made on or before the scheduled delivery date)

managers and only 30% of U.S. managers have explicit goals in their performance measures for reducing the number of products rejected for quality reasons.[3]

TIME AS A COMPETITIVE WEAPON

Companies increasingly view time as a key variable in competition.[4] We consider three areas in which time is important: new-product lead time, on-time performance, and customer response time.

Objective 4

Explain why managers view time as being strategically important

New-Product Lead Time

Successful new-product ventures are essential to most companies. Bringing new products to market faster than competitors enables a company to gain market share and to learn more quickly from customers about how to improve the product. **New-product lead time** (also called **time-to-market**) is the time from the initial concept of a new product to its market introduction.

Companies differ greatly in their new-product lead times. Consider the automobile industry. Japanese companies bring new products to market much faster than their U.S. or European competitors. The following data cover twenty-four new-car development projects by eight Japanese companies, three U.S. companies, and eight European companies:

	Average New-Product Lead Time
Japanese companies	43.3 months
U.S. companies	60.3 months
European companies	66.5 months

K. Clark and T. Fujimoto, "Lead Time in Automobile Product Development: Explaining the Japanese Advantage," Working Paper, Harvard Business School, 1989.

Industry observers argue that the Japanese gains in worldwide market share for automobiles are due, in part, to Japan's ability to bring high-quality new products to market faster than competitors. Not surprisingly, all automotive companies are now making new-product lead time a key variable in their planning processes.

Companies in other industries are also making reduction in new-product lead time a main concern. Three examples of dramatic changes made by companies in this area are:

		New-Product Lead Time	
Company	**Product**	**Before**	**After**
AT&T	Telephones	2 years	1 year
Navistar	Trucks	5 years	2.5 years
Hewlett-Packard	Computer printers	4.5 years	1.8 years

B. Dumaine, "How Managers Can Succeed Through Speed," *Fortune*, February 13, 1989.

To reduce new-product lead time by 50% or more usually requires major changes in operations. Central to these changes has been greater interaction among all the business functions at all stages of the new-product development process.

---

[3]S. Daniel and W. Reitsperger, "Management Control Systems for Quality: An Empirical Comparison of the U.S. and Japanese Electronics Industries," Working Paper, University of Hawaii, 1990.

[4]See G. Stalk and T. Hout, *Competing Against Time* (New York: Free Press, 1990).

On-Time Performance

When a customer orders a product or service, there is frequently an explicit time dimension to the transaction. For example, when a computer company orders semiconductor chips from a supplier, the contract will specify a quantity, cost, and delivery time. When a traveler purchases an airline ticket, the ticket specifies the type of seat, its cost, and the departure and arrival times. When a package is sent via an overnight courier service, the contract specifies a delivery time (say, by 10:30 A.M. the next morning) as well as a price.

Industry surveys report that superior on-time performance can provide a company with a competitive advantage. For example, the U.S. General Accounting Office reports the on-time performance of U.S. airlines. A flight is considered on-time when it departs and arrives within 15 minutes of the scheduled times. Over a recent 24-month period, on-time performance for fourteen airlines ranged from 74% to 85%. An airline that consistently averages higher on-time performance establishes with travelers and travel agents a time-based competitive advantage over airlines with lower on-time performance.

Customer-Response Time

The time to respond to customer requests is a key competitive factor in many industries. For example, a retail store able to deliver a washing machine of a given make and color in two days will probably attract more business than a store with a two-week delivery time. Similarly, a grocery store that averages 5 minutes per customer in a checkout line will probably attract more business than a store that averages 20-minutes per customer. The time a customer spends in line is critical in many other industries, including banks, car-rental agencies, and fast-food chains.

A **time driver** is any factor whose change causes a change in the speed with which an activity is undertaken. This activity could range from a watch company bringing a new product to market to a retail store responding to a customer order for a new television set. Companies that are able to identify the time drivers of their activities can seek ways to restructure their operations to become more competitive in the time dimension. For example, studies have found that new-product development projects with minimum overlap across their business functions have longer lead times than projects with sizable overlap across their business functions.[5] Companies can no longer afford to have research and development create a product in isolation, then have engineers design it, then have production managers determine how to manufacture it, then have marketing managers try to sell it, and then have customer-service managers try to fix it. Many companies are now encouraging early involvement of all business functions (such as manufacturing and customer service) in new-product development projects to reduce new-product lead times.

TIME-RELATED MANAGEMENT CONTROL INITIATIVES

Objective 5

Describe three ways that control systems can highlight time as a competitive weapon

Companies are experimenting with several initiatives in their management control systems to highlight or reinforce the importance of time as a competitive weapon.

[5] S. Wheelwright, "Time to Market in New Product Development," *ICL Technical Journal*, November 1989.

Capital Budgeting

Chapter 21 (pp. 686–88) discussed the breakeven-time capital-budgeting method. Breakeven time (BET) is the time from the start of a project (the initial idea date) to the time when the cumulative present value of the cash inflows of a project equals the present value of the total (to-market) cash outflows. Several companies are using breakeven time to decrease new-product lead times dramatically. For example, Hewlett-Packard has the following key managerial performance goal: a 50% reduction in the BET of a new product compared with the BET of comparable past products developed in that division. One manager observed: "It's an awesome challenge. . . . Hewlett-Packard hopes this BET benchmark will force its managers to redesign the product development process, getting people from all divisions to work together in flexible, fast-moving teams."

Choice of Cost Allocation (Application) Bases

Objective 6

Understand the motivation for selecting throughput time as a cost allocation base

Time-related variables are frequently used to apply indirect costs to products. For example, indirect manufacturing costs can be applied using direct labor hours or machine hours. Several firms have adopted throughput time as the base for applying indirect manufacturing costs to products in an attempt to speed up the manufacture of products. *Throughput time* is the time from the start of a product on the production line until it becomes a finished good.

Consider Zytec Corporation, a manufacturer of computer equipment. Indirect manufacturing costs were previously applied to products using direct labor hours. Changes in manufacturing and a reduction in the labor content of products led management to seek an alternative cost application base. Zytec's controller noted:

> We wanted to pick [a cost application base] that was meaningful to the people on the floor. We wanted the [chosen application base] to capture the essence of our drive for continuous improvement. In particular, we were convinced that the cost system could become a potent tool for behavior modification.[6]

Zytec adopted throughput time for applying indirect manufacturing costs to products in an effort to reduce the time taken to manufacture products.

The trend in many organizations is to increase the number of cost application bases (see p. 474). In contrast, several organizations have restricted the number of application bases in an effort to signal to managers which one or two variables top management views as critical. For example, Zytec considered using a larger number of cost application bases. The company, though, decided that the benefits of having managers focus on a single variable (throughput time) outweighed the costs of using multiple cost application bases. Note that these costs of using throughput time include having product-cost numbers that are less accurate than those that could be developed with a multiple cost-application base approach.[7]

Performance Measures

Companies that use time-based performance measures for evaluating managers or organization units send a signal that time is a key variable. Consider a metal-cutting machine company that manufactures over two thousand different types of

[6]R. Cooper and P. Turney, "Internally Focused Activity-Based Cost Systems," in *Performance Excellence in Manufacturing and Service Organizations*, ed. P. Turney (Sarasota, FL: American Accounting Association, 1990), p. 95.

[7]As an alternative, Zytec could have used multiple cost-application bases for product costing and a single variable for performance measurement. This approach has the attractive feature of sending a clear signal through the performance measurement system without compromising the accuracy of product cost numbers.

jobs each year. To promote faster throughput time for each job, time-based performance measures for the plant manager were introduced, including setup time for each new job, equipment repair time, and tooling design time. Several companies in the machining industry have reported dramatic reductions in setup times. For example, one German machining company reported a reduction in job setup times from 6.0 hours to 2.5 hours over a 12-month period.

The bonuses paid to regional managers of several overnight courier services include a component based on on-time delivery performance. Drivers for courier companies are given very clear signals that on-time delivery is their number-one priority each day.

COST AS A COMPETITIVE WEAPON

Many chapters in this book present cost analyses pertinent to firms using cost as a competitive weapon. This section covers two additional topics that illustrate ongoing developments in management accounting: (1) value-added versus non-value-added cost analysis and (2) customer-profitability analysis.[8]

Value-Added/Non-Value-Added Cost Analysis

Objective 7

Provide examples of value-added and non-value-added costs

The value-added/non-value-added classification of costs is being used by several companies aggressively seeking out ways to become more cost competitive. The classification focuses on whether a cost can be eliminated without the customer's perceiving a deterioration in the performance, function, or other quality of a product. By cutting a non-value-added cost, a company can price its product lower, with no apparent loss of quality in the customer's eyes.

Companies employing this classification for cost-management purposes find that the value-added and non-value-added categories are best viewed as ends of a spectrum with a large gray area in the middle. Consider how General Electric uses this approach to cost management.[9]

General Electric has used the value-added/non-value-added approach to cost management at a plant assembling medical equipment. GE implemented this approach in three steps:

Step 1: Identify the attributes of products that customers perceive to be valuable. For the GE plant, these attributes included quality, dependability, and price.

Step 2: Identify those activities that cause work in the operating process (such as the production line), and assess whether each activity adds value. Thirty activities were identified at GE. Operating personnel then classified each activity as value-added, non-value-added (waste), or in the gray area. Examples follow:

Category	Example
Value-added	Assembly time
Non-value-added	Rework time
Gray area	Work assignment

[8]An excellent coverage of many important topics in this general area is in C. Berliner and J. Brimson, *Cost Management for Today's Advanced Manufacturing: The CAM-I Conceptual Design* (Boston, MA: Harvard Business School Press, 1988)

[9]This discussion draws on a visit made by the authors to the General Electric plant and on H. T. Johnson, "Performance Measurement for Competitive Excellence," in *Measures for Manufacturing Excellence*, ed. R. S. Kaplan (Boston, MA: Harvard Business School Press, 1990).

EXHIBIT 29-3
Non-Value-Added Activities and Their Cost Drivers at a General Electric Medical Equipment Assembly Plan

Non-Value-Added Activity	Examples of Cost Drivers
1. Accumulated materials (materials and work in process inventories on shop floor)	Stock location (layout) Stocking procedures Assembly sequencing Number of separate part numbers in a product
2. Expediting of materials	Stock balance errors Ordering errors Variation in material yields Schedule changes
3. Movement of materials	Plant layout Handling equipment
4. Rework	Assembly errors Handling damages Supplier quality
5. Testing/verifying	Quality problems Training of personnel Design of products

The work force as a whole spent 35% of its time on value-added activities, 51% on non-value-added activities, and 14% on activities in the gray area.

Step 3: Identify and eliminate the causes of non-value-added activities. The first column in Exhibit 29-3 presents the five highest non-value-added activities identified by General Electric; the second column presents variables that operating personnel believed were the cost drivers of those activities. Management then devoted considerable effort to reducing the impact of the non-value-added cost drivers. For example, changes in plant layout reduced material movements. Similarly, changes in the design of products reduced the number of different parts in a product. The results of these and similar efforts over the next 12 months included a:

- 21% reduction in payroll cost per unit,
- 50% reduction in work in process inventory, and
- 50% reduction in defects detected in final testing.

The value-added/non-value-added classifications of individual activities developed at General Electric are contextual. Activities viewed as value-added at a medical instrument facility may be viewed as non-value-added at a home-appliance assembly plant. Also, as experience is gained managers may reclassify an activity. Adopting a value-added/non-valued-added classification of costs in an internal accounting system will probably require updating the system frequently.

Using Activity-Based Accounting for Customer-Profitability Analysis

The class of customer to emphasize and the price to charge for customized service are key strategic decisions facing managers. Management accountants are now giving increased attention to analyzing costs relevant to such decisions. Activity-based accounting (see pp. 150–57) can play a key role in customer-profitability analysis.

Objective 8

Describe how customers with equivalent sales dollars can differ in profitability

Consider the customer-profitability analysis conducted by Kanthal, a Swedish company that sells electric heating elements.[10] Customer-related selling costs constitute 34% of its cost structure. Until recently, Kanthal allocated these costs when conducting customer-profitability studies on the basis of the sales dollars of each customer. Management decided to reexamine customer profitability by analyzing the activities required to service customers. A detailed study of the resources used to service different types of customers revealed that two activities were important:

1. Number of orders placed. Each order had a large fixed cost, which did not vary with the quantity of items purchased. A customer who ordered 1,000 units per month in 10 orders of 100 units generated ten times the order cost as did a customer who ordered 1,000 units per month in a single order.
2. The availability of the ordered item "in stock." If the ordered item was not in stock, Kanthal incurred extra costs to produce it.

Kanthal made estimates of the per-order cost (say, $700), the cost of handling an in-stock item (say, $300), and the cost of handling an out-of-stock item (say, $2,300). Thus, the sales-related cost for an in-stock order was $1000 ($700 + $300), and the sales-related cost for an out-of-stock order was $3,000 ($700 + $2,300). (These cost figures were assumed to be purely variable.) An analysis was then made of the gross margin earned for each customer in the most recent year. Consider a customer who used 80 orders to purchase $890,000 of merchandise with a manufacturing cost of $680,000. The 80 orders comprised 30 orders for in-stock items and 50 orders for out-of-stock items. The gross margin on sales to this customer is $30,000:

Sales		$890,000
Costs:		
Manufacturing cost of goods sold	$680,000	
In-stock order costs (30 in-stock orders × $1,000)	30,000	
Out-of-stock order costs (50 out-of-stock orders × $3,000)	150,000	
Total costs		860,000
Gross margin		$ 30,000

Note how the customer's frequent purchases of out-of-stock items result in sizable sales-related costs.

Kanthal found that only 40% of its customers were profitable. Even more dramatic, 10% of Kanthal's customers lost 120% of its total gross margin. (That is, these 10% generated operating losses equal to 120% of Kanthal's total gross margin.) Two of the most unprofitable customers were in the top three in total sales volume. These two customers made many small orders of out-of-stock items.

Kanthal's use of activity-based accounting highlights how customers with equivalent sales dollars can differ dramatically in profitability. Kanthal can use this information to explore several approaches to becoming more cost competitive. For example, it could pursue the following three objectives:

- Reduce the fixed cost per order, the cost of handling an in-stock item, and the cost of handling an out-of-stock item
- Encourage customers to purchase using a lower number of orders
- Encourage customers to purchase heating elements that are regularly in-stock

[10]The Kanthal case draws on work by SAM, a Swedish consulting firm. It is described in "Kanthal" (Boston, MA: Harvard Business School, Case N9-189-129). An earlier discussion of ideas implicit in the Kanthal case appears in D. Longman and M. Schiff, *Practical Distribution Cost Analysis* (Homewood, IL: Irwin, 1955). For example, see their page 167. Instead of "activity-based accounting," they used "functional costing." Instead of "cost drivers," they used "control factor units."

Do not conclude that Kanthal should drop all customers showing a negative gross margin. For example, some customers may be willing to change their buying behavior away from numerous small orders of out-of-stock items. Other customers may be willing to pay extra to maintain their current buying patterns. Yet other customers may fall into the "unprofitable today–profitable tomorrow" category. For example, they may be expanding so rapidly that in 12 months' time they will be ordering in large volume. Kanthal may be willing to incur short-run losses on sales to these customers now in the expectation that they will be profitable customers in the longer run.

PROBLEM FOR SELF-STUDY

PROBLEM

Toulouse Engineering manufactures and sells contactors to electric companies. Tas Baitieri is examining individual customer profitability for Toulouse Engineering. He selects two representative customers. Summary data for 19_5 are (in thousands of francs):

	Total Sales	Manufacturing Cost of Goods Sold
Avignon Electric	F 860	F 694
Carcassonne Electric	F 490	F 417

Customer sales-related costs are currently allocated to individual customers using a rate of 15% of sales.

Baitieri attends an executive seminar run by SAM, a leading Swedish consulting firm on activity-based accounting. SAM presents an analysis of its engagement for Kanthal, a Swedish company selling electric heating elements. SAM reports that the two drivers of customer sales-related costs for Kanthal were the number of orders and whether an item was in-stock or out-of-stock.

Baitieri believes these two cost drivers also apply to Toulouse Engineering. He estimates the cost per order to be 0.6 thousand francs, the cost of handling an in-stock item to be 0.4 thousand francs, and the cost of handling an out-of-stock item to be 7.4 thousand francs. (Assume these cost figures are purely variable.) Thus, the sales-related cost for an in-stock order is 1.0 thousand francs (0.6 + 0.4) and the sales-related cost for an out-of-stock order is 8.0 thousand francs (0.6 + 7.4). He collects the following 19_5 information for the two representative customers:

	Number of In-Stock Orders	Number of Out-of-Stock Orders
Avignon Electric	8	20
Carcassonne Electric	10	4

Required

1. Compute the profitability of each customer using
   (a) The 15% of sales allocation method for customer sales-related costs
   (b) The activity-based accounting approach
   Comment on the results.
2. What should Toulouse Engineering do if it adopts the activity-based accounting approach to measuring individual customer profitability?

SOLUTION:

1(a)

	Avignon Electric	Carcassonne Electric
Sales	F 860.0	F 490.0
Costs:		
Manufacturing cost of goods sold	694.0	417.0
Customer sales-related costs (15% of sales)	129.0	73.5
Total costs	F 823.0	F 490.5
Gross margin	F 37.0	F (0.5)

1(b)

	Avignon Electric	Carcassonne Electric
Sales	F 860.0	F 490.0
Costs:		
Manufacturing cost of goods sold	694.0	417.0
In-stock costs (8,10 × F1.0)	8.0	10.0
Out-of-stock costs (20,4 × F8.0)	160.0	32.0
Total costs	F 862.0	F 459.0
Gross margin	F (2.0)	F 31.0

The two customers are ranked differently in terms of individual profitability under the two approaches. The sales franc allocation method shows Avignon Electric to be the more profitable customer. The activity-based accounting approach shows Carcassonne Electric to be the more profitable customer.

2. Actions Toulouse Engineering might take include the following:
   - Seek ways to reduce the cost per order, the cost of handling an in-stock item, and the cost of handling an out-of-stock item
   - Encourage each customer (especially Avignon Electric) to reduce the number of orders for out-of-stock items.

SUMMARY

Management accountants can increase their value to managers by remaining alert to the strategic areas in which companies compete and providing information to guide decisions in the areas of quality, time, and cost. Quality-related management control initiatives include cost of quality programs and quality-based performance measures. Time-related management control initiatives include breakeven time in capital budgeting and the use of time-based performance measures. Cost-related management control initiatives include value-added/non-value-added cost analysis and customer-profitability analysis. Cost-related management control initiatives include value added/no value added cost analysis and customer-profitability analysis.

Advances in strategic business thinking are expanding the ways management accounting can assist managers. It is likely that the rapid pace of experimentation in management accounting observed in recent years will continue in the 1990s.

TERMS TO LEARN

This chapter and the Glossary at the end of the book contain definitions of the following important terms:

appraisal costs *(p. 913)* cost of quality program *(913)*
external failure costs *(913)* internal failure costs *(913)*
new-product lead time *(915)* prevention costs *(913)* quality *(912)*
quality driver *(913)* time driver *(916)* time-to-market *(915)*

ASSIGNMENT MATERIAL

QUESTIONS

29-1 Give three examples of quality measures for a company assembling air-conditioning units.

29-2 Name two cost items classified as a prevention cost.

29-3 Distinguish between an *internal failure cost* and an *external failure cost.*

29-4 How is a cost of quality program different from a quality program based on a single business function?

29-5 In what three areas is time a key variable in competition?

29-6 Why should a company study quality drivers and time drivers?

29-7 "All firms are increasing the number of indirect cost application bases." Do you agree? Explain.

29-8 Give three examples of non-value-added costs that may be incurred at a plant assembling medical equipment.

29-9 Describe a three-step approach that can be used to implement a value-added/non-value-added approach to cost management.

29-10 "Why add customer-profitability analysis to accounting reports when we have more than enough difficulties handling product-profitability analysis?" How would you respond to this question?

EXERCISES AND PROBLEMS

29-11 Cost of quality program, nonfinancial quality measures. Baden Engineering manufactures automotive parts. A major customer has just given Baden an edict: "Improve quality or no more business." Hans Reichelstein, the controller of Baden Engineering, is given the task of developing a cost of quality program. He seeks your advice on classifying each of items *a–e* as (i) a prevention cost, (ii) an appraisal cost, (iii) an internal failure cost, or (iv) an external failure cost.

a. Cost of inspecting products on the production line by Baden quality inspectors
b. Payment of travel costs for a Baden Engineering customer representative to meet with a customer who detected defective products
c. Costs of reworking defective parts detected by Baden Engineering quality assurance group
d. Labor cost of product designer at Baden Engineering whose task is to design components that will not break under extreme temperature variations
e. Cost of automotive parts returned by customer

Required

1. Classify the five individual cost items into one of the four categories of prevention, appraisal, internal failure, or external failure.

2. Give two examples of nonfinancial performance measures Baden Engineering could monitor as part of a total quality control effort.

29-12 Cost of product quality program, classification of cost items. Home Quality Products (HQP) assembles and distributes household electrical products. The president of the company, with the enthusiastic backing of the board of directors, publicly announces a commitment by HQP to "excellence in product quality." A cost of product quality program is implemented to assist senior management in making decisions in this area.

Marci Young, an accountant at HQP, is charged with developing the cost accounting system to report on product quality. Individual cost items will be grouped into one of four categories: (i) prevention costs, (ii) appraisal costs, (iii) internal failure costs, and (iv) external failure costs.

Young receives the following cost items pertaining to HQP's steam-iron product line:

a. Labor and materials costs of reworking a batch of steam-iron handles at the Charlotte plant, $4,268.
b. Seminar costs for "Vendor Day," a program aiming to communicate to vendors the new quality requirements of HQP for all purchased components, $20,492.
c. Costs of conformance tests at the Charlotte plant to test the heating unit in steam irons, $8,409.
d. Replacement cost of 1,000 steam irons sold in the Pittsburgh area but returned because of a faulty water-spray unit, $28,628.
e. Costs of inspection tests at the Raleigh packaging plant to ensure that a complete set of warranty papers is included in each steam-iron box, $3,107.

Required

1. Classify the five individual cost items into one of the four categories of product quality cost (prevention, appraisal, internal failure, or external failure).
2. How might the cost of customer ill will created by the sale of defective products be estimated?
3. The production manager at the Charlotte plant is not enthusiastic about the extra work associated with HQP's cost of product quality program. He commented: "We are making detailed efforts to track quality costs at this plant. The time involved in adapting our reporting system to fit your format could be better spent on producing high-quality goods at a competitive cost." How should Young respond to this comment?

29-13 Measuring costs of product quality for manufacturer. Clean Shaven Inc. (CSI) produces electric shavers at its St. Louis, Missouri, plant. CSI, the sole North American manufacturer of electric shavers, has a 45% share of the North American market. Four manufacturers from Asia and Europe share the remainder of the market. Price and product quality are the two key areas in which companies compete in the electric shaver market.

CSI's manufacturing plant generally operates for one eight-hour shift, five days per week. Over 20,000 shavers can be produced on each eight-hour shift. The most important components in the shaver are the cutting system (made up of the cutter and the screen) and the motor. To manufacture the cutters, roles of steel are machine-stamped into rough blades that are then flattened and sharpened in lapping machines. In turn, these blades are fitted together to form the cutter unit that then passes through an ultrasonic cleaning system. The screens and the motors are purchased from outside suppliers. The cutters, screens, and motors pass through workstations where they are assembled along with electric cords and plastic molding into finished products. Finished products are packaged and sent to the finished goods warehouse for shipment.

Tory Kiam, the president of CSI, has a strong marketing focus. In recent years he has become concerned about the increasing quality level of the electric shavers sold by CSI's competitors. Two months ago, *Consumer Magazine* rated CSI's main brand of electric shaver as third in product quality of the five leading brands examined. A similar survey twelve months ago rated CSI's brand first in product quality. Kiam decided to devote more resources to product quality.

Kiam set up a task force that he headed to implement a formal quality improvement program. Part of this program involved developing a cost accounting system to track product quality costs. Mary Peterson was given responsibility for this project. She started with a system that several other consumer-oriented companies had used. Four categories of prod-

EXHIBIT 29-4

Cost of Quality Reporting by Clean Shaven Inc. for Eight Months of 19_2 (in thousands)

	January	February	March	April	May	June	July	August
PREVENTION								
Preventive machine maintenance	$ 166	$ 211	$ 196	$ 185	$ 209	$ 191	$ 211	$ 197
Seminars with suppliers	0	61	32	0	11	0	13	0
Product design reviews	202	166	167	159	179	185	192	189
	368	438	395	344	399	376	416	386
APPRAISAL								
Purchase inspection tests	67	72	69	31	24	25	23	21
Production line quality control	128	131	126	287	321	389	368	374
Final inspections and testing	157	161	154	191	327	317	347	326
	352	364	349	509	672	731	738	721
INTERNAL FAILURE								
Rework	647	623	631	651	550	496	421	396
Scrap	211	191	223	220	186	147	123	131
	858	814	854	871	736	643	544	527
EXTERNAL FAILURE								
Consumer repair claims	59	62	68	63	71	53	31	24
Product returns	235	247	221	262	251	122	116	87
	294	309	289	325	322	175	147	111
TOTAL COST OF QUALITY	$1,872	$ 1,925	$1,887	$ 2,049	$2,129	$1,925	$ 1,845	$ 1,745
TOTAL SALES	$8,508	$10,167	$8,986	$10,302	$8,406	$8,947	$11,206	$10,107

uct quality costs were distinguished: prevention, appraisal, internal failure, and external failure. Kiam's task force was set up in November 19_1. The following January (a low-production month), the product design staff held many seminars with all the employees of CSI. Starting in February, seminars were held with suppliers of CSI.

Mary Peterson's cost accounting project was implemented in January 19_2. She decided that for the first year (at least) she would restrict the cost of product quality system to items already recorded in the existing internal-reporting system. She viewed the task of collecting data already recorded in diverse parts of CSI and then presenting them in a single cost of quality format as challenging enough. At a later date she would consider recording opportunity costs that might be important when measuring the cost of product quality.

Exhibit 29-4 presents the monthly cost of product quality reports for the January 19_2 to August 19_2 period.

Required

1. What advantages might arise from developing a separate monthly cost of product quality report if the system is based only on data already recorded in the existing internal-reporting system?
2. What problems might Mary Peterson face in developing the cost of product quality system at CSI?
3. What trends are apparent in the monthly Cost of Product Quality reports for the January 19_2 to August 19_2 period? Comment on the uses management could make of the information in these reports.
4. What extensions might Mary Peterson consider if she is subsequently asked to develop a more comprehensive picture of the costs of product quality at CSI?

29-14 New-product proposals, life-cycle costing, breakeven time. (Solve Problem 21-22, p. 697, which is appropriate here.)

29-15 Product costing, effect of faster throughput time. Tulsa Caravans assembles prestige camping trailers at its state-of-the-art Tulsa plant. Demand for trailers is booming, and Tulsa Caravans has more demand than its assembly facility can satisfy. Direct manufacturing costs of its two highest-volume trailers in 19_8 are made up of only the direct materials costs:

	Ghengis Khan	Family Traveler
Direct materials	$6,000	$8,500

Indirect manufacturing costs (manufacturing overhead) are applied to each trailer on the basis of $80 per hour of throughput time. All manufacturing labor costs are included in indirect manufacturing costs.

In late 19_7 Kent Harris, production manager at Tulsa Caravans, made a major commitment to reducing the throughput time for each trailer. Average throughput time in hours per trailer for three selected months of 19_8 follows:

	Ghengis Khan	Family Traveler
January 19_8	60	75
June 19_8	47	61
December 19_8	39	47

Required

1. Compute the manufacturing costs of the Ghengis Khan and Family Traveler in January 19_8, June 19_8, and December 19_8.
2. The president of Tulsa Caravans believes that indirect manufacturing costs are driven by more variables than just throughput time. She seeks your advice on whether to revise the existing product-costing system to recognize multiple drivers of indirect manufacturing costs.

29-16 Effect of overhead cost application rate changes and product design changes on reported product costs. (Solve Problem 15-22, p. 517, which is appropriate here.)

29-17 Compensation linked with profitability, on-time delivery, and external quality performance measures. Pacific-Dunlop supplies tires to major automotive companies. It has two tire plants in North America—Detroit and Los Angeles. The quarterly bonus plan for each plant manager has three components:

1. Add 2% of operating income.
2. Add $10,000 if on-time delivery performance to the 10 major customers is 98% or better. If on-time performance is below 98%, add nothing.
3. Deduct 50% of cost of sales returns from 10 major customers.

These three components represent profitability performance, on-time delivery performance, and product quality performance.

Quarterly data for 19_7 pertaining to the Detroit and Los Angeles plants follow:

	Jan.–Mar.	Apr.–June	July–Sept.	Oct.–Dec.
DETROIT				
Operating income	$800,000	$850,000	$700,000	$900,000
On-time delivery	98.4%	98.6%	97.1%	97.9%
Cost of sales returns	$ 18,000	$ 26,000	$ 10,000	$ 25,000
LOS ANGELES				
Operating income	$1,600,000	$1,500,000	$1,800,000	$1,900,000
On-time delivery	95.6%	97.1%	97.9%	98.4%
Cost of sales returns	$ 35,000	$ 34,000	$ 28,000	$ 22,000

Required

1. Compute the bonus paid each quarter of 19_7 to the plant manager of the Detroit plant and to the plant manager of the Los Angeles plant.
2. Evaluate the three components of the bonus plan as measures of profitability, on-time delivery, and product quality.

29-18 Value-added versus non-valued-added cost classifications. Olivia Johns is manager of the Home Appliance plant of Newton Products. Johns decides to experiment with the value-added/non-value-added classification of costs in the accounting records. She selects the clothes dryer assembly product line as her test. She asks your advice on classifying items of labor time (and cost) in the plant:

a. Moving component parts from warehouse to assembly line
b. Assembling the tumbler unit
c. Expediting materials to the door-assembly area because of stock-balance error
d. Assembling the dial-presentation component
e. Inserting owner's manual and instruction guide in dryer package
f. Reworking faulty latches on doors
g. Testing the operating capabilities of the assembled unit
h. Packaging the clothes dryer in a breakage-resistant box

Required

1. What is the distinction between a value-added and a non-value-added cost?
2. Classify each of the eight labor time (and cost) items (a. to h.) as (i) a value-added cost, (ii) in the gray area, or (iii) a non-value-added cost.
3. How can Johns use your classifications in requirement 2?
4. Johns attends a conference where a well-known cost accounting writer expresses a great deal of cynicism about the value-added/non-value-added distinction. He calls it a "fad with a shelf life shorter than freshly cut roses." Johns has the chance to meet the cost accounting writer. What question should she pose to him about her proposed experiment with the value-added/non-value-added cost distinction?

29-19 Individual customer-profitability analysis (extension of Problem for Self-Study, p. 921) Tas Baitieri collects information on two customers in 19_5 (in thousands of francs):

	Total Sales	Manufacturing Cost of Goods Sold
Lyon Electric	F 746	F 582
Perpignan Electric	F 320	F 243

	Number of In-Stock Orders	Number of Out-of-Stock Orders
Lyon Electric	7	6
Perpignan Electric	2	12

Required

1. Compute the profitability of each customer using
   a. The 15% of sales franc allocation method for customer sales-related costs
   b. the activity-based accounting approach
   Comment on the results.
2. "Toulouse Engineering should discontinue sales to any customer showing a negative gross margin." Do you agree? Explain.

APPENDIX A

Recommended Readings

This literature on cost accounting and related areas is vast and varied. The footnotes in this book contain numerous citations to articles and books.

Advanced textbooks and casebooks related to topics covered in this book include:

ANTHONY, R., J. DEARDEN, and N. BEDFORD, *Management Control Systems,* 6th ed. Homewood, IL: Richard D. Irwin, 1989.

ANTHONY, R., and D. YOUNG, *Management Control in Nonprofit Organizations,* 4th ed. Homewood, IL: Richard D. Irwin, 1988.

FRECKA, T., ed., *Accounting for Manufacturing Productivity.* Wheeling, IL: Association for Manufacturing Excellence, 1988.

KAPLAN, R., and A. ATKINSON, *Advanced Management Accounting.* Englewood Cliffs, NJ: Prentice Hall, 1989.

MACIARIELLO, J., *Management Control Systems.* Englewood Cliffs, NJ: Prentice Hall, 1984.

MAGEE, R., *Advanced Managerial Accounting.* New York: Harper & Row, 1986.

RAMANATHAN, K., *Management Control in Nonprofit Organizations: Text and Cases.* New York: John Wiley, 1982.

ROTCH, W., and B. ALLEN, *Cases in Management Accounting and Control Systems,* 2d ed. Englewood Cliffs, NJ: Prentice Hall, 1990.

SHANK, J., and V. GOVINDARAJAN, *Strategic Cost Analysis.* Homewood, IL: Richard D. Irwin, 1989.

Books that explore specific topics in more detail include:

ALSTON, F., F. JOHNSON, M. WORTHINGTON, L. GOLDSMAN, and F. DEVITO, *Contracting with the Federal Government.* New York: John Wiley, 1984.

ATKINSON, A., *Intra-firm Cost and Resource Allocations: Theory and Practice.* Society of Management Accountants of Canada and Canadian Academic Accounting Association Research Monograph, 1987.

BEDINGFIELD, J., and L. ROSEN, *Government Contract Accounting.* Washington, DC: Federal Publications Inc., 1985.

COOPER, D., R. SCAPENS, and J. ARNOLD, *Management Accounting Research and Practice.* London: Institute of Cost and Management Accountants, 1983.

EUSKE, K., *Management Control: Planning, Control, Measurement and Evaluation.* Reading, MA: Addison-Wesley, 1984.

MANES, R., and C. CHENG, *The Marginal Approach to Joint Cost Allocation: Theory and Application.* Sarasota, FL: American Accounting Association, 1988.

MERCHANT, K., *Control in Business Organizations.* Boston: Pitman, 1984.

Books of readings, handbooks, and bibliographies related to cost or management accounting include:

ASHTON, R., ed., *The Evolution of Behavioral Accounting Research.* New York: Garland Publishing, 1984.

BELL, J., ed., *Accounting Control Systems: A Behavioral and Technical Integration.* New York: Markus Wiener, 1983.

BERRY, E., and G. HARWOOD, eds., *Governmental and Nonprofit Accounting: A Book of Readings.* Homewood, IL: Richard D. Irwin, 1984.

BULLOCH, J., D. KELLER, and L. VLASHO, eds., *Accountants' Cost Handbook,* 3d ed. New York: John Wiley, 1983.

CAPETTINI, R., and D. CLANCY, eds., *Cost Accounting, Robotics, and New Manufacturing Environment.* Sarasota, FL: American Accounting Association, 1987.

MONDEN, Y., and M. SAKURAI, eds., *Japanese Management Accounting.* Cambridge, MA: Productivity Press, 1989.

RAMANATHAN, K., and L. HEGSTAD, eds., *Readings in Management Control in Nonprofit Organizations.* New York: John Wiley, 1982.

ROSEN, L., ed., *Topics in Management Accounting.* Toronto: McGraw-Hill, 1984.

THOMAS, W., ed., *Readings in Cost Accounting, Budgeting, and Control,* 7th ed. Cincinnati: South-Western Publishing, 1988.

TURNEY, P., ed., *Performance Excellence in Manufacturing and Service Organizations.* Sarasota, FL: American Accounting Association, 1990.

ZIMMERMAN, V., ed., *Managerial Accounting: An Analysis of Current International Applications.* Urbana-Champaign: University of Illinois, 1984.

The Harvard Business School series in accounting and control offers important contributions to the cost accounting literature, including:

ANTHONY, R., *The Management Control Function.* Boston: Harvard Business School Press, 1988.

Berliner, C., and J. Brimson, eds., *Cost Management for Today's Advanced Manufacturing: The CAM-I Conceptual Design.* Boston: Harvard Business School Press, 1988.

Bruns, W., and R. Kaplan, eds., *Accounting and Management: Field Study Perspectives.* Boston: Harvard Business School Press, 1987.

Johnson, H., and R. Kaplan, *Relevance Lost: The Rise and Fall of Management Accounting.* Boston: Harvard Business School Press, 1987.

Kaplan, R., ed., *Measures for Manufacturing Excellence.* Boston: Harvard Business School Press, 1990.

Merchant, K. A., *Rewarding Results: Motivating Profit Center Managers.* Boston: Harvard Business School Press, 1989.

The Society of Management Accountants of Canada publishes a monograph series that presents coverage of many cost accounting topics, including:

Armitage, H., and A. Atkinson, *The Choice of Productivity Measures in Organizations: A Field Study of Practice in Seven Canadian Firms.* Hamilton, Ontario: Society of Management Accountants of Canada, 1990.

Sherman, H., *Service Organization Productivity Management.* Hamilton, Ontario: Society of Management Accountants of Canada, 1989.

The National Association of Accountants publishes monographs and books covering cost accounting topics, such as:

Bennett, R., J. Hendricks, D. Keys, and E. Rudnicki, *Cost Accounting for Factory Automation.* Montvale, NJ: National Association of Accountants, 1987.

Howell, R., and S. Soucy, *Factory 2000+: Management Accounting's Changing Role.* Montvale, NJ: National Association of Accountants, 1988.

McNair, C., W. Mosconi, and T. Norris, *Meeting the Technology Challenge: Cost Accounting in a JIT Environment.* Montvale, NJ: National Association of Accountants, 1988.

The Financial Executives Research Foundation publishes monographs and books concerning topics of interest to financial executives, such as:

Keating, P., and S. Jablonsky, *Changing Roles of Financial Management.* Morristown, NJ: Financial Executives Research Foundation, 1990.

The following are detailed annotated bibliographies of the cost and management accounting research literatures:

Clancy, D. *Annotated Management Accounting Readings.* Management Accounting Section of the American Accounting Association, 1986.

Deakin, E., M. Maher, and J. Cappel, *Contemporary Literature in Cost Accounting.* Homewood, IL: Richard D. Irwin, 1988.

Klemstine, C., and M. Maher, *Management Accounting Research: 1926–1983.* New York: Garland Publishing, 1984.

The *Journal of Cost Management for the Manufacturing Industry* contains numerous articles on modern management accounting. It is published by Warren, Gorham and Lamont, 210 South Street, Boston, MA 02111.

Two journals bearing on management accounting are published by sections of the American Accounting Association, 5717 Bessie Drive, Sarasota, FL 34233: *Journal of Management Accounting Research and Behavioral Research in Accounting.*

Professional associations that specialize in serving members with cost and management accounting interests include:

- *National Association of Accountants,* 10 Paragon Drive, P.O. Box 433, Montvale, NJ 07645. Publishes the *Management Accountant* journal.
- *Financial Executives Institute,* 10 Madison Avenue, P.O. Box 1938, Morristown, NJ 07960. Publishes *Financial Executive.*
- *Society of Cost Estimating and Analysis,* 101 South Whiting Street, Suite 313, Alexandria, VA 22304. Publishes the *Journal of Cost Analysis* and monographs related to cost estimation and price analysis in government and industry.
- *The Institute of Internal Auditors,* 249 Maitland Avenue, Altamonte Springs, FL 32701. Publishes *The Internal Auditor* journal. Also publishes monographs on topics related to internal control.
- *Society of Management Accountants of Canada,* 154 Main Street East, MPO Box 176, Hamilton, Ontario, L8N 3C3. Publishes the *CMA* magazine.
- *The Institute of Cost and Management Accountants,* 63 Portland Place, London, WIN 4AB. Publishes the *Management Accounting* journal. Also publishes monographs covering cost and managerial accounting topics.

In many countries, individuals with cost and management accounting interests belong to professional bodies that serve members with financial reporting and taxation, as well as cost and management accounting, interests. An example is the Australian Society of Accountants.

APPENDIX B

Notes on Compound Interest and Interest Tables

Interest is the cost of using money. It is the rental charge for funds, just as renting a building and equipment entails a rental charge. When the funds are used for a period of time, it is necessary to recognize interest as a cost of using the borrowed ("rented") funds. This requirement applies even if the funds represent ownership capital and if interest does not entail an outlay of cash. Why must interest be considered? Because the selection of one alternative automatically commits a given amount of funds that could otherwise be invested in some other alternative.

Interest is generally important, even when short-term projects are under consideration. Interest looms correspondingly larger when long-run plans are studied. The rate of interest has significant enough impact to influence decisions regarding borrowing and investing funds. For example, $100,000 invested now and compounded annually for ten years at 8% will accumulate to $215,900; at 20%, the $100,000 will accumulate to $619,200.

INTEREST TABLES

Many computer programs and pocket calculators are available that handle computations involving the time value of money. You may also turn to the following four basic tables to compute interest.

Table 1—Future Amount of $1

Table 1 shows how much $1 invested now will accumulate in a given number of periods at a given compounded interest rate per period. Consider investing $1,000 now for three years at 8% compound interest. A tabular presentation of how this $1,000 would accumulate to $1,259.70 follows:

Year	Interest per Year	Cumulative Interest Called Compound Interest	Total at End of Year
0	$ —	$ —	$1,000.00
1	80.00	80.00	1,080.00
2	86.40	166.40	1,166.40
3	93.30	259.70	1,259.70

This tabular presentation is a series of computations that could appear as follows:

$$S_1 = \$1{,}000(1.08)^1$$
$$S_2 = \$1{,}000(1.08)^2$$
$$S_3 = \$1{,}000(1.08)^3$$

The formula for the "amount of 1," often called the "future value of \$1" or "future amount of \$1," can be written:

$$S = P(1 + r)^n$$
$$S = \$1{,}000(1 + .08)^3 = \$1{,}259.70$$

S is the future value amount; P is the present value, \$1,000 in this case; r is the rate of interest; and n is the number of time periods.

Fortunately, tables make key computations readily available. A facility in selecting the *proper* table will minimize computations. Check the accuracy of the answer above using Table 1, page 934.

Table 2—Present Value of \$1

In the previous example, if \$1,000 compounded at 8% per year will accumulate to \$1,259.70 in three years, then \$1,000 must be the present value of \$1,259.70 due at the end of three years. The formula for the present value can be derived by reversing the process of *accumulation* (finding the future amount) that we just finished.

If

$$S = P(1 + r)^n$$

then

$$P = \frac{S}{(1 + r)^n}$$

$$P = \frac{\$1{,}259.70}{(1.08)^3} = \$1{,}000$$

Use Table 2, page 935, to check this calculation.

When accumulating, we advance or roll forward in time. The difference between our original amount and our accumulated amount is called *compound interest.* When discounting, we retreat or roll back in time. The difference between the future amount and the present value is called *compound discount.* Note the following formulas (where $P = \$1{,}000$):

$$\text{Compound interest} = P[(1 + r)^n - 1] = \$259.70$$

$$\text{Compound discount} = S\left[1 - \frac{1}{(1 + r)^n}\right] = \$259.70$$

Table 3—Amount of Annuity of \$1

An (ordinary) *annuity* is a series of equal payments (receipts) to be paid (or received) at the *end* of successive periods of equal length. Assume that \$1,000 is invested at the end of each of three years at 8%:

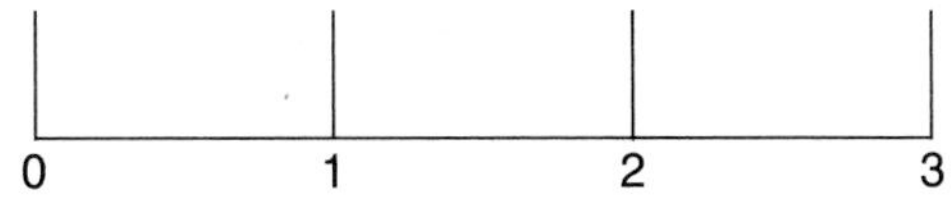

End of Year			Amount
1st payment	\$1,000.00 →	\$1,080.00 →	\$1,166.40, which is $\$1{,}000(1.08)^2$
2nd payment		\$1,000.00 →	1,080.00, which is $\$1{,}000(1.08)^1$
3rd payment			1,000.00
Accumulation (future amount)			\$3,246.40

The arithmetic shown above may be expressed algebraically as the amount of an ordinary annuity of \$1,000 for three years = $\$1{,}000(1 + r)^2 + \$1{,}000(1 + r)^1 + \$1{,}000$.

We can develop the general formula for S_n, the amount of an ordinary annuity of \$1, by using the example above as a basis:

1.	$S_n = 1 + (1 + r)^1 + (1 + r)^2$
2. Substitute:	$S_n = 1 + (1.08)^1 + (1.08)^2$
3. Multiply (2) by $(1 + r)$:	$(1.08)S_n = (1.08)^1 + (1.08)^2 + (1.08)^3$
4. Subtract (2) from (3): Note that all terms on the right-hand side are removed except $(1.08)^3$ in equation (3) and 1 in equation (2).	$1.08S_n - S_n = (1.08)^3 - 1$
5. Factor (4):	$S_n(1.08 - 1) = (1.08)^3 - 1$
6. Divide (5) by $(1.08 - 1)$:	$S_n = \frac{(1.08)^3 - 1}{1.08 - 1} = \frac{(1.08)^3 - 1}{.08}$
7. The general formula for the amount of an ordinary annuity of \$1 becomes:	$S_n = \frac{(1 + r)^n - 1}{r}$ or $\frac{\text{Compound interest}}{\text{Rate}}$

This formula is the basis for Table 3, page 936. Look at Table 3 or use the formula itself to check the calculations.

Table 4—Present Value of an Ordinary Annuity of \$1

Using the same example as for Table 3, we can show how the formula of P_n, *the present value of an ordinary annuity,* is developed.

End of Year			0	1	2	3
1st payment	$\frac{1,000}{(1.08)^1}$ =	\$ 926.14	←	\$1,000		
2nd payment	$\frac{1,000}{(1.08)^2}$ =	\$ 857.52	←	←	\$1,000	
3rd payment	$\frac{1,000}{(1.08)^3}$ =	\$ 794.00	←	←	←	\$1,000
Total present value		\$2,577.66				

For the general case, the present value of an ordinary annuity of \$1 may be expressed:

1.	$P_n = \frac{1}{1 + r} + \frac{1}{(1 + r)^2} + \frac{1}{(1 + r)^3}$
2. Substitute:	$P_n = \frac{1}{1.08} + \frac{1}{(1.08)^2} + \frac{1}{(1.08)^3}$
3. Multiply by $\frac{1}{1.08}$:	$P_n\frac{1}{1.08} = \frac{1}{(1.08)^2} + \frac{1}{(1.08)^3} + \frac{1}{(1.08)^4}$
4. Subtract (3) from (2):	$P_n - P_n\frac{1}{1.08} = \frac{1}{1.08} - \frac{1}{(1.08)^4}$
5. Factor:	$P_n\left(1 - \frac{1}{1.08}\right) = \frac{1}{1.08}\left[1 - \frac{1}{(1.08)^3}\right]$
6. or	$P_n\left(\frac{.08}{1.08}\right) = \frac{1}{1.08}\left[1 - \frac{1}{(1.08)^3}\right]$

7. Multiply by $\frac{1.08}{.08}$:

$$P_n = \frac{1}{.08}\left[1 - \frac{1}{(1.08)^3}\right]$$

The general formula for the present value of an annuity of $1.00 is:

$$P_n = \frac{1}{r}\left[1 - \frac{1}{(1 + r)^n}\right] = \frac{\text{Compound discount}}{\text{Rate}}$$

Solving,

$$P_n = \frac{.2062}{.08} = 2.577$$

The formula is the basis for Table 4, page 937. Check the answer in the table. The present value tables, Tables 2 and 4, are used most frequently in capital budgeting.

The tables for annuities are not essential. With Tables 1 and 2, compound interest and compound discount can readily be computed. It is simply a matter of dividing either of these by the rate to get values equivalent to those shown in Tables 3 and 4.

TABLE 1

Compound Amount of $1.00 (The Future Value of $1.00)
$S = P(1 + r)^n$. In this table $P = \$1.00$.

PERIODS	2%	4%	6%	8%	10%	12%	14%	16%	18%	20%	22%	24%	26%	28%	30%	32%	40%	PERIODS
1	1.020	1.040	1.060	1.080	1.100	1.120	1.140	1.160	1.180	1.200	1.220	1.240	1.260	1.280	1.300	1.320	1.400	1
2	1.040	1.082	1.124	1.166	1.210	1.254	1.300	1.346	1.392	1.440	1.488	1.538	1.588	1.638	1.690	1.742	1.960	2
3	1.061	1.125	1.191	1.260	1.331	1.405	1.482	1.561	1.643	1.728	1.816	1.907	2.000	2.097	2.197	2.300	2.744	3
4	1.082	1.170	1.262	1.360	1.464	1.574	1.689	1.811	1.939	2.074	2.215	2.364	2.520	2.684	2.856	3.036	3.842	4
5	1.104	1.217	1.338	1.469	1.611	1.762	1.925	2.100	2.288	2.488	2.703	2.932	3.176	3.436	3.713	4.007	5.378	5
6	1.126	1.265	1.419	1.587	1.772	1.974	2.195	2.436	2.700	2.986	3.297	3.635	4.002	4.398	4.827	5.290	7.530	6
7	1.149	1.316	1.504	1.714	1.949	2.211	2.502	2.826	3.185	3.583	4.023	4.508	5.042	5.629	6.275	6.983	10.541	7
8	1.172	1.369	1.594	1.851	2.144	2.476	2.853	3.278	3.759	4.300	4.908	5.590	6.353	7.206	8.157	9.217	14.758	8
9	1.195	1.423	1.689	1.999	2.358	2.773	3.252	3.803	4.435	5.160	5.987	6.931	8.005	9.223	10.604	12.166	20.661	9
10	1.219	1.480	1.791	2.159	2.594	3.106	3.707	4.411	5.234	6.192	7.305	8.594	10.086	11.806	13.786	16.060	28.925	10
11	1.243	1.539	1.898	2.332	2.853	3.479	4.226	5.117	6.176	7.430	8.912	10.657	12.708	15.112	17.922	21.199	40.496	11
12	1.268	1.601	2.012	2.518	3.138	3.896	4.818	5.936	7.288	8.916	10.872	13.215	16.012	19.343	23.298	27.983	56.694	12
13	1.294	1.665	2.133	2.720	3.452	4.363	5.492	6.886	8.599	10.699	13.264	16.386	20.175	24.759	30.288	36.937	79.371	13
14	1.319	1.732	2.261	2.937	3.797	4.887	6.261	7.988	10.147	12.839	16.182	20.319	25.421	31.691	39.374	48.757	111.120	14
15	1.346	1.801	2.397	3.172	4.177	5.474	7.138	9.266	11.974	15.407	19.742	25.196	32.030	40.565	51.186	64.359	155.568	15
16	1.373	1.873	2.540	3.426	4.595	6.130	8.137	10.748	14.129	18.488	24.086	31.243	40.358	51.923	66.542	84.954	217.795	16
17	1.400	1.948	2.693	3.700	5.054	6.866	9.276	12.468	16.672	22.186	29.384	38.741	50.851	66.461	86.504	112.139	304.913	17
18	1.428	2.026	2.854	3.996	5.560	7.690	10.575	14.463	19.673	26.623	35.849	48.039	64.072	85.071	112.455	148.024	426.879	18
19	1.457	2.107	3.026	4.316	6.116	8.613	12.056	16.777	23.214	31.948	43.736	59.568	80.731	108.890	146.192	195.391	597.630	19
20	1.486	2.191	3.207	4.661	6.727	9.646	13.743	19.461	27.393	38.338	53.358	73.864	101.721	139.380	190.050	257.916	836.683	20
21	1.516	2.279	3.400	5.034	7.400	10.804	15.668	22.574	32.324	46.005	65.096	91.592	128.169	178.406	247.065	340.449	1171.356	21
22	1.546	2.370	3.604	5.437	8.140	12.100	17.861	26.186	38.142	55.206	79.418	113.574	161.492	228.360	321.184	449.393	1639.898	22
23	1.577	2.465	3.820	5.871	8.954	13.552	20.362	30.376	45.008	66.247	96.889	140.831	203.480	292.300	417.539	593.199	2295.857	23
24	1.608	2.563	4.049	6.341	9.850	15.179	23.212	35.236	53.109	79.497	118.205	174.631	256.385	374.144	542.801	783.023	3214.200	24
25	1.641	2.666	4.292	6.848	10.835	17.000	26.462	40.874	62.669	95.396	144.210	216.542	323.045	478.905	705.641	1033.590	4499.880	25
26	1.673	2.772	4.549	7.396	11.918	19.040	30.167	47.414	73.949	114.475	175.936	268.512	407.037	612.998	917.333	1364.339	6299.831	26
27	1.707	2.883	4.822	7.988	13.110	21.325	34.390	55.000	87.260	137.371	214.642	332.955	512.867	784.638	1192.533	1800.927	8819.764	27
28	1.741	2.999	5.112	8.627	14.421	23.884	39.204	63.800	102.967	164.845	261.864	412.864	646.212	1004.336	1550.293	2377.224	12347.670	28
29	1.776	3.119	5.418	9.317	15.863	26.750	44.693	74.009	121.501	197.814	319.474	511.952	814.228	1285.550	2015.381	3137.935	17286.737	29
30	1.811	3.243	5.743	10.063	17.449	29.960	50.950	85.850	143.371	237.376	389.758	634.820	1025.927	1645.505	2619.996	4142.075	24201.432	30
35	2.000	3.946	7.686	14.785	28.102	52.800	98.100	180.314	327.997	590.668	1053.402	1861.054	3258.135	5653.911	9727.860	16599.217	130161.112	35
40	2.208	4.801	10.286	21.725	45.259	93.051	188.884	378.721	750.378	1469.772	2847.038	5455.913	10347.175	19426.689	36118.865	66520.767	700037.697	40

TABLE 2 *(Place a clip on this page for easy reference.)*

Present Value of $1.00.

$P = \frac{S}{(1 + r)^n}$. In this table S = $1.00.

PERIODS	2%	4%	6%	8%	10%	12%	14%	16%	18%	20%	22%	24%	26%	28%	30%	32%	40%	PERIODS
1	0.980	0.962	0.943	0.926	0.909	0.893	0.877	0.862	0.847	0.833	0.820	0.806	0.794	0.781	0.769	0.758	0.714	1
2	0.961	0.925	0.890	0.857	0.826	0.797	0.769	0.743	0.718	0.694	0.672	0.650	0.630	0.610	0.592	0.574	0.510	2
3	0.942	0.889	0.840	0.794	0.751	0.712	0.675	0.641	0.609	0.579	0.551	0.524	0.500	0.477	0.455	0.435	0.364	3
4	0.924	0.855	0.792	0.735	0.683	0.636	0.592	0.552	0.516	0.482	0.451	0.423	0.397	0.373	0.350	0.329	0.260	4
5	0.906	0.822	0.747	0.681	0.621	0.567	0.519	0.476	0.437	0.402	0.370	0.341	0.315	0.291	0.269	0.250	0.186	5
6	0.888	0.790	0.705	0.630	0.564	0.507	0.456	0.410	0.370	0.335	0.303	0.275	0.250	0.227	0.207	0.189	0.133	6
7	0.871	0.760	0.665	0.583	0.513	0.452	0.400	0.354	0.314	0.279	0.249	0.222	0.198	0.178	0.159	0.143	0.095	7
8	0.853	0.731	0.627	0.540	0.467	0.404	0.351	0.305	0.266	0.233	0.204	0.179	0.157	0.139	0.123	0.108	0.068	8
9	0.837	0.703	0.592	0.500	0.424	0.361	0.308	0.263	0.225	0.194	0.167	0.144	0.125	0.108	0.094	0.082	0.048	9
10	0.820	0.676	0.558	0.463	0.386	0.322	0.270	0.227	0.191	0.162	0.137	0.116	0.099	0.085	0.073	0.062	0.035	10
11	0.804	0.650	0.527	0.429	0.350	0.287	0.237	0.195	0.162	0.135	0.112	0.094	0.079	0.066	0.056	0.047	0.025	11
12	0.788	0.625	0.497	0.397	0.319	0.257	0.208	0.168	0.137	0.112	0.092	0.076	0.062	0.052	0.043	0.036	0.018	12
13	0.773	0.601	0.469	0.368	0.290	0.229	0.182	0.145	0.116	0.093	0.075	0.061	0.050	0.040	0.033	0.027	0.013	13
14	0.758	0.577	0.442	0.340	0.263	0.205	0.160	0.125	0.099	0.078	0.062	0.049	0.039	0.032	0.025	0.021	0.009	14
15	0.743	0.555	0.417	0.315	0.239	0.183	0.140	0.108	0.084	0.065	0.051	0.040	0.031	0.025	0.020	0.016	0.006	15
16	0.728	0.534	0.394	0.292	0.218	0.163	0.123	0.093	0.071	0.054	0.042	0.032	0.025	0.019	0.015	0.012	0.005	16
17	0.714	0.513	0.371	0.270	0.198	0.146	0.108	0.080	0.060	0.045	0.034	0.026	0.020	0.015	0.012	0.009	0.003	17
18	0.700	0.494	0.350	0.250	0.180	0.130	0.095	0.069	0.051	0.038	0.028	0.021	0.016	0.012	0.009	0.007	0.002	18
19	0.686	0.475	0.331	0.232	0.164	0.116	0.083	0.060	0.043	0.031	0.023	0.017	0.012	0.009	0.007	0.005	0.002	19
20	0.673	0.456	0.312	0.215	0.149	0.104	0.073	0.051	0.037	0.026	0.019	0.014	0.010	0.007	0.005	0.004	0.001	20
21	0.660	0.439	0.294	0.199	0.135	0.093	0.064	0.044	0.031	0.022	0.015	0.011	0.008	0.006	0.004	0.003	0.001	21
22	0.647	0.422	0.278	0.184	0.123	0.083	0.056	0.038	0.026	0.018	0.013	0.009	0.006	0.004	0.003	0.002	0.001	22
23	0.634	0.406	0.262	0.170	0.112	0.074	0.049	0.033	0.022	0.015	0.010	0.007	0.005	0.003	0.002	0.002	0.000	23
24	0.622	0.390	0.247	0.158	0.102	0.066	0.043	0.028	0.019	0.013	0.008	0.006	0.004	0.003	0.002	0.001	0.000	24
25	0.610	0.375	0.233	0.146	0.092	0.059	0.038	0.024	0.016	0.010	0.007	0.005	0.003	0.002	0.001	0.001	0.000	25
26	0.598	0.361	0.220	0.135	0.084	0.053	0.033	0.021	0.014	0.009	0.006	0.004	0.002	0.002	0.001	0.001	0.000	26
27	0.586	0.347	0.207	0.125	0.076	0.047	0.029	0.018	0.011	0.007	0.005	0.003	0.002	0.001	0.001	0.001	0.000	27
28	0.574	0.333	0.196	0.116	0.069	0.042	0.026	0.016	0.010	0.006	0.004	0.002	0.002	0.001	0.001	0.000	0.000	28
29	0.563	0.321	0.185	0.107	0.063	0.037	0.022	0.014	0.008	0.005	0.003	0.002	0.001	0.001	0.000	0.000	0.000	29
30	0.552	0.308	0.174	0.099	0.057	0.033	0.020	0.012	0.007	0.004	0.003	0.002	0.001	0.001	0.000	0.000	0.000	30
35	0.500	0.253	0.130	0.068	0.036	0.019	0.010	0.006	0.003	0.002	0.001	0.001	0.000	0.000	0.000	0.000	0.000	35
40	0.453	0.208	0.097	0.046	0.022	0.011	0.005	0.003	0.001	0.001	0.000	0.000	0.000	0.000	0.000	0.000	0.000	40

TABLE 3

Compound Amount of Annuity of $1.00 in Arrears* (Future Value of Annuity)

$$S_n = \frac{(1 + r)^n - 1}{r}$$

PERIODS	2%	4%	6%	8%	10%	12%	14%	16%	18%	20%	22%	24%	26%	28%	30%	32%	40%	PERIODS
1	1.000	1.000	1.000	1.000	1.000	1.000	1.000	1.000	1.000	1.000	1.000	1.000	1.000	1.000	1.000	1.000	1.000	1
2	2.020	2.040	2.060	2.080	2.100	2.120	2.140	2.160	2.180	2.200	2.220	2.240	2.260	2.280	2.300	2.320	2.400	2
3	3.060	3.122	3.184	3.246	3.310	3.374	3.440	3.506	3.572	3.640	3.708	3.778	3.848	3.918	3.990	4.062	4.360	3
4	4.122	4.246	4.375	4.506	4.641	4.779	4.921	5.066	5.215	5.368	5.524	5.684	5.848	6.016	6.187	6.362	7.104	4
5	5.204	5.416	5.637	5.867	6.105	6.353	6.610	6.877	7.154	7.442	7.740	8.048	8.368	8.700	9.043	9.398	10.946	5
6	6.308	6.633	6.975	7.336	7.716	8.115	8.536	8.977	9.442	9.930	10.442	10.980	11.544	12.136	12.756	13.406	16.324	6
7	7.434	7.898	8.394	8.923	9.487	10.089	10.730	11.414	12.142	12.916	13.740	14.615	15.546	16.534	17.583	18.696	23.853	7
8	8.583	9.214	9.897	10.637	11.436	12.300	13.233	14.240	15.327	16.499	17.762	19.123	20.588	22.163	23.858	25.678	34.395	8
9	9.755	10.583	11.491	12.488	13.579	14.776	16.085	17.519	19.086	20.799	22.670	24.712	26.940	29.369	32.015	34.895	49.153	9
10	10.950	12.006	13.181	14.487	15.937	17.549	19.337	21.321	23.521	25.959	28.657	31.643	34.945	38.593	42.619	47.062	69.814	10
11	12.169	13.486	14.972	16.645	18.531	20.655	23.045	25.733	28.755	32.150	35.962	40.238	45.031	50.398	56.405	63.122	98.739	11
12	13.412	15.026	16.870	18.977	21.384	24.133	27.271	30.850	34.931	39.581	44.874	50.895	57.739	65.510	74.327	84.320	139.235	12
13	14.680	16.627	18.882	21.495	24.523	28.029	32.089	36.786	42.219	48.497	55.746	64.110	73.751	84.853	97.625	112.303	195.929	13
14	15.974	18.292	21.015	24.215	27.975	32.393	37.581	43.672	50.818	59.196	69.010	80.496	93.926	109.612	127.913	149.240	275.300	14
15	17.293	20.024	23.276	27.152	31.772	37.280	43.842	51.660	60.965	72.035	85.192	100.815	119.347	141.303	167.286	197.997	386.420	15
16	18.639	21.825	25.673	30.324	35.950	42.753	50.980	60.925	72.939	87.442	104.935	126.011	151.377	181.868	218.472	262.356	541.988	16
17	20.012	23.698	28.213	33.750	40.545	48.884	59.118	71.673	87.068	105.931	129.020	157.253	191.735	233.791	285.014	347.309	759.784	17
18	21.412	25.645	30.906	37.450	45.599	55.750	68.394	84.141	103.740	128.117	158.405	195.994	242.585	300.252	371.518	459.449	1064.697	18
19	22.841	27.671	33.760	41.446	51.159	63.440	78.969	98.603	123.414	154.740	194.254	244.033	306.658	385.323	483.973	607.472	1491.576	19
20	24.297	29.778	36.786	45.762	57.275	72.052	91.025	115.380	146.628	186.688	237.989	303.601	387.389	494.213	630.165	802.863	2089.206	20
21	25.783	31.969	39.993	50.423	64.002	81.699	104.768	134.841	174.021	225.026	291.347	377.465	489.110	633.593	820.215	1060.779	2925.889	21
22	27.299	34.248	43.392	55.457	71.403	92.503	120.436	157.415	206.345	271.031	356.443	469.056	617.278	811.999	1067.280	1401.229	4097.245	22
23	28.845	36.618	46.996	60.893	79.543	104.603	138.297	183.601	244.487	326.237	435.861	582.630	778.771	1040.358	1388.464	1850.622	5737.142	23
24	30.422	39.083	50.816	66.765	88.497	118.155	158.659	213.978	289.494	392.484	532.750	723.461	982.251	1332.659	1806.003	2443.821	8032.999	24
25	32.030	41.646	54.865	73.106	98.347	133.334	181.871	249.214	342.603	471.981	650.955	898.092	1238.636	1706.803	2348.803	3226.844	11247.199	25
26	33.671	44.312	59.156	79.954	109.182	150.334	208.333	290.088	405.272	567.377	795.165	1114.634	1561.682	2185.708	3054.444	4260.434	15747.079	26
27	35.344	47.084	63.706	87.351	121.100	169.374	238.499	337.502	479.221	681.853	971.102	1383.146	1968.719	2798.706	3971.778	5624.772	22046.910	27
28	37.051	49.968	68.528	95.339	134.210	190.699	272.889	392.503	566.481	819.223	1185.744	1716.101	2481.586	3583.344	5164.311	7425.699	30866.674	28
29	38.792	52.966	73.640	103.966	148.631	214.583	312.094	456.303	669.447	984.068	1447.608	2128.965	3127.798	4587.680	6714.604	9802.923	43214.343	29
30	40.568	56.085	79.058	113.283	164.494	241.333	356.787	530.312	790.948	1181.882	1767.081	2640.916	3942.026	5873.231	8729.985	12940.859	60501.081	30
35	49.994	73.652	111.435	172.317	271.024	431.663	693.573	1120.713	1816.652	2948.341	4783.645	7750.225	12527.442	20188.966	32422.868	51869.427	325400.279	35
40	60.402	95.026	154.762	259.057	442.593	767.091	1342.025	2360.757	4163.213	7343.858	12936.535	22728.803	39792.982	69377.460	120392.883	207874.272	1750091.741	40

*Payments (or receipts) at the end of each period.

TABLE 4 *(Place a clip on this page for easy reference.)*

Present Value of Annuity of $1.00 in Arrears*

$$P_n = \frac{1}{r}\left[1 - \frac{1}{(1+r)^n}\right]$$

PERIODS	2%	4%	6%	8%	10%	12%	14%	16%	18%	20%	22%	24%	26%	28%	30%	32%	40%	PERIODS
1	0.980	0.962	0.943	0.926	0.909	0.893	0.877	0.862	0.847	0.833	0.820	0.806	0.794	0.781	0.769	0.758	0.714	1
2	1.942	1.886	1.833	1.783	1.736	1.690	1.647	1.605	1.566	1.528	1.492	1.457	1.424	1.392	1.361	1.331	1.224	2
3	2.884	2.775	2.673	2.577	2.487	2.402	2.322	2.246	2.174	2.106	2.042	1.981	1.923	1.868	1.816	1.766	1.589	3
4	3.808	3.630	3.465	3.312	3.170	3.037	2.914	2.798	2.690	2.589	2.494	2.404	2.320	2.241	2.166	2.096	1.849	4
5	4.713	4.452	4.212	3.993	3.791	3.605	3.433	3.274	3.127	2.991	2.864	2.745	2.635	2.532	2.436	2.345	2.035	5
6	5.601	5.242	4.917	4.623	4.355	4.111	3.889	3.685	3.498	3.326	3.167	3.020	2.885	2.759	2.643	2.534	2.168	6
7	6.472	6.002	5.582	5.206	4.868	4.564	4.288	4.039	3.812	3.605	3.416	3.242	3.083	2.937	2.802	2.677	2.263	7
8	7.325	6.733	6.210	5.747	5.335	4.968	4.639	4.344	4.078	3.837	3.619	3.421	3.241	3.076	2.925	2.786	2.331	8
9	8.162	7.435	6.802	6.247	5.759	5.328	4.946	4.607	4.303	4.031	3.786	3.566	3.366	3.184	3.019	2.868	2.379	9
10	8.983	8.111	7.360	6.710	6.145	5.650	5.216	4.833	4.494	4.192	3.923	3.682	3.465	3.269	3.092	2.930	2.414	10
11	9.787	8.760	7.887	7.139	6.495	5.938	5.453	5.029	4.656	4.327	4.035	3.776	3.543	3.335	3.147	2.978	2.438	11
12	10.575	9.385	8.384	7.536	6.814	6.194	5.660	5.197	4.793	4.439	4.127	3.851	3.606	3.387	3.190	3.013	2.456	12
13	11.348	9.986	8.853	7.904	7.103	6.424	5.842	5.342	4.910	4.533	4.203	3.912	3.656	3.427	3.223	3.040	2.469	13
14	12.106	10.563	9.295	8.244	7.367	6.628	6.002	5.468	5.008	4.611	4.265	3.962	3.695	3.459	3.249	3.061	2.478	14
15	12.849	11.118	9.712	8.559	7.606	6.811	6.142	5.575	5.092	4.675	4.315	4.001	3.726	3.483	3.268	3.076	2.484	15
16	13.578	11.652	10.106	8.851	7.824	6.974	6.265	5.668	5.162	4.730	4.357	4.033	3.751	3.503	3.283	3.088	2.489	16
17	14.292	12.166	10.477	9.122	8.022	7.120	6.373	5.749	5.222	4.775	4.391	4.059	3.771	3.518	3.295	3.097	2.492	17
18	14.992	12.659	10.828	9.372	8.201	7.250	6.467	5.818	5.273	4.812	4.419	4.080	3.786	3.529	3.304	3.104	2.494	18
19	15.678	13.134	11.158	9.604	8.365	7.366	6.550	5.877	5.316	4.843	4.442	4.097	3.799	3.539	3.311	3.109	2.496	19
20	16.351	13.590	11.470	9.818	8.514	7.469	6.623	5.929	5.353	4.870	4.460	4.110	3.808	3.546	3.316	3.113	2.497	20
21	17.011	14.029	11.764	10.017	8.649	7.562	6.687	5.973	5.384	4.891	4.476	4.121	3.816	3.551	3.320	3.116	2.498	21
22	17.658	14.451	12.042	10.201	8.772	7.645	6.743	6.011	5.410	4.909	4.488	4.130	3.822	3.556	3.323	3.118	2.498	22
23	18.292	14.857	12.303	10.371	8.883	7.718	6.792	6.044	5.432	4.925	4.499	4.137	3.827	3.559	3.325	3.120	2.499	23
24	18.914	15.247	12.550	10.529	8.985	7.784	6.835	6.073	5.451	4.937	4.507	4.143	3.831	3.562	3.327	3.121	2.499	24
25	19.523	15.622	12.783	10.675	9.077	7.843	6.873	6.097	5.467	4.948	4.514	4.147	3.834	3.564	3.329	3.122	2.499	25
26	20.121	15.983	13.003	10.810	9.161	7.896	6.906	6.118	5.480	4.956	4.520	4.151	3.837	3.566	3.330	3.123	2.500	26
27	20.707	16.330	13.211	10.935	9.237	7.943	6.935	6.136	5.492	4.964	4.524	4.154	3.839	3.567	3.331	3.123	2.500	27
28	21.281	16.663	13.406	11.051	9.307	7.984	6.961	6.152	5.502	4.970	4.528	4.157	3.840	3.568	3.331	3.124	2.500	28
29	21.844	16.984	13.591	11.158	9.370	8.022	6.983	6.166	5.510	4.975	4.531	4.159	3.841	3.569	3.332	3.124	2.500	29
30	22.396	17.292	13.765	11.258	9.427	8.055	7.003	6.177	5.517	4.979	4.534	4.160	3.842	3.569	3.332	3.124	2.500	30
35	24.999	18.665	14.498	11.655	9.644	8.176	7.070	6.215	5.539	4.992	4.541	4.164	3.845	3.571	3.333	3.125	2.500	35
40	27.355	19.793	15.046	11.925	9.779	8.244	7.105	6.233	5.548	4.997	4.544	4.166	3.846	3.571	3.333	3.125	2.500	40

*Payments (or receipts) at the end of each period.

APPENDIX C

Cost Accounting in Professional Examinations

This appendix describes the role of cost accounting in professional examinations. Cost/management accounting receives abundant attention in these examinations. A conscientious reader who has solved a representative sample of the problems at the end of the chapters will be well prepared for the professional examination questions dealing with cost accounting. This appendix aims to provide perspective, instill confidence, and encourage readers to take the examinations.

TYPES OF PROFESSIONAL EXAMINATIONS

CPA and CMA Designations

Many American readers may eventually take the Certified Public Accountant (CPA) examination or the Certified Management Accountant (CMA) examination. Similar examinations are conducted in other countries. Certification is important to professional accountants for many reasons, such as:

1. Recognition of achievement and technical competence by fellow accountants and by users of accounting services
2. Increased self-confidence in one's professional abilities
3. Membership in professional organizations offering programs of career-long education
4. Enhancement of career opportunities
5. Personal satisfaction

The CPA certificate is issued by individual states; it is necessary for obtaining a state's license to practice as a Certified Public Accountant. A prominent feature of public accounting is the use of independent (external) auditors to give assurance about the reliability of the financial statements supplied by managers. These auditors are called Certified Public Accountants in the United States and Chartered Accountants in many other English-speaking nations. The major U.S. professional association in the private sector that regulates the quality of external auditing is the American Institute of Certified Public Accountants (AICPA).[1]

[1]The AICPA also prepares the Uniform CPA Examination. Information regarding this examination can be obtained from the Director of Examinations, AICPA, 1211 Avenue of the Americas, New York, NY 10036. Information regarding *specific requirements* for taking the CPA Examination in a particular state is available from that state's Board of Public Accountancy. The *Journal of Accountancy* contains numerous advertisements for CPA review materials and courses.

The CMA designation is offered by the Institute of Certified Management Accountants. The ICMA is sponsored by the National Association of Accountants (NAA), the largest association of management accountants in the world.[2] The major objective of the CMA certification is to enhance the development of the management accounting profession. In particular, focus is placed on the modern role of the management accountant as an active contributor to and a participant in management. The CMA designation is gaining increased stature in the business community as a credential parallel to the CPA designation.[3]

The CMA examination consists of four parts taken during two days (16 hours):

- Part 1: Economics, finance, and management
- Part 2: Financial accounting and reporting
- Part 3: Management reporting, analysis, and behavioral issues
- Part 4: Decision analysis and information systems

Questions regarding ethical issues will appear on any part of the examination. A person who has successfully completed the U.S. CPA examination is exempt from Part 2.

The CMA certificate is not "in competition" with the CPA certificate. The two fields of management accounting and public accounting are compatible, not competitive. The CMA designation identifies the holder as having a required level of professional achievement in management accounting.

Taking Both Examinations

Students should plan to take both the CMA and CPA examinations. ***There are almost three times as many management accountants as there are public accountants. Many students work in public accounting immediately after graduation but spend most of their careers elsewhere.*** The material covered in both examinations is comparable in rigor and overlaps considerably. Therefore, candidates should try to prepare for both examinations at about the same time.

Students should take the CMA and CPA examinations as soon as possible after attaining their university degrees in accounting. At that point, if they conduct an appropriate review of their courses, students are likely to be as well prepared as they ever will be. The examinations are essentially academic in nature. They are aimed at determining whether a candidate is a qualified *entrant* into a profession rather than a full-fledged, experienced practitioner.

REVIEW FOR EXAMINATIONS

Cost/management accounting questions are prominent in the CMA examination. The CPA examination also includes such questions, although they are less extensive than questions regarding financial accounting, auditing, and business law. On

[2]The NAA has a wide range of activities driven by many committees. For example, the Management Accounting Practices Committee issues statements on both financial accounting and management accounting. The NAA also has an extensive continuing-education program.

[3]Information regarding the CMA program can be obtained from the Institute of Certified Management Accountants, 10 Paragon Drive, P.O. Box 405, Montvale, NJ 07645-0405. CMAs are required to be members of the National Association of Accountants. The periodical *Management Accounting* contains advertisements of CMA review materials and courses. A Certified Cost Estimator/Analyst (CCEA) program is administered by the Society of Cost Estimating and Analysis, 101 South Whiting Street, Suite 313, Alexandria, VA 22304. The society's primary purpose is to improve the effectiveness of cost estimation and price analysis.

the average, cost/managerial accounting represents 35%–40% of the CMA examination and 5% of the CPA examination. This book includes 53 questions and problems used in past CMA and CPA examinations.[4] In addition, a supplement to this book, *Student Guide and Review Manual* (John K. Harris and Dudley W. Curry [Englewood Cliffs, NJ: Prentice Hall, 1991]), contains over one hundred CMA and CPA questions and explanatory answers.

Careful study of appropriate topics in this book will give candidates sufficient background for succeeding in the cost accounting portions of the professional examinations. Chapters 2–11, 16–18, and 21–22 are particularly helpful for people taking the CPA examination. All the chapters are helpful for people taking the CMA examination.

CANADIAN PROFESSIONAL EXAMINATIONS

Three professional accounting designations are widely sought in Canada:

Designation	Sponsoring Organization
Certified Management Accountant (CMA)	Society of Management Accountants (SMA) 154 Main Street East, P.O. Box 176 M.P.O. Hamilton, Ontario L8N 3C3
Certified General Accountant (CGA)	Certified General Accountants' Association (CGA) 740-1176 West Georgia Street Vancouver, British Columbia V6E 4A2
Chartered Accountant (CA)	Canadian Institute of Chartered Accountants 150 Bloor Street West, Toronto

At least 50% of the Canadian CMA examination concentrates on cost/management accounting. The CGA and CA examinations also cover cost/management accounting, but to a lesser extent.

[4]These are designated in this book as "(CMA)" and "(CPA)." In addition, the problems designated "(SMA)" in this book are from the set of regular examinations (which are graduated by levels of difficulty) of the Society of Management Accountants of Canada, and the ones designated "(CGA)" are from the examinations of the Certified General Accountants' Association of Canada. Some of the professional examination questions in this book have been adapted to bring out particular points in the chapter for which they were used. Note that examinations are also given by the U.S. Institute of Internal Auditors; however, cost accounting is covered lightly in these examinations.

GLOSSARY

Abnormal spoilage. The spoilage that is not expected to arise under efficient operating conditions; it is not an inherent part of the selected production process. (593)

Absorption costing. Method of inventory costing that includes all direct manufacturing costs and all indirect manufacturing costs (both variable and fixed) as inventoriable costs. (289)

Accelerated depreciation. Any pattern of depreciation that writes off depreciable assets more quickly than does straight-line depreciation. (707)

Account analysis method. Cost estimation approach in which the cost accounts in the ledger are classified as variable, fixed, or mixed. (337)

Accounting rate of return. See *Accrual accounting rate of return.*

Accrual accounting rate of return. Accounting measure of income divided by an accounting measure of investment. Also called *accounting rate of return.* (683)

Activity accounting. See *Activity-based accounting.*

Activity-based accounting. A system of accounting that focuses on activities as the fundamental cost objects and uses the costs of these activities as building blocks for compiling the costs of other cost objects (such as a product or a department). Also called *activity-based costing* or *activity accounting.* (150)

Activity-based costing. See *Activity-based accounting.*

Activity variance. See *Production volume variance.*

Actual costing. A costing method that allocates actual direct costs to products and uses actual rates to apply indirect costs based on the actual inputs of the indirect cost application base. (261)

Actual costs. Amounts determined on the basis of costs incurred (historical costs), as distinguished from predicted or forecasted costs. (26)

Allowable cost. Cost that parties to a contract agree to include in the costs to be reimbursed. (507)

Applied factory overhead. Factory overhead allocated to products (or services), usually by means of some budgeted (predetermined) rate. (102)

Appraisal costs. Category of costs in a *cost of quality program* incurred in detecting which individual products of those produced do not conform to specification. (913)

Artificial cost. See *Complete reciprocated cost.*

Autocorrelation. See *Serial correlation.*

Autonomy. The degree of freedom to make decisions. (857)

Average cost. Computed by dividing some total cost (the numerator) by some denominator. (33)

Backflush costing. Standard costing system for product costing that focuses first on the output and then works backward to apply manufacturing costs to units sold and to inventories. Also called *delayed costing* or *post-deduct costing.* (627)

Bailout payback time. Capital budgeting method that measures the time when the cumulative cash operating savings plus the disposal price at the end of a particular year equal the original investment. (682)

Bias. Occurs in statistics when a random sample fails to represent a population because of some systematic error. (789)

Book value. Original cost of an asset minus accumulated depreciation. (337)

Breakeven point. Point of volume where total revenues and total costs (expenses) are equal; that is, there is neither profit nor loss. (60)

Breakeven time (BET). Capital budgeting method that measures the time taken from the start of a project (the initial idea date) to when the cumulative present value of the cash inflows of a project to equal the present value of the total cash outflows. (686)

Budget. Quantitative expression of plan of action and an aid to coordination and implementation. (5)

Budget variance. See *Flexible-budget variance.*

Budgeted factory-overhead rate. Rate derived by dividing the budgeted total overhead by the budgeted total volume of the application base. (104)

Budgeted volume. See *Master-budget volume.*

Business function cost. The sum of all costs (such as variable costs and fixed costs) in a particular business function. Business functions include research and development, product design, manufacturing, marketing, distribution, and customer service. (372)

Byproduct. Product that has a low sales value compared with the sales value of the main or joint product(s). (527)

CASB. See *Cost Accounting Standards Board.*

Capacity variance. See *Production volume variance.*

Capital budgeting. The making of long-term planning decisions for investments and their financing. (673)

Carrying costs. Costs arising when a business holds stocks of goods for sale. These costs consist of the opportunity cost of the investment tied up in inventory and the costs associated with storage. (741)

Cash budget. A schedule of expected cash receipts and disbursements. (194)

Cash cycle. See *Self-liquidating cycle.*

Certified Management Accountant (CMA). The professional designation for management accountants and financial executives. It is the internal accountant's counterpart of the CPA. (15)

Choice criterion. Objective that can be quantified. It provides a basis for choosing the best alternative action. Also called *objective function.* (647)

Clock card. Document used as a basis for determining individual employee earnings. (95)

Coefficient of determination. Measure of the extent to which the independent variable(s) in a regression model accounts for the variability in the dependent variable. Also called r^2. (786)

Committed costs. Costs that arise from having property, plant, equipment, and a functioning organization; little can be done in the short run to change committed costs. (432)

Common cost. Cost of operating a facility that is shared by two or more users. (473)

Complete reciprocated cost. The actual cost incurred by the service department plus a part of the costs of the other service departments that provide services to it; it is always larger than the actual cost. Also called *artificial cost* of the service department. (471)

Conference method. Cost estimation approach that develops cost estimates based on analysis and opinions gathered from various departments of an organization. (337)

Constant. Estimated component of total costs that, within the relevant range, does not vary with changes in the level of the cost driver. Also called *intercept.* (333)

Constant gross-margin percentage NRV method. Joint-cost allocation method that allocates joint costs so that the overall gross-margin percentage is identical for each individual product. (534)

Constant variance. Assumption of regression analysis that the standard deviation and variance of the residuals (u's) are constant for all values of the independent variable. (789)

Constraint. A mathematical inequality or equality that must be satisfied by the variables in a mathematical model. (765)

Continual improvement standard cost. A carefully predetermined cost that is successively reduced over succeeding time periods. Also called *moving cost reduction standard cost.* (232)

Contribution margin. Equal to sales minus all variable costs. (61)

Contribution-margin ratio. The total contribution margin divided by the total sales. (69)

Control. Action that implements the planning decision and performance evaluation that provides feedback of the results. (5)

Control chart. Graph of successive observations of a series in which each observation is plotted relative to specified points that present the expected distribution. Used in *statistical quality control.* (841)

Control-factor unit. Measure of workload used in work measurement. (437)

Controllability. The degree of influence that a specific manager has over costs, revenues, or other items in question. (190)

Controllable cost. Any cost that is primarily subject to the influence of a given *manager* of a given *responsibility center* for a given *time span.* (190)

Controller. The financial executive primarily responsible for both management accounting and financial accounting. (11)

Conversion costs. Consist of all manufacturing costs other than direct material costs. (37)

Correlation. The general relationship between two variables. (339)

Cost. A resource sacrificed or forgone to achieve a specific objective. (25)

Cost absorption. See *Cost application.*

Cost accounting. Part of management accounting plus a part of financial accounting—to the extent that cost accounting satisfies the requisites of external reporting. (4)

Cost Accounting Standards Board. Established in 1970 to develop cost accounting standards to be used by contractors doing business with the U.S. government. In 1988 established as an independent board within the Office of Federal Procurement Policy. Also called *CASB.* (505)

Cost accumulation. The collection of cost data in some organized way through an accounting system. (26)

Cost allocation. General term that refers to identifying accumulated costs with, or tracing accumulated costs to, cost objects such as departments, activities, or products. (26)

Cost allocation base. Factor that is the common denominator for systematically linking a cost or group of costs to a cost object such as a department or an activity. Where cost object is a product, the narrower term *cost application base* is often used. (103)

Cost application. Narrower term than cost allocation; it refers to the allocation of costs to products as distinguished from the allocation of costs to departments or activities. Also called *cost absorption.* (103)

Cost application base. Factor that is the common denominator for systematically relating a cost or a group of costs, such as factory overhead, to products. (103)

Cost-benefit approach. The primary criterion for choosing among alternative accounting systems or methods is how well they help achieve management goals in relation to their costs. (8)

Cost center. Responsibility center in which a manager is accountable for costs (expenses) only. (186)

Cost driver. Any factor whose change causes a change in the total cost of a related cost object. (28)

Cost estimation. Attempt to measure past cost relationships; an equation is formulated to describe these past relationships. (332)

Cost management. The performance by executives and others in the cost implications of their short-run and long-run planning and control functions. (4)

Cost object. Any activity or item for which a separate measurement of costs is desired. Also called *cost objective.* (25)

Cost objective. See *Cost object.*

Cost of capital. See *Required rate of return (RRR).*

Cost of quality program. Program that collects product-quality related costs incurred in different business functions of an organization and reports them in a comprehensive reporting format. (913)

Cost pool. A grouping of individual costs. (104)

Cost prediction. The forecasting of future costs. (332)

Cost-reimbursement contract. Contract reimbursement method in which the price the contractor receives is based on the actual costs the contractor incurs. (505)

Cross allocation method. See *Reciprocal allocation method.*

Cumulative average–time learning model. Learning curve model in which the cumulative average time per unit is reduced by a constant percentage each time the cumulative quantity of units produced is doubled. (347)

Current cost. The cost today of purchasing an asset currently held if identical assets can currently be purchased; it is the cost of purchasing the services provided by that asset if identical assets cannot currently be purchased. (889)

Currently attainable standards. *Standard costs* that are achievable by a specified level of effort and allow for normal spoilage, waste, and nonproductive time. (231)

Data mining. Search for variables in a regression model that is aimed solely at the maximization of r^2. (787)

Decentralization. The freedom to make decisions. Total decentralization in an organization means minimum constraints and maximum freedom for managers to make decisions at the lowest levels of an organization. (852)

Decision model. Formal method for making a choice, frequently involving quantitative analysis. (366)

Decision table. Summary of the contemplated actions, events, probabilities, and outcomes. Also called a *payoff table* or *payoff matrix.* (650)

Decision unit. The lowest unit in an organization for which a budget is prepared. (435)

Delayed costing. See *Backflush costing.*

Denominator activity. See *Denominator volume.*

Denominator level. See *Denominator volume.*

Denominator variance. See *Production volume variance.*

Denominator volume. The preselected production volume level used to set a budgeted fixed-factory-overhead rate for applying costs to inventory. Also called *denominator activity* or *denominator level.* (258)

Dependent variable. The variable to be predicted in a cost function (or regression equation). (339)

Differential cost. See *Incremental cost.*

Direct allocation method. Method of allocating service department costs that ignores any service rendered by one service department to another; it allocates each service department's total costs directly to the production departments. Also called *direct method.* (468)

Direct costing. See *Variable costing.*

Direct costs. Costs that can be identified specifically with or traced to a given cost object in an economically feasible way. (27)

Direct labor costs. The compensation of all labor that can be identified in an economically feasible way with a cost object. (35)

Direct labor mix variance. Variance used to analyze effect of changes in the mix of labor on direct labor cost. Computed as: [(Actual labor mix percentage − Standard labor mix percentage) × (Actual total units of labor inputs used)] × (Standard individual price per unit of labor input − Standard average price per unit of labor input). (833)

Direct labor yield variance. Variance used to analyze effect of changes in labor yield on labor cost. Computed as: (Actual units of labor inputs used − Standard units of labor inputs allowed for actual outputs) × (Standard average price per unit of labor input). (833)

Direct material mix variance. Variance used to analyze effect of changes in the mix of materials used on direct materials cost. Computed as: [(Actual material mix percentage − Standard material mix percentage) × (Actual total units of material inputs used)] × (Standard individual price per unit of material input − Standard average price per unit of material inputs). (830)

Direct materials cost. The acquisition costs of all materials that are identified as part of the cost object and that may be traced to the cost object in an economically feasible way. (35)

Direct materials inventory. Direct materials on hand and awaiting use in the production process. (39)

Direct material yield variance. Variance used to analyze effect of changes in material yield on materials cost. Computed as: (Actual units of material inputs used − Standard units of material inputs allowed for actual outputs) × (Standard average price per unit of material inputs). (830)

Direct method. See *Direct allocation method.*

Discounted cash flow (DCF). Measures the cash inflows and outflows of a project as if they occurred at a single point in time so that they can be compared in an appropriate way. (674)

Discount rate. See *Required rate of return (RRR).*

Discretionary costs. Costs that have two important features: (1) they arise from periodic (usually yearly) decisions regarding the maximum outlay to be incurred, and (2) they are not tied to a clear cause-and-effect relationship between inputs and outputs. Also called *managed costs* or *programmed costs.* (432)

Disturbance term. See *Residual term.*

Double-declining balance (DDB) depreciation. Form of accelerated depreciation in which first-year depreciation is twice the amount of straight-line depreciation when a zero terminal disposal price is assumed. (707)

Double-distribution allocation method. See *Reciprocal allocation method.*

Dual pricing. Transfer-pricing approach that uses two separate transfer-pricing methods to price each interdivision transaction. (863)

Dual-rate method. Method of allocating costs in which two cost functions are used. Typically, the two functions are a fixed-cost function and a variable-cost function. (464)

Dysfunctional decision making. See *Suboptimal decision making.*

Economic order quantity (EOQ). Decision model that calculates the optimal quantity of inventory to order. Simplest version of model incorporates only ordering costs and carrying costs into the calculation. (742)

Effectiveness. The attainment of a predetermined goal. The degree to which a predetermined goal (objective or target) is met. (217)

Efficiency. The relationship between inputs used and outputs achieved. The fewer the inputs used to obtain a given output, the greater the efficiency. (217)

Efficiency variance. The difference between the quantity of actual inputs used (such as yards of materials) and the quantity of inputs that *should have been used* (the flexible budget of any quantity of *units of good output achieved*), multiplied by the budgeted price. *Quantity variance* or *usage variance* are popular terms for a direct materials efficiency variance. (224)

Effort. Exertion toward a goal. (428)

Engineered costs. Costs that result specifically from a clear-cut, measured relationship between inputs and outputs. (432)

Environment. The set of uncontrollable factors that affect the success of a process. (6)

Equivalent units. Measure of the output in terms of the quantities of each of the factors of production that have been applied. (146)

Error term. See *Residual term.*

Estimated net realizable value (NRV) method. Joint-cost allocation method that allocates joint costs on the basis of their relative estimated net realizable value (expected final sales value in the ordinary course of business minus the expected separable costs of production and marketing). (533)

Events. Possible occurrences that can arise. (647)

Excess materials requisition. Form necessary to obtain any materials needed in excess of the standard amount allowed for the scheduled output. (234)

Excess present value index. The total present value of future net cash inflows divided by the total cash outflow. (722)

Executive compensation plan. One component of the reward system of a management control system. Includes the diverse set of short-run and long-run plans that detail the conditions under which, and the forms in which, compensation is to be made to executives of the organization. (894)

Expected annual activity. See *Master-budget volume.*

Expected annual capacity. See *Master-budget volume.*

Expected annual volume. See *Master-budget volume.*

Expected monetary value. *Expected value* measure where the outcomes are measured in monetary terms. (648)

Expected value. Weighted average of the outcomes with the probabilities of each outcome serving as the weights. (648)

Expected value of perfect information (EVPI). The difference between the expected value with perfect information and the highest expected value with existing information. (652)

Expected variances. Variances that are specified in the budget for cash planning. (232)

Experience curve. Function that shows how full costs per unit (including manufacturing, distribution, marketing, and so on) decline as units of output increase. Broader application of the *learning curve* concept. (347)

External failure costs. Category of incurred costs in a *cost of quality program* when a nonconforming product is detected *after* its shipment to customers. (913)

Factory burden. See *Indirect manufacturing costs.*

Factory overhead. See *Indirect manufacturing costs.*

Feedback. Consists of a comparison of the budget with actual results—that is, with historical information. (7)

Financial accounting. Focuses on how accounting can serve external decision makers. Heavily constrained by generally accepted accounting principles. Contrast with *management accounting.* (4)

Financial budget. That part of the master budget that comprises the capital budget, cash budget, budgeted balance sheet, and budgeted statement of cash flows. (176)

Financial planning model. Mathematical statement of the relationships among all the operating activities, financial activities, and other major internal and external factors that may affect decisions. (185)

Finished goods inventory. Goods fully completed but not yet sold. (39)

Fixed cost. Cost that does not change in total despite changes of a cost driver. (29)

Fixed-price contract. Contract reimbursement method in which the price the contractor receives is established at the outset. (505)

Flexible-budget variance. The difference between actual results and the flexible-budget amounts for the actual output achieved. Also called, for brevity, *budget variance.* (215)

Flexible-budget variance. The difference between actual results and the flexible-budget amounts for the actual output achieved. Also called, for brevity, *budget variance.* (215)

Full cost. The sum of all the costs in all the business functions (research and development, product design, manufacturing, marketing, distribution, and customer service). (373)

Functional authority. The right to command action laterally and downward with regard to a specific function or specialty. (10)

Goal congruence. Exists when individuals and groups work toward the organization goals that top management desires. (428)

Goods in process. See *Work in process inventory.*

Gross margin. The excess of sales over the cost of the goods sold. In a manufacturing company, the manufacturing cost of goods sold includes fixed indirect manufacturing (fixed manufacturing overhead) costs. Also called *gross profit.* (68)

Gross profit. See *Gross margin.*

High-low method. Method of estimating a cost function that uses only the highest and lowest values of the cost driver within the relevant range. (341)

Homogeneous cost pool. Cost pool in which each activity whose costs are included in it has the same or a similar cause-and-effect relationship between the cost driver and the costs at that activity. (463)

Hurdle rate. See *Required rate of return (RRR).*

Hybrid costing. Costing system with blends of characteristics from both job costing and process costing. (619)

Idle capacity variance. See *production volume variance.*

Idle time. Represents wages paid for unproductive time caused by machine breakdowns, material shortages, sloppy production scheduling, and the like. (42)

Imputed costs. Costs recognized in particular situations that are not regularly recognized by usual accounting procedures. (383)

Incremental approach. Approach to choosing among projects that includes only those cash outflows and inflows that differ between the two projects. (710)

Incremental common cost allocation method. Common cost allocation method that requires that one user be viewed as the primary party and the second user be viewed as the incremental party. (473)

Incremental cost. The difference in total cost between two alternatives. Also called *differential cost* or *net relevant cost.* (368)

Incremental unit–time learning model. Learning curve model in which the incremental unit time (the time needed to produce the last unit) is reduced by a constant percentage each time the cumulative quantity of units produced is doubled. (347)

Indirect costs. Costs that cannot be identified specifically with or traced to a given cost object in an economically feasible way. (27)

Indirect labor costs. All labor compensation other than direct labor compensation. (35)

Indirect manufacturing costs. All manufacturing costs that cannot be identified specifically with or traced to the cost object in an economically feasible way. Also called *factory overhead, factory burden, manufacturing overhead,* or *manufacturing expenses.* (35)

Industrial engineering method. Cost estimation approach that first analyzes the relationship between inputs and outputs in physical terms. Then the physical measures are transformed into standard or budgeted costs. Also called *work measurement method.* (337)

Inflation. Decline in the general purchasing power of the monetary unit. (715)

Intercept. See *Constant.*

Intermediate product. Product transferred from one segment of an organization to another segment of the same organization. (856)

Internal accounting. See *Management accounting.*

Internal failure costs. Category of costs in a *cost of quality program* incurred when a nonconforming product is detected *before* its shipment to customers. (913)

Internal rate-of-return method. Discounted cash-flow capital budgeting method. It is the rate of interest at which the present value of expected cash inflows from a project equals the present value of expected cash outflows of the project. Also called *time-adjusted rate of return.* (677)

Inventoriable costs. All costs of a product that are regarded as an asset for financial reporting under generally accepted accounting principles. (37)

Investment center. Responsibility center in which a manager is accountable for investments, revenues, and costs. (186)

Investment tax credit (ITC). Direct reduction of income taxes arising from the acquisition of depreciable assets. (706)

Job-cost record. The source document used by job-order costing to compile *product costs.* Also called *job-order* or *job-cost sheet.* (93)

Job-cost sheet. See *Job-cost record.*

Job-cost system. See *Job-order product costing system.*

Job-order. See *Job-cost record.*

Job-order costing system. See *Job-order product costing system.*

Job-order product costing system. Product costs are obtained by allocating costs to a specific unit or to a small batch of products or services that proceed through the production steps as a distinct identifiable job lot. Also called *job-order costing system, job-cost system, work-order system,* or *production-order system.* (93)

Joint cost. Cost of a single process that yields multiple products simultaneously. (527)

Joint products. Products that have a relatively high sales value and are not separately identifiable as individual products until the split-off point. (527)

Just-in-time (JIT) production. System in which each component on a production line is produced immediately as needed by the next step in the production line. (620)

Just-in-time (JIT) purchasing. Purchase of goods or materials such that delivery immediately precedes demand or use. (748)

Labor-paced manufacturing environment. Worker dexterity and productivity determine the speed of production. Machines function as tools that aid production workers. (499)

Learning curve. Function that shows how labor hours per unit decline as units of output increase. (347)

Life-cycle budgeting. Estimates the costs budgeted for each product from its initial research and development to its final customer servicing and support in the marketplace. (403)

Life-cycle costing. Tracks and accumulates the actual costs attributable to each product from its initial research and development to its final customer servicing and support in the marketplace. (403)

Line authority. Authority that is exerted downward over subordinates. (10)

Linear cost function. Function in which a single constant (a) and a single slope coefficient (b) describe (in an additive manner) the behavior of costs for all changes in the level of the cost driver. (333)

Linear programming (LP). Decision model that focuses on how best to allocate scarce resources to attain a chosen objective, such as maximization of operating costs. (763)

Machine-paced manufacturing environment. Machines conduct most (or all) phases of production, such as movement of materials to the production line, assembly and other activities on the production line, and shipment of finished goods to the delivery dock areas. (499)

Main product. Product with relatively high sales value obtained when a single process yields two or more products, only one with a relatively high sales value. (527)

Managed costs. See *Discretionary costs.*

Management accounting. Focuses on internal customers for accounting information and is the process of identification, measurement, accumulation, analysis, preparation, interpretation, and communication of information that assists executives in fulfilling organization goals. Also called *internal accounting.* Contrast with *financial accounting.* (4)

Management by exception. The practice of concentrating on areas that deserve attention and ignoring areas that are presumed to be running smoothly. (5)

Management by objectives (MBO). A subordinate and his or her superior jointly formulate the subordinate's set of goals and the plans for attaining those goals for a subsequent period. (191)

Management control system. A means of gathering data to aid and coordinate the process of making decisions throughout the organization. (6)

Manufacturing. The transformation of materials into other goods through the use of labor and factory facilities. (34)

Manufacturing expenses. See *Indirect manufacturing costs.*

Manufacturing overhead. See *Indirect manufacturing costs.*

Margin of safety. The excess of budgeted sales over the breakeven volume. (67)

Market share variance. Variance used to analyze effect of changes in market share on operating income. Computed as: (Actual market share percentage − Budgeted market share percentage) × (Actual industry sales volume in units) × (Budgeted average contribution margin per unit). (824)

Market size variance. Variance used to analyze effect of changes in market size on operating income. Computed as: (Budgeted market share percentage) × (Actual industry sales volume in units − Budgeted industry sales volume in units) × (Budgeted average contribution margin per unit). (824)

Master budget. Summarizes the objectives of all subunits of an organization—sales, production, research, marketing, customer service, and finance. (172)

Master-budget activity. See *Master-budget volume.*

Master-budget volume. Measure of capacity that is the anticipated level of capacity utilization for the coming year or other planning period (such as six months). Also called *budgeted volume, expected annual volume, expected annual capacity, expected annual activity,* or *master-budget activity.* (302)

Materials requirements planning (MRP). Planning system that focuses first on the amount and timing of finished goods demanded and then determines the demand for materials, components, and subassemblies at each of the prior stages of production. (751)

Materials requisitions. Forms used to charge departments and job-cost records for direct materials used. Also called *stores requisitions.* (94)

Matrix allocation method. See *Reciprocal allocation method.*

Merchandising. The marketing of goods without changing their basic form. (34)

Mixed cost. A cost function that has both fixed and variable elements. Also called *semivariable cost.* (334)

Modified Accelerated Cost Recovery System (MACRS). Federal tax regulation that classifies depreciable assets into one of several recovery periods, each of which has a designated pattern of allowable depreciation (double-declining balance, 150% declining balance, or straight-line). (711)

Motivation. Desire for a selected goal (the goal-congruence aspect) combined with the resulting drive or pursuit toward that goal (the effort aspect). (428)

Moving cost reduction standard cost. See *Continual improvement standard cost.*

Multicollinearity. Exists in regression analysis when two or more independent variables are highly correlated with each other. (794)

Multiple regression. Regression analysis where there are several independent variables. (783)

National Association of Accountants (NAA). The largest association of internal accountants in the United States. (15)

Negotiated static budget. A budget in which a fixed amount of costs is established through negotiations prior to the start of the budget period. (435)

Net income. Operating income plus nonoperating revenues minus nonoperating expenses minus income taxes. (60)

Net present value (NPV). Discounted cash-flow capital budgeting method of calculating the expected net monetary gain or loss from a project by discounting all expected future cash inflows and outflows to the present point in time, using the required rate of return. (675)

Net relevant cost. See *Incremental cost.*

New-product lead time. The time from the initial concept of a new product to its market introduction. Also called *time-to-market.* (915)

Nominal rate of interest. Rate of interest made up of two elements: (a) the real rate of interest and (b) an inflation element. (715)

Nonlinear cost function. Function in which a single constant (a) and a single slope coefficient (b) do not describe (in an additive manner) the behavior of costs for all changes in the level of the cost driver. (346)

Non-value-added cost. Cost incurred for any activity that can be eliminated without the customer's perceiving deterioration in the performance, function, or other quality of a product. (384)

Normal costing. A costing method that allocates actual direct costs to products and uses budgeted rates to apply indirect costs based on the actual inputs of the indirect cost application base. (261)

Normal spoilage. The spoilage that arises under efficient operating conditions; it is an inherent result of the particular process and thus is uncontrollable in the short run. (593)

Normal volume. Measure of capacity that is the level of capacity utilization (which is less than 100% of practical capacity) that will satisfy average consumer demand over a span of time (often five years) that includes seasonal, cyclical, and trend factors. (302)

Objective function. See *Choice criterion.*

Operating budget. The budgeted income statement and its supporting budgets. (176)

Operating cycle. See *Self-liquidating cycle.*

Operating department. See *Production department.*

Operating income. Revenues or sales from operations for the accounting period minus all operating costs or expenses, including cost of goods sold. (60)

Operation. A standardized method or technique that is repetitively performed regardless of the distinguishing features of the finished product. (622)

Operation costing. A hybrid costing system applied to batches of homogeneous products. Each batch uses the same resources of each activity to the same extent. (622)

Operations management. Management of resources to produce or deliver the products or services provided by an organization. (740)

Opportunity cost. The maximum available contribution to profit that is forgone (rejected) by using limited resources for a particular purpose. (376)

Ordering costs. Costs of preparing a purchase order. (741)

Ordinary incremental budget. A negotiated static budget that considers the previous period's budget and the actual result. (435)

Organization structure. An arrangement of lines of responsibility within the entity. (186)

Outcome. Measure of the predicted consequences in terms of the objective function of the various possible combinations of actions and events. Also called *payoffs.* (647)

Outlay cost. Cost that sooner or later requires a cash disbursement. (377)

Out-of-pocket costs. Current or near-future outlays made to meet costs incurred because of a specific decision. (383)

Overapplied overhead. Amount by which the applied overhead balance exceeds the incurred (actual) overhead balance. (108)

Overtime premium. Consists of the wages paid to all workers (for both direct labor and indirect labor) in *excess* of their straight-time wage rates. (42)

Parameter. A constant or a coefficient in a model or system of equations. (333)

Payback method. Capital budgeting method that measures the time it will take to recoup, in the form of cash inflow from operations, the total dollars invested in a project. (681)

Payoff matrix. See *Decision table.*

Payoff table. See *Decision table.*

Payoffs. See *Outcome.*

Perfectly competitive market. Market in which there is a homogeneous product with equivalent buying and selling prices and no individual buyers or sellers can affect those prices by their own actions. (860)

Performance plan. Long-run executive compensation plan that makes payment of a definite number of performance units (each assigned a fixed dollar value) contingent on the net income or ROI at the end of an extended time period (say, five years) being at a set level. (895)

Performance reports. Measure activities. These reports usually consist of comparisons of budgets with actual results. (5)

Period costs. All costs that are immediately expensed for financial reporting under generally accepted accounting principles. (37)

Periodic inventory method. Method of accounting for inventories that does not require a continuous record of inventory changes. (40)

Perpetual inventory method. Method of accounting for inventories that requires a continuous record of additions to and reductions in materials, work in process, and finished goods, thus measuring on a continuous basis not only these three inventories but also the cumulative cost of goods sold. (39)

Physical measure method. Joint-cost allocation method that allocates joint costs on the basis of their relative proportions at the split-off point, using a common physical measure such as weight or volume. (531)
Planning. Choosing goals, predicting potential results under various ways of achieving goals, and deciding how to attain the desired results. (5)
Post-deduct costing. See *Backflush costing.*
Practical capacity. Measure of capacity that is the maximum level at which the plant or department can operate efficiently. Practical capacity often allows for unavoidable operating interruptions such as repair time or waiting time. (301)
Predatory pricing. A pricing approach in which a business temporarily cuts prices in an effort to restrict supply and then raises prices rather than enlarge demand or meet competition. (411)
Present value. The asset measure based on DCF estimates. (888)
Prevention costs. Category of costs in a *cost of quality program* incurred in preventing the production of products that do not conform to specification. (913)
Previous department costs. See *Transferred-in costs.*
Price variance. The difference between actual unit prices and budgeted unit prices multiplied by the actual quantity of goods or services in question (that is, the actual quantity sold, purchased, or used). *Rate variance* is a popular term for a direct labor price variance. (224)
Prime costs. Prime costs are all the direct manufacturing costs. (37)
Priority incremental budget. A negotiated static budget similar to an ordinary incremental budget. However, priority incremental budgets also include a description of what incremental activities or changes would occur (1) if the budget were increased by, say, 10% and (2) if the budget were decreased by a similar percentage. (435)
Probability. The likelihood of occurrence of an event. (647)
Probability distribution. Describes the likelihood (or probability) of each of the mutually exclusive and collectively exhaustive set of events. (647)
Process. Collection of decisions or activities that should be aimed at some ends. (6)
Process-costing system. See *Process product-costing system.*
Process product-costing system. Product costs are obtained by allocating costs to masses of like units that usually proceed in continuous fashion through a series of uniform production steps. Also called *process-costing system.* (93)
Product-cost cross-subsidization. A source of undercosting or overcosting of products that arises when at least one miscosted product results in the miscosting of other products in the organization. (407)
Product costs. General term that denotes different costs allocated to products for different purposes. (44)
Product life cycle. The time from initial research and development to the time at which sales and support to customers are withdrawn. (403)
Product overcosting. A product-cost distortion that arises when a product consumes a relatively low level of resources but is reported to have a relatively high product cost. (406)
Product undercosting. A product-cost distortion that arises when a product consumes a relatively high level of resources but is reported to have a relatively low product cost. (406)
Production-cost report. Report of the units manufactured during a specified period and their costs. (562)
Production department. Department whose efforts are directed toward adding value to a product or service sold to a customer. Also called *operating department.* (462)
Production-order system. See *Job-order product costing system.*
Production volume variance. The difference between budgeted fixed overhead and applied fixed overhead. Also called *activity variance, capacity variance, denominator variance, idle capacity variance,* or *volume variance.* (263)
Profit center. Responsibility center in which a manager is accountable for revenues and costs. (186)
Pro forma statements. Term used to describe budgeted financial statements. (176)
Programmed costs. See *Discretionary costs.*
Project. Complex job that often takes months or years to complete and requires the work of many different departments, divisions, or subcontractors. (635)
Project cost variance. The actual cost of work performed to date (regardless of any budgets or schedules) minus the budgeted cost of work performed to date. (636)
Project schedule variance. Budgeted cost of work scheduled to date minus the budgeted cost of work performed to date. (636)
Proration. The spreading of underapplied or overapplied overhead among work in process and finished goods inventories and cost of goods sold. (109)
Purchase-ordering lead time. The time between the placement of an order and its delivery. (742)
P/V chart. A profit-volume graph showing the impact of changes in volume on operating income. (69)

Qualitative factors. Factors that cannot be measured in numerical terms. (369)
Quality. Quality of a product or service is its conformance with a prespecified (and often preannounced) standard. (741)
Quality driver. Any factor whose change causes a change in the quality level of a product or service. (913)
Quantitative factors. Factors that can be measured in numerical terms. (369)
Quantity variance. See *Efficiency variance.*

r^2. See *Coefficient of determination.*
Rate variance. See *Price variance.* Popular term for a direct labor price variance. (227)
Real rate of interest. Rate of interest made up of two elements: (a) a risk-free element—the "pure" rate of interest that is paid on long-term government bonds—and (b) a business-risk element—the "risk" premium above the pure rate that is demanded for undertaking risks. (715)
Reciprocal allocation method. Method of allocating service department costs that explicitly recognizes the mutual services rendered among all service departments when allocating service department costs to production departments. Also called the *cross allocation method, matrix allocation method,* or *double-distribution allocation method.* (470)
Regression analysis. Statistical model that measures the *average* amount of change in the dependent variable that is associated with a unit change in the amount of one or more independent variables (cost drivers in cost estimation settings). (340)
Relevant costs. Expected future costs that differ among alternative courses of action. (368)
Relevant range. The band of the cost driver in which a specific relationship between cost (revenue) and a cost (revenue) driver is valid. (30)

Relevant revenues. Expected future revenues that differ among alternative courses of action. (368)
Reorder point. The quantity level of the inventory that triggers a new order. (744)
Required rate of return (RRR). The minimum acceptable rate of return on an investment. Also called the *discount rate, hurdle rate,* or *cost of capital.* (675)
Residual income. Income minus an imputed interest charge for the investment base. (885)
Residual term. The deviation of the actual value of the dependent variable (y) from the regression line (y'). Also called *disturbance term* or *error term.* (789)
Responsibility accounting. System that measures the plans (by budgets) and the financial outcomes of actions (by historical records) of each responsibility center. (186)
Responsibility centers. Parts, segments, or subunits of an organization whose managers are accountable for specified sets of activities. (92)
Return on investment (ROI). Income divided by investment. (883)
Revenue center. Responsibility center in which a manager is accountable for revenues (sales) only. (186)
Revenue driver. Revenue driver is any factor whose change causes a change in the total revenue of a related product or service. (59)
Reworked units. Unacceptable units of production that are subsequently reworked and sold as acceptable finished goods. (592)
Risk averse. Risk attitude of a decision maker. There is more disutility from a loss of a given dollar amount than there is utility from a gain of the same dollar amount. (654)
Risk neutral. Risk attitude of a decision maker. Utility is directly proportional to monetary amounts. The decision maker weighs each dollar as a full dollar (no more, no less). (654)
Risk seeking. Risk attitude of a decision maker. There is more utility from a gain of a given dollar amount than there is disutility from the loss of the same dollar amount. (654)
ROI. See *Return on investment.*

Safety stock. The buffer inventory held as a cushion against unexpected increases in demand or lead time and unexpected unavailability of stock from suppliers. (744)
Sales mix. The relative combination of quantities of products that constitute total sales. (70)
Sales mix variance. The difference between the amount of contribution margin in the flexible budget based on actual sales volume at actual mix and that amount in the flexible budget based on actual sales volume at budgeted mix. (822)
Sales quantity variance. The difference between the amount of contribution margin in the flexible budget based on actual sales volume at budgeted sales mix and that amount in the static (master) budget. (822)
Sales value at split-off method. Joint-cost allocation method that allocates joint costs on the basis of each product's relative sales value at the split-off point. (530)
Sales volume variance. The difference between the flexible-budget amounts and the static (master) budget amounts. Unit selling prices, unit variable costs, and fixed costs are held constant. (215)
Scrap. Product that has a minimal (frequently zero) sales value compared with the sales value of the main or joint product(s). (527)
Segment. Any part of an organization for which a separate determination of revenues, costs, or both is desired. Also called *subunit.* (502)
Self-liquidating cycle. The movement from cash to inventories to receivables and back to cash. Also called *cash cycle, operating cycle,* or *working-capital cycle.* (196)
Semivariable cost. See *Mixed cost.*
Sensitivity analysis. A "what-if" technique that asks how a result will be changed if the original predicted data are not achieved or if an underlying assumption changes. (67)
Separable costs. Costs incurred beyond the split-off point that are identifiable with individual products. (527)
Sequential allocation method. See *Step-down allocation method.*
Sequential tracking. Any product costing method that is synchronized with the physical sequences of purchases and production. (627)
Serial correlation. Term used in regression analysis. Systematic pattern in the residuals such that knowledge of the residual in period t conveys information about the residual in period $t + 1$, $t + 2$, and so on. Also called *autocorrelation.* (790)
Service department. Department whose efforts are directed toward providing services to other (internal) departments (production departments and other service departments) in the organization. Also called *support department.* (462)
Setup time. The time required to prepare equipment and related resources for producing a specified number of finished units or operations. (235)
Simple regression. Regression analysis where only one independent variable is used. (783)
Simplex method. An iterative step-by-step procedure for determining the optimal solution to a linear programming problem. (767)
Single-rate method. Method of allocating costs in which a single-cost function is used. For example, no distinction is made between fixed-cost and variable-cost functions. (464)
Slope coefficient. The amount of change in total cost for each unit change in the cost driver within the relevant range. (333)
Specification analysis. Testing of the assumptions underlying regression analysis. Assumptions include linearity within the relevant range, constant variance of residuals, independence of residuals, and normality of residuals. (786)
Spending variance. Actual amount of overhead incurred minus the expected amount based on the flexible budget for actual inputs. (255)
Split-off point. Juncture in the process when products (in a joint-cost setting) become separately identifiable. (527)
Spoilage. Unacceptable units of production that are discarded and sold for disposal value. (592)
Staff authority. Authority to advise but not command others; it may be exercised laterally or upward. (10)
Stand-alone common cost allocation method. Common cost allocation method that allocates the common cost on the basis of each user's percentage of the total of the individual stand-alone costs. (473)
Standard bill of materials. Record of the standard quantities of materials required for manufacturing a finished unit. (234)
Standard costing. A cost method that allocates standard direct costs to products and uses budgeted rates to apply indirect costs based on the standard inputs allowed for actual output. (261)
Standard costs. Carefully predetermined costs that are usually expressed on a per-unit basis. They are costs that should be attained. (222)

Standard error of the estimate. See *Standard error of the residuals.*

Standard error of the residuals. A measure of the scatter of the actual observations about the regression line. Also called *standard error of the estimate.* (804)

Standard hours allowed. The number of standard hours that should have been used to obtain any given quantity of output (that is, actual goods produced or actual outputs achieved). Also called *standard hours earned, standard hours worked,* or, most accurately, *standard hours of input allowed for actual output produced.* (225)

Standard hours earned. See *Standard hours allowed.*

Standard hours of input allowed for actual output produced. See *Standard hours allowed.*

Standard hours worked. See *Standard hours allowed.*

States. See *Events.*

States of nature. See *Events.*

Static budget. A budget that is not adjusted or altered after it is drawn up, regardless of changes in volume, cost drivers, or other conditions during the budget period. (213)

Statistical quality control (SQC). Formal means of distinguishing between random variation and other sources of variation in an operating process. (841)

Step allocation method. See *Step-down allocation method.*

Step-down allocation method. Method of allocating service department costs that allows for partial recognition of services rendered by service departments to other service departments when allocating service department costs to production departments. Also called the *step allocation method* or *sequential allocation method.* (469)

Stockout. Arises when a customer demands a unit of stock and that unit is not readily available. (741)

Stores requisitions. See *Materials requisitions.*

Straight-line (SL) depreciation. Depreciation method in which an equal amount of depreciation is taken each year. (707)

Strategy. Broad term that usually means selection of overall objectives. (173)

Suboptimal decision making. Arises when a decision's benefit to one subunit is more than offset by the costs or loss of benefits to the organization as a whole. Also called *dysfunctional decision making.* (853)

Subunit. See *Segment.*

Sunk costs. Past costs that are unavoidable because they cannot be changed no matter what action is taken. (384)

Support department. See *Service department.*

Target cost. The estimated long-run cost of a product that will enable a company to enter or to remain in the market and compete profitably against its competitors. (409)

Theoretical capacity. Measure of capacity that assumes the production of output 100% of the time. Also called *maximum capacity* or *ideal capacity.* (301)

Throughput time. Time from when a product starts on the production line until when it becomes a finished good. (501)

Time-adjusted rate-of-return method. See *Internal rate-of-return method.*

Time driver. Any factor whose change causes a change in the speed with which an activity is undertaken. (916)

Time tickets. Forms used to charge departments and job-cost records for direct labor used. (94)

Time-to-market. See *New-product lead time.*

Time value of money. Takes into account the interest received (paid) from investing (borrowing) money for the time period under consideration. (672)

Total project approach. Approach to choosing among projects that includes all the cash outflows and inflows associated with each project. (710)

Transfer pricing. Price that one segment (subunit, department, division, and so on) of an organization charges for a product or service supplied to another segment of the same organization. (855)

Transferred-in costs. Costs incurred in a previous department that have been conveyed to a subsequent department. Also called *previous department costs.* (573)

Treasurer. The financial executive who is primarily responsible for obtaining investment capital and managing cash. (11)

Uncertainty. The possibility that an actual amount will deviate from an expected amount. (67)

Underapplied overhead. Amount by which the applied overhead balance is less than the incurred (actual) overhead balance. (108)

Usage variance. See *Efficiency variance.*

Utility function. A function that represents each outcome in terms of its utility value. (654)

Utility value. The value of a specific outcome to a particular decision maker. (654)

Utils. Quantifications of the utility value of given monetary amounts. (654)

Value-added cost. Cost incurred for any activity that cannot be eliminated without the customer's perceiving a deterioration in the performance, function, or other quality of a product. (384)

Value chain. The sequence of total business functions in which value is added to a firm's products or services. This sequence includes research and development, product design, manufacturing, marketing, distribution, and customer service. (113)

Variable budget. See *Flexible budget.*

Variable cost. Cost that does change in total in direct proportion to changes of a cost driver. (29)

Variable costing. Method of inventory costing that includes all direct manufacturing costs and variable indirect manufacturing costs as inventoriable costs; fixed indirect manufacturing costs are excluded from the inventoriable costs. Also called *direct costing.* (289)

Variable-cost ratio. The total variable costs divided by the total sales. (69)

Variances. Deviations of actual results from budget. (5)

Volume variance. See *Production volume variance.*

Work measurement. The careful analysis of a task, its size, the method used in its performance, and its efficiency. (436)

Work measurement method. See *Industrial engineering method.*

Working-capital cycle. See *Self-liquidating cycle.*

Work in process inventory. Goods undergoing the production process but not yet fully completed. Costs include the three major manufacturing costs (direct materials, direct labor, and factory overhead). Also called *work in progress* or *goods in process.* (39)

Work in progress. See *Work in process inventory.*

Work-order systems. See *Job-order product costing system.*

Zero-base budgeting (ZBB). Budgeting from the ground up, as though the budget were being prepared for the first time. Also called *ZBB.* (435)

Author Index

Company Index

Subject Index